郭柏灵院士

郭柏灵论文集

第十七卷（上）

Selected Papers of Guo Boling

Volume 17 (I)

郭柏灵　著

科学出版社

北京

内 容 简 介

《郭柏灵论文集第十七卷》收集的是郭柏灵先生发表于 2019 年度的主要科研论文，涉及的方程范围宽广，有确定性偏微分方程和随机偏微分方程，研究的问题包括适定性、爆破性、渐近性、孤立波等。

这些论文具有很高的学术价值，对偏微分方程、数学物理、非线性分析、计算数学等方向的科研工作者和研究生，是极好地参考著作。

图书在版编目（CIP）数据

郭柏灵论文集. 第 17 卷 / 郭柏灵著. -- 北京：科学出版社，2024.7.
-- ISBN 978-7-03-078966-2

I. O175-53

中国国家版本馆 CIP 数据核字第 20242RU899 号

责任编辑：李　欣 / 责任校对：彭珍珍
责任印制：张　伟 / 封面设计：陈　敬

科 学 出 版 社 出版
北京东黄城根北街 16 号
邮政编码：100717
http://www.sciencep.com

北京厚诚则铭印刷科技有限公司印刷
科学出版社发行　各地新华书店经销
*

2024 年 7 月第　一　版　开本：720×1000　1/16
2024 年 7 月第一次印刷　印张：33 1/4　插页：1
字数：670 000

定价: 398.00 元（全 2 册）
（如有印装质量问题，我社负责调换）

致 谢

《郭柏灵论文集》于 2006 年开始由华南理工大学出版社连续出版了十一卷，论文的收集截止到 2012 年。从第十二卷开始《郭柏灵论文集》将由科学出版社出版发行，从该卷起将收集郭柏灵院士于 2013 年以来先后发表的科研成果。在出版过程中科学出版社的同志，特别是陈玉琢老师进行了精心策划和细致编排，付出了辛勤的劳动。再次谨对他们的帮助表示衷心的感谢。

谭绍滨

2018 年 3 月于厦门大学

序

今年是恩师郭柏灵院士 70 寿辰，华南理工大学出版社决定出版《郭柏灵论文集》。郭老师的弟子，也就是我的师兄弟，推举我为文集作序。这使我深感荣幸。我于 1985 年考入北京应用物理与计算数学研究所，师从郭柏灵院士和周毓麟院士。研究生毕业后我留在研究所工作，继续跟随郭老师学习和研究偏微分方程理论。老师严谨的治学作风和对后学的精心培养与殷切期望，给我留下了深刻的印象，同时老师在科研上的刻苦精神也一直深深地印在我的脑海中。

郭老师 1936 年出于福建省龙岩市新罗区龙门镇，1953 年从福建省龙岩市第一中学考入复旦大学数学系，毕业后留校工作。1963 年，郭老师服从祖国的需要，从复旦大学调入北京应用物理与计算数学研究所，从事核武器研制中有关的数学、流体力学问题及其数值方法研究和数值计算工作。他全力以赴地做好了这项工作，为我国核武器的发展作出了积极的贡献。1978 年改革开放以后，他又在非线性发展方程数学理论及其数值方法领域开展研究工作，现为该所研究员、博士生导师，中国科学院院士。迄今他共发表学术论文 300 余篇、专著 9 部，1987 年获国家自然科学奖三等奖，1994 年和 1998 年两度获得国防科工委科技进步奖一等奖，为我国的国防建设与人才培养作出了巨大贡献。

郭老师的研究方向涉及数学的多个领域，其中包括非线性发展方程的数学理论及其数值解、孤立子理论、无穷维动力系统等，其研究工作的主要特点是紧密联系数学物理中提出的各种重要问题。他对力学及物理学等应用学科中出现的许多重要的非线性发展方程进行了系统深入的研究，其中对 Landau-Lifshitz 方程和 Benjamin-Ono 方程的大初值的整体可解性、解的唯一性、正则性、渐近行为以及爆破现象等建立了系统而深刻的数学理论。在无穷维动力系统方面，郭老师研究了一批重要的无穷维动力系统，建立了有关整体吸引子、惯性流形和近似惯性流形的存在性和分形维数精细估计等理论，提出了一种证明强紧吸引子的新方法，并利用离散化等方法进行理论分析和数值计算，展示了吸引子的结构和图像。下面我从这几个方面介绍郭老师的一些学术成就。

Landau-Lifshitz 方程 (又称铁磁链方程) 由于其结构的复杂性，特别是强耦合性和不带阻尼时的强退化性，在 20 世纪 80 年代之前国内外几乎没有从数学上进行理论研究的成果出现。最先进行研究的，当属周毓麟院士和郭柏灵老师，他们在 1982 年到 1986 年间，采用 Leray-Schauder 不动点定理、离散方法、Galerkin

方法证明了从一维到高维的各种边值问题整体弱解的存在性，比国外在 1992 年才出现的同类结果早了将近 10 年。

20 世纪 90 年代初期，周毓麟、郭柏灵和谭绍滨，郭柏灵和共敏纯得到了两个在国内外至今影响很大的经典结果。第一，通过差分方法结合粘性消去法，利用十分巧妙的先验估计，证明了一维 Landau-Lifshitz 方程光滑解的存在唯一性，对于一维问题给出了完整的答案，解决了长期悬而未决的难题。第二，系统分析了带阻尼的二维 Landau-Lifshitz 方程弱解的奇性，发现了 Landau-Lifshitz 方程与调和映照热流的联系，其弱解具有与调和映照热流完全相同的奇性。现在，国内外这方面的文章基本上都引用这个结果。调和映照的 Landau-Lifshitz 流的概念，即是源于此项结果。

20 世纪 90 年代中期，郭老师对于 Landau-Lifshitz 方程的长时间性态、Landau-Lifshitz 方程耦合 Maxwell 方程的弱解及光滑解的存在性问题进行了深入的研究，得到了一系列的成果。铁磁链方程的退化性以及缺少相应的线性化方程解的表达式，对研究解的长时间性态带来很大困难。郭老师的一系列成果克服了这些困难，证明了近似惯性流形的存在性、吸引子的存在性，给出了其 Hausdorff 和分形维数的上、下界的精细估计。此外，我们知道，与调和映照热流比较，高维铁磁链方程的研究至今还很不完善。其中最重要的是部分正则性问题，其难点在于单调不等式不成立，导致能量衰减估计方面的困难。另外一个是 Blow-up 解的存在性问题，至今没有解决；而对于调和映照热流来说，这样的问题的研究是比较成熟的。

对于高维问题，20 世纪 90 年代后期至今，郭老师和陈韵梅、丁时进、韩永前、杨士山一道，得出了许多成果，大大地推动了该领域的研究进程。第一，证明了二维问题的能量有限弱解的几乎光滑性及唯一性，这个结果类似于 Freire 关于调和映照热流的结果。第二，得到了高维 Landau-Lifshitz 方程初边值问题的奇点集合的 Hausdorff 维数和测度的估计。第三，得到了三维 Landau-Lifshitz-Maxwell 方程的奇点集合的 Hausdorff 维数和测度的估计。第四，得到了一些高维轴对称问题的整体光滑解和奇性解的精确表达式。郭老师还开创了一些新的研究领域。例如，关于一维非均匀铁磁链方程光滑解的存在唯一性结果后来被其他数学家引用并推广到一般流形上。第五，率先讨论了可压缩铁磁链方程测度值解的存在性。最近，在 Landau-Lifshitz 方程耦合非线性 Maxwell 方程与面，也取得了许多新的进展。

多年来，郭老师还对一大批非线性发展方程解的整体存在唯一性、有限时刻的爆破性、解的渐近性态等开展了广泛而深入的研究，受到国内外同行的广泛关注。研究的模型源于数学物理、水波、流体力学、相变等领域，如含磁场的 Schrödinger 方程组、Zakharov 方程、Schrödinger-Boussinesq 方程组、Schrödinger-KdV 方程组、长短波方程组、Maxwell 方程组、Davey-Stewartson 方程组、Klein-Gordon-Schrödinger 方程组、波动方程、广义 KdV 方程、Kadomtsev-Petviashvili (KP) 方程、Benjamin-Ono 方程、Newton-Boussinesq 方程、Cahn-Hilliard 方程、Ginzburg-Landau 方程等。其中不少耦合方程组都是郭老师得到了第一个结果，开创了研究的先河，对国内外同行的研究产生了深远的影响。

郭老师在无穷维动力系统方面也开展了广泛的研究，取得了丰硕的成果。对耗散非线性发展方程所决定的无穷维动力系统，研究了整体吸引子的存在性、分形维数估计、惯性流形、近似惯性流形、指数吸引子等问题。特别是在研究无界哉上耗散非线性发展方程的强紧整体吸引子存在性时所提出的化弱紧吸引子成为强紧吸引子的重要方法和技巧，颇受同行关注并广为利用。对五次非线性 Ginzburg-Landau 方程，郭老师利用空间离散化方法将无限维问题化为有限维问题，证明了该问题离散吸引子的存在性，并考虑五次 Ginzburg-Landau 方程的定态解、慢周期解、异宿轨道等的结构。利用有限维动力系统的理论和方法，结合数值计算得到具体的分形维数 (不超过 4) 和结构以及走向混沌、湍流的具体过程和图像，这是一种寻求整体吸引子细微结构新的探索和尝试，对其他方程的研究也是富有启发性的。1999 年以来，郭老师集中于近可积耗散的和 Hamilton 无穷维动力系统的结构性研究，利用孤立子理论、奇异摄动理论、Fenichel 纤维理论和无穷维 Melnikov 函数，对于具有小耗散的三次到五次非线性 Schrödinger 方程，证明了同宿轨道的不变性，并在有限维截断下证明了 Smale 马蹄的存在性，目前，正把这一方法应用于具小扰动的 Hamilton 系统的研究上。他对于非牛顿流无穷维动力系统也进行了系统深入的研究，建立了有关的数学理论，并把有关结果写成了专著。以上这些工作得到国际同行们的高度评价，被称为"有重大的国际影响""对无穷维动力系统理论有重要持久的贡献"。最近，郭老师及其合作者又证明了具耗散的 KdV 方程 L^2 整体吸引子的存在性，该结果也是引人注目的。

郭老师不仅自己辛勤地搞科研，还尽心尽力培养了大批的研究生 (硕士生、博士生、博士后)，据不完全统计，有 40 多人。他根据每个人不同的学习基础和特点，给予启发式的具体指导，其中的不少人已成为了该领域的学科带头人，有些人虽然开始时基础较差，经过培养，也得到了很大提高，成为了该方向的业务骨干。

《郭柏灵论文集》按照郭老师在不同时期所从事的研究领域，分成多卷出版。文集中所搜集的都是郭老师正式发表过的学术成果。把这些成果整理成集出版，不仅系统地反映了他的科研成就，更重要的是对于从事这方面学习、研究的学者无疑大有裨益。这本文集的出版得到了多方面的帮助与支持，特别要感谢华南理工大学校长李元元教授、华南理工大学出版社范家巧社长和华南理工大学数学科学院吴敏院长的支持。还要特别感谢华南理工大学的李用声教授、华南师范大学的丁时进教授、北京应用物理与计算数学研究所的苗长兴研究员等人在论文的搜集、选择与校对等工作中付出了辛勤的劳动。感谢华南理工大学出版社的编辑对文集的精心编排工作。

谭绍滨

2005 年 8 月于厦门大学

目 录

2019 年（上）

Long-time Asymptotic Behavior for an Extended Modified Korteweg-de Vries Equation ··· 1

Low Mach Number Limit of Strong Solutions to 3-D Full Navier-Stokes Equations with Dirichlet Boundary Condition ·· 49

Weak and Smooth Solutions to Incompressible Navier-Stokes-Landau-Lifshitz-Maxwell Equations ·· 81

Global Smooth Solution for the Compressible Landau-Lifshitz-Bloch Equation ······· 113

Global Weak Solution to the Quantum Navier-Stokes-Landau-Lifshitz Equations with Density-Dependent Viscosity ·· 126

Analytical Study of the Two-dimensional Time-fractional Navier-Stokes Equations ··· 162

On the Cauchy Problem for the Shallow-Water Model with the Coriolis Effect ······· 191

Unconditional Convergence of Linearized Implicit Finite Difference Method for the 2D/3D Gross-Pitaevskii Equation with Angular Momentum Rotation ················· 238

The Gerdjikov-Ivanov Type Derivative Nonlinear Schrödinger Equation: Long-time Dynamics of Nonzero Boundary Conditions ································· 265

On the Backward Uniqueness of the Stochastic Primitive Equations with Additive Noise ··· 305

Some Problems in Radiation Transport Fluid Mechanics and Quantum Fluid Mechanics ·· 329

Random Attractor for the 2D Stochastic Nematic Liquid Crystals Flows ············· 347

The Existence of Global Solutions for the Fourth-Order Nonlinear Schrödinger Equations ·· 386

The Suitable Weak Solution for the Cauchy Problem of the Double-Diffusive Convection System ··· 399

A Riemann-Hilbert Approach for the Modified Short Pulse Equation ··············· 424

Exponential Decay of Bénard Convection Problem with Surface Tension ············ 443

Interaction Behavior Associated with a Generalized (2+1)-Dimensional Hirota Bilinear
   Equation for Nonlinear Waves ··· 468
The Riemann-Hilbert Problem to Coupled Nonlinear Schrödinger Equation: Long-Time
   Dynamics on the Half-Line ··· 492

Long-time Asymptotic Behavior for an Extended Modified Korteweg-de Vries Equation*

Liu Nan (刘男), Guo Boling (郭柏灵), Wang Dengshan (王灯山), and Wang Yufeng (王玉风)

Abstract We investigate an integrable extended modified Korteweg–de Vries equation on the line with the initial value belonging to the Schwartz space. By performing the nonlinear steepest descent analysis of an associated matrix Riemann–Hilbert problem, we obtain the explicit leading-order asymptotics of the solution of this initial value problem as time t goes to infinity. For a special case $\alpha = 0$, we present the asymptotic formula of the solution to the extended modified Korteweg–de Vries equation in region $\mathcal{P} = \{(x,t) \in \mathbf{R}^2 | 0 < x \leqslant Mt^{\frac{1}{5}}, t \geqslant 3\}$ in terms of the solution of a fourth-order Painlevé II equation.

Keywords extended modified Korteweg-de Vries equation; Riemann–Hilbert problem; nonlinear steepest descent method; long-time asymptotics

1 Introduction

It is a well-known fact that the modified Korteweg–de Vries (mKdV) equation is a fundamental, completely integrable model in solitary waves theory, and is given in canonical form as

$$u_t + 6\sigma u^2 u_x + u_{xxx} = 0, \tag{1.1}$$

where $\sigma = \pm 1$, $u = u(x,t)$ is a real function with evolution variable t and transverse variable x. This equation gives rise to multiple soliton solutions

---

*Commun. Math. Sci., 2019, 17(7): 1877–1913. DOI:10.4310/CMS.2019.v17.n7.a6.

and multiple singular soliton solutions for $\sigma = +1$ and $\sigma = -1$, respectively. Moreover, the mKdV equation has significant applications in various physical contexts such as the generation of supercontinuum in optical fibres, acoustic waves in certain anharmonic lattices, nonlinear Alfvén waves propagating in plasma and fluid dynamics.

In this paper, we investigate an extended modified Korteweg-de Vries (emKdV) equation [26], which takes the form

$$u_t+\alpha(6u^2u_x+u_{xxx})+\beta(30u^4u_x+10u_x^3+40uu_xu_{xx}+10u^2u_{xxx}+u_{xxxxx}) = 0, \quad (1.2)$$

where $\alpha > 0$ and $\beta > 0$ stand for the third- and fifth-order dispersion coefficients matching the relevant nonlinear terms, respectively. Moreover, (1.2) also has certain applications for the description of nonlinear internal waves in a fluid stratified by both density and current [16, 25]. Equation (1.2) is integrable, and the infinitely many conservation laws have been constructed based on the Lax pair. Meanwhile, periodic and rational solutions have also been obtained by means of the N-fold Darboux transformation in a recent paper [26]. The Painlevé test and multi-soliton solutions via the simplified Hirota direct method for equation (1.2) have been recently studied in [28]. However, it is noted that the long-time asymptotics for the emKdV equation (1.2) on the line were not analyzed to the best of our knowledge.

In particular, the purpose of the present paper is to consider the initial-value problem (IVP) for the emKdV equation (1.2) on the line by a Riemann–Hilbert (RH) approach. Assuming that the initial data $u(x,0) = u_0(x)$ are smooth and decay sufficiently fast as $|x| \to \infty$, that is, $u_0(x) \in \mathcal{S}(\mathbf{R})$, one then can show that the solution $u(x,t)$ of the IVP for (1.2) can be represented in terms of the solution of a 2×2 matrix RH problem formulated in the complex k-plane with the jump matrices given in terms of two spectral functions $a(k)$, $b(k)$ obtained from the initial value $u_0(x)$. Then, this representation obtained allows us to apply the nonlinear steepest descent method for the associated RH problem and to obtain a detailed description of the leading term of the asymptotics of the solution for the Cauchy problem.

The nonlinear steepest descent method was first introduced in 1993 by Deift and Zhou [10], where they derived the long-time asymptotics for the IVP for the mKdV equation (1.1) with $\sigma = -1$. It then turns out to be very successful in analyzing the long-time asymptotics of IVPs for a large range of nonlinear integrable evolution equations in a rigorous and transparent form. Numerous new significant results about the asymptotics theory of initial-value and initial-boundary value problems for different completely integrable nonlinear equations were obtained based on the analysis of the corresponding RH problems [1-3, 6-8, 11, 15, 17-19, 21, 24, 27, 29].

Developing and extending the methods used in [10, 22], our goal here is to explore the long-time asymptotics of the solution $u(x,t)$ for the emKdV equation (1.2) on the line. Compared with other integrable equations, the long-time asymptotic analysis for (1.2) presents some distinctive features. For example, the spectral curve of the emKdV equation (1.2) is more involved, and it possesses four stationary points, which is different from that of the mKdV equation and Hirota equation considered in [10, 19], where the phase function has only two critical points. We note that in the case of the Camassa–Holm equation [4, 5], there is a sector $-\frac{1}{4} + C < c = \frac{x}{Nt} < 0$ where the corresponding phase function also has four stationary points. Moreover, in the case of the Degasperis–Procesi equation [6], depending on the range of x/t, one can also have four stationary points. However, our main asymptotic analysis still presents many particular pictures different from these literatures (see Sections 3 and 4). Therefore, the study of the long-time asymptotics for the IVP for (1.2) on the line is more interesting. Our main results of this paper are summarized by the following theorems.

Theorem 1.1 *Suppose that $u_0(x)$ lies in the Schwartz space $S(\mathbf{R})$ and be such that no discrete spectrum is present. Then, for any positive constant $\varepsilon > 0$, as $t \to \infty$, the solution $u(x,t)$ of the Cauchy problem for emKdV equation (1.2) on the line satisfies the following asymptotic formula*

$$u(x,t) = -\frac{u_{as}(x,t)}{\sqrt{t}} + O\left(\frac{\ln t}{t}\right), \quad t \to \infty, \ \xi = \frac{x}{t} \in \left(-\frac{9\alpha^2}{20\beta} + \varepsilon, -\varepsilon\right), \quad (1.3)$$

where the error term is uniform with respect to x in the given range, and the leading-order coefficient $u_{as}(x,t)$ is given by

$$u_{as}(x,t) = \sqrt{\frac{\nu(k_1)}{k_1(3\alpha - 40\beta k_1^2)}} \cos\left(16tk_1^3(8\beta k_1^2 - \alpha)\right.$$
$$\left. - \nu(k_1)\ln(16t(k_2 - k_1)^2(3\alpha k_1 - 40\beta k_1^3)) + \phi_a(\xi)\right)$$

$$+ \sqrt{\frac{\nu(k_2)}{k_2(40\beta k_2^2 - 3\alpha)}} \cos\left(16tk_2^3(8\beta k_2^2 - \alpha)\right.$$
$$\left. + \nu(k_2)\ln(16t(k_2-k_1)^2(40\beta k_2^3 - 3\alpha k_2)) + \phi_b(\xi)\right), \quad (1.4)$$

where

$$\phi_a(\xi) = -\frac{\pi}{4} - \arg r(k_1) + \arg \Gamma(i\nu(k_1)) + 2\nu(k_1)\ln\left(\frac{k_1+k_2}{2k_1}\right)$$
$$- \frac{1}{\pi}\int_{k_1}^{k_2} \ln\left(\frac{1+|r(s)|^2}{1+|r(k_1)|^2}\right)\left(\frac{1}{s-k_1} - \frac{1}{s+k_1}\right)ds,$$
$$\phi_b(\xi) = \frac{\pi}{4} - \arg r(k_2) - \arg \Gamma(i\nu(k_2)) + 2\nu(k_2)\ln\left(\frac{2k_2}{k_1+k_2}\right)$$
$$- \frac{1}{\pi}\int_{k_1}^{k_2} \ln\left(\frac{1+|r(s)|^2}{1+|r(k_2)|^2}\right)\left(\frac{1}{s-k_2} - \frac{1}{s+k_2}\right)ds,$$

and k_1, k_2, $\nu(k_1)$, $\nu(k_2)$ are defined by (3.6), (3.7), (3.43) and (3.45), respectively.

Remark 1.1 For $\xi < -\frac{9\alpha^2}{20\beta} - \varepsilon$, there are no real critical points for the phase function $\Phi(k)$. Thus, it is easy to proof that the solution $u(x,t)$ of emKdV equation (1.2) is rapidly decreasing as $t \to \infty$. However, for $\xi > \varepsilon$, there are two different real stationary points $\pm k_0 = \pm\sqrt{\frac{3\alpha}{40\beta}\left(1+\sqrt{1+\frac{20\beta\xi}{9\alpha^2}}\right)}$, this implies that it is possible to deform the RH problem through a series of transformations in exactly the same way as in the similarity region for the mKdV equation to find the asymptotics (one also can follow the strategy used in Section 3).

Theorem 1.2 Under the assumptions of Theorem 1.1, the solution $u(x,t)$ of equation (4.1), i.e., $\alpha = 0$ in emKdV equation (1.2), satisfies the following asymptotic formula as $t \to \infty$,

$$u(x,t) = \left(\frac{8}{5\beta t}\right)^{\frac{1}{5}} u_p\left(\frac{-x}{(20\beta t)^{\frac{1}{5}}}\right) + O(t^{-\frac{2}{5}}), \quad 0 < x \leqslant Mt^{\frac{1}{5}}, \quad (1.5)$$

where the formula holds uniformly with respect to x in the given range for any fixed $M > 1$ and the function $u_p(y)$ denotes the solution of the fourth order Painlevé II equation (A.5).

The organization of this paper is as follows. In Section 2, we show how the solution of emKdV equation (1.2) can be expressed in terms of the solution of a 2×2 matrix RH problem and give an auxiliary theorem which is useful for determining the long-time asymptotics. In Section 3, we derive the long-time asymptotic behavior of the solution of the emKdV equation (1.2) to prove our first main Theorem 1.1 in the physically interesting region. In Section 4, we present the asymptotic formula of the solution to a particular case $\alpha = 0$ of emKdV equation (1.2) in region $0 < x \leqslant Mt^{\frac{1}{5}}$. A few facts related to the RH problem associated with the fourth order Painlevé II equation are collected in the Appendix.

2 Preliminaries

2.1 Riemann–Hilbert formalism

The Lax pair of equation (1.2) is [26]

$$\Psi_x = X\Psi, \quad X = ik\sigma_3 + U,$$
$$\Psi_t = T\Psi, \quad T = (-16i\beta k^5 + 4i\alpha k^3)\sigma_3 + V \quad (2.1)$$

(namely, equation (1.2) is the compatibility condition $X_t - T_x + [X, T] = 0$ of equation (2.1)), where $\Psi(x, t; k)$ is a 2×2 matrix-valued function, $k \in \mathbf{C}$ is the spectral parameter and

$$\sigma_3 = \begin{pmatrix} 1 & 0 \\ 0 & -1 \end{pmatrix}, \quad U = \begin{pmatrix} 0 & u \\ -u & 0 \end{pmatrix}, \quad V = \begin{pmatrix} A & B \\ C & -A \end{pmatrix}, \quad (2.2)$$

$$A = 8i\beta k^3 u^2 - i(6\beta u^4 + 2\alpha u^2 + 4\beta u u_{xx} - 2\beta u_x^2)k,$$
$$B = -16\beta k^4 u + 8i\beta k^3 u_x + (8\beta u^3 + 4\alpha u + 4\beta u_{xx})k^2$$
$$\quad - i(12\beta u^2 u_x + 2\beta u_{xxx} + 2\alpha u_x)k$$
$$\quad - 6\beta u^5 - 2\alpha u^3 - 10\beta u^2 u_{xx} - 10\beta u u_x^2 - \beta u_{xxxx} - \alpha u_{xx},$$
$$C = -B + 16i\beta k^3 u_x - i(24\beta u^2 u_x + 4\alpha u_x + 4\beta u_{xxx})k.$$

Introducing a new eigenfunction $\mu(x, t; k)$ by

$$\mu(x, t; k) = \Psi(x, t; k)e^{-i[kx + (-16\beta k^5 + 4\alpha k^3)t]\sigma_3}, \quad (2.3)$$

we obtain the equivalent Lax pair

$$\mu_x - ik[\sigma_3, \mu] = U\mu,$$
$$\mu_t + i(16\beta k^5 - 4\alpha k^3)[\sigma_3, \mu] = V\mu. \quad (2.4)$$

We now consider the spectral analysis of the x-part of (2.4). Define two solutions μ_1 and μ_2 of the x-part of (2.4) by the following Volterra integral equations

$$\mu_1(x,t;k) = I + \int_{-\infty}^{x} e^{ik(x-x')\hat{\sigma}_3}[U(x',t)\mu_1(x',t;k)]dx', \tag{2.5}$$

$$\mu_2(x,t;k) = I - \int_{x}^{\infty} e^{ik(x-x')\hat{\sigma}_3}[U(x',t)\mu_2(x',t;k)]dx', \tag{2.6}$$

where $\hat{\sigma}_3$ acts on a 2×2 matrix X by $\hat{\sigma}_3 X = [\sigma_3, X]$, and $e^{\hat{\sigma}_3} = e^{\sigma_3}Xe^{-\sigma_3}$. We denote by $\mu^{(1)}$ and $\mu^{(2)}$ the columns of a 2×2 matrix $\mu = (\mu^{(1)}\ \mu^{(2)})$. Then it follows from (2.5)-(2.6) that for all (x,t):

(i) $\det \mu_j = 1$, $j = 1, 2$.

(ii) $\mu_2^{(1)}$ and $\mu_1^{(2)}$ are analytic and bounded in $\{k \in \mathbf{C}| \mathrm{Im} k > 0\}$, and $(\mu_2^{(1)}\ \mu_1^{(2)}) \to I$ as $k \to \infty$.

(iii) $\mu_1^{(1)}$ and $\mu_2^{(2)}$ are analytic and bounded in $\{k \in \mathbf{C}| \mathrm{Im} k < 0\}$, and $(\mu_1^{(1)}\ \mu_2^{(2)}) \to I$ as $k \to \infty$.

(iv) $\{\mu_j\}_1^2$ are continuous up to the real axis.

(v) Symmetry:

$$\overline{\mu_j(x,t;\bar{k})} = \mu_j(x,t;-k) = \sigma_2 \mu_j(x,t;k)\sigma_2, \tag{2.7}$$

where σ_2 is the second Pauli matrix,

$$\sigma_2 = \begin{pmatrix} 0 & -i \\ i & 0 \end{pmatrix}.$$

The symmetry relation (2.7) can be proved easily due to the symmetries of the matrix X:

$$\overline{X(x,t;\bar{k})} = X(x,t;-k) = \sigma_2 X(x,t;k)\sigma_2.$$

The solutions of the system of differential equation (2.4) must be related by a matrix independent of x and t, therefore,

$$\mu_1(x,t;k) = \mu_2(x,t;k)e^{i[kx+(-16\beta k^5 + 4\alpha k^3)t]\hat{\sigma}_3}s(k), \quad \det s(k) = 1,\ k \in \mathbf{R}. \tag{2.8}$$

Evaluation at $x \to \infty, t = 0$ gives

$$s(k) = \lim_{x\to\infty} e^{-ikx\hat{\sigma}_3}\mu_1(x,0;k), \tag{2.9}$$

that is,

$$s(k) = I + \int_{-\infty}^{\infty} e^{-ikx\hat{\sigma}_3}[U(x,0)\mu_1(x,0;k)]dx. \tag{2.10}$$

Due to the symmetry (2.7), the matrix-valued spectral function $s(k)$ can be defined in terms of two scalar spectral functions $a(k)$ and $b(k)$ by

$$s(k) = \begin{pmatrix} \bar{a}(k) & b(k) \\ -\bar{b}(k) & a(k) \end{pmatrix}, \qquad (2.11)$$

where $\bar{a}(k) = \overline{a(\bar{k})}$ and $\bar{b}(k) = \overline{b(\bar{k})}$ indicate the Schwartz conjugates. The spectral functions $a(k)$ and $b(k)$ can be determined by $u_0(x)$ through the solution of equation (2.10). On the other hand, $a(k)$ is analytic in the half-plane $\{k \in \mathbb{C} | \text{Im} k > 0\}$ and continuous in $\{k \in \mathbb{C} | \text{Im} k \geqslant 0\}$, and $a(k) \to 1$ as $k \to \infty$. Furthermore, $|a(k)|^2 + |b(k)|^2 = 1$ for $k \in \mathbb{R}$. Finally, $a(-k) = \bar{a}(k)$, $b(-k) = \bar{b}(k)$.

Assuming $u(x,t)$ be a solution of equation (1.2), the analytic properties of $\mu_j(x,t;k)$ stated above allow us to define a piecewise meromorphic, 2×2 matrix-valued function $M(x,t;k)$ by

$$M(x,t;k) = \begin{cases} \left(\dfrac{\mu_2^{(1)}(x,t;k)}{a(k)} \quad \mu_1^{(2)}(x,t;k) \right), & \text{Im} k > 0, \\ \left(\mu_1^{(1)}(x,t;k) \quad \dfrac{\mu_2^{(2)}(x,t;k)}{\bar{a}(k)} \right), & \text{Im} k < 0. \end{cases} \qquad (2.12)$$

Then, for each $x \in \mathbb{R}$ and $t \geqslant 0$, the boundary values $M_\pm(x,t;k)$ of M as k approaches $\mathbb{R}$ from the sides $\pm \text{Im} k > 0$ are related as follows:

$$M_+(x,t;k) = M_-(x,t;k) J(x,t;k), \qquad k \in \mathbb{R}, \qquad (2.13)$$

with

$$J(x,t;k) = \begin{pmatrix} 1 + |r(k)|^2 & \bar{r}(k) e^{-t\Phi(k)} \\ r(k) e^{t\Phi(k)} & 1 \end{pmatrix}, \qquad (2.14)$$

$$r(k) = \frac{\bar{b}(k)}{a(k)}, \quad \Phi(k) = 2i \left(-k\frac{x}{t} + 16\beta k^5 - 4\alpha k^3 \right).$$

In view of the properties of $\mu_j(x,t;k)$ and $s(k)$, $M(x,t;k)$ also satisfies the following properties:

(i) Behavior at $k = \infty$:

$$M(x,t;k) \to I \text{ as } k \to \infty. \qquad (2.15)$$

(ii) Symmetry:

$$\overline{M(x,t;\bar{k})} = M(x,t;-k) = \sigma_2 M(x,t;k)\sigma_2. \tag{2.16}$$

(iii) Residue conditions: Let $\{k_j\}_1^N$ be the set of zeros of $a(k)$. We assume these zeros are finite in number, simple, and no zero is real, then $M(x,t;k)$ satisfies the following residue conditions:

$$\operatorname{Res}_{k=k_j} M^{(1)}(x,t;k) = \frac{e^{t\Phi(k_j)}}{\dot{a}(k_j)b(k_j)} M^{(2)}(x,t;k_j) = i\chi_j e^{t\Phi(k_j)} M^{(2)}(x,t;k_j),$$

$$\operatorname{Res}_{k=\bar{k}_j} M^{(2)}(x,t;k) = -\frac{e^{-t\Phi(\bar{k}_j)}}{\dot{a}(k_j)\overline{b(k_j)}} M^{(1)}(x,t;\bar{k}_j) = i\bar{\chi}_j e^{-t\Phi(\bar{k}_j)} M^{(1)}(x,t;\bar{k}_j).$$
$$\tag{2.17}$$

Theorem 2.1 *Let $\{r(k), \{k_j, \chi_j\}_1^N\}$ be the spectral data determined by $u_0(x)$, and define $M(x,t;k)$ as the solution of the associated RH (2.13) with the jump matrix (2.14), the normalization condition (2.15), and the residue conditions (2.17). Then, $M(x,t;k)$ exists and is unique. Define $u(x,t)$ in terms of $M(x,t;k)$ by*

$$u(x,t) = -2i \lim_{k\to\infty} (kM(x,t;k))_{12}. \tag{2.18}$$

Then $u(x,t)$ solves the emKdV *equation (1.2). Furthermore, $u(x,0) = u_0(x)$.*

Proof In the case when $a(k)$ has no zeros, the existence and uniqueness of the solution of the above RH problem is a consequence of a 'vanishing lemma' for the associated RH problem with the vanishing condition at infinity $M(k) = O(1/k)$, $k \to \infty$ (see [23] since $J^\dagger(\bar{k}) = J(k)$). If $a(k)$ has zeros, the singular RH problem can be mapped to a regular one following the approach of [14]. Moreover, it follows from standard arguments using the dressing method [13] that if M solves the above RH problem and $u(x,t)$ is defined by (2.18), then $u(x,t)$ solves the emKdV equation (1.2). One observes that for $t = 0$, the RH problem reduces to that associated with $u_0(x)$, which yields $u(x,0) = u_0(x)$, owing to the uniqueness of the solution to the RH problem.

2.2 A model RH problem

After the formulation of the main RH problem, the main idea of the analysis of the long-time behavior is to reduce the original RH problem to a model RH problem which can be solved exactly. The following theorem turned out to be suitable for determining asymptotics of a class of RH problems which arise in the study of long-time asymptotics.

Let $X = X_1 \cup X_2 \cup X_3 \cup X_4 \subset \mathbf{C}$ be the cross defined by

$$X_1 = \{le^{\frac{i\pi}{4}} | 0 \leqslant l < \infty\}, \quad X_2 = \{le^{\frac{3i\pi}{4}} | 0 \leqslant l < \infty\},$$
$$X_3 = \{le^{-\frac{3i\pi}{4}} | 0 \leqslant l < \infty\}, \quad X_4 = \{le^{-\frac{i\pi}{4}} | 0 \leqslant l < \infty\}, \quad (2.19)$$

and oriented as in Fig.1. Define the function $\nu : \mathbf{C} \to (0, \infty)$ by $\nu(q) = \dfrac{1}{2\pi} \ln(1 + |q|^2)$. We consider the following RH problems parametrized by $q \in \mathbf{C}$,

$$\begin{cases} M_+^X(q, z) = M_-^X(q, z) J^X(q, z), & z \in X, \\ M^X(q, z) \to I, & z \to \infty, \end{cases} \quad (2.20)$$

where the jump matrix $J^X(q, z)$ is defined by

$$J^X(q, z) = \begin{cases} \begin{pmatrix} 1 & 0 \\ qe^{\frac{iz^2}{2}} z^{2i\nu(q)} & 1 \end{pmatrix}, & z \in X_1, \\[2ex] \begin{pmatrix} 1 & -\dfrac{\bar{q}}{1+|q|^2} e^{-\frac{iz^2}{2}} z^{-2i\nu(q)} \\ 0 & 1 \end{pmatrix}, & z \in X_2, \\[2ex] \begin{pmatrix} 1 & 0 \\ -\dfrac{q}{1+|q|^2} e^{\frac{iz^2}{2}} z^{2i\nu(q)} & 1 \end{pmatrix}, & z \in X_3, \\[2ex] \begin{pmatrix} 1 & \bar{q} e^{-\frac{iz^2}{2}} z^{-2i\nu(q)} \\ 0 & 1 \end{pmatrix}, & z \in X_4. \end{cases} \quad (2.21)$$

Then we have the following theorem.

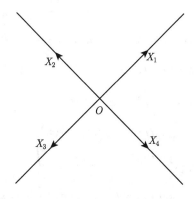

Fig. 1 The contour $X = X_1 \cup X_2 \cup X_3 \cup X_4$

Theorem 2.2 *The RH problem (2.20) has a unique solution $M^X(q,z)$ for each $q \in \mathbf{C}$. This solution satisfies*

$$M^X(q,z) = I - \frac{i}{z}\begin{pmatrix} 0 & \beta^X(q) \\ \overline{\beta^X(q)} & 0 \end{pmatrix} + O\left(\frac{q}{z^2}\right), \quad z \to \infty, \ q \in \mathbf{C}, \quad (2.22)$$

where the error term is uniform with respect to $\arg z \in [0, 2\pi]$ and the function $\beta^X(q)$ is given by

$$\beta^X(q) = \sqrt{\nu(q)}e^{i\left(\frac{\pi}{4} - \arg q - \arg \Gamma(i\nu(q))\right)}, \quad q \in \mathbf{C}, \quad (2.23)$$

where $\Gamma(\cdot)$ denotes the standard Gamma function. Moreover, for each compact subset $\mathcal{D}$ of $\mathbf{C}$,

$$\sup_{q \in \mathcal{D}} \sup_{z \in \mathbf{C} \setminus X} |M^X(q,z)| < \infty \quad (2.24)$$

and

$$\sup_{q \in \mathcal{D}} \sup_{z \in \mathbf{C} \setminus X} \frac{|M^X(q,z) - I|}{|q|} < \infty. \quad (2.25)$$

Proof For the proof of this theorem, we refer the readers to see [1, 10, 11].

3 Long-time asymptotics

In this section, we aim to transform the associated original RH problem (2.15) to a solvable RH problem and then find the explicitly asymptotic formula for the emKdV equation (1.2). In the following analysis, we suppose that $a(k) \neq 0$ for $\{k \in \mathbf{C} | \mathrm{Im}\, k \geqslant 0\}$ so that no discrete spectrum is present. Namely, we consider the following RH problem

$$\begin{cases} M_+(x,t;k) = M_-(x,t;k)J(x,t;k), & k \in \mathbf{R}, \\ M(x,t;k) \to I, & k \to \infty, \end{cases} \quad (3.1)$$

where the jump matrix $J(x,t;k)$ is defined by

$$J(x,t;k) = \begin{pmatrix} 1 + |r(k)|^2 & \bar{r}(k)e^{-t\Phi(k)} \\ r(k)e^{t\Phi(k)} & 1 \end{pmatrix}, \quad (3.2)$$

$$r(k) = \frac{\bar{b}(k)}{a(k)}, \quad \Phi(k) = 2i(-k\xi + 16\beta k^5 - 4\alpha k^3), \quad \xi = \frac{x}{t}.$$

In view of the symmetry relation in (2.7), we conclude that
$$r(-k) = \overline{r(\bar{k})}, \quad k \in \mathbf{R}. \tag{3.3}$$
Moreover, the relation between the solution $u(x,t)$ of the emKdV equation (1.2) and $M(x,t;k)$ is
$$u(x,t) = -2i \lim_{k\to\infty} (kM(x,t;k))_{12}. \tag{3.4}$$

The jump matrix J defined in (3.2) involves the exponentials $e^{\pm t\Phi}$, therefore, the sign structure of the quantity Re $\Phi(k)$ plays an important role in the following analysis. In particular, we suppose
$$-\frac{9\alpha^2}{20\beta} < \xi < 0. \tag{3.5}$$
It follows that there are four different real stationary points located at the points where $\frac{\partial \Phi}{\partial k} = 0$, namely, at
$$\pm k_1 = \pm \sqrt{\frac{3\alpha}{40\beta}\left(1 - \sqrt{1 + \frac{20\beta\xi}{9\alpha^2}}\right)}, \tag{3.6}$$
$$\pm k_2 = \pm \sqrt{\frac{3\alpha}{40\beta}\left(1 + \sqrt{1 + \frac{20\beta\xi}{9\alpha^2}}\right)}. \tag{3.7}$$
The signature table for Re$\Phi(k)$ is shown in Fig. 2.

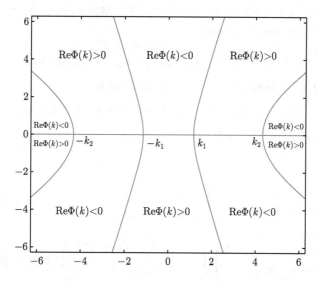

Fig. 2 The signature table for Re$\Phi(k)$ in the complex k-plane

Let $\varepsilon > 0$ be given constant. We restrict our attention here to the physically interesting region $\xi \in \mathcal{I} = \left(-\frac{9\alpha^2}{20\beta} + \varepsilon, -\varepsilon\right)$.

3.1 Transformations of the RH problem

One goes from the original RH problem (3.1) for M to the equivalent RH problem for the new function $M^{(1)}$ defined by

$$M^{(1)}(x,t;k) = M(x,t;k)\delta^{-\sigma_3}(k), \tag{3.8}$$

where the complex-valued function $\delta(k)$ is given by

$$\delta(k) = \exp\left\{\frac{1}{2\pi i}\left(\int_{-k_2}^{-k_1} + \int_{k_1}^{k_2}\right)\frac{\ln(1+|r(s)|^2)}{s-k}ds\right\}, \quad k \in \mathbf{C}\setminus([-k_2,-k_1]\cup[k_1,k_2]). \tag{3.9}$$

Lemma 3.1 *The function $\delta(k)$ has the following properties:*

(i) $\delta(k)$ *satisfies the following jump condition across the real axis oriented from $-\infty$ to ∞:*

$$\delta_+(k) = \delta_-(k)(1+|r(k)|^2), \quad k \in (-k_2,-k_1) \cup (k_1,k_2).$$

(ii) *As $k \to \infty$, $\delta(k)$ satisfies the asymptotic formula*

$$\delta(k) = 1 + O(k^{-1}), \quad k \to \infty. \tag{3.10}$$

(iii) $\delta(k)$ *and $\delta^{-1}(k)$ are bounded and analytic functions of $k \in \mathbf{C} \setminus ([-k_2,-k_1] \cup [k_1,k_2])$ with continuous boundary values on $(-k_2,-k_1) \cup (k_1,k_2)$.*

(iv) $\delta(k)$ *obeys the symmetry*

$$\delta(k) = \overline{\delta(\bar{k})}^{-1}, \quad k \in \mathbf{C} \setminus ([-k_2,-k_1] \cup [k_1,k_2]).$$

Then $M^{(1)}(x,t;k)$ satisfies the following RH problem

$$M_+^{(1)}(x,t;k) = M_-^{(1)}(x,t;k)J^{(1)}(x,t;k), \quad k \in \mathbf{R}, \tag{3.11}$$

with the jump matrix $J^{(1)} = \delta_-^{\sigma_3} J \delta_+^{-\sigma_3}$, namely,

$$J^{(1)}(x,t;k)$$

$$= \begin{cases} \begin{pmatrix} 1 & r_4(k)\delta^2(k)e^{-t\Phi(k)} \\ 0 & 1 \end{pmatrix}\begin{pmatrix} 1 & 0 \\ r_1(k)\delta^{-2}(k)e^{t\Phi(k)} & 1 \end{pmatrix}, & |k| > k_2, |k| < k_1, \\ \begin{pmatrix} 1 & 0 \\ r_3(k)\delta_-^{-2}(k)e^{t\Phi(k)} & 1 \end{pmatrix}\begin{pmatrix} 1 & r_2(k)\delta_+^2(k)e^{-t\Phi(k)} \\ 0 & 1 \end{pmatrix}, & k_1 < |k| < k_2, \end{cases} \tag{3.12}$$

where we define $\{r_j(k)\}_1^4$ by

$$r_1(k) = r(k), \qquad r_2(k) = \frac{\bar{r}(k)}{1+r(k)\bar{r}(k)}, \qquad (3.13)$$
$$r_3(k) = \frac{r(k)}{1+r(k)\bar{r}(k)}, \qquad r_4(k) = \bar{r}(k).$$

Before processing the next deformation, we first introduce analytic approximations of $\{r_j(k)\}_1^4$ following the idea of [22]. We define the open subsets $\{\Omega_j\}_1^4$, as displayed in Fig. 3, such that

$$\Omega_1 \cup \Omega_3 = \{k \in \mathbf{C} | \operatorname{Re}\Phi(k) < 0\},$$
$$\Omega_2 \cup \Omega_4 = \{k \in \mathbf{C} | \operatorname{Re}\Phi(k) > 0\}.$$

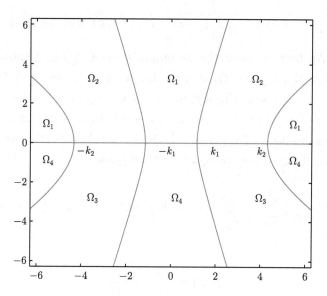

Fig. 3   The open sets $\{\Omega_j\}_1^4$ in the complex k-plane

Lemma 3.2 *There exist decompositions*

$$r_j(k) = \begin{cases} r_{j,a}(x,t,k) + r_{j,r}(x,t,k), & |k| > k_2,\ |k| < k_1,\ k \in \mathbf{R},\ j = 1, 4, \\ r_{j,a}(x,t,k) + r_{j,r}(x,t,k), & k_1 < |k| < k_2,\ k \in \mathbf{R},\ j = 2, 3, \end{cases}$$
(3.14)

where the functions $\{r_{j,a}, r_{j,r}\}_1^4$ have the following properties:

(1) For $\xi \in \mathcal{I}$ and each $t > 0$, $r_{j,a}(x,t,k)$ is defined and continuous for $k \in \bar{\Omega}_j$ and analytic for Ω_j, $j = 1, 2, 3, 4$.

(2) The functions $r_{1,a}$ and $r_{4,a}$ satisfy, for $\xi \in \mathcal{I}$, $t > 0$,

$$|r_{j,a}(x,t,k)| \leqslant \frac{C}{1+|k|^2} e^{\frac{t}{4}|\operatorname{Re}\Phi(k)|}, \quad k \in \bar{\Omega}_j \cap \{k \in \mathbf{C} | |\operatorname{Re}k| > k_2\}, \quad j = 1, 4, \tag{3.15}$$

where the constant C is independent of ξ, k, t.

(3) The L^1, L^2 and L^∞ norms of the functions $r_{1,r}(x,t,\cdot)$ and $r_{4,r}(x,t,\cdot)$ on $(-\infty, -k_2) \cup (k_2, \infty) \cup (-k_1, k_1)$ are $O(t^{-3/2})$ as $t \to \infty$ uniformly with respect to $\xi \in \mathcal{I}$.

(4) The L^1, L^2 and L^∞ norms of the functions $r_{2,r}(x,t,\cdot)$ and $r_{3,r}(x,t,\cdot)$ on $(-k_2, -k_1) \cup (k_1, k_2)$ are $O(t^{-3/2})$ as $t \to \infty$ uniformly with respect to $\xi \in \mathcal{I}$.

(5) For $j = 1, 2, 3, 4$, the following symmetries hold:

$$r_{j,a}(x,t,k) = \overline{r_{j,a}(x,t,-\bar{k})}, \quad r_{j,r}(x,t,k) = \overline{r_{j,r}(x,t,-\bar{k})}. \tag{3.16}$$

Proof We first consider the decomposition of $r_1(k)$. Denote $\Omega_1 = \Omega_1^1 \cup \Omega_1^2 \cup \Omega_1^3$, where Ω_1^1, Ω_1^2 and Ω_1^3 denote the parts of Ω_1 in $\{k \in \mathbf{C} | \operatorname{Re}k > k_2\}$, $\{k \in \mathbf{C} | \operatorname{Re}k < -k_2\}$ and the remaining part, respectively. We first derive a decomposition of $r_1(k)$ in Ω_1^1, and then extend it to Ω_1^2 by symmetry. Then, we derive a decomposition of $r_1(k)$ in Ω_1^3. Since $u_0(x) \in \mathcal{S}(\mathbf{R})$, this implies that $r_1(k) = r(k) \in \mathcal{S}(\mathbf{R})$. Then for $n = 0, 1, 2$, we have

$$r_1^{(n)}(k) = \frac{d^n}{dk^n}\left(\sum_{j=0}^{6} \frac{r_1^{(j)}(k_2)}{j!}(k-k_2)^j\right) + O((k-k_2)^{7-n}), \quad k \to k_2. \tag{3.17}$$

Let

$$f_0(k) = \sum_{j=5}^{11} \frac{a_j}{(k-i)^j}, \tag{3.18}$$

where $\{a_j\}_5^{11}$ are complex constants such that

$$f_0(k) = \sum_{j=0}^{6} \frac{r_1^{(j)}(k_2)}{j!}(k-k_2)^j + O((k-k_2)^7), \quad k \to k_2. \tag{3.19}$$

It is easy to verify that (3.19) imposes seven linearly independent conditions on the a_j, hence the coefficients a_j exist and are unique. Letting $f = r_1 - f_0$, it follows that

(i) $f_0(k)$ is a rational function of $k \in \mathbf{C}$ with no poles in $\bar{\Omega}_1^1$;

(ii) $f_0(k)$ coincides with $r_1(k)$ to six order at k_2, more precisely,

$$\frac{d^n}{dk^n}f(k) = \begin{cases} O((k-k_2)^{7-n}), & k \to k_2, \\ O(k^{-5-n}), & k \to \infty, \end{cases} \quad k \in \mathbf{R},\ n=0,1,2. \tag{3.20}$$

The decomposition of $r_1(k)$ can be derived as follows. The map $k \mapsto \phi = \phi(k)$ defined by

$$\phi(k) = -i\Phi(k) = 2(16\beta k^5 - 4\alpha k^3 - \xi k) \tag{3.21}$$

is a bijection $(k_2, \infty) \mapsto (-128\beta k_2^5 + 16\alpha k_2^3, \infty)$ (see Fig. 4), so we may define a function F by

$$F(\phi) = \begin{cases} \dfrac{(k-i)^3}{k-k_2} f(k), & \phi > -128\beta k_2^5 + 16\alpha k_2^3, \\ 0, & \phi \leqslant -128\beta k_2^5 + 16\alpha k_2^3. \end{cases} \tag{3.22}$$

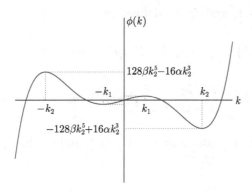

Fig. 4 The graph of the function $\phi(k)$ defined in (3.21)

Then,

$$F^{(n)}(\phi) = \left(\frac{1}{160\beta(k^2-k_2^2)\left(k^2+k_2^2-\dfrac{3\alpha}{20\beta}\right)} \frac{\partial}{\partial k} \right)^n \left(\frac{(k-i)^3}{k-k_2} f(k) \right),$$

$\phi > -128\beta k_2^5 + 16\alpha k_2^3$.

By (3.20), $F^{(n)}(\phi) = O(|\phi|^{-3/5})$ as $|\phi| \to \infty$ for $n=0,1,2$. In particular,

$$\left\| \frac{d^n F}{d\phi^n} \right\|_{L^2(\mathbf{R})} < \infty, \quad n=0,1,2, \tag{3.23}$$

that is, F belongs to $H^2(\mathbf{R})$. By the Fourier transform, $\hat{F}(s)$ defined by

$$\hat{F}(s) = \frac{1}{2\pi}\int_{\mathbf{R}} F(\phi)e^{-i\phi s}d\phi,$$

where

$$F(\phi) = \int_{\mathbf{R}} \hat{F}(s)e^{i\phi s}ds, \qquad (3.24)$$

it follows from Plancherel theorem that $\|s^2\hat{F}(s)\|_{L^2(\mathbf{R})} < \infty$. Equations (3.22) and (3.24) imply

$$f(k) = \frac{k-k_2}{(k-i)^3}\int_{\mathbf{R}} \hat{F}(s)e^{i\phi s}ds, \quad k > k_2. \qquad (3.25)$$

Writing
$$f(k) = f_a(x,t,k) + f_r(x,t,k), \quad t > 0, \ k > k_2,$$

where the functions f_a and f_r are defined by

$$f_a(x,t,k) = \frac{k-k_2}{(k-i)^3}\int_{-\frac{t}{4}}^{\infty} \hat{F}(s)e^{s\Phi(k)}ds, \quad t > 0, \ k \in \Omega_1^1,$$

$$f_r(x,t,k) = \frac{k-k_2}{(k-i)^3}\int_{-\infty}^{-\frac{t}{4}} \hat{F}(s)e^{s\Phi(k)}ds, \quad t > 0, \ k > k_2,$$

we infer that $f_a(x,t,\cdot)$ is continuous in $\bar{\Omega}_1^1$ and analytic in Ω_1^1. Moreover, we can get

$$|f_a(x,t,k)| \leq \frac{|k-k_2|}{|k-i|^3}\|\hat{F}(s)\|_{L^1(\mathbf{R})} \sup_{s \geq -\frac{t}{4}} e^{s\operatorname{Re}\Phi(k)}$$

$$\leq \frac{C|k-k_2|}{|k-i|^3}e^{\frac{t}{4}|\operatorname{Re}\Phi(k)|}, \quad t > 0, \ k \in \bar{\Omega}_1^1, \ \xi \in \mathcal{I}. \qquad (3.26)$$

Furthermore, we have

$$|f_r(x,t,k)| \leq \frac{|k-k_2|}{|k-i|^3}\int_{-\infty}^{-\frac{t}{4}} s^2|\hat{F}(s)|s^{-2}ds$$

$$\leq \frac{C}{1+|k|^2}\|s^2\hat{F}(s)\|_{L^2(\mathbf{R})}\sqrt{\int_{-\infty}^{-\frac{t}{4}} s^{-4}ds},$$

$$\leq \frac{C}{1+|k|^2}t^{-3/2}, \quad t > 0, \ k > k_2, \ \xi \in \mathcal{I}. \qquad (3.27)$$

Hence, the L^1, L^2, and L^∞ norms of f_r on (k_2, ∞) are $O(t^{-3/2})$. Letting

$$r_{1,a}(x,t,k) = f_0(k) + f_a(x,t,k), \qquad t > 0, \ k \in \bar{\Omega}_1^1,$$
$$r_{1,r}(x,t,k) = f_r(x,t,k), \qquad t > 0, \ k > k_2. \tag{3.28}$$

For $k < -k_2$, we use the symmetry (3.16) to extend this decomposition.

We next derive the decomposition of $r_1(k)$ for $-k_1 < k < k_1$. Following [10, 22], we split $r_1(k)$ into even and odd parts as follows:

$$r_1(k) = r_+(k^2) + kr_-(k^2), \quad k \in \mathbf{R}, \tag{3.29}$$

where $r_\pm : [0, \infty) \to \mathbf{C}$ are defined by

$$r_+(s) = \frac{r_1(\sqrt{s}) + r_1(-\sqrt{s})}{2}, \quad r_-(s) = \frac{r_1(\sqrt{s}) - r_1(-\sqrt{s})}{2\sqrt{s}}, \quad s \geqslant 0.$$

We write $r_1(k)$ as the following form of the Taylor series

$$r_1(k) = \sum_{j=0}^{10} q_j k^j + \frac{1}{10!} \int_0^k r_1^{(11)}(t)(k-t)^{10} dt, \quad q_j = \frac{r_1^{(j)}(0)}{j!}. \tag{3.30}$$

It then follows that

$$r_+(s) = \sum_{j=0}^{5} q_{2j} s^j + \frac{1}{2 \times 10!} \int_0^{\sqrt{s}} (r_1^{(11)}(t) - r_1^{(11)}(-t))(\sqrt{s} - t)^{10} dt,$$
$$r_-(s) = \sum_{j=0}^{4} q_{2j+1} s^j + \frac{1}{2 \times 10!\sqrt{s}} \int_0^{\sqrt{s}} (r_1^{(11)}(t) + r_1^{(11)}(-t))(\sqrt{s} - t)^{10} dt. \tag{3.31}$$

Letting $\{p_j^\pm\}_0^4$ denote the coefficients of the Taylor series representations

$$r_\pm(k^2) = \sum_{j=0}^{4} p_j^\pm (k^2 - k_1^2)^j + \frac{1}{4!} \int_{k_1^2}^{k^2} r_\pm^{(5)}(t)(k^2 - t)^4 dt,$$

we infer that the function $f_0(k)$ defined by

$$f_0(k) = \sum_{j=0}^{4} p_j^+ (k^2 - k_1^2)^j + k \sum_{j=0}^{4} p_j^- (k^2 - k_1^2)^j \tag{3.32}$$

has the following properties:

(i) $f_0(k)$ is a polynomial in $k \in \mathbf{C}$ whose coefficients are bounded.

(ii) The difference $f(k) = r_1(k) - f_0(k)$, which satisfies

$$\frac{d^n}{dk^n}f(k) \leqslant C|k^2 - k_1^2|^{5-n}, \quad -k_1 < k < k_1, \quad \xi \in \mathcal{I}, \, n = 0, 1, 2, \tag{3.33}$$

where C is independent of ξ, k. The decomposition of $r_1(k)$ for $-k_1 < k < k_1$ can now be derived as follows. Since the function $k \mapsto \phi$ defined in (3.21) is a bijection $(-k_1, k_1) \to (128\beta k_1^5 - 16\alpha k_1^3, -128\beta k_1^5 + 16\alpha k_1^3)$ (see Fig. 4), we may define a function $F(\phi)$ by

$$F(\phi) = \begin{cases} \dfrac{1}{k^2 - k_1^2} f(k), & |\phi| < -128\beta k_1^5 + 16\alpha k_1^3, \\ 0, & |\phi| \geqslant -128\beta k_1^5 + 16\alpha k_1^3. \end{cases} \tag{3.34}$$

Thus, we have

$$F^{(n)}(\phi) = \left(\frac{1}{160\beta(k^2 - k_1^2)\left(k^2 + k_1^2 - \dfrac{3\alpha}{20\beta}\right)}\frac{\partial}{\partial k}\right)^n \frac{f(k)}{k^2 - k_1^2},$$

$$|\phi| < -128\beta k_1^5 + 16\alpha k_1^3. \tag{3.35}$$

Equations (3.33) and (3.35) imply that

$$\left|\frac{d^n}{d\phi^n}F(\phi)\right| \leqslant C, \quad |\phi| < -128\beta k_1^5 + 16\alpha k_1^3, \quad n = 0, 1, 2.$$

Therefore, $F(\phi)$ satisfies (3.23). On the other hand, (3.24) and (3.34) imply

$$(k^2 - k_1^2)\int_{\mathbf{R}} \hat{F}(s)e^{s\Phi(k)}ds = \begin{cases} f(k), & |k| < k_1, \\ 0, & |k| \geqslant k_1. \end{cases} \tag{3.36}$$

Letting

$$f(k) = f_a(x, t, k) + f_r(x, t, k), \quad t > 0, \, |k| < k_1, \, \xi \in \mathcal{I},$$

where the functions f_a and f_r are defined by

$$f_a(x, t, k) = (k^2 - k_1^2)\int_{-\frac{t}{4}}^{\infty} \hat{F}(s)e^{s\Phi(k)}ds, \quad t > 0, \, k \in \Omega_1^3,$$

$$f_r(x, t, k) = (k^2 - k_1^2)\int_{-\infty}^{-\frac{t}{4}} \hat{F}(s)e^{s\Phi(k)}ds, \quad t > 0, \, |k| < k_1,$$

we infer that $f_a(x,t,\cdot)$ is continuous in $\bar{\Omega}_1^3$ and analytic in Ω_1^3. Moreover, we can get from (3.26) and (3.27) that

$$|f_a(x,t,k)| \leqslant C|k^2 - k_1^2|e^{\frac{t}{4}|\text{Re}\Phi(k)|}, \quad t>0, \ k \in \bar{\Omega}_1^3, \ \xi \in \mathcal{I},$$

$$|f_r(x,t,k)| \leqslant Ct^{-3/2}, \quad t>0, \ |k|<k_1, \ \xi \in \mathcal{I}.$$

It follows from

$$r_{1,a}(x,t,k) = f_0(k) + f_a(x,t,k), \quad r_{1,r}(x,t,k) = f_r(x,t,k)$$

that one gets a decomposition of $r_1(k)$ for $|k|<k_1$ with the properties listed in the statement of the lemma. Thus, we find a decomposition of $r_1(k)$ for $|k|>k_2$ and $|k|<k_1$ with the properties listed in the statement of the lemma. The decomposition of the function $r_3(k)$ can be obtained by a similar procedure as the decomposition of $r_1(k)$ for $|k|<k_1$.

Finally, the decompositions of $r_2(k)$ and $r_4(k)$ can be obtained from $r_3(k)$ and $r_1(k)$ by Schwartz conjugation.

The purpose of the next transformation is to deform the contour so that the jump matrix involves the exponential factor $e^{-t\Phi}$ on the parts of the contour where $\text{Re}\Phi$ is positive and the factor $e^{t\Phi}$ on the parts where $\text{Re}\Phi$ is negative. More precisely, we put

$$M^{(2)}(x,t;k) = M^{(1)}(x,t;k)G(k), \tag{3.37}$$

where

$$G(k) = \begin{cases} \begin{pmatrix} 1 & 0 \\ -r_{1,a}\delta^{-2}e^{t\Phi} & 1 \end{pmatrix}, & k \in D_1, \\ \begin{pmatrix} 1 & -r_{2,a}\delta^2 e^{-t\Phi} \\ 0 & 1 \end{pmatrix}, & k \in D_2, \\ \begin{pmatrix} 1 & 0 \\ r_{3,a}\delta^{-2}e^{t\Phi} & 1 \end{pmatrix}, & k \in D_3, \\ \begin{pmatrix} 1 & r_{4,a}\delta^2 e^{-t\Phi} \\ 0 & 1 \end{pmatrix}, & k \in D_4, \\ I, & k \in D_5 \cup D_6. \end{cases} \tag{3.38}$$

Then the matrix $M^{(2)}(x,t;k)$ satisfies the following RH problem

$$M_+^{(2)}(x,t;k) = M_-^{(2)}(x,t;k)J^{(2)}(x,t;k), \quad k \in \Sigma, \tag{3.39}$$

with the jump matrix $J^{(2)} = G_-^{-1}(k)J^{(1)}G_+(k)$ is given by

$$J_1^{(2)} = \begin{pmatrix} 1 & 0 \\ r_{1,a}\delta^{-2}e^{t\Phi} & 1 \end{pmatrix}, \quad J_2^{(2)} = \begin{pmatrix} 1 & -r_{2,a}\delta^2 e^{-t\Phi} \\ 0 & 1 \end{pmatrix},$$

$$J_3^{(2)} = \begin{pmatrix} 1 & 0 \\ -r_{3,a}\delta^{-2}e^{t\Phi} & 1 \end{pmatrix}, \quad J_4^{(2)} = \begin{pmatrix} 1 & r_{4,a}\delta^2 e^{-t\Phi} \\ 0 & 1 \end{pmatrix},$$

$$J_5^{(2)} = \begin{pmatrix} 1 & r_{4,r}\delta^2 e^{-t\Phi} \\ 0 & 1 \end{pmatrix} \begin{pmatrix} 1 & 0 \\ r_{1,r}\delta^{-2}e^{t\Phi} & 1 \end{pmatrix},$$

$$J_6^{(2)} = \begin{pmatrix} 1 & 0 \\ r_{3,r}\delta_-^{-2}e^{t\Phi} & 1 \end{pmatrix} \begin{pmatrix} 1 & r_{2,r}\delta_+^2 e^{-t\Phi} \\ 0 & 1 \end{pmatrix},$$

(3.40)

with $J_i^{(2)}$ denoting the restriction of $J^{(2)}$ to the contour Σ labeled by i in Fig. 5. It is easy to see that the jump matrix $J^{(2)}$ decays to identity matrix I as $t \to \infty$ everywhere except near the critical points $\pm k_1$ and $\pm k_2$. This implies that we only need to consider a neighborhood of the critical points $\pm k_1$ and $\pm k_2$ when we study the long-time asymptotics of $M^{(2)}(x,t;k)$ in terms of the corresponding RH problem.

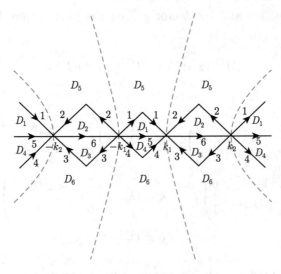

Fig. 5 The oriented contour Σ and the open sets $\{D_j\}_1^6$ in the complex k-plane

3.2 Local models near the critical points $\pm k_1$ and $\pm k_2$

We introduce the following scaling operators

$$S_{-k_2}: k \mapsto \frac{z}{4\sqrt{t(40\beta k_2^3 - 3\alpha k_2)}} - k_2,$$

$$S_{-k_1}: k \mapsto \frac{z}{4\sqrt{t(3\alpha k_1 - 40\beta k_1^3)}} - k_1,$$

$$S_{k_1}: k \mapsto \frac{z}{4\sqrt{t(3\alpha k_1 - 40\beta k_1^3)}} + k_1, \quad (3.41)$$

$$S_{k_2}: k \mapsto \frac{z}{4\sqrt{t(40\beta k_2^3 - 3\alpha k_2)}} + k_2.$$

Integrating by parts in formula (3.9) yields,

$$\delta(k) = \left(\frac{k-k_2}{k-k_1}\right)^{-i\nu(k_1)} \left(\frac{k+k_1}{k+k_2}\right)^{-i\nu(k_1)} e^{\chi_1(k)}$$
$$= \left(\frac{k-k_2}{k-k_1}\right)^{-i\nu(k_2)} \left(\frac{k+k_1}{k+k_2}\right)^{-i\nu(k_2)} e^{\chi_2(k)}, \quad (3.42)$$

where

$$\nu(k_1) = \frac{1}{2\pi} \ln(1 + |r(k_1)|^2) > 0, \quad (3.43)$$

$$\chi_1(k) = \frac{1}{2\pi i} \int_{k_1}^{k_2} \ln\left(\frac{1+|r(s)|^2}{1+|r(k_1)|^2}\right) \left(\frac{1}{s-k} - \frac{1}{s+k}\right) ds, \quad (3.44)$$

$$\nu(k_2) = \frac{1}{2\pi} \ln(1 + |r(k_2)|^2) > 0, \quad (3.45)$$

$$\chi_2(k) = \frac{1}{2\pi i} \int_{k_1}^{k_2} \ln\left(\frac{1+|r(s)|^2}{1+|r(k_2)|^2}\right) \left(\frac{1}{s-k} - \frac{1}{s+k}\right) ds. \quad (3.46)$$

Hence, we have

$$S_{-k_2}(\delta(k)e^{-\frac{t\Phi(k)}{2}}) = \delta^0_{-k_2}(z)\delta^1_{-k_2}(z),$$
$$S_{-k_1}(\delta(k)e^{-\frac{t\Phi(k)}{2}}) = \delta^0_{-k_1}(z)\delta^1_{-k_1}(z),$$
$$S_{k_1}(\delta(k)e^{-\frac{t\Phi(k)}{2}}) = \delta^0_{k_1}(z)\delta^1_{k_1}(z),$$
$$S_{k_2}(\delta(k)e^{-\frac{t\Phi(k)}{2}}) = \delta^0_{k_2}(z)\delta^1_{k_2}(z),$$

with

$$\delta^0_{-k_2}(z) = \left(16t(k_2-k_1)^2(40\beta k_2^3 - 3\alpha k_2)\right)^{-\frac{i\nu(k_2)}{2}} \left(\frac{2k_2}{k_1+k_2}\right)^{-i\nu(k_2)}$$

$$\times e^{\chi_2(-k_2)}e^{-8ik_2^3 t(8\beta k_2^2-\alpha)}, \tag{3.47}$$

$$\delta^1_{-k_2}(z) = (-z)^{i\nu(k_2)} e^{(\chi_2(z/\sqrt{16tk_2(40\beta k_2^2-3\alpha)})-k_2)-\chi_2(-k_2))}$$

$$\times \left(\frac{-z/\sqrt{16tk_2(40\beta k_2^2-3\alpha)} - k_1 + k_2}{k_2 - k_1} \right)^{-i\nu(k_2)}$$

$$\times \left(\frac{k_1 + k_2}{-z/\sqrt{16tk_2(40\beta k_2^2-3\alpha)} + k_1 + k_2} \right)^{-i\nu(k_2)}$$

$$\times \left(\frac{-z/\sqrt{16tk_2(40\beta k_2^2-3\alpha)} + 2k_2}{2k_2} \right)^{-i\nu(k_2)}$$

$$\times \exp\left(\frac{iz^2}{4}\left[1 - \frac{(40\beta k_2^2-\alpha)z}{4\sqrt{t}(40\beta k_2^3-3\alpha k_2)^{3/2}} + \frac{5\beta z^2}{4tk_2(40\beta k_2^2-3\alpha)^2} \right.\right.$$

$$\left.\left. - \frac{\beta z^3}{16t^{3/2}(40\beta k_2^3-3\alpha k_2)^{5/2}} \right] \right), \tag{3.48}$$

$$\delta^0_{-k_1}(z) = \left(16t(k_2-k_1)^2(3\alpha k_1 - 40\beta k_1^3) \right)^{\frac{i\nu(k_1)}{2}} \left(\frac{k_1+k_2}{2k_1} \right)^{-i\nu(k_1)}$$

$$\times e^{\chi_1(-k_1)} e^{-8ik_1^3 t(8\beta k_1^2-\alpha)}, \tag{3.49}$$

$$\delta^1_{-k_1}(z) = z^{-i\nu(k_1)} e^{(\chi_1(z/\sqrt{16tk_1(3\alpha-40\beta k_1^2)})-k_1)-\chi_1(-k_1))}$$

$$\times \left(\frac{k_2 - k_1}{z/\sqrt{16tk_1(3\alpha-40\beta k_1^2)} - k_1 + k_2} \right)^{-i\nu(k_1)}$$

$$\times \left(\frac{-z/\sqrt{16tk_1(3\alpha-40\beta k_1^2)} + k_1 + k_2}{k_1 + k_2} \right)^{-i\nu(k_1)}$$

$$\times \left(\frac{2k_1}{-z/\sqrt{16tk_1(3\alpha-40\beta k_1^2)} + 2k_1} \right)^{-i\nu(k_1)}$$

$$\times \exp\left(-\frac{iz^2}{4}\left[1 + \frac{(40\beta k_1^2-\alpha)z}{4\sqrt{t}(3\alpha k_1-40\beta k_1^3)^{3/2}} - \frac{5\beta z^2}{4tk_1(3\alpha-40\beta k_1^2)^2} \right.\right.$$

$$\left.\left. + \frac{\beta z^3}{16t^{3/2}(3\alpha k_1-40\beta k_1^3)^{5/2}} \right] \right), \tag{3.50}$$

$$\delta^0_{k_1}(z) = \left(16t(k_2-k_1)^2(3\alpha k_1 - 40\beta k_1^3) \right)^{-\frac{i\nu(k_1)}{2}} \left(\frac{2k_1}{k_1+k_2} \right)^{-i\nu(k_1)}$$

$$\times e^{\chi_1(k_1)} e^{8ik_1^3 t(8\beta k_1^2-\alpha)}, \tag{3.51}$$

$$\delta_{k_1}^1(z) = (-z)^{i\nu(k_1)} e^{(\chi_1(z/\sqrt{16tk_1(3\alpha-40\beta k_1^2)}+k_1)-\chi_1(k_1))}$$

$$\times \left(\frac{-z/\sqrt{16tk_1(3\alpha - 40\beta k_1^2)} - k_1 + k_2}{k_2 - k_1} \right)^{-i\nu(k_1)}$$

$$\times \left(\frac{k_1 + k_2}{z/\sqrt{16tk_1(3\alpha - 40\beta k_1^2)} + k_1 + k_2} \right)^{-i\nu(k_1)}$$

$$\times \left(\frac{z/\sqrt{16tk_1(3\alpha - 40\beta k_1^2)} + 2k_1}{2k_1} \right)^{-i\nu(k_1)}$$

$$\times \exp\left(\frac{iz^2}{4} \left[1 - \frac{(40\beta k_1^2 - \alpha)z}{4\sqrt{t}(3\alpha k_1 - 40\beta k_1^3)^{3/2}} - \frac{5\beta z^2}{4tk_1(3\alpha - 40\beta k_1^2)^2} \right. \right.$$

$$\left. \left. - \frac{\beta z^3}{16t^{3/2}(3\alpha k_1 - 40\beta k_1^3)^{5/2}} \right] \right), \tag{3.52}$$

$$\delta_{k_2}^0(z) = \left(16t(k_2 - k_1)^2 (40\beta k_2^3 - 3\alpha k_2) \right)^{\frac{i\nu(k_2)}{2}} \left(\frac{k_1 + k_2}{2k_2} \right)^{-i\nu(k_2)}$$

$$\times e^{\chi_2(k_2)} e^{8ik_2^3 t(8\beta k_2^2 - \alpha)}, \tag{3.53}$$

$$\delta_{k_2}^1(z) = z^{-i\nu(k_2)} e^{(\chi_2(z/\sqrt{16tk_2(40\beta k_2^2-3\alpha)}+k_2)-\chi_2(k_2))}$$

$$\times \left(\frac{k_2 - k_1}{z/\sqrt{16tk_2(40\beta k_2^2 - 3\alpha)} - k_1 + k_2} \right)^{-i\nu(k_2)}$$

$$\times \left(\frac{2k_2}{z/\sqrt{16tk_2(40\beta k_2^2 - 3\alpha)} + 2k_2} \right)^{-i\nu(k_2)}$$

$$\times \left(\frac{z/\sqrt{16tk_2(40\beta k_2^2 - 3\alpha)} + k_1 + k_2}{k_1 + k_2} \right)^{-i\nu(k_2)}$$

$$\times \exp\left(-\frac{iz^2}{4} \left[1 + \frac{(40\beta k_2^2 - \alpha)z}{4\sqrt{t}(40\beta k_2^3 - 3\alpha k_2)^{3/2}} + \frac{5\beta z^2}{4tk_2(40\beta k_2^2 - 3\alpha)^2} \right. \right.$$

$$\left. \left. + \frac{\beta z^3}{16t^{3/2}(40\beta k_2^3 - 3\alpha k_2)^{5/2}} \right] \right). \tag{3.54}$$

For $j = 1, 2$, let $D_\varepsilon(\pm k_j)$ denote the open disk of radius ε centered at $\pm k_j$ for a small $\varepsilon > 0$. Now we define

$$\tilde{M}(x, t; z) = M^{(2)}(x, t; k)(\delta_{-k_2}^0)^{\sigma_3}(z), \quad k \in D_\varepsilon(-k_2) \setminus \Sigma,$$

$$\check{M}(x, t; z) = M^{(2)}(x, t; k)(\delta_{-k_1}^0)^{\sigma_3}(z), \quad k \in D_\varepsilon(-k_1) \setminus \Sigma,$$

$$\check{\tilde{M}}(x,t;z) = M^{(2)}(x,t;k)(\delta^0_{k_1})^{\sigma_3}(z), \qquad k \in D_\varepsilon(k_1) \setminus \Sigma,$$
$$\tilde{M}(x,t;z) = M^{(2)}(x,t;k)(\delta^0_{k_2})^{\sigma_3}(z), \qquad k \in D_\varepsilon(k_2) \setminus \Sigma.$$

Then $\check{M}$, $\check{\tilde{M}}$, $\tilde{\tilde{M}}$, and $\tilde{M}$ are the sectionally analytic functions of z which satisfy

$$\tilde{\tilde{M}}_+(x,t;z) = \tilde{\tilde{M}}_-(x,t;z)\tilde{\tilde{J}}(x,t;z), \qquad k \in \mathcal{X}^\varepsilon_{-k_2},$$
$$\check{\tilde{M}}_+(x,t;z) = \check{\tilde{M}}_-(x,t;z)\check{\tilde{J}}(x,t;z), \qquad k \in \mathcal{X}^\varepsilon_{-k_1},$$
$$\check{M}_+(x,t;z) = \check{M}_-(x,t;z)\check{J}(x,t;z), \qquad k \in \mathcal{X}^\varepsilon_{k_1},$$
$$\tilde{M}_+(x,t;z) = \tilde{M}_-(x,t;z)\tilde{J}(x,t;z), \qquad k \in \mathcal{X}^\varepsilon_{k_2},$$

where $\mathcal{X}_{\pm k_j} = X \pm k_j$ denotes the cross X defined by (2.19) centered at $\pm k_j$ and $\mathcal{X}^\varepsilon_{\pm k_j} = \mathcal{X}_{\pm k_j} \cap D_\varepsilon(\pm k_j)$ for $j = 1, 2$. Moreover, the corresponding jump matrices are given by

$$\tilde{\tilde{J}}(x,t;z) = \begin{cases} \begin{pmatrix} 1 & r_{2,a}(\delta^1_{-k_2})^2 \\ 0 & 1 \end{pmatrix}, & k \in (\mathcal{X}^\varepsilon_{-k_2})_1, \\[2mm] \begin{pmatrix} 1 & 0 \\ -r_{1,a}(\delta^1_{-k_2})^{-2} & 1 \end{pmatrix}, & k \in (\mathcal{X}^\varepsilon_{-k_2})_2, \\[2mm] \begin{pmatrix} 1 & -r_{4,a}(\delta^1_{-k_2})^2 \\ 0 & 1 \end{pmatrix}, & k \in (\mathcal{X}^\varepsilon_{-k_2})_3, \\[2mm] \begin{pmatrix} 1 & 0 \\ r_{3,a}(\delta^1_{-k_2})^{-2} & 1 \end{pmatrix}, & k \in (\mathcal{X}^\varepsilon_{-k_2})_4, \end{cases} \quad (3.55)$$

$$\check{\tilde{J}}(x,t;z) = \begin{cases} \begin{pmatrix} 1 & 0 \\ r_{1,a}(\delta^1_{-k_1})^{-2} & 1 \end{pmatrix}, & k \in (\mathcal{X}^\varepsilon_{-k_1})_1, \\[2mm] \begin{pmatrix} 1 & -r_{2,a}(\delta^1_{-k_1})^2 \\ 0 & 1 \end{pmatrix}, & k \in (\mathcal{X}^\varepsilon_{-k_1})_2, \\[2mm] \begin{pmatrix} 1 & 0 \\ -r_{3,a}(\delta^1_{-k_1})^{-2} & 1 \end{pmatrix}, & k \in (\mathcal{X}^\varepsilon_{-k_1})_3, \\[2mm] \begin{pmatrix} 1 & r_{4,a}(\delta^1_{-k_1})^2 \\ 0 & 1 \end{pmatrix}, & k \in (\mathcal{X}^\varepsilon_{-k_1})_4, \end{cases} \quad (3.56)$$

$$\check{J}(x,t;z) = \begin{cases} \begin{pmatrix} 1 & r_{2,a}(\delta^1_{k_1})^2 \\ 0 & 1 \end{pmatrix}, & k \in (\mathcal{X}^\varepsilon_{k_1})_1, \\ \begin{pmatrix} 1 & 0 \\ -r_{1,a}(\delta^1_{k_1})^{-2} & 1 \end{pmatrix}, & k \in (\mathcal{X}^\varepsilon_{k_1})_2, \\ \begin{pmatrix} 1 & -r_{4,a}(\delta^1_{k_1})^2 \\ 0 & 1 \end{pmatrix}, & k \in (\mathcal{X}^\varepsilon_{k_1})_3, \\ \begin{pmatrix} 1 & 0 \\ r_{3,a}(\delta^1_{k_1})^{-2} & 1 \end{pmatrix}, & k \in (\mathcal{X}^\varepsilon_{k_1})_4, \end{cases} \tag{3.57}$$

and

$$\tilde{J}(x,t;z) = \begin{cases} \begin{pmatrix} 1 & 0 \\ r_{1,a}(\delta^1_{k_2})^{-2} & 1 \end{pmatrix}, & k \in (\mathcal{X}^\varepsilon_{k_2})_1, \\ \begin{pmatrix} 1 & -r_{2,a}(\delta^1_{k_2})^2 \\ 0 & 1 \end{pmatrix}, & k \in (\mathcal{X}^\varepsilon_{k_2})_2, \\ \begin{pmatrix} 1 & 0 \\ -r_{3,a}(\delta^1_{k_2})^{-2} & 1 \end{pmatrix}, & k \in (\mathcal{X}^\varepsilon_{k_2})_3, \\ \begin{pmatrix} 1 & r_{4,a}(\delta^1_{k_2})^2 \\ 0 & 1 \end{pmatrix}, & k \in (\mathcal{X}^\varepsilon_{k_2})_4. \end{cases} \tag{3.58}$$

For the jump matrix $\tilde{J}(x,t;z)$, define

$$q = r(k_2),$$

then for any fixed $z \in X$, we have $k(z) \to k_2$ as $t \to \infty$. Hence,

$$r_{1,a}(k) \to q, \quad r_{2,a}(k) \to \frac{\bar{q}}{1+|q|^2}, \quad \delta^1_{k_2} \to e^{-\frac{iz^2}{4}} z^{-i\nu(q)}.$$

This implies that the jump matrix $\tilde{J}$ tends to the matrix J^X defined in (2.21) for large t. In other words, the jumps of $M^{(2)}$ for k near k_2 approach those of the function $M^X(\delta^0_{k_2})^{-\sigma_3}$ as $t \to \infty$. Therefore, we can approximate $M^{(2)}$ in the neighborhood $D_\varepsilon(k_2)$ of k_2 by

$$M^{(k_2)}(x,t;k) = (\delta^0_{k_2})^{\sigma_3} M^X(q,z)(\delta^0_{k_2})^{-\sigma_3}, \tag{3.59}$$

where $M^X(q,z)$ is given by (2.22). On the other hand, according to the symmetry property (3.3), one can deduce that

$$\tilde{\tilde{J}}(x,t;z) \to \overline{J^X(q,-\bar{z})}, \quad \text{as } t \to \infty.$$

Thus, by uniqueness of the RH problem, we can approximate $M^{(2)}$ in the neighborhood $D_\varepsilon(-k_2)$ of $-k_2$ by

$$M^{(-k_2)}(x,t;k) = (\delta^0_{-k_2})^{\sigma_3} \overline{M^X(q,-\bar{z})} (\delta^0_{-k_2})^{-\sigma_3}. \tag{3.60}$$

For the case of $\check{J}(x,t;z)$, as $t \to \infty$, we find

$$r_{1,a}(k) \to r(k_1), \quad r_{2,a}(k) \to \frac{\overline{r(k_1)}}{1+|r(k_1)|^2}, \quad \delta^1_{k_1} \to e^{\frac{iz^2}{4}}(-z)^{i\nu(k_1)}.$$

This implies as $t \to \infty$ that

$$\check{J}(x,t;z) \to J^Y(p,z) = \begin{cases} \begin{pmatrix} 1 & \frac{\bar{p}}{1+|p|^2} e^{\frac{iz^2}{2}}(-z)^{2i\nu(p)} \\ 0 & 1 \end{pmatrix}, & z \in X_1, \\ \begin{pmatrix} 1 & 0 \\ -p e^{-\frac{iz^2}{2}}(-z)^{-2i\nu(p)} & 1 \end{pmatrix}, & z \in X_2, \\ \begin{pmatrix} 1 & -\bar{p} e^{\frac{iz^2}{2}}(-z)^{2i\nu(p)} \\ 0 & 1 \end{pmatrix}, & z \in X_3, \\ \begin{pmatrix} 1 & 0 \\ \frac{p}{1+|p|^2} e^{-\frac{iz^2}{2}}(-z)^{-2i\nu(p)} & 1 \end{pmatrix}, & z \in X_4, \end{cases}$$

if set

$$p = r(k_1).$$

It is easy to verify that

$$J^Y(p,z) = \overline{J^X(\bar{p},-\bar{z})},$$

which in turn implies that

$$M^Y(p,z) = \overline{M^X(\bar{p},-\bar{z})}, \tag{3.61}$$

where $M^Y(p,z)$ is the unique solution of the following RH problem

$$\begin{cases} M_+^Y(p,z) = M_-^Y(p,z) J^Y(p,z), & \text{for almost every } z \in X, \\ M^Y(p,z) \to I, & \text{as } z \to \infty. \end{cases}$$

Therefore, we find that

$$M^Y(p,z) = I - \frac{i}{z}\begin{pmatrix} 0 & \beta^Y(p) \\ \overline{\beta^Y(p)} & 0 \end{pmatrix} + O\left(\frac{p}{z^2}\right),$$

where

$$\beta^Y(p) = \sqrt{\nu(p)} e^{-i(\frac{\pi}{4} + \arg p + \arg \Gamma(-i\nu(p)))}.$$

As a consequence, we can approximate $M^{(2)}(x,t;k)$ in the neighborhood $D_\varepsilon(k_1)$ of k_1 by

$$M^{(k_1)}(x,t;k) = (\delta_{k_1}^0)^{\sigma_3} M^Y(p,z)(\delta_{k_1}^0)^{-\sigma_3}. \tag{3.62}$$

Using again (3.3), we can use

$$M^{(-k_1)}(x,t;k) = (\delta_{-k_1}^0)^{\sigma_3} \overline{M^Y(p,-\bar z)}(\delta_{-k_1}^0)^{-\sigma_3} \tag{3.63}$$

to approximate $M^{(2)}(x,t;k)$ in the neighborhood $D_\varepsilon(-k_1)$ of $-k_1$.

Lemma 3.3 *For each $t > 0$, $\xi \in \mathcal{I}$ and $j = 1,2$, the functions $M^{(\pm k_j)}(x,t;k)$ defined in (3.62), (3.63), (3.59) and (3.60) are analytic functions of $k \in D_\varepsilon(\pm k_j) \setminus \mathcal{X}_{\pm k_j}^\varepsilon$. Furthermore,*

$$|M^{(\pm k_j)}(x,t;k) - I| \leqslant C, \quad t > 3, \ \xi \in \mathcal{I}, \ k \in \overline{D_\varepsilon(\pm k_j)} \setminus \mathcal{X}_{\pm k_j}^\varepsilon, \ j = 1,2. \tag{3.64}$$

Across $\mathcal{X}_{\pm k_j}^\varepsilon$, $M^{(\pm k_j)}(x,t;k)$ satisfied the jump condition $M_+^{(\pm k_j)} = M_-^{(\pm k_j)} J^{(\pm k_j)}$, where the jump matrix $J^{(\pm k_j)}$ satisfy the following estimates for $1 \leqslant n \leqslant \infty$:

$$\|J^{(2)} - J^{(\pm k_j)}\|_{L^n(\mathcal{X}_{\pm k_j}^\varepsilon)} \leqslant C t^{-\frac{1}{2} - \frac{1}{2n}} \ln t, \quad t > 3, \ \xi \in \mathcal{I}, \ j = 1,2, \tag{3.65}$$

where $C > 0$ is a constant independent of t, ξ, k. Moreover, as $t \to \infty$,

$$\|(M^{(\pm k_j)})^{-1}(x,t;k) - I\|_{L^\infty(\partial D_\varepsilon(\pm k_j))} = O(t^{-1/2}), \tag{3.66}$$

and

$$\frac{1}{2\pi i} \int_{\partial D_\varepsilon(k_1)} ((M^{(k_1)})^{-1}(x,t;k) - I) dk = -\frac{(\delta_{k_1}^0)^{\hat\sigma_3} M_1^Y(\xi)}{4\sqrt{t k_1(3\alpha - 40\beta k_1^2)}} + O(t^{-1}), \tag{3.67}$$

$$\frac{1}{2\pi i}\int_{\partial D_\varepsilon(-k_1)}((M^{(-k_1)})^{-1}(x,t;k)-I)dk = \frac{(\delta^0_{-k_1})^{\hat\sigma_3}\overline{M_1^Y(\xi)}}{4\sqrt{tk_1(3\alpha-40\beta k_1^2)}}+O(t^{-1}),$$
(3.68)

$$\frac{1}{2\pi i}\int_{\partial D_\varepsilon(k_2)}((M^{(k_2)})^{-1}(x,t;k)-I)dk = -\frac{(\delta^0_{k_2})^{\hat\sigma_3}M_1^X(\xi)}{4\sqrt{tk_2(40\beta k_2^2-3\alpha)}}+O(t^{-1}),$$
(3.69)

$$\frac{1}{2\pi i}\int_{\partial D_\varepsilon(-k_2)}((M^{(-k_2)})^{-1}(x,t;k)-I)dk = \frac{(\delta^0_{-k_2})^{\hat\sigma_3}\overline{M_1^X(\xi)}}{4\sqrt{tk_2(40\beta k_2^2-3\alpha)}}+O(t^{-1}),$$
(3.70)

where $M_1^X(\xi)$ and $M_1^Y(\xi)$ are given by

$$M_1^X(\xi) = -i\begin{pmatrix} 0 & \beta^X(q) \\ \overline{\beta^X(q)} & 0 \end{pmatrix}, \quad M_1^Y(\xi) = -i\begin{pmatrix} 0 & \beta^Y(p) \\ \overline{\beta^Y(p)} & 0 \end{pmatrix}. \quad (3.71)$$

Proof We just consider the proof for the function $M^{(k_2)}(x,t;k)$, the others accordingly follow.

The analyticity of $M^{(k_2)}$ is obvious. Since $|\delta^0_{k_2}(z)|=1$, thus, the estimate (3.64) for $M^{(k_2)}$ follows from the definition of $M^{(k_2)}$ in (3.59) and the estimate (3.25). On the other hand, we have

$$J^{(2)}-J^{(k_2)}=(\delta^0_{k_2})^{\hat\sigma_3}(\tilde J-J^X),\quad k\in \mathcal{X}^\varepsilon_{k_2}.$$

However, proceeding the similar calculation as the Lemma 3.35 in [10] (also can see [9,29]), we have

$$\|\tilde J-J^X\|_{L^\infty((\mathcal{X}^\varepsilon_{k_2})_1)} \leqslant C|e^{\frac{i\gamma}{2}z^2}|t^{-1/2}\ln t,\quad 0<\gamma<\frac{1}{2},\ t>3,\ \xi\in\mathcal{I}, \quad (3.72)$$

for $k\in(\mathcal{X}^\varepsilon_{k_2})_1$, that is, $z=4\sqrt{tk_2(40\beta k_2^2-3\alpha)}se^{\frac{i\pi}{4}}$, $0\leqslant s\leqslant\varepsilon$. Thus,

$$\|\tilde J-J^X\|_{L^1((\mathcal{X}^\varepsilon_{k_2})_1)} \leqslant Ct^{-1}\ln t,\quad t>3,\ \xi\in\mathcal{I}. \quad (3.73)$$

By the general inequality $\|f\|_{L^n}\leqslant \|f\|_{L^\infty}^{1-1/n}\|f\|_{L^1}^{1/n}$, we find

$$\|\tilde J-J^X\|_{L^n((\mathcal{X}^\varepsilon_{k_2})_1)} \leqslant Ct^{-1/2-1/2n}\ln t,\quad t>3,\ \xi\in\mathcal{I}. \quad (3.74)$$

The norms on $(\mathcal{X}^\varepsilon_{k_2})_j$, $j=2,3,4$, are estimated in a similar way. Therefore, (3.65) follows.

If $k \in \partial D_\varepsilon(k_2)$, the variable $z = 4\sqrt{tk_2(40\beta k_2^2 - 3\alpha)}(k-k_2)$ tends to infinity as $t \to \infty$. It follows from (2.22) that

$$M^X(q,z) = I + \frac{M_1^X(\xi)}{4\sqrt{tk_2(40\beta k_2^2 - 3\alpha)}(k-k_2)} + O\left(\frac{q}{t}\right), \quad t \to \infty, \ k \in \partial D_\varepsilon(k_2),$$

where $M_1^X(\xi)$ is defined by (3.71). Since

$$M^{(k_2)}(x,t;k) = (\delta_{k_2}^0)^{\hat\sigma_3} M^X(q,z),$$

thus we have

$$(M^{(k_2)})^{-1}(x,t;k) - I = -\frac{(\delta_{k_2}^0)^{\hat\sigma_3} M_1^X(\xi)}{4\sqrt{tk_2(40\beta k_2^2 - 3\alpha)}(k-k_2)} + O\left(\frac{q}{t}\right), \quad k \in \partial D_\varepsilon(k_2). \tag{3.75}$$

The estimate (3.66) immediately follows from (3.75) and $|M_1^X| \leqslant C$. By Cauchy's formula and (3.75), we derive (3.69).

3.3 The final step

We now begin to establish the explicit long-time asymptotic formula for the emKdV equation (1.2) on the line. Define the approximate solution $M^{(app)}(x,t;k)$ by

$$M^{(app)} = \begin{cases} M^{(-k_2)}, & k \in D_\varepsilon(-k_2), \\ M^{(-k_1)}, & k \in D_\varepsilon(-k_1), \\ M^{(k_1)}, & k \in D_\varepsilon(k_1), \\ M^{(k_2)}, & k \in D_\varepsilon(k_2), \\ I, & \text{elsewhere.} \end{cases} \tag{3.76}$$

Let $\hat{M}(x,t;k)$ be

$$\hat{M} = M^{(2)}(M^{(app)})^{-1}, \tag{3.77}$$

then $\hat{M}(x,t;k)$ satisfies the following RH problem

$$\hat{M}_+(x,t;k) = \hat{M}_-(x,t;k)\hat{J}(x,t;k), \quad k \in \hat{\Sigma}, \tag{3.78}$$

where the jump contour $\hat{\Sigma} = \Sigma \cup \partial D_\varepsilon(-k_2) \cup \partial D_\varepsilon(-k_1) \cup \partial D_\varepsilon(k_1) \cup \partial D_\varepsilon(k_2)$ is depicted in Fig. 6, and the jump matrix $\hat{J}(x,t;k)$ is given by

$$\hat{J} = \begin{cases} M_-^{(app)} J^{(2)} (M_+^{(app)})^{-1}, & k \in \hat{\Sigma} \cap (D_\varepsilon(-k_2) \cup D_\varepsilon(-k_1) \cup D_\varepsilon(k_1) \cup D_\varepsilon(k_2)), \\ (M^{(app)})^{-1}, & k \in (\partial D_\varepsilon(-k_2) \cup \partial D_\varepsilon(-k_1) \cup \partial D_\varepsilon(k_1) \cup \partial D_\varepsilon(k_2)), \\ J^{(2)}, & k \in \hat{\Sigma} \setminus (\overline{D_\varepsilon(-k_2)} \cup \overline{D_\varepsilon(-k_1)} \cup \overline{D_\varepsilon(k_1)} \cup \overline{D_\varepsilon(k_2)}). \end{cases} \tag{3.79}$$

For convenience, we rewrite $\hat{\Sigma}$ as follows:
$$\hat{\Sigma} = \hat{\Sigma}_1 \cup \hat{\Sigma}_2 \cup \hat{\Sigma}_3 \cup \hat{\Sigma}_4,$$
where
$$\hat{\Sigma}_1 = \bigcup_1^4 \Sigma_j \setminus (D_\varepsilon(-k_2) \cup D_\varepsilon(-k_1) \cup D_\varepsilon(k_1) \cup D_\varepsilon(k_2)), \quad \hat{\Sigma}_2 = \bigcup_5^6 \Sigma_j,$$
$$\hat{\Sigma}_3 = \partial D_\varepsilon(-k_2) \cup \partial D_\varepsilon(-k_1) \cup \partial D_\varepsilon(k_1) \cup \partial D_\varepsilon(k_2),$$
$$\hat{\Sigma}_4 = \mathcal{X}^\varepsilon_{-k_2} \cup \mathcal{X}^\varepsilon_{-k_1} \cup \mathcal{X}^\varepsilon_{k_1} \cup \mathcal{X}^\varepsilon_{k_2},$$

and $\{\Sigma_j\}_1^6$ denoting the restriction of Σ to the contour labeled by j in Fig. 5. Then we have the following lemma if let $\hat{w} = \hat{J} - I$.

Lemma 3.4 *For $1 \leqslant n \leqslant \infty$, $t > 3$ and $\xi \in \mathcal{I}$, the following estimates hold:*

$$\|\hat{w}\|_{L^n(\hat{\Sigma}_1)} \leqslant Ce^{-ct}, \tag{3.80}$$

$$\|\hat{w}\|_{L^n(\hat{\Sigma}_2)} \leqslant Ct^{-3/2}, \tag{3.81}$$

$$\|\hat{w}\|_{L^n(\hat{\Sigma}_3)} \leqslant Ct^{-1/2}, \tag{3.82}$$

$$\|\hat{w}\|_{L^n(\hat{\Sigma}_4)} \leqslant Ct^{-\frac{1}{2}-\frac{1}{2n}} \ln t. \tag{3.83}$$

Proof For $k \in \Omega_1 \cap \{k \in \mathbf{C} | \text{Re}\, k > k_2\} \cap \hat{\Sigma}_1$, we have $-|\text{Re}\Phi(k)| \leqslant -c\varepsilon^2$. Since $\hat{w}$ only has a nonzero $r_{1,a}\delta^{-2}e^{t\Phi}$ in (21) entry, hence, for $t \geqslant 1$, by (3.15), we get

$$|\hat{w}_{21}| = |r_{1,a}\delta^{-2}e^{t\Phi}| \leqslant \frac{C}{1+|k|^2}e^{-\frac{3t}{4}|\text{Re}\Phi|} \leqslant Ce^{-c\varepsilon^2 t}.$$

In a similar way, the other estimates on $\hat{\Sigma}_1$ hold. This proves (3.80). Since the matrix $\hat{w}$ on $\hat{\Sigma}_2$ only involves the small remainder $r_{j,r}$ for $j = 1, \cdots, 4$, by Lemma 3.2, the estimate (3.81) follows. The inequality (3.82) is a consequence of (3.66), (3.76) and (3.79). For $k \in \mathcal{X}^\varepsilon_{\pm k_j}$, we find

$$\hat{w} = M_-^{(\pm k_j)}(J^{(2)} - J^{(\pm k_j)})(M_+^{(\pm k_j)})^{-1}, \quad j = 1, 2.$$

Therefore, it follows from (3.64) and (3.65) that the estimate (3.83) holds.

The estimates in Lemma 3.4 imply that

$$\begin{aligned}\|\hat{w}\|_{(L^1 \cap L^2)(\hat{\Sigma})} &\leqslant Ct^{-1/2}, \\ \|\hat{w}\|_{L^\infty(\hat{\Sigma})} &\leqslant Ct^{-1/2} \ln t,\end{aligned} \quad t > 3, \xi \in \mathcal{I}. \tag{3.84}$$

Let $\hat{C}$ denote the Cauchy operator associated with $\hat{\Sigma}$:

$$(\hat{C}f)(k) = \int_{\hat{\Sigma}} \frac{f(\zeta)}{\zeta - k} \frac{d\zeta}{2\pi i}, \quad k \in \mathbf{C} \setminus \hat{\Sigma}, \; f \in L^2(\hat{\Sigma}).$$

We denote the boundary values of $\hat{C}f$ from the left and right sides of $\hat{\Sigma}$ by $\hat{C}_+ f$ and $\hat{C}_- f$, respectively. Define the operator $\hat{C}_{\hat{w}} : L^2(\hat{\Sigma}) + L^\infty(\hat{\Sigma}) \to L^2(\hat{\Sigma})$ by $\hat{C}_{\hat{w}} f = \hat{C}_-(f\hat{w})$, that is, $\hat{C}_{\hat{w}}$ is defined by $\hat{C}_{\hat{w}}(f) = \hat{C}_+(f\hat{w}_-) + \hat{C}_-(f\hat{w}_+)$ where we have chosen, for simplicity, $\hat{w}_+ = \hat{w}$ and $\hat{w}_- = 0$. Then, by (3.84), we find

$$\|\hat{C}_{\hat{w}}\|_{B(L^2(\hat{\Sigma}))} \leqslant C\|\hat{w}\|_{L^\infty(\hat{\Sigma})} \leqslant Ct^{-1/2} \ln t, \tag{3.85}$$

where $B(L^2(\hat{\Sigma}))$ denotes the Banach space of bounded linear operators $L^2(\hat{\Sigma}) \to L^2(\hat{\Sigma})$. Therefore, there exists a $T > 0$ such that $I - \hat{C}_{\hat{w}} \in B(L^2(\hat{\Sigma}))$ is invertible for all $\xi \in \mathcal{I}$, $t > T$. Following this, we may define the 2×2 matrix-valued function $\hat{\mu}(x, t; k)$ whenever $t > T$ by

$$\hat{\mu} = I + \hat{C}_{\hat{w}} \hat{\mu}. \tag{3.86}$$

Then

$$\hat{M}(x, t; k) = I + \frac{1}{2\pi i} \int_{\hat{\Sigma}} \frac{(\hat{\mu}\hat{w})(x, t; \zeta)}{\zeta - k} d\zeta, \quad k \in \mathbf{C} \setminus \hat{\Sigma} \tag{3.87}$$

is the unique solution of the RH problem (3.78) for $t > T$. Moreover, using the Neumann series (see [22]), the function $\hat{\mu}(x, t; k)$ satisfies

$$\|\hat{\mu}(x, t; \cdot) - I\|_{L^2(\hat{\Sigma})} = O(t^{-1/2}), \quad t \to \infty, \; \xi \in \mathcal{I}. \tag{3.88}$$

It follows from (3.87) that

$$\lim_{k \to \infty} k(\hat{M}(x, t; k) - I) = -\frac{1}{2\pi i} \int_{\hat{\Sigma}} (\hat{\mu}\hat{w})(x, t; k) dk. \tag{3.89}$$

Using (3.80) and (3.88), we have

$$\int_{\hat{\Sigma}_1} (\hat{\mu}\hat{w})(x, t; k) dk = \int_{\hat{\Sigma}_1} \hat{w}(x, t; k) dk + \int_{\hat{\Sigma}_1} (\hat{\mu}(x, t; k) - I)\hat{w}(x, t; k) dk$$

$$\leqslant \|\hat{w}\|_{L^1(\hat{\Sigma}_1)} + \|\hat{\mu} - I\|_{L^2(\hat{\Sigma}_1)} \|\hat{w}\|_{L^2(\hat{\Sigma}_1)}$$

$$\leqslant Ce^{-ct}, \quad t \to \infty.$$

By (3.81) and (3.88), the contribution from $\hat{\Sigma}_2$ to the right-hand side of (3.89) is

$$O(\|\hat{w}\|_{L^1(\hat{\Sigma}_2)} + \|\hat{\mu} - I\|_{L^2(\hat{\Sigma}_2)}\|\hat{w}\|_{L^2(\hat{\Sigma}_2)}) = O(t^{-3/2}), \quad t \to \infty.$$

Similarly, by (3.83) and (3.88), the contribution from $\hat{\Sigma}_4$ to the right-hand side of (3.89) is

$$O(\|\hat{w}\|_{L^1(\hat{\Sigma}_4)} + \|\hat{\mu} - I\|_{L^2(\hat{\Sigma}_4)}\|\hat{w}\|_{L^2(\hat{\Sigma}_4)}) = O(t^{-1}\ln t), \quad t \to \infty.$$

Finally, by (3.67)-(3.70), (3.82) and (3.88), we can get

$$-\frac{1}{2\pi i}\int_{\hat{\Sigma}_3}(\hat{\mu}\hat{w})(x,t;k)dk$$
$$= -\frac{1}{2\pi i}\int_{\hat{\Sigma}_3}\hat{w}(x,t;k)dk - \frac{1}{2\pi i}\int_{\hat{\Sigma}_3}(\hat{\mu}(x,t;k) - I)\hat{w}(x,t;k)dk$$
$$= -\frac{1}{2\pi i}\int_{\partial D_\varepsilon(k_1)}\left((M^{(k_1)})^{-1}(x,t;k) - I\right)dk$$
$$\quad -\frac{1}{2\pi i}\int_{\partial D_\varepsilon(-k_1)}\left((M^{(-k_1)})^{-1}(x,t;k) - I\right)dk$$
$$\quad -\frac{1}{2\pi i}\int_{\partial D_\varepsilon(k_2)}\left((M^{(k_2)})^{-1}(x,t;k) - I\right)dk$$
$$\quad -\frac{1}{2\pi i}\int_{\partial D_\varepsilon(-k_2)}\left((M^{(-k_2)})^{-1}(x,t;k) - I\right)dk$$
$$\quad + O(\|\hat{\mu} - I\|_{L^2(\hat{\Sigma}_3)}\|\hat{w}\|_{L^2(\hat{\Sigma}_3)})$$
$$= \frac{(\delta_{k_1}^0)^{\hat{\sigma}_3}M_1^Y(\xi)}{4\sqrt{tk_1(3\alpha - 40\beta k_1^2)}} - \frac{(\delta_{-k_1}^0)^{\hat{\sigma}_3}\overline{M_1^Y(\xi)}}{4\sqrt{tk_1(3\alpha - 40\beta k_1^2)}}$$
$$\quad + \frac{(\delta_{k_2}^0)^{\hat{\sigma}_3}M_1^X(\xi)}{4\sqrt{tk_2(40\beta k_2^2 - 3\alpha)}} - \frac{(\delta_{-k_2}^0)^{\hat{\sigma}_3}\overline{M_1^X(\xi)}}{4\sqrt{tk_2(40\beta k_2^2 - 3\alpha)}} + O(t^{-1}), \quad t \to \infty.$$

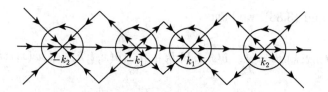

Fig. 6   The contour $\hat{\Sigma}$

Thus, we obtain the following important relation

$$\lim_{k \to \infty} k(\hat{M}(x,t;k) - I)$$

$$= \frac{(\delta_{k_1}^0)^{\hat{\sigma}_3} M_1^Y(\xi)}{4\sqrt{tk_1(3\alpha - 40\beta k_1^2)}} - \frac{(\delta_{-k_1}^0)^{\hat{\sigma}_3} \overline{M_1^Y(\xi)}}{4\sqrt{tk_1(3\alpha - 40\beta k_1^2)}}$$

$$+ \frac{(\delta_{k_2}^0)^{\hat{\sigma}_3} M_1^X(\xi)}{4\sqrt{tk_2(40\beta k_2^2 - 3\alpha)}} - \frac{(\delta_{-k_2}^0)^{\hat{\sigma}_3} \overline{M_1^X(\xi)}}{4\sqrt{tk_2(40\beta k_2^2 - 3\alpha)}} + O(t^{-1} \ln t), \quad t \to \infty.$$

Taking into account that (3.4), (3.8), (3.37) and (3.77), for sufficient large $k \in \mathbf{C} \setminus \hat{\Sigma}$, we get

$$u(x,t) = -2i \lim_{k \to \infty} (kM(x,t;k))_{12}$$

$$= -2i \lim_{k \to \infty} k(\hat{M}(x,t;k) - I)_{12}$$

$$= -\left(\frac{\beta^Y(p)(\delta_{k_1}^0)^2 + \overline{\beta^Y(p)}(\delta_{-k_1}^0)^2}{2\sqrt{tk_1(3\alpha - 40\beta k_1^2)}} + \frac{\beta^X(q)(\delta_{k_2}^0)^2 + \overline{\beta^X(q)}(\delta_{-k_2}^0)^2}{2\sqrt{tk_2(40\beta k_2^2 - 3\alpha)}} \right)$$

$$+ O\left(\frac{\ln t}{t}\right).$$

Using

$$\delta_{-k_1}^0 = \overline{\delta_{k_1}^0}, \quad \delta_{-k_2}^0 = \overline{\delta_{k_2}^0} \qquad (3.90)$$

as $\chi_j(-k_j) = -\chi_j(k_j) = \overline{\chi_j(k_j)}$, $j = 1,2$, and collecting the above computations, we obtain our main results as stated in the Theorem 1.1.

4 Asymptotics for a special case $\alpha = 0$

In this section, we consider the long-time asymptotics to the solutions for a particular case, namely, $\alpha = 0$ of emKdV equation (1.2),

$$u_t + \beta(30u^4 u_x + 10u_x^3 + 40uu_x u_{xx} + 10u^2 u_{xxx} + u_{xxxxx}) = 0. \qquad (4.1)$$

As we all know, the Hirota equation can be reduced to the complex-valued mKdV equation under the Galilean transformation. Thus, if we rewrite equation (1.2) as a complex-valued form

$$u_t + \alpha(6|u|^2 u_x + u_{xxx}) + \beta[30|u|^4 u_x + 10(u|u_x|^2 + |u|^2 u_{xx})_x + u_{xxxxx}] = 0, \quad (4.2)$$

similarly, the Galilean transformation can reduce (4.2) into a complex-valued form of (4.1), however, where we take u as a real-valued function. In fact, the fifth order KdV equation indeed can be excluded the third derivative term (see [12]), where its prolongation structure was considered.

As in Section 2, under the condition that $a(k) \neq 0$ for $\{k \in \mathbf{C} | \mathrm{Im}\, k \geqslant 0\}$, the corresponding RH problem associated with (4.1) is

$$\begin{cases} N_+(x,t;k) = N_-(x,t;k)v(x,t;k), & k \in \mathbf{R}, \\ N(x,t;k) \to I, & k \to \infty, \end{cases} \quad (4.3)$$

where the jump matrix $v(x,t;k)$ is defined by

$$v(x,t;k) = \begin{pmatrix} 1 + |r(k)|^2 & \bar{r}(k)e^{-t\theta(k)} \\ r(k)e^{t\theta(k)} & 1 \end{pmatrix},$$

$$r(k) = \frac{\bar{b}(k)}{a(k)}, \quad \theta(k) = 2i(16\beta k^5 - k\xi), \quad \xi = \frac{x}{t}. \quad (4.4)$$

Also, we have

$$r(-k) = \overline{r(\bar{k})}, \quad k \in \mathbf{R}. \quad (4.5)$$

Moreover, the relation between solution $u(x,t)$ of the equation (4.1) and $N(x,t;k)$ is

$$u(x,t) = -2i \lim_{k \to \infty} (kN(x,t;k))_{12}. \quad (4.6)$$

For this case, there are two real and pure imaginary critical points of $\theta(k)$ located at the points $\pm k_0$ and $\pm i k_0$, where

$$k_0 = \sqrt[4]{\frac{\xi}{80\beta}}. \quad (4.7)$$

Our aim in this section is to find the asymptotics of solution $u(x,t)$ to the equation (4.1) in region $0 < x \leqslant Mt^{\frac{1}{5}}$, where $M > 1$ is a constant. We see that as $t \to \infty$, the critical points $\pm k_0$ approach 0 at least as fast as $t^{-\frac{1}{5}}$, i.e., $k_0 \leqslant \left(\frac{M}{80\beta}\right)^{\frac{1}{4}} t^{-\frac{1}{5}}$. We will show that the asymptotics of the solution $u(x,t)$ in this region is given in terms of the solution of a fourth order Painlevé II equation.

Let $\Gamma \subset \mathbf{C}$ denote the contour $\Gamma = \mathbf{R} \cup \Gamma_1 \cup \Gamma_2$, where

$$\Gamma_1 = \{k_0 + le^{\frac{\pi i}{6}} | l \geqslant 0\} \cup \{-k_0 + le^{\frac{5\pi i}{6}} | l \geqslant 0\},$$
$$\Gamma_2 = \{k_0 + le^{-\frac{\pi i}{6}} | l \geqslant 0\} \cup \{-k_0 + le^{-\frac{5\pi i}{6}} | l \geqslant 0\},$$
(4.8)

and we orient Γ to the right. Let $\mathcal{V}$ and $\mathcal{V}^*$ denote the triangular domains shown in Fig. 7.

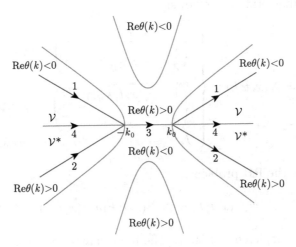

Fig. 7   The oriented contour Γ and the sets $\mathcal{V}$ and $\mathcal{V}^*$

Denote
$$\mathcal{P} = \{(x,t) \in \mathbf{R}^2 | 0 < x \leqslant Mt^{\frac{1}{5}}, t \geqslant 3\}.$$

By Lemma 3.2, we also have the following analytic approximation lemma for $r(k)$.

Lemma 4.1   *There exists a decomposition*
$$r(k) = r_a(x,t,k) + r_r(x,t,k), \quad k \in (-\infty, -k_0) \cup (k_0, \infty), \quad (4.9)$$

where the functions r_a and r_r satisfy the following properties:

(i) *For $(x,t) \in \mathcal{P}$, $r_a(x,t,k)$ is defined and continuous for $k \in \bar{\mathcal{V}}$ and analytic for $\mathcal{V}$.*

(ii) *The function r_a satisfies*
$$|r_a(x,t,k)| \leqslant \frac{C}{1+|k|^2} e^{\frac{t}{4}|\mathrm{Re}\theta(k)|}, \quad k \in \bar{\mathcal{V}}, \quad (4.10)$$

and
$$|r_a(x,t,k) - r(k_0)| \leqslant C|k - k_0| e^{\frac{t}{4}|\mathrm{Re}\theta(k)|}, \quad k \in \bar{\mathcal{V}}. \quad (4.11)$$

(iii) The L^1, L^2 and L^∞ norms of the function $r_r(x,t,\cdot)$ on $(-\infty, -k_0) \cup (k_0, \infty)$ are $O(t^{-3/2})$ as $t \to \infty$ uniformly with respect to $(x,t) \in \mathcal{P}$.

(iv) The following symmetries hold:
$$r_a(x,t,k) = \overline{r_a(x,t,-\bar{k})}, \quad r_r(x,t,k) = \overline{r_r(x,t,-\bar{k})}. \tag{4.12}$$

The first transform is as follows:
$$N^{(1)}(x,t;k) = N(x,t;k) \times \begin{cases} \begin{pmatrix} 1 & 0 \\ -r_a(x,t,k)e^{t\theta(k)} & 1 \end{pmatrix}, & k \in \mathcal{V}, \\ \begin{pmatrix} 1 & \overline{r_a(x,t,\bar{k})}e^{-t\theta(k)} \\ 0 & 1 \end{pmatrix}, & k \in \mathcal{V}^*, \\ I, & \text{elsewhere.} \end{cases} \tag{4.13}$$

Then, we obtain the RH problem
$$N_+^{(1)}(x,t;k) = N_-^{(1)}(x,t;k)v^{(1)}(x,t,k) \tag{4.14}$$

on the contour Γ depicted in Fig. 7. The jump matrix $v^{(1)}(x,t,k)$ is given by
$$v_1^{(1)} = \begin{pmatrix} 1 & 0 \\ r_a e^{t\theta} & 1 \end{pmatrix}, \quad v_2^{(1)} = \begin{pmatrix} 1 & \bar{r}_a e^{-t\theta} \\ 0 & 1 \end{pmatrix},$$
$$v_3^{(1)} = \begin{pmatrix} 1 & \bar{r} e^{-t\theta} \\ 0 & 1 \end{pmatrix} \begin{pmatrix} 1 & 0 \\ r e^{t\theta} & 1 \end{pmatrix}, \quad v_4^{(1)} = \begin{pmatrix} 1 & \bar{r}_r e^{-t\theta} \\ 0 & 1 \end{pmatrix} \begin{pmatrix} 1 & 0 \\ r_r e^{t\theta} & 1 \end{pmatrix},$$

where $v_i^{(1)}$ denotes the restriction of $v^{(1)}$ to the contour labeled by i in Fig. 7.

Let us introduce the new variables y and z by
$$y = \frac{-x}{(20\beta t)^{\frac{1}{5}}}, \quad z = (20\beta t)^{\frac{1}{5}} k, \tag{4.15}$$

such that
$$t\theta(k) = 2i\left(\frac{4}{5}z^5 + yz\right). \tag{4.16}$$

We now have $-C \leqslant y < 0$. Fix $\varepsilon > 0$ and let $D_\varepsilon(0) = \{k \in \mathbf{C} | |k| < \varepsilon\}$ denote the open disk of radius ε centered at the origin. Let $\Gamma^\varepsilon = (\Gamma \cap D_\varepsilon(0)) \setminus ((-\infty, -k_0) \cup (k_0, \infty))$. Let Z denote the contour defined in (B.1) with $z_0 = (20\beta t)^{\frac{1}{5}} k_0 = $

$\sqrt[4]{-y}/\sqrt{2}$. The map $k \mapsto z$ maps Γ^ε onto $Z \cap \{z \in \mathbb{C} | |z| < (20\beta t)^{\frac{1}{5}}\varepsilon\}$. We write $\Gamma^\varepsilon = \cup_{j=1}^3 \Gamma_j^\varepsilon$, where Γ_j^ε denotes the inverse image of $Z_j \cap \{z \in \mathbb{C} | |z| < (20\beta t)^{\frac{1}{5}}\varepsilon\}$ under this map, see Fig. 8.

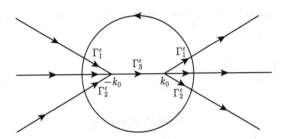

Fig. 8   The oriented contour $\hat{\Gamma}$ and Γ^ε

For large t and fixed z, the jump matrices $\{v_j^{(1)}\}_1^4$ tend to the matrix v^Z defined in (B.3) if we set $s = r(0)$. Thus, we expect that $N^{(1)}$ approaches the solution $N^0(x,t,k)$ defined by

$$N^0(x,t;k) := N^Z(s,y,z,z_0) \qquad (4.17)$$

for large t, where $N^Z(s,y,z,z_0)$ is the solution of the model RH problem for (B.2) with $z_0 = \sqrt[4]{-y}/\sqrt{2}$. Moreover, if $(x,t) \in \mathcal{P}$, then $(y,t,z_0) \in \mathbb{P}$, where $\mathbb{P}$ is the parameter subset defined in (B.4). Thus, Lemma B.1 ensures that N^0 is well-defined by (4.17).

Lemma 4.2   *For each $(x,t) \in \mathcal{P}$, the function $N^0(x,t;k)$ defined in (4.17) is an analytic function of $k \in D_\varepsilon(0) \setminus \Gamma^\varepsilon$ such that*

$$|N^0(x,t;k)| \leqslant C, \quad (x,t) \in \mathcal{P}, \ k \in D_\varepsilon(0) \setminus \Gamma^\varepsilon. \qquad (4.18)$$

Across Γ^ε, N^0 obeys the jump condition $N_+^0 = N_-^0 v^0$, where the jump matrix v^0 satisfies, for $1 \leqslant n \leqslant \infty$,

$$\|v^{(1)} - v^0\|_{L^n(\Gamma^\varepsilon)} \leqslant Ct^{-\frac{1}{5}(1+\frac{1}{n})}, \quad (x,t) \in \mathcal{P}. \qquad (4.19)$$

Furthermore, as $t \to \infty$, we have

$$\|N^0(x,t;k)^{-1} - I\|_{L^\infty(\partial D_\varepsilon(0))} = O(t^{-\frac{1}{5}}), \qquad (4.20)$$

and

$$\frac{1}{2\pi i} \int_{\partial D_\varepsilon(0)} (N^0(x,t;k)^{-1} - I)dk = -\frac{iN_1^0(y)}{(20\beta t)^{\frac{1}{5}}}, \qquad (4.21)$$

where
$$N_1^0(y) = \begin{pmatrix} -2\int_\infty^y u_p^2(y')dy' & u_p(y) \\ u_p(y) & 2\int_\infty^y u_p^2(y')dy' \end{pmatrix}. \tag{4.22}$$

Proof The analyticity and boundedness of N^0 are a consequence of Lemma B.1. Moreover,

$$v^{(1)} - v^0 = \begin{cases} \begin{pmatrix} 0 & 0 \\ (r_a(x,t,k) - r(0))e^{t\theta} & 0 \end{pmatrix}, & k \in \Gamma_1^\varepsilon, \\ \begin{pmatrix} 0 & (\overline{r_a(x,t,\bar{k})} - \overline{r(0)})e^{-t\theta} \\ 0 & 0 \end{pmatrix}, & k \in \Gamma_2^\varepsilon, \\ \begin{pmatrix} |r(k)|^2 - |r(0)|^2 & (\overline{r(k)} - \overline{r(0)})e^{-t\theta} \\ (r(k) - r(0))e^{t\theta} & 0 \end{pmatrix}, & k \in \Gamma_3^\varepsilon. \end{cases} \tag{4.23}$$

For $k = k_0 + le^{\frac{\pi i}{6}}$, $0 \leqslant l \leqslant \varepsilon$, we obtain

$$\mathrm{Re}\,\theta(k) = -16\beta l^2(l^3 + 5\sqrt{3}k_0 l^2 + 20k_0^2 l + 10\sqrt{3}k_0^3) \leqslant -16\beta|k-k_0|^5.$$

On the other hand, if $|k - k_0| \geqslant k_0$, then $|k - k_0| \geqslant |k|/2$, and hence

$$e^{-12\beta t|k-k_0|^5} \leqslant e^{-\frac{3}{8}\beta t|k|^5}.$$

If $|k - k_0| < k_0$, then $|k| \leqslant Ct^{-\frac{1}{5}}$, and so

$$e^{-12\beta t|k-k_0|^5} \leqslant 1 \leqslant Ce^{-\frac{3}{8}\beta t|k|^5}.$$

Thus, for $k = k_0 + le^{\frac{\pi i}{6}}$, $0 \leqslant l \leqslant \varepsilon$, we find

$$e^{-\frac{3}{4}t|\mathrm{Re}\,\theta|} \leqslant e^{-12\beta t|k-k_0|^5} \leqslant Ce^{-\frac{3}{8}\beta t|k|^5} \leqslant Ce^{-\frac{3}{160}|z|^5}. \tag{4.24}$$

As a consequence, we have

$$|v^{(1)} - v^0| \leqslant C|r_a(x,t,k) - r(k_0)|e^{t\mathrm{Re}\,\theta} + C|r(k_0) - r(0)|e^{t\mathrm{Re}\,\theta}$$
$$\leqslant C|k-k_0|e^{-\frac{3}{4}t|\mathrm{Re}\,\theta|} + Ck_0 e^{-t|\mathrm{Re}\,\theta|} \leqslant C|z|t^{-\frac{1}{5}}e^{-\frac{3}{160}|z|^5}. \tag{4.25}$$

A similar computation shows that (4.25) also holds for $k = -k_0 + le^{\frac{5\pi i}{6}}$, $0 \leqslant l \leqslant \varepsilon$. Consequently, writing $l = |z|$,

$$\|v^{(1)} - v^0\|_{L^\infty(\Gamma_1^\varepsilon)} \leqslant Ct^{-\frac{1}{5}}, \tag{4.26}$$

and
$$\|v^{(1)} - v^0\|_{L^1(\Gamma_1^\varepsilon)} \leqslant C \int_0^\infty lt^{-\frac{1}{5}} e^{-\frac{3}{160}l^5} \frac{dl}{t^{\frac{1}{5}}} \leqslant Ct^{-\frac{2}{5}}. \tag{4.27}$$

Using the general inequality $\|f\|_{L^n} \leqslant \|f\|_{L^\infty}^{1-1/n} \|f\|_{L^1}^{1/n}$, (4.19) holds for $k \in \Gamma_1^\varepsilon$. Similar estimates applying to Γ_j^ε, $j = 2, 3$ show that (4.19) holds.

The variable $z = (20\beta t)^{\frac{1}{5}} k$ satisfies $|z| = (20\beta t)^{\frac{1}{5}} \varepsilon$ if $|k| = \varepsilon$. Thus, equation (B.5) yields

$$N^0(x, t; k) = I + \frac{iN_1^0(y)}{(20\beta t)^{\frac{1}{5}} k} + O(t^{-\frac{2}{5}}), \quad k \in \partial D_\varepsilon(0), \ t \to \infty. \tag{4.28}$$

Thus, (4.20) and (4.21) follow from (4.28) and Cauchy's formula.

Let $\hat{\Gamma} = \Gamma \cup \partial D_\varepsilon(0)$ and assume that the boundary of $D_\varepsilon(0)$ is oriented counterclockwise, see Fig. 8. Define $\hat{N}(x, t; k)$ by

$$\hat{N}(x, t; k) = \begin{cases} N^{(1)}(x, t; k) N^0(x, t; k)^{-1}, & k \in D_\varepsilon(0), \\ N^{(1)}(x, t; k), & k \in \mathbf{C} \setminus \overline{D_\varepsilon(0)}, \end{cases} \tag{4.29}$$

then $\hat{N}(x, t; k)$ satisfies the following RH problem

$$\hat{N}_+(x, t; k) = \hat{N}_-(x, t; k) \hat{v}(x, t; k), \quad k \in \hat{\Gamma}, \tag{4.30}$$

where the jump contour $\hat{\Gamma} = \Gamma^\varepsilon \cup \partial D_\varepsilon(0) \cup (\mathbf{R} \setminus [-k_0, k_0]) \cup \hat{\Gamma}'$, $\hat{\Gamma}' = \Gamma \setminus (\mathbf{R} \cup \overline{D_\varepsilon(0)})$ is depicted in Fig. 8, and the jump matrix $\hat{v}(x, t; k)$ is given by

$$\hat{v} = \begin{cases} N_-^0 v^{(1)} (N_+^0)^{-1}, & k \in \hat{\Gamma} \cap D_\varepsilon(0), \\ (N^0)^{-1}, & k \in \partial D_\varepsilon(0), \\ v^{(1)}, & k \in \hat{\Gamma} \setminus \overline{D_\varepsilon(0)}. \end{cases} \tag{4.31}$$

Lemma 4.3 *Let $\hat{\omega} = \hat{v} - I$. For each $1 \leqslant n \leqslant \infty$, the following estimates hold:*

$$\|\hat{\omega}\|_{L^n(\partial D_\varepsilon(0))} \leqslant Ct^{-\frac{1}{5}}, \tag{4.32}$$

$$\|\hat{\omega}\|_{L^n(\Gamma^\varepsilon)} \leqslant Ct^{-\frac{1}{5}(1+\frac{1}{n})}, \tag{4.33}$$

$$\|\hat{\omega}\|_{L^n(\mathbf{R} \setminus [-k_0, k_0])} \leqslant Ct^{-\frac{3}{2}}, \tag{4.34}$$

$$\|\hat{\omega}\|_{L^n(\hat{\Gamma}')} \leqslant Ce^{-ct}. \tag{4.35}$$

Proof The estimate (4.32) follows from (4.20). For $k \in \Gamma^\varepsilon$, we have

$$\hat{\omega} = N_-^0(v^{(1)} - v^0)(N_+^0)^{-1}.$$

As a consequence, (4.18) and (4.19) imply (4.33). On $\mathbf{R} \setminus [-k_0, k_0]$, the jump matrix $\hat{\omega}$ only involves the small remainder r_r, so the estimate (4.34) holds as a consequence of Lemma 4.1 and (4.18). Finally, (4.35) follows from $e^{-t|\mathrm{Re}\theta|} \leqslant Ce^{-ct}$ uniformly on $\hat{\Gamma}'$.

As the discussion in Subsection 3.3, the estimates in Lemma 4.3 show that the RH problem (4.30) for $\hat{N}$ has a unique solution given by

$$\hat{N}(x,t;k) = I + \frac{1}{2\pi i} \int_{\hat{\Gamma}} (\hat{\mu}\hat{\omega})(x,t;s) \frac{ds}{s-k}, \qquad (4.36)$$

where $\hat{\mu} = I + (I - \hat{C}_{\hat{\omega}})^{-1} \hat{C}_{\hat{\omega}} I$, and $\hat{C}_{\hat{\omega}} f = \hat{C}_-(f\hat{\omega})$, $\hat{C}$ denote the Cauchy operator associated with $\hat{\Gamma}$. Moreover, the function $\hat{\mu}(x,t;k)$ satisfies

$$\|\hat{\mu}(x,t;\cdot) - I\|_{L^2(\hat{\Sigma})} = O(t^{-\frac{1}{5}}), \quad t \to \infty, \ (x,t) \in \mathcal{P}. \qquad (4.37)$$

It follows from (4.36) that

$$\lim_{k \to \infty} k(\hat{N}(x,t;k) - I) = -\frac{1}{2\pi i} \int_{\hat{\Gamma}} (\hat{\mu}\hat{\omega})(x,t;k) dk. \qquad (4.38)$$

By (4.21), (4.32) and (4.37), we can get

$$-\frac{1}{2\pi i} \int_{\partial D_\varepsilon(0)} (\hat{\mu}\hat{\omega})(x,t;k) dk$$

$$= -\frac{1}{2\pi i} \int_{\partial D_\varepsilon(0)} \hat{\omega}(x,t;k) dk - \frac{1}{2\pi i} \int_{\partial D_\varepsilon(0)} (\hat{\mu}(x,t;k) - I)\hat{\omega}(x,t;k) dk$$

$$= -\frac{1}{2\pi i} \int_{\partial D_\varepsilon(0)} \left((N^0)^{-1}(x,t;k) - I \right) dk + O(\|\hat{\mu} - I\|_{L^2(\partial D_\varepsilon(0))} \|\hat{w}\|_{L^2(\partial D_\varepsilon(0))})$$

$$= \frac{iN_1^0(y)}{(20\beta t)^{\frac{1}{5}}} + O(t^{-\frac{2}{5}}), \quad t \to \infty.$$

Using (4.33) and (4.37), we have

$$\int_{\Gamma^\varepsilon} (\hat{\mu}\hat{\omega})(x,t;k) dk = \int_{\Gamma^\varepsilon} \hat{\omega}(x,t;k) dk + \int_{\Gamma^\varepsilon} (\hat{\mu}(x,t;k) - I)\hat{\omega}(x,t;k) dk$$

$$\leqslant \|\hat{\omega}\|_{L^1(\Gamma^\varepsilon)} + \|\hat{\mu} - I\|_{L^2(\Gamma^\varepsilon)} \|\hat{\omega}\|_{L^2(\Gamma^\varepsilon)}$$

$$\leqslant Ct^{-\frac{2}{5}}, \quad t \to \infty.$$

By (3.34) and (3.37), the contribution from $\mathbf{R} \setminus [-k_0, k_0]$ to the right-hand side of (4.38) is $O(t^{-\frac{3}{2}})$, and similarly, by (3.35) and (3.37), the contribution from $\hat{\Gamma}'$ to the right-hand side of (4.38) is $O(e^{-ct})$, as $t \to \infty$. Thus, we obtain the following important relation

$$\lim_{k \to \infty} k(\hat{M}(x,t;k) - I) = \frac{iN_1^0(y)}{(20\beta t)^{\frac{1}{5}}} + O(t^{-\frac{2}{5}}), \quad t \to \infty. \tag{4.39}$$

Recalling the definition of $N_1^0(y)$ in (4.22) and the relation (4.6), we obtain our another main result stated in Theorem 1.2.

Remark 4.1 We did not directly consider the asymptotic behavior of the solution to equation (1.2) in region $\mathcal{P}$ because there is no suitable scale transformation to eliminate t from the coefficients of k^3 and k^5 in terms of the phase function $\Phi(k)$ given by (3.2) at the same time.

Appendix A. Fourth order Painlevé II RH problem

Let Y denote the contour $Y = Y_1 \cup Y_2$ oriented as in Fig. 9, where

$$Y_1 = \{le^{\frac{\pi i}{6}} | l \geq 0\} \cup \{le^{\frac{5\pi i}{6}} | l \geq 0\}, \quad Y_2 = \{le^{-\frac{\pi i}{6}} | l \geq 0\} \cup \{le^{-\frac{5\pi i}{6}} | l \geq 0\}.$$

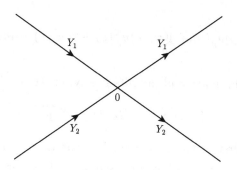

Fig. 9 The oriented contour Y

Lemma A.1 (fourth order Painlevé II RH problem) Let $s \in \mathbf{C}$ be a complex number. Then the following **RH** problems parametrized by $y \in \mathbf{R}, s \in \mathbf{C}$:

$$\begin{cases} N_+^Y(y,z) = N_-^Y(y,z) v^Y(y,z), & z \in Y, \\ N^Y(y,z) \to I, & z \to \infty, \end{cases} \tag{A.1}$$

where the jump matrix $v^Y(y,z)$ is defined by

$$v^Y(y,z) = \begin{cases} \begin{pmatrix} 1 & 0 \\ se^{2i(\frac{4}{5}z^5+yz)} & 1 \end{pmatrix}, & z \in Y_1, \\ \begin{pmatrix} 1 & \bar{s}e^{-2i(\frac{4}{5}z^5+yz)} \\ 0 & 1 \end{pmatrix}, & z \in Y_2 \end{cases} \quad (A.2)$$

has a unique solution $N^Y(y,z)$ for each $y \in \mathbf{R}$. Moreover, there exists a smooth functions $\{N_j^Y(y)\}^4$ of $y \in \mathbf{R}$ with decay as $y \to \infty$ such that

$$N^Y(y,z) = I + \sum_{j=1}^{4} \frac{N_j^Y(y)}{z^j} + O(z^{-5}), \quad z \to \infty, \quad (A.3)$$

uniformly for y in compact subsets of $\mathbf{R}$ and for $\arg z \in [0, 2\pi]$. The leading coefficient N_1^Y is given by

$$N_1^Y(y) = i \begin{pmatrix} -2\int_\infty^y u_p^2(y')dy' & u_p(y) \\ u_p(y) & 2\int_\infty^y u_p^2(y')dy' \end{pmatrix}, \quad (A.4)$$

where the real-valued function $u_p(y)$ satisfies the following fourth-order Painlevé II equation (see [20])

$$u_p''''(y) + 40u_p^2(y)u_p''(y) + 40u_p(y)u_p'^2(y) + 96u_p^5(y) + 4yu_p(y) = 0. \quad (A.5)$$

Proof The jump matrix v^Y admits the symmetries

$$v^Y(y,z) = (v^Y)^\dagger(y,\bar{z}) = \overline{v^Y(y,-\bar{z})}. \quad (A.6)$$

We infer from the first of these symmetries that the RH problem for $N^Y(y,z)$ admits a vanishing lemma, and as a consequence, there exists a unique solution $N^Y(y,z)$, which admits an expansion of the form (A.3). Assume that

$$(N^Y)^{-1}(y,z) = I + \sum_{j=1}^{4} \frac{\varphi_j(y)}{z^j} + O(z^{-5}), \quad (A.7)$$

a direct calculation shows that

$$\varphi_1 = -N_1^Y, \quad \varphi_2 = (N_1^Y)^2 - N_2^Y, \quad \varphi_3 = N_1^Y N_2^Y + N_2^Y N_1^Y - (N_1^Y)^3 - N_3^Y,$$
$$\varphi_4 = (N_1^Y)^4 + N_1^Y N_3^Y + N_3^Y N_1^Y - (N_1^Y)^2 N_2^Y - N_1^Y N_2^Y N_1^Y$$

$$-N_2^Y(N_1^Y)^2+(N_2^Y)^2-N_4^Y. \tag{A.8}$$

Let $\phi(y,z)=N^Y(y,z)e^{-i(\frac{4}{5}z^5+yz)\sigma_3}$. Then the function $\mathcal{Y}(y,z)$ defined by

$$\mathcal{Y}=\phi_y\phi^{-1}=(N_y^Y-izN^Y\sigma_3)(N^Y)^{-1} \tag{A.9}$$

is an entire function of z, hence according to (A.3), (A.7) and (A.8), we have

$$\mathcal{Y}(y,z)=-iz\sigma_3+i[\sigma_3,N_1^Y]. \tag{A.10}$$

Thus, we find

$$N_y^Y-izN^Y\sigma_3=\mathcal{Y}N^Y. \tag{A.11}$$

Substituting the expansion (A.3) into (A.11) and collecting terms with $O(z^{-n})$, one can get

$$\begin{aligned}N_{1y}^Y+i[\sigma_3,N_2^Y]&=i[\sigma_3,N_1^Y]N_1^Y,\\ N_{2y}^Y+i[\sigma_3,N_3^Y]&=i[\sigma_3,N_1^Y]N_2^Y,\\ N_{3y}^Y+i[\sigma_3,N_4^Y]&=i[\sigma_3,N_1^Y]N_3^Y.\end{aligned} \tag{A.12}$$

Accordingly, since

$$\mathcal{Z}=\phi_z\phi^{-1}=\left(N_z^Y-i(4z^4+y)N^Y\sigma_3\right)(N^Y)^{-1} \tag{A.13}$$

is entire, and thus we get

$$\mathcal{Z}=\mathcal{Z}_0+\mathcal{Z}_1z+\mathcal{Z}_2z^2+\mathcal{Z}_3z^3+\mathcal{Z}_4z^4. \tag{A.14}$$

Substituting the expansion (A.3) and (A.7) into (A.13), it follows from (A.8) and (A.12) that

$$\begin{aligned}\mathcal{Z}_4&=-4i\sigma_3,\quad \mathcal{Z}_3=4i[\sigma_3,N_1^Y],\quad \mathcal{Z}_2=-4N_{1y}^Y,\\ \mathcal{Z}_1&=-4N_{2y}^Y+4N_{1y}^YN_1^Y,\\ \mathcal{Z}_0&=-4N_{3y}^Y-4N_{1y}^Y(N_1^Y)^2+4N_{1y}^YN_2^Y+4N_{2y}^YN_1^Y-iy\sigma_3.\end{aligned} \tag{A.15}$$

We have shown that ϕ obeys the Lax pair equations

$$\begin{cases}\phi_y=\mathcal{Y}\phi,\\ \phi_z=\mathcal{Z}\phi,\end{cases} \tag{A.16}$$

where $\mathcal{Y}$ and $\mathcal{Z}$ are given by (A.10) and (A.14), respectively.

The symmetries (A.6) of the jump matrix $v^Y(y,z)$ implies that $N^Y(y,z)$ satisfies the symmetries

$$N^Y(y,z) = (N^Y)^\dagger(y,\bar{z})^{-1} = \sigma_2 N^Y(y,-z)\sigma_2. \tag{A.17}$$

In particular, the coefficients $N_1^Y(y)$, $N_2^Y(y)$ and $N_3^Y(y)$ satisfy

$$\begin{aligned}N_1^Y &= -(N_1^Y)^\dagger = -\sigma_2 N_1^Y \sigma_2,\\ N_2^Y &= \sigma_2 N_2^Y \sigma_2, \quad N_3^Y = -\sigma_2 N_3^Y \sigma_2.\end{aligned} \tag{A.18}$$

Therefore, we can write

$$N_1^Y(y) = \begin{pmatrix} \psi_1(y) & \psi_2(y) \\ \psi_2(y) & -\psi_1(y) \end{pmatrix},$$

$$N_2^Y(y) = \begin{pmatrix} f_1(y) & f_2(y) \\ -f_2(y) & f_1(y) \end{pmatrix}, \tag{A.19}$$

$$N_3^Y(y) = \begin{pmatrix} g_1(y) & g_2(y) \\ g_2(y) & -g_1(y) \end{pmatrix},$$

where $\{\psi_j(y), f_j(y), g_j(y)\}_1^2$ are complex-valued functions and $\psi_1(y), \psi_2(y) \in i\mathbf{R}$. Then, the compatibility condition

$$\mathcal{Y}_z - \mathcal{Z}_y + \mathcal{Y}\mathcal{Z} - \mathcal{Z}\mathcal{Y} = 0 \tag{A.20}$$

of the Lax pair (A.16) can then be rewritten as

$$-i\sigma_3 - \mathcal{Z}_{0y} + i[\sigma_3, N_1^Y]\mathcal{Z}_0 - i\mathcal{Z}_0[\sigma_3, N_1^Y] = 0, \tag{A.21}$$

since one can directly calculate the coefficients of z, z^2, z^3, and z^4 in (A.20) vanish identically. On the other hand, by substituting (A.19) into (A.12), we find

$$\begin{cases} \psi_1' = 2i\psi_2^2, \\ \psi_2' + 2if_2 = -2i\psi_1\psi_2, \\ f_1' = -2i\psi_2 f_2, \\ f_2' + 2ig_2 = 2i\psi_2 f_1, \\ g_1' = 2i\psi_2 g_2. \end{cases} \tag{A.22}$$

Substituting (A.19) into (A.21) and using the above relations, it follows from (12)-entry of (A.21) that

$$\psi_2'''' - 40\psi_2^2\psi_2'' - 40\psi_2\psi_2'^2 + 96\psi_2^5 + 4y\psi_2 = 0, \tag{A.23}$$

however, the (11)-entry of (A.21) vanish identically. If we set $\psi_2(y) = iu_p(y)$, then $u_p(y)$ satisfies the fourth order Painlevé II equation (A.5). The lemma follows.

Appendix B. Model RH problem for sector $\mathcal{P}$

Given a number $z_0 \geqslant 0$, let Z denote the contour $Z = Z_1 \cup Z_1 \cup Z_3$, where the line segments

$$Z_1 = \{z_0 + le^{\frac{\pi i}{6}} | l \geqslant 0\} \cup \{-z_0 + le^{\frac{5\pi i}{6}} | l \geqslant 0\},$$
$$Z_2 = \{z_0 + le^{-\frac{\pi i}{6}} | l \geqslant 0\} \cup \{-z_0 + le^{-\frac{5\pi i}{6}} | l \geqslant 0\}, \qquad (B.1)$$
$$Z_3 = \{l | -z_0 \leqslant l \leqslant z_0\}$$

are oriented as in Fig. 10. It turns out that the long-time asymptotics in the sector $\mathcal{P}$ is related to solution N^Z of the following family of RH problems parametrized by $y < 0, s \in \mathbf{C}, z_0 \geqslant 0$:

$$\begin{cases} N_+^Z(s, y, z, z_0) = N_-^Z(s, y, z, z_0) v^Z(s, y, z, z_0), & z \in Z, \\ N^Z(s, y, z, z_0) \to I, & z \to \infty, \end{cases} \qquad (B.2)$$

where the jump matrix $v^Z(s, y, z, z_0)$ is defined by

$$v^Z(s, y, z, z_0) = \begin{cases} \begin{pmatrix} 1 & 0 \\ se^{2i(\frac{4}{5}z^5 + yz)} & 1 \end{pmatrix}, & z \in Z_1, \\ \begin{pmatrix} 1 & \bar{s}e^{-2i(\frac{4}{5}z^5 + yz)} \\ 0 & 1 \end{pmatrix}, & z \in Z_2, \qquad (B.3) \\ \begin{pmatrix} 1 & \bar{s}e^{-2i(\frac{4}{5}z^5 + yz)} \\ 0 & 1 \end{pmatrix} \begin{pmatrix} 1 & 0 \\ se^{2i(\frac{4}{5}z^5 + yz)} & 1 \end{pmatrix}, & z \in Z_3. \end{cases}$$

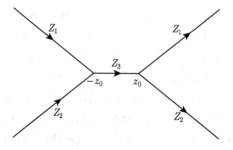

Fig. 10   The oriented contour Z

Lemma B.1 (model RH problem for sector $\mathcal{P}$) *Define the parameter subset*

$$\mathbb{P} = \{(y, t, z_0) \in \mathbf{R}^3 | -C_1 \leqslant y < 0, t \geqslant 3, \sqrt[4]{-y}/\sqrt{2} \leqslant z_0 \leqslant C_2\}, \tag{B.4}$$

where $C_1, C_2 > 0$ are constants. Then for $(y, t, z_0) \in \mathbb{P}$, the RH problem (B.2) *has a unique solution $N^Z(s, y, z, z_0)$ which satisfies*

$$N^Z(s, y, z, z_0) = I + \frac{i}{z}\begin{pmatrix} -2\int_\infty^y u_p^2(y')dy' & u_p(y) \\ u_p(y) & 2\int_\infty^y u_p^2(y')dy' \end{pmatrix} + O\left(\frac{1}{z^2}\right), \quad z \to \infty, \tag{B.5}$$

where $u_p(y)$ denotes the solution of the fourth order Painlevé II equation (A.5) *and $N^Z(s, y, z, z_0)$ is uniformly bounded for $z \in \mathbf{C} \setminus Z$. Furthermore, N^Z obeys the symmetries*

$$N^Z(s, y, z, z_0) = (N^Z)^\dagger(s, y, \bar{z}, z_0)^{-1} = \sigma_2 N^Z(s, y, -z, z_0)\sigma_2. \tag{B.6}$$

Proof Note that

$$\mathrm{Re}\left(2i\left(\frac{4}{5}z^5 + yz\right)\right) \leqslant l^2\left(-\frac{4}{5}l^3 - 4\sqrt{3}z_0 l^2 - 16z_0^2 l - 8\sqrt{3}z_0^3\right)$$

for all $z = z_0 + le^{\frac{\pi i}{6}}$ and $z = -z_0 + le^{\frac{5\pi i}{6}}$ with $l \geqslant 0, z_0 \geqslant 0$ and $-4z_0^4 \leqslant y < 0$. Thus, we have

$$|e^{2i(\frac{4}{5}z^5 + yz)}| \leqslant Ce^{-|z\pm z_0|^2(\frac{4}{5}|z\pm z_0|^3 + 4\sqrt{3}z_0|z\pm z_0|^2 + 16z_0^2|z\pm z_0| + 8\sqrt{3}z_0^3)}, \quad z \in Z_1.$$

Analogous estimates hold for $z \in Z_2$. However, $|e^{\pm 2i(\frac{4}{5}z^5 + yz)}| = 1$ for $z \in Z_3$, this shows that $v^Z \to I$ exponentially fast as $z \to \infty$.

The jump matrix v^Z obeys the same symmetries (A.6) as v^Y. In particular, v^Z is Hermitian and positive definite on $Z \cap \mathbf{R}$ and satisfies $v^Z(s, y, z, z_0) = (v^Z)^\dagger(s, y, z, z_0)$ on $Z \setminus \mathbf{R}$. This implies the existence of a vanishing lemma from which we deduce the unique existence of the solution N^Z. The symmetries (B.6) follow from the symmetries of v^Z. Moreover, the RH problem (B.2) for $N^Z(s, y, z, z_0)$ can be transformed into the RH problem (A.1) for $N^Y(y, z)$ up to a trivial contour deformation. Thus (B.5) follows from (A.3) and (A.4).

Acknowledgements N. Liu is supported by the China Postdoctoral Science Foundation under Grant no. 2019TQ0041.

References

[1] Arruda LK, J. Lenells J. Long-time asymptotics for the derivative nonlinear Schrödinger equation on the half-line [J]. Nonlinearity, 2017, 30(11): 4141–4172.

[2] Biondini G, Mantzavinos D. Long-time asymptotics for the focusing nonlinear Schrödinger equation with nonzero boundary conditions at infinity and asymptotic stage of modulational instability [J]. Commun. Pure Appl. Math., 2017, 70(12): 2300–2365.

[3] Boutet de Monvel A, Its A, Kotlyarov V. Long-time asymptotics for the focusing NLS equation with time-periodic boundary condition on the half-line [J]. Commun. Math. Phys., 2009, 290(2): 479–522.

[4] Boutet de Monvel A, Kostenko A, Shepelsky D, Teschl G. Long-time asymptotics for the Camassa–Holm equation [J]. SIAM J. Math. Anal., 2009, 41(4): 1559–1588.

[5] Boutet de Monvel A, D. Shepelsky D. Long-time asymptotics of the Camassa–Holm equation on the line [J]. in proceeding of the Conference on Integrable Systems, Random Matrices, and Applications: A conference in honor of Percy Deift's 60th birthday, Contemporary Mathematics, 2008, 458: 99–116.

[6] Boutet de Monvel A, Shepelsky D. A Riemann–Hilbert approach for the Degasperis-Procesi equation [J]. Nonlinearity, 2013, 26(7): 2081–2107.

[7] Boutet de Monvel A, Shepelsky D, Zielinski L. The short pulse equation by a Riemann–Hilbert approach [J]. Lett. Math. Phys., 2017, 107(7): 1345–1373.

[8] Buckingham R, Venakides S. Long-time asymptotics of the nonlinear Schrödinger equation shock problem [J]. Commun. Pure Appl. Math., 2007, 60(9): 1349–1414.

[9] Cheng Po-Jen, Venakides S, Zhou X. Long-time asymptotics for the pure radiation solution of the sine-Gordon equation [J]. Commun. Partial Differential Equations, 24(7-8)1195–1262, 1999.

[10] Deift P, Zhou X. A steepest descent method for oscillatory Riemann–Hilbert problems. Asymptotics for the MKdV equation [J]. Ann. Math., 1993, 137: 295–368.

[11] Deift P, Zhou X. Long-time behavior of the non-focusing nonlinear Schrödinger equation-a case study [M]. Lectures in Mathematical Sciences, University of Tokyo, Tokyo, 1995.

[12] Dodd RK, Gibbon J.D. The prolongation structure of a higher order Korteweg-de Vries equation [J]. Proc. R. Soc. Lond. A, 1978, 358(1694):287–296.

[13] Fokas AS. A unified approach to boundary value problems [M]. in: CBMS–NSF Regional Conference Series in Applied Mathematics. SIAM, 2008.

[14] Fokas AS, Its A. The linearization of the initial-boundary value problem of the nonlinear Schröinger equation [J]. SIAM J. Math. Anal., 1996, 27(3): 738–764.

[15] Fokas AS, Its A, Sung LY. The nonlinear Schröinger equation on the half-line [J]. Nonlinearity, 2005, 18(4): 1771–1822.

[16] Grimshaw R, Pelinovsky E, Poloukhina O. Higher-order Korteweg-de Vries models for internal solitary waves in a stratified shear flow with a free surface [J]. Nonlin. Processes

Geophys., 2002, 9(3/4): 221–235.

[17] Guo BL, Liu N. Long-time asymptotics for the Kundu–Eckhaus equation on the half-line [J]. J. Math. Phys., 2018, 59(6): 061505.

[18] Guo BL, Liu N, Wang YF. Long-time asymptotics for the Hirota equation on the half-line [J]. Nonlinear Anal. TMA, 2018, 174:118–140.

[19] Huang L, Xu J, Fan EG. Long-time asymptotic for the Hirota equation via nonlinear steepest descent method [J] Nonlinear Anal. Real World Appl., 2015, 26: 229–262.

[20] Kudryashov NA, Soukharev MB. Uniformization and transcendence of solutions for the first and second Painlevé hierarchies [J]. Phys. Lett. A, 1998, 237(4-5): 206–216.

[21] Lenells J. The nonlinear steepest descent method: asymptotics for initial-boundary value problems [J]. SIAM J. Math. Anal., 2016, 48(3): 2076–2118.

[22] Lenells J. The nonlinear steepest descent method for Riemann–Hilbert problems of low regularity [J]. Indiana Univ. Math. J., 2017, 66(4): 1287–1332.

[23] Lenells L, Fokas AS. On a novel integrable generalization of the nonlinear Schrödinger equation [J]. Nonlinearity, 2009, 22(1): 11–27.

[24] Liu N, Guo BL. Long-time asymptotics for the Sasa–Satsuma equation via nonlinear steepest descent method [J]. J. Math. Phys., 2019, 60(1): 011504.

[25] Pelinovskii E, Polukhina O, Lamb K. Nonlinear internal waves in the ocean stratified in density and current [J]. Oceanology, 2000, 40(6): 757–766.

[26] Wang X, Zhang J, Wang L. Conservation laws, periodic and rational solutions for an extended modified Korteweg-de Vries equation [J]. Nonlinear Dyn., 2018, 92(4): 1507–1516.

[27] Wang DS, Wang XL. Long-time asymptotics and the bright N-soliton solutions of the Kundu–Eckhaus equation via the Riemann–Hilbert approach [J]. Nonlinear Anal. Real World Appl., 2018, 41: 334–361.

[28] Wazwaz AM, Xu G. An extended modified KdV equation and its Painlevé integrability [J]. Nonlinear Dyn., 2016, 86(3): 1455–1460.

[29] Xu J, Fan EG. Long-time asymptotics for the Fokas–Lenells equation with decaying initial value problem: Without solitons [J]. J. Differential Equations, 2015, 259(3): 1098–1148.

Low Mach Number Limit of Strong Solutions to 3-D Full Navier-Stokes Equations with Dirichlet Boundary Condition*

Guo Boling (郭柏灵), Zeng Lan (曾兰), and Ni Guoxi (倪国喜)

Abstract In this paper, we consider the low Mach number limit of the full compressible Navier-Stokes equations in a three-dimensional bounded domain where the velocity field and the temperature satisfy the Dirichlet boundary conditions and the Neumann boundary condition, respectively. The uniform estimates in the Mach number for the strong solutions are derived in a short time interval, provided that the initial density and temperature are close to the constant states. Thus, the solutions of the fully compressible Navier-Stokes equations converge to the the isentropic incompressible Navier-Stokes equations, as the Mach number tends to zero.

Keywords incompressible limit; full Navier-Stokes equations; Dirichlet boundary condition

1 Introduction

In this paper, we consider the low Mach number limit of the full Navier-stokes equations in a bounded domain $\Omega \subset \mathbf{R}^3$. The motion of subsonic viscous fluids is described by the following non-dimensional Navier-Stokes equations:

$$\rho_t + \mathrm{div}(\rho u) = 0, \tag{1.1}$$

$$(\rho u)_t + \mathrm{div}(\rho u \otimes u) - \mathrm{div}(2\mu D(u) + \xi \mathrm{div} u I) + \frac{1}{\epsilon^2}\nabla P = 0, \tag{1.2}$$

$$(\rho e)_t + \mathrm{div}(\rho u e) + P \mathrm{div} u - \mathrm{div}(\kappa \nabla \mathcal{T}) = \epsilon^2(2\mu|D(u)|^2 + \xi(\mathrm{div} u^2)), \tag{1.3}$$

---

* Adv. Math. (China), 2019, 48(6): 667–691. DOI: 10.11845/sxjz.2018082b.

where $\rho, u, P, e, \mathcal{T}$ stand for the density, velocity, pressure, internal energy and temperature, respectively. The constants μ and λ are viscous coefficients with $\mu > 0$, $2\mu + 3\xi \geqslant 0$, ϵ is the Mach number, κ is the heat conductivity coefficient, and $D(u) = (\nabla u + \nabla u^t)/2$. Moreover, we assume that the fluid is a polytropic ideal gas, that is,

$$e = C_v \mathcal{T}, \quad P = R\rho\mathcal{T}, \tag{1.4}$$

where $C_v > 0$ is the specific heat at constant volume, and R is the generic gases constant. The ratio of specific heats is $\gamma = 1 + R/C_v$.

By a formal derivation, as ϵ tends to zero, the solutions of (1.1)-(1.4) will converge to the solution (ρ, u, π) of the following problem:

$$\rho_t + \mathrm{div}(\rho u) = 0, \tag{1.5}$$

$$(\rho u)_t + \mathrm{div}(\rho u \otimes u) - \mathrm{div}(2\mu D(u) + \xi \mathrm{div} u I) + \nabla \pi = 0, \tag{1.6}$$

$$\gamma \mathrm{div} u = \mathrm{div}\left[\kappa \nabla\left(\frac{1}{\rho}\right)\right]. \tag{1.7}$$

This low Mach number limit process is singular, due to the large parameter $1/\epsilon^2$ in the momentum equation (1.2). It is a challenging mathematical problem to obtain uniform estimates in Mach number, which are necessary for the strong convergence to the background flows.

In fluid mechanics, the Mach number is an important physical quantity to determine whether the fluid is compressible or incompressible. If the Mach number is small, the fluid should behave asymptotically like an incompressible one, provided velocity and viscosity are small. As a result, the low Mach number limit has attracted much attention in recent years. Especially, in the isentropic case, a great number of results have been obtained during the last four decades since the pioneering works of [8]. The readers may refer to [5], [6], [14], [16], [17] and the references therein for details.

In the non-isentropic case, the large pressure gradient of $O(1/\epsilon^2)$ in the momentum equations is related to the behavior of both the density and the temperature. Thus, the low Mach number analysis is much more complicated. Note that the uniform estimates with respect to the Mach number are established in both the whole space or periodic space, such as [1], [3], [9], [10], [13], [14]. However, in the bounded domains, the geometry of the bounded domain leads to the boundary effects of the acoustic waves, which will reflect against the boundary, thus increasing the complexity of the low Mach number analysis. From the viewpoint of mathematical analysis, the difficulty is that the techniques of Fourier transform in [13] or pseudo-differential operators in [1], cannot be employed in the case of bounded boundary. Thus, it prevents us from using the usual way to balance the singular differential operators in the whole space or periodic space.

Recently, progress for the cases in the bounded domain for the full Navier-Stokes-Fourier equations have been obtained in [4], [7], [14], [20], for instance. When the velocity field satisfies the slip boundary condition, Dou et.al. [4] obtained uniform estimates by estimating the divergence and vorticity of velocity, respectively, thanks to the slip boundary conditions. In the case of no-slip boundary condition for velocity field, Jiang and Ou in [15] considered the low Mach number limit of the non-isentropic Navier-Stokes equations with zero thermal conductivity in 3-D bounded domains.

In this paper, we attempt to extend the results in [4] and [15], that is, to establish the low Mach number limit within a short time interval for (1.1)-(1.4) when the density and temperature are small perturbations around constant states, with positive heat conductivity and no-slip boundary conditions. The method used in [4] does not work here for the no-slip Dirichlet boundary condition since integrations by parts usually fail in estimating the high-order spacial derivations and the information of the vertical direction of velocity on the boundary is not available. We will mainly use the method applied in [15], in which they applied the idea in the works of Valli and Zajaczkowski in [22], [21], to introduce the isothermal coordinates in local regions to estimate the higher-order spatial derivatives of the velocity u near the boundary.

The low Mach number fluid can be regarded as a perturbation near the background isentropic fluid, where the density and temperature are usually set to be constants. Hence, we introduce the density and temperature variations by σ and θ, respectively, as follows:

$$\rho = 1 + \epsilon\sigma, \quad \mathcal{T} = 1 + \epsilon\theta.$$

Then the non-dimensional system (1.1)-(1.4) can be rewritten as the following form:

$$\sigma_t + \operatorname{div}(\sigma u) + \frac{1}{\epsilon}\operatorname{div} u = 0, \qquad (1.8)$$

$$\rho(u_t + u \cdot \nabla u) + \frac{R}{\epsilon}(\nabla\sigma + \nabla\theta) + R\nabla(\sigma\theta) = \mu\Delta u + \lambda\nabla\operatorname{div} u, \qquad (1.9)$$

$$C_v\rho(\theta_t+u\cdot\nabla\theta)+R(\rho\theta+\sigma)\mathrm{div}u+\frac{R}{\epsilon}\mathrm{div}u = \kappa\Delta\theta+\epsilon(2\mu|D(u)|^2+\xi(\mathrm{div}u)^2), \quad (1.10)$$

where $\lambda = \mu + \xi$. Furthermore, we suppose the following initial and boundary conditions:

$$(\sigma,u,\theta)|_{t=0} = (\sigma_0,u_0,\theta_0)(x) \quad \text{in } \Omega, \quad (1.11)$$

$$u = 0, \quad \frac{\partial\theta}{\partial n} = 0 \quad \text{on } \partial\Omega \times (0,T), \quad (1.12)$$

where $\Omega \in \mathbf{R}^3$ is a simply connected, bounded domain with a smooth boundary $\partial\Omega$, n is the unit outer normal to $\partial\Omega$.

2 Main results

In this paper, we will denote the norm in $H^k(\Omega)$ (the usual Sobolev space) by $\|\cdot\|_{H^k}$ for $k \geqslant 0$, where $H^0(\Omega) = L^2(\Omega)$, and denote the norm in $L^q(0,T;H^k(\Omega))$ by $\|\cdot\|_{L_t^q(H^k)}$, for $1 \leqslant q \leqslant \infty$, $k \geqslant 0$. The positive constants C, C_i for $i = 1, 2\cdots$ below depend only on Ω, μ, λ and κ, but not on the time t and ϵ.

First, the local existence results for the problem (1.8)-(1.12) can be established in a similar way as in [22].

Theorem 2.1 (Local existence) Let $\epsilon \in (0,1]$ be a fixed constant. Suppose that the initial dataum $(\sigma_0^\epsilon, u_0^\epsilon, \theta_0^\epsilon)$ satisfies the following conditions,

$$(\sigma_0^\epsilon, u_0^\epsilon, \theta_0^\epsilon) \in H^2(\Omega), \quad (\sigma_t^\epsilon(0), u_t^\epsilon(0), \theta_t^\epsilon(0)) \in L^2(\Omega),$$

with $1 + \epsilon\sigma_0^\epsilon \geqslant m$ for some constant $m > 0$. Moreover, assume the following compatibility conditions are satisfied:

$$\partial_t^k u^\epsilon(0) = 0 \quad \text{on } \partial\Omega, \quad k = 0, 1, \quad (2.1)$$

$$\frac{\partial_t^k \theta^\epsilon(0)}{\partial n} = 0 \quad \text{on } \partial\Omega, \quad k = 0, 1. \quad (2.2)$$

Then, there exists a constant $T = T(\sigma_0^\epsilon, u_0^\epsilon, \theta_0^\epsilon, m, \epsilon) > 0$ such that the initial-boundary problem (1.8)-(1.12) admits a unique solution $(\sigma^\epsilon, u^\epsilon, \theta^\epsilon)$ satisfying that $1 + \epsilon\theta^\epsilon > 0$ in $\Omega \times (0,T)$, and

$$(\sigma^\epsilon, u^\epsilon, \theta^\epsilon) \in C([0,T], H^2), \quad (\sigma_t^\epsilon, u_t^\epsilon, \theta_t^\epsilon) \in C([0,T], L^2),$$

$$(u^\epsilon, \theta^\epsilon) \in L^2(0,T;H^3), \quad (u_t^\epsilon, \theta_t^\epsilon) \in L^2(0,T;H^1).$$

Remark 2.1 To simplify the statement, we have used $u_t^\epsilon(0)$ to denote the quantity $u_t^\epsilon|_{t=0}$ which can be obtained from (1.9), that is,

$$(1+\epsilon\sigma_0^\epsilon)(u_t^\epsilon(0) + u_0^\epsilon \cdot \nabla u_0^\epsilon) + \frac{R}{\epsilon}(\nabla\sigma_0^\epsilon + \nabla\theta_0^\epsilon) + R\nabla(\sigma_0^\epsilon\theta_0^\epsilon) = \mu\Delta u_0^\epsilon + \lambda\nabla\mathrm{div} u_0^\epsilon.$$

Similarly, "$\sigma_t^\epsilon(0)$" and "$\theta_t^\epsilon(0)$" are given by (1.8) and (1.10), respectively.

Then the main results of this paper are stated as follows, which show the uniform estimates of strong solutions to (1.8)-(1.12), and the corresponding low Mach number limit.

Theorem 2.2 (1) Assume that $(\sigma^\epsilon, u^\epsilon, \theta^\epsilon)$ is the solution obtained in Theorem 2.1, where the initial datum $(\sigma_0^\epsilon, u_0^\epsilon, \theta_0^\epsilon)$ satisfies

$$\|u_0^\epsilon\|_{H^1} + \|(\sigma_0^\epsilon, \theta_0^\epsilon)\|_{H^2} + \|(\sigma_t^\epsilon, u_t^\epsilon, \theta_t^\epsilon)(0)\|_{L^2} + \|(1+\epsilon\sigma_0^\epsilon)^{-1}\|_{L^\infty} \leqslant D_0,$$

and the compatibility conditions (2.1)-(2.2). Then there exist positive constants T_0 and D such that $(\sigma^\epsilon, u^\epsilon, \theta^\epsilon)$ satisfies the uniform estimates:

$$\sup_{0\leqslant t\leqslant T_0} \left(\|u^\epsilon\|_{H^1} + \|(\sigma^\epsilon,\theta^\epsilon)\|_{H^2} + \|(\sigma_t^\epsilon, u_t^\epsilon, \theta_t^\epsilon)\|_{L^2} + \|(1+\epsilon\sigma^\epsilon)^{-1}\|_{L^\infty}\right)$$

$$+ \left[\int_0^{T_0} \|(u^\epsilon,\theta^\epsilon)\|_{H^3}^2 dt\right]^{\frac{1}{2}} + \left[\int_0^{T_0} \|(u_t^\epsilon, \theta_t^\epsilon)\|_{H^1}^2 dt\right]^{\frac{1}{2}} \leqslant D, \quad (2.3)$$

where D_0, T_0 and D are constants independent of $\epsilon \in (0,1]$.

(2) As the Mach number $\epsilon \to 0$, $u^\epsilon \to v$ strongly in $C([0,T^0]; H^s)$ for any $0 \leqslant s < 2$. There exists a function $\pi(x,t)$, such that (v,π) solves the initial-boundary value problem of the isentropic incompressible Navier-Stokes equations:

$$\mathrm{div}\, v = 0, \quad \text{in } \Omega \times (0, T_0),$$

$$v_t + v \cdot \nabla v + \nabla \pi = \mu \Delta v, \quad \text{in } \Omega \times (0, T_0),$$

$$v = 0, \quad \text{on } \partial\Omega \times (0, T_0),$$

$$v|_{t=0} = v_0, \quad \text{in } \Omega,$$

where v_0 is the strong limit of $u^\epsilon(x,0)$ in H^s as $\epsilon \to 0$.

Definition 2.3

$$M^\epsilon(t) \equiv \sup_{0 \leqslant s \leqslant t} \left(\|u^\epsilon\|_{H^1} + \|(\sigma^\epsilon, \theta^\epsilon)\|_{H^2} + \|(\sigma_s^\epsilon, u_s^\epsilon, \theta_s^\epsilon)\|_{L^2} + \|(1+\epsilon\sigma^\epsilon)^{-1}\|_{L^\infty}\right)$$

$$+ \left[\int_0^t \|(u_s^\epsilon, \theta_s^\epsilon)\|_{H^1}^2 ds\right]^{\frac{1}{2}} + \left[\int_0^t \|(u^\epsilon, \theta^\epsilon)\|_{H^3}^2 ds\right]^{\frac{1}{2}},$$

$$M_0^\epsilon \equiv M^\epsilon(t=0).$$

Similarly, as in [1,4], it suffices to show that the following theorem to get the uniform estimates in (2.3).

Theorem 2.4 *Let T^ϵ be the maximal time of existence for the initial-boundary value problem (1.8)-(1.12) in the sense of Theorem 2.1. Then for any $t \in [0, T^\epsilon)$, we have*

$$M^\epsilon(t) \leqslant C_0(M_0^\epsilon) \exp[t^{1/8} C(M^\epsilon(t))],$$

for some given nondecreasing continuous functions $C_0(\cdot)$ and $C(\cdot)$.

For simplicity, we will drop the superscript ϵ of $\rho^\epsilon, \sigma^\epsilon, u^\epsilon, \theta^\epsilon$, and we will write $M^\epsilon(t)$, M_0^ϵ as M, M_0, respectively, for short.

This article is organized as follows. In Section 2, we state some elementary lemmas and calculus inequalities for the convenience of readers. In Section 3, uniform estimates are shown by employing a careful analysis of both temporal and spatial derivatives. In Section 4, we give the proofs of Theorems 1.2 and 1.4.

3 Preliminaries

In this section, we list some lemmas which will be frequently used throughout this paper.

Lemma 3.1 (Poincaré's inequality) *There exists a positive constant C such that*

$$\|u\|_{L^2(\Omega)} \leqslant C(\|\nabla u\|_{L^2(\Omega)} + \|u\|_{L^2(\partial\Omega)}),$$

for any $u \in H^1(\Omega)$.

Lemma 3.2 (See [2]) *Let Ω be a bounded domain in $\mathbf{R}^N$ with a smooth boundary $\partial\Omega$ and outward normal n. Then, for any $u \in H^1(\Omega)$, $s \geqslant 1$, there exists a constant $C > 0$ independent of u, such that*

$$\|u\|_{H^s(\Omega)} \leqslant C(\|\mathrm{div} u\|_{H^{s-1}(\Omega)} + \|\mathrm{curl} u\|_{H^{s-1}(\Omega)} + \|u \cdot n\|_{H^{s-\frac{1}{2}}(\partial\Omega)} + \|u\|_{H^{s-1}(\Omega)}),$$

where the vorticity $\mathrm{curl} u = (\partial_2 u_3 - \partial_3 u_2, \partial_3 u_1 - \partial_1 u_3, \partial_1 u_2 - \partial_2 u_1)^\mathrm{T}$.

Lemma 3.3 (See [11]) *Let $\Omega \subset \mathbf{R}^N$ be a bounded domain with C^k-boundary and u be a function in $W^{k,r}(\Omega)$ with $1 \leqslant r, q \leqslant \infty$. For any integer j with $0 \leqslant j < k$, and for any number a in the interval $[j/k, 1]$, set*

$$\frac{1}{p} = \frac{j}{N} + a\left(\frac{1}{r} - \frac{k}{N}\right) + (1-a)\frac{1}{q}.$$

If $k-j-N/r$ is not a nonnegative integer, then

$$\|D^j u\|_{L^p(\Omega)} \leqslant C\|u\|_{W^{k,r}(\Omega)}^a \|u\|_{L^q(\Omega)}^{1-a}. \qquad (3.1)$$

If $k-j-N/r$ is a nonnegative integer, then (3.1) only holds for $a = j/k$. The constant C depends only on Ω, r, q, k, j, a.

In what follows, we will use the fact that

$$\Delta u = \nabla \mathrm{div} u - \mathrm{curl}\mathrm{curl} u. \qquad (3.2)$$

4 Uniform estimates

4.1 Estimates for ρ

Note that ρ and its derivatives always appear as a coefficient of (σ, u, θ) and their derivatives. Thus, for convenience, we estimate ρ in terms of the initial data and u, by using the conservation of mass equation, namely the equation (1.1), and the proof can be found in [19]. Here, we only give the results.

Lemma 4.1 For any $0 \leqslant t \leqslant T^\epsilon$,

$$(\|\rho(t)\|_{H^2} + \|\rho_t(t)\|_{L^2} + \|\rho(t)^{-1}\|_{L^\infty})(t) \leqslant C_0(M_0)\exp(\sqrt{t}C(M)).$$

4.2 L^2-estimates for (σ, u, θ)

Lemma 4.2 For any $0 \leqslant t \leqslant \min\{T^\epsilon, 1\}$,

$$\|(\sigma, u, \theta)\|_{L^2}^2 + \|(u,\theta)\|_{L_t^2(H^1)}^2 ds \leqslant C_0(M_0)\exp(\sqrt{t}C(M)).$$

Proof Multiplying (1.8)-(1.10) by $R\sigma, u$ and θ, respectively, and integrating over $\Omega \times (0, t)$, we obtain

$$\frac{1}{2}\|(\sqrt{R}\sigma, \sqrt{\rho}u, \sqrt{C_v\rho}\theta)\|_{L^2}^2 + \int_0^t \|(\sqrt{\mu}\nabla u, \sqrt{\lambda}\mathrm{div}u, \sqrt{\kappa}\nabla\theta)\|_{L^2}^2 ds$$

$$= \int_0^t \int_\Omega [\mathrm{div} u(-R\rho\theta^2 - \frac{1}{2}R\sigma^2) + \epsilon\theta(2\mu|D(u)|^2 + \xi|\mathrm{div}u|^2)]dxds.$$

Using Lemma 4.1, we get

$$\|(\sigma, u, \theta)\|_{L^2}^2 + \|(u, \theta)\|_{L_t^2(H^1)}^2$$

$$\leqslant C_0(M_0) + C\int_0^t \|u\|_{H^1}(\|\rho\|_{H^2}\|\theta\|_{H^1}^2 + \|\sigma\|_{H^1}^2 + \|\theta\|_{H^2}\|u\|_{H^1}^2)ds$$

$$\leqslant C_0(M_0)\exp(\sqrt{t}C(M)).$$

4.3 Estimates for first-order derivatives of (σ, u, θ)

Next, we estimate the first-order spatial derivations.

Lemma 4.3 *For any* $0 \leqslant t \leqslant \min\{T^\epsilon, 1\}$,

$$\|(\nabla\sigma, \text{div}u, \nabla\theta)\|_{L^2}^2 + \|(\nabla\text{div}u, \Delta\theta)\|_{L_t^2(L^2)}^2 \leqslant C_0(M_0)\exp(\sqrt{t}C(M)).$$

Proof We integrate the product of $\nabla\text{div}u$ and (1.9) over $\Omega \times (0, t)$ to show

$$\frac{1}{2}\|\sqrt{\rho}\text{div}u\|_{L^2}^2 + (\mu+\lambda)\|\nabla\text{div}u\|_{L_t^2(L^2)}^2 - \frac{R}{\epsilon}\int_0^t\int_\Omega (\nabla\sigma + \nabla\theta)\cdot\text{div}u dxds$$

$$\leqslant C_0(M_0) + \int_0^t\int_\Omega \left(\frac{1}{2}\rho_t(\text{div}u)^2 - \nabla\rho\cdot u_t\text{div}u\right)dxds$$

$$+ \int_0^t\int_\Omega (\rho u\cdot\nabla u + R\nabla(\sigma\theta))\cdot\nabla\text{div}u dxds + \mu\int_0^t\int_\Omega \text{curlcurl}u\cdot\nabla\text{div}u dxds.$$

By using the Hölder inequality, (3.1) and Lemma 4.1, we can obtain

$$\frac{1}{2}\|\sqrt{\rho}\text{div}u\|_{L^2}^2 + (\mu+\lambda)\|\nabla\text{div}u\|_{L_t^2(L^2)}^2 - \frac{R}{\epsilon}\int_0^t\int_\Omega (\nabla\sigma + \nabla\theta)\cdot\text{div}u dxds$$

$$\leqslant C_0(M_0)\exp(\sqrt{t}C(M)). \tag{4.1}$$

To balance the singular term on the left-hand side of the above inequality, we apply ∇ to (1.8) and multiply the resulting equality by $R\nabla\sigma$ over $L^2(\Omega) \times (0, t)$ to derive

$$\frac{R}{2}\|\nabla\sigma(t)\|_{L^2}^2 + \frac{R}{\epsilon}\int_0^t\int_\Omega \nabla\text{div}u\cdot\nabla\sigma dxds$$

$$= \frac{R}{2}\|\nabla\sigma_0\|_{L^2}^2 - R\int_0^t\int_\Omega \nabla\text{div}(\sigma u)\cdot\nabla\sigma dxds$$

$$\leqslant C_0(M_0) + C\sqrt{t}\|\sigma\|_{L_t^\infty(H^2)}^2\int_0^t\|u\|_{H^3}^2 ds$$

$$\leqslant C_0(M_0)\exp(\sqrt{t}C(M)). \tag{4.2}$$

Similarly, from (1.9) and (1.10), we can deduce

$$\frac{C_v}{2}\|\sqrt{\rho}\nabla\theta(t)\|_{L^2}^2 + \kappa\|\Delta\theta\|_{L_t^2(L^2)}^2 + \frac{R}{\epsilon}\int_0^t\int_\Omega \nabla\text{div}u\cdot\nabla\theta dxds$$

$$\leqslant \frac{C_v}{2}\|\sqrt{\rho_0}\nabla\theta_0\|_{L^2}^2 + C\int_0^t [\|\rho_t\|_{L^2}\|\theta\|_{H^1}^2 + \|\rho\|_{H^2}\|\theta_t\|_{L^2}\|\nabla\theta\|_{H^1}$$

$$+ (\|\rho\|_{H^2}\|\theta\|_{H^2} + \|\sigma\|_{H^2})\|u\|_{H^2}\|\nabla\theta\|_{H^1} + \epsilon\|\nabla\|^2_{H^1}\|\nabla\theta\|_{H^1}]ds$$
$$\leqslant C_0(M_0)\exp(\sqrt{t}C(M)). \tag{4.3}$$

By collecting (4.1)-(4.3) and applying Lemma 4.1, we can obtain the results.

According to Lemma 3.2, we need to estimate $\|\mathrm{curl}u\|^2_{L^2}$.

Lemma 4.4 For any $0 \leqslant t \leqslant \min(T^\epsilon, 1)$,
$$\|\mathrm{curl}u\|^2_{L^2} \leqslant C_0(M_0)\exp(\sqrt{t}C(M)).$$

Proof Let $\omega = \mathrm{curl}u$. From (1.9), we easily derive the following equation for
$$\rho(\omega_t + u \cdot \nabla\omega) = \mu\Delta\omega + K, \tag{4.4}$$

where
$$K = -\left(\partial_2\rho u_{3t} - \partial_3\rho u_{2t}, \partial_3\rho u_{1t} - \partial_1\rho u_{3t}, \partial_1\rho u_{2t} - \partial_2\rho u_{1t}\right)$$
$$-\left(\partial_2(\rho u_i)\partial_i u_3 - \partial_3(\rho u_i)\partial_i u_2, \partial_3(\rho u_i)\partial_i u_1 - \partial_1(\rho u_i)\partial_i u_3,\right.$$
$$\left.\partial_1(\rho u_i)\partial_i u_2 - \partial_2(\rho u_i)\partial_i u_1\right).$$

Multiplying (4.4) by ω and integrating the resulting equation over $\Omega \times (0,t)$, we derive
$$\frac{1}{2}\|\sqrt{\rho}\mathrm{curl}u\|^2_{L^2} \leqslant C_0(M_0) + \int_0^t\int_\Omega K \cdot \omega dxds + \mu\int_0^t\int_\Omega \Delta\omega \cdot \omega dxds$$
$$\leqslant C_0(M_0) + \int_0^t \|\rho\|_{H^2}\|u\|_{H^1}(\|u_t\|_{H^1} + \|u\|^2_{H^2})ds$$
$$+ \int_0^t \|\Delta\omega\|_{L^2}\|\omega\|_{L^2}ds$$
$$\leqslant C_0(M_0) + \sqrt{t}C(M).$$

Next, we show the estimates for the first-order temporal derivatives $(\sigma_t, u_t, \theta_t)$. Differentiating (1.8)-(1.10) with respect to t, respectively, we have
$$\sigma_{tt} + \frac{1}{\epsilon}\mathrm{div}u_t = -\mathrm{div}(\sigma u)_t, \tag{4.5}$$
$$\rho(u_{tt} + u \cdot \nabla u_t) + \frac{R}{\epsilon}(\nabla\sigma_t + \nabla\theta_t) - 2\mu\mathrm{div}(D(u_t)) - \xi\nabla\mathrm{div}u_t$$

$$= -\rho_t u_t - (\rho u)_t \cdot \nabla u - R\nabla(\sigma\theta)_t, \tag{4.6}$$

$$C_v\rho(\theta_{tt} + u \cdot \nabla\theta_t) - \kappa\Delta\theta_t + \frac{R}{\epsilon}\text{div}u_t = \epsilon(2\mu|D(u)|^2 + \xi(\text{div}u)^2)_t$$
$$- C_v(\rho_t\theta_t + (\rho u)_t \cdot \nabla\theta) + R((\rho\theta + \sigma)\text{div}u)_t. \tag{4.7}$$

Lemma 4.5 For any $0 \leqslant t \leqslant \min(T^\epsilon, 1)$,

$$\|(\sigma_t, u_t, \theta_t)\|_{L^2}^2 + \|(u_t, \theta_t)\|_{L_t^2(H^1)}^2 \leqslant C_0(M_0)\exp(t^{\frac{1}{8}}C(M)).$$

Proof Multiplying (4.5), (4.6) and (4.7) by $R\sigma_t$, u_t, θ_t, respectively, and integrating over $\Omega \times (0, t)$, we reach

$$\frac{1}{2}\|(\sqrt{R}\sigma_t, \sqrt{\rho}, \sqrt{C_v\rho}\theta_t)\|_{L^2}^2 + \|(\sqrt{\mu}\nabla u_t, \sqrt{\mu+\xi}\text{div}u_t, \sqrt{\kappa}\nabla\theta_t)\|_{L_t^2(L^2)}^2$$
$$\leqslant C_0(M_0) - \int_0^t\int_\Omega R\sigma_t\text{div}(\sigma u)_t dxds - \int_0^t\int_\Omega u_t\Big[\rho_t u_t + (\rho u)_t \cdot \nabla u$$
$$+ R\nabla(\sigma\theta)_t\Big]dxds + \int_0^t\int_\Omega \theta_t\Big[\epsilon(2\mu|D(u)|^2 + \xi(\text{div}u)^2)_t$$
$$- C_v(\rho_t\theta_t + (\rho u)_t \cdot \nabla\theta) - R((\rho\theta+\sigma)\text{div}u)_t\Big]dxds$$
$$= C_0(M_0) + I_1 + I_2 + I_3, \tag{4.8}$$

where

$$|I_1| \leqslant \int_0^t\int_\Omega \sigma_t(\nabla\sigma_t u + \sigma_t\text{div}u + \nabla\sigma u_t + \sigma\text{div}u_t)dxds$$
$$\leqslant \int_0^t (\|\sigma_t\|_{H^1}^{\frac{3}{2}}\|\sigma_t\|_{L^2}^{\frac{1}{2}}\|u\|_{H^1} + \|u_t\|_{H^1}^{\frac{1}{2}}\|u_t\|_{L^2}^{\frac{1}{2}}\|\sigma_t\|_{L^2}\|\sigma\|_{H^2}$$
$$+ \|u_t\|_{H^1}\|\sigma_t\|_{L^2}\|\sigma\|_{H^2})ds$$
$$\leqslant t^{\frac{1}{4}}C(M). \tag{4.9}$$

$$|I_2| \leqslant \int_0^t (\|u_t\|_{L^4}^2\|\rho_t\|_{L^2} + \|u\|_{L^\infty}\|u_t\|_{L^6}\|\nabla u\|_{L^3}\|\rho_t\|_{L^2}$$
$$+ \|u_t\|_{L^2}\|\nabla\sigma_t\|_{L^2}\|\theta\|_{H^2} + \|u_t\|_{L^2}\|\nabla\theta_t\|_{L^2}\|\sigma\|_{H^2})ds$$
$$\leqslant \int_0^t (\|u_t\|_{H^1}^{\frac{3}{2}}\|u_t\|_{L^2}^{\frac{1}{2}}\|\rho_t\|_{L^2} + \|u\|_{H^3}^{\frac{3}{4}}\|u_t\|_{H^1}\|u\|_{H^1}^{\frac{5}{4}}\|\rho_t\|_{L^2}$$
$$+ \|u_t\|_{L^2}\|\sigma_t\|_{H^1}\|\theta\|_{H^2} + \|u_t\|_{L^2}\|\theta_t\|_{H^1}\|\sigma\|_{H^2})ds$$

$$\leq t^{\frac{1}{4}}C(M) + \left(\int_0^t \|u\|_{H^3}^{\frac{7}{4}}ds + \int_0^t \|u_t\|_{H^1}^{\frac{7}{4}}ds\right)\|u\|_{L_t^\infty(H^1)}^{\frac{5}{4}}\|\rho_t\|_{L_t^\infty(L^2)}$$

$$\leq t^{\frac{1}{8}}C(M). \tag{4.10}$$

$$|I_3| \leq \int_0^t \|\nabla u\|_{L^\infty}\|\theta_t\|_{L^2}\|u_t\|_{H^1}ds + \sqrt{t}C(M)$$

$$\leq \int_0^t \|u\|_{H^3}^{\frac{3}{4}}\|u\|_{H^1}^{\frac{1}{4}}\|\theta_t\|_{L^2}\|u_t\|_{H^1}ds + \sqrt{t}C(M)$$

$$\leq t^{\frac{1}{8}}C(M). \tag{4.11}$$

Thus, inserting the estimates (4.9)-(4.11) into (4.8) and applying Lemma 4.1 we can obtain the results.

4.4 Stokes problem

Lemma 4.6 *There exist positive constants such that*

$$\|u\|_{H^2}^2 + \frac{1}{\epsilon^2}\|(\nabla\sigma, \nabla\theta)\|_{L^2}^2$$

$$\leq C\|u\|_{H^3}\|u\|_{H^1} + C\|\rho\|_{H^2}^2(\|u_t\|_{L^2}^2 + \|u\|_{H^3}^{\frac{3}{2}}\|u\|_{H^1}^{\frac{7}{2}}) + C\|\sigma\|_{H^2}^2\|\theta\|_{H^2}^2, \tag{4.12}$$

$$\|u\|_{H^3}^2 + \frac{1}{\epsilon^2}\|(\nabla\sigma, \nabla\theta)\|_{H^1}^2$$

$$\leq C\|\nabla^2\mathrm{div}u\|_{L^2}^2 + C\|\rho\|_{H^2}^2(\|u_t\|_{H^1}^2 + \|u\|_{H^3}^{\frac{3}{2}}\|u\|_{H^1}^{\frac{5}{2}}) + C\|\sigma\|_{H^2}^2\|\theta\|_{H^2}^2. \tag{4.13}$$

Proof We rewrite (1.9) and (1.11) as an inhomogeneous Stokes problem to derive the desired bounds.

$$\begin{aligned}
-\mu\Delta u + \frac{R}{\epsilon}(\nabla\sigma + \nabla\theta) &= f \quad \text{in } \Omega, \\
\mathrm{div}u &= g \quad \text{in } \Omega, \\
u &= 0 \quad \text{on } \partial\Omega,
\end{aligned} \tag{4.14}$$

where $f = (\mu+\xi)\nabla\mathrm{div}u - \rho(u_t + u\cdot\nabla u) - R\nabla(\sigma\theta)$. We begin with the estimates of low order. By the usual estimates for the steady Stokes problem [12], we have

$$\|u\|_{H^2}^2 + \frac{1}{\epsilon^2}\|(\nabla\sigma, \nabla\theta)\|_{L^2}^2$$

$$\leq C(\|g\|_{H^1}^2 + \|f\|_{L^2}^2)$$

$$\leqslant C\|u\|_{H^3}\|u\|_{H^1} + C\|\rho\|_{H^2}^2(\|u_t\|_{L^2}^2 + \|u\|_{H^3}^{\frac{1}{2}}\|u\|_{H^1}^{\frac{7}{2}}) + C\|\sigma\|_{H^2}^2\|\theta\|_{H^2}^2,$$

and

$$\|u\|_{H^3}^2 + \frac{1}{\epsilon^2}\|(\nabla\sigma, \nabla\theta)\|_{H^1}^2$$

$$\leqslant C(\|g\|_{H^2}^2 + \|f\|_{H^1}^2)$$

$$\leqslant C\|\nabla^2 \text{div} u\|_{L^2}^2 + C\|\rho\|_{H^2}^2(\|u_t\|_{H^1}\|u_t\|_{L^2} + \|u\|_{H^3}\|u\|_{H^1}^3$$

$$+ \|u_t\|_{H^1}^2 + \|u\|_{H^3}^{\frac{3}{2}}\|u\|_{H^1}^{\frac{5}{2}} + \|\sigma\|_{H^2}^2\|\theta\|_{H^2}^2)$$

$$\leqslant C\|\nabla^2 \text{div} u\|_{L^2}^2 + C\|\rho\|_{H^2}^2(\|u_t\|_{H^1}^2 + \|u\|_{H^3}^{\frac{3}{2}}\|u\|_{H^1}^{\frac{5}{2}}) + C\|\sigma\|_{H^2}^2\|\theta\|_{H^2}^2.$$

Hence, we can obtain the results.

4.5 Estimates of high order differentials

We adapt the idea, due to Valli et al. [21, 22], by dividing it into the interior part and the part near the boundary.

4.5.1 Interior estimates

In fact, we will give an interior estimate of $\|\nabla^3 u\|_{L^2_t(L^2)}$ and $\|\nabla^3 \theta\|_{L^2}^2$. For simplicity, we will use the Einstein summation convention in what follows.

Lemma 4.7 *For any* $0 \leqslant t \leqslant \min\{T^\epsilon, 1\}$,

$$\frac{1}{2}\|\chi_0 \partial_{jk} u\|_{L^2}^2 + \frac{R}{2}\|\chi_0 \partial_{jk}\sigma\|_{L^2}^2 + \frac{C_v}{2}\|\chi_0 \partial_{jk}\theta\|_{L^2}^2$$

$$+ \|\chi_0 \nabla^3 u\|_{L^2_t(L^2)}^2 + \|\chi_0 \nabla^3 \theta\|_{L^2_t(L^2)}^2$$

$$\leqslant C_0(M_0)\exp(t^{\frac{1}{8}}C(M)) + \delta \int_0^t \frac{1}{\epsilon^2}\|(\nabla^2\sigma, \nabla^2\theta)\|_{L^2}^2 ds,$$

where $\chi_0 \in C_0^\infty(\Omega)$, δ *is a small enough positive constant.*

Proof We apply ∂_{jk} to (1.9) and multiply by $\chi_0^2 \partial_{jk} u$ to integrate over $(0,t) \times \Omega$, then we have

$$\frac{1}{2}\|\sqrt{\rho}\chi_0 \partial_{jk}u\|_{L^2}^2 + \mu \int_0^t \|\chi_0 \partial_{jk}\nabla u\|_{L^2}^2 ds$$

$$+ \frac{R}{\epsilon}\int_0^t \int_\Omega \chi_0^2(\partial_{jk}\nabla\theta + \partial_{jk}\nabla\sigma)\partial_{jk}u dxds$$

$$\leqslant C_0(M_0) + \frac{1}{2}\int_0^t \int_\Omega \chi_0^2 \rho_t |\partial_{jk}u|^2 dxds - 2\int_0^t \int_\Omega \chi_0[\mu\partial_{jk}\nabla u \cdot (\chi_0 \partial_{jk}u)$$

$$+ (\mu+\xi)\partial_{jk}\text{div}u\nabla\chi_0 \cdot \partial_{jk}u]dxds - R\int_0^t\int_\Omega \chi_0^2\partial_{jk}\nabla(\sigma\theta)\cdot\partial_{jk}udxds$$

$$-\int_0^t\int_\Omega \chi_0^2\partial_{jk}u\rho\partial_{jk}(u\cdot\nabla u)dxds - \int_0^t\int_\Omega \chi_0^2\partial_{jk}u[\partial_j\rho\partial_k(u_t+u\cdot\nabla u)$$

$$+\partial_k\rho\partial_j(u_t+u\cdot\nabla u)]dxds - \int_0^t\int_\Omega \chi_0^2\partial_{jk}u\partial_{jk}\rho(u_t+u\cdot\nabla u)dxds$$

$$=C_0(M_0)+\sum_{i=1}^6 J_i, \tag{4.15}$$

where

$$J_4 = -\int_0^t\int_\Omega \chi_0^2\partial_{jk}u_i\rho(\partial_ju_l\partial_{kl}u_i+\partial_ku_l\partial_{kl}u_i+\partial_{jk}u_l\partial_lu_i)dxds$$

$$-\int_0^t\int_\Omega \chi_0^2\partial_{jk}u_i\rho u_l\partial_{jkl}u_idxds$$

$$=-\int_0^t\int_\Omega \chi_0^2\partial_{jk}u_i\rho(\partial_ju_l\partial_{kl}u_i+\partial_ku_l\partial_{kl}u_i+\partial_{jk}u_l\partial_lu_i)dxds$$

$$+\int_0^t\int_\Omega \chi_0\partial_l\chi_0\rho u_l|\partial_{jk}u|^2 dxds - J_1.$$

Hence,

$$|J_1+J_4| \leq \int_0^t\int_\Omega \chi_0^2\partial_{jk}u_i\rho(\partial_ju_l\partial_{kl}u_i+\partial_ku_l\partial_{kl}u_i+\partial_{jk}u_l\partial_lu_i)dxds$$

$$+\int_0^t\int_\Omega \chi_0\partial_l\chi_0\rho u_l|\partial_{jk}u|^2 dxds$$

$$\leq C\int_0^t(\|\nabla^2 u\|_{L^2}\|\nabla^2 u\|_{L^3}\|\nabla u\|_{L^6}+\|\nabla^2 u\|_{L^4}^2\|u\|_{L^2})ds$$

$$\leq t^{\frac{1}{8}}C(M).$$

$$|J_2| \leq C\int_0^t \|\nabla^2 u\|_{L^2}\|u\|_{H^3}ds \leq t^{\frac{1}{4}}C(M).$$

$$|J_3| \leq R\int_0^t\int_\Omega |\partial_{jk}(\sigma\theta)\chi_0^2\partial_{jk}\text{div}u|+2|\partial_{jk}(\sigma\theta)\chi_0\nabla\chi_0\cdot\partial_{jk}u|dxds$$

$$\leq t^{\frac{1}{8}}C(M).$$

$$|J_5+J_6| \leq C\int_0^t \|\rho\|_{H^2}\|\nabla^2 u\|_{L^3}(\|u_t\|_{H^1}+\|\nabla u\|_{L^4}^2+\|\nabla^2 u\|_{L^2}\|u\|_{L^\infty})ds$$

$$\leq t^{\frac{1}{4}}C(M).$$

Substituting the estimates of $J_i(i = 1, \cdots, 6)$ into (4.15), we arrive at

$$\frac{1}{2}\|\chi_0\partial_{jk}u\|_{L^2}^2 + \mu\int_0^t \|\chi_0\partial_{jk}\nabla u\|_{L^2}^2 ds + \frac{R}{\epsilon}\int_0^t\int_\Omega \chi_0^2(\partial_{jk}\nabla\theta$$
$$+ \partial_{jk}\nabla\sigma)\cdot\partial_{jk}u dxds$$
$$\leqslant C_0(M_0)\exp(t^{\frac{1}{8}}C(M)). \tag{4.16}$$

To balance the singular term on the left-hand side of the above inequality, we apply ∂_{jk} to (1.8) and multiply the resulting equality by $R\chi_0^2\partial_{jk}\sigma$ over $L^2(\Omega) \times (0, t)$ to get

$$\frac{R}{2}\|\chi_0\partial_{jk}\sigma\|_{L^2}^2 + \frac{R}{\epsilon}\int_0^t\int_\Omega \chi_0^2\partial_{jk}\sigma\partial_{jk}\text{div}u dxds$$
$$= \frac{R}{2}\|\chi_0\partial_{jk}\sigma_0\|_{L^2}^2 - R\int_0^t\int_\Omega (\partial_j\sigma\partial_k\text{div}u + \partial_k\sigma\partial_j\text{div}u$$
$$+ \sigma\partial_{jk}\text{div}u + \partial_j\nabla\sigma\cdot\partial_k u + \partial_k\nabla\sigma\cdot\partial_j u + \nabla\sigma\cdot\partial_{jk}u)\chi_0^2\partial_{jk}\sigma dxds$$
$$+ \frac{R}{2}\int_0^t\int_\Omega \partial_{jk}\sigma\text{div}u\chi_0^2\partial_{jk}\sigma dxds + R\int_0^t\int_\Omega \partial_{jk}\sigma u\cdot\nabla\chi_0\chi_0\partial_{jk}\sigma dxds$$
$$\leqslant C_0(M_0) + C\int_0^t (\|\sigma\|_{H^2}\|u\|_{H^3} + \|\sigma\|_{H^2}^2\|u\|_{H^3}^{\frac{3}{4}}\|u\|_{H^1}^{\frac{1}{4}})ds$$
$$\leqslant C_0(M_0)\exp(t^{\frac{1}{4}}C(M)). \tag{4.17}$$

Similarly, applying ∂_{jk} to (1.10) and multiplying the resulting equality by $\chi_0^2\partial_{jk}\theta$ over $(\Omega) \times (0, t)$, we obtain

$$\frac{C_v}{2}\|\chi_0\sqrt{\rho_0}\partial_{jk}\theta\|_{L^2}^2 + \int_0^t \kappa\|\chi_0\partial_{jk}\nabla\theta\|_{L^2}^2 ds + \frac{R}{\epsilon}\int_t^0\int_\Omega \chi_0^2\partial_{jk}\theta\partial_{jk}\text{div}u dxds$$
$$= \frac{C_v}{2}\|\chi_0\sqrt{\rho_0}\partial_{jk}\theta_0\|_{L^2}^2 + \frac{C_v}{2}\int_0^t\int_\Omega \rho_t|\partial_{jk}\theta|^2 dxds$$
$$+ C_v\int_0^t\int_\Omega [\rho\partial_{jk}(u\cdot\nabla\theta)\cdot\chi_0^2\partial_{jk}\theta + \partial_{jk}\rho u\cdot\nabla\theta\chi_0^2\partial_{jk}\theta]dxds$$
$$+ C_v\int_0^t\int_\Omega [\partial_j\rho\partial_k(u\cdot\nabla\theta) + \partial_k\rho\partial_j(u\cdot\nabla\theta)]\cdot\chi_0^2\partial_{jk}\theta dxds$$
$$+ R\int_0^t\int_\Omega [\partial_{jk}(\rho\theta + \sigma)\text{div}u\cdot\chi_0^2\partial_{jk}\theta + (\rho\theta + \sigma)\partial_{jk}\text{div}u\cdot\chi_0^2\partial_{jk}\theta]dxds$$
$$+ R\int_0^t\int_\Omega [\partial_j(\rho\theta + \sigma)\partial_k\text{div}u + \partial_k(\rho\theta + \sigma)\partial_j\text{div}u]\cdot\chi_0^2\partial_{jk}\theta dxds$$

$$+ C_v \int_0^t \int_\Omega (\partial_{jk}\rho\theta_t + \partial_j\rho\partial_k\theta_t + \partial_k\rho\partial_j\theta_t) \cdot \chi_0^2 \partial_{jk}\theta dxds$$

$$+ \epsilon \int_0^t \int_\Omega \partial_{jk}(2\mu|D(u)|^2 + \xi(\text{div}u)^2) \cdot \chi_0^2 \partial_{jk}\theta dxds$$

$$= \frac{C_v}{2}\|\chi_0\sqrt{\rho_0}\partial_{jk}\theta_0\|_{L^2}^2 + \sum_{i=7}^{13} J_i, \qquad (4.18)$$

where

$$|J_7| \leqslant C\int_0^t \|\rho_t\|_{L^2}\|\nabla^2\|_{L^4}^2 ds \leqslant t^{\frac{1}{8}}C(M),$$

$$|J_8 + J_9| \leqslant C\int_0^t \|\rho\|_{H^2}(\|\nabla^2 u\|_{L^2}\|\nabla^2\theta\|_{L^3}\|\theta\|_{H^2}$$
$$+ \|u\|_{L^\infty}\|\nabla^2\theta\|_{L^2}\|\nabla^3\theta\|_{L^2})ds$$
$$\leqslant t^{\frac{3}{8}}C(M),$$
$$|J_{10} + J_{11}| \leqslant t^{\frac{3}{8}}C(M),$$

$$|J_{12}| \leqslant \int_0^t \|\rho\|_{H^2}\|\theta_t\|_{H^1}\|\nabla^2\theta\|_{L^3} ds \leqslant t^{\frac{1}{8}}C(M),$$

$$|J_{13}| \leqslant \int_0^t (\|\theta\|_{H^2}\|\nabla^3 u\|_{L^2}\|\nabla u\|_{L^\infty} + \|\nabla^2 u\|_{L^4}^2\|\theta\|_{H^2})ds$$
$$\leqslant t^{\frac{1}{8}}C(M).$$

Note that, the singular terms in (4.16)-(4.18) can be estimated as follow:

$$\frac{R}{\epsilon}\int_0^t \int_\Omega \chi_0^2(\partial_{jk}\nabla\theta + \partial_{jk}\nabla\sigma) \cdot \partial_{jk}u dxds$$

$$+ \frac{R}{\epsilon}\int_t^0 \int_\Omega \chi_0^2(\partial_{jk}\sigma + \partial_{jk}\theta)\partial_{jk}\text{div}u dxds$$

$$= -\frac{2R}{\epsilon}\int_0^t \int_\Omega \chi_0\nabla\chi_0(\partial_{jk}\sigma + \partial_{jk}\theta)\partial_{jk}u dxds$$

$$\leqslant \int_0^t \left(\delta\frac{1}{\epsilon^2}\|(\nabla^2\sigma, \nabla^2\theta)\|_{L^2}^2 + C_\delta\|\nabla^2 u\|_{L^2}^2\right) ds$$

$$\leqslant \delta\int_0^t \frac{1}{\epsilon^2}\|(\nabla^2\sigma, \nabla^2\theta)\|_{L^2}^2 ds + \sqrt{t}C(M),$$

where δ is small enough constant. Then we finish this lemma.

4.5.2 Boundary estimates

In order to derive the estimates near the boundary, we construct the local coordinates by the isothermal coordinates $\lambda(\psi,\varphi)$ (see [22]), and $\lambda(\psi,\varphi)$ satisfies

$$\lambda_\psi \cdot \lambda_\psi > 0, \quad \lambda_\varphi \cdot \lambda_\varphi > 0, \quad \text{and} \quad \lambda_\psi \cdot \lambda_\varphi = 0,$$

where $\lambda_\psi = \dfrac{\partial \lambda}{\partial \psi}, \lambda_\varphi = \dfrac{\partial \lambda}{\partial \varphi}$.

We cover the boundary $\partial\Omega$ by a finite number of bounded open sets $W^k \subset \mathbf{R}^3$, $k = 1, 2, \cdots, L$, such that for any $x \in W^k \bigcap \Omega$,

$$x = \lambda^k(\psi,\varphi) + rn(\lambda^k(\psi,\varphi)) = \Lambda^k(\psi,\varphi,r),$$

where n is the unit outer normal to $\partial\Omega$. For simplicity, in what follows, we omit the superscript k in each W^k. Then, we construct the orthonormal system corresponding to the local coordinates by

$$e_1 = \frac{\lambda_\psi}{|\lambda_\psi|}, \quad e_2 = \frac{\lambda_\varphi}{|\lambda_\varphi|}, \quad e_3 = n \circ \lambda = e_1 \times e_2,$$

where the notation '$\circ$' stands for the composition of operators.

By a straightforward calculation, we see that $J \in C^2$ and

$$J = \det \mathrm{Jac}\Lambda = (\Lambda_\psi \times \Lambda_\varphi) \cdot e_3$$
$$= |\lambda_\psi||\lambda_\varphi| + r(|\lambda_\psi|n_\varphi \cdot e_2 + |\lambda_\varphi|n_\psi \cdot e_1)$$
$$+ r^2[(n_\psi \cdot e_1)(n_\varphi \cdot e_2)(n_\varphi \cdot e_2) - (n_\psi \cdot e_2)(n_\varphi \cdot e_1)] > 0,$$

for sufficiently small $r > 0$. According to $\mathrm{Jac}\Lambda^{-1} \circ \Lambda = (\mathrm{Jac}\Lambda)^{-1}$, we can easily derive the following relations:

$$[\nabla(\Lambda^{-1})^1] \circ \Lambda = (1/J)(\Lambda_\varphi \times e_3),$$
$$[\nabla(\Lambda^{-1})^2] \circ \Lambda = (1/J)(e_3 \times \Lambda_\varphi),$$
$$[\nabla(\Lambda^{-1})^3] \circ \Lambda = (1/J)(\Lambda_\varphi \times \Lambda_\psi) = e_3.$$

Set $y := (y_1, y_2, y_3) := (\psi, \varphi, r)$, $a_{ij} = ((\mathrm{Jac}\Lambda)^{-1})_{ij}$. Then, $n = (a_{31}, a_{32}, a_{33})$, the tangential directions $\tau_i = (a_{i1}, a_{i2}, a_{i3}) (i = 1, 2)$, and

$$a_{ij}a_{3j} = 0, \text{ for } i = 1, 2; \quad a_{kj}a_{kj} = 1, \text{ for } j = 1, 2, 3. \tag{4.19}$$

Then, we denote by D_i the partial derivative with respect to y_i in local coordinates. To be precise, D_3 is the normal derivative and D_i for $i = 1, 2$ are the tangential derivatives in the original coordinates. Moreover, we have

$$\partial_{x_j} = a_{kj} D_k.$$

We rewrite (1.8)-(1.10) in $[0, T] \times \tilde{\Omega}$, $\tilde{\Omega} := \Lambda^{-1}(W \cap \Omega)$, as

$$\tilde{\sigma}_t + a_{kj} D_k(\tilde{\sigma} \tilde{u}_j) + \frac{1}{\epsilon} a_{kj} D_k \tilde{u}_j = 0, \tag{4.20}$$

$$\tilde{\rho}(\tilde{u}_{it} + \tilde{u}_j a_{kj} D_k \tilde{u}_i) + \frac{R}{\epsilon}(a_{ki} D_k \tilde{\sigma} + a_{ki} D_k \tilde{\theta}) + R a_{ki} D_k(\tilde{\sigma} \tilde{\theta})$$
$$= \mu a_{kj} D_k(a_{lj} D_l \tilde{u}_i) + \lambda a_{ki} D_k(a_{lj} D_l \tilde{u}_j), \tag{4.21}$$

$$C_v \tilde{\rho}(\tilde{\theta}_t + \tilde{u}_j a_{kj} D_k \tilde{\theta}) + R(\tilde{\rho}\tilde{\theta} + \tilde{\sigma}) a_{kj} D_k \tilde{u}_j + \frac{R}{\epsilon} a_{kj} D_k \tilde{u}_j$$
$$= \kappa a_{kj} D_k(a_{lj} D_l \tilde{\theta}) + \epsilon \left(\frac{\mu}{2}(a_{ki} D_k \tilde{u}_j + a_{kj} D_k \tilde{u}_i)^2 + \xi(a_{kj} D_k \tilde{u}_j)^2 \right), \tag{4.22}$$

where $\tilde{u}(t, y) := u(t, \Lambda(y))$, $\tilde{\theta}(t, y) := \theta(t, \Lambda(y))$, $\tilde{\sigma}(t, y) := \sigma(t, \Lambda(y))$ and $\tilde{\rho}(t, y) := \rho(t, \Lambda(y))$, and the initial and boundary conditions for $(\tilde{\sigma}, \tilde{u}, \tilde{\theta})$ are:

$$(\tilde{\sigma}, \tilde{u}, \tilde{\theta})|_{t=0} = (\tilde{\sigma}_0, \tilde{u}_0, \tilde{\theta}_0)(x) \quad \text{in } \tilde{\Omega}, \tag{4.23}$$

$$\tilde{u} = 0, \quad a_{ki} D_i \tilde{\theta} a_{3i} = 0, \quad \text{on } \partial \tilde{\Omega} \times (0, T). \tag{4.24}$$

Moreover, this localized system has the following properties (see [22]).

Proposition 4.1 $D_i(J a_{ij}) = 0 (j = 1, 2, 3)$; $\chi D_\tau \tilde{u} = 0$, $\chi D_\tau D_\zeta \tilde{u} = 0$ on $\partial \tilde{\Omega}$ in the tangential directions $\tau, \zeta = 1, 2$, where $\chi \in C_0^\infty(\Lambda^{-1}(W))$.

The following relations also hold:

$$\|D_y \tilde{u}\|_{L^p(\tilde{\Omega})} \leqslant C\|\nabla_x u\|_{L^p(\Omega)}, \quad \|D_y^2 \tilde{u}\|_{L^p(\tilde{\Omega})} \leqslant C\|\nabla_x u\|_{W^{1,p}(\Omega)},$$

where $1 \leqslant p \leqslant \infty$.

Estimate of the derivatives in the tangential directions.

Lemma 4.8 For any $0 \leqslant t \leqslant \min\{T^\epsilon, 1\}$, there exist positive constants μ_1 and κ_1, such that

$$\frac{1}{2} \int_{\tilde{\Omega}} J \chi^2 \tilde{\rho} |D_{\zeta \tau} \tilde{u}|^2 dy + \frac{R}{2} \int_{\tilde{\Omega}} J \chi^2 |D_{\zeta \tau} \tilde{\sigma}|^2 dy + \frac{C_v}{2} \int_{\tilde{\Omega}} J \chi^2 \tilde{\rho} |D_{\zeta \tau} \tilde{\theta}|^2 dy$$

$$+ \mu_1 \int_0^t \int_{\tilde{\Omega}} J\chi^2 |D_{y\zeta\tau}\tilde{u}|^2 dyds + \kappa_1 \int_0^t \int_{\tilde{\Omega}} J\chi^2 |D_{y\zeta\tau}\tilde{\theta}|^2 dyds$$

$$\leqslant C_0(M_0)\exp\left(t^{\frac{1}{8}}C(M)\right) + \delta \int_0^t \left(\frac{1}{\epsilon^2}\|(\nabla^2\sigma, \nabla^2\theta)\|_{L^2}^2 + \|u\|_{H^3}^2\right)ds.$$

Proof Applying $D_{\zeta\tau}$ to (4.21) and taking the product of the resulting equation and $J\chi^2 D_{\zeta\tau}\tilde{u}_i$ we obtain

$$\frac{1}{2}\int_{\tilde{\Omega}} J\chi^2\tilde{\rho}|D_{\zeta\tau}\tilde{u}|^2 dy + \frac{R}{\epsilon}\int_0^t\int_{\tilde{\Omega}} J\chi^2 D_{\zeta\tau}\tilde{u}_i D_{\zeta\tau}(a_{ki}D_k\tilde{\sigma} + a_{ki}D_k\tilde{\theta})dyds$$

$$+ \int_0^t\int_{\tilde{\Omega}} D_{k\zeta\tau}\tilde{u}_i a_{lj}(\mu J\chi^2 a_{kj}D_{l\zeta\tau}\tilde{u}_i + \lambda J\chi^2 a_{ki}D_{l\zeta\tau}\tilde{u}_j)dyds$$

$$= \frac{1}{2}\int_{\tilde{\Omega}} J\chi^2\tilde{\rho}_0|D_{\zeta\tau}\tilde{u}_0|^2 dy + \frac{1}{2}\int_0^t\int_{\tilde{\Omega}} D_k(J\chi^2 a_{kj})\tilde{\rho}\tilde{u}_j|D_{\zeta\tau}\tilde{u}_i|^2 dyds$$

$$- \int_0^t\int_{\tilde{\Omega}} J\chi^2 D_{\zeta\tau}\tilde{u}_i(D_\zeta\tilde{\rho}D_\tau\tilde{u}_{it} + D_\tau\tilde{\rho}D_\zeta\tilde{u}_{it} + D_{\zeta\tau}\tilde{\rho}\tilde{u}_{it})dyds$$

$$- \int_0^t\int_{\tilde{\Omega}} J\chi^2 D_{\zeta\tau}\tilde{u}_i[D_\zeta(\tilde{\rho}\tilde{u}_j a_{kj})D_{k\tau}\tilde{u}_i + D_\tau(\tilde{\rho}\tilde{u}_j a_{kj})D_{k\zeta}\tilde{u}_i$$

$$+ D_{\zeta\tau}(\tilde{\rho}\tilde{u}_j a_{kj})D_k\tilde{u}_i]dyds - R\int_0^t\int_{\tilde{\Omega}} J\chi^2 D_{\zeta\tau}\tilde{u}_i D_{\zeta\tau}(a_{k\tau}D_k(\tilde{\sigma}\tilde{\theta}))dyds$$

$$- \int_0^t\int_{\tilde{\Omega}}\{a_{lj}D_{\zeta\tau}\tilde{u}_i[\mu D_k(J\chi^2 a_{kj})D_{l\zeta\tau}\tilde{u}_i + \lambda D_k(J\chi^2 a_{ki})D_{l\zeta\tau}\tilde{u}_j]$$

$$+ D_{\zeta\tau}a_{lj}[\mu D_l\tilde{u}_i D_k(J\chi^2 a_{kj}D_{\zeta\tau}\tilde{u}_i) + \lambda D_l\tilde{u}_j D_k(J\chi^2 a_{ki}D_{\zeta\tau}\tilde{u}_i)]\}dyds$$

$$+ \int_0^t\int_{\tilde{\Omega}} J\chi^2 D_{\zeta\tau}\tilde{u}_i\{\mu[a_{kj}D_k(D_\zeta a_{lj}D_{l\tau}\tilde{u}_i + D_\tau a_{lj}D_{l\zeta}\tilde{u}_i)$$

$$+ D_\zeta a_{kj}D_{k\tau}(a_{lj}D_l\tilde{u}_i) + D_\tau a_{kj}D_{k\zeta}(a_{lj}D_l\tilde{u}_i)$$

$$+ D_{\zeta\tau}a_{kj}D_k(a_{lj}D_l\tilde{u}_i)] + \lambda[a_{ki}D_k(D_\zeta a_{lj}D_{l\tau}\tilde{u}_j + D_\tau a_{lj}D_{l\zeta}\tilde{u}_j)$$

$$+ D_\zeta a_{ki}D_{k\tau}(a_{lj}D_l\tilde{u}_j) + D_\tau a_{ki}D_{k\zeta}(a_{lj}D_l\tilde{u}_j) + D_{\zeta\tau}a_{ki}D_k(a_{lj}D_l\tilde{u}_j)]\}dyds$$

$$= \frac{1}{2}\int_{\tilde{\Omega}} J\chi^2\tilde{\rho}_0|D_{\zeta\tau}\tilde{u}_0|^2 dy + \sum_{i=1}^{6} K_i, \tag{4.25}$$

where

$$|K_1| \leqslant C\int_0^t \|\tilde{\rho}\|_{H^2}\|\nabla^2\tilde{u}\|_{L^2}^2\|\nabla^2\tilde{u}\|_{L^3}ds$$

$$\leqslant C \int_0^t \|\tilde{\rho}\|_{H^2} \|\tilde{u}\|_{H^3}^{\frac{7}{4}} \|\tilde{u}\|_{H^1}^{\frac{5}{4}} ds$$

$$\leqslant t^{\frac{1}{8}} C(M),$$

$$|K_2| \leqslant \int_0^t \|\tilde{\rho}\|_{H^2} \|\nabla^2 \tilde{u}\|_{L^3} \|\tilde{u}_t\|_{H^1} ds$$

$$\leqslant C \|\tilde{u}\|_{L_t^\infty(H^1)}^{\frac{1}{4}} \|\tilde{\rho}\|_{L_t^\infty(H^2)} \int_0^t \|\tilde{u}\|_{H^3}^{\frac{3}{4}} \|\tilde{u}_t\|_{H^1} ds$$

$$\leqslant C \|\tilde{u}\|_{L_t^\infty(H^1)}^{\frac{1}{4}} \|\tilde{\rho}\|_{L_t^\infty(H^2)} \left(\int_0^t \|\tilde{u}\|_{H^3}^{\frac{7}{4}} ds + \int_0^t \|\tilde{u}_t\|_{H^1}^{\frac{7}{4}} ds \right)$$

$$\leqslant t^{\frac{1}{8}} C(M),$$

similarly,

$$|K_3| \leqslant t^{\frac{3}{8}} C(M),$$

$$|K_5 + K_6| \leqslant t^{\frac{1}{4}} C(M),$$

by using the integrate by parts, we have

$$|K_4|$$

$$\leqslant \left| -R \int_0^t \int_{\tilde{\Omega}} J\chi^2 D_{\zeta\tau} \tilde{u}_i [D_{\zeta\tau} a_{ki} D_k(\tilde{\sigma}\tilde{\theta}) + D_\zeta a_{ki} D_{k\tau}(\tilde{\sigma}\tilde{\theta}) + D_\tau a_{ki} D_{k\zeta}(\tilde{\sigma}\tilde{\theta})] dyds \right.$$

$$\left. + R \int_0^t \int_{\tilde{\Omega}} [D_k(a_{ki} J\chi^2) D_{\zeta\tau}(\tilde{\sigma}\tilde{\theta}) D_{\zeta\tau}\tilde{u}_i + J\chi^2 a_{ki} D_{\zeta\tau}(\tilde{\sigma}\tilde{\theta}) D_{k\zeta\tau}\tilde{u}_i] dyds \right|$$

$$\leqslant t^{\frac{1}{4}} C(M).$$

It is obvious that $\{\sum_k a_{ki} a_{kj}\}$ is strictly positive definite, so we obtain

$$\mu \sum_{i,j} a_{kj} D_{k\zeta\tau} \tilde{u}_i a_{lj} D_{l\zeta\tau} \tilde{u}_i \geqslant \mu_1 |D_{y\zeta\tau}\tilde{u}|^2,$$

for some positive constant μ_1. Substituting the estimates of $K_i (i = 1, \cdots, 6)$ into (4.25), we obtain

$$\frac{1}{2} \int_{\tilde{\Omega}} J\chi^2 \tilde{\rho} |D_{\zeta\tau}\tilde{u}|^2 dy + \frac{R}{\epsilon} \int_0^t \int_{\tilde{\Omega}} J\chi^2 D_{\zeta\tau}\tilde{u}_i D_{\zeta\tau}(a_{ki} D_k \tilde{\sigma} + a_{ki} D_k \tilde{\theta}) dyds$$

$$+ C \int_0^t \int_{\tilde{\Omega}} J\chi^2 |D_{y\zeta\tau}\tilde{u}|^2 dyds$$

$$\leqslant C_0(M_0) \exp\left(t^{\frac{1}{8}} C(M)\right). \tag{4.26}$$

Then, we apply $D_{\zeta\tau}$ to (4.20) and integrate the product of the resulting equation and $RJ\chi^2 D_{\zeta\tau}\tilde{\sigma}$ to find

$$\frac{R}{2}\int_{\tilde{\Omega}} J\chi^2 |D_{\zeta\tau}\tilde{\sigma}|^2 dy + \frac{R}{\epsilon}\int_0^t\int_{\tilde{\Omega}} J\chi^2 D_{\zeta\tau}\tilde{\sigma} D_{\zeta\tau}(a_{kj}D_k\tilde{u}_j)dyds$$

$$=\frac{R}{2}\int_{\tilde{\Omega}} J\chi^2 |D_{\zeta\tau}\tilde{\sigma}_0|^2 dy - R\int_0^t\int_{\tilde{\Omega}} J\chi^2 D_{\zeta\tau}\tilde{\sigma} D_{\zeta\tau}(a_{kj}D_k(\tilde{\sigma}\tilde{u}_j))dyds$$

$$\leqslant C_0(M_0) + \Big| - R\int_0^t\int_{\tilde{\Omega}} J\chi^2 D_{\zeta\tau}\tilde{\sigma}[D_{\zeta\tau}a_{kj}D_k(\tilde{\sigma}\tilde{u}_j) + D_\zeta a_{kj}D_{k\tau}(\tilde{\sigma}\tilde{u}_j)$$

$$+ D_\tau a_{kj}D_{k\zeta}(\tilde{\sigma}\tilde{u}_j)]dyds - R\int_0^t\int_{\tilde{\Omega}} J\chi^2 D_{\zeta\tau}\tilde{\sigma} a_{kj}D_k(D_\zeta\tilde{\sigma}D_\tau\tilde{u}_j + D_\tau\tilde{\sigma}D_\zeta\tilde{u}_j$$

$$+ \tilde{\sigma}D_{\zeta\tau}u_j)dyds + \frac{R}{2}\int_0^t\int_{\tilde{\Omega}} D_k(J\chi^2 a_{kj}\tilde{u}_j)|D_{\zeta\tau}\tilde{\sigma}|^2 dyds\Big|$$

$$\leqslant C_0(M_0)\exp(t^{\frac{1}{4}}C(M)). \tag{4.27}$$

Similarly, applying $D_{\zeta\tau}$ to (4.22) and integrating the product of the resulting equation and $J\chi^2 D_{\zeta\tau}\tilde{\theta}$, we have

$$\frac{C_v}{2}\int_{\tilde{\Omega}} J\chi^2 \tilde{\rho}|D_{\zeta\tau}\tilde{\theta}|^2 dy + k\int_0^t\int_{\tilde{\Omega}} J\chi^2 a_{kj}D_{k\zeta\tau}\tilde{\theta} a_{lj}D_{l\zeta\tau}\tilde{\theta} dyds$$

$$+ \frac{R}{\epsilon}\int_0^t\int_{\tilde{\Omega}} J\chi^2 D_{\zeta\tau}\tilde{\theta} D_{\zeta\tau}(a_{kj}D_k\tilde{u}_j)dyds$$

$$= \frac{C_v}{2}\int_{\tilde{\Omega}} J\chi^2 \tilde{\rho}_0|D_{\zeta\tau}\tilde{\theta}_0|^2 dy - C_v\int_0^t\int_{\tilde{\Omega}} J\chi^2 D_{\zeta\tau}\tilde{\theta}(D_\zeta\tilde{\rho}D_\tau\tilde{\theta}_t$$

$$+ D_\tau\tilde{\rho}D_\zeta\tilde{\theta}_t + D_{\zeta\tau}\tilde{\rho}\tilde{\theta}_t)dyds + \frac{C_v}{2}\int_0^t\int_{\tilde{\Omega}} D_k(J\chi^2 a_{kj})\tilde{\rho}\tilde{u}_j|D_{\zeta\tau}\tilde{\theta}|^2 dyds$$

$$- C_v\int_0^t\int_{\tilde{\Omega}} J\chi^2 D_{\zeta\tau}\tilde{\theta}[D_\zeta(\tilde{\rho}\tilde{u}_j a_{kj})D_{k\tau}\tilde{\theta} + D_\tau(\tilde{\rho}\tilde{u}_j a_{kj})D_{k\zeta}\tilde{\theta}$$

$$+ D_{\zeta\tau}(\tilde{\rho}\tilde{u}_j a_{kj})D_k\tilde{\theta}]dyds - R\int_0^t\int_{\tilde{\Omega}} J\chi^2 D_{\zeta\tau}\tilde{\theta} D_{\zeta\tau}[(\tilde{\rho}\tilde{\theta}$$

$$+ \tilde{\sigma})a_{kj}D_k\tilde{u}_j]dyds - k\int_0^t\int_{\tilde{\Omega}} a_{lj}D_{\zeta\tau}\tilde{\theta} D_k(J\chi^2 a_{kj})D_{l\zeta\tau}\tilde{\theta} dyds$$

$$+ k\int_0^t\int_{\tilde{\Omega}} J\chi^2 D_{\zeta\tau}\tilde{\theta}[a_{kj}D_k(D_\zeta a_{lj}D_{l\tau}\tilde{\theta} + D_\tau a_{lj}D_{l\zeta}\tilde{\theta} + D_{\zeta\tau}a_{lj}D_l\tilde{\theta})$$

$$+ D_\zeta a_{kj}D_{k\tau}(a_{lj}D_l\tilde{\theta}) + D_\tau a_{kj}D_{k\zeta}(a_{lj}D_l\tilde{\theta}) + D_{\zeta\tau}a_{kj}D_k(a_{lj}D_l\tilde{\theta})]dyds$$

$$+ \epsilon \int_0^t \int_{\tilde{\Omega}} J\chi^2 D_{\zeta\tau}\tilde{\theta} D_{\zeta\tau} \left[\frac{\mu}{2}(a_{kj}D_k\tilde{u}_j + a_{kj}D_k\tilde{u}_i)^2 + \xi(a_{kj}D_k\tilde{u}_j)^2\right] dyds. \quad (4.28)$$

For similar estimates, we obtain

$$\frac{C_v}{2}\int_{\tilde{\Omega}} J\chi^2 \tilde{\rho}|D_{\zeta\tau}\tilde{\theta}|^2 dy + k_1 \int_0^t \int_{\tilde{\Omega}} J\chi^2 |D_{y\zeta\tau}\tilde{\theta}|^2 dyds$$

$$+ \frac{R}{\epsilon}\int_0^t \int_{\tilde{\Omega}} J\chi^2 D_{\zeta\tau}\tilde{\theta} D_{\zeta\tau}(a_{kj}D_k\tilde{u}_j) dyds$$

$$\leqslant C_0(M_0)\exp(t^{\frac{1}{8}}C(M)). \quad (4.29)$$

Note that, the singular terms in (4.26)-(4.27) and (4.29) can be estimated as follow:

$$\frac{R}{\epsilon}\int_0^t \int_{\tilde{\Omega}} J\chi^2 D_{\zeta\tau}\tilde{u}_i D_{\zeta\tau}(a_{ki}D_k\tilde{\sigma} + a_{ki}D_k\tilde{\theta}) dyds$$

$$+ \frac{R}{\epsilon}\int_0^t \int_{\tilde{\Omega}} J\chi^2(D_{\zeta\tau}\tilde{\sigma} + D_{\zeta\tau}\tilde{\theta})D_{\zeta\tau}(a_{kj}D_k\tilde{u}_j) dyds$$

$$\leqslant \int_0^t \frac{1}{\epsilon^2}\|(\nabla^2\sigma,\nabla^2\theta)\|_{L^2}^2 ds + \sqrt{t}C(M).$$

Finally, combing (4.26),(4.27),(4.29) and the above inequality, we can prove this lemma.

Estimate of the derivatives in the normal directions.

We will adapt an idea of Valli in [22] to estimate the derivatives in the normal direction. Multiplying (4.21) by a_{3i}, we have:

$$(\mu + \lambda)D_3(a_{lj}D_l\tilde{u}_j)$$

$$= \frac{R}{\epsilon}(D_3\tilde{\sigma} + D_3\tilde{\theta}) + \tilde{\rho}(\tilde{u}_{it} + \tilde{u}_j a_{kj} D_k\tilde{u}_i)a_{3i} + RD_3(\tilde{\sigma}\tilde{\theta})$$

$$+ \mu(D_3(a_{lj}D_l\tilde{u}_j) - a_{kj}a_{3i}D_k(a_{lj}D_l\tilde{u}_i))$$

$$= \frac{R}{\epsilon}(D_3\tilde{\sigma} + D_3\tilde{\theta}) + M_1 + M_2 + M_3. \quad (4.30)$$

Recalling that $a_{3j}a_{3j} = 1$ and $a_{3j}a_{ij} = 0$ $(i = 1, 2)$, M_3 can be rewritten as

$$\mu(D_3 a_{3j} D_3\tilde{u}_j + D_3 a_{\tau j} D_\tau\tilde{u}_j + a_{\tau j} D_{3\tau}\tilde{u}_j - a_{3j} D_3 a_{3j} a_{3i} D_3\tilde{u}_i$$

$$- a_{\tau j} a_{3i} D_\tau a_{lj} D_l\tilde{u}_i - a_{\tau j} a_{\zeta j} a_{3i} D_{\zeta\tau}\tilde{u}_i - a_{3j} a_{3i} D_3 a_{\tau j} D_\tau\tilde{u}_i), \quad \zeta, \tau = 1, 2.$$

Thus, we can see that the second-order normal derivative $D_{33}\tilde{u}$ is not involved on the right-hand side of (4.30). This is a critical observation in the estimates for derivatives.

Lemma 4.9 *For any $0 \leqslant t \leqslant \min\{T^\epsilon, 1\}$, there exists a positive constant κ_2, such that*

$$\frac{R}{2}\int_{\tilde{\Omega}} J\chi^2 |D_{\tau 3}\tilde{\sigma}|^2 dy + \frac{C_v}{2}\int_{\tilde{\Omega}} J\chi^2 \tilde{\rho}|D_{\tau 3}\tilde{\theta}|^2 dy$$
$$+ \frac{\mu+\lambda}{2}\int_0^t \int_{\tilde{\Omega}} J\chi^2 |D_{\tau 3}(a_{lj}D_j\tilde{u}_j)|^2 dyds + \kappa_2 \int_0^t \int_{\tilde{\Omega}} J\chi^2 |D_{y\tau 3}\tilde{\theta}|^2 dyds$$
$$\leqslant C_0(M_0)\exp(t^{\frac{1}{8}}C(M)) + C_1 \int_0^t \int_{\tilde{\Omega}} J\chi^2 |D_{y\zeta\tau}\tilde{u}|^2 dyds.$$

Proof We apply D_τ ($\tau = 1, 2$) to (4.30), and multiply by $J\chi^2 D_{\tau 3}(a_{lj}D_l\tilde{u}_j)$ in $L^2_{((0,t)\times\Omega)}$ to get

$$(\mu+\lambda)\int_0^t \int_{\tilde{\Omega}} J\chi^2 |D_{\tau 3}(a_{lj}D_l\tilde{u}_j)|^2 dyds$$
$$- \frac{R}{\epsilon}\int_0^t \int_{\tilde{\Omega}} J\chi^2 D_{\tau 3}(a_{lj}D_l\tilde{u}_j)(D_{\tau 3}\tilde{\sigma} + D_{\tau 3}\tilde{\theta}) dyds$$
$$= \int_0^t \int_{\tilde{\Omega}} J\chi^2 D_{\tau 3}(a_{lj}D_l\tilde{u}_j) D_\tau(M_1 + M_2 + M_3) dyds,$$

$$\left|\int_0^t \int_{\tilde{\Omega}} J\chi^2 D_{\tau 3}(a_{lj}D_l\tilde{u}_j) D_\tau M_1 dyds\right|$$
$$\leqslant \left|\int_0^t \int_{\tilde{\Omega}} J\chi^2 D_{\tau 3}(a_{lj}D_l\tilde{u}_j) D_\tau \tilde{\rho}(\tilde{u}_{it} + \tilde{u}_j a_{kj} D_k \tilde{u}_i) a_{3i} dyds\right|$$
$$+ \left|\int_0^t \int_{\tilde{\Omega}} J\chi^2 D_{\tau 3}(a_{lj}D_l\tilde{u}_j)\tilde{\rho} D_\tau(\tilde{u}_{it} + \tilde{u}_j a_{kj} D_k \tilde{u}_i) a_{3i} dyds\right|$$
$$+ \left|\int_0^t \int_{\tilde{\Omega}} J\chi^2 D_{\tau 3}(a_{lj}D_l\tilde{u}_j)\tilde{\rho}(\tilde{u}_{it} + \tilde{u}_j a_{kj} D_k \tilde{u}_i) D_\tau a_{3i} dyds\right|$$
$$\leqslant \delta \int_0^t \int_{\tilde{\Omega}} J\chi^2 |D_{\tau 3}(a_{lj}D_l\tilde{u}_j)|^2 dyds + C_\delta \|\tilde{\rho}\|^2_{L^\infty_t(H^2)} \|\tilde{u}_t\|_{L^\infty_t(L^2)} \int_0^t \|\tilde{u}_t\|_{H^1} ds$$
$$+ C_\delta \|\tilde{\rho}\|^2_{L^\infty_t(H^2)} \|\tilde{u}\|^{\frac{5}{2}}_{L^\infty_t(H^1)} \int_0^t \|\tilde{u}\|^{\frac{3}{2}}_{H^3} ds + C_\delta \|\tilde{\rho}\|^2_{L^\infty_t(H^2)} \int_0^t \|\tilde{u}_t\|^2_{H^1} ds$$
$$+ C_\delta \|\tilde{\rho}\|^2_{L^\infty_t(H^2)} \|\tilde{u}\|^{\frac{9}{4}}_{L^\infty_t(H^1)} \int_0^t \|\tilde{u}\|^{\frac{7}{8}}_{H^3} ds$$

$$\leqslant \delta \int_0^t \int_{\tilde{\Omega}} J\chi^2 |D_{\tau 3}(a_{lj} D_l \tilde{u}_j)|^2 dyds + C_0(M_0) \exp(t^{\frac{1}{8}} C(M)).$$

In the above inequality, we have applied the Lemma 4.5.

$$\left| \int_0^t \int_{\tilde{\Omega}} J\chi^2 D_{\tau 3}(a_{lj} D_l \tilde{u}_j) D_\tau M_2 dyds \right|$$

$$\leqslant \delta \int_0^t \int_{\tilde{\Omega}} J\chi^2 |D_{\tau 3}(a_{lj} D_l \tilde{u}_j)|^2 dyds + C_\delta \int_0^t \|\sigma\|_{H^2}^2 \|\theta\|_{H^2}^2 ds$$

$$\leqslant \delta \int_0^t \int_{\tilde{\Omega}} J\chi^2 |D_{\tau 3}(a_{lj} D_l \tilde{u}_j)|^2 dyds + tC(M),$$

similarly,

$$\left| \int_0^t \int_{\tilde{\Omega}} J\chi^2 D_{\tau 3}(a_{lj} D_l \tilde{u}_j) D_\tau M_3 dyds \right|$$

$$\leqslant \delta \int_0^t \int_{\tilde{\Omega}} J\chi^2 |D_{\tau 3}(a_{lj} D_l \tilde{u}_j)|^2 dyds + C_\delta \int_0^t \int_{\tilde{\Omega}} J\chi^2 |D_{y\varsigma\tau} \tilde{u}|^2 dyds$$

$$+ C_\delta \int_0^t \|\nabla^2 u\|_{L^2}^2 ds$$

$$\leqslant \delta \int_0^t \int_{\tilde{\Omega}} J\chi^2 |D_{\tau 3}(a_{lj} D_l \tilde{u}_j)|^2 dyds + C_\delta \int_0^t \int_{\tilde{\Omega}} J\chi^2 |D_{y\varsigma\tau} \tilde{u}|^2 dyds + \sqrt{t} C(M).$$

According to the above estimates, we have

$$\frac{\mu+\lambda}{2} \int_0^t \int_{\tilde{\Omega}} J\chi^2 |D_{\tau 3}(a_{lj} D_l \tilde{u}_j)|^2 dyds$$

$$- \frac{R}{\epsilon} \int_0^t \int_{\tilde{\Omega}} J\chi^2 D_{\tau 3}(a_{lj} D_l \tilde{u}_j)(D_{\tau 3}\tilde{\sigma} + D_{\tau 3}\tilde{\theta}) dyds$$

$$\leqslant C_0(M_0) \exp(t^{\frac{1}{4}} C(M)) + C_1 \int_0^t \int_{\tilde{\Omega}} J\chi^2 |D_{y\varsigma\tau} \tilde{u}|^2 dyds. \tag{4.31}$$

To eliminate the singular term in the above inequality, we apply $D_{\tau 3}$ ($\tau = 1, 2$) to (4.20) and (4.22), and then multiply the resulting equalities by $RJ\chi^2 D_{\tau 3}\tilde{\sigma}$ and $J\chi^2 D_{\tau 3}\tilde{\theta}$, respectively, over $\Omega \times (0, t)$ to get

$$\frac{R}{2} \int_{\tilde{\Omega}} J\chi^2 |D_{\tau 3}\tilde{\sigma}|^2 dy + \frac{R}{\epsilon} \int_0^t \int_{\tilde{\Omega}} J\chi^2 D_{\tau 3}\tilde{\sigma} D_{\tau 3}(a_{kj} D_k \tilde{u}_j) dyds$$

$$= \frac{R}{2} \int_{\tilde{\Omega}} J\chi^2 |D_{\tau 3}\tilde{\sigma}_0|^2 dy - R \int_0^t \int_{\tilde{\Omega}} J\chi^2 D_{\tau 3}\tilde{\sigma} D_{\tau 3}[a_{kj} D_k(\tilde{u}_j \tilde{\sigma})] dyds$$

$$\leqslant C_0(M_0)\exp(t^{\frac{1}{4}}C(M)) \tag{4.32}$$

and

$$\frac{C_v}{2}\int_{\tilde{\Omega}} J\chi^2\tilde{\rho}|D_{\tau 3}\tilde{\theta}|^2 dy + k\int_0^t\int_{\tilde{\Omega}} J\chi^2 a_{kj}D_{k\tau 3}\tilde{\theta}a_{lj}D_{l\tau 3}\tilde{\theta}dyds$$
$$+\frac{R}{\epsilon}\int_0^t\int_{\tilde{\Omega}} J\chi^2 D_{\tau 3}\tilde{\theta}D_{\tau 3}(a_{kj}D_k\tilde{u}_j)dyds$$
$$=\frac{C_v}{2}\int_{\tilde{\Omega}} J\chi^2\tilde{\rho}_0|D_{\tau 3}\tilde{\theta}_0|^2 dy - C_v\int_0^t\int_{\tilde{\Omega}} J\chi^2 D_{\tau 3}\tilde{\theta}(D_3\tilde{\rho}D_\tau\tilde{\theta}_t$$
$$+D_\tau\tilde{\rho}D_3\tilde{\theta}_t + D_{\tau 3}\tilde{\rho}\tilde{\theta}_t)dyds + \frac{C_v}{2}\int_0^t\int_{\tilde{\Omega}} D_k(J\chi^2 a_{kj})\tilde{\rho}\tilde{u}_j|D_{\tau 3}\tilde{\theta}|^2 dyds$$
$$-C_v\int_0^t\int_{\tilde{\Omega}} J\chi^2 D_{\tau 3}\tilde{\theta}[D_3(\tilde{\rho}\tilde{u}_j a_{kj})D_{k\tau}\tilde{\theta} + D_\tau(\tilde{\rho}\tilde{u}_j a_{kj})D_{k3}\tilde{\theta}$$
$$+D_{\tau 3}(\tilde{\rho}\tilde{u}_j a_{kj})D_k\tilde{\theta}]dyds - R\int_0^t\int_{\tilde{\Omega}} J\chi^2 D_{\tau 3}\tilde{\theta}D_{\tau 3}[(\tilde{\rho}\tilde{\theta}$$
$$+\tilde{\sigma})a_{kj}D_k\tilde{u}_j]dyds - k\int_0^t\int_{\tilde{\Omega}} a_{lj}D_{\tau 3}\tilde{\theta}D_k(J\chi^2 a_{kj})D_{l\tau 3}\tilde{\theta}dyds$$
$$+k\int_0^t\int_{\tilde{\Omega}} J\chi^2 D_{\tau 3}\tilde{\theta}[a_{kj}D_k(D_3 a_{lj}D_{l\tau}\tilde{\theta} + D_\tau a_{lj}D_{l3}\tilde{\theta} + D_{\tau 3}a_{lj}D_l\tilde{\theta})$$
$$+D_3 a_{kj}D_{k\tau}(a_{lj}D_l\tilde{\theta}) + D_\tau a_{kj}D_{k3}(a_{lj}D_l\tilde{\theta}) + D_{\tau 3}a_{kj}D_k(a_{lj}D_l\tilde{\theta})]dyds$$
$$+\epsilon\int_0^t\int_{\tilde{\Omega}} J\chi^2 D_{\tau 3}\tilde{\theta}D_{\tau 3}[\frac{\mu}{2}(a_{kj}D_k\tilde{u}_j + a_{kj}D_k\tilde{u}_i)^2 + \xi(a_{kj}D_k\tilde{u}_j)^2]dyds.$$

Similarly, as (4.28), we derive that the following inequality holds for some positive constant κ_2,

$$\frac{C_v}{2}\int_{\tilde{\Omega}} J\chi^2\tilde{\rho}|D_{\tau 3}\tilde{\theta}|^2 dy + k_2\int_0^t\int_{\tilde{\Omega}} J\chi^2|D_{y\tau 3}\tilde{\theta}|^2 dyds$$
$$+\frac{R}{\epsilon}\int_0^t\int_{\tilde{\Omega}} J\chi^2 D_{\tau 3}\tilde{\theta}D_{\tau 3}(a_{kj}D_k\tilde{u}_j)dyds.$$
$$\leqslant C_0(M_0)\exp(t^{\frac{1}{8}}C(M)). \tag{4.33}$$

Summarizing (4.31),(4.32) and (4.33), we prove this lemma.

Now, it suffices to estimate $\|D_{33}(a_{lj}D_l\tilde{u}_j)\|_{L^2((0,t)\times\tilde{\Omega})}$ to close the estimates for div$u$.

Lemma 4.10 For any $0 \leqslant t \leqslant \min\{T^\varepsilon, 1\}$, there exists a positive constant κ_3, such that

$$\frac{R}{2}\int_{\tilde{\Omega}} J\chi^2|D_{33}\tilde{\sigma}|^2 dy + \frac{C_v}{2}\int_{\tilde{\Omega}} J\chi^2 \tilde{\rho}|D_{33}\tilde{\theta}|^2 dy$$

$$+ \frac{\mu+\lambda}{2}\int_0^t\int_{\tilde{\Omega}} J\chi^2|D_{33}(a_{lj}D_l\tilde{u}_j)|^2 dyds + \kappa_3\int_0^t\int_{\tilde{\Omega}} J\chi^2|D_{y33}\tilde{\theta}|^2 dyds$$

$$\leqslant C_0(M_0)\exp(t^{\frac{1}{8}}C(M)) + C_2\int_0^t\int_{\tilde{\Omega}} J\chi^2|D_{33\tau}\tilde{u}|^2 dyds$$

$$+ C_3\int_0^t\int_{\tilde{\Omega}} J\chi^2|D_{3\zeta\tau}\tilde{u}|^2 dyds.$$

Proof Applying D_3 to (4.30), we can see that

$$(\mu+\lambda)D_{33}(a_{lj}D_l\tilde{u}_j) - \frac{R}{\varepsilon}(D_{33}\tilde{\sigma} + D_{33}\tilde{\theta})$$

$$= D_3[\tilde{\rho}a_{3i}(\tilde{u}_{it} + \tilde{u}_j a_{kj}D_k\tilde{u}_i)] + RD_{33}(\tilde{\sigma}\tilde{\theta})$$

$$+ O(1)(D_{33\tau}\tilde{u}_j + D_{3\zeta\tau}\tilde{u}_i + D_{3l}\tilde{u}_j + D_l\tilde{u}_i)$$

$$= \sum_{i=4}^{6} M_i.$$

Multiplying by $J\chi^2 D_{33}(a_{lj}D_l\tilde{u}_j)$ and integrating, we obtain

$$(\mu+\lambda)\int_0^t\int_{\tilde{\Omega}} J\chi^2|D_{33}(a_{lj}D_l\tilde{u}_j)|^2 dyds$$

$$-\frac{R}{\varepsilon}\int_0^t\int_{\tilde{\Omega}} J\chi^2 D_{33}(a_{lj}D_l\tilde{u}_j)(D_{33}\tilde{\sigma} + D_{33}\tilde{\theta}) dyds$$

$$= \int_0^t\int_{\tilde{\Omega}} J\chi^2 D_{33}(a_{lj}D_l\tilde{u}_j)(M_4 + M_5 + M_6) dyds.$$

Furthermore, we obtain

$$\left|\int_0^t\int_{\tilde{\Omega}} J\chi^2 D_{33}(a_{lj}D_l\tilde{u}_j)M_4 dyds\right|$$

$$= \left|\int_0^t\int_{\tilde{\Omega}} J\chi^2 D_{33}(a_{lj}D_l\tilde{u}_j)D_3\tilde{\rho}(\tilde{u}_{it} + \tilde{u}_j a_{kj}D_k\tilde{u}_i)a_{3i} dyds\right|$$

$$+ \left|\int_0^t\int_{\tilde{\Omega}} J\chi^2 D_{33}(a_{lj}D_l\tilde{u}_j)\tilde{\rho}D_3(\tilde{u}_{it} + \tilde{u}_j a_{kj}D_k\tilde{u}_i)a_{3i} dyds\right|$$

$$+\Big|\int_0^t\int_{\tilde\Omega}J\chi^2 D_{33}(a_{lj}D_l\tilde u_j)\tilde\rho(\tilde u_{it}+\tilde u_j a_{kj}D_k\tilde u_i)D_3 a_{3i}dyds\Big|$$

$$\leqslant \delta\int_0^t\int_{\tilde\Omega}J\chi^2|D_{33}(a_{lj}D_l\tilde u_j)|^2 dyds+C_\delta\|\tilde\rho\|^2_{L^\infty_t(H^2)}\|\tilde u_t\|_{L^\infty_t(L^2)}\int_0^t\|\tilde u_t\|_{H^1}ds$$

$$+C_\delta\|\tilde\rho\|^2_{L^\infty_t(H^2)}\|\tilde u\|^{\frac{5}{2}}_{L^\infty_t(H^1)}\int_0^t\|\tilde u\|^{\frac{3}{2}}_{H^3}ds+C_\delta\|\tilde\rho\|^2_{L^\infty_t(H^2)}\int_0^t\|\tilde u_t\|^2_{H^1}ds$$

$$+C_\delta\|\tilde\rho\|^2_{L^\infty_t(H^2)}\|\tilde u\|^{\frac{9}{4}}_{L^\infty_t(H^1)}\int_0^t\|\tilde u\|^{\frac{7}{8}}_{H^3}ds$$

$$\leqslant \delta\int_0^t\int_{\tilde\Omega}J\chi^2|D_{33}(a_{lj}D_l\tilde u_j)|^2 dyds+C_0(M_0)\exp(t^{\frac{1}{8}}C(M)),$$

where the result in Lemma 4.5 has been used.

$$\Big|\int_0^t\int_{\tilde\Omega}J\chi^2 D_{33}(a_{lj}D_l\tilde u_j)M_5 dyds\Big|$$

$$\leqslant \delta\int_0^t\int_{\tilde\Omega}J\chi^2|D_{33}(a_{lj}D_l\tilde u_j)|^2 dyds+C_\delta\int_0^t\|\sigma\|^2_{H^2}\|\theta\|^2_{H^2}ds$$

$$\leqslant \delta\int_0^t\int_{\tilde\Omega}J\chi^2|D_{33}(a_{lj}D_l\tilde u_j)|^2 dyds+tC(M),$$

and

$$\Big|\int_0^t\int_{\tilde\Omega}J\chi^2 D_{33}(a_{lj}D_l\tilde u_j)M_6 dyds\Big|$$

$$\leqslant \delta\int_0^t\int_{\tilde\Omega}J\chi^2|D_{33}(a_{lj}D_l\tilde u_j)|^2 dyds+C_\delta\int_0^t\int_{\tilde\Omega}J\chi^2|D_{33\tau}\tilde u|^2 dyds$$

$$+C_\delta\int_0^t\int_{\tilde\Omega}J\chi^2|D_{3\zeta\tau}\tilde u|^2 dyds+\sqrt{t}C(M).$$

From the above estimates, we can derive that

$$\frac{\mu+\lambda}{2}\int_0^t\int_{\tilde\Omega}J\chi^2|D_{33}(a_{lj}D_l\tilde u_j)|^2 dyds$$

$$-\frac{R}{\epsilon}\int_0^t\int_{\tilde\Omega}J\chi^2 D_{33}(a_{lj}D_l\tilde u_j)(D_{33}\tilde\sigma+D_{33}\tilde\theta)dyds$$

$$\leqslant C_2\int_0^t\int_{\tilde\Omega}J\chi^2|D_{33\tau}\tilde u|^2 dyds+C_3\int_0^t\int_{\tilde\Omega}J\chi^2|D_{3\zeta\tau}\tilde u|^2 dyds$$

$$+C_0(M_0)\exp(t^{\frac{1}{8}}C(M)). \tag{4.34}$$

In order to balance the singular term, similar to (4.32) and (4.33), we obtain

$$\frac{R}{2}\int_{\tilde{\Omega}} J\chi^2|D_{33}\tilde{\sigma}|^2 dy + \frac{R}{\epsilon}\int_0^t\int_{\tilde{\Omega}} J\chi^2 D_{33}(a_{kj}D_k\tilde{u}_j)D_{33}\tilde{\sigma}dyds$$

$$=\frac{R}{2}\int_{\tilde{\Omega}} J\chi^2|D_{33}\tilde{\sigma}_0|^2 dy - R\int_0^t\int_{\tilde{\Omega}} J\chi^2 D_{33}\tilde{\sigma}D_{33}[a_{kj}D_k(\tilde{u}_j\tilde{\sigma})]dyds$$

$$\leqslant C_0(M_0)\exp(t^{\frac{1}{4}}C(M)), \qquad (4.35)$$

and

$$\frac{C_v}{2}\int_{\tilde{\Omega}} J\chi^2\tilde{\rho}|D_{33}\tilde{\theta}|^2 dy + k_3\int_0^t\int_{\tilde{\Omega}} J\chi^2|D_{y33}\tilde{\theta}|^2 dyds$$

$$+\frac{R}{\epsilon}\int_0^t\int_{\tilde{\Omega}} J\chi^2 D_{33}\tilde{\theta}D_{33}(a_{kj}D_k\tilde{u}_j)dyds.$$

$$\leqslant C_0(M_0)\exp(t^{\frac{1}{8}}C(M)), \qquad (4.36)$$

where κ_3 is a positive constant.

Combining (4.34), (4.35) with (4.36), we can prove this lemma.

Lemma 4.11 *For any* $0 \leqslant t \leqslant \min\{T^\epsilon, 1\}$,

$$\int_0^t\int_{\tilde{\Omega}} J\chi^2|D_{33\tau}\tilde{u}|^2 dyds$$

$$\leqslant C_4\int_0^t\int_{\tilde{\Omega}} J\chi^2|D_{y\zeta\tau}\tilde{u}|^2 dyds + C_5\int_0^t\int_{\tilde{\Omega}} J\chi^2|D_{\tau 3}(a_{lj}D_l\tilde{u}_j)|^2 dyds$$

$$+ C_6\int_0^t \frac{1}{\epsilon^2}\|(\nabla\sigma, \nabla\theta)\|_{L^2}^2 ds + C_0(M_0)\exp(t^{\frac{1}{8}}C(M)).$$

Proof To estimate $\int_0^t\int_{\tilde{\Omega}} J\chi^2|D_{33\tau}\tilde{u}|^2 dyds$, we introduce an auxiliary Stokes problem in the original coordinates in the region near the boundary:

$$-\mu\Delta_x[(\chi D_\tau\tilde{u})\circ\Lambda^{-1}] + \frac{R}{\epsilon}\nabla_x(\chi D_\tau\tilde{\sigma} + \chi D_\tau\tilde{\sigma})\circ\Lambda^{-1}] = G_1 \quad \text{in } W\cap\Omega,$$

$$\text{div}[(\chi D_\tau\tilde{u})\circ\Lambda^{-1}] = G_2 \quad \text{in } W\cap\Omega,$$

$$(\chi D_\tau\tilde{u})\circ\Lambda^{-1} = 0 \quad \text{on } \partial(W\cap\Omega),$$

where

$$G_1^i = \chi D_\tau[\tilde{\rho}(\tilde{u}_{it} + \tilde{u}_j a_{kj}D_k\tilde{u}_i)] - \lambda\chi D_\tau[a_{ki}D_k(a_{lj}D_l\tilde{u}_j)]$$

$$+ R\chi D_\tau[a_{ki}D_k(\tilde{\sigma}\tilde{\theta})] + O(1)(D_l\tilde{u}_i + D_{kl}\tilde{u}_i + \frac{1}{\epsilon}D_k\tilde{\sigma} + \frac{1}{\epsilon}D_k\tilde{\theta}),$$

$$G_2^j = O(1)(D_\tau \tilde{u}_j + D_k \tilde{u}_j + D_{k\tau}\tilde{u}_j).$$

By the regularity theory of the Stokes problem in [12], we get

$$\int_\Omega |\nabla_x[(\chi D_\tau \tilde{u}) \circ \Lambda^{-1}]|^2 dx \leqslant C(\|G_2\|_{H^1(W\cap\Omega)}^2 + \|G_1\|_{L^2(W\cap\Omega)}^2).$$

According to the relations $a_{3j}a_{ij} = 0$ $(i = 1, 2)$, we have

$$D_{33\tau}\tilde{u} = \sum_{k,l=1}^{3}\left(\sum_{j=1}^{3} a_{kj}a_{lj}\right)D_{kl\tau}\tilde{u} - \sum_{1\leqslant k,l\leqslant 2}\sum_{j=1}^{3} a_{kj}a_{lj}D_{kl\tau}\tilde{u}.$$

Moreover, notice that

$$\int_\Omega |\nabla_x[(\chi D_\tau \tilde{u}) \circ \Lambda^{-1}]|^2 dx$$
$$= \int_{\tilde{\Omega}} J \Big| \sum_{j=1}^{3}\sum_{k=1}^{3} a_{kj} D_k \Big(\sum_{l=1}^{3} a_{lj} D_l(\chi D_\tau \tilde{u})\Big)\Big|^2 dy$$
$$\leqslant \int_{\tilde{\Omega}} J\chi^2 \Big|\sum_{j,k,l=1}^{3} a_{kj}a_{lj}D_{kl\tau}\tilde{u}\Big|^2 dy + O(1)\int_{\tilde{\Omega}}(|D_\tau \tilde{u}|^2 + |D_{y\tau}\tilde{u}|^2)dy.$$

Thus

$$\int_{\tilde{\Omega}} J\chi^2 |D_{33\tau}\tilde{u}|^2 dy \leqslant C(\|G_1\|_{L^2(W\cap\Omega)}^2 + \|G_2\|_{H^1(W\cap\Omega)}^2)$$
$$+ C\int_{\tilde{\Omega}} J\chi^2 |D_{\iota\zeta\tau}\tilde{u}|^2 dy + C\|\nabla^2 u\|_{L^2}^2 + C\|u\|_{H^1}^2, \quad (4.37)$$

where $\iota, \zeta, \tau = 1, 2$. By the representations of G_1 and G_2, it is not difficult to see that

$$\|G_1\|_{L^2(W\cap\Omega)}^2 \leqslant C\frac{1}{\epsilon^2}\|(\nabla\sigma, \nabla\theta)\|_{L^2}^2 + C\|\rho\|_{H^2}^2(\|u_t\|_{H^1}^2 + \|\tilde{u}\|_{H^3}^{\frac{3}{2}}\|\tilde{u}\|_{H^1}^{\frac{5}{2}})$$
$$+ C\int_{\tilde{\Omega}} J\chi^2 |D_{\tau k}(a_{lj}D_l\tilde{u}_j)|^2 dy + \|\sigma\|_{H^2}^2\|\theta\|_{H^2}^2, \quad (4.38)$$

and

$$\nabla_x G_2^j(\Lambda^{-1}(x)) = O(1)(D_l\tilde{u}_j + D_{kl}\tilde{u}_j + \chi D_{k\tau}(a_{lj}D_l\tilde{u}_j)).$$

Hence,
$$\|G_2\|_{H^1(W\cap\Omega)}^2 \leqslant C\int_{\tilde{\Omega}} J\chi^2 |D_{\tau k}(a_{lj}D_l\tilde{u}_j)|^2 dy + C\|\nabla^2 u\|_{L^2}^2 + C\|u\|_{H^1}^2. \quad (4.39)$$

Substituting (4.38), (4.39) into (4.37) and integrating over $(0,t) \times \tilde{\Omega}$, we obtain

$$\int_0^t \int_{\tilde{\Omega}} J\chi^2 |D_{33\tau}\tilde{u}|^2 dyds$$
$$\leqslant C_6 \int_0^t \frac{1}{\epsilon^2}\|(\nabla\sigma,\nabla\theta)\|_{L^2}^2 ds + C_5 \int_0^t \int_{\tilde{\Omega}} J\chi^2 |D_{\tau 3}(a_{lj}D_l\tilde{u}_j)|^2 dyds$$
$$+ C_4 \int_0^t \int_{\tilde{\Omega}} J\chi^2 |D_{y\varsigma\tau}\tilde{u}|^2 dyds + \int_0^t \|\rho\|_{H^2}^2(\|u_t\|_{H^1}^2 + \|\tilde{u}\|_{H^3}^{\frac{3}{2}}\|\tilde{u}\|_{H^1}^{\frac{5}{2}}) ds$$
$$+ \int_0^t \|\sigma\|_{H^2}^2 \|\theta\|_{H^2}^2 ds + C\int_0^t \|\nabla^2 u\|_{L^2}^2 ds + C\int_0^t \|u\|_{H^1}^2 ds.$$

Then, we prove this lemma.

Next, we give a conclusion of the boundary estimates. Denoting

$$\Phi_\chi(t) = \frac{C_8+1}{2\mu_1}\int_{\tilde{\Omega}} J\chi^2(\tilde{\rho}|D_{\varsigma\tau}\tilde{u}|^2 + R|D_{\varsigma\tau}\tilde{\sigma}|^2 + C_v\tilde{\rho}|D_{\varsigma\tau}\tilde{\theta}|^2) dy$$
$$+ \frac{1}{2}\int_{\tilde{\Omega}} J\chi^2(C_7 R|D_{\tau 3}\tilde{\sigma}|^2 + C_7 C_v\tilde{\rho}|D_{\tau 3}\tilde{\theta}|^2 + R|D_{33}\tilde{\sigma}|^2 + C_v\tilde{\rho}|D_{33}\tilde{\theta}|^2) dy,$$

where $C_7 = \dfrac{2}{\mu+\lambda}[(C_2+1)C_5+1]$, and

$$\Psi_\chi(t) = \int_0^t \int_{\tilde{\Omega}} J\chi^2 |D_{y\varsigma\tau}\tilde{u}|^2 dyds + \frac{\kappa_1(C_8+1)}{\mu_1}\int_0^t \int_{\tilde{\Omega}} J\chi^2 |D_{y\varsigma\tau}\tilde{\theta}|^2 dyds$$
$$+ \int_0^t \int_{\tilde{\Omega}} J\chi^2(|D_{\tau 3}(a_{lj}D_l\tilde{u}_j)|^2 + \kappa_2 C_7|D_{y\tau 3}\tilde{\theta}|^2 + |D_{33}(a_{lj}D_l\tilde{u}_j)|^2$$
$$+ \kappa_3|D_{y33}\tilde{\theta}|^2 + |D_{33\tau}\tilde{u}|^2) dyds.$$

We deduce from Lemma 4.8-Lemma 4.11 that

$$\Phi_\chi(t) + \Psi_\chi(t)$$
$$\leqslant C_0(M_0)\exp(t^{\frac{1}{8}}C(M)) + \delta\int_0^t \left(\frac{1}{\epsilon^2}\|(\nabla^2\sigma,\nabla^2\theta)\|_{L^2}^2 + \|u\|_{H^3}^2\right) ds$$
$$+ (1+C_2)C_6 \int_0^t \frac{1}{\epsilon^2}\|(\nabla\sigma,\nabla\theta)\|_{L^2}^2 ds, \quad (4.40)$$

where $C_8 = C_1C_7 + C_3 + (1+C_2)$ and δ is a small enough positive constant.

Conclusions for the estimates of $\|\nabla^2 \mathrm{div} u\|^2_{L^2_t(L^2)}$ *and* $\|\nabla^3 \theta\|^2_{L^2_t(L^2)}$.

Combining the interior estimates in the Lemma 4.7 with the estimates near the boundary (4.40) and transforming the local coordinates into usual ones, we have

Lemma 4.12 *For any* $0 \leqslant t \leqslant \min\{T^\epsilon, 1\}$,

$$\|\chi_0 \nabla^2 u\|^2_{L^2} + \int_{\tilde{\Omega}} J\chi^2 \tilde{\rho} |D_{\zeta_\tau} \tilde{u}|^2 dy + \|(\nabla^2 \sigma, \nabla^2 \theta)\|^2_{L^2}$$
$$+ \|\nabla^2 \mathrm{div} u\|^2_{L^2_t(L^2)} + \|\nabla^3 \theta\|^2_{L^2_t(L^2)}$$
$$\leqslant C_0(M_0) \exp(t^{\frac{1}{8}} C(M)) + \delta \int_0^t \left(\frac{1}{\epsilon^2} \|(\nabla^2 \sigma, \nabla^2 \theta)\|^2_{L^2} + \|u\|^2_{H^3} \right) ds$$
$$+ (1+C_2) C_6 \int_0^t \frac{1}{\epsilon^2} \|(\nabla \sigma, \nabla \theta)\|^2_{L^2} ds. \tag{4.41}$$

Finally, submitting (4.12), (4.13) into (4.41), we can close the high-order estimates of u and θ. For any $0 \leqslant t \leqslant \min\{T^\epsilon, 1\}$, we have

$$\|(\nabla^2 \sigma, \nabla^2 \theta)\|^2_{L^2} + \|(u,\theta)\|_{L^2_t(H^3)}$$
$$\leqslant C_0(M_0) \exp(t^{\frac{1}{8}} C(M)) + C \int_0^t \Big[\|u\|_{H^3} \|u\|_{H^1} + \|\rho\|^2_{H^2} (\|u_t\|^2_{L^2}$$
$$+ \|u\|^{\frac{1}{2}}_{H^3} \|u\|^{\frac{7}{2}}_{H^1}) + C\|\sigma\|^2_{H^2} \|\theta\|^2_{H^2} \Big] ds$$
$$+ \delta \int_0^t \Big[\|\nabla^2 \mathrm{div} u\|^2_{L^2} + \|\rho\|^2_{H^2} (\|u_t\|^2_{H^1} + \|u\|^{\frac{3}{2}}_{H^3} \|u\|^{\frac{5}{2}}_{H^1}) + \|\sigma\|^2_{H^2} \|\theta\|^2_{H^2} \Big] ds,$$

where δ is a small enough positive constant. According to lemma 4.5, we obtain, for any $0 \leqslant t \leqslant \min\{T^\epsilon, 1\}$ and $0 < \epsilon \leqslant 1$,

$$\|(\nabla^2 \sigma, \nabla^2 \theta)(t)\|^2_{L^2} + \|(u,\theta)\|_{L^2_t(H^3)} \leqslant C_0(M_0) \exp(t^{\frac{1}{8}} C(M)). \tag{4.42}$$

5 Proof of the main theorems

Proof of Theorem 2.4 Combining the estimates obtained in Lemma 4.1-Lemma 4.4 and (4.42), we can derive, for any $0 \leqslant t \leqslant \min\{T^\epsilon, 1\}$ and $0 < \epsilon \leqslant 1$,

$$\|(\sigma, u, \theta)(t)\|_{L^2} + \|(\nabla \sigma, \mathrm{div} u, \mathrm{curl} u, \nabla \theta)(t)\|_{L^2} + \|(\nabla^2 \sigma, \nabla^2 \theta)(t)\|_{L^2}$$
$$+ \|(\sigma_t, u_t, \theta_t)(t)\|_{L^2} + \|(u_t, \theta_t)\|^2_{L^2_t(H^1)} + \|(u,\theta)\|_{L^2_t(H^3)}$$

$$\leqslant C_0(M_0)\exp(t^{\frac{1}{8}}C(M)).$$

This completes the proof of Theorem 2.4.

Proof of Theorem 2.2 The proof is based on [1], [18]. Assume that Theorem 2.4 holds and $T^\epsilon < +\infty$ is the maximal lifetime of existence for the solution obtained in Theorem 2.1. Then, for any $0 \leqslant t \leqslant \min\{T^\epsilon, 1\}$, we have

$$M^\epsilon(t) \leqslant C_0(M_0^\epsilon)\exp(t^{\frac{1}{8}}C(M^\epsilon(t))), \tag{5.1}$$

where $M_0^\epsilon \leqslant D_0$ for $0 < \epsilon \leqslant 1$. In the sequence, we choose $D > C_0(D_0)$ and $T_1 \leqslant 1$ such that

$$C_0(D_0)\exp(T_1^{\frac{1}{8}}C(D)) < D. \tag{5.2}$$

Let $t < \min\{T^\epsilon, T_1\}$. Combining the inequalities (5.1) and (5.2) with the assumption $M^\epsilon(0) = M_0^\epsilon$, we obtain that $M^\epsilon(t) \neq D$. Moreover, we can assume without restriction that $D_0 \leqslant D$, so that $M^\epsilon(0) \leqslant D$. Since the function $M^\epsilon(t)$ is continuous, we obtain

$$M^\epsilon(t) \leqslant D \quad \text{for } t < \min\{T^\epsilon, T_1\} \text{ and } 0 < \epsilon \leqslant 1. \tag{5.3}$$

Then $T^\epsilon > T_1$ for $0 < \epsilon \leqslant 1$. Otherwise, by using the uniform estimates in (5.3) and applying Theorem 2.1 repeatedly, one can extend the time interval of existence to $[0, T_1]$, which contradicts the maximality of T^ϵ. Therefore, $M^\epsilon(t) \leqslant D$ for any $t \in [0, T_1]$ where T_1 is independent of $\epsilon \in (0, 1]$. This completes the proof of Theorem 2.2.

References

[1] Alazard T. Low Mach number limit of the full Navier-Stokes equations [J]. Arch. Ration. Mech. Anal., 2006, 180: 1–73.

[2] Bourguignon J. Brezis H. Remarks on the Euler equation [J]. J. Funct. Anal., 1974, 15: 341–363.

[3] Bresch D, Desjardins B, Grenier E, Lin C. Low Mach number limit of viscous polytropic flows: formal asymptotics in the periodic case [J]. Stud. Appl. Math., 2002, 109: 125–149.

[4] Dou C, Jiang S, Ou Y. Low Mach number limit of full Navier-Stokes equations in a 3D bounded domain [J]. J. Differential Equations, 2015, 258: 379–398.

[5] Danchin R. Zero Mach number limit for compressible flows with periodic boundary conditions [J]. Amer. J. Math., 2002, 124: 1153–1219.

[6] Desjardins B, Grenier E. Low Mach number limit of viscous compressible flows in the whole space [J]. Proc. R. Soc. Lond. Ser. A Math. Phys. Eng. Sci., 1999, 455: 2271–2279.

[7] Desjardins B, Grenier E, Lions P, Masmoudi N. Incompressible limit for solutions of the isentropic Navier- Stokes equations with Dirichlet boundary conditions [J]. J. Math. Pures Appl., 1999, 78: 461–471.

[8] Ebin D. Motion of a slightly compressible fluid [J]. Proc. Natl. Acad. Sci. USA, 1975, 72: 539–542.

[9] Feireisl E, Novotny A. The low Mach number limit for the full Navier-Stokes-Fourier system [J]. Arch. Ration. Mech. Anal., 2007, 186: 77–107.

[10] Feireisl E, Novotny A. Inviscid incompressible limits of the full Navier-Stokes-Fourier system [J]. Commun. Math. Phys., 2013, 321: 605–628.

[11] Friedman A. Partial Differential Equations [M]. New York: Dover Publications, 2008.

[12] Galdi G. An Introduction to the Mathematical Theory of the Navier-Stokes Equations. Vol. I. Linearized Steady Problems [M]. New York: Springer-Verlag, 1994.

[13] Hagstrom T, Lorenz J. On the stability of approximate solutions of hyperbolic-parabolic systems and the all-time existence of smooth, slightly compressible flows [J]. Indiana Univ. Math. J., 2002, 51: 1339–1387.

[14] Jiang S, Ju Q, Li F. Incompressible limit of the compressible magnetohydrodynamic equations with periodic boundary conditions [J]. Comm. Math. Phys., 2010, 297: 371–400.

[15] Jiang S, Ou Y. Incompressible limit of the non-isentropic Navier-Stokes equations with well-prepared initial data in three-dimensional bounded domains [J]. J. Math. Pures Appl., 2011, 96: 1–28.

[16] Klainerman S, Majda A. Singular perturbations of quasilinear hyperbolic systems with large parameters and the incompressible limit of compressible fluids [J]. Comm. Pure Appl. Math., 1981, 34: 481–524.

[17] Klainerman S, Majda A. Compressible and incompressible fluids [J]. Comm. Pure Appl. Math., 1982, 35: 629–653.

[18] Métivier G, Schchet S. The incompressible limit of the non-isentropic Euler equations[J]. Arch. Ration. Mech. Anal., 2001, 158: 61–90.

[19] Ou Y. Low Mach number limit of viscous polytropic fluid flows [J]. J. Differential Equations, 2011, 251: 2037–2065.

[20] Ren D, Ou Y. Incompressible limit of all-time solutions to 3-D full Navier-Stokes equations for perfect gas with well-prepared initial condition [J]. Z. Angew. Math. Phys., 2016, 67: 1–27.

[21] Valli A. Periodic and stationary solutions for compressible Navier-Stokes equations via a stability method [J]. Ann. Scuola Norm. Sup. Pisa Cl. Sci., 1983, 10: 607–647.

[22] Valli A, Zajaczkowski W. Navier-stokes for compressible fluid: global existence and qualitative properties of the solutions in the general case [J]. Commun. Math. Phys., 1986, 103: 259–296.

Weak and Smooth Solutions to Incompressible Navier-Stokes-Landau-Lifshitz-Maxwell Equations*

Guo Boling (郭柏灵), Liu Fengxia(刘峰霞)

Abstract This paper is concerned with Navier-Stokes-Landau-Lifshitz-Maxwell equations. In dimensions two and three, we use the Galerkin method to prove the existence of a weak solution. Then combining the a priori estimates and induction technique, we obtain the existence of the smooth solution.

Keywords weak solution; smooth solution; Navier-Stokes-Landau-Lifshitz-Maxwell equations

1 Introduction and main results

In this paper, we consider the following Navier-Stokes-Landau-Lifshitz-Maxwell system

$$u_t + (u \cdot \nabla)u + \nabla P - \nu \Delta u = -\lambda \nabla \cdot (\nabla d \odot \nabla d) - (E + u \times H), \quad (1.1)$$

$$d_t + (u \cdot \nabla)d = \gamma(-d \times (d \times \Delta d) + d \times \Delta d), \quad (1.2)$$

$$\nabla \cdot u = 0, \quad (1.3)$$

$$\frac{\partial E}{\partial t} - \nabla \times H = -\sigma E + u, \quad (1.4)$$

$$\frac{\partial (H + \beta d)}{\partial t} + \nabla \times E = 0, \quad (1.5)$$

$$\nabla \cdot E = 0, \quad (1.6)$$

---

* Frontier of Mathematics in China, 2019, 14(6), 29. DOI: https://doi.org/10.1007/s11464-019-0800-x.

$$\nabla \cdot (H + \beta d) = 0, \tag{1.7}$$

following initial and boundary conditions

$$u(x,0) = u_0(x), \quad \nabla \cdot u_0 = 0, \quad d(x,0) = d_0(x), \ x \in \Omega, \tag{1.8}$$

$$u(x,t) = d_t(x,t) = 0, \quad (x,t) \in (\partial\Omega \times (0,T)) \tag{1.9}$$

in $\Omega \times (0,T)$, for a bounded smooth domain Ω in $\mathbf{R}^3$ or in $\mathbf{R}^2$, $T \in (0,\infty)$. (1.1), (1.3) are the well-known density independent Navier-Stokes equations, while (1.2) is the Lauden-Lifshitz when $u \equiv 0$. Here $u(x,t) : \Omega \times (0,T) \to \mathbf{R}^n$ represents the velocity field of the flow, $d(x,t) : \Omega \times (0,T) \to S^n$, the unit sphere in $\mathbf{R}^{n+1}$ ($n = 1, 2$) is a unit vector that represents the macroscopic molecular orientation of the liquid crystal material. $P(x,t) : \Omega \times (0,+\infty) \to \mathbf{R}$ represents the pressure function, ν, λ, γ are positive constants that represent viscosity, the competition between kinetic energy and potential energy, the microscopic elastic relaxation time for the molecular orientation field. In this paper, we note

$$-(d|\nabla d|^2 + d \times \Delta d) = f(d). \tag{1.10}$$

In the system (1.1)-(1.5), the unusual term $\nabla \cdot (\nabla d \odot \nabla d)$ denotes the $n \times n$ matrix whose (i,j)-th entry is given by $d_{x_i} \cdot d_{x_j}$, for $1 \leqslant i, j \leqslant n$, "$\times$" denotes the vector outer product, $\sigma \geqslant 0$ denotes the constant conductivity and the constant β can be viewed as the magnetic permeability of free space.

For system (1.1)-(1.3), H. Kim [1] proved the following regularity criterion

$$u \in L^s(0,t; L^{p,\infty}(\mathbf{R}^3)) \ \text{ with } \ \frac{2}{s} + \frac{3}{p} = 1 \ \text{ and } \ 3 < p \leqslant \infty. \tag{1.11}$$

Fan [2] extends (1.11) to the multiplier space

$$\dot{X}_r := M(\dot{H}^r, L^2) = \left\{ f \Big| \|f\|_{\dot{X}_r} := \sup\left(\frac{\|fg\|_{L^2}}{\|g\|_{H^r}}\right) < \infty \right\}.$$

When the term $d \times \Delta d$ is omitted, the system (1.1)-(1.3) will be

$$u_t + u \cdot \nabla u + \nabla P - \nu \Delta u = -\lambda \nabla \cdot (\nabla d \odot \nabla d), \tag{1.12}$$

$$\nabla \cdot u = 0, \tag{1.13}$$

$$d_t + u \cdot \nabla d = \gamma(\Delta d + d|\nabla d|^2), \tag{1.14}$$

which is a simplified version of the Ericksen-Leslie model, which reduces to the Ossen-Frank model in the static case, for the hydrodynamics of nematic liquid crystals developed during the period 1958-1968 [3]- [5].

It is a macroscopic continuum description of the time evolution of the materials under the influce of both the flow field $u(x,t)$, and the macroscopic description of the microscopic orientation configurations $d(x,t)$ of rod-like liquid crystals. Roughly speaking, the system (1.12)-(1.14) is a coupling between the non-homogeneous Navier- Stokes equation and the transported flow of harmonic maps. It is probably the simplest mathematical model one can derive, without destroying the basic nonlinear structure, from the original equations in the continuum theory of nematic liquid crystals proposed by Ericksen in [7] and [9] and Leslie in [8]. In a series of papers, Lin [10]- [12] initiated the mathematical analysis of the system (1.12)-(1.14) in the 1990's.

More precisely, Lin considered in [11] the Leslie system of variable length, i.e., the Dirichlet energy $\frac{1}{2}\int_\Omega |\nabla d|^2 dx$ for $d: \Omega \to S^{n-1}$ is replaced by the Ginzburg-Landau energy $\int_\Omega \left(\frac{1}{2}|\nabla d|^2 + \frac{(1-|d|^2)^2}{4\varepsilon^2}\right)$ for $d: \Omega \to \mathbf{R}^n$ and proved existence of global classical and weak solutions in dimensions two or three. In [12], they proved the partial regularity theorem for suitable weak solutions, similar to the classical theorem by Caffarelli-Kohn-Nirenberg [14] for the Navier-Stokes equation. However, as pointed out in [11]- [12], both their estimates and arguments depend on ε, and it is a challenging problem to study the convergence as ε tends to zero.

The Ericksen-Leslie theory has successfully predicted several effects and analyzed many others rather well; see, for example, the survey articles by Leslie in [13]. Most analytical work, however, has been carried out under rather special assumptions concerning either the type of flow or the form of the solutions. There are also other reasons this system merits attention.

The coupling of this new Navier-Stokes-Landau-Lifshitz-Maxwell system can be derived from the full Maxwell system as follows

$$\frac{\partial B}{\partial t} = -\Delta \times E, \quad \frac{\partial D}{\partial t} + \sigma E = \nabla \times H, \tag{1.15}$$

where E and H are the electric and magnetic field, $\sigma > 0$ is the conductivity, D

and B are the electric and magnetic displacements defined by

$$D = \varepsilon_0 E, \quad B = \mu_0(H+d),$$

where ε_0 is the permittivity of free space, μ_0 is the magnetic permeability of free space, u is the velocity field of the flow, and E is the electric polarization. Substituting these definitions into (1.15), we may couple u, E, H and d by systems (1.1)-(1.7), for more information about (1.15), we refer to [15].

In the past few years, progress has been made in the analysis of the model (1.1)-(1.7) by overcoming the supercritical nonlinearity $|\nabla d|^2 d$. The existence of weak solutions was established in [16]. To our best knowledge, however, there are no results available on weak solutions to the multi-dimensional problem (1.1)-(1.7) with supercritical nonlinearity. Since Eq (1.1)-(1.7) is strongly coupled, it is not easy to obtain the weak solution by use of the theory of semigroups as Habib Ammari in [6]. We are going to use the Galerkin method here. We are interested in global weak or smooth solutions to the problem (1.1)-(1.9) in the domain $\Omega \times (0, T) = Q_T$.

This paper is written as follows. In section 2, we use the standard method to obtain the existence of a weak solution for the problem (1.1)-(1.9) in the domain $\Omega \times (0, T) = Q_T$, where $\Omega \subset \mathbf{R}^2$ or $\mathbf{R}^3$. In section 3, we obtain the global smooth solution by establishing a prior estimate and induction techniques.

C is a generic constant and may assume different values in different formulates.

For the sake of simplicity, we denote $\|\cdot\|_{L^p(\Omega)} = \|\cdot\|_p$, $p \geqslant 2$.

Denote $H^m(\Omega), m = 1, 2, \cdots$ be the Sobolev space of complex-valued functions with the norm

$$\|u\|_{H^m} = \Big(\int_\Omega \sum_{|\alpha| \leqslant m} |D^\alpha u|^2 dx\Big)^{\frac{1}{2}},$$

$H_0^1(\Omega) = $ closure of $C_0^\infty(\Omega, \mathbf{R}^n)$ in the norm $\Big(\int_\Omega |\nabla u|^2 dx\Big)^{\frac{1}{2}}$,

$H^{-1}(\Omega) = $ the dual of $H_0^1(\Omega)$, $V = C_0^\infty(\Omega, \mathbf{R}^n) \cap \{u : \nabla \cdot u = 0\}$,

$J = $ closure of V in $L^2(\Omega, \mathbf{R}^n)$, $K = $ closure of V in $H^1(\Omega, \mathbf{R}^n)$.

Definition 1.1 *A vector* $(u(x,t), d(x,t), E(x,t), H(x,t)) \in (L^\infty(0,T;L^2(\Omega)),$ $L^\infty(0,T;H^1(\Omega)), L^\infty(0,T;L^2(\Omega)), L^\infty(0,T;L^2(\Omega)))$ *is called a weak solution to problem* (1.1)-(1.9), *if for any vector valued test function* $\Psi(x,t) \in C^1(Q_T)$ *such* that $\Psi(x,T) = 0$, the following equalities hold

$$-\iint_{Q_T} u \cdot \Psi_t + \iint_{Q_T} \nabla u \cdot \nabla \Psi + \iint_{Q_T} u \cdot \nabla u \cdot \Psi + \iint_{Q_T} \nabla P \cdot \Psi$$
$$= \int_\Omega u_0 \Psi(x,0) - \lambda \iint_{Q_T} (\nabla d \odot \nabla d) \cdot \nabla \Psi - \iint_{Q_T} E \cdot \Psi - \iint_{Q_T} (u \times H) \cdot \Psi, \quad (1.16)$$

$$-\iint_{Q_T} d \cdot \Psi_t + \iint_{Q_T} \nabla u \cdot \nabla \Psi + \iint_{Q_T} u \cdot \nabla d \cdot \Psi$$
$$= \gamma \iint_{Q_T} (-d \times (d \times \Delta d) + d \times \Delta d) \cdot \Psi, \quad (1.17)$$

$$\iint_{Q_T} E\Psi_t e^{\sigma t} - \sigma \iint_{Q_T} e^{\sigma t} E \cdot \Psi + \iint_{Q_T} e^{\sigma t}(\nabla \times \Psi) \cdot H$$
$$+ \int_\Omega E_0 \cdot \Psi(x,0) + \int_\Omega u \cdot e^{\sigma t}\Psi = 0, \quad (1.18)$$

$$\iint_{Q_T} (H + \beta d)\Psi_t - \iint_{Q_T} (\nabla \times \Psi) \cdot E + \int_\Omega (H_0 + \beta d_0) \cdot \Psi(x,0) = 0. \quad (1.19)$$

Theorem 1.1 Assume $(u_0, d_0, E_0, H_0) \in (L^2(\Omega), H^1(\Omega), L^2(\Omega), L^2(\Omega))$. Then the problem (1.1)-(1.9) admits at least one global initial valued solution $(u(x,t), d(x,t), E(x,t), H(x,t))$ such that

$$u \in L^2(0,T;H^1(\Omega)) \cap L^\infty(0,T;L^2(\Omega)) \cap C(0,T;H^{-1}(\Omega)),$$

$$d \in L^2(0,T;H^2(\Omega)) \cap L^\infty(0,T;H^1(\Omega)) \cap C^{(0,\frac{1}{6})}(0,T;L^2(\Omega)),$$

$$E \in L^\infty(0,T;L^2(\Omega)) \cap C(0,T;H^{-1}(\Omega)),$$

$$H \in L^\infty(0,T;L^2(\Omega)) \cap C(0,T;H^{-1}(\Omega)).$$

Theorem 1.2 Assume $(u_0, d_0, E_0, H_0) \in (H^1(\Omega), H^2(\Omega), H^{m-1}(\Omega), H^{m-1}(\Omega))$ $(m \geqslant 2)$. Then there exists a unique smooth solution $(u(x,t), d(x,t), E(x,t), H(x,t))$ of the problem (1.1)-(1.9) such that

$$(u,d) \in (L^\infty(0,T;H^m(\Omega)))^2,$$

$$(E,H) \in (L^\infty(0,T;H^{m-1}(\Omega)))^2$$

and that either

$$\dim \Omega = 2$$

or
$$\dim \Omega = 3$$
and
$$\nu \geqslant \nu_0(u_0, d_0, E_0, H_0).$$

For various notations and definitions of function spaces in the statement of the above theorem, and throughout the paper, we refer to Temam and Ladyzhenskaya; see [17] and [18].

2 Weak solution to (1.1)-(1.9)

The sizes of the viscosity constants ν, λ, γ do not play important roles in our proof of Theorem 1.1, i.e., the global existence of weak solutions of the Cauchy boundary value problem (1.1)-(1.9). Since ν, λ, γ is not crucial in this section, we assume that $\nu = \lambda = \gamma = 1$ for simplicity.

Lemma 2.1 (the Gagliardo-Nirenberg inequality) Assume that $u \in L^q(\Omega)$, $D^m u \in L^r(\Omega)$, $\Omega \subseteq \mathbf{R}^n$, $1 \leqslant q, r \leqslant \infty$, $0 \leqslant j \leqslant m$. Let p and α satisfy
$$\frac{1}{p} = \frac{j}{n} + \alpha\left(\frac{1}{r} - \frac{m}{n}\right) + (1-\alpha)\frac{1}{q}; \quad \frac{j}{m} \leqslant \alpha \leqslant 1.$$

Then
$$\|D^j u\|_p \leqslant C(p, m, j, q, r)\|D^m u\|_r^\alpha \|u\|_q^{1-\alpha}, \tag{2.1}$$

where $C(p, m, j, q, r)$ is a positive constant.

Lemma 2.2 (the Gronwall's inequality)([20]) Let c be a constsnt, and $b(t)$, $u(t)$ be nonnegative continuous functions in the interval $[0, T]$ satisfying
$$u(t) \leqslant c + \int_0^t b(\tau)u(\tau)d\tau, \quad t \in [0, T].$$

Then $u(t)$ satisfies the estimate
$$u(t) \leqslant c \exp\left(\int_0^t b(\tau)d\tau\right), \quad t \in [0, T]. \tag{2.2}$$

Lemma 2.3 ([21]) Assume $X \subset E \subset Y$ are Banach spaces and $X \hookrightarrow\hookrightarrow E$. Then the following imbeddings are compact, if $1 \leqslant q \leqslant \infty$, or $1 < r \leqslant \infty$

(i) $L^q(0, T; X) \cap \left\{\varphi : \dfrac{\partial \varphi}{\partial t} \in L^1(0, T; Y)\right\} \hookrightarrow\hookrightarrow L^q(0, T; E);$ \hfill (2.3)

(ii) $L^\infty(0,T;X) \cap \left\{\varphi : \dfrac{\partial \varphi}{\partial t} \in L^r(0,T;Y)\right\} \hookrightarrow\hookrightarrow C(0,T;E).$ \hfill (2.4)

Lemma 2.4 *If Ω is a smooth, bounded domain in $\mathbf{R}^n$ and $u|_{\partial\Omega} = 0$, then the following are true:*

$$\|u\|_4^4 \leqslant 2\|\nabla u\|_2^2\|u\|_2^2, \quad \|u\|_4^8 \leqslant C\|\nabla u\|_2^5\|\nabla^2 u\|_2^3, \quad n=2, \tag{2.5}$$

$$\|u\|_4^4 \leqslant 4\|\nabla u\|_2^3\|u\|_2, \quad \|\nabla u\|_4^6 \leqslant C\|u\|_\infty^5\|\nabla^3 u\|_2, \quad n=3. \tag{2.6}$$

2.1 Global weak solutions to (1.1)-(1.9)

Firstly, we get the system of the ordinary differential equations (2.9)-(2.12) admits at least one continuously differentiable global solution. Secondly, we use the Galerkin method to obtain the existence of the weak solution to the problem (1.1)-(1.9).

Let $w_n(x)(n = 1, 2, 3 \cdots)$ be the unit eigenfunctions satisfying the equation $\Delta w_n + \lambda_n w_n = 0$, $w_n(x) = 0$, $x \in \partial\Omega$ and $\lambda_n (n = 1, 2, 3 \cdots)$ the corresponding eigenvalues different from each other.

Denote the approximate solution of the problem following from

$$u_m(x,t) = \sum_{s=1}^m \alpha_{sm}(t)w_s(x), \quad d_m(x,t) = \sum_{s=1}^m \beta_{sm}(t)w_s(x),$$

$$P_m(x,t) = \sum_{s=1}^m \zeta_{sm}(t)w_s(x),$$

$$E_m(x,t) = \sum_{s=1}^m \gamma_{sm}(t)w_s(x), \quad H_m(x,t) = \sum_{s=1}^m \delta_{sm}(t)w_s(x),$$

where $\alpha_{sm}(t), \beta_{sm}(t), \gamma_{sm}(t), \delta_{sm}(t)(t \in \mathbf{R}^+)(s = 1, 2, \cdots, m; m = 1, 2, \cdots)$ are n-dimensional vector-valued functions satisfying the following system of ordinary differential equations:

$$\int_\Omega u_{mt} w_s(x) dx \tag{2.7}$$

$$= \int_\Omega (-u_m \cdot \nabla u_m + \Delta u_m + \nabla P_m - \nabla \cdot (\nabla d_m \odot \nabla d_m) - E_m) w_s(x) \tag{2.8}$$

$$- (u_m \times H_m) w_s(x) dx, \tag{2.9}$$

$$\int_\Omega (d_{mt} + u_m \cdot \nabla d_m) w_s(x) dx = \int_\Omega (-d_m \times (d_m \times \Delta d_m) + d_m \times \Delta d_m) w_s(x) dx, \tag{2.10}$$

$$\int_\Omega E_{mt} w_s(x) dx = \int_\Omega (\nabla \times H_m + u_m - \sigma E_m) w_s(x) dx, \qquad (2.11)$$

$$\int_\Omega (H_{mt} + \beta d_{mt}) w_s(x) dx = -\int_\Omega (\nabla \times E_m) w_s(x) dx, \qquad (2.12)$$

$$\int_\Omega \nabla P_m w_s(x) dx = 0, \qquad (2.13)$$

and the initial condition

$$\int_\Omega u_m(x,0) w_s(x) dx = \int_\Omega u_0(x) w_s(x) dx, \qquad (2.14)$$

$$\int_\Omega d_m(x,0) w_s(x) dx = \int_\Omega d_0(x) w_s(x) dx, \qquad (2.15)$$

$$\int_\Omega E_m(x,0) w_s(x) dx = \int_\Omega E_0(x) w_s(x) dx, \qquad (2.16)$$

$$\int_\Omega H_m(x,0) w_s(x) dx = \int_\Omega H_0(x) w_s(x) dx. \qquad (2.17)$$

It follows from the standard theory on nonlinear ordinary differential equations that the problem (2.9)-(2.17) admits unique local solution. The following estimates can ensure the existence and uniqueness of the solution of (2.9)-(2.17) and also obtain the global solution to the problem (1.1)-(1.9).

Lemma 2.5 *Assume* $(u_{m0}, d_{m0}, E_{m0}, H_{m0}) \in (L^2(\Omega), H^1(\Omega), L^2(\Omega), L^2(\Omega))$. *Then, for the solutions of the initial value problem (2.9)-(2.17), we have the following estimates*

$$\sup_{0 \leqslant t \leqslant T} (\|u_m\|_2^2 + \|d_m\|_{H^1}^2 + \|E_m\|_2^2 + \|H_m\|_2^2) \leqslant C, \qquad (2.18)$$

$$\|u_m\|_{L^2(0,T;H^1(\Omega))} + \|d_m\|_{L^2(0,T;H^2(\Omega))} \leqslant C, \qquad (2.19)$$

where C is independent of m.

Proof Testing (2.11) by γ_{sm} and (2.12) by δ_{sm}, summing up the result for $s = 1, 2, \cdots, m$ and integrating by parts

$$\frac{1}{2}\frac{d}{dt} \int_\Omega (|E_m|^2 + |H_m|^2) dx + \sigma \int_\Omega |E_m|^2 dx + \beta \int_\Omega d_{mt} H_m dx = \int_\Omega u_m E_m dx. \qquad (2.20)$$

Testing (2.11) by $(\alpha_{sm} + \gamma_{sm})$, summing up the result for $s = 1, 2, \cdots, m$ and integrating by parts to deduce that

$$\frac{1}{2}\frac{d}{dt} \int_\Omega |E_m|^2 dx + \int_\Omega u_m E_{mt} dx = \int_\Omega (E_m + u_m)(-\sigma E_m + u_m) dx$$

$$+ \int_\Omega (\nabla \times H_m)(E_m + u_m) dx. \qquad (2.21)$$

Adding (2.20) and (2.21), then testing the result by α_0, we have

$$\frac{1}{2}\frac{d}{dt}\int_\Omega (2\alpha_0 |E_m|^2 + 2\alpha_0 |H_m|^2) dx + 2\beta\alpha_0 \int_\Omega d_{mt} H_m dx$$
$$+ \alpha_0 \int_\Omega u_m E_{mt} dx + \sigma\alpha_0 \int_\Omega |E_m|^2 dx$$
$$= \alpha_0 \int_\Omega u_m E_m dx + \alpha_0 \int_\Omega (\nabla \times H_m) u_m dx + \alpha_0 \int_\Omega (E_m + u_m)(-\sigma E_m + u_m) dx. \qquad (2.22)$$

Multiplying (2.9) by $\alpha_{sm}(t)$ and summing up the products for $s = 1, 2, \cdots m$ to deduce that

$$\frac{1}{2}\frac{d}{dt}\int_\Omega |u_m|^2 dx + \int_\Omega |\nabla u_m|^2 dx = -\int_\Omega (u_m \cdot \nabla) d_m \cdot \Delta d_m dx - \int_\Omega E_m u_m dx. \qquad (2.23)$$

Testing (2.10) by $-\lambda_m \beta_{sm}$, summing up the products for $s = 1, 2, \cdots m$ and then integrating by parts, then adding the above equality

$$\frac{1}{2}\frac{d}{dt}\int_\Omega (|u_m|^2 + |\nabla d_m|^2) dx + \int_\Omega (|\nabla u_m|^2 + |\Delta d_m|^2 |d_m|^2) dx$$
$$= \int_\Omega (d_m \cdot \Delta d_m)^2 dx - \int_\Omega E_m u_m dx \leqslant \int_\Omega |\Delta d_m|^2 |d_m|^2 dx - \int_\Omega E_m u_m dx. \qquad (2.24)$$

Combining (2.24) with (2.22) we get

$$\frac{1}{2}\frac{d}{dt}\int_\Omega (|u_m|^2 + |\nabla d_m|^2 + 2\alpha_0 |E_m|^2 + 2\alpha_0 |H_m|^2) dx + \int_\Omega |\nabla u_m|^2 dx$$
$$+ 2\beta\alpha_0 \int_\Omega d_{mt} H_m dx + \alpha_0 \int_\Omega E_{mt} u_m dx + \sigma\alpha_0 \int_\Omega |E_m|^2 dx$$
$$\leqslant \alpha_0 \int_\Omega (u_m - \sigma E_m)(E_m + u_m) dx + \alpha_0 \int_\Omega (\nabla \times H_m) u_m dx$$
$$+ (\alpha_0 - 1) \int_\Omega E_m u_m dx. \qquad (2.25)$$

In order to deal with the term $\int_\Omega d_{mt} H_m dx$, we multiply (2.12) by $(2\beta\alpha_0 - \beta)\beta_{sm}$ and sum up the product for $s = 1, 2, \cdots, m$ to obtain

$$(2\beta\alpha_0 - \beta) \int_\Omega d_m H_{mt} dx + \frac{\beta(2\beta\alpha_0 - \beta)}{2}\frac{d}{dt}\int_\Omega |d_m|^2 dx$$

$$+ (2\beta\alpha_0 - \beta)\int_\Omega (\nabla \times E_m)d_m dx = 0. \tag{2.26}$$

Using (2.21), adding (2.25) and (2.26) to obtain

$$\frac{1}{2}\frac{d}{dt}\int_\Omega (|u_m|^2 + |\nabla d_m|^2 + (2\alpha_0 + 1)|E_m|^2 + (2\alpha_0 + 1)|H_m|^2)dx$$
$$+ (2\beta\alpha_0 - \beta)\frac{d}{dt}\int_\Omega d_m H_m dx + \int_\Omega |\nabla u_m|^2 dx + \alpha_0\int_\Omega E_{mt}u_m dx$$
$$+ (\sigma\alpha_0 + \sigma)\int_\Omega |E_m|^2 dx$$
$$\leqslant \alpha_0\int_\Omega (u_m - \sigma E_m)(E_m + u_m)dx + \alpha_0\int_\Omega (\nabla \times H_m)u_m dx$$
$$- (2\beta\alpha_0 - \beta)\int_\Omega (\nabla \times E_m)d_m dx$$
$$- \frac{\beta(2\beta\alpha_0 - \beta)}{2}\frac{d}{dt}\int_\Omega |d_m|^2 dx + (\alpha_0 - 2)\int_\Omega E_m u_m dx. \tag{2.27}$$

Similarly, to deal with the term $\int_\Omega u_m E_{mt}dx$, we multiply (2.11) by $\alpha_0\alpha_{sm}$ and sum up the product for $s = 1, 2, \cdots, m$ to obtain

$$\alpha_0\int_\Omega E_{mt}u_m dx - \alpha_0\int_\Omega |u_m|^2 dx + \alpha_0\sigma\int_\Omega E_m u_m dx - \alpha_0\int_\Omega (\nabla \times H_m)\cdot u_m dx = 0. \tag{2.28}$$

Taking (2.28) into the inequality (2.27) and denote

$$\frac{1}{2}\frac{d}{dt}\int_\Omega (|u_m|^2 + |\nabla d_m|^2 + (2\alpha_0 + 1)|E_m|^2 + (2\alpha_0 + 1)|H_m|^2)dx + \alpha_0\int_\Omega |u_m|^2 dx$$
$$+ (2\beta\alpha_0 - \beta)\frac{d}{dt}\int_\Omega d_m H_m dx + \int_\Omega |\nabla u_m|^2 dx + \frac{\beta(2\beta\alpha_0 - \beta)}{2}\frac{d}{dt}\int_\Omega |d_m|^2 dx$$
$$\leqslant \alpha_0\int_\Omega (u_m - \sigma E_m)(E_m + u_m)dx + (2\beta\alpha_0 - \beta)\int_\Omega (\nabla \times E_m)d_m dx$$
$$- \sigma\alpha_0\int_\Omega |E_m|^2 dx + (\alpha_0\sigma + \alpha_0 - 2)\int_\Omega E_m u_m dx. \tag{2.29}$$

We will estimate each term of the right-hand sight of (2.29):

$$I_1 = \alpha_0\int_\Omega (u_m - \sigma E_m)(E_m + u_m)dx \leqslant \frac{\alpha_0}{2}\|u_m - \sigma E_m\|_2^2 + \frac{\alpha_0}{2}\|E_m + u_m\|_2^2,$$

$$I_2 = (2\beta\alpha_0 - \beta)\int_\Omega (\nabla \times d_m)E_m dx \leqslant \frac{2\beta\alpha_0 - \beta}{2}(\|\nabla \times d_m\|_2^2 + \|E_m\|_2^2),$$

$$I_3 = (\alpha_0\sigma + \alpha_0 - 2)\int_\Omega u_m E_m dx \leqslant \frac{\alpha_0\sigma + \alpha_0 - 2}{2}(\|u_m\|_2^2 + \|E_m\|_2^2).$$

So,

$$I_1 + I_2 + I_3 \leqslant \frac{\alpha_0}{2}\|u_m - \sigma E_m\|_2^2 + \frac{\alpha_0}{2}\|E_m + u_m\|_2^2$$
$$+ \left(\frac{2\beta\alpha_0 - \beta}{2} + \frac{\alpha_0\sigma + \alpha_0 - 2}{2}\right)\|E_m\|_2^2$$
$$+ \frac{\alpha_0\sigma + \alpha_0 - 2}{2}\|u_m\|_2^2 + \frac{2\beta\alpha_0 - \beta}{2}\|\nabla \times d_m\|_2^2$$
$$\leqslant \frac{2\beta\alpha_0 - \beta}{2}\|\nabla \times d_m\|_2^2 + \frac{\alpha_0\sigma + 5\alpha_0 - 2}{2}\|u_m\|_2^2$$
$$+ \frac{\alpha_0\sigma + 3\alpha_0 - 2 + 2\alpha_0\sigma^2 + 2\beta\alpha_0 - \beta}{2}\|E_m\|_2^2.$$

Taking the above inequalities into the inequality (2.29) we get

$$\frac{1}{2}\frac{d}{dt}\int_\Omega (|u_m|^2 + |\nabla d_m|^2 + 2\alpha_0|E_m|^2 + 2\alpha_0|H_m|^2)dx$$
$$+ (2\beta\alpha_0 - \beta)\frac{d}{dt}\int_\Omega d_m H_m dx + \int_\Omega |\nabla u_m|^2 dx + \frac{\beta(2\beta\alpha_0 - \beta)}{2}\frac{d}{dt}\int_\Omega |d_m|^2 dx$$
$$\leqslant \frac{\alpha_0\sigma + 5\alpha_0 - 2}{2}\|u_m\|_2^2 + \frac{\alpha_0\sigma + 3\alpha_0 - 2 + 2\alpha_0\sigma^2 + 2\beta\alpha_0 - \beta}{2}\|E_m\|_2^2\|E_m\|_2^2$$
$$+ \frac{2\beta\alpha_0 - \beta}{2}\|\nabla \times d_m\|_2^2. \tag{2.30}$$

Integrating the inequality (2.30) with the variant t to deduce that

$$\frac{1}{2}(\|u_m\|_2^2 + \|\nabla d_m\|_2^2 + 2\alpha_0\|E_m\|_2^2 + 2\alpha_0\|H_m\|_2^2)$$
$$- (2\beta\alpha_0 - \beta)\left(\frac{\|d_m\|_2^2}{2} + \frac{\|H_m\|_2^2}{2}\right) + \int_0^t \|\nabla u_m\|_2^2 dt + \frac{\beta(2\beta\alpha_0 - \beta)}{2}\|d_m\|_2^2$$
$$\leqslant \frac{\alpha_0\sigma + 5\alpha_0 - 2}{2}\int_0^t \|u_m\|_2^2 + \frac{2\beta\alpha_0 - \beta}{2}\int_0^t \|\nabla \times d_m\|_2^2 dt$$
$$+ \frac{\alpha_0\sigma + 3\alpha_0 - 2 + 2\alpha_0\sigma^2 + 2\beta\alpha_0 - \beta}{2}\int_0^t \|E_m\|_2^2 dt,$$

from the above inequality, we get

$$\beta > 1, \quad \max\left\{\frac{2}{5+\sigma}, \frac{2+\beta-2\beta\alpha_0}{\sigma+2\sigma^2+3}, \frac{1}{2}\right\} < \alpha_0 < \frac{1}{2(\beta-1)}.$$

Take
$$C_1 = \min\left\{1, 2\alpha_0, \frac{\beta + 2\alpha_0 - 2\beta\alpha_0}{2}, \frac{(\beta-1)(2\beta\alpha_0 - \beta)}{2}\right\},$$
$$C_2 = \max\left\{\frac{\alpha_0\sigma + 5\alpha_0 - 2}{2}, \frac{\alpha_0\sigma + 3\alpha_0 - 2 + 2\alpha_0\sigma^2 + 2\beta\alpha_0 - \beta}{2}, \frac{2\beta\alpha_0 - \beta}{2}\right\},$$
so
$$C_1(\|u_m\|_2^2 + \|\nabla d_m\|_2^2 + \|E_m\|_2^2 + \|H_m\|_2^2)$$
$$\leq C_2 \left(\int_0^t (\|u_m\|_2^2 + \|E_m\|_2^2 + \|\nabla \times d_m\|_2^2) dt\right),$$
by Gronwall's inequality, we have
$$\sup_{0 \leq t \leq T}(\|u_m\|_2^2 + \|d_m\|_{H^1}^2 + \|E_m\|_2^2 + \|H_m\|_2^2) \leq C,$$
where C is independent of m. □

Lemma 2.6 *Under the conditions of Lemma 2.5, for the solution (u_m, d_m, E_m, H_m) of the system (2.9)-(2.17), there exists $C > 0$ independent of m such that*
$$\int_0^t (\|u_{mt}\|_{H^{-2}}^2 + \|d_{mt}\|_{H^{-2}}^2 + \|E_{mt}\|_{H^{-2}}^2 + \|H_{mt}\|_{H^{-2}}^2) dt \leq C.$$

Proof For the function $\phi \in H_0^2(\Omega)$, ϕ can be represented by
$$\phi = \phi_m + \bar{\phi}_m, \quad \phi_m = \sum_{s=1}^m \eta_s w_s(x), \quad \bar{\phi}_m = \sum_{s=m+1}^\infty \eta_s w_s(x), \quad (2.31)$$
for $s \geq m+1$, we have $\int_\Omega d_{mt} w_s(x) dx = 0$. Firstly, we consider $\Omega \in \mathbf{R}^2$. By Lemma 2.5 and Ladyzhenskaya inequality [18], there holds
$$\left|\int_\Omega d_{mt}\phi(x) dx\right| = \left|\int_\Omega d_{mt}\phi_m(x) dx\right|$$
$$= \left|\int_\Omega (-u_m \cdot \nabla d_m + \Delta d_m + d_m|\nabla d_m|^2 + d_m \times \Delta d_m)\phi_m(x) dx\right|$$
$$\leq \|\phi_m\|_\infty \int_\Omega |u_m \cdot \nabla d_m| dx + \int_\Omega |d_m \cdot \Delta \phi_m| dx$$
$$+ \|\phi_m\|_\infty \int_\Omega d_m|\nabla d_m|^2 dx$$

$$+ \int_\Omega |\nabla d_m \times \nabla d_m| \phi_m(x) dx + \int_\Omega |d_m \times \nabla d_m| \nabla \phi_m(x) dx$$

$$\leqslant (\|u_m\|_2 \|\nabla d_m\|_2 + \|d_m\|_2 + \|d_m\|_2 \|\nabla d_m\|_4^2)(\|\phi_m\|_\infty + \|\phi_m\|_2)$$

$$+ (\|\nabla d_m\|_2^2 + \|d_m\|_2 \|\nabla d_m\|_2)(\|\phi_m\|_\infty + \|\phi_m\|_2)$$

$$\leqslant (\|u_m\|_2 \|\nabla d_m\|_2 + \|d_m\|_2 + C\|d_m\|_2 \|\nabla d_m\|_2 \|\Delta d_m\|_2$$

$$+ \|\Delta d_m\|_2 \|d_m\|_2) \|\phi\|_{H^2}$$

$$\leqslant C \|\Delta d_m\|_2 \|\phi\|_{H^2},$$

$$\left| \int_\Omega u_{mt} \phi(x) dx \right| = \left| \int_\Omega u_{mt} \phi_m(x) dx \right|$$

$$\leqslant C \Big(\int_\Omega |u_m \cdot \nabla u_m \cdot \phi_m| dx + \int_\Omega |\nabla u_m \cdot \nabla \phi_m| dx$$

$$+ \int_\Omega |\nabla \cdot (\nabla d_m \odot \nabla d_m) \cdot \phi_m| dx \Big) + \int_\Omega |E_m \phi_m| dx$$

$$+ \int_\Omega |(u_m \times H_m) \phi_m| dx)$$

$$\leqslant C(\|\nabla u_m\|_2 \|\nabla \phi_m\|_2 + \|u_m\|_2 \|\nabla u_m\|_2 \|\phi_m\|_\infty$$

$$+ \|E_m\|_2 \|\phi_m\|_2 + \|u_m\|_2 \|H_m\|_2 \|\phi_m\|_\infty$$

$$+ \|\nabla d_m\|_2 \|\Delta d_m\|_2 \|\phi_m\|_\infty),$$

so

$$\left| \int_\Omega u_{mt} \phi(x) dx \right| \leqslant C(\|\nabla u_m\|_2^2 + |\Delta d_m\|_2^2) \|\phi_m\|_{H^2}.$$

$$\left| \int_\Omega E_{mt} \phi(x) dx \right| = \left| \int_\Omega E_{mt} \phi_m(x) dx \right|$$

$$= |\int_\Omega -u_{mt} \phi_m - \sigma E_m \phi_m dx + (\nabla \times H_m) \phi_m dx$$

$$\leqslant C(\|E_m\|_2 \|\phi_m\|_2 + \|H_m\|_2 \|\nabla \phi_m\|_2 + \|u_m\|_2 \|\phi_m\|_2)$$

$$\leqslant C\|\phi_m\|_{H^2}(\|\nabla u_m\|_2^2 + |\Delta d_m\|_2^2),$$

$$\left| \int_\Omega H_{mt} \phi(x) dx \right| = \left| \int_\Omega H_{mt} \phi_m(x) dx \right| \leqslant C\|\phi_m\|_{H^2},$$

where the constant C is independent of m. When $\Omega \in \mathbf{R}^3$, by Lemma 2.4, the above estimate also holds. Finally, integrate the above estimates with respect to t, and the lemma is proved. □

Lemma 2.7 *Under the estimates of Lemma 2.5 for the solution (u_m, d_m, E_m, H_m) of the system (2.9)-(2.17), there exists $C > 0$ independent of m such that*

$$\|d_m(\cdot, t_1) - d_m(\cdot, t_2)\|_2 \leqslant C|t_1 - t_2|^{\frac{1}{6}}$$

$$u_m, E_m, H_m \in C(0, T; H^{-1}(\Omega)).$$

Proof By Sobolev's interpolation of negative order, there is

$$\|d_m(\cdot, t_1) - d_m(\cdot, t_2)\|_2 \leqslant C \|d_m(\cdot, t_1) - d_m(\cdot, t_2)\|_{H^{-2}}^{\frac{1}{3}} \|d_m(\cdot, t_1) - d_m(\cdot, t_2)\|_{H^1}^{\frac{2}{3}}$$

$$\leqslant C \| \int_{t_1}^{t_2} d_{mt} dt \|_{H^{-2}}^{\frac{1}{3}} \leqslant C \left(\int_{t_1}^{t_2} \|d_{mt}\|_{H^{-2}}^2 dt \right)^{\frac{1}{6}} \cdot |t_1 - t_2|^{\frac{1}{6}}$$

$$\leqslant C |t_1 - t_2|^{\frac{1}{6}},$$

on the other hand, it follows from Lemma 2.3 and

$$L^2(\Omega) \hookrightarrow H^{-1}(\Omega) \hookrightarrow H^{-2}(\Omega),$$

$$H_m \in L^\infty(0, T; L^2(\Omega)) \cap \left\{ \psi : \frac{\partial \psi}{\partial t} \in L^2(0, T; H^{-2}(\Omega)) \right\},$$

then

$$H_m \in C([0, T]; H^{-1}(\Omega)).$$

Similarly, we have

$$E_m \in C([0, T]; H^{-1}(\Omega)), \quad u_m \in C([0, T]; H^{-1}(\Omega)). \quad \square$$

In fact, it follows from (2.25)-(2.26) that the solution (u_m, d_m, E_m, H_m) of the system (2.9)-(2.17) doesn't blow up at finite from ODE problem, we have the following lemma.

Lemma 2.8 *Under the conditions of Lemma 2.5, the initial problem of the system of the ordinary differential equations (2.9)-(2.17) admits at least one continuously differentiable global solution*

$$\bar{\alpha}_{sm}(t), \bar{\beta}_{sm}(t), \bar{\gamma}_{sm}(t), \bar{\delta}_{sm}(t).$$

2.2 Existence of weak solution for the problem (1.1)-(1.9)

First of all, as in Definition 1.1, we may define the weak solution of the problem (1.1)-(1.9). In the proof of Theorem 1.1, we must use the following lemma, which is well-known to all.

Lemma 2.9 *If $u_n \to u$ strongly in $L^2(Q_T)$ and $v_n \to v$ weakly in $L^2(Q_T)$, then $u_n v_n \to uv$ weakly in $L^1(Q_T)$ and in the sense of distribution.*

Now we prove the existence of weak solutions for the problem (1.1)-(1.9) and the proof of Theorem 1.1.

Proof of Theorem 1.1 The uniform estimates for the approximate solution $u_m(x,t), d_m(x,t), E_m(x,t), H_m(x,t)$ in the above section yields that there is a subsequence of $u_m(x,t), d_m(x,t), E_m(x,t), H_m(x,t)$ such that

$$u_m(x,t) \to u(x,t) \text{ weakly } * \text{ in } L^\infty(0,T;J), \tag{2.32}$$

$$u_m(x,t) \to u(x,t) \text{ weakly } * \text{ in } L^2(0,T;K), \tag{2.33}$$

$$d_m(x,t) \to d(x,t) \text{ weakly } * \text{ in } L^\infty(0,T;H^1(\Omega)), \tag{2.34}$$

$$d_m(x,t) \to d(x,t) \text{ strongly in } L^2(0,T;H^1(\Omega)), \tag{2.35}$$

$$d_m(x,t) \to d(x,t) \text{ weakly } * \text{ in } L^2(0,T;H^2(\Omega)), \tag{2.36}$$

$$H_m(x,t) \to H(x,t) \text{ weakly } * \text{ in } L^\infty(0,T;L^2(\Omega)), \tag{2.37}$$

$$E_m(x,t) \to E(x,t) \text{ weakly } * \text{ in } L^\infty(0,T;L^2(\Omega)). \tag{2.38}$$

For any vector-value test function $\Psi(x,t) \in C^1(\bar{Q}_T)$, $\Psi(x,T) = 0$. We define an approximate sequence

$$\Psi_m(x,t) = \sum_{s=1}^{m} \eta_s(t) w_s(x),$$

where $\eta_s(t) = \int_\Omega \Psi(x,t) w_s(x) dx$, then

$$\Psi_m(x,t) \to \Psi(x,t) \text{ in } C^1(Q_T) \text{ and in } L^p(Q_T), \quad \forall p > 1. \tag{2.39}$$

Making the scalar product of $\eta_s(t)$ in (2.9), (2.10), (2.12) and $e^{\sigma t}\eta_s(t)$ in (2.11), summing up the result for $s = 1, 2, \cdots, m$ and integrating by parts we have

$$\iint_{Q_T} u_m \Psi_{mt} dx dt + \int_\Omega u_m(\cdot,0) \Psi_m(\cdot,0) dx$$

$$= \iint_{Q_T} (u_m \cdot \nabla u_m - \Delta u_m + \nabla P + \nabla \cdot (\nabla d_m \odot \nabla d_m) + E_m)) \Psi_m dx dt$$

$$+ (u_m \times H_m) \Psi_m dx dt$$

$$= \iint_{Q_T} [(u_m \cdot \nabla u_m) \Psi_m - \Delta u_m \Psi_m + \nabla \cdot (\nabla d_m \odot \nabla d_m) \Psi_m + \nabla P \Psi_m$$

$$+ (E_m + (u_m \times H_m)) \Psi_m] dx dt, \tag{2.40}$$

$$\iint_{Q_T} d_m \Psi_{mt} dx dt + \int_\Omega d_m(\cdot, 0) \Psi_m(\cdot, 0) dx$$

$$= \iint_{Q_T} [(u_m \cdot \nabla) d_m \Psi_m - (d_m \times \Delta d_m) \cdot (d_m \times \Psi_m) - (d_m \times \Delta d_m) \Psi_m] dx dt, \tag{2.41}$$

$$\iint_{Q_T} E_m \Psi_{mt} e^{\sigma t} dx dt + \iint_{Q_T} u_m \Psi_m e^{\sigma t} dx dt + \iint_{Q_T} (\nabla \times \Psi_m) \cdot H_m e^{\sigma t} dx dt$$

$$+ \int_\Omega E_m(\cdot, 0) \Psi_m(\cdot, 0) dx = 0, \tag{2.42}$$

$$\iint_{Q_T} (H_m + \beta d_m) \Psi_{mt} dx dt + \int_\Omega ((H_m(\cdot, 0) + \beta d_m(\cdot, 0)) \Psi_m(\cdot, 0) dx$$

$$- \iint_{Q_T} (\nabla \times \Psi_m) \cdot E_m dx dt = 0. \tag{2.43}$$

Now we are in the position to prove that $u(x,t), d(x,t), E(x,t), H(x,t)$ is a weak solution of (1.1)-(1.9). To this aim, one should set $m \to \infty$ in (2.40)-(2.43). From (2.32)-(2.39) and Lemma 2.9, it suffices to deal with the nonlinear terms in (2.40)-(2.43). By (2.32) and (2.33)

$$\iint_{Q_T} (\nabla u_m \cdot \nabla \Psi_m - \nabla u \cdot \nabla \Psi) dx dt$$

$$= \int_0^T \int_\Omega (\nabla u_m - \nabla u) \cdot \nabla \Psi_m dx dt + \nabla u \cdot (\nabla \Psi_m - \nabla \Psi) dx dt$$

$$\leq \int_0^T \|\nabla \Psi_m\|_2 \|\nabla u_m - \nabla u\|_2 dt + \int_0^T \|\nabla u\|_2 \|\nabla \Psi_m - \nabla \Psi\|_2 dt \to 0, \quad m \to \infty,$$

$$\iint_{Q_T} (u_m \cdot \nabla u_m - u \cdot \nabla u) dx dt$$

$$= \iint_{Q_T} [(u_m - u) \cdot \nabla u_m + u \cdot (\nabla u_m - \nabla u)] dx dt \to 0, \quad m \to \infty,$$

$$\iint_{Q_T} (u_m \times H_m - u \times H)\Phi dxdt$$

$$\leq \iint_{Q_T} [(H_m \times (u_m - u))\Phi + (u_m \times (H_m - H))\Phi]dxdt \to 0, \quad m \to \infty.$$

By Lemma 2.9, we only need to prove

$$(\nabla \cdot (\nabla d_m \odot \nabla d_m)) \rightharpoonup (\nabla \cdot (\nabla d \odot \nabla d)) \text{ weakly in } L^2(Q_T).$$

In fact, by (2.35), (2.36), for any $\Psi \in C^1(Q_T)$,

$$\left|\iint_{Q_T} (\nabla \cdot (\nabla d_m \odot \nabla d_m) - \nabla \cdot (\nabla d \odot \nabla d)) \cdot \Psi dxdt\right|$$

$$\leq \iint_{Q_T} |(\nabla |\nabla d_m|^2 - \nabla |\nabla d|^2) \cdot \Psi| dxdt$$

$$+ \iint_{Q_T} |(\Delta d_m \cdot \nabla d_m - \Delta d \cdot \nabla d) \cdot \Psi| dxdt$$

$$\leq \int_0^T \int_\Omega |(|\nabla d_m|^2 - |\nabla d|^2) \cdot \nabla \Psi| dxdt + \int_0^T \int_\Omega |\nabla d_m \cdot (\Delta d_m - \Delta d)\Psi| dxdt$$

$$+ \int_0^T \int_\Omega |\Delta d \cdot (\nabla d_m - \nabla d)\Psi| dxdt$$

$$\leq \|\nabla \Psi\|_{L^\infty(Q_T)} \|\nabla d_m - \nabla d\|_{L^2(Q_T)} + \int_0^T \|\nabla d_m\|_2 \|\Delta d_m - \Delta d\|_2 \|\Psi\|_\infty dt$$

$$+ \int_0^T \|\Delta d\|_2 \|\nabla d_m - \nabla d\|_2 \|\Psi\|_\infty dt = I_1^m + I_2^m + I_3^m.$$

(2.35) imply that $I_1^m, I_3^m \to 0$ as $m \to \infty$, the fact that ∇d_m is uniformly bounded in $L^2(Q_T)$ and (2.36) yield that $I_2^m \to 0$ as $m \to \infty$. We will prove

$$d_m \times \Delta d_m \to d \times \Delta d \text{ weakly in } L^2(Q_T).$$

By virtue of (2.35), (2.36), we have

$$\iint_{Q_T} |(d_m \times \Delta d_m - d \times \Delta d)\Psi| dxdt$$

$$= \iint_{Q_T} |[(d_m - d) \times \Delta d_m + d \times (\Delta d_m - \Delta d)]\Psi| dxdt$$

$$\leq \|\Psi\|_\infty \|d_m - d\|_{L^2(Q_T)} \|\Delta d_m\|_{L^2(Q_T)} + \int_0^T \|d\|_2 \|\Delta d_m - \Delta d\|_2 \|\Psi\|_\infty dt \to 0,$$

(2.44)

as $m \to \infty$. To prove the existence of the generalized solution remains to prove
$$\iint_{Q_T} (d_m \times \Delta d_m) \cdot (d_m \times \Psi_m) dx dt \to \iint_{Q_T} (d \times \Delta d) \cdot (d \times \Psi) dx dt.$$

In fact
$$\iint_{Q_T} (d_m \times \Delta d_m) \cdot (d_m \times \Psi_m) dx dt - \iint_{Q_T} (d \times \Delta d) \cdot (d \times \Psi) dx dt$$
$$= \iint_{Q_T} (d_m \times \Delta d_m - d \times \Delta d) \cdot (d \times \Psi) dx dt$$
$$+ \iint_{Q_T} d_m \times \Delta d_m \cdot (d_m \times \Psi_m - d \times \Psi) dx dt$$
$$= J_1^m + J_2^m.$$

It follows from (2.44) that $J_1^m \to 0$, moreover
$$J_2^m \leqslant \|d_m \times \Delta d_m\|_{L^2(Q_T)} \left(\iint_{Q_T} |d_m \times \Psi_m - d \times \Psi|^2 dx dt \right)^{\frac{1}{2}}$$
$$\leqslant C \left(\iint_{Q_T} |d_m \times (\Psi_m - \Psi) + (d_m - d) \times \Psi|^2 dx dt \right)^{\frac{1}{2}} \to 0.$$

Finally, from the above arguments, one may take $m \to \infty$ in (2.40)-(2.43) to obtain (u, d, E, H) is a global weak solution of (1.1)-(1.9), this completes the proof of Theorem 1.1. □

3 Global smooth solution for problem (1.1)-(1.9)

By applying the Banach compression mapping theorem and induction technique, we can obtain the smooth solution for the problem (1.1)-(1.9), then we get Theorem 1.2. In order to prove that there exists a global smooth solution for the problem (1.1)-(1.9), one need to establish a priori estimate.

In this section, we first consider the case $n = 2$.

Remark 3.1   We consider a classical solution (u, d, E, H) of the problem (1.1)-(1.9). In fact,
$$|d| \equiv 1 \quad \text{if} \quad |d_0| = 1. \tag{3.1}$$

Multiplying (1.2) *by* d *we obtain*
$$\partial_t |d|^2 + u \cdot \nabla |d|^2 = 0,$$

i.e.,
$$\partial_t(|d|^2 - 1) + u \cdot \nabla(|d|^2 - 1) = 0, \qquad (3.2)$$

multiplying (3.2) by $|d|^2 - 1$ and then integrating by parts over Ω to deduce

$$\frac{d}{dt} \int_\Omega (|d|^2 - 1)^2 dx = 0,$$

we immediately verify (3.1).

Due to the above Remak, the equation (1.2) equals to

$$d_t + (u \cdot \nabla)d = \gamma(|\nabla d|^2 d + \Delta d + d \times \Delta d). \qquad (3.3)$$

In the following, we will consider the problem (1.1), (3.3), (1.3)-(1.7).

Lemma 3.1 Assume $(u_0, d_0, E_0, H_0) \in (H^1(\Omega), H^2(\Omega), H^1(\Omega), H^1(\Omega))$. Then there exists the smooth solution $(u(x,t), d(x,t), E(x,t), H(x,t))$ of the problem (1.1)-(1.9) satisfying

$$\sup_{0 \leqslant t \leqslant T} (\|\Delta u\|_2^2 + \|\Delta d\|_2^2 + \|\nabla E\|_2^2 + \|\nabla H\|_2^2) \leqslant C, \qquad (3.4)$$

when $\dim \Omega = 2$.

Proof Testing (1.2) by $\Delta(\Delta d + H)$, by Young and Sobolev inequality we have

$$\frac{1}{2}\frac{d}{dt}\int_\Omega |\Delta d|^2 dx + \int_\Omega d_t \cdot \Delta H dx + \int_\Omega |\nabla \Delta d|^2$$

$$= \int_\Omega \nabla u \cdot \nabla d \cdot \nabla \Delta d\, dx + \int_\Omega u \cdot \nabla^2 d \cdot \nabla \Delta d\, dx + \int_\Omega \nabla u \cdot \nabla d \cdot \nabla H dx$$

$$- \int_\Omega \nabla d \cdot |\nabla d|^2 \cdot \nabla \Delta d\, dx - \int_\Omega d \cdot \nabla |\nabla d|^2 \cdot \nabla \Delta d\, dx - \int_\Omega \nabla d \cdot |\nabla d|^2 \cdot \nabla H dx$$

$$- \int_\Omega d \cdot \nabla |\nabla d|^2 \cdot \nabla H dx + \int_\Omega \Delta d \cdot \Delta H dx - \int_\Omega (\nabla d \times \Delta d) \cdot \nabla \Delta d\, dx$$

$$- \int_\Omega (\nabla d \times \Delta d) \cdot \nabla H dx - \int_\Omega (d \times \nabla \Delta d) \cdot \nabla H dx + \int_\Omega u \cdot \nabla^2 d \cdot \nabla H dx$$

$$\leqslant C(\Omega)(\|\nabla u\|_2 \|\nabla d\|_\infty \|\nabla \Delta d\|_2 + \|u\|_4 \|\Delta d\|_4 \|\nabla \Delta d\|_2 + \|\nabla u\|_4 \|\nabla d\|_4 \|\nabla H\|_2$$

$$+ \|u\|_4 \|\Delta d\|_4 \|\nabla H\|_2 + \|\nabla d\|_\infty \||\nabla d|^2\|_2 \|\nabla \Delta d\|_2 + \|d\|_4 \|\nabla |\nabla d|^2\|_4 \|\nabla \Delta d\|_2$$

$$+ \|\nabla d\|_\infty \||\nabla d|^2\|_2 \|\nabla H\|_2 + \|d\|_4 \|\nabla |\nabla d|^2\|_4 \|\nabla H\|_2 + \|\nabla d\|_4 \|\Delta d\|_4 \|\nabla \Delta d\|_2$$

$$+ \|d\|_\infty \|\nabla \Delta d\|_2 \|\nabla H\|_2 + \|\nabla d\|_4 \|\Delta d\|_4 \|\nabla H\|_2)$$

$$\leqslant \varepsilon \|\nabla \Delta d\|_2^2 + C(\|\nabla H\|_2^2 + \|\Delta d\|_2^2 + \|\Delta d\|_4^4 + \|\Delta u\|_4^4 + \|\Delta u\|_2^2).$$

Here we have used

$$\|\nabla d\|_\infty \||\nabla d|^2\|_2 \|\nabla \Delta d\|_2 \leqslant C(\Omega) \|\nabla d\|_2^{\frac{1}{2}} \|\nabla \Delta d\|_2^{\frac{1}{2}} \|\nabla d\|_4^2 \|\nabla \Delta d\|_2$$

$$\leqslant C(\Omega) \|\nabla \Delta d\|_2^{\frac{3}{2}} \|\nabla d\|_4^2$$

$$\leqslant C(\Omega)(\epsilon \|\nabla \Delta d\|_2^2 + \|\nabla d\|_4^8),$$

$$\|\nabla d\|_4^8 \leqslant C(\Omega) \|\nabla d\|_2^4 \|\Delta d\|_2^4 \leqslant C(\Omega) \|\Delta d\|_4^4,$$

$$\|d\|_\infty \|\nabla d\|_4 \|\Delta d\|_4 \|\nabla \Delta d\|_2 \leqslant C(\Omega)(\|\nabla d\|_4^2 \|\Delta d\|_4^2 + \epsilon \|\nabla \Delta d\|_2^2)$$

$$\leqslant C(\Omega)(\|\Delta d\|_4^4 + \epsilon \|\nabla \Delta d\|_2^2),$$

so,

$$\frac{1}{2}\frac{d}{dt}\int_\Omega |\Delta d|^2 dx + \int_\Omega |\nabla \Delta d|^2 + \int_\Omega d_t \cdot \Delta H dx$$

$$\leqslant \varepsilon \|\nabla \Delta d\|_2^2 + C(\|\nabla H\|_2^2 + \|\Delta d\|_2^2 + \|\Delta d\|_4^4 + \|\Delta u\|_4^4 + \|\Delta u\|_2^2)). \qquad (3.5)$$

Taking the same procedure to (1.1), we obtain

$$\frac{1}{2}\frac{d}{dt}\int_\Omega |\Delta u|^2 dx + \int_\Omega |\nabla \Delta u|^2 + \int_\Omega u_t \cdot \Delta E dx$$

$$\leqslant \varepsilon(\|\nabla \Delta d\|_2^2 + \|\nabla \Delta u\|_2^2) + C(\|\nabla H\|_2^2 + \|\nabla E\|_2^2 + \|\Delta d\|_2^2 + \|\Delta u\|_2^2 + \|\Delta u\|_4^4),$$

$$(3.6)$$

here we can choose $\varepsilon = \dfrac{1}{8}$.

Making the scalar product of ΔE with (1.4) and ΔH with (1.5) respectively and then integrating the resulting equation with respective to $x \in \Omega$, we obtain

$$\int_\Omega \Big[\frac{\partial E}{\partial t}\Delta E - (\nabla \times H)\cdot \Delta E + \sigma E \Delta E$$

$$+ \frac{\partial (H+\beta d)}{\partial t}\Delta H - u\Delta E + (\nabla \times E)\Delta H\Big]dx = 0,$$

i.e.,

$$-\frac{1}{2}\frac{d}{dt}\int_\Omega (|\nabla E|^2 + |\nabla H|^2)dx + \int_\Omega u\Delta E dx + \beta \int_\Omega d_t \Delta H dx - \sigma \int_\Omega |\nabla E|^2 dx = 0.$$

$$(3.7)$$

Different (1.1) with respect to t, and then making the scalar product with respect to u_t, we obtain

$$u_{tt} \cdot u_t + [(u_t \cdot \nabla)u + (u \cdot \nabla)u_t]u_t + \nabla P_t \cdot u_t - \Delta u_t \cdot u_t$$
$$= -[\nabla \cdot (\nabla d \odot \nabla d)]_t \cdot u_t - E_t u_t - (u \times H_t)u_t, \qquad (3.8)$$

integrate (3.8) over Ω, combining Sobolev imbedding theorem to deduce

$$\frac{1}{2}\frac{d}{dt}\int_\Omega |u_t|^2 dx + \int_\Omega |\nabla u_t|^2 dx$$
$$\leqslant \int_\Omega |\nabla d||\nabla d_t||\nabla u_t| dx + \int_\Omega |u \cdot \nabla u_t \cdot u_t| dx + \int_\Omega |(u_t \cdot \nabla)u \cdot u_t| dx$$
$$+ \int_\Omega (|E_t u_t| + |(u \times H_t) u_t|) dx$$
$$\leqslant \|\nabla d\|_4 \|\nabla d_t\|_4 \|\nabla u_t\|_2 + \|u\|_4 \|u_t\|_4 \|\nabla u_t\|_2 + \|u_t\|_4 \|\nabla u\|_2 \|u_t\|_4$$
$$+ \|E_t\|_2 \|u_t\|_2 + \|u\|_\infty \|u_t\|_2 \|H_t\|_2$$
$$\leqslant \varepsilon(\|\nabla u_t\|_2^2 + \|\nabla \Delta d\|_2^2)$$
$$+ C(\Omega)(\|u_t\|_2^2 + \|\nabla d_t\|_2^2 + \|\Delta u\|_2^2 + \|\nabla H\|_2^2 + \|\nabla E\|_2^2 + \|\nabla d_t\|_4^4$$
$$+ \|u_t\|_4^4 + \|\Delta d\|_2^2),$$

take $\varepsilon = \dfrac{1}{8}$, whence

$$\frac{1}{2}\frac{d}{dt}\int_\Omega |u_t|^2 dx + \int_\Omega |\nabla u_t|^2 dx$$
$$\leqslant \frac{1}{8}(\|\nabla u_t\|_2^2 + \|\nabla \Delta d\|_2^2)$$
$$+ C(\Omega)(\|u_t\|_2^2 + \|\nabla d_t\|_2^2 + \|\Delta u\|_2^2 + \|\nabla H\|_2^2 + \|\nabla E\|_2^2 + \|\nabla d_t\|_4^4$$
$$+ \|u_t\|_4^4 + \|\Delta d\|_2^2). \qquad (3.9)$$

Applying ∂_t to (1.2), multiplying by Δd_t, integrating by parts

$$\frac{1}{2}\frac{d}{dt}\int_\Omega |\nabla d_t|^2 dx + \int_\Omega |\Delta d_t|^2 dx$$
$$\leqslant \int_\Omega |(u_t \cdot \nabla)d \cdot \Delta d_t + (u \cdot \nabla)d_t \cdot \Delta d_t| dx + \int_\Omega |[d_t|\nabla d|^2 + d(|\nabla d|^2)_t] \cdot \Delta d_t| dx$$

$$+ \int_\Omega |(d_t \times \Delta d) \cdot \Delta d_t| dx$$
$$\leqslant \|u_t\|_4 \|\nabla d\|_4 \|\Delta d_t\|_2 + \|u\|_4 \|\nabla d_t\|_4 \|\Delta d_t\|_2 + \|d_t\|_\infty \|\nabla d\|_4^2 \|\Delta d_t\|_2$$
$$+ \|d\|_\infty \|\nabla d\|_4 \|\nabla d_t\|_4 \|\Delta d_t\|_2 + \|d_t\|_4 \|\Delta d\|_4 \|\Delta d_t\|_2$$
$$\leqslant C(\|\nabla d_t\|_2^2 + \|\Delta d\|_4^4 + \|\nabla d_t\|_4^4 + \|\Delta u\|_2^2) + \frac{1}{8}(\|\nabla u_t\|_2^2 + \|\Delta d_t\|_2^2 + \|\nabla \Delta d\|_2^2). \tag{3.10}$$

Whence (3.5)-(3.7), (3.9) and (3.10) imply (3.4) and
$$u_t \in L^\infty(0,T;L^2(\Omega)) \cap L^2(0,T;H^1(\Omega)),$$
$$\nabla d_t \in L^\infty(0,T;L^2(\Omega)) \cap L^2(0,T;H^1(\Omega)). \tag{3.11}$$

□

Lemma 3.2
$$u_t \in L^\infty(0,T;H^1(\Omega)) \cap L^2(0,T;H^2(\Omega)).$$

Proof Applying ∂_t to (1.1), multiplying by Δu_t, and integrating by parts, we get
$$\frac{d}{dt} \int_\Omega |\nabla u_t|^2 dx + \int_\Omega |\Delta u_t|^2 dx$$
$$\leqslant \int_\Omega [|u_t \cdot \nabla u \cdot \Delta u_t| + |u \cdot \nabla u_t \cdot \Delta u_t| + |\nabla d_t \cdot \Delta d \cdot \Delta u_t| + |E_t \cdot \Delta u_t|$$
$$+ |\nabla d \cdot \Delta d_t \cdot \Delta u_t| + |\nabla d|_t^2 \cdot d \cdot \Delta u_t| + |\nabla d|^2 \cdot d_t \cdot \Delta u_t| + |(u \times H_t) \cdot \Delta u_t|] dx$$
$$\leqslant \|\nabla u\|_4 \|u_t\|_4 \|\Delta u_t\|_2 + \|u\|_\infty \|\nabla u_t\|_2 \|\Delta u_t\|_2 + \|\nabla d_t\|_\infty \|\Delta d\|_2 \|\Delta u_t\|_2$$
$$+ \|\nabla d\|_\infty \|\Delta d_t\|_2 \|\Delta u_t\|_2 + \|\nabla d_t\|_4 \|\nabla d\|_4 \|d\|_\infty \|\Delta u_t\|_2 + \|\nabla d\|_4^2 \|d_t\|_\infty \|\Delta u_t\|_2$$
$$+ \|E_t\|_2 \|\Delta u_t\|_2 + \|u\|_\infty \|H_t\|_2 \|\Delta u_t\|_2$$
$$\leqslant C(\|u_t\|_2^2 + \|\nabla u_t\|_2^2 + \|\nabla d_t\|_2^2 + \|\Delta d_t\|_2^2 + \|E_t\|_2^2 + \|H_t\|_2^2) + \frac{1}{8} \|\Delta u_t\|_2^2,$$

together with Lemma 3.1, we have
$$\frac{d}{dt} \int_\Omega |\nabla u_t|^2 dx + \int_\Omega |\Delta u_t|^2 dx$$
$$\leqslant C(\|u_t\|_2^2 + \|\nabla u_t\|_2^2 + \|\nabla d_t\|_2^2 + \|\Delta d_t\|_2^2) + \frac{1}{8} \|\Delta u_t\|_2^2. \tag{3.12}$$

By Gronwall's inequality, (3.11) and (3.12), we deduce

$$\nabla u_t \in L^\infty(0,T; L^2(\Omega)) \cap L^2(0,T; H^1(\Omega)).\qquad \square$$

Lemma 3.3 *Assume* $(u_0, d_0, E_0, H_0) \in (H^1(\Omega), H^2(\Omega), H^{m-1}(\Omega), H^{m-1}(\Omega))$ $(m \geqslant 3)$. *Then there exists a unique smooth solution* $(u(x,t), d(x,t), E(x,t), H(x,t))$ *of the problem* (1.1)-(1.9) *satisfying*

$$\sup_{0 \leqslant t \leqslant T} \left(\|u\|_{H^m}^2 + \|d\|_{H^m}^2 + \|E\|_{H^{m-1}}^2 + \|H\|_{H^{m-1}}^2 \right) \leqslant C. \qquad (3.13)$$

Proof (existence part) This lemma will be proved by the induction for m. From Theorem 1.1 and Lemma 3.1, the estimate is held for $m = 1, 2$.

Now we assume that the estimate holds for $m = M \geqslant 3$, i.e.,

$$\sup_{0 \leqslant t \leqslant T} \left(\|u\|_{H^M}^2 + \|d\|_{H^M}^2 + \|E\|_{H^{M-1}}^2 + \|H\|_{H^{M-1}}^2 \right) \leqslant C. \qquad (3.14)$$

We will prove (3.14) is held for $m = M + 1$.

Making the scalar product of $\Delta^M(\Delta d + H)$ with (1.2) and integrating the resulting equation with respect to $x \in \Omega$, we have

$$\frac{1}{2}\frac{d}{dt}\|D^{M+1}d\|_2^2 + \int_\Omega D^M d_t \cdot D^M H - \int_\Omega |D^M \Delta d|^2 + \int_\Omega D^M u \cdot \nabla d \cdot D^M H$$
$$+ (-1)^M \int_\Omega u \cdot D^{M+1} d \cdot D^M H + \int_\Omega D^M u \cdot \nabla d \cdot D^M \Delta d$$
$$+ \int_\Omega u \cdot D^{M+1} d \cdot D^M \Delta d - \int_\Omega D^M \Delta d \cdot D^M H$$
$$= \int_\Omega D^M d \cdot |\nabla d|^2 \cdot D^M \Delta d + \int_\Omega d \cdot D^M(|\nabla d|^2) \cdot D^M \Delta d$$
$$+ \int_\Omega D^M d \cdot |\nabla d|^2 \cdot D^M H + \int_\Omega d \cdot D^M |\nabla d|^2 \cdot D^M H$$
$$+ \int_\Omega (D^M d \times \Delta d) \cdot D^M(\Delta d) + \int_\Omega (D^M d \times \Delta d) \cdot D^M H$$
$$+ \int_\Omega (D^{M+2} d \times d) \cdot D^M H, \qquad (3.15)$$

through Sobolev's inequality and Young's inequality, we get

$$\frac{1}{2}\frac{d}{dt}\|D^{M+1}d\|_2^2 + \|D^{M+2}d\|_2^2 + \int_\Omega D^M d_t \cdot D^M H dx$$

$$\leqslant \frac{1}{8}\|D^{M+2}d\|_2^2 + C(\|D^{M+1}d\|_2^2 + \|D^{M+1}u\|_2^2 + \|D^M H\|_2^2). \tag{3.16}$$

Similarly, Making the scalar product of $\Delta^M(\Delta u + E)$ with (1.1) and integrating the resulting equation with respect to $x \in \Omega$, we have

$$\frac{1}{2}\frac{d}{dt}\|D^{M+1}u\|_2^2 + \|D^{M+2}u\|_2^2 + \int_\Omega D^M u_t \cdot D^M E dx$$
$$\leqslant C(\|D^{M+1}d\|_2^2 + \|D^{M+1}u\|_2^2 + \|D^M E\|_2^2 + \|D^M H\|_2^2)$$
$$+ \frac{1}{8}(\|D^{M+2}u\|_2^2 + \|D^{M+2}d\|_2^2). \tag{3.17}$$

Making the scalar product of $\Delta^M E$ with (1.4) and $\Delta^M H$ with (1.5), respectively and then integrating the resulting equation with respective to $x \in \Omega$, we obtain

$$\int_\Omega \Big[\frac{\partial E}{\partial t}\Delta^M E - (\nabla \times H)\cdot \Delta^M E + \sigma E \cdot \Delta^M E + \frac{\partial(H+\beta d)}{\partial t}\cdot \Delta^M H$$
$$+ (\nabla \times E)\cdot \Delta^M H - u\Delta^M E\Big] = 0,$$

i.e.,

$$\frac{1}{2}\frac{d}{dt}(\|D^M E\|_2^2 + \|D^M H\|_2^2) - \int_\Omega D^M u \cdot D^M E + \beta \int_\Omega D^M d_t \cdot D^M H$$
$$+ \sigma \int_\Omega |D^M E|^2 = 0. \tag{3.18}$$

Next we will estimate the term $\int_\Omega D^M u_t dx$ and $\int_\Omega D^M d_t dx$.

Taking ∂_t of (1.1), multiplying by $\Delta^M u_t$, then integrate the resulting over Ω,

$$\frac{1}{2}\frac{d}{dt}\int_\Omega |D^M u_t|^2 dx + \int_\Omega |D^{M+1}u_t|^2 dx$$
$$\leqslant \|D^M u_t\|_2\|\nabla u\|_4\|D^M u_t\|_4 + \|u_t\|_4\|D^{M+1}u\|_2\|D^M u_t\|_4$$
$$+ \|\nabla u_t\|_2\|D^M u\|_4\|D^M u_t\|_4 + \|D^{M+1}u_t\|_2\|u\|_\infty\|D^M u_t\|_2$$
$$+ \|D^{M+1}d_t\|_2\|\Delta d\|_4\|D^M u_t\|_4 + \|\nabla d_t\|_4\|D^{M+2}d\|_2\|D^M u_t\|_4$$
$$+ \|D^{M+1}d\|_4\|\Delta d_t\|_2\|D^M u_t\|_4 + \|\nabla d\|_\infty\|D^{M+1}d_t\|_2\|D^{M+1}u_t\|_2$$
$$+ \|\Delta d\|_4\|D^{M+1}d_t\|_2\|D^M u_t\|_4 + \|D^{M+1}d\|_4\|\nabla d_t\|_2\|d\|_\infty\|D^M u_t\|_4$$

$$+ \|\nabla d\|_\infty \|D^{M+1}d_t\|_2 \|D^M u_t\|_2 + \|D^M |\nabla d|^2\|_2 \|d_t\|_\infty \|D^M u_t\|_2$$
$$+ \|\nabla d_t\|_2 \|D^M u_t\|_2 \|D^M d\|_\infty \|\nabla d\|_\infty + \||\nabla d|^2\|_\infty \|D^M u_t\|_2 \|D^M d_t\|_2$$
$$+ \|D^M E_t\|_2 \|D^M u_t\|_2 + \|D^M (u \times H)_t\|_2 \|D^M u_t\|_2$$
$$\leqslant C(\|D^M u_t\|_2^2 + \|D^M d_t\|_2^2 + \|D^M E\|_2^2 + \|D^M H\|_2^2 + \|D^M u\|_2^2)$$
$$+ \frac{1}{8}(\|D^{M+1} u_t\|_2^2 + \|D^{M+1} d_t\|_2^2).$$

So,
$$\frac{1}{2}\frac{d}{dt}\int_\Omega |D^M u_t|^2 dx + \int_\Omega |D^{M+1} u_t|^2 dx$$
$$\leqslant C(\|D^M u_t\|_2^2 + \|\Delta d_t\|_2^4 + \|\nabla d_t\|_2^4) + \frac{1}{8}\|D^{M+1} u_t\|_2^2 + \|D^{M+1} d_t\|_2^2). \tag{3.19}$$

Taking the similar procedure to (1.2), we get
$$\frac{1}{2}\frac{d}{dt}\int_\Omega |D^M d_t|^2 + \int_\Omega |D^{M+1} d_t|^2 \leqslant C(\|D^M d_t\|_2^2 + \|D^M u_t\|_2^2) + \frac{1}{8}\|D^{M+1} d_t\|_2^2. \tag{3.20}$$

Whence, together with Gronwall's inequality, (3.16)-(3.18) we deduce
$$\|D^M u_t\|_2 + \|D^M d_t\|_2 \leqslant C, \quad M \geqslant 2.$$
$$\sup_{0\leqslant t \leqslant T} (\|u\|_{H^{M+1}}^2 + \|d\|_{H^{M+1}}^2 + \|E\|_{H^M}^2 + \|H\|_{H^M}^2) \leqslant C.$$

(Uniqueness part) Next we will give the proof of the uniqueness of the solution. Suppose $(\bar{u}, \bar{d}, \bar{E}, \bar{H})$, $(\bar{\bar{u}}, \bar{\bar{d}}, \bar{\bar{E}}, \bar{\bar{H}})$ are two solutions of (1.1)-(1.9) as obtained in Lemma 3.1,
$$u^* = \bar{u} - \bar{\bar{u}}, \quad d^* = \bar{d} - \bar{\bar{d}}, \quad E^* = \bar{E} - \bar{\bar{E}}, \quad H^* = \bar{H} - \bar{\bar{H}}.$$

Then we have the following energy estimates of (u^*, d^*):
$$\frac{1}{2}\int_\Omega (|u^*|^2 + |\nabla d^*|^2) dx(T)$$
$$\leqslant \frac{1}{2}\int_\Omega (|u^*|^2 + |\nabla d^*|^2) dx(0) - \int_0^T \int_\Omega (|\nabla u^*|^2 + |\Delta d^*|^2) dx dt$$
$$- \int_0^T \int_\Omega (u^* \nabla \bar{u} u^* + \Delta \bar{d} \nabla d^* u^* + \bar{u} \nabla d^* \Delta d^* - (f(\bar{d}) - f(\bar{\bar{d}})) \Delta d^*) dx dt$$

$$-\int_0^T \int_\Omega (E^* u^* + (\bar{u} \times H^*)u^* + (u^* \times \bar{H})u^*)dxdt. \tag{3.21}$$

On the other hand, we deduce

$$E_t^* - \nabla \times H^* = -\sigma E^* + u^*, \tag{3.22}$$

$$H_t^* + \beta d_t^* + \nabla \times E^* = 0, \tag{3.23}$$

testing (3.22) and (3.23) with E^* and H^* respectively, we have

$$\int_\Omega (|E^*|^2 + |H^*|^2)dx(T) = \int_\Omega (|E^*|^2 + |H^*|^2)dx(0) - \sigma \int_0^T \int_\Omega |E^*|^2 dxdt$$
$$- \beta \int_0^T \int_\Omega d_t^* H^* dxdt + \int_0^T \int_\Omega u^* E^* dxdt. \tag{3.24}$$

By using Lemma 2.4 and the zero boundary condition, we have

$$\int_\Omega u^* \nabla \bar{u} u^* dx = \int_\Omega u^* \nabla \bar{\bar{u}} \bar{u} dx = \int_\Omega u^* \nabla u^* \bar{u} dx$$
$$\leqslant \|\nabla u^*\|_2 \|u^*\|_4 \|\bar{u}\|_4$$
$$\leqslant \varepsilon \|\nabla u^*\|_2^2 + C\|u^*\|_2^2,$$

here, ε is an arbitrary small number, and C is a constant. We also note that $\bar{u}$ is the good solution as in Lemma 3.1, so that $\|\bar{u}\|_4$ is bounded for all t.

The same argument also works for the other terms:

$$\int_\Omega \Delta \bar{d} \nabla d^* u^* dx \leqslant \|\nabla d^*\|_4 \|u^*\|_4 \|\Delta \bar{d}\|_2$$
$$\leqslant C\|\nabla d^*\|_2 \|u^*\|_2 + \varepsilon \|\Delta d^*\|_2 \|\nabla u^*\|_2$$
$$\leqslant C(\|\nabla d^*\|_2^2 + \|u^*\|_2^2) + \varepsilon(\|\Delta d^*\|_2^2 + \|\nabla u^*\|_2^2),$$

$$\int_\Omega \bar{u} \nabla d^* \Delta d^* dx \leqslant \|\Delta d^*\|_2 \|\bar{u}\|_4 \|\nabla d^*\|_4 \leqslant \varepsilon \|\Delta d^*\|_2^2 + C\|\nabla d^*\|_2^2,$$

$$\int_\Omega (f(\bar{d}) - f(\bar{\bar{d}}))\Delta d^* dx \leqslant \|\Delta d^*\|_2 \|f(\bar{d}) - f(\bar{\bar{d}})\|_2 \leqslant \varepsilon \|\Delta d^*\|_2^2 + C\|\nabla d^*\|_2^2,$$

where $f(d)$ is defined as in (1.10).

$$\int_\Omega d_t^* H^* dx \leqslant \|d_t^*\|_2 \|H^*\|_2$$

$$\leqslant (\|u^*\nabla \bar{d}\|_2 + \|\bar{u}\nabla d^*\|_2 + \|\Delta d^*\|_2 + \|f(\bar{d}) - f(\bar{\bar{d}})\|_2)\|H^*\|_2$$

$$\leqslant C(\|u^*\|_2^2 + \|H^*\|_2^2 + \|\nabla d^*\|_2^2) + \varepsilon\|\Delta d^*\|_2^2, \tag{3.25}$$

$$\int_\Omega (\bar{u} \times H^*) u^* dx \leqslant \|\bar{u}\|_\infty \|H^*\|_2 \|u^*\|_2 \leqslant C(\|H^*\|_2^2 + \|u^*\|_2^2). \tag{3.26}$$

Whence (3.21), (3.24)-(3.26) yields that

$$\frac{1}{2}\int_\Omega (|u^*|^2 + |\nabla d^*|^2 + |E^*|^2 + |H^*|^2) dx(T)$$
$$\leqslant \frac{1}{2}\int_\Omega (|u^*|^2 + |\nabla d^*|^2 + |E^*|^2 + |H^*|^2) dx(0)$$
$$+ C\int_0^T \int_\Omega (|u^*|^2 + |\nabla d^*|^2 + |E^*|^2 + |H^*|^2) dx dt,$$

since $u^*(0) = d^*(0) = E^*(0) = H^*(0) = 0$, and Gronwall's inequality yields

$$u^* = d^* = E^* = H^* = 0. \qquad \square$$

When the dimension of Ω is 3, we see that the size of ν plays a rather crucial role while the other viscosity constants λ, γ do not as long as λ, γ are positive constants. This fact can be seen from the following calculations. Thus we shall, for simplicity, assume that $\lambda = \gamma = 1$.

Denote

$$A^2(t) = \int_\Omega (|\Delta u|^2 + |\Delta d|^2 + |\nabla E|^2 + |\nabla H|^2) dx,$$

$$B^2(t) = \int_\Omega (|\nabla u|^2 + |\Delta d|^2) dx, \quad D^2(t) = \int_\Omega (|\Delta u|^2 + |\Delta d|^2) dx.$$

Lemma 3.4 *Assume* $(u_0, d_0, E_0, H_0) \in (H^1(\Omega), H^2(\Omega), H^1(\Omega), H^1(\Omega))$ ($\Omega \subseteq \mathbf{R}^3$). *Then there exists a unique smooth solution* $(u(x,t), d(x,t), E(x,t), H(x,t))$ *of the problem* (1.1)-(1.9) *such that*

$$(u, d) \in (L^\infty(0, T; H^2(\Omega)))^2,$$
$$(E, H) \in (L^\infty(0, T; H^1(\Omega)))^2,$$

when

$$\nu \geqslant \nu_0(\lambda, \gamma, u_0, d_0, E_0, H_0).$$

Proof First, we prove that
$$A^2(t) \in L^2(0,T).$$

Multiplying (1.1) and (1.2) with u and $\Delta d + |\nabla d|^2 d$ respectively,
$$\frac{1}{2}\frac{d}{dt}\int_\Omega (|u|^2 + |\nabla d|^2) dx = -\int_\Omega (|\nabla u|^2 + |\Delta d + |\nabla d|^2 d|^2 + E\cdot u) dx,$$

i.e.,
$$\sup_{0 \leqslant t \leqslant T} \int_\Omega (|u|^2 + |\nabla d|^2) dx + 2\int_0^T [\|\nabla u\|_2^2 + |\Delta d + |\nabla d|^2 d|^2] dt$$
$$+ \int_0^T \int_\Omega Eu\, dx\, dt \leqslant \int_\Omega (|u_0|^2 + |\nabla d_0|^2) dx. \tag{3.27}$$

By using Theorem 1.1, (3.27) and elliptic estimates, $\forall\, t \in [0, T-1]$, there is a $t_1 \in [t, t+1]$ such that
$$B^2(t_1) \leqslant 2M, \tag{3.28}$$

where
$$M = \|u_0\|_2^2 + \|\nabla d_0\|_2^2 + TC_0,$$

where
$$C_0 \geqslant \|E\|_2^2 + \|u\|_2^2.$$

Then, by a similar procedure, we deduce
$$D^2(t_1) \leqslant CM + \|\nabla E\|_2^2(t_1) + \|\nabla H\|_2^2(t_1). \tag{3.29}$$

By (3.7), (3.28), for the above t_1 we chose, there holds
$$\int_\Omega (|\nabla E|^2 + |\nabla H|^2) dx(t_1) \leqslant C(B^2(0) + \|\nabla E_0\|_2^2 + \|\nabla H_0\|_2^2). \tag{3.30}$$

Therefore
$$A^2(t_1) \leqslant 2M', \tag{3.31}$$

where
$$M' = \|u_0\|_2^2 + \|\nabla d_0\|_2^2 + C_0 T + \|\nabla E_0\|_2^2 + \|\nabla H_0\|_2^2.$$

We calculate
$$\frac{1}{2}\frac{d}{dt} A^2 = \int_\Omega [-(\Delta(\Delta u), u_t) + (\Delta d, \Delta d_t)] dx$$

$$+ \int_\Omega u_t \Delta E dx - \beta \int_\Omega \nabla d_t \cdot \nabla H dx - \sigma \int_\Omega |\nabla E|^2 dx$$

$$= -(\|\nabla \Delta u\|_2^2 + \|\nabla \Delta d\|_2^2) + \int_\Omega [u \nabla u \Delta(\Delta u) + \Delta(\Delta u) \nabla d \Delta d$$

$$+ \Delta d(\Delta d - u \nabla d) + \Delta d(-\Delta u \nabla d - u \nabla \Delta d - 2 \nabla u \cdot \nabla^2 d)] dx$$

$$- \int_\Omega E \cdot \Delta(\Delta u) dx - \int_\Omega (u \times H) \cdot \Delta(\Delta u) dx$$

$$+ \int_\Omega u_t \Delta E dx - \beta \int_\Omega \nabla d_t \cdot \nabla H dx - \sigma \int_\Omega |\nabla E|^2 dx. \qquad (3.32)$$

Now we can work with the right-hand side of (3.32) term by term, for simplicity, we just give some terms that may be difficulties for the calculation.

$$\int_\Omega u \nabla u \Delta(\Delta u) \leqslant \|\nabla u\|_4^2 \|\nabla \Delta u\|_2 + \|u\|_4 \|\Delta u\|_4 \|\nabla \Delta u\|_2$$

$$\leqslant C(\|\nabla u\|_2^{\frac{1}{2}} \|\Delta u\|_2^{\frac{3}{2}} \|\nabla \Delta u\|_2 + \|\Delta u\|_2^{\frac{1}{4}} \|\nabla \Delta u\|_2^{\frac{7}{4}})$$

$$\leqslant C(\|\nabla u\|_2^2 \|\nabla \Delta u\|_2^2 + \|\Delta u\|_2^2 \|\nabla \Delta u\|_2^2 + \|\nabla \Delta u\|_2^2),$$

$$\int_\Omega u \nabla u \Delta E dx \leqslant \|\Delta u\|_2^2 \|\nabla E\|_2^2 + \|\Delta u\|_2^2 + \|\nabla E\|_2^2 \|\nabla \Delta u\|_2^2,$$

$$\int_\Omega \Delta u \Delta E dx \leqslant C(\|\nabla \Delta u\|_2^2 + \|\nabla E\|_2^2),$$

$$- \int_\Omega (u \times H) \cdot \Delta(\Delta u) dx \leqslant C(\|\nabla H\|_2^2 + \|\Delta u\|_2^2 \|\nabla \Delta u\|_2^2 + \|\nabla \Delta u\|_2^2),$$

$$\int_\Omega (u \times H) \Delta E dx \leqslant C(\|\nabla H\|_2^2 + \|\Delta u\|_2^2 \|\nabla E\|_2^2 + \|\nabla E\|_2^2),$$

$$\int_\Omega u \cdot \nabla d \cdot \Delta \Delta d = - \int_\Omega \nabla u \cdot \nabla d \cdot \nabla \Delta d - \int_\Omega u \cdot \nabla^2 d \cdot \nabla \Delta d$$

$$\leqslant C(\|\nabla u\|_\infty \|\nabla d\|_2 \|\nabla \Delta d\|_2 + \|u\|_\infty \|\Delta d\|_2 \|\nabla \Delta d\|_2)$$

$$\leqslant C \left(\frac{1}{\nu} \|\nabla \Delta d\|_2^2 + \frac{1}{\nu} \|\nabla \Delta u\|_2^2 + \nu \|\Delta u\|_2^2 + \frac{1}{\nu} A^2(t) \|\nabla \Delta d\|_2^2 \right)$$

$$\leqslant C \left(\frac{1}{\nu} \|\nabla \Delta u\|_2^2 + \frac{A^2(t) + C}{\nu} \|\nabla \Delta d\|_2^2 + \nu A^2(t) \right),$$

$$\int_\Omega |\nabla d|^2 d \Delta \Delta d = - \int_\Omega \nabla |\nabla d|^2 d \nabla \Delta d - \int_\Omega |\nabla d|^2 \nabla d \nabla \Delta d,$$

using Lemma 2.4, we deduce

$$\int_\Omega (\nabla|\nabla d|^2) d\nabla\Delta d \leqslant \|d\|_\infty \|\nabla d\|_4 \|\Delta d\|_4 \|\nabla\Delta d\|_2$$

$$\leqslant C(\Omega)\left(\nu\|\nabla d\|_4^2\|\Delta d\|_4^2 + \frac{1}{\nu}\|\nabla\Delta d\|_2^2\right)$$

$$\leqslant C(\Omega)\left(\nu^3\|\Delta d\|_4^4 + \frac{1}{\nu}\|\nabla\Delta d\|_2^2 + \frac{1}{\nu}\|d\|_\infty^{\frac{10}{3}}\|\nabla^3 d\|_2^{\frac{2}{3}}\right)$$

$$\leqslant C(\Omega)\left(\nu^3\|\Delta d\|_4^4 + \frac{1}{\nu}\|\nabla\Delta d\|_2^2\right).$$

Take the above inequality into (3.32), we get

$$\frac{1}{2}\frac{d}{dt}A^2 \leqslant -\left(\nu - CA^2 - \frac{1}{\nu}\right)\|\nabla\Delta u\|_2^2 - \left(C - \frac{A^2(t)+C}{\nu}\right)\|\nabla\Delta d\|_2^2$$
$$+ (C + CA^2)A^2. \tag{3.33}$$

By setting $\tilde{A}^2 = A^2 + 1$, we have

$$\frac{1}{2}\frac{d}{dt}\tilde{A}^2 \leqslant -\left(\frac{\nu^2 - \tilde{A}^2 C}{\nu}\right)\|\nabla\Delta u\|_2^2 - \left(C - \frac{\tilde{A}^2(t)C}{\nu}\right)\|\nabla\Delta d\|_2^2 + (\tilde{A}^4 + C\tilde{A}^2). \tag{3.34}$$

Next, we shall assume that ν is so large that

$$\nu \geqslant 2C[\tilde{A}^4(0) + 1 + 4M']. \tag{3.35}$$

Then, initially there is some $T_0 > 0$ such that

$$\frac{\nu^2 - \tilde{A}^2 C}{\nu} \geqslant 0, \quad C - \frac{\tilde{A}^2(t)C}{\nu} \geqslant 0 \tag{3.36}$$

for all $t \in [0, T_0]$. We assume T_* is the largest such T_0, then we claim $T_* = T$.

To see this, we first show $T_* \geqslant \min\{1, T\}$. Indeed, by (3.33), (3.35) and (3.36) we have for all t with $0 \leqslant t \leqslant T_*$,

$$\tilde{A}^2(t) \leqslant \tilde{A}^2(0) + 2C\int_0^t (\tilde{A}^2(t) + \tilde{A}^4(t))dt \leqslant \tilde{A}^2(0) + 2C(M' + M'^2).$$

For $T > 1$, to see $T = T_*$, we use (3.31). In fact, there is a $t_* \in \left[T_* - \frac{1}{2}, T_*\right]$ such that we obtain that

$$\tilde{A}^2(t_*) \leqslant 4M'$$

and then the inequality (3.36) is valid at t_* in the strict sense by our choice of ν. We repeat the above reasoning with t replaced by $t - t_*$ to conclude a contradiction if $T < T_*$.

Then from the above computation, we have

$$(u,d) \in (L^\infty(0,T;H^2(\Omega)))^2, \quad (E,H) \in (L^\infty(0,T;H^1(\Omega)))^2$$

under the condition of (3.35).

The uniqueness can be proved exactly as in the 2D case. □

Lemma 3.5 *Assume* $(u_0, d_0, E_0, H_0) \in (H^1(\Omega), H^2(\Omega), H^{m-1}(\Omega), H^{m-1}(\Omega))$ ($m \geqslant 3$). *Then there exists a unique smooth solution* $(u(x,t), d(x,t), E(x,t), H(x,t))$ *of the problem* (1.1)-(1.9) *satisfying*

$$\sup_{0 \leqslant t \leqslant T} (\|u\|_{H^m}^2 + \|d\|_{H^m}^2 + \|E\|_{H^{m-1}}^2 + \|H\|_{H^{m-1}}^2) \leqslant C, \quad (3.37)$$

when

$$\nu \geqslant \nu_0(u_0, d_0, E_0, H_0).$$

Proof This lemma can be proved by the induction for m. The procedure is similar to Lemma 3.3 and Lemma 3.4, we omit it for simplicity. So Theorem 1.2 is proved. □

References

[1] H. Kim. A blow-up criterion for the nonhomogeneous incompressible Navier-Stokes equations. SIAM J. Math. Anal., 2006, 37: 1417-1434.
[2] Fan J, Gao H, Guo B. Regularity criteria for the Navier-Stokes-Landau-Lifshitz system [J]. Journal of Mathematical Analysis and Applications, 2009, 363(1): 29-37.
[3] Ericksen J L. Hydrostatic theory of liquid crystals[J]. Archive for Rational Mechanics and Analysis, 1962, 9(1): 371-378.
[4] Ericksen J. Conservation laws for liquid crystals. Trans. Soc. Rheol., 1961, 5: 22-34.
[5] Leslie F M. Some constitutive equations for liquid crystals[J]. Archive for Rational Mechanics and Analysis, 1968, 28(4): 265-283.
[6] Gennes P G D, Alben R. The Physics of Liquid Crystals[J]. Physics Today, 1975, 28(6): 54-55.
[7] Ericksen J. Continuum theory of nematic liquid crystals. Res. Mechanica., 1987, 22: 381-392.
[8] Leslie F M. Theory of flow phenomena in liquid crystals[J]. Advances in Liquid Crystals, G. Brown, ed.. New York: Academic Press, 1979, (5): 1-81.

[9] Ericksen J. Equilibrium theory of liquid crystals[J]. Advances in Liquid Crystals, G. Brown. ed.. New York: Academic Press, 1975, (2): 233-398.

[10] Lin F. A new proof of the Caffarelli-Kohn-Nirenberg theorem[J]. Communications on Pure and Applied Mathematics, 1998, 51(3): 241-257.

[11] Lin F, Liu C. Nonparabolic dissipative systems modeling the flow of liquid crystals[J]. Communications on Pure and Applied Mathematics, 2010, 48(5): 501-537.

[12] Lin F, Liu C. Partial Regularity of The Dynamic System Modeling The Flow of Liquid Cyrstals. DCDS, 1998, 2(1): 1-22.

[13] Ericksen J L, Kinderlehrer D. Theory and Applications of Liquid Crystals[J]. IMA Vol. 5, New York: Springer-Verlag, 1986.

[14] Caffarelli L, Kohn R, Nirenberg L. Partial regularity of suitable weak solutions of the Navier-Stokes equations[J]. Communications on Pure and Applied Mathematics, 2010, 35(6): 771-831.

[15] Greenberg J M, Maccamy R C, Coffman C V. On the long-time behavior of ferroelectric systems[J]. Physica D Nonlinear Phenomena, 1999, 134(3): 362-383.

[16] Lin F, Lin J, Wang C. Liquid Crystal Flows in Two Dimensions[J]. Archive for Rational Mechanics and Analysis, 2010, 197(1): 297-336.

[17] Temam R. Navier-Stokes Equations[M]. rev. ed.. Studies in Mathematics and its Applications 2, North-Holland, Amsterdam, 1977.

[18] Ladyzhenskaya O A, Ural'Tseva N N, Solonnikov N A. Linear and quasilinear elliptic equations[M]. Academic Press, 1968.

[19] Böttcher C. Theory of electric polarization. Elsevier, 1952.

[20] Feireisl E. Dynamics of Viscous Compressible Fluids. Oxford: Oxford University Press, 2004.

[21] Simon J. Nonhomogeneous viscous incompressible fluids: the existence of viscosity, density, and pressure. SIAM J. Math. Anal., 1990, 20: 1093-1117.

Global Smooth Solution for the Compressible Landau-Lifshitz-Bloch Equation*

Guo Boling (郭柏灵), Li Fangfang (李方方)

Abstract The Landau-Lifshitz-Bloch equation is often used to describe the micro-magnetic phenomenon under high temperature. In this paper, we establish the existence and uniqueness of a global smooth solution for the initial problem of the compressible Landau-Lifshitz-Bloch equation in dimension one.

Keywords Landau-Lifshitz-Bloch equation; compressible model; global solution

1 Introduction

As well known, the Landau–Lifshitz equation well describes the magnetization dynamics of ferromagnets at low temperatures. It is famous, and many important results have been obtained; see [7]. In order to describe the magnetization dynamics in a ferromagnetic body for a wide range of temperatures, Garanin [5,6] derived the Landau-Lifshitz-Bloch equation from statistical mechanics with the mean field approximation in the 1990s. The Landau-Lifshitz-Bloch equation is able to rule the time evolution of both the direction and the modulus of the vector Z and has been recently applied to simulations of FePtin [9, 14]. It generalizes the well-known Landau-Lifshitz equation.

The Landau–Lifshitz–Bloch equation is given as follows

$$Z_t = -\gamma Z \times H^{\text{eff}} + \frac{L_1}{|Z|^2}(Z \cdot H^{\text{eff}})Z - \frac{L_2}{|Z|^2} Z \times (Z \times H^{\text{eff}}), \qquad (1.1)$$

where $Z(x,t) = (Z_1(x,t), Z_2(x,t), Z_3(x,t))$ is a magnetization functional vector, γ, L_1, L_2 are constants, "$\times$" denotes the vector outer product, H^{eff} is the effective field. We can also rewrite (1.1) as follows

$$Z_t = -\gamma Z \times H^{\text{eff}} + \frac{\gamma a_\parallel}{|Z|^2}(H^{\text{eff}} \cdot Z)Z - \frac{\gamma a_\perp}{|Z|^2} Z \times (Z \times H^{\text{eff}})$$

with $\gamma a_\parallel = L_1$, $\gamma a_\perp = L_2$. Here $a_\parallel$ and $a_\perp$ are dimensionless damping parameters that depend on the temperature and are defined as follows [1]

$$a_\parallel(\theta) = \frac{2\theta}{3\theta_c}\lambda, \quad a_\perp(\theta) = \begin{cases} \lambda\left(1 - \dfrac{\theta}{3\theta_c}\right), & \text{if } \theta < \theta_c, \\ a_\parallel(\theta), & \text{if } \theta \geqslant \theta_c, \end{cases}$$

where $\lambda > 0$ is a constant.

The effective field H^{eff} is given by

$$H^{\text{eff}} = \Delta Z - \frac{1}{\chi_\parallel}\left(1 + \frac{3}{5}\frac{T}{T - T_c}|Z|^2\right)Z, \tag{1.2}$$

where $\chi_\parallel$ is the longitudinal susceptibility, if $L_1 = L_2$, (1.1) can be reduced as follows [10]

$$Z_t = k_1 \Delta Z + \gamma Z \times \Delta Z - k_2(1 + \mu|Z|^2)Z, \tag{1.3}$$

where the coeffcients $k_1, k_2, \gamma, \mu > 0$ and the existence of a global weak solution for the equation (1.3) has been obtained.

In 1982, Fivez derived the classical compressible Heisenberg chain equation

$$Z_t = (G(Z_x)Z \times Z_x)_x, \quad x \in \mathbf{R}, \tag{1.4}$$

where $G(\xi) = A + B|\xi|^2$ and $A, B > 0$ are constants. Magyari obtained the solutions of equation (1.4) with $B = 0$ in [13]. When $B = 0, A = g(x)$ is some given function, the existence and uniqueness of the smooth solution of (1.4) were obtained in [15] with $g(x) \equiv 1$ and $g(x) \neq$ constant in [3]. The existence of a measure-valued solution of the equation was established in [2].

In this paper, we will consider the following generalized compressible Landau–Lifshitz–Bloch equation

$$Z_t = k_1 Z_{xx} + (G(Z)Z \times Z_x)_x - k_2(1 + \mu|Z|^2)Z \tag{1.5}$$

with the initial value

$$Z(x, 0) = Z_0(x), \quad x \in \mathbf{R}, \tag{1.6}$$

where $G(\xi) = A + B|\xi|^q, q \geqslant 2$ and $A, B > 0$ are constants, which can be viewed as a generalization of the Landau–Lifshitz–Bloch equation and compressible Heisenberg chain equation; no one has discussed the compressible LLB equation at present, we will concentrate on the existence and uniqueness of global smooth solution for the initial problem (1.5)-(1.6). Developing and extending the methods of [4, 8, 11], we obtain the following theorem:

Theorem 1.1 Let the initial data $Z_0(x) \in H^m(m \geqslant 2)$, $k > 0, \mu > 0$, then the problem (1.5)-(1.6) admits a global smooth solution satisfying

$$\partial_t^j \partial_x^\alpha Z \in L^\infty([0,T]; L^2(\mathbf{R})), \quad \partial_t^j \partial_x^\beta Z \in L^2([0,T]; L^2(\mathbf{R})),$$

where $2j + |\alpha| \leqslant m$ and $2k + \beta \leqslant m + 1$.

The rest of this paper is as follows. In Section 2, the proof of smooth local solution of (1.5)-(1.6) is proved in $\mathbf{R}$, a priori uniform estimates in H^m is established, and the existence and uniqueness of the global smooth solution of (1.5)-(1.6) is proved in $\mathbf{R}$.

2 The proof of Theorem 1.1

From [12] it can be shown that there exists $T > 0$ and a smooth solution of problem (1.5)-(1.6) in $[0, T]$. Indeed it is easy to check that $e^{tk_1\Delta}$ is a analytic semigroup generated by $k_1\Delta$ in $L^2(\mathbf{R})$, let

$$X = \{Z | Z \in C([0,T]; H^m(\mathbf{R})), \quad t^\alpha Z \in C^\alpha([0,T]; H^m(\mathbf{R})), Z(0) = Z_0\}$$

and

$$Y = \{Z | Z \in X, \|Z\|_{C([0,T]; H^m(\mathbf{R}))} + [t^\alpha Z]_{C^\alpha([0,T]; H^m(\mathbf{R}))} \leqslant \delta\},$$

where $0 < \alpha < 1$ and $m \geqslant 2$. Define a nonlinear operator Γ on Y by $\Gamma(Z) = v$, where v is the solution of

$$v_t = k_1 v_{xx} + (G(Z)Z \times Z_x)_x - k(1 + \mu|Z|^2)Z.$$

By Theorem 4.3.5 of Reference [12] (pp.137-139), for every $Z \in Y, \Gamma(Z) \in C([0,T]; H^m(\mathbf{R}))$ and $t^\alpha \Gamma(Z) \in C^\alpha([0,T]; H^m(\mathbf{R}))$, then, using the same arguments as in the proof of Theorem 8.1.1, there exists $T > 0$ and $\delta > 0$ such that $\Gamma : Y \to Y$ is a contraction, there exists a unique smooth local solution of problem (1.5)-(1.6).

The following Gagliardo-Nirenberg inequality will be used many times.

Lemma 2.1(Gagliardo-Nirenberg inequality) Assume that $u \in L^q(\Omega), D^m u \in L^r(\Omega), \Omega \subset \mathbf{R}^n, 1 \leqslant q,r \leqslant \infty, 0 \leqslant j \leqslant m$. Then

$$\|D^j u\|_{L^p(\Omega)} \leqslant C(j,m;p,r,q)\|u\|_{W_r^m(\Omega)}^a \|u\|_{L^q(\Omega)}^{1-a}, \tag{2.1}$$

where $C(j,m;p,r,q)$ is a positive constant, and

$$\frac{1}{p} = \frac{j}{n} + a\left(\frac{1}{r} - \frac{m}{n}\right) + (1-a)\frac{1}{q}, \quad \frac{j}{m} \leqslant a \leqslant 1.$$

For simplicity, we denote

$$\|\cdot\|_{L^p(\mathbf{R}^1)} = \|\cdot\|_p, p \geqslant 2.$$

Lemma 2.2 Assume that the initial data $Z_0(x) \in H^m(m \geqslant 1)$, for the smooth solution of problem (1.5)-(1.6), we have

$$\|Z(\cdot,t)\|_2^2 + \int_0^t \|Z_x(\cdot,t)\|_2^2 dt \leqslant C, \tag{2.2}$$

$$\|Z_x(\cdot,t)\|_2^2 + \int_0^t \|Z_{xx}(\cdot,t)\|_2^2 dt \leqslant C. \tag{2.3}$$

Proof Taking the scalar product of Z with equation (1.5), and integrating the result over $\mathbf{R}$, we get

$$\frac{1}{2}\frac{d}{dt}\|Z(\cdot,t)\|_2^2 + k_1 \|Z_x(\cdot,t)\|_2^2 = -\int_{\mathbf{R}} G(Z) Z \times Z_x \cdot Z_x dx$$
$$- k_2 \int_{\mathbf{R}} (1+\mu|Z|^2)|Z|^2 dx,$$

then by the fact $A \times B \cdot B = 0, A, B$ are vectors, we find

$$\frac{1}{2}\frac{d}{dt}\|Z(\cdot,t)\|_2^2 + k_1 \|Z_x(\cdot,t)\|_2^2 + k_2 \int_{\mathbf{R}} (1+\mu|Z|^2)|Z|^2 dx = 0.$$

Since $k_1 > 0, k_2 > 0, \mu > 0$, we get

$$\|Z(\cdot,t)\|_2^2 + \int_0^t \|Z_x(\cdot,t)\|_2^2 dt \leqslant \|Z(\cdot,0)\|_2^2. \tag{2.4}$$

Taking the scalar product of $|Z|^{p-2}Z(p \geqslant 2)$ with equation (1.5), and integrating the result over $\mathbf{R}$, we get

$$\int_{\mathbf{R}} |Z|^{p-2} Z \cdot Z_t dx$$

$$= \int_{\mathbf{R}} |Z|^{p-2} Z \cdot Z_{xx} dx + \int_{\mathbf{R}} |Z|^{p-2} Z \cdot (G(Z)Z \times Z_x)_x dx$$

$$- k_2 \int_{\mathbf{R}} |Z|^{p-2} \cdot (1 + \mu|Z|^2) Z^2 dx$$

$$\leq - \int_{\mathbf{R}} |Z|^{p-2} Z_x \cdot Z_x dx - (p-2) \int_{\mathbf{R}} |Z|^{p-4} (Z \cdot Z_x)^2 dx$$

$$\leq 0,$$

so we have

$$\frac{1}{p}\frac{d}{dt}\|Z(\cdot,t)\|_{L^p}^p \leq 0$$

which implies

$$\|Z(\cdot,t)\|_{L^p} \leq \|Z_0(x)\|_{L^p}. \tag{2.5}$$

Let $p \to \infty$, we have

$$\|Z(\cdot,t)\|_{L^\infty} \leq \|Z_0(x)\|_{L^\infty}, \quad \forall t \geq 0. \tag{2.6}$$

Differentiating (1.5) with respect to x and multiplying it by Z_x, we have

$$Z_{xt} Z_x = k_1 Z_{xxx} Z_x + (G(Z)Z \times Z_x)_{xx} Z_x - k_2 ((1+\mu|Z|^2)Z)_x Z_x. \tag{2.7}$$

Integrating the result over $\mathbf{R}$, we find

$$\frac{1}{2}\frac{d}{dt}\|Z_x\|_2^2 + k_1 \|Z_{xx}\|_2^2$$

$$= -\int_{\mathbf{R}} (G(Z)Z \times Z_x)_x Z_{xx} dx - \int_{\mathbf{R}} k_2((1+\mu|Z|^2)Z)_x Z_x dx$$

$$= -\int_{\mathbf{R}} qB|Z|^{q-2}(Z \cdot Z_x) Z \times Z_x \cdot Z_{xx} dx$$

$$- \int_{\mathbf{R}} k_2 (2\mu Z^2 Z_x^2 + (1+\mu|Z|^2) Z_x^2) dx$$

$$\leq C\|Z\|_\infty^q \|Z_x\|_4^2 \|Z_{xx}\|_2 + C(\|Z\|_\infty^2 + 1)\|Z_x\|_2^2$$

$$\leq \frac{1}{2}\|Z_{xx}\|_2^2 + C(\|Z_x\|_2^4 + \|Z_x\|_2^2).$$

The generalized Gronwall's inequality says that if $f' = C(f \cdot g) + C, f \leq C \exp\left(\int_0^t g dt\right) + C$, by replacing f and g by $\|Z_x\|_2^2$ and $\|Z_x\|_2^2$ respectively, and the boundedness of $\int_0^t g dt$ from (2.4), we have the estimate (2.3). Thus the lemma is proved.

Lemma 2.3 *Assume that the initial data* $Z_0(x) \in H^m (m \geqslant 2)$, *for the smooth solution of problem* (1.5)-(1.6), *we have the following estimate*

$$\|Z_{xx}(\cdot,t)\|_2^2 + \int_0^t \|Z_{xxx}(\cdot,t)\|_2^2 dt \leqslant C, \tag{2.8}$$

where the constant C *depends on* $k_1, k_2, \mu, \|Z_0(x)\|_2^2$.

Proof Taking the scalar product of Z_{xxxx} with equation (1.5), and integrating the result over $\mathbf{R}$, we get

$$\frac{1}{2}\frac{d}{dt}\|Z_{xx}\|_2^2 - k_1 \int_{\mathbf{R}} Z_{xx} Z_{xxxx} dx$$
$$= \frac{1}{2}\frac{d}{dt}\|Z_{xx}\|_2^2 + \|Z_{xxx}\|_2^2$$
$$= -\int_{\mathbf{R}} (G(Z) Z \times Z_x)_{xx} Z_{xxx} dx - \int_{\mathbf{R}} k_2((1+\mu|Z|^2)Z) Z_{xxxx} dx, \tag{2.9}$$

where

$$\int_{\mathbf{R}} (G(Z) Z \times Z_x)_{xx} Z_{xxx} dx$$
$$= \int_{\mathbf{R}} Bq(q-2)|Z|^{q-4}(Z \cdot Z_x)^2 Z \times Z_x \cdot Z_{xxx} dx$$
$$+ \int_{\mathbf{R}} Bq|Z|^{q-2}(Z_x^2 + Z \cdot Z_{xx}) Z \times Z_x \cdot Z_{xxx} dx$$
$$+ \int_{\mathbf{R}} 2Bq|Z|^{q-2}(Z \cdot Z_x) Z \times Z_{xx} \cdot Z_{xxx} dx$$
$$+ \int_{\mathbf{R}} G(Z) Z_x \times Z_{xx} \cdot Z_{xxx} dx$$
$$\leqslant C\|Z\|_\infty^{q-1}\|Z_x\|_6^3\|Z_{xxx}\|_2 + C\|Z\|_\infty^q\|Z_x\|_4\|Z_x\|_4\|Z_{xxx}\|_2$$
$$+ C(1+\|Z\|_\infty^q)\|Z_x\|_4\|Z_{xx}\|_4\|Z_{xxx}\|_2.$$

By Gagliardo-Nirenlerg inequality, we get

$$\|Z_x\|_\infty \leqslant C\|Z_x\|_2^{\frac{3}{4}}\|Z_{xxx}\|_2^{\frac{1}{4}}, \quad \|Z_x\|_4 \leqslant C\|Z_x\|_2^{\frac{3}{4}}\|Z_{xx}\|_2^{\frac{1}{4}},$$
$$\|Z_x\|_6 \leqslant C\|Z_x\|_2^{\frac{5}{6}}\|Z_{xxx}\|_2^{\frac{1}{6}}, \quad \|Z_{xx}\|_4 \leqslant C\|Z_{xx}\|_2^{\frac{3}{4}}\|Z_{xxx}\|_2^{\frac{1}{4}},$$

so we have

$$\int_{\mathbf{R}} (G(Z)Z \times Z_x)_{xx} Z_{xxx} dx$$
$$\leqslant C\|Z\|_\infty^{q-1}\|Z_x\|_2^{\frac{5}{2}}\|Z_{xxx}\|_2^{\frac{3}{2}} + C\|Z\|_\infty^q \|Z_x\|_2^{\frac{3}{4}} \|Z_{xx}\|_2 \|Z_{xxx}\|_2^{\frac{5}{4}}$$
$$+ C(1+\|Z\|_\infty^q)\|Z_x\|_2^{\frac{3}{4}}\|Z_{xx}\|_2\|Z_{xxx}\|_2^{\frac{5}{4}}$$
$$\leqslant C(\|Z_{xx}\|_2^4) + \frac{1}{4}\|Z_{xxx}\|_2^2 + C.$$

By the Hölder inequality, it follows that

$$-\int_{\mathbf{R}} k_2((1+\mu|Z|^2)Z) Z_{xxxx} dx$$
$$= k_2 \int_{\mathbf{R}} ((1+\mu|Z|^2)Z)_x Z_{xxx} dx$$
$$= k_2 \int_{\mathbf{R}} (2\mu(Z \cdot Z_x)Z + (1+\mu|Z|^2)Z_x) Z_{xxx} dx$$
$$\leqslant C\|Z\|_\infty^2 \|Z_x\|_2 \|Z_{xxx}\|_2$$
$$\leqslant \frac{1}{4}\|Z_{xxx}\|_2^2 + C,$$

then

$$\frac{d}{dt}\|Z_{xx}\|_2^2 + \|Z_{xxx}\|_2^2 \leqslant C\|Z_{xx}\|_2^4 + C, \tag{2.10}$$

by Gronwall's inequality, the (2.8) holds.

Lemma 2.4 *Assume that the initial data $Z_0(x) \in H^m(m \geqslant 2)$, then for the smooth solution of problem (1.5)-(1.6), we have the following estimate*

$$\|Z_{xxx}(\cdot,t)\|_2^2 + \int_0^t \|Z_{xxxx}(\cdot,t)\|_2^2 dt \leqslant C, \tag{2.11}$$

where the constant C depends on $k_1, k_2, \mu, \|Z_0(x)\|_2^2$.

Proof Taking the scalar product of Z_{xxxxxx} with equation (1.5) and integrating the result over $\mathbf{R}$, we have

$$\frac{1}{2}\frac{d}{dt}\|Z_{xxx}\|_2^2 - k_1 \int_{\mathbf{R}} Z_{xxxxx} Z_{xxx} dx$$
$$= \int_{\mathbf{R}} (G(Z)Z \times Z_x)_{xxxx} Z_{xxx} dx - \int_{\mathbf{R}} k_2((1+\mu|Z|^2)Z)_{xxx} Z_{xxx} dx$$

$$= -\int_{\mathbf{R}} (G(Z)Z \times Z_x)_{xxx} Z_{xxxx} dx + \int_{\mathbf{R}} k_2((1+\mu|Z|^2)Z)_{xx} Z_{xxxx} dx, \quad (2.12)$$

where

$$\int_{\mathbf{R}} (G(Z)Z \times Z_x)_{xxx} Z_{xxxx} dx$$

$$= \int_{\mathbf{R}} \Big\{ 2G_{xxx}(Z)Z \times Z_x + 3G_{xx}(Z)Z \times Z_{xx} + 3G_x(Z)Z_x \times Z_{xx}$$

$$+ 3G_x(Z)Z \times Z_{xxx} + 2G(Z)Z_x \times Z_{xxx} + G(Z)Z \times Z_{xxxx} \Big\} Z_{xxxx} dx$$

$$= \int_{\mathbf{R}} \Big\{ Bq(q-2)(q-4)|Z|^{q-6}(Z \cdot Z_x)^3$$

$$+ 3Bq(q-2)|Z|^{q-4}(Z \cdot Z_x)(Z_x^2 + Z \cdot Z_{xx})$$

$$+ Bq|Z|^{q-2}(3Z_x \cdot Z_{xx} + Z \cdot Z_{xxxx}) \Big\} Z \times Z_x \cdot Z_{xxxx} dx$$

$$+ \int_{\mathbf{R}} \Big\{ Bq(q-2)|Z|^{q-4}(Z \cdot Z_x)^2$$

$$+ Bq|Z|^{q-2}(Z_x^2 + Z \cdot Z_{xx}) \Big\} Z \times Z_{xx} \cdot Z_{xxxx} dx$$

$$+ \int_{\mathbf{R}} 3qB|Z|^{q-2}(Z \cdot Z_x)Z_x \times Z_{xx} \cdot Z_{xxxx} dx$$

$$+ \int_{\mathbf{R}} 3qB|Z|^{q-2}(Z \cdot Z_x)Z \times Z_{xxx} \cdot Z_{xxxx} dx$$

$$+ \int_{\mathbf{R}} 2G(Z)Z_x \times Z_{xxx} \Big\} Z_{xxxx} dx$$

$$\leqslant C\|Z\|_{\infty}^{q-2}\|Z_x\|_{\infty}^4\|Z_{xxxx}\|_2 + C\|Z\|_{\infty}^{q-1}\|Z_x\|_{\infty}^2\|Z_{xx}\|_2\|Z_{xxxx}\|_2$$

$$+ C\|Z\|_{\infty}^q\|Z_x\|_{\infty}^2\|Z_{xxx}\|_2\|Z_{xxxx}\|_2$$

$$+ C(\|Z\|_{\infty}^{q-1}\|Z_x\|_{\infty}^2\|Z_{xx}\|_2 + \|Z\|_{\infty}^q\|Z_{xx}\|_4^2)\|Z_{xxxx}\|_2$$

$$+ C\|Z\|_{\infty}^q\|Z_x\|_{\infty}\|Z_{xx}\|_2\|Z_{xxxx}\|_2$$

$$+ C(1 + \|Z\|_{\infty}^q)\|Z_x\|_{\infty}\|Z_{xx}\|_2\|Z_{xxx}\|_2\|Z_{xxxx}\|_2.$$

By Gagliardo-Nirenlerg inequality, we get

$$\|Z_x\|_{\infty} \leqslant C\|Z_x\|_2^{\frac{3}{4}}\|Z_{xxx}\|_2^{\frac{1}{4}}, \quad \|Z_{xx}\|_4 \leqslant C\|Z_{xx}\|_2^{\frac{3}{4}}\|Z_{xxx}\|_2^{\frac{1}{4}},$$

thus,

$$\int_{\mathbf{R}} (G(Z)Z \times Z_x)_{xxx} Z_{xxxx} dx$$

$$\leqslant C\|Z_x\|_2^3 \|Z_{xxx}\|_2 \|Z_{xxxx}\|_2 + C\|Z_x\|_2^{\frac{3}{2}} \|Z_{xxx}\|_2^{\frac{1}{2}} \|Z_{xx}\|_2 \|Z_{xxxx}\|_2$$

$$+ C\|Z_x\|_2^{\frac{3}{2}} \|Z_{xxx}\|_2^{\frac{3}{2}} \|Z_{xxxx}\|_2 + C\|Z_x\|_2^{\frac{3}{4}} \|Z_{xxx}\|_2^{\frac{1}{4}} \|Z_{xx}\|_2 \|Z_{xxxx}\|_2$$

$$+ C\|Z_{xx}\|_2^{\frac{3}{4}} \|Z_{xxx}\|_2^{\frac{1}{4}} \|Z_{xxxx}\|_2 + C\|Z_x\|_2^{\frac{3}{4}} \|Z_{xxx}\|_2^{\frac{5}{4}} \|Z_{xx}\|_2 \|Z_{xxxx}\|_2$$

$$\leqslant \frac{1}{4} \|Z_{xxxx}\|_2^2 + C(\|Z_{xxx}\|_2^2 + \|Z_{xxx}\|_2^4).$$

By the Hölder inequality, it follows that

$$\int_{\mathbf{R}} k_2((1+\mu|Z|^2)Z)_{xx} Z_{xxxx} dx$$

$$= k_2 \int_{\mathbf{R}} (2\mu(Z_x \cdot Z_x)Z + 2\mu Z \cdot Z_{xx} \cdot Z$$

$$+ 5\mu Z \cdot Z_x \cdot Z_x + (1+\mu|Z|^2)Z_{xx}) Z_{xxxx} dx$$

$$\leqslant C\|Z\|_\infty \|Z_x\|_4^2 \|Z_{xxxx}\|_2 + C\|Z\|_\infty^2 \|Z_{xx}\|_2 \|Z_{xxxx}\|_2$$

$$\leqslant \frac{1}{4} \|Z_{xxx}\|_2^2 + C,$$

then,

$$\frac{d}{dt} \|Z_{xxx}\|_2^2 + \|Z_{xxxx}\|_2^2 \leqslant C(\|Z_{xxx}\|_2^2 + \|Z_{xxx}\|_2^4) + C, \quad (2.13)$$

by Gronwall's inequality, the (2.11) holds.

By induction, we also have

Lemma 2.5 *Assume that the initial data $Z_0(x) \in H^m (m \geqslant 2)$, then for the smooth solution of problem (1.5)-(1.6), we have the following estimate*

$$\sup_{0 \leqslant t \leqslant T} \|Z_{x^j}(\cdot, t)\|_2^2 + \int_0^t \|Z_{x^{j+1}}(\cdot, t)\|_2^2 dt \leqslant C, \quad (2.14)$$

where the constant C depends on $k_1, k_2, \mu, \|Z_0(x)\|_2^2, j \geqslant 3$.

Now we will deal with the uniqueness of the solution in problem (1.5)-(1.6). Assume that there exist two solutions Z, Y. Let $W = Z - Y$, then W satisfies the following equation

$$W_t = k_1 W_{xx} + (G(Z)W \times Z_x)_x + (B(R(Z,Y) \cdot W)Y \times Z_x)_x$$

$$+ (G(Y)Y \times W_x)_x - k_2(1+\mu|Z|^2)W - k\mu(Z+Y) \cdot WY, \qquad (2.15)$$

where $R(Z,Y) = \sum_{i=1}^{q-1} |Z|^i |Y|^{q-1-i}$, making the scalar product of W with equation (2.15) and then integrating the result over $\mathbf{R}$, we have

$$\frac{1}{2}\frac{d}{dt}\|W\|_2^2 + k_1\|W_x\|_2^2$$

$$= -\int_{\mathbf{R}} G(Z)W \times Z_x \cdot W_x dx - \int_{\mathbf{R}} B((R(Z,Y) \cdot W)Y \times Z_x \cdot W_x dx$$

$$+ \int_{\mathbf{R}} k(1+\mu|Z|^2)|W|^2 dx + k_2\mu \int_{\mathbf{R}} (Z+Y) \cdot WY \cdot W dx$$

$$\leqslant C\|W\|_\infty (1+\|Z\|_\infty^q)\|Z_x\|_2 \|W_x\|_2$$

$$+ C\|Y\|_\infty \|W\|_\infty (\|Z\|_\infty^{q-1} + \|Y\|_\infty^{q-1})\|Z_x\|_2 \|W_x\|_2$$

$$+ C\|Z\|_\infty^2 \|W\|_2^2 + C(\|Z\|_\infty + \|Y\|_\infty)\|Y\|_\infty \|W\|_2^2. \qquad (2.16)$$

By using Gagliardo-Nirenberg inequality, one has

$$\|W\|_{L^\infty} \leqslant C\|W\|_2^{\frac{3}{4}} \|W_{xx}\|_2^{\frac{1}{4}}. \qquad (2.17)$$

Applying the estimate (2.14) and inequality (2.17), we get

$$\frac{1}{2}\frac{d}{dt}\|W\|_2^2 + k_1\|W_x\|_2^2 \leqslant \frac{1}{4}\|W_{xx}\|_2^2 + C(\|W_x\|_2^2 + \|W\|_2^2). \qquad (2.18)$$

On the other hand, multiplying (2.15) by W_{xx} and then integrating the result over $\mathbf{R}$, we have

$$\frac{1}{2}\frac{d}{dt}\|W_x\|_2^2 + k_1\|W_{xx}\|_2^2$$

$$= \int_{\mathbf{R}} (G(Z)W \times Z_x)_x \cdot W_{xx} dx + \int_{\mathbf{R}} B((R(Z,Y) \cdot W)Y \times Z_x)_x \cdot W_{xx} dx$$

$$+ \int_{\mathbf{R}} (G(Y)Y \times W_x)_x \cdot W_{xx} dx + \int_{\mathbf{R}} k_2(1+\mu|Z|^2)W \cdot W_{xx} dx$$

$$+ k_2\mu \int_{\mathbf{R}} (Z+Y) \cdot WY \cdot W_{xx} dx$$

$$\leqslant \frac{1}{4}\|W_{xx}\|_2^2 + C(\|W_x\|_2^2 + \|W\|_2^2), \qquad (2.19)$$

where

$$\int_{\mathbf{R}} (G(Z)W \times Z_x)_x \cdot W_{xx} dx$$

$$\leqslant C(\|W\|_\infty \|Z\|_\infty^{q-1} \|Z_x\|_4^2 \|W_{xx}\|_2 + \|Z\|_\infty^q \|Z_x\|_4 \|W_x\|_4 \|W_{xx}\|_2$$
$$+ \|W\|_\infty \|Z\|_\infty^q \|Z_{xx}\|_2 \|W_{xx}\|_2),$$

similarly,

$$\int_{\mathbf{R}} B((R(Z,Y) \cdot W) Y \times Z_x)_x \cdot W_{xx} dx$$
$$= B\Big\{ \int_{\mathbf{R}} (R_x(Z,Y) \cdot W) Y \times Z_x \cdot W_{xx} dx + \int_{\mathbf{R}} (R(Z,Y) \cdot W_x) Y \times Z_x \cdot W_{xx} dx$$
$$+ \int_{\mathbf{R}} (R(Z,Y) \cdot W) Y_x \times Z_x \cdot W_{xx} dx$$
$$+ \int_{\mathbf{R}} (R(Z,Y) \cdot W) Y \times Z_{xx} \cdot W_{xx} dx \Big\}$$
$$\leqslant C \bar{R}(\|Z\|_\infty^i \|Y\|_\infty^{q-1-i}) \|W\|_\infty (\|Z_x\|_4^2 + \|Z_x\|_4 \|Y_x\|_4) \|W_{xx}\|_2$$
$$+ C \bar{R}(\|Z\|_\infty, \|Y\|_\infty) \|W\|_\infty \|Z_x\|_4 \|Y_x\|_4 \|W_{xx}\|_2$$
$$+ C \bar{R}(\|Z\|_\infty, \|Y\|_\infty) \|Z_x\|_4 \|W_x\|_4 \|W_{xx}\|_2$$
$$+ C \bar{R}(\|Z\|_\infty, \|Y\|_\infty) \|W\|_\infty \|Z_{xx}\|_2 \|W_{xx}\|_2,$$

where $\bar{R}(\|Z\|_\infty, \|Y\|_\infty) = \sum_{i=1}^{q-1} \|Z\|_\infty^i \|Y\|_\infty^{q-i}$, and

$$\int_{\mathbf{R}} (G(Y) Y \times W_x)_x \cdot W_{xx} dx$$
$$= \int_{\mathbf{R}} G_x(Y) Y \times W_x \cdot W_{xx} dx + \int_{\mathbf{R}} G(Y) Y_x \times W_x \cdot W_{xx} dx$$
$$+ \int_{\mathbf{R}} G(Y) Y \times W_{xx} \cdot W_{xx} dx$$
$$\leqslant C(\|Y\|_\infty \|Z\|_\infty^q \|Y_x\|_4 \|W_x\|_4 \|W_{xx}\|_2$$
$$+ C(1 + \|Y\|_\infty^{q-1}) \|Y_x\|_4 \|W_x\|_4 \|W_{xx}\|_2 + C \|Y\|_\infty^{q+1} \|W_{xx}\|_2^2,$$

then by (2.14) and the following embedding inequality

$$\|W\|_4 \leqslant C \|W\|_2^{\frac{3}{4}} \|W_x\|_2^{\frac{1}{4}},$$

we have

$$\frac{1}{2} \frac{d}{dt} \|W_x\|_2^2 + k_1 \|W_{xx}\|_2^2 \leqslant \frac{1}{4} \|W_{xx}\|_2^2 + C(\|W\|_2^2 + \|W_x\|_2^2). \qquad (2.20)$$

Combining (2.18) and (2.20), we can get

$$\frac{d}{dt}\{\|W\|_2^2 + \|W_x\|_2^2\} + \{\|W_x\|_2^2 + \|W_{xx}\|_2^2\} \leqslant C(\|W\|_2^2 + \|W_x\|_2^2),$$

and then by Gronwall's inequality, we get the uniqueness.

References

[1] Berti V., Fabrizio M., Giorgi C. A three-dimensional phase transition model in ferromagnetism: Existence and uniqueness[J]. J. Math. Anal. Appl. 355 (2009), 661-674.

[2] Ding S., Guo B., Su F. Measure-valued solution to the strongly degenerate compressible Heisenberg chain equations[J]. J. Math. Phys. 40 (1999), 1153-1162.

[3] Ding S., Guo B., Su F. Smooth solution for one-dimensional inhomogeneous Heisenberg chain equations[J]. Proc. Roy. Soc. Edinburgh Sect. A 129 (1999), 1171-1184.

[4] Gao J., Han L., and Huang Y. Solitary Waves for the Generalized Nonautonomous Dual-power Nonlinear Schrödinger Equations with Variable Coefficients[J]. Journal of Nonlinear Modeling and Analysis, 2019, 1(2), 251-260.

[5] Garanin D. Generalized equation of motion for a ferromagnet[J]. Phys. A Statistical Mechanics and Its Applications, 172 (1991), 470-491.

[6] Garanin D. V. V. lshtchenko, L.V. Panina, Dynamics of an ensemble of single-domain magnetic particles[J]. Teor. Mat. Fiz. 82 (1990), 169-179.

[7] Guo B., Ding S. Landau-Lifshitz Equations[M]. Frontiers of Research with the Chinese Academy of Sciences, vol.1, World Scientific Publishing Co. Pty. Ltd., Hackensack, NJ, 2008.

[8] Guo C., Guo B. The existence of global solutions for the fourth-order nonlinear Schrödinger equations[J]. J. Appl. Anal.Comput., 2019, 9(3),1183-1192.

[9] Kazantseva N., Hinzke D., NowakU.,Chantrell R. W., Atxitia U., Chubykalo-Fesenko O. Towards multiscale modeling of magnetic materials: simulations of FePt[J]. Phys. Rev. B 77(2008),184428.

[10] Le K. N. Weak solutions of the Landau-Lifshitz-Bloch equation[J]. J. Differential Equations, 261 (2016), 6699-6717.

[11] Long Q., Chen J., Yang G. Finite time blow-up and global existence of weak solutions for pseudo-parabolic equation with exponential nonlinearity[J]. J. Appl. Anal.Comput., 2018, 8(1),105-122.

[12] Lunardi A. Analytic semigroups and optimal regularity in parabolic problems-Analytic semigroups and optimal regularity in parabolic problems[M], Birkhä- user, 1995.

[13] Magyari E. Solitary waves along the compressible Heisenberg chain[J]. Phys. C. Solid State Phys. 15 (1982), 1159-1163.

[14] Ostler T., Ellis M., Hinzke D., Nowak U. Temperature dependent ferromagnetic resonance via the Landau – Lifshitz – Bloch equation: application to FePt[J]. Phys. Rev. B 90(2014), 094402.

[15] Zhou Y., Guo B., Tan S. Existence and uniqueness of smooth solution for system of ferromagnetic chain[J]. Sci. Sinica A 34 (1991), 157-166.

Global Weak Solution to the Quantum Navier-Stokes-Landau-Lifshitz Equations with Density-Dependent Viscosity*

Wang Guangwu (王光武), Guo Boling (郭柏灵)

Abstract In this paper, we investigate the global existence of the weak solutions to the quantum Navier-Stokes-Landau-Lifshitz equations with density dependent viscosity in two dimensional case. We research the model with singular pressure and the dispersive term. The main technique is using the uniform energy estimates and B-D entropy estimates to prove the convergence of the solutions to the approximate system. We also use some convergent theorems in Sobolev space.

Keywords Navier-Stokes-Landau-Lifshitz equations; global weak solutions; energy estimate; B-D entropy estimate

1 Introduction

In this paper, we will focus on the following quantum Navier-Stokes-Landau-Lifshitz equations with density-dependent viscosity(DQNSLL) in $(0, T) \times \Omega$:

$$\partial_t \rho + \mathrm{div}(\rho u) = 0, \tag{1.1}$$

$$\partial_t(\rho u) + \mathrm{div}(\rho u \otimes u) + \nabla P(\rho) = \nu \mathrm{div}(\rho D(u)) + \kappa \rho \nabla \left(\frac{\Delta \sqrt{\rho}}{\sqrt{\rho}}\right)$$

$$- \lambda \nabla \cdot \left(\nabla d \odot \nabla d - \frac{|\nabla d|^2}{2} I\right), \tag{1.2}$$

$$d_t + u \cdot \nabla d = \alpha_1 d \times \Delta d - \alpha_2 d \times (d \times \Delta d), \quad |d| = 1. \tag{1.3}$$

---

*Discrete Contin. Dyn. Syst. Ser. B, 2019, 24 (11): 6141-6166. Doi: 10.3934/dcdsb.2019133.

Here we denote by $\Omega = T^2 \subset R^2$ the two-dimensional periodic domain, that is, $\Omega = \{x = (x_1, x_2) | |x_i| < D; i = 1, 2\}$. ρ and u represent the density and the velocity field of the flow respectively. P is the pressure function. We only consider the isentropic case: $P = \rho^\gamma (\gamma > 1)$ for simplicity. $d : \Omega \to S^1$, the unit sphere in R^2, denotes the magnetization field. $D(u)$ stands for the symmetric part of the velocity gradient, namely $D(u) = \dfrac{\nabla u + \nabla^T u}{2}$. The positive constants ν and γ represent the viscosity constant and the competition between the kinetic energy and potential energy respectively. $\alpha_1 \geqslant 0$ is Gilbert damping coefficient. The constant α_2 is positive. κ is the scaled Planck constant. $\nabla \cdot$ denotes the divergence operator, and $\nabla d \odot \nabla d$ denotes the 2×2 matrix whose (i, j)-the entry is given by $\nabla_i d \cdot \nabla_j d$ for $1 \leqslant i, j \leqslant 2$. The expression $\dfrac{\Delta \sqrt{\rho}}{\sqrt{\rho}}$ can be interpreted as the quantum Bohm potential. λ and δ are parameters.

This model is coupled between the compressible quantum Navier-Stokes equations and Landau-Lifshitz equation which can be used to describe the dispersive theory of magnetization of ferromagnets with quantum effect.

If $\kappa = 0$, the model (1.1)-(1.3) is called the compressible Navier-Stokes-Landau-Lifshitz equation with density-dependent viscosity. Furthermore, if the viscosity is constant and the velocity satisfies the incompressible condition, the equation (1.1)-(1.3) becomes the inhomogeneous compressible Navier-Stokes-Landau-Lifshitz equation. These two models can be used to describe the dispersive theory of magnetization of ferromagnets. J.S. Fan, H.J. Gao and B.L. Guo [13] have studied the regularity criteria for the smooth solution to the inhomogeneous compressible Navier-Stokes-Landau-Lifshitz equation in Besov space and the multiplier spaces. We [39] have investigated the existence and the uniqueness of the weak solution to the inhomogeneous compressible Navier-Stokes-Landau-Lifshitz equation in two dimensions.

If ρ is a constant and $u = 0$, the system (1.1)-(1.3) is the Landau-Lifshitz equation, which can be used to describe the spin of the ferromagnets. There are lots of achievements about the Landau-Lifshitz equation having been made. In the 1960s-1970s, the traveling wave solution and the soliton solution have been researched [2,32]. Zhou Y-L, Sun H-S, Guo B-L [42]- [49] proved the existence of the global weak solutions to the initial value problems and initial boundary value problems for Landau-Lifshitz equations from one dimension to multi-dimensions.

Similar results by the penalty method were obtained by Alouges and Soyeur [3] in 1992. In 1991, Zhou Y-L, Guo B-L, and Tan S-B [41] got the existence and uniqueness of global smooth solution to one-dimensional Landau-Lifshitz equations with or without Gilbert damping by using a mobile frame on S^2 and a priori estimates. In 1993, Guo B-L, Hong M-C established in [17] the relations between two-dimensional Landau-Lifshitz equations and harmonic maps and applied the approaches studying harmonic maps to get the global existence and uniqueness of partially regular weak solutions. In 1998, Chen Y-M, Ding S-J, and Guo B-L [10] proved that all the finite energy weak solutions were globally smooth with the exception of finitely many singular points at most. The uniqueness was also given. In 2004, the partial regularity of the stationary weak solutions of higher dimensional Landau-Lifshitz equations was proved by X.G. Liu [28]. The Hausdorff dimensions and the Hausdorff measures of the singular set were also estimated. Similar results for lower dimensional Landau-Lifshitz equations were established by Moser R [31] by different methods. Wang C-Y [38] proved the partial regularity for the weak solutions of the initial value problems and initial boundary value problems on three and four-dimensional manifolds. In 2007, Ding S-J and Wang C-Y [11] rigorously proved that in three and four dimensions, some Dirichlet problems and Neumann problems for Landau-Lifshitz equations indeed admitted finite time blow-up solutions. A summary of the results of the Landau-Lifshitz equation can be referred to in [16].

If d is a constant vector, we call the equation (1.1)-(1.3) the compressible quantum Navier-Stokes equation with density dependent viscosity(DQNS). In 2010, Jüngel [24] proved the global existence of DQNS when the scaled Planck constant was bigger than the viscosity constant ($\hbar > \nu$), for or $\alpha > 3$(3-dimension) or $\alpha > 1$(2-dimension). Dong [12] and Jiang [23] extended that results to the case of the viscosity constant was equal to the scaled Plank constant ($\hbar = \nu$) and the viscosity constant was bigger than the scaled Planck constant ($\hbar < \nu$), respectively. We should notice that the definition of global weak solutions used in [12,23,24] followed the idea introduced in [8] by testing the momentum equation by $\rho\phi$ with a test function ϕ. Later, Antonelli and Spirito [4, 5] proved the existence of the global weak solutions to the DQNS for 2D, $\hbar < \nu$, $\alpha > 1$ or 3D, $\hbar^2 < \nu^2 < \frac{9}{8}\hbar^2$, $1 < \alpha < 3$. In 2015, Gisclon and Violet [15] investigated the existence of the weak solution to the DQNS equations with singular pressure, where the authors adopt some arguments from [40] to make use of the cold pressure for compactness in the case of $\hbar < \nu$, $\alpha \geqslant 1$, $1 \leqslant d \leqslant 3$ (d is the dimensional number). In 2016, A.F. Vasseur and C. Yu [37] researched the existence of the weak solution to the DQNS with damping term $(-r_0 u - r_1 \rho |u|^2 u)$.

If $\kappa = 0$ and d is a constant vector, the equation (1.1)-(1.3) is the usual compressible Navier-Stokes equation with density-dependent viscosity(DCNS). It is not difficult to find that, the DCNS is highly degenerated at the vacuum because the velocity cannot even be defined when the density vanishes. It is very difficult to deduce any estimate of the gradient on the velocity field due to the vacuum. However, there are lots of results about DCNS. D. Hoff proves the existence of the global weak solution to the 1-dimensional DCNS with large initial data [20] and discontinuous initial data [21]. When the initial density had the upper and lower bounds, the global existence of the weak solution to 1D DCNS was given by D. Hoff [22]. The research on the existence of a strong solution to 1D DCNS was established by B. Haspot [19]; A. Mellet and Vassure [29]. For the multi-dimensional case, the first tool of handling this difficult is due to Bresch, Desjardins, and Lin [8], where the authors developed a new mathematical entropy to show the structure of the diffusion terms providing some regularity for density. The result was extended for the case with an additional quadratic friction term $r\rho|u|u$, refer to Bresch and Desjardins [6, 7] and the resent results by Bresch, Desjardins, Zatorska [9] and by Zatorska [40]. Later, by obtaining a new a-priori estimate on smooth approximate solutions, Mellet, Vasseur [30] study the stability of DCNS without any additional drag term. Recently, Vasseur and Yu [36] have proved that there existed a global weak solution to DCNS by constructing a new approximate system allowing to derive the Mellet-Vasseur type inequality for the weak solutions; almost at the same time, Li and Xin gave another approach in [26].

In [18], we have obtained the global weak solutions to the finite weak solution to the viscous quantum Navier-Stokes-Landau-Lifshitz-Maxwell equation in two dimensions.

Since it is very difficult to get the B-D entropy for the system (1.1)-(1.3) directly, we replace the original system by adding an extra cold pressure and the damping term as introduced in [6, 15]. The main purpose in this paper is to prove the global existence of weak solutions to the following system, for $x \in \Omega$ and $t > 0$:

$$\partial_t \rho + \operatorname{div}(\rho u) = 0, \qquad (1.4)$$

$$\partial_t(\rho u) + \operatorname{div}(\rho u \otimes u) + \nabla P(\rho) + \nabla P_c(\rho) = \nu \operatorname{div}(\rho D(u)) + \kappa \rho \nabla \left(\frac{\Delta \sqrt{\rho}}{\sqrt{\rho}} \right)$$

$$- \lambda \nabla \cdot \left(\nabla d \odot \nabla d - \frac{|\nabla d|^2}{2} I \right) - \delta u, \qquad (1.5)$$

$$d_t + u \cdot \nabla d = \alpha_1 d \times \Delta d - \alpha_2 d \times (d \times \Delta d), \quad |d| = 1. \qquad (1.6)$$

Where P_c is a suitable increasing function satisfying

$$\lim_{\rho \to 0} P_c(\rho) = +\infty,$$

and called cold pressure. More precisely, we assume

$$P_c'(\rho) = \begin{cases} c\rho^{-4k-1}, & \rho \leqslant 1, \ k > 1, \\ \rho^{\gamma-1}, & \rho > 1, \ \gamma > 1 \end{cases} \qquad (1.7)$$

for some constant $c > 0$.

As we know, the cold pressure term will give us the higher integrability, a priori bounds crucial to proving the global regularity of the approximating solutions. However, this also brings some difficulties in the analysis, first of all that due to the presence of the coupled term between the magnetization field and the velocity, we can not let the cold pressure vanish directly.

For system (1.4)-(1.6), we impose the following initial conditions:

$$\rho|_{t=0} = \rho_0, \quad u|_{t=0} = u_0, \quad d|_{t=0} = d_0, \qquad (1.8)$$

which satisfy that

$$\rho_0 \geqslant 0, \quad |d_0(x)| = 1, \quad d_0(x) \in H^2(\Omega), \quad \inf_x d_0^2 > 0. \qquad (1.9)$$

Definition 1.1 *We call (ρ, u, d) the finite energy weak solution to the quantum Navier-Stokes-Landau-Lifshitz equation (1.4)-(1.6) with singular pressure, if the following is satisfied: for any $t \in [0, T]$,*

(1) *the initial data (1.7) and (1.8) hold in $\mathcal{D}'(\Omega)$,*

(2) *(1.4)-(1.6) holds in the sense of $\mathcal{D}'((0, T) \times \Omega)$, i.e. $\forall \varphi, \phi, \psi \in \mathcal{D}$,*

$$\int_\Omega \rho(T) \phi dx - \int_\Omega \rho_0 \phi dx - \int_0^T \int_\Omega \sqrt{\rho} \sqrt{\rho} u \cdot \nabla \phi dx dt = 0,$$

$$\int_\Omega \rho(T)u(T)\cdot\varphi dx - \int_\Omega \rho_0 u_0\cdot\varphi dx - \int_0^T\int_\Omega (\sqrt{\rho}u\otimes\sqrt{\rho}u):\nabla\varphi dxdt$$
$$-\int_0^T\int_\Omega \rho^\gamma \text{div}\varphi dxdt - \int_0^T\int_\Omega P_c(\rho)\text{div}\varphi dxdt + \int_0^T\int_\Omega \nu\rho Du:\nabla\varphi dxdt$$
$$=\int_0^T\int_\Omega 2\kappa\Delta\sqrt{\rho}\nabla\sqrt{\rho}\cdot\varphi dxdt + \int_0^T\int_\Omega \kappa\Delta\sqrt{\rho}\sqrt{\rho}\text{div}\varphi dxdt + \int_0^T\int_\Omega \delta u\varphi dxdt,$$

$$\int_\Omega d(T)\psi dx - \int_\Omega d_0\psi dx + \int_0^T\int_\Omega u\cdot\nabla d\cdot\psi dxdt$$
$$=\int_0^T\int_\Omega \alpha_1 d\times(d\times\Delta d)\cdot\psi dxdt - \int_0^T\int_\Omega \alpha_2 d\times(d\times\Delta d)\psi dxdt,$$

(3) *the following property of the weak solution is satisfied:*

$$\rho\in L^\infty([0,T];L^1(\Omega)\cap L^\gamma(\Omega)), \quad \nabla\sqrt{\rho}\in L^\infty([0,T];L^2(\Omega)),$$
$$\sqrt{\rho}u\in L^\infty([0,T];L^2(\Omega)), \quad \nabla\rho^{\frac{\gamma}{2}}\in L^2([0,T];L^2(\Omega)),$$
$$\sqrt{\rho}Du\in L^2([0,T];L^2(\Omega)), \quad \sqrt{\rho}\nabla u\in L^2([0,T];L^2(\Omega)),$$
$$\kappa^{\frac{1}{2}}\nabla^2\sqrt{\rho}\in L^2([0,T];L^2(\Omega)), \quad \kappa^{\frac{1}{4}}\nabla\rho^{\frac{1}{4}}\in L^4([0,T];L^4(\Omega)),$$
$$u\in L^2([0,T];L^2(\Omega)),$$

(4) *the weak solutions satisfy the energy inequality*

$$\int_\Omega \frac{1}{2}\rho|u|^2 + \frac{1}{\gamma-1}\rho^\gamma + H_c(\rho) + \kappa|\nabla\sqrt{\rho}|^2 + \frac{\lambda}{2}|\nabla d|^2$$
$$+\nu\int_0^T\int_\Omega \rho|Du|^2 dxdt + \int_0^T\int_\Omega \alpha_2\lambda|d\times\Delta d|^2 dxdt + \delta\int_0^T\int_\Omega |u|^2 dxdt$$
$$\leqslant C(\rho_0,u_0,d_0)<+\infty, \tag{1.10}$$

and the **B-D** *entropy*

$$\int_\Omega \frac{1}{2}\rho|u+\nu\nabla\log\rho|^2 + \frac{1}{\gamma-1}\rho^\gamma + H_c(\rho) + \kappa|\nabla\sqrt{\rho}|^2 + \frac{\lambda}{2}|\nabla d|^2 dx$$
$$+\frac{4}{\gamma}\nu\int_0^T\int_\Omega |\nabla\rho^{\frac{\gamma}{2}}|^2 dxdt + \int_0^T\int_\Omega H_c''(\rho)|\nabla\rho|^2 dxdt$$
$$+\nu\int_0^T\int_\Omega \frac{1}{4}\rho|\nabla u-\nabla^T u|^2 dxdt + \alpha_2\lambda\int_0^T\int_\Omega |d\times\Delta d|^2 dxdt$$

$$+ \kappa \int_0^T \int_\Omega \rho |\nabla^2 \log \rho|^2 dx dt + \delta \int_0^T \int_\Omega |u|^2 dx dt$$
$$\leqslant C(\rho_0, u_0, d_0) < +\infty, \tag{1.11}$$

almost everywhere for $t \in (0, T)$.

Theorem 1.1 Set $\Omega = T^3 \subset \mathbf{R}^3$, $\gamma > 1$. If the initial data satisfy the following condition:

$$\rho_0 \in L^\gamma(\Omega), \quad \rho_0 \geqslant 0, \quad \nabla\sqrt{\rho_0} \in L^2(\Omega),$$
$$m_0 \in L^1(\Omega), \quad m_0 = 0, \quad if \ \rho_0 = 0, \quad \frac{|m_0|^2}{\rho_0} \in L^1(\Omega), \tag{1.12}$$
$$-\log_- \rho_0 = -\log(\min\{\rho_0, 1\}) \in L^1(\Omega),$$

then there exists a finite energy weak solution (ρ, u, d) to the quantum Navier-Stokes-Landau-Lifshitz equations (1.4)-(1.6) for any $\gamma > 1$, any $T > 0$. The weak solution (ρ, u, d) also satisfies the energy inequality (1.10) and B-D entropy (1.11).

It is easy to see that (1.4)-(1.6) lacks compactness, which we need to get the estimate of L^2 or H^1-norm for u and the compactness for the pressure function. To deal with these difficulties, we will firstly research the following approximate system:

$$\partial_t \rho + \text{div}(\rho u) = \varepsilon \Delta \rho, \tag{1.13}$$

$$\partial_t(\rho u) + \text{div}(\rho u \otimes u) + \nabla P(\rho) + \nabla P_c(\rho)$$
$$= \varepsilon \nabla \rho \cdot \nabla u + \nu \text{div}(\rho D(u)) + \kappa \rho \nabla \left(\frac{\Delta\sqrt{\rho}}{\sqrt{\rho}}\right)$$
$$-\lambda \nabla \cdot \left(\nabla d \odot \nabla d - \frac{|\nabla d|^2}{2} I\right) - \delta u, \tag{1.14}$$

$$d_t + u \cdot \nabla d = \alpha_1 d \times \Delta d - \alpha_2 d \times (d \times \Delta d), \quad |d| = 1. \tag{1.15}$$

Then we want to send the parameter ε to 0. Finally, we obtain the desired weak solution to the original problem (1.4)-(1.6).

The arrangement of this paper is as follows. In Section 2, we give some notations, lemmas and theorems which will be used in the rest sections. In Section 3, we establish the local-in-time existence of weak solutions to the approximate system. Then we prove the global-in-time existence of finite energy weak solutions to the equations (1.1)-(1.5) by using the uniform energy estimate and the B-D entropy estimate in Section 4. The main techniques are the Faedo-Galerkin method and compactness theory.

2 Preliminaries

In this section, we first give some notations, lemmas, and theorems, which will be used in the following section. In this paper, we denote that C is the constant dependent of N and ε. $L^p([0,T];L^q(\Omega))(p,q \geqslant 1)$ is a space whose element is the p-integrable with respect to the time variable and q-integrable with respect to the space variable function. $W^{k,p}$ and H^s are the Sobolev spaces. $(H^s)^*$ is the dual space of H^s. Ω is a periodic domain in $\mathbf{R}^2$.

Lemma 2.1 (Gagliardo-Nirenberg inequality [34]) *Let $\Omega \subset \mathbf{R}^d (d \geqslant 1)$ be a bounded open set with $\partial\Omega \in C^{0,1}$, $m \in \mathbf{N}$, $1 \leqslant p,q,r \leqslant \infty$. Then there exists a constant $C > 0$ such that for all $u \in W^{m,p}(\Omega) \cap L^q(\Omega)$*

$$\|D^\alpha u\|_{L^r(\Omega)} \leqslant C\|u\|_{W^{m,p}(\Omega)}^\theta \|u\|_{L^q(\Omega)}^{1-\theta},$$

where $0 \leqslant |\alpha| \leqslant m-1$, $\theta = |\alpha|/m$, and $|\alpha| - d/r = \theta(m - d/p) - (1-\theta)d/q$. If $m - |\alpha| - d/p \notin \mathbf{N}_0$, then $\theta \in [|\alpha|/m, 1]$ is allowed.

Lemma 2.2 (Aubin-Lions Lemma [35]) *Assume $X \subset Y \subset Z$ are Banach spaces and $X \hookrightarrow\hookrightarrow Y$. Then the following embedding are compact:*

$$L^q([0,T];X) \cap \{\varphi : \partial_t\varphi \in L^1([0,T];Z)\} \hookrightarrow\hookrightarrow L^q([0,T];Y), \quad \forall 1 \leqslant q \leqslant \infty, \quad (2.1)$$

$$L^\infty([0,T];X) \cap \{\varphi : \partial_t\varphi \in L^r([0,T];Z)\} \hookrightarrow\hookrightarrow C([0,T];Y), \quad \forall 1 \leqslant r \leqslant \infty. \quad (2.2)$$

Lemma 2.3 (Egoroffs theorem [33]) *Let $f_n \to f$ a.e. in Ω a bounded measurable set in $\mathbf{R}^n$, with f finite a.e.. Then for any $\varepsilon > 0$ there exists a measurable subset $\Omega_\varepsilon \subset \Omega$ such that $|\Omega\backslash\Omega_\varepsilon| < \varepsilon$ and $f_n \to f$ uniformly in Ω_ε, moreover, if*

$$f_n \to f \quad a.e. \ in \ \Omega,$$

$$f_n \in L^p(\Omega) \ \ and \ uniformly \ bounded, for \ any \ 1 < p \leqslant +\infty,$$

then, we have

$$f_n \to f \quad strongly \ in \ L^s(\Omega), \quad for \ any \ s \in [1,p).$$

Lemma 2.4(Sobolev embedding theorem [1]) *Let U be a bounded open subset of $\mathbf{R}^n$, with a C^1 boundary. Assume $u \in W^{k,p}(U)$.*

(i) *If $kp < n$, then $u \in L^q(U)$, where $\dfrac{1}{q} = \dfrac{1}{p} - \dfrac{k}{n}$. We have in addition to the estimate* $\|u\|_{L^q(U)} \leqslant C\|u\|_{W^{k,p}(U)}$, *and* $W^{k,p}(\Omega) \hookrightarrow\hookrightarrow L^{q^*}(\Omega)(1 \leqslant q^* < q)$.

(ii) *If $kp = n$, then $u \in L^q(U)$, where $\forall 1 \leqslant q < \infty$, and $W^{k,p} \hookrightarrow\hookrightarrow L^q(\Omega)$.*

(iii) *If $kp > n$, then $u \in C^{k-[\frac{n}{p}]-1,\sigma}(\bar{U})$, where*

$$\sigma = \begin{cases} \left[\dfrac{n}{p}\right] + 1 - \dfrac{n}{p}, & \text{if } \dfrac{n}{p} \text{ is not an integer,} \\ \text{any positive number} < 1, & \text{if } \dfrac{n}{p} \text{ is an integer.} \end{cases}$$

We have in addition the estimate

$$\|u\|_{C^{k-[\frac{n}{p}]-1,\sigma}(\bar{U})} \leqslant C\|u\|_{W^{k,p}(U)},$$

and $W^{k,p} \hookrightarrow\hookrightarrow C^{k-[\frac{n}{p}]-1,\sigma}(\bar{U})$. Here the constant C depends only on k, p, n and U.

3 Local-in-time existence

In this section, we will show the local existence of weak solutions to the viscosity system (1.13)-(1.15) by the Faedo-Galerkin method. Let $T^* > 0$, and let $\{\omega_j\}$ be an orthogonal basis of $L^2(\Omega)$, which is also an orthogonal basis of $H^1(\Omega)$. Introduce the finite-dimensional space $X_N = span\{\omega_1, \cdots, \omega_N\}$, $n \in N$. Denote the approximate solutions of the problem (1.13)-(1.15) by u_N, d_N in the following form

$$u_N(x,t) = \sum_{j=1}^{N} \lambda_j(t) \omega_j(x).$$

For some functions $\lambda_i(t)$, the norm of v in $C^0([0, T^*]; X_N)$ can be formulated as

$$\|v\|_{C^0([0,T^*];X_N)} = \max_{t \in [0,T^*]} \sum_{i=1}^{N} |\lambda_i(t)|.$$

As a consequence, v can be bounded in $C^0([0, T^*]; C^k(\Omega))$ for any $k \in N$, and there exists a constant $C > 0$ depending on k such that

$$\|u_N\|_{C^0([0,T^*];C^k(\Omega))} \leqslant C\|u_N\|_{C^0([0,T^*];L^2(\Omega))}. \tag{3.1}$$

3.1 Solvability of the density equation

The approximate system is defined as follows. Let $\rho \in C^1([0,T^*];C^3(\Omega))$ be the classical solution to

$$\rho_t + \text{div}(\rho u) = \nu \Delta \rho, \quad \rho|_{t=0} = \rho_0(x). \tag{3.2}$$

The maximum principle provides the lower and upper bounds ([14], chapter 7.3)

$$\inf_{x \in \Omega} \rho_0(x) \exp\left(-\int_0^t \|\text{div}u\|_{L^\infty(\Omega)} ds\right) \leq \rho(x) \leq \sup_{x \in \Omega} \rho_0(x) \exp\left(\int_0^t \|\text{div}u\|_{L^\infty} ds\right).$$

Since we assumed that $\rho_0(x) > \bar{\rho} > 0$, $\rho(x)$ is strictly positive. In view of (3.1), for $\|u\|_{C^0([0,T^*];L^2(\Omega))} \leq C$, there exists constant $\underline{\rho}(C)$ and $\bar{\rho}(C)$ such that

$$0 < \underline{\rho}(C) \leq \rho(x,t) \leq \bar{\rho}(C).$$

We introduce the operator $S_1 : C^0([0,T^*];X_N) \to C^0([0,T^*];C^3(\Omega))$ by $S_1(u) = \rho$. Since the equation for ρ is linear, S_1 is Lipschitz continuous in the following sense:

$$\|S_1(u_1) - S_1(u_2)\|_{C^0([0,T^*];C^k(\Omega))} \leq C(N,k)\|u_1 - u_2\|_{C^0([0,T^*];L^2(\Omega))}. \tag{3.3}$$

3.2 Solvability of the magnetization field equation

Using the fact that

$$d \times (d \times \Delta d) = (d \cdot \Delta d)d - |d|^2 \Delta d \stackrel{|d|^2=1}{=} |\nabla d|^2 d - \Delta d. \tag{3.4}$$

We set $\varphi_j (j = 1, 2, \cdots)$ as the eigenvector of the Laplace operator, which satisfies the following equation

$$-\Delta \phi_j = \lambda_j \varphi_j, \quad j = 1, 2, \cdots. \tag{3.5}$$

We can set the approximate solution d_M of the system (3.4) as following

$$d_M = \sum_{j=1}^{M} \beta_j \varphi_j.$$

Then we can rewrite the equation (3.4) as following

$$\int_\Omega d_{Mt} \cdot \varphi_j + \int_\Omega u_N \cdot \nabla d_M \cdot \varphi_j dx$$

$$= \int_\Omega \Delta d_M \cdot \varphi_j dx + \int_\Omega |\nabla d_M|^2 d_M \cdot \varphi_j dx + \int_\Omega d_M \times \Delta d_M \cdot \varphi_j dx. \quad (3.6)$$

This is the ordinary differential equation about the coefficients β_j. By the existence of the ordinary differential equation we can easily get the existence of the local existence of the solution to the magnetization field.

3.3 Solvability of the velocity equation

Next we wish to solve (1.9) on the space X_N. To this end, for given $\rho = S_1(u)$, we are looking for functions $u_N^\gamma \in (C^0([0,T^*];X_N))$ such that

$$-\int_\Omega \rho_0 u_0 \cdot \phi(\cdot,0) dx$$
$$= \int_0^T \int_\Omega \Big(\rho u_N \cdot \phi_t + \rho(u_N \otimes u_N) : \nabla\phi + (P(\rho) + P_c(\rho))\text{div}\phi$$
$$- \frac{\hbar^2}{2}\frac{\Delta\sqrt{\rho}}{\sqrt{\rho}}\text{div}(\rho\phi) - \nu\nabla(\rho u_N) : \nabla\phi + \lambda\left(\nabla d_M \odot \nabla d_M - \frac{|\nabla d_M|^2}{2}I\right)\nabla\phi$$
$$+ (\varepsilon\nabla\rho u_N \phi + u_N \cdot \phi)\Big) dx dt, \quad (3.7)$$

for all $\phi \in C^1([0,T^*];X_N)$ such that $\phi(\cdot,T^*) = 0$. The reason is that we will apply Banach fixed point theorem to prove the local-in-time existence of solutions. The regularization yields the H^1 regularity of u_N^γ needed to conclude the global existence of solutions. For simplicity, we set $M \leqslant N$.

To solve (3.7), we follow ([27], Chapter 7.3.3) and introduce the following family of operators, given a function $\varrho \in L^1(\Omega)$ with $\varrho \geqslant \underline{\varrho} > 0$:

$$\mathcal{M}[\varrho] : X_N \to X_N^*, \quad \langle\mathcal{M}[\varrho]u,\omega\rangle = \int_\Omega \varrho u \cdot \omega, \quad u,\omega \in X_N.$$

These operators are symmetric and positive definite with the smallest eigenvalue

$$\int_{\|u\|_{L^2(\Omega)}=1} \langle\mathcal{M}[\varrho]u,u\rangle = \int_{\|u\|_{L^2(\Omega)}=1} \int_\Omega \varrho|u|^2 dx \geqslant \inf_{x\in\Omega} \varrho(x) \geqslant \underline{\varrho}.$$

Hence, since X_N is finite-dimensional, the operators are invertible with

$$\|\mathcal{M}^{-1}[\varrho]\|_{L(X_N^*,X_N)} \leqslant \underline{\varrho}^{-1},$$

where $L(X_N^*, X_N)$ is the set of bounded linear mappings from X_N^* to X_N. Moreover (see [27] Chapter 7.3.3), $\mathcal{M}^{-1}$ is Lipschitz continuous in the sense

$$\|\mathcal{M}^{-1}[\varrho_1] - \mathcal{M}^{-1}[\varrho_2]\|_{L(X_N^*,X_N)} \leqslant C(N,\underline{\varrho})\|\varrho_1 - \varrho_2\|_{L^1(\Omega)}, \quad (3.8)$$

for all $\varrho_1, \varrho_2 \in L^1(\Omega)$ such that $\varrho_1, \varrho_2 \geqslant \underline{\varrho} > 0$.

Now the integral equation (3.7) can be rephrased as an ordinary differential equation on the finite-dimensional space X_N:

$$\frac{d}{dt}(\mathcal{M}[\rho(t)]\omega_N) = \mathcal{N}[\rho, u, d_M, \omega_N], \quad t > 0, \quad \mathcal{M}[\rho_0]\omega_N(0) = \mathcal{M}[\rho_0]\omega_0, \quad (3.9)$$

where $\rho = S_1(u)$, and

$$\langle \mathcal{N}[\rho, u, d_M], \phi \rangle = \int_\Omega \Big(\rho(u \otimes \omega_N) : \nabla \phi + (P(\rho) + P_c(\rho))\mathrm{div}\phi - \frac{\hbar^2}{2}\frac{\Delta\sqrt{\rho}}{\sqrt{\rho}}\mathrm{div}(\rho\phi)$$

$$- \nu \nabla(\rho\omega_N) : \nabla\phi + \lambda(\nabla d_M \odot \nabla d_M - \frac{|\nabla d_M|^2}{2}I)\nabla\phi$$

$$+ (\delta \nabla \rho \omega_N \phi + \omega_N \cdot \phi) \Big) dx,$$

$\forall \phi \in X_N$. For operator $\mathcal{N}[\rho, u, d_M, \cdot]$, defined for every $t \in [0, T]$ as an operator form X_N to X_N^*, is continuous in time. Standard theory for systems of ordinary differential equations then provide the existence of a unique classical solution to (3.9), i.e., for a given $u \in C^0([0,T]; X_N)$, there exists a unique solution $u_N^\gamma \in C^1([0,T^*]; X_N)$ to (3.7).

Integrating (3.9) over $(0, t)$ yields the following nonlinear equation:

$$u_N(t) = \mathcal{M}^{-1} S_1(u_N)(t) \Big(\mathcal{M}[\rho_0] u_0 + \int_0^t \mathcal{N}[S_1(u_N), u_N, d_M, u_N(s)] ds \Big), \quad \text{in } X_N.$$

(3.10)

Since the operators S_1 and $\mathcal{M}$ are Lipschitz type, (3.10) can be solved by evoking the fixed point theorem of Banach on a short time interval $[0, T']$, where $T' \leqslant T^*$, in the space $C^0([0,T]; X_N)$. In fact, we have even $u_N \in C^1([0,T']; X_N)$. Then we can solve the equation (3.7). Thus, there exists a unique local-in-time solution (ρ_N, u_N, d_M) to (1.13)-(1.15).

4 Global existence

4.1 A priori estimates

In this subsection, we will derive the a priori estimates of the local solutions (ρ_N, u_N, d_N) to the approximate system (1.13)-(1.15).

Theorem 4.1(basic energy estimates) *Under the assumption of the Theorem 1.1, we have the following estimates:*

$$\int_\Omega \frac{1}{2}\rho_N|u_N|^2 + \frac{1}{\gamma-1}\rho_N^\gamma + H_c(\rho_N) + \frac{\lambda}{2}|\nabla d_N|^2 + \frac{\kappa}{2}|\nabla\sqrt{\rho_N}|dx$$
$$+ \int_0^T \int_\Omega \nu\rho_N|D(u_N)|^2 dxdt + \lambda\alpha_2 \int_0^T \int_\Omega |d_N \times \Delta d_N|^2 dxdt + \delta \int_0^T \int_\Omega |u_N|^2 dxdt$$
$$+ \int_0^T \int_\Omega \varepsilon\gamma\rho_N^{\gamma-2}|\nabla\rho_N|^2 + \varepsilon H_c''(\rho_N)dxdt + \int_0^T \int_\Omega \varepsilon\kappa\frac{1}{2}\rho_N|\nabla^2 \log\rho_N|^2 dxdt$$
$$\leqslant C. \tag{4.1}$$

Here $H_c''(\rho_N) = \dfrac{P_c'(\rho_N)}{\rho_N}$.

Proof Multiplying (1.14) by u_N and integrating by parts, we have

$$\int_\Omega (\rho_N u_N)_t \cdot u_N + \mathrm{div}(\rho_N u_N \otimes u_N) \cdot u_N dx$$
$$= \int_\Omega \rho_{Nt}|u_N|^2 + \rho_N \partial_t \left(\frac{|u_N|^2}{2}\right) + \mathrm{div}(\rho_N u_N)|u_N|^2 + \rho_N u_N \cdot \nabla\left(\frac{|u_N|^2}{2}\right) dx$$
$$= \int_\Omega \rho_{Nt}|u_N|^2 + \rho_N \partial_t\left(\frac{|u_N|^2}{2}\right) + \frac{1}{2}\mathrm{div}(\rho_N u_N)|u_N|^2 dx$$
$$= \partial_t \int_\Omega \frac{1}{2}\rho_N|u_N|^2 dx + \frac{1}{2}\int_\Omega \varepsilon\Delta\rho_N|u_N|^2 dx,$$

$$\int_\Omega \nabla\rho_N^\gamma \cdot u_N dx = \int_\Omega \frac{\gamma}{\gamma-1}\nabla\rho_N^{\gamma-1} \cdot (\rho_N u_N)dx$$
$$= \int_\Omega \frac{\gamma}{\gamma-1}\rho_N^{\gamma-1}\mathrm{div}(\rho_N u_N)dx = \frac{\gamma}{\gamma-1}\int_\Omega \rho_N^{\gamma-1}(\rho_{Nt} - \varepsilon\Delta\rho_N)dx$$
$$= \frac{1}{\gamma-1}\partial_t\int_\Omega \rho_N^\gamma dx + \varepsilon\int_\Omega \gamma\rho_N^{\gamma-2}|\nabla\rho_N|^2 dx.$$

From the equation (1.7), we know that

$$P_c(\rho_N) = \begin{cases} A_1 \dfrac{1}{-4k}\rho_N^{-4k}, & \rho \leqslant 1,\ k \geqslant 1, \\ \dfrac{1}{\gamma}\rho_N^\gamma, & \rho > 1,\ \gamma > 1. \end{cases}$$

Thus for $\rho_N > 1$, we can get

$$\int_\Omega \frac{1}{\gamma} \nabla \rho_N^\gamma \cdot u_N dx = \frac{1}{\gamma(\gamma-1)} \partial_t \int_\Omega \rho_N^\gamma dx + \varepsilon \int_\Omega \rho_N^{\gamma-2} |\nabla \rho_N|^2 dx$$

$$= \partial_t \int_\Omega H_c(\rho_N) dx + \varepsilon \int_\Omega H_c''(\rho_N) |\nabla \rho_N|^2 dx;$$

for $\rho_N \leqslant 1$, we can get

$$\int_\Omega \frac{1}{-4k} \nabla \rho_N^{-4k} dx = \int_\Omega \rho_N^{-4k-2} \nabla \rho_N \cdot (\rho_N u_N) dx$$

$$= -\int_\Omega \frac{1}{-4k-1} \rho_N^{-4k-1} \text{div}(\rho_N u_N) dx$$

$$= \int_\Omega \frac{1}{-4k-1} \rho_N^{-4k-1} (\rho_{Nt} - \varepsilon \Delta \rho_N) dx$$

$$= \frac{1}{4k(4k+1)} \partial_t \int_\Omega \rho_N^{-4k} dx + \varepsilon \int_\Omega \rho_N^{-4k-2} |\nabla \rho_N|^2 dx$$

$$= \partial_t \int_\Omega H_c(\rho_N) dx + \varepsilon \int_\Omega H_c''(\rho_N) |\nabla \rho_N|^2 dx,$$

$$\int_\Omega \rho_N \nabla \left(\frac{\Delta \sqrt{\rho_N}}{\sqrt{\rho_N}} \right) \cdot u_N dx$$

$$= -\int_\Omega \left(\frac{\Delta \sqrt{\rho_N}}{\sqrt{\rho_N}} \right) \text{div}(\rho_N u_N) dx$$

$$= \int_\Omega \Delta \sqrt{\rho_N} 2 \partial_t (\sqrt{\rho_N}) dx - \varepsilon \int_\Omega \frac{\Delta \sqrt{\rho_N}}{\sqrt{\rho_N}} \Delta \rho_N dx$$

$$= -\partial_t \int_\Omega |\nabla \sqrt{\rho_N}|^2 dx + \varepsilon \int_\Omega \rho_N \nabla \left(\frac{\Delta \sqrt{\rho_N}}{\sqrt{\rho_N}} \right) \cdot \nabla \log \rho_N dx$$

$$= -\partial_t \int_\Omega |\nabla \sqrt{\rho_N}|^2 dx + \frac{1}{2} \varepsilon \int_\Omega \text{div}(\rho_N \nabla^2 \log \rho_N) \cdot \nabla \log \rho_N dx$$

$$= -\partial_t \int_\Omega |\nabla \sqrt{\rho_N}|^2 dx - \frac{1}{2} \varepsilon \int_\Omega \rho_N |\nabla^2 \log \rho_N|^2 dx,$$

$$\int_\Omega \nu \text{div}(\rho_N D(u_N)) \cdot u_N dx = -\nu \int_\Omega \rho |D(u_N)|^2 dx,$$

$$\int_\Omega -\lambda \nabla \cdot \left(\nabla d_N \odot \nabla d_N - \frac{|\nabla d_N|^2}{2} I \right) \cdot u_N dx$$

$$= \lambda \int_\Omega \nabla d_N \odot \nabla d_N : \nabla u_N dx - \int_\Omega \frac{|\nabla d_N|^2}{2} \nabla u_N dx.$$

Multiplying (1.15) by Δd_N, it holds that

$$\int_\Omega d_{Nt} \cdot \Delta d_N dx = -\frac{1}{2}\partial_t \int_\Omega |\nabla d_N|^2 dx,$$

$$\int_\Omega (u_N \cdot \nabla d_N) \cdot \Delta d_N dx = \int_\Omega d_{Njj} u_N^i d_{Ni} dx$$

$$= \int_\Omega \left[(d_{Nj} u_N^i d_{Ni})_j - d_{Ni} \cdot d_{Nj} u_{Nj}^i - u_N^i \left(\frac{|\nabla d_N|^2}{2} \right)_i \right] dx$$

$$= -\int_\Omega d_{Ni} \cdot d_{Nj} u_{Nj}^i + u_{Ni}^i \left(\frac{|\nabla d_N|^2}{2} \right) dx$$

$$= -\int_\Omega \nabla d_N \odot \nabla d_N : \nabla u_N dx + \int_\Omega \frac{|\nabla d_N|^2}{2} \nabla u_N dx,$$

$$\alpha_1 \int_\Omega d_N \times (d_N \times \Delta d_N) \cdot \Delta d_N dx = -\alpha_1 \int_\Omega |d_N \times \Delta d_N|^2 dx,$$

$$\alpha_2 \int_\Omega d_N \times \Delta d_N \cdot \Delta d_N dx = 0.$$

Combining the above equations we have

$$\partial_t \int_\Omega \frac{1}{2}\rho_N |u_N|^2 + \frac{1}{\gamma-1}\rho_N^\gamma + H_c(\rho_N) + \frac{\kappa}{2}|\nabla\sqrt{\rho_N}| + \frac{\lambda}{2}|\nabla d_N|^2 dx$$
$$+ \int_\Omega \varepsilon\gamma\rho^{\gamma-2}|\nabla\rho_N|^2 + \varepsilon\int_\Omega H_c''(\rho_N)|\nabla\rho_N|^2 dx + \int_\Omega \varepsilon\kappa\frac{1}{2}\rho_N|\nabla^2 \log\rho_N|^2 dx$$
$$+ \int_\Omega \alpha_2\lambda |d_N \times \Delta d_N|^2 dx + \delta \int_\Omega |u_N|^2 dx \leqslant 0. \tag{4.2}$$

Then by the Gronwall inequality, we can obtain the equation (4.1). □

From Theorem 4.1 we have the following corollary:

Corollary 4.1 *Under the assumption of Theorem 1.1, we have the following estimates*:

$$\|\sqrt{\rho_N}\|_{L^\infty([0,T];H^1(\Omega))} \leqslant C, \tag{4.3}$$

$$\|\rho_N\|_{L^\infty([0,T];L^1(\Omega)\cap L^\gamma(\Omega))} + \|\sqrt{\varepsilon}\nabla\rho_N^{\frac{\gamma}{2}}\|_{L^2([0,T];L^2(\Omega))} \leqslant C, \tag{4.4}$$

$$\|\rho_N^{-4k}\|_{L^\infty([0,T];L^1(\Omega))} + \|\sqrt{\varepsilon}\nabla\rho_N^{-2k}\|_{L^2([0,T];L^2(\Omega))} \leqslant C, \quad \rho_N \leqslant 1, \tag{4.5}$$

$$\|\sqrt{\rho_N}u_N\|_{L^\infty(0,T);L^2(\Omega)} + \|\sqrt{\rho_N}\nabla u_N\|_{L^2([0,T];L^2(\Omega))} \leqslant C, \tag{4.6}$$

$$\|\nabla d_N\|_{L^\infty([0,T];L^2(\Omega))} \leqslant C, \tag{4.7}$$

$$\|d_N \times \Delta d_N\|_{L^2([0,T];L^2(\Omega))} \leqslant C, \tag{4.8}$$

$$\|\sqrt{\varepsilon}\sqrt{\rho_N}\nabla^2 \log \rho_N\|_{L^2([0,T];L^2(\Omega))} \leqslant C, \tag{4.9}$$

$$\delta\|u_N\|_{L^2([0,T];L^2(\Omega))} \leqslant C. \tag{4.10}$$

Proof Here we only prove the equations (4.5).

Since

$$P_c(\rho_N) = \begin{cases} A_1 \dfrac{1}{-4k}\rho_N^{-4k}, & \rho_N \leqslant 1,\ k \geqslant 1, \\ \dfrac{1}{\gamma}\rho_N^{\gamma}, & \rho_N > 1,\ \gamma > 1, \end{cases}$$

we have

$$H_c(\rho_N) = \begin{cases} A_1 \dfrac{1}{4k(4k+1)}\rho_N^{-4k}, & \rho_N \leqslant 1,\ k \geqslant 1, \\ \dfrac{1}{\gamma(\gamma-1)}\rho_N^{\gamma}, & \rho_N > 1,\ \gamma > 1. \end{cases}$$

From the equation (4.1), we can easily obtain

$$\int_\Omega \rho_N^\alpha dx + \int_0^T \int_\Omega |\nabla \rho_N^{\frac{\alpha}{2}}|^2 dx \leqslant C,$$

for $\rho_N \geqslant 1$, and for $\rho_N \leqslant 1$,

$$\int_\Omega \rho_N^{-4k} dx + \int_0^T \int_\Omega |\nabla \rho_N^{-2k}|^2 dx \leqslant C.$$

Thus we can get (4.5). □

The energy inequality (4.1) and Corollary 4.1 allows us to conclude some estimates.

Lemma 4.1 *For any positive function $\rho_N(x)$, we have*

$$\int_\Omega \rho_N |\nabla^2 \log \rho_N|^2 dx \geqslant \dfrac{1}{7}\int_\Omega |\nabla^2 \sqrt{\rho_N}|^2 dx \tag{4.11}$$

and

$$\int_\Omega \rho_N |\nabla^2 \log \rho_N|^2 dx \geqslant \dfrac{1}{8}\int_\Omega |\nabla \rho_N^{\frac{1}{4}}|^4 dx. \tag{4.12}$$

Proof This lemma was proved in [24, 37]. Here we state it again. We note

$$\sqrt{\rho_N} \cdot \nabla^2 \log \sqrt{\rho_N} = \sqrt{\rho_N} \cdot \nabla \left(\frac{\nabla \sqrt{\rho_N}}{\sqrt{\rho_N}} \right) = \nabla^2 \sqrt{\rho_N} - \frac{\nabla \sqrt{\rho_N} \otimes \nabla \sqrt{\rho_N}}{\sqrt{\rho_N}},$$

thus

$$\int_\Omega \rho_N |\nabla^2 \log \sqrt{\rho_N}|^2 dx = \int_\Omega |\nabla^2 \sqrt{\rho_N}|^2 dx + \int_\Omega |2\nabla \rho_N^{\frac{1}{4}}|^4 dx$$
$$- 2 \int_\Omega \nabla^2 \sqrt{\rho_N} \cdot \frac{\nabla \sqrt{\rho_N} \otimes \nabla \sqrt{\rho_N}}{\sqrt{\rho_N}} dx$$
$$= A + B - I.$$

For I, we control it as follows

$$I = 2 \int_\Omega \nabla^2 \sqrt{\rho_N} \cdot \frac{\nabla \sqrt{\rho_N} \otimes \nabla \sqrt{\rho_N}}{\sqrt{\rho_N}} dx$$
$$= -2 \int_\Omega \frac{|\nabla \sqrt{\rho_N}|^2}{\sqrt{\rho_N}} \Delta \log \sqrt{\rho_N} dx - 2 \int_\Omega \nabla^2 \sqrt{\rho} \cdot \frac{\nabla \sqrt{\rho_N} \otimes \nabla \sqrt{\rho_N}}{\sqrt{\rho_N}} dx.$$

Hence,

$$2I = -2 \int_\Omega \frac{|\nabla \sqrt{\rho_N}|^2}{\sqrt{\rho_N}} \Delta \log \sqrt{\rho_N} dx \leqslant 2\sqrt{3BD},$$

where $D = \int_\Omega \rho_N |\nabla^2 \log \rho|^2 dx$, and hence

$$A + B = D + I \leqslant (1 + 6)D + \frac{1}{8}B,$$

and hence,

$$\frac{1}{7}A + \frac{1}{8}B \leqslant D.$$

Thus we proved this lemma. □

By (4.9), we have

$$(\kappa\varepsilon)^{\frac{1}{2}} \|\sqrt{\rho_N}\|_{L^2([0,T];H^2(\Omega))} + (\kappa\varepsilon)^{\frac{1}{4}} \|\nabla \rho_N^{\frac{1}{4}}\|_{L^4([0,T];L^4(\Omega))} \leqslant C, \qquad (4.13)$$

where the constant $C > 0$ is independent of N.

We are able to deduce more regularity from the H^2 bound for $\sqrt{\rho_N}$.

Lemma 4.2 (space regularity for ρ_N and $\rho_N u_N$) *The following uniform estimates hold for some constant $C > 0$ independent on N and δ:*

$$\|\rho_N u_N\|_{L^2([0,T];W^{1,q}(\Omega))} \leqslant C, \tag{4.14}$$

$$\|\rho_N\|_{L^2([0,T];W^{2,q}(\Omega))} \leqslant C, \tag{4.15}$$

$$\|\rho_N\|_{L^{4\gamma/3+1}([0,T];L^{4\gamma/3+1}(\Omega))} \leqslant C, \tag{4.16}$$

where $1 < q < 2$.

Proof Since the space $H^2(\Omega)$ embeds continuously into $L^\infty(\Omega)$, showing that

$$\sqrt{\rho_N} \text{ is bounded in } L^2([0,T]; L^\infty(\Omega)).$$

Thus, in view of (4.3) and (4.6) $\rho_N u_N = \sqrt{\rho_N} \sqrt{\rho_N} u_N$ is uniformly bounded in $L^2([0,T]; L^2(\Omega))$. By (4.3) and (4.13), $\nabla \sqrt{\rho_N}$ is bounded in $L^2([0,T]; L^p(\Omega))(2 < p < \infty)$ and $\sqrt{\rho_N}$ is bounded in $L^\infty([0,T]; L^p(\Omega))(2 < p < \infty)$. This, together with (4.3), implies that

$$\nabla(\rho_N u_N) = 2\nabla\sqrt{\rho_N} \otimes (\sqrt{\rho_N} u_N) + \sqrt{\rho_N} \nabla u_N \sqrt{\rho_N}$$

is uniformly bounded in $L^2([0,T]; L^q(\Omega)) \left(q = \dfrac{2p}{p+2}\right)$, proving the first claim.

Since

$$\nabla^2 \rho_N = 2(\sqrt{\rho_N} \nabla^2 \sqrt{\rho_N} + \nabla\sqrt{\rho_N} \otimes \nabla\sqrt{\rho_N})$$

and $H^1(\Omega) \hookrightarrow\hookrightarrow L^{p'}(1 \leqslant p' < \infty)$, we have $\nabla^2 \rho_N \in L^2([0,T]; L^q(\Omega))$.

Then, the Gagliardo-Nirenberg inequality, with $\theta = 3/(4\gamma + 3)$ and $q_1 = 2(4\gamma + 3)/3$,

$$\|\sqrt{\rho_N}\|_{L^{q_1}([0,T];L^{q_1}(\Omega))}^{q_1} \leqslant C \int_0^T \|\sqrt{\rho_N}\|_{H^2(\Omega)}^{q_1\theta} \|\sqrt{\rho_N}\|_{L^{2\gamma}(\Omega)}^{q_1(1-\theta)} dt$$

$$\leqslant C \|\rho_N\|_{L^\infty([0,T];L^\gamma(\Omega))}^{q_1(1-\theta)} \int_0^T \|\sqrt{\rho_N}\|_{H^2(\Omega)}^2 dt \leqslant C,$$

shows that ρ_N is bounded in $L^{q_1/2}([0,T]; L^{q_1/2}(\Omega))$.

This finishes the proof. □

Lemma 4.3 *Under the assumption of Theorem 1.1, there yields that*

$$\|\rho_N^\gamma\|_{L^{\frac{5}{3}}([0,T];L^{\frac{5}{3}}(\Omega))} \leqslant C, \tag{4.17}$$

$$\|P_c(\rho_N)\|_{L^{\frac{5}{3}}([0,T];L^{\frac{5}{3}}(\Omega))} \leqslant C. \tag{4.18}$$

Proof Noticing that $\nabla \rho_N^{\frac{\gamma}{2}}$ is bounded in $L^2([0,T]; L^2(\Omega))$, using the Sobolev embedding theorem gives us ρ_N^γ is bounded in $L^1([0,T]; L^3(\Omega))$, then we apply the Hölder inequality to have

$$\|\rho_N^\gamma\|_{L^{\frac{5}{3}}([0,T]; L^{\frac{5}{3}}(\Omega))} \leqslant \|\rho_N^\gamma\|_{L^\infty([0,T]; L^1(\Omega))}^{\frac{2}{5}} \|\rho_N^\gamma\|_{L^1([0,T]; L^3(\Omega))}^{\frac{3}{5}} \leqslant C.$$

Similarly, we can get (4.18). □

From (4.8) we can get the estimate about $\|\Delta d_N\|_{L^2([0,T]; H^2(\Omega))}$.

Lemma 4.4
$$\|\Delta d_N\|_{L^2([0,T]; L^2(\Omega))} \leqslant C. \tag{4.19}$$

Proof Since

$$|d_N \times \Delta d_N|^2 = (d_N \cdot d_N)(\Delta d_N \cdot \Delta d_N) - (d_N \cdot \Delta d_N)^2,$$
$$= |\Delta d_N|^2 - |(|\nabla d_N|^2 d_N|^2,$$

we have

$$\int_0^T \int_\Omega |\Delta d_N|^2 dxdt$$
$$\leqslant \int_0^T \int_\Omega |d_N \times \Delta d_N|^2 + 2(|\nabla d_N|^2 d_N)^2 dxdt$$
$$\leqslant C_1 \varepsilon \int_0^T \int_\Omega |\Delta d_N|^2 dxdt + C_2 \int_0^T \|\nabla d_N\|_{L^4(\Omega)}^4 dt + C_3$$
$$\leqslant C_1 \varepsilon \int_0^T \int_\Omega |\Delta d_N|^2 dxdt + C_4 \int_0^T \int_\Omega |\nabla d_N|^4 dxdt + C.$$

From [25], we know that if $d_N : \mathbf{R}^2 \to S^1$, $\nabla d_N \in H^1(\mathbf{R}^2)$, $\|\nabla d_N\|_{L^2} \leqslant C$, $d_{N2} \geqslant \varepsilon_0$, then $\|\nabla d_N\|_{L^4}^4 \leqslant (1-\delta_0) \|\Delta d_N\|_{L^2}^2$.

Therefore, if $1 - C_1\varepsilon - C_4(1-\delta_0) > 0$, we can get (4.19). □

Lemma 4.5 (time regularity for ρ_N and ρu_N) *The following uniform estimates hold for $s > 2$:*

$$\|\partial_t \rho_N\|_{L^2([0,T]; L^{3/2}(\Omega))} \leqslant C, \tag{4.20}$$

$$\|\partial_t (\rho_N u_N)\|_{L^{4/3}([0,T]; (H^s(\Omega))^*)} \leqslant C. \tag{4.21}$$

Proof By (4.15)-(4.16), we find that $\partial_t \rho_N = -\text{div}(\rho_N u_N) - \nu \Delta \rho_N$ is uniformly bounded in $L^2([0,T]; L^{3/2}(\Omega))$, achieving the first claim.

The sequence $\rho_N u_N \otimes u_N$ is bounded in $L^\infty([0,T]; L^1(\Omega))$; hence, $\text{div}(\rho_N u_N \otimes u_N)$ is bounded in $L^\infty([0,T]; (W^{1,\infty}(\Omega))^*)$, and because of the continuous embedding of $H^s(\Omega)$ into $W^{1,\infty}(\Omega)$ for $s > 2$, and also in $L^\infty([0,T]; (H^s(\Omega))^*)$. The estimate

$$\int_0^T \int_\Omega \rho_N \nabla\left(\frac{\Delta\sqrt{\rho_N}}{\sqrt{\rho_N}}\right) \cdot \phi \, dx dt$$

$$= -\int_0^T \int_\Omega \Delta\sqrt{\rho_N}(2\nabla\sqrt{\rho_N}\cdot\phi + \sqrt{\rho_N}\text{div}\phi) dx dt$$

$$\leqslant \|\Delta\sqrt{\rho_N}\|_{L^2([0,T];L^2(\Omega))}(2\|\sqrt{\rho_N}\|_{L^\infty([0,T];L^2(\Omega))}\|\phi\|_{L^4([0,T];L^6(\Omega))})$$

$$+ \|\sqrt{\rho_N}\|_{L^\infty([0,T];L^4(\Omega))}\|\phi\|_{L^2([0,T];W^{1,4}(\Omega))}$$

$$\leqslant C\|\phi\|_{L^2([0,T];W^{1,4}(\Omega))},$$

for all $\phi \in L^2([0,T]; W^{1,4}(\Omega))$ proves that $\rho\Delta\sqrt{\rho_N}/\sqrt{\rho_N}$ is uniformly bounded in $L^2([0,T]; (W^{1,4}(\Omega))^*) \hookrightarrow L^2([0,T]; (H^s(\Omega))^*)$. In view of (4.17), ρ_N^γ is bounded in $L^{4/3}([0,T]; L^{4/3}(\Omega)) \hookrightarrow L^{4/3}([0,T]; (H^s(\Omega))^*)$. Furthermore, by (4.15), $\Delta(\rho_N u_N)$ is uniformly bounded in $L^2([0,T]; (W^{1,3}(\Omega))^*)$, and by (4.10), δu_N is bounded in $L^2([0,T]; (H^1(\Omega))^*)$. Therefore, using Corollary 4.1 and Lemma 4.3, we get that

$$(\rho_N u_N)_t = -\text{div}(\rho_N u_N \otimes u_N) - \nabla P(\rho_N) - \nabla P_c(\rho) + \nu\text{div}(\rho_N D(u_N))$$

$$-\lambda\nabla\cdot\left(\nabla d_N \odot \nabla d_N - \frac{|\nabla d_N|^2}{2}I\right) + \kappa\rho_N\nabla\left(\frac{\Delta\sqrt{\rho_N}}{\sqrt{\rho_N}}\right)$$

$$+ \varepsilon\nabla\rho_N \cdot \nabla u - \delta u_N$$

is uniformly bounded in $L^2([0,T]; (H^s(\Omega))^*)$. □

The $L^4([0,T]; W^{1,4}(\Omega))$ bound (4.13) on $\sqrt[4]{\rho_N}$ provides a uniform estimate for $\partial_t \sqrt{\rho_N}$.

Lemma 4.6 (time regularity for $\sqrt{\rho_N}$) *The following estimate holds:*

$$\|\partial_t \sqrt{\rho_N}\|_{L^2([0,T];(H^1(\Omega))^*)} \leqslant C. \tag{4.22}$$

Proof Dividing the mass equation (1.8) by $2\sqrt{\rho}$ gives

$$\partial_t\sqrt{\rho_N} = -\nabla\sqrt{\rho_N}\cdot u_N - \frac{1}{2}\sqrt{\rho_N}\text{div}u_N + \nu(\Delta\sqrt{\rho_N} + 4|\nabla\sqrt[4]{\rho_N}|^2)$$

$$= -\text{div}(\sqrt{\rho_N}u_N) + \frac{1}{2}\sqrt{\rho_N}\text{div}u_N + \nu(\Delta\sqrt{\rho_N} + 4|\nabla\sqrt[4]{\rho_N}|^2).$$

The first term on the right-hand side is bounded in $L^2([0,T];(H^1(\Omega))^*)$ by (4.11) and (4.12). The remaining terms are uniformly bounded in $L^2([0,T];L^2(\Omega))$; see (4.11), (4.12), (4.17). □

Lemma 4.7 *There holds that*
$$\|\partial_t d_N\|_{L^2([0,T];(H^2(\Omega))^*)} \leqslant C. \tag{4.23}$$

Proof Multiplying (1.10) by a test function φ, we have

$$\left|\int_0^T \int_\Omega \partial_t d_N \varphi dx dt\right|$$
$$= \left|-\int_\Omega u_N \cdot \nabla d_N \cdot \varphi + \alpha_1 d_N \times \Delta d_N \cdot \varphi - \alpha_2 d_N \times (d_N \times \Delta d_{\Delta_N}) \cdot \varphi dx\right|$$
$$\leqslant \|u_N\|_{L^2([0,T];L^2(\Omega))} \|\nabla d_N\|_{L^\infty([0,T];L^2(\Omega))} \|\varphi\|_{L^2([0,T];L^\infty(\Omega))}$$
$$+ \alpha_1 \|d_N\|_{L^\infty([0,T];L^4(\Omega))} \|\Delta d_N\|_{L^2([0,T];L^2(\Omega))} \|\varphi\|_{L^2([0,T];L^4(\Omega))}$$
$$+ \alpha_2 \|d_N\|_{L^\infty([0,T];L^4(\Omega))}^2 \|\Delta d_N\|_{L^2([0,T];L^2(\Omega))} \|\varphi\|_{L^2([0,T];L^\infty(\Omega))}.$$

Then we have
$$\|\partial_t d_N\|_{L^2((0,T);(H^2(\Omega))^*)} \leqslant C.$$

Thus we have completed the proof of this lemma. □

Lemma 4.8 *Under the assumption of Theorem 1.1, there holds that*
$$\left\|\nabla\left(\frac{1}{\sqrt{\rho_N}}\right)\right\|_{L^2([0,T];L^2(\Omega))} \leqslant C. \tag{4.24}$$

Proof Since $\nabla\sqrt{\rho_N} \in L^\infty([0,T];L^2(\Omega))$, and
$$\nabla\left(\frac{1}{\sqrt{\rho_N}}\right) = -\frac{1}{2}\frac{\nabla \rho_N}{\rho_N^{\frac{3}{2}}} = -\frac{\nabla\sqrt{\rho_N}}{\rho_N},$$

we have, if $\rho_N > 1$, $\nabla(1/\sqrt{\rho_N}) \in L^\infty([0,T];L^2(\Omega))$.

Now we take into account of the case $\rho_N \leqslant 1$. From $\int_0^T \int_\Omega H_c''(\rho_N)|\nabla \rho_N|^2 dx dt \leqslant C$, we get

$$\int_0^T \int_\Omega \frac{A_1}{4k^2}|\nabla \rho_N^{-2k}|^2 dx dt \leqslant C.$$

Make the connection between $\nabla(\rho_N^{-2k})$ and $\nabla \rho_N^{-1/2}$:

$$\nabla\left(\frac{1}{\sqrt{\rho_N}}\right) = \nabla\left(\frac{1}{\rho_N^{2k}}\rho_N^{2k-1/2}\right)$$

$$= \rho_N^{2k-1/2}\nabla\left(\frac{1}{\rho_N^{2k}}\right) + \frac{1}{\rho_N^{2k}}\nabla \rho_N^{2k-1/2}$$

$$= \rho_N^{2k-1/2}\nabla\left(\frac{1}{\rho_N^{2k}}\right) + (2k-1/2)\rho_N^{-2k}\nabla\rho_N\rho_N^{2k-3/2}$$

$$= \rho_N^{2k-1/2}\nabla\left(\frac{1}{\rho_N^{2k}}\right) + (2k-1/2)\nabla\rho_N\rho_N^{-3/2}$$

$$= \rho_N^{2k-1/2}\nabla\left(\frac{1}{\rho_N^{2k}}\right) - 2(2k-1/2)\nabla\left(\frac{1}{\sqrt{\rho_N}}\right),$$

we have

$$(1+4k-1)\nabla\left(\frac{1}{\sqrt{\rho_N}}\right) = \rho_N^{2k-1/2}\nabla\left(\frac{1}{\rho^{2k}}\right),$$

and

$$\left|\nabla\left(\frac{1}{\sqrt{\rho_N}}\right)\right|^2 = \frac{1}{16k^2}\rho^{4k-1}|\nabla\left(\frac{1}{\rho_N^{2k}}\right)|^2$$

$$= \frac{1}{16k^2}\rho_N^{4k-1}4k^2\frac{1}{A_1}H_c''(\rho_N)|\nabla\rho_N|^2$$

$$= \frac{1}{4A_1}\rho^{4k-1}H_c''(\rho_N)|\nabla\rho_N|^2.$$

Such as $\rho_N \leqslant 1$ and $\int_0^T \int_\Omega H_c''(\rho_N)|\nabla\rho_N|^2 dxdt < +\infty$, we have $\nabla\left(\frac{1}{\sqrt{\rho_N}}\right) \in L^2([0,T]; L^2(\Omega))$. □

4.2 The limit of $N \to 0$

In this subsection, we will show the limits of the solutions (ρ_N, u_N, d_N) as the global weak solutions to the approximate system (1.13)-(1.15) as N tends to infinite with other parameters fixed.

Convergence of the density and momentum.

Firstly, we should notice the following embedding formula:

$$W^{2,p}(\Omega) \hookrightarrow\hookrightarrow L^\infty(\Omega)(p > 3/2), \quad H^2(\Omega) \hookrightarrow\hookrightarrow H^1(\Omega), \quad W^{1,3/2}(\Omega) \hookrightarrow\hookrightarrow L^2(\Omega),$$
(4.25)

where $\hookrightarrow\hookrightarrow$ represents the compact embedding.

Combining the regularity (4.15) and (4.20), using the Aubin-Lions Lemma, we have the following convergence for density ρ_N. There exists a subsequence of ρ_N (we still denote itself) such that

$$\rho_N \to \rho \text{ strongly in } L^2([0,T]; L^\infty(\Omega)).$$

Similarly using the regularity (4.13) and (4.22) for $\sqrt{\rho_N}$ and the regularity (4.14) and (4.21) for $\rho_N u_N$, it yields that

$$\sqrt{\rho_n} \to \sqrt{\rho} \text{ weakly in } L^2([0,T]; H^2(\Omega)),$$

$$\sqrt{\rho_n} \to \sqrt{\rho} \text{ strongly in } L^2([0,T]; H^1(\Omega)),$$

$$\rho_N u_N \to J \text{ strongly in } L^2([0,T]; L^2(\Omega)).$$

The estimate (4.10) on u_N provides further the existence of a subsequence (not relabeled) such that, as $N \to \infty$,

$$u_N \rightharpoonup u \text{ weakly in } L^2([0,T]; L^2(\Omega)).$$

Then, since $\rho_n u_N$ converges weakly to ρu in $L^1([0,T]; L^6(\Omega))$, we infer that $J = \rho u$.

We are now in the position to let $N \to \infty$ in the approximate system (3.1), (3.6), (3.8) with $\rho = \rho_N$, $u = u_N$, $d = d_N$. Clearly, the limit $N \to \infty$ shows immediately that n solves

$$\rho_t + \text{div}(\rho u) = \nu \Delta \rho.$$

Next we consider the weak formulation (3.7) term by term. The strong convergence of $\rho_N u_N$ in $L^2([0,T]; L^2(\Omega))$ and the weak convergence of u_N in $L^2([0,T]; L^2(\Omega))$ leads to

$$\rho_N u_N \otimes u_N \rightharpoonup \rho u \otimes u \text{ weakly in } L^1([0,T]; L^1(\Omega)).$$

Furthermore, in view of (4.21) (up to a subsequence),

$$\nabla(\rho_n u_N) \rightharpoonup \nabla(\rho u) \text{ weakly in } L^2([0,T]; L^{3/2}(\Omega)).$$

From the $\sqrt{\rho_N} \to \rho$ strongly in $L^2([0,T]; H^1(\Omega))$, we know $\rho_N \to \rho$ a.e., then we have $\rho_N^\gamma \to \rho^\gamma$ a.e.. Thus combing the Egoroffs theorem and $\rho_N^\gamma \in L^{\frac{5}{3}}([0,T]; L^{\frac{5}{3}}(\Omega))$, we can get

$$\rho_N^\gamma \to \rho^\gamma, \text{ strongly in } L^1([0,T]; L^1(\Omega)).$$

Similarly, we also can get the convergence of ρ_N^{-4k} for $\rho_N \leqslant 1$:

$$\rho_N^{-4k} \to \rho^{-4k}, \quad \text{strongly in } L^1([0,T]; L^1(\Omega)).$$

Finally, the above convergence results show that the limit $N \to \infty$ of

$$\int_\Omega \frac{\Delta\sqrt{\rho_N}}{\sqrt{\rho_N}} \operatorname{div}(\rho_N \phi) dx = \int_\Omega \Delta\sqrt{\rho_N}(2\nabla\sqrt{\rho_N} \cdot \phi + \sqrt{\rho_N}\operatorname{div}\phi) dx$$

equals, for sufficiently smooth test functions,

$$\int_\Omega \Delta\sqrt{\rho}(2\nabla\sqrt{\rho} \cdot \phi + \sqrt{\rho}\operatorname{div}\phi) dx.$$

Next we devote to proof the convergence of the nonlinear diffusion terms. Let $\varphi \in C_{per}^\infty([0,T];\Omega)$, with $C_{per}^\infty([0,T];\Omega)$ defined by

$$C_{per}^\infty([0,T];\Omega) = \{\phi \in C^\infty([0,T];\Omega) | \phi \text{ is periodic in } x\}.$$

$$\int_0^T \int_\Omega \operatorname{div}(\rho_N Du_N)\varphi dxdt$$

$$= -\frac{1}{2}\int_0^T \int_\Omega \partial_i \rho_N \nabla u_N \nabla\varphi dxdt - \frac{1}{2}\int_0^T \int_\Omega \rho_N \nabla^T u_N \nabla\varphi dxdt$$

$$= -\frac{1}{2}\int_0^T \int_\Omega \partial_i(\rho_N u_{Nj})\partial_i\varphi_j dxdt + \frac{1}{2}\int_0^T \int_\Omega \partial_i\rho_N u_{Nj}\partial_i\varphi_j dxdt$$

$$- \frac{1}{2}\int_0^T \int_\Omega \partial_j(\rho_N u_{Ni})\partial_i\varphi_j dxdt + \frac{1}{2}\int_0^T \int_\Omega \partial_j\rho_N u_{Ni}\partial_i\varphi_j dxdt$$

$$= \frac{1}{2}\int_\Omega (\rho_N u_{Nj})\partial_{ii}\varphi_j dxdt + \frac{1}{2}\int_0^T \int_\Omega \partial_i\rho_N u_{Nj}\partial_i\varphi_j dxdt$$

$$+ \frac{1}{2}\int_\Omega (\rho_N u_{Ni})\partial_{ji}\varphi_j dxdt + \frac{1}{2}\int_0^T \int_\Omega \partial_j\rho_N u_{Ni}\partial_i\varphi_j dxdt,$$

since $\rho_N \to \rho$ strongly in $L^2([0,T]; L^\infty(\Omega))$, $\rho_N u_N \to \rho u$ strongly in $L^2([0,T]; L^2(\Omega))$, $u_N \rightharpoonup u$ weakly in $L^2([0,T]; L^2(\Omega))$, so we have

$$\int_0^T \int_\Omega (\rho_N u_{Nj})\partial_{ii}\varphi_j dxdt \to \int_0^T \int_\Omega (\rho u_j)\partial_{ii}\varphi_j dxdt,$$

$$\int_0^T \int_\Omega \partial_i\rho_N u_{Nj}\partial_i\varphi_j dxdt \to \int_0^T \int_\Omega \partial_i\rho u_j \partial_i\varphi_j dxdt,$$

$$\int_0^T \int_\Omega (\rho_N u_{Ni})\partial_{ji}\varphi_j dxdt \to \int_0^T \int_\Omega (\rho u_i)\partial_{ji}\varphi_j dxdt,$$

$$\int_0^T \int_\Omega \partial_j \rho_N u_{Ni} \partial_i \varphi_j dxdt \to \int_0^T \int_\Omega \partial_j \rho u_i \partial_i \varphi_j dxdt.$$

Using Corollary 4.1, Lemma 4.3, Lemma 4.6 and Lemma 4.7, we can get the convergence of d_N and B_N:

$$d_N \to d, \quad \text{strongly in } L^2([0,T]; H^1(\Omega)) \cap C([0,T]; L^2(\Omega)),$$

$$d_N \rightharpoonup d, \quad \text{weakly in } L^2([0,T]; H^2(\Omega)),$$

$$d_N \rightharpoonup d, \quad \text{weakly } * \text{ in } L^\infty([0,T]; H^1(\Omega)).$$

Therefore we prove the limit of (ρ_N, u_N, d_N) is the weak solution to the equations (1.8)-(1.12). Moreover, the limit function (ρ, u, d) satisfies the following equality, for all test functions $\forall \phi(x, T) = 0 \in \mathcal{D}(\Omega)$:

$$-\int_\Omega \rho_0(x) u_0(x) \cdot \varphi(x, 0) dx$$
$$= \int_0^T \int_\Omega \Big(\rho u \varphi_t + \rho(u \otimes u_N) : \nabla \varphi + P(\rho) \mathrm{div}\phi + P_c(\rho) \mathrm{div}\varphi$$
$$- \frac{\hbar^2}{2} \frac{\Delta\sqrt{\rho}}{\sqrt{\rho}} \mathrm{div}(\rho\varphi) - \nu \nabla(\rho u) : \nabla\varphi + \lambda \Big(\nabla d \odot \nabla d - \frac{|\nabla d|^2}{2} I \Big) \nabla\varphi - \delta u \cdot \varphi \Big) dxdt,$$
(4.26)

$$-\int_\Omega d_0(x) \psi(x, 0) dx - \int_0^T \int_\Omega d\psi_t dxdt + \int_0^T \int_\Omega u \cdot \nabla d \psi dxdt$$
$$= \alpha_1 \int_0^T \int_\Omega d \times \Delta d \psi dxdt - \alpha_2 \int_0^T \int_\Omega d \times (d \times \Delta d) \psi dxdt.$$

Furthermore the weak solutions also satisfy the energy inequality:

$$\int_\Omega \frac{1}{2} \rho |u|^2 + \frac{1}{\gamma - 1} \rho^\gamma + H_c(\rho) + \frac{\lambda}{2} |\nabla d|^2 + \frac{\kappa}{2} |\nabla \sqrt{\rho}| dx$$
$$+ \int_0^T \int_\Omega \nu \rho |D(u)|^2 dxdt + \lambda \alpha_2 \int_0^T \int_\Omega |d \times \Delta d|^2 dxdt + \delta \int_0^T \int_\Omega |u|^2 dxdt$$
$$+ \int_0^T \int_\Omega \varepsilon \gamma \rho^{\gamma-2} |\nabla \rho|^2 + \varepsilon H_c''(\rho) dxdt + \int_0^T \int_\Omega \varepsilon \kappa \frac{1}{2} \rho |\nabla^2 \log \rho|^2 dxdt \leqslant C. \quad (4.27)$$

4.3 The limit of $\varepsilon \to 0$

From the a priori estimates in Theorem 4.1 we know that we can not directly send ε tends to 0. Therefore, we need to derive new estimates (the B-D entropy) which is uniformly in ε.

Set $(\rho_\varepsilon, u_\varepsilon, d_\varepsilon)$ are the weak solutions to the approximate system (1.13)-(1.15) which is constructed in the above subsection. Then we can derive the following estimates.

Theorem 4.2 (B-D entropy) *Under the assumption of Theorem 1.1, there holds that*

$$\frac{d}{dt}\int_\Omega \frac{1}{2}\rho_\varepsilon |u_\varepsilon + \nu\nabla \log \rho_\varepsilon|^2 + \frac{1}{\gamma-1}\rho_\varepsilon^\gamma + H_c(\rho_\varepsilon) + \frac{\lambda}{2}|\nabla d_\varepsilon|^2 dx$$

$$+ \int_\Omega \frac{4}{\gamma}(\nu+\varepsilon)|\nabla \rho_\varepsilon^{\frac{\gamma}{2}}|^2 dx + \int_\Omega (\nu+\varepsilon) H_c''(\rho_\varepsilon)|\nabla \rho_\varepsilon|^2 dx$$

$$+ \nu\int_\Omega \frac{1}{4}\rho_\varepsilon|\nabla u_\varepsilon - \nabla^T u_\varepsilon|^2 dx + \frac{1}{2}\kappa(\nu+\varepsilon)\int_\Omega \rho_\varepsilon|\nabla^2 \log \rho_\varepsilon|^2 dx + \delta^2\int_\Omega |u_\varepsilon|^2 dx$$

$$+ \lambda\alpha_2 \int_\Omega |d_\varepsilon \times \Delta d_\varepsilon|^2 dx$$

$$= -\frac{1}{2}\int_\Omega \varepsilon\Delta\rho_\varepsilon|u_\varepsilon + \nu\nabla \log \rho_\varepsilon|^2 dx - \delta\nu\int_\Omega u_\varepsilon\dot\nabla \log \rho_\varepsilon dx + \nu\int_\Omega \Delta d_\varepsilon \nabla d_\varepsilon \nabla \log \rho_\varepsilon dx$$

$$= R_1 + R_2 + R_3. \tag{4.28}$$

Proof Firstly using the mass equation (1.13) we can derive a equation about $\rho_\varepsilon \nabla \log \rho_\varepsilon = \nabla\rho_\varepsilon$:

$$\partial_t(\rho_\varepsilon\nabla \log \rho_\varepsilon) + \mathrm{div}(\rho_\varepsilon u_\varepsilon \otimes \nabla \log \rho_\varepsilon) + \mathrm{div}(\rho_\varepsilon\nabla^T u_\varepsilon) = \varepsilon\Delta(\rho_\varepsilon\nabla \log \rho_\varepsilon). \tag{4.29}$$

Then we obtain

$$\partial_t(\rho_\varepsilon(u_\varepsilon+\nu\nabla \log \rho_\varepsilon)) + \mathrm{div}(\rho_\varepsilon u_\varepsilon \otimes (u_\varepsilon + \nu\nabla \log \rho_\varepsilon)) + \nu\mathrm{div}(\rho_\varepsilon\nabla^T u_\varepsilon) + \nabla\rho_\varepsilon^\gamma$$

$$+ \nabla P_c(\rho_\varepsilon) - \nu\mathrm{div}(\rho_\varepsilon Du_\varepsilon) + \varepsilon\nabla\rho_\varepsilon \cdot \nabla u_\varepsilon = -\lambda\nabla\cdot\left(\nabla d_\varepsilon \odot \nabla d_\varepsilon - \frac{|\nabla d_\varepsilon|^2}{2}\right) - \delta u_\varepsilon$$

$$+ \kappa\rho_\varepsilon\nabla\left(\frac{\Delta\sqrt{\rho_\varepsilon}}{\sqrt{\rho_\varepsilon}}\right) + \nu\varepsilon\Delta(\rho_\varepsilon \log \rho_\varepsilon).$$

Next we will using the following facts:

$$\int_\Omega \partial_t(\rho_\varepsilon(u_\varepsilon+\nu\nabla \log \rho_\varepsilon))\cdot(u_\varepsilon+\nu\nabla \log \rho_\varepsilon)dx$$

$$= \int_\Omega \partial_t\rho_\varepsilon|u_\varepsilon+\nu\nabla \log \rho_\varepsilon|^2 dx + \int_\Omega \rho_\varepsilon\partial_t\left(\frac{1}{2}|u_\varepsilon+\nu\nabla \log \rho_\varepsilon|^2\right)dx$$

$$= \int_\Omega \frac{1}{2}\partial_t\rho_\varepsilon|u_\varepsilon+\nu\nabla \log \rho_\varepsilon|^2 dx + \partial_t\int_\Omega \frac{1}{2}\rho_\varepsilon|u_\varepsilon+\nu\nabla \log \rho_\varepsilon|^2 dx$$

$$= \int_\Omega \frac{1}{2}(\varepsilon\Delta\rho_\varepsilon - \mathrm{div}(\rho_\varepsilon u_\varepsilon))|u_\varepsilon + \nu\nabla\log\rho_\varepsilon|^2 dx + \partial_t \int_\Omega \frac{1}{2}\rho_\varepsilon |u_\varepsilon + \nu\nabla\log\rho_\varepsilon|^2 dx,$$

$$\int_\Omega \mathrm{div}(\rho_\varepsilon u_\varepsilon \otimes (u_\varepsilon + \nu\nabla\log\rho_\varepsilon)) \cdot (u_\varepsilon + \nu\nabla\log\rho_\varepsilon) dx$$

$$= \int_\Omega \mathrm{div}(\rho_\varepsilon u_\varepsilon)|u_\varepsilon + \nu\nabla\log\rho_\varepsilon|^2 + \rho_\varepsilon u_\varepsilon \cdot \nabla \frac{|u_\varepsilon + \nu\nabla\log\rho_\varepsilon|^2}{2} dx$$

$$= \int_\Omega \frac{1}{2}\mathrm{div}(\rho_\varepsilon u_\varepsilon)|u_\varepsilon + \nu\nabla\log\rho_\varepsilon|^2 dx,$$

$$\int_\Omega \nabla\rho_\varepsilon^\gamma (u_\varepsilon + \nu\nabla\log\rho_\varepsilon) dx = \int_\Omega \gamma\rho_\varepsilon^{\gamma-2}\nabla\rho_\varepsilon(\rho_\varepsilon u_\varepsilon) + \nu\gamma\rho_\varepsilon^{\gamma-2}|\nabla\rho_\varepsilon|^2 dx$$

$$= \int_\Omega -\frac{\gamma}{\gamma-1}\rho_\varepsilon^{\gamma-1}\mathrm{div}(\rho_\varepsilon u_\varepsilon) dx + \nu\gamma \int_\Omega \rho_\varepsilon^{\gamma-2}|\nabla\rho_\varepsilon|^2 dx$$

$$= \int_\Omega \frac{\gamma}{\gamma-1}\rho_\varepsilon^{\gamma-1}(\rho_{\varepsilon t} - \varepsilon\Delta\rho_\varepsilon) dx + \frac{4\nu}{\gamma}\int_\Omega |\nabla\rho_\varepsilon^{\frac{\gamma}{2}}|^2 dx$$

$$= \partial_t \int_\Omega \frac{1}{\gamma-1}\rho_\varepsilon^\gamma dx + \frac{4}{\gamma}(\nu+\varepsilon)\int_\Omega |\nabla\rho_\varepsilon^{\frac{\gamma}{2}}|^2 dx,$$

$$\int_\Omega \frac{1}{-4k}\nabla\rho_\varepsilon^{-4k}(u_\varepsilon + \nu\nabla\log\rho_\varepsilon) dx$$

$$= \int_\Omega \frac{1}{-4k-1}\nabla\rho_\varepsilon^{-4k-1}(\rho_\varepsilon u_\varepsilon) dx + \nu \int_\Omega \rho_\varepsilon^{-4k-2}|\nabla\rho_\varepsilon|^2 dx$$

$$= \int_\Omega \frac{1}{4k+1}\rho_\varepsilon^{-4k-1}\mathrm{div}(\rho_\varepsilon u_\varepsilon) dx + \nu \int_\Omega H_c''(\rho_\varepsilon)|\nabla\rho_\varepsilon|^2 dx$$

$$= \frac{1}{4k+1}\int_\Omega \rho_\varepsilon^{-4k-1}(-\rho_{\varepsilon t} + \varepsilon\Delta\rho_\varepsilon) dx + \nu \int_\Omega H_c''(\rho_\varepsilon)|\nabla\rho_\varepsilon|^2 dx$$

$$= \partial_t \int_\Omega H_c(\rho_\varepsilon) dx + (\nu+\varepsilon)\int_\Omega H_c''(\rho_\varepsilon)|\nabla\rho_\varepsilon|^2 dx,$$

$$\int_\Omega [\mathrm{div}(\rho_\varepsilon \nabla^T u_\varepsilon) - \mathrm{div}(\rho_\varepsilon D(u_\varepsilon))](u_\varepsilon + \nu\nabla\log\rho_\varepsilon) dx$$

$$= -\int_\Omega \frac{1}{2}\mathrm{div}(\rho_\varepsilon(\nabla u_\varepsilon - \nabla^T u_\varepsilon))(u_\varepsilon + \nu\nabla\log\rho_\varepsilon) dx$$

$$= \frac{1}{4}\int_\Omega \rho_\varepsilon |\nabla u_\varepsilon - \nabla^T u_\varepsilon|^2 dx + \nu\frac{1}{2}(\nabla u_\varepsilon - \nabla^T u_\varepsilon)\rho_\varepsilon\nabla^2\log\rho_\varepsilon dx$$

$$= \frac{1}{4}\int_\Omega \rho_\varepsilon |\nabla u_\varepsilon - \nabla^T u_\varepsilon|^2 dx,$$

$$-\int_\Omega \delta u_\varepsilon \cdot (u_\varepsilon + \nu \nabla \log \rho_\varepsilon) dx$$

$$= -\int_\Omega \delta |u_\varepsilon|^2 dx - \delta\nu \int_\Omega u_\varepsilon \cdot \nabla \log \rho_\varepsilon dx,$$

$$\int_\Omega \kappa \rho_\varepsilon \nabla \left(\frac{\Delta \sqrt{\rho_\varepsilon}}{\sqrt{\rho_\varepsilon}}\right) \cdot (u_\varepsilon + \nu \nabla \log \rho_\varepsilon) dx$$

$$= -\kappa \int_\Omega \operatorname{div}(\rho_\varepsilon u_\varepsilon) \frac{\Delta \sqrt{\rho_\varepsilon}}{\sqrt{\rho_\varepsilon}} dx + \kappa\nu \int_\Omega \rho_\varepsilon \nabla\left(\frac{\Delta \sqrt{\rho_\varepsilon}}{\sqrt{\rho_\varepsilon}}\right) \cdot \nabla \log \rho_\varepsilon dx$$

$$= \kappa \int_\Omega (\rho_{\varepsilon t} - \Delta \rho_\varepsilon) \frac{\Delta \sqrt{\rho_\varepsilon}}{\sqrt{\rho_\varepsilon}} dx + \kappa\nu \frac{1}{2} \int_\Omega \operatorname{div}(\rho_\varepsilon \nabla^2 \log \rho_\varepsilon) \cdot \nabla \log \rho_\varepsilon dx$$

$$= -\kappa \frac{d}{dt} \int_\Omega |\nabla \sqrt{\rho_\varepsilon}|^2 dx - \kappa \frac{1}{2}(\nu + \varepsilon) \int_\Omega \rho_\varepsilon |\nabla^2 \log \rho_\varepsilon|^2 dx,$$

$$\int_\Omega \nabla \cdot \left(\nabla d_\varepsilon \odot \nabla d_\varepsilon - \frac{|\nabla d_\varepsilon|^2}{2} I\right)(u_\varepsilon + \nu \nabla \log \rho_\varepsilon) dx$$

$$= \int_\Omega \Delta d_\varepsilon \nabla d_\varepsilon \cdot u_\varepsilon + \nu \Delta d_\varepsilon \nabla d_\varepsilon \nabla \log \rho_\varepsilon dx.$$

Similar to the subsection 4.1, multiplying (1.15) by Δd_ε, then summing the above results we can get the equation (4.27). □

Then integrating (4.27) respect to t, we have

$$\int_\Omega \frac{1}{2} \rho_\varepsilon |u_\varepsilon + \nu \nabla \log \rho_\varepsilon|^2 + \frac{1}{\gamma - 1} \rho_\varepsilon^\gamma + H_c(\rho_\varepsilon) + \frac{\lambda}{2} |\nabla d_\varepsilon|^2 dx$$

$$+ \int_0^T \int_\Omega \frac{4}{\gamma}(\nu + \varepsilon) |\nabla \rho_\varepsilon^{\frac{\gamma}{2}}|^2 dxdt + \int_0^T \int_\Omega (\nu + \varepsilon) H_c''(\rho_\varepsilon) |\nabla \rho_\varepsilon|^2 dxdt$$

$$+ \frac{\nu}{4} \int_0^T \int_\Omega \rho_\varepsilon |\nabla u_\varepsilon - \nabla^T u_\varepsilon|^2 dxdt + \int_0^T \int_\Omega \kappa(\nu + \varepsilon) \rho_\varepsilon |\nabla^2 \log \rho_\varepsilon|^2 dxdt$$

$$+ \delta^2 \int_0^T \int_\Omega |u_\varepsilon|^2 dxdt + \lambda \alpha_2 \int_0^T \int_\Omega |d_\varepsilon \times \Delta d_\varepsilon|^2 dxdt$$

$$= \int_0^T R_1 dt + \int_0^T R_2 dt + \int_0^T R_3 dt. \qquad (4.30)$$

Next we estimate $\int_0^T R_1 dt, \int_0^T R_2 dt, \int_0^T R_3 dt$ one by one.

Since $\sqrt{\rho_\varepsilon} \in L^\infty([0,T]; L^2(\Omega)) \cap L^2([0,T]; L^{p_1}(\Omega))$, using the interpolation

theorem we can obtain
$$\|\sqrt{\rho_\varepsilon}\|_{L^{\frac{8k}{2k-1}}([0,T];L^{\frac{24k}{6k-1}}(\Omega))} \leqslant \frac{\kappa(\nu+\varepsilon)}{64}\int_0^T\int_\Omega \|\nabla^2\sqrt{\rho_\varepsilon}\|^2 dxdt + C.$$

Therefore we can estimate $\int_0^T R_3 dt$ as following.

$$\left|\int_0^T R_1 dt\right| = \left|-\frac{1}{2}\int_0^T\int_\Omega \varepsilon\Delta\rho_\varepsilon |u_\varepsilon + \nu\nabla\log\rho_\varepsilon|^2 dxdt\right|$$

$$\leqslant \left|\frac{1}{4}\int_0^T\int_\Omega \rho_\varepsilon |u_\varepsilon + \nu\nabla\log\rho_\varepsilon|^2 dxdt\right| + C_1\left|\int_0^T\int_\Omega \left(\frac{\Delta\rho_\varepsilon}{\sqrt{\rho_\varepsilon}}\right)^2 dxdt\right|$$

$$\leqslant \left|\frac{1}{4}\int_0^T\int_\Omega \rho_\varepsilon |u_\varepsilon + \nu\nabla\log\rho_\varepsilon|^2 dxdt\right|$$

$$+ C_1\left|\int_0^T\int_\Omega (\Delta\sqrt{\rho_\varepsilon} + 8(\nabla\rho_\varepsilon^{\frac{1}{4}})^2)^2 dxdt\right|$$

$$\leqslant \frac{1}{4}\int_0^T\int_\Omega \rho_\varepsilon |u_\varepsilon + \nu\nabla\log\rho_\varepsilon|^2 dxdt + \frac{\kappa(\nu+\varepsilon)}{16}\int_0^T\int_\Omega \|\nabla^2\sqrt{\rho_\varepsilon}\|^2 dxdt$$

$$+ \frac{\kappa(\nu+\varepsilon)}{16}\int_0^T\int_\Omega \|\nabla\rho_\varepsilon^{\frac{1}{4}}\|^4 dxdt + C, \tag{4.31}$$

$$\left|\int_0^T R_3 dt\right| = \left|\nu\int_0^T\int_\Omega \Delta d_\varepsilon \nabla d_\varepsilon \nabla\log\rho_\varepsilon dxdt\right|$$

$$\leqslant \|\Delta d_\varepsilon\|_{L^2([0,T];L^2(\Omega))} \|\nabla d_\varepsilon\|_{L^4([0,T];L^4(\Omega))} \|\nabla\sqrt{\rho_\varepsilon}\|_{L^{\frac{8k}{2k-1}}([0,T];L^{\frac{24k}{6k-1}}(\Omega))}$$

$$\cdot \left\|\frac{1}{\sqrt{\rho_\varepsilon}}\right\|_{L^{8k}([0,T];L^{24k}(\Omega))}$$

$$\leqslant \frac{\kappa(\nu+\varepsilon)}{16}\int_0^T\int_\Omega \|\nabla^2\sqrt{\rho_\varepsilon}\|^2 dxdt + C, \tag{4.32}$$

$$\int_0^T R_2 dt = -\int_0^T \nu u_\varepsilon \cdot \frac{\nabla\rho_\varepsilon}{\rho_\varepsilon} dxdt$$

$$= \int_0^T\int_\Omega \rho_\varepsilon^{-1}(\rho_{\varepsilon t} + \rho_\varepsilon \text{div} u_\varepsilon - \varepsilon\Delta\rho_\varepsilon) dxdt$$

$$= \int_0^T\int_\Omega \partial_t(\log\rho_\varepsilon) dxdt - \varepsilon\int_0^T\int_\Omega \frac{\Delta\rho_\varepsilon}{\rho_\varepsilon} dxdt. \tag{4.33}$$

Since ρ_ε is bounded in $L^\infty([0,T];L^\gamma(\Omega))$, we have

$$\int_\Omega \log_+ \rho_\varepsilon dx \leqslant C, \quad \text{where } \log_+ g = \log\max(g,1).$$

On the other hand, it holds that

$$\left|\varepsilon \int_0^T \int_\Omega \frac{\Delta\rho_\varepsilon}{\rho_\varepsilon} dxdt\right| = \left|\varepsilon \int_0^T \int_\Omega \frac{\Delta\rho_\varepsilon}{\sqrt{\rho_\varepsilon}} \frac{1}{\sqrt{\rho_\varepsilon}} dxdt\right|$$

$$\leqslant \left\|\sqrt{\varepsilon}\frac{\Delta\rho_\varepsilon}{\sqrt{\rho_\varepsilon}}\right\|_{L^2([0,T];L^2(\Omega))} \left\|\sqrt{\varepsilon}\frac{1}{\sqrt{\rho_\varepsilon}}\right\|_{L^2([0,T];L^2(\Omega))}$$

$$\leqslant \|\sqrt{\varepsilon}\Delta\sqrt{\rho_\varepsilon}\|_{L^2([0,T];L^2(\Omega))} \left\|\sqrt{\varepsilon}\frac{1}{\sqrt{\rho_\varepsilon}}\right\|_{L^2([0,T];L^2(\Omega))}$$

$$+ \left\|8\sqrt{\varepsilon}(\nabla\rho_\varepsilon^{\frac{1}{4}})^2\right\|_{L^2([0,T];L^2(\Omega))} \left\|\sqrt{\varepsilon}\frac{1}{\sqrt{\rho_\varepsilon}}\right\|_{L^2([0,T];L^2(\Omega))}$$

$$\leqslant \frac{(\nu+\varepsilon)\kappa}{16}\|\nabla^2\sqrt{\rho_\varepsilon}\|_{L^2([0,T];L^2(\Omega))}$$

$$+ \frac{(\nu+\varepsilon)\kappa}{16}\|\nabla\rho_\varepsilon^{\frac{1}{4}}\|_{L^2([0,T];L^2(\Omega))}$$

$$+ \frac{\varepsilon}{16}\int_0^T\int_\Omega (\nu+\varepsilon)H_c''(\rho_\varepsilon)|\nabla\rho_\varepsilon|^2 dxdt + C.$$

Thus we can get the following inequality:

$$\int_\Omega \frac{1}{2}\rho_\varepsilon|u_\varepsilon + \nu\nabla\log\rho_\varepsilon|^2$$

$$+ \frac{1}{\gamma-1}\rho_\varepsilon^\gamma + H_c(\rho_\varepsilon) + \frac{\lambda}{2}|\nabla d_\varepsilon|^2 dx + \int_0^T\int_\Omega \frac{4}{\gamma}(\nu+\varepsilon)|\nabla\rho_\varepsilon^{\frac{\gamma}{2}}|^2 dxdt$$

$$+ \int_0^T\int_\Omega \frac{(\nu+\varepsilon)}{8}H_c''(\rho_\varepsilon)|\nabla\rho_\varepsilon|^2 dxdt + \frac{\nu}{4}\int_0^T\int_\Omega \rho_\varepsilon|\nabla u_\varepsilon - \nabla^T u_\varepsilon|^2 dxdt$$

$$+ \int_0^T\int_\Omega \frac{\kappa(\nu+\varepsilon)}{16}|\nabla^2\rho_\varepsilon|^2 dxdt + \int_0^T\int_\Omega \frac{\kappa(\nu+\varepsilon)}{16}|\nabla\rho_\varepsilon^{\frac{1}{4}}|^4 dxdt$$

$$+ \delta^2\int_0^T\int_\Omega |u_\varepsilon|^2 dxdt + \lambda\alpha_2 \int_0^T\int_\Omega |d_\varepsilon \times \Delta d_\varepsilon|^2 dxdt \leqslant C. \quad (4.34)$$

Combining the energy estimates and B-D entropy estimates, we can get the following corollary uniformly about ε.

Corollary 4.2 *There holds that*

$$\|\sqrt{\rho_\varepsilon}\|_{L^\infty([0,T];H^1(\Omega))} \leqslant C, \tag{4.35}$$

$$\|\rho_\varepsilon\|_{L^\infty([0,T];L^1(\Omega)\cap L^\gamma(\Omega))} + \|\nabla\rho_\varepsilon^{\frac{\gamma}{2}}\|_{L^2([0,T];L^2(\Omega))} \leqslant C, \tag{4.36}$$

$$\|\rho_\varepsilon^{-4k}\|_{L^\infty([0,T];L^1(\Omega))} + \|\nabla\rho_\varepsilon^{-2k}\|_{L^2([0,T];L^2(\Omega))} \leqslant C, \quad \rho_\varepsilon \leqslant 1, \tag{4.37}$$

$$\|\sqrt{\rho_\varepsilon}u_\varepsilon\|_{L^\infty(0,T);L^2(\Omega))} + \|\sqrt{\rho_\varepsilon}\nabla u_\varepsilon\|_{L^2([0,T];L^2(\Omega))} \leqslant C, \tag{4.38}$$

$$\|\nabla d_\varepsilon\|_{L^\infty([0,T];L^2(\Omega))} + \|d_\varepsilon \times \Delta d_\varepsilon\|_{L^2([0,T];L^2(\Omega))} \leqslant C, \tag{4.39}$$

$$\delta\|u_\varepsilon\|_{L^2([0,T];L^2(\Omega))} \leqslant C, \tag{4.40}$$

$$\|\sqrt{\kappa}\nabla^2\sqrt{\rho_\varepsilon}\|_{L^2([0,T];L^2(\Omega))} + \|\kappa^{\frac{1}{4}}\nabla\rho_\varepsilon^{\frac{1}{4}}\|_{L^4([0,T];L^4(\Omega))} \leqslant C, \tag{4.41}$$

$$\|\rho_\varepsilon u_\varepsilon\|_{L^2([0,T];W^{1,q}(\Omega))} + \|\partial_t(\rho_\varepsilon u_\varepsilon)\|_{L^{\frac{4}{3}}([0,T];(H^s(\Omega))^*)} \leqslant C, \tag{4.42}$$

$$\|\rho_\varepsilon\|_{L^2([0,T];W^{2,q}(\Omega))} + \|\rho_\varepsilon\|_{L^{4\gamma/3+1}([0,T];L^{4\gamma/3+1}(\Omega))} \leqslant C, \tag{4.43}$$

$$\|\rho_\varepsilon^\gamma\|_{L^{\frac{5}{3}}([0,T];L^{\frac{5}{3}}(\Omega))} + \|P_c(\rho_\varepsilon)\|_{L^{\frac{5}{3}}([0,T];L^{\frac{5}{3}}(\Omega))} \leqslant C, \tag{4.44}$$

$$\|\Delta d_\varepsilon\|_{L^2([0,T];L^2(\Omega))} + \|\partial_t d_\varepsilon\|_{L^2([0,T];(H^2(\Omega))^*)} \leqslant C, \tag{4.45}$$

$$\|\partial_t\rho_\varepsilon\|_{L^2([0,T];L^{3/2}(\Omega))} + \|\partial_t\sqrt{\rho_\varepsilon}\|_{L^2([0,T];(H^1(\Omega))^*)} + \left\|\nabla\left(\frac{1}{\sqrt{\rho_\varepsilon}}\right)\right\|_{L^2([0,T];L^2(\Omega))} \leqslant C. \tag{4.46}$$

Lemma 4.9 *Under the assumption of Theorem 4.1, we have*

$$\|\nabla u_\varepsilon\|_{L^{p_2}([0,T];L^{q_2}(\Omega))} \leqslant C, \tag{4.47}$$

where $p_2 = \dfrac{8k}{4k+1}$ *and* $q_2 = \dfrac{24k}{12k+1}$.

Proof Note $\nabla u_\varepsilon = \dfrac{1}{\sqrt{\rho_\varepsilon}}\sqrt{\rho_\varepsilon}\nabla u$. Since $\sqrt{\rho_\varepsilon}\nabla u_\varepsilon \in L^2([0,T];L^2(\Omega))$, it remains to obtain an estimate for $1/\sqrt{\rho_\varepsilon}$ in the appropriate space. Using the estimate of the singular pressure, we have

$$\|\rho_\varepsilon^{-2k}\|^2_{L^2([0,T];L^2(\Omega))} = \int_0^T \left(\int_\Omega \left(\frac{1}{\sqrt{\rho_\varepsilon}}\right)^{24k} dx\right)^{\frac{8k}{24k}} dt,$$

we obtain

$$\left\|\frac{1}{\sqrt{\rho_\varepsilon}}\right\|_{L^{8k}([0,T];L^{24k}(\Omega))} \leqslant C,$$

with C independent of ε.

As previously said, since $\nabla u_\varepsilon = \dfrac{1}{\sqrt{\rho_\varepsilon}}\sqrt{\rho_\varepsilon}\nabla u.$, using the Hölder inequality, we have the result. □

Using the Sobolev embedding, a direct consequence of Lemma 4.9 is the following one.

Lemma 4.10 *Under the assumption of Theorem 4.1, there exists a constant C independent of ε, such that*

$$\|u_\varepsilon\|_{L^{p_2}([0,T];L^{q_3}(\Omega))} \leqslant C, \tag{4.48}$$

with $p_2 = \dfrac{8k}{4k+1}$ and $q_3 = \dfrac{3q_2}{3-q_2} = \dfrac{24k}{4k+1}$.

Using the above estimate we can get the convergence

$\rho_\varepsilon \to \rho$ weakly in $L^2([0,T];L^\infty(\Omega))$,

$\sqrt{\rho_\varepsilon} \to \sqrt{\rho}$ weakly in $L^2([0,T];H^2(\Omega))$,

$\sqrt{\rho_\varepsilon} \to \sqrt{\rho}$ strongly in $L^2([0,T];H^1(\Omega))$,

$\rho_\varepsilon u_\varepsilon \to \rho u$ strongly in $L^2([0,T];L^2(\Omega))$,

$u_\varepsilon \rightharpoonup u$ weakly in $L^2([0,T];L^2(\Omega)) \cap L^{p_2}([0,T];L^{q_3}(\Omega)) \cap L^{p_2}([0,T];W^{1,q_2}(\Omega))$,

$\rho_\varepsilon^\gamma \to \rho^\gamma$ strongly in $L^1([0,T];L^1(\Omega))$,

$\rho_\varepsilon^{-4k} \to \rho^{-k}$ strongly in $L^1([0,T];L^1(\Omega))$,

$d_\varepsilon \to d$ strongly in $L^2([0,T];H^1(\Omega)) \cap C([0,T];L^2(\Omega))$,

$d_\varepsilon \rightharpoonup d$ weakly in $L^2([0,T];H^2(\Omega))$,

$d_\varepsilon \rightharpoonup d$ weak $*$ in $L^\infty([0,T];H^1(\Omega))$.

We only need to prove the convergence of the term about $\varepsilon\Delta\rho_\varepsilon$ and $\varepsilon\nabla\rho_\varepsilon \cdot \nabla u_\varepsilon$.

In fact, for any test function $\phi \in \mathcal{D}'$, we have

$$\int_0^T \int_\Omega \varepsilon\rho_\varepsilon \cdot \varphi dx dt = \int_0^T \int_\Omega \varepsilon\rho_\varepsilon \Delta\varphi dx dt \to 0, \quad \text{as } \varepsilon \to 0,$$

and

$$\left| \int_0^T \int_\Omega \varepsilon \nabla \rho_\varepsilon \cdot \nabla u_\varepsilon \cdot \varphi dxdt \right|$$

$$= \left| \int_0^T \int_\Omega 2\varepsilon \nabla \sqrt{\rho_\varepsilon} \cdot \sqrt{\rho_\varepsilon} \nabla u_\varepsilon \cdot \varphi dxdt \right|$$

$$\leqslant \varepsilon \|\nabla \sqrt{\rho_\varepsilon}\|_{L^2([0,T];L^4(\Omega))} \cdot \|\sqrt{\rho_\varepsilon} \nabla u_\varepsilon\|_{L^2([0,T];L^2(\Omega))} \|\varphi\|_{L^\infty([0,T];L^4(\Omega))}$$

$$\to 0, \ as \ \varepsilon \to 0.$$

Thus we can get that when $\varepsilon \to 0$, the approximate system

$$\partial_t \rho + \mathrm{div}(\rho u) = 0, \qquad (4.49)$$

$$\partial_t(\rho u) + \mathrm{div}(\rho u \otimes u) + \nabla P(\rho) + \nabla P_c(\rho) = \nu \mathrm{div}(\rho D(u)) + \kappa \rho \nabla \left(\frac{\Delta \sqrt{\rho}}{\sqrt{\rho}} \right)$$

$$-\lambda \nabla \cdot \left(\nabla d \odot \nabla d - \frac{|\nabla d|^2}{2} I \right) - \delta u, \qquad (4.50)$$

$$d_t + u \cdot \nabla d = \alpha_1 d \times \Delta d - \alpha_2 d \times (d \times \Delta d), \quad |d| = 1, \qquad (4.51)$$

exists weak solutions.

References

[1] Adams R-A, Fournier J-F, Sobolev spaces. Second edition[M], Academic Press, 2003.

[2] Akhiezer A-I, Yakhtar V-G, Peletminskii S-V, Spin Waves[M], North-Holland, 1968.

[3] Alouges F, Soyeur A, On global weak solutions for Landau-Lifshitz equations: Existence and nonuniqueness[J], Nonlinear Anal. TMA, 1992, 18: 1071-1084.

[4] Antonelli P, Spirito S, Global existence of finite energy weak solutions of quantum Navier-Stokes equations[J], Arch. Rational Mech. Anal., 2017, 225: 1161-1199.

[5] Antonelli P, Spirito S, On the compactness of finite energy weak solutions to the quantum Navier-Stokes equations[J], Journal of Hyperbolic Differential Equations, 2018, 15(01): 133-147.

[6] Bresch D, Desjardins B, Existence of global weak solutions for 2D viscous shallow water equations and convergence to the quasi-geostrophic model[J], Comm. Math. Phys., 2003, 238(1-3): 211-233.

[7] Bresch D, Desjardins B, On the construction of approximate solutions for the 2D viscous shallow water model and for compressible Navier-Stokes models[J], J. Math. Pure. Appl., 2006, 86(4):362-368.

[8] Bresch D, Desjardins B, Lin C-K, On some compressible fluid models: Korteweg, lubrication, and shallow water systems[J], Comm. Partial Differential Equations., 2003, 28(3-4):843-868.

[9] Bresh D, Desjardins B, Zatorska E, Two-velocity hydrodynamics in fluid mechanics: Part II Existence of global κ-entropy solutions to the compressible Navier-Stokes system with degenerate viscosities[J], J. Math. Pure Appl., 2015, 104: 801-836.

[10] Chen Y-M, Ding S-J, Guo B-L, Partial regularity for two-dimensional Landau-Lifshitz equations[J], Acta Math. Sinica, Eng. Ser., 1998, 14: 423-432.

[11] Ding S-J, Wang C-Y, Finite time singularity of Landau-Lifshitz-Gilbert equations[J], Int. Math. Res. Notices, 2007(4): 2007.

[12] Dong J-W, A note on barotropic compressible quantum Navier-Stokes equations[J], Nonlinear Analysis: TMA, 2010, 73: 854-856.

[13] Fan J-S, Gao H-J, Guo B-L, Regularity critera for the Navier-Stokes-Landau-Lifshitz system[J], J. Math. Anal. Appl., 2010, 363:29-37.

[14] Feireisl E, Dynamics of viscous compressible fluid[M], Oxford lecture series in Mathematics and its applications, vol. 26. Oxford Science Publications, The Clarendon Press, Oxford University Press, New York, 2004.

[15] Gisclon M, Lacroix-Violet I, About the barotropic compressible quantum Navier-Stokes equations[J], Nonlinear Analysis: TMA, 2015, 128:106-121.

[16] Guo B-L, Ding S-J, Landau-Lifshitz equations[M], World Scientific, 2008.

[17] Guo B-L, Hong M-C, The Landau-Lifshitz equation of the ferromagnetic spin chain and harmonic maps[J], Cal. Var. Partial Diff. Eqns., 1993, 1(3):311-334.

[18] Guo B-L, Wang G-W, Global finite energy weak solution to the viscous quantum Navier-Stokes-Landau-Lifshitz-Maxwell equation in 2-dimension[J], Annl. Appl. Math., 2016, 32(2):111-132.

[19] Haspot B, Existence of global strong solution for the compressible Navier-Stokes equations with degenerate viscosity coefficients in 1D[J], arXiv:1411.5503, 2014.

[20] Hoff D, Global existence for 1D, compressible, isentropic Navier-Stokes equations with large initial data[J], Trans. Amer. Math. Soc., 1987, 303(1):169-169.

[21] Hoff D, Global well-posedness of the cauchy problem for the Navier-Stokes equations of nonisentropic flow with discontinuous initial data[J], J. Diff. Eqns.,1992, 95:33-74.

[22] Hoff D, Global solutions of the equations of one-dimensional, compressible flow with large data and forces, and with differing end states[J], Z. Ange. Math. Phys., 1998, 49(5):774-785.

[23] Jiang F, A remark on weak solutions to the barotropic compressible quantum Navier-Stokes equations[J], Nonlinear Anlaysis: RWA, 2011, 12: 1733-1735.

[24] Jüngel A, Global weak solution to compressible Navier-Stokes equations for quantum fluids[J], SIAM J. Math. Anal., 2010, 42(3): 1025-1045.

[25] Lei Z, Li D, Zhang X-Y, Remarks of global wellposedness of liquid crystal flows and heat flow of harmonic maps in two dimensions[J], Proceedings of American Mathematical

Society, 2012, 142(11): 3801-3810.

[26] Li J, Xin Z-P, Global existence of weak solutions to the barotropic compressible Navier-Stokes flows with degenerate viscosities[J], arXiv:1504.06826v2, 2015.

[27] Lions P-L, Mathematical topics in fluid mechanics[M], Vol. 2. Compressible models. Oxford Lecture Series in Mathematics and its Applications, vol.10. Oxford science publications, the Clarendon Press, Oxford University Press, New York, 1998.

[28] Liu X-G, Partial regularity for the Landau-Lifshitz system[J], Cal. Var. Partial Diff. Eqns., 2004, 20(2): 153-173.

[29] Mellet A, Vasseur A, Existence and uniqueness of global strong solutions for one-dimensional compressible Navier-Stokes equations[J], SIAM J. Math. Anal., 2007, 39(4): 1344-1365.

[30] Mellet A, Vasseur A, On the barotropic compressible Navier-Stokes equations[J], Comm. Partial Differential Equations, 2007, 32(1-3): 431-452.

[31] Moser R, Partial regularity for the Landau-Lifshitz equation in small dimensions[J], MPI (Leipzig) preprint, 2002.

[32] Nakamura K, Sasada T, Soliton and wave trains in ferromagnets[J], Phys. Lett. A, 1974, 48: 321-322.

[33] Novotný A, Straškraba I, Introduction to the mathematical theory of compressible flow[M], Oxford Lecture Series in Mathematics and its Applications, 27. Oxford University Press, Oxford, 2004.

[34] Nirenberg L, On elliptic partial differential equations[J]. Ann. Scuola Norm. Sup. Pisa, 1959, 13(3):115-162.

[35] Simon J, Compact sets in the space $L^p([0,T];B)$[J], Ann. Mat. Pura Appl., 1987, 146(4): 65-96.

[36] Vasseur A-F, Yu C, Existence of global weak solutions for 3D degenerate compressible Navier-Stokes equations[J], Invent. Math., 2016, 206(3): 935-974.

[37] Vasseur A-F, Yu C, Global weak solutions to the compressible quantum Navier-Stokes equations with damping[J], SIAM J. Math. Anal., 2016, 48(2):1489-1511.

[38] Wang C-Y, On Landau-Lifshitz equation in dimensions at most four[J], Indiana University Mathematics Journal, 2006, 55(5): 1615-1644.

[39] Wang G-W, Guo B-L, Existence and uniqueness of the weak solution to the incompressible Navier-Stokes-Landau-Lifshitz model in 2-dimension[J], Acta Mathematica Scientia, 2017, 37(5): 1361-1372.

[40] Zatorska E, On the flow of chemically reacting gaseous mixture[J], J. Diff. Eqns., 2012, 253(12): 3471-3500.

[41] Zhou Y-L, Guo B-L, Tan S-B, Existence and uniqueness of smooth solution for system of ferromagnetic chain[J], Sci. China Ser. A, 1991, 34(3): 257-266.

[42] Zhou Y-L, Sun H-S, Guo B-L, Existence of weak solution for boundary problems of systems of ferromagnetic chain[J], Sci. Sin. A, 1981, 27: 779-811.

[43] Zhou Y-L, Sun H-S, Guo B-L, On the solvability of the initial value problem for the

quasilinear degenerate parabolic system: $\vec{Z}_t = \vec{Z} \times \vec{Z}_{xx} + \vec{f}(x, t, \vec{Z})$[J], Proc. Symp., 1982, 3: 713-732.

[44] Zhou Y-L, Sun H-S, Guo B-L, Finite difference solutions of the boundary problems for systems of ferromagnetic chain[J], J. Comp. Math., 1983, 1: 294-302.

[45] Zhou Y-L, Sun H-S, Guo B-L, Existence of weak solution for boundary problems of ferromagnetic chain[J], Sci. Sin. A, 1984, 27: 799-811.

[46] Zhou Y-L, Sun H-S, Guo B-L, The weak solution of homogeneous boundary value problem for the system of ferromagnetic chain with several variables[J], Sci. Sin. A, 1986, 4: 337-349.

[47] Zhou Y-L, Sun H-S, Guo B-L, Some boundary problems of the spin system and the system of ferro magnetic chain I: Nonlinear boundary problems[J], Acta Math. Sci., 1986, 6: 321-337.

[48] Zhou Y-L, Sun H-S, Guo B-L, Some boundary problems of the spin system and the system of ferromagnetic chain II: Mixed problems and others[J], Acta Math. Sci., 1987, 7: 121-132.

[49] Zhou Y-L, Sun H-S, Guo B-L, Weak solution systems of ferromagnetic chain with several variables[J], Science in China A, 1987, 30: 1251-1266.

Analytical Study of the Two-dimensional Time-fractional Navier-Stokes Equations*

Shao Jing (邵晶), Guo Boling (郭柏灵), and Duan Lingling (段玲玲)

Abstract In this paper, the two-dimensional (2D) Holf-Cole transformation with mass conservation in the frame of the conformable derivative is developed, and then by introducing some exact solutions that satisfy linear differential equations and using the symbolic computation method, four exact solutions of 2D-nonlinear Navier-Stokes equations (NSEs) with the conformable time-fractional derivative are established. Some physical properties of the exact solutions are described preliminarily. Our results are the first ones on analytical study for the 2D time-fractional NSEs.

Keywords 2D Navier-Stokes equations; conformable fractional derivative; Holf-Cole transformation; exact solution

1 Introduction

Nonlinear Navier-Stokes equations (NSEs) have been addressed extensively because of their demonstrated applications in hydromechanics, aeronautical sciences, meteorology, and other science branches. In recent literatures only a small number of exact solutions of the NSEs are reported, concerning the pulsating dean flow in a channel with porous walls, the steady flow in an annulus with the porous wall, the stagnation flow on a plate with anisotropic slip and so on (see [10], [13], [15], [17], [18], [21], [22], [24], [25], [27] and references therein).

In this paper, we study the following two dimensional (2D)-nonlinear NSEs with the conformable time-fractional derivative

$$\mathbf{D}_t^\alpha \vec{u} + \vec{u} \cdot \nabla \vec{u} = -\nabla p + \nu \nabla^2 \vec{u}, \tag{1.1}$$

---

*J. Appl. Anal. Comput., 2019, 9(5): 1999–2022. Doi:10.11948/20190065.

$$\nabla \cdot \vec{u} = 0, \qquad (1.2)$$

where $\mathbf{D}_t^\alpha(\cdot)$ is the conformable fractional derivative, $0 < \alpha \leqslant 1$, $\vec{u} = \vec{u}(t,x,y)$, $p = p(t,x,y)$, ν and ∇ are the fluid velocity field, the fluid pressure, the viscosity and the gradient, respectively. The variables x and y form a Cartesian coordinate system; u_1, u_2 are the components of $\vec{u}$.

Notice that the problem (1.1) reduces to the classical NSEs for $\alpha = 1$:

$$\vec{u}_t + \vec{u} \cdot \nabla \vec{u} = -\nabla p + \nu \nabla^2 \vec{u}. \qquad (1.3)$$

In 1977, Takeo Saitoh [15] considered the full NSEs (1.3) by a numerical scheme with a high degree of accuracy. A fortunate exact solution was produced for flow in a porous pipe in [18]. By Chebyshev expansion methods, H. C. Ku et al. [10] studied the solutions of the steady 2D-NSEs in both the vorticity-stream function and the vorticity-velocity formulation. In 1998, G. Profilo et al. [13] solved the 2D-NSEs by the symmetry approach. The pseudo-spectral solutions of the 2D incompressible NSEs on a disk with no-slip boundary conditions were studied in [21]. The analyticity of solutions for randomly forced 2D-NSEs with periodic boundary conditions was discussed in [17]. By similarity transform, C. Y. Wang [24] investigated the flow due to a stretching flat boundary with partial slip. Analytical solutions of the equations of motion of a Newtonian fluid for the fully developed laminar flow between two concentric cylinders were presented by S. Tsangaris in the literature [22]. The numerical methods for solving the 2D-NSEs have been investigated by many authors [3], [6]- [8], [16], [26].

Fractional calculus has attracted much attention from mathematicians, physicists, biology, chemistry, engineering and other areas of applications in recent decades. Various types of definitions of fractional derivative are given, such as Grünwald-Letnikov, Riemann-Liouville and Caputo's fractional derivatives [5,12]. Most of them are defined via fractional integrals, thus they have nonlocal properties. The theory of conformable fractional calculus is a new topic of research which is introduced by Khalil et al. [9] in 2014. This new fractional derivative is a well-behaved definition, which depends on the basic limit definition of the derivative, and has governed much attention in recent years. In 2015, Abdeljawad studied fractional versions of the chain rule, exponential functions, Gronwall inequality, integration by parts (see [1] for details). Atangana et al. [2] investigated some new properties of this derivative, such as Taylor power series expansions, the conformable partial derivative, the conformable gradient, the conformable divergence theorem, and so on. The fractional Newtonian mechanics and the fractional version of the calculus of variations were introduced, and the fractional Euler-Lagrange equation was constructed in [4]. D. Zhao and M. Luo [28] generalized the conformable derivative and gave the physical and geometrical interpretation of the generalized conformable derivative in 2017.

There is a considerable interest in the study of time fractional Navier-Stokes equations (TFNSEs). Most of them are 1D-TFNSEs. There are also some analytical methods available for solving the TFNSEs. The homotopy analysis method is used to obtain an approximate solution of the nonlinear 1D-TFNSEs by introducing the Caputo's fractional derivative, see Ragab, Hemida, Mohamed and Salam [14]. In [29], Y. Zhou and L. Peng established the existence criterion of weak solutions of the 1D-TFNSEs by means of Galerkin approximations in the case that the dimension $n \leqslant 4$, which can be used to simulate anomalous diffusion in fractal media. Moreover, L. Peng, A. Debbouche and Y. Zhou investigated the existence and Faedo-Galerkin approximations of solutions for 1D-TFNSEs with Caputo derivative operators in the paper [11]. In 2018, G. Zou et al. [30] solved the numerical solution of 1D-TFNSEs by applying a composite idea of semi-discrete finite difference approximation in time and Galerkin finite element method in space with Caputo derivative of order $0 < \alpha < 1$.

However, to our best knowledge, there is no result on the exact solution of the 2D-TFNSEs, especially the result with the conformable fractional derivative operator. In 2007, C. Wu et al. [23] considered the following 2D-NSEs

$$\begin{cases} u_t + uu_x + vu_y + p_x = \dfrac{1}{\mathrm{Re}}(u_{xx} + u_{yy}), \\ v_t + uv_x + vv_y + p_y = \dfrac{1}{\mathrm{Re}}(v_{xx} + v_{yy}), \end{cases}$$

and the continuity equation is

$$u_x + v_y = 0,$$

where x and y form a Cartesian coordinate system; the variables u and v are the components of the fluid vector $\vec{V}$ in the x and y directions, respectively; Re is Reynolds number; variable p is the fluid pressure; $u_t \triangleq \dfrac{\partial u(t,x,y)}{\partial t}$, $u_{xx} \triangleq \dfrac{\partial^2 u(t,x,y)}{\partial x^2}$. Three exact solutions of the 2D-NSEs are presented by using the method of 2D Hopf-Cole transformation with mass conservation.

Followed the above references, the main contribution of our paper is to provide some exact solutions of 2D-TFNSEs in the frame of conformable derivatives and to discuss some interesting physical properties of these exact solutions. The rest of the paper is arranged as follows. In section 2, we give some definitions, properties of conformable fractional operators and the procedure of the Jacobi elliptic function expansion method. In section 3, the 2D Holf-Cole transformation with mass conservation in the frame of the conformable derivative is developed. And then in section 4, by introducing some 2D exact solutions that satisfy linear differential equations and using the symbolic computation method in Refs. [19], [20], four exact solutions of 2D-TFNSEs are established. In order to reveal some relevant physical aspects of the obtained results, some figures are presented for various parameters by using the analytical solutions obtained in section 3. In section 5, we give some comments on our paper.

2 Basic definitions and tools

To address our main result, here we represent the definitions, symbols and known properties of conformable fractional operators which will be used in the remainder of this paper.

Definition 2.1 *The (left) conformable fractional derivative starting from t_0 of a function $f : [t_0, \infty) \to \mathbf{R}$ of order α with $0 < \alpha \leqslant 1$ is defined by*

$$\left({}_{t_0}\mathbf{D}_t^\alpha f\right)(t) = \lim_{\varepsilon \to 0} \frac{f(t + \varepsilon(t - t_0)^{1-\alpha}) - f(t)}{\varepsilon}.$$

If $\left({}_{t_0}\mathbf{D}_t^\alpha f\right)(t)$ exists at $t \geqslant t_0$, we say f is α-differentiable at point t, and if $\left({}_{t_0}\mathbf{D}_t^\alpha f\right)(t)$ exists on (t_0, t_1), then $\left({}_{t_0}\mathbf{D}_t^\alpha f\right)(t_0) = \lim_{t \to t_0^+} \left({}_{t_0}\mathbf{D}_t^\alpha f\right)(t)$.

Definition 2.2 *Let $\alpha \in (0, 1]$. Then the left conformable fractional integral of order α starting at t_0 is defined by*

$$\left({}_{t_0}\mathbf{I}_t^\alpha f\right)(t) = \int_{t_0}^t (x - t_0)^{\alpha-1} f(x) dx.$$

If $t_0 = 0$, $\mathbf{D}_t^\alpha f \triangleq \left({}_{t_0}\mathbf{D}_t^\alpha f\right)(t)$, $\mathbf{I}_t^\alpha f \triangleq \left({}_{t_0}\mathbf{I}_t^\alpha f\right)(t)$. We note that for $\alpha \in (0, 1]$, the definition of conformable fractional integral is the same as Riemann-Liouville fractional integral up to a constant multiplier.

In the higher order case, we can generalize to the following definitions.

Definition 2.3 Let $\alpha \in (n, n+1]$. The (left) conformable fractional derivative starting from t_0 of a function $f : [t_0, \infty) \to \mathbf{R}$ of order α, where $f^{(n)}(t)$ exists, is defined by

$$\left(_{t_0}\mathbf{D}_t^\alpha f\right)(t) = \left(_{t_0}\mathbf{D}_t^{\alpha-n} f^{(n)}\right)(t).$$

Definition 2.4 Let $\alpha \in (n, n+1]$. The (left) conformable fractional integral of order α starting at t_0 is defined by

$$\left(_{t_0}\mathbf{I}_t^\alpha f\right)(t) =_{t_0} \mathbf{I}_t^{n+1}\left((t-t_0)^{\alpha-n-1} f\right)(t) = \frac{1}{n!}\int_{t_0}^t (t-x)^n (x-t_0)^{\alpha-n-1} f(x)dx.$$

Lemma 2.1 (see [9]) If $\alpha \in (n, n+1]$ and $f : [t_0, \infty) \to \mathbf{R}$ is an $(n+1)$ times differentiable function for $t > t_0$. Then, for all $t > t_0$, we have

$$_{t_0}\mathbf{I}_t^\alpha \, _{t_0}\mathbf{D}_t^\alpha(f)(t) = f(t) - \sum_{k=0}^n \frac{f^{(k)}(t_0)(t-t_0)^k}{k!}. \tag{2.1}$$

Lemma 2.2 (see [1]) Let $\alpha \in (0, 1]$ and suppose f, g are α-differentiable at point $t > 0$. Then

1. $_{t_0}\mathbf{D}_t^\alpha(af+bg) = a \cdot_{t_0}\mathbf{D}_t^\alpha(f) + b \cdot_{t_0}\mathbf{D}_t^\alpha(g)$ for all real constant a, b;
2. $_{t_0}\mathbf{D}_t^\alpha(fg) = f \cdot_{t_0}\mathbf{D}_t^\alpha(g) + g \cdot_{t_0}\mathbf{D}_t^\alpha(f)$;
3. $_{t_0}\mathbf{D}_t^\alpha(t^p) = pt^{p-\alpha}$, for all p;
4. $_{t_0}\mathbf{D}_t^\alpha\left(\frac{f}{g}\right) = \frac{g \cdot_{t_0}\mathbf{D}_t^\alpha(f) - f \cdot_{t_0}\mathbf{D}_t^\alpha(g)}{g^2}$;
5. $_{t_0}\mathbf{D}_t^\alpha(c) = 0$, where c is a constant;
6. $\left(_{t_0}\mathbf{D}_t^\alpha f\right)(t) = (t-t_0)^{1-\alpha} f'(t)$.

Now we describe the procedure of the Jacobi elliptic function expansion method. Given a nonlinear wave equation

$$F\left(u, \mathbf{D}_t^\alpha u, \frac{\partial u}{\partial x}, \frac{\partial u}{\partial y}, \mathbf{D}_t^{2\alpha} u, \frac{\partial^2 u}{\partial x^2}, \frac{\partial^2 u}{\partial y^2}, \cdots\right) = 0, \tag{2.2}$$

where $\mathbf{D}_t^{n\alpha}(\cdot) = \underbrace{\mathbf{D}_t^\alpha \cdots \mathbf{D}_t^\alpha}_{n}(\cdot), n \in N$. Transforming (2.2), applying the chain rule [1] and letting $t_0 = 0$,

$$u = u(\xi), \xi = l\frac{t^\alpha}{\alpha} + mx + ny,$$

where l, m, and n are arbitrary constants,

$$\mathbf{D}_t^\alpha(\cdot) = l\frac{d(\cdot)}{d\xi},\ \frac{\partial(\cdot)}{\partial x} = m\frac{d(\cdot)}{d\xi},\ \frac{\partial(\cdot)}{\partial y} = n\frac{d(\cdot)}{d\xi},\cdots \quad (2.3)$$

yields an ordinary differential equation (ODE) for $u(\xi)$,

$$O(u, u', u'', u''', \cdots),$$

where the prime denotes the derivation with respect to ξ.

Example 2.1 2D conformable time fractional heat-conduction equation.

Using the Jacobi elliptic function expansion method, let us consider the conformable time-fractional heat conduction equation

$$\mathbf{D}_t^\alpha W(t,x,y) = \nu\left(W_{xx} + W_{yy}\right).$$

Suppose that $W(t,x,y) = W(\xi)$, $\xi = x + y - l\frac{t^\alpha}{\alpha}$, where l is constant, we get

$$W_{\xi\xi} + \frac{l}{2\nu}W_\xi = 0. \quad (2.4)$$

It is easy to get a simple solution of (2.4). That is

$$W(\xi) = \frac{2C\nu}{l} + C^* \exp\left\{-\frac{l}{2\nu}\xi\right\},$$

where C and C^* are constants. Hence we get

$$W(t,x,y) = \frac{2C\nu}{l} + C^* \exp\left\{-\frac{l}{2\nu}\left(x + y - l\frac{t^\alpha}{\alpha}\right)\right\}.$$

3 2D Hopf-Cole transformation with mass conservation

3.1 Expression of 2D Hopf-Cole transformation

The conformable nonlinear 2D-TFNSEs are of the form

$$\begin{cases} \mathbf{D}_t^\alpha u_1 + u_1 u_{1x} + u_2 u_{1y} + p_x = \nu\left(u_{1xx} + u_{1yy}\right), \\ \mathbf{D}_t^\alpha u_2 + u_1 u_{2x} + u_2 u_{2y} + p_y = \nu\left(u_{2xx} + u_{2yy}\right), \end{cases} \quad (3.1)$$

and

$$u_{1x} + u_{2y} = 0. \quad (3.2)$$

We recall that the stream function ψ is defined by

$$u_1 = -\frac{\partial \psi}{\partial y}, u_2 = \frac{\partial \psi}{\partial x}. \tag{3.3}$$

Substituting (3.3) into (3.2), we get

$$\begin{aligned} u_1 u_{1x} + u_2 u_{1y} &= -(\psi_x \psi_y)_y + (\psi_y^2)_x, \\ u_1 u_{2x} + u_2 u_{2y} &= -(\psi_x \psi_y)_x + (\psi_x^2)_y. \end{aligned} \tag{3.4}$$

Using (3.3) and (3.4) in (3.1), we have

$$\begin{aligned} -\left[\mathbf{D}_t^\alpha \psi + \frac{1}{2}\left(\psi_x^2 + \psi_y^2\right) - \nu\left(\psi_{xx} + \psi_{yy}\right)\right]_y + \frac{1}{2}\left[(\psi_x - \psi_y)^2\right]_y + p_x + (\psi_y^2)_x &= 0, \\ \left[\mathbf{D}_t^\alpha \psi + \frac{1}{2}\left(\psi_x^2 + \psi_y^2\right) - \nu\left(\psi_{xx} + \psi_{yy}\right)\right]_x - \frac{1}{2}\left[(\psi_x + \psi_y)^2\right]_x + p_y + (\psi_x^2)_y &= 0. \end{aligned} \tag{3.5}$$

Letting

$$\begin{aligned} p_x &= -\frac{1}{2}\left[(\psi_x - \psi_y)^2\right]_y - (\psi_y^2)_x, \\ p_y &= \frac{1}{2}\left[(\psi_x + \psi_y)^2\right]_x - (\psi_x^2)_y, \end{aligned} \tag{3.6}$$

and the integral constant being zero, we get the following simple expression via performing integration for x and y directions,

$$\mathbf{D}_t^\alpha \psi + \frac{1}{2}\left(\psi_x^2 + \psi_y^2\right) - \nu\left(\psi_{xx} + \psi_{yy}\right) = 0. \tag{3.7}$$

We introduce a variable W now which satisfies the following linear differential equation

$$\mathbf{D}_t^\alpha W - \nu(W_{xx} + W_{yy}) = 0, \tag{3.8}$$

and we assume that

$$W = W(t, x, y) = F(\psi), \tag{3.9}$$

then we can structure a transformation between variables F and ψ. Substituting (3.9) into (3.8) and using Lemma 2.2, we have

$$\mathbf{D}_t^\alpha \psi - \frac{\nu F_{\psi\psi}}{F_\psi}\left(\psi_x^2 + \psi_y^2\right) - \nu\left(\psi_{xx} + \psi_{yy}\right) = 0. \tag{3.10}$$

Comparing (3.7) with (3.10), we can see that there is a relation

$$-\frac{1}{2} = \nu \frac{F_{\psi\psi}}{F_{\psi}}. \tag{3.11}$$

Making all integral constants zero in the course of solving (3.11), then the following transformation between variables F and ψ can be obtained

$$\psi = -2\nu \ln\left(-\frac{1}{2\nu}W\right). \tag{3.12}$$

Substituting (3.12) into (3.3) and (3.7), we get

$$u_1 = 2\nu \frac{W_y}{W}, \; u_2 = -2\nu \frac{W_x}{W}; \tag{3.13}$$

$$p_x = -2\nu^2 \left(\frac{(W_y - W_x)^2}{W^2}\right)_y - 4\nu^2 \left(\frac{W_y^2}{W^2}\right)_x$$

$$= -\frac{1}{2}\left[(u_1 + u_2)^2\right]_y - (u_1^2)_x,$$

$$p_y = 2\nu^2 \left(\frac{(W_x + W_y)^2}{W^2}\right)_x - 4\nu^2 \left(\frac{W_x^2}{W^2}\right)_y$$

$$= \frac{1}{2}\left[(u_2 - u_1)^2\right]_x - (u_2^2)_y. \tag{3.14}$$

Furthermore, W can be solved in (3.8), and then u_1, u_2 and p_x, p_y can be obtained via solving (3.13) ∼ (3.14). Substituting u_1, u_2 and p_x, p_y into (3.1) ∼ (3.2), we get

$$2\nu \mathbf{D}_t^\alpha (\ln F_\psi)_y + 4\nu^2 (\ln F_\psi)_y (\ln F_\psi)_{yx} + 4\nu^2 (\ln F_\psi)_x (\ln F_\psi)_{xy} + p_x$$
$$= 2\nu^2 (\ln F_\psi)_{yxx} + 2\nu^2 (\ln F_\psi)_{yyy},$$

$$2\nu \mathbf{D}_t^\alpha (\ln F_\psi)_x + 4\nu^2 (\ln F_\psi)_y (\ln F_\psi)_{xx} - 4\nu^2 (\ln F_\psi)_x (\ln F_\psi)_{xy} + p_y$$
$$= 2\nu^2 (\ln F_\psi)_{xxx} + 2\nu^2 (\ln F_\psi)_{xyy}.$$

Using the symbolic computation method in Refs. [19], [20] to calculate these variables, we know that u_1, u_2 and p_x, p_y satisfy (3.1) ∼ (3.2). Thus (3.13) ∼ (3.14) make up a 2D Hopf-Cole transformation of the conformable nonlinear 2D-TFNSEs.

If u_1, u_2 and p_x, p_y are structured by using some trigonometric functions and exponential functions as in section 4 case 4, then p_x and p_y do not satisfy (3.14). In this case, u_1 and u_2 have been obtained by (3.13) under the precondition that W satisfies (3.8), and then p_x and p_y can be obtained by the following equations

$$p_x = \nu(u_{1xx} + u_{1yy}) - \mathbf{D}_t^\alpha u_1 - u_1 u_{1x} - u_2 u_{1y},$$
$$p_y = \nu(u_{2xx} + u_{2yy}) - \mathbf{D}_t^\alpha u_2 - u_1 u_{2x} - u_2 u_{2y}. \tag{3.15}$$

3.2 Modified expression of 2D Hopf-Cole transformation

Followed the procedure of 2D Hopf-Cole transformation, we obtain (3.4) $\sim$ (3.7). Introducing a new variable V that satisfies the following linear differential equation

$$\mathbf{D}_t^\alpha V + h(t, x, y) - \nu(V_{xx} + V_{yy}) = 0, \tag{3.16}$$

and assuming

$$V = C_0 + V_1(t) H(t, x, y) = C_0 + V_1(t) H(\psi), \tag{3.17}$$

here $h(t, x, y)$ may be structured as follows

$$h(t, x, y) = -\mathbf{D}_t^\alpha (V_1(t)) H(t, x, y), \tag{3.18}$$

where $V_1(t)$ is a point source varying with time and C_0 is a constant.

Using the similar discussion as 2D Hopf-Cole transformation, we can obtain a modified expression of 2D Hopf-Cole transformation with mass conservation as follows

$$\psi = -2\nu \ln\left(-\frac{1}{2\nu} V\right);$$

$$u_1 = 2\nu \frac{V_y}{V}, \quad u_2 = -2\nu \frac{V_x}{V};$$

$$p_x = -2\nu^2 \left(\frac{(W_y - W_x)^2}{W^2}\right)_y - 4\nu^2 \left(\frac{W_y^2}{W^2}\right)_x + A_1$$

$$= -\frac{1}{2}\left[(u_1 + u_2)^2\right]_y - (u_1^2)_x + A_1,$$

$$p_y = 2\nu^2 \left(\frac{(W_x + W_y)^2}{W^2}\right)_x - 4\nu^2 \left(\frac{W_x^2}{W^2}\right)_y + A_2$$

$$= \frac{1}{2}\left[(u_2 - u_1)^2\right]_x - (u_2^2)_y + A_2, \tag{3.19}$$

where
$$A_1 = -\mathbf{D}_t^\alpha [V_1(t)] \cdot \frac{u_1}{V_1(t)V(t)} \tag{3.20}$$
and
$$A_2 = -\mathbf{D}_t^\alpha [V_1(t)] \cdot \frac{u_2}{V_1(t)V(t)}. \tag{3.21}$$

If $V_1(t) = 1$ in (3.17), then (3.17) reduces to (3.9). By the symbolic computation method, we learn that $H(t,x,y)$ is the same as the solution of $W(t,x,y)$, and $V_1(t)$ is a differentiable function, $h(t,x,y)$ can structure the $V(x,y,t)$ that satisfies (3.16).

If the point source is a constant, then A_1 and A_2 are zeros in (3.20) and (3.21), thus, (3.19) reduces to (3.13) and (3.14).

3.3 Properties of p_x and p_y

Note that since some exact solutions of 2D-NSEs in this paper are vortex solutions, they can be expressed in two orthogonal coordinate systems reciprocally. We introduce a new coordinate system (X, Y) and two new variables U_1 and U_2 which are defined by
$$x = -Y, \quad y = X, \quad u_1 = -U_2, \quad u_2 = U_1. \tag{3.22}$$
Under the new coordinate system (3.22), the expressions in (3.19) are changed into
$$\begin{aligned} p_Y &= \frac{1}{2}\left[(U_2 - U_1)^2\right]_X - (U_2^2)_Y - \mathbf{D}_t^\alpha [V_1(t)] \cdot \frac{U_2}{V_1(t)V(t)}, \\ p_X &= -\frac{1}{2}\left[(U_1 + U_2)^2\right]_Y - (U_1^2)_X - \mathbf{D}_t^\alpha [V_1(t)] \cdot \frac{U_1}{V_1(t)V(t)}. \end{aligned} \tag{3.23}$$
Substituting (3.22) into (3.15), we get
$$p_Y = \nu\left(U_{2XX} + U_{2YY}\right) - \mathbf{D}_t^\alpha U_2 - U_1 U_{2X} - U_2 U_{2Y},$$
$$p_X = \nu\left(U_{1XX} + U_{1YY}\right) - \mathbf{D}_t^\alpha U_1 - U_1 U_{1X} - U_2 U_{1Y}. \tag{3.24}$$
Comparing (3.23), (3.24) with (3.19), (3.15), we get the following conclusions.
(i) The expressions of p_x, p_y and p_X, p_Y will be exchanged reciprocally.
(ii) p_x and p_y are orthogonal symmetry reciprocally, and
$$\begin{aligned} \int p_x dx &= p(u_1, u_2, x, y, t), \\ \int p_y dy &= \int p_X dX = p(U_1, U_2, X, Y, t). \end{aligned} \tag{3.25}$$

(iii) $p(u_1, u_2, x, y, t) = p(U_1, U_2, X, Y, t)$.

(iv) The fluid velocity field $\vec{u}$, the fluid pressure p in some vortex solutions are orthogonal symmetric distribution for origin ($x = 0$, $y = 0$). Thus p_x and p_y are compatible to each other.

4 Statement of the problem and its exact solutions on infinity plane

In this section, we consider some exact solutions of 2D-TFNSEs in many cases.

Case 1. An exact solution on infinity plane

Introducing an exact solution that satisfies (3.8) of the form

$$W(t, x, y) = \frac{2C\nu}{l} + C^* E_1 \tag{4.1}$$

and substituting (4.1) into (3.12) $\sim$ (3.14), we have

$$\psi = -2\nu \ln\left(-\frac{C}{l} - C^* E_1\right), \quad C < 0, C^* < 0;$$

$$u_1 = \frac{-C^* l^2 E_1}{C + lC^* E_1}, \quad u_2 = \frac{C^* l^2 E_1}{C + lC^* E_1};$$

$$p_x = \frac{C^{*2} C l^5 E_1^2}{\nu(C + lC^* E_1)^3}, \quad p_y = \frac{\nu C C^{*2} l^5 E_1^2 - 2C^{*2} C l^5 E_1^4}{\nu(C + lC^* E_1)^3}, \tag{4.2}$$

where ψ is the stream function and $E_1 = e^{-\frac{1}{2\nu}(x+y-l\frac{t^\alpha}{\alpha})}$. Substituting u_1, u_2 and p_x, p_y into (3.1) $\sim$ (3.2) and using the symbolic computation method to calculate these variables, respectively, we can learn that the variables satisfy (3.1) $\sim$ (3.2). Hence u_1, u_2 and p_x, p_y in (4.2) constitute an exact solution of conformable time-fractional 2D-NSEs.

In the first case, the coefficients of (4.2) are $\nu = 0.1$, $l = 1$, $C = -1$, $C^* = -2$. Then we can find some interesting physical behavior of this exact solution with the fractional order $\alpha = \frac{1}{2}$ at the time $t = 1$, $t = 10$ and $t = 100$, as shown in Figs.1~3.

(i) From Fig.1, we see that the vectorial distribution of the fluid velocity vector $\vec{u}$ is strip region and it is increasing as t increasing. The flown line distribution of $\vec{u}$ is a series of parallel lines. Particularly, the flown line distribution of $\vec{u}$ in the region $X \times Y \in [-20, 20] \times [-20, 20]$, the flown line distribution of $\vec{u}$ is more intensive than in the rest of the region.

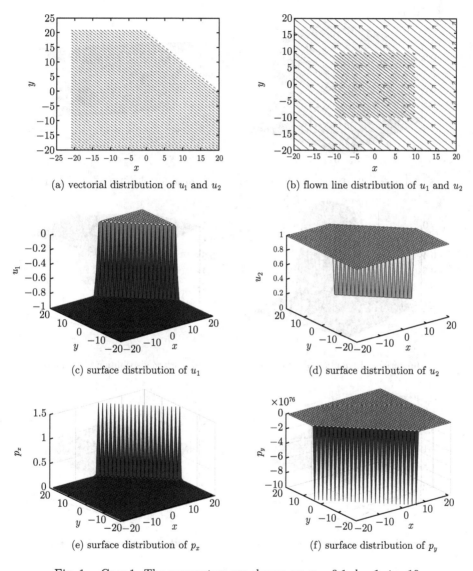

Fig. 1　Case 1. The parameters are chosen as: $\nu = 0.1$, $l = 1$, $t = 10$, $C = -1$, $C^* = -2$ and $\alpha = \dfrac{1}{2}$

(ii) Fig.2 shows that the surface distribution of u_1 and u_2 change with $\alpha = \dfrac{1}{2}$ at the time $t = 1$, $t = 10$, and $t = 100$. It can be seen clearly that the valued field of u_1 is $[-1, 0]$ and the surface distribution of u_1 decreases gradually to $u_1 = 1$

as the increasing of t. And then the valued field of u_2 is $[0,1]$ and the surface distribution of u_2 increases gradually to $u_2 = 1$ with the increasing of t.

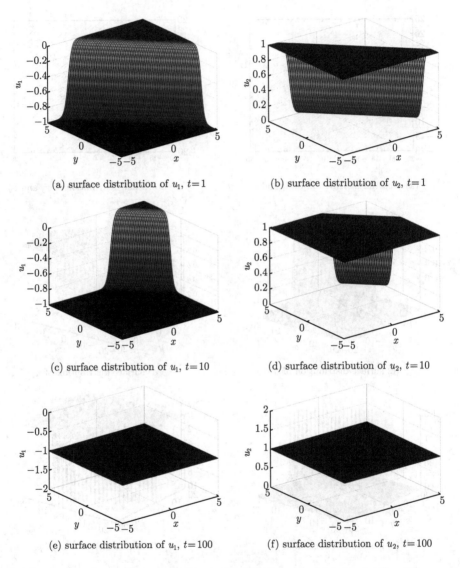

Fig. 2 Case 1. The parameters are chosen as: $\nu = 0.1$, $l = 1$, $C = -1$, $C^* = -2$ and $\alpha = \dfrac{1}{2}$

(iii) Fig.3 depicts the surface distribution of p_x and p_y for three different values of t. It is clear that the surface distribution of p_x decreases from $+\infty$ to 0, and the surface distribution of p_y increases from $-\infty$ to 0. Hence we get that as time goes on, the surface distribution of p_x and p_y will be on the plane $p_x = 0$ and $p_y = 0$ eventually.

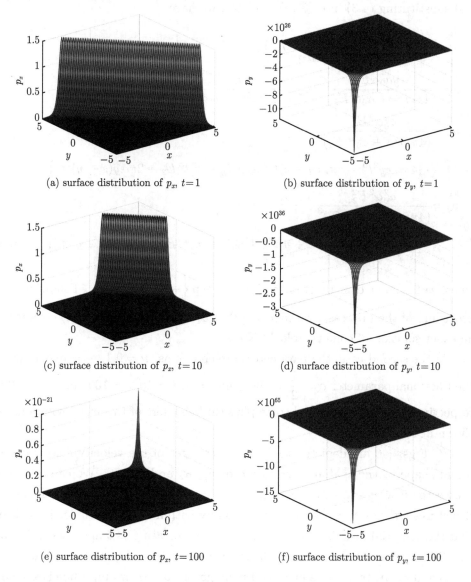

Fig. 3 Case 1. The parameters are chosen as: $\nu = 0.1$, $l = 1$, $C = -1$, $C^* = -2$ and $\alpha = \dfrac{1}{2}$

Case 2. An exact solution under action of a point source on infinity plane
Introducing an exact solution that satisfies (3.8) of the form

$$W(t,x,y) = 1 + \frac{a_0}{4\pi\nu t^\alpha} E_2 \qquad (4.3)$$

and substituting (4.3) into (3.12) ∼ (3.14), we have

$$\psi = -2\nu \ln\left(-\frac{1}{2\nu} - \frac{a_0}{8\pi\nu^2 t^\alpha} E_2\right);$$

$$u_1 = \frac{-y a_0 \alpha E_2}{4\pi\nu t^{2\alpha} + a_0 t^\alpha E_2}, \quad u_2 = \frac{x a_0 \alpha E_2}{4\pi\nu t^{2\alpha} + a_0 t^\alpha E_2};$$

$$p_x = \frac{a_0^2 \alpha^2 E_2^2}{(4\pi\nu t^{2\alpha} + a_0 t^\alpha E_2)^3}$$
$$\cdot \left[4\pi\alpha x y^2 t^\alpha - 4\pi\nu(y-x)t^{2\alpha} - a_0(y-x)t^\alpha E_2 - 2\pi\alpha y(x-y)^2 t^\alpha\right],$$

$$p_y = \frac{a_0^2 \alpha^2 E_2^2}{(4\pi\nu t^{2\alpha} + a_0 t^\alpha E_2)^3}$$
$$\cdot \left[4\pi\alpha x^2 y t^\alpha + 4\pi\nu(x+y)t^{2\alpha} + a_0(x+y)t^\alpha E_2 - 2\pi\alpha x(x+y)^2 t^\alpha\right], \qquad (4.4)$$

where ψ is the stream function and $E_2 = \exp\left\{-\dfrac{\alpha(x^2+y^2)}{4t^\alpha \nu}\right\}$. Following the procedure of the first case, we conclude that u_1, u_2 and p_x, p_y in (4.4) constitute an exact solution of conformable TFNSEs.

In the second case, the parameters are chosen as: $\nu = 0.1$, $a_0 = \exp(1)$ with the fractional parameter $\alpha = \dfrac{1}{2}$ at the time $t = 1$, $t = 10$, $t = 100$ and $t = 10000$, respectively. Then we can find some physical behaviors of this exact solution, as shown in Figs.4∼8.

(i) Figs.4∼6 are depicted to show the changes of the velocity field and the initial shape of the spatial distribution of u_1, u_2 and p_x, p_y. We can see that the vector $\vec{u}$ whirls around origin $(x=0, y=0)$ only and the stream function ψ pictures a series of concentric circles merely within a certain circle. It is clear that the vectorial distribution of u_1 and u_2 are increasing as t increases in Fig.6.

(ii) Figs.7 and 8 demonstrate the initial steep shape of spatial distribution of u_1, u_2 and p_x, p_y tend to be more and more gentle and their amplitude is smaller with the increasing of t.

(iii) If we choose $\nu = \dfrac{1}{\text{Re}}$, where Re is Reynolds number, then we get that the change in ν value may lead to change in shape of the spatial distribution of u_1, u_2 and p_x, p_y. With the increase in ν value, their shape tends to be more and

Analytical Study of the Two-dimensional Time-fractional Navier-Stokes Equations

more gentle.

Choosing the fractional parameter $\alpha = 1$ and $\nu = \dfrac{1}{\text{Re}}$ in (4.3), we can get the same exact solution as in Ref. [23].

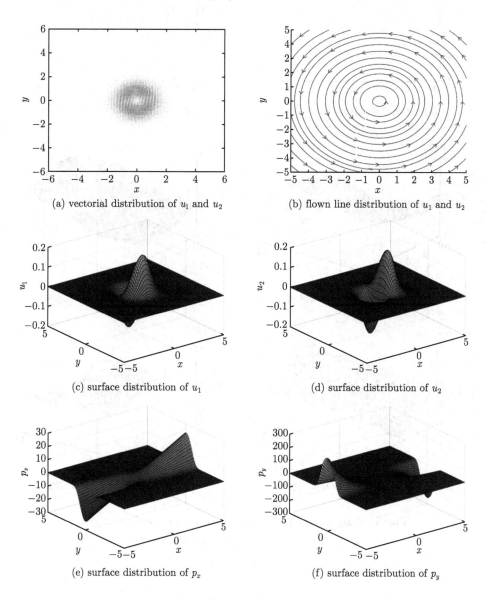

Fig. 4 Case 2. The parameters are chosen as: $\nu = 0.1$, $t = 1$, $x \times y \in [-5,5] \times [-5,5]$, $a_0 = \exp(1)$ and $\alpha = \dfrac{1}{2}$

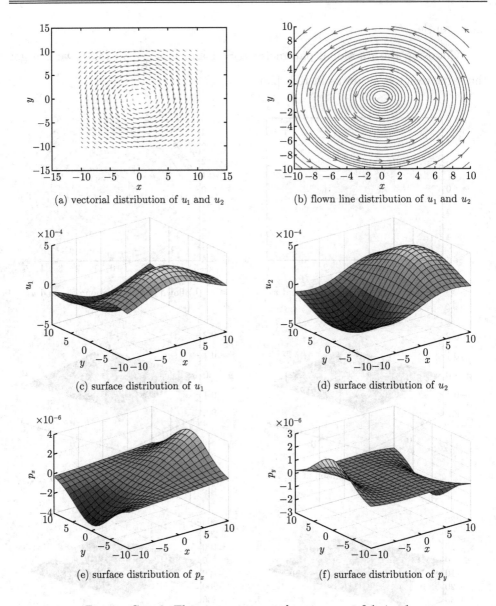

(a) vectorial distribution of u_1 and u_2

(b) flown line distribution of u_1 and u_2

(c) surface distribution of u_1

(d) surface distribution of u_2

(e) surface distribution of p_x

(f) surface distribution of p_y

Fig. 5 Case 2. The parameters are chosen as: $\nu = 0.1$, $t = 1$, $x \times y \in [-10, 10] \times [-10, 10]$, $a_0 = \exp(1)$ and $\alpha = \dfrac{1}{2}$

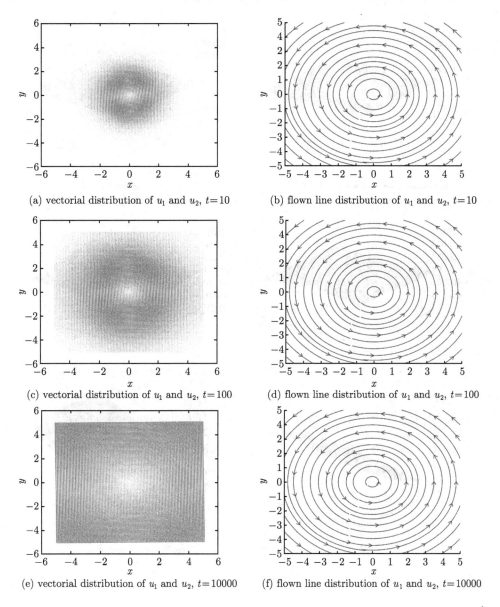

Fig. 6 Case 2. The parameters are chosen as: $\nu = 0.1$, $t = 10000$, $a_0 = \exp(1)$ and $\alpha = \dfrac{1}{2}$

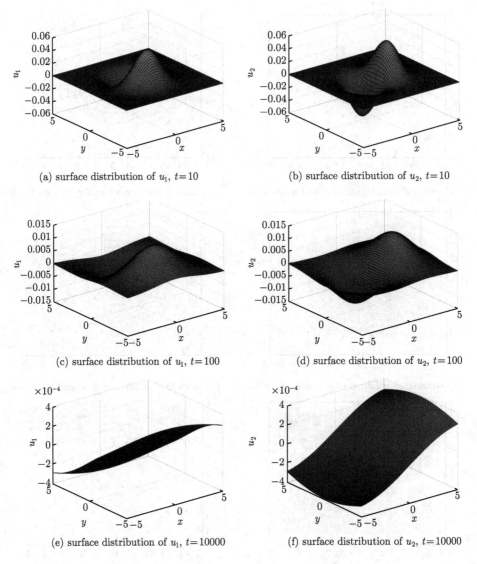

(a) surface distribution of u_1, $t=10$

(b) surface distribution of u_2, $t=10$

(c) surface distribution of u_1, $t=100$

(d) surface distribution of u_2, $t=100$

(e) surface distribution of u_1, $t=10000$

(f) surface distribution of u_2, $t=10000$

Fig. 7 Case 2. The parameters are chosen as: $\nu = 0.1$, $t = 10000$, $a_0 = \exp(1)$ and $\alpha = \dfrac{1}{2}$

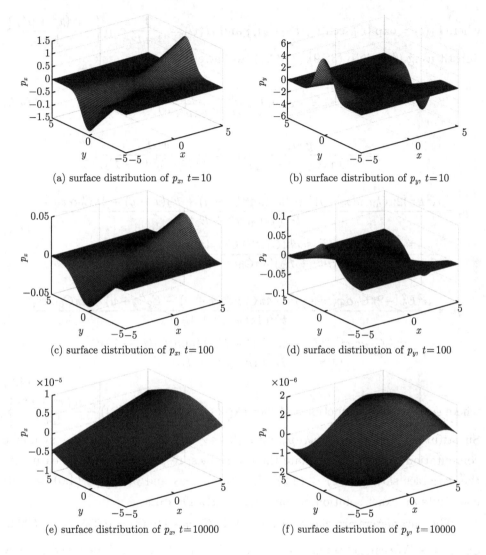

Fig. 8 Case 2. The parameters are chosen as: $\nu = 0.1$, $t = 10000$, $a_0 = \exp(1)$ and $\alpha = \dfrac{1}{2}$

Case 3. An exact solution under action of a point source varying with time on infinity plane

Introducing an exact solution that satisfies (3.16) of the form

$$V(t,x,y) = C_0 + V_1(t)H(\psi)$$

$$= C_0 + \exp\{C_0\operatorname{sech}(C_1(t-C_2))\}\frac{1}{4\pi\nu t^\alpha}\exp\left\{-\frac{\alpha(x^2+y^2)}{4\nu t^\alpha}\right\}, \quad (4.5)$$

where $V_1(t) = \exp\{C_0\operatorname{sech}(C_1(t-C_2))\}$ and $H(\psi) = \dfrac{1}{4\pi\nu t^\alpha}\exp\left\{-\dfrac{\alpha(x^2+y^2)}{4\nu t^\alpha}\right\}$. Substituting (4.5) into (3.19) $\sim$ (3.21), we have

$$\psi = -2\nu \ln\left(-\frac{C_0}{2\nu} - \frac{E_2}{8\pi\nu^2 t^\alpha}\right); \tag{4.6}$$

$$u_1 = \frac{-y\alpha E_2}{t^\alpha(4\pi C_0 \nu t^\alpha + E_2)}, \quad u_2 = \frac{x\alpha E_2}{t^\alpha(4\pi C_0 \nu t^\alpha + E_2)}; \tag{4.7}$$

$$\begin{aligned}p_x = &\frac{\alpha^2 E_2^2[2\pi C_0\alpha y(x-y)^2 + 4\pi C_0\nu t^\alpha(x-y) + E_2(x-y) + 4\pi C_0\alpha xy^2]}{t^{2\alpha}(4\pi C_0\nu t^\alpha + E_2)^3} \\ &- \frac{4\pi C_0 C_1 \nu\alpha y t^{1-\alpha} E_2 \operatorname{sech}(C_1(t-C_2))\tanh(C_1(t-C_2))}{(4\pi\nu t^\alpha + E_2)(4\pi C_0\nu t^\alpha + E_2)},\end{aligned} \tag{4.8}$$

$$\begin{aligned}p_y = &\frac{\alpha^2 E_2^2[-2\pi C_0\alpha x(x+y)^2 - 4\pi C_0\nu t^\alpha(x+y) - E_2(x+y) + 4\pi C_0\alpha x^2 y]}{t^{2\alpha}(4\pi C_0\nu t^\alpha + E_2)^3} \\ &+ \frac{4\pi C_0 C_1 \nu\alpha x t^{1-\alpha} E_2 \operatorname{sech}(C_1(t-C_2))\tanh(C_1(t-C_2))}{(4\pi\nu t^\alpha + E_2)(4\pi C_0\nu t^\alpha + E_2)},\end{aligned} \tag{4.9}$$

where ψ is the stream function and $E_2 = \exp\left\{C_0\operatorname{sech}(C_1(t-C_2)) - \dfrac{\alpha(x^2+y^2)}{4\nu t^\alpha}\right\}$. Substituting u_1, u_2 and p_x, p_y into (3.1) $\sim$ (3.2) and using the symbolic computation method to calculate these variables, respectively, we can learn that the variables satisfy (3.1) $\sim$ (3.2), thus the u_1, u_2 and p_x, p_y in (4.6) $\sim$ (4.9) constitute an exact solution of conformable 2D-TFNSEs.

The parameters are chosen as: $\nu = 0.1$, $C_0 = 10.0$, $C_1 = 0.005$, $C_2 = 500.0$, and $\alpha = \dfrac{1}{4}$. We can get some properties, as shown in Fig. 9. It is depicted to show the changes of the velocity field and the initial shape of the spatial distribution of u_1, u_2 and p_x, p_y, respectively. We can see that the vector $\vec{u}$ whirls around origin $(x=0, y=0)$. Especially, the flown line distribution of $\vec{u}$ in the region $X \times Y \in [-5,5] \times [-5,5]$, is more intensive than that in the rest of the region. The initial steep shape of spatial distribution of u_1, u_2 and p_x, p_y tend to be more and more gentle and their amplitude to be smaller with the increase in time t.

Choosing the fractional parameter $\alpha = 1$, $C_0 = 1$ and $\nu = \dfrac{1}{\operatorname{Re}}$ in (4.5), we can get the same exact solution as in Ref. [23].

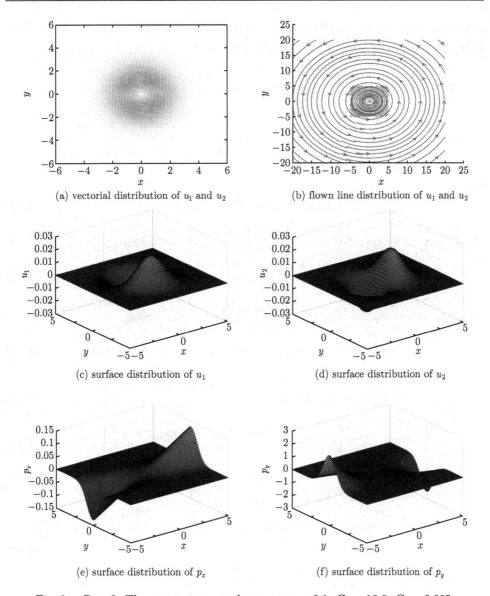

Fig. 9 Case 3. The parameters are chosen as: $\nu = 0.1$, $C_0 = 10.0$, $C_1 = 0.005$, $C_2 = 500.0$, and $\alpha = \dfrac{1}{4}$.

Case 4. An exact solution with respect to initial boundary value problem in foursquare region

Introducing an exact solution that satisfies (3.8) of the form

$$W(t,x,y) = 1 + a_0 E_3 \frac{l^2}{\pi^2} \cos\frac{\pi x}{l} \cos\frac{\pi y}{l}, \tag{4.10}$$

where a_0 is an initial boundary value, and l is a side length of foursquare region. Substituting (4.10) into (3.12) $\sim$ (3.14), we have

$$\psi = -2\nu \ln\left(-\frac{1}{2\nu} - \frac{a_0 l^2 E_3}{2\nu\pi^2}\cos\frac{\pi x}{l}\cos\frac{\pi y}{l}\right);$$

$$u_1 = \frac{2\nu\pi a_0 l E_3 \cos\frac{\pi}{l}x \sin\frac{\pi}{l}y}{\pi^2 + a_0 l^2 E_3 \cos\frac{\pi}{l}x \cos\frac{\pi}{l}y}, \quad u_2 = \frac{2\nu\pi a_0 l E_3 \sin\frac{\pi}{l}x \cos\frac{\pi}{l}y}{\pi^2 + a_0 l^2 E_3 \cos\frac{\pi}{l}x \cos\frac{\pi}{l}y}. \quad (4.11)$$

Let

$$B(t,x,y) = \pi^2 + a_0 l^2 E_3 \cos\frac{\pi}{l}x \sin\frac{\pi}{l}y, \quad (4.12)$$

$$B_1(t,x,y) = \nu(u_{1xx} + u_{1yy}) = -\frac{4\nu^2\pi^3 a_0 E_3}{lB^3}\cdot\left(\pi^4 \cos\frac{\pi}{l}x \sin\frac{\pi}{l}y\right.$$
$$\left.+\pi^2 a_0 l^2 E_3 \sin^2\frac{\pi}{l}x \sin\frac{\pi}{l}y \cos\frac{\pi}{l}y - a_0^2 l^4 E_3^2 \cos^3\frac{\pi}{l}x \sin\frac{\pi}{l}x\right), \quad (4.13)$$

$$B_2(t,x,y) = \nu(u_{2xx} + u_{2yy}) = -\frac{4\nu^2\pi^3 a_0 E_3}{lB^3}\cdot\left(\pi^4 \sin\frac{\pi}{l}x \cos\frac{\pi}{l}y\right.$$
$$\left.-\pi^2 a_0 l^2 E_3 \sin\frac{\pi}{l}x \sin\frac{\pi}{l}x \sin^2\frac{\pi}{l}y - a_0^2 l^4 E_3^2 \cos^3\frac{\pi}{l}y \sin\frac{\pi}{l}x\right), \quad (4.14)$$

$$B_3(t,x,y) = \mathbf{D}_t^\alpha u_1 = \frac{-4\nu^2\pi^5 a_0 E_3 \cos\frac{\pi}{l}x \sin\frac{\pi}{l}y}{lB^2} \quad (4.15)$$

$$B_4(t,x,y) = \mathbf{D}_t^\alpha u_2 = \frac{-4\nu^2\pi^5 a_0 E_3 \sin\frac{\pi}{l}x \cos\frac{\pi}{l}y}{lB^2} \quad (4.16)$$

$$B_5(t,x,y) = u_1 u_{1x} + u_2 u_{1y} = \frac{4\nu^2\pi^3 a_0^2 l E_3^2}{B^3}\left(-\pi^2 \sin\frac{\pi}{l}x \cos\frac{\pi}{l}x \sin^2\frac{\pi}{l}y\right.$$
$$\left.+\pi^2 \sin\frac{\pi}{l}x \cos\frac{\pi}{l}x \cos^2\frac{\pi}{l}y + a_0 l^2 E_3 \sin\frac{\pi}{l}x \cos\frac{\pi}{l}y \cos^2\frac{\pi}{l}x\right), \quad (4.17)$$

$$B_6(t,x,y) = u_1 u_{2x} + u_2 u_{2y} = \frac{4\nu^2\pi^3 a_0^2 l E_3^2}{B^3}\left(\pi^2 \cos^2\frac{\pi}{l}x \sin\frac{\pi}{l}y \cos\frac{\pi}{l}y\right.$$
$$\left.+a_0 l^2 E_3 \cos\frac{\pi}{l}x \sin\frac{\pi}{l}y \cos^2\frac{\pi}{l}y - \pi^2 \sin^2\frac{\pi}{l}x \sin\frac{\pi}{l}y \cos\frac{\pi}{l}y\right), \quad (4.18)$$

where $E_3 = \exp\left\{-\frac{2\nu\pi^2 t^\alpha}{\alpha l^2}\right\}$. Then we get

$$\begin{aligned} p_x &= B_1(t,x,y) - B_3(t,x,y) - B_5(t,x,y), \\ p_y &= B_2(t,x,y) - B_4(t,x,y) - B_6(t,x,y). \end{aligned} \quad (4.19)$$

Following the procedure of the first case, we conclude that u_1, u_2 and p_x, p_y in (4.4) constitute an exact solution of the conformable 2D-TFNSEs.

The exact solution for 2D-TFNSEs with respect to the initial boundary value problem in foursquare region is obtained using Hopf-Cole transform. For a complete study and for possible comparisons, we present the parameters by making $\nu = 0.01$, $t = 10$, $a_0 = \exp(3)$, $\alpha = \dfrac{1}{4}$ and $X \times Y \in [-5, 5] \times [-5, 5]$ or $X \times Y \in [-20, 20] \times [-20, 20]$. Some physical properties are found as shown in Fig. 10 and Fig. 11.

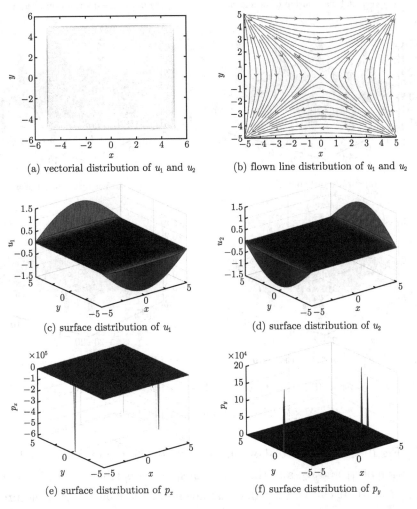

(a) vectorial distribution of u_1 and u_2

(b) flown line distribution of u_1 and u_2

(c) surface distribution of u_1

(d) surface distribution of u_2

(e) surface distribution of p_x

(f) surface distribution of p_y

Fig. 10 Case 4. The parameters are chosen as: $\nu = 0.01$, $t = 10$, $a_0 = \exp(3)$, $X \times Y \in [-5, 5] \times [-5, 5]$ and $\alpha = \dfrac{1}{4}$

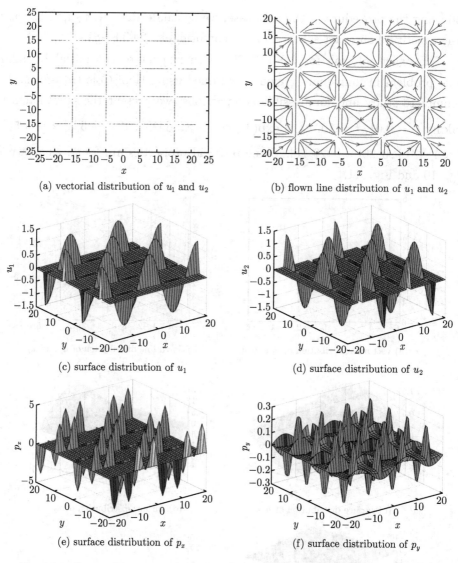

Fig. 11 Case 4. The parameters are chosen as: $\nu = 0.01$, $t = 10$, $a_0 = \exp(3)$, $X \times Y \in [-20, 20] \times [-20, 20]$ and $\alpha = \dfrac{1}{4}$

(i) Fig.10 is depicted to show the changes of the vector field, the flown line distribution of u_1 and u_2, and the shapes of spatial distribution of u_1, u_2 and p_x, p_y when $t = 10$. It can be seen that the flown line distribution of $\vec{u}$ pictures two families of hyperbolas.

(ii) In Fig. 11, it is clearly seen that u_1 and u_2 are the space periodic solutions, and their region of periodic is $X \times Y \in [-20, 20] \times [-20, 20]$.

(iii) If the parameters are chosen as Fig. 10, the shape of spatial distribution of u_1, u_2 and p_x, p_y remain the same as Fig. 10 with the increasing of t. These figures are omitted here for reasons of limited space.

5 Conclusions

The 2D-TFNSEs are considered in this paper for the first time. These new fluid models have a series of physical properties. Many conclusions are obtained as follows:

(i) Compared to 2D-NSEs, it is hard to solve the exact solutions of 2D-TFNSEs due to the time fractional derivative. With mass conservations (3.2) and the stream function ψ (3.3), 2D Hopf-Cole transformations are established by using the properties of conformable fractional operators and introducing a new variable W that satisfies the linear differential equations (3.8). If W can be found, it is possible to structure the corresponding exact solutions of TFNSEs. Next we must use the symbolic computation method to test the obtained exact solutions according to the point of Chiping Wu et al. in Refs. [23].

(ii) Exact solution in Case 1 provides that the velocity of the fluid tends to the constant value $\vec{u} = (-1, 1)$ gradually as the fluid pressure p varying from strong to weak with t increasing. The flown line distribution of $\vec{u}$ is a series of parallel lines. Especially the flown line distribution of $\vec{u}$ in the region $X \times Y \in [-20, 20] \times [-20, 20]$, is more intensive than that in the rest of the region. It is seen clearly that the valued field of u_1 is $[-1, 0]$ and the surface distribution of u_1 decreases gradually to $u_1 = 1$ with the increasing of t. And then the valued field of u_2 is $[0, 1]$ and the surface distribution of u_2 gradually decreases to $u_2 = 1$ with the increase in time t. The surface distribution of p_x and p_y for three different values of t. It is clear that the surface distribution of p_x decreases from $+\infty$ to $p_x = 0$, and the surface distribution of p_y increases from $-\infty$ to $p_x = 0$.

(iii) Exact solution in Case 2 describes the gradual decreasing in the vortex on the infinity plane due to the influence of turbulent diffusion. Fig. 7 and Fig. 8 demonstrate the rotating vortex will gradually flatten out as the fluid pressure becoming weaker as time goes on. We can see that the vector $\vec{u}$ whirls around origin $(x = 0, y = 0)$ only and stream function ψ pictures a series of concentric circles merely within a certain circle. It is clear that the vectorial distribution of u_1 and u_2 are increasing as t increases. The initial steep shape of the spatial distribution of u_1, u_2 and p_x, p_y tend to be more and more gentle and their amplitude to be smaller with the increasing of t. If we choose $\nu = \dfrac{1}{\text{Re}}$, then we get that the change in ν value may lead to change in shape of spatial distribution of u_1, u_2 and p_x, p_y. With the increasing of ν value, their shape tend to be more and more gentle. In other words, the higher of the density becomes, the gentler of fluid motion.

(iv) Exact solution in Case 3 provides a description of 2D vortex moving from weak to strong and then to weak on infinity plane. The vector $\vec{u}$ in this case whirls around origin $(x = 0, y = 0)$, too. Especially, the flown line distribution of $\vec{u}$ in the region $X \times Y \in [-5, 5] \times [-5, 5]$, is more intensive than that in the rest of the region. The initial steep shape of spatial distribution of u_1, u_2 and p_x, p_y tend to be more and more gentle and their amplitude to be smaller with the increasing of t.

(v) The exact solution in Case 4 describes the circumfluence obvious characteristic of periodic changes within foursquare region because of the influence of turbulent diffusion, see Fig. 11. We can see that the flown line distribution of $\vec{u}$ pictures two families of hyperbolas. It is clearly seen that u_1 and u_2 are the space periodic solutions, and their region of periodic is $X \times Y \in [-5, 5] \times [-5, 5]$. If the parameters are chosen as Fig. 10, the shape of spatial distribution of u_1, u_2 and p_x, p_y remain the same in Fig. 10 with the increasing of t.

Acknowledgements The authors are grateful to the anonymous referees for their useful suggestions which improve the contents of this article.

References

[1] Abdeljawad T. On conformable fractional calculus[J]. Comput. Appl. Math., 2015, 279: 57-66.

[2] Atangana A, Baleanu D, Alsaedi A. New properties of conformable derivative[J]. Open Math., 2015, 13: 889-898.

[3] Bruneau C H, Tancogne S. Far field boundary conditions for incompressible flows computation[J]. J. Appl. Anal. Comput., 2018, 8(3): 690-709.

[4] Chung W. Fractional Newton mechanics with conformable fractional derivative[J]. Comput. Appl. Math., 2015, 290: 150-158.

[5] Diethelm K. The analysis of fractional differential equations[M]. Springer, Berlin, 2010.

[6] He Y, Sun W. Stability and convergence of the Crank-Nicolson/Adams-Bashforth scheme for the time-dependent Navier-Stokes equations[J]. SIAM J. Numer. Anal., 2007, 45(2): 837-869.

[7] Ingram R. A new linearly extrapolated Crank-Nicolson time-stepping scheme for the Navier-Stokes equations[J]. Math. Comp., 2013, 82(284): 1953-1973.

[8] Kaya S, Rivière B. A discontinuous subgrid eddy viscosity method for the time-dependent Navier-Stokes equations[J]. SIAM J. Numer. Anal., 2005, 43(4): 1572-1595.

[9] Khalil R, Alhorani M, Yousef A, Sababheh M. A new definition of fractional derivative[J]. Comput. Appl. Math., 2014, 65-70.

[10] Ku H C, Hatziavramidis D. Solutions of the two-dimensional Navier-Stokes equations by Chebyshev expansion methods[J]. Computer & Fluids, 1985, 13(1): 99-113.

[11] Peng L, Debbouche A, Zhou Y. Existence and approximations of solutions for time-fractional Navier-Stokes equations[J]. Math. Meth. Appl. Sci., 2018, 4: 1-12.

[12] Podlubny I. Fractional differential equations[M]. Academic Press, San Diego, 1999.

[13] Profilo G, Soliani G, Tebaldi C. Some exact solutions of the two-dimensional Navier-Stokes equations[J]. Int. J. Engng Sci., 1998, 36(4): 459-471.

[14] Ragab A A, Hemida K M, Mohamed M S, Abd El Salam M A. Solution of time fractional Navier-Stokes equation by using Homotopy analysis method[J]. Gen. Math. Notes, 2012, 13(2): 13-21.

[15] Saitoh T. A numerical method for two-dimensional Navier-Stokes equation by multipoint finite differences[J]. International J. Numer. Meth. Engin., 1977, 11: 1439-1454.

[16] Shan L, Hou Y. A fully discrete stabilized finite element method for the time-dependent Navier-Stokes equations[J]. Appl. Math. Comput., 2009, 215(1): 85-99.

[17] Shirikyan A. Analyticity of solutions for randomly forced two-dimensional Navier-Stokes equations[J]. Russian Math. Surveys, 2002, 54(4): 785-799.

[18] Terrill R M. An exact solution for flow in a porous pipe[J]. Appl. Math. Phys., 1982, 33(4): 547-552.

[19] Tian B, Gao Y, Spherical Kadomtsev-Petviashvili equation and nebulous for dustion-acoustic waves with symbolic computation[J]. Physics Letters A, 2005, 340: 243-250.

[20] Tian B, Gao Y. On the solitonic structures of the cylindrical dust-acoustic and dustion-acoustic waves with symbolic computation[J]. Physics Letters A, 2005, 340: 449-455.

[21] Torres D J, Coutsias E A. Pseudospectral solution of the two-dimensional Navier-Stokes equations in a disk[J]. SIAM J. Sci. Comput., 1999, 21(1): 378-403.

[22] Tsangaris S, Kondaxakis D, Vlachakis N W. Exact solution of the Navier-Stokes equations for the pulsating Dean flow in a channel with porous walls[J]. International J. Engin. Sci., 2006, 44: 1498-1509.

[23] Wu C, Ji Z, Zhang Y, Hao J, Jin X. Some new exact solutions for the two-dimensional Navier-Stokes equations[J]. Physics Letters A, 2007, 371: 438-452.

[24] Wang C Y. Flow due to a stretching boundary with partial slip-an exact solution of the Navier-Stokes equations[J]. Chemical Engineering Science, 2002, 57: 3745-3747.

[25] Zhang Z, Ouyang X, Yang X. Refined a priori estimates for the axisymmetric Navier-Stokes equations[J]. J. Appl. Anal. Comput., 2017, 7: 554-558.

[26] Zhang Y, Li Y, An R. Two-level iteration penalty and variational multiscale method for steady incompressible flows[J]. J. Appl. Anal. Comput., 2016, 6: 607-627.

[27] Zhao Z. Exact solutions of a class of second-order nonlocal boundary value problems and applications[J]. Appl. Math. Comput., 2009, 215(5): 1926-1936.

[28] Zhao D, Luo M. General conformable fractional derivative and its physical interpretation[J]. Calcolo, 2017. https://DOI 10.1007/s10092-017-0213-8.

[29] Zhou Y, Peng L. Weak solutions of the time fractional Navier-Stokes equations and optimal control[J]. Comput. Math. Appl., 2017, 73(6): 1016-1027.

[30] Zou G, Zhou Y, Ahmad B, Alsaedi A. Finite difference/element method for time-fractional Navier-Stokes equations[J]. Appl. Comput. Math., 2018, arXiv:1802.09779v1[math. NA].

On the Cauchy Problem for the Shallow-Water Model with the Coriolis Effect*

Mi Yongsheng(米永生), Liu Yue(刘跃), Luo Ting(罗婷),

and Guo Boling (郭柏灵)

Abstract In this paper, we are concerned with an asymptotic model for wave propagation in shallow water with the effect of the Coriolis force. We first establish the local well-posedness in a range of the Besov spaces, as well as the local well-posedness in the critical space. Then, we study the Gevrey regularity of the shallow-water model by using a generalized Cauchy-Kovalevsky theorem, which implies that the shallow-water model admits analytical solutions locally in time and globally in space. Moreover, we obtain a precise lower bound of the lifespan and the continuity of the solution. Finally, working with moderate weight functions commonly used in time-frequency analysis, some persistence results to the shallow-water model are illustrated..

Keywords the Besov spaces; the shallow-water model; the local well-posedness; the Coriolis effect

1 Introduction

In this paper, we consider the following shallow-water model with the coriolis effect (see [64] for the derivation of the model)

$$\begin{cases} u_t - u_{txx} + uu_x + 4u_xu_{xx} + uu_{xxx} + \eta_x + 2\Omega\eta_t = 0, & t > 0, x \in \mathbf{R}, \\ \eta_t + ((1+\eta)u)_x = 0, & t > 0, x \in \mathbf{R}, \\ u(0,x) = u_0(x), \quad \eta(0,x) = \eta_0(x), & x \in \mathbf{R}, \end{cases} \quad (1.1)$$

---

* J. Differential Equations, 2019, 267(11): 6370–6408. DOI:10.1016/j.jde.2019.06.023.

where u is connected with the average of horizontal velocity, η is related to free surface elevation from equilibrium with the boundary condition $u \to 0$ and $\eta \to 0$ when $|x| \to \infty$, Ω is a dimensionless parameter describing the strength of the Coriolis effect ($\Omega > 0$).

Recently, some new model equations of long-crested shallow-water waves propagating in the equatorial ocean regions with the Coriolis effect due to the Earth's rotation have attracted more and more attention. It is noted that the Coriolis effect at the equator manifests itself in forcing azimuthal wave propagation [23]. While the flow field has a three-dimensional character, the meridional and vertical velocity components are very small, thus it is reasonable to consider the two-dimensional flows in order to capture the main dynamics [24, 25, 60]. Moreover, when the interest is focused on the interaction of waves with the equatorial undercurrent, i.e., the flows at low latitude, we adopted the f-plane approximation [22]. The following equations are derived using the setting mentioned above.

One of the model equations with the Coriolis effect is the rotation-Camassa-Holm (R-CH) equation

$$\begin{cases} u_t - \beta\mu u_{txx} + cu_x + 3\alpha\varepsilon uu_x - \beta_0\mu u_{xx} + \omega_1\varepsilon^2 u^2 u_x + \omega_2\varepsilon^3 u^3 u_x \\ \qquad = \alpha\beta\varepsilon\mu(2u_x u_{xx} + uu_{xxx}), \end{cases} \quad (1.2)$$

where the parameter Ω is the constant rotational frequency due to the Coriolis effect. The other constants appearing in the equation are defined by

$$c = \sqrt{1+\Omega^2} - \Omega, \quad \alpha = \frac{c^2}{1+c^2}, \quad \beta_0 = \frac{c(c^4+6c^2-1)}{6(c^2+1)^2}, \quad \beta = \frac{3c^4+8c^2-1}{6(c^2+1)^2},$$

$$\omega_1 = \frac{-3c(c^2-1)(c^2-2)}{2(1+c^2)^3}, \quad \text{and} \quad \omega_2 = \frac{(c^2-1)^2(c^2-2)(8c^2-1)}{2(1+c^2)^5}$$

satisfying $c \to 1, \beta \to 5/12, \beta_0 \to 1/4, \omega_1, \omega_2 \to 0$ and $\alpha \to 1/2$ when $\Omega \to 0$. The rotation-Camassa-Holm (R-CH) equation (1.2) was derived in [9, 42] as a model equation that describes the motion of the fluid with the Coriolis effect from the incompressible shallow water in the equatorial region. This model equation has nonlocal cubic and even quartic nonlinearities and a formal Hamiltonian structure. In [9, 42, 43], the effects of the Coriolis force caused by the Earth's rotation and nonlocal higher nonlinearities on blow-up criteria and wave-breaking phenomena are also investigated. Furthermore, they demonstrate nonexistence of the Camassa-Holm-type peaked solution and classify various localized traveling wave solutions to the equation (1.2) depending on the range of the rotation parameter. The global existence and uniqueness of the energy conservative weak solutions in the energy space H^1 to equation (1.2) was established in [70].

Recently, the new modeling of equatorial shallow-water waves with the Coriolis effect

$$\begin{cases} m_t + (\rho u)_x + \sigma(2mu_x + um_x) + 3(1-\sigma)uu_x + \frac{1}{2}(\rho^2)_x + 2\Omega(\rho^2 u)_x \\ \quad -8\Omega(\rho u)_x + 4\Omega u_x + 4\beta_2(u^3)_x + 24\beta_1(u^2\rho(\rho-1)) = 0, \\ \rho_t + (\rho u)_x + 2\Omega u u_x \rho + 8\beta_1(u^3)_x \rho = 0 \end{cases} \quad (1.3)$$

was derived [57], where

$$c = \sqrt{1+\Omega^2} - \Omega, \quad \beta_1 = \frac{5c^2 - 11}{48c^2}, \quad \beta_2 = \frac{2c^4 - 2c^2 + 8}{8c^3} \quad \text{and} \quad m = u - u_{xx}.$$

The model equation which is analogous to the Green-Naghdi equations with the second-order approximation of the Camassa-Holm scaling captures stronger nonlinear effects than the classical dispersive integrable equations like the Korteweg-de Vries and two-component Camassa-Holm system. The local well-posedness of the Cauchy problem is then established by the linear transport theory and wave-breaking phenomenon is investigated based on the method of characteristics and the Riccati-type differential inequality. Finally, the condition of permanent waves is demonstrated by the analyzing competition between the slope of an average of the horizontal velocity component and the free surface component.

The third member of the model equations with the Coriolis effect is the following rotation-two-component Camassa-Holm (R2CH) system

$$\begin{cases} u_t - u_{txx} - Au_x + 3uu_x = \sigma(2u_x u_{xx} + uu_{xxx}) - \mu u_{xxx} \\ \quad -(1-2\Omega A)\rho\rho_x + 2\Omega\rho(\rho u)_x, \\ \rho_x + u\rho_x = -\rho u_x, \end{cases} \quad (1.4)$$

where A characterizes a linear underlying shear flow, the real dimensionless constant σ provides the competition, or balance, in fluid convection between nonlinear steepening and amplification due to stretching, μ is a nondimensional parameter and characterizes the constant rotational speed of the Earth. Recently, Liu, etc. [35] derived the system (1.4) in the spirit of Ivanov's asymptotic perturbation analysis for the governing equations of two-dimensional rotational gravity water waves [61]. The authors mainly established the precise blow-up mechanism, and it is shown that the system (1.4) can only have singularities that corresponded to wave breaking. Moreover, some initial conditions are provided to guarantee that the wave breaking phenomena occurs in finite time. In [8], the authors showed that there are solitary waves with singularities, like peakons and cuspons, depending on the values of the rotating parameter and the balance index σ, they also proved that horizontally symmetric weak solutions of this model must be traveling waves. Further study of the system (1.4) on the line and circle can be found [63, 68, 72, 73].

In fact, the system (1.4) has a significant relationship with several models describing the motion of waves at the free surface of shallow water under the influence of gravity. If we consider the system (1.4) without effect of the Earth's rotation, i.e., $\Omega = 0$, it becomes the generalized DGH system [10, 45]

$$\begin{cases} u_t - u_{txx} - Au_x + 3uu_x = \sigma(2u_xu_{xx} + uu_{xxx}) - \mu u_{xxx} - \rho\rho_x, \\ \rho_x + (u\rho)_x = 0. \end{cases} \quad (1.5)$$

When $\sigma = 1$ and $\mu = 0$, the system (1.5) recovers the standard two-component integrable Camassa-Holm system

$$\begin{cases} u_t - u_{txx} - Au_x + 3uu_x - 2u_xu_{xx} - uu_{xxx} + \rho\rho_x = 0, \\ \rho_t + (\rho u)_x = 0, \end{cases} \quad (1.6)$$

where $u(t, x)$ describes the horizontal velocity of the fluid and $\rho(t, x)$ is in connection with the horizontal deviation of the surface from equilibrium, all measured in dimensionless units. Moreover, u and ρ satisfy the boundary conditions: $u \to 0$ and $\rho \to 1$ as $|x| \to \infty$. The system can be identified with the first negative flow of the AKNS hierarchy and possesses the interesting peakon and multi-kink solutions [11]. Moreover, it is connected with the time dependent Schrödinger spectral problem [11]. Popowicz [69] observes that the system is related to the bosonic sector of an $N = 2$ supersymmetric extension of the classical Camassa-Holm equation. Recently, many researchers have paid extensive attention to it. In [34], Escher et al. establish the local well-posedness and present the precise blow-up scenarios and several blow-up results of strong solutions to the system (1.6) on the line. In [21], Constantin and Ivanov investigate the global existence and blow-up phenomena of strong solutions of the system (1.6) on the line. Later, Guan and Yin [39] obtain a new global existence result for strong solutions to the system (1.6) and get several blow-up results, which improve the recent results in [21]. Recently, they study the global existence of weak solutions to the system (1.6) [40]. In [46], Henry studies the infinite propagation speed for the system (1.6). Gui and Liu [41] establish the local well-posedness for the system (1.6) in a range of the Besov spaces, they also derive a wave breaking mechanism for strong solutions. Mustafa [66] gives a simple proof of existence for the smooth travelling waves for the system (1.6). Hu and Yin [58,59] study the blow-up phenomena and the global existence of the system (1.6) on the circle.

It is worth pointing out that the system (1.6) can be regarded as a generalization of the famous Camassa-Holm (CH) equation in the case of $\rho = 0$,

$$u_t - u_{txx} + 3uu_x = 2u_x u_{xx} + uu_{xxx} \tag{1.7}$$

modeling the unidirectional propagation of shallow water waves over a flat bottom, $u(t,x)$ stands for the fluid velocity at time t in the spatial direction x. It is a well-known integrable equation describing the velocity dynamics of shallow water waves. This equation spontaneously exhibits emergence of singular solutions from smooth initial conditions. It has a bi-Hamilton structure [36] and is completely integrable [6,12]. In particular, it possesses an infinity of conservation laws and is solvable by its corresponding inverse scattering transform. After the birth of the Camassa-Holm equation, many works have been carried out to probe its dynamic properties. Such as, Camassa-Holm equation has traveling wave solutions of the form $ce^{-|x-ct|}$, called peakons, which describes an essential feature of the travelling waves of the largest amplitude (see [13, 18, 19, 30]). It is shown in [14, 20, 28] that the inverse spectral or scattering approach is a powerful tool to handle the Camassa-Holm equation and analyze its dynamics. It is worthwhile to mention that the Camassa-Holm equation gives rise to geodesic flow of a certain invariant metric on the Bott-Virasoro group [26,65], and this geometric illustration leads to a proof that the Least Action Principle holds. It is shown in [16] that the blow-up occurs in the form of breaking waves, namely, the solution remains bounded but its slope becomes unbounded in finite time. Moreover, the Camassa-Holm equation has global conservative solutions [4, 55] and dissipative solutions [5, 56]. Other methods to handle the problems relating to various dynamic properties of the Camassa-Holm equation and other shallow water equations, the reader is referred to [2, 15-17, 27, 29, 31, 38, 47-53] and the references therein.

Recently, Liu, etc. derived the system (1.1) in the spirit of Ivanov's asymptotic perturbation analysis for the governing equations of two-dimensional rotational gravity water waves [61]. The authors mainly established the local well-posedness in the Sobolev Space and wave breaking phenomena for the system (1.1). Moreover, the nonexistence of the Camassa-Holm-type peaked solution and classification of various traveling-wave solutions to the system (1.1) are also established.

Motivated by the references cited above, we first establish the local well-posedness in a range of the Besov spaces $B_{p,r}^s(\mathbf{R}) \times B_{p,r}^{s-1}(\mathbf{R})$ with $s > \max\{1 + 1/p, 3/2\}$, based on the Littlewood-Paley theory and the transport theory, the local well-posedness in the critical space $B_{2,1}^{3/2}(\mathbf{R}) \times B_{2,1}^{1/2}(\mathbf{R})$ is also established. Then, we study the gevrey regularity of the system (1.1) by using a generalized Cauchy-Kovalevsky theorem, which implies that system (1.1) admits analytical solutions locally in time and globally in space. Moreover, we obtain a precise lower bound of the lifespan and the continuity of the solution. Finally, working with moderate weight functions that are commonly used in time-frequency analysis, some persistent results to the equation are illustrated.

The remainder of the paper is organized as follows. In section 2, we recall some facts on the Besov space and the transport theory and establish the local well-posedness of solutions to (1.1) in Besov space. Section 3 is devoted to the local well-posedness of solutions to the system (1.1) in the critical space. In section 4, we investigate the Gevrey regularity and analyticity of the system (1.1). In Section 5, some fundamentals concerning moderate weight functions and the functional analytic setting for the system (1.1) are presented and persistence results for the system (1.1) to its supersymmetric extension are given.

2 Local well-posedness in the Besov spaces

In this section, we shall establish local well-posedness of the initial value problem (1.1) in the Besov spaces. First, for the convenience of the readers, we recall some facts on the Littlewood-Paley decomposition and some useful lemmas.

Notation $\mathcal{S}$ stands for the Schwartz space of smooth functions over $\mathbf{R}^d$ whose derivatives of all order decay at infinity. The set $\mathcal{S}'$ of temperate distributions is the dual set of $\mathcal{S}$ for the usual pairing. We denote the norm of the Lebesgue space $L^p(\mathbf{R})$ by $\|\cdot\|_{L^p}$ with $1 \leqslant p \leqslant \infty$, and the norm in the Sobolev space $H^s(\mathbf{R})$ with $s \in \mathbf{R}$ by $\|\cdot\|_{H^s}$.

Proposition 2.1(littlewood-Paley decomposition [7]) *Let* $\mathcal{B} \doteq \left\{\xi \in \mathbf{R}^d, |\xi| \leqslant \frac{4}{3}\right\}$ *and* $\mathcal{C} \doteq \left\{\xi \in \mathbf{R}^d, \frac{4}{3} \leqslant |\xi| \leqslant \frac{8}{3}\right\}$. *There exist two radial functions* $\chi \in C_c^\infty(\mathcal{B})$ *and* $\varphi \in C_c^\infty(\mathcal{C})$ *such that*

$$\chi(\xi) + \sum_{q \geqslant 0} \varphi(2^{-q}\xi) = 1, \quad \forall \xi \in \mathbf{R}^d,$$

$$|q - q'| \geqslant 2 \Rightarrow \operatorname{Supp} \varphi(2^{-q}\cdot) \cap \operatorname{Supp} \varphi(2^{-q'}\cdot) = \varnothing,$$

$$q \geqslant 1 \Rightarrow \operatorname{Supp} \chi(\cdot) \cap \operatorname{Supp} \varphi(2^{-q'}\cdot) = \varnothing,$$

$$\frac{1}{3} \leqslant \chi(\xi)^2 + \sum_{q \geqslant 0} \varphi(2^{-q}\xi)^2 \leqslant 1, \quad \forall \xi \in \mathbf{R}^d.$$

Furthermore, let $h \doteq \mathcal{F}^{-1}\varphi$ *and* $\tilde{h} \doteq \mathcal{F}^{-1}\chi$. *Then for all* $f \in \mathcal{S}'(\mathbf{R}^d)$, *the dyadic operators* Δ_q *and* S_q *can be defined as follows*

$$\Delta_q f \doteq \varphi(2^{-q}D)f = 2^{qd} \int_{\mathbf{R}^d} h(2^q y) f(x - y) \, dy \text{ for } q \geqslant 0,$$

$$S_q f \doteq \chi(2^{-q}D)f = \sum_{-1 \leqslant k \leqslant q-1} \Delta_k = 2^{qd} \int_{\mathbf{R}^d} \tilde{h}(2^q y) f(x - y) \, dy,$$

$$\Delta_{-1} f \doteq S_0 f \text{ and } \Delta_q f \doteq 0 \text{ for } q \leqslant -2.$$

Hence,

$$f = \sum_{q \geqslant 0} \Delta_q f \text{ in } \mathcal{S}'(\mathbf{R}^d),$$

where the right-hand side is called the nonhomogeneous Littlewood-Paley decomposition of f.

Lemma 2.1 (Bernstein's inequality [33]) *Let* $\mathcal{B}$ *be a ball with center 0 in* $\mathbf{R}^d$ *and* $\mathcal{C}$ *a ring with center 0 in* $\mathbf{R}^d$. *A constant* C *exists so that, for any positive real number* λ, *any nonnegative integer* k, *any smooth homogeneous function* σ *of degree* m *and any couple of real numbers* (a, b) *with* $b \geqslant a \geqslant 1$, *there hold*

$$\operatorname{Supp} \hat{u} \subset \lambda \mathcal{B} \Rightarrow \sup_{|\alpha|=k} \|\partial^\alpha u\|_{L^a} \leqslant C^{k+1} \lambda^{k+d(\frac{1}{a} - \frac{1}{b})} \|u\|_{L^a},$$

$$\text{Supp } \hat{u} \subset \lambda \mathcal{C} \Rightarrow C^{-k-1}\lambda^k \|u\|_{L^a} \leqslant \sup_{|\alpha|=k} \|\partial^\alpha u\|_{L^a} \leqslant C^{k+1}\lambda^k \|u\|_{L^a},$$

$$\text{Supp } \hat{u} \subset \lambda \mathcal{C} \Rightarrow \|\sigma(D)u\|_{L^b} \leqslant C_{\sigma,m}\lambda^{m+d(\frac{1}{a}-\frac{1}{b})}\|u\|_{L^a},$$

for any function $u \in L^a$.

Definition 2.1 (Besov space) Let $s \in \mathbf{R}, 1 \leqslant p, r \leqslant \infty$. The inhomogenous Besov space $B^s_{p,r}(\mathbf{R}^d)$ ($B^s_{p,r}$ for short) is defined by

$$B^s_{p,r} \doteq \{f \in \mathcal{S}'(\mathbf{R}^d); \|f\|_{B^s_{p,r}} < \infty\},$$

where

$$\|f\|_{B^s_{p,r}} \doteq \begin{cases} \left(\sum_{q \in \mathbf{Z}} 2^{qsr}\|\Delta_q f\|^r_{L_p}\right)^{\frac{1}{r}}, & \text{for } r < \infty, \\ \sup_{q \in \mathbf{Z}} 2^{qs}\|\Delta_q f\|_{L_p}, & \text{for } r = \infty. \end{cases}$$

If $s = \infty, B^\infty_{p,r} \doteq \bigcap_{s \in \mathbf{R}} B^s_{p,r}$.

Proposition 2.2 (see [33]) Suppose that $s \in \mathbf{R}, 1 \leqslant p, r, p_i, r_i \leqslant \infty (i = 1, 2)$. We have

(1) *Topological properties*: $B^s_{p,r}$ is a Banach space which is continuously embedded in $\mathcal{S}'$.

(2) *Density*: C^∞_c is dense in $B^s_{p,r} \Leftrightarrow 1 \leqslant p, r < \infty$.

(3) *Embedding*: $B^s_{p_1,r_1} \hookrightarrow B^{s-n(\frac{1}{p_1}-\frac{1}{p_2})}_{p_2,r_2}$, if $p_1 \leqslant p_2$ and $r_1 \leqslant r_2$. $B^{s_2}_{p,r_2} \hookrightarrow B^{s_1}_{p,r_1}$ locally compact, if $s_1 < s_2$.

(4) *Algebraic properties*: $\forall s > 0, B^s_{p,r} \cap L^\infty$ is an algebra. Moreover, $B^s_{p,r}$ is an algebra, provided that $s > \frac{n}{p}$ or $s \geqslant \frac{n}{p}$ and $r = 1$.

(5) *Complex interpolation*:

$$\|u\|_{B^{\theta s_1+(1-\theta)s_2}_{p,r}} \leqslant C\|u\|^\theta_{B^{s_1}_{p,r}}\|u\|^{1-\theta}_{B^{s_2}_{p,r}}, \quad \forall u \in B^{s_1}_{p,r} \cap B^{s_2}_{p,r}, \quad \forall \theta \in [0,1].$$

(6) *Fatou lemma*: If $(u_n)_{n \in \mathbf{N}}$ is bounded in $B^s_{p,r}$ and $u_n \to u$ in $\mathcal{S}'$, then $u \in B^s_{p,r}$ and

$$\|u\|_{B^s_{p,r}} \leqslant \liminf_{n \to \infty} \|u_n\|_{B^s_{p,r}}.$$

(7) Let $m \in \mathbf{R}$ and f be an S^m-multiplier (i.e., $f : \mathbf{R}^d \to \mathbf{R}$ is smooth and satisfies that $\forall \alpha \in \mathbf{N}^d$, there exists a constant C_α, s.t. $|\partial^\alpha f(\xi)| \leqslant C_\alpha(1+|\xi|^{m-|\alpha|})$ for all $\xi \in \mathbf{R}^d$). Then the operator $f(D)$ is continuous from $B^s_{p,r}$ to $B^{s-m}_{p,r}$.

Now we state some useful results in the transport equation theory, which are crucial to the proofs of our main theorems later.

Lemma 2.2 (see [32, 33]) *Suppose that $(p,r) \in [1, +\infty]^2$ and $s > -\frac{d}{p}$. Let v be a vector field such that ∇v belongs to $L^1([0,T]; B_{p,r}^{s-1})$ if $s > 1 + \frac{d}{p}$ or to $L^1([0,T]; B_{p,r}^{\frac{d}{p}} \cap L^\infty)$ otherwise. Suppose also that $f_0 \in B_{p,r}^s$, $F \in L^1([0,T]; B_{p,r}^s)$ and that $f \in L^\infty(L^1([0,T]; B_{p,r}^s) \cap C([0,T]; \mathcal{S}')$ solves the d-dimensional linear transport equations*

$$\text{(T)} \quad \begin{cases} \partial_t f + v \cdot \nabla f = F, \\ f|_{t=0} = f_0. \end{cases}$$

Then there exists a constant C depending only on s, p and d such that the following statements hold:

(1) *If $r = 1$ or $s \neq 1 + \frac{d}{p}$, then*

$$\|f\|_{B_{p,r}^s} \leq \|f_0\|_{B_{p,r}^s} + \int_0^t \|F(\tau)\|_{B_{p,r}^s} d\tau + C \int_0^t V'(\tau) \|f(\tau)\|_{B_{p,r}^s} d\tau,$$

or

$$\|f\|_{B_{p,r}^s} \leq e^{CV(t)} \left(\|f_0\|_{B_{p,r}^s} + \int_0^t e^{-CV(\tau)} \|F(\tau)\|_{B_{p,r}^s} d\tau \right) \quad (2.1)$$

hold, where $V(t) = \int_0^t \|\nabla v(\tau)\|_{B_{p,r}^{\frac{d}{p}} \cap L^\infty} d\tau$ if $s < 1 + \frac{d}{p}$ and $V(t) = \int_0^t \|\nabla v(\tau)\|_{B_{p,r}^{s-1}} d\tau$ else.

(2) *If $s \leq 1 + \frac{d}{p}$ and $\nabla f_0 \in L^\infty$, $\nabla f \in L^\infty([0,T] \times \mathbf{R}^d)$ and $\nabla F \in L^1([0,T]; L^\infty)$, then*

$$\|f\|_{B_{p,r}^s} + \|\nabla f\|_{L^\infty}$$
$$\leq e^{CV(t)} \left(\|f_0\|_{B_{p,r}^s} + \|\nabla f_0\|_{L^\infty} + \int_0^t e^{-CV(\tau)} \left(\|F(\tau)\|_{B_{p,r}^s} + \|\nabla F(\tau)\|_{L^\infty} \right) d\tau \right)$$

with $V(t) = \int_0^t \|\nabla v(\tau)\|_{B_{p,r}^{\frac{d}{p}} \cap L^\infty} d\tau$.

(3) *If $f = v$, then for all $s > 0$, the estimate (2.1) holds with*

$$V(t) = \int_0^t \|\nabla v(\tau)\|_{L^\infty} d\tau.$$

(4) If $r < +\infty$, then $f \in C([0,T]; B_{p,r}^s)$. If $r = +\infty$, then $f \in C([0,T]; B_{p,r}^{s'})$ for all $s' < s$.

Lemma 2.3 (existence and uniqueness see [32,33]) Let $(p, p_1, r) \in [1, +\infty]^3$ and $s > -d \min\left\{\frac{1}{p_1}, \frac{1}{p'}\right\}$ with $p' \doteq \left(1 - \frac{1}{p}\right)^{-1}$. Assume that $f_0 \in B_{p,r}^s$, $F \in L^1([0,T]; B_{p,r}^s)$. Let v be a time dependent vector field such that $v \in L^\rho([0,T]; B_{\infty,\infty}^{-M})$ for some $\rho > 1, M > 0$ and $\nabla v \in L^1([0,T]; B_{p_1,r}^{\frac{d}{p_1}} \cap L^\infty)$ if $s < 1 + \frac{d}{p_1}$ and $\nabla v \in L^1([0,T]; B_{p_1,r}^{s-1})$ if $s > 1 + \frac{d}{1p}$ or $s = 1 + \frac{d}{p_1}$ and $r = 1$. Then the transport equations (T) has a unique solution $f \in L^\infty([0,T]; B_{p,r}^s) \cap (\cap_{s' < s} C[0,T]; B_{p,r}^{s'})$ and the inequalities in Lemma 2.2 hold. Moreover, $r < \infty$, then we have $f \in C[0,T]; B_{p,1}^s)$.

Lemma 2.4 (1-D Moser-type estimates [32,33]) Assume that $1 \leqslant p, r \leqslant +\infty$, the following estimates hold:

(i) For $s > 0$,
$$\|fg\|_{B_{p,r}^s} \leqslant C(\|f\|_{B_{p,r}^s} \|g\|_{L^\infty} + \|g\|_{B_{p,r}^s} \|f\|_{L^\infty});$$

(ii) $\forall s_1 \leqslant \frac{1}{p} < s_2 \left(s_2 \geqslant \frac{1}{p} \text{ if } r = 1\right)$ and $s_1 + s_2 > 0$, we have
$$\|fg\|_{B_{p,r}^{s_1}} \leqslant C\|f\|_{B_{p,r}^{s_1}} \|g\|_{B_{p,r}^{s_2}};$$

(iii) In Sobolev spaces $H^s = B_{2,2}^s$, we have for $s > 0$,
$$\|f\partial_x g\|_{H^s} \leqslant C(\|f\|_{H^{s+1}}\|g\|_{L^\infty} + \|\partial_x g\|_{H^s}\|f\|_{L^\infty}),$$

where C is a positive constant independent of f and g.

Lemma 2.5 (interpolation inequality [32,33]) (1) Let $s_2 > s_1, \theta \in (0,1)$, we have
$$\|u\|_{B_{p,1}^{\theta s_1 + (1-\theta)s_2}} \leqslant \frac{C}{s_2 - s_1}\left(\frac{1}{\theta} + \frac{1}{1-\theta}\right)\|u\|_{B_{p,\infty}^{s_1}}^\theta \|u\|_{B_{p,\infty}^{s_2}}^{1-\theta}.$$

(2) $\forall s \in \mathbf{R}, \epsilon > 0$ and $1 \leqslant p \leqslant \infty$, there exists a constant $C > 0$ such that
$$\|u\|_{B_{p,1}^s}^\theta \leqslant C \frac{\epsilon + 1}{\epsilon} \|u\|_{B_{p,\infty}^s}^\theta \left(1 + \ln \frac{\|u\|_{B_{p,\infty}^{s+\epsilon}}^\theta}{\|u\|_{B_{p,\infty}^s}^\theta}\right).$$

Definition 2.2 For $T > 0$, $s \in \mathbf{R}$ and $1 \leq p \leq +\infty$, we set

$$E^s_{p,r}(T) \triangleq C([0,T]; B^s_{p,r}) \cap C^1([0,T]; B^{s-1}_{p,r}) \quad \text{if} \quad r < +\infty,$$

$$E^s_{p,\infty}(T) \triangleq L^\infty([0,T]; B^s_{p,\infty}) \cap \text{Lip}([0,T]; B^{s-1}_{p,\infty})$$

and $E^s_{p,r} \triangleq \bigcap_{T>0} E^s_{p,r}(T).$

Note that if $g(x) = \frac{1}{2}e^{-|x|}$, $x \in \mathbf{R}$, then $(1-\partial_x^2)^{-1} f = g * f$ for all $f \in L^2(\mathbf{R})$, where $*$ denotes the spatial convolution. Using this identity, and applying the pseudo-differential operator $(1 - \partial_x^2)^{-1}$ to the system (1.1), one can rewrite the system (1.1) as a quasi-linear nonlocal evolution system of hyperbolic type

$$\begin{cases} u_t - uu_x = -\partial_x(1-\partial_x^2)^{-1}\left(u^2 + \frac{1}{2}u_x^2 + \eta - 2\Omega((1+\eta)u)\right), & t > 0, x \in \mathbf{R}, \\ \eta_t + u\eta_x = -(1+\eta)u_x, & t > 0, x \in \mathbf{R}, \\ u(0,x) = u_0(x), \quad \eta(0,x) = \eta_0(x), & x \in \mathbf{R}. \end{cases}$$
(2.2)

If we denote $P(D)$ as the Fourier integral operator with the Fourier multiplier $-i\xi(1+\xi^2)^{-1}$, then the system (2.2) becomes

$$\begin{cases} u_t - uu_x = P(D)\left(u^2 + \frac{1}{2}u_x^2 + \eta - 2\Omega((1+\eta)u)\right), & t > 0, x \in \mathbf{R}, \\ \eta_t + u\eta_x = (1+\eta)u_x, & t > 0, x \in \mathbf{R}, \\ u(0,x) = u_0(x), \quad \eta(0,x) = \eta_0(x), & x \in \mathbf{R}. \end{cases} \quad (2.3)$$

We are now ready to state the first main result of the paper.

Theorem 2.1 Let $p, r \in [1, \infty]$ and $s > \max\left\{\frac{3}{2}, 1+\frac{1}{p}\right\}$. Assume that $(u_0, \eta_0) \in B^s_{p,r} \times B^{s-1}_{p,r}$. There exists a time $T > 0$ such that the initial-value problem (1.1) or (2.3) has a unique solution $(u, \eta) \in E^s_{p,r}(T) \times E^{s-1}_{p,r}(T)$ and the map $(u_0, \eta_0) \mapsto (u, \eta)$ is continuous from a neighborhood of (u_0, η_0) in $B^s_{p,r} \times B^{s-1}_{p,r}$ into

$$C([0,T]; B^{s'}_{p,r}) \cap C^1([0,T]; B^{s'-1}_{p,r}) \times C([0,T]; B^{s'-1}_{p,r}) \cap C^1([0,T]; B^{s'-2}_{p,r})$$

for every $s' < s$ when $r = \infty$ and $s' = s$ whereas $r < \infty$.

Remark 2.1 When $p = r = 2$, the Besov space $B^s_{p,r}(\mathbf{R})$ coincides with the Sobolev space $H^s(\mathbf{R})$. Theorem 2.1 implies that under the assumption $(u_0, \eta_0) \in$

$H^s(\mathbf{R}) \times H^{s-1}(\mathbf{R})$ with $s > \frac{3}{2}$, we can obtain the local well-posedness for system (1.1), which improves the Theorem 3.1 in [64].

In the following, we denote $C > 0$ a generic constant only depending on p, r, s. Uniqueness and continuity with respect to the initial data are an immediate consequence of the following result.

Proposition 2.3 *Let $1 \leqslant p, r \leqslant +\infty$ and $s > \max\left\{\frac{3}{2}, 1+\frac{1}{p}\right\}$. Suppose that $(u^{(i)}; \eta^{(i)}) \in \left(L^\infty([0,T]; B_{p,r}^s) \cap C([0,T]; \mathcal{S}')\right) \times \left(L^\infty([0,T]; B_{p,r}^{s-1}) \cap C([0,T]; \mathcal{S}')\right)$ $(i = 1, 2)$ be two given solutions of the initial-value problem (1.1) or (2.3) with the initial data $(u_0^{(i)}; \eta_0^{(i)}) \in B_{p,r}^s \times B_{p,r}^{s-1} (i = 1, 2)$. Then for every $t \in [0; T]$, we have*

$$\|u^{(1)}(t) - u^{(2)}(t)\|_{B_{p,r}^{s-1}} + \|\eta^{(1)}(t) - \eta^{(2)}(t)\|_{B_{p,r}^{s-2}}$$
$$\leqslant \left(\|u_0^{(1)} - u_0^{(2)}\|_{B_{p,r}^{s-1}} + \|\eta_0^{(1)} - \eta_0^{(2)}\|_{B_{p,r}^{s-2}}\right)$$
$$\times e^{C \int_0^t \left(\|u^{(1)}(\tau)\|_{B_{p,r}^s} + \|u^{(2)}(\tau)\|_{B_{p,r}^s} + \|\eta^{(1)}(\tau)\|_{B_{p,r}^{s-1}} + \|\eta^{(2)}(\tau)\|_{B_{p,r}^{s-1}} + 1\right) d\tau}. \quad (2.4)$$

Proof Denote $u^{(12)} = u^{(2)} - u^{(1)}$, $\eta^{(12)} = \eta^{(2)} - \eta^{(1)}$. It is obvious that

$$u^{(12)} \in L^\infty([0,T]; B_{p,r}^s) \cap C([0,T]; \mathcal{S}'), \eta^{(12)} \in L^\infty([0,T]; B_{p,r}^{s-1}) \cap C([0,T]; \mathcal{S}')$$

which implies that $(u^{(12)}, \eta^{(12)}) \in C([0,T]; B_{p,r}^{s-1}) \times C([0,T]; B_{p,r}^{s-2})$ and $(u^{(12)}, \eta^{(12)})$ solves the transport equations

$$\begin{cases} u_t^{(12)} - u^{(1)} u_x^{(12)} = F, \\ \eta_t^{(12)} + u^{(1)} \eta_x^{(12)} = G, \\ u^{(12)}|_{t=0} = u_0^{(12)} := u_0^{(2)} - u_0^{(1)}, \\ \eta^{(12)}|_{t=0} = \eta_0^{(12)} := \eta_0^{(2)} - \eta_0^{(1)}, \end{cases} \quad (2.5)$$

with

$$F = u^{(12)} u_x^{(2)}$$
$$+ P(D)\left(u^{(12)}\left(u^{(1)} + u^{(2)} - 2\Omega \eta^{(1)} - 2\Omega\right) + \frac{1}{2} u_x^{(12)}(u_x^{(1)} + u_x^{(2)})\right.$$
$$\left. + \eta^{(12)}(1 - 2\Omega u^{(2)})\right),$$

$$G = -u^{(12)}\eta_x^{(2)} - (\eta^{(12)}u_x^{(1)} + \eta^{(2)}u_x^{(12)} + u_x^{(12)}).$$

According to Lemma 2.2, we have

$$e^{-C\int_0^t \|\partial_x u^{(1)}(\tau')\|_{B^{s-1}_{p,r}} d\tau'} \|u^{(12)}(t)\|_{B^{s-1}_{p,r}}$$

$$\leqslant \|u_0^{(12)}\|_{B^{s-1}_{p,r}} + C\int_0^t e^{-C\int_0^\tau \|\partial_x u^{(1)}(\tau')\|_{B^{s-1}_{p,r}} d\tau'} \|F\|_{B^{s-1}_{p,r}} d\tau, \quad (2.6)$$

and

$$e^{-C\int_0^t \|\partial_x u^{(1)}(\tau')\|_{B^{s-1}_{p,r}} d\tau'} \|\eta^{(12)}(t)\|_{B^{s-2}_{p,r}}$$

$$\leqslant \|\eta_0^{(12)}\|_{B^{s-2}_{p,r}} + C\int_0^t e^{-C\int_0^\tau \|\partial_x u^{(1)}(\tau')\|_{B^{s-1}_{p,r}} d\tau'} \|G\|_{B^{s-2}_{p,r}} d\tau. \quad (2.7)$$

For $s > \max\left\{1 + \frac{1}{p}, \frac{3}{2}\right\}$, $B^{s-1}_{p,r} \subset L^\infty$ is an algebra, $P(\varepsilon)$ is S^{-1} multiplies, according to Proposition 2.1 and Lemma 2.4, so we have

$$\|F\|_{B^{s-1}_{p,r}}$$

$$= \left\| P(D)\left(u^{(12)}(u^{(1)} + u^{(2)} - 2\Omega\eta^{(1)}) + \frac{1}{2}u_x^{(12)}(u_x^{(1)} + u_x^{(2)}) + \eta^{(12)}(1 - 2\Omega u^{(2)})\right) \right\|_{B^{s-1}_{p,r}}$$

$$+ \|u^{(12)}u_x^{(2)}\|_{B^{s-1}_{p,r}}$$

$$= C\|u^{(12)}\|_{B^{s-1}_{p,r}}\left(\|u^{(1)}\|_{B^{s-1}_{p,r}} + \|u^{(2)}\|_{B^{s-1}_{p,r}} + \|\eta^{(1)}\|_{B^{s-1}_{p,r}} + 1\right)$$

$$+ C\|u_x^{(12)}\|_{B^{s-2}_{p,r}}\left(\|u_x^{(1)}\|_{B^{s-1}_{p,r}} + \|u_x^{(2)}\|_{B^{s-1}_{p,r}}\right) + \|\eta^{(12)}\|_{B^{s-2}_{p,r}}(\|u^{(2)}\|_{B^{s-1}_{p,r}} + 1)$$

$$+ C\|u^{(12)}\|_{B^{s-1}_{p,r}}\|u_x^{(2)}\|_{B^{s-1}_{p,r}}$$

$$\leqslant C\left(\|u^{(12)}\|_{B^{s-1}_{p,r}} + \|\eta^{(12)}\|_{B^{s-2}_{p,r}}\right)$$

$$\times \left(\|u^{(1)}\|_{B^s_{p,r}} + \|u^{(2)}\|_{B^s_{p,r}} + \|\eta^{(1)}\|_{B^{s-1}_{p,r}} + \|\eta^{(2)}\|_{B^{s-1}_{p,r}} + 1\right), \quad (2.8)$$

$$\|G\|_{B^{s-2}_{p,r}}$$

$$= \left\| -u^{(12)}\eta_x^{(2)} - (\eta^{(12)}u_x^{(1)} + \eta^{(2)}u_x^{(12)} + u_x^{(12)}) \right\|_{B^{s-2}_{p,r}}$$

$$= \|u^{(12)}\eta_x^{(2)}\|_{B^{s-2}_{p,r}} + \|\eta^{(12)}u_x^{(1)}\|_{B^{s-2}_{p,r}} + \|\eta^{(2)}u_x^{(12)}\|_{B^{s-2}_{p,r}} + \|u_x^{(12)}\|_{B^{s-2}_{p,r}}$$

$$= \|u^{(12)}\|_{B^{s-1}_{p,r}}\|\eta_x^{(2)}\|_{B^{s-2}_{p,r}} + \|\eta^{(12)}\|_{B^{s-1}_{p,r}}\|u_x^{(1)}\|_{B^{s-1}_{p,r}}$$

$$+ \|\eta^{(2)}\|_{B^{s-2}_{p,r}}\|u_x^{(12)}\|_{B^{s-1}_{p,r}} + \|u_x^{(12)}\|_{B^{s-2}_{p,r}}$$

$$\leqslant C \left(\|u^{(12)}\|_{B_{p,r}^{s-1}} + \|\eta^{(12)}\|_{B_{p,r}^{s-2}} \right)$$
$$\times \left(\|u^{(1)}\|_{B_{p,r}^{s}} + \|u^{(2)}\|_{B_{p,r}^{s}} + \|\eta^{(1)}\|_{B_{p,r}^{s-1}} + \|\eta^{(2)}\|_{B_{p,r}^{s-1}} + 1 \right). \tag{2.9}$$

Therefore, by inserting the above estimates to (2.6)-(2.9), we obtain

$$e^{-C\int_0^t \|\partial_x u^{(1)}(\tau')\|_{B_{p,r}^{s-1}} d\tau'} \left(\|u^{(12)}(t)\|_{B_{p,r}^{s-1}} + \|\eta^{(12)}(t)\|_{B_{p,r}^{s-2}} \right)$$
$$\leqslant \|u_0^{(12)}\|_{B_{p,r}^{s-1}} + \|\eta_0^{(12)}\|_{B_{p,r}^{s-2}}$$
$$+ C\int_0^t e^{-C\int_0^\tau \|\partial_x u^{(1)}(\tau')\|_{B_{p,r}^{s-2}} d\tau'} \left(\|u^{(12)}\|_{B_{p,r}^{s-1}} + \|\eta^{(12)}\|_{B_{p,r}^{s-2}} \right)$$
$$\times \left(\|u^{(1)}\|_{B_{p,r}^{s}} + \|u^{(2)}\|_{B_{p,r}^{s}} + \|\eta^{(1)}\|_{B_{p,r}^{s-1}} + \|\eta^{(2)}\|_{B_{p,r}^{s-1}} + 1 \right) d\tau.$$

Hence, applying the Gronwall inequality, we reach (2.4).

Now let us start the proof of Theorem 2.1, which is motivated by the proof of the local existence theorem about the Camassa-Holm equation in [32]. Firstly, we shall use the classical Friedrichs regularization method to construct the approximate solutions to the Cauchy problem problem (2.3).

Lemma 2.6 *Assume that $u^{(0)} = \eta^{(0)} = 0$. Let $1 \leqslant p, r \leqslant +\infty$, $s > \max\left\{\frac{3}{2}, 1 + \frac{1}{p}\right\}$ and $(u_0, \eta_0) \in B_{p,r}^s \times B_{p,r}^{s-1}$. Then there exists a sequence of smooth functions $(u^{(l)}, \eta^{(l)})_{l \in \mathbf{N}} \in C(\mathbf{R}^+; B_{p,r}^\infty)^2$ solving the following linear transport equation by induction*

$$\begin{cases} \left(\partial_t - u^{(l)} \partial_x \right) u^{(l+1)} \\ \quad = P(D) \left((u^{(l)})^2 + \frac{1}{2}(u_x^{(l)})^2 + \eta^{(l)} - 2\Omega(1+\eta^{(l)})u^{(l)} \right), & dt > 0, \ x \in \mathbf{R}, \\ \left(\partial_t + u^{(l)} \partial_x \right) \eta^{(l+1)} = -\eta^{(l)} \partial_x u^{(l)} - \partial_x u^{(l)}, & t > 0, \ x \in \mathbf{R}, \\ u^{(l+1)}(x, 0) = u_0^{(l+1)}(x) = S_{l+1} u_0, \quad \eta^{(l+1)}(x, 0) = \eta_0^{(l+1)}(x) = S_{l+1} \eta_0, & x \in \mathbf{R}. \end{cases} \tag{2.10}$$

Moreover, there is a positive T such that the solutions satisfying the following properties:

(i) $(u^{(l)}, \eta^{(l)})_{l \in \mathbf{N}}$ *is uniformly bounded in $E_{p,r}^s(T) \times E_{p,r}^{s-1}(T)$.*

(ii) $(u^{(l)}, \eta^{(l)})_{l \in \mathbf{N}}$ *is a Cauchy sequence in $C([0,T]; B_{p,r}^{s-1}) \times C([0,T]; B_{p,r}^{s-2})$.*

Proof Since all the data $S_{n+1} u_0$ and $S_{n+1} \eta_0$ belong to $B_{p,r}^\infty$, Lemma 2.3 enables us to show by induction that for all $l \in \mathbf{N}$, the equation (2.10) has a global solution which belongs to $C(\mathbf{R}^+; B_{p,r}^\infty)^2$. Thanks to Lemma 2.2 and the proof of Proposition 2.3, we have the following inequality for all $l \in \mathbf{N}$,

$$e^{-C\int_0^t \|u^{(l)}(\tau')\|_{B_{p,r}^s} d\tau'} \|u^{(l+1)}(t)\|_{B_{p,r}^s}$$

$$\leqslant \|u_0\|_{B_{p,r}^s} + \frac{C}{2}\int_0^t e^{-C\int_0^\tau \|u^{(l)}(\tau')\|_{B_{p,r}^s} d\tau'}$$

$$\times \left(\|u^{(l)}\|_{B_{p,r}^s}^2 + \|\eta^{(l)}\|_{B_{p,r}^{s-1}}^2 + \|u^{(l)}\|_{B_{p,r}^s}\|\eta^{(l)}\|_{B_{p,r}^{s-1}} + \|\eta^{(l)}\|_{B_{p,r}^{s-1}} + \|u^{(l)}\|_{B_{p,r}^{s-1}}\right) d\tau,$$

and

$$e^{-C\int_0^t \|u^{(l)}(\tau')\|_{B_{p,r}^s} d\tau'} \|\eta^{(l+1)}(t)\|_{B_{p,r}^s}$$

$$\leqslant \|\eta_0\|_{B_{p,r}^s} + \frac{C}{2}\int_0^t e^{-C\int_0^\tau \|u^{(l)}(\tau')\|_{B_{p,r}^s} d\tau'} \left(\|u^{(l)}\|_{B_{p,r}^s}\|\eta^{(l)}\|_{B_{p,r}^{s-1}} + \|u^{(l)}\|_{B_{p,r}^s}\right) d\tau.$$

Hence, we have

$$e^{-C\int_0^t \|\partial_x u^{(l)}(\tau')\|_{B_{p,r}^s} d\tau'} \left(\|u^{(l+1)}(t)\|_{B_{p,r}^s} + \|\eta^{(l+1)}(t)\|_{B_{p,r}^s}\right)$$

$$\leqslant \|u_0\|_{B_{p,r}^s} + \|\eta_0\|_{B_{p,r}^{s-1}} + \frac{C}{2}\int_0^t e^{-C\int_0^\tau \|u^{(l)}(\tau')\|_{B_{p,r}^s} d\tau'}$$

$$\times \left(\|u^{(l)}\|_{B_{p,r}^s} + \|\eta^{(l)}\|_{B_{p,r}^{s-1}}\right)\left(\|u^{(l)}\|_{B_{p,r}^s} + \|\eta^{(l)}\|_{B_{p,r}^{s-1}} + 1\right) d\tau. \qquad (2.11)$$

Let us choose a $T > 0$ such that

$$T < \min\left\{\frac{1}{4C\left(\|u_0\|_{B_{p,r}^s} + \|\eta_0\|_{B_{p,r}^{s-1}}\right)}, \frac{1}{2C}\right\}$$

and suppose by induction that for all $t \in [0, T]$,

$$\|u^{(l)}(t)\|_{B_{p,r}^s} + \|\eta^{(l)}(t)\|_{B_{p,r}^{s-1}} \leqslant \frac{2\left(\|u_0\|_{B_{p,r}^s} + \|v_0\|_{B_{p,r}^s}\right)}{1 - 4C\left(\|u_0\|_{B_{p,r}^s} + \|v_0\|_{B_{p,r}^s}\right)t}. \qquad (2.12)$$

And then inserting (2.12) into (2.11) leads to

$$\sqrt{1 - 4C\left(\|u_0\|_{B_{p,r}^s} + \|\eta_0\|_{B_{p,r}^{s-1}}\right)t} \left(\|u^{(l+1)}(t)\|_{B_{p,r}^s} + \|\eta^{(l+1)}(t)\|_{B_{p,r}^{s-1}}\right)$$

$$\leqslant \|u_0\|_{B_{p,r}^s} + \|\eta_0\|_{B_{p,r}^{s-1}} + \int_0^t \frac{2C\left(\|u_0\|_{B_{p,r}^s} + \|\eta_0\|_{B_{p,r}^{s-1}}\right)^2 d\tau}{\left(1 - 4C\left(\|u_0\|_{B_{p,r}^s} + \|\eta_0\|_{B_{p,r}^{s-1}}\right)\tau\right)^{\frac{3}{2}}}$$

$$+ \int_0^t \frac{C\left(\|u_0\|_{B_{p,r}^s} + \|\eta_0\|_{B_{p,r}^{s-1}}\right) d\tau}{\left(1 - 4C\left(\|u_0\|_{B_{p,r}^s} + \|\eta_0\|_{B_{p,r}^{s-1}}\right)\tau\right)^{\frac{1}{2}}}$$

$$\leqslant \frac{\|u_0\|_{B_{p,r}^s} + \|\eta_0\|_{B_{p,r}^{s-1}}}{\sqrt{1 - 4C\left(\|u_0\|_{B_{p,r}^s} + \|\eta_0\|_{B_{p,r}^{s-1}}\right)t}}$$

$$+ \frac{1}{2}\left(1 - \sqrt{1 - 4C\left(\|u_0\|_{B_{p,r}^s} + \|\eta_0\|_{B_{p,r}^{s-1}}\right)t}\right)$$

$$\leqslant \frac{2\left(\|u_0\|_{B_{p,r}^s} + \|\eta_0\|_{B_{p,r}^{s-1}}\right)}{\sqrt{1 - 4C\left(\|u_0\|_{B_{p,r}^s} + \|\eta_0\|_{B_{p,r}^{s-1}}\right)t}}. \tag{2.13}$$

Hence, one can see that

$$\|u^{(l+1)}(t)\|_{B_{p,r}^s} + \|\eta^{(l+1)}(t)\|_{B_{p,r}^s} \leqslant \frac{2\left(\|u_0\|_{B_{p,r}^s} + \|v_0\|_{B_{p,r}^s}\right)}{1 - 4C\left(\|u_0\|_{B_{p,r}^s} + \|v_0\|_{B_{p,r}^s}\right)t},$$

which implies that $(u^{(l)}, \eta^{(l)})_{l \in \mathbf{N}}$ is uniformly bounded in

$$C([0;T]; B_{p,r}^s) \times C([0;T]; B_{p,r}^{s-1}).$$

Using the fact that $B_{p,r}^{s-1}$ with $s > 1 + \dfrac{1}{p}$ is an algebra, together with the Moser-type estimates (see Lemma 2.4), one can see that

$$-u^{(l)}\partial_x u^{(l+1)}, \quad P(D)\left((u^{(l)})^2 + \frac{1}{2}(u_x^{(l)})^2 + \eta^{(l)} - 2\Omega(1 + \eta^{(l)})u^{(l)}\right)$$

are uniformly bounded in $C([0;T]; B_{p,r}^{s-1})$, and

$$\eta^{(l)}\partial_x \eta^{(l+1)}, -\eta^{(l)}\partial_x u^{(l)} - \partial_x u^{(l)}$$

are uniformly bounded in $C([0;T]; B_{p,r}^{s-2})$. Hence, using the equation (2.10), we have

$$(\partial_x u^{(l+1)}, \partial_x \eta^{(l+1)})_{l \in \mathbf{N}} \in C([0;T]; B_{p,r}^{s-1}) \times C([0;T]; B_{p,r}^{s-2})$$

uniformly bounded, which yields that the sequence $(u^{(l)}, \eta^{(l)})_{\in \mathbf{N}}$ is uniformly bounded in $E_{p,r}^s(T) \times E_{p,r}^{s-1}(T)$.

Now, it suffices to show that $(u^{(l)}, \eta^{(l)})_{l \in \mathbf{N}}$ is a Cauchy sequence in $C([0;T]; B_{p,r}^{s-1}) \times C([0;T]; B_{p,r}^{s-2})$. In fact, for all $l, k \in \mathbf{N}$, from (2.10), we have

$$\begin{cases} (\partial_t - u^{(l+k)}\partial_x)(u^{(l+k+1)} - u^{(l+1)}) = F', \\ (\partial_t + u^{(l+k)}\partial_x)(\eta^{(l+k+1)} - \eta^{(l+1)}) = G', \end{cases} \quad (2.14)$$

with

$$F' = (u^{(l+k)} - u^{(l)})\partial_x u^{(l+1)} + P(D)\Big((u^{(l+k)} - u^{(l)})(u^{(l)} + u^{(l+k)} - 2\Omega\eta^{(l)} - 2\Omega)$$

$$+ \frac{1}{2}(u_x^{(l+k)} - u_x^{(l)})(u_x^{(l)} + u_x^{(l+k)}) + (\eta^{(l+k)} - \eta^{(l)})(1 - 2\Omega u^{(l+k)})\Big),$$

$$G' = -(u^{(l+k)} - u^{(l)})\partial_x \eta^{(l+1)} - ((\eta^{(l+k)} - \eta^{(l)})u_x^{(l)}$$

$$- (u_x^{(l+k)} - u_x^{(l)})\eta_x^{(l+k)} - (u_x^{(l+k)} - u_x^{(l)})).$$

Similar to the proof of Proposition 2.3, then for every $t \in [0, T]$, we obtain

$$e^{-C\int_0^t \|u^{(k+l)}(\tau')\|_{B_{p,r}^s} d\tau'} \|u^{(k+l+1)} - u^{(l+1)}\|_{B_{p,r}^{s-1}}$$

$$\leqslant \|u_0^{(k+l+1)} - u_0^{(l+1)}\|_{B_{p,r}^{s-1}} + C \int_0^t e^{-C\int_0^\tau \|(u^{(l+k)})(\tau')\|_{B_{p,r}^s} d\tau'}$$

$$\times \Big(\|u^{(k+l)} - u^{(l)}\|_{B_{p,r}^{s-1}}$$

$$\times (\|u^{(l)}\|_{B_{p,r}^s} + \|u^{(l+1)}\|_{B_{p,r}^s} + \|u^{(l+k)}\|_{B_{p,r}^s} + \|\eta^{(l)}\|_{B_{p,r}^{s-1}} + 1)$$

$$+ \|\eta^{(k+l)} - \eta^{(l)}\|_{B_{p,r}^{s-2}}(\|u^{(l+k)}\|_{B_{p,r}^{s-1}} + 1)\Big) d\tau,$$

and

$$e^{-C\int_0^t \|u^{(k+l)}(\tau')\|_{B_{p,r}^s} d\tau'} \|\eta^{(k+l+1)} - \eta^{(l+1)}\|_{B_{p,r}^{s-2}}$$

$$\leqslant \|u_0^{(k+l+1)} - u_0^{(l+1)}\|_{B_{p,r}^{s-2}} + C \int_0^t e^{-C\int_0^\tau \|(u^{(l+k)})(\tau')\|_{B_{p,r}^s} d\tau'}$$

$$\times \Big(\|u^{(k+l)} - u^{(l)}\|_{B_{p,r}^{s-1}}(\|\eta^{(l+1)}\|_{B_{p,r}^{s-1}} + \|\eta^{(l+k)}\|_{B_{p,r}^{s-1}} + 1)$$

$$+ \|\eta^{(k+l)} - \eta^{(l)}\|_{B_{p,r}^{s-2}} \|u^{(l+1)}\|_{B_{p,r}^s} \Big) d\tau.$$

Since $(u^{(l)}, \eta^{(l)})_{l \in \mathbf{N}}$ is uniformly bounded in $E_{p,r}^s(T) \times E_{p,r}^{s-1}(T)$ and

$$u_0^{(l+k+1)} - u_0^{(l+1)} = S_{k+l+1}u_0 - S_{l+1}u_0 = \sum_{q=l+1}^{l+k} \Delta_q u_0,$$

$$\eta_0^{(l+k+1)} - \eta_0^{(l+1)} = S_{k+l+1}\eta_0 - S_{l+1}\eta_0 = \sum_{q=l+1}^{l+k} \Delta_q \eta_0,$$

we get a constant C_T independent of l, k such that for all $t \in [0, T]$,

$$\|(u^{(k+l+1)} - u^{(l+1)})(t)\|_{B_{p,r}^{s-1}} + \|(\eta^{(k+l+1)} - \eta^{(l+1)})(t)\|_{B_{p,r}^{s-2}}$$
$$\leqslant C_T \left(2^{-l} + \int_0^t \left(\|(u^{(k+l)} - u^{(l)})(\tau)\|_{B_{p,r}^{s-1}} + \|(\eta^{(k+l)} - \eta^{(l)})(\tau)\|_{B_{p,r}^{s-2}} \right) d\tau \right).$$

Arguing by induction with respect to the index l, one can easily prove that

$$\|(u^{(k+l+1)} - u^{(l+1)})(t)\|_{L_T^\infty(B_{p,r}^{s-1})} + \|(\eta^{(k+l+1)} - \eta^{(l+1)})(t)\|_{L_T^\infty(B_{p,r}^{s-2})}$$
$$\leqslant \frac{(TC_T)^{l+1}}{(l+1)!}(\|u^k\|_{L_T^\infty(B_{p,r}^s)} + \|\eta^k\|_{L_T^\infty(B_{p,r}^{s-1})}) + C_T \sum_{q=0}^{l} 2^{q-l} \frac{(TC_T)^q}{q!}.$$

As $\|u^{(k)}\|_{L_T^\infty(B_{p,r}^{s-1})}, \|\eta^{(k)}\|_{L_T^\infty(B_{p,r}^{s-2})}$ and C are bounded independently of k, there exists constant C_T' independent of l, k such that

$$\|(u^{(k+l+1)} - u^{(l+1)})(t)\|_{L_T^\infty(B_{p,r}^{s-1})} + \|(\eta^{(k+l+1)} - \eta^{(l+1)})(t)\|_{L_T^\infty(B_{p,r}^{s-1})} \leqslant C_T' 2^{-n}.$$

Thus $(u^{(l)}, \eta^{(l)})_{l \in \mathbf{N}}$ is a Cauchy sequence in $C([0,T]; B_{p,r}^{s-1}) \times ([0,T]; B_{p,r}^{s-2})$.

Proof of Theorem 2.1 Thanks to Lemma 2.6, we obtain that $(u^{(l)}, \eta^{(l)})_{l \in \mathbf{N}}$ is a Cauchy sequence in $C([0,T]; B_{p,r}^{s-1}) \times C([0,T]; B_{p,r}^{s-2})$, so it converges to some function $(u, \eta) \in C([0,T]; B_{p,r}^{s-1}) \times C([0,T]; B_{p,r}^{s-2})$. We now have to check that (u, η) belongs to $E_{p,r}^s(T) \times E_{p,r}^{s-1}(T)$ and solves the Cauchy problem (1.1) or (2.3). Since $(u^{(l)}, \eta^{(l)})_{l \in \mathbf{N}}$ is uniformly bounded in $L^\infty([0,T]; B_{p,r}^s) \times L^\infty([0,T]; B_{p,r}^{s-1})$ according to Lemma 2.6, the Fatou property for the Besov spaces (Proposition 2.2) guarantees that (u, η) also belongs to

$$L^\infty([0,T]; B_{p,r}^s) \times L^\infty([0,T]; B_{p,r}^{s-1}).$$

On the other hand, as $(u^{(l)}, \eta^{(l)})_{l \in \mathbf{N}}$ converges to (u, η) in $C([0,T]; B_{p,r}^{s-1}) \times C([0,T]; B_{p,r}^{s-2})$, an interpolation argument ensures that the convergence holds in

$C([0,T]; B_{p,r}^{s'}) \times C([0,T]; B_{p,r}^{s'-1})$, for any $s' < s$. It is then easy to pass to the limit in the equation (2.10) and to conclude that (u, η) is indeed a solution to the Cauchy problem (1.1) or (2.3). Thanks to the fact that (u, η) belongs to $L^\infty([0,T]; B_{p,r}^s) \times L^\infty([0,T]; B_{p,r}^{s-1})$, the right-hand side of the equation

$$u_t - uu_x = P(D)\left(u^2 + \frac{1}{2}u_x^2 + \eta - 2\Omega(1+\eta)u\right)$$

belongs to $L^\infty([0,T]; B_{p,r}^s)$, and the right-hand side of the equation

$$\eta_t + u\eta_x = -\eta u_x - u_x$$

belongs to $L^\infty([0,T]; B_{p,r}^{s-1})$ In particular, for the case $r < \infty$, Lemma 2.3 enables us to conclude that $(u, \eta) \in C([0,T]; B_{p,r}^{s'}) \times C([0,T]; B_{p,r}^{s'-1})$ for any $s' < s$. Finally, using the equation again, we see that $(\partial_t u, \partial_t \eta) \in C([0,T]; B_{p,r}^{s-1}) \times C([0,T]; B_{p,r}^{s-2})$ if $r < \infty$, and in $L^\infty([0,T]; B_{p,r}^{s-1}) \times L^\infty([0,T]; B_{p,r}^{s-2})$ otherwise. Therefore, (u, η) belongs to $E_{p,r}^s(T) \times E_{p,r}^{s-1}(T)$. Moreover, the standard use of a sequence of viscosity approximate solutions $(u_\varepsilon, v_\varepsilon)_{\varepsilon>0}$ for the Cauchy problem (1.1) or (2.3) which converges uniformly in $C([0,T]; B_{p,r}^s) \cap C^1([0,T]; B_{p,r}^{s-1}) \times C([0,T]; B_{p,r}^s) \cap C^1([0,T]; B_{p,r}^{s-2})$ gives the continuity of the solution (u, η) in $E_{p,r}^s \times E_{p,r}^{s-1}$. The proof of Theorem 2.1 is completed.

3 The well-posedness in the critical space

This section aims to investigate the local well-posedness of the system (1.1) or (2.3) in the critical Besov space $B_{2,1}^{\frac{3}{2}} \times B_{2,1}^{\frac{1}{2}}$ and it is shown that the data-to-solution mapping is Hölder continuous. The uniqueness and existence of the solution in the critical space is guaranteed by the following result.

Theorem 3.1 Let $(u_0, \eta_0) \in B_{2,1}^{\frac{3}{2}} \times B_{2,1}^{\frac{1}{2}}$ the system (1.1) or (2.3) admits a unique solution $(u, \eta) \in E_{2,1}^{\frac{3}{2}}(T) \times E_{2,1}^{\frac{1}{2}}(T)$ for some positive T depends only on the initial data, and there exists a constant C such that if

$$T = \frac{1}{8C\left(\|u_0\|_{B_{2,1}^{\frac{3}{2}}} + \|\eta_0\|_{B_{2,1}^{\frac{1}{2}}}\right)},$$

then

$$\|u\|_{L^\infty(0,T;B_{2,1}^{\frac{3}{2}})} + \|\eta\|_{L^\infty(0,T;B_{2,1}^{\frac{1}{2}})} \leqslant 2\left(\|u\|_{B_{2,1}^{\frac{3}{2}}} + \|\eta\|_{B_{2,1}^{\frac{1}{2}}}\right). \quad (3.1)$$

Moreover, the data-to-solution mapping $(u_0, \eta_0) \mapsto (u, \eta) : B_{2,1}^{\frac{3}{2}} \times B_{2,1}^{\frac{1}{2}} \mapsto B_{2,1}^s \times B_{2,1}^{s-1}$ is Hölder continuous for any $s \in \left(\frac{1}{2}, \frac{3}{2}\right)$. Namely, let $(u^{(i)}, \eta^{(i)})$ be solutions to the system (1.1) or (2.3) with respect to $(u_0^{(i)}, \eta_0^{(i)})$, $i = 1, 2$, then we have

$$\sup_{t \in [0,T]} \|u^{(1)}(t) - u^{(2)}(t)\|_{B_{2,1}^s} + \sup_{t \in [0,T]} \|\eta^{(1)}(t) - \eta^{(2)}(t)\|_{B_{2,1}^{s-1}}$$

$$\leqslant C \left(\|u_0^{(1)}(t) - u_0^{(2)(t)}\|_{B_{2,1}^{\frac{3}{2}}} + \|\eta_0^{(1)}(t) - \eta_0^{(2)(t)}\|_{B_{2,1}^{\frac{3}{2}}} \right)^{\theta \exp\{-Ct\}}, \qquad (3.2)$$

$\forall \theta = \frac{3}{2} - s \in (0, 1)$.

Proof The proof will be divided into several steps.

Step 1 Consider the system (2.10) in Section 2 and assume that $(u^{(l)}, \eta^{(l)}) \in L^\infty(0, T; B_{2,1}^{\frac{3}{2}} \times B_{2,1}^{\frac{1}{2}})$ Since $B_{2,1}^{\frac{3}{2}} \hookrightarrow B_{2,1}^{\frac{1}{2}}$ and $B_{2,1}^{\frac{1}{2}}$ is a Banach algebra, it is easy to verify that the right-hand side of the system (2.10) belongs to $L_{loc}^\infty(\mathbf{R}^+; B_{2,1}^{\frac{3}{2}})$ and $L_{loc}^\infty(\mathbf{R}^+; B_{2,1}^{\frac{1}{2}})$, respectively. Taking advantage of the Lemma 2.3 ensures that the system (2.10) has a global solution in $E_{2,1}^{\frac{3}{2}}(T)$ for any given $T > 0$.

Step 2 Similar to the proof of Theorem 2.1, we now consider two cases one can find $T > 0$ such that

$$T < \frac{1}{4C \left(\|u_0\|_{B_{2,1}^{\frac{3}{2}}} + \|\eta_0\|_{B_{2,1}^{\frac{1}{2}}} \right)},$$

and

$$\|u^{(l)}(t)\|_{B_{2,1}^{\frac{3}{2}}} + \|\eta^{(l)}(t)\|_{B_{2,1}^{\frac{1}{2}}}$$

$$\leqslant \frac{2 \left(\|u_0\|_{B_{2,1}^{\frac{3}{2}}} + \|\eta_0\|_{B_{2,1}^{\frac{1}{2}}} \right)}{1 - 4C \left(\|u_0\|_{B_{2,1}^{\frac{3}{2}}} + \|\eta_0\|_{B_{2,1}^{\frac{1}{2}}} \right) t}$$

$$\leqslant \frac{2 \left(\|u_0\|_{B_{2,1}^{\frac{3}{2}}} + \|\eta_0\|_{B_{2,1}^{\frac{1}{2}}} \right)}{1 - 4C \left(\|u_0\|_{B_{2,1}^{\frac{3}{2}}} + \|\eta_0\|_{B_{2,1}^{\frac{1}{2}}} \right) t} \equiv M, \quad \forall t \in [0, T]. \qquad (3.3)$$

This shows that the sequence $(u^{(l)}, \eta^{(l)})_{l \in N}$ is uniformly bounded in $L^\infty(0, T; B_{2,1}^{\frac{3}{2}} \times B_{2,1}^{\frac{1}{2}})$. By using the system (2.10), we can verify that the sequence $(u^{(l)}, \eta^{(l)})_{l \in N}$ is uniformly bounded in $E_{2,1}^{\frac{3}{2}}$. Especially, the estimate (3.1) can be easily obtained by taking

$$T = \frac{1}{8C\left(\|u_0\|_{B_{2,1}^{\frac{3}{2}}} + \|\eta_0\|_{B_{2,1}^{\frac{1}{2}}}\right)}.$$

Step 3 We are going to show that $(u^{(l)}, \eta^{(l)})_{l\in\mathbb{N}}$ is a Cauchy sequence in $C([0;T]; B_{2,\infty}^{1/2} \times B_{2,\infty}^{-1/2})$. Applying the Lemma 2.2 to (2.14), it follows from the uniformly boundedness of $(u^{(l)}, \eta^{(l)})_{l\in\mathbb{N}}$ that

$$\|u^{(l+k+1)}(t) - u^{(l+1)}(t)\|_{B_{2,\infty}^{\frac{1}{2}}}$$

$$\leqslant C\left(\|S_{l+k+1}u_0 - S_{l+1}u_0\|_{B_{2,\infty}^{\frac{1}{2}}} + C\int_0^t \|F'\|_{B_{2,\infty}^{\frac{1}{2}}}\,d\tau\right). \quad (3.4)$$

Noting that $P(D) \in S^{-1}$, by virtue of the embedding $B_{2,1}^{-\frac{1}{2}} \hookrightarrow B_{2,1}^{-\frac{3}{2}}$ and $B_{2,1}^{\frac{1}{2}} \hookrightarrow B_{2,\infty}^{\frac{3}{2}} \cap L^{\infty}$, the second term on the right-hand side of (3.4) can be estimated as

$$\|F'\|_{B_{2,\infty}^{\frac{1}{2}}} \leqslant C\Big(\|(u^{(l+k)} - u^{(l)})\partial_x u^{(l+1)}\|_{B_{2,1}^{\frac{1}{2}}} + \|(u^{(l+k)} - u^{(l)})(u^{(l)} + u^{(l+k)})\|_{B_{2,1}^{\frac{1}{2}}} + 1)$$

$$+ \|(u^{(l+k)} - u^{(l)})\eta^{(l)}\|_{B_{2,\infty}^{-\frac{1}{2}}} + \|(u_x^{(l+k)} - u_x^{(l)})(u_x^{(l)} + u_x^{(l+k)})\|_{B_{2,\infty}^{-\frac{1}{2}}}$$

$$+ \|\eta^{(l+k)} - \eta^{(l)}\|_{B_{2,\infty}^{-\frac{1}{2}}} + \|(\eta^{(l+k)} - \eta^{(l)})u^{(l+k)}\|_{B_{2,\infty}^{-\frac{1}{2}}}\Big). \quad (3.5)$$

By using the Banach algebra property of $B_{2,1}^{\frac{1}{2}}$ and the embedding $B_{2,1}^{\frac{1}{2}} \hookrightarrow B_{2,\infty}^{\frac{3}{2}} \cap L^{\infty}$, one can estimate the terms on the right-hand side of (3.5) by

$$\|(u^{(l+k)} - u^{(l)})\partial_x u^{(l+1)}\|_{B_{2,1}^{\frac{1}{2}}} + \|(u^{(l+k)} - u^{(l)})(u^{(l)} + u^{(l+k)})\|_{B_{2,1}^{\frac{1}{2}}} + 1)$$

$$\leqslant \|u^{(l+k)} - u^{(l)}\|_{B_{2,1}^{\frac{1}{2}}} \|u^{(l+1)}\|_{B_{2,1}^{\frac{3}{2}}} + \|u^{(l+k)} - u^{(l)}\|_{B_{2,1}^{\frac{1}{2}}} \|u^{(l)} + u^{(l+k)}\|_{B_{2,1}^{\frac{1}{2}}}$$

$$+ \|u^{(l+k)} - u^{(l)}\|_{B_{2,1}^{\frac{1}{2}}}$$

$$\leqslant C\|u^{(l+k)} - u^{(l)}\|_{B_{2,1}^{\frac{1}{2}}},$$

since $\|u^{(l)}(t)\|_{L^{\infty}(0,T;B_{2,1}^{\frac{3}{2}})} \leqslant M$ is uniformly bounded.

Moreover, by using the Moser-type estimate (Lemma (2.4)), we have

$$\|(u^{(l+k)} - u^{(l)})\eta^{(l)}\|_{B_{2,\infty}^{-\frac{1}{2}}} + \|(u_x^{(l+k)} - u_x^{(l)})(u_x^{(l)} + u_x^{(l+k)})\|_{B_{2,\infty}^{-\frac{1}{2}}}$$

$$+ \|\eta^{(l+k)} - \eta^{(l)}\|_{B_{2,\infty}^{-\frac{1}{2}}} + \|(\eta^{(l+k)} - \eta^{(l)})u^{(l+k)}\|_{B_{2,\infty}^{-\frac{1}{2}}},$$

$$\leqslant C \left(\|u^{(l+k)} - u^{(l)}\|_{B_{2,1}^{\frac{1}{2}}} + \|\eta^{(l+k)} - \eta^{(l)}\|_{B_{2,\infty}^{-\frac{1}{2}}} \right).$$

It is not difficult to verify that

$$\|u_0^{(l+k+1)}(t) - u_0^{(l+1)}(t)\|_{B_{2,\infty}^{\frac{1}{2}}} = \|S_{l+k+1}u_0 - S_{l+1}u_0\|_{B_{2,\infty}^{\frac{1}{2}}} \leqslant C 2^{-l} \|u_0\|_{B_{2,1}^{\frac{3}{2}}},$$

$$\|\eta_0^{(l+k+1)}(t) - \eta_0^{(l+1)}(t)\|_{B_{2,\infty}^{-\frac{1}{2}}} = \|S_{l+k+1}\eta_0 - S_{l+1}\eta_0\|_{B_{2,\infty}^{-\frac{1}{2}}} \leqslant C 2^{-l} \|u_0\|_{B_{2,1}^{\frac{1}{2}}}. \quad (3.6)$$

Plugging the above estimates into (3.5), one can conclude from (3.4) that

$$\|u^{(l+k+1)}(t) - u^{(l+1)}(t)\|_{B_{2,\infty}^{\frac{1}{2}}}$$

$$\leqslant C \left(2^{-l} \|u_0\|_{B_{2,1}^{\frac{3}{2}}} + \int_0^t \left(\|u^{(l+k)} - u^{(l)}\|_{B_{2,1}^{\frac{1}{2}}} + \|\eta^{(l+k)} - \eta^{(l)}\|_{B_{2,\infty}^{-\frac{1}{2}}} \right) d\tau \right). \quad (3.7)$$

On the other hand, by the same taken for the second equation in (2.10), one can deduce that

$$\|\eta^{(l+k+1)}(t) - \eta^{(l+1)}(t)\|_{B_{2,\infty}^{-\frac{1}{2}}}$$

$$\leqslant C \left(2^{-l} \|\eta_0\|_{B_{2,1}^{\frac{1}{2}}} + \int_0^t \left(\|u^{(l+k)} - u^{(l)}\|_{B_{2,1}^{\frac{1}{2}}} + \|\eta^{(l+k)} - \eta^{(l)}\|_{B_{2,\infty}^{-\frac{1}{2}}} \right) d\tau \right) \quad (3.8)$$

Setting

$$H^{k,l} = \|u^{(l+k)}(t) - u^{(l)}(t)\|_{B_{2,\infty}^{\frac{1}{2}}} + \|\eta^{(l+k)}(t) - \eta^{(l)}(t)\|_{B_{2,\infty}^{-\frac{1}{2}}}.$$

It follows from (3.7) and (3.8) that

$$H^{k,l+1} \leqslant C \left(2^{-l} + \int_0^t \left(H^{k,l} + \|u^{(l+k)} - u^{(l)}\|_{B_{2,1}^{\frac{1}{2}}} \right) d\tau \right). \quad (3.9)$$

However, by means of the Log-type interpolation inequality (see Lemma 2.5), we have

$$\|u^{(k+l)} - u^{(l)}\|_{B_{2,1}^{\frac{1}{2}}} \leqslant C \|u^{(k+l)} - u^{(l)}\|_{B_{2,\infty}^{\frac{1}{2}}} \ln \left(e + \frac{\|u^{(k+l)} - u^{(l)}\|_{B_{2,\infty}^{\frac{3}{2}}}}{\|u^{(k+l)} - u^{(l)}\|_{B_{2,\infty}^{\frac{1}{2}}}} \right)$$

$$\leqslant C\|u^{(k+l)} - u^{(l)}\|_{B_{2,\infty}^{\frac{1}{2}}} \ln\left(e + \frac{2M}{\|u^{(k+l)} - u^{(l)}\|_{B_{2,\infty}^{\frac{1}{2}}}}\right)$$

$$\leqslant CH^{k,l}(t) \ln\left(e + \frac{2M}{H^{k,l}(t)}\right),$$

since $f(x) = x \ln\left(e + \frac{C}{x}\right)$ $(C > 0)$ s a nondecreasing function, which combined with (3.9) yields that

$$H^{k,l+1}(t) \leqslant C2^{-l} + C \int_0^t H^{k,l}(t) \ln\left(e + \frac{2M}{H^{k,l}(t)}\right) d\tau.$$

Define $\overline{H}^l(t) = \sup_{\tau \in [0,t], k \in \mathbb{N}} H^{k,l}(\tau)$, the previous inequality implies that

$$\overline{H}^{l+1}(t) \leqslant C2^{-l} + C \int_0^t \overline{H}^l(t) \ln\left(e + \frac{2M}{\overline{H}^l(t)}\right) d\tau. \tag{3.10}$$

Let $\overline{H} = \limsup_{l \in \mathbb{N}} \overline{H}^{l+1}(t)$. By taking the superior limit on both sides of the inequality (3.10) with respect to l, and using the Fatou's Lemma and again the monotone property of function $f(x) = x \ln\left(e + \frac{C}{x}\right)$ $(C > 0)$. We have

$$\overline{H}(t) \leqslant C \int_0^t \limsup_{l \in \mathbb{N}} \overline{H}^l(t) \ln\left(e + \frac{2M}{\limsup_{l \in \mathbb{N}} \overline{H}^l(t)}\right) d\tau$$

$$= \int_0^t \overline{H}(t) \ln\left(e + \frac{2M}{\overline{H}(t)}\right) d\tau. \tag{3.11}$$

An application of the Osgood's Lemma to the above inequality implies that $\overline{H}(t) \equiv 0$, which shows that the sequence $(u^{(l)}, \eta^{(l)})_{l \in \mathbb{N}}$ is a Cauchy sequence

$$C(0,T; B_{2,\infty}^{\frac{1}{2}} \times B_{2,\infty}^{-\frac{1}{2}}).$$

Hence there exists a function (u,η) such that $(u^{(l)}, \eta^{(l)}) \to (u,\eta)$ in $C(0,T; B_{2,\infty}^{\frac{1}{2}} \times B_{2,\infty}^{-\frac{1}{2}})$ as $n \to \infty$ Then by taking the similar argument as those in the proof of Theorem 2.1, one can conclude that $(u,\eta) \in E_{2,1}^{\frac{3}{2}}(T) \times E_{2,1}^{\frac{1}{2}}(T)$ is indeed a solution for the system (2.10).

Step 4 In order to prove the uniqueness of the solution, let $(u^{(i)}, \eta^{(i)})$, $i = 1, 2$ be two solutions corresponding to the initial data $(u_0^{(i)}, \eta_0^{(i)})$, $i = 1, 2$ Consider the system (2.5) with respect to the difference between two solutions. Using the Lemma 2.2 and following the similar computations as we did in Step 3, we obtain

$$\|u^{(12)}(t)\|_{B_{2,\infty}^{\frac{1}{2}}} + \|\eta^{(12)}(t)\|_{B_{2,\infty}^{-\frac{1}{2}}}$$
$$\leqslant C\left(\|u_0^{(12)}\|_{B_{2,1}^{\frac{1}{2}}} + \|\eta_0^{(12)}\|_{B_{2,1}^{\frac{1}{2}}} + \int_0^t \left(\|u^{(12)}(t)\|_{B_{2,1}^{\frac{1}{2}}} + \|\eta^{(12)}(t)\|_{B_{2,\infty}^{-\frac{1}{2}}}\right) d\tau\right). \quad (3.12)$$

Applying the Lemma 2.5, we have

$$\|u^{(12)}(t)\|_{B_{2,1}^{\frac{1}{2}}} \leqslant \|u^{(12)}(t)\|_{B_{2,\infty}^{\frac{1}{2}}} \ln\left(e + \frac{2M}{\|u^{(12)}(t)\|_{B_{2,\infty}^{\frac{1}{2}}}}\right). \quad (3.13)$$

Inserting the above inequality into (3.12), and noting that $\ln(e + \frac{a}{x}) \leqslant \ln(e + a)(1 - \ln x)$ with $x > 0$ and $a > 0$, we get

$$\frac{\|u^{(12)}(t)\|_{B_{2,\infty}^{\frac{1}{2}}} + \|\eta^{(12)}(t)\|_{B_{2,\infty}^{-\frac{1}{2}}}}{2M}$$
$$\leqslant C \frac{\|u_0^{(12)}\|_{B_{2,\infty}^{\frac{1}{2}}} + \|\eta_0^{(12)}\|_{B_{2,\infty}^{-\frac{1}{2}}}}{2M} + C\int_0^t \frac{\|u^{(12)}(\tau)\|_{B_{2,\infty}^{\frac{1}{2}}} + \|\eta^{(12)}(\tau)\|_{B_{2,\infty}^{-\frac{1}{2}}}}{2M}$$
$$\times \ln\left(e + \frac{2M}{\|u^{(12)}(\tau)\|_{B_{2,\infty}^{\frac{1}{2}}} + \|\eta^{(12)}(\tau)\|_{B_{2,\infty}^{-\frac{1}{2}}}}\right) d\tau$$
$$\leqslant C \frac{\|u_0^{(12)}\|_{B_{2,\infty}^{\frac{1}{2}}} + \|\eta_0^{(12)}\|_{B_{2,\infty}^{-\frac{1}{2}}}}{2M} + C\int_0^t \frac{\|u^{(12)}(\tau)\|_{B_{2,\infty}^{\frac{1}{2}}} + \|\eta^{(12)}(\tau)\|_{B_{2,\infty}^{-\frac{1}{2}}}}{2M}$$
$$\times \left(1 - \ln\frac{\|u^{(12)}(\tau)\|_{B_{2,\infty}^{\frac{1}{2}}} + \|\eta^{(12)}(\tau)\|_{B_{2,\infty}^{-\frac{1}{2}}}}{2M}\right) d\tau. \quad (3.14)$$

By taking

$$c = C\frac{\|u_0^{(12)}\|_{B_{2,\infty}^{\frac{1}{2}}} + \|\eta_0^{(12)}\|_{B_{2,\infty}^{-\frac{1}{2}}}}{2M},$$

$\beta(x) = x(1-\ln x)$ and $\gamma(t) = C$, it then follows from the Osgood's Lemma that

$$\frac{\|u^{(12)}(t)\|_{B_{2,\infty}^{\frac{1}{2}}} + \|\eta^{(12)}(t)\|_{B_{2,\infty}^{-\frac{1}{2}}}}{2M} \leq \left(C\frac{\|u_0^{(12)}\|_{B_{2,\infty}^{\frac{1}{2}}} + \|\eta_0^{(12)}\|_{B_{2,\infty}^{-\frac{1}{2}}}}{2M}\right)^{\exp\{-Ct\}},$$

which is equivalent to

$$\|u^{(12)}(t)\|_{B_{2,\infty}^{\frac{1}{2}}} + \|\eta^{(12)}(t)\|_{B_{2,\infty}^{-\frac{1}{2}}} \leq C\left(\|u_0^{(12)}\|_{B_{2,1}^{\frac{3}{2}}} + \|\eta_0^{(12)}\|_{B_{2,1}^{\frac{1}{2}}}\right)^{\exp\{-Ct\}}.$$

By using the interpolation inequality and the uniform boundedness of the solutions, we have

$$\|u^{(12)}(t)\|_{B_{2,1}^{s}} + \|\eta^{(12)}(t)\|_{B_{2,1}^{s-1}}$$

$$\leq C\left(\|u^{(12)}(t)\|_{B_{2,\infty}^{\frac{3}{2}}}^{1-\theta}\|u^{(12)}(t)\|_{B_{2,\infty}^{\frac{1}{2}}}^{\theta} + \|\eta^{(12)}(t)\|_{B_{2,\infty}^{-\frac{1}{2}}}^{\theta}\|\eta^{(12)}(t)\|_{B_{2,\infty}^{\frac{1}{2}}}^{1-\theta}\right)$$

$$\leq C\left(\left(\|u^{(12)}(t)\|_{B_{2,\infty}^{\frac{3}{2}}} + \|\eta^{(12)}(t)\|_{B_{2,\infty}^{\frac{1}{2}}}\right)^{1-\theta}\left(\|u^{(12)}(t)\|_{B_{2,\infty}^{\frac{1}{2}}}^{\theta} + \|\eta^{(12)}(t)\|_{B_{2,\infty}^{-\frac{1}{2}}}\right)^{\theta}\right)$$

$$\leq C\left(\|u_0^{(12)}\|_{B_{2,1}^{\frac{3}{2}}} + \|\eta_0^{(12)}\|_{B_{2,1}^{\frac{1}{2}}}\right)^{\exp\{-Ct\}}, \tag{3.15}$$

for any $\theta = \frac{3}{2} - s \in (0,1)$, which implies that the data-to-solution mapping is Holder continuous from $B_{2,1}^{\frac{3}{2}} \times B_{2,1}^{\frac{1}{2}}$ into $B_{2,1}^{s} \times B_{2,1}^{s-1}$ for any $s \in \left(\frac{1}{2}, \frac{3}{2}\right)$. This completes the proof of Theorem 3.1.

4 Gevery regularity and analyticity

In this section, we show that the system (2.2) is locally well-posed in the Gevery-Sobolev spaces in the sense of Hardamard. Especially, the system (2.2) admits unique analytic solutions locally in time and globally in space. Moreover, we obtain a lower bound of the lifespan.

To begin with, we recall the classical definition of the Gevrey class.

Definition 4.1 *A function $g \in C^{\infty}(\mathbf{R}^n)$ is said to be in the Gevrey class $G^{\tau}(\mathbf{R}^n)$ for some $\tau > 0$, if there exists positive constants C and $\mathbf{R}$ such that*

$$|D^{\alpha}G(x)| \leq C\frac{(k!)^{\tau}}{R^k}, \quad |\alpha| = k, \quad \alpha \in \mathbf{N}^n, \quad x \in \mathbf{T}^n,$$

where $\mathbf{R}$ represents the radius of Gevrey-class regularity of the function f.

Remark 4.1 The class $G^1(\mathbf{R}^n)$ is equal to the space of real-analytic functions on $\mathbf{R}^n$. If $0 < \tau < 1$, then f is called an ultra-analytic function. Moreover, the functions in $G^\tau(\mathbf{R}^n)$ with $\tau > 1$ is a smooth but not analytic function.

The following definition, which was introduced by Foias and Temam [37] to study the analyticity of solution to the Navier-Stokes equation, is an equivalent characterization of the Gevrey-class.

Definition 4.2 ([37]) For all $\tau \geqslant 1$ and $s \geqslant 0$, the Gevrey-class $G^\tau(\mathbf{R}^n)$ is given by
$$\bigcup_{\delta > 0} D\left((-\Delta)^s e^{\delta(-\Delta)^{\frac{1}{\tau}}}\right),$$
where
$$\left\|(-\Delta)^s e^{\delta(-\Delta)^{\frac{1}{2\tau}}} f\right\|_{L^2}^2 (2\pi)^n \sum_{k \in \mathbf{Z}^n} |k|^{2s} e^{2\delta |k|^{\frac{1}{\tau}}} |\widehat{f}(k)|^2,$$
where $\widehat{f}$ is the Fourier coefficient of f in $\mathbf{T}^n$.

Combining the above two definitions, we now introduce the framework of the Gevery class spaces $G^\tau(\mathbf{R}^n)$, which is the working space in the present paper.

Definition 4.3 Given τ, δ and $s \in \mathbf{R}$, the Sobolev-Gevrey spaces $G^\delta_{\tau,s}(\mathbf{R}^n)$ is defined by
$$G^\delta_{\tau,s}(\mathbf{R}^n) = \left\{ f \in C^\infty(\mathbf{R}^n); \|f\|_{G^\tau_{\tau,s}} = \left(\int_{\mathbf{R}^n} (1+|\xi|^2)^s e^{2\delta |\xi|^{\frac{1}{\tau}}} |\widehat{f}(\xi)|^2 d\xi\right)^{\frac{1}{2}} \leqslant \infty \right\}.$$

The following two lemmas on the property of $G^\delta_{\tau,s}(\mathbf{R})$ can be found in [67].

Lemma 4.1 Let s be a real number and τ. Assume that $0 < \delta' < \delta$. Then
$$\|\partial_x f\|_{G^{\delta'}_{\tau,s}(\mathbf{R})} \leqslant \frac{e^{-\tau} \tau^\tau}{(\delta - \delta')^\tau} \|f\|_{G^\delta_{\tau,s}(\mathbf{R})}.$$

Moreover, for $s \in \mathbf{R}, \tau, \delta > 0$ and $f \in G^\delta_{\tau,s-1}$ we have
$$\|(1-\partial_x^2)^{-1} f\|_{G^\delta_{\tau,s}(\mathbf{R})} = \|f\|_{G^\delta_{\tau,s-2}(\mathbf{R})} \leqslant \|f\|_{G^\delta_{\tau,s}(\mathbf{R})}$$
$$\|\partial_x (1-\partial_x^2)^{-1} f\|_{G^\delta_{\tau,s}(\mathbf{R})} \leqslant \frac{1}{2} \|f\|_{G^\delta_{\tau,s}(\mathbf{R})}$$
$$\|\partial_x (1-\partial_x^2)^{-1} f\|_{G^\delta_{\tau,s}(\mathbf{R})} \leqslant \|f\|_{G^\delta_{\tau,s-1}(\mathbf{R})}.$$

Lemma 4.2 (1) Let $s > \dfrac{1}{2}$ and $\tau \geqslant 1$ and $\delta > 0$. The space $G_{\tau,s}^\delta$ is a Banach algebra, and there exist constants $C_s; C_s' > 0$ such that $\forall u, v \in G_{\tau,s}^\delta$,

$$\|uv\|_{G_{\tau,s}^\delta} \leqslant C_s \|u\|_{G_{\tau,s}^\delta} \|v\|_{G_{\tau,s}^\delta} \quad \text{and} \quad \|uv\|_{G_{\tau,s-1}^\delta} \leqslant C_s' \|u\|_{G_{\tau,s-1}^\delta} \|v\|_{G_{\tau,s}^\delta}.$$

(2) Let $0 < \delta' < \delta, 0 < \tau' < \tau$ and $s' < s$, we have

$$G_{\tau,s}^\delta \hookrightarrow G_{\tau,s}^{\delta'}, \quad G_{\tau',s}^\delta \hookrightarrow G_{\tau,s}^\delta, \quad G_{\tau,s'}^\delta \hookrightarrow G_{\tau,s}^\delta.$$

Proposition 4.1 ([67]) Let $\{X_\delta\}_{0<\delta<1}$ be a scale of decreasing Banach spaces, namely for any $\delta' < \delta$ we have $X_\delta \subset X_{\delta'}$ and $\|\cdot\|_{\delta'} \leqslant \|\cdot\|_\delta$. Consider the Cauchy problem

$$\begin{cases} \dfrac{du}{dt} = F(t, u(t)), \\ u|_{t=0} = u_0. \end{cases} \tag{4.1}$$

Let $T, R, \tau \geqslant 1$. For given $u_0 \in X_1$, assume that F satisfies the following conditions:

(1) If for $0 < \delta' < \delta < 1$ the function $t \mapsto u(t)$ is holomorphic in $|t| < T$ and continuous on $|t| \leqslant T$ with values in X_δ and

$$\sup_{|t| \leqslant T} \|u(t)\|_\delta < R,$$

then $t \mapsto F(t, u(t))$ is a holomorphic function on $|t| < T$ with values in $X_{s'}$.

(2) For any $0 < \delta' < \delta < 1$ and any $u, v \in \overline{B(u_0, H)} \subset X_\delta$, there exists a positive constant L depending on u_0 and R such that

$$\sup_{|t|<T} \|F(t,u) - F(t,v)\|_{\delta'} \leqslant \dfrac{L}{(\delta - \delta')^\tau} \|u - v\|_\delta.$$

(3) There exists $M > 0$ such that for any $0 < \delta < 1$,

$$\sup_{|t|<T} \|F(t, u_0)\|_\delta \leqslant \dfrac{M}{(1-\delta)^\tau}.$$

Then there exists a time $T_0 \in (0, T)$ and a unique function $u(t)$, which for every $\delta \in (0,1)$ is holomorphic in $|t| < \dfrac{T_0(1-\delta)^\tau}{2^\tau - 1}$ with values in X_δ.

Remark 4.2 In fact, $T_0 = \min\left\{\dfrac{1}{2^{2\tau+4}L}, \dfrac{(2^\tau-1)R}{(2^\tau-1)2^{2\tau+3}LR+M}\right\}$, which gives a lower bound of the lifespan.

Theorem 4.1 Let $\tau \geqslant 1$ and $s > \dfrac{3}{2}$. Assume that $(u,\eta) \in G^1_{\tau,s} \times G^1_{\tau,s-1}$. Then for every $0 < \delta < 1$, there exists a $T_0 > 0$ such that the system (2.2) has a unique solution (u,η) which is holomorphic in $|t| < \dfrac{T_0(1-\delta)^\tau}{2^\tau - 1}$ with values in $G^\delta_{\tau,s}(\mathbf{R}) \times G^\delta_{\tau,s-1}(\mathbf{R})$. Moreover, $T_0 \approx 1/\left(\|u\|_{G^\delta_{\tau,s}}(\mathbf{R}) + \|\eta\|_{G^\delta_{\tau,s-1}}(\mathbf{R})\right)$.

Proof Define $F_1 \equiv -uu_x - \partial_x(1-\partial_x^2)\left(u^2 + \dfrac{1}{2}u_x^2 + \eta - 2\Omega(1+\eta)u\right)$, $F_2 \equiv -u\eta_x - \eta u_x - u_x$ $U \equiv (u,\eta), F \equiv (F_1, F_2)$. For our needs, let us rewrite the system (2.2) in the form of (4.1), i.e.,

$$\begin{cases} \dfrac{dU}{dt} = F(t, U(t)), \\ U(0) = U_0(u_0, \eta_0). \end{cases} \tag{4.2}$$

For any $\tau \geqslant 1$ and $s > \dfrac{3}{2}$ $X_\delta = G^\delta_{\tau,s} \times G^\delta_{\tau,s-1}$ and

$$\|U\|_{X_\delta} = \|u\|_{G^\delta_{\tau,s}} + \|\eta\|_{G^\delta_{\tau,s-1}}, \quad \forall u \in G^\delta_{\tau,s}, \eta \in G^\delta_{\tau,s-1}. \tag{4.3}$$

Thanks to Lemma 4.2, we have $\{X_\delta\}_{0<\delta<1}$ is a decreasing scale of Banach spaces. Let $C_s > 0$ be the constant given in Lemma 2.3. According to Lemmas 4.1-4.2, we have for any

$$\left\|\partial_x(1-\partial_x^2)^{-1}\left(u^2 + \dfrac{1}{2}u_x^2 + \eta - 2\Omega(1+\eta)u\right)\right\|_{G^{\delta'}_{\tau,s}}$$

$$\leqslant \left\|u^2 + \dfrac{1}{2}u_x^2 + \eta - 2\Omega(1+\eta)u)\right\|_{G^{\delta'}_{\tau,s-1}}$$

$$\leqslant C_s \|u\|^2_{G^{\delta'}_{\tau,s}} + \|u_x\|^2_{G^{\delta'}_{\tau,s-1}} + \|\eta\|_{G^{\delta'}_{\tau,s-1}} + 2\Omega\|u\|_{G^{\delta'}_{\tau,s}} + 2\Omega C_s \|u\|_{G^{\delta'}_{\tau,s}}\|\eta\|_{G^{\delta'}_{\tau,s-1}}$$

$$\leqslant C_s \|u\|^2_{G^{\delta'}_{\tau,s}} + \dfrac{e^{-\tau}\tau^\tau C_s}{(\delta-\delta')^\tau}\|u\|^2_{G^\delta_{\tau,s-1}} + \|\eta\|_{G^{\delta'}_{\tau,s-1}} + 2\Omega\|u\|_{G^{\delta'}_{\tau,s}}$$

$$+ 2\Omega C_s \|u\|_{G^{\delta'}_{\tau,s}}\|\eta\|_{G^{\delta'}_{\tau,s-1}}$$

$$\leqslant \left(1 + C_s + 2\Omega C_s + \dfrac{e^{-\tau}\tau^\tau C_s}{(\delta-\delta')^\tau}\right)\|U\|^2_{X_\delta} + (1+2\Omega)\|U\|_{X_\delta} \tag{4.4}$$

and

$$\|uu_x\|_{G^\delta_{\tau,s}} \leq \frac{e^{-\tau}\tau^\tau C_s}{(\delta-\delta')^\tau}\|u\|^2_{G^\delta_{\tau,s}} \leq \frac{e^{-\tau}\tau^\tau C_s}{(\delta-\delta')^\tau}\|U\|^2_{X_\delta}. \quad (4.5)$$

Therefore, by combing (4.4) and (4.5), we get the estimate for $F_1(U)$,

$$\|F_1(U)\|_{G^\delta_{\tau,s}} \leq \left(1+C_s+2\Omega C_s+\frac{2e^{-\tau}\tau^\tau C_s}{(\delta-\delta')^\tau}\right)\|U\|^2_{X_\delta}+(1+2\Omega)\|U\|_{X_\delta}. \quad (4.6)$$

Similarly, we have

$$\|F_2(U)\|_{G^\delta_{\tau,s}} \leq \frac{e^{-\tau}\tau^\tau C_s}{(\delta-\delta')^\tau}\|U\|^2_{X_\delta}+\|U\|_{X_\delta}. \quad (4.7)$$

It thus follows from (4.6)-(4.7) that

$$\|F(U)\|_{G^\delta_{\tau,s}} = \|F_1(U)\|_{G^\delta_{\tau,s}} + \|F_2(U)\|_{G^\delta_{\tau,s}}$$

$$\leq \left(1+C_s+2\Omega C_s+\frac{3e^{-\tau}\tau^\tau C_s}{(\delta-\delta')^\tau}\right)\|U\|^2_{X_\delta}+(2+2\Omega)\|U\|_{X_\delta}$$

$$\leq \frac{1}{(\delta-\delta')^\tau}\left(M_1\|U\|^2_{X_\delta}+M_2\|U\|_{X_\delta}\right), \quad (4.8)$$

with $M_1 = 1+C_s+2\Omega C_s+3e^{-\tau}\tau^\tau C_s$, $M_2 = 2+2\Omega$.

By taking the similar argument, it is not difficult to obtain that, for any $\delta \in (0,1)$

$$\|F(U)\|_{G^\delta_{\tau,s}} \leq \frac{1}{(\delta-\delta')^\tau}\left(M_1\|U_0\|^2_{X_\delta}+M_2\|U_0\|_{X_\delta}\right) \cong \frac{M}{(\delta-\delta')^\tau}. \quad (4.9)$$

In order to prove our desired result, it suffices to show that F satisfies the condition (3) of Proposition 4.1.

Assume that $U_1, U_2 \in \overline{B(U_0,R)} \subset X_\delta$. Taking advantage of Lemmas 4.1-4.2, we get and the fact that the space $G^\delta_{\tau,s-1}$ is a Banach algebra for $\frac{3}{2}$, we deduce that

$$\|F_1(U_1)-F_1(U_2)\|_{G^{\delta'}_{\tau,s}}$$

$$\leq \|u_1\partial_x u_1 - u_2\partial_x u_2\|_{G^{\delta'}_{\tau,s}} + \|u_1^2-u_2^2\|_{G^{\delta'}_{\tau,s-1}} + \frac{1}{2}\|(\partial_x u_1)^2-(\partial_x u_2)^2\|_{G^{\delta'}_{\tau,s-1}}$$

$$+ \|\eta_1-\eta_2\|_{G^{\delta'}_{\tau,s-1}} + 2\Omega\|u_1-u_2\|_{G^{\delta'}_{\tau,s-1}} + 2\Omega\|\eta_1 u_1 - \eta_1 u_2\|_{G^{\delta'}_{\tau,s-1}}$$

$$\leqslant \frac{1}{2} \|\partial_x(u_1-u_2)\|_{G_{\tau,s}^{\delta'}} + C_s \|u_1-u_2\|_{G_{\tau,s-1}^{\delta'}} \|u_1+u_2\|_{G_{\tau,s-1}^{\delta'}} + \|\eta_1-\eta_2\|_{G_{\tau,s-1}^{\delta'}}$$

$$+ \frac{C_s}{2} \|\partial_x(u_1^2-u_2^2)\|_{G_{\tau,s-1}^{\delta'}} \|\partial_x(u_1+u_2)\|_{G_{\tau,s-1}^{\delta'}} + 2\Omega \|u_1-u_2\|_{G_{\tau,s-1}^{\delta'}}$$

$$+ 2\Omega \|\eta_1 u_1 - \eta_1 u_2\|_{G_{\tau,s-1}^{\delta'}}$$

$$\leqslant \frac{e^{-\tau}\tau^\tau C_s}{2(\delta-\delta')^\tau} 2C_s(R+\|u_0\|_{G_{\tau,s}^\delta}) + 2C_s(R+\|u_0\|_{G_{\tau,s}^\delta}) \|u_1-u_2\|_{G_{\tau,s-1}^{\delta'}}$$

$$+ \|\eta_1-\eta_2\|_{G_{\tau,s-1}^{\delta'}} + C_s(R+\|u_0\|_{G_{\tau,s}^\delta}) \|u_1-u_2\|_{G_{\tau,s-1}^{\delta'}} + 2\Omega \|u_1-u_2\|_{G_{\tau,s-1}^{\delta'}}$$

$$+ 2\Omega C_s(R+\|\eta_0\|_{G_{\tau,s-1}^\delta}) \|u_1-u_2\|_{G_{\tau,s-1}^{\delta'}} + C_s(R+\|u_0\|_{G_{\tau,s}^\delta}) \|\eta_1-\eta_2\|_{G_{\tau,s-1}^{\delta'}}$$

$$\leqslant \left(\frac{e^{-\tau}\tau^\tau C_s}{2(\delta-\delta')^\tau} 2C_s(R+\|u_0\|_{G_{\tau,s}^\delta}) + 3C_s \left(R+\|u_0\|_{G_{\tau,s}^\delta} + 2\Omega C_s\left(R+\|\eta_0\|_{G_{\tau,s-1}^\delta}\right)\right)\right)$$

$$\times \|u_1-u_2\|_{G_{\tau,s}^{\delta'}} + C_s(1+R+\|u_0\|_{G_{\tau,s}^\delta}) \|\eta_1-\eta_2\|_{G_{\tau,s}^{\delta'}}$$

$$\leqslant \frac{J_1}{(\delta-\delta')^\tau} \|U_1-U_2\|_{X_\delta}, \tag{4.10}$$

where $J_1 = e^{-\tau}\tau^\tau C_s + 4C_s R + 6\Omega C_s^2 + C_s + (C_s + 3C_s + 6\Omega C_s + e^{-\tau}\tau^\tau C_s)\|U_0\|_{X_1}$. Similarly, we have

$$\|F_2(U_1) - F_2(U_2)\|_{G_{\tau,s-1}^{\delta'}} \leqslant \frac{J_2}{(\delta-\delta')^\tau} \|U_1-U_2\|_{X_\delta}, \tag{4.11}$$

where $J_2 = e^{-\tau}\tau^\tau C_s(1+R+\|U_0\|_{X_1})$. It follows that from (4.10) and (4.11) that

$$\|F(U_1) - F(U_2)\|_{X_{\delta'}} = \|F_1(U_1) - F_1(U_2)\|_{G_{\tau,s}^{\delta'}} + \|F_2(U_1) - F_2(U_2)\|_{G_{\tau,s-1}^{\delta'}}$$

$$\leqslant \frac{J_1+J_2}{(\delta-\delta')^\tau} \|U_1-U_2\|_{X_\delta}, \tag{4.12}$$

where the constants J_1, J_1 only depend on the radius R and the initial data u_0, η_0. F satisfies the condition (2) of Proposition 4.1 with $L = J_1 + J_2 = e^{-\tau}\tau^\tau C_s(1+2R) + 4C_s R + 6\Omega C_s^2 + C_s + (e^{-\tau}\tau^\tau C_s + 4C_s + 6\Omega C_s)\|U_0\|_{X_1}$. Moreover, $|t| < \frac{T_0(1-\delta)^\tau}{2^\tau - 1}$ with values in X_δ by setting $R = \|U_0\|_{X_1}$ we see that $L = J_1 + J_2 = e^{-\tau}\tau^\tau C_s + 6\Omega C_s^2 + C_s + (3e^{-\tau}\tau^\tau C_s + 8C_s + 6\Omega C_s)\|U_0\|_{X_1}$ and $M \leqslant 2^{2\tau+3}LR$. Then, we get that

$$T_0 = \frac{1}{2^{\delta+4} 2 e^{-\tau}\tau^\tau C_s + 6\Omega C_s^2 + C_s + (3e^{-\tau}\tau^\tau C_s + 8C_s + 6\Omega C_s)\|U_0\|_{X_1}}.$$

To prove the well-posedness of the system in the sense of Hadamard, following the idea in [67], we shall first introduce a new Banach space E_α with $\alpha > 0$.

Definition 4.4 Let $\tau \geqslant 1$ and $\alpha > 0$. We denote by E_α the Banach space of all functions $U(t)$ such that for every $0 < \delta < 1$ and $|t| < \dfrac{\alpha(1-\delta)^\tau}{2^\tau - 1}$, the functions $U(t)$ are holomorphic and continuous functions of t with values in X_δ. The space E_α is equipped with the norm

$$\|U\|_{E_\alpha} \equiv \sup_{|t| < \frac{\alpha(1-\delta)^\tau}{2^\tau-1},\, 0<\delta<1} \left\{ \|U\|_{X_\delta} (1-\delta)^\tau \left(1 - \frac{|t|}{\alpha(1-\delta)^\tau}\right)^{\frac{1}{2}} \right\}, \quad (4.13)$$

where the norm in space X_δ is defined by (4.3).

Lemma 4.3 ([67]) Let $\tau \geqslant 1$. For every $\alpha > 0$ and $U \in E_\alpha$, $0 < \delta < 1$ and $0 \leqslant t < \dfrac{\alpha(1-\delta)^\tau}{2^\tau - 1}$, we have

$$\int_0^t \frac{\|U(\omega)\|_{X_{\delta_\alpha(\omega)}}}{(\delta_\alpha(\omega) - \delta)^\tau} d\omega \leqslant \frac{\alpha 2^{2\tau+3} \|U\|_{\delta_\alpha}}{(1-\delta)^\tau} \left(\frac{\alpha(1-\delta)^\tau}{\alpha(1-\delta)^\tau - t} \right)^{\frac{1}{2}},$$

where

$$\delta_\alpha(\omega)$$

$$= \frac{1}{2}(1+\delta) + \left(\frac{1}{2}\right)^{2+\frac{1}{\tau}} \left\{ \left[(1-\delta)^\tau - \frac{t}{\alpha}\right]^{\frac{1}{\tau}} - \left[(1-\delta)^\tau + (2^{\tau+1} - 1)\frac{t}{\alpha}\right]^{\frac{1}{\tau}} \right\} \in (\delta, 1).$$

Inspired by [67], we give the definition of what means the data-to-solution mapping is continuous.

Definition 4.5 Let $\delta \geqslant 1$ and $s > \dfrac{3}{2}$. We say that the solution mapping for the system (2.2) is continuous, if for any $U_0^\infty = (u_0^\infty, \eta_0^\infty) \in X_1$, there exists $T > 0$ such that for any $U_0^n = (u_0^n, \eta_0^n) \in X_1$ satisfying $\|U_0^n - U_0^\infty\|_{X_1} \to 0$ the corresponding solutions U^n satisfying $\|U_0^n - U_0^\infty\|_{E_T} \to 0$, $n \to \infty$ where the norm E_T is defined in (4.13) with $\alpha = T$.

Theorem 4.2 Let $\tau \geqslant 1$ and $s > \dfrac{3}{2}$. Given initial data $U_0 \in X_1$, there exists a time $\widetilde{T}$ such that the data-to-solution mapping $U_0 \mapsto U : X_1 \mapsto E_{\widetilde{T}}$ is continuous.

Proof The existence and uniqueness of the solution have been proved in Theorem 4.1, and it follows that the lifespan for the solutions with data U_0^n and

U_0^∞ are given by $T^n = \dfrac{1}{2^{2\tau+4} L^n}$ and $T^\infty = \dfrac{1}{2^{2\tau+4} L^\infty}$ respectively, where

$$L^n = e^{-\tau}\tau^\tau C_s + 6\Omega C_s^2 + C_s + \left(3e^{-\tau}\tau^\tau C_s + 8C_s + 6\Omega C_s\right) \|U_0^n\|_{X_1},$$

and

$$L^\infty = e^{-\tau}\tau^\tau C_s + 6\Omega C_s^2 + C_s + \left(3e^{-\tau}\tau^\tau C_s + 8C_s + 6\Omega C_s\right) \|U_0^\infty\|_{X_1}.$$

Since $U_0^n \to U_0^\infty$ in X_1 as $n \to \infty$, there must be a interge $N > 0$ such that

$$\|U_0^n\|_{X_1} \leqslant \|U_0^\infty\|_{X_1} + 1, \quad \forall n > N. \tag{4.14}$$

For each $n > N$, there is a constant $\overline{T} < \max\left\{\widetilde{T}^n, \widetilde{T}^\infty\right\}$ such that, for any $|t| < \dfrac{\overline{T}(1-\delta)^\tau}{2^\tau - 1}$, the solutions to the Cauchy problem (4.2) satisfying

$$U^n(t) = U_0^n + \int_0^t F(U^n(\nu))\, d\nu, \quad \text{and} \quad U^\infty(t) = U_0^\infty + \int_0^t F(U^\infty(\nu))\, d\nu, \tag{4.15}$$

where the functional $F(U^n)$ is the same as (4.10). For instance, one can take $\overline{T} = 2e^{-\tau}\tau^\tau C_s + 6\Omega C_s^2 + C_s + \left(2e^{-\tau}\tau^\tau C_s + 8C_s + 6\Omega C_s\right)\left(\|U_0^\infty\|_{X_1} + 1\right)$. It follows from (4.15) that for each integer $n > N$ and $|t| < \dfrac{\overline{T}(1-\delta)^\tau}{2^\tau - 1}$, we have

$$\|U^n(t) - U^\infty(t)\|_{X_\delta} = \|U_0^n - U_0^\infty\|_{X_\delta} + \int_0^t \|F(U^n(\nu)) - F(U^\infty(\nu))\|_{X_\delta}\, d\nu. \tag{4.16}$$

By applying the Lemma 4.3, for $0 \leqslant t < \dfrac{\overline{T}(1-\delta)^\tau}{2^\tau - 1}$ we have $\delta < \delta_{\overline{T}}(\omega) < 1$. Taking the similar argument in Theorem 4.1, one can obtain

$$\|F^n - F^\infty\|_{X_\delta} \leqslant \dfrac{L}{(\delta_{\overline{T}}(\omega) - \delta)^\tau} \|U^n - U^\infty\|_{X_{\overline{T}}(\omega)}, \tag{4.17}$$

where L is a constant similar to that in the Theorem 4.1, i.e., $L = e^{-\tau}\tau^\tau C_s(2 + R) + 4C_s R + 6\Omega C_s^2 + C_s + \left(e^{-\tau}\tau^\tau C_s + 4C_s + 6\Omega C_s\right) \|U_0\|_{X_1}$. On the other hand, by virtue of the Lemma 4.3 with $\alpha = \overline{T}$, we have $\delta_{\overline{T}}(\omega) \in (\delta, 1)$ and

$$\int_0^t \dfrac{\|U^n(\omega) - U^\infty(\omega)\|_{X_{\overline{T}}(\omega)}}{(\delta_{\overline{T}}(\omega) - \delta)^\tau}\, d\omega \leqslant \dfrac{\overline{T} 2^{2\tau+3} \|U^n(t) - U^\infty(t)\|_{E_{\overline{T}}}}{(1-\delta)^\tau} \left(\dfrac{\overline{T}(1-\delta)^\tau}{\overline{T}(1-\delta)^\tau - t}\right)^{\frac{1}{2}}. \tag{4.18}$$

Then it follows from (4.16)-(4.18) that

$$\|U^n(t) - U^\infty(t)\|_{X_\delta}$$
$$\leqslant \|U_0^n(t) - U_0^\infty(t)\|_{X_\delta} + \frac{L\overline{T}2^{2\tau+3}\|U^n(t) - U^\infty(t)\|_{E_{\overline{T}}}}{(1-\delta)^\tau}\left(\frac{\overline{T}(1-\delta)^\tau}{\overline{T}(1-\delta)^\tau - t}\right)^{\frac{1}{2}}. \tag{4.19}$$

Simple calculation yields that $L\overline{T}2^{2\tau+3} < \frac{1}{2}$ we get from the above inequality that

$$(1-\delta)^\tau \left(\frac{\overline{T}(1-\delta)^\tau}{\overline{T}(1-\delta)^\tau - t}\right)^{\frac{1}{2}} \|U^n(t) - U^\infty(t)\|_{X_\delta}$$
$$\leqslant \frac{1}{2}\|U^n(t) - U^\infty(t)\|_{E_{\overline{T}}} + (1-\delta)^\tau \left(\frac{\overline{T}(1-\delta)^\tau}{\overline{T}(1-\delta)^\tau - t}\right)^{\frac{1}{2}} \|U_0^n - U_0^\infty\|_{X_\delta}$$
$$\leqslant \frac{1}{2}\|U^n(t) - U^\infty(t)\|_{E_{\overline{T}}} + (1-\delta)^\tau \|U_0^n - U_0^\infty\|_{X_1}. \tag{4.20}$$

By taking the supremum both sides over $0 < \delta < 1, 0 < t < \frac{\overline{T}(1-\delta)^\tau}{2^\tau - 1}$, we deduce from the definition of the space $E_{\overline{T}}$ that

$$\|U^n(t) - U^\infty(t)\|_{E_{\overline{T}}}$$
$$\leqslant \frac{1}{2}\|U^n(t) - U^\infty(t)\|_{E_{\overline{T}}} + (1-\delta)^\tau \|U_0^n - U_0^\infty\|_{X_1}, \quad \forall n > N. \tag{4.21}$$

Hence, we have

$$\|U^n(t) - U^\infty(t)\|_{E_{\overline{T}}} \leqslant 2(1-\delta)^\tau \|U_0^n - U_0^\infty\|_{X_1} \to 0, \text{ as } n \to \infty. \tag{4.22}$$

Therefore, the proof of Theorem 4.2 is completed.

5 Persistence properties of solutions

In this section, we shall discuss the persistence properties for the system (2.3) in weighted L^p spaces. More specifically, we intend to find a large class of weight functions such that

$$\sup_{t \in [0,T]} \left(\|\phi u(t)\|_{L^p} + \|\phi u_x(t)\|_{L^p} + \|\phi u_{xx}(t)\|_{L^p} + \|\phi \eta(t)\|_{L^p} + \|\phi \eta_x(t)\|_{L^p} \right) < \infty,$$

where $\|\cdot\|_{L^p}$ denotes the usual L^p norm. This way we obtain a persistence result on solutions (u,η) to the system (2.3) in the weight L^p spaces $L^{p,\phi} := L^p(\mathbf{R}, \phi^p\, dx)$. As a consequence and an application we determine the spatial asymptotic behavior of certain solutions to the system (2.3).

For the convenience of the readers, we present some standard definitions. In general a weight function is simply a non-negative function. A weight function $v : \mathbf{R}^n \to \mathbf{R}$ is called sub-multiplicative if $v(x+y) \leqslant v(x)v(y)$, for all $x, y \in \mathbf{R}^n$. Given a sub-multiplicative function v, a positive function ϕ is v-moderate if and only if $\exists C_0 > 0 : \phi(x+y) \leqslant C_0 v(x) \phi(y)$, for all $x, y \in \mathbf{R}^n$. If ϕ is v-moderate for some sub-multiplicative function v, then we say that ϕ is moderate. This is the usual terminology in time-frequency analysis papers [1,44]. Let us recall the most standard examples of such weights. Let

$$\phi(x) = \phi_{a,b,c,d}(x) = e^{a|x|^b}(1+|x|^c)\ln(e+|x|)^d.$$

We have (see [3]) the following conditions:

(i) For $a, c, d \geqslant 0$ and $0 \leqslant b \leqslant 1$ such a weight is sub-multiplicative.

(ii) If $a, c, d \in \mathbf{R}$ and $0 \leqslant b \leqslant 1$, then ϕ is moderate. More precisely, $\phi_{a,b,c,d}$ is $\phi_{\alpha,\beta,\gamma,\delta}$-moderate for $|a| \leqslant \alpha, |b| \leqslant \beta, |c| \leqslant \gamma$ and $|d| \leqslant \delta$. The elementary properties of sub-multiplicative and moderate weights can be found in [3].

We say that $\phi : \mathbf{R} \to \mathbf{R}$ is an admissible weight function for the system (2.3) if it is a locally absolutely continuous function such that for some $A > 0$ and a.e. $x \in \mathbf{R}, |\phi'(x)| \leqslant A|\phi(x)|$, and ϕ is v-moderate for a sub-multiplicative weight function v satisfying $\inf_{\mathbf{R}} v > 0$ and

$$\int_{\mathbf{R}} \frac{v(x)}{e^{|x|}} \, dx < \infty. \tag{5.1}$$

It is easily checked that the functions $\phi_{a,b,c,d}$ are admissible for the system (2.3) if $a \geqslant 0, 0 \leqslant b \leqslant 1$ and $ab < 1$; see [3] for further details.

Theorem 5.1 *Let $T > 0, s > 2$, and $2 \leqslant p < \infty$. Let also $(u, \eta) \in C([0,T], H^s \times H^{s-1}) \cap C^1([0,T], H^{s-1} \times H^{s-2})$ be a strong solution of the system (2.3), starting from $z_0 = (u_0, \eta_0)$ so that $\phi u_0, \phi u_{0x}, \phi u_{0xx}, \phi \eta_0, \phi \eta_{0x} \in L^p(\mathbf{R})$ for an admissible weight function ϕ of the system (2.3). Let*

$$M \equiv \sup_{t \in [0,T]} \left(\|\phi u(t)\|_{L^\infty} + \|\phi u_x(t)\|_{L^\infty} + \|\phi u_{xx}(t)\|_{L^\infty} + \|\phi \eta(t)\|_{L^\infty} + \|\phi \eta_x(t)\|_{L^\infty} \right).$$

Then there is a constant $C > 0$ depending only on the weight ϕ such that

$$\|\phi u(t)\|_{L^p} + \|\phi u_x(t)\|_{L^p} + \|\phi u_{xx}(t)\|_{L^p} + \|\phi \eta(t)\|_{L^p} + \|\phi \eta_x(t)\|_{L^p}$$
$$\leq e^{CMt} \left(\|\phi u_0\|_{L^p} + \|\phi u_{x0}\|_{L^p} + \|\phi u_{xx0}\|_{L^p} + \|\phi \eta_0\|_{L^p} + \|\phi \eta_{0x}\|_{L^p} \right)$$

for all $t \in [0, T]$

Remark 5.1 (1) *Power weights:* Take $\phi = \phi_{0,0,c,0}$ with $c > 0$, and choose $p = \infty$. In this case Theorem 5.1 states that the that the algebraic decay of the initial datum $z_0 = (u_0, \eta_0)$

$$|u_0(x)| + |u_{0x}(x)| + |u_{0xx}(x)| + |\eta_0(x)| + |\eta_{0x}(x)| \leq C(1 + |x|)^{-c},$$

for all $x \in \mathbf{R}$, is preserved by the solution $z = (u, \eta)$ with $z(0) = z_0$ on $[0, T)$, i.e.,

$$|u(x,t)| + |u_x(x,t)| + |u_{xx}(x,t)| + |\eta(x,t)| + |\eta_x(x,t)| \leq C'(1 + |x|)^{-c}$$

for all $(x, t) \in \mathbf{R} \times [0, T)$, where $C, C' > 0$ are constants. Theorem 5.1 thus generalizes the main result of Ni and Zhou [71] on algebraic decay rates of strong solutions to the Camassa-Holm equation.

(2) *Exponential weights:* Choose $\phi = \phi_{a,1,0,0}$ if $x \geq 0$ and $\phi(x) = 1$, if $x \leq 0$ with $0 \leq a < 1$. It is easy to see that such a weight is an admissible weight function for the system (2.3). Let further $p = \infty$ in Theorem 5.1. Then one deduces that the system (2.3) preserves the pointwise decay $O(e^{-ax})$ as $x \to +\infty$ for any $t > 0$. Similarly, we have persistence of the decay $O(e^{-ax})$ as $x \to -\infty$. Some similar results on the persistence of strong solutions of the Camassa-Holm equation (or the two-component Camassa-Holm system) can be found in [54, 62].

Clearly, the limit case $\phi = \phi_{1,1,c,d}$ is not covered by Theorem 5.1. In the following theorem however we may choose the weight $\phi = \phi_{1,1,c,d}$ with $c < 0, d \in \mathbf{R}$, and $\frac{1}{|c|} < p \leq \infty$, or more generally when $(1 + |\cdot|)^c \ln(e + |\cdot|)^d \in L^p(\mathbf{R})$. See Theorem 5.2 below, which covers the case of such fast growing weights. In other words, we want to establish a variant of Theorem 5.1 that can be applied to some v-moderate weights ϕ for which condition (5.1) does not hold. Instead of assuming (5.1), we now put the weaker condition

$$ve^{-|\cdot|} \in L^p(\mathbf{R}), \tag{5.2}$$

where $2 \leq p \leq \infty$.

Theorem 5.2 Let $2 \leqslant p \leqslant \infty$ and ϕ be a locally absolutely continuous and v-moderate weight function satisfying condition (5.2) instead of (5.1). Assume that $\phi u_0, \phi u_{0x}, \phi u_{0xx}, \phi \eta_0, \phi \eta_{0x} \in L^p(\mathbf{R})$ and $\phi^{\frac{1}{2}} u_0, \phi^{\frac{1}{2}} u_{0x}, \phi^{\frac{1}{2}} u_{0xx}, \phi^{\frac{1}{2}} \eta_0, \phi^{\frac{1}{2}} \eta_{0x} \in L^2(\mathbf{R})$. For $s > 3$. Let also $z = (u, \eta) \in C([0,T], H^s \times H^{s-1}) \cap C^1([0,T], H^{s-1} \times H^{s-2})$ be a strong solution of the system (2.3), starting from $z_0 = (u_0, \eta_0)$. Then

$$\sup_{t \in [0,T]} (\|\phi u(t)\|_{L^p} + \|\phi u_x(t)\|_{L^p} + \|\phi u_{xx}(t)\|_{L^p} + \|\phi \eta(t)\|_{L^p} + \|\phi \eta_x(t)\|_{L^p})$$

and

$$\sup_{t \in [0,T]} \left(\|\phi^{\frac{1}{2}} u(t)\|_{L^2} + \|\phi^{\frac{1}{2}} u_x(t)\|_{L^2} + \|\phi^{\frac{1}{2}} u_{xx}(t)\|_{L^2} + \|\phi^{\frac{1}{2}} \eta(t)\|_{L^2} + \|\phi^{\frac{1}{2}} \eta_x(t)\|_{L^2} \right)$$

are finite.

Remark 5.2 For the particular choice $c = d = 0$ and $p = \infty$, we conclude from

$$|u_0(x)| + |u_{0x}(x)| + |u_{0xx}(x)| + |\eta_0(x)| + |\eta_{0x}(x)| \leqslant Ce^{-|x|},$$

for all $x \in \mathbf{R}$, that the unique solution $z = (u, \eta) \in C([0,T]; H^s \times H^{s-1})$ of the system (2.3) with $z_0 = (u_0, \eta_0)$ satisfies

$$|u(x,t)| + |u_x(x,t)| + |u_{xx}(x,t)| + |\eta(x,t)| + |\eta_x(x,t)| \leqslant C'e^{-|x|}$$

on $R \times [0, T]$.

In the following Theorem 5.3, we compute the spatial asymptotic profiles of solutions with exponential decay. As a further consequence we may infer that the peakon-like decay $O\left(e^{-|x|}\right)$ mentioned above is the fastest possible decay for a nontrivial solution $z = (u, \eta)$ of the system (2.3) to propagate.

Theorem 5.3 Let $\phi = e^{\frac{|x|}{2}} (1+|x|)^{\frac{1}{2}} \ln(e+|x|)^d$, for some $d > \dfrac{1}{2}$. For $s > 3$, let $z_0 = (u_0, \eta_0) \in C([0,T]; H^s \times H^{s-1})$ be nonzero and assume that

$$\sup_{x \in \mathbf{R}} \{\phi(x) (|u_0(x)| + |u_{0x}(x)| + |u_{0xx}(x)| + |\eta_0(x)| + |\eta_{0x}(x)|)\} < \infty. \qquad (5.3)$$

Then condition (5.3) is conserved for the solution

$$z = (u, \eta) \in C([0,T], H^s \times H^{s-1}) \times C^1([0,T], H^{s-1} \times H^{s-2})$$

of the system (2.3) starting from $z_0 = (u_0, \eta_0)$, and we have the asymptotic behavior

$$z(t) = \begin{cases} u_0(x) + e^{-x}t \begin{pmatrix} \Phi^+(t) + \epsilon_1^+(x,t) \\ \epsilon_2^+(x,t) \end{pmatrix}, & \text{with } \lim_{x \to +\infty} \epsilon_{1,2}^+(x,t) = 0, \\ u_0(x) - e^{x}t \begin{pmatrix} \Phi^-(t) + \epsilon_1^-(x,t) \\ \epsilon_2^-(x,t) \end{pmatrix}, & \text{with } \lim_{x \to -\infty} \epsilon_{1,2}^-(x,t) = 0 \end{cases}$$
(5.4)

for all $t \in [0,T]$, and $c_1 \leqslant \Phi^{\pm}(t) \leqslant c_2$ with $c_1, c_2 > 0$ independent of t.

Proof of Theorem 5.1 Let $\Psi(u,\eta) = u^2 + \frac{1}{2}u_x^2 + \eta - 2\Omega(1+\eta)u$ Assume that ϕ is v-moderate satisfying the above conditions. Our first observation is that the first equation in the system (2.3) can be rewritten as

$$u_t - uu_x - \partial_x G * \Psi(u,\eta) = 0, \tag{5.5}$$

with the kernel $G = \frac{1}{2}e^{-|x|}$.

On the other hand, from the assumption

$$(u,\eta) \in C([0,T], H^s) \times C([0,T], H^{s-1}), \quad s > 3,$$

we get

$$M \equiv \sup_{t \in [0,T]} (\|u(t)\|_\infty + \|\partial_x u(t)\|_\infty + \|\partial_{xx} u(t)\|_\infty + \|\eta(t)\|_\infty + \|\partial_x \eta(t)\|_\infty) < \infty.$$

For any $N \in \mathbf{Z}^+$ let us consider the N-truncations of $\phi(x)$: $f(x) = f_N(x) = \min\{\phi, N\}$. Then $f : \mathbf{R} \to \mathbf{R}$ is a locally absolutely continuous function such that $\|f\|_\infty \leqslant N$, $|f'(x)| \leqslant A|f(x)|$ a.e. on $\mathbf{R}$. In addition, if $C_1 = \max\{C_0, \alpha^{-1}\}$, where $\alpha = \inf_{x \in \mathbf{R}} v(x) > 0$, then $f(x+y) \leqslant C_1 v(x) f(y)), \forall x, y \in \mathbf{R}$. Moreover, as shown in [3], the N-truncations f of a v-moderate weight ϕ are uniformly v-moderate with respect to N.

We start considering the case $2 \leqslant p < \infty$. Multiply the first equation in (5.5) by $f|uf|^{p-2}(uf)$ and integrate to obtain

$$\int_{\mathbf{R}} |uf|^{p-2}(uf)(\partial_t uf)\,dx - \int_{\mathbf{R}} |uf|^p \partial_x u\,dx - \int_{\mathbf{R}} |uf|^{p-2}(uf)(f\partial_x G * \Psi(u,\eta)) = 0.$$
(5.6)

Note that the estimates

$$\int_{\mathbf{R}} |uf|^{p-2}(uf)(\partial_t uf)\,dx = \frac{1}{p}\frac{d}{dt}\|uf\|_{L^p}^p = \|uf\|_{L^p}^{p-1}\frac{d}{dt}\|uf\|_{L^p},$$

and

$$\left|2\int_{\mathbf{R}}|uf|^p \partial_x u\, dx\right| \leqslant 2\|\partial_x u\|_{L^\infty}\|uf\|_{L^p}^p \leqslant 2M\|uf\|_{L^p}^p,$$

are true. Moreover, we get

$$\left|\int_{\mathbf{R}}|uf|^{p-2}(uf)\left(f\partial_x G * \Psi(u,\eta)\right)\right|$$
$$\leqslant \|uf\|_{L^p}^{p-1}\|f\partial_x G * \Psi(u,\eta)\|_{L^p}$$
$$\leqslant C\|uf\|_{L^p}^{p-1}\left(\|\partial_x G\|_{L^1}\|(u^2 + \frac{1}{2}u_x^2 + \eta - 2\Omega(1+\eta)u)f\|_{L^p}\right)$$
$$\leqslant CM\|uf\|_{L^p}^{p-1}\left(\|uf\|_{L^p} + \|f\partial_x u\|_{L^p} + \|f\eta\|_{L^p}\right).$$

In the first inequality we used Hölder's inequality, and in the second inequality we applied Propositions 3.1 and 3.2 in [3], and in the last we used condition (5.1). Here, C depends only on v and ϕ. Form (5.6) we can obtain

$$\frac{d}{dt}\|uf\|_{L^p} \leqslant C_1 M(\|uf\|_{L^p} + \|f\partial_x u\|_{L^p} + \|f\eta\|_{L^p}). \tag{5.7}$$

Next, we will give estimates on $u_x f$. Differentiating (5.5) with respect to x-variable, next multiplying by f produces the equation

$$\partial_t[(\partial_x u)f] - uf\partial_x^2 u - f(\partial_x u)^2 - f\partial_x^2 G * (u^2 + \frac{1}{2}u_x^2 + \eta - 2\Omega(1+\eta)u) = 0.$$

Multiply this equation by $|f\partial_x u|^{p-2}(f\partial_x u)$ with $p \in \mathbf{Z}^+$, integrate the result in the x-variable, and note that

$$\int_{\mathbf{R}}|f\partial_x u|^{p-2}(f\partial_x u)f\partial_t[\partial_x u]\, dx = \|f\partial_x u\|_{L^p}^{p-1}\frac{d}{dt}\|f\partial_x u\|_{L^p},$$

and

$$\int_{\mathbf{R}}|f\partial_x u|^{p-2}(f\partial_x u)f\partial_x^2 G * (u^2 + \frac{1}{2}u_x^2 + \eta - 2\Omega(1+\eta)u)\, dx$$
$$\leqslant \|f\partial_x u\|_{L^p}^{p-1}\|f\partial_x^2 G * (u^2 + \frac{1}{2}u_x^2 + \eta - 2\Omega(1+\eta)u)\|_{L^p}$$
$$\leqslant CM\|f\partial_x u\|_{L^p}^{p-1}\left(\|uf\|_{L^p} + \|f\partial_x u\|_{L^p} + \|\eta f\|_{L^p}\right),$$

and

$$\left|\int_{\mathbb{R}} |f\partial_x u|^{p-2}(f\partial_x u)f(\partial_x u)^2\, dx\right| \leq \|\partial_x uf\|_{L^p}^{p-1}\|u_x fu_x\|_{L^p} \leq M\|u_x f\|_{L^p}^{p-1}\|u_x f\|_{L^p}.$$

For the second order derivative term, we have

$$\left|\int_{\mathbb{R}} |f\partial_x u|^{p-2}(f\partial_x u)uf\partial_x^2 u\, dx\right| = \left|\int_{\mathbb{R}} |f\partial_x u|^{p-2}(f\partial_x u)[\partial_x(u_x f)u_x f_x]\, dx\right|$$

$$\leq M\|u_x f\|_{L^p}^{p-1}\|uf\|_{L^p}.$$

Thus, it follows that

$$\frac{d}{dt}\|f\partial_x u\|_{L^p} \leq CM(\|uf\|_{L^p} + \|f\partial_x u\|_{L^p} + \|\eta f\|_{L^p}). \tag{5.8}$$

Multiplying the equation $u_{txx} - uu_{xxx} - 3u_x u_{xx} + \partial_x G * \Psi(u, \eta) - \partial_x \Psi(u, \eta) = 0$ with $|fu_{xx}|^{p-2}(fu_{xx})$ yields

$$\frac{1}{p}\frac{d}{dt}\|fu_{xx}\|_{L^p}^p - \int_{\mathbb{R}} uu_{xx}u_{xxx}f^2|fu_{xx}|^{p-2}\, dx - 3\int_{\mathbb{R}} u_x u_{xx}^2 f^2|fu_{xx}|^{p-2}\, dx$$

$$+ \int_{\mathbb{R}} f(\partial_x G * \Psi(u, \eta))|fu_{xx}|^{p-2} fu_{xx}\, dx - \int_{\mathbb{R}} f(\partial_x \Psi(u, \eta))|fu_{xx}|^{p-2} fu_{xx}\, dx. \tag{5.9}$$

We obtain

$$\int_{\mathbb{R}} f(\partial_x G * \Phi(u, \eta))|fu_{xx}|^{p-2} fu_{xx}\, dx$$

$$\leq \|f(\partial_x G * \Phi(u, \eta))\|_{L^p}\|fu_{xx}\|_{L^p}^{p-1}$$

$$\leq CM(\|fu\|_{L^p} + \|fu_x\|_{L^p} + \|f\eta\|_{L^p})\|fu_{xx}\|_{L^p}^{p-1}, \tag{5.10}$$

$$\int_{\mathbb{R}} f\partial_x \Phi(u, \eta)|fu_{xx}|^{p-2} fu_{xx}\, dx$$

$$\leq \|f(\partial_x G * \Phi(u, \eta))\|_{L^p}\|fu_{xx}\|_{L^p}^{p-1}$$

$$\leq CM(\|fu\|_{L^p} + \|fu_x\|_{L^p} + \|f\eta\|_{L^p})\|fu_{xx}\|_{L^p}^{p-1}, \tag{5.11}$$

and

$$3\int_{\mathbb{R}} u_x u_{xx}^2 f^2|fu_{xx}|^{p-2}\, dx \leq CM\|fu_{xx}\|_{L^p}^p. \tag{5.12}$$

The integral $\int_{\mathbf{R}} uu_{xx}u_{xxx}f^2|fu_{xx}|^{p-2}\,dx$ may be rewritten as

$$\int_{\mathbf{R}} uu_{xx}u_{xxx}f^2|fu_{xx}|^{p-2}\,dx$$
$$= \int_{\mathbf{R}} [\partial_x(fu_{xx} - f_x u_{xx})]uu_{xx}f|fu_{xx}|^{p-2}\,dx = \Gamma_1 - \Gamma_2. \qquad (5.13)$$

As $|f_x| \leqslant C|f|$, we have $\Gamma_2 \leqslant CM\|u_{xx}f\|_{L^p}^p$. For any function $g : \mathbf{R} \times [0, T] \to \mathbf{R}$ which is weakly differentiable in the first argument one has

$$\frac{\partial}{\partial x}\left(\frac{|g(x,t)|^p}{p}\right) = |g(x,t)|^{p-2}g(x,t)g_x(x,t). \qquad (5.14)$$

Since $s > 3$ and f is locally absolutely continuous, Eq. (5.14) holds a.e. for $g = fu_{xx}$ and we have

$$\Gamma_1 = \int_{\mathbf{R}} u\partial_x\left(\frac{|fu_{xx}|^p}{p}\right)dx = -\frac{1}{p}u_x|fu_{xx}|^p\,dx, \qquad (5.15)$$

the boundary terms vanish when performing integration by parts in view of Sobolev's embedding theorem. Hence $|\Gamma_1| \leqslant \dfrac{M}{p}\|fu_{xx}\|_{L^p}^p$. We have shown that

$$\frac{d}{dt}\|fu_{xx}\|_{L^p} \leqslant CM(\|uf\|_{L^p} + \|u_xf\|_{L^p} + \|u_{xx}f\|_{L^p} + \|\eta f\|_{L^p}). \qquad (5.16)$$

We now multiply the equation $\eta_t + u\eta_x + \eta u_x + u_x = 0$ with $f|f\eta|^{p-2}f\eta$ and integrate to obtain the identity

$$\frac{1}{p}\frac{d}{dt}\|f\eta\|_{L^p}^p + \int_{\mathbf{R}}\eta\eta_x uf^2|f\eta|^{p-2}\,dx + \int_{\mathbf{R}}(f\eta)^2 u_x|f\eta|^{p-2}\,dx + \int_{\mathbf{R}}\eta u_x f^2|f\eta|^{p-2}\,dx$$
$$=0. \qquad (5.17)$$

Note that the estimates

$$\int_{\mathbf{R}}\eta\eta_x uf^2|f\eta|^{p-2}\,dx \leqslant M\|\eta_x f\|_{L^p}\|\eta f\|_{L^p}^{p-1},$$
$$\int_{\mathbf{R}}(f\eta)^2 u_x|f\eta|^{p-2}\,dx \leqslant M\|\eta f\|_{L^p}^p, \qquad (5.18)$$
$$\int_{\mathbf{R}}\eta u_x f^2|f\eta|^{p-2}\,dx \leqslant M\|u_x f\|_{L^p}\|\eta f\|_{L^p}^{p-1},$$

are true. This yields

$$\frac{d}{dt}\|f\eta\|_{L^p} \leqslant M(\|f\eta\|_{L^p} + \|f\eta_x\|_{L^p} + \|fu_x\|_{L^p}). \tag{5.19}$$

Multiply the equation $\eta_{tx}+u\eta_{xx}+2\eta_x u_x+u_{xx}=0$ with $f|f\eta_x|^{p-2}f\eta_x$ and integrate to obtain the identity

$$\frac{1}{p}\frac{d}{dt}\|f\eta_x\|_{L^p}^p + \int_{\mathbf{R}} \eta_x\eta_{xx}uf^2|f\eta_x|^{p-2}\,dx + 2\int_{\mathbf{R}} \eta_x^2 u_x f^2|f\eta_x|^{p-2}\,dx$$
$$+ \int_{\mathbf{R}} \eta\eta_x u_{xx}f^2|f\eta_x|^{p-2}\,dx + \int_{\mathbf{R}} \eta_x u_{xx}f^2|f\eta_x|^{p-2}\,dx = 0. \tag{5.20}$$

For the integrals on the left-hand side of (5.20) we have

$$2\int_{\mathbf{R}} \eta_x^2 u_x f^2|f\eta_x|^{p-2}\,dx \leqslant CM\|\eta_x f\|_{L^p}^p,$$

$$\int_{\mathbf{R}} \eta\eta_x u_{xx}f^2|f\eta_x|^{p-2}\,dx \leqslant CM\|\eta f\|_{L^p}^p \|\eta_x f\|_{L^p}^p, \tag{5.21}$$

$$\int_{\mathbf{R}} \eta_x u_{xx}f^2|f\eta_x|^{p-2}\,dx \leqslant CM\|u_{xx}f\|_{L^p}^p \|\eta_x f\|_{L^p}^{p-1}.$$

For the first integral on the left-hand side of (5.20), we have

$$\int_{\mathbf{R}} \eta_x\eta_{xx}uf^2|f\eta_x|^{p-2}\,dx$$
$$= \int_{\mathbf{R}} [\partial_x(f\eta_x) - f\eta_x]u|f\eta_x|^{p-2}f\eta_x f\,dx$$
$$= \int_{\mathbf{R}} u\partial_x\left(\frac{|f\eta_x|^p}{p}\right)dx - \int_{\mathbf{R}} f\eta_x u|f\eta_x|^{p-2}f\eta_x f\,dx$$
$$\leqslant CM(\|fu_x\|_{L^p}^p + \|fu\|_{L^p}^p).$$

In the last inequality, we used $f_x \leqslant Cf(x)$ for a.e. x. Combining (5.7), (5.8), (5.16), (5.19), and (5.20), there exists a constant C, such that

$$\frac{d}{dt}\left(\|fu\|_{L^p} + \|fu_x\|_{L^p} + \|fu_{xx}\|_{L^p} + \|f\eta\|_{L^p} + \|f\eta_x\|_{L^p}\right)$$
$$\leqslant CM\left(\|fu\|_{L^p} + \|fu_x\|_{L^p} + \|fu_{xx}\|_{L^p} + \|f\eta\|_{L^p} + \|f\eta_x\|_{L^p}\right),$$

so that, by Gronwall's Lemma,

$$\|fu(t)\|_{L^p} + \|fu_x(t)\|_{L^p} + \|fu_{xx}(t)\|_{L^p} + \|f\eta(t)\|_{L^p} + \|f\eta_x(t)\|_{L^p}$$

$$\leqslant (\|fu_0\|_{L^p} + \|fu_{0x}\|_{L^p} + \|fu_{0xx}\|_{L^p} + \|f\eta_0\|_{L^p} + \|f\eta_{0x}\|_{L^p}) e^{CMt}, \quad \forall t \in [0, T).$$

Since $f(x) = f_N(x) \to \phi(x)$ as $N \to \infty$ for a.e. $x \in \mathbf{R}$. Recalling that $u_0\phi, u_{0x}\phi,$ $u_{0xx}\phi, \eta_0\phi, \eta_{0x}\phi \in L^p(\mathbf{R})$, the assertion of the theorem follows for the case $p \in [2, \infty)$. Since $\|\cdot\|_\infty = \lim\limits_{p\to\infty} \|\cdot\|_p$ it is clear that the theorem also applies for $p = \infty$.

Proof of Theorem 5.2 As explained in [3], the function $\phi^{\frac{1}{2}}$ is a $v^{\frac{1}{2}}$-moderate weight such that $\left(\phi^{\frac{1}{2}}\right)'(x) \leqslant \dfrac{A}{2} \phi^{\frac{1}{2}}(x)$. Moreover, $\inf\limits_{\mathbf{R}} v^{\frac{1}{2}} > 0$. Then Theorem 5.1 applies with $p = 2$ to the weight $\phi^{\frac{1}{2}}$ yielding

$$\|\phi^{\frac{1}{2}} u(t)\|_{L^2} + \|\phi^{\frac{1}{2}} u_x(t)\|_{L^2} + \|\phi^{\frac{1}{2}} u_{xx}(t)\|_{L^2} + \|\phi^{\frac{1}{2}} \eta(t)\|_{L^2} + \|\phi^{\frac{1}{2}} \eta_x(t)\|_{L^2}$$
$$\leqslant \left(\|\phi^{\frac{1}{2}} u_0\|_{L^2} + \|\phi^{\frac{1}{2}} u_{0x}\|_{L^2} + \|fu_{0xx}\|_{L^2} + \|\phi^{\frac{1}{2}} \eta_0\|_{L^2} + \|\phi^{\frac{1}{2}} \eta_{0x}\|_{L^2}\right) e^{CMt}. \quad (5.22)$$

Let f be as in the proof of Theorem 5.1. Then (5.22) holds equally with Ψ replaced by f. By the definition of $\Psi(u, \eta)$ and (5.22), there is a constant C depending only on ϕ and the initial data such that $\|f\Psi(u, \eta)\|_{L^p} \leqslant Ce^{2CMt}$. Using (5.6), the weighted Young inequality and $ve^{-|\cdot|} \in L_p(\mathbf{R})$ we conclude that $\|f(\partial_x G * \Psi(u, \eta))\|_{L^p} \leqslant Ce^{2CMt}$ and that

$$\|f(\partial_x^2 G * \Psi(u, \eta))\|_{L^p} \leqslant \|f(G * \Psi(u, \eta))\|_{L^p} + \|f\Psi(u, \eta)\|_{L^p}$$
$$\leqslant Ce^{2CMt} + M\left(\|fu\|_{L^p} + \|fu_x\|_{L^p} + \|f\eta\|_{L^p}\right).$$

Using (5.7), (5.8), and (5.16) this yields

$$\frac{d}{dt}\left(\|fu\|_{L^p} + \|fu_x\|_{L^p} + \|fu_{xx}\|_{L^p} + \|f\eta\|_{L^p} + \|f\eta_x\|_{L^p}\right)$$
$$\leqslant CM\left(\|fu\|_{L^p} + \|fu_x\|_{L^p} + \|fu_{xx}\|_{L^p} + \|f\eta\|_{L^p} + \|f\eta_x\|_{L^p}\right) + Ce^{2CMt}$$

and the constants C, depend only on ϕ and the initial data. Integrating this equation and letting $N \to \infty$, we obtain the result of the corollary for $p \in [2, \infty)$. The case $p = \infty$ is again obtained from a standard limit argument. This achieves the proof.

Proof of Theorem 5.3 Conservation of (5.3) follows from Theorem 5.1 with $p = \infty$ and $\phi(x) = e^{|x|/(2)}(1 + |x|)^{1/2} \ln(e + |x|)^d$. Integration of (5.5) yields

$$\begin{pmatrix} u \\ \eta \end{pmatrix} = \begin{pmatrix} u_0 \\ \eta_0 \end{pmatrix} - \begin{pmatrix} \int_0^t \partial_x G * \Psi(x, s)\, ds + \int_0^t (uu_x)(x, s)\, ds \\ \int_0^t (\eta u_x + \eta_x u + u_x)(x, s)\, ds \end{pmatrix}.$$

We now use that

$$\left|\int_0^t (\eta u_x)(x,s)\,ds\right|, \left|\int_0^t (\eta u_x(x,s)\,ds\right|, \left|\int_0^t u_x(x,s)\,ds\right|$$
$$\leqslant Cte^{-|x|}(1+|x|)^{-1}\ln(e+|x|)^{-2d}$$

and let

$$\Phi^\pm = \frac{1}{2}\int_{-\infty}^\infty e^{\pm y}h(y,t)\,dy, \quad h(x,t) = \frac{1}{t}\int_0^t \Psi(x,s)\,ds.$$

By the arguments in [3], this achieves the asymptotic representation of (u,η).

Acknowledgements The authors would like to thank the referees for constructive suggestions and comments. The work of Mi is partially supported by NSF of China (11671055), partially supported by NSF of Chongqing (cstc2018jcyjAX0273), partially supported by Key project of science and technology research program of Chongqing Education Commission (KJZD-K20180140). The work of Liu is supported in part by the Simons Foundation (grant-499875). The work of Luo is partially supported by China Postdoctoral Science Foundation (2018M641271).

References

[1] Aldroubi A, Gröchenig K. Nonuniform sampling and reconstruction in shift-invariant spaces [J]. SIAM Rev., 2001, 43: 585–620.
[2] Beals R, Sattinger D, Szmigielski J. Multi-peakons and a theorem of Stieltjes [J]. Inverse Problems, 1999, 15: 1–4.
[3] Brandolese L. Breakdown for the Camassa-Holm equation using decay criteria and persistence in weighted spaces [J]. Int. Math. Res. Not., IMRN 2012, 22: 5161–5181.
[4] Bressan A, Constantin A. Global conservative solutions of the Camassa-Holm equation [J]. Arch. Ration. Mech. Anal., 2007, 183: 215–239.
[5] Bressan A, Constantin A. Global dissipative solutions of the Camassa-Holm equation [J]. Anal. Appl., 2007, 5: 1–27.
[6] Camassa R, Holm R. An integrable shallow water equation with peaked solitons [J]. Phys. Rev. Lett., 1993, 71: 1661–1664.
[7] Chemin J. Localization in Fourier space and Navier-Stokes system, Phase Space Analysis of Partial Differential Equations. Proceedings, CRM series, Pisa, 2004, 53–136.
[8] Chen M, Fan L, Gao H, Liu Y. Breaking waves and solitary waves to the rotation-two component Camassa-Holm system [J]. SIAM J. Math. Anal., 2017, 49: 3573–3602.

[9] Chen M, Gui G, Liu Y. On a shallow-water approximation to the Green-Naghdi equations with the Coriolis effect [J]. Adv. Math., 2018, 340: 106–137.

[10] Chen M, Liu Y. Wave breaking and global existence for a generalized two-component Camassa-Holm system [J]. Int. Math. Res. Not., 2011, 6: 1381–1416.

[11] Chen M, Liu S, Zhang Y. A 2-component generalization of the Camassa-Holm equation and its solutions [J]. Lett. Math. Phys., 2006, 75: 1–5.

[12] Constantin A. On the scattering problem for the Camassa-Holm equation [J]. Proc. R. Soc. Lond. A, 2001, 457: 953–970.

[13] Constantin A. The trajectories of particles in Stokes waves [J]. Invent. Math., 2006, 166: 23–535.

[14] Constantin A. On the inverse spectral problem for the Camassa-Holm equation [J]. J. Funct. Anal., 1998, 155: 352–363.

[15] Constantin A. Global existence of solutions and breaking waves for a shallow water equation: a geometric approach [J]. Ann. Inst. Fourier (Grenoble), 2000, 50: 321–362.

[16] Constantin A, Escher J. Wave breaking for nonlinear nonlocal shallow water equations [J]. Acta Math., 1998, 181: 229–243.

[17] Constantin A, Escher J. Global existence and blow-up for a shallow water equation [J]. Ann. Sc. Norm. Super. Pisa, 1998, 26: 303–328.

[18] Constantin A, Escher J. Particle trajectories in solitary water waves [J]. Bull. Amer. Math. Soc., 2007, 44: 423–431.

[19] Constantin A, Escher J. Analyticity of periodic traveling free surface water waves with vorticity [J]. Ann. of Math., 2011, 173: 559–568.

[20] Constantin A, Gerdjikov V, Ivanov R I. Inverse scattering transform for the Camassa-Holm equation [J]. Inverse Problems, 2006, 22: 2197–2207.

[21] Constantin A, Ivanov R. On an integrable two-component Camassa-Holm shallow water system [J]. Phys. Lett. A, 2008, 372: 7129–7132.

[22] Constantin A, Johnson R S. The dynamics of waves interacting with the equatorial undercurrent[J]. Geophys. Astrophys. Fluid Dyn. 2015, 109: 311–358.

[23] Constantin A, Johnson R S. An exact, steady, purely azimuthal equatorial flow with a free surface [J]. J. Phys. Oceanogr. 2016, 46: 1935–1945.

[24] Constantin A, Johnson R S. A nonlinear, three-dimensional model for ocean flows, motivated by some observations of the Pacific Equatorial Undercurrent and thermocline[J]. Phys. Fluids, 2017, 29: 056604.

[25] Constantin A, Johnson R S. On the nonlinear, three-dimensional structure of equatorial oceanic flows [J]. J. Phys. Oceanogr., 2019, https://doi.org/10.1175/JPO-D-19-0079.1, in press.

[26] Constantin A, Kappeler T, Kolev B, Topalov T. On Geodesic exponential maps of the Virasoro group [J]. Ann. Global Anal. Geom., 2007, 31: 155–180.

[27] Constantin A, Lannes D. The hydro-dynamical relevance of the Camassa-Holm and Degasperis-Procesi equations [J]. Arch. Ration. Mech. Anal., 2009, 193: 165–186.

[28] Constantin A, McKean H P. A shallow water equation on the circle [J]. Comm. Pure Appl. Math., 1999, 52: 949–982.

[29] Constantin A, Molinet L. Global weak solutions for a shallow water equation [J]. Comm. Math. Phys., 2000, 211: 45–61.

[30] Constantin A, Strauss W. Stability of peakons, Comm. Pure Appl. Math., 2000, 53: 603-610.

[31] Constantin A, Strauss W. Stability of the Camassa-Holm solitons [J]. J. Nonlinear Sci., 2002, 12: 415–422.

[32] Danchin R. A few remarks on the Camassa-Holm equation [J]. Differential Integral Equations, 2001, 14: 953–988.

[33] Danchin R. Fourier analysis methods for PDEs [M]. Lecture Notes, 14 November, 2003.

[34] Escher J, Lechtenfeld O, Yin Z. Well-posedness and blow-up phenomena for the 2-component Camassa-Holm equation [J]. Discrete Contin. Dyn. Syst., 2007, 19: 493–413.

[35] Fan L, Gao H, Liu Y. On the rotation-two-component Camassa-Holm system modeling the equatorial water waves [J]. Adv. Math., 2016, 291: 59–89.

[36] Fokas A, Fuchssteiner B. Symplectic structures, their Backlund transformation and hereditary symmetries [J]. Physica D, 1981, 4: 47–66.

[37] Foias C, Temam R. Gevrey class regularity for the solutions of the Navier-Stokes equations [J]. J. Funct. Anal., 1989, 87: 359–369.

[38] Fu Y, Gu G, Liu Y, Qu Z. On the Cauchy problem for the integrable Camassa-Holm type equation with cubic nonlinearity [J]. J. Differential Equations, 2013, 255: 1905–1938.

[39] Guan C, Yin Z. Global existence and blow-up phenomena for an integrable two-component Camassa-Holm shallow water system [J]. J. Differential Equations, 2010, 248: 2003–2014.

[40] Guan C, Yin Z. Global weak solutions for a two-component Camassa-Holm shallow water system [J]. J. Funct. Anal., 2011, 260: 1132–1154.

[41] Gui G, Liu Y. On the Cauchy problem for the two-component Camassa-Holm system [J]. Math. Z., 2011, 268: 45–66.

[42] Gui G, Liu Y, Sun J. A nonlocal shallow-water model arising from the full water waves with the Coriolis effect [J]. J. Math. Fluid Mech., 2019, 21: 1–27.

[43] Gui G, Liu Y, Luo T. Model equations and traveling waves solutions for shallow-water eaves with the Coriolis effect [J]. J. Nonlinear Sci., 2019, 29: 993–1039.

[44] Gröchenig K. Weight functions in time-frequency analysis, in: Pseudo-differential operators: partial differential equations and time frequency analysis [J]. Fields Inst. Commun., 2007, 52: 343–366.

[45] Han Y, Guo F, Gao H. On solitary waves and wave-breaking phenomena for a generalized two-component integrable Dullin-Gottwald-Holm system [J]. J. Nonlinear Sci., 2013, 23: 617–656.

[46] Henry D. Infinite propagation speed for a two component Camassa-Holm equation [J]. Discrete Contin. Dyn. Syst. Ser. B Appl. Algor., 2009, 12: 597–506.

[47] Himonas A, Kenig C. Non-uniform dependence on initial data for the Camassa-Holm equation on the line [J]. Differential Integral Equations, 2009, 22: 201-224.

[48] Himonas A, Kenig C, Misiolek G. Non-uniform dependence for the periodic Camassa-Holm equation [J]. Comm. Partial Differential Equations, 2010, 35: 1145–1162.

[49] Himonas A, Misiolek G. The cauchy problem for an integrable shallowwater equation [J]. Differential Integral Equations, 2001, 14: 821–831.

[50] Himonas A, Misiolek G. High-frequency smooth solutions and well-posedness of the Camassa-Holm equation [J]. Int. Math. Res. Not., 2005, 51: 3135–3151.

[51] Himonas A, Misiolek G. Non-uniform dependence on initial data of solutions to the Euler equations of hydrodynamics [J]. Comm. Math. Phys., 2010, 296: 285–301.

[52] Himonas A, Misiolek G. Analyticity of the Cauchy problem for an integrable evolution equation [J]. Math. Ann., 2003, 327: 575–584.

[53] Himonas A, Misiolek G, Ponce G. Non-uniform continuity in H^1 of the solution map of the Camassa-Holm equation [J]. Asian J. Math., 2007, 11: 141–150.

[54] Himonas A A, Misiolek G, Ponce G, Zhou Y. Persistence properties and unique continuation of solutions of the Camassa-Holm equation [J]. Comm. Math. Phys., 2007, 271: 511–522.

[55] Holden H, Raynaud X. Global conservative solutions of the Camassa-Holm equations-a Lagrangianpoiny of view [J]. Comm. Partial Differential Equations, 2007, 32: 1511–1549.

[56] Holden H, Raynaud X. Dissipative solutions for the Camassa-Holm equation [J]. Discrete Contin. Dyn. Syst., 2009, 24: 1047–1112.

[57] Hu T, Liu Y. On the modeling of equatorial shallow-water waves with the Coriolis effect [J]. Physica D: Nonli. Phenomena, 2019, 391: 87–110.

[58] Hu Q, Yin Z. Global existence and blowup phenomena for the periodic 2-component Camassa-Holm equation [J]. Monatsh. Math., 2012, 165: 217–235.

[59] Hu Q, Yin Z. Well-posedness and blowup phenomena for the periodic 2-component Camassa-Holm equation [J]. Proc. R. Soc. Edinburgh Sect. A, 2011, 141: 93–107.

[60] Ionescu-Kruse D. A three-dimensional autonomous nonlinear dynamical system modelling equatorial ocean flows [J]. J. Differential Equations, 2018, 264: 4650–4668.

[61] Ivanov R. Two-component integrable systems modelling shallow water waves: the constant vorticity case [J]. Wave Motion, 2009, 46: 389–396.

[62] Kohlmann M. The two-component Camassa-Holm system in weighted L_p spaces [J]. Z. Angew. Math. Mech., 2014, 94: 264–272.

[63] Liu J. Blow-up phenomena for the rotation-two-component Camassa-Holm system [J]. arXiv:1812.10075.

[64] Luo T, Liu Y, Mi Y, Moon B. On a shallow-water model with the Coriolis effect [J]. J. Differential Equations, 2019, 267: 3232–3270.

[65] Misiolek G A. Shallow water equation as a geodesic flow on the Bott-Virasoro group [J]. J. Geom. Phys., 1998, 24: 203–208.

[66] Mustafa O. On smooth travelling waves of an integrable two-component Camassa-Holm shallow water system [J]. Wave Motion, 2009, 46: 397–402.

[67] Luo W, Yin Z. Gevrey regularity and analyticity for Camassa-Holm type systems [J]. Ann. Sc. Norm. Super. Pisa Cl. Sci., 2018, 18: 1061–1079.

[68] Moon B. On the wave-breaking phenomena and global existence for the periodic rotation-twocomponent Camassa Holm system [J]. J. Math. Anal. Appl., 2017, 451: 84–101.

[69] Popowicz Z. A 2-component or $N = 2$ supersymmetric Camassa-Holm equation [J]. Phys. Lett. A, 2006, 354: 110–114.

[70] Tu T, Liu T, Mu C. Existence and uniqueness of the global conservative weak solutions to the rotation-Camassa-Holm equation [J]. J. Differential Equations, 2019, 266: 4864–4900.

[71] Ni L, Zhou Y. A new asymptotic behavior of solutions to the Camassa-Holm equation [J]. Proc. Amer. Math. Soc., 2012, 140: 607–614.

[72] Zhang Y. Wave breaking and global existence for the periodic rotation-Camassa-Holm system [J]. Discrete Contin. Dyn. Syst., 2017, 37: 2243–2257.

[73] Zhang L, Liu B. Well-posedness, blow-up criteria and gevrey regularity for a rotation-twocomponent Camassa-Holm system [J]. Discrete Contin. Dyn. Syst., 2018, 38: 2655–2685.

Unconditional Convergence of Linearized Implicit Finite Difference Method for the 2D/3D Gross-Pitaevskii Equation with Angular Momentum Rotation*

Wang Tingchun (王廷春), Guo Boling (郭柏灵)

Abstract This paper is concerned with the time-step condition of the linearized implicit finite difference method for solving the Gross-Pitaevskii equation with an angular momentum rotation term. Unlike the existing studies in literature where the cut-off function technique was used to establish the error estimates under some conditions of the time-step size, this paper introduces an induction argument and a 'lifting' technique as well as some useful inequalities to build the optimal maximum error estimate without any constraints on the time-step size. The analysis method can be directly extended to the general nonlinear Schrödinger-type equations in two- and three-dimensions and other linear implicit finite difference schemes. As a by-product, this paper defines a new type of energy functional of the grid functions by using a recursive relation to prove that the proposed scheme preserves well the total mass and energy in the discrete sense. Several numerical results are reported to verify the error estimates and conservation laws.

Keywords Gross-Pitaevskii equation; angular momentum; rotation; finite difference method; mass and energy conservation; unconditional and optimal error estimate

---

* Sci. China Math., 2018, 62: 1669–1686. DOI: 10.1007/s11425-016-9212-1.

1 Introduction

After proper non-dimensionalization and dimension reduction [5], a rotating Bose-Einstein condensate (BEC) [9, 18] can be well modeled by the following Gross-Pitaevskii equation (GPE) with an angular momentum rotation term (AMR) in dimensionless form

$$i\partial_t \psi = \left[-\frac{1}{2}\Delta + V - \gamma L_z + \beta |\psi|^2\right]\psi, \quad \boldsymbol{x} \in \Omega \subset \mathbf{R}^d, \ t > 0. \qquad (1.1)$$

Here, t is time, $\boldsymbol{x} = (x, y)$ in two dimensions ($d = 2$) and respectively $\boldsymbol{x} = (x, y, z)$ in three dimensions ($d = 3$) are the Cartesian coordinates, $\psi = \psi(\boldsymbol{x}, t)$ is the unknown complex-valued wave function, $V = V(\boldsymbol{x})$ is a real-valued function corresponding to the external trapping potential and it is chosen as a harmonic potential (i.e., a quadratic polynomial) in most experiments, γ is a dimensionless constant corresponding to the angular speed of the laser beam in experiment, β is a dimensionless constant characterizing the interaction (positive for repulsive interaction and negative for attractive interaction) between particles in the rotating BEC, Ω is a bounded computational domain, and L_z is the z-component of the angular momentum defined as

$$L_z = -i(x\partial y - y\partial x) = -i\partial_\theta,$$

where (r, θ) and (r, θ, z) are the polar coordinates in two dimensions (2D) and cylindrical coordinates in three dimensions (3D), respectively.

In this paper, we numerically study the GPE (1.1) in a bounded domain with the homogeneous Dirichlet boundary condition

$$\psi(\boldsymbol{x}, t) = 0, \quad \boldsymbol{x} \in \partial\Omega, \ t > 0, \qquad (1.2)$$

and the given initial condition

$$\psi(\boldsymbol{x}, 0) = \varphi(\boldsymbol{x}), \quad \boldsymbol{x} \in \overline{\Omega} = \Omega \cup \partial\Omega, \qquad (1.3)$$

where $\varphi = \varphi(\boldsymbol{x})$ is a given smooth complex-valued function.

The initial-boundary value problem (1.1)-(1.3) preserves the total mass

$$Q(\psi(\cdot, t)) := \|\psi(\cdot, t)\|^2 = \int_\Omega |\psi(\boldsymbol{x}, t)|^2 \, d\boldsymbol{x} \equiv Q(\psi(\cdot, 0)) = Q(\varphi), \quad t \geq 0, \qquad (1.4)$$

and energy

$$E(\psi(\cdot,t)) := \int_\Omega \left[\frac{1}{2}|\nabla\psi|^2 + V(\boldsymbol{x})|\psi|^2 + \frac{\beta}{2}|\psi|^4 - \gamma\overline{\psi}L_z\psi\right] d\boldsymbol{x} \equiv E(\varphi), \quad t \geqslant 0, \tag{1.5}$$

where $\overline{f}$ represents the conjugate of the function f.

Many results including mathematical studies and numerical simulations have been carried out in the literature to study the GPE with AMR. Along the mathematical front, for the derivation, well-posedness and dynamical properties of the GPE with AMR, we refer to [14, 15, 17] and the references therein. Along the numerical front, numerous efforts have been devoted to the development of the efficient and accurate numerical methods including the time-splitting pseudospectral method [6, 7, 20], time-splitting alternating direction implicit method [8], finite difference method [5] for the GPE with AMR. Of course, each numerical method has its advantages and disadvantages. For numerical comparisons between different numerical methods for GPE without AMR (i.e., $\gamma = 0$), we refer to [2, 3] and references therein. Mostly, completely implicit methods are unconditionally stable. However, at each time step, one has to solve a system of nonlinear equations. Though explicit methods are much easier in the practical computation, they suffer severe constraints of the time-step for convergence or stability. Linearized (semi-)implicit numerical methods become the most popular and widely used ones, at each time step, the methods only require the solution of one or several linear systems. However, the time-step size condition (or constraints on the grid ratios) is a key issue in analysis and computation.

The error estimates of finite difference methods for solving the one-dimensional nonlinear Schrödinger(NLS)-type equations including the GPE without AMR have been established in the literature (see [10, 11, 21] and references therein). The techniques used in establishing the error estimates are strongly dependent on the energy conservation laws and the Sobolev embedding inequality in one dimension, so they can not be extended to high-dimensional cases and the non-conservative schemes. However, the GPE with AMR is either in two dimensions or in three dimensions [5-8]. Bao and Cai[5] introduce the cut-off function technique used in [1, 4, 12, 13, 19] together with standard energy method to give the first result of the error estimates of finite difference method for solving the GPE with AMR. However, their results require a time-step condition, i.e., $\tau \leqslant O(h)$, and their error estimates in some cases are not optimal, e.g., their error estimates in the discrete H^1-norm is merely $O(h^{3/2} + \tau^{3/2})$ when $\gamma \neq 0$, as is not consistent with their numerical results which show that the convergence rate in the discrete H^1-norm is of $O(h^2 + \tau^2)$. In [16], the authors establish the maximum error estimates of a conservative finite difference scheme for two- and three-dimensional linear Schrödinger equation without any constraints on the time-step size, but their analysis can not be extended to the GPE with AMR, due to the existence of the nonlinear term and the AMR term. In [22], the authors established the unconditional L^2 convergence of a conservative and compact finite difference scheme for solving the nonlinear Schrödinger equation in two dimensions, however, their results require that the initial value is small enough for the focusing case. Though in [23], the authors establish the optimal error estimates in the maximum norm of several finite difference sechemes for solving the coupled GPE without AMR, their error estimates still require a condition of the time-step size, i.e., $\tau \leqslant O(h)$. Moreover, for two- or three-dimensional cases, L^2- or H^1-error estimates do not provide immediate insight on the phase error occurring during time evolution. As we measure computation errors in practice, error estimates in the grid-independent maximum norm are preferable in numerical analysis.

Thus, the purpose of this investigation is to design a new linearized semi-implicit finite difference scheme for solving the GPE with AMR and establish the optimal error estimate of the proposed scheme in maximum norm without any constraints on the time-step size. Besides the standard energy method, we introduce an induction argument as well as a 'lifting' technique (i.e., by taking the discrete L^2-norm of both sides of the 'error' equations, the estimate of the higher order difference quotient of the error functions can be obtained from the estimate of the lower order difference quotient of the error functions and local truncation error functions) in analyzing the convergence and stability of the proposed schemes. Our analysis method can be directly extended to the general NLS/GP equations in 2D/3D and other linearized implicit finite difference schemes including the one given [5]. Another interest is that we define a new type of energy functional of the grid functions by using a recursive relation and prove that the new scheme preserves the total mass and energy in the discrete sense. By using this method one can find that there exists numerous energy-preserving finite difference schemes for solving the NLS-type equations.

The remainder of this paper is organized as follows. In Section 2, we give a linearized semi-implicit finite difference scheme, prove that the new scheme preserves the total mass and energy in the discrete sense. In Section 3, we establish the optimal error estimate of the proposed scheme in the maximum norm with no requirement of the time-step size. In Section 4, some numerical results are reported to test the theoretical analysis. Finally, some concise conclusions are drawn in Section 5.

2 Finite difference scheme and conservation laws

In this section, we propose a linearized finite difference scheme for the GPE with AMR in 2D on a rectangle $\Omega = (x_L, x_R) \times (y_L, y_R)$ and in 3D on a cube $\Omega = (x_L, x_R) \times (y_L, y_R) \times (z_L, z_R)$, respectively, and prove that the new scheme also preserves the total mass and energy in the discrete sense.

For simplicity, by the similar processing in [5], we here only consider the numerical method in two dimensions, i.e., $d = 2$ and $\Omega = (x_L, x_R) \times (y_L, y_R)$ in (1.1). Extension to 3D is straightforward, and the error estimates in the maximum norm are the same in 2D and 3D. That is, we consider the GPE with AMR

$$i\partial_t \psi = \left[-\frac{1}{2}\Delta + V(x,y) - \gamma L_z + \beta |\psi|^2\right]\psi, \quad (x,y) \in \Omega, \ t > 0, \quad (2.1)$$

with the homogeneous Dirichlet boundary condition

$$\psi(x, y, t) = 0, \quad (x, y) \in \partial\Omega, \ t > 0, \quad (2.2)$$

and initial condition

$$\psi(x, y, 0) = \varphi(x, y), \quad (x, y) \in \overline{\Omega}. \quad (2.3)$$

2.1 Numerical methods

For a positive integer N, choose time-step $\tau = T/N$ and denote time steps $t_n = n\tau$, $n = 0, 1, 2, \cdots, N$, where $0 < T < T_{\max}$ with $T_{\max}$ the maximal existing time of the solution; choose mesh sizes $h_1 = (x_R - x_L)/J$, $h_2 = (y_R - y_L)/K$ with two positive integers J and K, denote $h = \max\{h_1, h_2\}$, $h_{\min} = \min\{h_1, h_2\}$ and grid points $(x_j, y_k) = (x_L + jh_1, y_L + kh_2)$, $j = 0, 1, \cdots, J$, $k = 0, 1, \cdots, K$. Denote the index sets

$$\mathcal{T}_h^0 := \{(j,k)|\, j = 0, 1, 2, \cdots, J,\ k = 0, 1, 2, \cdots, K\},$$

$$\mathcal{T}_h := \{(j,k)|\, j = 1, 2, 3, \cdots, J-1,\ k = 1, 2, 3, \cdots, K-1\},$$

and three grid sets

$$\overline{\Omega}_h := \{(x_j, y_k)|\, (j,k) \in \mathcal{T}_h^0\}, \quad \Omega_h := \overline{\Omega}_h \cap \Omega, \quad \partial\Omega_h := \overline{\Omega}_h \cap \partial\Omega.$$

For simplicity, we define a space of grid functions as

$$X_h := \{u = (u_{jk})_{(j,k) \in \mathcal{T}_h^0}|\, u_{jk} = 0,\ \text{when}\ (j,k) \notin \mathcal{T}_h\} \subseteq \mathbb{C}^{(J+1) \times (K+1)}.$$

Let ψ_{jk}^n and ϕ_{jk}^n be the numerical approximation and the exact value of ψ at the point (x_j, y_k, t_n) for $(j,k) \in \mathcal{T}_h, n = 0, 1, 2, \cdots, N$, respectively. Denote $\psi^n \in \mathbb{C}^{(J+1) \times (K+1)}$ and $\phi^n \in \mathbb{C}^{(J+1) \times (K+1)}$ be the numerical vector solution and the exact vector solution at time $t = t_n$. For a grid function $u^n \in X_h$, we introduce the following finite difference quotient operators:

$$\delta_t^+ u_{jk}^n = \frac{1}{\tau}\left(u_{jk}^{n+1} - u_{jk}^n\right), \quad \delta_x^+ u_{jk}^n = \frac{1}{h_1}\left(u_{j+1k}^n - u_{jk}^n\right),$$

$$\delta_x u_{jk}^n = \frac{1}{2h_1}\left(u_{j+1k}^n - u_{j-1k}^n\right), \quad \delta_x^2 u_{jk}^n = \frac{1}{h_1^2}\left(u_{j-1k}^n - 2u_{jk}^n + u_{j+1k}^n\right).$$

Difference quotient operators $\delta_y^+ u_{jk}^n, \delta_y u_{jk}^n, \delta_y^2 u_{jk}^n$ are defined similarly. Besides, we introduce the discrete version of the rotation operator, the gradient operator and Laplacian operator as follows

$$L_z^h u_{jk}^n = -i(x_j \delta_y - y_k \delta_x) u_{jk}^n, \quad \nabla_h u_{jk}^n = (\delta_x^+ u_{jk}^n, \delta_y^+ u_{jk}^n)^T, \quad \Delta_h u_{jk}^n = \delta_x^2 u_{jk}^n + \delta_y^2 u_{jk}^n.$$

We define discrete inner products and discrete norms over X_h as

$$\langle u, v \rangle := h_1 h_2 \sum_{j=1}^{J-1} \sum_{k=1}^{K-1} u_{jk} \overline{v_{jk}}, \quad \|u\| := \langle u, u \rangle^{\frac{1}{2}}, \quad \langle u, v \rangle_0 := h_1 h_2 \sum_{j=0}^{J-1} \sum_{k=0}^{K-1} u_{jk} \overline{v_{jk}},$$

$$|u|_1 := \left[\langle \delta_x^+ u_{jk}, \delta_x^+ u_{jk} \rangle_0 + \langle \delta_y^+ u_{jk}, \delta_y^+ u_{jk} \rangle_0\right]^{\frac{1}{2}}, \quad \|\|u\|\|_1 := \left(\|u\|^2 + |u|_1^2\right)^{\frac{1}{2}},$$

$$|u|_2 := \langle \Delta_h u, \Delta_h u \rangle^{\frac{1}{2}}, \quad \|u\|_\infty := \max_{(j,k) \in \mathcal{T}_h} |u_{jk}|, \quad \|\|u\|\|_2 := \left(\|\|u\|\|_1^2 + |u|_2^2\right)^{\frac{1}{2}},$$

where $|u|_k$ and $|||u|||_k$ $(k=1,2)$ are respectively Sobolev's semi-norms and norms for the grid function $u \in X_h$. Let $H^k(\Omega_h)$ denote the space of complex-valued or real-valued discrete functions with the Sobolev's norm $|||\cdot|||_k$, $(k=1,2)$. Let $H_0^1(\Omega_h)$ be the subspace of the space $H^1(\Omega_h)$ satisfying the homogeneous Dirichlet boundary condition. Throughout the paper, we denote C as a generic positive constant which may be dependent on the regularity of exact solution and the given data but independent of the discrete parameters. In particular, we choose regular h_1 and h_2 such that $h \leqslant Ch_{\min}$.

Now, by using a local extrapolation technique to discretize the coefficient of the nonlinear term in temporal direction and adopting the centered finite difference technique to approximate the spatial derivatives of the wave function, we give the following linearized semi-implicit finite difference (LEFD) scheme

$$i\delta_t^+ \psi_{jk}^n = \frac{1}{2}\left[-\frac{1}{2}\Delta_h + V_{jk} - \gamma L_z^h + \frac{\beta}{2}(3|\psi_{jk}^n|^2 - |\psi_{jk}^{n-1}|^2)\right]$$
$$\times (\psi_{jk}^n + \psi_{jk}^{n+1}), \quad (j,k) \in \mathcal{T}_h, \ n=1,2,\cdots,N-1, \qquad (2.4)$$

$$\psi_{jk}^0 = \varphi(x_j, y_k), \quad (j,k) \in \mathcal{T}_h^0, \qquad (2.5)$$

$$\psi^n \in X_h, \quad n=1,2,\cdots,N, \qquad (2.6)$$

where $V_{jk} = V(x_j, y_k)$. As a three-level scheme, the LEFD scheme can not start by itself, the numerical solution at the first step can be computed by any second or higher order time integrator, e.g., the following predictor-corrector method

$$i\delta_t^* \psi_{jk}^0 = \frac{1}{2}\left(-\frac{1}{2}\Delta_h + V_{jk} - \gamma L_z^h\right)\left(\psi_{jk}^0 + \psi_{jk}^{1/2}\right) + \beta\left|\psi_{jk}^0\right|^2 \psi_{jk}^0, \quad (j,k) \in \mathcal{T}_h, \qquad (2.7)$$

$$i\delta_t^+ \psi_{jk}^0 = \frac{1}{2}\left(-\frac{1}{2}\Delta_h + V_{jk} - \gamma L_z^h + \beta\left|\psi_{jk}^{1/2}\right|^2\right)\left(\psi_{jk}^0 + \psi_{jk}^1\right), \quad (j,k) \in \mathcal{T}_h, \qquad (2.8)$$

where $\delta_t^* \psi_{jk}^0 = \dfrac{\psi_{jk}^{1/2} - \psi_{jk}^0}{\tau/2}$, $\psi_{jk}^{1/2}$ is the approximation of $\psi(x_j, y_k, t_{\frac{1}{2}})$. For the LEFD scheme, at each time step, only a system of linear algebraic equations is to be solved, which is much less expensive in the practical computation than that of nonlinear finite difference schemes such as the conservative Crank-Nicolson finite difference (CNFD) scheme given in [5].

2.2 Conservation laws in the discrete sense

In this subsection, we aim to prove that the LEFD scheme preserves the total mass and energy in the discrete sense.

Before giving the energy functional of the grid functions in X_h, we use a recursive relation to define a functional $\langle F^n, 1\rangle$ on X_h for $n = 1, 2, \cdots, N$ as

$$\langle F^n, 1\rangle = h_1 h_2 \sum_{j=1}^{J-1} \sum_{k=1}^{K-1} F_{jk}^n, \tag{2.9}$$

where $F^n \in X_h$ with components

$$F_{jk}^{n+1} := F_{jk}^n + \frac{\beta}{2}(3|\psi_{jk}^n|^2 - |\psi_{jk}^{n-1}|^2)(|\psi_{jk}^{n+1}|^2 - |\psi_{jk}^n|^2), \ (j,k) \in \mathcal{T}_h, \ 1 \leqslant n < N, \tag{2.10}$$

$$F_{jk}^1 := \beta |\psi_{jk}^{1/2}|^2 |\psi_{jk}^1|^2, \quad F_{jk}^0 := \beta |\psi_{jk}^0|^2 |\psi_{jk}^{1/2}|^2, \ (j,k) \in \mathcal{T}_h. \tag{2.11}$$

In the next analysis, we need the following lemmas,

Lemma 2.1 ([5]) *For any $u, w \in H_0^1(\Omega_h) \cap H^2(\Omega_h)$, there are*

$$\langle \delta_x u, w\rangle = -\langle u, \delta_x w\rangle, \quad \langle \delta_y u, w\rangle = -\langle u, \delta_y w\rangle, \quad \langle \delta_x^2 u, w\rangle = -\langle \delta_x^+ u, \delta_x^+ w\rangle_0,$$

$$\langle \delta_y^2 u, w\rangle = -\langle \delta_y^+ u, \delta_y^+ w\rangle_0, \quad \|u\| \leqslant C\|\nabla_h u\|, \quad \langle L_z^h u, w\rangle = \langle u, L_z^h w\rangle,$$

$$\frac{1}{2}(1 - \gamma^2/\mu^2) \|\nabla_h u\|^2 \leqslant \mathcal{E}(u) \leqslant C\|\nabla_h u\|^2,$$

where μ is given in (3.1) and $\mathcal{E}(u) = \frac{1}{2}\|\nabla_h u\|^2 + h_1 h_2 \sum_{j=1}^{J-1}\sum_{k=1}^{K-1}[V_{jk}|u_{jk}|^2 - \gamma \overline{u_{jk}} L_z^h u_{jk}]$.

Lemma 2.2 *For any grid function $u, w \in H_0^1(\Omega_h) \cap H^2(\Omega_h)$, there are*

$$\langle L_z^h u, \Delta_h w\rangle = \langle \Delta_h u, L_z^h w\rangle, \tag{2.12}$$

$$\text{Re}(L_z^h(u+w), \Delta_h(u-w)) = \langle L_z^h u, \Delta_h u\rangle - \langle L_z^h w, \Delta_h w\rangle, \tag{2.13}$$

where $\text{Re}(s)$ represents taking the real part of s.

Proof For any grid function $u, w \in H_0^1(\Omega_h) \cap H^2(\Omega_h)$, we have

$$\langle L_z^h u, \Delta_h w\rangle = h_1 h_2 \sum_{j=1}^{J-1} \sum_{k=1}^{K-1} \Delta_h \overline{w_{jk}} L_z^h u_{jk}$$

$$= -ih_1 h_2 \sum_{j=1}^{J-1}\sum_{k=1}^{K-1} (\delta_x^2 \overline{w_{jk}} + \delta_y^2 \overline{w_{jk}})(x_j \delta_y u_{jk} - y_k \delta_x u_{jk})$$

$$= ih_1 h_2 \sum_{j=1}^{J-1}\sum_{k=1}^{K-1} [x_j \delta_y \overline{w_{jk}} \delta_x^2 u_{jk} + 2\delta_y \overline{w_{jk}} \delta_x u_{jk} - y_k \delta_x \overline{w_{jk}} \delta_x^2 u_{jk}]$$

$$+ ih_1 h_2 \sum_{j=1}^{J-1}\sum_{k=1}^{K-1} [x_j \delta_y \overline{w_{jk}} \delta_y^2 u_{jk} - 2\delta_x \overline{w_{jk}} \delta_y u_{jk} - y_k \delta_x \overline{w_{jk}} \delta_y^2 u_{jk}]$$

$$= h_1 h_2 \sum_{j=1}^{J-1}\sum_{k=1}^{K-1} i(x_j \delta_y \overline{w_{jk}} - y_k \delta_x \overline{w_{jk}})(\delta_x^2 u_{jk} + \delta_y^2 u_{jk})$$

$$+ 2ih_1 h_2 \sum_{j=1}^{J-1}\sum_{k=1}^{K-1} [\delta_y \overline{w_{jk}} \delta_x u_{jk} - \delta_x \overline{w_{jk}} \delta_y u_{jk}]$$

$$= \langle \Delta_h u, L_z^h w \rangle, \tag{2.14}$$

where $w \in H_0^1(\Omega_h) \cap H^2(\Omega_h)$. This gives

$$\langle L_z^h u, \Delta_h u \rangle = \langle \Delta_h u, L_z^h u \rangle, \tag{2.15}$$

and

$$\text{Re}\langle L_z^h(u+w), \Delta_h(u-w)\rangle$$
$$= \text{Re}\langle L_z^h u, \Delta_h u\rangle - \text{Re}\langle L_z^h w, \Delta_h u\rangle + \text{Re}\langle L_z^h w, \Delta_h u\rangle - \text{Re}\langle L_z^h u, \Delta_h w\rangle$$
$$= \text{Re}\langle L_z^h u, \Delta_h u\rangle - \text{Re}\langle L_z^h w, \Delta_h u\rangle + \text{Re}\langle L_z^h w, \Delta_h u\rangle - \text{Re}\langle \Delta_h u, L_z^h w\rangle$$
$$= \langle L_z^h u, \Delta_h u\rangle - \langle L_z^h w, \Delta_h w\rangle. \tag{2.16}$$

Combining (2.14)-(2.16) gives (2.12)-(2.13).

Now we give the conservative results satisfied by the LEFD scheme as follows,

Lemma 2.3 *The LEFD scheme preserves the total mass and energy in the discrete sense, i.e.,*

$$Q^n := \|\psi^n\|^2 \equiv Q^0, \quad n = 0, 1, 2, \cdots, N, \tag{2.17}$$

$$E^n := \frac{1}{2}\|\nabla_h \psi^n\|^2 + h_1 h_2 \sum_{j=1}^{J-1}\sum_{k=1}^{K-1}\left[V_{jk}|\psi_{jk}^n|^2 - \gamma \overline{\psi_{jk}^n} L_z^h \psi_{jk}^n\right]$$

$$+ \langle F^n, 1\rangle \equiv E^0, \quad n = 0, 1, 2, \cdots, N. \tag{2.18}$$

Proof Computing the inner product of (2.4) with $\psi^n + \psi^{n+1}$ and taking the imaginary part give

$$\frac{1}{\tau}(\|\psi^{n+1}\|^2 - \|\psi^n\|^2) = 0, \quad n = 1, 2, \cdots, N-1, \tag{2.19}$$

where Lemma 2.1 was used. Similarly, computing the inner product of (2.8) with $\psi^0 + \psi^1$ and taking the imaginary part give

$$\frac{1}{\tau}(\|\psi^1\|^2 - \|\psi^0\|^2) = 0. \tag{2.20}$$

Combining (2.19) and (2.20) immediately gives (2.17).

Computing the inner product of (2.4) with $2(\psi^{n+1} - \psi^n)$ and taking the real part give

$$\frac{1}{2}(\|\nabla_h \psi^{n+1}\|^2 - \|\nabla_h \psi^n\|^2) + h_1 h_2 \sum_{j=1}^{J-1} \sum_{k=1}^{K-1} V_{jk}(|\psi_{jk}^{n+1}|^2 - |\psi_{jk}^n|^2)$$

$$- \gamma h_1 h_2 \sum_{j=1}^{J-1} \sum_{k=1}^{K-1} \left(\overline{\psi_{jk}^{n+1}} L_z^h \psi_{jk}^{n+1} - \overline{\psi_{jk}^n} L_z^h \psi_{jk}^n \right)$$

$$+ \frac{\beta}{2} h_1 h_2 \sum_{j=1}^{J-1} \sum_{j=1}^{K-1} (3|\psi_{jk}^n|^2 - |\psi_{jk}^{n-1}|^2)(|\psi_{jk}^{n+1}|^2 - |\psi_{jk}^n|^2) = 0,$$

$$n = 1, 2, \cdots, N-1, \tag{2.21}$$

where Lemma 2.1 was used. It follows from (2.10) that

$$\frac{\beta}{2} h_1 h_2 \sum_{j=1}^{J-1} \sum_{k=1}^{K-1} (3|\psi_{jk}^n|^2 - |\psi_{jk}^{n-1}|^2)(|\psi_{jk}^{n+1}|^2 - |\psi_{jk}^n|^2)$$

$$= \langle F^{n+1}, 1 \rangle - \langle F^n, 1 \rangle, \quad n = 1, 2, \cdots, N-1. \tag{2.22}$$

Plugging (2.22) into (2.21) gives

$$\frac{1}{2}(\|\nabla_h \psi^{n+1}\|^2 - \|\nabla_h \psi^n\|^2) + h_1 h_2 \sum_{j=1}^{J-1} \sum_{k=1}^{K-1} V_{jk}(|\psi_{jk}^{n+1}|^2 - |\psi_{jk}^n|^2)$$

$$- \gamma h_1 h_2 \sum_{j=1}^{J-1} \sum_{k=1}^{K-1} \left[\overline{\psi_{jk}^{n+1}} L_z^h \psi_{jk}^{n+1} - \overline{\psi_{jk}^n} L_z^h \psi_{jk}^n \right] + \langle F^{n+1} - F^n, 1 \rangle = 0,$$

$$n = 1, 2, \cdots, N-1. \tag{2.23}$$

Similarly, computing the inner product of (2.8) with $2(\psi^1 - \psi^0)$ and taking the real part give

$$\frac{1}{2}(\|\nabla_h \psi^1\|^2 - \|\nabla_h \psi^0\|^2) + h_1 h_2 \sum_{j=1}^{J-1}\sum_{k=1}^{K-1} V_{jk}(|\psi_{jk}^1|^2 - |\psi_{jk}^0|^2)$$

$$- \gamma h_1 h_2 \sum_{j=1}^{J-1}\sum_{k=1}^{K-1} \left(\overline{\psi_{jk}^1} L_z^h \psi_{jk}^1 - \overline{\psi_{jk}^0} L_z^h \psi_{jk}^0\right)$$

$$+ \beta h_1 h_2 \sum_{j=1}^{J-1}\sum_{j=1}^{K-1} |\psi_{jk}^{1/2}|^2(|\psi_{jk}^1|^2 - |\psi_{jk}^0|^2) = 0. \qquad (2.24)$$

It follows from (2.11) that

$$\beta h_1 h_2 \sum_{j=1}^{J-1}\sum_{j=1}^{K-1} |\psi_{jk}^{1/2}|^2(|\psi_{jk}^1|^2 - |\psi_{jk}^0|^2) = \langle F^1, 1\rangle - \langle F^0, 1\rangle. \qquad (2.25)$$

Plugging (2.25) into (2.24) gives

$$\frac{1}{2}(\|\nabla_h \psi^1\|^2 - \|\nabla_h \psi^0\|^2) + h_1 h_2 \sum_{j=1}^{J-1}\sum_{k=1}^{K-1} [V_{jk}(|\psi_{jk}^1|^2 - |\psi_{jk}^0|^2)]$$

$$- \gamma h_1 h_2 \sum_{j=1}^{J-1}\sum_{k=1}^{K-1} \left[\overline{\psi_{jk}^1} L_z^h \psi_{jk}^1 - \overline{\psi_{jk}^0} L_z^h \psi_{jk}^0\right] + \langle F^1 - F^0, 1\rangle = 0,$$

$$n = 1, 2, \cdots, N-1. \qquad (2.26)$$

Combining (2.23) and (2.26) immediately gives (2.18).

3 Error estimate of the LEFD scheme

In this section, we establish the optimal error estimate of the numerical solution without any restrictions on the time-step size. To do this, we need some assumptions and useful lemmas.

(A) Assumption on the external trapping potential $V(\boldsymbol{x})$ and the rotation speed γ: there exists a constant $\mu > 0$ such that

$$V(\boldsymbol{x}) \in C(\Omega), \quad V(\boldsymbol{x}) \geqslant \frac{1}{2}\mu^2|\boldsymbol{x}|^2, \quad \forall \boldsymbol{x} \in \Omega, \quad |\gamma| < \mu; \qquad (3.1)$$

and

(B) Assumption on the exact solution ψ:

$$\psi \in C^3([0,T]; L^\infty(\Omega)) \cap C^2([0,T]; W^{2,\infty}(\Omega)) \cap C^0([0,T]; W^{4,\infty}(\Omega) \cap H_0^1(\Omega)). \tag{3.2}$$

Lemma 3.1 *For any grid function $w \in H^2(\Omega_h) \cap H_0^1(\Omega_h)$ in d ($d = 2, 3$) dimensions, we have*

$$|w|_1 \leqslant \|w\|^{\frac{1}{2}} |w|_2^{\frac{1}{2}}.$$

Proof For any grid function $w \in H^2(\Omega_h) \cap H_0^1(\Omega_h)$, by using Lemma 2.1, Hölder's inequality and the definitions of $|\cdot|_1$ and $|\cdot|_2$, we have

$$|w|_1 = \langle \nabla_h w, \nabla_h w \rangle_0^{\frac{1}{2}} = (-\langle w, \Delta_h w \rangle)^{\frac{1}{2}} \leqslant \|w\|^{\frac{1}{2}} |w|_2^{\frac{1}{2}}.$$

This completes the proof.

Lemma 3.2 ([24, 25]) *For any grid function $w \in H^2(\Omega_h) \cap H_0^1(\Omega_h)$ in d ($d = 2, 3$) dimensions, we have*

$$\|w\|_\infty \leqslant C|w|_2. \tag{3.3}$$

Lemma 3.3 *For any grid function $w \in H^2(\Omega_h) \cap H_0^1(\Omega_h)$ in d ($d = 2, 3$) dimensions, we have*

$$|w|_2 \leqslant Ch^{-2}\|w\|. \tag{3.4}$$

Proof The results can be obtained by using direct computations, and we omit the details here for brevity.

Define the local truncation errors $\eta^*, \eta^n \in X_h$ of the LEFD scheme for $n = 0, 1, 2, \cdots, N-1$ as

$$\eta_{jk}^n := i\delta_t^+ \phi_{jk}^n + \frac{1}{2}\left[\frac{1}{2}\Delta_h - V_{jk} + \gamma L_z^h - \frac{\beta}{2}(3|\phi_{jk}^n|^2 - |\phi_{jk}^{n-1}|^2)\right]$$
$$\times \left(\phi_{jk}^n + \phi_{jk}^{n+1}\right), \quad (j,k) \in \mathcal{T}_h, \tag{3.5}$$

$$\eta_{jk}^0 := i\delta_t^+ \phi_{jk}^0 + \frac{1}{2}\left[\frac{1}{2}\Delta_h - V_{jk} + \gamma L_z^h - \beta\left|\phi_{jk}^{1/2}\right|^2\right](\phi_{jk}^0 + \phi_{jk}^1), \quad (j,k) \in \mathcal{T}_h, \tag{3.6}$$

$$\eta_{jk}^* := i\delta_t^* \phi_{jk}^0 + \frac{1}{2}\left[\frac{1}{2}\Delta_h - V_{jk} + \gamma L_z^h\right]\left(\phi_{jk}^0 + \phi_{jk}^{1/2}\right) + \beta|\phi_{jk}^0|^2 \phi_{jk}^0, \quad (j,k) \in \mathcal{T}_h, \tag{3.7}$$

where $\phi_{jk}^{1/2} = \psi(x_j, y_k, t_{\frac{1}{2}})$, $\phi_{jk}^n = \psi(x_j, y_k, t_n)$ for $n = 0, 1, 2, \cdots, N$. By using Taylor's expansion and noticing the boundary conditions, we can obtain the following lemma.

Lemma 3.4 (local truncation error) *Under assumptions (A) and (B), we have*

$$\|\eta^*\| \leqslant C(\tau + h^2), \quad \|\eta^n\| \leqslant C(\tau^2 + h^2), \quad n = 0, 1, 2, \cdots, N-1. \tag{3.8}$$

Define the "error" functions $e^{1/2}, e^n \in X_h$ for $n = 0, 1, 2, \cdots, N$ as

$$e_{jk}^{1/2} = \phi_{jk}^{1/2} - \psi_{jk}^{1/2}, \quad e_{jk}^n = \phi_{jk}^n - \psi_{jk}^n, \quad (j, k) \in \mathcal{T}_h^0. \tag{3.9}$$

Then for the error estimates of the LEFD scheme proposed in this paper, we have

Lemma 3.5 *Under assumptions (A) and (B), there exist $h_0 > 0$ and $\tau_0 > 0$ sufficiently small, when $0 < h \leqslant h_0$ and $0 < \tau \leqslant \tau_0$, we have the following optimal error estimate for the LEFD scheme*

$$\|e^{\frac{1}{2}}\| \leqslant C\tau(\tau + h^2), \quad |\psi^{\frac{1}{2}}|_2 \leqslant C, \quad \|\psi^{\frac{1}{2}}\|_\infty \leqslant C,$$

$$\|e^n\| \leqslant C(h^2 + \tau^2), \quad |\psi^n|_2 \leqslant C, \quad \|\psi^n\|_\infty \leqslant C, \quad n = 1, 2, \cdots, N. \tag{3.10}$$

Proof In order to avoid the difficulty in obtaining the *a priori* estimate of the numerical solution, we here adopt an induction argument to prove the error estimate (3.60). Subtracting (2.7) and (2.8) from (3.6) and (3.7), respectively, one can obtain

$$i\delta_t^+ e_{jk}^0 + \frac{1}{2}\left[\frac{1}{2}\Delta_h - V_{jk} + \gamma L_z^h\right](e_{jk}^0 + e_{jk}^1) - \xi_{jk}^{1/2} = \eta_{jk}^0, \quad (j,k) \in \mathcal{T}_h, \tag{3.11}$$

$$i\delta_t^* e_{jk}^0 + \frac{1}{2}\left[\frac{1}{2}\Delta_h - V_{jk} + \gamma L_z^h\right](e_{jk}^0 + e_{jk}^{1/2}) = \eta_{jk}^*, \quad (j,k) \in \mathcal{T}_h, \tag{3.12}$$

where $\xi^{1/2} \in X_h$ is defined as

$$\xi_{jk}^{1/2} := \frac{\beta}{2}\left|\phi_{jk}^{1/2}\right|^2(\phi_{jk}^0 + \phi_{jk}^1) - \frac{\beta}{2}\left|\psi_{jk}^{1/2}\right|^2(\psi_{jk}^0 + \psi_{jk}^1)$$

$$= \frac{\beta}{2}\left(\phi_{jk}^{1/2}\overline{e_{jk}^{1/2}} + e_{jk}^{1/2}\overline{\psi_{jk}^{1/2}}\right)(\phi_{jk}^0 + \phi_{jk}^1) + \frac{\beta}{2}\left|\psi_{jk}^{1/2}\right|^2(e_{jk}^0 + e_{jk}^1)$$

$$= \frac{\beta}{2}\left(\phi_{jk}^{1/2}\overline{e_{jk}^{1/2}} + e_{jk}^{1/2}\overline{\psi_{jk}^{1/2}}\right)(\phi_{jk}^0 + \phi_{jk}^1) + \frac{\beta}{2}\left|\psi_{jk}^{1/2}\right|^2 e_{jk}^1, \quad (j,k) \in \mathcal{T}_h. \tag{3.13}$$

Noticing $e_{jk}^0 = 0$ for $(j,k) \in \mathcal{T}_h^0$, one can rewrite (3.11) and (3.12) as follows

$$\frac{i}{\tau}e_{jk}^1 + \frac{1}{2}\left[\frac{1}{2}\Delta_h - V_{jk} + \gamma L_z^h\right]e_{jk}^1 - \xi_{jk}^{1/2} = \eta_{jk}^0, \quad (j,k) \in \mathcal{T}_h, \tag{3.14}$$

$$\frac{2i}{\tau}e_{jk}^{1/2} + \frac{1}{2}\left[\frac{1}{2}\Delta_h - V_{jk} + \gamma L_z^h\right]e_{jk}^{1/2} = \eta_{jk}^*, \quad (j,k) \in \mathcal{T}_h. \tag{3.15}$$

Computing the inner product of (3.15) with $e^{1/2}$ and taking the imaginary part give

$$\|e^{1/2}\|^2 = \frac{\tau}{2}\mathrm{Im}\langle\eta^*, e^{1/2}\rangle \leqslant C\tau^2\|\eta^*\|^2 + \frac{1}{2}\|e^{1/2}\|^2 \leqslant C\tau^2(\tau+h^2)^2 + \frac{1}{2}\|e^{1/2}\|^2, \tag{3.16}$$

where Lemma 3.4 was used. This gives

$$\|e^{1/2}\| \leqslant C\tau(\tau+h^2). \tag{3.17}$$

Then, by using the 'lifting' technique, we obtain from (3.15) that

$$|e^{1/2}|_2 = \left(h_1 h_2 \sum_{j=1}^{J-1}\sum_{k=1}^{K-1}\left|\Delta_h e_{jk}^{1/2}\right|^2\right)^{\frac{1}{2}}$$

$$= \left(h_1 h_2 \sum_{j=1}^{J-1}\sum_{k=1}^{K-1}\left|-\frac{8i}{\tau}e_{jk}^{1/2} + 2V_{jk}e_{jk}^{1/2} - 2\gamma L_z^h e_{jk}^{1/2} + 4\eta_{jk}^*\right|^2\right)^{\frac{1}{2}}$$

$$\leqslant C(\tau^{-1}\|e^{1/2}\| + \|e^{1/2}\| + |e^{1/2}|_1 + \|\eta^*\|)$$

$$\leqslant C(\tau^{-1}\|e^{1/2}\| + \|e^{1/2}\| + \|e^{1/2}\|^{\frac{1}{2}}|e^{1/2}|_2^{\frac{1}{2}} + \|\eta^*\|)$$

$$\leqslant C(\tau^{-1}\|e^{1/2}\| + \|e^{1/2}\| + \|\eta^*\|) + \frac{1}{2}|e^{1/2}|_2, \tag{3.18}$$

where (3.17), Cauchy's inequality, Lemma 3.1 and Lemma 3.4 were used. This means that, without any constraints on the time-step size, there is

$$|e^{1/2}|_2 \leqslant C(\tau+h^2) \leqslant C, \quad |\psi^{1/2}|_2 \leqslant |\phi^{1/2}|_2 + |e^{1/2}|_2 \leqslant C. \tag{3.19}$$

Consequently, by using Lemma 3.2, we obtain

$$\|e^{1/2}\|_\infty \leqslant |e^{1/2}|_2 \leqslant C, \quad \|\psi^{1/2}\|_\infty \leqslant |\psi^{1/2}|_2 \leqslant C. \tag{3.20}$$

By using (3.17) and (3.20), one can obtain from (3.13) that

$$\|\xi^{\frac{1}{2}}\| \leqslant C(\|e^{\frac{1}{2}}\| + \|e^1\|) \leqslant C(h^2+\tau^2) + C\|e^1\|. \tag{3.21}$$

Computing the inner product of (3.14) with e^1 and taking the imaginary part give

$$||e^1||^2 = -\tau \text{Im}\langle \xi^{\frac{1}{2}}, e^1\rangle + \tau \text{Im}\langle \eta^0, e^1\rangle \leqslant C\tau(h^2+\tau^2)^2 + C\tau||e^1||^2. \tag{3.22}$$

Hence, if τ is small enough, there is

$$||e^1|| \leqslant C(h^2+\tau^2). \tag{3.23}$$

By using the lifting technique, we obtain from (3.14) that

$$|e^1|_2 \leqslant C(\tau^{-1}||e^1|| + ||e^1|| + |e^1|_1 + ||\eta^0||)$$
$$\leqslant C(\tau^{-1}||e^1|| + ||e^1|| + ||e^1||^{\frac{1}{2}}|e^1|_2^{\frac{1}{2}} + ||\eta^0||)$$
$$\leqslant C(\tau^{-1}||e^1|| + ||e^1|| + ||\eta^0||) + \frac{1}{2}|e^1|_2, \tag{3.24}$$

where (3.23), Cauchy's inequality, Lemma 3.1 and Lemma 3.4 were used. This gives

$$|e^1|_2 \leqslant C(\tau^{-1}||e^1|| + ||e^1|| + ||\eta^0||) \leqslant C\tau^{-1}(h^2+\tau^2). \tag{3.25}$$

On the other hand, by using Lemma 3.3, one can obtain

$$|e^1|_2 \leqslant Ch^{-2}||e^1|| \leqslant Ch^{-2}(h^2+\tau^2). \tag{3.26}$$

Combining (3.24) and (3.26) implies that, without any constraints on the time step size, it is always true to have

$$|e^1|_2 \leqslant C, \quad |\psi^1|_2 \leqslant |\phi^1|_2 + |e^1|_2 \leqslant C. \tag{3.27}$$

This together with Lemma 3.2 gives

$$||e^1||_\infty \leqslant C|e^1|_2 \leqslant C, \quad ||\psi^1||_\infty \leqslant ||\phi^1||_\infty + ||e^1||_\infty \leqslant C. \tag{3.28}$$

We now suppose that (3.60) holds for $n = 1, 2, \cdots, m$ with $1 \leqslant m \leqslant N-1$, i.e.,

$$||e^n|| \leqslant C(h^2+\tau^2), \quad |\psi^n|_2 \leqslant C, \quad n = 1, 2, \cdots, m. \tag{3.29}$$

This together with Lemma 3.2 gives

$$||\psi^n||_\infty \leqslant C|\psi^n|_2 \leqslant C, \quad n = 1, 2, \cdots, m. \tag{3.30}$$

Next, we are going to prove that (3.60) also holds for $n = m+1$. Subtracting (2.4) from (3.5) gives the following 'error' equation:

$$i\delta_t^+ e_{jk}^n + \frac{1}{2}\left[\frac{1}{2}\Delta_h - V_{jk} + \gamma L_z^h\right](e_{jk}^n + e_{jk}^{n+1}) - \xi_{jk}^{n+1} = \eta_{jk}^n,$$

$$(j,k) \in \mathcal{T}_h, \quad n = 1, 2, \cdots, N-1, \tag{3.31}$$

where

$$\xi_{jk}^{n+1} = \frac{\beta}{4}[(3|\phi_{jk}^n|^2 - |\phi_{jk}^{n-1}|^2)(\phi_{jk}^n + \phi_{jk}^{n+1}) - (3|\psi_{jk}^n|^2 - |\psi_{jk}^{n-1}|^2)(\psi_{jk}^n + \psi_{jk}^{n+1})]$$

$$= \frac{\beta}{4}[3(e_{jk}^n \overline{\phi_{jk}^n} + \psi_{jk}^n \overline{e_{jk}^n}) - (e_{jk}^{n-1}\overline{\phi_{jk}^{n-1}} + \psi_{jk}^{n-1}\overline{e_{jk}^{n-1}})](\phi_{jk}^n + \phi_{jk}^{n+1})$$

$$+ \frac{\beta}{4}(3|\psi_{jk}^n|^2 - |\psi_{jk}^{n-1}|^2)(e_{jk}^n + e_{jk}^{n+1}), \quad (j,k) \in \mathcal{T}_h, \quad n = 1, 2, \cdots, N-1.$$

$$\tag{3.32}$$

By using Cauchy's inequality, assumption (B), (3.30), we obtain that, when $1 \leqslant n \leqslant m$, there are

$$\|\xi^{n+1}\| \leqslant C(\|e^n\| + \|e^{n-1}\| + \|e^{n+1}\|). \tag{3.33}$$

Computing the inner product of (3.31) with $\tau(e^n + e^{n+1})$, then taking the imaginary part, we obtain

$$\|e^{n+1}\|^2 - \|e^n\|^2 = \tau\text{Im}\langle \xi^{n+1}, e^n + e^{n+1}\rangle + \tau\text{Im}\langle \eta^n, e^n + e^{n+1}\rangle, \tag{3.34}$$

where Lemma 2.1 was used. Summing (3.34) together over n from 1 to m gives

$$\|e^{m+1}\|^2 - \|e^1\|^2 = \tau\sum_{n=1}^{m}\text{Im}\langle \xi^{n+1}, e^n + e^{n+1}\rangle + \tau\sum_{n=1}^{m}\text{Im}\langle \eta^n, e^n + e^{n+1}\rangle. \tag{3.35}$$

On the estimates of the last two terms of (3.35), by using (3.30), (3.33) and Cauchy's inequality, we have

$$|\text{Im}\langle \xi^{n+1}, e^n + e^{n+1}\rangle| \leqslant \|\xi^{n+1}\|^2 + \|e^n\|^2 + \|e^{n+1}\|^2$$

$$\leqslant C(\|e^{n-1}\|^2 + \|e^n\|^2 + \|e^{n+1}\|^2), \quad n = 1, 2, \cdots, m,$$

$$\tag{3.36}$$

$$|\text{Im}\langle \eta^n, e^n + e^{n+1}\rangle| \leqslant \|\eta^n\|^2 + \frac{1}{2}\|e^n\|^2 + \frac{1}{2}\|e^{n+1}\|^2, \quad n = 1, 2, \cdots, m. \tag{3.37}$$

Substituting (3.36)-(3.37) into (3.35), we obtain

$$\|e^{m+1}\|^2 \leqslant \|e^1\|^2 + \tau\sum_{n=1}^{m}\|\eta^n\|^2 + \tau\sum_{n=1}^{m+1}\|e^n\|^2 \leqslant C(h^2+\tau^2)^2 + \tau\sum_{n=1}^{m+1}\|e^n\|^2. \tag{3.38}$$

By using Gronwall's inequality, we obtain from (3.38) that, for a sufficiently small τ, there is

$$\|e^{m+1}\| \leqslant C(h^2+\tau^2). \tag{3.39}$$

This together with (3.29) gives

$$\|\delta_t^+ e^n\| \leqslant C\tau^{-1}(h^2+\tau^2), \quad n=1,2,\cdots,m. \tag{3.40}$$

By using the lifting technique, (3.29)-(3.30), (3.33), (3.39), Lemma 3.1, Lemma 3.4 and Cauchy's inequality, we obtain from (3.31) that

$$|e^n + e^{n+1}|_2$$
$$\leqslant C(\|\delta_t^+ e^n\| + \|e^n + e^{n+1}\| + |e^n + e^{n+1}|_1 + \|\xi^{n+1}\| + \|\eta^n\|)$$
$$\leqslant C(\|\delta_t^+ e^n\| + \|e^n + e^{n+1}\| + \|e^n + e^{n+1}\|^{\frac{1}{2}}|e^n + e^{n+1}|_2^{\frac{1}{2}} + \|\xi^{n+1}\| + \|\eta^n\|)$$
$$\leqslant C(\|\delta_t^+ e^n\| + \|e^n\| + \|e^{n+1}\| + \|\xi^{n+1}\| + \|\eta^n\|) + \frac{1}{2}|e^n + e^{n+1}|_2$$
$$\leqslant C\tau^{-1}(h^2+\tau^2) + \frac{1}{2}|e^n + e^{n+1}|_2, \quad n=1,2,\cdots,m. \tag{3.41}$$

This gives

$$|e^{n+1}|_2 - |e^n|_2 \leqslant |e^n + e^{n+1}|_2 \leqslant C\tau^{-1}(h^2+\tau^2), \quad n=1,2,\cdots,m. \tag{3.42}$$

Summing (3.42) together for n from 1 to m gives

$$|e^{m+1}|_2 \leqslant |e^1|_2 + Cm\tau^{-1}(h^2+\tau^2) \leqslant C\tau^{-2}(h^2+\tau^2). \tag{3.43}$$

On the other hand, by using Lemma 3.3, one can obtain

$$|e^{m+1}|_2 \leqslant Ch^{-2}\|e^{m+1}\| \leqslant Ch^{-2}(h^2+\tau^2). \tag{3.44}$$

Hence, without any constraints on the time-step size, it is always true to have

$$|e^{m+1}|_2 \leqslant C, \quad |\psi^{m+1}|_2 \leqslant |\phi^{m+1}|_2 + |e^{m+1}|_2 \leqslant C. \tag{3.45}$$

This together with Lemma 3.2 gives

$$||e^{m+1}||_\infty \leq |e^{m+1}|_2 \leq C, \quad ||\psi^{m+1}||_\infty \leq |\psi^{m+1}|_2 \leq C. \tag{3.46}$$

This completes the induction argument and consequently completes the proof.
If the exact solution satisfies higher regularity in time, i.e.,

(C) Assumption of the exact solution ψ:

$$\psi \in C^4([0,T]; L^\infty(\Omega)) \cap C^3([0,T]; W^{2,\infty}(\Omega)) \cap C^1([0,T]; W^{4,\infty}(\Omega) \cap H_0^1(\Omega)).$$

We have the following result:

Lemma 3.6 *Under assumptions (A) and (C), there exist $h_0 > 0$ and $\tau_0 > 0$ sufficiently small, when $0 < h \leq h_0$ and $0 < \tau \leq \tau_0$, we have the following optimal error estimate for the LEFD scheme,*

$$|e^n|_1 \leq C(h^2 + \tau^2), \quad n = 1, 2, \cdots, N. \tag{3.47}$$

Proof Under Assumptions (A) and (C), by using Taylor's expansion, one can obtain an estimate of the local truncation error as follows

$$||\delta_t^+ \eta^n|| \leq C(h^2 + \tau^2), \quad n = 1, 2, \cdots, N-2. \tag{3.48}$$

By using Cauchy's inequality, assumption (C) and Lemma 3.5, one can obtain from (3.13) that

$$||\xi^{\frac{1}{2}}|| \leq C(||e^{\frac{1}{2}}|| + ||e^1||) \leq C(h^2 + \tau^2), \tag{3.49}$$

$$||\xi^{n+1}|| \leq C(||e^n|| + ||e^{n-1}|| + ||e^{n+1}||) \leq C(h^2 + \tau^2), \quad n = 1, 2, \cdots, N-1, \tag{3.50}$$

$$|\xi^{n+1}|_1 \leq C(|e^n|_1 + |e^{n-1}|_1 + |e^{n+1}|_1) + C(h^2 + \tau^2), \quad n = 1, 2, \cdots, N-1. \tag{3.51}$$

Computing the inner product of (3.14) with $-2e^1$ and taking the real part give

$$\mathcal{E}(e^1) = 2\text{Re}\langle \xi^{\frac{1}{2}}, e^1 \rangle - 2\text{Re}\langle \eta^0, e^1 \rangle \leq C(||\xi^{\frac{1}{2}}||^2 + ||e^1||^2 + ||\eta^0||^2) \leq C(h^2 + \tau^2)^2. \tag{3.52}$$

This together with Lemma 2.1 gives

$$|e^1|_1 \leq C(h^2 + \tau^2). \tag{3.53}$$

Computing the inner product of (3.31) with $2(e^{n+1} - e^n)$ and taking the real part yield

$$\mathcal{E}(e^{n+1}) - \mathcal{E}(e^n) = -2\tau \mathrm{Re}\langle \xi^{n+1}, \delta_t^+ e^n \rangle - 2\mathrm{Re}\langle \eta^n, e^{n+1} - e^n \rangle, \quad n = 1, 2, \cdots, N-1. \tag{3.54}$$

Summing the above equation over n from 1 to m, then replacing m by n, one can obtain

$$\mathcal{E}(e^{n+1}) - \mathcal{E}(e^1) = -2\tau \sum_{l=1}^{n} \mathrm{Re}\langle \xi^{l+1}, \delta_t^+ e^l \rangle - 2\tau \sum_{l=1}^{n} \mathrm{Re}\langle \eta^l, \delta_t^+ e^l \rangle, \quad n = 1, 2, \cdots, N-1. \tag{3.55}$$

By using summation by parts, Cauchy's inequality, Lemma 3.4, Lemma 3.5, (3.48), (3.50) and (3.51), one can obtain

$$\begin{aligned}
\left|2\tau \sum_{l=1}^{n} \mathrm{Re}\langle \eta^l, \delta_t^+ e^l \rangle\right| &= \left|-2\tau \sum_{l=2}^{n} \mathrm{Re}\langle \delta_t^+ \eta^{l-1}, e^l \rangle + 2\mathrm{Re}\langle \eta^n, e^{n+1} \rangle - 2\mathrm{Re}\langle \eta^1, e^1 \rangle\right| \\
&\leqslant 2\tau \sum_{l=2}^{n} |\langle \delta_t^+ \eta^{l-1}, e^l \rangle| + 2|\langle \eta^n, e^{n+1} \rangle| + |\langle \eta^1, e^1 \rangle| \\
&\leqslant \tau \sum_{l=2}^{n} (\|\delta_t^+ \eta^{l-1}\|^2 + \|e^l\|^2) + \|\eta^n\|^2 + \|e^{n+1}\|^2 + \|\eta^1\|^2 + \|e^1\|^2 \\
&\leqslant C(\tau^2 + h^2)^2, \quad n = 1, 2, \cdots, N-1. \tag{3.56}
\end{aligned}$$

$$\begin{aligned}
|\mathrm{Re}\langle \xi^{l+1}, \delta_t^+ e^l \rangle| &= \left|\mathrm{Re}\left\langle \xi^{l+1}, \frac{i}{2}\left[\frac{1}{2}\Delta_h - V + \gamma L_z^h\right](e^l + e^{l+1}) - i\xi^l - i\eta^l \right\rangle\right| \\
&= \left|\mathrm{Re}\left\langle \xi^{l+1}, \frac{i}{2}\left[\frac{1}{2}\Delta_h - V + \gamma L_z^h\right](e^l + e^{l+1}) - i\eta^l \right\rangle\right| \\
&= \left|\frac{1}{4}\mathrm{Im}\langle \xi^{l+1}, \Delta_h(e^l + e^{l+1})\rangle - \frac{1}{2}\mathrm{Im}\langle \xi^{l+1}, V(e^l + e^{l+1})\rangle \right. \\
&\quad \left. + \frac{\gamma}{2}\mathrm{Im}\langle \xi^{l+1}, L_z^h(e^l + e^{l+1})\rangle - \mathrm{Im}\langle \xi^{l+1}, \eta^l \rangle\right| \\
&= \left|-\frac{1}{4}\mathrm{Im}\langle \nabla_h \xi^{l+1}, \nabla_h(e^l + e^{l+1})\rangle_0 - \frac{1}{2}\mathrm{Im}\langle \xi^{l+1}, V(e^l + e^{l+1})\rangle \right. \\
&\quad \left. + \frac{\gamma}{2}\mathrm{Im}\langle \xi^{l+1}, L_z^h(e^l + e^{l+1})\rangle - \mathrm{Im}\langle \xi^{l+1}, \eta^l \rangle\right| \\
&\leqslant \frac{1}{4}|\langle \nabla_h \xi^{l+1}, \nabla_h(e^l + e^{l+1})\rangle_0| + \frac{1}{2}|\langle \xi^{l+1}, V(e^l + e^{l+1})\rangle|
\end{aligned}$$

$$+\frac{|\gamma|}{2}|\langle\xi^{l+1}, L_z^h(e^l+e^{l+1})\rangle|+|\langle\xi^l,\eta^l\rangle|$$

$$\leqslant \frac{1}{8}|\xi^{l+1}|_1^2+\frac{1}{4}|e^l|_1^2+\frac{1}{4}|e^{l+1}|_1^2+\frac{1}{4}\|\xi^{l+1}\|^2$$

$$+\frac{1}{2}\|V\|_\infty(\|e^l\|^2+\|e^{l+1}\|^2)+\frac{|\gamma|}{4}\|\xi^{l+1}\|^2$$

$$+\frac{|\gamma|}{2}(\|L_z^h e^l\|^2+\|L_z^h e^{l+1}\|^2)+\frac{1}{2}\|\xi^{l+1}\|^2+\frac{1}{2}\|\eta^l\|^2$$

$$\leqslant C(h^2+\tau^2)^2+C(|e^{l-1}|_1^2+|e^l|_1^2+|e^{l+1}|_1^2). \tag{3.57}$$

Substituting (3.52), (3.56) and (3.57) into (3.55) and using Lemma 2.1 give

$$|e^{n+1}|_1^2 \leqslant C(h^2+\tau^2)^2+C\tau\sum_{l=1}^{n+1}|e^l|_1^2. \tag{3.58}$$

Then by using Gronwall's inequality, one can obtain that, when τ is small enough, there is

$$|e^n|_1 \leqslant C(h^2+\tau^2), \quad n=1,2,\cdots,N. \tag{3.59}$$

This completes the induction argument and consequently completes the proof.

Following Lemma 3.5 and Lemma 3.6, we are ready to give and prove our main error estimate result of the LEFD scheme.

Theorem 3.1 *Under assumptions (A) and (C), there exist $h_0 > 0$ and $\tau_0 > 0$ sufficiently small, when $0 < h \leqslant h_0$ and $0 < \tau \leqslant \tau_0$, we have the following optimal error estimate for the LEFD scheme,*

$$\|e^n\|_\infty \leqslant C(h^2+\tau^2), \quad n=1,2,\cdots,N. \tag{3.60}$$

Proof By using Cauchy's inequality, assumption (C), Lemmas 3.5 and Lemma 3.6, one can obtain from (3.13) that

$$|\xi^{n+1}|_2 \leqslant C(|e^n|_2+|e^{n-1}|_2+|e^{n+1}|_2)+C(h^2+\tau^2), \quad n=1,2,\cdots,N-1. \tag{3.61}$$

Computing the inner product of (3.14) with $\Delta_h e^1$ and taking the real part give

$$\frac{1}{4}|e^1|_2^2 = \frac{1}{2}\mathrm{Re}\langle Ve^1,\Delta_h e^1\rangle-\frac{\gamma}{2}\mathrm{Re}\langle L_z^h e^1,\Delta_h e^1\rangle-\mathrm{Re}\langle\xi^{\frac{1}{2}},\Delta_h e^1\rangle+\mathrm{Re}\langle\eta^0,\Delta_h e^1\rangle$$

$$\leqslant C(\|e^1\|^2+\|L_z^h e^1\|^2+\|\xi^{\frac{1}{2}}\|^2+\|e^1\|^2+\|\eta^0\|^2)+\frac{1}{8}|e^1|_2^2$$

$$\leqslant C(h^2+\tau^2)^2 + \frac{1}{8}|e^1|_2^2, \tag{3.62}$$

where Cauchy's inequality, Lemma 3.4, Lemma 3.5 and Lemma 3.6 were used. This gives

$$|e^1|_2 \leqslant C(h^2+\tau^2). \tag{3.63}$$

Computing the inner product of (3.31) with $2(\Delta_h e^{n+1} - \Delta_h e^n)$ and taking the real part yield

$$\frac{1}{2}|e^{n+1}|_2^2 + \gamma h_1 h_2 \sum_{j=1}^{J-1}\sum_{k=1}^{K-1} \Delta_h \overline{e_{jk}^{n+1}} L_z^h e_{jk}^{n+1}$$

$$-\frac{1}{2}|e^n|_2^2 - \gamma h_1 h_2 \sum_{j=1}^{J-1}\sum_{k=1}^{K-1} \Delta_h \overline{e_{jk}^{n}} L_z^h e_{jk}^{n}$$

$$=\tau h_1 h_2 \sum_{j=1}^{J-1}\sum_{k=1}^{K-1} \mathrm{Re}[v_{jk}(e_{jk}^n + e_{jk}^{n+1})\Delta_h \delta_t^+ \overline{e_{jk}^n}] + 2\tau \mathrm{Re}\langle \xi^{n+1}, \delta_t^+ \Delta_h e^n\rangle$$

$$+2\tau \mathrm{Re}\langle \eta^n, \delta_t^+ \Delta_h e^n\rangle, \quad n=1,2,\cdots,N-1, \tag{3.64}$$

where Lemma 2.1 and Lemma 2.2 were used. Summing the above equation over n from 1 to m, then replacing m by n, one can obtain

$$\frac{1}{2}|e^{n+1}|_2^2 + \gamma h_1 h_2 \sum_{j=1}^{J-1}\sum_{k=1}^{K-1} \Delta_h \overline{e_{jk}^{n+1}} L_z^h e_{jk}^{n+1}$$

$$=\frac{1}{2}|e^1|_2^2 + \gamma h_1 h_2 \sum_{j=1}^{J-1}\sum_{k=1}^{K-1} \Delta_h \overline{e_{jk}^{1}} L_z^h e_{jk}^{1}$$

$$+\tau\sum_{l=1}^{n} h_1 h_2 \sum_{j=1}^{J-1}\sum_{k=1}^{K-1} \mathrm{Re}[v_{jk}(e_{jk}^l + e_{jk}^{l+1})\Delta_h \delta_t^+ \overline{e_{jk}^l}]$$

$$+2\tau\sum_{l=1}^{n}\mathrm{Re}\langle \xi^{l+1}, \Delta_h \delta_t^+ e^l\rangle + 2\tau\sum_{l=1}^{n}\mathrm{Re}\langle \eta^l, \delta_t^+ \Delta_h e^l\rangle,$$

$$n=1,2,\cdots,N-1. \tag{3.65}$$

By using summation by parts, Cauchy's inequality, Lemma 3.4, Lemma 3.5, Lemma 3.6, (3.48), (3.50)-(3.51) and (3.61), one can obtain

$$\Big|2\tau\sum_{l=1}^{n}\mathrm{Re}\langle \eta^l, \delta_t^+ \Delta_h e^l\rangle\Big|$$

$$=\left|-2\tau\sum_{l=2}^{n}\operatorname{Re}\langle\delta_t^+\eta^{l-1},\Delta_h e^l\rangle+2\operatorname{Re}\langle\eta^n,\Delta_h e^{n+1}\rangle-2\operatorname{Re}\langle\eta^1,\Delta_h e^1\rangle\right|$$

$$\leqslant\tau\sum_{l=2}^{n}(\|\delta_t^+\eta^{l-1}\|^2+|e^l|_2^2)+2\|\eta^n\|^2+\frac{1}{8}|e^{n+1}|_2^2+\|\eta^1\|^2+|e^1|_2^2$$

$$\leqslant\tau\sum_{l=2}^{n}|e^l|_2^2+\frac{1}{8}|e^{n+1}|_2^2+C(\tau^2+h^2)^2, \tag{3.66}$$

$$|\operatorname{Re}\langle\xi^{l+1},\delta_t^+\Delta_h e^l\rangle|$$

$$=\left|\operatorname{Re}\langle\Delta_h\xi^{l+1},\frac{i}{2}\left[\frac{1}{2}\Delta_h-V+\gamma L_z^h\right](e^l+e^{l+1})-i\xi^l-i\eta^l\rangle\right|$$

$$=\left|\operatorname{Re}\langle\Delta_h\xi^{l+1},\frac{i}{2}\left[\frac{1}{2}\Delta_h-V+\gamma L_z^h\right](e^l+e^{l+1})-i\eta^l\rangle\right|$$

$$=\left|\frac{1}{4}\operatorname{Im}\langle\Delta_h\xi^{l+1},\Delta_h(e^l+e^{l+1})\rangle-\frac{1}{2}\operatorname{Im}\langle\Delta_h\xi^{l+1},V(e^l+e^{l+1})\rangle\right.$$

$$\left.+\frac{\gamma}{2}\operatorname{Im}\langle\Delta_h\xi^{l+1},L_z^h(e^l+e^{l+1})\rangle-\operatorname{Im}\langle\Delta_h\xi^{l+1},\eta^l\rangle\right|$$

$$\leqslant\frac{1}{4}|\langle\Delta_h\xi^{l+1},\Delta_h(e^l+e^{l+1})\rangle|+\frac{1}{2}|\langle\Delta_h\xi^{l+1},V(e^l+e^{l+1})\rangle|$$

$$+\frac{|\gamma|}{2}|\langle\Delta_h\xi^{l+1},L_z^h(e^l+e^{l+1})\rangle|+|\langle\Delta_h\xi^l,\eta^l\rangle|$$

$$\leqslant\frac{1}{8}|\xi^{l+1}|_2^2+\frac{1}{4}|e^l|_2^2+\frac{1}{4}|e^{l+1}|_2^2+\frac{1}{4}|\xi^{l+1}|_2^2$$

$$+\frac{1}{2}\|V\|_\infty(\|e^l\|^2+\|e^{l+1}\|^2)+\frac{|\gamma|}{4}|\xi^{l+1}|_2^2$$

$$+\frac{|\gamma|}{2}(\|L_z^h e^l\|^2+\|L_z^h e^{l+1}\|^2)+\frac{1}{2}|\xi^{l+1}|_2^2+\frac{1}{2}\|\eta^l\|^2$$

$$\leqslant C(h^2+\tau^2)^2+C(|e^{l-1}|_2^2+|e^l|_2^2+|e^{l+1}|_2^2), \tag{3.67}$$

$$h_1 h_2\sum_{j=1}^{J-1}\sum_{k=1}^{K-1}\operatorname{Re}[v_{jk}(e_{jk}^l+e_{jk}^{l+1})\Delta_h\delta_t^+\overline{e_{jk}^l}]$$

$$\leqslant h_1 h_2\sum_{j=1}^{J-1}\sum_{k=1}^{K-1}\operatorname{Re}[\Delta_h(v_{jk}(e_{jk}^l+e_{jk}^{l+1}))\delta_t^+\overline{e_{jk}^l}]$$

$$\leqslant C(h^2+\tau^2)^2+C(|e^{l-1}|_2^2+|e^l|_2^2+|e^{l+1}|_2^2), \tag{3.68}$$

$$\frac{1}{2}|e^1|_2^2 + \gamma h_1 h_2 \sum_{j=1}^{J-1}\sum_{k=1}^{K-1} \Delta_h \overline{e_{jk}^1} L_z^h e_{jk}^1 \leqslant C(|e^1|_2^2 + \|L_z^h e_{jk}^1\|^2) \leqslant C(h^2+\tau^2)^2, \tag{3.69}$$

$$|\gamma h_1 h_2 \sum_{j=1}^{J-1}\sum_{k=1}^{K-1} \Delta_h \overline{e_{jk}^{n+1}} L_z^h e_{jk}^{n+1}| \leqslant \frac{1}{8}|e^{n+1}|_2^2 + C(\tau^2+h^2)^2. \tag{3.70}$$

Substituting (3.66)-(3.70) into (3.65) gives

$$|e^{n+1}|_2^2 \leqslant C(h^2+\tau^2)^2 + C\tau \sum_{l=1}^{n+1} |e^l|_2^2. \tag{3.71}$$

Then by using Gronwall's inequality, one can obtain that, when τ is small enough, there is

$$|e^n|_2 \leqslant C(h^2+\tau^2), \quad n=1,2,\cdots,N. \tag{3.72}$$

This completes the proof.

Similarly, one can prove the following stability result of the LEFD scheme.

Theorem 3.2 *Under assumptions (A) and (C), there exist $h_0 > 0$ and $\tau_0 > 0$ sufficiently small, when $0 < h \leqslant h_0$ and $0 < \tau \leqslant \tau_0$, the LEFD scheme is unconditionally stable with respect to the initial data in the maximum norm.*

4 Numerical experiment

In this section, we report several numerical results of the LEFD scheme for the problem (1.1)-(1.3) to test the error estimate and the conservation laws in the discrete sense.

We take $\Omega = [-8,8] \times [-8,8]$, $V(\boldsymbol{x}) = \frac{1}{2}(x^2+y^2)$ in (1.1) and $\psi_0 = \frac{2}{\sqrt{\pi}}(x+iy)e^{-8(x^2+y^2)}$ in (1.2). For comparison, the numerical 'exact' solution ψ_e is obtained by the LEFD scheme with a very fine mesh and a small time step, e.g. $h = 0.001$ and $\tau = 0.0001$. Let $e(h,\tau)$ denote the error $\|e^N\|_\infty$ with time step τ and mesh-size h. Denote

$$\text{rate1} = \frac{\log(e(h_1,\tau)/e(h_2,\tau))}{\log(h_1/h_2)}, \quad \text{rate2} = \frac{\log(e(h,\tau_1)/e(h,\tau_2))}{\log(\tau_1/\tau_2)}. \tag{4.1}$$

Thanks to the good stability, we measure the temporal error and the spatial error of the proposed numerical methods separately. To test the spatial error test, we choose different h, β, γ and use $\tau = 0.0001$ which is small enough to ignore the temporal error. To test the temporal error test, we choose different τ, β, γ and fix $h = 0.001$ which is small enough to ignore the spatial error. The error is presented at time $t = 0.6$ and measured under the maximum norm with exactly the same form as given in the theoretical estimate (3.60). Data in Tables 1-2 show rate1 ≈ 2, rate2 ≈ 2, which verifies our theoretical analysis in this paper.

Table 1 Spatial errors $\|e^N\|_\infty$ of the LEFD scheme in computing the example with $\tau = 0.0001$ and different γ, β

	$h_0 = 0.16$	$h = h_0/2$	$h = h_0/4$	$h = h_0/8$
$\gamma = 0, \beta = -1$	4.1492e-3	1.0531e-3	2.5918e-4	6.4238e-5
	rate1	1.98	2.02	2.01
$\gamma = 0, \beta = 1$	4.1569e-3	1.0558e-3	2.5986e-4	6.4403e-5
	rate1	1.98	2.02	2.01
$\gamma = 0.5, \beta = -1$	4.1285e-3	1.0472e-3	2.5781e-4	6.3872e-5
	rate1	1.98	2.02	2.01
$\gamma = 0.5, \beta = 1$	4.1362e-3	1.0499e-3	2.5846e-4	6.4039e-5
	rate1	1.98	2.02	2.01
$\gamma = 0.9, \beta = -1$	4.0848e-3	1.0343e-3	2.5463e-4	6.3109e-5
	rate1	1.98	2.02	2.01
$\gamma = 0.9, \beta = 1$	4.0921e-3	1.0371e-3	2.5529e-4	6.3271e-5
	rate1	1.98	2.02	2.01

Table 2 Temporal errors $\|e^N\|_\infty$ of the LEFD scheme in computing the example with $h = 0.001$ and different γ, β

	$\tau_0 = 0.06$	$\tau = \tau_0/2$	$\tau = \tau_0/4$	$\tau = \tau_0/8$
$\gamma = 0, \beta = -1$	1.8264e-2	5.1794e-3	1.2725e-3	3.0571e-4
	rate2	1.82	2.03	2.05
$\gamma = 0, \beta = 1$	1.8278e-2	5.1870e-3	1.2743e-3	3.0616e-4
	rate2	1.82	2.03	2.05
$\gamma = 0.5, \beta = -1$	1.7802e-2	5.0697e-3	1.2429e-3	2.9842e-4
	rate2	1.82	2.03	2.05
$\gamma = 0.5, \beta = 1$	1.7815e-2	5.0771e-3	1.2448e-3	2.9889e-4
	rate2	1.82	2.03	2.05
$\gamma = 0.9, \beta = -1$	1.7769e-2	5.0111e-3	1.2251e-3	2.9417e-4
	rate2	1.82	2.03	2.05
$\gamma = 0.9, \beta = 1$	1.7791e-2	5.0184e-3	1.2270e-3	2.9466e-4
	rate2	1.82	2.03	2.05

In order to test the conservation laws possessed by the LEFD scheme, we draw the total mass and energy in the discrete level in Fig. 1. We see from the figure that the LEFD scheme preserves the total mass and energy in the discrete sense very well. This verifies the results given in Lemma 2.3.

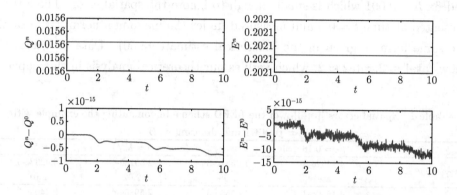

Fig. 1 Total mass and energy computed by LEFD with $\gamma = 0.5, \beta = 1, h = 0.25, \tau = 0.01, T = 10$

5 Conclusion

In this paper, we propose a new linearized semi-implicit finite difference scheme for solving the GPE with AMR. By using a recursive relation to introduce a new kind of energy functional of the discrete functions, we prove that the proposed linearized scheme preserves both the total mass and the total energy in the discrete sense. The main contribution of this paper is that, without imposing any constraints on the time-step size, we establish the maximum error estimate of the numerical solution. The convergence rate of the numerical solution is proved to be of $O(\tau^2 + h^2)$ which is consistent with the local truncation error of the proposed scheme, this means that the error estimate is optimal. Our analysis techniques and the error estimates can be directly extended to other linearized finite difference schemes for solving the general NLS/GP equations in 2D/3D.

Acknowledgements This work was supported by the National Natural Science Foundation of China (Grant Nos. 11571181,11731014), the Natural Science Foundation of Jiangsu Province (Grant No. BK20171454) and Qing Lan Project.

References

[1] Akrivis G, Dougalis V, Karakashian O. On fully discrete Galerkin methods of second order temporal accuracy for the nonlinear Schrödinger equation[J]. Numer. Math., 1991, 59: 31-53.

[2] Antoine X, Bao W, Besse C. Computational methods for the dynamics of the nonlinear Schrödinger/Gross-Pitaevskii equations[J]. Comput. Phys. Comm., 2013, 184(12): 2621-2633.

[3] Bao W, Cai Y. Mathematical theory and numerical methods for Bose-Einstein condensation[J]. Kinet. Relat. Mod., 2013, 6(1): 1-135.

[4] Bao W, Cai Y. Uniform error estimates of finite difference methods for the nonlinear schrödinger equation with wave operator[J]. SIAM J. Numer. Anal., 2012, 50(2): 492-521.

[5] Bao W, Cai Y. Optimal error estimates of finite difference methods for the Gross-Pitaevskii equation with angular momentum rotation[J]. Math. Comput., 2013, 82(281): 99-128.

[6] Bao W, Du Q, Zhang Y. Dynamics of rotating Bose-Einstein condensates and its efficient and accurate numerical computation[J]. SIAM J. Appl. Math., 2006, 66(3): 758-786.

[7] Bao W, Li H, Shen J. A generalized-Laguerre-Fourier-Hermite pseudospectral method for computing the dynamics of rotating Bose-Einstein condensates[J]. SIAM J. Sci. Comput., 2009, 31(5): 3685-3711.

[8] Bao W, Wang H. An efficient and spectrally accurate numerical method for computing dynamics of rotating Bose-Einstein condensates[J]. J. Comput. Phys., 2006, 217(2): 612-626.

[9] Bao W, Wang H, Markowich P A. Ground, symmetric and central vortex states in rotating Bose-Einstein condensates[J]. Comm. Math. Sci., 2005, 3(1): 57-88.

[10] Chang Q, Guo B, Jiang H. Finite difference method for generalized Zakharov equations[J]. Math. Comp., 1995, 64(210): 537-553.

[11] Chang Q, Jia E, Sun W. Difference schemes for solving the generalized nonlinear Schrödinger equation[J]. J. Comput. Phys., 1999, 148(2): 397-415.

[12] Chippada S, Dawson C N, Martinez M L, Wheeler M F. Finite element approximations to the system of shallow water equations, Part II: Discrete time a priori error estimates[M]. SIAM J. Numer. Anal., 1999, 36: 226-250.

[13] Dawson C N, Martinez M L. A characteristic-Galerkin approximation to a system of shallow water equations[J]. Numer. Math., 2000, 86(2): 239-256.

[14] Hao C, Hsiao L, Li H. Global well posedness for the Gross-Pitaevskii equation with an angular momentum rotational term in three dimensions[J]. J. Math. Phys., 2007, 48(10): article 102105.

[15] Hao C, Hsiao L, Li H. Global well posedness for the Gross-Pitaevskii equation with an angular momentum rotational term[J]. Math. Meth. Appl. Sci., 2008, 31(6): 655-664.

[16] Liao H, Sun Z. Error estimate of fourth-order compact scheme for linear Schrödinger equations[J]. SIAM J. Numer. Anal., 2010, 47(6) 4381-4401.

[17] Lieb E H, Seiringer R. Derivation of the Gross-Pitaevskii equation for rotating Bose gases[J]. Commun. Math. Phys., 2006, 264(2): 505-537.

[18] Pitaevskii L P, Stringary S. Bose-Einstein condensation[M]. Clarendon Press, 2003.

[19] Thomée V. Galerkin Finite Element Methods for Parabolic Problems[M]. Berlin: Springer-Verlag, 1997.

[20] Wang H. A time-splitting spectral method for coupled Gross-Pitaevskii equations with applications to rotating Bose-Einstein condensates[J]. J. Comput. Appl. Math., 2007, 205(1): 88-104.

[21] Wang T. Optimal Point-Wise Error Estimate of a Compact Difference Scheme for the Coupled Gross-Pitaevskii Equations in One Dimension[J]. J. Sci. Comput., 2014, 59(1): 158-186.

[22] Wang T, Guo B, Xu Q. Fourth-order compact and energy conservative difference schemes for the nonlinear Schrödinger equation in two dimensions[J]. J. Comput. Phys., 2013, 243: 382-399.

[23] Wang T, Zhao X. Optimal l^∞ error estimates of finite difference methods for the coupled Gross-Pitaevskii equations in high dimensions[J]. Sci. China Math., 2014, 57(10): 2189-2214.

[24] Zhang F, Han B. The finite difference method for dissipative Klein-Gordon-Schrödinger equations in three dimensions[J]. J. Comput. Math., 2010, 28(6): 879-900.

[25] Zhou Y. Application of discrete functional analysis to the finite difference methods[M]. International Academic Publishers, Beijing, 1990.

The Gerdjikov-Ivanov Type Derivative Nonlinear Schrödinger Equation: Long-Time Dynamics of Nonzero Boundary Conditions[*]

Guo Boling（郭柏灵）, and Liu Nan（刘男）

Abstract We consider the Gerdjikov–Ivanov type derivative nonlinear Schrödinger equation

$$iq_t + q_{xx} - iq^2\bar{q}_x + \frac{1}{2}(|q|^4 - q_0^4)q = 0$$

on the line. The initial value $q(x,0)$ is given and satisfies the symmetric, nonzero boundary conditions at infinity, that is, $q(x,0) \to q_\pm$ as $x \to \pm\infty$, and $|q_\pm| = q_0 > 0$. The goal of this paper is to study the asymptotic behavior of the solution of this initial-value problem as $t \to \infty$. The main tool is the asymptotic analysis of an associated matrix Riemann–Hilbert problem by using the steepest descent method and the so-called g-function mechanism. We show that the solution $q(x,t)$ of this initial value problem has a different asymptotic behavior in different regions of the xt-plane. In the regions $x < -2\sqrt{2}q_0^2 t$ and $x > 2\sqrt{2}q_0^2 t$, the solution takes the form of a plane wave. In the region $-2\sqrt{2}q_0^2 t < x < 2\sqrt{2}q_0^2 t$, the solution takes the form of a modulated elliptic wave.

Keywords Gerdjikov–Ivanov type derivative nonlinear Schrödinger equation; nonlinear steepest descent method; Riemann–Hilbert problem; long-time asymptotics

1 Introduction

The celebrated nonlinear Schrödinger (NLS) equation has been recognized as a ubiquitous mathematical model among many integrable systems, which

---

[*] Math. Methods Appl. Sci., 2019, 42(14): 4839–4861. DOI:10.1002/mma.5698.

governs weakly nonlinear and dispersive wave packets in one-dimensional physical systems. It plays an important role in nonlinear optics [15], water waves [3], and Bose-Einstein condensates [29]. Another integrable system of NLS type, the derivative-type NLS equation

$$iu_t + u_{xx} - iu^2 \bar{u}_x + \frac{1}{2}|u|^4 u = 0 \tag{1.1}$$

is derived by Gerdjikov–Ivanov [20], which is called the GI equation. Here and below, the bar refers to the complex conjugate. The GI equation can be regarded as an extension of the NLS when certain higher-order nonlinear effects are taken into account. In recent years, there has been much work on the GI equation, such as its Darboux transformation and Hamiltonian structures [16, 17], the algebra-geometric solutions [11], and the rogue wave and breather solution [35]. Particularly, the long-time asymptotic behavior of the solution to the GI equation (1.1) was established in [31, 34] by using the nonlinear steepest descent method.

In particular, via the transformation $u(x,t) = q(x,t)e^{\frac{1}{2}iq_0^4 t}$, equation (1.1) is trivially converted into

$$iq_t + q_{xx} - iq^2 \bar{q}_x + \frac{1}{2}(|q|^4 - q_0^4)q = 0. \tag{1.2}$$

Our purpose in the present work is devoted to the study of the long-time asymptotics of equation (1.2) on the line with symmetric, nonzero boundary conditions at infinity, that is,

$$\lim_{x \to \pm\infty} q(x,0) = q_\pm. \tag{1.3}$$

Hereafter, $q_\pm$ are complex constants and $|q_\pm| = q_0 > 0$. This form of equation (1.2) has the advantage that the solutions of (1.2) which satisfy (1.3) are asymptotically time independent as $x \to \infty$. The problems with nonzero boundary conditions at infinity of the kind (1.3) have already been studied. For example, the inverse scattering transform (IST) and the long-time asymptotics for the focusing NLS equation with nonzero boundary conditions at infinity were developed recently in [5] and [7], respectively. Furthermore, for the multicomponent case, the initial value problem for the focusing Manakov system with nonzero boundary conditions at infinity is solved in [27] by developing an appropriate IST. The three-component defocusing NLS equation with nonzero boundary conditions was analyzed in [6] by the theory of IST. On the other hand, for the asymmetric non-zero boundary conditions (i.e., when the limiting values of the solution at space infinities have different non-zero moduli), the IST for the focusing and defocusing NLS equation were formulated in [14] and [4], respectively.

Our present work was motivated by the long-time asymptotic analysis for the focusing NLS equation developed in [7]. Our goal here is to compute the long-time asymptotics for the GI-type derivative NLS equation (1.2) with the initial value condition satisfied (1.3). The main tool is the asymptotic analysis of an associated matrix Riemann–Hilbert (RH) problem by using the steepest descent method and the so-called g-function mechanism [12]. The well-known nonlinear steepest descent method introduced by Deift and Zhou in [13] provides a detailed rigorous proof to calculate the large-time asymptotic behaviors of the integrable nonlinear evolution equations. This approach has been successfully applied in determining asymptotic formulas for the initial value problems of a number of integrable systems associated with 2×2 matrix sprectral problems including the mKdV equation [13], the KdV equation [21], the Hirota equation [24], the derivative NLS equation [30], the Fokas–Lenells equation [33] and the Kundu–Eckhaus equation [32]. For the 3×3 matrix spectral problem, the large-time asymptotic behavior for the coupled NLS equations was obtained in [19] via nonlinear steepest descent. Moreover, there are also many beautiful results about the study of asymptotics of solutions of the initial value problems with shock-type oscillating initial data [10], nondecaying step-like initial data [9, 26, 34] and the initial-boundary value problems with t-periodic boundary condition [8, 31] for various integrable equations. Recently, Lenells also has derived some interesting asymptotic formulas for the solution of integrable equations on the half-line [2, 28] by using the steepest descent method. We also have done some meaningful work about determining the long-time asymptotics for integrable equations on the half-line, see [22, 23].

The organization of the paper is as follows. In Section 2, we formulate the main Riemann–Hilbert problem to solve the initial value problem for the GI-type derivative NLS equation (1.2) with nonzero boundary conditions. We then use this RH problem to compute the long-time asymptotic behavior of the solution in different regions of the xt-plane in Section 3.

2 The Riemann–Hilbert problem

In this section, we will use the approach proposed in [7] to formulate the main RH problem, which allows us to give a representation of the solution for the equation (1.2).

2.1 Eigenfunctions

The integrability of equation (1.2) follows from its Lax pair representation

$$\Psi_x = X\Psi, \quad \Psi_t = T\Psi, \tag{2.1}$$

where

$$X = ik^2\sigma_3 + ikQ - \frac{i}{2}Q^2\sigma_3, \tag{2.2}$$

$$T = -2ik^4\sigma_3 - 2ik^3Q + ik^2Q^2\sigma_3 + kQ_x\sigma_3 + \frac{i}{4}(Q^4 - q_0^4 I)\sigma_3 + \frac{1}{2}(Q_xQ - QQ_x), \tag{2.3}$$

$\Psi(x,t;k)$ is a vector or a 2×2 matrix-valued function and $k \in \mathbf{C}$ is the spectral parameter, and

$$\sigma_3 = \begin{pmatrix} 1 & 0 \\ 0 & -1 \end{pmatrix}, \quad Q = Q(x,t) = \begin{pmatrix} 0 & q(x,t) \\ -\bar{q}(x,t) & 0 \end{pmatrix}. \tag{2.4}$$

Let $X_\pm = \lim_{x\to\pm\infty} X(x,t;k)$ and $T_\pm = \lim_{x\to\pm\infty} T(x,t;k)$. It is straightforward to see that the eigenvector matrix of $X_\pm$ can be written as

$$E_\pm = \begin{pmatrix} 1 & \dfrac{\lambda - \left(k^2 + \frac{1}{2}q_0^2\right)}{k\bar{q}_\pm} \\ \dfrac{\lambda - \left(k^2 + \frac{1}{2}q_0^2\right)}{kq_\pm} & 1 \end{pmatrix}, \tag{2.5}$$

while the corresponding eigenvalues $\pm i\lambda$ are defined by

$$\lambda(k) = \left(k^4 + \frac{1}{4}q_0^4\right)^{\frac{1}{2}}. \tag{2.6}$$

The branch cut for $\lambda(k)$ is taken along the segment

$$\varsigma \cup \bar{\varsigma} = \left\{ k \in \mathbf{C} \mid k_1^2 - k_2^2 = 0, \ |k_1 k_2| \leqslant \frac{1}{4}q_0^2 \right\}, \tag{2.7}$$

where $\varsigma = \left\{ k \in \mathbf{C} \mid k_1^2 - k_2^2 = 0,\ |k_1 k_2| \leqslant \frac{1}{4} q_0^2,\ \mathrm{Im}\, k^2 > 0 \right\}$, $k_1 = \mathrm{Re}\, k$, $k_2 = \mathrm{Im}\, k$.

On the other hand, one obtains that $T_\pm = -2k^2 X_\pm$, we seek simultaneous solution $\Psi_\pm$ of Lax pair (2.1) such that

$$\Psi_\pm(x,t;k) = E_\pm(k) e^{i\Phi(x,t;k)\sigma_3}(1+o(1)), \quad x \to \pm\infty, \tag{2.8}$$

where

$$\Phi(x,t;k) = \lambda(x - 2k^2 t). \tag{2.9}$$

We also find that for any $t \geqslant 0$, $\Psi_\pm(x,t;k)$ remain bounded as $x \to \pm\infty$ if and only if $k \in \Sigma_0 = \mathbf{R} \cup i\mathbf{R} \cup \varsigma \cup \bar{\varsigma}$. The contour Σ_0 can be reduced to the contour $\Sigma = \mathbf{R} \cup \gamma \cup \bar{\gamma}$ in the k^2-plane as shown in Fig. 1. Introducing a new eigenfunction $\mu(x,t;k)$ by

$$\mu_\pm(x,t;k) = \Psi_\pm(x,t;k) e^{-i\Phi(x,t;k)\sigma_3}, \tag{2.10}$$

we obtain that

$$\mu_\pm(x,t;k) = E_\pm(k) + o(1), \quad x \to \pm\infty,\ k \in \Sigma_0. \tag{2.11}$$

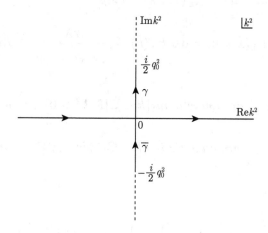

Fig. 1   The oriented contour $\Sigma = \mathbf{R} \cup \gamma \cup \bar{\gamma}$ in k^2-plane

Using the method of variation of parameters, we get the following Volterra integral equations for $\mu_\pm$,

$$\mu_+(x,t;k) = E_+(k) + \int_{+\infty}^{x} E_+(k) e^{i\lambda(x-y)\sigma_3} E_+^{-1}(k)(\Delta X_+ \mu_+)(y,t;k) e^{-i\lambda(x-y)\sigma_3}\, dy, \tag{2.12}$$

$$\mu_-(x,t;k) = E_-(k) + \int_{-\infty}^{x} E_-(k)e^{i\lambda(x-y)\sigma_3} E_-^{-1}(k)(\triangle X_- \mu_-)(y,t;k)e^{-i\lambda(x-y)\sigma_3} dy, \tag{2.13}$$

where $\triangle X_\pm(x,t;k) = X(x,t;k) - X_\pm = ik(Q-Q_\pm) - \dfrac{i}{2}(Q^2 + q_0^2 I)\sigma_3$ and $Q_\pm = \lim_{x\to\pm\infty} Q(x,t)$. Assuming $(q(x,t) - q_\pm) \in L^{1,1}(\mathbf{R}_x^\pm)$, and $(|q(x,t)|^2 - q_0^2) \in L^{1,1}(\mathbf{R}_x^\pm)$, where

$$L^{1,1}(\mathbf{R}^\pm) = \left\{ f : \mathbf{R} \to \mathbf{C} \Big| \int_{\mathbf{R}^\pm} (1+|x|)|f(x)|dx < \infty \right\},$$

the analysis of the Neumann series for the integral equations (2.12) and (2.13) allows one to prove existence and uniqueness of the eigenfunctions $\mu_\pm$ for all $k \in \Sigma_0$ (the detailed proof can be founded [4,5]). We denote by $\mu_\pm^{(1)}(x,t;k)$ and $\mu_\pm^{(2)}(x,t;k)$ the columns of $\mu_\pm(x,t;k)$. Then we have the following lemma.

Lemma 2.1 *For all (x,t), the matrices $\mu_\pm(x,t;k)$ have the following properties:*

(i) *The determinants of $\mu_\pm(x,t;k)$ satisfy*

$$\det \mu_\pm(x,t;k) = \det E_\pm(k) = \frac{2\lambda}{\lambda + k^2 + \frac{1}{2}q_0^2} \triangleq d(k). \tag{2.14}$$

(ii) $\mu_+^{(1)}$ *and* $\mu_-^{(2)}$ *are analytic in* $\{k \in \mathbf{C} | Imk^2 > 0\} \setminus \gamma$, *and* $(\mu_+^{(1)}\ \mu_-^{(2)}) \to I$ *as* $k \to \infty$.

(iii) $\mu_-^{(1)}$ *and* $\mu_+^{(2)}$ *are analytic in* $\{k \in \mathbf{C} | Imk^2 < 0\} \setminus \bar{\gamma}$, *and* $(\mu_-^{(1)}\ \mu_+^{(2)}) \to I$ *as* $k \to \infty$.

(iv) *Symmetry:*

$$\overline{\mu_\pm(x,t;\bar{k})} = \begin{pmatrix} 0 & 1 \\ 1 & 0 \end{pmatrix} \mu_\pm(x,t;k) \begin{pmatrix} 0 & 1 \\ 1 & 0 \end{pmatrix}, \tag{2.15}$$

$$\mu_\pm(x,t;-k) = \sigma_3 \mu_\pm(x,t;k)\sigma_3. \tag{2.16}$$

(v) *Moreover,*

$$\mu_\pm(x,t;k) = I + \frac{\tilde{\mu}(x,t)}{k} + O(k^{-1}), \quad k \to \infty,$$

where

$$[\sigma_3, \tilde{\mu}(x,t)] = \begin{pmatrix} 0 & -q(x,t) \\ \bar{q}(x,t) & 0 \end{pmatrix}.$$

Remark 2.1 The definition (2.6) of λ implies that d is nonzero and non-singular for all $k^2 \in \Sigma^* = \Sigma \setminus \left\{ \pm \frac{i}{2} q_0^2 \right\}$. The symmetry relations (2.15) and (2.16) can be proved easily due to the symmetries of the matrix $X(x,t;k)$:

$$\overline{X(x,t;\bar{k})} = \begin{pmatrix} 0 & 1 \\ 1 & 0 \end{pmatrix} X(x,t;k) \begin{pmatrix} 0 & 1 \\ 1 & 0 \end{pmatrix}, \quad X(x,t;-k) = \sigma_3 X(x,t;k) \sigma_3.$$

The large k asymptotics of $\mu_\pm(x,t;k)$ can be obtained from the x-part of Lax pair (2.1) and (2.10) as well as the asymptotics of $\lambda(k)$, that is, $\lambda(k) = k^2 + O(k^{-2})$ as $k \to \infty$.

Moreover, any two solutions of (2.1) are related, thus we define the spectral matrix $s(k)$ by

$$\mu_-(x,t;k) = \mu_+(x,t;k) e^{i\Phi(x,t;k)\sigma_3} s(k) e^{-i\Phi(x,t;k)\sigma_3}, \quad k^2 \in \Sigma^*. \tag{2.17}$$

In particular, it follows from (2.14) that

$$\det s(k) = 1. \tag{2.18}$$

Due to the symmetry relation (2.15), the matrix-valued spectral functions $s(k)$ can be defined in terms of two scalar spectral functions, $a(k)$ and $b(k)$ by

$$s(k) = \begin{pmatrix} \bar{a}(k) & b(k) \\ \bar{b}(k) & a(k) \end{pmatrix}, \quad a(k)\bar{a}(k) - b(k)\bar{b}(k) = 1, \tag{2.19}$$

where $\bar{a}(k) = \overline{a(\bar{k})}$ and $\bar{b}(k) = \overline{b(\bar{k})}$ indicate the Schwartz conjugates. Meanwhile, by the symmetry relation (2.16), one can infer that

$$a(k) = a(-k), \quad b(k) = -b(-k). \tag{2.20}$$

Finally, equations (2.10), (2.17) and (2.19) imply

$$a(k) = \frac{1}{d(k)} \mathrm{wr}(\Psi_+^{(1)}(x,t;k), \Psi_-^{(2)}(x,t;k)), \tag{2.21}$$

$$b(k) = \frac{1}{d(k)} \text{wr}(\Psi_-^{(2)}(x,t;k), \Psi_+^{(2)}(x,t;k)), \tag{2.22}$$

$$\bar{a}(k) = \frac{1}{d(k)} \text{wr}(\Psi_-^{(1)}(x,t;k), \Psi_+^{(2)}(x,t;k)), \tag{2.23}$$

$$\bar{b}(k) = \frac{1}{d(k)} \text{wr}(\Psi_+^{(1)}(x,t;k), \Psi_-^{(1)}(x,t;k)), \tag{2.24}$$

where wr denotes the Wronskian determinant. Thus, we can find that $a(k)$ is analytic in $\{k \in \mathbf{C} | \text{Im} k^2 > 0\} \setminus \gamma$.

The jump discontinuity of λ across $\gamma \cup \bar{\gamma}$ induces a corresponding jump for the eigenfunctions and scattering data.

Lemma 2.2 *The columns of fundamental solutions $\Psi(x,t;k)$ and scattering data $a(k), b(k)$ satisfy the following jump conditions across $\gamma \cup \bar{\gamma}$:*

$$(\Psi_+^{(1)})^+(x,t;k) = -\frac{\lambda + k^2 + \frac{1}{2}q_0^2}{kq_+} \Psi_+^{(2)}(x,t;k),$$

$$(\Psi_-^{(2)})^+(x,t;k) = -\frac{\lambda + k^2 + \frac{1}{2}q_0^2}{k\bar{q}_-} \Psi_-^{(1)}(x,t;k), \qquad k^2 \in \gamma, \tag{2.25}$$

$$(\Psi_-^{(1)})^+(x,t;k) = -\frac{\lambda + k^2 + \frac{1}{2}q_0^2}{kq_-} \Psi_-^{(2)}(x,t;k),$$

$$(\Psi_+^{(2)})^+(x,t;k) = -\frac{\lambda + k^2 + \frac{1}{2}q_0^2}{k\bar{q}_+} \Psi_+^{(1)}(x,t;k), \qquad k^2 \in \bar{\gamma}, \tag{2.26}$$

and

$$a^+(k) = \frac{q_-}{q_+} \bar{a}(k), \quad k^2 \in \gamma. \tag{2.27}$$

Proof It is noted that the pairs $\{\Psi_-^{(1)}, \Psi_-^{(2)}\}$ and $\{\Psi_+^{(1)}, \Psi_+^{(2)}\}$ are both fundamental sets of solutions of x-part Lax pair (2.1). Thus, the limit $(\Psi_+^{(1)})^+(x,t;k)$ satisfies

$$(\Psi_+^{(1)})^+(x,t;k) = \beta_1 \Psi_+^{(1)}(x,t;k) + \beta_2 \Psi_+^{(2)}(x,t;k), \quad k^2 \in \gamma,$$

where β_1, β_2 are independent of x. Letting $x \to +\infty$ on both sides of the above equation, we get from (2.5) and (2.8) that

$$\beta_1 = 0, \quad \beta_2 = -\frac{\lambda + k^2 + \frac{1}{2}q_0^2}{kq_+},$$

which yields the first relation of (2.25). The others follow the similar arguments. Combining (2.21) with (2.25), (2.23), we can easily obtain (2.27).

2.2 The Riemann–Hilbert problem and the solution of the Cauchy problem

The scattering relation (2.17) can be rewritten in the form of conjugation of boundary values of a piecewise analytic matrix-valued function on a contour in the complex k-plane. The final form is

$$M^+(x,t;k) = M^-(x,t;k)J(x,t;k), \quad k^2 \in \Sigma, \tag{2.28}$$

where $M^\pm(x,t;k)$ denote the boundary values of $M(x,t;k)$ according to a chosen orientation of Σ, see Fig. 1. Indeed, define the matrix $M(x,t;k)$ as follows:

$$M(x,t;k) = \begin{cases} \left(\dfrac{\mu_+^{(1)}}{ad}, \mu_-^{(2)}\right) = \left(\dfrac{\Psi_+^{(1)}}{ad}, \Psi_-^{(2)}\right) e^{-i\Phi\sigma_3}, & \mathrm{Im}\, k^2 > 0 \setminus \gamma, \\ \left(\mu_-^{(1)}, \dfrac{\mu_+^{(2)}}{\bar{a}d}\right) = \left(\Psi_-^{(1)}, \dfrac{\Psi_+^{(2)}}{\bar{a}d}\right) e^{-i\Phi\sigma_3}, & \mathrm{Im}\, k^2 < 0 \setminus \bar\gamma, \end{cases} \tag{2.29}$$

where the dependence on the variables x, t, k on the right-hand side has been suppressed for brevity. Then the boundary values $M^\pm(x,t;k)$ relative to Σ are related by (2.28) with the jump matrix J is given by

$$J(x,t;k)$$

$$= \begin{cases} \begin{pmatrix} \dfrac{1}{d(k)}[1 - r(k)\bar r(k)] & -\bar r(k) e^{2i\Phi(x,t;k)} \\ r(k) e^{-2i\Phi(x,t;k)} & d(k) \end{pmatrix}, & k^2 \in \mathbf{R}, \\[2em] \begin{pmatrix} -\dfrac{\lambda - (k^2 + \tfrac12 q_0^2)}{kq_-}\bar r(k) e^{2i\Phi(x,t;k)} & -\dfrac{2\lambda}{k\bar q_-} \\ \dfrac{k\bar q_-}{2\lambda}[1 - r(k)\bar r(k)] & \dfrac{\lambda + k^2 + \tfrac12 q_0^2}{k\bar q_-} r(k) e^{-2i\Phi(x,t;k)} \end{pmatrix}, & k^2 \in \gamma, \\[2em] \begin{pmatrix} \dfrac{\lambda + k^2 + \tfrac12 q_0^2}{kq_-}\bar r(k) e^{2i\Phi(x,t;k)} & \dfrac{kq_-}{2\lambda}[1 - r(k)\bar r(k)] \\ -\dfrac{2\lambda}{kq_-} & -\dfrac{\lambda - (k^2 + \tfrac12 q_0^2)}{k\bar q_-} r(k) e^{-2i\Phi(x,t;k)} \end{pmatrix}, & k^2 \in \bar\gamma, \end{cases}$$

$$\tag{2.30}$$

where
$$r(k) = -\frac{\bar{b}(k)}{a(k)}. \qquad (2.31)$$

We note that the jump of M across $\gamma \cup \bar{\gamma}$ is obtained according to the jump conditions in Lemma 2.2. Finally, it follows from Lemma 2.1 and the relationship (2.29) between M and $\mu_\pm$, it can be shown that M admits the following large k asymptotic expansion

$$M(x,t;k) = I + \frac{M_1(x,t)}{k} + O(k^{-2}), \quad k \to \infty. \qquad (2.32)$$

Thus, the solution $q(x,t)$ of the GI-type derivative NLS equation (1.2) can be expressed in terms of the solution of the basic RH problem as follows:

$$q(x,t) = -2(M_1(x,t))_{12} = -2 \lim_{k \to \infty} (kM(x,t;k))_{12}, \qquad (2.33)$$

where M is the solution of the following RH problem:

Suppose that $a(k) \neq 0$ for $\{\mathrm{Im}\,k^2 > 0\} \cup \Sigma$. Determine a 2×2 matrix-valued function $M(x,t;k)$ such that

- $M(x,t;k)$ is a sectionally meromorphic function in $\mathbf{C} \setminus \{k^2 \in \Sigma\}$;
- $M(x,t;k)$ satisfies the jump condition in (2.28) with the jump matrix given by (2.30);
- $M(x,t;k)$ has the following asymptotics:

$$M(x,t;k) = I + O(k^{-1}), \quad k \to \infty. \qquad (2.34)$$

Although it is possible to perform this analysis directly in the complex k-plane, the symmetry of the spectral functions $a(k)$, $b(k)$ in (2.20) suggest that it is convenient to introduce a new spectral variable z by $z = k^2$. This change of spectral parameter appeared already in [25] and was further employed in [2,34]. One advantage of working with z is that we can more easily analyze the signature table of the real part for the new phase function in order to find the long-time asymptotics.

The symmetry relation (2.16) implies that the solution $M(x,t;k)$ obeys the symmetry

$$M(x,t;k) = \sigma_3 M(x,t;-k)\sigma_3, \quad k \in \mathbf{C} \setminus \Sigma_0. \qquad (2.35)$$

Hence, letting $z = k^2$, we can define $m(x,t;z)$ by the equation

$$m(x,t;z) = \begin{pmatrix} 1 & 0 \\ \frac{1}{2}\bar{q} & 1 \end{pmatrix} k^{-\frac{\hat{\sigma}_3}{2}} M(x,t;k), \quad z \in \mathbf{C} \setminus \Sigma. \qquad (2.36)$$

The factor $k^{-\frac{\hat{\sigma}_3}{2}}$ is included in (2.36) in order to make the right-hand side an even function of k and $k^{-\frac{\hat{\sigma}_3}{2}}$ acts on a 2×2 matrix A by $k^{-\frac{\sigma_3}{2}} A k^{\frac{\sigma_3}{2}}$; the matrix involving $\bar{q}$ is included to ensure that $m \to I$ as $z \to \infty$. We define the new spectral function $\rho(z)$ by

$$\rho(z) = \frac{r(k)}{k}, \quad z \in \mathbf{R}. \qquad (2.37)$$

Then the Riemann–Hilbert problem for $M(x,t;k)$ can be rewritten in terms of $m(x,t;z)$ as

$$\begin{aligned} m^+(x,t;z) &= m^-(x,t;z) J_1^{(0)}(x,t;z), & z \in \mathbf{R}, \\ m^+(x,t;z) &= m^-(x,t;z) J_2^{(0)}(x,t;z), & z \in \gamma, \\ m^+(x,t;z) &= m^-(x,t;z) J_3^{(0)}(x,t;z), & z \in \bar{\gamma}, \end{aligned} \qquad (2.38)$$

with the normalization condition

$$m(x,t;z) = I + O(z^{-1}), \quad z \to \infty, \qquad (2.39)$$

where

$$J_1^{(0)}(x,t;z) = \begin{pmatrix} \frac{1}{d(z)}[1 - z\rho(z)\bar{\rho}(z)] & -\bar{\rho}(z)e^{2it\theta(\xi;z)} \\ z\rho(z)e^{-2it\theta(\xi;z)} & d(z) \end{pmatrix},$$

$$J_2^{(0)}(x,t;z) = \begin{pmatrix} -\dfrac{\lambda - (z + \frac{1}{2}q_0^2)}{q_-} \bar{\rho}(z)e^{2it\theta(\xi;z)} & -\dfrac{2\lambda}{z\bar{q}_-} \\ \dfrac{z\bar{q}_-}{2\lambda}[1 - z\rho(z)\bar{\rho}(z)] & \dfrac{\lambda + z + \frac{1}{2}q_0^2}{\bar{q}_-}\rho(z)e^{-2it\theta(\xi;z)} \end{pmatrix}, \quad (2.40)$$

$$J_3^{(0)}(x,t;z) = \begin{pmatrix} \dfrac{\lambda + z + \frac{1}{2}q_0^2}{q_-}\bar{\rho}(z)e^{2it\theta(\xi;z)} & \dfrac{q_-}{2\lambda}[1 - z\rho(z)\bar{\rho}(z)] \\ -\dfrac{2\lambda}{q_-} & -\dfrac{\lambda - (z + \frac{1}{2}q_0^2)}{\bar{q}_-}\rho(z)e^{-2it\theta(\xi;z)} \end{pmatrix},$$

and

$$\lambda(z) = \left(z^2 + \frac{1}{4}q_0^4\right)^{\frac{1}{2}}, \quad d(z) = \frac{2\lambda}{\lambda + z + \frac{1}{2}q_0^2}, \quad (2.41)$$

$$\theta(\xi; z) = \frac{\Phi(x, t; k)}{t} = \lambda(\xi - 2z), \quad \xi = \frac{x}{t}. \quad (2.42)$$

In terms of $m(x, t; z)$, (2.33) can be expressed as

$$q(x, t) = -2 \lim_{z \to \infty} (zm(x, t; z))_{12}. \quad (2.43)$$

3 Long-time asymptotics

In this section, we compute the long-time asymptotic behavior of the solution $q(x,t)$ of the GI-type derivative NLS equation (1.2), as obtained by equation (2.43) by using the Deift-Zhou nonlinear steepest descent method to perform the asymptotic analysis of oscillating RH problems [7-10, 13]. The key point of this method is the choice of contour deformations, which depends crucially on the sign structure of the quantity $\text{Re}(i\theta)$. We note that

$$\theta(\xi; z) = \sqrt{z^2 + \frac{1}{4}q_0^4}(\xi - 2z). \quad (3.1)$$

Thus, we get

$$\frac{d\theta(\xi; z)}{dz} = \frac{-4z^2 + \xi z - \frac{1}{2}q_0^4}{\lambda}, \quad (3.2)$$

which implies that θ has two stationary points

$$z_{\pm} = \frac{1}{8}(\xi \pm \sqrt{\xi^2 - 8q_0^2}). \quad (3.3)$$

Letting $z = z_1 + iz_2$, we have

$$\theta^2(\xi; z) = \left(z_1^2 - z_2^2 + \frac{1}{4}q_0^4 + 2iz_1z_2\right)(4z_1^2 - 4z_2^2 + \xi^2 - 4\xi z_1 + i(8z_1z_2 - 4\xi z_2)).$$

Hence, according to

$$\{\text{Im}\theta(\xi; z) = 0\} = \{\text{Im}\theta^2(\xi; z) = 0\} \cap \{\text{Re}\theta^2(\xi; z) \geqslant 0\},$$

we can find the curves of the sets $\{\text{Im}\theta(\xi; z) = 0\}$. In fact, the sign structure of $\text{Re}(i\theta)$ in the complex z-plane is shown in Fig. 2.

Remark 3.1 It is noted that $|\xi| > 2\sqrt{2}q_0^2$, that is, $x < -2\sqrt{2}q_0^2 t$ or $x > 2\sqrt{2}q_0^2 t$, the function θ has two real stationary points. We will show that this sector of the xt-plane correspond to plane wave regions, whereas the sectors where $|\xi| < 2\sqrt{2}q_0^2$ correspond to modulated elliptic wave regions.

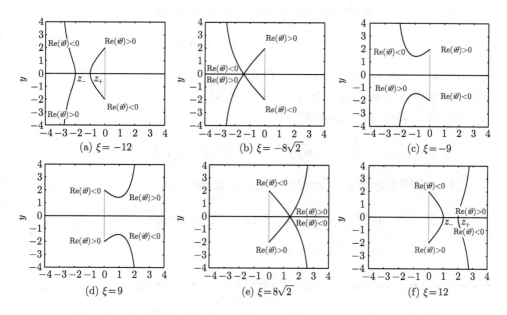

Fig. 2 The signature table for $\text{Re}(i\theta)$ in the complex z-plane for various values of ξ and $q_0 = 2$

3.1 The plane wave region I: $x < -2\sqrt{2}q_0^2 t$

For $x < -2\sqrt{2}q_0^2 t$, that is, $\xi < -2\sqrt{2}q_0^2$, $\theta(\xi; z)$ has two real stationary points $z_\pm$ given by (3.3). This implies that one can introduce a complex-valued function $\delta(z)$ to transform the original RH problem for $m(x,t;z)$ to the RH problem for the new function $m^{(1)}(x,t;z)$ by

$$m^{(1)}(x,t;z) = m(x,t;z)\delta^{-\sigma_3}(z), \qquad (3.4)$$

where

$$\delta(z) = \exp\left\{\frac{1}{2\pi i}\int_{-\infty}^{z_-}\frac{\ln[1-s\rho(s)\bar{\rho}(s)]}{s-z}ds\right\}, \quad z \in \mathbf{C}\setminus(-\infty, z_-]. \qquad (3.5)$$

Lemma 3.1 *The function $\delta(z)$ has the following properties:*

(i) $\delta(z)$ satisfies the following jump condition across the real axis oriented in Fig. 1:
$$\delta^+(z) = \delta^-(z)[1 - z\rho(z)\bar{\rho}(z)], \quad z \in (-\infty, z_-).$$

(ii) As $z \to \infty$, $\delta(z)$ satisfies the asymptotic formula
$$\delta(z) = 1 + O(z^{-1}), \quad z \to \infty. \tag{3.6}$$

(iii) $\delta(z)$ and $\delta^{-1}(z)$ are bounded and analytic functions of $z \in \mathbf{C} \setminus (-\infty, z_-]$ with continuous boundary values on $(-\infty, z_-)$.

(iv) $\delta(z)$ obeys the symmetry
$$\delta(z) = \overline{\delta(\bar{z})}^{-1}, \quad z \in \mathbf{C} \setminus (-\infty, z_-].$$

Then $m^{(1)}(x, t; z)$ satisfies the jump condition
$$m^{(1)+}(x, t; z) = m^{(1)-}(x, t; z) J^{(1)}(x, t; z),$$
$$z \in \Sigma^{(1)} = \Sigma = \mathbf{R} \cup \gamma \cup \bar{\gamma}, \tag{3.7}$$

where the jump matrix $J^{(1)} = (\delta^-)^{\sigma_3} J^{(0)} (\delta^+)^{-\sigma_3}$ is given by

$$J_1^{(1)} = \begin{pmatrix} d^{-\frac{1}{2}} & 0 \\ \dfrac{z\rho e^{-2it\theta}(\delta^-)^{-2}}{1 - z\rho\bar{\rho}} d^{\frac{1}{2}} & d^{\frac{1}{2}} \end{pmatrix} \begin{pmatrix} d^{-\frac{1}{2}} & -\dfrac{\bar{\rho} e^{2it\theta}(\delta^+)^2}{1 - z\rho\bar{\rho}} d^{\frac{1}{2}} \\ 0 & d^{\frac{1}{2}} \end{pmatrix}, \quad z \in (-\infty, z_-),$$

$$J_2^{(1)} = \begin{pmatrix} d^{-\frac{1}{2}} & -\bar{\rho} e^{2it\theta} \delta^2 d^{-\frac{1}{2}} \\ 0 & d^{\frac{1}{2}} \end{pmatrix} \begin{pmatrix} d^{-\frac{1}{2}} & 0 \\ z\rho e^{-2it\theta} \delta^{-2} d^{-\frac{1}{2}} & d^{\frac{1}{2}} \end{pmatrix}, \quad z \in (z_-, \infty),$$

$$J_3^{(1)} = \begin{pmatrix} -\dfrac{\lambda - (z + \frac{1}{2}q_0^2)}{q_-} \bar{\rho} e^{2it\theta} & -\dfrac{2\lambda \delta^2}{z\bar{q}_-} \\ \dfrac{z\bar{q}_- \delta^{-2}}{2\lambda}[1 - z\rho\bar{\rho}] & \dfrac{\lambda + z + \frac{1}{2}q_0^2}{\bar{q}_-} \rho e^{-2it\theta} \end{pmatrix}, \quad z \in \gamma,$$

$$J_4^{(1)} = \begin{pmatrix} \dfrac{\lambda + z + \frac{1}{2}q_0^2}{q_-} \bar{\rho} e^{2it\theta} & \dfrac{q_- \delta^2}{2\lambda}[1 - z\rho\bar{\rho}] \\ -\dfrac{2\lambda \delta^{-2}}{q_-} & -\dfrac{\lambda - (z + \frac{1}{2}q_0^2)}{\bar{q}_-} \rho e^{-2it\theta} \end{pmatrix}, \quad z \in \bar{\gamma}.$$

$$\tag{3.8}$$

The new jump matrix $J^{(1)}(x,t;z)$ can be analytically extended from $\Sigma^{(1)}$. This leads to the next transformation:

$$m^{(2)}(x,t;z) = m^{(1)}(x,t;z)H(z), \qquad (3.9)$$

where $H(z)$ is defined by

$$H(z) = \begin{cases} \begin{pmatrix} d^{\frac{1}{2}} & 0 \\ -z\rho e^{-2it\theta}\delta^{-2}d^{-\frac{1}{2}} & d^{-\frac{1}{2}} \end{pmatrix}, & z \in D_1, \\[1em] \begin{pmatrix} d^{\frac{1}{2}} & \dfrac{\bar{\rho}e^{2it\theta}\delta^2}{1-z\rho\bar{\rho}}d^{\frac{1}{2}} \\ 0 & d^{-\frac{1}{2}} \end{pmatrix}, & z \in D_2, \\[1em] \begin{pmatrix} d^{-\frac{1}{2}} & 0 \\ \dfrac{z\rho e^{-2it\theta}\delta^{-2}}{1-z\rho\bar{\rho}}d^{\frac{1}{2}} & d^{\frac{1}{2}} \end{pmatrix}, & z \in D_3, \\[1em] \begin{pmatrix} d^{-\frac{1}{2}} & -\bar{\rho}e^{2it\theta}\delta^2 d^{-\frac{1}{2}} \\ 0 & d^{\frac{1}{2}} \end{pmatrix}, & z \in D_4, \\[1em] I, & z \in D_5 \cup D_6. \end{cases}$$

The domains $\{D_j\}_1^6$ are shown on Fig. 3. Then the new function $m^{(2)}(x,t;z)$ solves the following equivalent RH problem:

$$m^{(2)+}(x,t;z) = m^{(2)-}(x,t;z)J^{(2)}(x,t;z),$$
$$z \in \Sigma^{(2)} = L_1 \cup L_2 \cup L_3 \cup L_4 \cup \gamma \cup \bar{\gamma}, \qquad (3.10)$$

where the curves $L_1, \cdots, L_4$ can be chosen freely, only respecting that they pass through z_-, do not cross $\gamma \cup \bar{\gamma}$, go to ∞, and lie entirely in domains with the appropriate sign of $\text{Re}(i\theta)$, and the jump matrix $J^{(2)} = (H^-)^{-1}(z)J^{(1)}H^+(z)$ is given by

$$J_1^{(2)} = \begin{pmatrix} d^{-\frac{1}{2}} & 0 \\ z\rho e^{-2it\theta}\delta^{-2}d^{-\frac{1}{2}} & d^{\frac{1}{2}} \end{pmatrix}, \quad z \in L_1,$$

$$J_2^{(2)} = \begin{pmatrix} d^{-\frac{1}{2}} & -\dfrac{\bar{\rho}e^{2it\theta}\delta^2}{1-z\rho\bar{\rho}}d^{\frac{1}{2}} \\ 0 & d^{\frac{1}{2}} \end{pmatrix}, \quad z \in L_2,$$

$$J_3^{(2)} = \begin{pmatrix} d^{-\frac{1}{2}} & 0 \\ \dfrac{z\rho e^{-2it\theta}\delta^{-2}}{1-z\rho\bar{\rho}} d^{\frac{1}{2}} & d^{\frac{1}{2}} \end{pmatrix}, \quad z \in L_3,$$

$$J_4^{(2)} = \begin{pmatrix} d^{-\frac{1}{2}} & -\bar{\rho} e^{2it\theta}\delta^2 d^{-\frac{1}{2}} \\ 0 & d^{\frac{1}{2}} \end{pmatrix}, \quad z \in L_4,$$

$$J_5^{(2)} = \begin{pmatrix} 0 & \dfrac{q_-\delta^2}{i\sqrt{z}q_0} \\ -\dfrac{i\sqrt{z}q_0}{q_-\delta^2} & 0 \end{pmatrix}, \quad z \in \gamma \cup \bar{\gamma}. \qquad (3.11)$$

We have used the jump condition

$$\rho^+(z) = \frac{\bar{q}_-}{q_-}\bar{\rho}(z) \qquad (3.12)$$

for $\rho(z)$ across $\gamma \cup \bar{\gamma}$ to derive the jump matrix $J_5^{(2)}$.

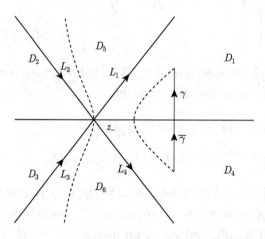

Fig. 3 The oriented contour $\Sigma^{(2)}$ and the open sets $\{D_j\}_1^6$ in the complex z-plane

The next deformation aims to remove the function d from the jump matrices (3.11) so that the jumps along the contours L_j eventually tend to the identity matrix as $t \to \infty$. This can be accomplished by letting

$$m^{(3)}(x,t;z) = m^{(2)}(x,t;z)\hat{H}(z), \qquad (3.13)$$

where $\hat{H}(z)$ is defined by

$$\hat{H}(z) = \begin{cases} I, & z \in D_j, \ j = 1, 2, 3, 4, \\ d^{\frac{\sigma_3}{2}}, & z \in D_5, \\ d^{-\frac{\sigma_3}{2}}, & z \in D_6. \end{cases}$$

Then it follows that $m^{(3)}(x,t;z)$ satisfies the following jump conditions

$$m^{(3)+}(x,t;z) = m^{(3)-}(x,t;z) J^{(3)}(x,t;z), \quad z \in \Sigma^{(3)} = \Sigma^{(2)}, \tag{3.14}$$

and the jump matrix $J^{(3)}$ is given by

$$J_1^{(3)} = \begin{pmatrix} 1 & 0 \\ z\rho e^{-2it\theta} \delta^{-2} & 1 \end{pmatrix}, \quad z \in L_1,$$

$$J_2^{(3)} = \begin{pmatrix} 1 & -\dfrac{\bar{\rho} e^{2it\theta} \delta^2}{1 - z\rho\bar{\rho}} \\ 0 & 1 \end{pmatrix}, \quad z \in L_2,$$

$$J_3^{(3)} = \begin{pmatrix} 1 & 0 \\ \dfrac{z\rho e^{-2it\theta} \delta^{-2}}{1 - z\rho\bar{\rho}} & 1 \end{pmatrix}, \quad z \in L_3, \tag{3.15}$$

$$J_4^{(3)} = \begin{pmatrix} 1 & -\bar{\rho} e^{2it\theta} \delta^2 \\ 0 & 1 \end{pmatrix}, \quad z \in L_4,$$

$$J_5^{(3)} = J_5^{(2)} = \begin{pmatrix} 0 & \dfrac{q_- \delta^2}{i\sqrt{z} q_0} \\ -\dfrac{i\sqrt{z} q_0}{q_- \delta^2} & 0 \end{pmatrix}, \quad z \in \gamma \cup \bar{\gamma}.$$

Obviously, for $z \in L_j, j = 1, \cdots, 4$, the jump matrix $J^{(3)}(x,t;z)$ decays to identity matrix I as $t \to \infty$ exponentially fast and uniformly outside any neighborhood of $z = z_-$. In order to arrive at a RH problem whose jump matrix does not depend on z, we introduce a factorization involving a scalar function $F(z)$ to be defined:

$$J_5^{(3)}(x,t;z) = \begin{pmatrix} (F^-)^{-1}(z) & 0 \\ 0 & F^-(z) \end{pmatrix} \begin{pmatrix} 0 & \dfrac{2}{\bar{q}_-} \\ -\dfrac{\bar{q}_-}{2} & 0 \end{pmatrix} \begin{pmatrix} F^+(z) & 0 \\ 0 & (F^+)^{-1}(z) \end{pmatrix}$$

$$\tag{3.16}$$

in such a way that the boundary values $F^{\pm}(z)$ of $F(z)$ along the two sides of $\gamma \cup \bar{\gamma}$ satisfy

$$F^+(z)F^-(z) = \frac{2i\sqrt{z}}{q_0}\delta^{-2}(z), \quad z \in \gamma \cup \bar{\gamma}. \tag{3.17}$$

Indeed, once (3.16) is satisfied, one can absorb the diagonal factors into a new piecewise analytic function whose jump across $\gamma \cup \bar{\gamma}$ is only the constant middle factor in (3.16).

The function $F(z)$ can be determined explicitly as follows. Dividing condition (3.17) by $\lambda^+(z)$, we deduce that

$$\left[\frac{\ln F(z)}{\lambda(z)}\right]^+ - \left[\frac{\ln F(z)}{\lambda(z)}\right]^- = \frac{\ln\left[\frac{2i\sqrt{z}}{q_0}\delta^{-2}(z)\right]}{\lambda^+(z)}, \quad z \in \gamma \cup \bar{\gamma}, \tag{3.18}$$

and

$$\frac{\ln F(z)}{\lambda(z)} = O(z^{-1}), \quad z \to \infty.$$

Plemelj's formula [1] then yields $F(z)$ in the explicit form

$$F(z) = \exp\left\{\frac{\lambda(z)}{2\pi i}\int_{\gamma \cup \bar{\gamma}} \frac{\ln\left[\frac{2i\sqrt{s}}{q_0}\delta^{-2}(s)\right]}{s-z}\frac{ds}{\lambda^+(s)}\right\}, \quad z \notin \gamma \cup \bar{\gamma}. \tag{3.19}$$

As $z \to \infty$, we find that

$$F(\infty) = e^{i\phi(\xi)}, \tag{3.20}$$

where

$$\phi(\xi) = \frac{1}{2\pi}\int_{\gamma \cup \bar{\gamma}}\left[\frac{\ln\left[\frac{2i\sqrt{s}}{q_0}\right]}{\lambda^+(s)} + \frac{i}{\pi\lambda^+(s)}\int_{-\infty}^{z-}\frac{\ln[1-u\rho(u)\bar{\rho}(u)]}{u-s}du\right]ds. \tag{3.21}$$

The factorization (3.16) suggests the final transformation

$$m^{(4)}(x,t;z) = F^{\sigma_3}(\infty)m^{(3)}(x,t;z)F^{-\sigma_3}(z). \tag{3.22}$$

Then we have

$$m^{(4)+}(x,t;z) = m^{(4)-}(x,t;z)J^{(4)}(x,t;z), \tag{3.23}$$

for $z \in \Sigma^{(4)} = \Sigma^{(3)} = L_1 \cup L_2 \cup L_3 \cup L_4 \cup \gamma \cup \bar{\gamma}$ (Fig.3). Meanwhile, the jump matrix $J^{(4)} = (F^-)^{\sigma_3} J^{(3)} (F^+)^{-\sigma_3}$ satisfies:

- for $z \in \gamma \cup \bar{\gamma}$, jump matrix $J^{(4)}(x, t; z)$ is a constant

$$J^{(4)}(x, t; z) = J^{\mathrm{mod}} = \begin{pmatrix} 0 & \frac{2}{\bar{q}_-} \\ -\frac{\bar{q}_-}{2} & 0 \end{pmatrix};$$

- for $z \in L_j$, $j = 1, \cdots, 4$, jump matrix $J^{(4)}(x, t; z)$ decays to the identity

$$J^{(4)}(x, t; z) = I + O(e^{-\epsilon t})$$

uniformly outside any neighborhood of $z = z_-$.

Finally, one can write $m^{(4)}$ in the form

$$m^{(4)}(x, t; z) = m^{\mathrm{err}}(x, t; z) m^{\mathrm{mod}}(x, t; z), \qquad (3.24)$$

where $m^{\mathrm{mod}}(x, t; z)$ solves the model problem:

$$m^{\mathrm{mod}+}(x, t; z) = m^{\mathrm{mod}-}(x, t; z) J^{\mathrm{mod}}, \quad z \in \gamma \cup \bar{\gamma} \qquad (3.25)$$

with constant jump matrix

$$J^{\mathrm{mod}} = \begin{pmatrix} 0 & \frac{2}{\bar{q}_-} \\ -\frac{\bar{q}_-}{2} & 0 \end{pmatrix},$$

and

$$m^{\mathrm{err}}(x, t; z) = I + O(t^{-\frac{1}{2}}). \qquad (3.26)$$

The last estimate (3.26) can be justified by considering the parametrix associated with the RH problem for $m^{(4)}(x, t; z)$, see [7, 10]. The error of order $O(t^{-1/2})$ comes from the contribution of the jump near z_-.

Define

$$\nu(z) = \left(\frac{z - \frac{i}{2} q_0^2}{z + \frac{i}{2} q_0^2} \right)^{\frac{1}{4}}. \qquad (3.27)$$

Then, we have

$$\nu^+(z) = i \nu^-(z)$$

on $\gamma \cup \bar{\gamma}$, and $\nu(z)$ admits the large-z expansion

$$\nu(z) = 1 - \frac{iq_0^2}{4z} + O(z^{-2}), \quad z \to \infty.$$

As for the model RH problem, its solution thus can be given explicitly in terms of $\nu(z)$:

$$m^{\mathrm{mod}}(x,t;z) = \frac{1}{2}\begin{pmatrix} \nu(z) + \frac{1}{\nu(z)} & \frac{2}{i\bar{q}_-}\left(\nu(z) - \frac{1}{\nu(z)}\right) \\ \frac{i\bar{q}_-}{2}\left(\nu(z) - \frac{1}{\nu(z)}\right) & \nu(z) + \frac{1}{\nu(z)} \end{pmatrix}. \quad (3.28)$$

Then, going back to the determination of $q(x,t)$ in terms of the solution of the basic RH problem, we have

$$q(x,t) = -2\lim_{z\to\infty}(zm(x,t;z))_{12}$$
$$= -2\lim_{z\to\infty}(zm^{(4)}(x,t;z))_{12}F^{-2}(\infty) + O(t^{-\frac{1}{2}})$$
$$= -2\lim_{z\to\infty}(zm^{\mathrm{mod}}(x,t;z))_{12}F^{-2}(\infty) + O(t^{-\frac{1}{2}}). \quad (3.29)$$

Taking into account that $-2\lim_{z\to\infty}(zm^{\mathrm{mod}}(x,t;z))_{12} = q_-$, and $F^{-2}(\infty) = e^{-2i\phi(\xi)}$, we arrive at the following theorem.

Theorem 3.1 *In the region $x < -2\sqrt{2}q_0^2 t$, as $t \to \infty$, the asymptotics of the solution $q(x,t)$ of the initial value problem for the GI-type derivative NLS equation (1.2) takes the form of a plane wave:*

$$q(x,t) = q_- e^{-2i\phi(\xi)} + O(t^{-\frac{1}{2}}), \quad t \to \infty, \quad (3.30)$$

where $\phi(\xi)$ is defined by (3.21).

Remark 3.2 Let $\xi \to -\infty$, then $z_- \to -\infty$. Thus, $\phi(\xi) \to \phi = \frac{1}{2\pi}\int_{\gamma\cup\bar{\gamma}}\frac{\ln\left[\frac{2i\sqrt{s}}{q_0}\right]}{\lambda^+(s)}ds$, and then the asymptotic formulae (3.30) reduces to $q(x,t) = q_- e^{-2i\phi}$. This is correspondence to our initial condition up to a phase shift as $x \to -\infty$.

3.2 The modulated elliptic wave region I: $-2\sqrt{2}q_0^2 t < x < 0$

In this subsection we compute the leading-order long-time asymptotics of the solution of the GI-type derivative NLS equation (1.2) in the region $-2\sqrt{2}q_0^2 t <$

$x < 0$. Recalling the sign structure of $\text{Re}(i\theta)$ in this region depicted in Fig. 2, the main difference is the absence of real stationary points compared with the plane wave region I. This implies that it is not possible anymore to use the previous factorizations and deformations to lift the contours of the real z-axis in such a way that the corresponding jump matrices remain bounded as $t \to \infty$. Therefore, developing and extending the ideas used in [7,10], we introduce a new g-function $g(z)$ appropriate for the region under consideration, which has a new real stationary point $z_0 \in \mathbf{R}^-$.

Let us perform the same transformations

$$m(x,t;z) \rightsquigarrow m^{(1)}(x,t;z) \rightsquigarrow m^{(2)}(x,t;z) \rightsquigarrow m^{(3)}(x,t;z)$$

as in Subsection 3.1 for the plane wave region I but with $\delta(z)$ characterized by z_0 instead of the point z_-, that is,

$$\delta(z) = \exp\left\{\frac{1}{2\pi i}\int_{-\infty}^{z_0} \frac{\ln[1-s\rho(s)\bar{\rho}(s)]}{s-z}ds\right\}, \quad z \in \mathbf{C}\setminus(-\infty, z_0]. \quad (3.31)$$

In this subsection, we will use the notation L_5 to denote $\gamma \cup \bar{\gamma}$ for simplicity. It then follows that the function $m^{(3)}(x,t;z)$ is analytic in $\mathbf{C}\setminus(\cup_{j=1}^{5} L_j)$ and satisfies the jump conditions

$$m^{(3)+}(x,t;z) = m^{(3)-}(x,t;z)J_j^{(3)}(x,t;z), \quad z \in L_j, \ j=1,\cdots,5, \quad (3.32)$$

with the normalization condition

$$m^{(3)}(x,t;z) = I + O(z^{-1}), \quad z \to \infty. \quad (3.33)$$

The jump contours L_j is shown in Fig. 4, and the jump matrices $J_j^{(3)}$ are given by equations (3.15) and $\delta(z)$ defined by (3.31).

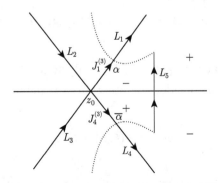

Fig. 4   The oriented contours $\{L_j\}_1^5$ in the complex z-plane

We now encounter a new phenomenon that is not present in the plane wave region I, namely, the jumps $J_1^{(3)}$ and $J_4^{(3)}$ grow exponentially with t along the segments $[z_0, \alpha]$ and $[z_0, \bar\alpha]$ as shown in Fig. 4, respectively. In order to overcome this difficulty, we employ the new factorizations for these jumps:

$$J_1^{(3)} = J_8^{(3)} J_6^{(3)} J_8^{(3)}, \quad J_4^{(3)} = J_9^{(3)} J_7^{(3)} J_9^{(3)}, \qquad (3.34)$$

where

$$J_6^{(3)} = \begin{pmatrix} 0 & -\frac{\delta^2}{z\rho} e^{2it\theta} \\ \frac{z\rho}{\delta^2} e^{-2it\theta} & 0 \end{pmatrix}, \quad J_7^{(3)} = \begin{pmatrix} 0 & -\bar\rho \delta^2 e^{2it\theta} \\ \frac{e^{-2it\theta}}{\bar\rho \delta^2} & 0 \end{pmatrix},$$

$$J_8^{(3)} = \begin{pmatrix} 1 & \frac{\delta^2}{z\rho} e^{2it\theta} \\ 0 & 1 \end{pmatrix}, \quad J_9^{(3)} = \begin{pmatrix} 1 & 0 \\ -\frac{e^{-2it\theta}}{\bar\rho \delta^2} & 1 \end{pmatrix}.$$

These factorizations allow for the deformation of the segments $[z_0, \alpha]$ and $[z_0, \bar\alpha]$ of Fig. 4 into the contours shown in Fig. 5, where $J_j^{(3)}$ ($j = 6, \cdots, 9$) denotes the restriction of $J^{(3)}$ to the contours labeled by L_j in Fig. 5. We next employ the g-function mechanism, more precisely, let

$$m^{(4)}(x, t; z) = e^{iG(\infty)t\sigma_3} m^{(3)}(x, t; z) e^{-iG(z)t\sigma_3} \qquad (3.35)$$

for a function $G(z)$ that is required to be analytic in $\mathbf{C} \setminus (L_5 \cup L_6 \cup L_7)$. In fact, it is more convenient to consider the function $g(z)$ define by

$$g(z) = \theta(\xi; z) + G(z). \qquad (3.36)$$

Then, one can infer that $g(z)$ is analytic in $\mathbf{C} \setminus (L_5 \cup L_6 \cup L_7)$ and has jump discontinuities across L_5 and $L_6 \cup L_7$. Furthermore, the jump conditions of function $m^{(4)}(x, t; z)$ read

$$J_1^{(4)} = \begin{pmatrix} 1 & 0 \\ z\rho\delta^{-2}e^{-2igt} & 1 \end{pmatrix}, \quad J_2^{(4)} = \begin{pmatrix} 1 & -\dfrac{\bar{\rho}\delta^2}{1-z\rho\bar{\rho}}e^{2igt} \\ 0 & 1 \end{pmatrix},$$

$$J_3^{(4)} = \begin{pmatrix} 1 & 0 \\ \dfrac{z\rho\delta^{-2}}{1-z\rho\bar{\rho}}e^{-2igt} & 0 \end{pmatrix}, \quad J_4^{(4)} = \begin{pmatrix} 1 & -\bar{\rho}\delta^2 e^{2igt} \\ 0 & 1 \end{pmatrix},$$

$$J_5^{(4)} = \begin{pmatrix} 0 & \dfrac{q_-\delta^2}{i\sqrt{z}q_0}e^{i(g^++g^-)t} \\ -\dfrac{i\sqrt{z}q_0}{q_-\delta^2}e^{-i(g^++g^-)t} & 0 \end{pmatrix},$$

$$J_6^{(4)} = \begin{pmatrix} 0 & -\dfrac{\delta^2}{z\rho}e^{i(g^++g^-)t} \\ \dfrac{z\rho}{\delta^2}e^{-i(g^++g^-)t} & 0 \end{pmatrix},$$

$$J_7^{(4)} = \begin{pmatrix} 0 & -\bar{\rho}\delta^2 e^{i(g^++g^-)t} \\ \dfrac{e^{-i(g^++g^-)t}}{\bar{\rho}\delta^2} & 0 \end{pmatrix},$$

$$J_8^{(4)} = \begin{pmatrix} 1 & \dfrac{\delta^2}{z\rho}e^{2igt} \\ 0 & 1 \end{pmatrix}, \quad J_9^{(4)} = \begin{pmatrix} 1 & 0 \\ -\dfrac{e^{-2igt}}{\bar{\rho}\delta^2} & 1 \end{pmatrix}.$$

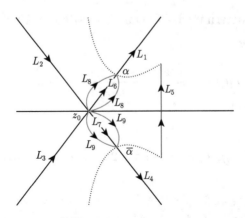

Fig. 5 The new oriented contours $\{L_j\}_1^9$ in the complex z-plane

We next turn to determine z_0, $\alpha = \alpha_1 + i\alpha_2$ and $g(z)$ such that all these jump matrices remain bounded as $t \to \infty$ by using the methods in [7,9]. Let

$$B = L_6 \cup (-L_7).$$

Define function $w(z)$ with branch cuts L_5 and B by

$$w(z) = \sqrt{(z^2 + \frac{1}{4}q_0^4)(z-\alpha)(z-\bar{\alpha})}. \tag{3.37}$$

We fix the branch cut B to be oriented upwards, then $w(z) = w^-(z) = -w^+(z)$ for $z \in L_5 \cup B$. The $w(z)$ gives rise to a genus-1 Riemann surface $\mathcal{X}$ with sheets $\mathcal{X}_1$, $\mathcal{X}_2$ and a basis $\{a, b\}$ of cycles defined as follows: the b-cycle is a closed, anticlockwise contour around the branch cut L_5 that remains entirely on the first sheet $\mathcal{X}_1$ of the Riemann surface; the a-cycle consists of an anticlockwise contour that starts on the left side of B, then goes B from the right while on the first sheet $\mathcal{X}_1$, and finally returns to the starting point via the second sheet $\mathcal{X}_2$.

Next, let $g(z)$ be given by the sum of two Abelian integrals:

$$g(z) = \frac{1}{2}\left(\int_{-\frac{i}{2}q_0^2}^{z} + \int_{\frac{i}{2}q_0^2}^{z}\right) dg(s),$$

where the Abelian differential $dg(z)$ is defined by

$$dg(z) = -4\frac{(z-z_0)(z-\alpha)(z-\bar{\alpha})}{w(z)}dz, \tag{3.38}$$

and z_0, α are to be determined. We can also write the Abelian differential $dg(z)$ in the following form:

$$dg(z) = -4\frac{z^3 + c_2 z^2 + c_1 z + c_0}{w(z)}dz,$$

then, we have

$$c_2 = -(2\alpha_1 + z_0), \quad c_1 = \alpha_1^2 + \alpha_2^2 + 2\alpha_1 z_0, \quad c_0 = -z_0(\alpha_1^2 + \alpha_2^2). \tag{3.39}$$

We then have

$$g(z) = -2\left(\int_{-\frac{i}{2}q_0^2}^{z} + \int_{\frac{i}{2}q_0^2}^{z}\right)\frac{s^3 + c_2 s^2 + c_1 s + c_0}{w(s)}ds. \tag{3.40}$$

In order to preserve the sign structure in the transition from θ to g, we require that $g(z)$ has the same large-z behavior as the function $\theta(\xi; z)$:

$$g(z) = -2z^2 + \xi z + O(1), \quad z \to \infty.$$

This condition implies:

$$c_1 = \frac{1}{2}\alpha_2^2 + \frac{1}{4}\xi\alpha_1 + \frac{1}{8}q_0^4,$$

$$c_2 = -\frac{\xi}{4} - \alpha_1.$$
(3.41)

Observe that

$$2\left(\int_{-\frac{i}{2}q_0^2}^{z} + \int_{\frac{i}{2}q_0^2}^{z}\right)\left(s - \frac{\xi}{4}\right) = 2z^2 - \xi z + \frac{1}{2}q_0^4.$$

Thus, the large-z asymptotics of $g(z)$ can be specified as

$$g(z) = -2z^2 + \xi z + g_\infty + O(z^{-1}), \quad z \to \infty,$$
(3.42)

where

$$g_\infty = -2\left(\int_{-\frac{i}{2}q_0^2}^{\infty} + \int_{\frac{i}{2}q_0^2}^{\infty}\right)\left[\frac{s^3 + c_2 s^2 + c_1 s + c_0}{w(s)} - \left(s - \frac{\xi}{4}\right)\right]ds - \frac{1}{2}q_0^4. \quad (3.43)$$

Moreover, by combining (3.39) and (3.41), we get

$$\alpha_1 = \frac{\xi}{4} - z_0, \quad \alpha_2^2 = 2z_0^2 - \frac{1}{2}\xi z_0 + \frac{1}{4}q_0^4.$$
(3.44)

It thus remains to determine z_0. We do so by analyzing the behavior of g near α. In a neighborhood of α, it follows from (3.38) that

$$\frac{dg}{dz} = \frac{-4(\alpha - \bar{\alpha})^{\frac{1}{2}}(\alpha - z_0)}{(\alpha^2 + \frac{1}{4}q_0^4)^{\frac{1}{2}}}\left[(z-\alpha)^{\frac{1}{2}} + \left(\frac{1}{\alpha - z_0} + \frac{1}{2(\alpha - \bar{\alpha})} - \frac{\alpha}{\alpha^2 + \frac{1}{4}q_0^4}\right)(z-\alpha)^{\frac{3}{2}}\right.$$

$$\left. + O((z-\alpha)^{\frac{5}{2}})\right], \quad z \to \alpha.$$

Thus, we obtain the expansion

$$g(z) = g(\alpha) - \frac{4(\alpha - \bar{\alpha})^{\frac{1}{2}}(\alpha - z_0)}{(\alpha^2 + \frac{1}{4}q_0^4)^{\frac{1}{2}}}\left[\frac{2}{3}(z-\alpha)^{\frac{3}{2}}\right.$$

$$+ \frac{2}{5}\left(\frac{1}{\alpha-z_0} + \frac{1}{2(\alpha-\bar{\alpha})} - \frac{\alpha}{\alpha^2 + \frac{1}{4}q_0^4}\right)(z-\alpha)^{\frac{5}{2}} + O((z-\alpha)^{\frac{7}{2}})\Big], \quad z \to \alpha,$$

where

$$g(\alpha) = -2\left(\int_{-\frac{i}{2}q_0^2}^{\alpha} + \int_{\frac{i}{2}q_0^2}^{\bar{\alpha}}\right)\frac{s^3 + c_2 s^2 + c_1 s + c_0}{w(s)}ds - 2\int_{\bar{\alpha}}^{\alpha}\frac{s^3 + c_2 s^2 + c_1 s + c_0}{w(s)}ds. \tag{3.45}$$

In order to obtain the desired sign structure as shown in Fig. 6, the leading-order term of the expansion of $\mathrm{Re}(ig)$ near α should be of $O((z-\alpha)^{\frac{3}{2}})$. Thus, we must have $\mathrm{Im} g(\alpha) = 0$. Hence, we get

$$\int_{\bar{\alpha}}^{\alpha} \frac{s^3 + c_2 s^2 + c_1 s + c_0}{w(s)}ds = 0. \tag{3.46}$$

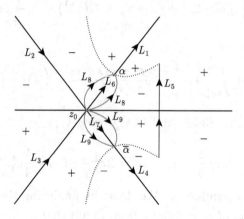

Fig. 6   The sign structure of $\mathrm{Re}(ig)$ in the complex z-plane

According to the discussion in [7], one can rewrite the condition (3.46) as the following forms

$$\int_{-\frac{i}{2}q_0^2}^{\frac{i}{2}q_0^2} \frac{s^3 + c_2 s^2 + c_1 s + c_0}{w(s)}ds = \int_{-\frac{i}{2}q_0^2}^{\frac{i}{2}q_0^2} \sqrt{\frac{(s-\alpha_1)^2 + \alpha_2^2}{s^2 + \frac{1}{4}q_0^4}}(s-z_0)ds = 0. \tag{3.47}$$

Finally, substituting (3.44) into (3.47), we find

$$\int_{-\frac{i}{2}q_0^2}^{\frac{i}{2}q_0^2} \sqrt{\frac{\left(s - \frac{\xi}{4} + z_0\right)^2 + 2z_0^2 - \frac{1}{2}\xi z_0 + \frac{1}{4}q_0^4}{s^2 + \frac{1}{4}q_0^4}} (s - z_0)ds = 0. \qquad (3.48)$$

It is enough to determine the z_0 from equation (3.48), and hence α, function $g(z)$.

Lemma 3.2 For all $\xi \in (-2\sqrt{2}q_0^2, 0)$, the integral equation (3.48) admits a unique solution $z_0 = z_0(\xi)$.

Proof The strategy of the proof follows from [7,9]. Let

$$x = -\frac{\xi}{4q_0^2}, \quad y = \frac{4z_0 - \xi}{2q_0^2}.$$

Then, equation (3.48) turns into

$$\mathcal{F}(x, y) = \int_{-1}^{1} \sqrt{\frac{(i\tau + y)^2 + 2y^2 - 4xy + 1}{1 - \tau^2}} (i\tau + 2x - y)d\tau = 0, \qquad (3.49)$$

which is considered for $(x, y) \in [0, \sqrt{2}/2] \times [0, \sqrt{2}]$. It is easy to check that $\mathcal{F}(0, 0) = \mathcal{F}(\sqrt{2}/2, \sqrt{2}/2) = 0$, $\mathcal{F}_y(0, 0) < 0$. Moreover, $\mathcal{F}_y(x, y) < 0$ for $(x, y) \neq (\sqrt{2}/2, \sqrt{2}/2)$. Therefore, by the implicit function theorem, (3.49) determines a unique function $y = y(x)$ for any $x \in [0, \sqrt{2}/2]$ such that $y(0) = 0$ and $y(\sqrt{2}/2) = \sqrt{2}/2$. That is, for any $\xi \in [-2\sqrt{2}q_0^2, 0]$ there exists a unique solution $z_0(\xi)$ of the integral equation (3.48) with $z_0(-2\sqrt{2}q_0^2) = -(\sqrt{2}/4)q_0^2$.

We have now specified a g-function $g(z)$ which has an appropriate signature table as in Fig. 6. Moreover, across the branch cuts L_5 and B, $g(z)$ satisfies the following jump conditions

$$\begin{aligned} g^+(z) + g^-(z) &= 0, \quad z \in L_5, \\ g^+(z) + g^-(z) &= \Omega, \quad z \in L_6 \cup L_7, \end{aligned} \qquad (3.50)$$

where the real constant Ω is defined by

$$\Omega = -4\left(\int_{-\frac{i}{2}q_0^2}^{\alpha} + \int_{\frac{i}{2}q_0^2}^{\bar{\alpha}}\right)\frac{(z - z_0)(z - \alpha)(z - \bar{\alpha})}{w(z)}dz. \qquad (3.51)$$

In the following, we continue to deform the RH problem. By the jumps (3.50) of $g(z)$, we infer that the jump matrices of $m^{(4)}$ can be rewritten as

$$J_1^{(4)} = \begin{pmatrix} 1 & 0 \\ z\rho\delta^{-2}e^{-2igt} & 1 \end{pmatrix}, \quad J_2^{(4)} = \begin{pmatrix} 1 & -\frac{\bar{\rho}\delta^2}{1 - z\rho\bar{\rho}}e^{2igt} \\ 0 & 1 \end{pmatrix},$$

$$J_3^{(4)} = \begin{pmatrix} 1 & 0 \\ \dfrac{z\rho\delta^{-2}}{1-z\rho\bar\rho}e^{-2igt} & 0 \end{pmatrix}, \quad J_4^{(4)} = \begin{pmatrix} 1 & -\bar\rho\delta^2 e^{2igt} \\ 0 & 1 \end{pmatrix},$$

$$J_5^{(4)} = \begin{pmatrix} 0 & \dfrac{q_-\delta^2}{i\sqrt{z}q_0} \\ -\dfrac{i\sqrt{z}q_0}{q_-\delta^2} & 0 \end{pmatrix}, \quad J_6^{(4)} = \begin{pmatrix} 0 & -\dfrac{\delta^2}{z\rho}e^{i\Omega t} \\ \dfrac{z\rho}{\delta^2}e^{-i\Omega t} & 0 \end{pmatrix},$$

$$J_7^{(4)} = \begin{pmatrix} 0 & -\bar\rho\delta^2 e^{i\Omega t} \\ \dfrac{e^{-i\Omega t}}{\bar\rho\delta^2} & 0 \end{pmatrix}, \quad J_8^{(4)} = \begin{pmatrix} 1 & \dfrac{\delta^2}{z\rho}e^{2igt} \\ 0 & 1 \end{pmatrix},$$

$$J_9^{(4)} = \begin{pmatrix} 1 & 0 \\ -\dfrac{e^{-2igt}}{\bar\rho\delta^2} & 1 \end{pmatrix},$$

where for $j = 1,\cdots,9$ we denote by $J_j^{(4)}$ the jump associated with the contour L_j. Furthermore, the normalization condition of $m^{(4)}$ is

$$m^{(4)}(x,t;z) = I + O(z^{-1}), \quad z \to \infty. \tag{3.52}$$

Recalling (3.36), (3.42), (3.43) and

$$\theta(\xi;z) = -2z^2 + \xi z - \frac{1}{4}q_0^4 + O(z^{-1}), \quad z \to \infty,$$

we find that the real constant $G(\infty)$ involved in the (3.35) is equal to

$$G(\infty) = -2\left(\int_{-\frac{i}{2}q_0^2}^{\infty} + \int_{\frac{i}{2}q_0^2}^{\infty}\right)\left[\frac{s^3 + c_2 s^2 + c_1 s + c_0}{w(s)} - \left(s - \frac{\xi}{4}\right)\right]ds - \frac{1}{4}q_0^4. \tag{3.53}$$

Our final task is to eliminate the dependence on z from the jump matrices $J_j^{(4)}$ across the branch cuts L_j for $j = 5, 6, 7$. To achieve this goal, let

$$m^{(5)}(x,t;z) = e^{-i\hat g(\infty)\sigma_3} m^{(4)}(x,t;z) e^{i\hat g(z)\sigma_3}. \tag{3.54}$$

The function $\hat g(z)$ is analytic in $\mathbf{C}\setminus\bigcup_{j=5}^{7} L_j$ with jumps

$$\hat g^+(z) + \hat g^-(z) = i\ln\left(\frac{2i\sqrt{z}}{q_0\delta^2}\right), \quad z \in L_5,$$

$$\hat g^+(z) + \hat g^-(z) = i\ln\left(\frac{z\rho}{\delta^2}\right) + \omega + \pi, \quad z \in L_6, \tag{3.55}$$

$$\hat{g}^+(z) + \hat{g}^-(z) = -i\ln(\bar{\rho}\delta^2) + \omega, \qquad z \in L_7,$$

with the function $\delta(z)$ defined by (3.31) and the real constant ω determined by

$$\omega = i\frac{\int_{L_5}\frac{\ln\left[\frac{q_0\delta^2(s)}{2i\sqrt{s}}\right]}{w(s)}ds + \int_{L_6}\frac{\ln\left[\frac{\delta^2(s)}{s\rho(s)}\right]+i\pi}{w(s)}ds - \int_{L_7}\frac{\ln[\bar{\rho}(s)\delta^2(s)]}{w(s)}ds}{\left(\int_{L_6}-\int_{L_7}\right)\frac{ds}{w(s)}}. \qquad (3.56)$$

It follows from the Plemelj's formulae that

$$\hat{g}(z) = \frac{w(z)}{2\pi}\left\{\int_{L_5}\frac{\ln\left[\frac{q_0\delta^2(s)}{2i\sqrt{s}}\right]}{w(s)(s-z)}ds + \int_{L_6}\frac{\ln\left[\frac{\delta^2(s)}{s\rho(s)}\right]+i(\omega+\pi)}{w(s)(s-z)}ds \right.$$
$$\left. - \int_{L_7}\frac{\ln[\bar{\rho}(s)\delta^2(s)]+i\omega}{w(s)(s-z)}ds\right\}. \qquad (3.57)$$

The definition (3.56) of ω ensures that

$$\hat{g}(z) = \hat{g}(\infty) + O(z^{-1}), \qquad z \to \infty,$$

where the real constant $\hat{g}(\infty)$ is given by

$$\hat{g}(\infty) = -\frac{1}{2\pi}\left\{\int_{L_5}\frac{\ln\left[\frac{q_0\delta^2(s)}{2i\sqrt{s}}\right]}{w(s)}sds + \int_{L_6}\frac{\ln\left[\frac{\delta^2(s)}{s\rho(s)}\right]+i(\omega+\pi)}{w(s)}sds\right.$$
$$\left. - \int_{L_7}\frac{\ln[\bar{\rho}(s)\delta^2(s)]+i\omega}{w(s)}sds\right\}. \qquad (3.58)$$

Finally, we can obtain the following RH problem for $m^{(5)}(x,t;z)$:

$$m^{(5)+}(x,t;z) = m^{(5)-}(x,t;z)J^{(5)}(x,t;z), \qquad z \in L_j, \qquad (3.59)$$

with the jump matrices given by

$$J_1^{(5)} = \begin{pmatrix} 1 & 0 \\ z\rho\delta^{-2}e^{-2i(gt-\hat{g})} & 1 \end{pmatrix}, \quad J_2^{(5)} = \begin{pmatrix} 1 & -\dfrac{\bar{\rho}\delta^2}{1-z\rho\bar{\rho}}e^{2i(gt-\hat{g})} \\ 0 & 1 \end{pmatrix},$$

$$J_3^{(5)} = \begin{pmatrix} 1 & 0 \\ \dfrac{z\rho\delta^{-2}}{1-z\rho\bar{\rho}}e^{-2i(gt-\hat{g})} & 0 \end{pmatrix}, \quad J_4^{(5)} = \begin{pmatrix} 1 & -\bar{\rho}\delta^2 e^{2i(gt-\hat{g})} \\ 0 & 1 \end{pmatrix},$$

$$J_5^{(5)} = \begin{pmatrix} 0 & \dfrac{2}{\bar{q}_-} \\ -\dfrac{\bar{q}_-}{2} & 0 \end{pmatrix}, \quad J_6^{(5)} = \begin{pmatrix} 0 & e^{i(\Omega t - \omega)} \\ -e^{-i(\Omega t - \omega)} & 0 \end{pmatrix},$$

$$J_7^{(5)} = \begin{pmatrix} 0 & -e^{i(\Omega t - \omega)} \\ e^{-i(\Omega t - \omega)} & 0 \end{pmatrix}, \quad J_8^{(5)} = \begin{pmatrix} 1 & \dfrac{\delta^2}{z\rho}e^{2i(gt-\hat{g})} \\ 0 & 1 \end{pmatrix},$$

$$J_9^{(5)} = \begin{pmatrix} 1 & 0 \\ -\dfrac{e^{-2i(gt-\hat{g})}}{\bar{\rho}\delta^2} & 1 \end{pmatrix},$$

and the normalization condition

$$m^{(5)}(x,t;z) = I + O(z^{-1}), \quad z \to \infty. \tag{3.60}$$

In other words, for the jump matrix $J^{(5)}(x,t;z)$, we have

$$J^{(5)}(x,t;z) = \begin{cases} J^{\mathrm{mod}}, & z \in L_5 \cup B, \\ I + O(e^{-\epsilon t}), & z \in L_j, \, j = 1, \cdots, 4, 8, 9, \end{cases}$$

where $B = L_6 \cup (-L_7)$ and

$$J^{\mathrm{mod}} = \begin{cases} \begin{pmatrix} 0 & \dfrac{2}{\bar{q}_-} \\ -\dfrac{\bar{q}_-}{2} & 0 \end{pmatrix}, & z \in L_5, \\ \begin{pmatrix} 0 & e^{i(\Omega t - \omega)} \\ -e^{-i(\Omega t - \omega)} & 0 \end{pmatrix}, & z \in B. \end{cases} \tag{3.61}$$

Thus, we arrive at the following model problem $\mathrm{RH}^{\mathrm{mod}}$:

$$\begin{aligned} m^{\mathrm{mod}+}(x,t;z) &= m^{\mathrm{mod}-}(x,t;z) J^{\mathrm{mod}}, \quad z \in L_5 \cup B, \\ m^{\mathrm{mod}}(x,t;z) &= I + O(z^{-1}), \quad z \to \infty. \end{aligned} \tag{3.62}$$

The solution of RH^{mod} approximates $m^{(5)}(x,t;z)$ as follows (see the parametrix analysis in [10])

$$m^{(5)}(x,t;z) = (I + O(t^{-\frac{1}{2}}))m^{mod}(x,t;z). \qquad (3.63)$$

The model RH problem (3.62) can be solved in terms of elliptic theta functions. Let us consider the Abelian differential

$$du = \frac{c}{w(z)}dz, \quad c = \left(\oint_b \frac{1}{w(z)}dz\right)^{-1},$$

which is normalized so that $\oint_b du = 1$ and has Riemann period τ defined by

$$\tau = \oint_a du.$$

Note that $c \in i\mathbf{R}$. It can be shown that $\tau \in i\mathbf{R}^+$ (see [18]). Define

$$U(z) = \int_{\frac{i}{2}q_0^2}^{z} du. \qquad (3.64)$$

Then the following relations are valid:

$$\begin{aligned} U^+(z) + U^-(z) &= n, & n \in \mathbb{Z}, \ z \in L_5, \\ U^+(z) + U^-(z) &= -\tau + n, & n \in \mathbb{Z}, \ z \in B. \end{aligned} \qquad (3.65)$$

Next, we define a new function $\nu(z)$ by

$$\nu(z) = \left(\frac{(z - \frac{i}{2}q_0^2)(z - \alpha)}{(z + \frac{i}{2}q_0^2)(z - \bar{\alpha})}\right)^{\frac{1}{4}}, \qquad (3.66)$$

which has the same jump discontinuity across both L_5 and B, namely,

$$\nu^+(z) = i\nu^-(z), \quad z \in L_5 \cup B,$$

and admits the large-z asymptotic behavior

$$\nu(z) = 1 + \frac{q_0^2 + 2\alpha_2}{4iz} + O(z^{-2}), \quad z \to \infty.$$

The last ingredient is the theta function with $i\tau < 0$:

$$\theta_3(z) = \sum_{l \in \mathbb{Z}} e^{i\pi\tau l^2 + 2i\pi l z},$$

which has the following properties

$$\theta_3(-z) = \theta_3(z), \quad \theta_3(z+n) = \theta_3(z), \quad \theta(z+\tau) = e^{-i\pi\tau - 2i\pi z}\theta_3(z). \tag{3.67}$$

Now we define the 2×2 matrix-valued function $\Theta(z) = \Theta(x, t; z)$ with entries:

$$\Theta_{11}(z) = \frac{1}{2}\left[\nu(z) + \frac{1}{\nu(z)}\right] \frac{\theta_3\left(U(z) + U_0 - \frac{\Omega t}{2\pi} + \frac{\omega}{2\pi} + \frac{i\ln\left(\frac{\bar{q}_-}{2}\right)}{2\pi}\right)}{\sqrt{\frac{2}{\bar{q}_-}}\theta_3(U(z) + U_0)},$$

$$\Theta_{12}(z) = \frac{1}{2}i\left[\nu(z) - \frac{1}{\nu(z)}\right] \frac{\theta_3\left(-U(z) + U_0 - \frac{\Omega t}{2\pi} + \frac{\omega}{2\pi} + \frac{i\ln\left(\frac{\bar{q}_-}{2}\right)}{2\pi}\right)}{-\sqrt{\frac{\bar{q}_-}{2}}\theta_3(-U(z) + U_0)},$$

$$\Theta_{21}(z) = \frac{1}{2}i\left[\nu(z) - \frac{1}{\nu(z)}\right] \frac{\theta_3\left(U(z) - U_0 - \frac{\Omega t}{2\pi} + \frac{\omega}{2\pi} + \frac{i\ln\left(\frac{\bar{q}_-}{2}\right)}{2\pi}\right)}{-\sqrt{\frac{2}{\bar{q}_-}}\theta_3(U(z) - U_0)},$$

$$\Theta_{22}(z) = \frac{1}{2}\left[\nu(z) + \frac{1}{\nu(z)}\right] \frac{\theta_3\left(-U(z) - U_0 - \frac{\Omega t}{2\pi} + \frac{\omega}{2\pi} + \frac{i\ln\left(\frac{\bar{q}_-}{2}\right)}{2\pi}\right)}{-\sqrt{\frac{\bar{q}_-}{2}}\theta_3(-U(z) - U_0)},$$

where

$$U_0 = U(z_*) + \frac{1}{2}(1+\tau), \quad z_* = \frac{q_0^2 \alpha_1}{q_0^2 + 2\alpha_2}. \tag{3.68}$$

Then the solution of the model RH problem (3.62) is given by

$$m^{\mathrm{mod}}(x, t; z) = \Theta^{-1}(x, t; \infty)\Theta(x, t; z). \tag{3.69}$$

Taking into account (3.35), (3.54) and (3.63), we get

$$q(x,t) = -2 \lim_{z \to \infty} (zm(x,t;z))_{12}$$
$$= -2 \lim_{z \to \infty} (zm^{(4)}(x,t;z))_{12} e^{-2iG(\infty)t} + O(t^{-\frac{1}{2}})$$
$$= -2 \lim_{z \to \infty} (zm^{(5)}(x,t;z))_{12} e^{2i(\hat{g}(\infty)-G(\infty))t} + O(t^{-\frac{1}{2}})$$
$$= -2 \lim_{z \to \infty} (zm^{\text{mod}}(x,t;z))_{12} e^{2i(\hat{g}(\infty)-G(\infty))t} + O(t^{-\frac{1}{2}}). \tag{3.70}$$

However, we have

$$-2 \lim_{z \to \infty} (zm^{\text{mod}}(x,t;z))_{12}$$

$$= \left(q_- + \frac{2\alpha_2}{\bar{q}_-} \right) \frac{\theta_3(-U(\infty) + U_0 - \frac{\Omega t}{2\pi} + \frac{\omega}{2\pi} + \frac{i \ln\left(\frac{\bar{q}_-}{2}\right)}{2\pi}) \theta_3(U(\infty) + U_0)}{\theta_3(U(\infty) + U_0 - \frac{\Omega t}{2\pi} + \frac{\omega}{2\pi} + \frac{i \ln\left(\frac{\bar{q}_-}{2}\right)}{2\pi}) \theta_3(-U(\infty) + U_0)},$$

where

$$U(\infty) = \int_{\frac{i}{2}q_0^2}^{\infty} du. \tag{3.71}$$

Taking into account (3.53) and (3.58), we get the asymptotics of the solution in the region $-2\sqrt{2}q_0^2 < \xi < 0$.

Theorem 3.2 *In the region $-2\sqrt{2}q_0^2 t < x < 0$, the asymptotics of the solution $q(x,t)$ of the initial value problem for the GI-type derivative NLS equation (1.2) takes, as $t \to \infty$, the form of a modulated elliptic wave:*

$$q(x,t) = \left(q_- + \frac{2\alpha_2}{\bar{q}_-} \right) \frac{\theta_3(-U(\infty) + U_0 - \frac{\Omega t}{2\pi} + \frac{\omega}{2\pi} + \frac{i \ln\left(\frac{\bar{q}_-}{2}\right)}{2\pi}) \theta_3(U(\infty) + U_0)}{\theta_3(U(\infty) + U_0 - \frac{\Omega t}{2\pi} + \frac{\omega}{2\pi} + \frac{i \ln\left(\frac{\bar{q}_-}{2}\right)}{2\pi}) \theta_3(-U(\infty) + U_0)}$$
$$\times e^{2i(\hat{g}(\infty) - G(\infty))t} + O(t^{-\frac{1}{2}}), \quad t \to \infty, \tag{3.72}$$

where the constants α_2, Ω, ω, $U(\infty)$, U_0, $G(\infty)$ and $\hat{g}(\infty)$ are given by the equations (3.44), (3.51), (3.56), (3.71), (3.68), (3.53) and (3.58), respectively.

3.3 The plane wave and modulated elliptic wave regions II

In the rest of the present paper, we devoted to discuss the long-time asymptotics of the solution of the initial value problem for GI-type derivative NLS equation (1.2) in the plane wave region II: $x > 2\sqrt{2}q_0^2 t$ and modulated elliptic wave region II: $0 < x < 2\sqrt{2}q_0^2 t$.

We first consider the region $x > 2\sqrt{2}q_0^2 t$. In this case we first rescale the RH problem (2.38) for $m(x,t;z)$ as follows, which is motivated by the ideas used in [7]. Let

$$\tilde{m}(x,t;z) = \begin{cases} m(x,t;z)A(k), & z \in \mathbf{C}^+ \setminus \gamma, \\ m(x,t;z)\bar{A}^{-1}(k), & z \in \mathbf{C}_- \setminus \bar{\gamma}, \end{cases} \tag{3.73}$$

where

$$A(k) = \begin{pmatrix} a(k) & 0 \\ 0 & a^{-1}(k) \end{pmatrix}.$$

Then, $\tilde{m}(x,t;z)$ is analytic in $\mathbf{C} \setminus \Sigma$ and satisfies the following jump conditions

$$\begin{aligned} \tilde{m}^+(x,t;z) &= \tilde{m}^-(x,t;z)\tilde{J}_1^{(0)}(x,t;z), & z \in \mathbf{R}, \\ \tilde{m}^+(x,t;z) &= \tilde{m}^-(x,t;z)\tilde{J}_2^{(0)}(x,t;z), & z \in \gamma, \\ \tilde{m}^+(x,t;z) &= \tilde{m}^-(x,t;z)\tilde{J}_3^{(0)}(x,t;z), & z \in \bar{\gamma}, \end{aligned} \tag{3.74}$$

with the normalization condition

$$\tilde{m}(x,t;z) = I + O(z^{-1}), \quad z \to \infty, \tag{3.75}$$

where

$$\tilde{J}_1^{(0)}(x,t;z) = \begin{pmatrix} \dfrac{1}{d(z)} & -\bar{\varrho}(z)e^{2it\theta(\xi;z)} \\ z\varrho(z)e^{-2it\theta(\xi;z)} & d(z)[1 - z\varrho(z)\bar{\varrho}(z)] \end{pmatrix},$$

$$\tilde{J}_2^{(0)}(x,t;z) = \begin{pmatrix} -\dfrac{\lambda - (z + \frac{1}{2}q_0^2)}{q_+}\bar{\varrho}(z)e^{2it\theta(\xi;z)} & -\dfrac{2\lambda}{z\bar{q}_+}[1 - z\varrho(z)\bar{\varrho}(z)] \\ \dfrac{z\bar{q}_+}{2\lambda} & \dfrac{\lambda + z + \frac{1}{2}q_0^2}{\bar{q}_+}\varrho(z)e^{-2it\theta(\xi;z)} \end{pmatrix},$$

$$\tilde{J}_3^{(0)}(x,t;z) = \begin{pmatrix} \dfrac{\lambda + z + \dfrac{1}{2}q_0^2}{q_+}\bar{\varrho}(z)e^{2it\theta(\xi;z)} & \dfrac{q_+}{2\lambda} \\ -\dfrac{2\lambda}{q_+}[1 - z\varrho(z)\bar{\varrho}(z)] & -\dfrac{\lambda - (z + \dfrac{1}{2}q_0^2)}{\bar{q}_+}\varrho(z)e^{-2it\theta(\xi;z)} \end{pmatrix},$$

and

$$\varrho(z) = -\dfrac{\bar{b}(k)}{k\bar{a}(k)}.$$

Then, we have

$$\varrho^+(z) = \dfrac{\bar{q}_+}{q_+}\bar{\varrho}(z), \quad z \in \gamma \cup \bar{\gamma}. \tag{3.76}$$

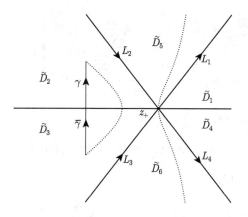

Fig. 7 The oriented contour $\Sigma^{(2)}$ and the open sets $\{\tilde{D}_j\}_1^6$ in the plane wave region II

As the discussions in plane wave region I, the first deformation for $\tilde{m}(x,t;z)$ is

$$\tilde{m}^{(1)}(x,t;z) = \tilde{m}(x,t;z)\delta^{-\sigma_3}(z), \tag{3.77}$$

where

$$\delta(z) = \exp\left\{\dfrac{i}{2\pi}\int_{z_+}^{\infty}\dfrac{\ln[1 - s\varrho(s)\bar{\varrho}(s)]}{s - z}ds\right\}, \quad z \in \mathbb{C} \setminus [z_+, \infty). \tag{3.78}$$

Then $\tilde{m}^{(1)}(x,t;z)$ satisfies the following RH problem

$$\tilde{m}^{(1)+}(x,t;z) = \tilde{m}^{(1)-}(x,t;z)\tilde{J}^{(1)}(x,t;z), \quad z \in \Sigma, \tag{3.79}$$

where the jump matrix is given by

$$\tilde{J}_1^{(1)} = \begin{pmatrix} d^{-\frac{1}{2}} & 0 \\ z\varrho e^{-2it\theta}\delta^{-2}d^{\frac{1}{2}} & d^{\frac{1}{2}} \end{pmatrix} \begin{pmatrix} d^{-\frac{1}{2}} & -\bar{\varrho}e^{2it\theta}\delta^2 d^{\frac{1}{2}} \\ 0 & d^{\frac{1}{2}} \end{pmatrix}, \quad z \in (-\infty, z_+),$$

$$\tilde{J}_2^{(1)} = \begin{pmatrix} d^{-\frac{1}{2}} & -\dfrac{\bar{\varrho}e^{2it\theta}(\delta^-)^2}{1-z\varrho\bar{\varrho}}d^{-\frac{1}{2}} \\ 0 & d^{\frac{1}{2}} \end{pmatrix} \begin{pmatrix} d^{-\frac{1}{2}} & 0 \\ \dfrac{z\varrho e^{-2it\theta}(\delta^+)^{-2}}{1-z\varrho\bar{\varrho}}d^{-\frac{1}{2}} & d^{\frac{1}{2}} \end{pmatrix}, \quad z \in (z_+, \infty),$$

$$\tilde{J}_3^{(1)} = \begin{pmatrix} -\dfrac{\lambda-(z+\frac{1}{2}q_0^2)}{q_+}\bar{\varrho}e^{2it\theta} & -\dfrac{2\lambda\delta^2}{z\bar{q}_+}[1-z\varrho\bar{\varrho}] \\ \dfrac{z\bar{q}_+\delta^{-2}}{2\lambda} & \dfrac{\lambda+z+\frac{1}{2}q_0^2}{\bar{q}_+}\varrho e^{-2it\theta} \end{pmatrix}, \quad z \in \gamma,$$

$$\tilde{J}_4^{(1)} = \begin{pmatrix} \dfrac{\lambda+z+\frac{1}{2}q_0^2}{q_+}\bar{\varrho}e^{2it\theta} & \dfrac{q_+\delta^2}{2\lambda} \\ -\dfrac{2\lambda\delta^{-2}}{q_+}[1-z\varrho\bar{\varrho}] & -\dfrac{\lambda-(z+\frac{1}{2}q_0^2)}{\bar{q}_+}\varrho e^{-2it\theta} \end{pmatrix}, \quad z \in \bar{\gamma}.$$

The next transformation is

$$\tilde{m}^{(2)}(x,t;z) = \tilde{m}^{(1)}(x,t;z)\tilde{H}(z), \tag{3.80}$$

where $\tilde{H}(z)$ is defined by

$$\tilde{H}(z) = \begin{cases} \begin{pmatrix} d^{\frac{1}{2}} & 0 \\ -\dfrac{z\varrho e^{-2it\theta}\delta^{-2}}{1-z\varrho\bar{\varrho}}d^{-\frac{1}{2}} & d^{-\frac{1}{2}} \end{pmatrix}, & z \in \tilde{D}_1, \\[2ex] \begin{pmatrix} d^{\frac{1}{2}} & \bar{\varrho}e^{2it\theta}\delta^2 d^{\frac{1}{2}} \\ 0 & d^{-\frac{1}{2}} \end{pmatrix}, & z \in \tilde{D}_2, \\[2ex] \begin{pmatrix} d^{-\frac{1}{2}} & 0 \\ z\varrho e^{-2it\theta}\delta^{-2}d^{\frac{1}{2}} & d^{\frac{1}{2}} \end{pmatrix}, & z \in \tilde{D}_3, \\[2ex] \begin{pmatrix} d^{-\frac{1}{2}} & -\dfrac{\bar{\varrho}e^{2it\theta}\delta^2}{1-z\varrho\bar{\varrho}}d^{-\frac{1}{2}} \\ 0 & d^{\frac{1}{2}} \end{pmatrix}, & z \in \tilde{D}_4, \\[2ex] I, & z \in \tilde{D}_5 \cup \tilde{D}_6, \end{cases}$$

and the domains $\{\tilde{D}_j\}_1^6$ are shown in Fig. 7. Then the function $\tilde{m}^{(2)}(x,t;z)$ solves the following equivalent RH problem:

$$\tilde{m}^{(2)+}(x,t;z) = \tilde{m}^{(2)-}(x,t;z)\tilde{J}^{(2)}(x,t;z),$$
$$z \in \Sigma^{(2)} = L_1 \cup L_2 \cup L_3 \cup L_4 \cup \gamma \cup \bar{\gamma},$$
(3.81)

where the contours L_j are shown in Fig. 7 and the jump matrix $\tilde{J}^{(2)}$ is given by

$$\tilde{J}_1^{(2)} = \begin{pmatrix} d^{-\frac{1}{2}} & 0 \\ \frac{z\varrho e^{-2it\theta}(\delta^+)^{-2}}{1-z\varrho\bar{\varrho}}d^{-\frac{1}{2}} & d^{\frac{1}{2}} \end{pmatrix}, \quad z \in L_1,$$

$$\tilde{J}_2^{(2)} = \begin{pmatrix} d^{-\frac{1}{2}} & -\bar{\varrho}e^{2it\theta}\delta^2 d^{\frac{1}{2}} \\ 0 & d^{\frac{1}{2}} \end{pmatrix}, \quad z \in L_2,$$

$$\tilde{J}_3^{(2)} = \begin{pmatrix} d^{-\frac{1}{2}} & 0 \\ z\varrho e^{-2it\theta}\delta^{-2}d^{\frac{1}{2}} & d^{\frac{1}{2}} \end{pmatrix}, \quad z \in L_3,$$

$$\tilde{J}_4^{(2)} = \begin{pmatrix} d^{-\frac{1}{2}} & -\frac{\bar{\varrho}e^{2it\theta}(\delta^-)^2}{1-z\varrho\bar{\varrho}}d^{-\frac{1}{2}} \\ 0 & d^{\frac{1}{2}} \end{pmatrix}, \quad z \in L_4,$$

$$\tilde{J}_5^{(2)} = \begin{pmatrix} 0 & \frac{q_+\delta^2}{i\sqrt{z}q_0} \\ -\frac{i\sqrt{z}q_0}{q_+\delta^2} & 0 \end{pmatrix}, \quad z \in \gamma \cup \bar{\gamma}.$$

Then, performing the similar deformation and analysis as in Subsection 3.1, we can get a model RH problem that can be solved explicitly and obtain the long-time asymptotics for the solution $q(x,t)$. We just give the main results here.

Theorem 3.3 *In the region $x > 2\sqrt{2}q_0^2 t$, the asymptotics of the solution $q(x,t)$ as $t \to \infty$ of the initial value problem for the GI-type derivative NLS equation (1.2) takes the form of a plane wave:*

$$q(x,t) = q_+ e^{2i\tilde{\phi}(\xi)} + O(t^{-\frac{1}{2}}), \quad t \to \infty,$$
(3.82)

where $\tilde{\phi}(\xi)$ is given by

$$\tilde{\phi}(\xi) = \frac{1}{2\pi}\int_{\gamma\cup\bar{\gamma}}\left[\frac{\ln\left[\frac{2i\sqrt{s}}{q_0}\right]}{\lambda^+(s)} + \frac{1}{i\pi\lambda^+(s)}\int_{z_+}^{\infty}\frac{\ln[1-u\varrho(u)\bar{\varrho}(u)]}{u-s}du\right]ds.$$

For the modulated elliptic wave region II: $0 < x < 2\sqrt{2}q_0^2 t$, the asymptotic analysis is entirely analogous to Subsection 3.2 after the rescaling of the RH problem for $m(x,t;z)$ as stated above. We omit the detailed computation here.

Acknowledgements This work was supported in part by the National Natural Science Foundation of China under grants 11731014, 11571254 and 11471099. The authors would like to thank the anonymous referees for their careful reading and useful comments.

References

[1] Ablowitz MJ, Fokas AS. Complex Variables: Introduction and Applications [M]. 2nd ed. Cambridge: Cambridge University Press; 2003.

[2] Arruda LK, Lenells J. Long-time asymptotics for the derivative nonlinear Schrödinger equation on the half-line [J]. Nonlinearity, 2017, 30: 4141–4172.

[3] Benney DJ, Newell AC. The propagation of nonlinear wave envelopes [J]. Stud. Appl. Math., 1967, 46: 133–139.

[4] Biondini G, Fagerstrom E, Prinari B. Inverse scattering transform for the defocusing nonlinear Schrödinger equation with fully asymmetric non-zero boundary conditions [J]. Phys. D, 2016, 333: 117–136.

[5] Biondini G, G. Kovačič G. Inverse scattering transform for the focusing nonlinear Schrödinger equation with nonzero boundary conditions [J]. J. Math. Phys., 2014, 55: 31506.

[6] Biondini G, Kraus D, Prinari B. The three-component defocusing nonlinear Schröinger equation with nonzero boundary conditions [J]. Comm. Math. Phys., 2016, 348: 475–533.

[7] Biondini G, Mantzavinos D. Long-time asymptotics for the focusing nonlinear Schrödinger equation with nonzero boundary conditions at infinity and asymptotic stage of modulational instability [J]. Comm. Pure Appl. Math., 2017, 70: 2300–2365.

[8] Boutet de Monvel A, Its A, Kotlyarov V. Long-time asymptotics for the focusing NLS equation with time-periodic boundary condition on the half-line [J]. Comm. Math. Phys., 2009, 290: 479–522.

[9] Boutet de Monvel A, Kotlyarov VP, Shepelsky D. Focusing NLS equation: long-time dynamics of step-like initial data [J]. Int. Math. Res. Not., 2011, 2011: 1613–1653.

[10] Buckingham R, Venakides S. Long-time asymptotics of the nonlinear Schrödinger equation shock problem [J]. Comm. Pure. Appl. Math., 2007, 60: 1349–1414.

[11] Dai H, Fan E, separation V. Variable separation and algebro-geometric solutions of the Gerdjikov–Ivanov equation [J]. Chaos, Solitons Fractals, 2004, 22: 93–101.

[12] Deift P, Kriecherbauer T, McLaughlin KT-R, Venakides S, Zhou X. Uniform asymptotics for polynomials orthogonal with respect to varying exponential weights

and applications to universality questions in random matrix theory [J]. Comm. Pure Appl. Math., 1999, 52: 1335-1425.
[13] Deift P, Zhou X. A steepest descent method for oscillatory Riemann–Hilbert problems. Asymptotics for the MKdV equation [J]. Ann Math., 1993, 137: 295-368.
[14] Demontis F, Prinari B, van der Mee C, Vitale F. The inverse scattering transform for the focusing nonlinear Schrödinger equation with asymmetric boundary conditions [J]. J. Math. Phys. 2014, 55: 101505.
[15] Desyatnikov AS, Kivshar YS, Torner L. Optical vortices and vortex solitons [J]. Prog Opt. 2005, 47: 291-391.
[16] Fan E. Integrable evolution systems based on Gerdjikov–Ivanov equations, bi-Hamiltonian structure, finite-dimensional integrable systems and N-fold Darboux transformation [J]. J. Math. Phys., 2000, 41: 7769-7782.
[17] Fan E. transformation D. Soliton-like solutions for the Gerdjikov–Ivanov equation [J]. J. Phys. A Math. Theor., 2000, 33: 6925-6933.
[18] Farkas H, Kra I. Riemann surfaces [M]. Graduate Texts in Mathematics. New York: Springer. 2nd ed.; 1992: 71.
[19] Geng X, Liu H. The nonlinear steepest descent method to long-time asymptotics of the coupled nonlinear Schrödinger equation [J]. J. Nonlinear Sci., 2018, 28: 739-763.
[20] Gerdjikov V, Ivanov I. A quadratic pencil of general type and nonlinear evolution equations. II. Hierarchies of Hamiltonian structures [J]. Bulg J Phys., 1983, 10: 130-143.
[21] Grunert K, Teschl G. Long-time asymptotics for the Korteweg–de Vries equation via nonlinear steepest descent [J]. Math. Phys. Anal. Geom., 2009, 12: 287-324.
[22] Guo B, Liu N. Long-time asymptotics for the Kundu–Eckhaus equation on the half-line [J]. J. Math. Phys., 2018, 59: 61505.
[23] Guo B, Liu N, Wang Y. Long-time asymptotics for the Hirota equation on the half-line [J]. Nonlinear Anal. TMA, 2018, 174: 118-140.
[24] Huang L, Xu J, Fan E. Long-time asymptotic for the Hirota equation via nonlinear steepest descent method [J]. Nonlinear Anal. RWA, 2015, 26: 229-262.
[25] Kaup DJ, Newell AC. An exact solution for a derivative nonlinear Schrödinger equation [J]. J. Math. Phys., 1978, 19: 789-801.
[26] Kotlyarov V, Minakov A. Riemann–Hilbert problem to the modified Korteveg–de Vries equation: long-time dynamics of the steplike initial data [J]. J. Math. Phys., 2010, 51: 093506.
[27] Kraus D, Biondini G, Kovačič G. The focusing Manakov system with nonzero boundary conditions [J]. Nonlinearity, 2015, 28: 3101-3151.
[28] Lenells J. The nonlinear steepest descent method: asymptotics for initial-boundary value problems [J]. SIAM J. Math. Anal., 2016, 48: 2076-2118.
[29] Liang Z, Zhang Z, Liu W. Dynamics of a bright soliton in Bose–Einstein condensates with time-dependent atomic scattering length in an expulsive parabolic potential [J].

Phys. Rev. Lett., 2005, 94: 50402.

[30] Liu J, Perry PA, Sulem C. Long-time behavior of solutions to the derivative nonlinear Schrödinger equation for soliton-free initial data [J]. Ann. Inst. H. Poincaré Anal. Non Linéaire, 2018, 35: 217–265.

[31] Tian S, Zhang T. Long-time asymptotic behavior for the Gerdjikov–Ivanov type of derivative nonlinear Schrödinger equation with time-periodic boundary condition [J]. Proc. Amer. Math. Soc., 2018, 146: 1713–1729.

[32] Wang D, Wang X. Long-time asymptotics and the bright N-soliton solutions of the Kundu–Eckhaus equation via the Riemann–Hilbert approach [J]. Nonlinear Anal. RWA, 2018, 41: 334–361.

[33] Xu J, Fan E. Long-time asymptotics for the Fokas–Lenells equationwith decaying initial value problem: without solitons [J]. J. Differ. Equations, 2015, 259: 1098–1148.

[34] Xu J, Fan E, Chen Y. Long-time asymptotic for the derivative nonlinear Schrödinger equation with step-like initial value [J]. Math. Phys. Anal. Geom., 2013, 16: 253–288.

[35] Xu S, He J. The rogue wave and breather solution of the Gerdjikov–Ivanov equation [J]. J. Math. Phys., 2012, 53: 63507.

On the Backward Uniqueness of the Stochastic Primitive Equations with Additive Noise*

Guo Boling (郭柏灵), Zhou Guoli (周国立)

Abstract The previous works focus on the uniqueness for the initial-value problems of stochastic primitive equations. Uniqueness for the initial-value problems means that if the two initial conditions are the same, then the two solutions coincide with each other. However there is no work to answer what will happen to the solutions if the two initial conditions are different. This problem for the stochastic three dimensional primitive equations is addressed by the backward uniqueness established in this article. The backward uniqueness means that if two solutions intersect at time $t > 0$, then they are equal everywhere on the interval $(0, t)$. In other words, given two different initial-value conditions, the corresponding two solutions will never cross in the future. Hence this article can be viewed as a further study of the dependence of the solutions on the initial data.

Keywords stochastic primitive equations; backward uniqueness

1 Introduction

The paper is concerned with the primitive equations (PEs) in a bounded domain with Wiener noises. To outline its content in detail, we introduce a bounded domain $\mho = M \times \left(-\frac{1}{2}, \frac{1}{2}\right)$ with $M = (0,1) \times (0,1)$ and consider the following 3D stochastic PEs of Geophysical Fluid Dynamics.

$$\partial_t \boldsymbol{v} + (\boldsymbol{v} \cdot \nabla_H)\boldsymbol{v} + w\partial_z \boldsymbol{v} + f\boldsymbol{v}^\perp + \nabla_H p - \Delta \boldsymbol{v} = \dot{W}_1, \qquad (1.1)$$

$$\partial_z p + \theta = 0, \qquad (1.2)$$

---

* Discrete Conti. Dyn. Systems-Ser. B, 2018, 23, 4305–4327.

$$\nabla_H \cdot \boldsymbol{v} + \partial_z w = 0, \qquad (1.3)$$

$$\partial_t \theta + \boldsymbol{v} \cdot \nabla_H \theta + w \partial_z \theta - w - \Delta \theta = Q + \dot{W}_2. \qquad (1.4)$$

The unknowns for the 3D stochastic viscous PEs are the fluid velocity field $(\boldsymbol{v}, w) = (v_1, v_2, w) \in \mathbf{R}^3$ with $\boldsymbol{v} = (v_1, v_2)$ and $\boldsymbol{v}^\perp = (-v_2, v_1)$ being horizontal, the temperature θ and the pressure p. $f = f_0(\beta + y)$ is the given Coriolis parameter, Q is a given heat source which is periodic in x, y, z and is odd in z. In this paper we use the notations $\nabla_H = (\partial_x, \partial_y)$ and $\Delta = \partial_x^2 + \partial_y^2 + \partial_z^2$ to denote the horizontal gradient and the three dimensional Laplacian operators. Here, we take $\dot{W}_i(t, x, y, z), i = 1, 2$, the informal derivative for the Wiener process W_i given below.

We supplement (1.1) – (1.4) with the following boundary and initial conditions:

$$\boldsymbol{v}, w, p, \text{ and } \theta \text{ are periodic in } x, y, z, \qquad (1.5)$$

$$\boldsymbol{v} \text{ and } p \text{ are even in } z, w \text{ and } \theta \text{ are odd in } z, \qquad (1.6)$$

$$(\boldsymbol{v}, \theta)|_{t=0} = (\boldsymbol{v}_0, \theta_0). \qquad (1.7)$$

The primitive equations are the basic model used in the study of climate and weather prediction, which can be used to describe the motion of the atmosphere when the hydrostatic assumption is enforced (see [16, 23, 24] and the references therein). This model has been intensively investigated because of the interests stemmed from physics and mathematics. As far as we know, their mathematical study was initiated by Lions, Temam and Wang (see e.g. [29-32]). For example, the existence of global weak solutions for the primitive equations was established in [30]. Guillén-González et al. obtained the global existence of strong solutions to the primitive equations with small initial data in [19]. The local existence of strong solutions to the primitive equations under the small depth hypothesis was established by Hu et al. in [25]. Cao and Titi developed a beautiful approach to dealing with the L^6-norm of the fluctuation $\tilde{v}$ of horizontal velocity and obtained the global well-posedness for the 3D viscous primitive equations in [9]. Subsequently, in [26], Kukavica and Ziane developed a different method to handle non-rectangular domains and boundary conditions with physical reality. For the global well-posedness of 3D primitive equations with partial dissipation, we refer the reader to some papers, e.g. [5-8, 10].

Along with the great successful developments of deterministic primitive equations, the random case has also been developed rapidly. In [17], Guo and Huang studied the long-time behavior of the strong solution to the stochastic primitive equations with additive stochastic forcing. When the noise is multiplicative, Gao and Sun [20] proved the global existence and uniqueness for the strong solution. Moreover, when the noise tends to zero, Gao and Sun [21] established the large deviation principle for this stochostic system. In [22], Gao and Sun studied the long-time behavior of stochastic PEs when the velocity is perturbed by an additive noise. Debussche, Glatt-Holtz, Temam and Ziane are concerned with the global well-posedness of strong solutions when the primitive equations are driven by multiplicative stochastic forcing in [12]. Under the periodic conditions, Glatt-Holtz, Kukavica, Vicol and Ziane considered the existence of invariant measure for the 3D PEs in [18]. The uniqueness of the invariant measure and large deviations for the 3D stochastic primitive equations were obtained by Dong, Zhai and Zhang in [14, 15] under the periodic boundary conditions. Some analytical properties of weak solutions of 3D stochastic primitive equations with periodic boundary conditions were obtained in [13], in which the martingale problem associated with this model is shown to have a family of solutions satisfying the Markov property.

Our main goal of this article is to establish the backward uniqueness of the stochastic three dimensional primitive equations. This kind of uniqueness is different from uniqueness for the initial-value problems. Backward uniqueness means that given two solutions if these solutions intersect at some time t, they must be equal everywhere before t. In other words, if the two solutions start with two different initial states, then the two solutions will never cross. Therefore, this article can be viewed as a further study about the dependence of the solutions to stochastic primitive equations on the initial data. To prove our result, we adopted the logarithmic energy method from [33]. The idea of the method is try to estimate $\log |U|_2^2$ where U is the difference between the two solutions. Exactly speaking, we try to obtain some kind of estimates like

$$\log |U(t)|_2^2 \geqslant \log |U(t_0)|_2^2,$$

where t represents the present time and t_0 is the initial time. The above inequality implies that if the two solutions cross at present time t, they will equal at any previous time t_0 (see Theorem 3.4). To show the inequality, the difficulty lies in the establishment of the bound of $\frac{\|U(t)\|_1}{|U(t)|_2}$ on the finite interval, where $|\cdot|_2$ and $\|\cdot\|_1$ are norms in Lebesgue space L^2 and Sobolev space H^1 respectively. The *a priori* estimates of $\frac{\|U(t)\|_1}{|U(t)|_2}$ heavily rely on the delicate and careful argument on the special geometric structure of stochastic primitive equations (see Theorem 3.3).

The remaining of this paper is organized as follows. Notations and some results are recalled in section 2. In section 3, the regularity of the strong solutions to stochastic PEs is improved (see Theorem 3.1 and Theorem 3.2). Then the bound of $\frac{\|U\|_1^2}{|U|_2^2}$ is obtained in Theorem 3.3 by virtue of Theorem 3.1 and Theorem 3.2. The main result the backward uniqueness for the 3D stochastic PEs is established in Theorem 3.4. As usual, constants C may change from one line to the next, unless, we give a special declaration ; we denote by $C(a)$ a constant which depends on some parameter a.

2 Preliminaries

For $1 \leqslant p \leqslant \infty$, let $L^p(\mho)$ be the usual Lebesgue spaces with the norm $|\cdot|_p$, when $p = 2$, we denote by $\langle \cdot, \cdot \rangle$ the inner product in $L^2(\mho)$. For a positive integer m, we denote by $(H^{m,p}(\mho), \|\cdot\|_{m,p})$ the usual Sobolev spaces, see([1]). When $p = 2$, we denote by $(H^m(\mho), \|\cdot\|_m)$ with inner product $\langle , \rangle_{H^m}$.

In the following we will need terminology to describe periodic boundary conditions. Let $m \in \mathbb{N}$. Then we say that a smooth function $f : \bar{\mho} \to \mathbf{R}$ is periodic of order m on $\partial \mho$ if

$$\frac{\partial^\alpha f}{\partial x^\alpha}(0, y, z) = \frac{\partial^\alpha f}{\partial x^\alpha}(1, y, z) \text{ and } \frac{\partial^\alpha f}{\partial y^\alpha}(x, 0, z) = \frac{\partial^\alpha f}{\partial y^\alpha}(x, 1, z),$$

$$\frac{\partial^\alpha f}{\partial z^\alpha}(x, y, -\frac{1}{2}) = \frac{\partial^\alpha f}{\partial z^\alpha}(x, y, \frac{1}{2}) \tag{2.1}$$

for all $\alpha = 0, \cdots, m$. We assume space periodicity with period $\mho$, meaning that all functions are taken to satisfy:

$$f(x, y, z, t) = f(x+1, y, z, t) = f(x, y+1, z, t) = f(x, y, z+1, t),$$

when extended to $\mathbf{R}^3$. Define

$$\dot{H}^m_{per}(\mho) := \{f \in H^m(\mho) | f \text{ is periodic of order } m-1$$

on $\partial \mho$ and satisfy $\int_\mho f d\mho = 0\}$. (2.2)

We introduce our working space in the following

$$V_1 = \left\{ v \in (\dot{H}^1_{per}(\mho))^2, v \text{ even in } z, \int_{-\frac{1}{2}}^{\frac{1}{2}} \nabla_H \cdot v dz = 0 \right\},$$

$$V_2 = \left\{ \theta \in \dot{H}^1_{per}(\mho), \theta \text{ odd in } z, \right\},$$

$\mathbb{H} =$ closure of V_1 in $(\dot{L}^2(\mho))^2$,
$\mathbb{H}^1 =$ closure of V_1 in $(\dot{H}^1_{per}(\mho))^2$,
$\mathbb{H}^2 =$ closure of $V_1 \cap (\dot{H}^2_{per}(\mho))^2$ in $(\dot{H}^2_{per}(\mho))^2$,
$\mathbb{H}^3 =$ closure of $V_1 \cap (\dot{H}^3_{per}(\mho))^2$ in $(\dot{H}^3_{per}(\mho))^2$,
$\mathbf{H} =$ closure of V_2 in $\dot{L}^2(\mho)$,
$\mathbf{H}^1 =$ closure of V_2 in $\dot{H}^1_{per}(\mho)$,
$\mathbf{H}^2 =$ closure of $V_2 \cap \dot{H}^2_{per}(\mho)$ in $\dot{H}^2_{per}(\mho)$,
$\mathbf{H}^3 =$ closure of $V_2 \cap \dot{H}^3_{per}(\mho)$ in $\dot{H}^3_{per}(\mho)$. (2.3)

We denote the dual space of $\mathbb{H}^m$ and $\mathbf{H}^m$ by $\mathbb{H}^{-m}$ and $\mathbf{H}^{-m}$ respectively. W_1 and W_2 are standard Wiener processes in $\mathbb{H}$ and $\mathbf{H}$ defined below and have the form of

$$W_j(t) := \Sigma_{i=1}^\infty \lambda_i^{\frac{1}{2}} e_{i,j} B_i(t), \quad j = 1, 2,$$

where $((B_i(t))_{t \in \mathbb{R}})_{i \in \mathbb{N}}$ be a sequence of one-dimensional, independent, identically distributed Brownian motions defined on the complete probability space $(\Omega, \mathcal{F}, \mathbb{P})$, $(e_{i,1})_{i \in \mathbb{N}}$ and $(e_{i,2})_{i \in \mathbb{N}}$ are orthonormal basis in $\mathbb{H}$ and $\mathbf{H}$ respectively, $(\lambda_i)_{i \in \mathbb{N}}$ is a convergent sequence of positive numbers which ensure W_1 and W_2 are standard Wiener processes in $\mathbb{H}$ and $\mathbf{H}$ respectively.

Let $U_i := (v_i, \theta_i)$ be the horizontal velocity and temperature with $i = 1, 2, 3$. We equip $V := \mathbb{H}^1 \times \mathbf{H}^1$ with the inner product

$$\langle U_1, U_2 \rangle_V := \int_\mho (\nabla v_1 \cdot \nabla v_2 + \nabla \theta_1 \cdot \nabla \theta_2) d\mho,$$

where $\nabla = (\partial_x, \partial_y, \partial_z)$. Subsequently, the norm of V is defined by $\|U_i\|_1 = \langle U_i, U_i \rangle_V^{\frac{1}{2}}$. We define the inner product of $H := (L^2(\mho))^3$ by

$$\langle U_1, U_2 \rangle_H := \langle v_1, v_2 \rangle + \langle \theta_1, \theta_2 \rangle = \int_\mho (v_1 \cdot v_2 + \theta_1 \cdot \theta_2) d\mho.$$

Define $a : V \times V \to \mathbf{R}$ bilinear, continuous, coercive

$$a(U_1, U_2) = \langle U_1, U_2 \rangle_V, \tag{2.4}$$

$b : V \times V \times \mathbb{H}^2 \to \mathbf{R}$ trilinear:

$$\begin{aligned}b(U_1, U_2, U_3) &= \langle \boldsymbol{v}_1 \cdot \nabla_H \boldsymbol{v}_2, \boldsymbol{v}_3 \rangle + \langle w(\boldsymbol{v}_1)\partial_z \boldsymbol{v}_2, \boldsymbol{v}_3 \rangle \\ &\quad + \langle \boldsymbol{v}_1 \cdot \nabla_H \theta_2, \theta_3 \rangle + \langle w(\boldsymbol{v}_1)\partial_z \theta_2, \theta_3 \rangle,\end{aligned} \tag{2.5}$$

where $w(f) = -\int_{-\frac{1}{2}}^{z} \nabla_H \cdot f(x, y, z') dz', f \in \mathbf{R}^2$. Obviously, we have $b(U_1, U_2, U_3) = -b(U_1, U_3, U_2)$ and $b(U_1, U_2, U_2) = 0$. Define $e : V \times V \to \mathbf{R}$ bilinear, continuous:

$$e(U_1, U_2) = \langle \boldsymbol{v}_1^\perp, \boldsymbol{v}_2 \rangle + \langle \theta_1, w(\boldsymbol{v}_2) \rangle - \langle \theta_2, w(\boldsymbol{v}_1) \rangle. \tag{2.6}$$

Obviously, $e(U_1, U_1) = 0$. Denote by V' the dual space of V. In the following, we introduce the operators A, B and E.

A is a linear continuous from V into V', given by

$$\langle AU_1, U_2 \rangle = a(U_1, U_2), \quad \forall\, U_1, U_2 \in V.$$

B is a bilinear continuous from $V \times V$ into $\mathbb{H}^{-2}$, defined by

$$\langle B(U_1, U_2), U_3 \rangle = b(U_1, U_2, U_3), \quad \forall\, U_1, U_2 \in V, \quad \forall\, U_3 \in \mathbb{H}^2 \times \mathbf{H}^2.$$

E is a linear continuous from V into V' satisfying

$$\langle EU_1, U_2 \rangle = e(U_1, U_2), \quad \forall\, U_1, U_2 \in V.$$

Define the linear operator $A_1 : \mathbb{H}^1 \mapsto \mathbb{H}^{-1}$, as follows:

$$\langle A_1 \boldsymbol{v}_1, \boldsymbol{v}_2 \rangle = \langle \nabla \boldsymbol{v}_1, \nabla \boldsymbol{v}_2 \rangle, \quad \forall\, \boldsymbol{v}_1, \boldsymbol{v}_2 \in \mathbb{H}^1.$$

Define the linear operator $A_2 : \mathbf{H}^1 \mapsto \mathbf{H}^{-1}$, as follows:

$$\langle A_2 \theta_1, \theta_2 \rangle = \langle \nabla \theta_1, \nabla \theta_2 \rangle, \quad \forall\, \theta_1, \theta_2 \in \mathbf{H}^1.$$

Since the operator $A_i, i = 1, 2$ is positive selfadjoint with compact resolvent, by the classical spectral theorems there exists a sequence $\{\alpha_{i,j}\}_{j \in \mathbb{N}}$ of eigenvalues of A_i such that

$$0 < \alpha_{i,1} \leqslant \alpha_{i,2} \leqslant \cdots, \quad \alpha_{i,j} \to \infty$$

corresponding to the eigenvectors $e_{i,j}$. Assume

$$\Sigma_{j=1}^{\infty} \lambda_j \alpha_{i,j}^3 < \infty. \tag{2.7}$$

For arbitrary constant $T > 0, i = 1, 2$ and $j \in \mathbb{N}$, we define

$$z_i^j(t) = \Sigma_{n=1}^{j} \sqrt{\lambda_n} \int_0^t e^{-A_i(t-s)} e_{i,n} dB_n(s), \quad t \in [0, T]$$

and

$$z_i(t) = \Sigma_{n=1}^{\infty} \sqrt{\lambda_n} \int_0^t e^{-A_i(t-s)} e_{i,n} dB_n(s), \quad t \in [0, T].$$

Obviously,

$$z_1^j(w) \in C([0,T]; \mathbb{H}^3) \text{ and } z_2^j(w) \in C([0,T]; \mathbb{H}^3), \quad \mathbb{P}\text{-a.e. } w \in \Omega.$$

For $k \in \mathbb{N}$ and $k > j$, In view of an infinite dimensional version of Burkholder-Davis-Gundy type of inequality for stochastic convolutions (see Theorem 1.2.6 in [27] and references), we have

$$E \sup_{t \in [0,T]} \|A_i^{3/2}(z_i^j - z_i^k)\|_{L^2}^2$$

$$\leqslant CT \Sigma_{n=j+1}^{k} \lambda_n \alpha_{i,n}^3 \to 0, \quad \text{as } j \to \infty.$$

Therefore

$$z_1(w) \in C([0,T]; \mathbb{H}^3), \text{ and } z_2(w) \in C([0,T]; \mathbb{H}^3), \quad \mathbb{P}\text{-a.e. } w \in \Omega. \tag{2.8}$$

Obviously, $(z_1(t))_{t \in [0,T]}$ is the unique solution which is periodic in x, y, z and even in z to the following initial condition problem defined on $\mho$

$$dz_1 - \Delta z_1 dt = dW_1,$$

$$z_1|_{t=0} = 0.$$

Similarly, $(z_2(t))_{t \in [0,T]}$ is the unique solution which is periodic in x, y, z and odd in z to the following initial condition problem defined on $\mho$

$$dz_2 - \Delta z_2 dt = dW_2,$$

$$z_2|_{t=0} = 0.$$

To simplify the notations we set

$$\int_{\mho} \cdot := \int_{\mho} \cdot d\mho.$$

Definition 2.1 Given $T > 0$, we say a continuous $\mathcal{V}$-valued $(\mathcal{F}_t) = (\sigma(W_j^H(s), s \in [0, t]), j = 1, 2)$ adapted random field $(U(., t))_{t \in [0, T]} = (\boldsymbol{v}(., t), \theta(., t))_{t \in [0, T]}$ defined on $(\Omega, \mathcal{F}, \mathbb{P})$ is a strong solution to problem $(1.1) - (1.7)$ if the following two conditions hold:

(1) For $(\boldsymbol{v}_0, \theta_0) \in \mathbb{H}^1 \times \mathbf{H}^1$, we have $U \in C([0, T]; \mathbb{H}^1 \times \mathbf{H}^1) \cap L^2([0, T]; \mathbb{H}^2 \times \mathbf{H}^2)$ a.s..

(2) The integral relation

$$\int_\mho \boldsymbol{v}(t) \cdot \phi_1$$
$$- \int_0^t ds \int_\mho \left\{ [(\boldsymbol{v} \cdot \nabla)\phi_1 + w(\boldsymbol{v})\partial_z \phi_1] \boldsymbol{v} - \left[(fk \times \boldsymbol{v}) \cdot \phi_1 + \left(\int_{-\frac{1}{2}}^z \theta dz' \right) \nabla \cdot \phi_1 \right] \right\}$$
$$+ \int_0^t ds \int_\mho \boldsymbol{v} \cdot A_1 \phi_1 = \int_\mho \boldsymbol{v}_0 \cdot \phi_1 + \int_\mho W_1(t, w) \cdot \phi_1,$$
$$\int_\mho \theta(t) \phi_2 - \int_0^t ds \int_\mho \{[(\boldsymbol{v} \cdot \nabla)\phi_2 + w(\boldsymbol{v})\partial_z \phi_2]\theta - \theta A_2 \phi_2\} = \int_\mho \theta_0 \phi_2$$
$$+ \int_0^t ds \int_\mho Q \phi_2 + \int_\mho W_2(t, w) \cdot \phi_2,$$

hold a.s. for all $t \in [0, T]$ and $\phi = (\phi_1, \phi_2) \in D(A_1) \times D(A_2)$.

Denote by $\boldsymbol{u} = \boldsymbol{v} - \boldsymbol{z}_1$ and $\vartheta = \theta - z_2$. Consider the primitive equations driven by a stochastic forcing in a Cartesian system. It is easy to see that if $(\boldsymbol{u}, \vartheta)$ is the unique strong solution of the following system $(2.16) - (2.21)$, then equivalently $(\boldsymbol{v}, \theta)$ is the unique strong solution of $(1.1) - (1.7)$.

$$\partial_t \boldsymbol{u} - \Delta \boldsymbol{u} + [(\boldsymbol{u} + \boldsymbol{z}_1) \cdot \nabla_H](\boldsymbol{u} + \boldsymbol{z}_1) + w(\boldsymbol{u} + \boldsymbol{z}_1)\partial_z(\boldsymbol{u} + \boldsymbol{z}_1)$$
$$+ f(\boldsymbol{u} + \boldsymbol{z}_1)^\perp + \nabla_H p_s - \int_{-\frac{1}{2}}^z \nabla_H \vartheta dz' = 0; \tag{2.9}$$

$$\partial_t \vartheta - \Delta \vartheta + [(\boldsymbol{u} + \boldsymbol{z}_1) \cdot \nabla_H](\vartheta + z_2) + w(\boldsymbol{u} + \boldsymbol{z}_1)\partial_z(\vartheta + z_2) - w(\boldsymbol{u} + \boldsymbol{z}_1) = 0; \tag{2.10}$$

$\boldsymbol{u}$, and ϑ are periodic in x, y, z; $\tag{2.11}$

$\boldsymbol{u}$ is even in z, ϑ is odd in z; $\tag{2.12}$

$$\int_{-1}^0 \nabla_H \cdot \boldsymbol{u} dz = 0; \tag{2.13}$$

$$(u\big|_{t=0}, \vartheta\big|_{t=0}) = (v_0, \theta_0). \qquad (2.14)$$

Definition 2.2 Let $z_j, j = 1, 2$, are defined above, $(v_0, \theta_0) \in \mathbb{H}^1 \times \mathbf{H}^1$ and T be a fixed positive time. For P-a.e.$\omega \in \Omega$, (u, ϑ) is called a strong solution of the system $(2.16) - (2.21)$ on the time interval $[0, T]$ if it satisfies $(2.16) - (2.17)$ in the weak sense such that

$$u \in C([0,T]; \mathbb{H}^1) \cap L^2([0,T]; \mathbb{H}^2),$$
$$\vartheta \in C([0,T]; \mathbf{H}^1) \cap L^2([0,T]; \mathbf{H}^2).$$

In our paper, we will frequently use the following inequalities, so we state them in the following. For their proof, one can refer to [9].

Lemma 2.1 Let $v \in (H^2(\mho))^2, \mu \in (H^1(\mho))^2$ and $\nu \in (L^2(\mho))^2$. Then, there exists a positive constant c independent of v, μ and ν such that

$$\left|\left\langle \left(\int_\xi^1 \nabla_H \cdot v(t; \theta, \phi, \xi')d\xi'\right)\mu, \nu\right\rangle\right|$$
$$\leqslant c|\nabla_H \cdot v|_2^{\frac{1}{2}}(|\nabla_H \cdot v|_2^{\frac{1}{2}} + |\Delta v|_2^{\frac{1}{2}})|\mu|_2^{\frac{1}{2}}(|\nabla_H \mu|_2^{\frac{1}{2}} + |\Delta_H \mu|_2^{\frac{1}{2}})|\nu|_2.$$

Lemma 2.2 Let $v \in (H^1(\mho))^2, \mu \in (H^1(\mho))^2$ and $\nu \in (H^1(\mho))^2$. Then, there exists a positive constant c independent of v, μ and ν such that

$$\left|\left\langle \left(\int_\xi^1 \nabla_H \cdot v(t; \theta, \phi, \xi')d\xi'\right)\mu, \nu\right\rangle\right|$$
$$\leqslant c|\nabla_H \cdot v|_2|\mu|_2^{\frac{1}{2}}(|\mu|_2^{\frac{1}{2}} + |\nabla_H \mu|_2^{\frac{1}{2}})|\nu|_2^{\frac{1}{2}}(|\nu|_2^{\frac{1}{2}} + |\nabla_H \nu|_2^{\frac{1}{2}}).$$

In [17], one of the authors of this article obtained the global well-posedness for the stochastic 3D primitive equations.

Theorem 2.1 For $Q \in L^2(\mho)$, $(v_0, \theta_0) \in \mathbb{H}^1 \times \mathbf{H}^1$. There exists a unique strong solution (v, θ) to $(1.1) - (1.7)$ or equivalently there exists a unique strong solution (u, ϑ) to $(2.16) - (2.21)$.

In order to study the backward uniqueness of 3D stochastic primitive equations, we need to improve the regularity of the strong solution. The following Lemma, a special case of a general result of Lions and Magenes [28], will help us to achieve our goal.

Lemma 2.3 Let V, H, V' be three Hilbert spaces such that $V \subset H = H \subset V'$, where H' and V' are the dual spaces of H and V respectively. Suppose

$u \in L^2(0,T;V)$ and $u' \in L^2(0,T;V')$. Then u is almost everywhere equal to a function continuous from $[0,T]$ into H.

3 Backward uniqueness for the strong solutions of the three dimensional primitive equations

Theorem 3.1 Given $(\boldsymbol{v}_0, \theta_0) \in \mathbb{H}^1 \times \mathbf{H}^1$ and $(\partial_z \boldsymbol{v}_0, \partial_z \theta_0) \in \mathbb{H}^1 \times \mathbf{H}^1$, let $(\boldsymbol{u}, \vartheta)$ be the unique strong solution to (2.16) − (2.21), then $\boldsymbol{u}_z \in C([0,T]; \mathbb{H}^1) \cap L^2([0,T]; \mathbb{H}^2)$ and $\vartheta_z \in C([0,T]; \mathbf{H}^1) \cap L^2([0,T]; \mathbf{H}^2)$.

Proof Taking the derivative of (2.16) with respect to z yields

$$\partial_t \boldsymbol{u}_z - \Delta \boldsymbol{u}_z + [(\boldsymbol{u} + \boldsymbol{z}_1) \cdot \nabla_H](\boldsymbol{u}_z + \partial_z \boldsymbol{z}_1) + [(\boldsymbol{u}_z + \partial_z \boldsymbol{z}_1) \cdot \nabla_H](\boldsymbol{u} + \boldsymbol{z}_1)$$
$$- (\boldsymbol{u}_z + \partial_z \boldsymbol{z}_1) \nabla_H \cdot (\boldsymbol{u} + \boldsymbol{z}_1) + w(\boldsymbol{u} + \boldsymbol{z}_1)(\boldsymbol{u}_{zz} + \partial_{zz} \boldsymbol{z}_1)$$
$$+ fk \times (\boldsymbol{u}_z + \partial_z \boldsymbol{z}_1) - \nabla_H \vartheta = 0. \tag{3.1}$$

Multiplying $-\Delta \boldsymbol{u}_z$ and integrating over $\mho$, we have

$$\frac{1}{2}\partial_t \|\boldsymbol{u}_z\|_1^2 + |\Delta \boldsymbol{u}_z|_2^2$$
$$= -\langle [(\boldsymbol{u} + \boldsymbol{z}_1) \cdot \nabla_H](\boldsymbol{u}_z + \partial_z \boldsymbol{z}_1), -\Delta \boldsymbol{u}_z \rangle$$
$$- \langle [(\boldsymbol{u}_z + \partial_z \boldsymbol{z}_1) \cdot \nabla_H](\boldsymbol{u} + \boldsymbol{z}_1), -\Delta \boldsymbol{u}_z \rangle$$
$$+ \langle (\boldsymbol{u}_z + \partial_z \boldsymbol{z}_1) \nabla_H \cdot (\boldsymbol{u} + \boldsymbol{z}_1), -\Delta \boldsymbol{u}_z \rangle$$
$$- \langle w(\boldsymbol{u} + \boldsymbol{z}_1)(\boldsymbol{u}_{zz} + \partial_{zz} \boldsymbol{z}_1), -\Delta \boldsymbol{u}_z \rangle$$
$$- \langle fk \times (\boldsymbol{u}_z + \partial_z \boldsymbol{z}_1), -\Delta \boldsymbol{u}_z \rangle + \langle \nabla_H \vartheta, -\Delta \boldsymbol{u}_z \rangle$$
$$= \Sigma_{k=1}^6 I_k. \tag{3.2}$$

By the Hölder inequality, the interpolation inequality and Young's inequality we have

$$I_1 \leqslant |\Delta \boldsymbol{u}_z|_2 |\nabla_H(\boldsymbol{u}_z + \partial_z \boldsymbol{z}_1)|_3 |\boldsymbol{u} + \boldsymbol{z}_1|_6$$
$$\leqslant C|\Delta \boldsymbol{u}_z|_2 |\nabla_H(\boldsymbol{u}_z + \partial_z \boldsymbol{z}_1)|_2^{1/2} |\Delta(\boldsymbol{u}_z + \partial_z \boldsymbol{z}_1)|_2^{1/2} |\boldsymbol{u} + \boldsymbol{z}_1|_6$$
$$\leqslant \varepsilon |\Delta \boldsymbol{u}_z|_2^2 + C\|\boldsymbol{z}_1\|_3^2 + C|\nabla_H(\boldsymbol{u}_z + \partial_z \boldsymbol{z}_1)|_2^2 |\boldsymbol{u} + \boldsymbol{z}_1|_6^4$$
$$\leqslant \varepsilon |\Delta \boldsymbol{u}_z|_2^2 + C\|\boldsymbol{z}_1\|_3^2 + C(\|\boldsymbol{u}\|_2^2 + \|\boldsymbol{z}_1\|_2^2)(\|\boldsymbol{u}\|_1^4 + \|\boldsymbol{z}_1\|_1^4).$$

In view of the Hölder inequality, the interpolation inequality, Sobolev imbedding theorem and Young's inequality we obtain

$$I_2 \leqslant |\Delta u_z|_2 |\nabla_H(u+z_1)|_4 |u_z + \partial_z z_1|_4$$
$$\leqslant C|\Delta u_z|_2 |\nabla_H(u+z_1)|_2^{1/4} \|u+z_1\|_2^{3/4} |u_z + \partial_z z_1|_2^{1/4} \|u_z + \partial_z z_1\|_1^{3/4}$$
$$\leqslant C|\Delta u_z|_2 \|u+z_1\|_1^{1/2} \|u+z_1\|_2^{3/4} \|u_z + \partial_z z_1\|_1^{3/4}$$
$$\leqslant \varepsilon|\Delta u_z|_2^2 + C\|u+z_1\|_1 \|u+z_1\|_2^{3/2} \|u_z + \partial_z z_1\|_1^{3/2}$$
$$\leqslant \varepsilon|\Delta u_z|_2^2 + C\|u+z_1\|_1^4 + C\|u+z_1\|_2^2 \|u_z + \partial_z z_1\|_1^2.$$

Similarly,

$$I_3 \leqslant |\Delta u_z|_2 |\nabla_H \cdot (u+z_1)|_3 |u_z + \partial_z z_1|_6$$
$$\leqslant |\Delta u_z|_2 \|u+z_1\|_2 \|u_z + \partial_z z_1\|_1$$
$$\leqslant \varepsilon|\Delta u_z|_2^2 + C\|u+z_1\|_2^2 \|u_z + \partial_z z_1\|_1^2.$$

By Lemma 2.1 and Young's inequality we have

$$I_4 \leqslant |\Delta u_z|_2 |u_{zz} + \partial_{zz} z_1|_2^{1/2} |\nabla_H u_{zz} + \nabla_H \partial_{zz} z_1|_2^{1/2}$$
$$\times \|u+z_1\|_1^{1/2} \|u+z_1\|_2^{1/2}$$
$$\leqslant \varepsilon|\Delta u_z|_2^2 + \varepsilon|\nabla_H u_{zz} + \nabla_H \partial_{zz} z_1|_2^2$$
$$+ C|u_{zz} + \partial_{zz} z_1|_2^2 \|u+z_1\|_1^2 \|u+z_1\|_2^2$$
$$\leqslant \varepsilon|\Delta u_z|_2^2 + C\|z_1\|_3^2 + C\|u_z\|_1^2 \|u+z_1\|_1^2 \|u+z_1\|_2^2$$
$$+ C\|z_1\|_2^2 \|u+z_1\|_1^2 \|u+z_1\|_2^2.$$

Thanks to Hölder inequality and Young's inequality, we obtain

$$I_5 + I_6 \leqslant \varepsilon|\Delta u_z|_2^2 + C\|u+z_1\|_1^2 + C\|\vartheta\|_1^2.$$

Based on (3.23), estimates from I_1 to I_6, the Gronwall inequality and the Theorem 2.1 we conclude that

$$u_z \in L^\infty([0,T]; \mathbb{H}^1) \cap L^2([0,T]; \mathbb{H}^2). \tag{3.3}$$

Multiplying (3.22) by $\eta \in \mathbb{H}^1$, integrating with respect to space variable yields

$$\langle \partial_t A_1^{1/2} u_z, \eta \rangle$$

$$=\langle \partial_t \boldsymbol{u}_z, A_1^{1/2}\eta\rangle$$

$$=\langle \Delta \boldsymbol{u}_z, A_1^{1/2}\eta\rangle - \langle [(\boldsymbol{u}+\boldsymbol{z}_1)\cdot \nabla_H](\boldsymbol{u}_z+\partial_z\boldsymbol{z}_1), A_1^{1/2}\eta\rangle$$

$$- \langle [(\boldsymbol{u}_z+\partial_z\boldsymbol{z}_1)\cdot \nabla_H](\boldsymbol{u}+\boldsymbol{z}_1), A_1^{1/2}\eta\rangle$$

$$+ \langle (\boldsymbol{u}_z+\partial_z\boldsymbol{z}_1)\nabla_H\cdot(\boldsymbol{u}+\boldsymbol{z}_1), A_1^{1/2}\eta\rangle - \langle w(\boldsymbol{u}+\boldsymbol{z}_1)(\boldsymbol{u}_{zz}+\partial_{zz}\boldsymbol{z}_1), A_1^{1/2}\eta\rangle$$

$$- \langle f k\times (\boldsymbol{u}_z+\partial_z\boldsymbol{z}_1)-\nabla_H\vartheta, A_1^{1/2}\eta\rangle. \tag{3.4}$$

By the Hölder inequality, Lemma 2.1 and Sobolev imbedding theorem we have

$$\|\partial_t A_1^{1/2}\boldsymbol{u}_z\|_{\mathbb{H}_1^{-1}}$$

$$\leqslant C\|\boldsymbol{u}_z\|_2 + C|\nabla_H(\boldsymbol{u}_z+\partial_z\boldsymbol{z}_1)|_3|\boldsymbol{u}+\boldsymbol{z}_1|_6$$

$$+ C|\boldsymbol{u}_z+\partial_z\boldsymbol{z}_1|_4|\nabla_H(\boldsymbol{u}+\boldsymbol{z}_1)|_4$$

$$+ C|\boldsymbol{u}_{zz}+\partial_{zz}\boldsymbol{z}_1|_2^{1/2}\|\boldsymbol{u}_z+\partial_z\boldsymbol{z}_1\|_2^{1/2}\|\boldsymbol{u}+\boldsymbol{z}_1\|_1^{1/2}\|\boldsymbol{u}+\boldsymbol{z}_1\|_2^{1/2}$$

$$+ C|\boldsymbol{u}_z+\partial_z\boldsymbol{z}_1|_2 + C|\nabla_H\vartheta|_2$$

$$\leqslant C\|\boldsymbol{u}_z\|_2 + C\|\boldsymbol{u}_z+\partial_z\boldsymbol{z}_1\|_2\|\boldsymbol{u}+\boldsymbol{z}_1\|_1 + C\|\boldsymbol{u}_z+\partial_z\boldsymbol{z}_1\|_1\|\boldsymbol{u}+\boldsymbol{z}_1\|_2$$

$$+ C\|\boldsymbol{u}_z+\partial_z\boldsymbol{z}_1\|_1^{1/2}\|\boldsymbol{u}_z+\partial_z\boldsymbol{z}_1\|_2^{1/2}\|\boldsymbol{u}+\boldsymbol{z}_1\|_1^{1/2}\|\boldsymbol{u}+\boldsymbol{z}_1\|_2^{1/2}$$

$$+ C|\boldsymbol{u}_z+\partial_z\boldsymbol{z}_1|_2 + C\|\vartheta\|_1, \tag{3.5}$$

which implies that

$$\partial_t A_1^{1/2}\boldsymbol{u}_z \in L^2([0,T]; \mathbb{H}^{-1}). \tag{3.6}$$

Therefore in view of Lemma 2.3, we have

$$\boldsymbol{u}_z \in C([0,T]; \mathbb{H}^1). \tag{3.7}$$

Taking the derivative of (2.17) with respective to z yields

$$\partial_t\vartheta_z - \Delta\vartheta_z + [(\boldsymbol{u}_z+\partial_z\boldsymbol{z}_1)\cdot\nabla_H](\vartheta+z_2)$$

$$+ [(\boldsymbol{u}+\boldsymbol{z}_1)\cdot\nabla_H](\vartheta_z+\partial_z z_2) - (\nabla_H\cdot\boldsymbol{u}+\nabla_H\cdot\boldsymbol{z}_1)(\vartheta_z+\partial_z z_2)$$

$$+ w(\boldsymbol{u}+\boldsymbol{z}_1)(\vartheta_{zz}+\partial_{zz}z_2) + \nabla_H\cdot\boldsymbol{u}+\nabla_H\cdot\boldsymbol{z}_1 = 0. \tag{3.8}$$

Multiplying $-\Delta\vartheta_z$ and integrating over $\mho$ yields,

$$\frac{1}{2}\partial_t\|\vartheta_z\|_1^2 + |\Delta\vartheta_z|_2^2$$

$$= -\langle[(u_z + \partial_z z_1) \cdot \nabla_H](\vartheta + z_2), -\Delta\vartheta_z\rangle$$
$$- \langle[(u + z_1) \cdot \nabla_H](\vartheta_z + \partial_z z_2), -\Delta\vartheta_z\rangle$$
$$+ \langle(\nabla_H \cdot u + \nabla_H \cdot z_1)(\vartheta_z + \partial_z z_2), -\Delta\vartheta_z\rangle$$
$$- \langle w(u + z_1)(\vartheta_{zz} + \partial_{zz} z_2), -\Delta\vartheta_z\rangle$$
$$- \langle(\nabla_H \cdot u + \nabla_H \cdot z_1), -\Delta\vartheta_z\rangle = \Sigma_{k=1}^{5} J_k. \tag{3.9}$$

By the Hölder inequality, the Sobolev imbedding theorem and Young's inequality we have
$$J_1 \leq \varepsilon|\Delta\vartheta_z|_2^2 + C|\nabla_H(\vartheta + z_2)|_4^2 |u_z + \partial_z z_1|_4^2$$
$$\leq \varepsilon|\Delta\vartheta_z|_2^2 + C\|\vartheta + z_2\|_2^2 \|u_z + \partial_z z_1\|_1^2.$$

In view of the Sobolev imbedding theorem, the Hölder inequality and Young's inequality we obtain
$$J_2 \leq |u + z_1|_\infty \|\vartheta_z + \partial_z z_2\|_1 |\Delta\vartheta_z|_2$$
$$\leq \varepsilon|\Delta\vartheta_z|_2 + c\|u + z_1\|_2^2 \|\vartheta_z + \partial_z z_2\|_1^2.$$

Taking a similar argument as in J_1 yields,
$$J_3 \leq \varepsilon|\Delta\vartheta_z|_2^2 + C\|\vartheta_z + \partial_z z_2\|_1^2 \|u + z_1\|_2^2.$$

According to Lemma 2.1 and Young's inequality,
$$J_4 \leq |\Delta\vartheta_z|_2 |\nabla_H(\vartheta_{zz} + \partial_{zz} z_2)|_2^{\frac{1}{2}} |\vartheta_{zz} + \partial_{zz} z_2|_2^{\frac{1}{2}} \|u + z_1\|_1^{\frac{1}{2}} \|u + z_1\|_1^{\frac{1}{2}}$$
$$\leq \varepsilon|\Delta\vartheta_z|_2^2 + C\|z_2\|_2^2 + C|\vartheta_{zz} + \partial_{zz} z_2|_2^2 \|u + z_1\|_1^2 \|u + z_1\|_2^2$$
$$\leq \varepsilon|\Delta\vartheta_z|_2^2 + C\|z_2\|_2^2 + C\|\vartheta_z + \partial_z z_2\|_1^2 \|u + z_1\|_1^2 \|u + z_1\|_2^2.$$

By Young's inequality, we obtain
$$J_5 \leq \varepsilon|\Delta\vartheta_z|_2^2 + C\|u\|_1^2 + C\|z_1\|_1^2.$$

By (3.29) and estimates of $J_1 - J_5$, we obtain
$$\partial_t \|\vartheta_z\|_1^2 + |\Delta\vartheta_z|_2^2 \leq C\|\vartheta + z_2\|_2^2 \|u_z + \partial_z z_1\|_1^2$$
$$+ C(1 + \|u + z_1\|_2^2) \|u + z_1\|_2^2 \|\vartheta_z + \partial_z z_2\|_1^2$$
$$+ C\|u\|_1^2 + C\|z_1\|_1^2. \tag{3.10}$$

Based on (3.30), the Gronwall inequality and the Theorem 2.1 we conclude that
$$\vartheta_z \in L^\infty([0,T]; \mathbf{H}^1) \cap L^2([0,T]; \mathbf{H}^2). \tag{3.11}$$

Multiplying (3.29) by $\xi \in \mathbf{H}^1$, integrating with respect to space variable yields

$$\begin{aligned}\langle \partial_t A_2^{1/2}\vartheta_z, \xi \rangle &= \langle \partial_t \vartheta_z, A_2^{1/2}\xi \rangle \\ &= \langle \Delta \vartheta_z, A_2^{1/2}\xi \rangle - \langle [(\boldsymbol{u}_z + \partial_z \boldsymbol{z}_1) \cdot \nabla_H](\vartheta + z_2), A_2^{1/2}\xi \rangle \\ &\quad - \langle [(\boldsymbol{u} + \boldsymbol{z}_1) \cdot \nabla_H](\vartheta_z + \partial_z z_2), A_2^{1/2}\xi \rangle \\ &\quad + \langle (\nabla_H \cdot \boldsymbol{u} + \nabla_H \cdot \boldsymbol{z}_1)(\vartheta_z + \partial_z z_2), A_2^{1/2}\xi \rangle \\ &\quad - \langle w(\boldsymbol{u} + \boldsymbol{z}_1)(\vartheta_{zz} + \partial_{zz} z_2), A_2^{1/2}\xi \rangle \\ &\quad - \langle \nabla_H \cdot \boldsymbol{u} + \nabla_H \cdot \boldsymbol{z}_1, A_2^{1/2}\xi \rangle. \end{aligned} \tag{3.12}$$

By the Hölder inequality and the Sobolev inequality, we obtain

$$\|\partial_t A_2^{1/2}\vartheta_z\|_{\mathbf{H}^{-1}} \leqslant c|\Delta\vartheta|_2 + C\|\boldsymbol{u}_z + \partial_z \boldsymbol{z}_1\|_1 \|\vartheta + z_2\|_2 \\ + C|\boldsymbol{u} + \boldsymbol{z}_1|_\infty \|\vartheta_z + \partial_z z_2\|_1 + C\|\boldsymbol{u} + \boldsymbol{z}_1\|_2(\|\vartheta_z + \partial_z z_2\|_1 + 1). \tag{3.13}$$

Hence, by (3.24), (3.32) and Theorem 2.1, we infer
$$\partial_t A_2^{1/2}\vartheta_z \in L^2([0,T]; \mathbf{H}^{-1}),$$

which combined with (3.32) and Lemma 2.3 implies
$$\vartheta_z \in C([0,T]; \mathbf{H}^1). \tag{3.14}$$

Finally, the conclusions of this theorem follow from (3.24), (3.28), (3.32) and (3.35). $\square$

Theorem 3.2 *Given* $(\boldsymbol{v}_0, \theta_0) \in \mathbb{H}^2 \times \mathbf{H}^2$, *let* $(\boldsymbol{u}, \vartheta)$ *be the unique strong solution to* (2.16) − (2.21), *then* $(\boldsymbol{u}, \vartheta) \in C([0,T]; \mathbb{H}^2 \times \mathbf{H}^2) \cap L^2([0,T]; \mathbb{H}^3 \times \mathbf{H}^3)$.

Proof Taking the inner product of (2.16) with $A_1^2 \boldsymbol{u}$ in $\mathbb{H}$ yields

$$\frac{1}{2}\partial_t |A_1 \boldsymbol{u}|_2^2 + |A_1^{3/2}\boldsymbol{u}|_2^2 = -\int_\mathcal{U} [(\boldsymbol{u}+Z_1)\cdot \nabla_H](\boldsymbol{u}+Z_1)\cdot A_1^2 \boldsymbol{u} \\ - \int_\mathcal{U} w(\boldsymbol{u}+\boldsymbol{z}_1)\partial_z(\boldsymbol{u}+\boldsymbol{z}_1)\cdot A_1^2\boldsymbol{u} - \int_\mathcal{U} f(\boldsymbol{u}+\boldsymbol{z}_1)^\perp \cdot A_1^2 \boldsymbol{u} \\ - \int_\mathcal{U} \nabla_H p_s \cdot A_1^2\boldsymbol{u} + \int_\mathcal{U}\Big(\int_{-1}^z \nabla_H \vartheta dz'\Big)\cdot A_1^2 \boldsymbol{u} = \sum_{k=1}^5 J_k. \tag{3.15}$$

By integration by parts and Young's inequality, we have

$$J_3 = \int_U f(u+z_1)^\perp \cdot A_1^2 u = \int_U fz_1^\perp \cdot A_1^2 u \leq \varepsilon |A_1^{3/2} u|_2^2 + C\|z_1\|_1^2$$

and

$$J_4 = \int_U \nabla_H p_s \cdot A_1^2 u = 0.$$

Similarly,

$$J_5 = \int_U \left(\int_{-1}^z \nabla_H \vartheta dz' \right) \cdot A_1^2 u \leq \varepsilon |A_1^{3/2} u|_2^2 + C\|\vartheta\|_2^2.$$

Since by integration by parts,

$$J_1 + J_2$$
$$= \int_U (u+z_1) \cdot \nabla_H (A_1(u+z_1)) \cdot A_1 u + \int_U w(u+z_1) \partial_z (A_1(u+z_1)) \cdot A_1 u$$
$$+ \int_U \left(\nabla(u+z_1) : \nabla_H \nabla(u+z_1) \right) \cdot A_1 u + \int_U \left(\nabla w(u+z_1) : \partial_z \nabla(u+z_1) \right) \cdot A_1 u$$
$$+ \int_U A_1(u+z_1) \cdot \nabla_H (u+z_1) A_1 u + \int_U A_1 w(u+z_1) \partial_z (u+z_1) A_1 u. \quad (3.16)$$

In the following, we will estimate the terms on the right-hand side of (3.37) respectively. Obviously, by integration by parts we have

$$\int_U u \cdot \nabla_H (A_1 u) \cdot (A_1 u) + \int_U w(u) \partial_z (A_1 u) \cdot (A_1 u) = 0. \quad (3.17)$$

In view of the Hölder inequality, the Sobolev inequality and Young's inequality we obtain

$$\int_U z_1 \cdot \nabla_H A_1(u+z_1) \cdot A_1 u + \int_U w(z_1) \partial_z A_1(u+z_1) \cdot A_1 u$$
$$\leq |z_1|_\infty |A_1^{3/2}(u+z_1)|_2 |A_1 u|_2 + \|z_1\|_3 |A_1^{3/2}(u+z_1)|_2 |A_1 u|_2$$
$$\leq \varepsilon |A_1^{3/2} u|_2^2 + C\|z_1\|_3^2 \|u\|_2 + C\|z_1\|_3^2 \|u\|_2^2. \quad (3.18)$$

Thanks to the Hölder inequality, the Sobolev inequality and Young's inequality we have

$$\int_U u \cdot (\nabla_H A_1 z_1) \cdot A_1 u + \int_U w(u) \partial_z (A_1 z_1) \cdot A_1 u$$

$$\leqslant C|\boldsymbol{u}|_\infty |A_1\boldsymbol{u}|_2\|\boldsymbol{z}_1\|_3 + \|\boldsymbol{z}_1\|_3 |A_1\boldsymbol{u}|_2^{1/2}|A_1^{3/2}\boldsymbol{u}|_2^{1/2}\|\boldsymbol{u}\|_1^{1/2}|A_1\boldsymbol{u}|_2^{1/2}$$
$$\leqslant \varepsilon |A_1^{3/2}\boldsymbol{u}|_2^2 + C\|\boldsymbol{z}_1\|_3^2\|\boldsymbol{u}\|_2^2 + C\|\boldsymbol{z}_1\|_3^4\|\boldsymbol{u}\|_1^2 + C\|\boldsymbol{u}\|_2^2. \tag{3.19}$$

With the aid of the Hölder inequality, the interpolation inequality and Young's inequality we have

$$\int_\mho [\nabla(\boldsymbol{u}+\boldsymbol{z}_1)\cdot \nabla_H] : \nabla(\boldsymbol{u}+\boldsymbol{z}_1)\cdot A_1\boldsymbol{u}$$
$$\leqslant |A_1\boldsymbol{u}|_2|\nabla_H\nabla(\boldsymbol{u}+\boldsymbol{z}_1)|_4|\nabla(\boldsymbol{u}+\boldsymbol{z}_1)|_4$$
$$\leqslant |A_1\boldsymbol{u}|_2|A_1(\boldsymbol{u}+\boldsymbol{z}_1)|_2^{1/4}|A_1^{3/2}(\boldsymbol{u}+\boldsymbol{z}_1)|_2^{3/4}|\nabla(\boldsymbol{u}+\boldsymbol{z}_1)|_2^{1/4}\|\boldsymbol{u}+\boldsymbol{z}_1\|_2^{3/4}$$
$$\leqslant \varepsilon |A_1^{3/2}(\boldsymbol{u}+\boldsymbol{z}_1)|_2^2 + C|A_1\boldsymbol{u}|_2^{8/5}|A_1(\boldsymbol{u}+\boldsymbol{z}_1)|_2^{2/5}|\nabla(\boldsymbol{u}+\boldsymbol{z}_1)|_2^{2/5}\|\boldsymbol{u}+\boldsymbol{z}_1\|_2^{6/5}$$
$$\leqslant \varepsilon |A_1^{3/2}\boldsymbol{u}|_2^2 + C\|\boldsymbol{z}_1\|_3^2 + C|A_1\boldsymbol{u}|_2^{8/5}\|\boldsymbol{u}+\boldsymbol{z}_1\|_1^{8/5}\|\boldsymbol{u}+\boldsymbol{z}_1\|_2^{2/5}$$
$$\leqslant \varepsilon |A_1^{3/2}\boldsymbol{u}|_2^2 + C\|\boldsymbol{z}_1\|_3^2 + C\|\boldsymbol{u}+\boldsymbol{z}_1\|_1^2 + C|A_1\boldsymbol{u}|_2^2\|\boldsymbol{u}+\boldsymbol{z}_1\|_2^2. \tag{3.20}$$

Thanks to Lemma 2.2 and Young's inequality we obtain

$$\int_\mho \left(\nabla w(\boldsymbol{u}+\boldsymbol{z}_1) : \partial_z\nabla(\boldsymbol{u}+\boldsymbol{z}_1)\right)\cdot A_1\boldsymbol{u}$$
$$\leqslant |A_1(\boldsymbol{u}+\boldsymbol{z}_1)|_2|\partial_z\nabla(\boldsymbol{u}+\boldsymbol{z}_1)|_2^{1/2}|\partial_z A_1(\boldsymbol{u}+\boldsymbol{z}_1)|_2^{1/2}|A_1\boldsymbol{u}|_2^{1/2}|A_1^{3/2}\boldsymbol{u}|_2^{1/2}$$
$$\leqslant \varepsilon |A_1^{3/2}\boldsymbol{u}|_2^2 + C|A_1(\boldsymbol{u}+\boldsymbol{z}_1)|_2^{4/3}|\partial_z\nabla(\boldsymbol{u}+\boldsymbol{z}_1)|_2^{2/3}$$
$$\times |\partial_z A_1(\boldsymbol{u}+\boldsymbol{z}_1)|_2^{2/3}|A_1\boldsymbol{u}|_2^{2/3}$$
$$\leqslant \varepsilon |A_1^{3/2}\boldsymbol{u}|_2^2 + C|A_1\boldsymbol{u}|_2^2\|\partial_z\boldsymbol{u}+\partial_z\boldsymbol{z}_1\|_1^{2/3}\cdot \|\partial_z\boldsymbol{u}+\partial_z\boldsymbol{z}_1\|_2^{2/3}$$
$$+C\|\boldsymbol{z}_1\|_2^{4/3}\|\boldsymbol{u}\|_2^{2/3}\|\partial_z\boldsymbol{u}+\partial_z\boldsymbol{z}_1\|_1^{2/3}\cdot \|\partial_z\boldsymbol{u}+\partial_z\boldsymbol{z}_1\|_2^{2/3}$$
$$\leqslant \varepsilon |A_1^{3/2}\boldsymbol{u}|_2^2 + C|A_1\boldsymbol{u}|_2^2\|\partial_z\boldsymbol{u}+\partial_z\boldsymbol{z}_1\|_1^{2/3}\cdot \|\partial_z\boldsymbol{u}+\partial_z\boldsymbol{z}_1\|_2^{2/3}$$
$$+C|A_1\boldsymbol{u}|_2^2 + C\|\boldsymbol{z}_1\|_2^2\|\partial_z\boldsymbol{u}+\partial_z\boldsymbol{z}_1\|_1\|\partial_z\boldsymbol{u}+\partial_z\boldsymbol{z}_1\|_2. \tag{3.21}$$

By the Hölder inequality, the interpolation inequality and Young's inequality we have

$$\int_\mho \left(A_1(\boldsymbol{u}+\boldsymbol{z}_1)\right)\cdot \nabla_H(\boldsymbol{u}+\boldsymbol{z}_1)\cdot A_1\boldsymbol{u}$$
$$\leqslant |A_1(\boldsymbol{u}+\boldsymbol{z}_1)|_2|\nabla_H(\boldsymbol{u}+\boldsymbol{z}_1)|_4|A_1\boldsymbol{u}|_4$$
$$\leqslant C|A_1(\boldsymbol{u}+\boldsymbol{z}_1)|_2|\nabla_H(\boldsymbol{u}+\boldsymbol{z}_1)|_4|A_1\boldsymbol{u}|_2^{1/4}|A_1|^{\frac{3}{2}}\boldsymbol{u}|_2^{3/4}$$
$$\leqslant \varepsilon |A_1^{3/2}\boldsymbol{u}|_2^2 + C|A_1(\boldsymbol{u}+\boldsymbol{z}_1)|_2^{8/5}|\nabla(\boldsymbol{u}+\boldsymbol{z}_1)|_2^{8/5}|A_1\boldsymbol{u}|_2^{2/5}$$
$$\leqslant \varepsilon |A_1^{3/2}\boldsymbol{u}|_2^2 + C(|A_1\boldsymbol{u}|_2^2 + \|\boldsymbol{z}_1\|_2^2)\|\boldsymbol{u}+\boldsymbol{z}_1\|_2^{6/5}\|\boldsymbol{u}+\boldsymbol{z}_1\|_1^{2/5}$$

$$\leqslant \varepsilon |A_1^{3/2} u|_2^2 + C(|A_1 u|_2^{16/5} + \|z_1\|_2^{16/5})\|u+z_1\|_1^{2/5}$$
$$\leqslant \varepsilon |A_1^{3/2} u|_2^2 + C|A_1 u|_2^2 \|u\|_2^2 + C\|z_1\|_2^4 + C\|u+z_1\|_2^4. \tag{3.22}$$

In view of Lemma 2.2 and Young's inequality we obtain

$$\int_\mho A_1 w(u+z_1) \cdot \partial_z(u+z_1)(A_1 u)$$
$$\leqslant C|A_1^{3/2}(u+z_1)|_2 |A_1 u|_2^{1/2} |A_1^{3/2} u|_2^{1/2} |\nabla(u+z_1)|_2^{1/2} |A_1(u+z_1)|_2^{1/2}$$
$$\leqslant \varepsilon |A_1^{3/2}(u+z_1)|_2^2 + C|A_1 u|_2^2 \|u+z_1\|_1^2 \|u+z_1\|_2^2. \tag{3.23}$$

From (3.37) − (3.44) we obtain the estimates of $J_1 + J_2$ that

$$J_1 + J_2 \leqslant C(\|z_1\|_3^4 + 1)(\|u\|_2^2 + 1)$$
$$+ C(\|z_1\|_2^2 + |A_1 u|_2^2)(\|u\|_1^2 + \|z_1\|_1^2 + 1)(\|u\|_2^2 + \|z_1\|_2^2 + 1)$$
$$+ C\|z_1\|_2^2(\|u_z\|_1 + \|\partial_z z_1\|_1)(\|u_z\|_2 + \|\partial_z z_1\|_2)$$

Combining the estimates of $J_1 - J_5$ and (3.36) as well as the Gronwall inequality yields

$$u \in L^\infty([0,T];\mathbb{H}^2) \cap L^2([0,T];\mathbb{H}^3).$$

Following the same steps as in (3.25) and (3.26) with minor revisions, we can further prove

$$u \in C([0,T];\mathbb{H}^2) \cap L^2([0,T];\mathbb{H}^3). \tag{3.24}$$

Taking inner product of (2.17) with $A_2^2 \vartheta$ in $\mathbf{H}$ yields,

$$\frac{1}{2}\partial_t |A_2 \vartheta|_2^2 + |A_2^{3/2} \vartheta|_2^2 = -\int_\mho [(u+z_1) \cdot \nabla_H](\vartheta+z_2) A_2^2 \vartheta$$
$$-\int_\mho w(u+z_1)\partial_z(\vartheta+z_2) A_2^2 \vartheta + \int_\mho w(u+z_1) A_2^2 \theta$$
$$=: K_1 + K_2 + K_3. \tag{3.25}$$

Taking a similar argument as in (3.37) yields,

$$K_1 + K_2$$
$$= -\int_\mho [A_2(u+z_1) \cdot \nabla_H](\theta+z_2) A_2 \theta - \int_\mho w(A_2(u+z_1))\partial_z(\vartheta+z_2) A_2 \vartheta$$

$$-\int_{\mho}[(u+z_1)\cdot\nabla_H](A_2\vartheta+A_2z_1)A_2\theta - \int_{\mho}w(u+z_1)\partial_z(A_2\vartheta+A_2z_2)A_2\theta$$
$$+\int_{\mho}\nabla(u+z_1)\nabla_H\nabla(\vartheta+z_2)A_2\vartheta + \int_{\mho}\nabla w(u+z_1)\partial_z(\nabla\vartheta+\nabla z_2)A_2\theta. \quad (3.26)$$

In the following, we will estimate each term on the right-hand side of (3.47) respectively. By integration by parts,

$$-\int_{\mho}u\cdot\nabla_H A_2\vartheta A_2\vartheta - \int_{\mho}w(u)\partial_z A_2\vartheta A_2\vartheta = 0. \quad (3.27)$$

In view of the Hölder inequality, Lemma 2.1 and Young's inequality, we have

$$-\int_{\mho}u\cdot\nabla_H A_2 z_2 A_2\vartheta - \int_{\mho}w(u)\partial_z A_2 z_2 A_2\vartheta$$
$$-\int_{\mho}z_1\cdot\nabla_H A_2(\vartheta+z_2)A_2\vartheta - \int_{\mho}w(z_1)\partial_z(A_2\vartheta+A_2z_2)A_2\vartheta$$
$$\leqslant C|u|_\infty\|z_2\|_3|A_2\vartheta|_2 + C\|z_2\|_3|A_2\vartheta|_2^{\frac{1}{2}}|A_2^{3/2}\vartheta|_2^{\frac{1}{2}}\|u\|_1^{\frac{1}{2}}\|u\|_2^{\frac{1}{2}}$$
$$+C\|z_1\|_3(|A_2^{3/2}\vartheta|_2 + \|z_2\|_3)|A_2\vartheta|_2$$
$$\leqslant \varepsilon|A_2^{3/2}\vartheta|_2^2 + C\|u\|_2^2 + C(\|z_1\|_3^4 + \|z_2\|_3^4 + 1)(|A_2\vartheta|_2^2 + 1). \quad (3.28)$$

With the aid of the Hölder inequality and Young's inequality, we deduce

$$-\int_{\mho}A_2(u+z_1)\cdot\nabla_H(\vartheta+z_2)A_2\vartheta$$
$$\leqslant|\nabla_H(\vartheta+z_2)|_\infty|A_2(u+z_1)|_2|A_2\vartheta|_2$$
$$\leqslant \varepsilon|A_2^{3/2}\vartheta|_2^2 + C\|z_2\|_3^2 + C(\|u\|_2^2 + \|z_1\|_2^2)|A_2\vartheta|_2^2. \quad (3.29)$$

By virtue of Lemma 2.1 and Young's inequality, we obtain

$$-\int_{\mho}w(A_2 u+A_2 z_1)\partial_z(\vartheta+z_2)A_2\vartheta$$
$$\leqslant C|A_2 u+A_2 z_1|_2^{1/2}\|u+z_1\|_3^{1/2}\|\vartheta+z_2\|_1^{1/2}\|\vartheta+z_2\|_2^{1/2}|A_2^{3/2}\vartheta|_2$$
$$\leqslant \varepsilon|A_2^{3/2}\vartheta|_2^2 + C\|u+z_1\|_2^2\|u+z_1\|_3^2$$
$$+C\|\vartheta+z_2\|_1^2\|\vartheta+z_2\|_2^2. \quad (3.30)$$

With the aid of the Hölder inequality, the interpolation inequality and Young's inequality we have

$$\int_{\mho}\nabla(u+z_1)\nabla_H\nabla(\vartheta+z_2)A_2\vartheta$$

$$\leqslant |A_2\vartheta|_2 |\nabla(\boldsymbol{u}+\boldsymbol{z}_1)|_4 |\nabla^2(\vartheta+z_2)|_4$$
$$\leqslant C|A_2\vartheta|_2 \|\boldsymbol{u}+\boldsymbol{z}_1\|_1^{1/4} \|\boldsymbol{u}+\boldsymbol{z}_1\|_2^{3/4} \|\vartheta+z_2\|_2^{1/4} |A_2^{3/2}(\vartheta+z_2)|_2^{3/4}$$
$$\leqslant \varepsilon |A_2^{3/2}\vartheta|_2^2 + C\|z_2\|_3^2 + C\|\vartheta+z_2\|_2^2 + C|A_2\vartheta|_2^2 \|\boldsymbol{u}+\boldsymbol{z}_1\|_2^2. \tag{3.31}$$

Taking a similar argument as in (3.42) yields,

$$\int_{\mho} \nabla w(\boldsymbol{u}+\boldsymbol{z}_1)\partial_z(\nabla\vartheta+\nabla z_2)A_2\vartheta$$
$$\leqslant |A_2\vartheta|_2 \|\vartheta+z_2\|_2^{1/2} |A_2^{3/2}(\vartheta+z_2)|_2^{1/2} \|\boldsymbol{u}+\boldsymbol{z}_1\|_2^{1/2} \|\boldsymbol{u}+\boldsymbol{z}_1\|_3^{1/2}$$
$$\leqslant \varepsilon |A_2^{3/2}\vartheta|_2^2 + C\|z_2\|_3^2 + C|A_2\vartheta|_2^2$$
$$+C(|A_2\vartheta|_2^2 + \|z_2\|_2^2)\|\boldsymbol{u}+\boldsymbol{z}_1\|_2^2 \|\boldsymbol{u}+\boldsymbol{z}_1\|_3^2. \tag{3.32}$$

In view of Young's inequality we get

$$K_3 \leqslant \varepsilon |A_2^{3/2}\vartheta|_2^2 + C\|\boldsymbol{u}+\boldsymbol{z}_1\|_2^2. \tag{3.33}$$

With the aid of (3.46) – (3.54), we obtain

$$\frac{1}{2}\partial_t |A_2\vartheta|_2^2 + |A_2^{3/2}\vartheta|_2^2$$
$$\leqslant C(1 + \|z_1\|_3^4 + \|z_2\|_3^4 + \|\boldsymbol{u}\|_2^2 + \|\vartheta\|_1^2)(|A_2\vartheta|_2^2 + 1)$$
$$+C(\|\boldsymbol{u}\|_2^2 + \|z_1\|_2^2)(\|\boldsymbol{u}\|_3^2 + \|z_1\|_3^2)(|A_2\vartheta|_2^2 + \|z_2\|_2^2 + 1).$$

By (2.15), (3.45) and the Gronwall inequality, we have

$$\vartheta \in L^\infty([0,T]; \mathbf{H}^2) \cap L^2([0,T]; \mathbf{H}^3).$$

Following the same steps as in (3.33) – (3.35) with minor revisions, we can further prove

$$\vartheta \in C([0,T]; \mathbf{H}^2) \cap L^2([0,T]; \mathbf{H}^3). \tag{3.34}$$

Then conclusions of Theorem 3.2 follow from (3.45) and (3.55). □

Theorem 3.3 Let $G \in L^2([0,T]; \mathbb{H}^1 \times \mathbf{H}^1), U_0 \in \mathbb{H}^1 \times \mathbf{H}^1$. Assume U be the solution to the linear primitive equations:

$$U'(t) + AU(t) + EU(t) = G,$$
$$U(0) = U_0.$$

For all time such that $U(t) \neq 0$, define

$$g(t) = \frac{\langle (A+E)U(t), U(t) \rangle}{|U(t)|_2^2}.$$

Then

$$g'(t) \leqslant \frac{|G(t)|_2^2}{|U(t)|_2^2}.$$

Proof Taking a similar argument as in Theorem 2.1, we can show that U satisfies

$$U \in C([0,T]; \mathbb{H}^1 \times \mathbf{H}^1) \cap L^2([0,T]; \mathbb{H}^2 \times \mathbf{H}^2).$$

Since $\langle EU(t), U(t) \rangle = 0$, we have

$$g(t) = \frac{\langle (A+E)U(t), U(t) \rangle}{|U(t)|_2^2} = \frac{\langle AU(t), U(t) \rangle}{|U(t)|_2^2}.$$

Here, as $|U(t)|_2$ is continuous on $[0, T]$, g is defined on the open set of $(0, T)$ which satisfies $|U(t)|_2 > 0$. The following calculations are carried out formally. To give a rigorous proof, we should perform the following calculations on the Galerkin approximation sequence and then extend the result to the infinite dimensional case by Aubin-Lions Lemma. Taking the derivative of g with respect to time t yields,

$$g'(t) = \frac{2\langle AU'(t), U(t) \rangle}{|U(t)|_2^2} - \frac{2\langle AU(t), U(t) \rangle \langle U'(t), U(t) \rangle}{|U(t)|_2^4}.$$

Since by integration by parts,

$$\langle AU'(t), U(t) \rangle = \langle G - AU(t) - EU(t), AU(t) \rangle$$
$$= \langle G - AU(t), AU(t) \rangle = \langle G, AU(t) \rangle - \langle AU(t), AU(t) \rangle$$

and

$$\langle AU(t), U(t) \rangle \langle U'(t), U(t) \rangle = \langle AU(t), U(t) \rangle \langle G - AU(t) - EU(t), U(t) \rangle$$
$$= \langle AU(t), U(t) \rangle \langle G - AU(t), U(t) \rangle,$$

we have

$$g'(t) = \frac{2\langle G, AU(t) \rangle - 2\langle AU(t), AU(t) \rangle}{|U(t)|_2^2}$$

$$-\frac{2\langle AU(t), U(t)\rangle\langle G - AU(t), U(t)\rangle}{|U(t)|_2^4}$$

$$= \frac{2\langle G, AU(t)\rangle - 2\langle AU(t), AU(t)\rangle}{|U(t)|_2^2}$$

$$+ \frac{2\langle AU(t), U(t)\rangle^2 - 2\langle G, U(t)\rangle\langle AU(t), U(t)\rangle}{|U(t)|_2^4}$$

$$= \frac{2\langle G, AU(t)\rangle - 2\langle AU(t), AU(t)\rangle}{|U(t)|_2^2}$$

$$+ \frac{2|\langle AU(t), U(t)\rangle - \frac{1}{2}\langle G, U(t)\rangle|^2}{|U(t)|_2^4} - 2\frac{\frac{1}{4}\langle G, U(t)\rangle^2}{|U(t)|_2^4}$$

$$\leqslant \frac{2\langle G, AU(t)\rangle - 2\langle AU(t), AU(t)\rangle}{|U(t)|_2^2}$$

$$+ \frac{2|AU(t) - \frac{1}{2}G|_2^2}{|U(t)|_2^2} + \frac{1}{2}\frac{|G|_2^2}{|U(t)|_2^2}$$

$$\leqslant \frac{|G|_2^2}{|U(t)|_2^2}.$$

$\square$

Theorem 3.4 *Given $(v_1(0), \theta_1(0))$ and $(v_2(0), \theta_2(0)) \in \mathbb{H}^1 \times \mathbf{H}^1$, let (v_1, θ_1) and (v_2, θ_2) be the two corresponding strong solutions to stochastic primitive equations with (2.14) holds. If we assume $v_1(t) = v_2(t)$ and $\theta_1(t) = \theta_2(t)$ for some $t > 0$, a.s., then*

$$\mathbb{P}(v_1(s) = v_2(s), \theta_1(s) = \theta_2(s), \quad \forall s \in [0, t]) = 1.$$

Proof Let $v = v_1 - v_2, \theta = \theta_1 - \theta_2, p_1 - p_2 = p$. Then (v, p, θ) satisfies

$$\partial_t v - \Delta v + v \cdot \nabla_H v_1 + v_2 \cdot \nabla_H v$$
$$+ w(v)\partial_z v_1 + w(v_2)\partial_z v + f v^\perp + \nabla_H p_s - \int_{-\frac{1}{2}}^z \nabla_H \theta dz' = 0, \quad (3.35)$$

$$\nabla_H \cdot v + \partial_z w(v) = 0, \quad (3.36)$$

$$\partial_t \theta - \Delta\theta + v \cdot \nabla_H \theta_1 + v_2 \cdot \nabla_H \theta + w(v)\partial_z \theta_1 + w(v_2)\partial_z \theta - w(v) = 0, \quad (3.37)$$

where $p = p_s - \int_{-1}^z \theta dz'$. For $t > 0$, define

$$\tilde{G}U = B(U, U_1) + B(U_2, U),$$

where $U_i = (\boldsymbol{v}_i, \theta_i), i = 1, 2$. Then, rewriting the above equations in the abstract form

$$U' + AU + EU + \tilde{G}U = 0. \tag{3.38}$$

Taking the inner product of (3.56) and (3.58) in $\mathbb{H}$ and $\mathbf{H}$ with $\boldsymbol{v}$ and ϑ respectively and adding the two equations yields,

$$\frac{1}{2}\partial_t(|\boldsymbol{v}|_2^2 + |\theta|_2^2) + (|\nabla\boldsymbol{v}|_2^2 + |\nabla\theta|_2^2)$$
$$= -\langle \boldsymbol{v} \cdot \nabla_H \boldsymbol{v}_1 + w(\boldsymbol{v})\partial_z \boldsymbol{v}_1, \boldsymbol{v}\rangle - \langle \boldsymbol{v} \cdot \nabla_H \theta_1 + w(\boldsymbol{v})\partial_z \theta_1, \theta\rangle := \langle \tilde{G}U, U\rangle. \tag{3.39}$$

In view of Theorem 3.3, we have

$$g'(t) \leqslant \frac{|\tilde{G}U|_2^2}{|U(t)|_2^2} \leqslant \frac{C\|U\|_1^2 \|U_1\|_3^2}{|U|_2^2} = Cg(t)\|U_1\|_3^2,$$

by the Gronwall inequality, we have

$$g(t) \leqslant g(t_0) e^{C\int_{t_0}^{t} \|U_1\|_3^2 ds} \leqslant C. \tag{3.40}$$

In the following we try to estimates $\log |U(t)|_2^2$

$$\frac{d \log |U(t)|_2^2}{dt} = \frac{-2(|\nabla \boldsymbol{v}|_2^2 + |\nabla \vartheta|_2^2) - 2\langle G(t)U, U\rangle}{|U(t)|_2^2}$$
$$\geqslant -2g(t) - 2\frac{\|U(t)\|_1 \|U_1(t)\|_3}{|U(t)|_2}$$
$$\geqslant -2g(t) - 2g^{\frac{1}{2}}(t)\|U_1(t)\|_3$$
$$\geqslant -C - C\|U_1(t)\|_3. \tag{3.41}$$

Since $U_1 \in L^2([0,T]; \mathbb{H}^3 \times \mathbf{H}^3)$, for arbitrary $t_0 \in [0,t)$, integrating (3.62) over $[t_0, t]$ yields

$$\log |U(t)|_2^2 \geqslant C(t - t_0) + \log |U(t_0)|_2^2, \quad \forall t \in [t_0, t]. \tag{3.42}$$

Hence, from (3.63) we see that If $\boldsymbol{v}_1(t) = \boldsymbol{v}_2(t)$ and $\theta_1(t) = \theta_2(t)$ for some $t > 0$ a.s., which is equivalent to $U(t) = 0$ a.s., then $U(t_0) = 0$ a.s.. Then the result of this theorem follows. □

References

[1] R. A. Adams, Sobolev Spaces, Academic Press, New York, 1975.

[2] H. Crauel, Markov measures for random dynamical systems, Stochastics Stochastics Rep. 3(1991), 153–173.

[3] H. Crauel, A. Debussche, F. Flandoli, Random attractors, J. Dynam. Differential Equations 9 (1997), 307–341.

[4] H. Crauel, F. Flandoli, Attractors for random dynamical systems, Probab. Theory Relat. Fields. 100(1994), 365–393.

[5] C. Cao, S. Ibrahim, K. Nakanishi, E.S. Titi, Finite-time blowup for the inviscid primitive equations of oceanic and atmospheric dynamics, Comm. Math. Phys. 337(2015), 473–482.

[6] C. Cao, J. Li, E.S. Titi, Global well-posedness of strong solutions to the 3D primitive equations with horizontal eddy diffusivity, J. Differential Equations 257 (2014), 4108–4132.

[7] C. Cao, J. Li, E.S. Titi, Local and global well-posedness of strong solutions to the 3D primitive equations with vertical eddy diffusivity, Arch. Ration. Mech. Anal. 214 (2014), 35–76.

[8] C. Cao, J. Li, E.S. Titi, Global well-posedness of the three-dimensional primitive equations with only horizontal viscosity and diffusion, Communications on Pure and Applied Mathematics Vol. LXIX(2016), 1492–1531.

[9] C. Cao, E.S. Titi, Global well-posedness of the three-dimensional viscous primitive equations of large scale ocean and atmosphere dynamics, Ann. of Math. 166(2007), 245–267.

[10] C. Cao, E.S. Titi, Global well-posedness of the 3D primitive equations with partial vertical turbulence mixing heat diffusion, Comm. Math. Phys. 310 (2012), 537–568.

[11] G. Da Prato, and J. Zabczyk, Stochastic equations in infinite dimensions, Cambridge University Press, Cambridge, 1992.

[12] A. Debussche, N. Glatt-Holtz, R. Temam, M. Ziane, Global existence and regularity for the 3D stochastic primitive equations of the ocean and atmosphere with multiplicative white noise, Nonlinearity 25(2012), 2093–2118.

[13] Z. Dong, R. Zhang, Markov selection and W-strong Feller for 3D stochastic primitive equations, Science China Mathematics 60(2017), 1873–1900.

[14] Z. Dong, J. Zhai, R. Zhang, Large deviation principles for 3D stochastic primitive equations, J. Differential Equations 263(2017), 3110-3146.

[15] Z. Dong, J. Zhai, R. Zhang, Exponential mixing for 3D stochastic primitive equations of the large scale ocean, Available at arXiv: 1506.08514.

[16] A. E. Gill, Atmosphere-ocean dynamics, International Geophysics Series, Vol. 30, Academic Press, San Diego, 1982.

[17] B. Guo and D. Huang, 3d stochastic primitive equations of the large-scale ocean: global well- posedness and attractors, Commun. Math. Phys. 286(2009), 697–723.

[18] N. Glatt-Holtz, I. Kukavica, V. Vicol, M. Ziane, Existence and regularity of invariant measures for the three dimensional stochastic primitive equations, J. Math. Phys. 55(2014), 051504.

[19] F. Guillén-González, N. Masmoudi, M.A. Rodríguez-Bellido, Anisotropic estimates and strong solutions for the primitive equations, Diff. Int. Equ. 14(2001), 1381–1408.

[20] Hongjun Gao, Chengfeng Sun, Well-posedness and large deviations for the stochastic primitive equations in two space dimensions, COMMUN. MATH. SCI. 10(2012), 575–593.

[21] Hongjun Gao, Chengfeng Sun, Well-posedness of stochastic primitive equations with multiplicative noise in three dimensions, Disc. and Cont. Dyn. Sys. B. 21(2016), 3053–3073.

[22] Hongjun Gao, Chengfeng Sun, Hausdorff dimension of random attractor for stochastic Navier-Stokes-Voight equations and primitive equations. Dyn Partial Differ Equ. 7(2010), 307–326.

[23] G. J. Haltiner, Numerical weather prediction, J.W. Wiley & Sons, New York, 1971.

[24] G. J. Haltiner and R. T. Williams, Numerical prediction and dynamic meteorology, John Wiley & Sons, New York, 1980.

[25] C. Hu, R. Temam, M. Ziane, The primimitive equations of the large scale ocean under the small depth hypothesis, Disc. and Cont. Dyn. Sys. 9(2003), 97–131.

[26] I. Kukavica and M. Ziane, On the regularity of the primitive equations of the ocean, Nonlinearity 20(2007), 2739–2753.

[27] K. Liu, Stability of stochastic differential equations in infinite dimensions, Springer Verlag, New York, 2004.

[28] J. Lions and B. Magenes, "Nonhomogeneous Boundary Value Problems and Applications," Springer-Verlag, New York, 1972.

[29] J.L. Lions, R. Temam and S. Wang, New formulations of the primitive equations of atmosphere and applications, Nonlinearity 5(1992), 237–288.

[30] J.L. Lions, R. Temam and S. Wang, On the equations of the large scale ocean, Nonlinearity 5(1992), 1007–1053.

[31] J.L. Lions, R. Temam and S. Wang, Models of the coupled atmosphere and ocean($CAOI$), Computational Mechanics Advance 1(1993), 1–54.

[32] J.L. Lions, R. Temam and S. Wang, Mathematical theory for the coupled atmosphere-ocean models ($CAOIII$), J. Math. Pures Appl. 74(1995), 105–163.

[33] M.Petcu, On the backward uniqueness of the primitive equations, J.Math.Pures Appl. 87(2007), 275–289.

Some Problems in Radiation Transport Fluid Mechanics and Quantum Fluid Mechanics*

Guo Boling（郭柏灵）, and Wu Jun（巫军）

Abstract We introduce the radiation transport equations, the radiation fluid mechanics equations and the fluid mechanics equations with quantum effects. We obtain the unique global weak solution for the radiation transport fluid mechanics equations under certain initial and boundary values. In addition, we also obtain that the periodic region problem of the compressible N-S equations with quantum effect has weak solutions under some conditions.

Keywords radiation transport equation; radiation fluid mechanics equations; fluid mechanics equations with quantum effects

1 Radiation transport equation and radiation fluid mechanics equations

A radiation transport equation is as follows

$$\frac{1}{c}\frac{\partial I(v,\Omega)}{\partial t}+\Omega\cdot\nabla I(v,\Omega) = S(v)-\sigma_a(v)I(v,\Omega)$$
$$+\int_0^\infty dv'\int_{S^{n-1}}\left[\frac{v}{v'}\sigma_s(v'\to v,\Omega'\cdot\Omega)I(v',\Omega')\right.$$
$$\left.-\sigma_s(v\to v',\Omega\cdot\Omega')I(v,\Omega)\right]d\Omega', \qquad (1.1)$$

where $I(v,\Omega) = I(x,t,v,\Omega)$ is radiation intensity, $S(v)$ is production rate of photons, $\sigma_a(v)$ is absorption rate, $\sigma_s(v)$ is scattering rate. In general, $\sigma_a = O(\rho^\alpha \theta^{-\beta})$, $\alpha > 0, \beta > 0$, where ρ is the density of matter, θ is the temperature of

---

* Annals of Applied Mathematics, 2019, 35(2): 111–125.

matter, and the radiation intensity of scattering out is

$$\int_0^\infty dv' \int_{S^{n-1}} \sigma_s(v \to v', \Omega \cdot \Omega') I(v, \Omega) \, d\Omega',$$

the radiation intensity of scattering in is

$$\int_0^\infty dv' \int_{S^{n-1}} \sigma_s(v' \to v, \Omega' \cdot \Omega) I(v', \Omega') \, d\Omega',$$

where S^{n-1} is a unit sphere in $\mathbf{R}^{n-1}$.

Define the absorption coefficient and compton scattering nucleus

$$\sigma_a(v) = c_1 \rho \theta^{-\frac{1}{2}} \exp\left[-\frac{c_2}{\theta^{\frac{1}{2}}} \left(\frac{v - v_0}{v_0}\right)^2\right],$$

$$\sigma_s(v \to v', \xi) = \frac{c_3 \rho (1 + \xi^2)}{[1 + \gamma(1 - \xi)]^2} \times \left\{1 + \frac{\gamma^{2(1-\xi)^2}}{(1 + \xi^2)[1 + \gamma(1 - \xi)]}\right\}$$

$$\times \delta\left(v' - \frac{v}{1 + \gamma(1 - \xi)}\right),$$

where $\gamma = c_4 v, \xi = \Omega \cdot \Omega', c_i$ is positive constant, v_0 is frequency. We now define the radiation energy density, radiation flux and radiation pressure as follows, respectively.

$$\begin{cases} E_r = \dfrac{1}{c} \displaystyle\int_0^\infty dv \int_{S^{n-1}} I(v, \Omega) \, d\Omega, \\ F_r = \displaystyle\int_0^\infty dv \int_{S^{n-1}} \Omega I(v, \Omega) \, d\Omega, \\ P_r = \dfrac{1}{c} \displaystyle\int_0^\infty dv \int_{S^{n-1}} \Omega \otimes \Omega I(v, \Omega) \, d\Omega. \end{cases} \quad (1.2)$$

The radiation transport fluid mechanics equations are

$$\begin{cases} \rho_t + \mathrm{div}(\rho u) = 0, \\ \left(\rho u + \dfrac{1}{c^2} F_r\right)_t + \nabla(\rho u \otimes u + P_m + P_r) = \mathrm{div} S, \\ \left(\dfrac{1}{2} \rho u^2 + E_m + E_r\right)_t + \nabla\left[\left(\dfrac{1}{2} \rho u^2 + E_m + P_m\right) u + F_r\right] = \mathrm{div}(Su + k \nabla \theta), \end{cases}$$

$$(1.3)$$

where $\rho, u, P_m = P_m(\rho, \theta), E_m = E_m(\rho, \theta)$ and θ are the density, speed, pressure, internal energy and temperature of fluid, respectively. $k = k(\rho, \theta)$ is the thermal conductivity of fluid, S is viscous tensor

$$S = \lambda(\text{div})I + \mu(\nabla u + (\nabla u)^T),$$

where λ, μ are viscous coefficients with $2\lambda + \mu > 0$.

The radiant transport equation through the absorption of photons and scattering interaction is

$$\frac{1}{c}\frac{\partial I(v,\Omega)}{\partial t} + \Omega \cdot \nabla I(v,\Omega)$$
$$= S(v)(1 + \frac{c^2 I(v,\Omega)}{2hv^3}) - \sigma_a(v)I(v,\Omega)$$
$$+ \int_0^\infty dv' \int_{S^{n-1}} \left[\sigma_s(v' \to v, \Omega' \cdot \Omega) I(v',\Omega')\left(1 + \frac{c^2 I(v,\Omega)}{2hv^3}\right)\right.$$
$$\left.-\sigma_s(v \to v', \Omega \cdot \Omega')I(v,\Omega)\left(1 + \frac{c^2 I(v',\Omega')}{2hv^3}\right)\right]d\Omega'.$$

From $S = \sigma_a B, \sigma = \sigma'_a(1 + \frac{c^2 I(v,\Omega)}{2hv^3})$, then we can get

$$\frac{1}{c}\frac{\partial I(v,\Omega)}{\partial t} + \Omega \cdot \nabla I(v,\Omega)$$
$$= \sigma'_a [B(v) - I(v,\Omega)]$$
$$+ \int_0^\infty dv' \int_{S^{n-1}} \left[\sigma_s(v' \to v, \Omega' \cdot \Omega)I(v',\Omega')\left(1 + \frac{c^2 I(v,\Omega)}{2hv^3}\right)\right.$$
$$\left.-\sigma_s(v \to v', \Omega \cdot \Omega')I(v,\Omega)\left(1 + \frac{c^2 I(v',\Omega')}{2hv^3}\right)\right]d\Omega'. \quad (1.4)$$

The radiation transport fluid mechanics system is formed by (1.4) and (1.2), where $B(v) = 2hv^3 c^{-2}(e^{\frac{hv}{\kappa\theta}} - 1)^{-1}$, h is plank constant, B is plank function, κ is Boltzmann constant.

Diffusion approximation

$$I(x,t,v,\Omega) = \frac{1}{4\pi}I_0(x,t,v) + \frac{3}{4\pi}\Omega I_1(x,t,v) \quad (1.5)$$

can be integrated as

$$I_0(x,t,v) = \int_{4\pi} d\Omega I(x,t,v,\Omega).$$

Multiplying (1.5) by Ω and integrating it, we obtain

$$I_1(x,t,v) = \int_{4\pi} d\Omega \, \Omega I(x,t,v,\Omega).$$

Putting the expressions of (1.5) and I_0, I_1 into (1.4), multiplying by Ω and integrating, we can omit $c^{-1}\partial_t I_1$ and assume that the scattering kernel is diagonal,

$$I_1(x,t,v) = -D(x,t,v)\nabla I_0(x,t,v). \quad (Fick\ law)$$

Putting (1.5) into (1.4), and integrating with respect to Ω and with the Fick law, we have

$$\frac{1}{c}\frac{\partial I(v,\Omega)}{\partial t} + \text{div}(\nabla I_0(v))$$
$$=\sigma_a'[4\pi B(v) - I_0(v)] - \sigma_s(v)I_0$$
$$+ \frac{C^2 I_0(v)}{8\pi h}\int_0^\infty dv' \left[\frac{\sigma_{s0}(v' \to v)}{v^2 v'} - \frac{\sigma_{s0}(v \to v')}{v'^3}\right]I_0. \qquad (1.6)$$

We call (1.6) as diffusion approximation of (1.4).

By balanced diffusion approximation, we can set $I_0(x,t,v) \sim I_0(v)$, then (1.6) can be represented by approximate radiation field

$$I_0(v) = 4\pi B = \frac{8\pi h v^3}{c^2}(e^{\frac{hv}{\kappa\theta}} - 1)^{-1},$$

and obtain

$$E_r = a\theta^4, \quad F_r = -\frac{ac}{3\sigma_\Omega}\nabla \theta^4, \quad P_r = \frac{a}{3}\theta^4, \qquad (1.7)$$

where $\sigma_\Omega \sim a\theta^m \rho^{-m}$ is Rosseland average.

Consider the boundary condition

$$I(x_s,v,\Omega,t) = \Gamma(x_s,v,\Omega,t), \quad n\cdot\Omega < 0, \qquad (1.8)$$

where we denote the known function, one point of the surface and the normal vector of this point as Γ, x_s, n. For the vacuum conditions of free surfaces of (1.8),

$$I(x_s,v,\Omega,t) = 0, \quad n\cdot\Omega < 0. \qquad (1.9)$$

2 Some related results

(1) Consider the systems

$$\begin{cases} \eta_t = v_x, \\ v_t = \sigma_x - \eta(S_F)_R, \\ \left(e + \frac{1}{2}v^2\right)_t = (\sigma v - q)_x - \eta(S_E) - R, \end{cases} \quad (2.1)$$

and the transport equation

$$I_t + \eta^{-1}(c\omega - v)i_x = cS. \quad (2.2)$$

with boundary conditions

$$v|_{x=0} = v|_{x=M} = 0, \quad q|_{x=0} = q|_{x=M} = 0, \quad (2.3)$$

$$I|_{x=0} = 0, \quad \omega \in (0,1), \quad I|_{x=M} = 0, \quad \omega \in (-1,0), \quad (2.4)$$

$$\eta|_{t=0} = \eta^0(x), \quad v|_{t=0} = v^0(x), \quad \theta|_{t=0} = \theta^0(x), \quad x \in \Omega,$$
$$I|_{t=0} = I^0(x,v,\omega), \quad \text{on } \Omega \times O, \quad (2.5)$$

where

$$(S_F)_R = \frac{1}{c}\int_{-1}^{1}\int_0^\infty \omega S(y,\tau,v,\omega)\,dv\,d\omega, \quad (S_E)_R = \int_{-1}^{1}\int_0^\infty S(y,\tau,v,\omega)\,dv\,d\omega,$$

$$S = S_{ae} + S_s,$$

$$S_{ae}(x,t,\Omega,v) = -\frac{v_0}{v}\sigma_a I(x,t,\Omega,v) + \left(\frac{v_0}{v}\right)^2 \sigma_a B(x,t,\Omega,v),$$

$$\eta = \frac{1}{\rho}, \quad \Omega = (0,M), \quad q = -\kappa\frac{\theta_x}{\eta}, \quad \sigma = -p + \mu\frac{v_x}{\eta}, \quad O = (0,\infty) \times S^1.$$

Theorem 2.1 *Under the initial condition*

$$\begin{cases} \eta^0 > 0, \quad \Omega, \eta^0 \in L^1(\Omega), \\ v^0 \in L^2(\Omega), \quad v_x^0 \in L^2(\Omega), \\ \theta^0 \in L^2(\Omega), \quad \inf_\Omega \theta^0 > 0, \quad I^0 \in L^1(\Omega \times O), \end{cases} \quad (2.6)$$

suppose that e, p, k satisfy some growth conditions, then there exists a unique global weak solution

$$\begin{cases} \eta \in L^\infty(Q_T), \quad \eta_t \in L^\infty([0,T], L^2(\Omega)), \\ v \in L^\infty([0,T], \quad L^4(\Omega)), v_t \in L^\infty([0,T], L^2(\Omega)), v_x \in L^\infty([0,T], L^2(\Omega)), \\ \sigma_x \in L^\infty([0,T], L^2(\Omega)), \theta \in L^\infty([0,T], L^q(\Omega)), \theta_x \in L^\infty([0,T], L^2(\Omega)), \\ I \in L^\infty([0,T], L^1(\Omega \times O)), \end{cases}$$

where $Q_T = \Omega \times (0,T)$.

The weak solution is

$$\eta(x,t) = \eta^0(x) + \int_0^t v_x \, ds,$$

$x \in \Omega, t > 0$, and for any test function $\phi \in L^2([0,T], H^1(\Omega)), \phi_t \in L^1([0,T], L^2(\Omega), \phi(\cdot,T) = 0$, which satisfy

$$\begin{cases} \int_Q \phi_t v + \phi_x p - \dfrac{\mu \phi_x}{\eta} v_x \, dx \, dt = \int_Q \phi(0,x) v^0(x) \, dx, \\ \int_Q \phi_t \left(e + \dfrac{1}{2} v^2\right) + \phi_x (\sigma v - q) \, dx \, dt = \int_Q \phi(0,x) \left(e + \dfrac{1}{2}(v^0)^2(x)\right) dx, \end{cases} \quad (2.7)$$

also for any test function $\psi \in L^2([0,T], H^1(\Omega \times O)), \psi_t \in L^1([0,T], L^2(\Omega \times O)), \phi(\cdot,T,\cdot,\cdot) = 0$, then we obtain

$$\int_{Q \times O} [\psi_t \eta I + \psi_x (v - c\omega)I + \psi \eta S] \, dv \, d\omega \, dx \, dt = \int_{Q \times O} \psi(0,x) \eta^0(x) I^0(x) \, dv \, d\omega \, dx. \tag{2.8}$$

For the system

$$\begin{cases} \eta_t = v_x, \\ v_t = \sigma_x - \eta(S_F)_R, \\ \left(e + \dfrac{1}{2} v^2\right)_t = (\sigma v - q)_x, \\ I_t + \eta^{-1}(c\omega - v)I_x = cS, \end{cases} \tag{2.9}$$

where $q = -\kappa \dfrac{\theta_x}{\eta}, \sigma = -p + \mu \dfrac{v_x}{\eta}, v \in R_+, Q = \Omega \times R_+, \Omega = (0,M), \omega \in (-1,1)$.

$$S(x,t,v,\omega) = \sigma_s(v;\eta,\theta)[\tilde{I}(x,t,v,\omega) - I(x,t,v,\omega)],$$

$$\tilde{I}(x,t,v) = \frac{1}{2}\int_{-1}^{1} I(x,t,v,\omega)\,d\omega, \quad \sigma_s > 0,$$

$$(S_F)_R = \frac{1}{c}\int_{-1}^{1}\int_{0}^{\infty} \omega S(x,t,v,\omega)\,dv\,d\omega.$$

Theorem 2.2 *Assume that the initial values of the systems (2.9) satisfy the following conditions*

$$\begin{cases} \eta^0 > 0, \Omega, \eta^0 \in L^1(\Omega), v^0 \in L^2(\Omega), v_x^0 \in L^2(\Omega), \\ \theta^0 \in L^2(\Omega), \inf_\Omega \theta^0 > 0, I^0 \in L^1(\Omega \times R_+ \times [-1,1]), \end{cases} \quad (2.10)$$

with boundary value conditions

$$\begin{cases} v|_{x=0} = v|_{x=M} = 0, q_{x=0} = q_{x=M} = 0, \\ I|_{x=0} = 0, \omega \in (0,1), I|_{x=M} = 0, \omega \in (-1,0), \\ I_\infty|_{x=0} = 0, \omega \in (0,1), I_\infty|_{x=M} = 0, \omega \in (-1,0), \end{cases} \quad (2.11)$$

and the initial condition

$$\eta|_{t=0} = \eta^0(x), v|_{t=0} = v^0(x), \theta|_{t=0} = \theta^0(x), \quad x \in \Omega, \quad (2.12)$$

then this system exists a unique global weak solution.

Theorem 2.3 *Under the above conditions, the solution of equation (2.9) tends to constant solution at $t \to \infty$. ($\eta_\infty, v_\infty = 0, \theta_\infty, I_\infty = 0$). We have the following estimate*

$$\|\eta - \eta_\infty\|_{L^2(\Omega)} + \|\theta - \theta_\infty\|_{L^2(\Omega)} + \|v\|_{L^2(\Omega)} + \|I - I_\infty\|_{L^2(\Omega)} \leq Ke^{-\Gamma t}, \quad (2.13)$$

where $\Gamma > 0, t \geq t_\infty$.

(2) We consider the following radiation transport fluid mechanic equation

$$\frac{1}{c}\frac{\partial I(v,\Omega)}{\partial t} + \Omega \cdot \nabla I(v,\Omega) = S(v) - \sigma_a(v)I(v,\Omega)$$

$$+ \int_0^\infty dv' \int_{S^{N-1}} \left[\frac{v}{v'}\sigma_s(v' \to v, \Omega' \cdot \Omega)I(v',\Omega') - \sigma_s(v \to v', \Omega \cdot \Omega')I(v,\Omega)\right]d\Omega',$$

$$(2.14)$$

where $I(v,\Omega) = I(x,t,v,\Omega), S^{N-1}$ is unit ball of S^N, $S(v) = S(x,t,v)$ is energy production rate. Consider

$$\begin{cases} \dfrac{\partial \rho}{\partial t} + \nabla \cdot (\rho u) = 0, \\ \dfrac{\partial}{\partial t}\left(\rho u + \dfrac{1}{c^2} F_r\right) + \nabla P_m + \nabla \cdot (\rho u \otimes u + P_r) = 0, \\ \dfrac{\partial}{\partial t}\left(\dfrac{1}{2}\rho u^2 + E_m + E_r\right) + \nabla \cdot \left[\left(\dfrac{1}{2}\rho u^2 + E_m + P_m\right)u + F_r\right] = 0 \end{cases} \quad (2.15)$$

with $E_m = c_v \rho \theta, P_m = R\rho\theta \equiv c_v(r-1)\rho\theta$.

We can rewrite (2.14) and (2.15) as

$$\frac{1}{c}\frac{\partial I(v,\Omega)}{\partial t} + \Omega \cdot \nabla I(v,\Omega) = S(v) - \sigma(v)I(v,\Omega) \\ + \int_0^\infty \int_{S^{N-1}} \frac{v}{v'}\sigma_s(v' \to v, \Omega' \cdot \Omega) I(v',\Omega')\, d\Omega'\, dv, \quad (2.16)$$

where $\sigma(v) = \sigma_a(v) + \sigma_s(v)$.

$$\sigma_s(v) = \int_0^\infty \int_{S^{N-1}} \sigma_s(v \to v', \Omega \cdot \Omega')\, d\Omega'\, dv'.$$

Denote $V = (\rho, u_1, \cdots, u_N, I), \tilde{A}_j(V) = (\tilde{a}_{mn})$, where $\tilde{a}_{ij} = u_j, \tilde{a}_{1,j+1} = \rho, \tilde{a}_{j+1,1} = k\dfrac{\theta}{\rho}, \tilde{a}_{j+1,N+2} = R, \tilde{a}_{N+2,j+1} = \dfrac{k\theta}{cv}$, $j = 1, \cdots, N$, and for other cases, $\tilde{a}_{ij} = 0$.

Set $G(v,I) = (g_0, g_1, \cdots, g_{N+1})^t, g_0 = 0, \quad j = 1, \cdots, N,$

$$g_j = -\frac{1}{c\rho}\left\{\int_0^\infty \int_{S^{N-1}} \Omega_j[S(v) - \sigma(v)I(v,\Omega)]\, d\Omega \\ + \int_0^\infty dv \int_{S^{N-1}} d\Omega \int_0^\infty dv' \int_{S^{N-1}} \Omega_j \frac{v}{v'}\sigma_s(v' \to v, \Omega' \cdot \Omega)I(v',\Omega')\, d\Omega'\right\},$$

$$g_{N+1} = \frac{1}{cc_v\rho}\left\{\int_0^\infty \int_{S^{N-1}} u \cdot \Omega[S(v) - \sigma(v)I(v,\Omega)]\, d\Omega \\ + \int_0^\infty dv \int_{S^{N-1}} d\Omega \int_0^\infty dv' \int_{S^{N-1}} u \cdot \Omega \frac{v}{v'}\sigma_s(v' \to v, \Omega' \cdot \Omega)I(v',\Omega')\, d\Omega'\right\}$$

$$-\frac{1}{c_{v\rho}}\left\{\int_0^\infty \int_{S^{N-1}} (S(v)-\sigma(v)I(v,\Omega))\,d\Omega\right.$$

$$\left. +\int_0^\infty dv \int_{S^{N-1}} d\Omega \int_0^\infty dv' \int_{S^{N-1}} \Omega_j \frac{v}{v'}\sigma_s(v'\to v,\Omega'\cdot\Omega)I(v',\Omega')\,d\Omega'\right\}.$$

Then we obtain

$$\frac{\partial V}{\partial t}+\sum_{j=1}^N \tilde{A}_j(V)\frac{\partial V}{\partial x_j}=G(V,I). \qquad (2.17)$$

We now consider the initial value of (2.16) and (2.17)

$$I(x,0,v,\Omega)=I_0(x,v,\Omega),\quad V(x,c)=V_0(x),\quad x\in R^N. \qquad (2.18)$$

Theorem 2.4 Assume that $s>\dfrac{N}{2}+1$, and

(A1) $S\in L^\infty(0,T;L^2(0,\infty;H^s(R^N)))$,

(A2) $\max_{(v,\Omega)\in(0,\infty)\times S^{N-1}}\|\sigma(\cdot,t,v,\Omega,\rho,\theta)-\bar\sigma\|_s-c\|\rho-\bar\rho\|_s\|\theta-\bar\theta\|_s$, $t>0$,

(A3)

$$\int_0^\infty dv\int_{S^{N-1}}\left(\int_0^\infty dv'\int_{S^{N-1}}\frac{v^2}{v'^2}\|\sigma_s(\cdot,t,v'\to v,\Omega'\cdot\Omega,\rho,\theta)\bar\sigma_s\|_s^2\,d\Omega'\right)^\lambda d\Omega$$

$$\leqslant C\Big(\|\rho-\bar\rho\|_s,\|\theta-\bar\theta\|_s\Big),\quad t\geqslant 0,(\lambda=\frac{1}{2},\ \text{or}\ \lambda=1).$$

where $(\rho-\bar\rho,\theta-\bar\theta)\in H^s(R^N)$, $M_1\leqslant \rho,\theta\leqslant M_2$ and

$$(V_0,I_0)\in G=\Big\{(V,I)|(\rho-\bar\rho,\theta-\bar\theta)\in H^s(R^N),$$

$$I(x,v,\Omega)\in L^2((0,\infty)\times S^{N-1},H^s(R^N)),$$

$$M_3\leqslant \int_0^\infty \int_{S^{N-1}} g\,dv\,d\Omega)\leqslant M_4\Big\}.$$

$\bar\rho,\bar\theta,M_i$ are positive constant, $\bar\sigma\equiv\sigma(x,t,v,\Omega,\bar\rho,\bar\theta)$, then we can choose $T>0$ such that there exists a unique smooth solution of (2.6)-(2.8), and $V\in C^1(R^N\times[0,T])$, $I\in C^1(R^N\times[0,T],(0,\infty)\times S^{N-1})$, $(V,I)\in G$, $\bar G_1\subset G$.

Remark 2.1 (A1), (A2) *can be satisfied,*

$$\sigma_a(v) = c_1\rho\theta^{-1}\exp\left[-\frac{c_2}{\theta^{\frac{1}{2}}}\left(\frac{v-v_0}{v}\right)^2\right],$$

$$\sigma_s(v \to v', \xi) = \frac{c_3\rho(1+\xi^2)}{[1+\gamma(1-\xi)]^2} \times \left\{1 + \frac{\gamma^2(1-\xi)^2}{(1+\xi^2)[1+\gamma(1-\xi)]}\right\} \sigma\left(v' - \frac{v}{1+\gamma(1-\xi)}\right),$$

where $\gamma = c_4 v, \xi = \Omega \cdot \Omega', c_i (i = 1, \cdots, 4)$ *is positive constant,* v_0 *is fixed frequency.*

Consider equations

$$\begin{cases} \dfrac{1}{c}\dfrac{\partial I}{\partial t} + \Omega \cdot \nabla I = \sigma'_a(v)(\bar{B}(v) - I), \\ A_0(v)\dfrac{\partial V}{\partial t} + \sum_{j=1}^{3} A_j(v)\dfrac{\partial V}{\partial x_j} = F(V, I) \end{cases} \tag{2.19}$$

with the initial value

$$I(x, 0, v, \Omega) = I_0(x, v, \Omega), V(X, 0) = V^0(x) = (\rho^0(x), u^0(x), \theta^0(x)),$$

$$A_0(v) = \begin{pmatrix} R\rho^{-1}\theta^2 & 0 & 0 \\ 0 & \rho\theta I_3 & 0 \\ 0 & 0 & c_v\rho \end{pmatrix}.$$

Symmetric matrix $A(v) = A_0 B_j(v), B_j(v) = (b_{mn})_{5\times 5}, b_{ij} = u_j, b_{1,j+1} = \rho, b_{j+1,1} = \dfrac{R\theta}{\rho}, b_{j+1,5} = R, b_{5,j+1} = \dfrac{R\theta}{cv}, j = 1, 2, 3$, the others are zeros,

$$F(V, I) = A_0(v) G(V, I) = (f_0, f_1, \cdots, f_{N+1})^t,$$

$$f_0 = 0, f_j = -\frac{\theta}{c}\int_0^\infty \int_{S^2} \Omega_j \sigma'_a(v)(\bar{B} - I) d\Omega, \quad j = 1, 2, 3,$$

$$f_4 = \frac{1}{c}\int_0^\infty dv \int_{S^2} u \cdot \Omega \sigma'_a(v)(\bar{B} - I) d\Omega - \int_0^\infty dv \int_{S^2} \sigma'_a(v)(\bar{B} - I) d\Omega.$$

Set

$$\begin{cases} I^0(x, v, \Omega) = \bar{B}(v), \quad V^0(x) = V \equiv (\bar{\rho}, 0, \bar{\theta}), \quad |x| \geqslant R_0, \\ \sigma'_a(v) > 0, \quad \rho^0(x) > 0, \quad \theta^0(x) > 0. \end{cases} \tag{2.20}$$

Denote
$$D(t) = \{x \in R^3 : |x| \geqslant R_0 + \beta t, \beta = (R^2 \bar{\theta} c_v^{-1} + R\bar{\theta})^{\frac{1}{2}}\},$$
$$E = \{(x,t) : x \in D(t), 0 \leqslant t \leqslant T\}.$$

Theorem 2.5 (finite speed of propagation) *Suppose $\beta \geqslant 1$ and (2.20) holds. If the solution of (2.19) satisfies $(V, I) \in C^1$, then $(V, I) \equiv (\bar{V}, \bar{I}) \in E$.*

Theorem 2.6 (blow up) *Suppose (2.20) holds, there exists solution $(\rho, u, \theta, I) \in C^1$ of (2.19), $0 \leqslant t \leqslant T$, where T is the maximum existence interval. If*

$$\begin{cases} 1 < \gamma < \dfrac{3}{5}, I^0(x, v, \Omega) \geqslant \bar{B}, m(0) \geqslant 0, \displaystyle\int_{R^3}(E_m^0 - \bar{E}_m)\,dx \geqslant 0, \\ F(0) \geqslant \dfrac{16}{3}\pi\beta R_0^4 \left\{\dfrac{2}{5-3\gamma}\max \rho^0(x) + \dfrac{1}{c^3}\max \displaystyle\int_0^\infty dv \int_{S^2}(I^0 - \bar{B})\,d\Omega \right\}, \end{cases} \quad (2.21)$$

then T is finite, $m(0) = \displaystyle\int_{S^2}(\rho_0 - \bar{\rho})\,dx$.

3 Plasma two-fluid with quantum effect equations

(1) Plasma two-fluid with quantum effect equations

$$\begin{cases} \dfrac{\partial n_e}{\partial t} + \dfrac{\partial (n_e u_e)}{\partial x} = 0, \\ \dfrac{\partial n_i}{\partial t} + \dfrac{\partial (n_i u_i)}{\partial x} = 0, \\ \dfrac{\partial u_e}{\partial t} + u_e \dfrac{\partial u_e}{\partial x} = \dfrac{e}{m_e}\dfrac{\partial \phi}{\partial x} - \dfrac{1}{m_e n_e}\dfrac{\partial P}{\partial x} + \dfrac{\hbar^2}{2m_e^2}\dfrac{\partial}{\partial x}\left(\dfrac{\partial^2 \sqrt{n_e}}{\partial x^2}/\sqrt{n_e}\right), \\ \dfrac{\partial u_i}{\partial t} + u_i \dfrac{\partial u_i}{\partial x} = -\dfrac{e}{m}\dfrac{\partial \phi}{\partial x}, \\ \dfrac{\partial^2 \phi}{\partial x^2} = \dfrac{e}{\varepsilon_0}(n_e - n_i). \end{cases} \quad (3.1)$$

where $n_e, n_i, u_{ei}, \hbar, \varepsilon_0, P = P(n_e)$ are electron number density, ion number density, electron ion fluid velocity, Planck constant, vacuum constant, and electronic pressure, respectively.

(2) Quantum KdV equation

Expand this equation according to the equilibrium

$$\begin{cases} n_i = 1 + \varepsilon n_{i1} + \varepsilon^2 n_{i2} + \cdots, \\ u_i = \varepsilon u_{i1} + \varepsilon^2 u_{i2} + \cdots, \\ n_e = \varepsilon n_{e1} + \varepsilon^2 n_{e2} + \cdots, \\ \phi = \varepsilon n_{e1} + \dfrac{\varepsilon^2}{2}(n_{e1}^2 + 2n_{e2}) - \dfrac{H^2}{4}\dfrac{\partial^2}{\partial x^2}\left(\varepsilon n_{e1} + \varepsilon^2\left(n_{e2} - \dfrac{n_{e2}^2}{\delta}\right)\right) \\ \qquad + \dfrac{H^2}{8}\varepsilon^2 n_{e1}\dfrac{\partial^2 n_{e1}}{\partial x^2} + \cdots. \end{cases} \quad (3.2)$$

Suppose $\xi = \varepsilon^{\frac{1}{2}}(x-t), \tau = \varepsilon^{\frac{3}{2}}t, n_{e1} = m_{01} = u_{i1} = U(\xi, \tau)$, then

$$\frac{\partial U}{\partial \tau} + 2U\frac{\partial U}{\partial \xi} + \frac{1}{2}\left(1 - \frac{H^2}{4}\right)\frac{\partial^3 U}{\partial \xi^3} = 0, \quad (3.3)$$

where $H \ne 2, U = U(\xi - c\tau)$.

If $H < 2$,

$$U = \frac{3c}{2}\operatorname{sech}^2\left(\sqrt{\frac{c}{2}}\frac{(\xi - c\tau)}{(1 - H^2/4)^{1/2}}\right). \quad (3.4)$$

If $H > 2$,

$$U = e - \frac{3c}{2}\operatorname{sech}^2\left(\sqrt{\frac{c}{2}}\frac{(\xi - e^2)}{(H^2/4 - 1)^{1/2}}\right). \quad (3.5)$$

(3) Quantum electromagnetic fluid mechanics equations

$$\begin{cases} \dfrac{\partial n_e}{\partial t} + \nabla \cdot (n_e u_e) = 0, \\[4pt] \dfrac{\partial n_i}{\partial t} + \nabla \cdot (n_i u_i) = 0, \\[4pt] \dfrac{\partial u_e}{\partial t} + u_e \nabla u_e = -\dfrac{\nabla \rho_e}{m_e n_e} - \dfrac{e}{m}(E + u_e \times B) + \dfrac{\hbar^2}{2m_e^2}\nabla\left(\dfrac{\nabla^2 \sqrt{n_e}}{\sqrt{n_e}}\right) - V_{ei}(u_e - u_i), \\[4pt] \dfrac{\partial u_i}{\partial t} + u_i \nabla u_i = -\dfrac{\nabla \rho_i}{m_i n_i} + \dfrac{e}{m}(E + u \times B) + \dfrac{\hbar^2}{2m_i^2}\nabla\left(\dfrac{\nabla^2 \sqrt{n_i}}{\sqrt{n_i}}\right) - V_{ie}(u_i - u_e), \\[4pt] \nabla \cdot E = \dfrac{\rho}{\varepsilon_0}, \\[4pt] \nabla \cdot B = 0, \\[4pt] \nabla \times E = -\dfrac{\partial B}{\partial t}, \\[4pt] \nabla \times D = \mu_0 J + \mu_0 \varepsilon_0 \dfrac{\partial E}{\partial t}, \end{cases}$$

$$(3.6)$$

where $\rho = e(n_i - n_e)$, $J = e(n_i u_i - n_e u_e)$.

Using
$$\bar{\rho}_m = m_e n_e + m_i n_i, \quad \bar{U} = \frac{m_e n_e U_e + m_i n_i U_i}{m_e n_e + m_i n_i},$$

and the dimensionless quantity method, we can simplify the model as follows,

$$\begin{cases} \dfrac{\partial \rho_m}{\partial t} + \nabla \cdot (\rho_m \cdot U) = 0, \\ \rho_m \left(\dfrac{\partial U}{\partial t} + U \nabla U \right) = -\dfrac{V_s^2}{V_A^2} \nabla \rho_m + (\nabla \times B) \times B + \dfrac{H^2}{2} \rho_m \nabla \left(\dfrac{\nabla^2 (\rho_m)^{1/2}}{\rho_m^{1/2}} \right), \\ \dfrac{\partial B}{\partial t} = \nabla \times (U \times B), \end{cases} \quad (3.7)$$

where $H = \dfrac{\hbar \Omega_i}{\sqrt{m_e m_i}} \sqrt{A}$, $V_A = (B_0^2/(u_0 \rho_0))^{1/2}$, $\Omega_i = eB_0/m_i$.

(4) One-dimensional and three-dimensional quantum Zakharov equations.
One-dimensional quantum Zakharov equations are

$$\begin{cases} i\dfrac{\partial E}{\partial t} + \dfrac{\partial^2 E}{\partial x^2} - H^2 \dfrac{\partial^4 E}{\partial x^4} = nE, \\ \dfrac{\partial^2 n}{\partial t^2} - \dfrac{\partial^2 n}{\partial u^2} + H^2 \dfrac{\partial^4 n}{\partial x^4} = \dfrac{\partial^2 |E|^2}{\partial x^2}. \end{cases} \quad (3.8)$$

One-dimensional quantum nonlinear Schrödinger equation is

$$i\dfrac{\partial E}{\partial t} + \dfrac{\partial^2 E}{\partial x^2} + |E|^2 E = H^2 \left(\dfrac{\partial^4 E}{\partial x^4} - E \dfrac{\partial^2 |E|^2}{\partial x^2} \right).$$

Three-dimensional quantum Zakharov system is

$$\begin{cases} i\dfrac{\partial \varepsilon}{\partial t} - \dfrac{5e^2}{3v_{F_e}^2} \nabla \times (\nabla \times \varepsilon) + \nabla(\nabla \cdot \varepsilon) = n\varepsilon + H\nabla[\nabla^2(\nabla \cdot \varepsilon)], \\ \dfrac{\partial^2 n}{\partial t^2} - \nabla^2 n - \nabla^2(|\varepsilon|^2) + H\nabla^4 n = 0, \end{cases} \quad (3.9)$$

where
$$H = \dfrac{m_e}{m_i} \left(\dfrac{5\hbar w}{3k_B T_{F_e}} \right)^2.$$

$$\begin{cases} \dfrac{\partial \alpha}{\partial t} = n + |\varepsilon|^2 - H\nabla^2 n, \\ \dfrac{\partial n}{\partial t} = \nabla^2 \alpha, \end{cases} \quad (3.10)$$

where

$$N = \int |\varepsilon|^2\, dr, \quad P_i = \int \left[\frac{i}{2}\left(\varepsilon_j \frac{\partial \varepsilon_j^*}{\partial r_i} - \varepsilon_j^* \frac{\partial \varepsilon_j}{\partial r_i}\right) - n\frac{\partial \alpha}{\partial r_i}\right] dr,$$

$$H = \int \left[n|\varepsilon|^2 + \frac{5c^2}{3V_{F_e}^2}|\nabla\times\varepsilon|^2 + |\nabla\cdot\varepsilon|^2 + H|\nabla(\nabla\cdot\varepsilon)|^2\right.$$
$$\left. + \frac{1}{2}(n^2 + H\nabla\eta)^2 + |\nabla\alpha|^2\right] dv.$$

Consider two-dimensional and three-dimensional QVNLLS

$$i\frac{\partial \varepsilon}{\partial t} + \nabla(\nabla\cdot\varepsilon) - \frac{5c^2}{3V_{F_e}^2}(\nabla\times(\nabla\times\varepsilon)) + |\varepsilon|^2\varepsilon = H\nabla(\nabla^2(\nabla\cdot\varepsilon)) - H\varepsilon\nabla^2(|\varepsilon|^2). \quad (3.11)$$

Two-dimensional variational solution is

$$\varepsilon = \left(\frac{W}{\pi}\right)^{1/2} \frac{1}{\sigma} \exp\left(-\frac{\rho^2}{2\sigma^2}\right) \exp(i(\theta + h\rho^2))(\cos\phi, \sin\phi, 0). \quad (3.12)$$

Three-dimensional variational solution is

$$\varepsilon = \left(\frac{W}{(\sqrt{\pi}\sigma)^3}\right)^{1/2} \exp\left[-\frac{r^2}{2\sigma^2} + e(\theta + hr^2)\right](\cos\phi\sin\theta, \sin\phi\sin\theta, \cos\theta). \quad (3.13)$$

4 Some results

(1) Consider the periodic region problem of the compressible N-S equation with the following quantum effects

$$\begin{cases} n_t + \mathrm{div}(nu) = 0, \quad x \in T^d, t > 0, \\ (nu)_t + \mathrm{div}(nu\otimes u) + \nabla P(n) - 2\varepsilon^2 n\nabla\left(\dfrac{\Delta\sqrt{n}}{\sqrt{n}}\right) - nf = 2v\mathrm{div}(nD(u)), \\ n(\cdot,0) = n_0, (nu)(\cdot,0) = n_0 u_0, \quad x \in T^d. \end{cases} \quad (4.1)$$

where $u \otimes u$ is the matrix of component $u_i u_j$, $D(u) = \dfrac{1}{2}(\nabla u + (\nabla u)^{\mathrm{T}})$, $d \leqslant 3$, $p(n) = n^\gamma$, $\gamma \geqslant 1$, $v > 0$.

From $(4.1)_1, (4.1)_2$, we obtain

$$E_\varepsilon(n,u) = \int_{T^d} \frac{n}{2}|u|^2 + H(n) + 2\varepsilon^2|\nabla\sqrt{n}|^2 \, dx, \quad (4.2)$$

where $H(n) = \dfrac{n^\gamma}{\gamma - 1}, \gamma > 1. H(n) = n(\log n - 1), \gamma = 1$.

Formally, for $f = 0$,

$$\frac{dE_\varepsilon}{dt}(n,u) + v\int_{T^d} n|\nabla u|^2 = 0. \quad (4.3)$$

In order to solve the compactness of $\nabla\sqrt{n}$, we introduce

$$\omega = u + v\nabla \log n, \quad (4.4)$$

then we can rewrite (4.1) as

$$\begin{cases} n_t + \mathrm{div}(n\omega) = v\Delta n, & x \in T^d, t > 0, \\ (n\omega)_t + \mathrm{div}(n\omega \otimes \omega) + \nabla P(n) - 2\varepsilon_0^2 n\nabla\left(\dfrac{\Delta\sqrt{n}}{\sqrt{n}}\right) - nf = v\Delta(n\omega), & (4.5) \\ n(\cdot,0) = n_0, (n\omega)(\cdot,0) = n_0\omega_0, & x \in T^d, \end{cases}$$

where $\omega_0 = u_0 + v\nabla\log n_0, \varepsilon_0^2 = \varepsilon^2 - v^2$. If $\varepsilon > v, f = 0$, then

$$\frac{dE_{\varepsilon_0}}{dt}(n,\omega) + v\int_{T^d}(n|\nabla u|^2 + H'(n)|\nabla u|^2 + \varepsilon_0^2 n|\nabla^2 \log n|^2) \, dx = 0. \quad (4.6)$$

Theorem 4.1 Assume that $d \leqslant 3, T > 0, \varepsilon_0, v > 0, p(n) = n^\gamma, \gamma > 3(d = 3), \gamma \geqslant 1(d = 2), f \in L^\infty(0,T;L^\infty(T^d))$ such that $n_0 \geqslant 0, E_{\varepsilon_0}(n_0,\omega_0)$ is finite, then there exists a weak solution of (4.5),

$$\begin{cases} \sqrt{n} \in L^\infty(0,T;H^1(T^d)) \cap L^2(0,T;H^2(T^d)), & n \geqslant 0, x \in T^d, \\ n \in H^1(0,T;L^2(T^d)) \cap L^\infty(0,T;L^\gamma(T^d)) \cap L^2(0,T;W^{1,3}(T^d)), \\ \sqrt{n}\omega \in L^\infty([0,T],L^2(T^d)), n\omega \in L^2(0,T;W^{1,\frac{3}{2}}(T^d)), n|\nabla\omega| \in L^2(0,T;L^2(T^d)), \end{cases}$$

which satisfies point for point in smooth experimental function ϕ with $\phi(\cdot,t) = 0$ satisfying

$$-\int_{T^d} n_0^2 u_0^2 \phi(\cdot,t) \, dx = \int_0^T (n^2\omega\phi_t - n^2\mathrm{div}(\omega)\phi - v(n\omega \otimes \nabla n)) : \nabla\phi,$$

$$+ n\omega \otimes n\omega : \nabla\phi + \frac{\gamma}{\gamma+1} n^{\gamma+1} \mathrm{div}\phi$$

$$- 2\varepsilon_0^2 \Delta \sqrt{n}(2\sqrt{n}\nabla n\phi + n^{\frac{3}{2}} \mathrm{div}\phi),$$

$$- n^2 f\phi - \upsilon \nabla(n\omega) : (n\nabla\phi + 2(\nabla\phi \otimes \phi)) \, dx \, dt. \qquad (4.7)$$

Corollary 4.1 *Assume that* $d \leqslant 3, T > 0, \varepsilon, \upsilon > 0, \varepsilon > \upsilon, p(n) = n^\gamma, \gamma > 3 (d=3), \gamma \geqslant 1 (d=2), f \in L^\infty(0,T; L^\infty(T^d))$ *such that* $n_0 > 0, E_\varepsilon(n_0, u_0 + \upsilon \nabla \log n_0) < \infty$, *then there exists a solution of* (4.1),

$$\sqrt{n}u \in L^\infty([0,T], L^2(T^d)), nu \in L^2(0,T; W^{1,\frac{3}{2}}(T^d)), n|\nabla u| \in L^2(0,T; L^2(T^d))$$

satisfying (4.1), *and for the experimental function* ϕ, $\phi(\cdot, t) = 0$ *satisfying*

$$-\int_{T^a} n_0^2 u_0^2 \cdot \phi(\cdot, 0) \, dx = \int_0^T \int_{T^a} (n^2 u - \phi_t - n^2 \mathrm{div}(u)u)$$
$$+ nu \otimes u\nabla\phi + \frac{\gamma}{\gamma+1} n^{\gamma+1} \mathrm{div}\phi - 2\varepsilon^2 \Delta n(2\sqrt{n} \cdot \nabla n \cdot \phi + n^{\frac{3}{2}} \mathrm{div}\phi)$$
$$- n^2 f \cdot \phi - \upsilon n D(u) \cdot (n\nabla\phi + \nabla n \otimes \phi) \, dx \, dt. \qquad (4.8)$$

(2) Consider the periodic initial value problem of the N-S equation with the following quantum effects

$$\begin{cases} n_t + \mathrm{div}(nu) = 0, \quad x \in T^d, t > 0, \\ (nu)_t + \mathrm{div}(nu \otimes u) + \nabla P(n) - \frac{\varepsilon^2}{6} n\nabla \left(\frac{\Delta\sqrt{n}}{\sqrt{n}}\right) - n\nabla V = 2\gamma \mathrm{div}(nD(u)), \\ n(\cdot, 0) = n_0, (nu)(\cdot, 0) = n_0 u_0, \quad x \in T^d, \end{cases}$$
$$(4.9)$$

where $P(n) = n^\gamma, n \geqslant 1, \mu(n) = \alpha n$, or $\mu(n) = \alpha$.

By $(4.9)_1, (4.9)_2$, we can get

$$E_{\varepsilon^2}(n,u) = \int_{T^d} \left(\frac{n}{2}|u|^2 + H(n) + \frac{\varepsilon^2}{\gamma}|\nabla\sqrt{n}|^2\right) dx, \qquad (4.10)$$

where $H(n) = n^\gamma/(\gamma-1), \gamma \neq 1; H(n) = n(\log n - 1), \gamma = 1$. If $\nabla V = 0$, then

$$\frac{dE_{\varepsilon^2(n,u)}}{dt} + \alpha \int_{T^d} n|D(u)|^2 \, dx = 0.$$

We introduce $w = u + \alpha \nabla \log \log n$,

$$\begin{cases} n_t + \text{div}(nu) = \alpha \Delta n, \\ (nu)_t + \text{div}(nw \otimes w) + \nabla P(n) - \dfrac{\varepsilon^2}{6} n\nabla \left(\dfrac{\Delta \sqrt{n}}{\sqrt{n}}\right) - n\nabla v = \alpha \Delta(nw), \end{cases} \quad (4.11)$$

where $w_0 = u_0 + \alpha \nabla \log n_0, \varepsilon_0 = \varepsilon^2 - 12\alpha^2$. If $\varepsilon^2 > 12\alpha^2$ and $\nabla v = 0$, then

$$\frac{dE_{\varepsilon^2}(n,u)}{dt} + \alpha \int_{T^d} (n|\nabla w|^2 + H'(n)|\nabla n|^2 + \frac{\varepsilon_0}{12} n |\nabla^2 \log n|^2) \, dx = 0.$$

Using the inequality

$$\int_{T^d} |\nabla_2 \sqrt{n}|^2 \, dx \leqslant C \int_{T^d} n |\nabla^2 \log n|^2 \, dx,$$

estimated by $L^2_{loc}(0,t; H^2(T^d))$ of $\sqrt{n}$. From the maximum principle of (4.10), if $n_0 > 0$, then $n > 0, w$ is smooth.

Theorem 4.2 *Assume that $d \leqslant 3, \alpha > 0, p(n) = n^\gamma, \gamma > 3(d=3), \gamma > 1(d=2), \nabla V \in C^\infty(0,\infty; L^\infty(T^d)), (n_0, u_0)$ such that $m_0 \geqslant 0, E_q(n_0, u_0 + \alpha \nabla \log n_0) < \infty$, then there exists a weak solution of (4.5) and*

$\sqrt{n} \in L^\infty_{loc}(0,\infty; H^1(T^d)) \cap L^2_{loc}(0,\infty; H^1(T^d)), n \geqslant 0, x \in L^\infty_{loc}(0,\infty; H^1(T^d))$,

$n \in H^1_{loc}(0,\infty; L^2(T^d)) \cap L^\infty_{loc}(0,\infty; L^\gamma(T^d)) \cap L^2_{loc}(0,\infty; W^{1,3}(T^d))$,

$\sqrt{n}u \in L^\infty_{loc}(0,\infty; L^2(T^d)), nu \in L^2_{loc}(0,\infty; W^{1,3}(T^d))$,

$n|\nabla u| \in L^2_{loc}(0,\infty; L^2(T^d))$.

References

[1] Mihalas D, Mihalas B. W. Foundations of radiation hydrodynamics[M]. New York: Oxford University Press, 1984.
[2] Chandrasekhar S. Radiative transfer[M]. New York: Dover Publications, 1960.
[3] Pomraning G. C. The equation of radiation hydrodynamics[M]. New York: Pergamon Press, 1973.
[4] Mihalas D. Radiation hydrodynamics[M]. Berlin: Springer, 1998.
[5] Thomas L. H. The radiation field in a fluid in motion[J]. Q. J. Math., 1930, 1: 239-251.
[6] Castor J. I. Radiative transfer in spherically symmetric flows[J]. Astrophys. J., 1972, 178: 779-792.

[7] Mihalas D. Solution of the comoving-frame equation of transfer in spherically symmetric flows[J]. Astrophys. J. 1980, 237: 574-589.

[8] Apruzese J. P, Whitney J. P, Davis J, Thornhill J. W. The physics of radiation transport in dense plasmas[J]. Phys. Plasmas., 2002, 9: 2411-2419.

[9] Duderstadt J. J, Martin W. R. Transport theory[M]. New York: John Wiley Sons. Inc, 1979.

[10] Gardner C. L. The quantum hydrodynamic model for semiconductor devices[J]. SIAM J. Appl. Math., 1994, 54: 409-427.

[11] Gamba I. M, Jüngel A, Vasseur A. Global existence of solutions to one-dimensional viscous quantum hydrodynamic equations[J]. J. Differential Equations, 2009, 247: 3117-3135.

[12] Zhang X, Jiang S. Local existence and finite-time blow-up in multidimensional radiation hydrodynamics[J]. J. Math. Fluid Mech., 2007, 9: 543-564.

Random Attractor for the 2D Stochastic Nematic Liquid Crystals Flows*

Guo Boling (郭柏灵), Han Yongqian (韩永前),
and Zhou Guoli (周国立)

Abstract We consider the long-time behavior for stochastic 2D nematic liquid crystals flows with the velocity field perturbed by an additive noise. The presence of the noises destroys the basic balance law of the nematic liquid crystals flows, so we can not follow the standard argument to obtain uniform a priori estimates for the stochastic flow even in the weak solution space under non-periodic boundary conditions. To overcome the difficulty we use a new technique some kind of logarithmic energy estimates to obtain the uniform estimates which improve the previous result for the orientation field that grows exponentially w.r.t.time t. Considering the existence of a random attractor, the common method is to derive uniform a priori estimates in functional space which is more regular than the solution space. We can follow the common method to prove the existence of a random attractor in the weak solution space. However, if we consider the existence of a random attractor in the strong solution space, it is very difficult and very complicated for such a highly non-linear stochastic system with no basic balance law and non-periodic boundary conditions. Here, we use compactness arguments of the stochastic flow and regularity of the solutions to the stochastic model to obtain the existence of the random attractor in the strong solution space, which implies the support of the invariant measure lies in a more regular space. As far as we know, it is the first article to attack the long-time behavior of stochastic nematic liquid crystals.

Keywords nematic liquid crystals; random attractor; invariant measure

---

* Commun. Pure Appl. Math. 2019, 18, 2349-2376.

1 Introduction

The paper is concerned with the following stochastic hydrodynamical model for the flow of nematic liquid crystals in $\mathbf{D} \times \mathbf{R}^+$, where $\mathbf{D} \subset \mathbf{R}^2$ is a bounded domain with smooth boundary Γ.

$$v_t + [(v \cdot \nabla)v - \mu \Delta v + \nabla p] + \lambda \nabla \cdot (\nabla d \odot \nabla d) = \dot{W}, \tag{1.1}$$

$$\nabla \cdot v(t) = 0, \tag{1.2}$$

$$d_t + [(v \cdot \nabla)d] - \gamma(\Delta d - f(d)) = 0. \tag{1.3}$$

The unknowns for the 2D stochastic hydrodynamical model are the fluid velocity field $v = (v^1, v^2) \in \mathbf{R}^2$, the averaged macroscopic/continuum molecular orientations $d = (d^1, d^2, d^3) \in \mathbf{R}^3$ and the scalar function $p(x,t)$ representing the pressure (including both the hydrostatic and the induced elastic part from the orientation field). The positive constants ν, λ and γ stand for viscosity, the competition between kinetic energy and potential energy, and macroscopic elastic relaxation time (Deborah number) for the molecular orientation field. Without loss of generality, we assume $\mu = \lambda = \gamma = 1$. W is a standard Wiener process in $\mathbf{H}$ defined below and has the form of

$$W(t) := \sum_{i=1}^{\infty} \lambda_i^{\frac{1}{2}} e_i B_i(t),$$

where $((B_i(t))_{t \in \mathbf{R}})_{i \in \mathbb{N}}$ is a sequence of one-dimensional, independent, identically distributed Brownian motions defined on the complete probability space $(\Omega, \mathcal{F}, \mathbb{P})$, $(e_i)_{i \in \mathbb{N}}$ is an orthonormal basis in $\mathbf{H}$, $(\lambda_i)_{i \in \mathbb{N}}$ is a convergent sequence of positive numbers which ensure W is a standard Wiener process in $\mathbf{H}$. Here $f : \mathbf{R}^2 \to \mathbf{R}^2$ is a general polynomial function whose details will be given later. The symbol $\nabla d \odot \nabla d$ denotes the 2×2 matrix whose (i,j)-th entry is given by

$$[\nabla d \odot \nabla d]_{i,j} = \sum_{k=1}^{3} \partial_{x_i} d^{(k)} \partial_{x_j} d^{(k)}, \quad i,j = 1,2.$$

In this paper, we consider the following initial boundary conditions for the stochastic nematic liquid crystals equations. The boundary conditions are

$$v(t,x) = 0, \quad d(x,t) = d_0(x), \quad \text{for } (x,t) \in \Gamma \times \mathbf{R}^+. \tag{1.4}$$

The initial conditions are

$$v|_{t=0} = v_0(x) \text{ with } \nabla \cdot v_0 = 0, \quad d|_{t=0} = d_0(x), \quad \text{for } x \in D, \qquad (1.5)$$

where n is the outward unit normal vector to Γ. In [21], F.H. Lin proposed a corresponding deterministic model of (1.1) – (1.3) as a simplified system of the original Ericksen-Leslie system (see [12,20]). By Ericksen-Leslie's hydrodynamical theory of the liquid crystals, the simplified system describing the orientation as well as the macroscopic motion reads as follows

$$dv + [(v \cdot \nabla)v - \mu\Delta v + \nabla p]dt = \lambda \nabla \cdot (\nabla d \odot \nabla d)dt, \qquad (1.6)$$

$$\nabla \cdot v(t) = 0, \qquad (1.7)$$

$$\partial_t d + [(v \cdot \nabla)d] = \gamma(\Delta d(t) + |\nabla d|^2 d), \quad |d| = 1. \qquad (1.8)$$

In order to avoid the nonlinear gradient in (1.8), one usually uses the Ginzburg-Landau approximation to relax the constraint $d = 1$. The corresponding approximate energy is

$$\int_D [\frac{1}{2}|\nabla d|^2 + \frac{1}{4\eta^2}(|d|^2 - 1)^2]dx \qquad (1.9)$$

where η is a positive constant. Then one arrives at the approximation system (1.6) – (1.8) with $f(d)$ and $F(d)$ given by

$$f(d) = \frac{1}{\eta^2}(|d|^2 - 1)d \text{ and } F(d) = \frac{1}{4\eta^2}(|d|^2 - 1)^2. \qquad (1.10)$$

In this work, we consider a more general polynomial function $f(d)$ which contains as a special case the (1.10). We define a function $\tilde{f}: [0, \infty) \to \mathbf{R}$ by

$$\tilde{f}(x) = \sum_{k=0}^{N} a_k x^k, \quad x \in \mathbf{R}_+, \qquad (1.11)$$

where $a_N > 0$ and $a_k \in \mathbf{R}$, $k = 0, 1, 2, \cdots, N - 1$. Let $f: \mathbf{R}^3 \to \mathbf{R}^3$ given by

$$f(d) = \tilde{f}(|d|^2)d. \qquad (1.12)$$

Denote by $F: \mathbf{R}^3 \to \mathbf{R}$ the Fréchet differentiable map such that for any $d \in \mathbf{R}^3$ and $\xi \in \mathbf{R}^3$

$$F'(d)[\xi] = f(d) \cdot \xi. \qquad (1.13)$$

Set $\tilde{F}$ to be an antiderivative of $\tilde{f}$ such that $\tilde{F}(0) = 0$. Then

$$\tilde{F}(x) = \sum_{k=0}^{N} \frac{a_k}{k+1} x^{k+1}, \quad x \in \mathbf{R}_+.$$

The Ericksen–Leslie system is well suited for describing many special flows for the materials, especially for those with small molecules, and is widely accepted in the engineering and mathematical communities studying liquid crystals. System (1.1)– (1.3) with $f(d)$ given by (1.10) can be possibly viewed as the simplest mathematical model, which keeps the most important mathematical structure as well as most of the essential difficulties of the original Ericksen–Leslie system (see [22]). This deterministic system with Dirichlet boundary conditions has been studied in a series of work not only theoretically (see [22], [23]) but also numerically (see [27], [28]).

The introduction of stochastic processes in nematic liquid crystals flows is aimed at accounting for a number of uncertainties and errors.

The state of the nematic liquid crystals is strongly dependent on the state of the environment. In natural systems external noise is often quite large. At an instability point the system is sensitive even to infinitesimally small perturbations and the role of external noise has to be investigated in the vicinity of a transition point. An experimental investigation also showed that, in the case of electrohydrodynamic instabilities, the average value of the voltage necessary for the transition is shifted to higher and higher values as the intensity of the external noise is increased i.e., the amplitude of the voltage fluctuations, is increased. Further study showed that the average voltage necessary to induce the transition to turbulent behavior, increases with the variance of the voltage fluctuations (see [16]). For more details one can also refer to [17, 33, 34].

Rheological predictions of the behavior of complex fluids like these, often start with the derivation of macroscopic, approximate equations for quantities of interest using various closure approximations. The difficulty in obtaining accurate closures has motivated the extensive, in recent years, use of direct simulations, either of the PDE governing the orientation distribution function, or of the equivalent stochastic differential equation, via "Brownian dynamics" simulations. The latter has the advantage that they are amenable to use with models with many internal degrees of freedom (as opposed to the PDE approach in which the "curse of dimensionality" precludes realistic computation). For more details one can see [13, 18, 26, 31, 32].

Despite the developments in the deterministic case, the theory for the stochastic nematic liquid crystals remains underdeveloped. To the best of our knowledge, there are few works on the stochastic nematic liquid crystals. In the papers [5, 6], Z.Brzezniak, E.Hausenblas and P.Razafimandimby have considered the model perturbed by multiplicative Gaussian noise and have proved the global well-posedness for the weak solution and strong solution in 2-D case. When the noise is jump and the dimension is two, Z.Brzezniak, U.Manna and A.A. Panda in [7] obtained the same result as the case of Gaussian noise. A weak martingale solution is also established for the three dimensional stochastic nematic liquid crystals with jump noise in [7].

One natural problem arising from this global existence result is the dynamical behavior of the 2D stochastic system.

As it is seen in the system (1.1) – (1.5), only the velocity field is disturbed by the noise. Why don't we consider the problem that both velocity field and orientation field are perturbed? This is because of the particular geometric structure of the nematic liquid crystals equations (see [22] for the *basic balance law*). Specifically, to obtain the energy estimates of velocity field in $\mathbf{H}$ we should estimate orientation field in the more regular space $\mathbb{H}^1$ at the same time. As we see, if (1.3) is also perturbed by the additive noise, by introducing two known Ornstein-Uhlenbeck processes z and z_1 with enough regularity and letting $\boldsymbol{u} = \boldsymbol{v} - \boldsymbol{z}$ and $\boldsymbol{d} - \boldsymbol{z}_1 = \theta$, (1.1) – (1.5) is equivalent to the following system,

$$\boldsymbol{u}_t + [(\boldsymbol{u}+\boldsymbol{z})\cdot\nabla(\boldsymbol{u}+\boldsymbol{z}) - \mu\Delta\boldsymbol{u} + \nabla p] + \lambda\nabla\cdot((\nabla\theta+\nabla\boldsymbol{z}_1)\odot(\nabla\theta+\nabla\boldsymbol{z}_1)) = 0,$$

$$\nabla\cdot\boldsymbol{u} = 0,$$

$$\theta_t + \boldsymbol{v}\cdot\nabla(\theta+\boldsymbol{z}_1) - \gamma\Delta\theta + \gamma f(\theta+\boldsymbol{z}_1) = 0,$$

where $\boldsymbol{u}$ and θ satisfy the following initial boundary conditions

$$\boldsymbol{u}(x,t)=0,\ \theta(x,t)=\boldsymbol{d}_0(x),\quad \forall (x,t)\in\Gamma\times\mathbf{R}_+,$$

$$\boldsymbol{u}|_{t=0} = \boldsymbol{v}_0(x) \text{ with } \nabla\cdot\boldsymbol{v}_0 = 0,\ \theta(x)|_{t=0} = \boldsymbol{d}_0(x),\quad \forall x\in\mathbf{D}.$$

For simplicity, we assume f is given by (1.10). Then we will arrive at some kind of the following estimates

$$\frac{1}{2}\frac{(|\theta|_{L^2}^2 + |\nabla\theta|_{L^2}^2 + |\boldsymbol{u}|_{L^2}^2)}{dt} + (|\nabla\theta|_{L^2}^2 + |\Delta\theta|_{L^2}^2 + |\nabla\boldsymbol{u}|_{L^2}^2 + |\theta|_{L^4}^4) \leqslant c|\theta|_{L^6}^6 + \cdots,$$

or

$$\frac{1}{2}\frac{(|\nabla\theta|_{\mathbb{L}^2}^2 + \int_D (|\theta|^2 - 1)^2 dx + |\boldsymbol{u}|_{\mathbb{L}^2}^2)}{dt} + (|\nabla \boldsymbol{u}|_{\mathbb{L}^2}^2 + |\Delta\theta - f(\theta)|_{\mathbb{L}^2}^2)$$
$$\leqslant c\int_D (|\theta|^2 - 1)^2 |\theta|_2^2 dx + \cdots,$$

where $|\cdot|_{\mathbb{L}^m}$ is the norm of the usual Lebesgue space $\mathbb{L}^m, m > 1$ and c is a positive constant that is bigger than one. From the above inequalities, we see that it is difficult to obtain the energy estimates of $(\boldsymbol{u}, \theta)$ or $(\boldsymbol{v}, \boldsymbol{d})$ if both velocity field and orientation field are perturbed by the noises. Therefore, in this article, we only consider the long-time behavior when the velocity field is perturbed.

The understanding of the asymptotic behavior of dynamical systems is one of the most important problems of modern mathematical physics. For the deterministic nematic liquid crystals equations, the existence of a global attractor for the strong solution is established (see [15,30]). For the corresponding stochastic model, there is no result about the existence of a random attractor. One reason is the absence of the *basic balance law* which results in the failure of applying the method of deterministic model to the stochastic model.

In this article, we firstly improve the bounds for the solutions to (1.1) – (1.5). These bounds are uniform with respect to present time and initial time (see lemma 3.1). These estimates of the uniform boundedness improve previous bounds obtained in [5-7], in which the bounds of $\boldsymbol{d}$ grow exponentially with respect to present time or initial time. In obtaining these time-uniform *a priori* estimates for orientation field $\boldsymbol{d}$ (for example in space $(L^2(D))^3$), the power of $\boldsymbol{d}$ from the nonlinear $f(\boldsymbol{d})$ will be much bigger (see (1.3)) than two. Using the Gronwall inequality (a standard argument) we will obtain that the solutions have exponential growth which is not sufficient to ensure the existence of random attractor. To overcome the difficulty, our ideal is that taking advantage of the property of the logarithmic function we reduce the power from nonlinear term. Roughly speaking, for a positive $f(t), t \in \mathbf{R}_+$, if we want to consider its uniform boundedness with respect to time t, we just need to estimate $\ln(1 + f(t))$. If $\ln(1 + f(t))$ is uniformly bounded with respect to t, so is $f(t)$. Using this new technique we obtain the uniform estimates for orientation field $\boldsymbol{d}$ (see Lemma 3.1) which opens a way to study the long-time behavior of stochastic nematic liquid crystals.

Our main goal of this article is to show the existence of a random attractor in the strong solution space $\mathbf{V} \times \mathbb{H}^2$. Although, we also obtain the existence of a random attractor in the weak solution space $\mathbf{H} \times \mathbb{H}^1$. As we know, the sufficient condition for ensuring the existence of a random attractor in $\mathbf{V} \times \mathbb{H}^2$ is to obtain an absorbing ball that is compact in $\mathbf{V} \times \mathbb{H}^2$. One method is to derive uniform a priori estimates in the functional space $\mathbf{H}^2 \times \mathbb{H}^3$. However, it is very difficult and very complicated for such *highly non-linear* stochastic system with *non-periodic boundary conditions*. Here, we use compactness arguments of the stochastic flow and regularity of the solutions to construct a compact absorbing ball in the strong solution space $\mathbf{V} \times \mathbb{H}^2$. For simplicity, we will outline the method of proving the existence of a random attractor in the weak solution space in detail. The method of proving the existence of random attractor in the strong solution space is the same as the case of weak solution space, but with more complicated calculus. We complete the proof of the existence of random attractor in the weak solution space by four steps. Firstly, using Lemma 3.1, we obtain the absorbing ball in the weak solution $\mathbf{H} \times \mathbb{H}^1$ (see Proposition 3.1). Secondly, we will verify the two a priori estimates of Aubin-Lions compact lemma to obtain a convergent subsequence of $(\boldsymbol{v}, \boldsymbol{d})$ which converges almost everywhere with respect to time $t \in [s,T], -\infty < s < T < \infty$ (see Proposition 3.3 and Proposition 3.4). Thirdly, by showing the continuity of the weak solution in the space $\mathbf{H} \times \mathbb{H}^1$ with respect to time t and with respect to the initial data $(\boldsymbol{v}_0, \boldsymbol{d}_0)$, we prove that the weak solution operators $S(t,s;\omega)$ are indeed a stochastic dynamic system and are almost surely compact in $\mathbf{H} \times \mathbb{H}^1$ for all (t,s) satisfying $-\infty < s < t < \infty$ (see Proposition 3.5). Finally, in Proposition 3.6 using the compact solution operator to act on the the absorbing ball yields a new set which is a compact and absorbing in $\mathbf{H} \times \mathbb{H}^1$. The existence of the random attractor in the weak solution space $\mathbf{H} \times \mathbb{H}^1$ follows directly from Proposition 3.6.

Remark 1.1 *Concerning the regularity of weak solution stated in the third step above, we show the following regularity in Corollary 3.1 for the weak solution to* (1.1) − (1.5)

$$\boldsymbol{v} \in C([0,T]; \mathbf{H}) \text{ and } \boldsymbol{d} \in C([0,T]; \mathbb{H}^1),$$

which is not available before. To show the existence of random attractor in the strong solution space $\mathbf{V} \times \mathbb{H}^2$, *the regularity for the strong solutions to* (1.1)−(1.5) *in* [5,6] *is not enough for our purpose. In* [5,6], *the strong solutions are Lipschitz* continuous in the space $\mathbf{H} \times \mathbb{H}^1$ with respect to initial data. To improve the regularity of the strong solutions, in Proposition 4.4, we obtain the continuity of the strong solution in $\mathbf{V} \times \mathbb{H}^2$ with respect to initial data. The new difficulty involved here in proving the continuity in space $\mathbf{V} \times \mathbb{H}^2$ with respect to initial data is the high non-linearity of the stochastic system. We overcome this difficulty by careful and delicate a priori estimates.

From the proof of the existence of random attractor in the strong solution space $\mathbf{V} \times \mathbb{H}^2$, one will see some advantages of our method over the method in [9, 10]. As we will see if the uniform a priori estimates for the stochastic flow to (1.1) − (1.5) in the strong solution space $\mathbf{V} \times \mathbb{H}^2$ hold, our method shows not only the existence of random attractor in the weak solution space $\mathbf{H} \times \mathbb{H}^1$ but also the existence of random attractor in the strong solution space $\mathbf{V} \times \mathbb{H}^2$. This improves the regularity of the support of the invariant measure which lies not only in $\mathbf{H} \times \mathbb{H}^1$ but also in $\mathbf{V} \times \mathbb{H}^2$ (see Remark 4.2). We should point out that support of the invariant measure in $\mathbf{V} \times \mathbb{H}^2$ may not be easily proved using the Krylov-Bogoliubov method. It is a new result for this stochastic model and may open a way to study the long-time behavior of stochastic nematic liquid crystals.

Recently, we noted these nice works [3, 4, 19] which study the existence of random attractor for stochastic hydrodynamic equations including stochastic Navier-Stokes equations defined on unbounded domains with an irregular noise. As we know, the study of long-term behavior for stochatic partial differential equations(SPDEs) defined on unbounded domains is difficult. Fortunately, a general frame which can be applied to prove the existence of random attractor for SPDEs defined on both the bounded case and the unbounded case is established in [3]. Furthermore, in [4], as an application of the beautiful notation 'asymptotic compactness random dynamical system' given in [4], the existence of invariant measure is obtained for stochastic Navier-Stokes equations defined on unbounded domains. This is the first result for stochastic hydrodynamic equations defined on unbounded domains. Finally, a natural question arises, could the method in this article be applied to SPDEs on unbounded domains? We think we need some work and we will extend the method to the unbounded domains in the next step.

The remaining of this paper is organized as follows. In section 2, we state some preliminaries and recall some results. The existence of a random attractor in weak solution space is presented in section 3, and section 4 is for the existence of a random attractor in the strong solution space. As usual, constants c may change from one line to the next, unless, we give a special declaration ; we denote a constant by $c(a)$, which depends on some parameter a.

2 Preliminaries

For $1 \leqslant p \leqslant \infty$, let $L^p(\mathbf{D})$ be the usual Lebesgue spaces with the norm $|\cdot|_p$. For a positive integer m, we denote by $(H^{m,p}(\mathbf{D}), \|\cdot\|_{m,p})$ the usual Sobolev spaces, see([1]). When $p = 2$, we denote by $(H^m(\mathbf{D}), \|\cdot\|_m)$ with inner product $\langle,\rangle_{H^m}$. Let

$$\mathcal{V} = \{v \in (C_0^\infty(\mho))^2 \ : \ \nabla \cdot v = 0\}.$$

We denote by $\mathbf{H}, \mathbf{V}$, and $\mathbf{H}^m$ be the closure spaces of $\mathcal{V}$ in $(L^2(\mathbf{D}))^2, (H^1(\mathbf{D}))^2$, and $(H^m(\mathbf{D}))^2$ respectively for positive integer $m \geqslant 2$. And set $|\cdot|_2$ and $\langle,\rangle$ to be the norm and inner product of $\mathbf{H}$ respectively. The notation $\langle,\rangle$ is also used to denote the inner product in $(L^2(\mathbf{D}))^2$. By the Poincaré inequality, there exists a constant c such that for any $v \in \mathbf{V}$, we have $\|v\|_1 \leqslant c|\nabla v|_2$. Without confusion, we let $\|\cdot\|_1$ and $\langle,\rangle_\mathbf{V}$ stand for the norm and the inner product in $\mathbf{V}$, respectively, where $\langle,\rangle_\mathbf{V}$ is defined by

$$\langle v_1, v_2\rangle_\mathbf{V} := \int_D \nabla v_1 \cdot \nabla v_2 dx, \quad \text{for } v_1, v_2 \in \mathbf{V}.$$

Denote by $\mathbf{V}'$ the dual space of $\mathbf{V}$. And define the linear operator $A_1 : \mathbf{V} \mapsto \mathbf{V}'$, as the following:

$$\langle A_1 v_1, v_2\rangle = \langle v_1, v_2\rangle_\mathbf{V}, \quad \text{for } v_1, v_2 \in \mathbf{V}.$$

Since the operator A_1 is positive selfadjoint with compact resolvent, by the classical spectral theorems, there exists a sequence $\{\alpha_j\}_{j \in \mathbb{N}}$ of eigenvalues of A_1 such that

$$0 < \alpha_1 \leqslant \alpha_2 \leqslant \cdots, \quad \alpha_j \to \infty$$

corresponding to the eigenvectors e_j which consist of an orthonormal basis of $\mathbf{H}$. Assume

$$\sum_{i=1}^\infty \lambda_i \alpha_i^2 < \infty. \tag{2.1}$$

For arbitrary constant $T > 0$ and $j \in \mathbb{N}$, we define

$$z^j(t) = \sum_{n=1}^{j} \sqrt{\lambda_n} \int_0^t e^{-A_1(t-s)} e_n dB_n(s), \quad t \in [0, T]$$

and

$$z(t) = \sum_{n=1}^{\infty} \sqrt{\lambda_n} \int_0^t e^{-A_1(t-s)} e_n dB_n(s), \quad t \in [0, T].$$

Obviously,

$$z^j(w) \in C([0, T]; \mathbf{H}^2), \quad \mathbb{P}\text{-a.e. } \omega \in \Omega.$$

For $k \in \mathbb{N}$ and $k > j$, in view of an infinite dimensional version of Burkholder-Davis-Gundy type of inequality for stochastic convolutions (see [11, 25] and references therein), we have

$$E \sup_{t \in [0,T]} \|A_1(z^j - z^k)\|_2^2$$

$$\leqslant CT \sum_{n=j+1}^{k} \lambda_n \alpha_n^2 \to 0, \quad as\ j \to \infty.$$

Therefore

$$z(w) \in C([0, T]; \mathbf{H}^2), \quad \mathbb{P}\text{-a.e. } \omega \in \Omega. \tag{2.2}$$

Obviously, z is the unique solution to the problem below

$$dz - \Delta z = dW_1 \tag{2.3}$$

$$z(x, t) = 0, \quad \forall (x, t) \in \Gamma \times \mathbf{R}_+, \tag{2.4}$$

$$z(x)|_{t=0} = 0, \quad \forall x \in \mathbf{D}. \tag{2.5}$$

Let $\mathbb{H}^m = (H^m(\mathbf{D}))^3, m = 0, 1, 2, \cdots$. When $m = 0$, set $\mathbb{H} = \mathbb{H}^0 = (L^2(\mathbf{D}))^3$ for simplicity. Denote the dual space of $\mathbb{H}^m$ by $\mathbb{H}^{-m}$. Then, similarly, we define the linear operator $A_2 : \mathbb{H}^1 \mapsto \mathbb{H}^{-1}$ as

$$\langle A_2 \mathbf{d}_1, \mathbf{d}_2 \rangle = \langle \mathbf{d}_1, \mathbf{d}_2 \rangle_{H^1}, \quad \text{for } \mathbf{d}_1, \mathbf{d}_2 \in \mathbb{H}^1.$$

Similarly, if we assume

$$\sum_{i=1}^{\infty} \lambda_i \alpha_i^3 < \infty, \tag{2.6}$$

we have that
$$z(\omega) \in C([0,T]; \mathbf{H}^3), \quad \text{P-a.e.} \omega \in \Omega.$$

Let $D(A_1) := \{\eta \in \mathbf{H}, A_1\eta \in \mathbb{H}\}$ and $D(A_2) := \{\theta \in \mathbb{H}^1, A_2\theta \in \mathbb{H}\}$. Because A_1^{-1} and A_2^{-1} are self-adjoint compact operators in $\mathbf{H}$ and $\mathbb{H}$ respectively, thanks to the classic spectral theory, we can define the power A_i^s for any $s \in \mathbf{R}$. Then $D(A_i)' = D(A_i^{-1})$ is the dual space of $D(A_i)$. Furthermore, we have the compact embedding relationship

$$D(A_1) \subset \mathbf{V} \subset \mathbf{H} \subset \mathbf{V}' \subset D(A_1)',$$

and

$$\langle \cdot, \cdot \rangle_{\mathbf{V}} = \langle A_1 \cdot, \cdot \rangle = \langle A_1^{\frac{1}{2}} \cdot, A_1^{\frac{1}{2}} \cdot \rangle.$$

Similarly,

$$D(A_2) \subset \mathbb{H}^1 \subset \mathbb{H} \subset \mathbb{H}^{-1} \subset D(A_2)',$$

and

$$\langle \cdot, \cdot \rangle_{\mathbb{H}^1} = \langle A_2 \cdot, \cdot \rangle = \langle A_2^{\frac{1}{2}} \cdot, A_2^{\frac{1}{2}} \cdot \rangle.$$

Definition 2.1 *We say a continuous $\mathbf{H} \times \mathbb{H}^1$ valued $(\mathcal{F}_t) = (\sigma(W(s), s \in [0,t]))$ adapted random field $(\boldsymbol{v}(.,t), \mathbf{B}(.,t))_{t \in [0,T]}$ defined on $(\Omega, \mathcal{F}, \mathbb{P})$ is a weak solution to problem $(1.1)-(1.5)$ if for $(\boldsymbol{v}_0, \boldsymbol{d}_0) \in \mathbf{H} \times \mathbb{H}^1$ the following conditions hold:*

$$\boldsymbol{v} \in C([0,T]; \mathbf{H}) \cap L^2([0,T]; \mathbf{V}),$$
$$\boldsymbol{d} \in C([0,T]; \mathbb{H}^1) \cap L^2([0,T]; \mathbb{H}^2),$$

and the integral relation

$$\langle \boldsymbol{v}(t), v \rangle + \int_0^t \langle A_1 \boldsymbol{v}(s), v \rangle ds + \int_0^t \langle \boldsymbol{v}(s) \cdot \nabla \boldsymbol{v}(s), v \rangle ds$$
$$+ \int_0^t \langle \nabla \cdot (\nabla \boldsymbol{d}(s) \odot \nabla \boldsymbol{d}(s)), v \rangle ds = \langle \boldsymbol{v}_0, v \rangle + \langle W(t), v \rangle,$$

$$\langle \boldsymbol{d}(t), d \rangle + \int_0^t \langle A_2 \boldsymbol{d}(s), d \rangle ds + \int_0^t \langle \boldsymbol{v}(s) \cdot \nabla \boldsymbol{d}(s), d \rangle ds$$
$$= \langle \boldsymbol{d}_0, d \rangle - \int_0^t \langle f(\boldsymbol{d}(s)), d \rangle ds,$$

hold a.s. for all $t \in [0,T]$ and $(v,d) \in \mathbf{V} \times \mathbb{H}$.

Definition 2.2 We say a continuous $\mathbf{V} \times \mathbb{H}^2$ valued $(\mathcal{F}_t) = (\sigma(W(s), s \in [0,t]))$ adapted random field $(\boldsymbol{v}(.,t), \mathbf{B}(.,t))_{t \in [0,T]}$ defined on $(\Omega, \mathcal{F}, \mathbb{P})$ is a strong solution to problem $(1.1) - (1.5)$ if for $(\boldsymbol{v}_0, \boldsymbol{d}_0) \in \mathbf{V} \times \mathbb{H}^2$ the following conditions hold:

$$\boldsymbol{v} \in C([0,T]; \mathbf{V}) \cap L^2([0,T]; \mathbf{H}^2),$$
$$\boldsymbol{d} \in C([0,T]; \mathbb{H}^2) \cap L^2([0,T]; \mathbb{H}^3),$$

and the integral relation

$$\boldsymbol{v}(t) + \int_0^t A_1 \boldsymbol{v}(s) ds + \int_0^t \boldsymbol{v}(s) \cdot \nabla \boldsymbol{v}(s) ds$$
$$+ \int_0^t \nabla \cdot (\nabla \boldsymbol{d}(s) \odot \nabla \boldsymbol{d}(s)) ds = \boldsymbol{v}_0 + W(t),$$
$$\boldsymbol{d}(t) + \int_0^t A_2 \boldsymbol{d}(s) ds + \int_0^t \boldsymbol{v}(s) \cdot \nabla \boldsymbol{d}(s) ds = \boldsymbol{d}_0 - \int_0^t f(\boldsymbol{d}(s)) ds,$$

hold a.s. for all $t \in [0,T]$.

Taking a similar argument as in [6] with minor revisions, we can obtain the following result concerning the global well-posedness of $(1.1) - (1.5)$.

Theorem 2.1 Let $(\boldsymbol{v}_0, \boldsymbol{d}_0) \in \mathbf{H} \times \mathbb{H}^1$ (or $(\boldsymbol{v}_0, \boldsymbol{d}_0) \in \mathbf{V} \times \mathbb{H}^2$). Assume conditions (2.14) (or (2.19)) hold. Then there exists a unique weak (or a unique strong) solution $(\boldsymbol{v}, \boldsymbol{d})$ of the system $(1.1) - (1.5)$ on the interval $[0,T]$, which is Lipschitz continuous with respect to the initial data in $\mathbf{H} \times \mathbb{H}^1$.

Remark 2.1 (1) In Theorem 3.2 of [5], it is proved that the weak solution $(\boldsymbol{v}, \boldsymbol{d})$ to $(1.1) - (1.5)$ satisfies

$$\boldsymbol{v} \in C([0,T]; \mathbf{V}^{-\beta}) \text{ and } \boldsymbol{d} \in C([0,T]; \mathbb{X}_{\beta - \frac{1}{2}}), \quad a.s., \quad \beta \in \left(0, \frac{1}{2}\right).$$

In Corollary 3.1, we improve the regularity and obtain that

$$\boldsymbol{v} \in C([0,T]; \mathbf{H}), \text{ and } \boldsymbol{d} \in C([0,T]; \mathbb{H}^1), \quad a.s.,$$

which ensures the solution operator is a time continuous stochastic dynamical system. And using this continuity, we can prove that, the solution operator is compact in $\mathbf{H} \times \mathbb{H}^1$, almost surely.

(2) In [5], it is proved that the strong solution (see Definition 2.2) is continuous with respect to initial data in $\mathbf{H} \times \mathbb{H}^1$. This kind of continuous

dependence on the initial data is not enough for us to prove the existence of random attractor in the strong solution space. We will show in Proposition 4.4 that the strong solution to (1.1) – (1.5) is continuous with respect to initial data in $\mathbf{V} \times \mathbb{H}^2$.

In the following, we recall the notations and results in stochastic dynamical systems which will be used to prove the main results of this article.

Let (X, d) be a Polish space and $(\tilde{\Omega}, \tilde{\mathcal{F}}, \tilde{\mathbb{P}})$ be a probability space, where $\tilde{\Omega}$ is the two-sided Wiener space $C_0(\mathbf{R}; X)$ of continuous functions with values in X, equal to 0 at $t = 0$. We consider a family of mappings $S(t, s; \omega) : X \to X$, $-\infty < s \leqslant t < \infty$, parametrized by $\omega \in \tilde{\Omega}$, satisfying for $\tilde{\mathbb{P}}$-a.e. ω the following properties (i)-(iv):

(i) $S(t, r; \omega)S(r, s; \omega)x = S(t, s; \omega)x$ for all $s \leqslant r \leqslant t$ and $x \in X$;

(ii) $S(t, s; \omega)$ is continuous in X, for all $s \leqslant t$;

(iii) for all $s < t$ and $x \in X$, the mapping

$$\omega \mapsto S(t, s; \omega)x$$

is measurable from $(\tilde{\Omega}, \tilde{\mathcal{F}})$ to $(X, \mathcal{B}(X))$ where $\mathcal{B}(X)$ is the Borel-σ- algebra of X;

(iv) for all $t, x \in X$, the mapping $s \mapsto S(t, s; \omega)$ is right continuous at any point.

A set valued map $K : \tilde{\Omega} \to 2^X$ taking values in the closed subsets of X is said to be measurable if for each $x \in X$ the map $\omega \mapsto d(x, K(\omega))$ is measurable, where $d(A, B) = \sup\{\inf\{d(x, y) : y \in B\} : x \in A\}$ for $A, B \in 2^X, A, B \neq \emptyset$; and $d(x, B) = d(\{x\}, B)$. Since $d(A, B) = 0$ if and only if $A \subset B$, d is not a metric. A closed set valued measurable map $K : \tilde{\Omega} \to 2^X$ is named a random closed set.

Given $t \in \mathbf{R}$ and $\omega \in \tilde{\Omega}$, $K(t, \omega) \subset X$ is called an attracting set at time t if, for all bounded sets $B \subset X$,

$$d(S(t, s; \omega)B, K(t, \omega)) \to 0, \quad \text{provided } s \to -\infty.$$

Moreover, if for all bounded sets $B \subset X$, there exists $t_B(\omega)$ such that for all $s \leqslant t_B(\omega)$,

$$S(t, s; \omega)B \subset K(t, \omega),$$

we say $K(t, \omega)$ is an absorbing set at time t.

Let $\{\vartheta_t : \tilde{\Omega} \to \tilde{\Omega}\}, t \in T, T = \mathbf{R}$, be a family of measure preserving transformations of the probability space $(\tilde{\Omega}, \tilde{\mathcal{F}}, \tilde{\mathbb{P}})$ such that for all $s < t$ and $\omega \in \tilde{\Omega}$,

(a) $(t, \omega) \to \vartheta_t \omega$ is measurable;
(b) $\vartheta_t(\omega)(s) = \omega(t+s) - \omega(t)$;
(c) $S(t, s; \omega)x = S(t-s, 0; \vartheta_s \omega)x$.

Thus $(\vartheta_t)_{t \in T}$ is a flow, and $((\tilde{\Omega}, \tilde{\mathcal{F}}, \tilde{\mathbb{P}}), (\vartheta_t)_{t \in T})$ is a measurable dynamical system.

Definition 2.3 *Given a bounded set $B \subset X$, the set*

$$\mathcal{A}(B, t, \omega) = \bigcap_{T \leqslant t} \overline{\bigcup_{s \leqslant T} S(t, s, \omega) B}$$

is said to be the Ω-limit set of B at time t. Obviously, if we denote $\mathcal{A}(B, 0, \omega) = \mathcal{A}(B, \omega)$, we have $\mathcal{A}(B, t, \omega) = \mathcal{A}(B, \vartheta_t \omega)$.

We may identify

$$\mathcal{A}(B, t, \omega) = \{x \in X : \text{there exist } s_n \to -\infty \text{ and } x_n \in B \\ \text{such that } \lim_{n \to \infty} S(t, s_n, \omega) x_n = x\}.$$

Furthermore, if there exists a compact attracting set $K(t, \omega)$ at time t, it is not difficult to check that $\mathcal{A}(B, t, \omega)$ is a nonempty compact subset of X and $\mathcal{A}(B, t, \omega) \subset K(t, \omega)$.

Definition 2.4 *If, for all $t \in \mathbf{R}$ and $\omega \in \tilde{\Omega}$, the random closed set $\omega \to \mathcal{A}(t, \omega)$ satisfying the following properties:*

(1) $\mathcal{A}(t, \omega)$ is a nonempty compact subset of X,

(2) $\mathcal{A}(t, \omega)$ is the minimal closed attracting set, i.e., if $\tilde{\mathcal{A}}(t, \omega)$ is another closed attracting set, then $\mathcal{A}(t, \omega) \subset \tilde{\mathcal{A}}(t, \omega)$,

(3) it is invariant, in the sense that, for all $s \leqslant t$,

$$S(t, s; \omega) \mathcal{A}(s, \omega) = \mathcal{A}(t, \omega),$$

$\mathcal{A}(t, \omega)$ *is called the random attractor.*

Let

$$\mathcal{A}(\omega) = \mathcal{A}(0, \omega).$$

Then the invariance property writes

$$S(t, s; \omega) \mathcal{A}(\vartheta_s \omega) = \mathcal{A}(\vartheta_t \omega).$$

To prove the existence of the random attractor, we will use the following sufficient condition given in [9]. For convenience, we cite it here.

Theorem 2.2 Let $(S(t,s;\omega))_{t \geqslant s, \omega \in \tilde{\Omega}}$ be a stochastic dynamical system satisfying (i), (ii), (iii) and (iv). Assume that there exists a group $\vartheta_t, t \in \mathbf{R}$, of measure preserving mappings such that condition (c) holds and that, for $\tilde{\mathbb{P}}$-a.e. ω, there exists a compact attracting set $K(\omega)$ at time 0. For $\tilde{\mathbb{P}}$-a.e. ω, we set

$$\mathcal{A}(\omega) = \overline{\bigcup_{B \subset X} \mathcal{A}(B, \omega)}$$

where the union is taken over all the bounded subsets of X. Then we have for $\tilde{\mathbb{P}}$-a.e. $\omega \in \tilde{\Omega}$.

(1) $\mathcal{A}(\omega)$ is a nonempty compact subset of X, and if X is connected, it is a connected subset of $K(\omega)$.

(2) The family $\mathcal{A}(\omega)$, $\omega \in \Omega$, is measurable.

(3) $\mathcal{A}(\omega)$ is invariant in the sense that

$$S(t,s;\omega)\mathcal{A}(\vartheta_s\omega) = \mathcal{A}(\vartheta_t\omega), \quad s \leqslant t.$$

(4) It attracts all bounded sets from $-\infty$: for bounded $B \subset X$ and $\omega \in \tilde{\Omega}$

$$d(S(t,s;\omega)B, \mathcal{A}(\vartheta_t\omega)) \to 0, \quad \text{when } s \to -\infty.$$

Moreover, it is the minimal closed set with this property: if $\tilde{\mathcal{A}}(\vartheta_t\omega)$ is a closed attracting set, then $\mathcal{A}(\vartheta_t\omega) \subset \tilde{\mathcal{A}}(\vartheta_t\omega)$.

(5) For any bounded set $B \subset X$, $d(S(t,s;\omega)B, \mathcal{A}(\vartheta_t\omega)) \to 0$ in probability when $t \to \infty$.

And if the time shift $\vartheta_t, t \in \mathbf{R}$ is ergodic.

(6) there exists a bounded set $B \subset X$ such that

$$\mathcal{A}(\omega) = \mathcal{A}(B, \omega).$$

(7) $\mathcal{A}(\omega)$ is the largest compact measurable set which is invariant in sense of Definition 2.4.

In this article, we will use the two lemmas below to prove our main results. The first lemma is Aubin-Lions Lemma whose proof can be found in [2, 24].

Lemma 2.1 Let B_0, B, B_1 be Banach spaces such that B_0, B_1 are reflexive and $B_0 \overset{c}{\subset} B \subset B_1$. Define, for $0 < T < \infty$,

$$X := \left\{ h \,\middle|\, h \in L^2([0,T]; B_0), \frac{dh}{dt} \in L^2([0,T]; B_1) \right\}.$$

Then X is a Banach space equipped with the norm $|h|_{L^2([0,T];B_0)} + |h'|_{L^2([0,T];B_1)}$. Moreover,
$$X \overset{c}{\subset} L^2([0,T];B).$$

The following lemma, a special case of a general result of Lions and Magenes [29], will help us to show the continuity of the solution to stochastic nematic liquid crystals with respect to time.

Lemma 2.2 Let V, H, V' be three Hilbert spaces such that $V \subset H = Hc \subset V'$, where H' and V' are the dual spaces of H and V respectively. Suppose $u \in L^2([0,T];V)$ and $u' \in L^2([0,T];V')$. Then u is almost everywhere equal to a function continuous from $[0,T]$ into H.

3 Existence of random attractor in $\mathbf{H} \times \mathbb{H}^1$

One of our main results in this article is to prove:

Theorem 3.1 Let $v_0 \in \mathbf{H}, d_0 \in \mathbb{H}^1$ in (1.4) and $f(d)$ is given by (1.12). Assume (2.14) hold. Then the solution operator $(S(t,s;\omega))_{t \geqslant s, \omega \in \tilde{\Omega}}$ of (1.1) – (1.5): $S(t,s;\omega)(v_s, d_s) = (v(t), d(t))$ has properties (i) – (iv) of Theorem 2.2 and possesses a compact absorbing ball $\mathcal{B}(0,\omega)$ in $\mathbf{H} \times \mathbb{H}^1$ at time 0. Furthermore, for $\tilde{\mathbb{P}}$-a.e. ω, the set
$$\mathcal{A}(\omega) = \overline{\bigcup_{B \subset \mathbf{H} \times \mathbb{H}^1} \mathcal{A}(B,\omega)},$$
where the union is taken over all the bounded subsets of $\mathbf{H} \times \mathbb{H}^1$ is the random attractor of (1.1) – (1.5) and possesses the properties (1) – (7) of Theorem 2.2 with space X replaced by space $\mathbf{H} \times \mathbb{H}^1$.

Proof The results of this theorem follow directly from Proposition 3.6 and Theorem 2.2. □

The rest of this section is to find a compact absorbing ball for (1.1) – (1.5) in $\mathbf{H} \times \mathbb{H}^1$. We will achieve our goal in six steps. In subsection 3.1, we use *a new technique* logarithmic energy estimates to obtain the uniform a priori estimates in $(L^{4N+2}(D))^3$ which is very important to study the long-time behavior of the stochastic nematic liquid system (see Lemma 3.1 and the proof of Proposition 3.1). Then in subsection 3.2, we obtain the absorbing ball for the solution to (1.1) – (1.5) in the space $\mathbf{H} \times \mathbb{H}^1$ in Proposition 3.1. As the third step, we prove the solution operator is a stochastic dynamical system in subsection 3.3.

In the next subsection, by Proposition 3.3 and Proposition 3.4 we verified two a priori estimates of the Aubin-Lions lemma which is used to obtain a convergent subsequence of the solutions to (1.1)−(1.5). In subsection 3.5, taking advantage of the convergent subsequence and the continuity of the solutions with respect to initial data in $\mathbf{H} \times \mathbb{H}^1$, we prove the solution operator $S(t, s; \omega)$ is almost surely compact from $\mathbf{H} \times \mathbb{H}^1$ to $\mathbf{H} \times \mathbb{H}^1$ for all $s, t \in \mathbf{R}, s < t$ (see Proposition 3.5). Finally, using the existence of absorbing ball and compactness of the solution operator, we construct a compact absorbing ball in Proposition 3.6.

To study the long time behavior of (1.1) − (1.5), we introduce a modified stochastic convolution. Let $t \in \mathbf{R}$ and $\beta \in \mathbf{R}_+$. For simplicity, we still define

$$z(t) := \int_{-\infty}^{t} e^{-(t-s)(A_1+\beta)} dW(s). \qquad (3.1)$$

Then by (2.2), we have $z(\omega) \in C([0, T]; \mathbf{H}^2), \mathbb{P}$-a.e. and satisfies the linear equation

$$dz = (-A_1 z - \beta z)dt + dW,$$
$$z(x, t) = 0, \quad \forall (x, t) \in \Gamma \times \mathbf{R},$$
$$z(t_0) = z_{t_0},$$

where $z_{t_0} = \int_{-\infty}^{t_0} e^{s(A_1+\beta)} dW(s)$. Let $(v_{t_0}, d_{t_0}) \in \mathbf{H} \times \mathbb{H}^1$, then in view of Theorem 2.1, (v, d) is the unique global weak solution to (1.1) − (1.5) on $[t_0, \infty)$ with $v(t_0) = v_{t_0}$ and $d(t_0) = d_0(x)$. Making the classic change $(v, d) = (u + z, d)$, then (u, d) satisfies the following system (3.2) − (3.6).

$$du + ((u + z) \cdot \nabla(u + z) + \nabla p - \Delta u)dt + \nabla \cdot (\nabla d \odot \nabla d)dt = \beta z, \qquad (3.2)$$
$$\nabla \cdot u = 0, \qquad (3.3)$$
$$d_t + [(u + z) \cdot \nabla d] - (\Delta d - f(d)) = 0, \qquad (3.4)$$
$$u(x, t) = 0, \; d(x, t) = d_0(x), \; (x, t) \in \Gamma \times [t_0, \infty), \qquad (3.5)$$
$$u|_{t=t_0} = v_{t_0}(x) \text{ with } \nabla \cdot v_{t_0} = 0, \quad d|_{t=t_0} = d_0(x), \; x \in D. \qquad (3.6)$$

3.1 Absorbing ball of d in $(L^{4N+2}(\mathbf{D}))^3$

Lemma 3.1 *Denote by $d(t, \omega; t_0, d_0)$ the weak solution to (1.3) with $d(t_0) = d_0$, then it is almost surely uniformly bounded w.r.t. time t and initial time t_0 in*

$(L^{4N+2}(\mathbf{D}))^3$ provided the initial data is bounded in $\mathbf{H} \times \mathbb{H}^1$, i.e.,

$$\sup_{(t,t_0)\in\{(t,t_0)|t\in\mathbf{R}, t_0 \leqslant t\}} |\boldsymbol{d}(t,\omega;t_0,\boldsymbol{d}_0)|^{4N+2}_{4N+2} < \infty, \quad \mathbb{P}\text{-}a.e..$$

Furthermore,

$$|f(\boldsymbol{d}(t,\omega;t_0,\boldsymbol{d}_0))|_2^2 \text{ and } \int_D \tilde{F}(|\boldsymbol{d}(t,\omega;t_0,\boldsymbol{d}_0)|^2)dx$$

are uniformly bounded w. r. t. t and t_0. (3.7)

Remark 3.1 The estimates of this lemma improved bounds for the solutions to (1.1) − (1.3). These bounds are uniform with respect to time t and initial time t_0. These estimates of the uniform boundedness improve bounds obtained in Proposition C.1 of [5] and bounds in Proposition 5.4 of [7]. These uniform bounds also allow us to obtain the absorbing balls for the solution in various function spaces(see (3.19)). The radii of these absorbing balls are independent of the initial data. This opens the way for finding the random attractor in the weak solution space $\mathbf{H} \times \mathbb{H}^1$ and strong solutions space $\mathbf{V} \times \mathbb{H}^2$. Maybe this lemma is a basic result to study the long-time behavior of stochastic nematic liquid crystals i.e. the existence of random attractor and ergodicity for this stochastic system.

Proof Taking inner product with (3.5) in $\mathbb{H}$ with $|\boldsymbol{d}|^{4N}\boldsymbol{d}$ we have

$$|\boldsymbol{d}(t)|^{4N+2}_{4N+2} = |\boldsymbol{d}(t_0)|^{4N+2}_{4N+2} - (4N+2)\int_{t_0}^{t} \langle |\boldsymbol{d}|^{4N}\boldsymbol{d}, -\Delta\boldsymbol{d} + (\boldsymbol{u}+\boldsymbol{z})\cdot\nabla\boldsymbol{d} + f(\boldsymbol{d})\rangle ds$$

$$= |\boldsymbol{d}(t_0)|^{4N+2}_{4N+2} - (4N+2)(4N+1)\int_{t_0}^{t}\int_D |\boldsymbol{d}|^{4N}|\nabla\boldsymbol{d}|^2 dx ds$$

$$-(4N+2)\int_{t_0}^{t}\langle |\boldsymbol{d}|^{4N}\boldsymbol{d}, f(\boldsymbol{d})\rangle ds,$$

which implies that

$$d|\boldsymbol{d}(t)|^{4N+2}_{4N+2} + (4N+2)(4N+1)$$
$$\cdot \int_{\mathbf{D}} |\boldsymbol{d}|^{4N}|\nabla\boldsymbol{d}|^2 dx + (4N+2)\langle |\boldsymbol{d}|^{4N}\boldsymbol{d}, f(\boldsymbol{d})\rangle = 0. \quad (3.8)$$

By Young's inequality, for a small positive constant ε there exists a positive constant c such that

$$|a_k||\boldsymbol{d}|^{2k+4N+2} \leqslant \frac{\varepsilon}{N}|\boldsymbol{d}|^{6N+2} + \frac{c}{N}|\boldsymbol{d}|^{4N+2}, \quad k=0,1,\cdots,N-1.$$

Therefore,

$$\langle |d|^{4N}d, f(d)\rangle = \sum_{k=0}^{N} a_k |d|^{2k+4N+2}$$
$$\geq -\varepsilon |d|^{6N+2} - c|d|^{4N+2} + a_N |d|^{6N+2}. \tag{3.9}$$

Combing (3.8) and (3.9) yields

$$d|d(t)|_{4N+2}^{4N+2} + c|d(t)|_{6N+2}^{6N+2} dt \leq c|d(t)|_{4N+2}^{4N+2} dt,$$

which implies

$$d(|d(t)|_{4N+2}^{4N+2} + 1) + c(|d(t)|_{6N+2}^{6N+2} + 1)dt \leq c(|d(t)|_{4N+2}^{4N+2} + 1)dt. \tag{3.10}$$

Diving $(|d(t)|_{4N+2}^{4N+2} + 1)$ on both sides of (3.29) yields,

$$\frac{d}{dt}\ln(|d(t)|_{4N+2}^{4N+2} + 1) + c(|d(t)|_{4N+2}^{4N+2} dt + 1)^{\frac{3N+1}{2N+1}} \leq c. \tag{3.11}$$

Since

$$\ln(1+|x|) \leq 1 + |x| \leq (1+|x|)^{\frac{3N+1}{2N+1}}, \quad \text{for all } x \in \mathbf{R}, \tag{3.12}$$

By (3.11) – (3.12) we have

$$\frac{d}{dt}\ln(|d(t)|_{4N+2}^{4N+2} + 1) + c\ln(|d(t)|_{4N+2}^{4N+2} + 1) \leq c. \tag{3.13}$$

Let $y(t) = \ln(|d(t)|_{4N+2}^{4N+2} + 1)$. Then multiplying e^{ct} on both sides yields

$$\frac{d}{dt}(y(t)e^{ct}) \leq ce^{ct},$$

which implies

$$y(t) \leq y(t_0)e^{-c(t-t_0)} + \int_{t_0}^{t} e^{-c(t-s)} ds. \tag{3.14}$$

By (3.33), $y(t)$ is uniformly bounded with respect to time t and initial time t_0 provided the initial data $y(t_0)$ is bounded. Therefore, this in turn implies the uniform boundedness of d in $(L^{4N+2}(\mathbf{D}))^3$ with respect to time t and initial time t_0. (3.26) follows by the uniform estimate of d with respect to time and initial time in $(L^{4N+2}(\mathbf{D}))^3$. □

3.2 Absorbing ball of (u, d) in $\mathbf{H} \times \mathbb{H}^1$

Proposition 3.1 *There exists an absorbing ball for the weak solution (v, d) to $(1.1) - (1.5)$ at any time $t(\in \mathbf{R})$ in $\mathbf{H} \times \mathbb{H}^1$.*

Proof Taking the inner product of (3.4) with $\Delta d - f(d)$ yields

$$\frac{1}{2}\frac{d(|\nabla d|_2^2 + \int_D \tilde{F}(|d|^2)dx)}{dt} + |\Delta d - f(d)|_2^2 = \langle (u+z) \cdot \nabla d, \Delta d - f(d) \rangle. \quad (3.15)$$

Taking inner product of (3.2) with u in $\mathbf{H}$ yields,

$$\frac{1}{2}\frac{d|u|_2^2}{dt} + |\nabla u|_2^2 = \langle u \cdot \nabla z + z \cdot \nabla z, u \rangle - \langle \nabla \cdot (\nabla d \odot \nabla d), u \rangle + \beta \langle z, u \rangle. \quad (3.16)$$

Since by integration by parts and boundary conditions (3.5),

$$\langle u \cdot \nabla d, \Delta d \rangle = \int_D u^i \partial_{x_i} d^k \partial_{x_j x_j} d^k$$
$$= -\int_D \partial_{x_j} u^i \partial_{x_i} d^k \partial_{x_j} d^k d\mathbf{D} - \int_D u^i \partial_{x_i x_j} d^k \partial_{x_j} d^k d\mathbf{D}$$
$$= -\int_D \partial_{x_j} u^i \partial_{x_i} d^k \partial_{x_j} d^k d\mathbf{D},$$

and

$$-\langle \nabla \cdot (\nabla d \odot \nabla d), u \rangle = -\int_D \partial_{x_i}(\partial_{x_i} d^k \partial_{x_j} d^k) u^j dD$$
$$= \int_D \partial_{x_i} d^k \partial_{x_j} d^k \partial_{x_i} u^j dD,$$

combing (3.15) and (3.16) together yields

$$\frac{1}{2}\frac{d(|u|_2^2 + |\nabla d|_2^2 + \int_D \tilde{F}(|d|^2)dx)}{dt} + |\nabla u|_2^2 + |\Delta d - f(d)|_2^2$$
$$= \langle u \cdot \nabla z + z \cdot \nabla z, u \rangle + \langle z \cdot \nabla d, \Delta d \rangle$$
$$\leqslant c|\nabla z|_\infty |u|_2^2 + c|z|_\infty^2 |u|_2^2 + c|\nabla z|_2^2 + \varepsilon|\Delta d|_2^2 + c|z|_\infty^2 |\nabla d|_2^2. \quad (3.17)$$

Since by the Poincaré inequality and the Minkovski inequality

$$c|\nabla d|_2^2 \leqslant \frac{1}{2}|\Delta d|_2^2 \leqslant |\Delta d - f(d)|_2^2 + |f(d)|_2^2, \quad (3.18)$$

then in view of the Poincaré inequality, the Hölder inequality, the Sobolev imbedding theorem, (3.17) − (3.18) and Lemma 3.1, we have

$$\frac{1}{2}\frac{d(|u|_2^2 + |\nabla d|_2^2 + \int_D \tilde{F}(|d|^2)dx)}{dt}$$
$$+ c|\nabla u|_2^2 + c|\Delta d|_2^2 + c(1 - \|z\|_1^2)\int_D \tilde{F}(|d|^2)dx$$
$$\leqslant c|f(d)|_2^2 + c(1 - \|z\|_1^2)\int_D \tilde{F}(|d|^2)dx + c|z|_\infty^2|\nabla d|_2^2$$
$$+ |\nabla z|_\infty |u|_2^2 + |z|_\infty^2 |u|_2^2 + |\nabla z|_2^2$$
$$\leqslant c(\|z\|_1 + \|z\|_1^2)(|u|_2^2 + |\nabla d|_2^2) + c + c\|z\|_1^2. \tag{3.19}$$

Let $g(t) = |u(t)|_2^2 + |\nabla d(t)|_2^2 + \int_D \tilde{F}(|d(t)|^2)dx$, with $t \in \mathbf{R}$. Then by (3.38) we have

$$\frac{d}{dt}g(t) + c(1 - \|z\|_1^2)g(t) \leqslant c + c\|z\|_1^2.$$

Therefore,

$$\frac{d\left(g(t)e^{\int_{t_0}^t c(1-\|z(s)\|_1^2)ds}\right)}{dt} \leqslant c(1 + \|z\|_1^2)e^{\int_{t_0}^t c(1-\|z(s)\|_1^2)},$$

which implies

$$g(t) \leqslant g(t_0)e^{-\int_{t_0}^t c(1-\|z(s)\|_1^2)ds} + c\int_{t_0}^t (\|z\|_1^2 + 1)e^{-\int_s^t c(1-\|z(u)\|_1^2)du}ds. \tag{3.20}$$

Since, the process $z(t)$ (see (3.20)) is stationary and ergodic and $\|z(t)\|_1$ has polynomial growth when $t \to -\infty$, following the classic arguments (see [9, 11] and other references), (3.20) gives us the desired uniform estimate which yields an absorbing ball for (u, d) in $\mathbf{H} \times \mathbb{H}^1$. Following the standard arguments, from (3.19), we can show that for any constant $r(> 0)$, $t(\in \mathbf{R})$ and given bounded initial data (v_{t_0}, d_0)

$$\int_{t-r}^t (|\nabla u(s)|_2^2 + |\Delta d(s)|_2^2)ds$$

is uniformly bounded w.r.t. initial time $t_0 (\leqslant t - r)$. \hfill (3.21)

□

Remark 3.2 As we see from (3.19) the uniform bounds for $\tilde{F}(d)$ obtained in Lemma 3.1 play a vital role to obtain the absorbing ball for (v, d) in $\mathbf{H} \times \mathbb{H}^1$.

3.3 The solution operator $S(t, s; \omega)$ to (1.1) – (1.5) is a stochastic dynamical system

In this part, we need to show the solution operator to (1.1) – (1.5) is indeed a stochastic dynamical system. For this purpose, we will introduce some notations below. Define $\hat{e}_j = (e_j, 0, 0, 0)^{\mathrm{T}}, j \in \mathbb{N}$, where the $(e_j)_{j \in \mathbb{N}}$ is an orthonormal basis of $\mathbf{H}$ (see section 2). Let

$$\mathbf{W} := \sum_{i=1}^{\infty} \lambda_i^{\frac{1}{2}} \hat{e}_j B_i(t), \ t \in \mathbf{R},$$

where λ_i and B_i are illustrated in the introduction. Then $\mathbf{W}$ is the two-sided $\mathbf{H} \times \mathbb{H}^1$-valued Wiener process, which has a version ω in $C_0(\mathbf{R}, \mathbf{H} \times \mathbb{H}^1) := \Omega$, the space of continuous functions which are zero at zero. In what follows, we consider a canonical version of $\mathbf{W}$ given by the probability space $(C_0(\mathbf{R}, \mathbf{H} \times \mathbb{H}^1), B(C_0(\mathbf{R}, \mathbf{H} \times \mathbb{H}^1)), \mathbb{P})$ where $\mathbb{P}$ is the Wiener-measure generated by $\mathbf{W}$. Then

$$\mathbf{W}(t, \omega) = \omega(t), \quad t \in \mathbf{R}, \ \omega \in \Omega. \tag{3.22}$$

On this probability space, we can also introduce the shift

$$\vartheta_s \omega(t) = \omega(t+s) - \omega(s), \quad s, t \in \mathbf{R}.$$

By Theorem 2.1, for $s \in \mathbf{R}$ and $(v_s, d_s) \in \mathbf{H} \times \mathbb{H}^1$ a.s., there exists a weak unique solution defined on $[s, \infty)$ to (1.1) – (1.5), $(v(t, \omega), d(t, \omega))$ such that

$$(v(s, \omega), d(s, \omega)) = (v_s, d_s), \quad \mathbb{P}\text{-}a.s.. \tag{3.23}$$

Define the solution operator $(S(t, s; \omega))_{t \geqslant s, \omega \in \Omega}$ to (1.1) – (1.5) by

$$S(t, s; \omega)(v_s, d_s) = (u(t, \omega) + z(t, \omega), d(t, \omega)), \tag{3.24}$$

where (u, d) is the unique weak solution to (3.2) – (3.6).

Proposition 3.2 The solution operator $(S(t, s; \omega))_{t \geqslant s, \omega \in \Omega}$ to (1.1) – (1.5) and $\vartheta_t, t \in \mathbf{R}$, satisfy properties (i) – (iv) and (a) – (c) in the preliminaries, respectively.

Proof Denote

$$H := \{(v,d) \in (L^2(\mathbf{D}))^2 \times \mathbb{H}^1 : \nabla \cdot v = 0 \text{ in } \mathbf{D}, v(t,x) = 0, d(t,x)$$
$$= d_0(x), \forall (x,t) \in \Gamma \times \mathbf{R}_+\}.$$

Define P_H to be the Leray type projection operator from $(L^2(\mathbf{D}))^2 \times \mathbb{H}^1$ onto H. The principal linear portion of (1.1) – (1.5) is defined by

$$AU = P_H(A_1 v, A_2 d)^\mathrm{T}, \qquad (3.25)$$

where for a matrix C, C^T stands for the transpose of C and $U = (v,d) \in D(A) := D(A_1) \times D(A_2)$. Take, for $U \in D(A_1) \times \mathbb{H}^3$ we deifne

$$B_1(U) = P_H(v \cdot \nabla v, f(d))^\mathrm{T}, \quad B_2(U) = P_H(\nabla \cdot (\nabla d \odot \nabla d), v \cdot \nabla d)^\mathrm{T}. \quad (3.26)$$

Let $B(U) := B_1(U) + B_2(U)$. Collecting the operators defined above we reformulate (1.1) – (1.5) as the following abstract evolution system

$$dU + (AU + BU)dt = d\mathbf{W}, \quad U(0) = U_0. \qquad (3.27)$$

Obviously, for $U_0 \in \mathbf{H} \times \mathbb{H}^1$ or $U_0 \in \mathbf{V} \times \mathbb{H}^2$, in view of Theorem 2.1, there exists a unique weak solution or a unique strong solution to (3.27). The definitions of weak solution and strong solution to (3.27) are similar to Definition 2.1 and Definition 2.2, respectively. Moreover, for $U_0 \in \mathbf{H} \times \mathbb{H}^1$ or $U_0 \in \mathbf{V} \times \mathbb{H}^2$, the solution U is Lipschitz continuous with respect to U_0 in $\mathbf{H} \times \mathbb{H}^1$ or $\mathbf{V} \times \mathbb{H}^2$, respectively. For the latter case, we will give a proof in Proposition 4.4.

Then the conclusions of this proposition follow from the global existence, uniqueness and regularity of the solutions to (3.27). □

3.4 Two a priori estaimes for the Aubin-Lions Lemma

To establish that the solution operator to (1.1) – (1.5) is compact in $\mathbf{H} \times \mathbb{H}^1$, the first step is to use the Aubin-Lions lemma to obtain a convergent subsequence of (v,d) or equivalently a convergent subsequence of (u,d) which converges almost everywhere with respect to time $t \in [s,T]$, $s, T \in \mathbf{R}$ and $s < T$, in $\mathbf{H} \times \mathbb{H}^1$. Then we have to verify the two a priori estimates of Aubin-Lions lemma. The first one is to obtain a priori estimates about $(u, A_2^{\frac{1}{2}} d)$ in $L^2([s,T]; \mathbf{V} \times \mathbb{H}^1)$, which is proved in the following Proposition 3.3. The other one is to obtain a priori estimates of $\left(\dfrac{d}{dt} u, \dfrac{d}{dt} A_2^{\frac{1}{2}} d\right)$ in $L^2([s,T]; \mathbf{H}^{-1} \times \mathbb{H}^{-1})$, which is obtained in Proposition 3.4.

Proposition 3.3 Let $B = \{(v_0, d_0) \in \mathbf{H} \times \mathbb{H}^1 | |v_0|_2 + \|d_0\|_1 \leqslant M\}$ for some positive constant M, then $\mathbb{P}$-a.s.

$$\sup_{(v_0, d_0) \in B} \Big(\int_0^T \|u(t)\|_1^2 dt + \int_0^T \|d(t)\|_2^2 dt \Big) < \infty. \tag{3.28}$$

Proof Proposition 3.3 follows directly from Lemma 3.1 and (3.19). □

Proposition 3.4 Let $B = \{(v_0, d_0) \in \mathbf{H} \times \mathbb{H}^1 | |v_0|_2 + \|d_0\|_1 \leqslant M\}$ for some positive constant M, then $\mathbb{P}$-a.s.

$$\sup_{(v_0, d_0) \in B} \Big(\int_0^T \|\frac{d}{dt} u(t)\|_{-1}^2 dt + \int_0^T \|\frac{d}{dt} A_2^{\frac{1}{2}} d(t)\|_{-1}^2 dt \Big) < \infty. \tag{3.29}$$

Proof For $\eta \in \mathbb{H}^1$, taking inner product of (3.4) with $A_2^{\frac{1}{2}} \eta$ in $\mathbb{H}$ yields,

$$\Big\langle \frac{d A_2^{\frac{1}{2}} d}{dt}, \eta \Big\rangle = \Big\langle \frac{dd}{dt}, A_2^{\frac{1}{2}} \eta \Big\rangle$$

$$= \langle \Delta d, A_2^{\frac{1}{2}} \eta \rangle - \langle u \cdot \nabla d, A_2^{\frac{1}{2}} \eta \rangle - \langle f(d), A_2^{\frac{1}{2}} \eta \rangle$$

$$\leqslant |\Delta d|_2 \|\eta\|_1 + |u|_4 |\nabla d|_4 \|\eta\|_1 + |f(d)|_2 \|\eta\|_1$$

$$\leqslant |\Delta d|_2 \|\eta\|_1 + c|u|_2^{\frac{1}{2}} \|u\|_1^{\frac{1}{2}} |\nabla d|_2^{\frac{1}{2}} |\Delta d|_2^{\frac{1}{2}} \|\eta\|_1 + |f(d)|_2 \|\eta\|_1. \tag{3.30}$$

By (3.30) and the Hölder inequality, we have

$$\Big| \frac{d A_2^{\frac{1}{2}} d}{dt} \Big|_{\mathbb{H}^{-1}}^2 \leqslant c|\Delta d|_2^2 + c|u|_2^2 \|u\|_1^2 |\nabla d|_2^2 + c|f(d)|_2^2, \tag{3.31}$$

where $|\cdot|_{\mathbb{H}^{-m}}$ is the norm of Sobolev space $\mathbb{H}^{-m}$ which is the dual space of $\mathbb{H}^m$ for $m \in \mathbb{N}$. Therefore, from Theorem 2.1 and (3.49) we conclude

$$\frac{d A_2^{\frac{1}{2}} d}{dt} \text{ is bounded in } L^2([0, T]; \mathbb{H}^{-1}). \tag{3.32}$$

For $\xi \in \mathbf{H}^1$, taking the inner product of (3.21) with ξ in $\mathbf{H}$ yields,

$$\Big\langle \frac{du}{dt}, \xi \Big\rangle = -\langle u \cdot \nabla u, \xi \rangle - \langle u \cdot \nabla z, \xi \rangle$$

$$- \langle z \cdot \nabla u, \xi \rangle - \langle z \cdot \nabla z, \xi \rangle - \langle \Delta u, \xi \rangle + \int_D \partial_{x_i} d^k \partial_{x_j} d^k \partial_{x_i} \xi^j. \tag{3.33}$$

By the incompressible property of the fluid (see (1.2)), the boundary condition (1.4) and integration by parts we obtain

$$-\langle u \cdot \nabla u, \xi \rangle = \langle u \cdot \nabla \xi, u \rangle \leqslant |\nabla \xi|_2 |u|_4^2 \leqslant c\|\xi\|_1 |u|_2 \|u\|_1. \qquad (3.34)$$

In view of (3.52) and (3.53) we have

$$\left\langle \frac{du}{dt}, \xi \right\rangle \leqslant c|u|_2 \|u\|_1 \|\xi\|_1 + |u|_2 |\nabla z|_4 |\xi|_4 + |\nabla u|_2 |z|_4 |\xi|_4$$
$$+ |\nabla z|_2 |z|_4 |\xi|_4 + |\nabla u|_2 |\nabla \xi|_2 + c|\nabla d|_4^2 \|\xi\|_1$$
$$\leqslant c|u|_2 \|u\|_1 \|\xi\|_1 + c|u|_2 \|z\|_2 \|\xi\|_1 + c\|u\|_1 \|z\|_1 \|\xi\|_1$$
$$+ c\|z\|_1^2 \|\xi\|_1 + c\|u\|_1 \|\xi\|_1 + c\|d\|_1 \|d\|_2 \|\xi\|_1,$$

which implies

$$\left| \frac{du}{dt} \right|^2_{\mathbf{H}^{-1}} \leqslant c|u|_2^2 \|u\|_1^2 + \|u\|_1^2 \|z\|_2^2 + c\|z\|_1^2 + c\|u\|_1^2 + c\|d\|_1^2 \|d\|_2^2, \qquad (3.35)$$

where $|\cdot|_{\mathbf{H}^{-1}}$ is the norm of the Sobolev space $\mathbf{H}^{-1}$ which is the dual space of $\mathbf{V}$. Then Theorem 2.1, (2.15) and (3.35) imply

$$\frac{du}{dt} \text{ is bounded in } L^2([0,T]; \mathbf{H}^{-1}). \qquad (3.36)$$

Therefore, (3.29) is followed by (3.32) and (3.36). □

3.5 The solution $S(t,s;\omega)$ operator to (1.1) − (1.5) is impact in $\mathbf{H} \times \mathbb{H}^1$

By virtue of Proposition 3.3, Proposition 3.4 and Lemma 2.2 with $V = \mathbf{V}$ or $\mathbb{H}^1$, $H = \mathbf{H}$ or $\mathbb{H}$ and $V' = \mathbf{H}^{-1}$ or $\mathbb{H}^{-1}$ we infer that

Corollary 3.1

$$v \in C([0,T]; \mathbf{H}), \quad \text{and } d \in C([0,T]; \mathbb{H}^1), \quad \text{for arbitrary } T > 0, \text{ a.s..}$$

Then we will use Aubin-Lions lemma and Corollary 3.1 to show the following compactness result for the solution operators to (1.1) − (1.5).

Proposition 3.5 *For $\omega \in \Omega$, $S(t,s;\omega)$ is compact from $\mathbf{H} \times \mathbb{H}^1$ to $\mathbf{H} \times \mathbb{H}^1$, for all $s,t \in \mathbf{R}$ and $s \leqslant t$.*

Proof In Proposition 3.1, we have obtained an absorbing ball for $(S(t,s;\omega))_{t\geqslant s, \omega\in\Omega}$ at any time $t \in \mathbf{R}$. We denote by $B(s, r(\omega))$, the absorbing ball at time s with center $0 \in \mathbf{H} \times \mathbb{H}^1$ and radius $r(\omega)$. Denote by $\mathcal{B}$ a bounded subset $\mathbf{H} \times \mathbb{H}^1$ and set $\mathcal{C}_T$ as a subset of the function space:

$$\mathcal{C}_T := \left\{ \left(\boldsymbol{v}, A_2^{\frac{1}{2}} \boldsymbol{d}\right) \middle| (\boldsymbol{v}(s), \boldsymbol{d}(s)) \in \mathcal{B}, (\boldsymbol{v}(t), \boldsymbol{d}(t)) \right.$$
$$\left. = S(t, s; \omega)(\boldsymbol{v}(s), \boldsymbol{d}(s)), t \in [s, T], s \leqslant T \right\}.$$

Since both embedding $\mathbf{H}^1 \subset \mathbf{H}$ and embedding $\mathbb{H}^1 \subset \mathbb{H}$ are compact, then the embedding $\mathbf{H}^1 \times \mathbb{H}^1 \subset \mathbf{H} \times \mathbb{H}$ is also compact. For arbitrary $(\boldsymbol{v}(s), \boldsymbol{d}(s)) \in \mathcal{B}$, by Proposition 3.2 and Proposition 3.3 we infer that

$$(\boldsymbol{u}, A_2^{\frac{1}{2}} \boldsymbol{d}) \text{ is bounded in } L^2([s, T]; \mathbf{H}^1 \times \mathbb{H}^1)$$

and

$$(\partial_t \boldsymbol{u}, \partial_t A_2^{\frac{1}{2}} \boldsymbol{d}) \text{ is bounded in } L^2([s, T]; \mathbf{H}^{-1} \times \mathbb{H}^{-1}).$$

Therefore, by Lemma 2.1, with

$$B_0 = \mathbf{H}^1 \times \mathbb{H}^1, \quad B = \mathbf{H} \times \mathbb{H}, \quad B_1 = \mathbf{H}^{-1} \times \mathbb{H}^{-1},$$

$\mathcal{C}_T$ is compact in $L^2([s, T]; \mathbf{H} \times \mathbb{H})$. In order to show that for any fixed $t \in (s, T], \omega \in \tilde{\Omega}, S(t, s; \omega)$ is a compact operator in $\mathbf{H} \times \mathbb{H}^1$, we take any bounded sequences $\{(\boldsymbol{v}_{0,n}, \boldsymbol{d}_{0,n})\}_{n \in \mathbb{N}} \subset \mathcal{B}$ and we want to extract, for any fixed $t \in (s, T]$ and $\omega \in \Omega$, a convergent subsequence from $\{S(t, s; \omega)(\boldsymbol{v}_{0,n}, \boldsymbol{d}_{0,n})\}$. Since $\{(\boldsymbol{v}, A_2^{\frac{1}{2}} \boldsymbol{d})\} \subset \mathcal{C}_T$, by Lemma 2.1, there is a function $(\boldsymbol{v}_*, \boldsymbol{d}_*)$:

$$(\boldsymbol{v}_*, \boldsymbol{d}_*) \in L^2([s, T]; \mathbf{H} \times \mathbb{H}^1),$$

and a subsequence of $\{S(t, s; \omega)(\boldsymbol{v}_{0,n}, \boldsymbol{d}_{0,n})\}_{n \in \mathbb{N}}$, still denoted by $\{S(t, s; \omega)(\boldsymbol{v}_{0,n}, \boldsymbol{d}_{0,n})\}_{n \in \mathbb{N}}$ for simplicity, such that

$$\lim_{n \to \infty} \int_s^T \|S(t, s; \omega)(\boldsymbol{v}_{0,n}, \boldsymbol{d}_{0,n}) - (\boldsymbol{v}_*(t), \boldsymbol{d}_*(t))\|_{\mathbf{H} \times \mathbb{H}^1}^2 dt = 0, \qquad (3.37)$$

where $\| \cdot \|_{\mathbf{H} \times \mathbb{H}^1}$ denotes the norm of the product space $\mathbf{H} \times \mathbb{H}^1$. By measure theory, convergence in mean square implies almost sure convergence of a subsequence. Therefore, it follows from (3.37) that there exists a subsequence of $\{S(t, s; \omega)(\boldsymbol{v}_{0,n}, \boldsymbol{d}_{0,n})\}_{n \in \mathbb{N}}$, still denoted by $\{S(t, s; \omega)(\boldsymbol{v}_{0,n}, \boldsymbol{d}_{0,n})\}_{n \in \mathbb{N}}$ for simplicity, such that

$$\lim_{n \to \infty} \|S(t, s; \omega)(\boldsymbol{v}_{0,n}, \boldsymbol{d}_{0,n}) - (\boldsymbol{v}_*(t), \boldsymbol{d}_*(t))\|_{\mathbf{H} \times \mathbb{H}^1} = 0, \quad a.e. \ t \in (s, T]. \qquad (3.38)$$

Fix any $t \in (s, T]$, by (3.38), we can select a $t_2 \in (s, t)$ such that

$$\lim_{n \to \infty} \|S(t_2, s, \omega)(\boldsymbol{v}_{0,n}, \boldsymbol{d}_{0,n}) - (\boldsymbol{v}_*(t_2), \boldsymbol{d}_*(t_2))\|_{\mathbf{H} \times \mathbb{H}^1} = 0.$$

Then by the continuity of the map $S(t, t_2; \omega)$ in $\mathbf{H} \times \mathbb{H}^1$ with respect to the initial value, we have

$$S(t, s; \omega)(\boldsymbol{v}_{0,n}, \boldsymbol{d}_{0,n}) = S(t, t_2; \omega) S(t_2, s; \omega)(\boldsymbol{v}_{0,n}, \boldsymbol{d}_{0,n})$$
$$\to S(t, t_2; \omega)(\boldsymbol{v}_*(t_2), \boldsymbol{d}_*(t_2)), \quad \text{in } \mathbf{H} \times \mathbb{H}^1.$$

Hence, for any $t \in (s, T]$, $\{S(t, s; \omega)(\boldsymbol{v}_{0,n}, \boldsymbol{d}_{0,n})\}_{n \in \mathbb{N}}$ contains a subsequence which is convergent in $\mathbf{H} \times \mathbb{H}^1$, which implies that for any fixed $t \in (s, T], \omega \in \Omega$, $S(t, s; \omega)$ is a compact operator in $\mathbf{H} \times \mathbb{H}^1$. □

3.6 The existence of compact absorbing ball in $\mathbf{H} \times \mathbb{H}^1$

Proposition 3.6 *There exists a compact absorbing ball at any time $t \in \mathbf{R}$ for the stochastic dynamical system (1.1) – (1.5) in $\mathbf{H} \times \mathbb{H}^1$.*

Proof Using the notations given in Proposition 3.5, for $s < t$, let $\mathcal{B}(t, \omega) = \overline{S(t, s; \omega) B(s, r(\omega))}$ be the closed set of $S(t, s; \omega) B(s, r(\omega))$ in $\mathbf{H} \times \mathbb{H}^1$. Then, by the above arguments, we infer $\mathcal{B}(t, \omega)$ is a random compact set in $\mathbf{H} \times \mathbb{H}^1$ for each ω. More precisely, $\mathcal{B}(t, \omega)$ is a compact absorbing set in $\mathbf{H} \times \mathbb{H}^1$ at time $t \in \mathbf{R}$. Indeed, for $(\boldsymbol{v}_{0,n}, \boldsymbol{d}_{0,n}) \in \mathcal{B}$, there exists $s_0(\mathcal{B}) \in \mathbf{R}_-$ such that if $s_0 \leqslant s(\mathcal{B})$, we have

$$S(t, s_0; \omega)(\boldsymbol{v}_{0,n}, \boldsymbol{d}_{0,n}) = S(t, s; \omega) S(s, s_0; \omega)(\boldsymbol{v}_{0,n}, \boldsymbol{d}_{0,n}) \subset S(t, s; \omega) B(s, r(\omega))$$
$$\subset \mathcal{B}(t, \omega). \qquad \Box$$

Remark 3.3 By Proposition 4.5 in [10], the existence of a random attractor as constructed in the proof of Theorem 3.1 implies the existence of an invariant Markov measure $\mu. \in \mathcal{P}_\Omega(\mathbf{H} \times \mathbb{H}^1)$ for φ (see Definition 4.1 in [10]). In view of [8] there exists an invariant measure μ for the Markovian semigroup $\mathbb{P}_t[f(S(t, 0, x))]$ satisfying

$$\mu(B) = \int_\Omega \mu_\omega(B) \mathbb{P}(d\omega) \quad \text{and} \quad \mu(\mathcal{A}(\omega)) = 1, \quad \mathbb{P}\text{-a.e.},$$

where $B \in \mathcal{B}(\mathbf{H} \times \mathbb{H}^1)$ which is a Borel σ-algebra on $\mathbf{H} \times \mathbb{H}^1$. If the invariant measure for $\mathbb{P}_t$ is unique, then the invariant Markov measure μ. for φ is unique

and given by
$$\mu_\omega = \lim_{t\to\infty} \varphi(t, \vartheta_{-t}\omega)\mu.$$

4 Existence of random attractor in $\mathbf{V} \times \mathbb{H}^2$

The main goal of this section is to show the existence of random attractor in $\mathbf{V} \times \mathbb{H}^2$ which will be proved in four steps. Firstly, in subsection 4.1, we obtain the absorbing ball in $\mathbf{V} \times \mathbb{H}^2$ for the system $(1.1) - (1.5)$ in Proposition 4.1. Then, by careful and delicate a priori estimates, we verify the two a priori estimates of Aubin-Lions Lemma in Proposition 4.2 and Proposition 4.3 of subsection 4.2. Thirdly, in the next subsection, we establish the continuity of the strong solution in $\mathbf{V} \times \mathbb{H}^2$ with respect to time and initial data, improving the corresponding continuity of the strong solutions in [5]. Finally, in the last section, with Aubin-Lions Lemma and regularity of strong solutions to $(1.1) - (1.5)$, following the same line of Proposition 3.5 and Proposition 3.6, the compact property of the solution operators in $\mathbf{V} \times \mathbb{H}^2$ and the existence of a compact absorbing ball in $\mathbf{V} \times \mathbb{H}^2$ hold, see Proposition 4.5 and Proposition 4.6 respectively. Theorem 4.1 one of our main results of this article is the direct result of Theorem 2.2 and Proposition 4.6.

Theorem 4.1 *Let $v_0 \in \mathbf{V}, d_0 \in \mathbb{H}^2$ in (1.4) and $f(d)$ is given by (1.12). Assume (2.15) hold. Then the solution operator $(S(t, s; \omega))_{t \geqslant s, \omega \in \tilde{\Omega}}$ of $(1.1) - (1.5) : S(t, s; \omega)(\boldsymbol{v}_s, \boldsymbol{d}_s) = (\boldsymbol{v}(t), \boldsymbol{d}(t))$ has properties (i) − (iv) of Theorem 2.2 and possesses a compact absorbing ball $\mathcal{B}(0, \omega)$ in $\mathbf{V} \times \mathbb{H}^2$ at time 0. Furthermore, for $\tilde{\mathbb{P}}$-a.e. ω, the set*
$$\mathcal{A}(\omega) = \overline{\bigcup_{B \subset \mathbf{V} \times \mathbb{H}^2} \mathcal{A}(B, \omega)},$$

where the union is taken over all the bounded subsets of $\mathbf{V} \times \mathbb{H}^2$ is the random attractor of $(1.1) - (1.5)$ and possesses the properties $(1) - (7)$ of Theorem 2.2 with space X replaced by space $\mathbf{V} \times \mathbb{H}^2$.

4.1 Absorbing ball of the strong solution to $(1.1) - (1.5)$

Proposition 4.1 *There exists an absorbing ball in $\mathbf{V} \times \mathbb{H}^2$ at time -1 for the strong solutions $(\boldsymbol{v}, \boldsymbol{d})$ to $(1.1) - (1.5)$.*

Remark 4.1 *In fact, we can prove that there exists an absorbing ball in $\mathbf{V} \times \mathbb{H}^2$ at any time. But here, to prove the Theorem 4.1, we only need the* existence of an absorbing ball in $\mathbf{V} \times \mathbb{H}^2$ at some time t, i.e., $t = -1$.

Proof Taking the inner product of (3.14) with $\Delta \boldsymbol{u}$ in $\mathbf{H}$ yields

$$\frac{d}{dt}|\nabla \boldsymbol{u}|_2^2 + |\Delta \boldsymbol{u}|_2^2 - \langle (\boldsymbol{u}+\boldsymbol{z}) \cdot \nabla(\boldsymbol{u}+\boldsymbol{z}), \Delta \boldsymbol{u}\rangle$$
$$= \langle \nabla \cdot (\nabla \boldsymbol{d} \odot \nabla \boldsymbol{d}), \Delta \boldsymbol{u}\rangle + \beta \langle \boldsymbol{z}, \Delta \boldsymbol{u}\rangle. \tag{4.1}$$

Since it is given in [22] that

$$\langle \nabla \cdot (\nabla \boldsymbol{d} \odot \nabla \boldsymbol{d}), \Delta \boldsymbol{u}\rangle = \langle \nabla \boldsymbol{d} \Delta \boldsymbol{d}, \Delta \boldsymbol{u}\rangle + \langle \nabla(\frac{|\nabla \boldsymbol{d}|^2}{2}), \Delta \boldsymbol{u}\rangle$$
$$= \langle \nabla \boldsymbol{d} \Delta \boldsymbol{d}, \Delta \boldsymbol{u}\rangle$$
$$= \langle \nabla \boldsymbol{d}(\Delta \boldsymbol{d} - f(\boldsymbol{d})), \Delta \boldsymbol{u}\rangle + \langle \nabla \boldsymbol{d} f(\boldsymbol{d}), \Delta \boldsymbol{u}\rangle, \tag{4.2}$$

where the second equality is followed by (3.3). Taking Fréchet derivative with respect to $|\Delta \boldsymbol{d} - f(\boldsymbol{d})|_2^2$ we have

$$\frac{d}{dt}|\Delta \boldsymbol{d} - f(\boldsymbol{d})|_2^2$$
$$= 2\langle \Delta \boldsymbol{d} - f(\boldsymbol{d}), \Delta \boldsymbol{d}_t - f'(\boldsymbol{d})\boldsymbol{d}_t\rangle$$
$$= -2|\nabla(\Delta \boldsymbol{d} - f(\boldsymbol{d}))|_2^2 + \int_D (\Delta \boldsymbol{d} - f(\boldsymbol{d}))(-f'(\boldsymbol{d}))(\Delta \boldsymbol{d} - f(\boldsymbol{d}) - (\boldsymbol{u}+\boldsymbol{z}) \cdot \nabla \boldsymbol{d})dx$$
$$+ 2\int_D (\Delta \boldsymbol{d} - f(\boldsymbol{d}))(-\Delta(\boldsymbol{u}+\boldsymbol{z})\nabla \boldsymbol{d} - (\boldsymbol{u}+\boldsymbol{z})\nabla(\Delta \boldsymbol{d}) - 2\nabla(\boldsymbol{u}+\boldsymbol{z})(\nabla^2 \boldsymbol{d}))dx. \tag{4.3}$$

By integration by parts, we obtain

$$\int_D (\Delta \boldsymbol{d} - f(\boldsymbol{d}))(\boldsymbol{u}+\boldsymbol{z})\nabla(\Delta \boldsymbol{d})dx$$
$$= \int_D (\Delta \boldsymbol{d} - f(\boldsymbol{d}))(\boldsymbol{u}+\boldsymbol{z})\nabla(\Delta \boldsymbol{d} - f(\boldsymbol{d}))dx$$
$$+ \int_D (\Delta \boldsymbol{d} - f(\boldsymbol{d}))(\boldsymbol{u}+\boldsymbol{z})\nabla f(\boldsymbol{d})dx$$
$$= \int_D (\Delta \boldsymbol{d} - f(\boldsymbol{d}))(\boldsymbol{u}+\boldsymbol{z})\nabla f(\boldsymbol{d})dx$$
$$= \int_D (\Delta \boldsymbol{d} - f(\boldsymbol{d}))(\boldsymbol{u}+\boldsymbol{z})f'(\boldsymbol{d})\nabla \boldsymbol{d} dx \tag{4.4}$$

and
$$\int_D (\Delta d - f(d))\nabla(u+z)\nabla^2 d\, dx = -\int_D (\Delta d - f(d))_{x_j}(u_{x_k}^j + z_{x_k}^j)d_{x_k} dx. \quad (4.5)$$

In view of (4.1)-(4.5), we have

$$\frac{1}{2}\frac{d}{dt}(|\nabla u|_2^2 + |\Delta d - f(d)|_2^2) + |\Delta u|_2^2 + |\nabla(\Delta d - f(d))|_2^2$$
$$= \langle (u+z)\cdot\nabla(u+z), \Delta u\rangle + \langle \nabla d, f(d)\Delta u\rangle$$
$$+\langle \nabla d(\Delta d - f(d)), \Delta u\rangle + \beta\langle z, \Delta u\rangle - \int_D (\Delta d - f(d))f'(d)(\Delta d - f(d))dx$$
$$+\int_D (\Delta d - f(d))f'(d)(u+z)\cdot\nabla d\, dx - \int_D (\Delta d - f(d))\Delta(u+z)\nabla d\, dx$$
$$-\int_D (\Delta d - f(d))f'(d)(u+z)\cdot\nabla d\, dx$$
$$-2\int_D (\Delta d - f(d))_{x_j}(u+z)_{x_k}^j d_{x_k} dx$$
$$= \langle (u+z)\cdot\nabla(u+z), \Delta u\rangle + \beta\langle z, \Delta u\rangle + \langle \nabla d, f(d)\Delta u\rangle$$
$$-\int_D (\Delta d - f(d))f'(d)(\Delta d - f(d))dx$$
$$-\int_D (\Delta d - f(d))\Delta z\nabla d\, dx$$
$$-2\int_D (\Delta d - f(d))_{x_j}(u_{x_k}^j + z_{x_k}^j)d_{x_k} dx. \quad (4.6)$$

By the Hölder inequality, the interpolation inequality and Young's inequality we have

$$\langle (u+z)\cdot\nabla(u+z), \Delta u\rangle$$
$$\leqslant c|\Delta u|_2|\Delta(u+z)|_2^{\frac{1}{2}}|\nabla(u+z)|_2^{\frac{1}{2}}|u+z|_2^{\frac{1}{2}}|\nabla(u+z)|_2^{\frac{1}{2}}$$
$$\leqslant \varepsilon|\Delta u|_2^2 + c|\Delta z|_2^2 + c|u+z|_2^2|\nabla(u+z)|_2^4$$
$$\leqslant \varepsilon|\Delta u|_2^2 + c|\Delta z|_2^2 + c|u+z|_2^2(|\nabla u|_2^4 + |\nabla z|_2^4). \quad (4.7)$$

In view of Young's inequality we have

$$\beta\langle z, \Delta u\rangle \leqslant \varepsilon|\Delta u|_2^2 + c|z|_2^2. \quad (4.8)$$

By integration by parts,

$$\langle (\nabla d)f(d), \Delta u\rangle = 0. \quad (4.9)$$

By the Hölder inequality and the Sobolev inequality,

$$\int_D (\Delta d - f(d)) f'(d)(\Delta d - f(d)) dx$$
$$\leqslant c|f'(d)|_\infty |\Delta d - f(d)|_2^2$$
$$\leqslant c(\|d\|_1^{2N} + 1)|\Delta d - f(d)|_2^2. \tag{4.10}$$

Taking a similar argument as in (4.64) yields,

$$-\int_D (\Delta d - f(d))\Delta z \nabla d\, dx$$
$$\leqslant |\Delta d - f(d)|_2 |\Delta z|_4 |\nabla d|_4$$
$$\leqslant |\Delta d - f(d)|_2 |\Delta z|_2^{\frac{1}{2}} |\nabla \Delta z|_2^{\frac{1}{2}} |\nabla d|_2^{\frac{1}{2}} |\Delta d|_2^{\frac{1}{2}}$$
$$\leqslant c|\Delta d - f(d)|_2^2 + c|\Delta d|_2^2 + c\|z\|_3^4 |\nabla d|_2^2$$
$$\leqslant c|\Delta d - f(d)|_2^2 + c|f(d)|_2^2 + c\|z\|_3^4 |\nabla d|_2^2. \tag{4.11}$$

Similarly,

$$-\int_D (\Delta d - f(d))_{x_j}(u_{x_k}^j + z_{x_k}^j) d_{x_k} dx$$
$$\leqslant |\nabla(\Delta d - f(d))|_2 |\nabla(u+z)|_4 |\nabla d|_4$$
$$\leqslant \varepsilon |\nabla(\Delta d - f(d))|_2^2 + c|\nabla(u+z)|_2 |\Delta(u+z)|_2 |\nabla d|_2 |\Delta d|_2$$
$$\leqslant \varepsilon |\nabla(\Delta d - f(d))|_2^2 + \varepsilon |\Delta u|_2^2 + \varepsilon |\Delta z|_2^2$$
$$+ c|u+z|_2^2 |\nabla d|_2^2 |f(d)|_2^2 + c|u+z|_2^2 |\nabla d|_2^2 |\Delta d - f(d)|_2^2. \tag{4.12}$$

By virtue of (4.6) − (4.12), we reach

$$\frac{d}{dt}(|\nabla u|_2^2 + |\Delta d - f(d)|_2^2) + |\Delta u|_2^2 + |\nabla(\Delta d - f(d))|_2^2$$
$$\leqslant c|\Delta z|_2^2 + c|u+z|_2^2 |\nabla(u+z)|_2^4 + c(\|d\|_1^{2N} + 1)|\Delta d - f(d)|_2^2$$
$$+ c\|z\|_3^4 |\nabla d|_2^2 + c(|u+z|_2^2 |\nabla d|_2^2 + 1)(|f(d)|_2^2 + |\Delta d - f(d)|_2^2). \tag{4.13}$$

Integrating (4.13) on interval $[t_0, -1]$ yields,

$$|\nabla u(-1)|_2^2 + |\Delta d(-1) - f(d)(-1)|_2^2$$
$$\leqslant |\nabla u(t_0)|_2^2 + |\Delta d(t_0) - f(d)(t_0)|_2^2$$
$$+ c\int_{t_0}^{-1} (|\Delta z|_2^2 + c|u+z|_2^2(|\nabla u|_2^4 + |\nabla z|_2^4) + c(\|d\|_1^{4N+2} + 1)|\Delta d - f(d)|_2^2) ds$$

$$+c\int_{t_0}^{-1}[\|z\|_3^4|\nabla d|_2^2+c(|u+z|_2^2|\nabla d|_2^2+1)(|f(d)|_2^2+|\Delta d-f(d)|_2^2)]ds. \quad (4.14)$$

By Gronwall inequality, we obtain

$$|\nabla u(-1)|_2^2+|\Delta d(-1)-f(d)(-1)|_2^2$$
$$\leqslant \Big(|\nabla u(t_0)|_2^2+|\Delta d(t_0)-f(d)(t_0)|_2^2$$
$$+\int_{t_0}^{-1}(|\Delta z|_2^2+c|u+z|_2^2|\nabla z|_2^4$$
$$+\|z\|_3^4|\nabla d|_2^2+c(|u+z|_2^2|\nabla d|_2^2+1)|f(d)|_2^2)ds\Big)$$
$$\times e^{c\int_{t_0}^{-1}(1+\|d\|_1^{2N}+|u+z|_2^2(|\nabla d|_2^2+|\nabla u|_2^2)ds}. \quad (4.15)$$

Taking integration of (4.72) over $[-2,-1]$ yields,

$$|\nabla u(-1)|_2^2+|\Delta d(-1)-f(d)(-1)|_2^2$$
$$\leqslant \Big(\int_{-2}^{-1}(|\nabla u(t_0)|_2^2+|\Delta d(t_0)-f(d)(t_0)|_2^2)dt_0$$
$$+\int_{-2}^{-1}\int_{t_0}^{-1}(|\Delta z|_2^2+c|u+z|_2^2|\nabla z|_2^4$$
$$+\|z\|_3^4|\nabla d|_2^2+c(|u+z|_2^2|\nabla d|_2^2+1)|f(d)|_2^2)dsdt_0\Big)$$
$$\times e^{c\int_{-2}^{-1}(1+\|d\|_1^{2N}+|u+z|_2^2(|\nabla d|_2^2+|\nabla u|_2^2)ds}. \quad (4.16)$$

Since we have shown in section 3 that (u,d) is uniformly bounded in $\mathbf{H}\times\mathbb{H}^1$, with respect to initial time, then (3.40) and (4.73) imply the uniform boundedness of $(\nabla u(-1),\Delta d(-1))$ in $\mathbf{H}\times\mathbb{H}$. Therefore, we obtain the absorbing ball for (u,d) in $\mathbf{V}\times\mathbb{H}^2$ at time $t=-1$. □

4.2 Two a priori estimates for the Aubin–Lions Lemma

Proposition 4.2 Let $B=\{(v_0,d_0)\in\mathbf{V}\times\mathbb{H}^2|\|v_0\|_1+\|d_0\|_2\leqslant M\}$ for some positive constant M, then $\mathbb{P}$-a.s.

$$\sup_{(v_0,d_0)\in B}\Big(\int_0^T\|u(t)\|_2^2dt+\int_0^T\|d(t)\|_3^2dt\Big)<\infty. \quad (4.17)$$

Proof Proposition 4.2 follows directly from Theorem 2.1 and (4.13). □

Proposition 4.3 Let $B = \{(v_0, d_0) \in \mathbf{H} \times \mathbb{H}^1 | \|v_0\|_1 + \|d_0\|_2 \leq M\}$ for some positive constant M, then $\mathbb{P}$-a.s.

$$\sup_{(v_0,d_0)\in B} \left(\int_0^T \|\frac{d}{dt}u(t)\|_{-1}^2 dt + \int_0^T \|\frac{d}{dt}A_2 d(t)\|_{-1}^2 dt \right) < \infty. \quad (4.18)$$

Proof For $\eta \in \mathbb{H}^2$, taking the inner product of (3.4) with $A_2\eta$ in $\mathbb{H}$ yields,

$$\left\langle \frac{d}{dt}A_2 d, \eta \right\rangle = \left\langle \frac{dd}{dt}, A_2\eta \right\rangle$$

$$= \langle \Delta d, A_2\eta \rangle - \langle (u+z) \cdot \nabla d, A_2\eta \rangle - \langle f(d), A_2\eta \rangle$$

$$\leq \|\Delta d\|_1 \|\eta\|_1 + |\nabla(u+z)|_4 |\nabla d|_4 \|\eta\|_1$$

$$+ |(u+z)|_4 |\nabla^2 d|_4 \|\eta\|_1 + |\nabla f(d)|_2 \|\eta\|_1$$

$$\leq \|d\|_3 \|\eta\|_1 + c\|u+z\|_2 \|d\|_2 \|\eta\|_1$$

$$+ \|u+z\|_1 \|d\|_3 \|\eta\|_1 + (1 + \|d\|_1^{2N+1}) \|\eta\|_1. \quad (4.19)$$

Given the bounded initial data $(v_0, d_0) \in \mathbb{H}^1 \times \mathbb{H}^2$, by (4.17) we obtain that

$$\frac{d}{dt}A_2 d \text{ is bounded in } L^2([0, T]; \mathbb{H}^{-1}). \quad (4.20)$$

For $\xi \in \mathbf{H}^1$, taking the inner product of (3.2) with ξ in $\mathbf{H}$ yields,

$$\left\langle \frac{dA_1^{\frac{1}{2}}u}{dt}, \xi \right\rangle = \left\langle \frac{du}{dt}, A_1^{\frac{1}{2}}\xi \right\rangle$$

$$= -\langle u \cdot \nabla u, A_1^{\frac{1}{2}}\xi \rangle - \langle u \cdot \nabla z, A_1^{\frac{1}{2}}\xi \rangle$$

$$- \langle z \cdot \nabla u, A_1^{\frac{1}{2}}\xi \rangle - \langle z \cdot \nabla z, A_1^{\frac{1}{2}}\xi \rangle - \langle \Delta u, A_1^{\frac{1}{2}}\xi \rangle$$

$$+ \int_D \partial_{x_i} d^k \partial_{x_j} d^k \partial_{x_i} A_1^{\frac{1}{2}} \xi^j$$

$$\leq |\nabla u|_4 |u|_4 \|\xi\|_1 + |u|_4 |\nabla z|_4 \|\xi\|_1 + |\nabla u|_4 |z|_4 \|\xi\|_1$$

$$+ |\nabla z|_4 |z|_4 \|\xi\|_1 + |\Delta u|_2 \|\xi\|_1 + c\|d\|_2 \|d\|_3 \|\xi\|_1$$

$$\leq c\|u\|_1 \|u\|_2 \|\xi\|_1 + \|u\|_1 \|z\|_2 \|\xi\|_1 + \|z\|_1 \|u\|_2 \|\xi\|_1$$

$$+ \|z\|_2 \|z\|_1 \|\xi\|_1 + \|u\|_2 \|\xi\|_1 + c\|d\|_2 \|d\|_3 \|\xi\|_1. \quad (4.21)$$

Given the bounded initial data $(v_0, d_0) \in \mathbb{H}^1 \times \mathbb{H}^2$, by (4.19) we obtain that

$$\frac{d}{dt}A_1^{\frac{1}{2}}u \text{ is bounded in } L^2([0, T]; \mathbf{H}^{-1}). \quad (4.22)$$

□

4.3 Continuity of the strong solutions in $\mathbf{V} \times \mathbb{H}^2$ w. r. t. time and initial data

By virtue of Proposition 4.2, Proposition 4.3 and Lemma 2.2 with $V = \mathbf{V}$ or $\mathbb{H}^1$, $H = \mathbf{H}$ or $\mathbb{H}$ and $V' = \mathbf{H}^{-1}$ or $\mathbb{H}^{-1}$ we infer that

Corollary 4.1

$\boldsymbol{v} \in C([0,T]; \mathbf{V})$, and $\boldsymbol{d} \in C([0,T]; \mathbb{H}^2)$, for arbitrary $T > 0$, a.s..

Proposition 4.4 Given $(\boldsymbol{v}_0, \boldsymbol{d}_0) \in \mathbf{V} \times \mathbb{H}^2$, the unique strong solution $(\boldsymbol{v}, \boldsymbol{d})$ to $(1.1) - (1.5)$ is Lipschitz continuous in $\mathbf{V} \times \mathbb{H}^2$ with respect to the initial data $(\boldsymbol{v}_0, \boldsymbol{d}_0)$.

Proof Let $(\boldsymbol{v}_1, p_1, \boldsymbol{d}_1)$ and $(\boldsymbol{v}_2, p_1, \boldsymbol{d}_2)$ be two strong solutions to $(1.1)-(1.5)$ with the initial data $(\boldsymbol{v}_{0,1}, \boldsymbol{d}_{0,1})$ and $(\boldsymbol{v}_{0,2}, \boldsymbol{d}_{0,2})$ in $\mathbf{V} \times \mathbb{H}^2$ respectively. Denote $\boldsymbol{v} := \boldsymbol{v}_1 - \boldsymbol{v}_2$, $\boldsymbol{d} := \boldsymbol{d}_1 - \boldsymbol{d}_2$ and $p := p_1 - p_2$. Then $(\boldsymbol{v}, p, \boldsymbol{d}))$ satisfies the following equations

$$\boldsymbol{v}_t + \boldsymbol{v} \cdot \nabla \boldsymbol{v}_1 + \boldsymbol{v}_2 \cdot \nabla \boldsymbol{v} - \Delta \boldsymbol{v} + \nabla p$$
$$+ \nabla \cdot (\nabla \boldsymbol{d}_1 \odot \nabla \boldsymbol{d})_1 - \nabla \cdot (\nabla \boldsymbol{d}_2 \odot \nabla \boldsymbol{d})_2 = 0, \qquad (4.23)$$

$$\nabla \cdot \boldsymbol{v}(t) = 0, \qquad (4.24)$$

$$\boldsymbol{d}_t + \boldsymbol{v} \cdot \nabla \boldsymbol{d}_1 + \boldsymbol{v}_2 \cdot \nabla \boldsymbol{d} - \Delta \boldsymbol{d} + f(\boldsymbol{d}_1) - f(\boldsymbol{d}_2) = 0, \qquad (4.25)$$

$$\boldsymbol{v}(x,t) = 0, \ \boldsymbol{d}(x,t) = \boldsymbol{d}_0(x) := \boldsymbol{d}_{1,0}(x) - \boldsymbol{d}_{2,0}(x), \ (x,t) \in \Gamma \times [t_0, \infty), \qquad (4.26)$$

$$\boldsymbol{v}|_{t=0} = \boldsymbol{v}_0(x) = \boldsymbol{v}_{0,1}(x) - \boldsymbol{v}_{0,2}(x) \text{ with } \nabla \cdot \boldsymbol{v}_0 = 0, \ \boldsymbol{d}|_{t=0} = \boldsymbol{d}_0(x), \quad x \in \mathbf{D}. \qquad (4.27)$$

Taking the inner product of (4.23) with $-\Delta \boldsymbol{v}$ in $\mathbf{H}$ yields,

$$\frac{1}{2}\frac{d|\nabla \boldsymbol{v}|_2^2}{dt} + |\Delta \boldsymbol{v}|_2^2$$
$$= \langle \boldsymbol{v} \cdot \nabla \boldsymbol{v}_1, \Delta \boldsymbol{v}\rangle + \langle \boldsymbol{v}_2 \cdot \nabla \boldsymbol{v}, \Delta \boldsymbol{v}\rangle$$
$$+ \langle \nabla \cdot (\nabla \boldsymbol{d}_1 \odot \nabla \boldsymbol{d}_1) - \nabla \cdot (\nabla \boldsymbol{d}_1 \odot \nabla \boldsymbol{d}_1), \Delta \boldsymbol{v}\rangle \qquad (4.28)$$

Taking a similar argument as in (4.2) yields,

$$\langle \nabla \cdot (\nabla \boldsymbol{d}_1 \odot \nabla \boldsymbol{d}_1) - \nabla \cdot (\nabla \boldsymbol{d}_1 \odot \nabla \boldsymbol{d}_1), \Delta \boldsymbol{v}\rangle = \langle \nabla \boldsymbol{d}_1 \Delta \boldsymbol{d}_1 - \nabla \boldsymbol{d}_2 \Delta \boldsymbol{d}_2, \Delta \boldsymbol{v}\rangle. \quad (4.29)$$

By the Hölder inequality, the interpolation inequality and Young's inequality, we have

$$\frac{1}{2}\frac{d|\nabla \boldsymbol{v}|_2^2}{dt} + |\Delta \boldsymbol{v}|_2^2$$

$$\leqslant |\Delta v|_2 |\nabla v_1|_4 |v|_4 + |\Delta v|_2 |\nabla v|_4 |v_2|_4$$
$$+ (|\nabla d_1|_4 |\Delta d_1|_4 + |\nabla d_2|_4 |\Delta d_2|_4) |\Delta v|_2$$
$$\leqslant \varepsilon |\Delta v|_2^2 + c |\nabla v_1|_4^2 |v|_4^2 + c |\nabla v|_4^2 |v_2|_4^2$$
$$+ c(|\Delta d_1|_2^2 |\nabla \Delta d_1|_2^2 + |\Delta d_2|_2^2 |\nabla \Delta d_2|_2^2)$$
$$\leqslant \varepsilon |\Delta v|_2^2 + c\|v_1\|_2^2 |\nabla v|_2^2 + c(|\Delta v|_2 + |\nabla v|_2) |\nabla v|_2 \|v_2\|_1^2$$
$$+ c(|\Delta d_1|_2^2 |\nabla \Delta d_1|_2^2 + |\Delta d_2|_2^2 |\nabla \Delta d_2|_2^2)$$
$$\leqslant \varepsilon |\Delta v|_2^2 + c(\|v_1\|_2^2 + \|v_2\|_1^2 + \|v_2\|_1^4) |\nabla v|_2^2$$
$$+ c(|\Delta d_1|_2^2 |\nabla \Delta d_1|_2^2 + |\Delta d_2|_2^2 |\nabla \Delta d_2|_2^2). \tag{4.30}$$

Taking a similar argument as in (4.60) yields,

$$\frac{1}{2} \frac{d|\Delta d|_2^2}{dt} = \langle \Delta d, \Delta d_t \rangle$$
$$= \langle \Delta d, \Delta^2 d \rangle + \langle \Delta v \nabla d_1, \Delta d \rangle + \langle v \nabla \Delta d_1, \Delta d \rangle + 2 \int_D \nabla v \nabla^2 d_1 \Delta d dx$$
$$+ \langle \Delta v_2 \nabla d, \Delta d \rangle + \langle v_2 \nabla \Delta d, \Delta d \rangle + 2 \int_D \nabla v_2 \nabla^2 d \Delta d dx$$
$$+ \langle \Delta (f(d_1) - f(d_2)), \Delta d \rangle. \tag{4.31}$$

Since by integration by parts, the Hölder inequality and the interpolation inequality we have

$$\langle \Delta (f(d_1) - f(d_2)), \Delta d \rangle$$
$$= \langle \nabla (f(d_1) - f(d_2)), \nabla \Delta d \rangle$$
$$\leqslant c |\nabla \Delta d|_2 (|\nabla f(d_1)|_2^2 + |\nabla f(d_2)|_2^2)$$
$$\leqslant \varepsilon |\nabla \Delta d|_2^2 + c\Big((|d_1|_\infty^{4N} + 1)\|d_1\|_1^2 + (|d_2|_\infty^{4N} + 1)\|d_2\|_1^2\Big)$$
$$\leqslant \varepsilon |\nabla \Delta d|_2^2 + c(\|d_1\|_1^{4N+2} + \|d_1\|_1^2 + \|d_2\|_1^{4N+2} + \|d_2\|_1^2). \tag{4.32}$$

By the Sobolev imbedding theorem and Young's inequality we have

$$2 \int_D \nabla v \nabla^2 d_1 \Delta d dx + 2 \int_D \nabla v_2 \nabla^2 d \Delta d dx$$
$$\leqslant c |\nabla v|_\infty \|d_1\|_2 |\Delta d|_2 + c |\nabla v_2|_\infty |\Delta d|_2^2$$
$$\leqslant \varepsilon (|\Delta v|_2^2 + |\nabla v|_2^2) + c(\|d_1\|_2^2 + \|v_2\|_2) |\Delta d|_2^2. \tag{4.33}$$

In view of (4.31) − (4.33), the Hölder inequality, the interpolation inequality and Young's inequality we obtain

$$\frac{1}{2}\frac{d|\Delta d|_2^2}{dt} + |\nabla\Delta d|_2^2$$
$$\leqslant |\Delta v|_2|\nabla d_1|_4|\Delta d|_4 + |\nabla\Delta d_1|_2|v|_4|\Delta d|_4 + |\Delta v_2|_2|\nabla d|_4|\Delta d|_4$$
$$+ |\nabla d|_2|\Delta d|_4|v_2|_4 + \varepsilon|\nabla\Delta d|_2^2$$
$$+ c(\|d_1\|_1^{4N+2} + \|d_1\|_1^2 + \|d_2\|_1^{4N+2} + \|d_2\|_1^2)$$
$$+ \varepsilon(|\Delta v|_2^2 + |\nabla v|_2^2) + c(\|d_1\|_2^2 + \|v_2\|_2)|\Delta d|_2^2$$
$$\leqslant c|\Delta v|_2|\Delta d_1|_2|\Delta d|_2^{\frac{1}{2}}|\nabla\Delta d|_2^{\frac{1}{2}} + c|\nabla\Delta d_1|_2|\nabla v|_2|\Delta d|_2^{\frac{1}{2}}|\nabla\Delta d|_2^{\frac{1}{2}}$$
$$+ c|\Delta v_2|_2|\Delta d|_2|\nabla\Delta d|_2 + c|\Delta d|_2^{\frac{3}{2}}|\Delta d|_2^{\frac{1}{2}}|\nabla v_2|_2 + \varepsilon|\nabla\Delta d|_2^2$$
$$+ c(\|d_1\|_1^{4N+2} + \|d_1\|_1^2 + \|d_2\|_1^{4N+2} + \|d_2\|_1^2)$$
$$+ \varepsilon(|\Delta v|_2^2 + |\nabla v|_2^2) + c(\|d_1\|_2^2 + \|v_2\|_2)|\Delta d|_2^2$$
$$\leqslant \varepsilon|\nabla\Delta d|_2^2 + \varepsilon|\Delta v|_2^2 + c|\Delta d_1|_2^4|\Delta d|_2^2 + c|\nabla\Delta d_1|_2^2|\nabla v|_2^2$$
$$+ c|d|_2^2 + c|\Delta d|_2^2 + c|\Delta v_2|_2^2|\Delta d|_2^2 + c|\Delta d|_2^2|\nabla v_2|_2^4$$
$$+ c(\|d_1\|_1^{4N+2} + \|d_1\|_1^2 + \|d_2\|_1^{4N+2} + \|d_2\|_1^2)$$
$$+ \varepsilon(|\Delta v|_2^2 + |\nabla v|_2^2) + c(\|d_1\|_2^2 + \|v_2\|_2)|\Delta d|_2^2. \tag{4.34}$$

To prove the continuity of the strong solution in $\mathbf{V} \times \mathbb{H}^2$ w.r.t. the initial data we introduce some notations

$$h(t) = |\nabla v|_2^2 + |\Delta d(t)|_2^2,$$
$$\xi(t) = 1 + \|v_1\|_2^2 + \|v_2\|_1^4 + \|v_2\|_2^2 + \|d_1\|_2^2 + |\nabla\Delta d_1|_2^2,$$
$$\eta(t) = (1 + \|d_1\|_1^{4N+2} + \|d_2\|_1^{4N+2} + |\Delta d_1|_2^2|\nabla\Delta d_1|_2^2 + |\Delta d_2|_2^2|\nabla\Delta d_2|_2^2),$$

where $t \geqslant 0$. In view of (4.30), (4.33), Theorem 2.1 and the Gronwall inequality, we obtain that

$$h(t) \leqslant c(\|v_0\|_1^2 + \|d_0\|_2^2) \int_0^T j(t)dt e^{c\int_0^T \xi(t)dt},$$

which implies that the strong solution to (1.1) − (1.5) is Lipschitz continuous in $\mathbf{V} \times \mathbb{H}^2$ w.r.t. initial data. □

4.4 The existence of compact absorbing Ball in $\mathbf{V} \times \mathbb{H}^2$

With the above a priori estimates and results, repeating the arguments of proposition 3.5 and proposition 3.6 with minor revisions, we have the following conclusions.

Proposition 4.5 *For $\omega \in \Omega$, $S(t,s;\omega)$ is compact from $\mathbf{V} \times \mathbb{H}^2$ to $\mathbf{V} \times \mathbb{H}^2$, for all $s,t \in \mathbf{R}$ and $s \leqslant t$.*

Proposition 4.6 *There exists a compact absorbing ball in $\mathbf{V} \times \mathbb{H}^2$ at any time $t \in \mathbf{R}$ for the stochastic dynamical system (1.1) − (1.5).*

Remark 4.2 *By Proposition 4.5 in [10], the existence of a random attractor as constructed in the proof of Theorem 4.1 implies the existence of an invariant Markov measure $\mu. \in \mathscr{P}_\Omega(\mathbf{V} \times \mathbb{H}^2)$ for φ (see Definition 4.1 in [10]). In view of [8] there exists an invariant measure μ for the Markovian semigroup $\mathbb{P}_t[f(S(t,0,x))]$ satisfying*

$$\mu(B) = \int_\Omega \mu_\omega(B)\mathbb{P}(d\omega), \quad \text{and } \mu(\mathcal{A}(\omega)) = 1, \ \mathbb{P}\text{-a.e.},$$

where $\mathcal{A}$ is given by Theorem 4.1 and $B \in \mathscr{B}(\mathbf{V} \times \mathbb{H}^2)$ which is a Borel σ-algebra on $\mathbf{V} \times \mathbb{H}^2$. If the invariant measure for $\mathbb{P}_t$ is unique, then the invariant Markov measure $\mu.$ for φ is unique and given by

$$\mu_\omega = \lim_{t \to \infty} \varphi(t, \vartheta_{-t}\omega)\mu.$$

Acknowledgements We deeply appreciate the valuable discussions with Professor Z. Brzeźniak for many days, and we also are deeply grateful to Professor Daiwen Huang and Dr Rangrang Zhang for their valuable suggestions.

References

[1] R. A. Adams, Sobolev Spaces, Academic Press, New York, 1975.

[2] J. Aubin, Un théorème de compacité, C. R. Acad. Sci. Paris, 256 (1963), 5042–5044.

[3] Z. Brzeźniak, T. Caraballo, J.A. Langa, Y. Li, G. Lukaszewicz, J. Real, Random attractors for stochastic 2D-Navier-Stokes equations in some unbounded domains. J. Differential Equations 255 (2013), no. 11, 3897–3919.

[4] Z. Brzeźniak, Y. Li, Asymptotic compactness and absorbing sets for 2D stochastic Navier-Stokes equations on some unbounded domains, Trans. Amer. Math. Soc. 358 (2006), no. 12, 5587–5629.

[5] Z. Brzeźniak, E. Hausenblas, and P. Razafimandimby, Some results on a system of Nonlinear SPDEs with Multiplicative Noise arising in the Dynamics of Nematic Liquid Crystals. arxiv: 1310.8641v3 [math.PR]2016.

[6] Z. Brzeźniak, E. Hausenblas, and P. Razafimandimby, Stochastic Nonparabolic dissipative systems modeling the flow of Liquid Crystals: Strong solution, RIMS Kokyuroku Proceeding of RIMS Symposium on Mathematical Analysis of Incompressible Flow, 1875(2014), 41–72.

[7] Z. Brzeźniak, U. Manna and P. Akash Ashirbad, Existence of weak martingale solution of nematic liquid crystals driven by pure jump noise in the Marcus canonical form, J. Differential Equations (2018), https://doi.org/10.1016/j.jde.2018.11.001

[8] H. Crauel, Markov measures for random dynamical systems, Stochastics Stochastics Rep. 37(1991), 153–173.

[9] H. Crauel, A. Debussche, F. Flandoli, Random attractors, J. Dynam. Differential Equations, 9 (1997), 307–341.

[10] H. Crauel, F. Flandoli, Attractors for random dynamical systems, Probab. Theory Relat. Fields. 100(1994), 365–393.

[11] G. Da Prato, J. Zabczyk, Stochastic Equations in Infinite Dimensions, Encyclopedia of Mathematics and its Applications vol 44, 1992.

[12] J. Ericksen, Conservation laws for liquid crystals, Trans. Soc. Rheol. 5 (1961), 22–34.

[13] N. Gershenfeld, The Nature of Mathematical Modeling ,Cambridge Press, Cambridge, 1999.

[14] B. Guo and D. Huang, 3d stochastic primitive equations of the large-scale ocean: global wellposedness and attractors, Commun. Math. Phys. 286(2009), 697–723.

[15] M. Grasselli and H. Wu , Finite-dimensional global attractor for a system modeling the 2D nematic liquid crystal flow, Z. Angew. Math. Phys. 62 (2011), 979–992.

[16] W. Horsthemke, C. R. Doering, R. Lefever and A. S. Chi, Effect of external-field fluctuations on instabilities in nematic liquid crystals, Phys. Rev. A, 31(1985), 1123–1135.

[17] W. Horsthemke and R. Lefever, Noise-Induced Transitions. Theory and Applications in Physics, Chemistry, and Biology. Springer Series in Synergetics, 15. Springer-Verlag, Berlin, 1984.

[18] R. M. Jendrejack, J. J. de Pablo and M. D. Graham, A method for multiscale simulation of flowing complex fluids, J. Non-Newtonian Fluid Mech. (108) 2002, 123–142.

[19] Y. Li, Z. Brzeźniak, J.Z. Zhou, Conceptual analysis and random attractor for dissipative random dynamical systems, Acta Math. Sci. Ser. B (Engl. Ed.) 28 (2008), no. 2, 253–268.

[20] F. M. Leslie, Theory of flow phenomena in liquid crystals, in "Advances in Liquid Crystals" (eds. G. Brown), Academic Press, 4 (1979), 1–81.

[21] F.H. Lin, Nonlinear theory of defects in nematic liquid crystals: Phase transitions and flow phenomena, Comm. Pure Appl. Math. XLII (1989), 789–814.

[22] F.-H. Lin and C. Liu, Nonparabolic dissipative systems modeling the flow of Liquid Crystals, Communications on Pure and Applied Mathematics, Vol. XLVIII(1995), 501–537 .

[23] F.-H. Lin and C. Liu, Partial regularities of the nonlinear dissipative systems modeling the flow of liquid crystals, Discrete Contin. Dyn. Syst., 2 (1996), 1–23.

[24] J. Lions, Quelques Méthode de Résolution des Problèmes aux Limites Non Linéaires,

Dunod, Paris, 1969.
[25] K. Liu, Stability of stochastic differential equations in infinite dimensions, Springer Verlag, New York, 2004.
[26] M. Laso and H. C. Öttinger, Calculation of viscoelastic flow using molecular models: the connffessit approach, J. Non-Newtonian Fluid Mech. 1 (1993), 1–20.
[27] C. Liu and N. J. Walkington, Approximation of liquid crystal flows, SIAM J. Numer. Anal. 37 (2000), 725–741.
[28] C. Liu and N. J. Walkington, Mixed methods for the approximation of liquid crystal flows, M2AN Math. Model. Numer. Anal. 36 (2002), 205–222.
[29] J. Lions and B. Magenes, Nonhomogeneous Boundary Value Problems and Applications, Springer-Verlag, New York, 1972.
[30] S. Shkoller, Well-posedness and global attractors for liquid crystals on Riemannian manifolds, Communications in Partial Differential Equations, 27(2002), 1103–1137.
[31] M. Somasi and B. Khomami, Linear stability and dynamics of viscoelastic flows using time-dependent stochastic simulation techniques, Journal of Non-Newtonian Fluid Mechanics, 93(2000), 339–362.
[32] M. Somasi and B. Khomami, A new approach for studying the hydrodynamic stability of fluids with microstructure, Phys Fluids, 13(2001), 1811–1814.
[33] M. San Miguel, Nematic liquid crystals in a stochastic magnetic field: Spatial correlations, Phys. Rev. A 32(1985), 3811–3813, 1985.
[34] F. Sagués and M. San Miguel. Dynamics of Fréedericksz transition in a fluctuating magnetic field. Phys. Rev. A. 32(3): 1843–1851, 1985.
[35] Guoli Zhou, Random attractor of the 3D viscous primitive equations driven by fractional noises, arXiv:1604.05376.

The Existence of Global Solutions for the Fourth-Order Nonlinear Schrödinger Equations*

Guo Chunxiao (郭春晓), and Guo Boling (郭柏灵)

Abstract In this paper, the problem of a class of multidimensional fourth-order nonlinear Schrödinger equation including the derivatives of the unknown function in the nonlinear term is studied, and the existence of global weak solutions of nonlinear Schrödinger equation is proved by the Galerkin method according to the different values of λ.

Keywords fourth-order nonlinear Schrödinger equation; galerkin method; initial boundary value problem

1 Introduction

In this paper, we consider the following multidimensional fourth-order nonlinear Schrödinger equation (1.1) with initial boundary values (1.2) and (1.3)

$$iu_t + \Delta^2 u + \lambda \sum_{j=1}^n \frac{\partial}{\partial x_j}\left(\left|\frac{\partial u}{\partial x_j}\right|^{p-2}\frac{\partial u}{\partial x_j}\right) - \alpha \Delta |u|^2 u + \beta(x) f(|u|^2) u = 0, \quad (1.1)$$

$$u|_{t=0} = u_0(x), \qquad x \in \Omega, \quad (1.2)$$

$$u|_{\partial\Omega} = 0, \quad \Delta u|_{\partial\Omega} = 0, \quad t \geq 0, \quad (1.3)$$

where Ω is a bounded smooth domain, $u(x,t) = (u_1(x,t), u_2(x,t), \cdots, u_n(x,t))$ is a complex-valued vector function, $f(s)$ is a real-valued function satisfying $|f(s)| \leq As^q + B$, with $A, B > 0$, and $q \geq 0$, $\alpha > 0$ is a real number and $\beta(x)$ is a

---

*J. Appl. Anal. Comput, 2019, 9(3): 1–10. Doi: 10.11948/2156-907X.20190095

given real-valued bounded function. The initial data $u_0(x)$ is a given complex-valued function, and the boundary condition is taken to be the Dirichlet boundary condition for both u and Δu on the boundary $\partial \Omega$.

The nonlinear Schrödinger (NLS) equation arises from the study of nonlinear wave propagation in dispersive and inhomogeneous media, such as plasma phenomena and non-uniform dielectric media [8]. In recent years, many studies have been devoted to the nonlinear Schrödinger equations with a variety of nonlinearities. Some methods, theoretical, numerical or analytical, have been used to deal with these problems. The fourth-order physical models occur in various areas of physics, including nonlinear optics, plasma physics, superconductivity and quantum mechanics [3,5,6]. Several exact solutions were obtained for the fourth-order nonlinear Schrödinger equation with cubic and power law nonlinearities in [11]. Karpman investigated the stability of the soliton solutions to the fourth-order nonlinear Schrödinger equations with arbitrary power nonlinearities in different space dimensions in [7]. Guo and Cui obtained the global existence of solutions for a fourth-order nonlinear Schrödinger equation in [2]. Huo and Jia in [4] considered the Cauchy problem for the fourth-order nonlinear Schrödinger equation related to the vortex filament. In this paper, we study the global existence of weak solutions for the initial boundary problem of the more general nonlinear Schrödinger equation. In particular, we will establish global existence of weak solutions for equations (1.1)–(1.3), which is different from that considered in [2] in several aspects. Also, the present results extend the results of [1] in the following aspects. Firstly, compared to the condition $n < 4$ in [1], we can prove that weak solutions exist for $n < 6$. Secondly, the emphasis of this paper will be the different values of λ. We define

$$E(t) \equiv \frac{1}{2}\|\Delta u\|^2 - \frac{\lambda}{p}\sum_{j=1}^{n}\int_{\Omega}|\frac{\partial u}{\partial x_j}|^p\,dx + \frac{\alpha}{4}\|\nabla|u|^2\|^2 + \frac{1}{2}\int_{\Omega}\beta(x)F(|u(t)|^2)\,dx,$$

where $F(s) = \int_0^s f(\tau)d\tau$. When $\lambda < 0$, the second term in energy $E(t)$ is positive, and we can obtain some useful estimates. But when $\lambda > 0$, we need some mathematical techniques to deal with this energy in order to obtain some estimates. One can refer to Lemma 2.2 and Lemma 2.3 for the details.

Let $L^p(\Omega)$ be the p-times integrable function space on Ω with norm denoted by $\|\cdot\|_{L^p}$. Of course, $\|\cdot\|_{L^2}$ and $\|\cdot\|$ coincide. Let H_0^s be the completion of $C_0^\infty(\Omega)$

in the H^s norm, and $W_0^{s,p}$ be the completion of $C_0^\infty(\Omega)$ in the $W^{s,p}$ norm. We use $\|\cdot\|_s$ to denote the H^s Sobolev-norm for $s \in R$, and use $\|\cdot\|_{W^{s,p}}$ to denote the $W^{s,p}$ norm. The letter C will denote some positive constant, which may change from one line to another.

Definition 1.1 *A complex-valued vector function $u(x,t)$ is called a global weak solution of equations (1.1)–(1.3), if for any $T > 0$, $u(x,t) \in L^\infty(0,T; H_0^2(\Omega) \cap W_0^{1,p}(\Omega)) \cap W^{1,\infty}(0,T; H^{-2}(\Omega) \cap W^{-1,p'}(\Omega))$, $p \geqslant 2$, and it satisfies:*

$$-i \int_0^T (u, v_t)\, dt + \int_0^T (\Delta u, \Delta v)\, dt - \lambda \int_0^T \int_\Omega \left(\sum_{j=1}^n |\frac{\partial u}{\partial x_j}|^{p-2} \frac{\partial u}{\partial x_j} \frac{\partial \overline{v}}{\partial x_j} \right) dx\, dt$$

$$- \alpha \int_0^T (\Delta |u|^2 u, v)\, dt + \int_0^T (\beta(x) f(|u|^2) u, v)\, dt - i(u_0(x), v(0)) = 0, \qquad (1.4)$$

for all $v(x,t) \in C^1(0,T; L^2(\Omega)) \cap C(0,T; H_0^2(\Omega) \cap W_0^{1,p}(\Omega))$, and $v(x,T) \equiv 0$.

The main result of this paper is as follows:

Theorem 1.1 *Assume that $n < 6$, $u_0 \in H_0^2(\Omega) \cap W_0^{1,p}(\Omega)$, and $\int_\Omega \beta(x) F(|u_0(x)|^2)\, dx < \infty$, $|\beta(x)| \leqslant M$. If either (P_1) or (P_2) below is satisfied,*

(P_1) $\lambda < 0$, and $0 \leqslant q < \min\left\{\dfrac{4}{n}, \dfrac{p}{2} - 1 + \dfrac{p}{n}\right\}$, $p \geqslant 2$,

(P_2) $\lambda > 0$, $0 \leqslant q < \dfrac{4}{n}$, and $2 \leqslant p < \dfrac{2n+8}{n+2}$,

then there exists a global weak solution for the equations (1.1)–(1.3).

This paper is organized as follows. In section 2, we give some *a prior* estimates according to the different value of λ; In section 3, we obtain the existence of weak solutions by the Galerkin method.

2 A priori estimates

Lemma 2.1 *Assume that $u_0(x) \in L^2(\Omega)$, $\beta(x)$ and $f(s)$ are real-valued functions, and $u(x,t)$ is the solution of the equations (1.1)–(1.3). Then $\|u(t)\|^2 = \|u_0\|^2$.*

Proof Taking the inner product of the equation (1.1) with u, we have

$$i(u_t, u) + (\Delta^2 u, u) + \lambda \left(\sum_{j=1}^n \frac{\partial}{\partial x_j} \left(|\frac{\partial u}{\partial x_j}|^{p-2} \frac{\partial u}{\partial x_j} \right), u \right)$$

$$-\alpha(\Delta|u|^2 u, u) + (\beta(x)f(|u|^2)u, u) = 0. \qquad (2.1)$$

Notice that

$$(\Delta^2 u, u) = \int_\Omega \Delta^2 u \cdot \bar{u}\, dx = \int_\Omega \Delta u \cdot \Delta \bar{u}\, dx + \int_{\partial\Omega}\left(\frac{\partial \Delta u}{\partial n}\cdot \bar{u} - \frac{\partial \bar{u}}{\partial n}\cdot \Delta u\right) dS = \|\Delta u\|^2,$$

where we used the boundary condition (1.3).

For the third term on the left-hand side of equation (2.1), applying the integration by parts, we have

$$\lambda\Big(\sum_{j=1}^n \frac{\partial}{\partial x_j}\Big(\Big|\frac{\partial u}{\partial x_j}\Big|^{p-2}\frac{\partial u}{\partial x_j}\Big), u\Big) = \lambda\int_\Omega \sum_{j=1}^n \frac{\partial}{\partial x_j}\Big(\Big|\frac{\partial u}{\partial x_j}\Big|^{p-2}\frac{\partial u}{\partial x_j}\Big)\bar{u}\, dx$$

$$= -\lambda\sum_{j=1}^n \int_\Omega \Big|\frac{\partial u}{\partial x_j}\Big|^{p-2}\frac{\partial u}{\partial x_j}\frac{\partial \bar{u}}{\partial x_j}\, dx$$

$$= -\lambda\sum_{j=1}^n \int_\Omega \Big|\frac{\partial u}{\partial x_j}\Big|^{p}\, dx. \qquad (2.2)$$

We can easily obtain the following estimates

$$-\alpha(\Delta|u|^2 u, u) = -\alpha\int_\Omega \Delta|u|^2 u\bar{u}\, dx = -\alpha\int_\Omega \Delta|u|^2 |u|^2\, dx, \qquad (2.3)$$

and

$$(\beta(x)f(|u|^2)u, u) = \int_\Omega \beta(x)f(|u|^2)|u|^2\, dx, \qquad (2.4)$$

where $\beta(x), f(s)$ are real-valued functions.

Hence, taking the imaginary part of the equation (2.1), we can obtain $\|u(t)\|^2 = \|u_0\|^2$.

Lemma 2.2 Let $u_0 \in H_0^2(\Omega) \cap W_0^{1,p}(\Omega)$, $\int_\Omega \beta(x)F(|u_0(x)|^2)\, dx < \infty$, and $u(x,t)$ be the solution of (1.1)–(1.3). Then we have

$$E(t) \equiv \frac{1}{2}\|\Delta u\|^2 - \frac{\lambda}{p}\sum_{j=1}^n \int_\Omega \Big|\frac{\partial u}{\partial x_j}\Big|^p dx + \frac{\alpha}{4}\|\nabla|u|^2\|^2 + \frac{1}{2}\int_\Omega \beta(x)F(|u(t)|^2)\, dx = E(0),$$

where $F(s) = \int_0^s f(\tau)\, d\tau$.

Proof Taking the inner product of (1.1) with u_t, we have

$$i(u_t, u_t) + (\Delta^2 u, u_t) + \lambda\bigg(\sum_{j=1}^{n} \frac{\partial}{\partial x_j}\big(|\frac{\partial u}{\partial x_j}|^{p-2}\frac{\partial u}{\partial x_j}\big), u_t\bigg)$$
$$- \alpha(\Delta|u|^2 u, u_t) + (\beta(x)f(|u|^2)u, u_t) = 0. \tag{2.5}$$

Next, we take the real part and estimate the terms one by one. First, for the second term on the left-hand side of equation (2.5), we get

$$\text{Re}(\Delta^2 u, u_t) = \text{Re}\int_\Omega \Delta u \cdot \Delta \overline{u_t}\, dx + \text{Re}\int_{\partial\Omega}\big(\overline{u_t}\frac{\partial \Delta u}{\partial n} - \Delta u \frac{\partial \overline{u_t}}{\partial n}\big)dS = \frac{1}{2}\frac{d}{dt}\|\Delta u\|^2. \tag{2.6}$$

Applying the integration by parts, we obtain

$$\text{Re}\lambda\bigg(\sum_{j=1}^{n}\frac{\partial}{\partial x_j}\big(|\frac{\partial u}{\partial x_j}|^{p-2}\frac{\partial u}{\partial x_j}\big), u_t\bigg)$$
$$= -\lambda\text{Re}\sum_{j=1}^{n}\int_\Omega \big(|\frac{\partial u}{\partial x_j}|^{p-2}\frac{\partial u}{\partial x_j}\big)\cdot\frac{\partial \overline{u_t}}{\partial x_j}\, dx$$
$$= -\frac{\lambda}{p}\frac{d}{dt}\sum_{j=1}^{n}\int_\Omega |\frac{\partial u}{\partial x_j}|^p\, dx, \tag{2.7}$$

$$-\alpha\text{Re}(\Delta|u|^2 u, u_t) = \frac{\alpha}{2}\text{Re}\int_\Omega \nabla|u|^2\cdot\frac{d}{dt}\nabla|u|^2\, dx - \frac{\alpha}{2}\text{Re}\int_{\partial\Omega}\frac{\partial|u|^2}{\partial n}\cdot\frac{d}{dt}|u|^2\, dx$$
$$= \frac{\alpha}{4}\frac{d}{dt}\|\nabla|u|^2\|^2, \tag{2.8}$$

$$\text{Re}(\beta(x)f(|u|^2)u, u_t) = \frac{1}{2}\int_\Omega \beta(x)f(|u|^2)\frac{d}{dt}|u|^2\, dx$$
$$= \frac{1}{2}\frac{d}{dt}\int_\Omega \beta(x)F(|u|^2)\, dx. \tag{2.9}$$

From (2.6)–(2.9), we deduce that

$$\frac{d}{dt}\bigg[\frac{1}{2}\|\Delta u\|^2 - \frac{\lambda}{p}\sum_{j=1}^{n}\int_\Omega |\frac{\partial u}{\partial x_j}|^p\, dx + \frac{\alpha}{4}\|\nabla|u|^2\|^2 + \frac{1}{2}\int_\Omega \beta(x)F(|u(t)|^2)\, dx\bigg] = 0.$$

The proof is complete.

Lemma 2.3 Assume that $|f(s)| \leqslant As^q + B$, $A, B > 0$. Under the assumptions in Lemma 2.2, if either

Case 1 when $\lambda < 0$ and $0 \leqslant q < \min\left\{\dfrac{4}{n}, \dfrac{p}{2} - 1 + \dfrac{p}{n}\right\}$, $p \geqslant 2$, or

Case 2 when $\lambda > 0$, $0 \leqslant q < \dfrac{4}{n}$, and $2 \leqslant p < \dfrac{2n+8}{n+2}$,

then we have

$$\|\Delta u\|^2 + \sum_{j=1}^{n} \int_\Omega \left|\frac{\partial u}{\partial x_j}\right|^p dx \leqslant C, \quad \int_\Omega \beta(x) F(|u(t)|^2) \, dx \leqslant C.$$

Proof From the assumed conditions, we have

$$\left|\int_\Omega \beta(x) F(|u(t)|^2) \, dx\right| \leqslant C\left(\int_\Omega |u(t)|^{2q+2} \, dx + \int_\Omega |u(t)|^2 \, dx\right),$$

where $F(s) = \displaystyle\int_0^s f(\tau) \, d\tau$.

According to Lemma 2.1, the second term on the right-hand side of above inequality is bounded. Then

$$\left|\int_\Omega \beta(x) F(|u(t)|^2) \, dx\right| \leqslant C\left(\int_\Omega |u(t)|^{2q+2} \, dx + 1\right). \tag{2.10}$$

By the Gagliardo-Nirenberg inequality, we get

$$\|u\|_{L^{2q+2}}^{2q+2} \leqslant C\|u\|^{\frac{4+q(4-n)}{2}} \|D^2 u\|^{\frac{qn}{2}}.$$

Noticing that $nq < 4$, combining Lemma 2.1 and the ϵ-Young inequality, we have

$$\|u\|_{L^{2q+2}}^{2q+2} \leqslant C\|\Delta u\|^{\frac{qn}{2}} \leqslant \frac{1}{4}\|\Delta u\|^2 + C. \tag{2.11}$$

Using the Gagliardo-Nirenberg inequality again,

$$\|u\|_{L^{2q+2}}^{2q+2} \leqslant C\|Du\|_{L^p}^{\frac{2npq}{np-2n+2p}} \|u\|^{\frac{4q(p-n)+2(np-2n+2p)}{np-2n+2p}}.$$

Combing with Lemma 2.1, when $q < \dfrac{p}{2} - 1 + \dfrac{p}{n}$, we have

$$\|u\|_{L^{2q+2}}^{2q+2} \leqslant C\|Du\|_{L^p}^{\frac{2npq}{np-2n+2p}} \leqslant \frac{|\lambda|}{2p}\|Du\|_{L^p}^p + C. \tag{2.12}$$

Case 1 $\lambda < 0$, the second term in $E(t)$ is positive. Then from Lemma 2.2, we have

$$\frac{1}{4}\|\Delta u\|^2 + \frac{\alpha}{4}\|\nabla |u|^2\|^2 - \frac{\lambda}{p}\sum_{j=1}^{n}\int_{\Omega}|\frac{\partial u}{\partial x_j}|^p\,dx \leqslant C.$$

Then

$$\|\Delta u\|^2 + \sum_{j=1}^{n}\int_{\Omega}|\frac{\partial u}{\partial x_j}|^p\,dx \leqslant C.$$

By (2.11), we have

$$\int_{\Omega}\beta(x)F(|u(t)|^2)\,dx \leqslant C.$$

In conclusion, when $0 \leqslant q < \min\{\frac{4}{n}, \frac{p}{2} - 1 + \frac{p}{n}\}$ and $p \geqslant 2$,

$$\|\Delta u\|^2 + \sum_{j=1}^{n}\int_{\Omega}|\frac{\partial u}{\partial x_j}|^p\,dx \leqslant C,$$

$$\int_{\Omega}\beta(x)F(|u(x,t)|^2)\,dx \leqslant C.$$

Case 2 $\lambda > 0$, and $(n-2)p < 2n$, if $\frac{2p-2n+np}{4} < 2$. Then by using the Gagliardo-Nirenberg inequality and the ϵ-Young inequality, we have

$$\frac{\lambda}{p}\sum_{j=1}^{n}\int_{\Omega}|\frac{\partial u}{\partial x_j}|^p\,dx \leqslant \frac{\lambda}{p}C\|Du\|_{L^p}^p \leqslant \frac{\lambda}{p}C\|u\|^{\frac{np+2p-2n}{4}}\|\Delta u\|^{\frac{np+2p-2n}{4}}$$

$$\leqslant \frac{\lambda}{p}C\|\Delta u\|^{\frac{2p-2n+np}{4}}$$

$$\leqslant \frac{1}{8}\|\Delta u\|^2 + C. \tag{2.13}$$

In order to satisfy the conditions $(n-2)p < 2n$ and $(n+2)p < 8+2n$, we only need to have $p < \frac{2n+8}{n+2}$. In particular, when $2 \leqslant p < \frac{2n+8}{n+2}$, from (2.11), we have

$$\frac{1}{8}\|\Delta u\|^2 + \frac{\alpha}{4}\|\nabla |u|^2\|^2 \leqslant C, \quad \int_{\Omega}\beta(x)F(|u(t)|^2)\,dx \leqslant C.$$

The proof is complete.

Lemma 2.4 *Under the assumptions in Lemma 2.3, if $n < 6$, then the following estimate holds*

$$\|u_t\|_{H^{-2}(\Omega)\cap W^{-1,p'}(\Omega)} \leq C.$$

Proof $\forall v \in H_0^2(\Omega) \cap W_0^{1,p}(\Omega)$, taking the inner product of the equation (1.1) with v, we have

$$i(u_t, v) + (\Delta^2 u, v) + \lambda \Big(\sum_{j=1}^{n} \frac{\partial}{\partial x_j} \Big(\Big|\frac{\partial u}{\partial x_j}\Big|^{p-2} \frac{\partial u}{\partial x_j} \Big), v \Big)$$
$$- \alpha(\Delta|u|^2 u, v) + (\beta(x)f(|u|^2)u, v) = 0. \tag{2.14}$$

For the second term on the left-hand side of (2.14), we have

$$(\Delta^2 u, v) = \int_\Omega \Delta^2 u \bar{v}\, dx = \int_\Omega \Delta u \Delta \bar{v}\, dx + \int_{\partial\Omega} \Big(\frac{\partial \Delta u}{\partial n} \cdot \bar{v} - \Delta u \cdot \frac{\partial \bar{v}}{\partial n}\Big) dS$$
$$\leq \|\Delta u\|\|\Delta v\| \leq C\|v\|_2.$$

For the third term, combing the conclusion of Lemma 2.3 and the Hölder inequality, we obtain

$$\Big|\lambda \sum_{j=1}^{n} \frac{\partial}{\partial x_j}\Big(\Big|\frac{\partial u}{\partial x_j}\Big|^{p-2} \frac{\partial u}{\partial x_j}\Big)\bar{v}\, dx\Big| \leq |\lambda| \sum_{j=1}^{n} \Big|\frac{\partial u}{\partial x_j}\Big|^{p-1} \Big|\frac{\partial \bar{v}}{\partial x_j}\Big|\, dx$$
$$\leq |\lambda| \sum_{j=1}^{n} \Big[\Big(\int_\Omega \Big|\frac{\partial u}{\partial x_j}\Big|^p dx\Big)^{\frac{p-1}{p}} \Big(\int_\Omega \Big|\frac{\partial \bar{v}}{\partial x_j}\Big|^p dx\Big)^{\frac{1}{p}}\Big]$$
$$\leq C\|v\|_{W_0^{1,p}}. \tag{2.15}$$

By the integration by parts, we deduce

$$|(\Delta|u|^2 u, v)| = \Big|\int_\Omega (\Delta|u|^2 u) \cdot \bar{v}\, dx\Big|$$
$$\leq 2\Big[\int_\Omega |u||\nabla u|^2|\bar{v}|\, dx + \int_\Omega |u|^2|\nabla u||\nabla \bar{v}|\, dx\Big]. \tag{2.16}$$

We estimate the terms in (2.16) respectively. Using the Hölder inequality

$$\int_\Omega |u||\nabla u|^2 |\bar{v}|\, dx \leq \|u\|_{L^6}\|\nabla u\|_{L^3}^2 \|v\|_{L^6},$$

and the Gagliardo-Nirenberg inequality for $n < 6$,
$$\|u\|_{L^6} \leqslant C\|u\|^{\frac{6-n}{6}}\|\Delta u\|^{\frac{n}{6}},$$
$$\|\nabla u\|_{L^3}^2 \leqslant C\|u\|^{\frac{6-n}{6}}\|\Delta u\|^{\frac{6+n}{6}}.$$

Thus, for the first term in (2.16), we have
$$\int_\Omega |u||\nabla u|^2|\bar{v}|\,dx \leqslant C\|u\|^{\frac{6-n}{3}}\|\Delta u\|^{\frac{n+3}{3}}\|v\|^{\frac{6-n}{6}}\|\Delta v\|^{\frac{n}{6}} \leqslant C.$$

Similarly, by the Hölder inequality, we obtain
$$\int_\Omega |u|^2|\nabla u||\nabla \bar{v}|\,dx \leqslant C\|u\|_{L^6}^2\|\nabla u\|_{L^3}\|\nabla v\|_{L^3}.$$

Using the Gagliardo-Nirenberg inequality for $n < 6$,
$$\|u\|_{L^6}^2 \leqslant C\|u\|^{\frac{6-n}{3}}\|\Delta u\|^{\frac{n}{3}},$$
$$\|\nabla u\|_{L^3} \leqslant C\|u\|^{\frac{6-n}{12}}\|\Delta u\|^{\frac{6+n}{12}}.$$

Then we have
$$\int_\Omega |u|^2|\nabla u||\nabla \bar{v}|\,dx \leqslant C\|u\|^{\frac{30-5n}{12}}\|\Delta u\|^{\frac{6+5n}{12}}\|v\|^{\frac{6-n}{12}}\|\Delta v\|^{\frac{6+n}{12}} \leqslant C.$$

Finally, we need to estimate the term $|(\beta(x)f(|u|^2)u, v)|$. By the assumed conditions, we get
$$|(\beta(x)f(|u|^2)u, v)| \leqslant C\left|\int_\Omega (A|u|^{2q} + B)u\bar{v}\,dx\right|$$
$$\leqslant C\int_\Omega |u|^{2q+1}|\bar{v}|\,dx + C\|u\|\|v\|.$$

By the Hölder inequality, the estimates (2.11) and (2.12), combining with Lemma 2.3, we have
$$\int_\Omega |u|^{2q+1}|\bar{v}|\,dx \leqslant \|u\|_{L^{2q+2}}^{2q+1}\|v\|_{L^{2q+2}} \leqslant C. \tag{2.17}$$

In conclusion, from (2.14)-(2.17), we obtain $\|u_t\|_{H^{-2}(\Omega) \cap W^{-1,p'}(\Omega)} \leqslant C$.

3 Existence of global solutions

In this section, we give the proof of Theorem 1.1.

The proof of Theorem 1.1 Utilizing the estimates from Lemmas 2.2–2.4, we have

$$\|\Delta u\| + \|u\|_{W_0^{1,p}(\Omega)} + \|u_t\|_{H^{-2}(\Omega)\cap W^{-1,p'}(\Omega)} \leqslant C.$$

Next we will use the Galerkin method to prove the existence of solutions. Assume $\{w_j\}_{j=1}^{\infty}$ is a complete basis in $H_0^2(\Omega)$, and let

$$u_m(x,t) = \sum_{j=1}^{m} (u, w_j) w_j.$$

Then $u_m(x,t)$ is the solution of approximate problems corresponding to equations (1.1)–(1.3). Considering the initial problem of the following ordinary differential equations:

$$i(u_{mt}, w_i) + (\Delta u_m, \Delta w_i) - \lambda \int_{\Omega} \sum_{j=1}^{n} |\frac{\partial u_m}{\partial x_j}|^{p-2} \frac{\partial u_m}{\partial x_j} \frac{\partial \overline{w}_i}{\partial x_j} dx$$

$$- \alpha(\Delta |u_m|^2 u_m, w_i) + (\beta(x) f(|u_m|^2) u_m, w_i) = 0,$$

$$u_m(0) = u_{0m}(x).$$

Assuming $u_{0m}(x) \to u_0(x)$ in $H_0^2 \cap W_0^{1,p}$ as $m \to \infty$, combining the estimates of Lemma 2.2 to Lemma 2.4, we have

$$\|\Delta u_m\| + \|u_m\|_{W_0^{1,p}(\Omega)} + \|u_{mt}\|_{H^{-2}(\Omega)\cap W^{-1,p'}(\Omega)} \leqslant C.$$

Consequently, there exits a subsequence of $\{u_m(x,t)\}$ (still denoted by $\{u_m(x,t)\}$) such that

$$u_m(x,t) \to u(x,t) \text{ weakly-} * \text{ in } L^{\infty}(0,T; H_0^2(\Omega));$$

$$u_m(x,t) \to u(x,t) \text{ weakly-} * \text{ in } L^{\infty}(0,T; W_0^{1,p}(\Omega));$$

$$u_{mt}(x,t) \to u_t(x,t) \text{ weakly-} * \text{ in } L^{\infty}(0,T; W^{-1,p'}(\Omega) \cap H^{-2}(\Omega)).$$

We obtain from [10] that there exists a subsequence of $\{u_m(x,t)\}$ (not relabeled) such that $u_m(x,t) \to u(x,t)$ strongly in $L^2(0,T; L^2(\Omega))$, $Du_m(x,t) \to Du(x,t)$ strongly in $L^2(0,T; L^2(\Omega))$. Then there exits a further subsequence of $\{u_m(x,t)\}$

(not relabeled again) such that $u_m(x,t) \to u(x,t)$, $Du_m(x,t) \to Du(x,t)$, for almost every $(x,t) \in \Omega \times [0,T]$.

Next, we need to prove that $u(x,t)$ satisfies equation (1.4). The main difficulty is the convergence of nonlinear terms. First, since $\|u_m\|_{W_0^{1,p}(\Omega)} \leq C$, $\sum_{j=1}^n |\frac{\partial u_m}{\partial x_j}|^{p-2} \frac{\partial u_m}{\partial x_j}|$ is uniformly bounded in $L^\infty(0,T; L^{\frac{p}{p-1}}(\Omega))$. Notice that $Du_m(x,t) \to Du(x,t)$ a.e. in $\Omega \times [0,T]$, we have the following result from the Lemma in [9]

$$\sum_{j=1}^n |\frac{\partial u_m}{\partial x_j}|^{p-2} \frac{\partial u_m}{\partial x_j} \to \sum_{j=1}^n |\frac{\partial u}{\partial x_j}|^{p-2} \frac{\partial u}{\partial x_j} \quad \text{weakly in } \Omega \times [0,T].$$

Furthermore,

$$\lambda \int_0^T \int_\Omega \sum_{j=1}^n |\frac{\partial u_m}{\partial x_j}|^{p-2} \frac{\partial u_m}{\partial x_j} \frac{\partial \overline{v}}{\partial x_j} \, dxdt \to \lambda \int_0^T \int_\Omega \sum_{j=1}^n |\frac{\partial u}{\partial x_j}|^{p-2} \frac{\partial u}{\partial x_j} \frac{\partial \overline{v}}{\partial x_j} \, dxdt.$$

Next we prove $\int_0^T (\Delta |u_m|^2 u_m, v) \, dt \to \int_0^T (\Delta |u|^2 u, v) \, dt$, and

$$(\Delta |u_m|^2 u_m, v) - (\Delta |u|^2 u, v) = (\Delta(|u_m|^2 - |u|^2) u_m, v) + (\Delta |u|^2 (u_m - u), v).$$

We consider the convergence one by one.

$$-(\Delta(|u_m|^2 - |u|^2)u_m, v) = -\int_\Omega \Delta(|u_m|^2 - |u|^2) u_m \overline{v} \, dx$$

$$= \int_\Omega \nabla(u_m \overline{u_m} - u\overline{u})(\nabla u_m \overline{v} + u_m \nabla \overline{v}) \, dx$$

$$= \int_\Omega \nabla[(u_m - u)\overline{u_m} + u(\overline{u_m} - \overline{u})](\nabla u_m \overline{v} + u_m \nabla \overline{v}) \, dx$$

$$= \int_\Omega \nabla(u_m - u)\overline{u_m}(\nabla u_m \overline{v} + u_m \nabla \overline{v}) \, dx + \int_\Omega (u_m - u)\nabla \overline{u_m}(\nabla u_m \overline{v} + u_m \nabla \overline{v}) \, dx$$

$$+ \int_\Omega \nabla u(\overline{u_m} - \overline{u})(\nabla u_m \overline{v} + u_m \nabla \overline{v}) \, dx + \int_\Omega u \nabla(\overline{u_m} - \overline{u})(\nabla u_m \overline{v} + u_m \nabla \overline{v}) \, dx$$

$$= I_1 + I_2 + I_3 + I_4.$$

By the Hölder inequality and Gagliardo-Nirenberg inequality, we get

$$I_1 \leq \|\nabla(u_m - u)\|_{L^3} \|u_m\|_{L^6} (\|v\|_{L^6} \|\nabla u_m\|_{L^3} + \|\nabla v\|_{L^3} \|u_m\|_{L^6});$$

$$I_2 \leqslant \|u_m - u\|_{L^6}\|\nabla u_m\|_{L^3}(\|v\|_{L^6}\|\nabla u_m\|_{L^3} + \|\nabla v\|_{L^3}\|u_m\|_{L^6});$$

$$I_3 \leqslant \|\nabla u\|_{L^3}\|u_m - u\|_{L^6}(\|\nabla v\|_{L^3}\|u_m\|_{L^6} + \|v\|_{L^6}\|\nabla u_m\|_{L^3});$$

$$I_4 \leqslant \|u\|_{L^6}\|\nabla(u_m - u)\|_{L^3}(\|\nabla v\|_{L^3}\|u_m\|_{L^6} + \|v\|_{L^6}\|\nabla u_m\|_{L^3}).$$

Furthermore, by the Gagliardo-Nirenberg inequality

$$\|\nabla(u_m - u)\|_{L^3} \leqslant C\|u_m - u\|^{\frac{6-n}{12}}\|\Delta(u_m - u)\|^{\frac{6+n}{12}},$$

$$\|u_m - u\|_{L^6} \leqslant C\|u_m - u\|^{\frac{6-n}{6}}\|\Delta(u_m - u)\|^{\frac{n}{6}},$$

we have for $n < 6$, $\int_0^T (\Delta(|u_m|^2 - |u|^2)u_m, v)\, dt \to 0$, as $m \to \infty$.

$$(\Delta|u|^2(u_m - u), v) = \int_\Omega \nabla|u|^2 \nabla((u_m - u)\bar{v})\, dx$$

$$\leqslant 2\int_\Omega (|u||\nabla u||\nabla(u_m - u)\bar{v}| + |u||\nabla u||(u_m - u)\nabla\bar{v}|)\, dx.$$

Similarly, we have $\int_0^T (\Delta|u|^2(u_m - u), v)\, dt \to 0$, $m \to \infty$. Then we get

$$\int_0^T (\Delta|u_m|^2 u_m, v)\, dt \to \int_0^T (\Delta|u|^2 u, v)\, dt.$$

At last, we show that

$$\int_0^T (\beta(x)f(|u_m|^2)u_m, v)\, dt \to \int_0^T (\beta(x)f(|u|^2)u, v)\, dt.$$

Since $\|u_m\|_{L^{2q+2}}$ is bounded, and

$$|\beta(x)f(|u_m|^2)u_m| \leqslant C(|u_m|^{2q+1} + |u_m|),$$

then we can obtain that $\beta(x)f(|u_m|^2)u_m$ is uniformly bounded in $L^\infty(0,T; L^{2q+2/2q+1})$, and combining the result $u_m(x,t) \to u(x,t)$ a.e.. Thus we have

$$\beta(x)f(|u_m|^2)u_m \to \beta(x)f(|u|^2)u \quad \text{weakly}.$$

Finally, we obtain

$$\int_0^T \int_\Omega \beta(x)f(|u_m|^2)u_m \bar{v}\, dxdt \to \int_0^T \int_\Omega \beta(x)f(|u|^2)u\bar{v}\, dxdt.$$

The proof is complete.

References

[1] Guo B. Global solution for a class of multidimensional fourth-order nonlinear Schrödinger equations [J]. Northeast Math., 1987, 3: 399–409.

[2] Guo C, Cui S. Global existence of solutions for a fourth-order nonlinear Schrödinger equation [J]. Appl. Math. Lett., 2006, 19: 706–711.

[3] Hirota R. Direct Methods in Soliton Theory [M]. Springer, Berlin, 1980.

[4] Huo Z, Jia Y. The Cauchy problem for the fourth-order nonlinear Schrödinger equation related to the vortex filament [J]. J. Differ. Equ., 2005, 214: 1–35.

[5] Ilan B, Fibich G, Papanicolaou G. Self-focusing with fourth-order dispersion [J]. SIAM J. Appl. Math., 2002, 62: 1437–1462.

[6] Karpman V I, Shagalov A G. Stability of solitons described by nonlinear Schrödinger type equation with higher-order dispersion [J]. Phys. D., 2000, 144: 194–210.

[7] Karpman V I. Stabilization of soliton instabilities by higher-order dispersion: Fourth-order nonlinear Schrödinger-type equations [J]. Phys. Rev. E., 1996, 53: 1336–1339.

[8] Laedke E W, Spatschek K H, Stenflo L. Evolution theorem for a class of perturbed envelope soliton solution [J]. J. Math. Phys., 1983, 24: 2764–2769.

[9] Lions J L, Magenes E. Non-homogeneous Boundary Value Problems and Applications [M]. New York, Berlin, Heidelberg: Springer-Verlag, 1972.

[10] Lions J L. Quelques Methodes de Resolution des Problemes aux Limites Non Lineaires [M]. Dunod, Paris, 1969.

[11] Wazwaz A M. Exact Solutions for the Fourth Order Nonlinear Schrödinger Equations with Cubic and Power Law Nonlinearities [M]. Math. and Comput. Model., 2006, 43: 802–808.

The Suitable Weak Solution for the Cauchy Problem of the Double-Diffusive Convection System*

Chen Fei (陈菲), and Guo Boling (郭柏灵)

Abstract In this paper, we consider the existence of the suitable weak solutions for the Cauchy problem of the double-diffusive convection system in $\mathbf{R}^3$. We establish the global existence of the suitable weak solutions of the double-diffusive convection system, especially, we obtain some generalized energy estimates which will play a very important role in studying the partial regularity of the suitable weak solutions of the double-diffusive convection system.

Keywords suitable weak solution; double-diffusive convection; Cauchy problem

1 Introduction

In this paper, we study the global suitable weak solution for the Cauchy problem of the non-dimensional double-diffusive convection system (see [1-3])

$$\begin{cases} \dfrac{\partial U}{\partial t} + U \cdot \nabla U - \mu \Delta U + \mu \nabla P - \mu(\lambda \theta - \eta S)\mathbf{K} = 0, & \text{in } Q_T, \\ \dfrac{\partial \theta}{\partial t} + U \cdot \nabla \theta - \Delta \theta - w = 0, & \text{in } Q_T, \\ \dfrac{\partial S}{\partial t} + U \cdot \nabla S - \tau \Delta S - w = 0, & \text{in } Q_T, \\ \nabla \cdot U = 0, & \text{in } Q_T, \\ (U, \theta, S)(x,t)|_{t=0} = (U_0(x), \theta_0(x), S_0(x)), & \text{on } \mathbf{R}^3, \end{cases} \quad (1.1)$$

---

* Appl. Anal., 2019, 98(9): 1724−1740. Doi: https://doi.org/10.1080/00036811.2018.1441995.

where $Q_T \equiv \mathbf{R}^3 \times (0,T)$, $U = (u,v,w)$ stands for the velocity vector of the fluid, θ the temperature and S the salinity. The non-dimensional parameters μ means the Prandtl number, τ the Lewis number, λ the thermal Rayleigh number and η the solute Rayleigh number. $\mu, \lambda, \eta, \tau > 0$. $\mathbf{K}$ is the vertical unit vector.

Double-diffusion convection is named by the convection motions which occur in a fluid when density variations appear caused by two different components with different rates of diffusion. After a long time since, in 1857, Jevons [4] first discovered the phenomenon of the double-diffusion convection in experiments, it was rediscovered as an "oceanographic curiosity" by Stommel-Arons-Blanchard [5], Veronis [3], Baines-Gill [6] and so on. Since the strong physical background such as in astrophysics, engineering and geology a large number of studies for the double-diffusive convection have been carried out (see e.g. [7-11] and the references therein).

There are also a lot of studies on the double-diffusive convection system in mathematics (see [1, 2, 12-19] and the references therein). Especially, there are some results about the well-posedness of the strong solution for the double-diffusive convection system based on Brinkman-Forchheimer equation (see e.g. [14-16, 19]). For example, Terasawa-Ôtani [19] and Ôtani-Uchida [14] established the unique global solution with Dirichlet and Neumann boundary conditions, respectively, in which the global unique H^1-solution was established for 3D bounded domains. Then, in 2016, Ôtani-Uchida [16] studied the case in unbounded domains (see [15] for the periodic solutions). However, all of the results in [14-16, 19], are under the assumption that the Brinkman-Forchheimer equation was linearized (originally, it has a convection term):

$$\partial_t \boldsymbol{u} = \nu \Delta \boldsymbol{u} - a\boldsymbol{u} - \nabla P + \boldsymbol{g}T + \boldsymbol{h}C + f_1. \tag{1.2}$$

In (1.2), it ignores the convection term, which, however, will cause a crucial difficulty in the study of the global existence just like that of the Navier-Stokes equations. Therefore, considering the system (1.1) which has a convection term in the first equation, Piniewski [17] proved the global existence, uniqueness and continuous dependence on initial data of weak solution in two dimensions for the bounded domain. Attractor bifurcation of 3D double-diffusive convection was considered by Hsia-Ma-Wang [1] in a parallel domain, in which they claimed the global weak solution and the strong solution to exist, but there was not detailed proof about them. Then, Chen et al. [20] got the global unique solution in $H_\sigma^N \times H^N \times H^N(\mathbf{R}^3)$ for any $N \geqslant 1$ with L^2 norm of the initial data being small. Moreover, in [21], they also considered the case in the bounded domain.

As is well known, the weak solutions are the only solutions that exist for all time without any smallness restriction on the initial data for 3D Navier-Stokes equations. However, the uniqueness of the weak solution is still open. Therefore, the properties of the solutions abtained by different methods could not coincide. The concept of suitable weak solution (with useful additional properties, which are (very likely) those having a physical meaning) for the Navier-Stokes equations was first introduced by Scheffer [22] in studying the very interesting result on the Hausdorff dimension of the singular set for suitable weak solution of the Navier-Stokes equations. Then, Caffarelli, Kohn and Nirenberg [23] proved the singular set for the suitable weak solution of the Navier-Stokes equations has one-dimensional Hausdorff measure zero. Considering the very important role of the existence for the suitable weak solution in studying the partial regularity for that of the Navier-Stokes equations, here, motivated by [22-25], we consider the global existence of the suitable weak solution of the double-diffusive convection system.

2 Definition and the main result

Throughout this paper, we use the following notations. For simplicity, we denote
$$\int f = \int_{\mathbf{R}^3} f \, dx, \qquad \iint f = \iint_{Q_T} f \, dx dt.$$
For $1 \leqslant p \leqslant \infty$, and $m \in \mathbf{N}$, the Sobolev spaces are defined in a standard manner.
$$L^p \triangleq L^p(\mathbf{R}^3), \quad H^m \triangleq W^{m,2}(\mathbf{R}^3), \quad \mathbf{L}^P \triangleq (L^P)^3, \quad \mathbf{H}^m \triangleq (H^m)^3.$$
And we define
$$\mathbf{L}_\sigma^2 \triangleq \{f \in \mathbf{L}^2; \nabla \cdot f = 0\},$$
$$\mathbf{H}_\sigma^1 \triangleq \{f \in \mathbf{H}^1; \nabla \cdot f = 0\},$$
$$\mathbf{H}_\sigma^2 \triangleq \{f \in \mathbf{H}^2; \nabla \cdot f = 0\},$$
where $\nabla \cdot f$ means the divergence of f in the distribution sense. Denote the dual of the space M by M^{-1}. Then we find
$$\mathbf{H}_\sigma^2 \hookrightarrow \mathbf{H}_\sigma^1 \hookrightarrow \mathbf{L}_\sigma^2 \hookrightarrow \mathbf{H}_\sigma^{-1} \hookrightarrow \mathbf{H}_\sigma^{-2},$$

$$H^2 \hookrightarrow H^1 \hookrightarrow L^2 \hookrightarrow H^{-1} \hookrightarrow H^{-2},$$

with continuous and locally dense embedding. Denote $C_w([0,T];X)$ the subspace of $L^\infty(0,T;X)$ of functions $f(t)$ which are weakly continuous on $[0,T]$ with the value in X, where X is a Banach space.

Definition 2.1 *Define $\{U, \theta, S, P\}$ a suitable weak solution of the problem (1.1) with the initial data $U_0 \in \mathbf{L}_\sigma^2$, $\theta_0 \in L^2$, $S_0 \in L^2$ if the following conditions are satisfied:*

(1) (*Integrability*)

$$U \in L^2(0,T;\mathbf{H}_\sigma^1) \cap C_w([0,T];\mathbf{L}_\sigma^2), \quad U_t \in L^{\frac{4}{3}}(0,T;\mathbf{H}^{-1}),$$

$$\theta \in L^2(0,T;H^1) \cap C_w([0,T];L^2), \quad \theta_t \in L^{\frac{4}{3}}(0,T;H^{-1}),$$

$$S \in L^2(0,T;H^1) \cap C_w([0,T];L^2), \quad S_t \in L^{\frac{4}{3}}(0,T;H^{-1}),$$

$$P \in L^{\frac{4}{3}}(0,T;L^2+L^6). \tag{2.1}$$

(2) (*Equations*)

(i) For any $\varphi \in (C^\infty(Q_T))^3$ with compact support respect to the space variables, $\nabla \cdot \varphi = 0$ and $\varphi(\cdot, T) = 0$,

$$\iint \left\{ -U \frac{\partial \varphi}{\partial t} + \mu \nabla U \cdot \nabla \varphi - U \cdot (U \cdot \nabla)\varphi \right\} = \iint \mu(\lambda \theta - \eta S)\mathbf{K} \cdot \varphi + \int U_0 \varphi(\cdot, 0). \tag{2.2}$$

(ii) For any $\psi \in C^\infty(Q_T)$ with compact support respect to the space variables and $\psi(\cdot, T) = 0$,

$$\iint \left\{ -\theta \frac{\partial \psi}{\partial t} + \nabla \theta \cdot \nabla \psi - \theta(U \cdot \nabla)\psi \right\} = \iint w\psi + \int \theta_0 \psi(\cdot, 0). \tag{2.3}$$

(iii) For any $\Psi \in C^\infty(Q_T)$ with compact support respect to the space variables and $\Psi(\cdot, T) = 0$,

$$\iint \left\{ -S \frac{\partial \Psi}{\partial t} + \tau \nabla S \cdot \nabla \Psi - S(U \cdot \nabla)\Psi \right\} = \iint w\Psi + \int S_0 \Psi(\cdot, 0). \tag{2.4}$$

(3) (*Generalized energy inequalities*)

For any smooth function $\phi = \phi(x,t) \geqslant 0$ with compact support respect to the space variables and for any $t \in [0,T]$,

$$\int |U(\cdot, t)|^2 \phi + 2\mu \iint |\nabla U|^2 \phi$$

$$\leqslant \int |U_0|^2 \phi(\cdot,0) + \iint |U|^2 \left(\frac{\partial \phi}{\partial t} + \mu \Delta \phi\right) + \iint (|U|^2 + 2\mu P) U \cdot \nabla \phi$$
$$+ 2\mu \iint (\lambda \theta - \eta S) w\phi, \qquad (2.5)$$

$$\int |\theta(\cdot,t)|^2 \phi + 2 \iint |\nabla \theta|^2 \phi$$
$$\leqslant \int |\theta_0|^2 \phi(\cdot,0) + \iint |\theta|^2 \left(\frac{\partial \phi}{\partial t} + \Delta \phi\right) + \iint |\theta|^2 U \cdot \nabla \phi + 2 \iint w\theta\phi, \qquad (2.6)$$

and

$$\int |S(\cdot,t)|^2 \phi + 2\tau \iint |\nabla S|^2 \phi$$
$$\leqslant \int |S_0|^2 \phi(\cdot,0) + \iint |S|^2 \left(\frac{\partial \phi}{\partial t} + \tau \Delta \phi\right) + \iint |S|^2 U \cdot \nabla \phi + 2 \iint wS\phi. \qquad (2.7)$$

(4) (*Energy inequalities*)

$$\|U\|^2_{L^\infty(0,t;\mathbf{L}^2_\sigma)} + \|\theta\|^2_{L^\infty(0,t;L^2)} + \|S\|^2_{L^\infty(0,t;L^2)} + \|\nabla U\|^2_{L^2(0,t;L^2)} + \|\nabla \theta\|^2_{L^2(0,t;L^2)} \qquad (2.8)$$
$$+ \|\nabla S\|^2_{L^2(0,t;L^2)} \leqslant C(\lambda,\eta,\mu,\tau,T)\{\|U_0\|^2_{\mathbf{L}^2_\sigma} + \|\theta_0\|^2_{L^2} + \|S_0\|^2_{L^2}\}, \quad \text{for all } t \in [0,T].$$

One can now state the main result of the paper:

Theorem 2.1 *If the initial data (U_0, θ_0, S_0) satisfy $U_0 \in \mathbf{L}^2_\sigma$, $\theta_0 \in L^2$, $S_0 \in L^2$, then there exists at least one suitable weak solution of the problem (1.1).*

Remark 2.1 (1) *Since the lack of the uniqueness of the weak solution, the properties of the weak solution obtained by different methods may not coincide. We are not sure if the generalized energy inequalities which are very important in studying the partial regularity of the suitable weak solution can be obtained by other methods (see e.g. [23, 24]).*

(2) *It needs to emphasize that our final objective is to get the partial regularity of the suitable weak solution to the double-diffusive convection system. The result of this paper is an important staged one.*

(3) *Owing to the temperature, the salinity and the velocity field interacting with each other, we can not get the strong energy inequality just like that of the Navier-Stokes equations which shows that the kinetic energy is a decreasing function of time, for almost all time $t \in [0,T]$. Here we only get (2.8).*

The rest of the paper unfolds as follows. In the next section, we introduce some lemmas, which will play a crucial role in the proof of Theorem 2.1. Then, we divide the proof of Theorem 2.1 into two steps: In Section 4, based on Lemma 3.3, we shall deal with the global existence and uniqueness of the strong solution of an approximating system with a stronger dissipative term; In the last section, by the compactness theory, letting $\varepsilon \to 0$, we finish the proof of Theorem 2.1.

3 Preliminaries

In the sequel, we will frequently use the following interpolation inequality.

Lemma 3.1 *For $2 \leqslant p \leqslant \infty$ and $0 \leqslant \beta, \beta_1 \leqslant \beta_2$; when $p = \infty$, we require further that $\beta_1 \leqslant \beta + 1$ and $\beta_2 \geqslant \beta + 2$. Then we have that for any $f \in C_0^\infty(\mathbf{R}^3)$,*

$$\|\nabla^\beta f\|_{L^p} \leqslant \|\nabla^{\beta_1} f\|_{L^2}^{1-\gamma} \|\nabla^{\beta_2} f\|_{L^2}^\gamma,$$

where $0 \leqslant \gamma \leqslant 1$ and β satisfy

$$\frac{1}{p} - \frac{\beta}{3} = (1-\gamma)\left(\frac{1}{2} - \frac{\beta_1}{3}\right) + \gamma\left(\frac{1}{2} - \frac{\beta_2}{3}\right).$$

In part 4.2, we'll use the following Aubin-Lions-Simon lemma (Theorem II.5.16, in [26]).

Lemma 3.2 *Let $B_0 \subset B_1 \subset B_2$ be three Banach spaces. We assume that the embedding of B_1 in B_2 is continuous and that the embedding of B_0 in B_1 is compact. Let p, r satisfy $1 \leqslant p, r \leqslant +\infty$. For $T > 0$, we define*

$$E_{p,r} = \left\{v \in L^p(0,T,B_0), \frac{dv}{dt} \in L^r(0,T,B_2)\right\}.$$

(1) *If $p < +\infty$, then the embedding of $E_{p,r}$ in $L^p(0,T,B_1)$ is compact.*

(2) *If $p = +\infty$ and $r > 1$, then the embedding of $E_{p,r}$ in $C^0([0,T],B_1)$ is compact.*

Then, let's introduce the well known results on linear evolution equations (see, [24, 25, 27]).

Lemma 3.3 *If $U_0 \in \mathbf{H}_\sigma^2$, $\theta_0 \in H^2$, $S_0 \in H^2$, $G_1 \in L^2(0,T;\mathbf{L}_\sigma^2)$, $G_2 \in L^2(0,T;L^2)$, $G_3 \in L^2(0,T;L^2)$, then for the following problems:*

$$\begin{cases} \dfrac{\partial U}{\partial t} + \varepsilon\Delta^2 U - \mu\Delta U = G_1, \\ \dfrac{\partial \theta}{\partial t} + \varepsilon\Delta^2\theta - \Delta\theta = G_2, \\ \dfrac{\partial S}{\partial t} + \varepsilon\Delta^2 S - \tau\Delta S = G_3, \\ \nabla \cdot U = 0, \\ (U,\theta,S)|_{t=0} = (U_0,\theta_0,S_0), \end{cases} \quad (3.1)$$

there exists a unique solution $(U,\theta\, S)$, which satisfies

$$\begin{cases} U \in L^2(0,T;\mathbf{H}^4) \cap C([0,T]; q\mathbf{H}_\sigma^2), & U_t \in L^2(0,T;\mathbf{L}_\sigma^2), \\ \theta \in L^2(0,T;H^4) \cap C([0,T]; H^2), & \theta_t \in L^2(0,T;L^2), \\ S \in L^2(0,T;H^4) \cap C([0,T]; H^2), & S_t \in L^2(0,T;L^2). \end{cases} \quad (3.2)$$

4 The proof of Theorem 2.1

4.1 Approximating problem

Instead of directly dealing with the system (1.1), we first study the approximating system (4.1) which has a stronger dissipative term so that it's much easier to get a globally unique solution with large initial data. Then, in part 4.2, we'll let $\varepsilon \to 0$ to recover the initial system (1.1). Now, for any $0 < T < \infty$, we assert the following theorem.

Theorem 4.1 *If $\varepsilon > 0$ and the initial data $(U_{0,\varepsilon}, \theta_{0,\varepsilon}, S_{0,\varepsilon}) \in \mathbf{H}_\sigma^2 \times H^2 \times H^2$, then there exists a global unique strong solution $(U_\varepsilon, \theta_\varepsilon, S_\varepsilon)$ of the approximating system*

$$\begin{cases} \dfrac{\partial U_\varepsilon}{\partial t} + \varepsilon\Delta^2 U_\varepsilon - \mu\Delta U_\varepsilon + U_\varepsilon \cdot \nabla U_\varepsilon = -\mu\nabla P_\varepsilon + \mu(\lambda\theta_\varepsilon - \eta S_\varepsilon)\mathbf{K}, & \text{in } Q_T, \\ \dfrac{\partial \theta_\varepsilon}{\partial t} + \varepsilon\Delta^2\theta_\varepsilon - \Delta\theta_\varepsilon + U_\varepsilon \cdot \nabla\theta_\varepsilon = w_\varepsilon, & \text{in } Q_T, \\ \dfrac{\partial S_\varepsilon}{\partial t} + \varepsilon\Delta^2 S_\varepsilon - \tau\Delta S_\varepsilon + U_\varepsilon \cdot \nabla S_\varepsilon = w_\varepsilon, & \text{in } Q_T, \\ \nabla \cdot U_\varepsilon = 0, & \text{in } Q_T, \\ (U_\varepsilon,\theta_\varepsilon,S_\varepsilon)(x,t)|_{t=0} = (U_{0,\varepsilon}(x), \theta_{0,\varepsilon}(x), S_{0,\varepsilon}(x)), & \text{on } \mathbf{R}^3, \end{cases} \quad (4.1)$$

and the solution $(U_\varepsilon, \theta_\varepsilon, S_\varepsilon)$ satisfies

$$\begin{cases} U_\varepsilon \in L^2(0,T;\mathbf{H}^4 \cap \mathbf{H}^2_\sigma) \cap C([0,T];\mathbf{L}^2_\sigma) \cap L^\infty(0,T;\mathbf{H}^2_\sigma), \\ \theta_\varepsilon \in L^2(0,T;H^4) \cap C([0,T];L^2) \cap L^\infty(0,T;H^2), \\ S_\varepsilon \in L^2(0,T;H^4) \cap C([0,T];L^2) \cap L^\infty(0,T;H^2), \\ \nabla P_\varepsilon \in L^2(0,T;\mathbf{L}^2), \end{cases} \quad (4.2)$$

$$\|U_\varepsilon\|^2_{C([0,T];\mathbf{L}^2_\sigma)} + 2\mu\|\nabla U_\varepsilon\|^2_{L^2(0,T;\mathbf{L}^2)} + 2\varepsilon\|\Delta U_\varepsilon\|^2_{L^2(0,T;\mathbf{L}^2)}$$
$$+ \|\theta_\varepsilon\|^2_{C([0,T];L^2)} + 2\|\nabla \theta_\varepsilon\|^2_{L^2(0,T;\mathbf{L}^2)} + 2\varepsilon\|\Delta \theta_\varepsilon\|^2_{L^2(0,T;L^2)}$$
$$+ \|S_\varepsilon\|^2_{C([0,T];L^2)} + 2\tau\|\nabla S_\varepsilon\|^2_{L^2(0,T;\mathbf{L}^2)} + 2\varepsilon\|\Delta S_\varepsilon\|^2_{L^2(0,T;L^2)}$$
$$\leqslant C(T)(\|U_{0,\varepsilon}\|^2_{\mathbf{L}^2_\sigma} + \|\theta_{0,\varepsilon}\|^2_{L^2} + \|S_{0,\varepsilon}\|^2_{L^2}), \quad (4.3)$$

and the generalized energy estimates (4.31)-(4.33) holds.

Remark 4.1 The constant C in (4.3) depends on $T, \eta, \lambda, \mu, \tau$, but does not depend on ε. Here and after, we use $C(T)$ to illustrate the dependence on T.

In order to use the results of the linear evolution equations Lemma 3.3 and the fixed point theorem to deal with the approximating system (4.1), we need to get rid of the influence of the unknown pressure term. Let's introduce the orthogonal projection $\mathbf{P}$ that projects $\mathbf{L}^2$ into the closed subspace $\mathbf{L}^2_\sigma$, which leads to for all $f \in \mathbf{L}^2$, there exists a unique orthogonal decomposition $f = f_1 + \nabla f_2$, where $f_1 \in H \triangleq \mathbf{L}^2_\sigma$ and $\nabla f_2 \in H^\perp \triangleq \{\nabla g; g \in L^2_{\text{loc}}, \nabla g \in \mathbf{L}^2\}$, which satisfies:

$$\begin{cases} \hat{f}_1 = \hat{f} - |\xi|^{-2}(\xi \cdot \hat{f})\xi, \\ \hat{f}_2 = |\xi|^{-2}\xi \cdot \hat{f}. \end{cases}$$

By acting the orthogonal projection $\mathbf{P}$ on the momentum equation of (4.1), we consider the following system (for simplicity, here we ignore the subscript ε):

$$\begin{cases} \dfrac{\partial U}{\partial t} + \varepsilon\Delta^2 U - \mu\Delta U + \mathbf{P}[U\cdot\nabla U] = \mathbf{P}[\mu(\lambda\theta - \eta S)\mathbf{K}], & \text{in } Q_T, \\ \dfrac{\partial \theta}{\partial t} + \varepsilon\Delta^2\theta - \Delta\theta + U\cdot\nabla\theta = w, & \text{in } Q_T, \\ \dfrac{\partial S}{\partial t} + \varepsilon\Delta^2 S - \tau\Delta S + U\cdot\nabla S = w, & \text{in } Q_T, \\ \nabla\cdot U = 0, & \text{in } Q_T, \\ (U,\theta,S)(x,t)|_{t=0} = (U_{0,\varepsilon}(x), \theta_{0,\varepsilon}(x), S_{0,\varepsilon}(x)), & \text{on } \mathbf{R}^3. \end{cases} \quad (4.4)$$

Now, we have the following proposition:

Proposition 4.1 *If the initial data* $(U_{0,\varepsilon}, \theta_{0,\varepsilon}, S_{0,\varepsilon}) \in \mathbf{H}_\sigma^2 \times H^2 \times H^2$, *then there exists a global unique strong solution* (U, θ, S) *of system* (4.4) *satisfying* (3.2).

Proof Step 1 We prove that there exists a unique local solution of system (4.4) satisfying (3.2). Now, we consider the following system:

$$\begin{cases} \dfrac{\partial U}{\partial t} + \varepsilon\Delta^2 U - \mu\Delta U = -\mathbf{P}[\bar{U}\cdot\nabla\bar{U}] + \mathbf{P}[\mu(\lambda\bar{\theta} - \eta\bar{S})\mathbf{K}], & \text{in } Q_T, \\ \dfrac{\partial \theta}{\partial t} + \varepsilon\Delta^2\theta - \Delta\theta = -\bar{U}\cdot\nabla\bar{\theta} + \bar{w}, & \text{in } Q_T, \\ \dfrac{\partial S}{\partial t} + \varepsilon\Delta^2 S - \tau\Delta S = -\bar{U}\cdot\nabla\bar{S} + \bar{w}, & \text{in } Q_T, \\ \nabla\cdot U = 0, & \text{in } Q_T, \\ (U,\theta,S)(x,t)|_{t=0} = (U_{0,\varepsilon}(x), \theta_{0,\varepsilon}(x), S_{0,\varepsilon}(x)), & \text{on } \mathbf{R}^3, \end{cases} \quad (4.5)$$

where $\bar{U} \in C([0,T]; \mathbf{H}_\sigma^1) \cap L^2(0,T; \mathbf{H}^2)$, $\bar{\theta}, \bar{S} \in C([0,T]; H^1) \cap L^2(0,T; H^2)$.

By Hölder's inequality and Sobolev inequality, we find that

$$\|\mathbf{P}[\bar{U}\cdot\nabla\bar{U}]\|_{\mathbf{L}_\sigma^2}^2 \leqslant \|\bar{U}\cdot\nabla\bar{U}\|_{L^2}^2 \leqslant \|\bar{U}\|_{L^6}^2\|\nabla\bar{U}\|_{L^2}\|\nabla\bar{U}\|_{L^6} \leqslant C\|\bar{U}\|_{\mathbf{H}^1}^3\|\bar{U}\|_{\mathbf{H}^2},$$

$$\|\bar{U}\cdot\nabla\bar{\theta}\|_{L^2}^2 \leqslant \|\bar{U}\|_{L^6}^2\|\nabla\bar{\theta}\|_{L^2}\|\nabla\bar{\theta}\|_{L^6} \leqslant C\|\bar{U}\|_{\mathbf{H}^1}^2\|\bar{\theta}\|_{H^1}\|\bar{\theta}\|_{H^2},$$

$$\|\bar{U}\cdot\nabla\bar{S}\|_{L^2}^2 \leqslant \|\bar{U}\|_{L^6}^2\|\nabla\bar{S}\|_{L^2}\|\nabla\bar{S}\|_{L^6} \leqslant C\|\bar{U}\|_{\mathbf{H}^1}^2\|\bar{S}\|_{H^1}\|\bar{S}\|_{H^2}. \quad (4.6)$$

Therefore, it follows from Lemma 3.3 that for any

$$\bar{U} \in C([0,T]; \mathbf{H}_\sigma^1) \cap L^2(0,T; \mathbf{H}_\sigma^2), \quad \bar{\theta}, \bar{S} \in C([0,T]; H^1) \cap L^2(0,T; H^2),$$

the system (4.5) has a unique solution (U, θ, S) satisfying (3.2). Now we can define a map J: $J(\bar{U}, \bar{\theta}, \bar{S}) = (U, \theta, S)$, and denote $\mathbf{M}$ the set

$$\Big\{(\bar{U}, \bar{\theta}, \bar{S}); \ \bar{U} \in C([0,t]; \mathbf{H}_\sigma^1) \cap L^2(0,t; \mathbf{H}_\sigma^2), \ \bar{\theta}, \bar{S} \in C([0,t]; H^1) \cap L^2(0,t; H^2),$$

$$(\bar{U}, \bar{\theta}, \bar{S})(x, 0) = (\bar{U}_{0,\varepsilon}(x), \bar{\theta}_{0,\varepsilon}(x), \bar{S}_{0,\varepsilon}(x)),$$

$$\|\bar{U}\|_{C([0,t]; \mathbf{H}_\sigma^1)}, \|\bar{U}\|_{L^2(0,t; \mathbf{H}_\sigma^2)}, \|\bar{\theta}\|_{C([0,t]; H^1)}, \|\bar{\theta}\|_{L^2(0,t; H^2)}, \|\bar{S}\|_{C([0,t]; H^1)}, \|\bar{S}\|_{L^2(0,t; H^2)}$$

$$\leqslant B\Big\},$$

where B and t will be determined later.

Now, we start to prove that J is a contraction map on $\mathbf{M}$, provided t is small enough. Firstly, it needs to demonstrate $(U, \theta, S) = J(\bar{U}, \bar{\theta}, \bar{S}) \in \mathbf{M}$, for any $(\bar{U}, \bar{\theta}, \bar{S}) \in \mathbf{M}$.

Multiplying $(4.5)_1$ by U, then integrating over Q_t, we get by Hölder's inequality and Young's inequality that

$$\frac{1}{2} \int (|U(x,t)|^2 - |U_{0,\varepsilon}(x)|^2) + \varepsilon \iint |\Delta U|^2 + \mu \iint |\nabla U|^2$$

$$\leqslant \frac{1}{2} \iint \Big(\big|\mathbf{P}[\bar{U} \cdot \nabla \bar{U}]\big|^2 + \big|\mathbf{P}[\mu(\lambda \bar{\theta} - \eta \bar{S})\mathbf{K}]\big|^2\Big) + \frac{1}{2} \iint |U|^2,$$

Noting that for any $(\bar{U}, \bar{\theta}, \bar{S}) \in \mathbf{M}$, by (4.6) and Hölder's inequality, one has

$$\|\mathbf{P}[\bar{U} \cdot \nabla \bar{U}]\|_{L^2(0,t; \mathbf{L}_\sigma^2)}^2 \leqslant C t^{\frac{1}{2}} \|\bar{U}\|_{L^\infty(0,t; \mathbf{H}^1)}^3 \|\bar{U}\|_{L^2(0,t; \mathbf{H}^2)} \leqslant C B^4 t^{1/2},$$

$$\|\mathbf{P}[\mu(\lambda \bar{\theta} - \eta \bar{S})\mathbf{K}]\|_{L^2(0,t; L^2)}^2 \leqslant C \|(\bar{\theta}, \bar{S})\|_{L^2(0,t; L^2)}^2 \leqslant C B^2 t, \tag{4.7}$$

where the constant C may depend on μ, λ, η, τ. Moreover, noting that $t \in (0, T]$ is finite, it follows from Gronwall's inequality and (4.7) that

$$\int |U(x,t)|^2 + \varepsilon \iint |\Delta U|^2 + \mu \iint |\nabla U|^2$$

$$\leqslant C \int |U_{0,\varepsilon}(x)|^2 + C \left(B^4 t^{1/2} + B^2 t\right). \tag{4.8}$$

Multiplying $(4.5)_1$ by ΔU, then integrating over Q_t, we obtain

$$\int (|\nabla U(x,t)|^2 - |\nabla U_{0,\varepsilon}(x)|^2) + \varepsilon \iint |\nabla \cdot \Delta U|^2 + \mu \iint |\Delta U|^2$$

$$\leqslant C \iint \Big(\big|\mathbf{P}[\bar{U} \nabla \bar{U}]\big|^2 + \big|\mathbf{P}[\mu(\lambda \bar{\theta} - \eta \bar{S})\mathbf{K}]\big|^2\Big) + \frac{\mu}{2} \iint |\Delta U|^2,$$

then by (4.7), one has

$$\int |\nabla U(x,t)|^2 + \varepsilon \iint |\nabla \Delta U|^2 + \mu \iint |\Delta U|^2$$
$$\leqslant C \int |\nabla U_{0,\varepsilon}(x)|^2 + C\left(B^4 t^{1/2} + B^2 t\right). \tag{4.9}$$

Similarly, according to

$$\|\bar{U} \cdot \nabla \bar{\theta}\|^2_{L^2(0,t;L^2)} + \|\bar{U} \cdot \nabla \bar{S}\|^2_{L^2(0,t;L^2)} + \|\bar{w}\|^2_{L^2(0,t;L^2)}$$
$$\leqslant C t^{\frac{1}{2}} \|\bar{U}\|^2_{L^\infty(0,t;\mathbf{H}^1)} \|(\bar\theta,\bar S)\|_{L^\infty(0,t;H^1)} \|(\bar\theta,\bar S)\|_{L^2(0,t;H^2)} + CB^2 t$$
$$\leqslant C(B^4 t^{1/2} + B^2 t),$$

we have

$$\int |\theta(x,t)|^2 + \varepsilon \iint |\Delta\theta|^2 + \iint |\nabla\theta|^2 \leqslant C\int |\theta_{0,\varepsilon}(x)|^2 + C\left(B^4 t^{1/2} + B^2 t\right), \tag{4.10}$$

$$\int |\nabla\theta(x,t)|^2 + \varepsilon \iint |\nabla\Delta\theta|^2 + \iint |\Delta\theta|^2$$
$$\leqslant \int |\nabla\theta_{0,\varepsilon}(x)|^2 + C\left(B^4 t^{1/2} + B^2 t\right), \tag{4.11}$$

and

$$\int |S(x,t)|^2 + \varepsilon \iint |\Delta S|^2 + \tau \iint |\nabla S|^2$$
$$\leqslant C \int |S_{0,\varepsilon}(x)|^2 + C\left(B^4 t^{1/2} + B^2 t\right), \tag{4.12}$$

$$\int |\nabla S(x,t)|^2 + \varepsilon \iint |\nabla\Delta S|^2 + \tau \iint |\Delta S|^2$$
$$\leqslant \int |\nabla S_{0,\varepsilon}(x)|^2 + C\left(B^4 t^{1/2} + B^2 t\right). \tag{4.13}$$

Summing up (4.8)–(4.13), one has

$$\|U\|^2_{C(0,t;\mathbf{H}^1_\sigma)} + \|U\|^2_{L^2(0,t;\mathbf{H}^2_\sigma)} + \|\theta\|^2_{C(0,t;H^1)}$$
$$+ \|\theta\|^2_{L^2(0,t;H^2)} + \|S\|^2_{C(0,t;H^1)} + \|S\|^2_{L^2(0,t;H^2)}$$
$$\leqslant C_0 \left\{ \|U_{0,\varepsilon}\|^2_{\mathbf{H}^1_\sigma} + \|\theta_{0,\varepsilon}\|^2_{H^1} + \|S_{0,\varepsilon}\|^2_{H^1} + B^4 t^{1/2} + B^2 t \right\}. \tag{4.14}$$

By choosing
$$B^2 = 3C_0 \left\{ \|U_{0,\varepsilon}\|_{\mathbf{H}_\sigma^1}^2 + \|\theta_{0,\varepsilon}\|_{H^1}^2 + \|S_{0,\varepsilon}\|_{H^1}^2 \right\}$$

and $t \leqslant T_1 \triangleq \min\left\{T, \dfrac{1}{(3C_0 B^2)^2}, \dfrac{1}{3C_0}\right\}$, we get $J(\mathbf{M}) \subset \mathbf{M}$.

Then, we will certify that J is a contraction mapping. Let

$$(\bar{U}_1, \bar{\theta}_1, \bar{S}_1),\ (\bar{U}_2, \bar{\theta}_2, \bar{S}_2) \in \mathbf{M},$$

and the corresponding solutions

$$(U_1, \theta_1, S_1) = J(\bar{U}_1, \bar{\theta}_1, \bar{S}_1), \quad (U_2, \theta_2, S_2) = J(\bar{U}_2, \bar{\theta}_2, \bar{S}_2).$$

Define $(\delta U, \delta\theta, \delta S) = (U_1 - U_2, \theta_1 - \theta_2, S_1 - S_2)$. Then

$$\begin{cases} \dfrac{\partial \delta U}{\partial t} + \varepsilon \Delta^2 \delta U - \mu \Delta \delta U = -\mathbf{P}[\bar{U}_1 \cdot \nabla \bar{U}_1 - \bar{U}_2 \cdot \nabla \bar{U}_2] \\ \qquad\qquad\qquad\qquad\qquad\qquad + \mu \mathbf{P}[(\lambda(\bar{\theta}_1 - \bar{\theta}_2) - \eta(\bar{S}_1 - \bar{S}_2))\mathbf{K}], \\ \dfrac{\partial \delta\theta}{\partial t} + \varepsilon \Delta^2 \delta\theta - \Delta \delta\theta = -[\bar{U}_1 \cdot \nabla \bar{\theta}_1 - \bar{U}_2 \cdot \nabla \bar{\theta}_2] + (\bar{w}_1 - \bar{w}_2), \\ \dfrac{\partial \delta S}{\partial t} + \varepsilon \Delta^2 \delta S - \Delta \delta S = -[\bar{U}_1 \cdot \nabla \bar{S}_1 - \bar{U}_2 \cdot \nabla \bar{S}_2] + (\bar{w}_1 - \bar{w}_2), \\ \nabla \cdot \delta U = \nabla \cdot U_1 = \nabla \cdot U_2 = 0, \\ (\delta U, \delta\theta, \delta S)(x, t)|_{t=0} = (0, 0, 0). \end{cases} \quad (4.15)$$

Just like the inequalities (4.6), we can get

$$\|\mathbf{P}[\bar{U}_1 \cdot \nabla \bar{U}_1 - \bar{U}_2 \cdot \nabla \bar{U}_2]\|_{\mathbf{L}_\sigma^2}^2 \leqslant C \|\bar{U}_1 - \bar{U}_2\|_{\mathbf{H}^1}^2 \|\bar{U}_1\|_{\mathbf{H}^1} \|\bar{U}_1\|_{\mathbf{H}^2}$$
$$+ C\|\bar{U}_2\|_{\mathbf{H}^1}^2 \|\bar{U}_1 - \bar{U}_2\|_{\mathbf{H}^1} \|\bar{U}_1 - \bar{U}_2\|_{\mathbf{H}^2},$$

$$\|\bar{U}_1 \cdot \nabla \bar{\theta}_1 - \bar{U}_2 \cdot \nabla \bar{\theta}_2\|_{L^2}^2 \leqslant C \|\bar{U}_1 - \bar{U}_2\|_{\mathbf{H}^1}^2 \|\bar{\theta}_1\|_{H^1} \|\bar{\theta}_1\|_{H^2}$$
$$+ C\|\bar{U}_2\|_{\mathbf{H}^1}^2 \|\bar{\theta}_1 - \bar{\theta}_2\|_{H^1} \|\bar{\theta}_1 - \bar{\theta}_2\|_{H^2},$$

$$\|\bar{U}_1 \cdot \nabla \bar{S}_1 - \bar{U}_2 \cdot \nabla \bar{S}_2\|_{L^2}^2 \leqslant C \|\bar{U}_1 - \bar{U}_2\|_{\mathbf{H}^1}^2 \|\bar{S}_1\|_{H^1} \|\bar{S}_1\|_{H^2}$$
$$+ C\|\bar{U}_2\|_{\mathbf{H}^1}^2 \|\bar{S}_1 - \bar{S}_2\|_{H^1} \|\bar{S}_1 - \bar{S}_2\|_{H^2},$$

which leads to

$$\|\mathbf{P}[\bar{U}_1 \cdot \nabla \bar{U}_1 - \bar{U}_2 \cdot \nabla \bar{U}_2]\|^2_{L^2(0,t;\mathbf{L}^2_\sigma)}$$
$$\leqslant CB^2 t^{1/2} \left\{ \|\bar{U}_1 - \bar{U}_2\|^2_{C([0,t];\mathbf{H}^1_\sigma)} + \|\bar{U}_1 - \bar{U}_2\|_{C([0,t];\mathbf{H}^1_\sigma)} \|\bar{U}_1 - \bar{U}_2\|_{L^2(0,t;\mathbf{H}^2_\sigma)} \right\}, \quad (4.16)$$

$$\|\bar{U}_1 \cdot \nabla \bar{\theta}_1 - \bar{U}_2 \cdot \nabla \bar{\theta}_2\|^2_{L^2(0,t;L^2)}$$
$$\leqslant CB^2 t^{1/2} \left\{ \|\bar{U}_1 - \bar{U}_2\|^2_{C([0,t];\mathbf{H}^1_\sigma)} + \|\bar{\theta}_1 - \bar{\theta}_2\|_{C([0,t];H^1)} \|\bar{\theta}_1 - \bar{\theta}_2\|_{L^2(0,t;H^2)} \right\}, \quad (4.17)$$

$$\|\bar{U}_1 \cdot \nabla \bar{S}_1 - \bar{U}_2 \cdot \nabla \bar{S}_2\|^2_{L^2(0,t;L^2)}$$
$$\leqslant CB^2 t^{1/2} \left\{ \|\bar{U}_1 - \bar{U}_2\|^2_{C([0,t];\mathbf{H}^1_\sigma)} + \|\bar{S}_1 - \bar{S}_2\|_{C([0,t];H^1)} \|\bar{S}_1 - \bar{S}_2\|_{L^2(0,t;H^2)} \right\}. \quad (4.18)$$

And

$$\|\mu \mathbf{P}[(\lambda(\bar{\theta}_1 - \bar{\theta}_2) - \eta(\bar{S}_1 - \bar{S}_2))\mathbf{K}]\|^2_{L^2(0,t;L^2)} + \|\bar{w}_1 - \bar{w}_2\|^2_{L^2(0,t;L^2)}$$
$$\leqslant Ct \left\{ \|\bar{\theta}_1 - \bar{\theta}_2\|^2_{C([0,t];L^2)} + \|\bar{S}_1 - \bar{S}_2\|^2_{C([0,t];L^2)} + \|\bar{w}_1 - \bar{w}_2\|^2_{C([0,t];L^2)} \right\}. \quad (4.19)$$

According to (4.15)–(4.19), we have

$$\|\delta U\|^2_{C([0,t];\mathbf{H}^1_\sigma)} + \|\delta U\|^2_{L^2(0,t;\mathbf{H}^2_\sigma)} + \|(\delta\theta, \delta S)\|^2_{C([0,t];H^1)} + \|(\delta\theta, \delta S)\|^2_{L^2(0,t;H^2)}$$
$$\leqslant \|\mathbf{P}[\bar{U}_1 \cdot \nabla \bar{U}_1 - \bar{U}_2 \cdot \nabla \bar{U}_2]\|^2_{L^2(0,t;\mathbf{L}^2_\sigma)} + \|\bar{U}_1 \cdot \nabla \bar{\theta}_1 - \bar{U}_2 \cdot \nabla \bar{\theta}_2\|^2_{L^2(0,t;L^2)}$$
$$+ \|\bar{w}_1 - \bar{w}_2\|^2_{L^2(0,t;L^2)} + \|\bar{U}_1 \cdot \nabla \bar{S}_1 - \bar{U}_2 \cdot \nabla \bar{S}_2\|^2_{L^2(0,t;L^2)}$$
$$+ \|\mu \mathbf{P}[(\lambda(\bar{\theta}_1 - \bar{\theta}_2) - \eta(\bar{S}_1 - \bar{S}_2))\mathbf{K}]\|^2_{L^2(0,t;L^2)}$$
$$\leqslant CB^2 t^{1/2} \left\{ \|\bar{U}_1 - \bar{U}_2\|^2_{C([0,t];\mathbf{H}^1_\sigma)} + \|\bar{U}_1 - \bar{U}_2\|^2_{L^2(0,t;\mathbf{H}^2_\sigma)} + \|\bar{\theta}_1 - \bar{\theta}_2\|^2_{C([0,t];H^1)} \right.$$
$$\left. + \|\bar{\theta}_1 - \bar{\theta}_2\|^2_{L^2(0,t;H^2)} + \|\bar{S}_1 - \bar{S}_2\|^2_{C([0,t];H^1)} + \|\bar{S}_1 - \bar{S}_2\|^2_{L^2(0,t;H^2)} \right\}$$
$$+ Ct \left\{ \|\bar{\theta}_1 - \bar{\theta}_2\|^2_{C([0,t];L^2)} + \|\bar{S}_1 - \bar{S}_2\|^2_{C([0,t];L^2)} + \|\bar{w}_1 - \bar{w}_2\|^2_{C([0,t];L^2)} \right\}$$
$$\leqslant \zeta \left\{ \|\bar{U}_1 - \bar{U}_2\|^2_{C([0,t];\mathbf{H}^1_\sigma)} + \|\bar{U}_1 - \bar{U}_2\|^2_{L^2(0,t;\mathbf{H}^2_\sigma)} + \|\bar{\theta}_1 - \bar{\theta}_2\|^2_{C([0,t];H^1)} \right.$$
$$\left. + \|\bar{\theta}_1 - \bar{\theta}_2\|^2_{L^2(0,t;H^2)} + \|\bar{S}_1 - \bar{S}_2\|^2_{C([0,t];H^1)} + \|\bar{S}_1 - \bar{S}_2\|^2_{L^2(0,t;H^2)} \right\}, \quad (4.20)$$

where $0 < \zeta < 1$, by choosing $t \leqslant T_* \triangleq \min\left\{T_1, \dfrac{1}{4C}, \dfrac{1}{(2CB^2)^2}\right\}$. So, we conclude that J is a contraction mapping on **M**. According to the principle of contraction mapping, we obtain that exists a unique local solution for the system (4.4) on $[0, T_*]$.

Step 2 We prove the unique local solution of the system (4.4) to be a global one.

In order to finish the proof of Proposition 4.1, we only need to get the global *a priori* uniform bound for $\|U(t)\|_{\mathbf{H}^2_\sigma} + \|\theta(t)\|_{H^2} + \|S(t)\|_{H^2}$.

Lemma 4.1
$$\|U(t)\|^2_{\mathbf{H}^2_\sigma} + \|\theta(t)\|^2_{H^2} + \|S(t)\|^2_{H^2} \leqslant C(T), \quad for \quad t \in [0, T],$$
where the bound $C(T)$ also depends on ε.

Remark 4.2 From (4.21), (4.27), (4.30), we find that the bound of the $L^\infty(0, T; L^2)$ norm of (U, θ, S) is independent of ε, while those of $(\nabla U, \nabla \theta, \nabla S)$ and $(\Delta U, \Delta \theta, \Delta S)$ depend on ε.

Proof Multiplying $(4.4)_1, (4.4)_2, (4.4)_3$ by U, θ, S and integrating over Q_t, respectively, integrating by parts, and noting $\nabla \cdot U = 0$, one has

$$\frac{1}{2}\frac{d}{dt}\int |U|^2 + \varepsilon \int |\Delta U|^2 + \mu \int |\nabla U|^2 \leqslant \frac{1}{2}\int |\mathbf{P}[\mu(\lambda\theta - \eta S)\mathbf{K}]|^2 + \frac{1}{2}\int |U|^2,$$

$$\frac{1}{2}\frac{d}{dt}\int |\theta|^2 + \varepsilon \int |\Delta\theta|^2 + \int |\nabla\theta|^2 \leqslant \frac{1}{2}\int |w|^2 + \frac{1}{2}\int |\theta|^2,$$

$$\frac{1}{2}\frac{d}{dt}\int |S|^2 + \varepsilon \int |\Delta S|^2 + \tau \int |\nabla S|^2 \leqslant \frac{1}{2}\int |w|^2 + \frac{1}{2}\int |S|^2,$$

which leads to

$$\frac{d}{dt}\int (|U|^2 + |\theta|^2 + |S|^2) + \varepsilon \int (|\Delta U|^2 + |\Delta\theta|^2 + |\Delta S|^2)$$
$$+ \int (\mu|\nabla U|^2 + |\nabla\theta|^2 + \tau|\nabla S|^2)$$
$$\leqslant C \int (|U|^2 + |\theta|^2 + |S|^2),$$

so, we obtain by Gronwall's inequality that

$$\int (|U(t)|^2 + |\theta(t)|^2 + |S(t)|^2) + \varepsilon \iint (|\Delta U|^2 + |\Delta\theta|^2 + |\Delta S|^2)$$

$$+ \iint (\mu|\nabla U|^2 + |\nabla\theta|^2 + \tau|\nabla S|^2)$$
$$\leqslant C(T) \int (|U_{0,\varepsilon}|^2 + |\theta_{0,\varepsilon}|^2 + |S_{0,\varepsilon}|^2) \quad \text{for} \quad t \in [0, T], \tag{4.21}$$

where the constant $C(T)$ is independent of ε.

Multiplying $(4.4)_1, (4.4)_2, (4.4)_3$ by $\Delta U, \Delta\theta, \Delta S$ and integrating over Q_t, respectively, it follows from integrating by parts that

$$\frac{1}{2}\frac{d}{dt}\int |\nabla U|^2 + \varepsilon \int |\nabla \Delta U|^2 + \mu \int |\Delta U|^2$$
$$= \int \mathbf{P}[U \cdot \nabla U] \cdot \Delta U - \mu \int \mathbf{P}[(\lambda\theta - \eta S)\mathbf{K}] \cdot \Delta U,$$
$$\frac{1}{2}\frac{d}{dt}\int |\nabla\theta|^2 + \varepsilon \int |\nabla \Delta\theta|^2 + \int |\Delta\theta|^2 = \int [U \cdot \nabla\theta]\Delta\theta - \int w\Delta\theta,$$
$$\frac{1}{2}\frac{d}{dt}\int |\nabla S|^2 + \varepsilon \int |\nabla \Delta S|^2 + \tau \int |\Delta S|^2 = \int [U \cdot \nabla S]\Delta S - \int w\Delta S. \tag{4.22}$$

Integrating by parts and noting $\nabla \cdot U = 0$, we get by Hölder's inequality and Young's inequality that

$$\int [U \cdot \nabla\theta]\Delta\theta = -\int [(\nabla U \cdot \nabla)\theta] \cdot \nabla\theta - \int (U \cdot \nabla)\nabla\theta \cdot \nabla\theta$$
$$= -\int [(\nabla U \cdot \nabla)\theta] \cdot \nabla\theta + \frac{1}{2}\int \nabla \cdot U |\nabla\theta|^2$$
$$\leqslant \left(\int |\nabla U|^3\right)^{1/3} \left(\int |\nabla\theta|^3\right)^{2/3}$$
$$\leqslant \int |\nabla U|^3 + \int |\nabla\theta|^3, \tag{4.23}$$

similarly, one has

$$\int \mathbf{P}[U \cdot \nabla U] \cdot \Delta U \leqslant \int |\nabla U|^3,$$
$$\int [U \cdot \nabla S]\Delta S \leqslant \int |\nabla U|^3 + |\nabla S|^3. \tag{4.24}$$

Then by Lemma 3.1 and Young's inequality, we obtain that

$$\int (|\nabla U|^3 + |\nabla\theta|^3 + |\nabla S|^3) \leqslant C(\varepsilon)\left(\int (|U|^2 + |\theta|^2 + |S|^2)\right)^3$$

$$+ \frac{\varepsilon}{2} \int \left(|\nabla \Delta U|^2 + |\nabla \Delta \theta|^2 + |\nabla \Delta S|^2 \right). \quad (4.25)$$

By Hölder's inequality and Young's inequality, we have

$$\mu \int \mathbf{P}[(\lambda \theta - \eta S)\mathbf{K}] \cdot \Delta U + \int w \Delta \theta + \int w \Delta S$$
$$\leqslant C \int \left(|U|^2 + |\theta|^2 + |S|^2 \right) + \frac{1}{2} \int \left(\mu |\Delta U|^2 + |\Delta \theta|^2 + \tau |\Delta S|^2 \right). \quad (4.26)$$

By substituting (4.23)-(4.26) into the summation of the equations in (4.22), we achieve at

$$\frac{d}{dt} \int \left(|\nabla U|^2 + |\nabla \theta|^2 + |\nabla S|^2 \right) + \varepsilon \int \left(|\nabla \Delta U|^2 + |\nabla \Delta \theta|^2 + |\nabla \Delta S|^2 \right)$$
$$+ \int \left(\mu |\Delta U|^2 + |\Delta \theta|^2 + \tau |\Delta S|^2 \right)$$
$$\leqslant C(\varepsilon) \left(\int \left(|U|^2 + |\theta|^2 + |S|^2 \right) \right)^3 + C \int \left(|U|^2 + |\theta|^2 + |S|^2 \right),$$

and by (4.21), we get that

$$\int \left(|\nabla U(t)|^2 + |\nabla \theta(t)|^2 + |\nabla S(t)|^2 \right) + \varepsilon \iint \left(|\nabla \Delta U|^2 + |\nabla \Delta \theta|^2 + |\nabla \Delta S|^2 \right)$$
$$+ \iint \left(\mu |\Delta U|^2 + |\Delta \theta|^2 + \tau |\Delta S|^2 \right) \leqslant C(T), \quad \text{for} \quad t \in [0, T], \quad (4.27)$$

where the constant $C(T)$ depends on ε.

Multiplying $(4.4)_1, (4.4)_2, (4.4)_3$ by $\Delta^2 U, \Delta^2 \theta, \Delta^2 S$ and integrating over Q_t, respectively, integrating by parts, we obtain

$$\frac{1}{2} \frac{d}{dt} \int |\Delta U|^2 + \varepsilon \int |\Delta^2 U|^2 + \mu \int |\nabla \Delta U|^2$$
$$= - \int \mathbf{P}[U \cdot \nabla U] \cdot \Delta^2 U + \mu \int \mathbf{P}[(\lambda \theta - \eta S)\mathbf{K}] \cdot \Delta^2 U,$$
$$\frac{1}{2} \frac{d}{dt} \int |\Delta \theta|^2 + \varepsilon \int |\Delta^2 \theta|^2 + \int |\nabla \Delta \theta|^2 = - \int (U \cdot \nabla \theta) \Delta^2 \theta + \int w \Delta^2 \theta,$$
$$\frac{1}{2} \frac{d}{dt} \int |\Delta S|^2 + \varepsilon \int |\Delta^2 S|^2 + \tau \int |\nabla \Delta S|^2 = - \int (U \cdot \nabla S) \Delta^2 S + \int w \Delta^2 S.$$
$$(4.28)$$

According to Hölder's inequality and Young's inequality, one has

$$\frac{d}{dt}\int(|\Delta U|^2+|\Delta\theta|^2+|\Delta S|^2)+\varepsilon\int(|\Delta^2 U|^2+|\Delta^2\theta|^2+|\Delta^2 S|^2)$$
$$+\int(\mu|\nabla\Delta U|^2+|\nabla\Delta\theta|^2+\tau|\nabla\Delta S|^2)$$
$$\leqslant C(\varepsilon)\int(|\mathbf{P}[U\cdot\nabla U]|^2+|U\cdot\nabla\theta|^2+|U\cdot\nabla S|^2)$$
$$+C(\varepsilon)\int(|\theta|^2+|S|^2+|w|^2). \tag{4.29}$$

Noting that

$$\int|U\cdot\nabla\theta|^2\leqslant\|U\|_{L^6}^2\|\nabla\theta\|_{L^3}^2\leqslant C\|\nabla U\|_{L^2}^2\|\nabla\theta\|_{L^2}\|\Delta\theta\|_{L^2}$$
$$\leqslant C\|\nabla U\|_{L^2}^4\|\nabla\theta\|_{L^2}^2+\|\Delta\theta\|_{L^2}^2,$$
$$\int|\mathbf{P}[U\cdot\nabla U]|^2\leqslant C\|\nabla U\|_{L^2}^6+\|\Delta U\|_{L^2}^2,$$
$$\int|U\cdot\nabla S|^2\leqslant C\|\nabla U\|_{L^2}^4\|\nabla S\|_{L^2}^2+\|\Delta S\|_{L^2}^2,$$

and using (4.21), (4.27), the inequality (4.29) changes into

$$\int(|\Delta U(t)|^2+|\Delta\theta(t)|^2+|\Delta S(t)|^2)+\varepsilon\iint(|\Delta^2 U|^2+|\Delta^2\theta|^2+|\Delta^2 S|^2)$$
$$+\iint(\mu|\nabla\Delta U|^2+|\nabla\Delta\theta|^2+\tau|\nabla\Delta S|^2)\leqslant C(T),\quad\text{for}\quad t\in[0,T], \tag{4.30}$$

where the constant $C(T)$ depends on ε. So the proof of Lemma 4.1 has been finished.

Then, we succeed in proving Proposition 4.1. Now we finish the proof of Theorem 4.1. Let $(U_\varepsilon,\theta_\varepsilon,S_\varepsilon)$ be the solution of system (4.4). Define

$$\Omega_\varepsilon=\frac{\partial U_\varepsilon}{\partial t}+\varepsilon\Delta^2 U_\varepsilon+U_\varepsilon\cdot\nabla U_\varepsilon-\mu\Delta U_\varepsilon-\mu[\lambda\theta_\varepsilon-\eta S_\varepsilon]\mathbf{K},$$

which and $(4.4)_1$ show $\mathbf{P}\Omega_\varepsilon=0$, and (3.2) leads to $\Omega_\varepsilon\in L^2(0,T;L^2)$. So, there exists a unique ∇P_ε satisfying $\Omega_\varepsilon=-\nabla P_\varepsilon$ such that (4.1) and (4.2) hold. The energy estimate (4.3) can be obtained just like the proof of (4.21).

Assume $\phi(x,t)$ be a nonnegative C^∞ function with compact support respect to the space variables for any $t\in[0,T]$, then, by taking the scalar product of

$(4.1)_1, (4.1)_2, (4.1)_3$ with $U_\varepsilon\phi$, $\theta_\varepsilon\phi$, $S_\varepsilon\phi$, respectively, and integrating on Q_t, we find that

$$\int |U_\varepsilon(\cdot,t)|^2\phi(\cdot,t) + 2\mu \iint |\nabla U_\varepsilon|^2\phi + 2\varepsilon \iint |\Delta U_\varepsilon|^2\phi$$

$$\leqslant \int |U_{0,\varepsilon}|^2\phi(\cdot,0) + \iint |U_\varepsilon|^2\left(\frac{\partial\phi}{\partial t} + \mu\Delta\phi\right) + \iint \left(|U_\varepsilon|^2 + 2\mu P_\varepsilon\right) U_\varepsilon \cdot \nabla\phi$$

$$+ 2\mu \iint (\lambda\theta_\varepsilon - \eta S_\varepsilon)w_\varepsilon\phi + 4\varepsilon\|\nabla\phi\|_{L^\infty(Q_t)}\|\nabla U_\varepsilon\|_{L^2(Q_t)}\|\Delta U_\varepsilon\|_{L^2(Q_t)}$$

$$+ 2\varepsilon\|\Delta\phi\|_{L^\infty(Q_t)}\|U_\varepsilon\|_{L^2(Q_t)}\|\Delta U_\varepsilon\|_{L^2(Q_t)}, \tag{4.31}$$

$$\int |\theta_\varepsilon(\cdot,t)|^2\phi(\cdot,t) + 2 \iint |\nabla\theta_\varepsilon|^2\phi + 2\varepsilon \iint |\Delta\theta_\varepsilon|^2\phi$$

$$\leqslant \int |\theta_{0,\varepsilon}|^2\phi(\cdot,0) + \iint |\theta_\varepsilon|^2\left(\frac{\partial\phi}{\partial t} + \Delta\phi\right) + \iint |\theta_\varepsilon|^2 U \cdot \nabla\phi + 2\iint w_\varepsilon\theta_\varepsilon\phi$$

$$+ 4\varepsilon\|\nabla\phi\|_{L^\infty(Q_t)}\|\nabla\theta_\varepsilon\|_{L^2(Q_t)}\|\Delta\theta_\varepsilon\|_{L^2(Q_t)}$$

$$+ 2\varepsilon\|\Delta\phi\|_{L^\infty(Q_t)}\|\theta_\varepsilon\|_{L^2(Q_t)}\|\Delta\theta_\varepsilon\|_{L^2(Q_t)}, \tag{4.32}$$

and

$$\int |S_\varepsilon(\cdot,t)|^2\phi(\cdot,t) + 2\tau \iint |\nabla S|^2\phi + 2\varepsilon \iint |\Delta S_\varepsilon|^2\phi$$

$$\leqslant \int |S_{0,\varepsilon}|^2\phi(\cdot,0) + \iint |S_\varepsilon|^2\left(\frac{\partial\phi}{\partial t} + \tau\Delta\phi\right) + \iint |S_\varepsilon|^2 U \cdot \nabla\phi + 2\iint w_\varepsilon S_\varepsilon\phi$$

$$+ 4\varepsilon\|\nabla\phi\|_{L^\infty(Q_t)}\|\nabla S_\varepsilon\|_{L^2(Q_t)}\|\Delta S_\varepsilon\|_{L^2(Q_t)}$$

$$+ 2\varepsilon\|\Delta\phi\|_{L^\infty(Q_t)}\|S_\varepsilon\|_{L^2(Q_t)}\|\Delta S_\varepsilon\|_{L^2(Q_t)}. \tag{4.33}$$

4.2 The limiting problem

Now, we let $\varepsilon \to 0$ to prove Theorem 2.1. Let (U_0, θ_0, S_0) satisfy $U_0 \in \mathbf{L}^2_\sigma$, $\theta_0 \in L^2$, $S_0 \in L^2$, and choose sequences $U_{0,\varepsilon} \in \mathbf{H}^2_\sigma$, $\theta_{0,\varepsilon} \in H^2$, $S_{0,\varepsilon} \in H^2$, such that

$$\lim_{\varepsilon\to 0} U_{0,\varepsilon} = U_0, \text{ in } \mathbf{L}^2_\sigma, \quad \lim_{\varepsilon\to 0}\theta_{0,\varepsilon} = \theta_0, \text{ in } L^2, \quad \lim_{\varepsilon\to 0} S_{0,\varepsilon} = S_0, \text{ in } L^2. \tag{4.34}$$

Denote the solution of (4.1) by $(U_\varepsilon, \theta_\varepsilon, S_\varepsilon, \nabla P_\varepsilon)$. Then by the energy estimate (4.3), we achieve at

$$\lim_{\varepsilon\to 0}\varepsilon\|\Delta U_\varepsilon\|_{L^2(0,T;\mathbf{L}^2)} = \lim_{\varepsilon\to 0}\sqrt{\varepsilon}\left(\sqrt{\varepsilon}\|\Delta U_\varepsilon\|_{L^2(0,T;\mathbf{L}^2)}\right) = 0,$$

$$\lim_{\varepsilon \to 0} \varepsilon \|\Delta \theta_\varepsilon\|_{L^2(0,T;L^2)} = 0, \ \lim_{\varepsilon \to 0} \varepsilon \|\Delta S_\varepsilon\|_{L^2(0,T;L^2)} = 0,$$

$$\lim_{\varepsilon \to 0} \varepsilon \|\Delta^2 U_\varepsilon\|_{L^2(0,T;\mathbf{H}^{-2})} = 0, \ \lim_{\varepsilon \to 0} \varepsilon \|\Delta^2 \theta_\varepsilon\|_{L^2(0,T;H^{-2})} = 0,$$

$$\lim_{\varepsilon \to 0} \varepsilon \|\Delta^2 S_\varepsilon\|_{L^2(0,T;H^{-2})} = 0.$$

Since we find that for

$$U_\varepsilon \in C([0,T];\mathbf{L}^2_\sigma) \cap L^2(0,T;\mathbf{H}^1_\sigma), \theta_\varepsilon \in C([0,T];L^2) \cap L^2(0,T;H^1),$$

$$\begin{aligned}
\|U_\varepsilon \cdot \nabla \theta_\varepsilon\|_{H^{-2}} &\leqslant \sup_{\|h\|_{H^2} \leqslant 1} \langle U_\varepsilon \cdot \nabla \theta_\varepsilon, h \rangle \\
&\leqslant \sup_{\|h\|_{H^2} \leqslant 1} \int |U_\varepsilon| |\theta_\varepsilon| |\nabla h| \\
&\leqslant \sup_{\|h\|_{H^2} \leqslant 1} \|U_\varepsilon\|_{L^3} \|\theta_\varepsilon\|_{L^2} \|h\|_{H^2} \\
&\leqslant C \|U_\varepsilon\|_{L^2}^{1/2} \|\nabla U_\varepsilon\|_{L^2}^{1/2} \|\theta_\varepsilon\|_{L^2},
\end{aligned}$$

so, we conclude that

$$\|U_\varepsilon \cdot \nabla \theta_\varepsilon\|_{L^2(0,T;H^{-2})}$$
$$\leqslant C \|U_\varepsilon\|_{C([0,T];\mathbf{L}^2_\sigma)}^{1/2} \|U_\varepsilon\|_{L^2(0,T;\mathbf{H}^1_\sigma)}^{1/2} \|\theta_\varepsilon\|_{C([0,T];L^2)}^{1/2} \|\theta_\varepsilon\|_{L^2(0,T;H^1)}^{1/2}, \tag{4.35}$$

similarly, for $S_\varepsilon \in C([0,T];L^2) \cap L^2(0,T;H^1)$, there hold

$$\|U_\varepsilon \cdot \nabla S_\varepsilon\|_{L^2(0,T;H^{-2})}$$
$$\leqslant C \|U_\varepsilon\|_{C([0,T];\mathbf{L}^2_\sigma)}^{1/2} \|U_\varepsilon\|_{L^2(0,T;\mathbf{H}^1_\sigma)}^{1/2} \|S_\varepsilon\|_{C([0,T];L^2)}^{1/2} \|S_\varepsilon\|_{L^2(0,T;H^1)}^{1/2},$$

$$\|\mathbf{P}[U_\varepsilon \cdot \nabla U_\varepsilon]\|_{L^2(0,T;\mathbf{H}^{-2}_\sigma)} \leqslant C \|U_\varepsilon\|_{C([0,T];\mathbf{L}^2_\sigma)} \|U_\varepsilon\|_{L^2(0,T;\mathbf{H}^1_\sigma)}. \tag{4.36}$$

Then, by (4.35), (4.36) and (4.3), one has

$$\|\mathbf{P}[U_\varepsilon \cdot \nabla U_\varepsilon]\|_{L^2(0,T;\mathbf{H}^{-2}_\sigma)} + \|U_\varepsilon \cdot \nabla \theta_\varepsilon\|_{L^2(0,T;H^{-2})} + \|U_\varepsilon \cdot \nabla S_\varepsilon\|_{L^2(0,T;H^{-2})}$$
$$\leqslant C(T)(\|U_{0,\varepsilon}\|_{\mathbf{L}^2_\sigma}^2 + \|\theta_{0,\varepsilon}\|_{L^2}^2 + \|S_{0,\varepsilon}\|_{L^2}^2). \tag{4.37}$$

Since

$$\begin{aligned}
\|\Delta^2 U_\varepsilon\|_{\mathbf{H}^{-2}_\sigma} &\leqslant \sup_{\|h\|_{H^2} \leqslant 1} \langle \Delta^2 U_\varepsilon, h \rangle \\
&\leqslant \sup_{\|h\|_{H^2} \leqslant 1} \int |\Delta U_\varepsilon| |\Delta h|
\end{aligned}$$

$$\leqslant \sup_{\|h\|_{H^2}\leqslant 1} \|\Delta U_\varepsilon\|_{L^2}\|\Delta h\|_{L^2}$$

$$\leqslant \|\Delta U_\varepsilon\|_{L^2},$$

which and (4.3) lead to

$$\varepsilon\|\Delta^2 U_\varepsilon\|_{L^2(0,T;\mathbf{H}_\sigma^{-2})} \leqslant \sqrt{\varepsilon}\sqrt{\varepsilon}\|\Delta^2 U_\varepsilon\|_{L^2(0,T;\mathbf{H}_\sigma^{-2})}$$

$$\leqslant \sqrt{\varepsilon}\sqrt{\varepsilon}\|\Delta U_\varepsilon\|_{L^2(0,T;\mathbf{L}^2)}$$

$$\leqslant C(T)\sqrt{\varepsilon}(\|U_{0,\varepsilon}\|_{\mathbf{L}_\sigma^2} + \|\theta_{0,\varepsilon}\|_{L^2} + \|S_{0,\varepsilon}\|_{L^2}). \tag{4.38}$$

From (4.3), (4.4), (4.37) and (4.38), we have

$$\|(U_\varepsilon)_t\|_{L^2(0,T;\mathbf{H}_\sigma^{-2})} + \|(\theta_\varepsilon)_t\|_{L^2(0,T;H^{-2})} + \|(S_\varepsilon)_t\|_{L^2(0,T;H^{-2})}$$

$$\leqslant C(T)(1+\sqrt{\varepsilon})(\|U_{0,\varepsilon}\|_{\mathbf{L}_\sigma^2} + \|\theta_{0,\varepsilon}\|_{L^2} + \|S_{0,\varepsilon}\|_{L^2})$$

$$+ C(T)(\|U_{0,\varepsilon}\|_{\mathbf{L}_\sigma^2}^2 + \|\theta_{0,\varepsilon}\|_{L^2}^2 + \|S_{0,\varepsilon}\|_{L^2}^2). \tag{4.39}$$

For any $r > 0$, $B_r = \{x \in \mathbf{R}^3; |x| < r\}$, (4.39) and (4.3) show that

$$\|U_\varepsilon\|_{L^2(0,T;\mathbf{H}^1(B_r))} + \|\theta_\varepsilon\|_{L^2(0,T;H^1(B_r))} + \|S_\varepsilon\|_{L^2(0,T;H^1(B_r))}$$

$$+ \|(U_\varepsilon)_t\|_{L^2(0,T;\mathbf{H}_\sigma^{-2}(B_r))} + \|(\theta_\varepsilon)_t\|_{L^2(0,T;H^{-2}(B_r))} + \|(S_\varepsilon)_t\|_{L^2(0,T;B^{-2}(B_r))} \leqslant C. \tag{4.40}$$

From (4.40) and lemma 3.2, we find there exist subsequences such that for any $r > 0$,

$$U_\varepsilon \to U, \text{ in } L^2(0,T;\mathbf{L}^2(B_r)), \quad \theta_\varepsilon \to \theta, \text{ in } L^2(0,T;L^2(B_r)),$$

$$S_\varepsilon \to S, \text{ in } L^2(0,T;L^2(B_r)), \tag{4.41}$$

$U_\varepsilon \to U$ weakly in $L^2(0,T;\mathbf{H}_\sigma^1)$, weakly star in $L^\infty(0,T;\mathbf{L}_\sigma^2)$, a.e. in Q_T,

$\theta_\varepsilon \to \theta$, weakly in $L^2(0,T;H^1)$, weakly star in $L^\infty(0,T;L^2)$, a.e. in Q_T,

$S_\varepsilon \to S$, weakly in $L^2(0,T;H^1)$, weakly star in $L^\infty(0,T;L^2)$, a.e. in Q_T. \tag{4.42}

Moreover,

$$(U_\varepsilon)_t \to U_t \text{ weakly in } L^2(0,T;\mathbf{H}_\sigma^{-2}),$$

$$(\theta_\varepsilon)_t \to \theta_t \text{ weakly in } L^2(0,T;H^{-2}),$$

$$(S_\varepsilon)_t \to S_t \text{ weakly in } L^2(0,T;H^{-2}). \tag{4.43}$$

Since $L^\infty(0,T;L^2) \cap L^2(0,T;L^6) \hookrightarrow L^{\frac{8}{3}}(0,T;L^4)$, we obtain

$$\|U_\varepsilon\|_{L^{\frac{8}{3}}(0,T;\mathbf{L}^4)} \leqslant C, \quad \|\theta_\varepsilon\|_{L^{\frac{8}{3}}(0,T;L^4)} \leqslant C, \quad \|S_\varepsilon\|_{L^{\frac{8}{3}}(0,T;L^4)} \leqslant C, \tag{4.44}$$

then

$$\|U_\varepsilon^i U_\varepsilon^j\|_{L^{\frac{4}{3}}(0,T;L^2)} \leqslant C, \quad \|U_\varepsilon^i \theta_\varepsilon\|_{L^{\frac{4}{3}}(0,T;L^2)} \leqslant C, \quad \|U_\varepsilon^i S_\varepsilon\|_{L^{\frac{4}{3}}(0,T;L^2)} \leqslant C, \tag{4.45}$$

which and (4.41),(4.42) lead to

$$U_\varepsilon^i U_\varepsilon^j \to U^i U^j \text{ weakly in } L^{\frac{4}{3}}(0,T;L^2),$$

$$U_\varepsilon^i \theta_\varepsilon \to U^i \theta \text{ weakly in } L^{\frac{4}{3}}(0,T;L^2),$$

$$U_\varepsilon^i S_\varepsilon \to U^i S \text{ weakly in } L^{\frac{4}{3}}(0,T;L^2), \tag{4.46}$$

where we have used Lemma 1.3 in [28], Chapter I.

And noting $\nabla \cdot U = 0$, such that

$$U^i \frac{\partial}{\partial x_i} U^j = \frac{\partial}{\partial x_i}(U^i U^j), \quad U^i \frac{\partial}{\partial x_i} \theta = \frac{\partial}{\partial x_i}(U^i \theta), \quad U^i \frac{\partial}{\partial x_i} S = \frac{\partial}{\partial x_i}(U^i S),$$

we get

$$U_\varepsilon \cdot \nabla U_\varepsilon \to U \cdot \nabla U \text{ weakly in } L^{\frac{4}{3}}(0,T;\mathbf{H}^{-1}),$$

$$U_\varepsilon \cdot \nabla \theta_\varepsilon \to U \cdot \nabla \theta \text{ weakly in } L^{\frac{4}{3}}(0,T;H^{-1}),$$

$$U_\varepsilon \cdot \nabla S_\varepsilon \to U \cdot \nabla S \text{ weakly in } L^{\frac{4}{3}}(0,T;H^{-1}). \tag{4.47}$$

Acting the divergence operator $\nabla\cdot$ on $(4.1)_1$, one has

$$-\mu \Delta P_\varepsilon = \nabla \cdot (U_\varepsilon \cdot \nabla U_\varepsilon) - \mu \nabla \cdot [(\lambda \theta_\varepsilon - \eta S_\varepsilon)\mathbf{K}]. \tag{4.48}$$

We decompose P_ε as:

$$P_\varepsilon = P_\varepsilon^{(1)} + P_\varepsilon^{(2)}, \tag{4.49}$$

with

$$-\mu \Delta P_\varepsilon^{(1)} = \sum_{i,j=1}^{3} \frac{\partial^2}{\partial x_i \partial x_j}(U_\varepsilon^i U_\varepsilon^j),$$

$$-\mu\Delta P_\varepsilon^{(2)} = -\mu \sum_{i=1}^{3} \frac{\partial}{\partial x_i}[(\lambda\theta_\varepsilon - \eta S_\varepsilon)\mathbf{K}]^i. \tag{4.50}$$

From (4.44) and the Fourier transform

$$\mu|\xi|^2 \mathcal{F}(P_\varepsilon^{(1)}) = -\sum_{i,j=1}^{3} \xi_i \xi_j \mathcal{F}(U_\varepsilon^i U_\varepsilon^j),$$

we get

$$\|P_\varepsilon^{(1)}\|_{L^{\frac{4}{3}}(0,T;L^2)} + \|\nabla P_\varepsilon^{(1)}\|_{L^{\frac{4}{3}}(0,T;H^{-1})} \leqslant C, \tag{4.51}$$

and from (4.45) and the Fourier transform $\mu|\xi|\mathcal{F}(P_\varepsilon^{(2)}) = -i|\xi|^{-1}\mathbf{K}\cdot\xi\mathcal{F}(\lambda\theta_\varepsilon - \eta S_\varepsilon)$, we have

$$\|P_\varepsilon^{(2)}\|_{L^2(0,T;L^6)} + \|\nabla P_\varepsilon^{(2)}\|_{L^2(0,T;L^2)} \leqslant C, \tag{4.52}$$

we know from (4.49), (4.51) and (4.52) that

$$P_\varepsilon \to P \text{ weakly in } L^{\frac{4}{3}}(0,T; L^2 + L^6),$$
$$\nabla P_\varepsilon \to \nabla P \text{ weakly in } L^{\frac{4}{3}}(0,T; H^{-1}). \tag{4.53}$$

So, we get

$$(U_\varepsilon)_t \to U_t \text{ weakly in } L^{\frac{4}{3}}(0,T;\mathbf{H}^{-1}) + L^2(0,T;\mathbf{H}^{-2}),$$
$$(\theta_\varepsilon)_t \to \theta_t \text{ weakly in } L^{\frac{4}{3}}(0,T;H^{-1}) + L^2(0,T;H^{-2}),$$
$$(S_\varepsilon)_t \to S_t \text{ weakly in } L^{\frac{4}{3}}(0,T;H^{-1}) + L^2(0,T;H^{-2}), \tag{4.54}$$

and $U_t \in L^{4/3}(0,T;\mathbf{H}^{-1})$, $\theta_t, S_t \in L^{4/3}(0,T;H^{-1})$. Then by a usual way, one has $U \in C_w(0,T;\mathbf{L}_\sigma^2)$, $\theta, S \in C_w(0,T;L^2)$. Now we can pass to the limit in (4.1) and (4.3) to obtain that $(U,\theta,S,\nabla P)$ satisfy (2.1)–(2.4) and (2.8).

At last, we show that (2.5)–(2.7) hold. According to (4.41), there exists a subsequence $U_\varepsilon(t) \to U(t)$, strongly in $L^2(B_r)$ a.e. $t \in [0,T]$, for any $r > 0$. Then, we get

$$\int |U_\varepsilon(\cdot,t)|^2 \phi(\cdot,t) \to \int |U(\cdot,t)|^2 \phi(\cdot,t) \qquad \text{for a.e. } t \in [0,T].$$

By (4.34), (4.41), (4.42), (4.45), (4.3), one has

$$\mu \iint |\nabla U|^2 \phi \leqslant \lim_{\varepsilon \to 0} \mu \iint |\nabla U_\varepsilon|^2 \phi,$$

$$\varepsilon \iint |\Delta U_\varepsilon|^2 \phi \geqslant 0,$$

$$\int |U_{0,\varepsilon}|^2 \phi(\cdot,0) \to \int |U_0|^2 \phi(\cdot,0),$$

$$\iint |U_\varepsilon|^2 \left(\frac{\partial \phi}{\partial t} + \mu \Delta \phi \right) \to \iint |U|^2 \left(\frac{\partial \phi}{\partial t} + \mu \Delta \phi \right),$$

$$\mu \iint (\lambda \theta_\varepsilon - \eta S_\varepsilon) w_\varepsilon \phi \to \mu \iint (\lambda \theta - \eta S) w \phi,$$

$$4\varepsilon \|\nabla \phi\|_{L^\infty(Q_t)} \|\nabla U_\varepsilon\|_{L^2(Q_t)} \|\Delta U_\varepsilon\|_{L^2(Q_t)} \to 0,$$

$$2\varepsilon \|\Delta \phi\|_{L^\infty(Q_t)} \|U_\varepsilon\|_{L^2(Q_t)} \|\Delta U_\varepsilon\|_{L^2(Q_t)} \to 0.$$

By Lemma 3.1 and U_ε being bounded in $L^\infty(0,T;L^2) \cap L^2(0,T;H^1)$, we know that

$$\iint |U_\varepsilon|^{\frac{10}{3}} \leqslant C,$$

which and (4.41) lead to

$$\iint |U_\varepsilon|^2 U_\varepsilon \cdot \nabla \phi \to \iint |U|^2 U \cdot \nabla \phi.$$

Noting for any $r > 0$,

$$\|U_\varepsilon - U\|_{L^4(0,T;L^2(B_r))} \leqslant \|U_\varepsilon - U\|_{L^\infty(0,T;L^2(B_r))}^{1/2} \|U_\varepsilon - U\|_{L^2(0,T;L^2(B_r))}^{1/2} \to 0, \quad \text{as } \varepsilon \to 0,$$

and by (4.53), we achieve at

$$2\mu \iint P_\varepsilon U_\varepsilon \cdot \nabla \phi \to 2\mu \iint PU \cdot \nabla \phi.$$

Obviously, we can get similar results for θ_ε and S_ε. Now, we have proved that (2.5)–(2.7) hold for almost everywhere $t \in [0,T]$. Then, by the weak continuity of U, θ, S with respect to t, we can conclude the weak continuity of $U(t)\sqrt{\phi(t)}$, $\theta(t)\sqrt{\phi(t)}$, $S(t)\sqrt{\phi(t)}$ in $[0,T]$ with values in $L^2(\mathbf{R}_x^3)$, such that

$$\int |U(t)|^2 \phi(t) \leqslant \liminf_{t_n \to t} \int |U(t_n)|^2 \phi(t_n) \, dx,$$

$$\int |\theta(t)|^2 \phi(t) \leqslant \liminf_{t_n \to t} \int |\theta(t_n)|^2 \phi(t_n) \,\mathrm{d}x,$$

$$\int |S(t)|^2 \phi(t) \leqslant \liminf_{t_n \to t} \int |S(t_n)|^2 \phi(t_n) \,\mathrm{d}x,$$

which means that (2.5)–(2.7) hold for any $t \in [0, T]$. Here we have completed the proof of Theorem 2.1.

Acknowledgements This work is supported by NNSF of China under grant number 11731014, 11571254, and by China Postdoctoral Science Foundation under grant number 2017M620688.

References

[1] Hsia C H, Ma T, Wang S. Attractor bifurcation of three-dimensional double-diffusive convection [J]. Z. Anal. Anwend., 2008, 27: 233–252.

[2] Joseph D D. Uniqueness criteria for the conduction-diffusion solution of the Boussinesq equations [J]. Arch. Rational. Mech. Anal., 1969, 35: 169–177.

[3] Veronis G. On finite amplitude instability in the thermohaline convection [J]. J. Marine Res., 1965, 23: 1–17.

[4] Jevons W S. On the cirrous form of cloud [J]. London, Edinburgh, Dublin Philos. Mag. J. Sci. Ser. 4, 1857, 14: 22–35.

[5] Stommel H, Arons A, Blanchard D. An oceanographical curiosity: the perpetual salt fountain [J]. Deep-Sea Res., 1956, 3: 152–153.

[6] Baines P G, Gill A. On thermohaline convection with linear gradients [J]. J. Fluid Mech., 1969, 37: 289–306.

[7] Brandt A, Fernando H J S. Double-Diffusive Convection [M]. Washington, Amer. Geophysical Union, 1995.

[8] Huppert H E, Turner J S. Double-diffusive convection [J]. J. Fluid Mech., 1981, 106: 299–329.

[9] Nield D A, Bejan A. Convection in Porous Medium [M]. New York, third edition, Springer, 2006.

[10] Radko T, Ball J, Colosi J, Flanagan J. Double-diffusive convection in a stochastic shear [J]. J. Phys. Oceanogr., 2015, 45: 3155–3167.

[11] Radko T, Flanagan J D, Stellmach S, Timmermans M L. Double-Diffusive Recipes. part II: Layer-Merging Events [J]. J. Phys. Oceanogr., 2014, 44: 1285–1305.

[12] Joseph D D. Global stability of the conduction-diffusion solution [J]. Arch. Rational. Mech. Anal., 1970, 36: 285–292.

[13] Lions J L, Temam R, Wang S H. On the equations of the large-scale ocean [J]. Nonlinearity, 1992, 5: 1007–1053.

[14] Ôtani M, Uchida S. Global sovability of some double-diffusive convection system coupled with Brinkman-Forchheimer equations [J]. Lib. Math., 2013, 33: 79–107.

[15] Ôtani M, Uchida S. The existence of periodic solutions of some double-diffusive convection system based on Brinkman-Forchheimer equations [J]. Adv. Math. Sci. Appl., 2013, 23: 77–92.

[16] Ôtani M, Uchida S. Global sovability for double-diffusive convection system based on Brinkman-Forchheimer equation in general domains [J]. Osaka J. Math., 2016, 53: 855–872.

[17] Piniewski M. Asymptotic dynamics in double-diffusive convection [J]. Appl. Math., 2008, 35: 223–245.

[18] Shir C C, Joseph D D. Convective instability in a temperature and concentration field [J]. Arch. Rational. Mech. Anal., 1968, 30: 38–80.

[19] Terasawa K, Ôtani M. Global solvability of double-diffusive convection systems based upon Brinkman-Forchheimer equations [J]. GAKUTO Internat. Ser. Math. Sci. Appl., 2010, 32: 505–515.

[20] Chen F, Guo B L, Zeng L. The well-posedness for the Cauchy problem of the double-diffusive convection system [J]. J. Math. Phys., 2019, 60: 011511.

[21] Chen F, Guo B L, Zeng L. The well-posedness of the double-diffusive convection system in a bounded domain [J]. Math. Meth. Appl. Sci., 2018, 41: 4327–4336.

[22] Scheffer V. Hausdorff measure and the Navier-Stokes equations [J]. Comm. Math. Phys., 1977, 55: 97–112.

[23] Caffarelli L, Kohn R, Nirenberg L. Partial regularity of suitable weak solutions of the Navier-Stokes equations [J]. Communs. Pure Appl. Math, 1982, 35: 771–831.

[24] Beirão Da Veiga H. On the suitable weak solutions to the Navier-Stokes equations in the whole space [J]. J. Math. pures Appl., 1985, 64: 77–86.

[25] Guo B L, Yuan G W. On the suitable weak solutions for the Cauchy problem of the Boussinesq equations [J]. Nonlinear Anal., 1996, 26: 1367–1385.

[26] Boyer F, Fabrice P. Mathematical Tools for the Study of the Incompressible Navier-Stokes Equations and Related Models [M]. New York, Springer Science+Bussiness Media, 2013.

[27] Lions J L, Magenes E. Problémes aux Limities non Homogénes et Applications [M]. Paris, Vol. I and II, Dunod, 1969.

[28] Lions J L. Quelques Méthodes de Resolution de Problemes aux Limits Non-linéairies [M]. Paris, Dunod, 1969.

A Riemann-Hilbert Approach for the Modified Short Pulse Equation*

Boling Guo(郭柏灵), and Nan Liu(刘男)

Abstract We present a Riemann–Hilbert approach for the modified short pulse equation

$$q_{xt} = q + \frac{1}{2}q(q^2)_{xx}$$

on the line. This approach allows us to give a representation of the solution to the Cauchy problem, which can be efficiently used in studying its long-time behaviour, and also to describe the soliton solutions.

Keywords modified short pulse equation; Riemann–Hilbert problem; long-time asymptotics; soliton solutions

1 Introduction

The so-called short pulse (SP) equation

$$q_{xt} = q + \frac{1}{6}(q^3)_{xx} \qquad (1.1)$$

was proposed by Schäfer and Wayne [7, 20] to describe the propagation of ultra-short optical pulses in nonlinear media, where $q(x,t)$ is a real-valued function and represents the magnitude of the electric field. The SP equation represents an alternative approach in contrast with the slowly varying envelope approximation which leads to the nonlinear Schrödinger (NLS) equation. In fact, in the regime of ultra-short pulses where the width of the optical pulse is in the order of femtosecond (10^{-15} s), the NLS equation becomes less accurate, while the SP

---

* Appl. Anal., 2019, 98:9 1646–1659. DOI:10.1080/00036811.2018.1437418.

equation provides an increasingly better approximation to the corresponding solution of the Maxwell equations [7]. Apart from the context of nonlinear optics, the SP equation has also been derived as an integrable differential equation associated with pseudospherical surfaces [16]. Thus, it is expected that the SP equation and its generalization will play more and more important roles in applications.

The SP equation has been shown to be completely integrable possessing a Lax pair representation and bi-Hamiltonian structure [6,17]. The loop-soliton solutions as well as an exact nonsingular solitary wave solutions of the SP equation were found in [14,18]. Periodic solutions regarding the SP equation were discussed in [15] in detail. It is interesting to notice that Boutet de Monvel et al. have recently developed a Riemann–Hilbert approach to the Cauchy problem on the line for the SP equation (1.1) in [2], and further analyzed the long-time behavior of the solution as well as retrieved the soliton solutions, which motivates our following analysis.

Recently, a modified short pulse (mSP) equation

$$q_{xt} = q + \frac{1}{2}q(q^2)_{xx} \tag{1.2}$$

was studied by Sakovich [19], in which the Lax pair, bi-Hamiltonian structure and the soliton solutions were provided. However, the soliton solutions were obtained by using the connection between the mSP equation and the sine-Gordon equation. The mSP equation is a direct reduction ($p = q$) of a coupled short pulse equation [10] proposed by Feng

$$p_{xt} = p + \frac{1}{6}(p^3)_{xx} + \frac{1}{2}q^2 p_{xx}, \quad q_{xt} = q + \frac{1}{6}(q^3)_{xx} + \frac{1}{2}p^2 q_{xx}.$$

Equation (1.2) is (at least, formally) integrable and it possesses a Lax pair representation [19]

$$\psi_x = X\psi, \quad \psi_t = T\psi, \tag{1.3}$$

with

$$X = \begin{pmatrix} \lambda(1-q_x^2) & 2\lambda q_x \\ 2\lambda q_x & -\lambda(1-q_x^2) \end{pmatrix}, \tag{1.4}$$

$$T = \begin{pmatrix} \lambda q^2(1-q_x^2) + \frac{1}{4\lambda} & 2\lambda q^2 q_x - q \\ 2\lambda q^2 q_x + q & -\lambda q^2(1-q_x^2) - \frac{1}{4\lambda} \end{pmatrix}, \tag{1.5}$$

and $\lambda \in \mathbf{C}$ is the spectral parameter.

Our purpose in the present paper is to propose a Riemann–Hilbert approach to the Cauchy problem for the mSP equation

$$q_{xt} = q + \frac{1}{2}q(q^2)_{xx}, \quad x \in (-\infty, +\infty),\ t > 0, \tag{1.6}$$

$$q(x,0) = q_0(x), \quad x \in (-\infty, +\infty), \tag{1.7}$$

on the line. We will assume that $q_0(x)$ is smooth and decays sufficiently fast at $\pm\infty$, that is,

$$q_0(x) \to 0, \quad x \to \pm\infty, \tag{1.8}$$

and we seek a solution $q(x,t)$ that decays as $x \to \pm\infty$ for all $t > 0$:

$$q(x,t) \to 0, \quad x \to \pm\infty. \tag{1.9}$$

The rest of the paper is organized as follows. In Section 2, we present appropriate Lax pairs associated with the mSP equation, whose dedicated solutions are used to formulate a matrix Riemann–Hilbert problem suitable for solving the Cauchy problem (1.6)-(1.7). Then we give a representation of the solution of the Cauchy problem in terms of the solution of an associated Riemann–Hilbert problem. The representation obtained allows us to apply the nonlinear steepest descent method for oscillatory Riemann–Hilbert problems and to obtain a detailed description for the long-time asymptotics of the solution of the Cauchy problem in Section 3. In Section 4, we discuss the construction of soliton solutions using the formalism of the Riemann–Hilbert problem.

2 Riemann–Hilbert formalism

In this section, we aim to transform the Lax pair (1.3) to a suitable form that can formulate a Riemann–Hilbert problem.

2.1 Lax pairs and eigenfunctions

Introducing a new spectral parameter $k = i\lambda$, it is noted that X and T have singularities in the extended complex k-plane at $k = 0$ and at $k = \infty$. In order to control the behavior of solutions to (1.3) as functions of the spectral parameter k, it is convenient to transform the Lax pair equations (1.3) to a certain form, which is standard for establishing analytic properties of solutions near the singular points with respect to the spectral parameter of the Lax pair equations. We will follow a strategy similar to that adopted for the SP equation [2] and the short-wave model for the Camassa–Holm equation [3].

In order to have a good control on the behavior of eigenfunctions at $k = \infty$, we transform the Lax pair (1.3) to the following form. Letting

$$P = \frac{1}{\sqrt{1+q_x^2}} \begin{pmatrix} 1 & q_x \\ -q_x & 1 \end{pmatrix}, \tag{2.1}$$

we can see that

$$PXP^{-1} = -ik(1+q_x^2)\sigma_3.$$

Then setting $\Psi = P\psi$ reduces (1.3) to

$$\Psi_x + ik(1+q_x^2)\sigma_3 \Psi = U\Psi, \tag{2.2}$$

$$\Psi_t + \left[ikq^2(1+q_x^2) + \frac{1}{4ik}\right]\sigma_3 \Psi = V\Psi, \tag{2.3}$$

where

$$U = -PP_x^{-1} = \frac{q_{xx}}{1+q_x^2} \begin{pmatrix} 0 & 1 \\ -1 & 0 \end{pmatrix}, \tag{2.4}$$

$$V = -\frac{q_x}{2ik(1+q_x^2)} \begin{pmatrix} -q_x & -1 \\ -1 & q_x \end{pmatrix} + \frac{q^2 q_{xx}}{1+q_x^2} \begin{pmatrix} 0 & 1 \\ -1 & 0 \end{pmatrix}. \tag{2.5}$$

Since $U = U(x,t;k)$ and $V = V(x,t;k)$ are bounded at $k = \infty$, moreover, the diagonal part of U vanishes identically while the diagonal part of V is $O(1/k)$ as $k \to \infty$, thus, the Lax pair (2.2)-(2.5) is appropriate for controlling the behavior of its solutions for large k.

The left-hand side of (2.2) and (2.3) suggest introducing the function

$$Q(x,t;k) = ik\left(x - \int_x^\infty q_\xi^2(\xi,t)d\xi\right) + \frac{t}{4ik}, \tag{2.6}$$

such that

$$Q_x = ik(1+q_x^2), \quad Q_t = ikq^2(1+q_x^2) + \frac{1}{4ik},$$

because the conservation law for the mSP equation

$$(1+q_x^2)_t = (q^2(1+q_x^2))_x. \tag{2.7}$$

Therefore, setting $\mu = \Psi e^{Q(x,t;k)\sigma_3}$, we have

$$\mu_x + Q_x[\sigma_3, \mu] = U\mu, \tag{2.8}$$

$$\mu_t + Q_t[\sigma_3, \mu] = V\mu. \tag{2.9}$$

We define two eigenfunctions $\{\mu_j\}_1^2$ of equation (2.8)-(2.9) by the following Volterra integral equations

$$\mu_j(x,t;k) = I + \int_{\Gamma_j} e^{(Q(x',t';k)-Q(x,t;k))\hat{\sigma}_3}(U(x',t')dx' + V(x',t';k)dt')\mu_j(x',t';k),$$

$$j = 1, 2, \tag{2.10}$$

where the contours $\{\Gamma_j\}_1^2$ denote the smooth curves from (x_j, t_j) to (x,t), and $(x_1, t_1) = (-\infty, t)$, $(x_2, t_2) = (+\infty, t)$, and $\hat{\sigma}_3$ denotes the operator which acts on a 2×2 matrix A by $\hat{\sigma}_3 A = [\sigma_3, A]$, then $e^{\hat{\sigma}_3 A} = e^{\sigma_3} A e^{-\sigma_3}$. That is

$$\mu_1(x,t;k) = I + \int_{-\infty}^{x} e^{-ik\int_{x'}^{x}(q_\xi^2(\xi,t)+1)d\xi \hat{\sigma}_3} U(x',t)\mu_1(x',t;k)dx', \tag{2.11}$$

$$\mu_2(x,t;k) = I - \int_{x}^{+\infty} e^{ik\int_{x}^{x'}(q_\xi^2(\xi,t)+1)d\xi \hat{\sigma}_3} U(x',t)\mu_2(x',t;k)dx'. \tag{2.12}$$

We denote by $\mu^{(1)}$ and $\mu^{(2)}$ the columns of a 2×2 matrix $\mu = (\mu^{(1)} \; \mu^{(2)})$. Then it follows from (2.11)-(2.12) that for all (x,t):

(i) $\det \mu_j = 1$, $j = 1, 2$.

(ii) $\mu_1^{(1)}$ and $\mu_2^{(2)}$ are analytic and bounded in $\{k | \text{Im}\, k > 0\}$, and $(\mu_1^{(1)} \; \mu_2^{(2)}) \to I$ as $k \to \infty$.

(iii) $\mu_2^{(1)}$ and $\mu_1^{(2)}$ are analytic and bounded in $\{k | \text{Im}\, k < 0\}$, and $(\mu_2^{(1)} \; \mu_1^{(2)}) \to I$ as $k \to \infty$.

(iv) $\{\mu_j\}_1^2$ are continuous up to the real axis.

(v) Symmetry:

$$\overline{\mu_j(x,t;\bar{k})} = \mu_j(x,t;-k) = \begin{pmatrix} 0 & 1 \\ -1 & 0 \end{pmatrix} \mu_j(x,t;k) \begin{pmatrix} 0 & -1 \\ 1 & 0 \end{pmatrix}, \quad \text{for } j = 1,2. \tag{2.13}$$

The symmetry relation (2.13) can be proved easily due to the symmetries of the matrix $\hat{U} \triangleq U - ik(1 + q_x^2)\sigma_3$:

$$\overline{\hat{U}(x,t;\bar{k})} = \hat{U}(x,t;-k) = \begin{pmatrix} 0 & 1 \\ -1 & 0 \end{pmatrix} \hat{U}(x,t;k) \begin{pmatrix} 0 & -1 \\ 1 & 0 \end{pmatrix}.$$

The solutions of the system of differential equation (2.8)-(2.9) must be related by a matrix independent of x and t, therefore,

$$\mu_2(x,t;k) = \mu_1(x,t;k) e^{-Q(x,t;k)\hat{\sigma}_3} s(k), \quad k \in \mathbf{R}. \tag{2.14}$$

Due to the symmetry (2.13), the matrix-valued spectral functions $s(k)$ can be defined in terms of two scalar spectral functions, $a(k)$ and $b(k)$ by

$$s(k) = \begin{pmatrix} \overline{a(k)} & b(k) \\ -\overline{b(k)} & a(k) \end{pmatrix}. \tag{2.15}$$

The spectral functions $a(k)$ and $b(k)$ can be determined by $q_0(x)$ through the solutions of equations (2.11)-(2.12) at $t=0$ with q replaced by $q(x,0)$. On the other hand, $a(k)$ is analytic in the half-plane $\{k|\mathrm{Im}\,k > 0\}$ and continuous in $\{k|\mathrm{Im}\,k \geqslant 0\}$, and $a(k) \to 1$ as $k \to \infty$. Furthermore, $|a(k)|^2 + |b(k)|^2 = 1$, for $k \in \mathbf{R}$.

In order to control the behavior of the eigenfunctions at $k=0$, it is convenient to rewrite the Lax pair (1.3) in the following form

$$\psi_x + ik\sigma_3\psi = U_0\psi, \tag{2.16}$$

$$\psi_t + \frac{1}{4ik}\sigma_3\psi = V_0\psi, \tag{2.17}$$

where

$$U_0 = -ik \begin{pmatrix} -q_x^2 & 2q_x \\ 2q_x & q_x^2 \end{pmatrix}, \tag{2.18}$$

$$V_0 = -ikq^2 \begin{pmatrix} 1-q_x^2 & 2q_x \\ 2q_x & q_x^2-1 \end{pmatrix} + q \begin{pmatrix} 0 & -1 \\ 1 & 0 \end{pmatrix}. \tag{2.19}$$

Since $U_0 = U_0(x,t;k)$ and $V_0 = V_0(x,t;k)$ are bounded at $k=0$ and $U_0 \to 0$, $V_0 \to 0$ as $|x| \to \infty$, the Lax pair (2.16)-(2.19) is appropriate for controlling the behavior of its solutions as $k \to 0$. Moreover, it is important that $U_0(x,t;0) = 0$ for all (x,t).

Introducing a new eigenfunction $\nu(x,t;k)$ by

$$\nu = \psi e^{(ikx+\frac{t}{4ik})\sigma_3}, \tag{2.20}$$

we obtain the equivalent Lax pair

$$\nu_x + ik[\sigma_3, \nu] = U_0\nu, \tag{2.21}$$

$$\nu_t + \frac{1}{4ik}[\sigma_3, \nu] = V_0\nu. \tag{2.22}$$

The eigenfunctions $\{\nu_j\}_1^2$ of (2.21)-(2.22) are determined, similarly to above, as the solutions of associated Volterra integral equations:

$$\nu_1(x,t;k) = I + \int_{-\infty}^{x} e^{-ik(x-x')\hat{\sigma}_3} U_0(x',t;k)\nu_1(x',t;k)dx', \tag{2.23}$$

$$\nu_2(x,t;k) = I - \int_{x}^{+\infty} e^{ik(x'-x)\hat{\sigma}_3} U_0(x',t;k)\nu_2(x',t;k)dx'. \tag{2.24}$$

An important consequence of (2.23) and (2.24) is that the expansions in powers of k of ν_j as $k \to 0$, are given by

$$\nu_j(x,t;k) = I - ik \begin{pmatrix} \int_x^\infty q_\xi^2(\xi,t)d\xi & 2q \\ 2q & -\int_x^\infty q_\xi^2(\xi,t)d\xi \end{pmatrix} + O(k^2), \quad j = 1, 2. \tag{2.25}$$

Now we note that the eigenfunctions μ and ν, being related to the same Lax pair (1.3), must be related to each other as

$$\mu_j(x,t;k) = P(x,t)\nu_j(x,t;k)e^{-(ikx+\frac{t}{4ik})\sigma_3} C_j(k) e^{Q(x,t;k)\sigma_3}, \tag{2.26}$$

with $C_j(k)$, $j = 1, 2$ independent of x and t. Evaluating (2.26) as $x \to \pm\infty$ gives $C_2(k) = I$ and $C_1(k) = e^{ik\alpha\sigma_3}$, where

$$\alpha = \int_{-\infty}^{\infty} q_\xi^2(\xi,t)d\xi$$

is a constant independent of t due to the conservation law (2.7).

Combining (2.25) with (2.26), we get the following expansions in powers of k of μ_j as $k \to 0$,

$$\mu_1(x,t;k) = P(x,t)\Bigg(I - ik\Bigg[\begin{pmatrix} \int_x^\infty q_\xi^2(\xi,t)d\xi & 2q \\ 2q & -\int_x^\infty q_\xi^2(\xi,t)d\xi \end{pmatrix} \\ - \int_{-\infty}^{x} q_\xi^2(\xi,t)d\xi\,\sigma_3\Bigg] + O(k^2)\Bigg), \tag{2.27}$$

$$\mu_2(x,t;k) = P(x,t)\left(I - ik\left[\left(\begin{array}{cc} \int_x^\infty q_\xi^2(\xi,t)d\xi & 2q \\ 2q & -\int_x^\infty q_\xi^2(\xi,t)d\xi \end{array}\right)\right.\right.$$

$$\left.\left. + \int_x^\infty q_\xi^2(\xi,t)d\xi\,\sigma_3\right] + O(k^2)\right). \tag{2.28}$$

Recalling the relation between μ_1 and μ_2 in (2.14), using the expansions in (2.27) and (2.28), we obtain

$$a(k) = 1 + ik\alpha + O(k^2), \quad k \to 0. \tag{2.29}$$

2.2 Riemann-Hilbert formulation

Assuming $q(x,t)$ be a solution of equation (1.6), the analytic properties of $\mu_j(x,t;k)$ stated above allow us to define a piecewise meromorphic, 2×2 matrix-valued function $M(x,t;k)$ by

$$M(x,t;k) = \begin{cases} \left(\dfrac{\mu_1^{(1)}(x,t;k)}{a(k)}\ \ \mu_2^{(2)}(x,t;k)\right), & \mathrm{Im}\,k > 0, \\[2mm] \left(\mu_2^{(1)}(x,t;k)\ \ \dfrac{\mu_1^{(2)}(x,t;k)}{\overline{a(\bar k)}}\right), & \mathrm{Im}\,k < 0. \end{cases} \tag{2.30}$$

Then, for each $x \in \mathbf{R}$ and $t \geqslant 0$, the boundary values $M_\pm(x,t;k)$ of M as k approaches $\mathbf{R}$ from the sides $\pm\mathrm{Im}\,k > 0$ are related as follows:

$$M_+(x,t;k) = M_-(x,t;k)J(x,t;k) = M_-(x,t;k)e^{-Q(x,t;k)\hat\sigma_3}J_0(k), \quad k \in \mathbf{R}, \tag{2.31}$$

with

$$J_0(k) = \begin{pmatrix} 1 + |r(k)|^2 & \overline{r(k)} \\ r(k) & 1 \end{pmatrix}, \quad r(k) = \frac{b(k)}{a(k)}. \tag{2.32}$$

In view of the properties of $\mu_j(x,t;k)$ and $s(k)$, $M(x,t;k)$ also satisfies the following properties:

(i) Behavior at $k = \infty$:

$$M(x,t;k) \to I \text{ as } k \to \infty. \tag{2.33}$$

(ii) Symmetry:

$$\overline{M(x,t;\bar k)} = M(x,t;-k) = \begin{pmatrix} 0 & 1 \\ -1 & 0 \end{pmatrix} M(x,t;k) \begin{pmatrix} 0 & -1 \\ 1 & 0 \end{pmatrix}. \tag{2.34}$$

(iii) Residue conditions: Let $\{k_j\}_1^N$ be the set of zeros of $a(k)$. We assume these zeros are finite in number, simple and no zero is real, then $M(x,t;k)$ satisfies the following residue conditions:

$$\operatorname{Res}_{k=k_j} M^{(1)}(x,t;k) = \frac{e^{2Q(x,t;k_j)}}{\dot{a}(k_j)b(k_j)} M^{(2)}(x,t;k_j) = i\chi_j e^{2Q(x,t;k_j)} M^{(2)}(x,t;k_j),$$

$$\operatorname{Res}_{k=\bar{k}_j} M^{(2)}(x,t;k) = -\frac{e^{-2Q(x,t;\bar{k}_j)}}{\dot{\bar{a}}(\bar{k}_j)\overline{b(\bar{k}_j)}} M^{(1)}(x,t;\bar{k}_j) = i\bar{\chi}_j e^{-2Q(x,t;\bar{k}_j)} M^{(1)}(x,t;\bar{k}_j).$$

(2.35)

(iv) Behavior at $k = 0$: An important characteristic property of M is its behavior as $k \to 0$. Indeed, substituting (2.27)-(2.28) and (2.29) into (2.30) gives

$$M(x,t;k) = P(x,t)\left(I - ik\left[\begin{pmatrix} \int_x^\infty q_\xi^2(\xi,t)d\xi & 2q \\ 2q & -\int_x^\infty q_\xi^2(\xi,t)d\xi \end{pmatrix}\right.\right.$$

$$\left.\left.+ \int_x^\infty q_\xi^2(\xi,t)d\xi \sigma_3\right] + O(k^2)\right), \quad k \to 0.$$

(2.36)

To solve the above Riemann–Hilbert problem, as in the case of Camassa–Holm equation [1] and SP equation, one faces the problem that the determination of the jump matrix, which is $e^{-Q}J_0(k)e^Q$, involves not only objects uniquely determined by the initial data $q(x,0)$ (the functions $a(k)$ and $b(k)$ involved in $J_0(k)$ and the constants involved in the residue conditions), but also $Q = Q(x,t;k)$, which is obviously not determined by $q(x,0)$, since it involves $q(x,t)$ for $t \geqslant 0$ which is unknown.

To remedy this, we introduce a new scale by

$$y(x,t) = x - \int_x^\infty q_\xi^2(\xi,t)d\xi.$$

(2.37)

This makes the jump matrix explicitly dependent on the parameters (y,t):

$$J(x,t;k) = \hat{J}(y,t;k) = e^{-(iky+\frac{t}{4ik})\hat{\sigma}_3} J_0(k).$$

(2.38)

Introducing

$$\hat{M}(y,t;k) = M(x,t;k),$$

then the jump condition takes the form

$$\hat{M}_+(y,t;k) = \hat{M}_-(y,t;k)\hat{J}(y,t;k) = \hat{M}_-(y,t;k)e^{-(iky+\frac{t}{4ik})\hat{\sigma}_3}J_0(k), \quad k \in \mathbf{R}, \tag{2.39}$$

with $J_0(k)$ as in (2.32). Accordingly, the residue conditions (2.35) take the following form

$$\begin{aligned}\operatorname{Res}_{k=k_j}\hat{M}^{(1)}(y,t;k) &= i\chi_j e^{2(ik_j y+\frac{t}{4ik_j})}\hat{M}^{(2)}(y,t;k_j), \\ \operatorname{Res}_{k=\bar{k}_j}\hat{M}^{(2)}(y,t;k) &= i\bar{\chi}_j e^{-2(i\bar{k}_j y+\frac{t}{4i\bar{k}_j})}\hat{M}^{(1)}(y,t;\bar{k}_j).\end{aligned} \tag{2.40}$$

Theorem 2.1 *Suppose that $q(x,t)$ is a solution of the Cauchy problem (1.6)-(1.7) for the mSP equation. Let $\{r(k), \{k_j, \chi_j\}_1^N\}$ be the spectral data determined by $q_0(x)$, and define $\hat{M}(y,t;k)$ as the solution of the associated Riemann–Hilbert (2.39) with (2.32), (2.40) and the normalization condition*

$$\hat{M}(y,t;k) \to I, \quad \text{as } k \to \infty. \tag{2.41}$$

Then, $\hat{M}(y,t;k)$ exists and is unique. Evaluating $\hat{M}(y,t;k)$ as $k \to 0$, we get a representation for the solution $q(x,t)$ of the Cauchy problem (1.6)-(1.7):

$$q(x,t) = \hat{q}(y(x,t),t) \tag{2.42}$$

with

$$x(y,t) = y + \beta(y,t), \tag{2.43}$$

$$\hat{q}(y,t) = \gamma(y,t), \tag{2.44}$$

where

$$\begin{pmatrix} \beta & \gamma \\ \gamma & -\beta \end{pmatrix}(y,t) = \lim_{k \to 0}\frac{i}{2k}\left(\hat{M}^{-1}(y,t;0)\hat{M}(y,t;k) - I\right). \tag{2.45}$$

Proof The structures of the above Riemann–Hilbert problem are the same as in the case of the focusing nonlinear Schrödinger (NLS) equation (only the dependence on the parameters (y,t) is different). Therefore, its unique solvability is a consequence of the same 'vanishing lemma' as for the NLS equation [11,12]. In view of (2.36), the representation formulas (2.42)-(2.45) can be obtained by

$$M(x,t;k) = P(x,t)\left(I - 2ik\begin{pmatrix} x-y & q \\ q & y-x \end{pmatrix} + O(k^2)\right), \quad k \to 0. \tag{2.46}$$

2.3 From the Riemann–Hilbert problem to a solution of the mSP equation

Theorem 2.1 gives a representation of the solution of the Cauchy problem (1.6)-(1.7) under assumption of existence of a solution q to this problem. On the other hand, we should check that the representation result actually satisfies the mSP equation and the initial condition. In fact, we have the following theorem.

Theorem 2.2 *Let $\hat{M}(y,t;k)$ be the solution of the associated Riemann–Hilbert problem (2.32), (2.39)-(2.41). Define $\beta(y,t)$, $\gamma(y,t)$ by (2.45) and introduce $x(y,t)$ and $\hat{q}(y,t)$ as in (2.43) and (2.44). Set*

$$\begin{pmatrix} f(y,t) & g(y,t) \\ -g(y,t) & f(y,t) \end{pmatrix} = \hat{M}(y,t;0),$$

$$\hat{p}(y,t) = \frac{2}{f^2(y,t) - g^2(y,t) + 1}, \quad \hat{w}(y,t) = \frac{2f(y,t)g(y,t)}{f^2(y,t) - g^2(y,t) + 1}, \tag{2.47}$$

then the following equations hold:

$$x_y = \frac{1}{\hat{p}}, \quad \hat{q}_y = \frac{\hat{w}}{\hat{p}}, \quad \hat{p}_t = 2\hat{p}\hat{w}\hat{q}. \tag{2.48}$$

Furthermore, if we let $q(x,t) = \hat{q}(y(x,t),t)$, $p(x,t) = \hat{p}(y(x,t),t)$, then $q(x,t)$ and $p(x,t)$ satisfy the system of equations

$$p_t = (q^2 p)_x, \quad p = 1 + q_x^2, \tag{2.49}$$

which is equivalent to equation (1.6). In particular, $q(x,t)$ satisfies the initial condition (1.7).

Proof Define

$$\Phi(y,t;k) = \hat{M}(y,t;k)e^{-(iky + \frac{t}{4ik})\sigma_3}.$$

Assuming $\hat{M}(y,t;k)$ admits the following expansion

$$\hat{M}(y,t;k) = I + \frac{M_1(y,t)}{ik} + O\left(\frac{1}{k^2}\right), \quad k \to \infty. \tag{2.50}$$

Then, we can get

$$(\Phi_y \Phi^{-1})(y,t;k) = -ik\sigma_3 + [\sigma_3, M_1(y,t)] + O\left(\frac{1}{k}\right), \quad k \to \infty. \tag{2.51}$$

Since $(\Phi_y \Phi^{-1})(y,t;k) + ik\sigma_3$ has neither jumps no singularities and is bounded in $k \in \mathbb{C}$, hence,

$$(\Phi_y \Phi^{-1})(y,t;k) = -ik\sigma_3 + [\sigma_3, M_1(y,t)]. \tag{2.52}$$

Meanwhile, as $k \to 0$, from (2.45), $\Phi(y,t;k)$ has the expansion

$$\Phi(y,t;k) = P_0(y,t)\left(I - 2ikP_1(y,t) + O(k^2)\right)e^{-(iky+\frac{t}{4ik})\sigma_3}, \tag{2.53}$$

where

$$P_0 = \begin{pmatrix} f & g \\ -g & f \end{pmatrix}, \quad P_1 = \begin{pmatrix} \beta & \gamma \\ \gamma & -\beta \end{pmatrix}.$$

Therefore, we can get

$$(\Phi_y \Phi^{-1})(y,t;k) = (P_0)_y P_0^{-1} - 2ikP_0\left((P_1)_y + \frac{1}{2}\sigma_3\right)P_0^{-1} + O(k^2), \quad k \to 0. \tag{2.54}$$

Comparing with (2.52), it follows that

$$2(P_1)_y = P_0^{-1}\sigma_3 P_0 - \sigma_3 = \begin{pmatrix} f^2 - g^2 - 1 & 2fg \\ 2fg & g^2 - f^2 + 1 \end{pmatrix}, \tag{2.55}$$

which implies

$$x_y = 1 + \beta_y = \frac{1}{2}(f^2 - g^2 + 1) = \frac{1}{\hat{p}}, \quad \hat{q}_y = \gamma_y = fg = \frac{\hat{w}}{\hat{p}}. \tag{2.56}$$

On the other hand,

$$(\Phi_t \Phi^{-1})(y,t;k) = O\left(\frac{1}{k}\right), \quad k \to \infty.$$

However, as $k \to 0$, we have

$$(\Phi_t \Phi^{-1})(y,t;k) = \left((P_0)_t + \frac{1}{2}P_0[P_1, \sigma_3]\right)P_0^{-1} + \frac{i}{4k}P_0\sigma_3 P_0^{-1} + O(k). \tag{2.57}$$

Thus, by Liouville's theorem,

$$(P_0)_t = -\frac{1}{2}P_0[P_1, \sigma_3] = \begin{pmatrix} -g\gamma & f\gamma \\ -f\gamma & -g\gamma \end{pmatrix}. \tag{2.58}$$

In view of the definitions of $\hat{p}$, $\hat{w}$ and $\hat{q}$, we immediately get

$$\hat{p}_t = 2\hat{p}\hat{w}\hat{q}. \tag{2.59}$$

Thus, we complete the proof of (2.48).

In the following, we will derive the relation (2.49). First, we notice that $\det \hat{M}(x,t;k) = 1$, thus, $f^2 + g^2 = 1$. Then, we have $\hat{p} = 1 + \hat{w}^2$. From (2.48), we can obtain $y_x(x,t) = p(x,t)$ and $q_x(x,t) = \hat{q}_y(y(x,t),t)y_x(x,t) = w$, where $w(x,t) = \hat{w}(y(x,t),t)$. Hence,

$$p = 1 + q_x^2. \tag{2.60}$$

In order to derive the first equation in (2.49), we can use (2.59) and (2.56) to get

$$\left(\frac{1}{\hat{p}}\right)_t = -\frac{\hat{p}_t}{\hat{p}^2} = -\frac{2\hat{p}\hat{w}\hat{q}}{\hat{p}^2} = -2\hat{q}\hat{q}_y = -(\hat{q}^2)_y. \tag{2.61}$$

Therefore, recalling $x_y = \dfrac{1}{\hat{p}}$ in (2.48), we can calculate $x_t(y,t)$ as follows

$$x_t(y,t) = \frac{\partial}{\partial t}\left(\int_\infty^y \frac{1}{\hat{p}(\xi,t)}d\xi\right) = -\hat{q}^2(y,t). \tag{2.62}$$

Substituting this into the identity $\hat{p}_t = p_x x_t + p_t$ and using (2.59), we get

$$p_t = 2\hat{p}\hat{w}\hat{q} + \hat{q}^2 p_x = 2qq_x p + q^2 p_x = (q^2 p_x)_x. \tag{2.63}$$

In order to verify the initial condition, one observes that for $t = 0$, the Riemann–Hilbert problem reduces to that associated with $q_0(x)$, which yields $q(x,0) = q_0(x)$, owing to the uniqueness of the solution of the Riemann–Hilbert problem.

3 Long-time asymptotics

The representation of the solution q to the Cauchy problem (1.6)-(1.7) in terms of an associated Riemann–Hilbert problem allows applying the nonlinear steepest descent method [8, 9] (the wide application of this method to the other integrable equations can be found in [4, 5, 13]) for obtaining a detailed long-time asymptotics of q, whose key ingredient is the deformation of the original Riemann–Hilbert problem in accordance with the 'signature table' for the phase function θ involved in the jump matrix $\hat{J}$ written in the form

$$\hat{J}(y,t;k) = e^{-it\theta(\hat{\xi},k)\hat{\sigma}_3} J_0(k) \tag{3.1}$$

with

$$\theta(\hat{\xi}, k) = \hat{\xi}k - \frac{1}{4k}, \quad \hat{\xi} = \frac{y}{t}. \tag{3.2}$$

Since the structure of the jump matrix is similar, in many aspects, to that in the case of the short pulse equation, the long time analysis shares many common features with that for the SP equation [2].

Theorem 3.1 *Let $q(x,t)$ be the solution of the Cauchy problem (1.6)-(1.7) with initial datum $q_0(x) \to 0$ as $x \to \pm\infty$. Assume that the spectral function $a(k)$ has no zeros in the upper half-plane and let ε be any small positive number. Then the behavior of $q(x,t)$ as $t \to \infty$ is as follows.*

- *In the domain $x/t > \varepsilon$, $q(x,t)$ tends to 0 with fast decay.*
- *In the domain $x/t < -\varepsilon$, $q(x,t)$ exhibits decaying, of order $O(t^{-\frac{1}{2}})$, modulated oscillations:*

$$q(x,t) = \sqrt{\frac{h(\kappa)}{\kappa t}} \cos\left(\frac{t}{\kappa} + h(\kappa)\log\frac{4t}{\kappa} + \phi(\kappa)\right), \tag{3.3}$$

where

$$\kappa = \frac{1}{2}\sqrt{\frac{t}{|x|}}, \quad h(\kappa) = -\frac{1}{2\pi}\log(1 + |r(\kappa)|^2), \tag{3.4}$$

$$\phi(\kappa) = -\frac{\pi}{4} - \arg(r(\kappa)) - \arg\Gamma(ih(\kappa)) + \frac{1}{\pi}\int_{\mathbb{R}\setminus[-\kappa,\kappa]} \log|k-s| d\log(1+|r(s)|^2)$$
$$+ \frac{\kappa}{\pi}\int_{\kappa}^{\infty} \frac{\log(1+|r(s)|^2)}{s^2} ds, \tag{3.5}$$

and Γ is the Euler Gamma function.

Proof The proof is analogous to [2], we omit the details.

Remark 3.1 The sectors of different asymptotic behavior match, as $\varepsilon \to 0$, through the fast decay. Indeed, as $x/t \to 0-$, $\kappa \to \infty$ and $h(\kappa) \to 0$ and thus the amplitude in (3.3) decays faster.

4 Solitons

In the context of the inverse scattering transform method, nonreal zeros of the scattering coefficient $a(k)$ correspond to eigenvalues of the 'x-equation' of the Lax pair and consequently to multi-soliton solutions of the underlying nonlinear equation. In the framework of the Riemann–Hilbert method, pure soliton solutions correspond to solving a meromorphic Riemann–Hilbert problem with trivial jump conditions ($J = I$) and with nontrivial residue conditions at zeros of $a(k)$, thus the solution of the Riemann–Hilbert problem, being a rational function of the spectral parameter, reduces to solving a system of linear algebraic equations only. Hence, this construction leads to explicit formulas for multi-solitons.

We suppress the y and t dependence for clarity. From the first equation of (2.40), we get

$$\hat{M}^{(1)}(k) = \begin{pmatrix} 1 \\ 0 \end{pmatrix} + \sum_{j=1}^{N} \frac{i\chi_j e^{2(ik_j y + \frac{t}{4ik_j})}}{k - k_j} \hat{M}^{(2)}(k_j). \tag{4.1}$$

Taking into account the symmetry (2.34), (4.1) can be rewritten as

$$\begin{pmatrix} \overline{\hat{M}_{22}(\bar{k})} \\ -\overline{\hat{M}_{12}(\bar{k})} \end{pmatrix} = \begin{pmatrix} 1 \\ 0 \end{pmatrix} + \sum_{j=1}^{N} \frac{i\chi_j e^{2(ik_j y + \frac{t}{4ik_j})}}{k - k_j} \begin{pmatrix} \hat{M}_{12}(k_j) \\ \hat{M}_{22}(k_j) \end{pmatrix}. \tag{4.2}$$

Evaluation at $\bar{k}_m$, $m = 1, \cdots, N$ yields

$$\begin{pmatrix} \overline{\hat{M}_{22}(k_m)} \\ -\overline{\hat{M}_{12}(k_m)} \end{pmatrix} = \begin{pmatrix} 1 \\ 0 \end{pmatrix} + \sum_{j=1}^{N} \frac{i\chi_j e^{2(ik_j y + \frac{t}{4ik_j})}}{\bar{k}_m - k_j} \begin{pmatrix} \hat{M}_{12}(k_j) \\ \hat{M}_{22}(k_j) \end{pmatrix}. \tag{4.3}$$

Solving this algebraic system for $\hat{M}_{12}(k_j)$, $\hat{M}_{22}(k_j)$, $j = 1, \cdots, N$, and substituting the solution into (4.1) yields an explicit expression for $\hat{M}^{(1)}(k)$. Then the second residue condition of (2.40) is the consequence of symmetry. Thus, we have

$$\hat{M}(k) = \begin{pmatrix} 1 & 0 \\ 0 & 1 \end{pmatrix} + \sum_{j=1}^{N} \begin{pmatrix} \frac{\Delta_j}{k - k_j} \hat{M}_{12}(k_j) & -\frac{\bar{\Delta}_j}{k - \bar{k}_j} \hat{M}_{11}(\bar{k}_j) \\ \frac{\Delta_j}{k - k_j} \hat{M}_{22}(k_j) & -\frac{\bar{\Delta}_j}{k - \bar{k}_j} \hat{M}_{21}(\bar{k}_j) \end{pmatrix}, \tag{4.4}$$

where $\Delta_j = i\chi_j e^{2(ik_j y + \frac{t}{4ik_j})}$. We are interested in the expansion of $\hat{M}(k)$ as $k \to 0$, so (4.4) leads to the following approximation

$$\hat{M}(k) = I + \sum_{j=1}^{N} \begin{pmatrix} -\frac{\Delta_j}{k_j}\hat{M}_{12}(k_j) & \frac{\bar{\Delta}_j}{\bar{k}_j}\hat{M}_{11}(\bar{k}_j) \\ -\frac{\Delta_j}{k_j}\hat{M}_{22}(k_j) & \frac{\bar{\Delta}_j}{\bar{k}_j}\hat{M}_{21}(\bar{k}_j) \end{pmatrix}$$

$$+ k\sum_{j=1}^{N} \begin{pmatrix} -\frac{\Delta_j}{k_j^2}\hat{M}_{12}(k_j) & \frac{\bar{\Delta}_j}{\bar{k}_j^2}\hat{M}_{11}(\bar{k}_j) \\ -\frac{\Delta_j}{k_j^2}\hat{M}_{22}(k_j) & \frac{\bar{\Delta}_j}{\bar{k}_j^2}\hat{M}_{21}(\bar{k}_j) \end{pmatrix} + O(k^2). \quad (4.5)$$

In particular,

$$\hat{M}(0) = I + \sum_{j=1}^{N} \begin{pmatrix} -\frac{\Delta_j}{k_j}\hat{M}_{12}(k_j) & \frac{\bar{\Delta}_j}{\bar{k}_j}\hat{M}_{11}(\bar{k}_j) \\ -\frac{\Delta_j}{k_j}\hat{M}_{22}(k_j) & \frac{\bar{\Delta}_j}{\bar{k}_j}\hat{M}_{21}(\bar{k}_j) \end{pmatrix}. \quad (4.6)$$

Recalling (2.45), one can find

$$-2i\hat{M}(0)\begin{pmatrix} \beta & \gamma \\ \gamma & -\beta \end{pmatrix} = \sum_{j=1}^{N} \begin{pmatrix} -\frac{\Delta_j}{k_j^2}\hat{M}_{12}(k_j) & \frac{\bar{\Delta}_j}{\bar{k}_j^2}\hat{M}_{11}(\bar{k}_j) \\ -\frac{\Delta_j}{k_j^2}\hat{M}_{22}(k_j) & \frac{\bar{\Delta}_j}{\bar{k}_j^2}\hat{M}_{21}(\bar{k}_j) \end{pmatrix}. \quad (4.7)$$

Directly solving this algebraic system for β, γ, and combining with (2.43)-(2.44), we can obtain the N-soliton solutions of equation (1.6).

In the following, we will derive explicit formulas for the one-soliton solutions. Assume $N = 1$ so that the zero of $a(k)$ is $k_1 = i\rho$, $\rho > 0$. By the symmetry condition (2.34), we have $\overline{\hat{M}(-\bar{k})} = \hat{M}(k)$. Therefore, $\Delta_1 = i\chi_1 e^{2\left(ik_1 y + \frac{t}{4ik_1}\right)} = i\operatorname{sign}(\chi_1)e^{-2\rho y - \frac{t}{2\rho} + \log|\chi_1|} := i\Delta$. From (4.3), we get

$$\overline{\hat{M}_{22}(i\rho)} = 1 + \frac{i\Delta}{-2i\rho}\hat{M}_{12}(i\rho), \quad (4.8)$$

$$-\overline{\hat{M}_{12}(i\rho)} = \frac{i\Delta}{-2i\rho}\hat{M}_{22}(i\rho). \quad (4.9)$$

Thus,

$$\hat{M}_{12}(i\rho) = \frac{2\rho\Delta}{4\rho^2 + \Delta^2}, \quad \hat{M}_{22}(i\rho) = \frac{4\rho^2}{4\rho^2 + \Delta^2}. \quad (4.10)$$

By the symmetry relation, we immediately find

$$\hat{M}_{21}(-i\rho) = \frac{-2\rho\Delta}{4\rho^2 + \Delta^2}, \quad \hat{M}_{11}(-i\rho) = \frac{4\rho^2}{4\rho^2 + \Delta^2}. \qquad (4.11)$$

Substituting (4.10) and (4.11) into (4.7), one can obtain

$$-\begin{pmatrix} \frac{4\rho^2 - \Delta^2}{4\rho^2 + \Delta^2} & \frac{4\rho\Delta}{4\rho^2 + \Delta^2} \\ \frac{-4\rho\Delta}{4\rho^2 + \Delta^2} & \frac{4\rho^2 - \Delta^2}{4\rho^2 + \Delta^2} \end{pmatrix}^{-1} \begin{pmatrix} \frac{\Delta^2}{\rho(4\rho^2 + \Delta^2)} & \frac{2\Delta}{4\rho^2 + \Delta^2} \\ \frac{2\Delta}{4\rho^2 + \Delta^2} & -\frac{\Delta^2}{\rho(4\rho^2 + \Delta^2)} \end{pmatrix} = \begin{pmatrix} \beta & \gamma \\ \gamma & -\beta \end{pmatrix}. \qquad (4.12)$$

Finally, using (2.43) and (2.44), we arrive at the one-soliton solutions $q(x,t)$ of the mSP equation (1.6) as follows:

$$q(x,t) = \hat{q}(y(x,t),t), \quad \hat{q}(y,t) = -\frac{2\Delta}{4\rho^2 + \Delta^2}, \qquad (4.13)$$

$$x(y,t) = y + \frac{\Delta^2}{\rho(4\rho^2 + \Delta^2)}, \quad \Delta = \text{sign}(\chi_1) e^{-2\rho y - \frac{t}{2\rho} + \log|\chi_1|}, \qquad (4.14)$$

where $\rho > 0$ and $\chi_1 \in \mathbf{R}$ are the soliton parameters.

Particularly, introducing

$$\eta(y,t) = 2\rho\left(y + \frac{t}{4\rho^2} - \frac{1}{2\rho}\log\frac{|\chi_1|}{2\rho}\right),$$

we can rewrite the soliton formulas (4.13)-(4.14) as

$$x = y + \frac{1}{2\rho}(1 - \tanh\eta(y,t)), \quad \hat{q}(y,t) = -\frac{1}{2\rho}\frac{\text{sign}(\chi_1)}{\cosh\eta(y,t)}. \qquad (4.15)$$

If we select $\rho = \frac{1}{2}$, $\chi_1 = -1$, (4.15) implies

$$x = y - \tanh(y+t) + 1, \quad \hat{q}(y,t) = 1/\cosh(y+t), \qquad (4.16)$$

this expression coincides with formula (44) in [19] for the soliton solution of the mSP equation, the additional constant in (4.16) provides that $x - y \to 0$ as $y \to +\infty$.

Acknowledgements This work was supported in part by the National Natural Science Foundation of China [grant number 11731014], [grant number 11471099].

References

[1] Boutet de Monvel A, Shepelsky D. Riemann–Hilbert approach for the Camassa–Holm equation on the line [J]. C. R. Math. Acad. Sci. Paris., 2006, 343: 627–632.

[2] Boutet de Monvel A, Shepelsky D, Zielinski L. The short pulse equation by a Riemann–Hilbert approach [J]. Lett. Math. Phys., 2017, 107: 1345–1373.

[3] Boutet de Monvel A, Shepelsky D, Zielinski L. The short-wave model for the Camassa–Holm equation: a Riemann–Hilbert approach [J]. Inverse Probl., 2011, 27: 105006.

[4] Boutet de Monvel A, Shepelsky D. The Ostrovsky–Vakhnenko equation by a Riemann–Hilbert approach [J]. J. Phys. A Math. Theor., 2015, 48: 035204.

[5] Boutet de Monvel A, Shepelsky D. A Riemann–Hilbert approach for the Degasperis–Procesi equation [J]. Nonlinearity, 2013, 26: 2081–2107.

[6] Brunelli JC. The bi-Hamiltonian structure of the short pulse equation [J]. Phys. Lett. A. 2006, 353: 475–478.

[7] Chung Y, Jones CKRT, Schäfer T, et al. Ultra-short pulses in linear and nonlinear media [J]. Nonlinearity, 2005, 18: 1351–1374.

[8] Deift P, Zhou X. A steepest descent method for oscillatory Riemann–Hilbert problems. Asymptotics for the MKdV equation [J]. Ann. Math., 1993, 137: 295–368.

[9] Deift P, Venakides S, Zhou X. New results in small dispersion KdV by an extension of the steepest descent method for Riemann–Hilbert problems [J]. Int. Math. Res. Notices., 1997, 6: 286–299.

[10] Feng B. An integrable coupled short pulse equation [J]. J. Phys. A Math. Theor., 2012, 45: 085202.

[11] Faddeev LD, Takhtajan LA. Hamiltonian methods in the theory of solitons [M]. Springer Series in SovietMathematics. Berlin: Springer; 1987.

[12] Fokas AS, Its AR. The linearization of the initial-boundary value problem of the nonlinear Schröinger equation [J]. SIAM J. Math. Anal. 1996, 27: 738–764.

[13] Huang L, Xu J, Fan E. Long-time asymptotic for the Hirota equation via nonlinear steepest descent method [J]. Nonlinear Anal Real World Appl., 2015, 26: 229–262.

[14] Kuetche VK, Bouetou TB, Kofane TC. On two-loop soliton solution of the Schäfer–Wayne short-pulse equation using Hirotas method and Hodnett–Moloney approach [J]. J. Phys. Soc. Japan., 2007, 76: 024004.

[15] Matsuno Y. Periodic solutions of the short pulse model equation [J]. J. Math. Phys., 2008, 49: 073508.

[16] Robelo ML. On equations which describe pseudospherical surfaces [J]. Stud. Appl. Math., 1989, 81: 221–248.

[17] Sakovich A, Sakovich S. The short pulse equation is integrable [J]. J. Phys. Soc. Japan., 2005, 74: 239–241.

[18] Sakovich A, Sakovich S. Solitary wave solutions of the short pulse equation [J]. J. Phys. A Math. Gen., 2006, 39: L361–L367.

[19] Sakovich S. Transformation and integrability of a generalized short pulse equation [J]. Commun. Nonlinear Sci. Numer. Simulat., 2016, 39: 21–28.

[20] Schäfer T, Wayne CE. Propagation of ultra-short optical pulses in cubic nonlinear media [J]. Physica D, 2004, 196: 90–105.

Exponential Decay of Bénard Convection Problem with Surface Tension*

Guo Boling (郭柏灵), Xie Binqiang (解斌强), and Zeng Lan (曾兰)

Abstract We consider the dynamics of an Boussinesq approximation Bénard convection fluid evolving in a three-dimensional domain bounded below by a fixed flatten boundary and above by a free moving surface. The domain is horizontally periodic and the effect of the surface tension is on the free surface. By developing a priori estimates for the model, we prove the exponential decay of solutions in the framework of high regularity.

Keywords exponential decay; Bénard convection; high regularity

1 Introduction

1.1 Formulation of the problem in Eulerian problem

We consider the Bénard convection problem in a shallow horizontal layer of a fluid heated from below evolving in a moving domain

$$\Omega(t) = \{y \in \Sigma \times R | -b < y_3 < \eta(y_1, y_2, t)\},$$

where we assume that $\Sigma = (L_1 \mathbb{T}) \times (L_2 \mathbb{T})$ for $\mathbb{T} = \mathbf{R}/\mathbb{Z}$ and $L_1, L_2 > 0$ periodicity lengths. The depth of the lower boundary $b > 0$ is assumed to be fixed constant, but the upper boundary is a free surface that is the graph of the unknown function $\eta : \Sigma \times \mathbf{R}_+ \to \mathbf{R}$. We will write $\Sigma(t) = \{y_3 = \eta(y_1, y_2, t)\}$ for the free surface of the fluid and $\Sigma_b = \{y_3 = -b\}$ for the fixed bottom surface of the fluid. Assuming the Boussinesq approximation, we obtain the basic hydrodynamic equations governing Bénard convection as follows:

$$\partial_t u + u \cdot \nabla u + \frac{1}{\rho_0} \nabla p = \mu \Delta u + g\alpha\theta \mathbf{e}_{y_3}, \quad \text{in } \Omega(t),$$

---

* J. Differential. Equations, 2019, 267(4): 2261-2283. Doi:10.1016/j.jde.2019.03.017.

$$\mathrm{div} u = 0, \quad \mathrm{in}\ \Omega(t),$$
$$\partial_t \theta + u \cdot \nabla \theta = \kappa \Delta \theta, \quad \mathrm{in}\ \Omega(t),$$
$$u|_{t=0} = u_0(y_1, y_2, y_3), \quad \theta|_{t=0} = \theta_0(y_1, y_2, y_3).$$

Here, $u = (u_1, u_2, u_3)$ is the velocity field of the fluid satisfying $\mathrm{div} u = 0$, p the pressure, $g > 0$ the strength of gravity, $\mu > 0$ the kinematic viscosity, α the thermal expansion coefficient, e_{y_3} the unit upward vector, θ the temperature field of the fluid, κ the thermal diffusively coefficient, and ρ_0 is the density at the temperature T_0. Notice that, we have made the shift of actual pressure $p = \bar{p} + gy_3 - p_{atm}$ with the constant atmosphere pressure p_{atm}.

The boundary condition is
$$\partial_t \eta = u_3 - u_1 \partial_{y_1} \eta - u_2 \partial_{y_2} \eta, \quad \mathrm{on}\ \Gamma(t),$$
$$(pI - \mu \mathbb{D}(u))n = g\eta n + \sigma H n, \quad \mathrm{on}\ \Gamma(t),$$
$$n \cdot \nabla \theta + Bi\theta = -1, \quad \mathrm{on}\ \Gamma(t),$$
$$u|_{y_3=-1} = 0, \theta|_{y_3=-1} = 0, \quad \mathrm{on}\ \Gamma_b.$$

Here, I the 3×3 identity matrix, $\mathbb{D}(u)_{ij} = \partial_i u_j + \partial_j u_i$ the symmetric gradient of u, $\mathcal{N}$ the upper normal vector of the free boundary $y_3 = \eta$, $n = \mathcal{N}/|\mathcal{N}|$ the unit upward vector of the free surface $y_3 = \eta$ where $\mathcal{N} = (-\partial_1 \eta, -\partial_2 \eta, 1)$ is the upward normal vector of the free surface $y_3 = \eta$ and $|\mathcal{N}| = \sqrt{(\partial_1 \eta)^2 + (\partial_2 \eta)^2 + 1}$, $Bi \geqslant 0$ the Biot number, σ the constant coefficient of surface tension, H is the mean curvature of the free surface and is given by

$$H(\eta) = \mathrm{div}_\star \left(\frac{\nabla_\star \eta}{\sqrt{1 + |\nabla_\star \eta|^2}} \right). \tag{1.1}$$

Here $\mathrm{div}_\star, \nabla_\star$ denotes the horizontal differential operator(along with writing $x_\star = (x_1, x_2)$).

We will assume always assume the natural condition that there exists a positive number δ_0 such that $1 + \eta_0 \geqslant \delta_0 > 0$ on $\Gamma(0)$, which means that the initial free surface is strictly separated from the bottom. And without loss of generality, we may assume that $\rho_0 = \mu = \kappa = \alpha = g = Bi = 1$, i.e., we will consider the equations

$$\begin{cases} \partial_t u + u \cdot \nabla u + \nabla p - \Delta u - \theta e_{y_3} = 0, & \text{in } \Omega(t) \\ \operatorname{div} u = 0, & \text{in } \Omega(t) \\ \partial_t \theta + u \cdot \nabla \theta - \Delta \theta = 0, & \text{in } \Omega(t) \\ (pI - \mathbb{D}u)n = \eta n + \sigma H n, & \text{on } \Gamma(t), \\ \nabla \theta \cdot n + \theta = -1, & \text{on } \Gamma(t), \\ \partial_t \eta + u_1 \partial_1 \eta + u_2 \partial_2 \eta = u_3, & \text{on } \Gamma(t), \\ u = 0, \ \theta = 0, & \text{on } \Gamma_b, \\ u|_{t=0} = u_0, \theta|_{t=0} = \theta_0, \eta|_{t=0} = \eta_0. \end{cases} \quad (1.2)$$

We assume that the initial surface function η_0 satisfies the 'zero average' condition

$$\frac{1}{L_1 L_2} \int_\Sigma \eta_0 = 0. \quad (1.3)$$

Notice that for sufficiently regular solutions to the periodic problem, the condition (1.3) persists in time since $\partial_t \eta = u \cdot \nu \sqrt{1 + |\nabla_* \eta|^2}$:

$$\frac{d}{dt} \int_\Sigma \eta = \int_\Sigma \partial_t \eta = \int_{\Gamma(t)} u \cdot \nu = \int_{\Omega(t)} \operatorname{div} u = 0.$$

1.2 Reformulation of equations

In order to work in a fixed domain, we use a flattening transformation introduce by Beale [1], [2], also see [6-8]. We consider the fixed equilibrium domain

$$\Omega := \{x \in \Sigma \times \mathbf{R} | -b < x_3 < 0\},$$

for which we will write the coordinates as $x \in \Omega$. We will think of Σ as the upper boundary of Ω, and we will write $\Sigma_b = \{x_3 = -b\}$ for the lower boundary. We continue to view η as a function on $\Sigma \times R^+$. We then define

$$\bar{\eta} := \mathcal{P}\eta = \text{ harmonic extension of } \eta \text{ into the lower half space,}$$

where $\mathcal{P}$ is as defined by

$$\mathcal{P}\eta(x) = \sum_{n \in (L_1^{-1}\mathbb{Z}) \times (L_2^{-1}\mathbb{Z})} e^{2\pi i n \cdot x_*} e^{2\pi |n| x_3} \hat{\eta}(n),$$

where we have written
$$\hat{\eta}(n) = \int_\Sigma \eta(x_\star) \frac{e^{-2\pi i n \cdot x_\star}}{L_1 L_2} dx_\star.$$

The harmonic extension $\bar{\eta}$ allows us to flatten the coordinate domain via the mapping
$$\Omega \ni x \mapsto (x_1, x_2, x_3 + \bar{\eta}(x,t)(1 + x_3/b)) = \Phi(x,t) = (y_1, y_2, y_3) \in \Omega(t). \quad (1.4)$$

Note that $\Phi(\Sigma, t) = \{y_3 = \eta(y_1, y_2, t)\} = \Phi(t)$ and $\Phi(\cdot, t)|_{\Sigma_b} = Id_{\Sigma_b}$, i.e., Φ maps Σ to the free surface and keeps the lower surface fixed. We have

$$\nabla\Phi = \begin{pmatrix} 1 & 0 & 0 \\ 0 & 1 & 0 \\ A & B & J \end{pmatrix} \quad \text{and} \quad \mathcal{A} := (\nabla\Phi^{-1})^{\mathrm{T}} = \begin{pmatrix} 1 & 0 & -AK \\ 0 & 1 & -BK \\ 0 & 0 & K \end{pmatrix},$$

for
$$A = \partial_1 \bar{\eta} \tilde{b} - (x_3 \bar{\eta} \partial_1 b)/b^2, \quad B = \partial_2 \bar{\eta} \tilde{b} - (x_3 \bar{\eta} \partial_2 b)/b^2,$$
$$J = 1 + \bar{\eta}/b + \partial_3 \bar{\eta} \tilde{b}, \quad K = J^{-1},$$
$$\tilde{b} = (1 + x_3/b).$$

Here $J = \det \nabla \Phi$ is the Jacobian of the coordinate transformation.

Now we define the transformed quantities as
$$u(t,x) := u(t, \Phi(t,x)), \quad p(t,x) := p(t, \Phi(t,x)), \quad \theta(t,x) := \theta(t, \Phi(t,x)).$$

In the new coordinates, we rewrite (1.2) as
$$\begin{cases} \partial_t u - \partial_t \bar{\eta} \tilde{b} K \partial_3 u + u \cdot \nabla_\mathcal{A} u + \nabla_\mathcal{A} p - \Delta_\mathcal{A} u - \theta \nabla_\mathcal{A} y_3 = 0, & \text{in } \Omega \\ \operatorname{div}_\mathcal{A} u = 0, & \text{in } \Omega \\ \partial_t \theta - \partial_t \bar{\eta} \tilde{b} K \partial_3 \theta + u \cdot \nabla_\mathcal{A} \theta - \Delta_\mathcal{A} \theta = 0, & \text{in } \Omega \\ (pI - \mathbb{D}_\mathcal{A} u)\mathcal{N} = \eta \mathcal{N} + \sigma H \mathcal{N}, & \text{on } \Sigma, \\ \nabla_\mathcal{A} \theta \cdot \mathcal{N} + \theta |\mathcal{N}| = -|\mathcal{N}|, & \text{on } \Sigma, \\ \partial_t \eta + u_1 \partial_1 \eta + u_2 \partial_2 \eta = u_3, & \text{on } \Sigma, \\ u = 0, \ \theta = 0, & \text{on } \Sigma_b. \end{cases} \quad (1.5)$$

Here we have written the differential operators $\nabla_{\mathcal{A}}$, $\text{div}_{\mathcal{A}}$, and $\Delta_{\mathcal{A}}$ with their actions given by $(\nabla_{\mathcal{A}} f)_i := \mathcal{A}_{ij}\partial_j f$, $\text{div}_{\mathcal{A}} X = \mathcal{A}_{ij}\partial_j X_i$, and $\Delta_{\mathcal{A}} f = \text{div}_{\mathcal{A}} \nabla_{\mathcal{A}} f$ for approximate f and X; for $u \cdot \nabla_{\mathcal{A}} u$ we mean $(u \cdot \nabla_{\mathcal{A}} u)_i := u_j \mathcal{A}_{jk} \partial_k u_i$. We have also written $(\mathbb{D}_{\mathcal{A}} u)_{ij} = \sum_k (\mathcal{A}_{ik} \partial_k u_j + \mathcal{A}_{jk} \partial_k u_i)$. Also, $\mathcal{N} = (-\nabla_\star \eta, 1)$ denotes the non-unit normal on $\Gamma(t)$.

The Bénard convection problem was firstly observed from the experiments by Bénard [3]. Later on, Rayleigh [5] gives the linearized stability of the Bénard convection model in the fixed slab $\{0 < x_3 < 1\}$. It was the experiment by Block [4] that identified surface tension as the cause of patterns observed by Bénard in very thin layers with a free surface. For small Rayleigh and Marangoni numbers, T.Ioraha [9] proved the existence of exponentially decaying solutions in the class of small initial data. Similar results can be seen in [11], [12], [13]. For the viscous surface wave problem with surface tension case, the existence and decay of global in time solutions with free boundary surface were proved by T.Nishida, Y.Teramoto and H. Yoshihara [10]. They all utilized the framework of [1] and [2] in the Lagrangian coordinates. In this paper, we will use the flattening coordinate method [6] and high regularity framework [8] to prove the exponential decay of solutions for the Bénard convection problem.

The paper is organized as follows. In section 2, we define the energy and dissipations, we also state our main result. In section 3, we develop basic energy-dissipative estimates. In section 4, we provide the estimates for the nonlinearities. In section 5, we enhanced the estimates with elliptic estimates. In section 6, we complete our a priori estimates and prove our main results.

2 Main results

In order to state our main results, we first define the energy and dissipation functionals that we shall use in our analysis. We define the energy via

$$\mathcal{E} := \|u\|_{H^2(\Omega)}^2 + \|\partial_t u\|_{H^0(\Omega)}^2 + \|p\|_{H^1(\Omega)}^2 + \|\eta\|_{H^3(\Sigma)}^2 + \|\partial_t \eta\|_{H^{\frac{3}{2}}(\Sigma)}^2 + \|\partial_t^2 \eta\|_{H^{-\frac{1}{2}}(\Omega)}^2$$

$$+ \|\theta\|_{H^2(\Omega)}^2 + \|\partial_t \theta\|_{H^0(\Omega)},$$

and define the dissipation as

$$\mathcal{D} := \|u\|_{H^3(\Omega)}^2 + \|\partial_t u\|_{H^1(\Omega)}^2 + \|p\|_{H^2(\Omega)}^2 + \|\eta\|_{H^{\frac{7}{2}}(\Sigma)}^2 + \|\partial_t \eta\|_{H^{\frac{5}{2}}(\Sigma)}^2 + \|\partial_t^2 \eta\|_{H^{\frac{1}{2}}(\Omega)}^2$$

$$+ \|\theta\|_{H^3(\Omega)}^2 + \|\partial_t \theta\|_{H^1(\Omega)}.$$

Here the spaces H^s denote the usual L^2-based Sobolev spaces of order s. For simplicity, we will write $\|\cdot\|_s$ for $H^s(\Omega)$ norm and $\|\cdot\|_{\Sigma,s}$ for $H^s(\Sigma)$ norms.

We now state our a priori estimates for the solution to (1.5).

Theorem 2.1   *Suppose that (u, p, η, θ) solves (1.5) on the temporal interval $[0, T]$. Let $\mathcal{E}$ and $\mathcal{D}$ be as defined in (2.1) and (2.1). Then there exists a universal constant $0 < \delta_*$(independent of T) such that if*

$$\sup_{0 \leqslant t \leqslant T} \mathcal{E}(t) \leqslant \delta_* \quad \text{and} \quad \int_0^T \mathcal{D}(t)dt < \infty,$$

then

$$\sup_{0 \leqslant t \leqslant T} e^{\lambda t} \mathcal{E}(t) + \int_0^T \mathcal{D}(t)dt \lesssim \mathcal{E}(0),$$

for all $t \in [0, T]$, where $\lambda > 0$ is a universal constant.

In order to prove the existence of global exponential decaying solutions, we couple a priori estimates with a local existence result. In [14], Zheng has constructed local-in-time solutions of the form (1.5) without surface tension. Similarly, in the surface tension case, we can also prove a local existence result. We will simply state the result that one can prove by modifying the known methods [6], [14] in straightforward ways.

Given the initial data u_0, η_0, θ_0, we need to construct the initial data $\partial_t u(\cdot, 0)$, $\partial_t \eta(\cdot, 0)$, $\partial_t \theta(0)$, and $p(\cdot, 0)$. To construct these we require a compatibility condition for the data. To state this properly, we define the orthogonal projection onto the tangent space of the surface $\Gamma(0) = \{x_3 = \eta_0(x_*)\}$ according to

$$\Pi_0 v = v - (v \cdot \mathcal{N}_0)\mathcal{N}_0|\mathcal{N}_0|^{-2}$$

for $\mathcal{N}_0 = (-\partial_1 \eta_0, \partial_2 \eta_0, 1)$. Then the compatibility conditions for the data read

$$\begin{cases} \Pi_0(D_{\mathcal{A}_0} u_0 \mathcal{N}_0) = 0, & \text{on } \Sigma, \\ \text{div}_{\mathcal{A}_0} u_0 = 0, & \text{in } \Omega, \\ u_0 = 0, & \text{on } \Sigma_b, \end{cases} \quad (2.1)$$

where here $\mathcal{A}_0$ and Γ_0 are determined by η_0. To state the local result, we will also need to define $\mathcal{H}_1 := \{u \in H^1(\Omega) | u|_{\Sigma_b} = 0\}$ and

$$\mathcal{X}_T = \{u \in L^2([0, T]; \mathcal{H}_1) | \text{div}_{\mathcal{A}(t)} u(t) = 0 \text{ for a.e. } t\}. \quad (2.2)$$

Having stated the compatibility conditions, we can now state the result of local existence.

Theorem 2.2 *Let $u_0 \in H^2(\Omega)$, $\eta_0 \in H^3(\Sigma)$, and $\theta_0 \in H^2(\Omega)$, and assume that η_0 satisfy the zero average condition (1.3). Further, assume that the initial data satisfy the compatibility conditions of (2.1). Let $T > 0$. Then there exists a universal constant $\kappa > 0$ such that if*

$$\|u_0\|_{H^2(\Omega)}^2 + \|\eta_0\|_{H^3(\Omega)}^2 + \|\theta_0\|_{H^2(\Omega)} \leqslant \kappa,$$

then there exists a unique strong solution (u, p, η, θ) to (1.5) on the temporal interval $[0, T]$ satisfying the estimate

$$\sup_{0 \leqslant t \leqslant T} \mathcal{E}(t) + \int_0^T \mathcal{D}(t) dt + \|\partial_t^2 u\|_{(\mathcal{X}_T)^*}^2 \lesssim \mathcal{E}(0). \tag{2.3}$$

Moreover, η is such that the mapping $\Phi(\cdot, t)$, defined by (1.4), is a C^1 diffeomorphism for each $t \in [0, T]$.

Coupled a priori estimates Theorem 2.1 with local existence Theorem 2.2, we may deduce a global existence and exponential decay result.

Theorem 2.3 *Let $u_0 \in H^2(\Omega)$, $\eta_0 \in H^3(\Sigma)$, and $\theta_0 \in H^2(\Omega)$, and assume that η_0 satisfy the zero average condition (1.3). Further assume that the initial data satisfy the compatibility conditions of (2.1). Let $T > 0$. Then there exists a universal constant $\kappa > 0$ such that if*

$$\|u_0\|_{H^2(\Omega)}^2 + \|\eta_0\|_{H^3(\Omega)}^2 + \|\theta_0\|_{H^2(\Omega)} \leqslant \kappa, \tag{2.4}$$

then there exists a unique strong solution (u, p, η, θ) to (1.5) on the temporal interval $[0, T]$ satisfying the estimate

$$\sup_{0 \leqslant t \leqslant T} e^{\lambda t} \mathcal{E}(t) + \int_0^T \mathcal{D}(t) dt \lesssim \mathcal{E}(0), \tag{2.5}$$

3 Energy-dissipation equations

In this section, we show two forms of the energy-dissipation for solutions to (1.5). The one form is the geometric form which is ideal for estimating temporal derivatives. The other form is the perturbed linear form which is ideal for estimating horizonal spatial derivatives and elliptic regularity.

3.1 Geometric form

In controlling the interaction between the highest time derivative pressure and velocity, the perturbed linear form will fail. Thus, we adopt the geometric form which is a linear formulation of (1.5). We assume that u and η are given and that $\mathcal{A}, \mathcal{N}, \mathcal{J}$, etc. are given in terms of η as in (1.5). Consider the following system for (v, q, ζ, h):

$$\begin{cases} \partial_t v - \partial_t \widetilde{\eta} b K \partial_3 v + u \cdot \nabla_{\mathcal{A}} v + \nabla_{\mathcal{A}} q - \Delta_{\mathcal{A}} v - \vartheta \nabla_{\mathcal{A}} y_3 = F^1, & \text{in } \Omega, \\ \text{div}_{\mathcal{A}} v = F^2, & \text{in } \Omega, \\ \partial_t \vartheta - \partial_t \widetilde{\eta} b K \partial_3 \vartheta + u \cdot \nabla_{\mathcal{A}} \vartheta - \Delta_{\mathcal{A}} \vartheta = F^3, & \text{in } \Omega \\ (qI - \mathbb{D}_{\mathcal{A}} v)\mathcal{N} = \zeta \mathcal{N} - \sigma \Delta_{\star} \mathcal{N} + F^4, & \text{on } \Sigma, \\ \nabla_{\mathcal{A}} \vartheta \cdot \mathcal{N} + \vartheta |\mathcal{N}| = F^5, & \text{on } \Sigma, \\ \partial_t \zeta - v \cdot \mathcal{N} = F^6, & \text{on } \Sigma, \\ v = \vartheta = 0, & \text{on } \Sigma_b. \end{cases} \quad (3.1)$$

We now record the energy-dissipation equality associated to solutions to (3.1).

Proposition 3.1 *Let u and η be given and solve (1.5). If (v, q, ζ, ϑ) solve (3.1), then*

$$\frac{d}{dt}\left(\int_\Omega \frac{|v|^2}{2} J + \int_\Sigma \frac{|\zeta|^2}{2} + \int_\Sigma \sigma \frac{|\nabla_\star \zeta|^2}{2}\right) + \int_\Omega \frac{|\mathbb{D}_{\mathcal{A}} v|^2}{2} J = \int_\Omega (v \cdot F^1 + q \cdot F^2) J$$
$$+ \int_\Sigma -v \cdot F^4 + \int_\Sigma (\zeta - \sigma \Delta_\star \zeta) F^6 + \int_\Omega \vartheta \nabla_{\mathcal{A}} y_3 \cdot v J, \qquad (3.2)$$

and

$$\frac{d}{dt} \int_\Omega \frac{|\vartheta|^2}{2} J + \int_\Omega |\nabla_{\mathcal{A}} \vartheta|^2 J + \int_\Sigma |\vartheta|^2 |\mathcal{N}| = \int_\Omega \vartheta \cdot F^3 J + \int_\Sigma \vartheta \cdot F^5. \qquad (3.3)$$

Proof We take the product of the first equation in (3.1) with Jv and integrate over Ω to find that

$$I + II = III,$$

for

$$I = \int_\Omega \partial_t v_i J v_i - \partial_t \widetilde{\eta} b \partial_3 v_i v_i + u_j \mathcal{A}_{jk} \partial_k v_i J v_i,$$

$$II = \int_\Omega \mathcal{A}_{jk}\partial_k S_{ij}(v,q)Jv_i, \quad III = \int_\Omega F^1 \cdot vJ.$$

A simple computation shows that

$$I = \frac{d}{dt}\int_\Omega \frac{|v|^2}{2} J.$$

To handle the term II we first integrate

$$II = \int_\Omega -\mathcal{A}_{jk}S_{ij}(v,q)J\partial_k v_i + \int_\Sigma J\mathcal{A}_{j3}S_{ij}(v,q)v_i$$

$$= \int_\Omega -q\mathcal{A}_{jk}\partial_k v_i J + J\frac{|\mathbb{D}_\mathcal{A}v|^2}{2} + \int_\Sigma S_{ij}(v,q)\mathcal{N}_j v_i$$

$$= \int_\Omega -qJF^2 + J\frac{|\mathbb{D}_\mathcal{A}v|^2}{2} + \int_\Sigma (\zeta\mathcal{N} - \sigma\Delta_\star\mathcal{N})\cdot v + F^4 \cdot v.$$

For the fourth equation in (3.1) we may compute

$$\int_\Sigma (\zeta\mathcal{N} - \sigma\Delta_\star\mathcal{N})\cdot v = \int_\Sigma (\zeta - \sigma\Delta_\star)(\partial_t\zeta - F^6)$$

$$= \frac{d}{dt}\left(\int_\Sigma \frac{|\zeta|^2}{2} + \int_\Sigma \sigma\frac{|\nabla_\star\zeta|^2}{2}\right) - \int_\Sigma (\zeta - \sigma\Delta_\star\zeta)F^6.$$

Similarly, multiplying the third equation in (3.1) with $J\vartheta$ and integrating over Ω, we have

$$IV + V = VI,$$

for

$$IV = \int_\Omega \partial_t\vartheta J\vartheta - \partial_t\tilde{\eta}\tilde{b}\partial_3\vartheta\vartheta + u_j\mathcal{A}_{jk}\partial_k\vartheta J\vartheta,$$

$$V = -\int_\Omega \Delta_\mathcal{A}\vartheta J\vartheta, \quad VI = \int_\Omega F^3\vartheta J.$$

A simple computation shows that

$$IV = \frac{d}{dt}\int_\Omega \frac{|\vartheta|^2}{2} J.$$

To handle the term V we first integrate

$$V = \int_\Omega |\nabla_\mathcal{A}\vartheta|^2 J - \int_\Sigma \nabla_\mathcal{A}\vartheta \cdot \mathcal{N}\vartheta$$

$$= \int_\Omega |\nabla_{\mathcal{A}}\vartheta|^2 J + \int_\Sigma |\vartheta|^2 |\mathcal{N}| - F^5 \vartheta.$$

We will employ the form (3.1) to study the temporal derivative of solutions to (1.5). That is, we will employ ∂_t to (1.5) to deduce that $(v, q, \zeta, \vartheta) = (\partial_t u, \partial_t p, \partial_t \eta, \partial_t \theta)$ satisfy (3.1) for certain terms F^i. Below we record the form of these forcing terms F^i, $i = 1, 2, 3, 4, 5, 6$ for this particular problem.

We have that $F^1 = \sum_{i=1}^{5} F^{1,i}$, for

$$\begin{aligned}
F_i^{1,1} &:= \partial_t(\partial_t \overline{\eta} \tilde{b} K) \partial_3 u \\
F_i^{1,2} &:= -\partial_t(u_j \mathcal{A}_{jk}) \partial_k u_i + \partial_t \mathcal{A}_{ik} \partial_k p, \\
F_i^{1,3} &:= \partial_t \mathcal{A}_{jk} \partial_k (\mathcal{A}_{im} \partial_m u_j + \mathcal{A}_{jm} \partial_m u_i), \\
F_i^{1,4} &:= \mathcal{A}_{jk} \partial_k (\partial_t \mathcal{A}_{im} \partial_m u_j + \partial_t \mathcal{A}_{jm} \partial_m u_i), \\
F_i^{1,5} &:= \theta \partial_t (\nabla_{\mathcal{A}} y_3),
\end{aligned} \tag{3.4}$$

$$F^2 := -\partial_t \mathcal{A}_{ij} \partial_j u_i, \tag{3.5}$$

$$\begin{aligned}
F^{3,1} &:= \partial_t(\partial_t \overline{\eta} \tilde{b} K) \partial_3 \theta \\
F^{3,2} &:= -\partial_t(u_j \mathcal{A}_{jk}) \partial_k \theta, \\
F^{3,3} &:= \partial_t \mathcal{A}_{il} \partial_l \mathcal{A}_{im} \partial_m \theta, \\
F^{3,4} &:= \mathcal{A}_{il} \partial_l \partial_t \mathcal{A}_{im} \partial_m \theta,
\end{aligned} \tag{3.6}$$

$$F^4 = F^{4,1} + F^{4,2},$$

where for $i = 1, 2, 3$ we have

$$F_i^{4,1} := (\eta - p) \partial_t \mathcal{N}_i + (\mathcal{A}_{ik} \partial_k u_j + \mathcal{A}_{jk} \partial_k u_i) \partial_t \mathcal{N}_j + (\partial_t \mathcal{A}_{ik} \partial_k u_j + \partial_t \mathcal{A}_{jk} \partial_k u_i) \mathcal{N}_j,$$

$$F_i^{4,2} := -(\sigma \partial_t H - \sigma \partial_t \Delta_* \eta) \mathcal{N}_i - \sigma H \partial_t \mathcal{N}_i, \tag{3.7}$$

$$F^5 := -\partial_t |\mathcal{N}| - \theta \partial_t |\mathcal{N}| - \nabla_{\partial_t \mathcal{A}} \theta \cdot \mathcal{N} - \nabla_{\mathcal{A}} \theta \cdot \partial_t |\mathcal{N}|, \tag{3.8}$$

$$F^6 := \partial_t D\eta \cdot u. \tag{3.9}$$

3.2 Perturbed Linear form

Next, we consider an alternate way of linearizing (1.5) that eliminates the $\mathcal{A}$ coefficients in favor of constant coefficients. This is advantageous for applying elliptic regularity results and is the context in which we will derive estimates horizontal spatial derivatives. We may rewrite (1.5) as

$$\begin{cases} \partial_t u + \nabla p - \Delta u - \theta e_3 = G^1, & \text{in } \Omega, \\ \text{div} u = G^2, & \text{in } \Omega, \\ \partial_t \theta - \Delta \theta = G^3, & \text{in } \Omega, \\ (pI - \mathbb{D}u - \eta I + \sigma \Delta_\star \eta)e_3 = G^4, & \text{on } \Sigma, \\ \nabla \theta \cdot e_3 + \theta = G^5, & \text{on } \Sigma, \\ \partial_t \eta - u_3 = G^6, & \text{on } \Sigma. \end{cases} \quad (3.10)$$

Here we have written the nonlinear terms G^i for $i = 1, \cdots, 5$ as follows. We write $G^1 := G^{1,1} + G^{1,2} + G^{1,3} + G^{1,4} + G^{1,5} + G^{1,6}$, for

$$G_i^{1,1} := (\delta_{ij} - \mathcal{A}_{ij})\partial_j p$$
$$G_i^{1,2} := u_j \mathcal{A}_{jk} \partial_k u_i,$$
$$G_i^{1,3} := [K^2(1 + A^2 + B^2) - 1]\partial_{33} u_i - 2AK\partial_{13} u_i - 2BK\partial_{23} u_i,$$
$$G_i^{1,4} := [-K^3(1 + A^2 + B^2)\partial_3 J + AK^2(\partial_1 J + \partial_3 A) + BK^2(\partial_2 J + \partial_3 B)]\partial_3 u_i,$$
$$G_i^{1,5} := \partial_t \bar{\eta}(1 + x_3/b)K\partial_3 u_i,$$
$$G_i^{1,6} := \theta \nabla_\mathcal{A} y_3 - \theta e_3, \quad (3.11)$$

$$G^2 := AK\partial_3 u_1 + BK\partial_3 u_2 + (1-K)\partial_3 u_3, \quad (3.12)$$

$$G^3 = G^{3,1} + G^{3,2} + G^{3,3},$$

for

$$G_i^{3,1} := u_j \mathcal{A}_{jk} \partial_k \theta,$$
$$G_i^{3,2} := [K^2(1 + A^2 + B^2) - 1]\partial_{33}\theta - 2AK\partial_{13} u_i - 2BK\partial_{23}\theta,$$
$$G_i^{3,3} := [-K^3(1 + A^2 + B^2)\partial_3 J + AK^2(\partial_1 J + \partial_3 A)$$
$$\qquad + BK^2(\partial_2 J + \partial_3 B)]\partial_3 \theta, \quad (3.13)$$

$$G^4 = G^{4,1} + G^{4,2},$$

for

$$G^{4,1} := \partial_1 \eta \begin{pmatrix} p - \eta - 2(\partial_1 u_1 - AK\partial_3 u_1) \\ -\partial_2 u_1 - \partial_1 u_2 + BK\partial_3 u_1 + AK\partial_3 u_2 \\ -\partial_1 u_3 - K\partial_3 u_1 + AK\partial_3 u_3 \end{pmatrix}$$

$$+\partial_2\eta\begin{pmatrix} -\partial_2 u_1 - \partial_1 u_2 + BK\partial_3 u_1 + AK\partial_3 u_2 \\ p - \eta - 2(\partial_2 u_2 - BK\partial_3 u_2) \\ -\partial_2 u_3 - K\partial_3 u_2 + BK\partial_3 u_3 \end{pmatrix} + \begin{pmatrix} (K-1)\partial_3 u_1 + AK\partial_3 u_3 \\ (K-1)\partial_3 u_2 + BK\partial_3 u_3 \\ 2(K-1)\partial_3 u_3 \end{pmatrix},$$

$$G^{4,2} := \sigma(H - \Delta_\star \eta)\mathcal{N} + \sigma\Delta_\star(\mathcal{N} - e_3), \tag{3.14}$$

$$G^5 := -|\mathcal{N}| - (\nabla_\mathcal{A}\theta \cdot \mathcal{N} + \theta|\mathcal{N}|) + \nabla\theta \cdot e_3 + \theta, \tag{3.15}$$

$$G^6 := D\eta \cdot u. \tag{3.16}$$

Next we consider the energy-dissipation evolution equation for solutions to problem of the form (3.10).

Proposition 3.2 *Suppose* (v, q, ζ, ϑ) *solve*

$$\begin{cases} \partial_t v + \nabla q - \Delta v - \vartheta e_3 = \Phi^1, & \text{in } \Omega, \\ \operatorname{div} v = \Phi^2, & \text{in } \Omega, \\ \partial_t \vartheta - \Delta \vartheta = \Phi^3, & \text{in } \Omega, \\ (qI - \mathbb{D}v - \zeta I + \sigma\Delta_\star \zeta)e_3 = \Phi^4, & \text{on } \Sigma, \\ \nabla\vartheta \cdot e_3 + \vartheta = \Phi^5, & \text{on } \Sigma, \\ \partial_t \zeta - v_3 = \Phi^6, & \text{on } \Sigma, \\ v = \theta = 0, & \text{on } \Sigma_b. \end{cases} \tag{3.17}$$

Then

$$\frac{d}{dt}\left(\int_\Omega \frac{|v|^2}{2} + \int_\Sigma \frac{|\zeta|^2}{2} + \int_\Sigma \sigma\frac{|\nabla_\star \zeta|^2}{2}\right) + \int_\Omega \frac{|\mathbb{D}v|^2}{2}$$
$$= \int_\Omega v \cdot \Phi^1 + q \cdot \Phi^2 + \int_\Sigma -v \cdot \Phi^4 + \int_\Sigma (\zeta - \sigma\Delta_\star\zeta)\Phi^6 + \int_\Omega \vartheta e_3 \cdot v, \tag{3.18}$$

and

$$\frac{d}{dt}\int_\Omega \frac{|\vartheta|^2}{2} + \int_\Omega |\nabla\vartheta|^2 + \int_\Sigma |\vartheta|^2 = \int_\Omega \vartheta \cdot \Phi^3 + \int_\Sigma \vartheta \cdot \Phi^5. \tag{3.19}$$

Proof From the first equation in (3.17) we compute

$$\partial_t v_i + \partial_i q - \Delta v_i - \vartheta e_3 - \partial_i \Phi^2 = \Phi^1_i - \partial_i \Phi^2.$$

By the usual energy estimates, we may compute

$$\frac{d}{dt}\int_\Omega \frac{|v|^2}{2} + \int_\Omega \frac{|Dv|^2}{2} + \underbrace{\int_\Sigma v_3\zeta}_{I} + \underbrace{\int_\Sigma -\sigma v_3 \Delta_*\zeta}_{II} = \int_\Omega v\cdot\Phi^1 + q\Phi^2 - v\cdot\nabla\Phi^4 + \int_\Omega \vartheta e_3\cdot v,$$

we compute I by integrating by parts and using (3.17):

$$I = \int_\Sigma \zeta\partial_t\zeta - \zeta\Phi^6 = \frac{d}{dt}\int_\Sigma \frac{|\zeta|^2}{2} - \int_\Sigma \zeta\Phi^6.$$

For II, we compute

$$II = -\int_\Sigma (\partial_t\zeta - \Phi^6)\Delta_*\zeta = \sigma\frac{d}{dt}\int_\Sigma \frac{|\nabla_*\zeta|^2}{2} + \sigma\int_\Sigma \Phi^6\Delta_*\zeta.$$

Similarly, from the third equation in (3.17) and usual energy estimates, we compute

$$\frac{d}{dt}\int_\Omega |\vartheta|^2 + \int_\Omega |\nabla\vartheta|^2 + \int_\Sigma |\vartheta|^2 = \int_\Omega \Phi^3\vartheta + \int_\Sigma \vartheta\Phi^5.$$

4 Estimates of the nonlinearities

In this section, we record estimates for the nonlinearities that appear in (3.1) and (3.10). Throughout this section we will repeatedly use the estimates of Lemmas B.1 and B.2 in [15] to estimate $\bar\eta$, as well as Lemma B.3 in [15] to estimate various nonlinearities. Before doing these, we firstly give a lemma for moving the appearance of J and $\mathcal{A}$ factors.

4.1 Useful L^∞ estimates

Lemma 4.1 *There exists a universal $0 < \delta < 1$ so that if $\|\eta\|_{5/2}^2 \leq \delta$, then the following hold.*

(1) We have the estimate

$$\|J-1\|_{L^\infty}^2 + \|A\|_{L^\infty}^2 + \|B\|_{L^\infty}^2 \leq \frac{1}{2}, \quad \text{and} \quad \|K\|_{L^\infty}^2 + \|\mathcal{A}\|_{L^\infty}^2 \lesssim 1.$$

(2) The map Θ defined by (1.4) is a diffeomorphism.

(3) There exists a universal constant $C > 0$ such that for all $v \in H^1(\Omega)$ such that $v = 0$ on Σ_b we have

$$\int_\Omega |Dv|^2 \leq \int_\Omega J|D_\mathcal{A}v|^2 + C\sqrt{\mathcal{E}}\int_\Omega |Dv|^2.$$

Proof The proof of this lemma can be found in [8].

4.2 Nonlinearities in (3.1)

Our goal now is to estimate the nonlinear terms F^i for $i = 1, \cdots, 6$, as defined in (3.4)-(3.9). These estimates will be used principally to estimate the interaction terms on the right side of (3.2) and (3.3).

Theorem 4.1 Let $F^1, \cdots, F^5$ be as defined in (3.4)-(3.9). Let $\mathcal{E}$ and $\mathcal{D}$ be as defined in (2.1) and (2.1). Suppose that $\mathcal{E} \leqslant \delta$, where $\delta \in (0,1)$ is the universal constant given in Lemma 4.1, and that $D < \infty$. Then,

$$\|F^1\|_0 + \|F^2\|_0 + \|F^3\|_0 + \|F^4\|_{\Sigma,0} + \|F^5\|_{\Sigma,0} + \|F^6\|_{\Sigma,0} \lesssim \sqrt{\mathcal{E}}\sqrt{\mathcal{D}}, \quad (4.1)$$

$$\left|\int_\Omega \partial_t p F^2 J - \frac{d}{dt}\int_\Omega p F^2 J\right| \lesssim \sqrt{\mathcal{E}}\mathcal{D}, \quad \text{and} \quad \left|\int_\Omega p F^2 J\right| \lesssim \mathcal{E}^{\frac{3}{2}}. \quad (4.2)$$

Proof We divide the proof into several steps. Throughout the lemmas, we will employ Holder's inequality, Sobolev embeddings, trace theory, and Lemma 4.1.

Step1 F^1 estimates. The estimates

$$\|F^{1,1}\|_0 + \|F^{1,2}\|_0 + \|F^{1,3}\|_0 + \|F^{1,4}\|_0 \lesssim \sqrt{\mathcal{E}}\sqrt{\mathcal{D}},$$

is proved in [8]. We only bound the last term $F^{1,5}$ in F^1 via

$$\|F^{1,5}\|_0 \lesssim \|\theta\|_{L^\infty(\Omega)}\|\partial_t \nabla\overline{\theta}\|_0\|\partial_i y_3\|_{L^\infty(\Omega)} \lesssim \sqrt{\mathcal{E}}\sqrt{\mathcal{D}}.$$

Next, we control the second term F^2 as follows

$$\|F^2\|_0 \lesssim \|\partial_t \nabla\overline{\eta}\|_0 \|\nabla u\|_{L^\infty(\Omega)} \lesssim \sqrt{\mathcal{E}}\sqrt{\mathcal{D}},$$

and the term F^5 can be estimated as

$$\|F^5\|_{0,\Sigma} \lesssim \|\partial_t \nabla\eta\|_{0,\Sigma} + \|\theta\|_{L^\infty(\Sigma)}\|\partial_t \nabla\eta\|_{0,\Sigma} + \|\partial_t \nabla\eta\|_{L^4,\Sigma}\|\nabla\theta\|_{L^4,\Sigma}\|\nabla\eta\|_{L^\infty,\Sigma}$$

$$\lesssim \|\partial_t\eta\|_{H^{\frac{5}{2}}(\Sigma)} + \|\theta\|_{H^2(\Omega)}\|\partial_t\eta\|_{H^{\frac{5}{2}}(\Sigma)} + \|\partial_t\eta\|_{H^{\frac{5}{2}}(\Sigma)}\|\theta\|_{H^2(\Omega)}\|\eta\|_{H^3(\Sigma)}$$

$$\lesssim (1 + \sqrt{\mathcal{E}} + \mathcal{E})\sqrt{\mathcal{D}}.$$

Similar to the F^1 term, whereas F^3, F^4, F^6 term can be handled as follows

$$\|F^3\|_0 + \|F^4\|_{\Sigma,0} + \|F^6\|_{\Sigma,0} \lesssim \sqrt{\mathcal{E}}\sqrt{\mathcal{D}}.$$

Whereas (4.2) is proved in [15].

4.3 Nonlinearities in (3.10)

Now we turn our attention to the nonlinear terms G^i for $i = 1, \cdots, 5$, as defined in (3.9)-(3.13).

Theorem 4.2 *Let $G^1, \cdots, G^5$ be as defined in (3.11)-(3.16). Let $\mathcal{E}$ and $\mathcal{D}$ be as defined in (2.1) and (2.1). Suppose that $\mathcal{E} \leqslant \delta$, where $\delta \in (0,1)$ is the universal constant given in Lemma 4.1, and that $\mathcal{D} < \infty$. Then,*

$$\|G^1\|_1 + \|G^2\|_2 + \|G^3\|_1 + \|G^4\|_{\Sigma, \frac{3}{2}} + \|G^5\|_{\Sigma, \frac{3}{2}} + \|G^6\|_{\Sigma, \frac{5}{2}} \lesssim \sqrt{\mathcal{E}}\sqrt{\mathcal{D}}, \qquad (4.3)$$

and

$$\|G^1\|_0 + \|G^2\|_1 + \|G^3\|_0 + \|G^4\|_{\Sigma, \frac{1}{2}} + \|G^5\|_{\Sigma, \frac{1}{2}} + \|G^6\|_{\Sigma, \frac{3}{2}} \lesssim \mathcal{E}. \qquad (4.4)$$

Proof Here we only estimate the term G^5, the other terms are similar to [15], so we omit it.

$$\|G^5\|_{\Sigma, \frac{3}{2}} \lesssim \|\nabla \eta\|_{\Sigma, \frac{3}{2}} + \|\nabla \overline{\eta}\|_{\Sigma, \frac{3}{2}} \|\nabla \theta\|_{\Sigma, \frac{3}{2}} \|\nabla \eta\|_{\Sigma, \frac{3}{2}} + \|\theta\|_{\Sigma, \frac{3}{2}} \|\nabla \eta\|_{\Sigma, \frac{3}{2}}$$
$$+ \|\nabla \theta\|_{\Sigma, \frac{3}{2}} + \|\theta\|_{\Sigma, \frac{3}{2}}$$
$$\lesssim \sqrt{\mathcal{D}} + \sqrt{\mathcal{E}}\sqrt{\mathcal{D}} \lesssim \sqrt{\mathcal{E}}\sqrt{\mathcal{D}},$$

$$\|G^5\|_{\Sigma, \frac{1}{2}} \lesssim \|\nabla \eta\|_{\Sigma, \frac{1}{2}} + \|\nabla \overline{\eta}\|_{\Sigma, \frac{3}{2}} \|\nabla \theta\|_{\Sigma, \frac{1}{2}} \|\nabla \eta\|_{\Sigma, \frac{3}{2}} + \|\theta\|_{\Sigma, \frac{1}{2}} \|\nabla \eta\|_{\Sigma, \frac{3}{2}}$$
$$+ \|\nabla \theta\|_{\Sigma, \frac{1}{2}} + \|\theta\|_{\Sigma, \frac{1}{2}}$$
$$\lesssim \mathcal{E}.$$

5 A priori estimates

In this section, we combine energy-dissipation estimates with various elliptic estimates and estimate the nonlinearities in order to deduce a system of a priori estimates.

5.1 Energy-dissipation estimates

In order to state our energy-dissipation estimates, we must first introduce some notation. Recall that for a multi-index $\alpha = (\alpha_0, \alpha_1, \alpha_2) \in \mathbb{N}^{1+2}$ we write $|\alpha| = 2\alpha_0 + \alpha_1 + \alpha_2$ and $\partial^\alpha = \partial_{\alpha_0}^t \partial_1^{\alpha_1} \partial_2^{\alpha_2}$. For $\alpha \in \mathbb{N}^{1+2}$, set

$$\overline{\mathcal{E}}_\alpha := \int_\Omega \frac{1}{2} |\partial^\alpha u|^2 + \int_\Sigma \frac{1}{2} |\partial^\alpha \eta|^2 + \frac{\sigma}{2} |\nabla_* \partial^\alpha \eta|^2 + \int_\Omega \frac{1}{2} |\partial^\alpha \theta|^2,$$

$$\overline{\mathcal{D}}_\alpha := \int_\Omega \frac{1}{2}|\mathbb{D}\partial^\alpha u|^2 + \int_\Omega |\nabla \partial^\alpha \theta|^2 + \int_\Sigma |\partial^\alpha \theta|^2 |\mathcal{N}|. \tag{5.1}$$

We then define
$$\overline{\mathcal{E}} := \sum_{\alpha \leqslant 2} \overline{\mathcal{E}}_\alpha \quad \text{and} \quad \overline{\mathcal{D}} := \sum_{\alpha \leqslant 2} \overline{\mathcal{D}}_\alpha. \tag{5.2}$$

We will also need to use the functional
$$\mathcal{F} := \int_\Omega pF^2 J. \tag{5.3}$$

Our next result encodes the energy-dissipation inequality associated with $\overline{\mathcal{E}}$ and $\overline{\mathcal{D}}$.

Theorem 5.1 *Suppose that* (u, p, η, θ) *solves* (1.5) *on the temporal interval* $[0, T]$. *Let* $\mathcal{E}$ *and* $\mathcal{D}$ *be as defined in* (2.1) *and* (2.1), *and suppose that*

$$\sup_{0 \leqslant t \leqslant T} \mathcal{E}(t) \leqslant \delta, \quad \text{and} \quad \int_0^T \mathcal{D}(t) dt < \infty,$$

where $\delta \in (0, 1)$ *is the universal constant given in Lemma 4.1. Let* $\overline{\mathcal{E}}$ *and* $\overline{\mathcal{D}}$ *be given by* (5.2) *and* $\mathcal{F}$ *be given by* (5.3). *Then*

$$\frac{d}{dt}(\overline{\mathcal{E}} - \mathcal{F}) + \overline{\mathcal{D}} \lesssim \sqrt{\mathcal{E}}\mathcal{D}, \tag{5.4}$$

for all $t \in [0, T]$.

Proof Let $\alpha \in \mathbb{N}^{1+2}$ with $|\alpha| \leqslant 2$. We apply ∂^α to (1.5) to derive an equation for $(\partial^\alpha u, \partial^\alpha p, \partial^\alpha \eta, \partial^\alpha \theta)$. We will consider the form of this equation in different ways depending on α.

Suppose that $|\alpha| = 2$ and $\alpha_0 = 1$, i.e. that $\partial^\alpha = \partial_t$. Then, $v = \partial_t u$, $q = \partial_t p$, $\zeta = \partial_t \eta$ and $\vartheta = \partial_t \theta$ satisfy (3.1) with $F^1, \cdots, F^5$ as given in (3.4)-(3.9). According to Proposition 3.1 we have that

$$\frac{d}{dt}\left(\int_\Omega \frac{|\partial_t u|^2}{2}J + \int_\Sigma \frac{|\partial_t \eta|^2}{2} + \int_\Sigma \sigma \frac{|\nabla_\star \partial_t \eta|^2}{2} + \int_\Omega \frac{|\partial_t \vartheta|^2}{2}J\right) + \int_\Omega \frac{|\mathbb{D}_A \partial_t u|^2}{2}J$$
$$+ \int_\Omega |\nabla_A \partial_t \theta|^2 J + \int_\Sigma |\partial_t \theta|^2 |\mathcal{N}|$$
$$= \int_\Omega (\partial_t u \cdot F^1 + \partial_t p F^2 + \partial_t \theta \cdot F^3) J$$
$$+ \int_\Sigma (-\partial_t u \cdot F^4 + \partial_t \theta F^5) + \int_\Sigma (\partial_t \eta - \sigma \Delta_\star \partial_t \eta) F^6 + \int_\Omega \vartheta \nabla_A y_3 \cdot vJ.$$

We then write

$$\int_\Omega \partial_t p F^2 J = \frac{d}{dt}\int_\Omega p F^2 J - \int_\Omega p(\partial_t F^2 J + F^2 J),$$

collect the temporal derivative terms, and apply the estimates (4.1)-(4.3) of Theorem 4.2, the estimates of Lemma 4.1, and the usual trace estimates to derive that

$$\frac{d}{dt}(\overline{\mathcal{E}}_{(1,0,0,0)} - \mathcal{F}) + \overline{\mathcal{D}}_{(1,0,0,0)} \lesssim \sqrt{\mathcal{E}}\mathcal{D}, \qquad (5.5)$$

where $\overline{\mathcal{E}}_{(1,0,0,0)}$ and $\overline{\mathcal{D}}_{(1,0,0,0)}$ are as defined in (5.1).

Next, we consider $\alpha \in \mathbb{N}^{1+2}$ with $\alpha_0 = 0$, i.e. no temporal derivatives. In this case we view (u, p, η, θ) in terms of (3.10), which then means that $(v, q, \zeta, \vartheta) = (\partial^\alpha u, \partial^\alpha p, \partial^\alpha \eta, \partial^\alpha \theta)$ satisfy (3.17) with $\Phi^i = \partial^\alpha G^i$ for $i = 1,\cdots,5$, where the nonlinearities G^i are as defined in (3.11)-(3.16). We may then apply Proposition 3.2 to see that for $|\alpha| \leqslant 2$ and $\alpha_0 = 0$, we have the identity

$$\frac{d}{dt}\overline{\mathcal{E}}_\alpha + \overline{\mathcal{D}}_\alpha = \int_\Omega (\partial^\alpha u \cdot \partial^\alpha G^1 + \partial^\alpha p \partial^\alpha G^2 + \partial^\alpha \theta \cdot \partial^\alpha G^3) + \int_\Omega \partial^\alpha \theta e_3 \cdot \partial^\alpha u$$
$$+ \int_\Sigma (-\partial^\alpha u \cdot \partial^\alpha G^4 + \partial^\alpha \theta \cdot \partial^\alpha G^5) + \partial^\alpha \eta \partial^\alpha G^6 - \sigma \partial^\alpha G^6 \Delta_* \partial^\alpha \eta. \quad (5.6)$$

When $|\alpha| = 2$ and $\alpha_0 = 0$, we write $\partial^\alpha = \partial^\beta \alpha^\omega$ for $|\beta| = |\omega| = 1$. We then integrate by parts in the G^1, G^6 terms in (5.6) to estimate

$$RHS \text{ of } (5.6)$$
$$= \int_\Omega (-\partial^{\alpha+\beta} u \cdot \partial^\omega G^1 + \partial^\alpha p \partial^\alpha G^2 + \partial^{\alpha+\beta}\theta \cdot \partial^\omega G^3) + \int_\Omega \partial^\alpha \theta e_3 \cdot \partial^\alpha u$$
$$+ \int_\Sigma (-\partial^\alpha u \cdot \partial^\alpha G^4 + \partial^\alpha \theta \cdot \partial^\alpha G^5) + \partial^\omega \eta \partial^{\alpha+\beta} G^6 - \sigma \partial^{\alpha+\beta} G^6 \Delta_* \partial^\omega \eta,$$
$$\lesssim \|u\|_3 \|G^1\|_1 + \|p\|_2 \|G^2\|_2 + \|\theta\|_3 \|G^1\|_1 + \|\theta\|_2 \|u\|_2 + \|D^2 u\|_{\Sigma,\frac{1}{2}} \|D^2 G^4\|_{\Sigma,-\frac{1}{2}}$$
$$+ \|D^2\theta\|_{\Sigma,\frac{1}{2}} \|D^2 G^5\|_{\Sigma,-\frac{1}{2}} + \|D^3 G^6\|_{\Sigma,-\frac{1}{2}} [\|D\eta\|_{\Sigma,\frac{1}{2}} + \|D^3\eta\|_{\Sigma,\frac{1}{2}}]$$
$$\lesssim \sqrt{\mathcal{D}}(\|G^1\|_1 + \|G^2\|_2 + \|G^3\|_1 + \|\theta\|_2 + \|G^4\|_{\Sigma,\frac{3}{2}} + \|G^5\|_{\Sigma,\frac{3}{2}} + \|G^6\|_{\Sigma,\frac{5}{2}}).$$

The estimate (4.3) of Theorem 4.3 then tells us that

$$RHS \text{ of } (5.6) \lesssim \sqrt{\mathcal{E}}\mathcal{D},$$

and so we find that for $\overline{\mathcal{E}}_\alpha$ and $\overline{\mathcal{D}}_\alpha$ as in (5.1) we have the inequality

$$\frac{d}{dt} \sum_{|\alpha|=2,\alpha_0=0} \overline{\mathcal{E}}_\alpha + \sum_{|\alpha|=2,\alpha_0=0} \overline{\mathcal{D}}_\alpha \lesssim \sqrt{\mathcal{E}}\mathcal{D}. \tag{5.7}$$

On the other hand, if $|\alpha| < 2$, then we must have that $\alpha_0 = 0$, and we can directly apply Theorem 4.3 to see that

$$RHS \text{ of } (5.6) \lesssim \sqrt{\mathcal{E}}\mathcal{D}.$$

From this we deduce that

$$\frac{d}{dt} \sum_{|\alpha|=2,\alpha_0=0} \overline{\mathcal{E}}_\alpha + \sum_{|\alpha|=2,\alpha_0=0} \overline{\mathcal{D}}_\alpha \lesssim \sqrt{\mathcal{E}}\mathcal{D}. \tag{5.8}$$

Now, to deduce (5.4) we simply sum (5.5),(5.6) and (5.8).

5.2 Enhanced energy estimates

From the energy-dissipative estimates of Theorem 5.1, we have control of $\overline{\mathcal{E}}$ and $\overline{\mathcal{D}}$. Our goal now is to show that these can be used to control $\mathcal{E}$ and $\mathcal{D}$ up to some error terms that we will be able to guarantee are small. Here we focus on the estimate for the energies, $\overline{\mathcal{E}}$, and $\mathcal{E}$.

Theorem 5.2 *Let $\mathcal{E}$ be as defined in (2.1). Suppose that $\mathcal{E} \leqslant \delta$, where $\delta \in (0,1)$ is the universal constant given in Lemma 4.1. Then*

$$\mathcal{E} \lesssim \overline{\mathcal{E}} + \mathcal{E}^2. \tag{5.9}$$

Proof According to the definitions of $\overline{\mathcal{E}}$ and $\mathcal{E}$, in order to prove (5.9), it suffices to prove that

$$\|u\|_2^2 + \|p\|_1^2 + \|\theta\|_2^2 + \|\partial_t \eta\|_{\Sigma,\frac{3}{2}}^2 + \|\partial_t^2 \eta\|_{\Sigma,-\frac{1}{2}}^2 \lesssim \overline{\mathcal{E}} + \mathcal{E}^2. \tag{5.10}$$

For estimating u and p, we recall the standard Stokes estimates: for $r \geqslant 0$,

$$\|u\|_r + \|p\|_{r-1} \lesssim \|\phi\|_{r-2} + \|\psi\|_{r-1} + \|\alpha\|_{\Sigma,r-\frac{3}{2}}, \tag{5.11}$$

if

$$\begin{cases} -\Delta u + \nabla p = \phi \in H^{r-2}(\Omega), \\ \operatorname{div} v = \psi \in H^{r-1}(\Omega), \\ (pI - \mathbb{D}u)e_3 = \alpha \in H^{r-\frac{3}{2}}(\Sigma), \\ u|_{\Sigma_b} = 0. \end{cases}$$

Now, according to (3.10), we have that

$$\begin{cases} -\Delta u + \nabla p = -\partial_t u + G^1, \\ \operatorname{div} v = G^2, \\ (pI - \mathbb{D}u)e_3 = (\eta I + \sigma \Delta_\star \eta)e_3 + G^4, \\ u = 0, \quad \text{on } \Sigma_b, \end{cases}$$

and hence we may apply (5.11) and the estimate (4.4) of Theorem 4.3 to see that

$$\|u\|_2 + \|p\|_1 \lesssim \|\partial_t u\|_0 + \|G^1\|_0 + \|G^2\|_1 + \|(\eta I + \sigma \Delta_\star \eta)e_3\|_{\Sigma,\frac{1}{2}} + \|G^4\|_{\Sigma,\frac{1}{2}},$$
$$\lesssim \sqrt{\mathcal{E}} + \|G^1\|_0 + \|G^2\|_1 + \|G^4\|_{\Sigma,\frac{1}{2}},$$
$$\lesssim \sqrt{\mathcal{E}} + \mathcal{E}.$$

From this, we deduce that the u, p estimates in (5.10) hold.

Similarly, for estimating θ, we have

$$\begin{cases} -\Delta \theta = -\partial_t \theta + G^3, \\ \nabla \theta \cdot e_3 + \theta = G^5, \\ \theta = 0, \quad \text{on } \Sigma_b, \end{cases}$$

and hence we deduce that

$$\|\theta\|_2 \lesssim \|\partial_t \theta\|_0 + \|G^3\|_0 + \|\theta\|_{\Sigma,\frac{1}{2}} + \|G^5\|_{\Sigma,\frac{1}{2}},$$
$$\lesssim \sqrt{\mathcal{E}} + \|G^3\|_0 + \|G^5\|_{\Sigma,\frac{1}{2}},$$
$$\lesssim \sqrt{\mathcal{E}} + \mathcal{E}.$$

From this we deduce that the θ estimate in (5.10) holds.

To estimate the $\partial_t \eta$ term in (5.10) we use the fourth equation of (3.10) in conjunction with the estimate (4.4) of Theorem 4.3 and the usual trace estimates to see that

$$\|\partial_t \eta\|_{\Sigma,\frac{3}{2}} \lesssim \|u_3\|_{\Sigma,\frac{3}{2}} + \|G^4\|_{\Sigma,\frac{3}{2}} \lesssim \|u\|_2 + \mathcal{E} \lesssim \sqrt{\mathcal{E}} + \mathcal{E}.$$

From this we deduce that the $\partial_t \eta$ estimate in (5.10) holds.

It remains only to estimate the $\partial_t^2 \eta$ term in (5.10). For this, we apply a temporal derivative to the fourth equation of (3.10) and integrate against a function $\phi \in H^{\frac{1}{2}}(\Sigma)$ to see that

$$\int_\Sigma \partial_t^2 \eta \phi dx_\star = \int_\Sigma \partial_t u_3 \phi dx_\star + \int_\Sigma \partial_t G^4 \phi dx_\star.$$

Choose an extension $E\phi \in H^1(\Omega)$ with $E\phi|_\Sigma = \phi, E\phi|_{\Sigma_b} = \phi$, and $\|E\phi\|_1 \lesssim \|\phi\|_{\Sigma,\frac{1}{2}}$. Then

$$\int_\Sigma \partial_t u_3 \phi dx_\star = \int_\Sigma \partial_t u \cdot \nabla_x E\phi + \int_\Sigma \partial_t G^2 E\phi \leqslant (\|\partial_t u\|_0 + \|\partial_t G^2\|_0)\|\phi\|_{\Sigma,\frac{1}{2}},$$

and again, Theorem 4.3 implies that

$$\|\partial_t^2 \eta\|_{\Sigma,\frac{1}{2}} \lesssim \|\partial_t u\|_0 + \|\partial_t G^2\|_0 + \|\partial_t G^4\|_{\Sigma,-\frac{1}{2}} \lesssim \sqrt{\mathcal{E}} + \mathcal{E}.$$

From this, we deduce that the $\partial^2 \eta$ estimate in (5.10) holds.

5.3 Enhanced dissipate estimates

We now complete Theorem 5.2 by proving a corresponding result for the dissipation.

Theorem 5.3 *Let $\mathcal{E}$ and $\mathcal{D}$ be as defined in (2.1) and (2.1). Suppose that $\mathcal{E} \leqslant \delta$, where $\delta \in (0,1)$ is the universal constant given in Lemma 4.1 and suppose that $\mathcal{D} < \infty$. Then*

$$\mathcal{D} \lesssim \overline{\mathcal{D}} + \mathcal{E}\mathcal{D}. \qquad (5.12)$$

Proof For the θ dissipative estimate, we directly use the Stokes elliptic estimates to the following equations

$$\begin{cases} -\Delta \theta = -\partial_t \theta + G^3, \\ \nabla \theta \cdot e_3 + \theta = G^5, \\ \theta = 0, \quad \text{on } \Sigma_b, \end{cases}$$

and hence we derive that

$$\|\theta\|_3 \lesssim \|\partial_t \theta\|_1 + \|G^3\|_1 + \|\theta\|_{\Sigma,\frac{5}{2}} + \|G^5\|_{\Sigma,\frac{5}{2}}$$
$$\lesssim \sqrt{\overline{\mathcal{D}}} + \|G^3\|_1 + \|G^5\|_{\Sigma,\frac{5}{2}},$$

$$\lesssim \sqrt{\mathcal{D}} + \sqrt{\mathcal{E}}\sqrt{\mathcal{D}}. \tag{5.13}$$

Recall the Stokes elliptic estimates for the Stokes problem with Dirichlet boundary conditions: for $r \geq 2$,

$$\|u\|_r + \|\nabla p\|_{r-2} \lesssim \|f\|_{r-2} + \|h\|_{r-1} + \|\psi_1\|_{\Sigma, r-\frac{1}{2}} + \|\psi_2\|_{\Sigma_b, r-\frac{1}{2}}, \tag{5.14}$$

if

$$\begin{cases} -\Delta u + \nabla p = f, & \text{in } \Omega, \\ \operatorname{div} v = h & \text{in } \Omega, \\ u = \psi_1 & \text{on } \Sigma, \\ u = \psi_2 & \text{on } \Sigma_b. \end{cases}$$

We know that

$$\|u\|_1 + \|\nabla_* u\|_1 + \|\nabla_*^2\|_1 \lesssim \sqrt{\mathcal{D}},$$

and so trace theory provides us with the estimate

$$\|u\|_{\Sigma, \frac{5}{2}} \lesssim \sqrt{\mathcal{D}}.$$

We also have that $\|\partial_t u\|_1 \lesssim \sqrt{\mathcal{D}}$, and Theorem 4.3 tells that

$$\|G^1\|_1 + \|G^2\|_2 \lesssim \sqrt{\mathcal{E}}\sqrt{\mathcal{D}},$$

we may thus apply (5.14) with $r = 3$ and $f = -\partial_t u + G^1$, $h = G^2$, $\psi_1 = u|_\Sigma$, and $\psi_2 = 0$ to obtain

$$\|u\|_3 + \|\nabla p\|_1 \lesssim \|-\partial_t u + G^1\|_1 + \|G^2\|_2 + \|u\|_{\Sigma, \frac{5}{2}} \lesssim \sqrt{\mathcal{D}} + \sqrt{\mathcal{E}}\sqrt{\mathcal{D}}. \tag{5.15}$$

We now turn to the η estimates. By using the third and fourth equation of (3.10) and elliptic estimates, we can deduce ([15])

$$\|\eta\|_{\Sigma, \frac{7}{2}} \lesssim \|\eta\|_{\Sigma, 0} + \|\nabla_* \eta\|_{\Sigma, \frac{5}{2}} \lesssim \|\nabla_* \eta\|_{\Sigma, \frac{5}{2}} \lesssim \sqrt{\mathcal{D}} + \sqrt{\mathcal{E}}\sqrt{\mathcal{D}}, \tag{5.16}$$

and

$$\|\partial_t \eta\|_{\Sigma, \frac{5}{2}} \lesssim \|u_3\|_{\Sigma, \frac{5}{2}} + \|G^4\|_{\Sigma, \frac{5}{2}} \lesssim \|u\|_3 + \|G^4\|_{\Sigma, \frac{5}{2}} \lesssim \sqrt{\mathcal{D}} + \sqrt{\mathcal{E}}\sqrt{\mathcal{D}}, \tag{5.17}$$

and

$$\|\partial_t^2 \eta\|_{\Sigma,\frac{1}{2}} \lesssim \|\partial_t u_3\|_{\Sigma,\frac{1}{2}} + \|\partial_t G^4\|_{\Sigma,\frac{1}{2}} \lesssim \|\partial_t u\|_1 + \|\partial_t G^4\|_{\Sigma,\frac{1}{2}} \lesssim \sqrt{\mathcal{D}} + \sqrt{\mathcal{E}}\sqrt{\mathcal{D}}. \quad (5.18)$$

Now we complete the estimate of the pressure by obtaining a bound for $\|p\|_0$. To this end we combine the estimates (5.15) and (5.16) with the Stokes estimate of (5.11) with $\phi = -\partial_t u + G^1, \psi = G^2$, and $\alpha = (\eta I - \sigma \Delta_\star \eta)e_3 + G^3 e_3$ to bound

$$\|u\|_3 + \|p\|_2 \lesssim \| -\partial_t u + G^1\|_1 + \|G^2\|_2 + \|(\eta I - \sigma \delta_\star \eta)e_3\|_{\Sigma,\frac{3}{2}},$$
$$\lesssim \|\partial_t u\|_1 + \|G^1\|_1 + \|G^2\|_2 + \|\eta\|_{\Sigma,\frac{7}{2}},$$
$$\lesssim \sqrt{\mathcal{D}} + \sqrt{\mathcal{E}}\sqrt{\mathcal{D}}.$$

Thus,
$$\|p\|_2 \lesssim \sqrt{\mathcal{D}} + + \sqrt{\mathcal{E}}\sqrt{\mathcal{D}}. \quad (5.19)$$

Finally, to deduce (5.12) we sum the squares of the estimates (5.14)–(5.19).

6 Proof of main results

6.1 Boundedness and decay

We now combine the estimates of the previous section in order to deduce our primary a priori estimates for solutions. It shows that under a smallness condition on the energy and a finiteness condition for the integrated dissipation, the energy decays exponentially and the dissipation integral is bounded by the initial data.

Theorem 6.1 *Suppose that (u, p, η, θ) solves (1.5) on the temporal interval $[0, T]$. Let $\mathcal{E}$ and $\mathcal{D}$ be as defined in (2.1) and (2.1). Then there exists universal constant $0 < \delta_\star < \delta$, where $\delta \in (0,1)$ is the universal constant given in lemma 4.1, such that if*

$$\sup_{0 \leqslant t \leqslant T} \mathcal{E}(t) \leqslant \delta_\star \text{ and } \int_0^T \mathcal{D}(t)dt < \infty,$$

then
$$\sup_{0 \leqslant t \leqslant T} e^{\lambda t}\mathcal{E}(t) + \int_0^T \mathcal{D}(t)dt \lesssim \mathcal{E}(0), \quad (6.1)$$

for all $t \in [0, T]$, where $\lambda > 0$ is a universal constant.

Proof The details of this proof can be seen in [15], but we omit it here.

6.2 Global well-posedness

We now couple to the local well-posedness to produce global-in-time solutions that decay to equilibrium exponentially fast.

Proof of Theorem 2.3 First note that given u_0, η_0 and θ_0, in the local existence result, Theorem 2, we construct the remaining data $\partial_t u(\cdot, 0)$, $\partial_t \eta(\cdot, 0)$ and $\partial_t \theta(\cdot, 0)$ in such a way that

$$\mathcal{E}(0) \leqslant C_0(\|u_0\|^2_{H^2(\Omega)} + \|\eta_0\|^2_{H^3(\Sigma)} + \|\theta_0\|^2_{H^2(\Omega)}), \tag{6.2}$$

for some universal constant $C_0 > 0$.

Let $T = 1$ and choose $\delta_\star > 0$ as in Theorem 6.1. Choose $\kappa > 0$ as in Theorem 2 and let $C_1 > 0$ denote the universal constant appearing on the right side of (2.3). Also, let $C_2 > 0$ be the universal constant appearing on the right side of (6.1) and $\lambda > 0$ be the constant appearing on the left. Set

$$\kappa_\star = \frac{1}{(1+C_0)(1+C_1)(1+C_2)} \min\{\kappa, \delta_\star\},$$

and assume that (2.4) is satisfied with $\kappa_\star$.

Due to (6.2), the unique solution on $[0,1]$ produced by Theorem 2 then satisfies

$$\sup_{0 \leqslant t \leqslant 1} \mathcal{E}(t) + \int_0^T \mathcal{D}(t) + \|\partial_t^2 u\|^2_{\mathcal{X}^*} \leqslant C_1 \mathcal{E}(0) \leqslant C_0 C_1 \kappa_\star \leqslant \delta_\star.$$

Consequently, we may apply Theorem 6.1 to see that

$$\sup_{0 \leqslant t \leqslant 1} e^{\lambda t} \mathcal{E}(t) + \int_0^1 \mathcal{D}(t) \leqslant C_2 \mathcal{E}(0) \leqslant C_0 C_2 \kappa_\star,$$

which in particular means that

$$\mathcal{E}(1) \leqslant e^{-\lambda} C_0 C_2 \kappa_\star \leqslant \kappa. \tag{6.3}$$

Due to (6.3), we may apply Theorem 2 with initial data $u(\cdot, 1), \eta(\cdot, 1)$, etc, to uniquely extend the solution to $[1,2]$ in such that a way that

$$\sup_{1 \leqslant t \leqslant 2} \mathcal{E}(t) + \int_1^2 \mathcal{D}(t) + \int_1^2 \|\partial_t^{2N+1} u\|^2_{\mathcal{X}^*} \leqslant C_1 \mathcal{E}(1) \leqslant e^{-\lambda} C_0 C_1 C_2 \kappa_\star \leqslant \delta_\star.$$

We may then apply the a priori estimate of Theorem 6.1 to see that

$$\sup_{0\leqslant t\leqslant 2} e^{\lambda t}\mathcal{E}(t) + \int_0^2 \mathcal{D}(t) \leqslant C_2\mathcal{E}(0) \leqslant C_0 C_2 \kappa_\star,$$

and hence that

$$\mathcal{E}(2) \leqslant e^{-2\lambda} C_0 C_2 \kappa_\star.$$

We may continue iterating the above argument to ultimately deduce that the solution exists on $[0,\infty)$ and obey the estimate (2.5).

References

[1] J. T. Beale, The initial value problem for the Navier-Stokes equations equations with a free surface, Comm. Pure Appl. Math., 34(1981) 359-392.

[2] J. T. Beale, Large-time regularity of viscous surface waves, Arch. Rational Mech. Anal. 84(1984), 307-352.

[3] H. Bénard, Les tourbillons cellulaires dans une nappe liquide, Revue Génard. Sci. Pure Appl. 11 (1990), 126 1-1271, 1309-1328.

[4] M. J. Block, Surface tension as the cause of Bénard cell and surface deformation in a liquid film, Nature 178 (1956), 650-651.

[5] P. G. Drazin and W. H. Reid, Hydrodynamic stability, 2nd., Cambridge University Press, Cambridge, 2004.

[6] Y. Guo, I.Tice, Local well-poseness of the viscous surface wave problem without surface tension, Anal PDE., 6(2003), 287-369.

[7] Y. Guo, I.Tice, Decay of viscous surface waves without surface tension in horizontally infinite domains, Anal PDE., 6(2003), 1429-1533.

[8] Y. Guo, I.Tice, Almost exponential decay of periodic viscous surface waves without surface tension Arch. Rational Mech. Anal., 207(2013), 459-531.

[9] K. Iohara, Bénard-Marangoni convection with a deformable surface, J. Math. Kyoto Univ., 38(1998), 255-270.

[10] T. Nishida, Y. Teramoto, H. Yoshihara. Global in time behavior of viscous surface waves: horizontal periodic motion. J. Math. Kyoto Univ. 44 (2004), no. 2, 271-323.

[11] T. Nishida and Y. Teramoto, On the linearized system arising in the study of Bénard-Marangoni convection, Kyoto Conference on the Navier-Stokes Equations and their applications, RIMS Kokyuroku Bessatsu B1, (2007), 271-286.

[12] T. Nishida and Y. Teramoto, Bifurcation theorems for the model system of Bénard-Marangoni convection, J. Math. Fluid Mech. 11(2009), 383-406

[13] T. Nishida and Y. Teramoto, Pattern formations in heat convection problems, Chin. Ann. Math. 30B(6), (2009), 769-784.

[14] Y. R. Zeng, Local well-posed for the Bénard convection without surface tension. Commun. Math. Sci. 15(4), (2017) 903-956.

[15] C. Kim and I. Tice, Dynamics and stability of Surfactant-driven surface waves. SIAM J. MATH. ANAL. 49(2), (2017), 1295-1332.

Interaction Behavior Associated with a Generalized (2+1)-Dimensional Hirota Bilinear Equation for Nonlinear Waves*

Hua Yanfei(花艳菲), Guo Boling(郭柏灵),
Ma Wenxiu(马文秀), and Lü Xing (吕兴)

Abstract In this paper, we focus on the interaction behavior associated with a generalized (2+1)-dimensional Hirota bilinear equation. With symbolic computation, two types of interaction solutions including lump-kink and lump-soliton ones are derived through mixing two positive quadratic functions with an exponential function, or two positive quadratic functions with a hyperbolic cosine function in the bilinear equation. The completely non-elastic interaction between a lump and a stripe is presented, which shows the lump is drowned or shallowed by the stripe. The interaction between lump and soliton is also given, where the lump moves from one branch to the other branch of the soliton. These phenomena exhibit the dynamics of nonlinear waves and the solutions are useful for the study on interaction behavior of nonlinear waves in shallow water, plasma, nonlinear optics and Bose-Einstein condensates.

Keywords Hirota bilinear method; lump solution; interaction solution; symbolic computation

1 Introduction

As is known, solitons possess many special characteristics of nonlinear waves[1] and are widely used to describe the nonlinear phenomena in such fields as shallow water[2,3], plasma[4-7], nonlinear optics[8], and Bose-Einstein

---

* Appl. Math. Model, 2019, 74: 184–198. DOI: https://doi.org/10.1016/j.apm.2019.04.044.

condensates[9]. Due to its vital application, the theoretical analysis on soliton solutions to nonlinear evolution equations is of great importance [1,10,11]. Many effective methods have been employed to solve nonlinear evolution equations, for example, the Hirota bilinear method[12], the exp-function method[13,14], the homotopy perturbation method[15], the variational iteration method[16], the Adomian decomposition method[16] and the Galerkin method[17,18]. As a type of rational solution, lump solutions are different from soliton solutions. Lump solutions are localized in all directions in the space. In 2006, lump solutions were studied with the variable separation method[19]. In 2015, lump solutions were constructed to the KP equation by substituting the positive quadratic function to the bilinear equation[20]. Moreover, lump or multi-lump solutions to the Boussinesq[21], the KPI equation[20,22,23], the BKP equation[24] and the potential-YTSF equation[25] have been obtained.

Recently, the interaction between lump solutions and soliton solutions has attracted more and more attention(see, Ref. [26, 27] and references therein). Interaction behavior of nonlinear waves appearing in many different systems in nature can be described and illustrated with the interaction solutions[28]. Interaction solutions are valuable in analyzing the nonlinear dynamics of waves in shallow water and can be used for forecasting the appearance of rogue waves[26,28,29]. It generally hold that the rogue waves turn at the interaction location of a lump with a two-soliton wave[28,29].

Lump dynamics has been studied for the following (2+1)-dimensional nonlinear evolution equation[30]

$$u_{yt} - u_{xxxy} - 3(u_x u_y)_x - 3u_{xx} + 3u_{yy} = 0, \tag{1.1}$$

which enjoys the Hirota bilinear form as

$$(D_t D_y - D_x^3 D_y - 3D_x^2 + 3D_y^2)f \cdot f = 0, \tag{1.2}$$

through the dependent variable transformation $u = 2\big[\ln f(x,y,t)\big]_x$, where the D-operator[12] is defined by

$$D_x^m D_y^n D_t^p (f \cdot g)$$
$$= \Big(\frac{\partial}{\partial x} - \frac{\partial}{\partial x'}\Big)^m \Big(\frac{\partial}{\partial y} - \frac{\partial}{\partial y'}\Big)^n \Big(\frac{\partial}{\partial t} - \frac{\partial}{\partial t'}\Big)^p f(x,y,t)g(x',y',t')\Big|_{x'=x,y'=y,t'=t}.$$

The coefficients of each term in some nonlinear evolution equations reflex different physical meaning or background, such as medium inhomogeneity, different boundary conditions, or different external force[2,3,6,7]. In order to show the influence of the coefficients on the wave interaction and investigate some more general cases, we will focus on the interaction behavior associated with a generalized (2+1)-dimensional Hirota bilinear equation as

$$(D_t D_y + c_1 D_x^3 D_y + c_2 D_y^2) f \cdot f$$
$$= 2[f_{yt}f - f_y f_t + c_1(f_{xxxy}f - 3f_{xxy}f_x + 3f_{xy}f_{xx} - f_y f_{xxx}) + c_2(f_{yy}f - f_y^2)]$$
$$= 0, \qquad (1.3)$$

which is linked with the following equation

$$u_{yt} + c_1 \left[u_{xxxy} + 3(2u_x u_y + u_{xy}u) + 3u_{xx} \int_{-\infty}^{x} u_y \, dx' \right] + c_2 u_{yy} = 0, \qquad (1.4)$$

through the dependent variable transformation $u = 2[\ln f(x,y,t)]_{xx}$, where c_1 and c_2 are arbitrary real constants.

With symbolic computation, two types of interaction solutions including lump-kink and lump-soliton ones will be derived to Eq. (1.4). The outline of this paper is as follows: In Section 2, we will analyze the interaction between a lump and a stripe by considering a mixed solution of two positive quadratic functions with an exponential function. The dynamic behaviors of the interaction solutions will be exhibited. In Section 3, we will discuss the interaction between a lump and a two-soliton by considering a mixed solution of two positive quadratic functions with a hyperbolic cosine function, and generate two cases of interaction solutions. We will display the propagation behaviors with some figures and study the dynamics with limitation analysis of these solutions. The last section is our concluding remarks.

2 Interaction solutions of lump-kink type

In this section, we will focus on computing interaction solutions between lump and stripe to Eq. (1.4) by making a combination of two positive quadratic functions with an exponential function as

$$f = g^2 + h^2 + ke^l + a_9, \qquad (2.1)$$

where three wave variables are defined by

$$g = a_1 x + a_2 y + a_3 t + a_4,$$

$$h = a_5 x + a_6 y + a_7 t + a_8,$$

$$l = k_1 x + k_2 y + k_3 t,$$

while a_i $(1 \leqslant i \leqslant 9)$ and k_j $(1 \leqslant j \leqslant 3)$ are real constants to be determined, and $k > 0$ is a real constant.

Case 1

$$\left\{ a_1 = -\frac{a_5 a_6}{a_2}, a_2 = a_2, a_3 = -a_2 c_2, a_4 = a_4, a_5 = a_5, a_6 = a_6, a_7 = -a_6 c_2, \right.$$

$$\left. a_8 = a_8, a_9 = a_9, k_1 = k_1, k_2 = 0, k_3 = c_1 k_1^3 \right\},$$

which needs to satisfy the condition

$$a_2 \neq 0, \tag{2.2}$$

to make the corresponding solution f be well-defined, the condition

$$a_9 > 0, \tag{2.3}$$

to guarantee the positiveness of f, the condition

$$a_5 \neq 0, \tag{2.4}$$

to realize the localization of u in all directions in the (x, y)-plane, and the condition

$$k_1 \neq 0, \tag{2.5}$$

to ensure the interaction solutions be obtained.

Case 2

$$\left\{ a_1 = a_1, a_2 = 0, a_3 = 0, a_4 = a_4, a_5 = 0, a_6 = a_6, a_7 = a_7, \right.$$

$$\left. a_8 = a_8, a_9 = a_9, k_1 = k_1, k_2 = 0, k_3 = c_1 k_1^3 \right\},$$

which needs to satisfy the conditions

$$a_1 a_6 \neq 0, \quad a_9 > 0, \quad k_1 \neq 0. \tag{2.6}$$

Case 3

$$\left\{a_1 = -\frac{a_5 a_6}{a_2}, a_2 = a_2, a_3 = -a_2 c_2, a_4 = a_4, a_5 = a_5, a_6 = a_6, a_7 = -a_6 c_2,\right.$$

$$\left. a_8 = a_8, a_9 = a_9, k_1 = 0, k_2 = k_2, k_3 = -c_2 k_2\right\},$$

which needs to satisfy the conditions

$$a_2 \neq 0, \quad a_9 > 0, \quad a_5 \neq 0, \quad k_2 \neq 0. \tag{2.7}$$

Case 4

$$\left\{a_1 = a_1, a_2 = 0, a_3 = 0, a_4 = a_4, a_5 = 0, a_6 = a_6, a_7 = a_7,\right.$$

$$\left. a_8 = a_8, a_9 = a_9, k_1 = 0, k_2 = k_2, k_3 = \frac{a_7 k_2}{a_6}\right\},$$

which needs to satisfy the conditions

$$a_1 a_6 \neq 0, \quad a_9 > 0, \quad k_2 \neq 0. \tag{2.8}$$

Substituting the four cases of parameters into the function f, we can get the interaction solutions to Eq. (1.4).

For any fixed value of t, the extremum points of the lump can be obtained. The extremum point of the lump locates at

$$\left(x = \frac{a_2 a_7 t_0 - a_3 a_6 t_0 + a_2 a_8 - a_4 a_6}{a_1 a_6 - a_2 a_5}, y = -\frac{a_1 a_7 t_0 - a_3 a_5 t_0 + a_1 a_8 - a_4 a_5}{a_1 a_6 - a_2 a_5}\right), \tag{2.9}$$

where the maximum of the amplitude of the lump is attained as $4(a_1^2 + a_5^2)/a_9$.

Substituting suitable values of a_i ($1 \leqslant i \leqslant 9$), k, k_j ($1 \leqslant j \leqslant 3$) and c_p ($p = 1, 2$) into the resulting solutions, we get various exact interaction solutions to Eq. (1.4).

The parameters are arbitrary but need to satisfy the corresponding conditions, that is, Constraints (2.2) to (2.8). To exhibit the interaction process clearly, we have to choose suitable parameters for the simulation because the parameters determine the location and height of the waves. After trying many times, we take $a_4 = a_8 = 0$ and choose $a_1 = 2, a_2 = -15, a_3 = 3, a_4 = 0, a_5 = -10, a_6 = -3, a_7 = 3/5, a_8 = 0, a_9 = 120, k = 120, k_1 = 1, k_2 = 0, k_3 = 1/10, c_1 = -1/10, c_2 = 1/5$ in Case 1, and $a_1 = 2, a_2 = 0, a_3 = 0, a_4 = 0, a_5 = 0, a_6 = 5, a_7 = 40/9, a_8 = 0, a_9 = 10, k = 120, k_1 = 1/2, k_2 = 0, k_3 = 1/8, c_1 = -1, c_2 = -8/9$ in Case 2 to simulate the interaction.

As an example, Figs. 1 and 2 show, respectively, the interaction phenomena between a lump and a stripe with the parameters given above.

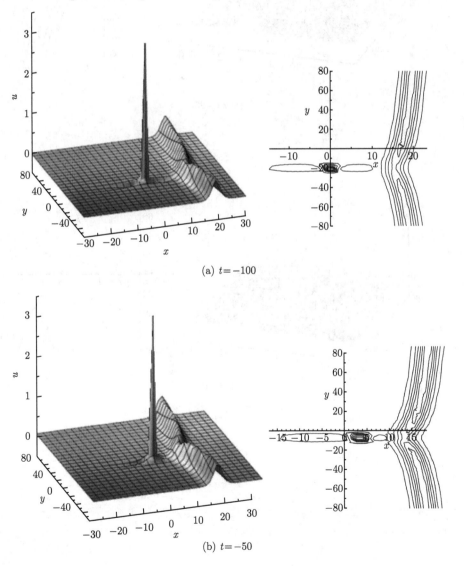

(a) $t=-100$

(b) $t=-50$

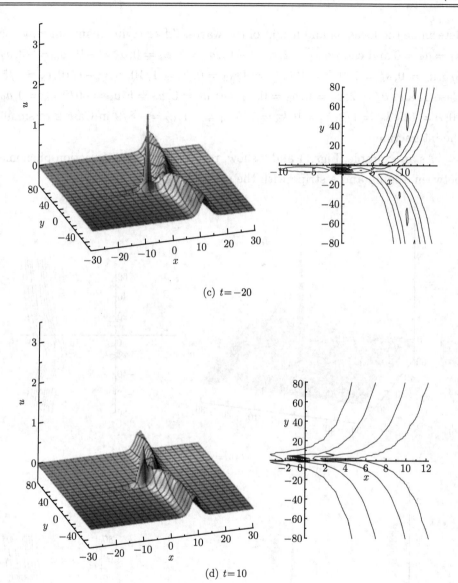

(c) $t=-20$

(d) $t=10$

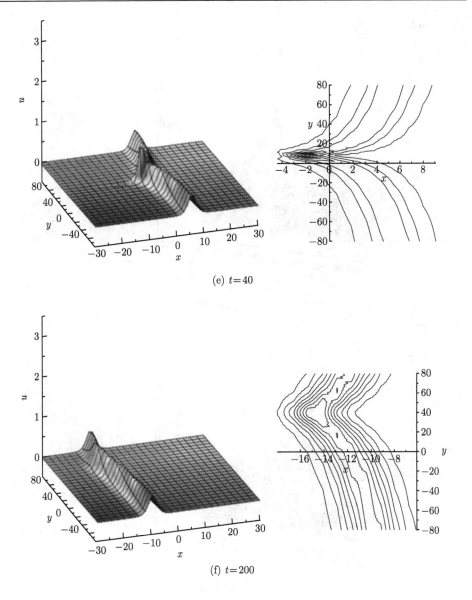

(e) $t=40$

(f) $t=200$

Fig. 1 Interaction behavior between a lump and a stripe in Case 1: 3d plots (left) and contour plots (right)

According to the expressions of the functions f, g, h and l, the asymptotic property of the lump and stripe waves can be analyzed. When $t \to -\infty$ and $k_3 > 0$, we have

$$\lim_{t \to -\infty} f = \lim_{t \to -\infty} (g^2 + h^2 + ke^l + a_9) = g^2 + h^2 + a_9,$$

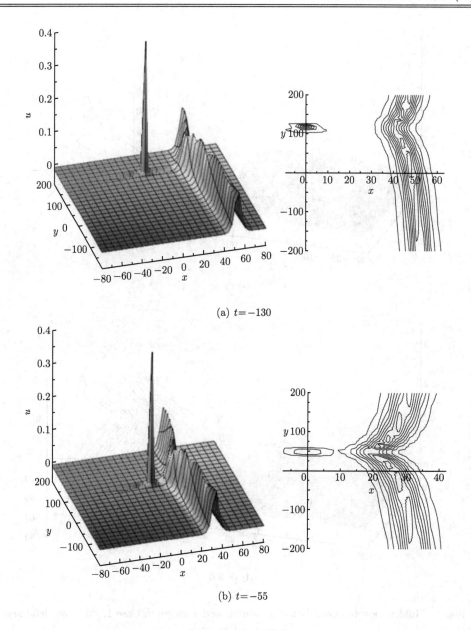

(a) $t=-130$

(b) $t=-55$

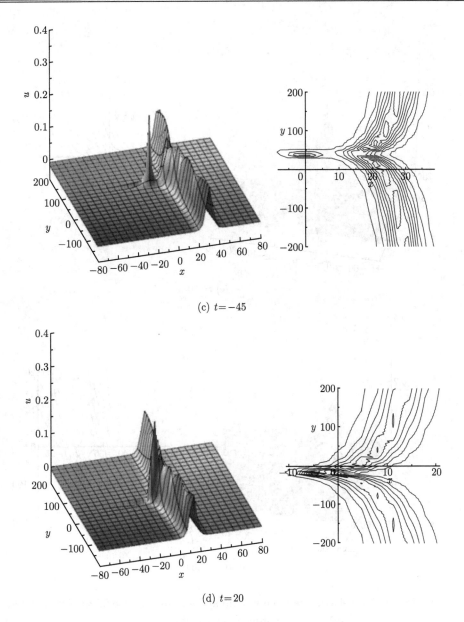

(c) $t=-45$

(d) $t=20$

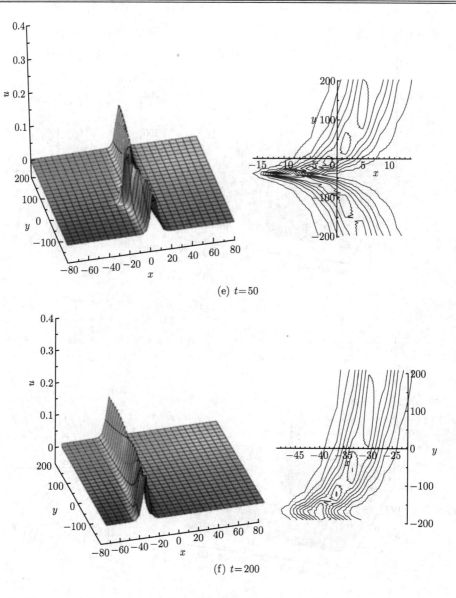

Fig. 2 Interaction behavior between a lump and a stripe in Case 2: 3d plots (left) and contour plots (right)

which means only lump wave exists and the stripe wave disappears. When $t \to -\infty$ and $k_3 = 0$, both lump and the stripe waves exist, but the lump wave plays the dominant role. When $t \to -\infty$ and $k_3 < 0$,

$$\lim_{t \to -\infty} ke^l = +\infty, \quad \lim_{t \to -\infty} g^2 = +\infty, \quad \lim_{t \to -\infty} h^2 = +\infty,$$

we need to compare g^2, h^2 and ke^l. It is proved that

$$\lim_{t \to \pm\infty} \frac{g^2}{h^2} = \frac{a_3^2}{a_7^2} = \text{constant}, \qquad (2.10)$$

so g^2 and h^2 have the same order, we just compare g^2 and ke^l. Hereby, both lump and the stripe waves exist, but the stripe wave plays the dominant role in terms of the following result

$$\lim_{t \to -\infty} \frac{g^2}{ke^l} = 0.$$

Similar analysis can be given correspondingly to the case of $t \to +\infty$. We conclude that when $t \to -\infty$ and $k_3 > 0$, as well as $t \to +\infty$ and $k_3 < 0$, only the lump wave exists, while when $t \to \pm\infty$ and $k_3 = 0$, both lump wave and stripe wave exist, but the lump wave plays the dominant role. When $t \to -\infty$ and $k_3 < 0$, as well as $t \to +\infty$ and $k_3 > 0$, both waves exist, but the lump is swallowed or drowned by the stripe.

It is also clear that

$$\lim_{t \to \pm\infty} u = \lim_{t \to \pm\infty} 2(\ln f)_{xx}$$

$$= \lim_{t \to \pm\infty} \left[\frac{2(2a_1^2 + 2a_5^2 + kk_1^2 e^l)}{g^2 + h^2 + ke^l + a_9} - \frac{2(2a_1 g + 2a_5 h + kk_1 e^l)^2}{(g^2 + h^2 + ke^l + a_9)^2} \right]$$

$$= 0,$$

where $k > 0$ is an arbitrary constant.

When choosing the parameters given in Case 1, the solution of u can be rewritten as

$$u = \frac{2(120e^{x+\frac{1}{10}t} + 208)}{(2x - 15y + 3t)^2 + \left(-10x - 3y + \frac{3}{5}t\right)^2 + 120e^{x+\frac{1}{10}t} + 120}$$

$$- \frac{2(120e^{x+\frac{1}{10}t} + 208x)^2}{\left((2x - 15y + 3t)^2 + \left(-10x - 3y + \frac{3}{5}t\right)^2 + 120e^{x+\frac{1}{10}t} + 120\right)^2},$$

which shows the stripe wave has the speed $v_s = -1/10$ along the x-axis and the speed along the y-axis is zero, while the lump wave has the speed $v_l = 1/5$

along the y-axis and the speed along the x-axis is zero. Because the stripe wave is exponentially localized in certain direction and its speed along the x-axis is faster than the lump wave, as the lump moves from the negative direction of y-axis to the positive direction, the stripe will finally catch up with the lump and interact with it. After the collision, the lump is swallowed or drowned by the stripe and the waves have a common speed.

As can be seen in both Figs. 1 and 2, when $t = -100$ in Fig. 1(a) and when $t = -130$ in Fig. 2(a), the wave consists of two separate parts: the lump wave and the stripe wave. Then the lump moves closely to the stripe and begins to interact with the stripe. The two waves collide over a period and the amplitudes, shapes and velocities of both waves change. When $t = 200$ in Figs. 1(f) and 2(f), the lump is swallowed or drowned by the stripe and the amplitude of the stripe turns higher than its original one, which presents the completely non-elastic interaction between the two different waves. This kind of interaction solution can be used in the fields of shallow water waves, plasma, nonlinear optics, Bose-Einstein condensates and so on[26, 31].

3 Interaction solutions of lump-soliton type

In this section, we will pay attention to the interaction solutions between lump and soliton to Eq. (1.4) by making a combinations of two positive quadratic functions and a hyperbolic cosine function. We suppose f is in the form of

$$f = g^2 + h^2 + \cosh(l) + a_9, \tag{3.1}$$

and three wave variables are defined by

$$g = a_1 x + a_2 y + a_3 t + a_4,$$
$$h = a_5 x + a_6 y + a_7 t + a_8,$$
$$l = k_1 x + k_2 y + k_3 t,$$

where a_i ($1 \leqslant i \leqslant 9$) and k_j ($1 \leqslant j \leqslant 3$) are real constants to be determined, $k > 0$ is a real parameter.

With symbolic computation, we obtain two cases of parameters:

Case 1

$$\left\{ a_1 = -\frac{a_5 a_6}{a_2}, a_2 = a_2, a_3 = -a_2 c_2, a_4 = a_4, a_5 = a_5, a_6 = a_6, a_7 = -a_6 c_2, \right.$$

$$a_8 = a_8, a_9 = a_9, k_1 = k_1, k_2 = 0, k_3 = -c_1 k_1^3 \Big\},$$

which needs to satisfy the conditions

$$a_2 \neq 0, \quad a_9 > 0, \quad a_5 \neq 0, \quad k_1 \neq 0. \tag{3.2}$$

Case 2

$$\Big\{ a_1 = a_1, a_2 = 0, a_3 = 0, a_4 = a_4, a_5 = 0, a_6 = a_6, a_7 = a_7,$$

$$a_8 = a_8, a_9 = a_9, k_1 = k_1, k_2 = 0, k_3 = -c_1 k_1^3 \Big\},$$

which needs to satisfy the conditions

$$a_1 a_6 \neq 0, \quad a_9 > 0, \quad k_1 \neq 0. \tag{3.3}$$

Substituting the two cases of parameters into the function f, we can get the interaction solutions to Eq. (1.4) through the transformation $u = 2[\ln f(x, y, t)]_{xx}$.

Figs.3 and 4 illustrate, respectively, the interaction phenomena between a lump and a two-soliton wave with the parameters $a_1 = -1/2, a_2 = 3, a_3 = 3, a_4 = 0, a_5 = 3/2, a_6 = 1, a_7 = 1, a_8 = 0, a_9 = 1, k_1 = 1, k_2 = 0, k_3 = 1, c_1 = -1, c_2 = -1$ in Case 1, and $a_1 = 3/2, a_2 = 0, a_3 = 0, a_4 = 0, a_5 = 0, a_6 = 2, a_7 = -2/3, a_8 = 0, a_9 = 1, k_1 = 4/5, k_2 = 0, k_3 = 64/125, c_1 = -1, c_2 = 1/3$ in Case 2.

For any fixed value of t, the extremum points of the lump can be obtained. The extremum point of the lump locates at

$$\left(x = \frac{a_2 a_7 t_0 - a_3 a_6 t_0 + a_2 a_8 - a_4 a_6}{a_1 a_6 - a_2 a_5}, y = -\frac{a_1 a_7 t_0 - a_3 a_5 t_0 + a_1 a_8 - a_4 a_5}{a_1 a_6 - a_2 a_5} \right), \tag{3.4}$$

where the maximum of the amplitude of the lump is attained as $4(a_1^2 + a_5^2)/a_9$.

When $t \to -\infty$ and $k_3 > 0$, we have

$$\lim_{t \to -\infty} \cosh(l) = \lim_{t \to -\infty} \frac{e^l + e^{-l}}{2} = \lim_{t \to -\infty} \frac{e^{-l}}{2} = +\infty,$$

$$\lim_{t\to -\infty} g^2 = +\infty, \quad \lim_{t\to -\infty} h^2 = +\infty.$$

We have proved that when $t \to \pm\infty$, g^2 and h^2 have the same order, and we need to compare g^2 and $\cosh(l)$ by virtue of

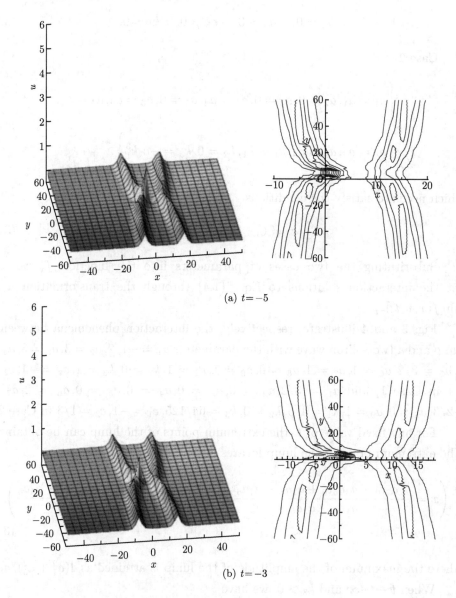

(a) $t=-5$

(b) $t=-3$

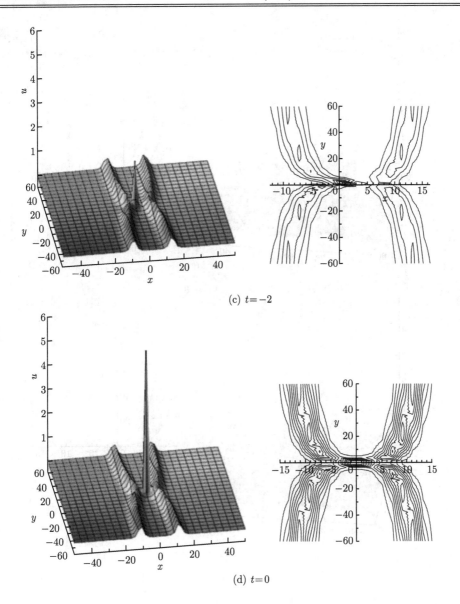

Fig. 3 Interaction behavior between a lump and a two-soliton wave in Case 1: 3d plots (left) and contour plots (right)

$$\lim_{t\to-\infty}\frac{g^2}{\cosh(l)}=2\lim_{t\to-\infty}\frac{g^2}{e^{-l}}=0, \quad (3.5)$$

which means both waves exist, but the soliton dominates the waves.

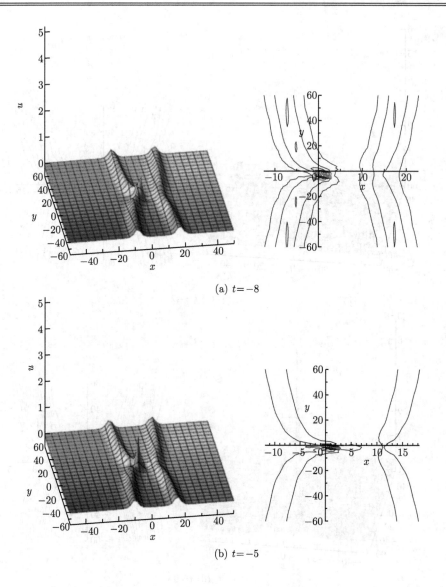

(a) $t=-8$

(b) $t=-5$

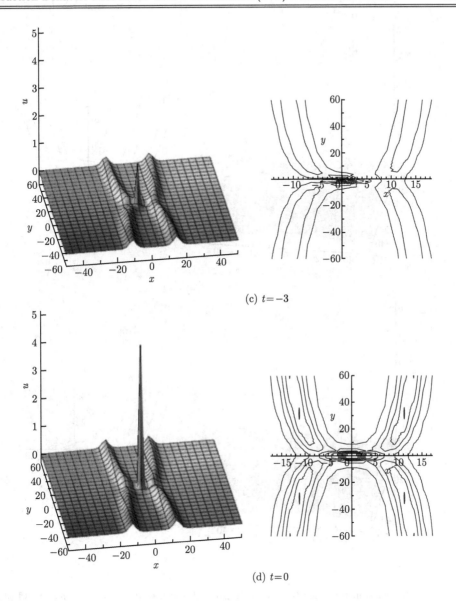

(c) $t=-3$

(d) $t=0$

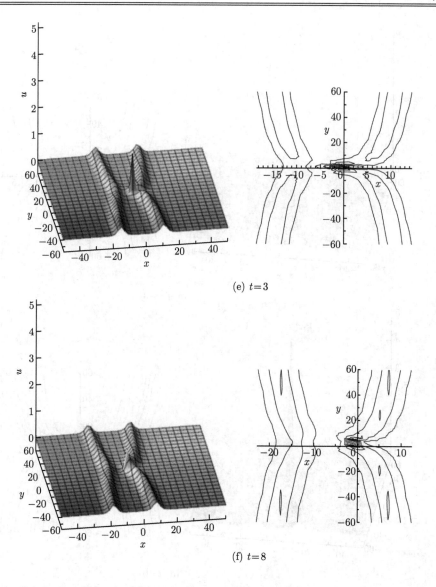

(e) $t=3$

(f) $t=8$

Fig. 4 Interaction behavior between a lump and a two-soliton wave in Case 2: 3d plots (left) and contour plots (right)

When $t \to -\infty$ and $k_3 = 0$, both waves exist, but the lump wave plays the dominant role.

When $t \to -\infty$ and $k_3 < 0$, the results

$$\lim_{t \to -\infty} \cosh(l) = \lim_{t \to -\infty} \frac{e^l}{2} = +\infty, \quad \lim_{t \to -\infty} g^2 = +\infty, \quad \lim_{t \to -\infty} h^2 = +\infty,$$

$$\lim_{t\to-\infty}\frac{g^2}{\cosh(l)}=2\lim_{t\to-\infty}\frac{g^2}{e^l}=0,$$

shows the soliton plays the dominant role.

Based on the similar analysis when $t\to+\infty$, we conclude that when $t\to\pm\infty$ and $k_3\neq 0$, both the lump and soliton waves exist, but the soliton plays the dominant role, while when $k_3=0$, the lump dominates the waves.

It is also clear that

$$\lim_{t\to\pm\infty}u=\lim_{t\to\pm\infty}2(\ln f)_{xx}$$
$$=\lim_{t\to\pm\infty}\left[\frac{2(2a_1^2+2a_5^2+k_1^2\cosh(l))(g^2+h^2+\cosh(l)+a_9)}{(g^2+h^2+\cosh(l)+a_9)^2}\right.$$
$$\left.-\frac{2(2a_1g+2a_5h+k_1\sinh(l))^2}{(g^2+h^2+\cosh(l)+a_9)^2}\right]$$
$$=0. \tag{3.6}$$

When choosing the parameters given in Case 1, the solution of u can be rewritten as

$$u=\frac{2(\cosh(x+t)+5)}{\left(-\frac{1}{2}x+3y+3t\right)^2+\left(\frac{3}{2}x+y+t\right)^2+\cosh(x+t)+1}$$
$$-\frac{2(\sinh(x+t)+5x)^2}{\left(\left(-\frac{1}{2}x+3y+3t\right)^2+\left(\frac{3}{2}x+y+t\right)^2+\cosh(x+t)+1\right)^2}.$$

The characteristic lines of the soliton wave are

$$l_1:x+t+b_1=0,$$
$$l_2:x+t+b_2=0, \tag{3.7}$$

where $b_1\neq b_2$, b_1 and b_2 are two constants. It is easy to find that two branches of the soliton wave are parallel in (x,t)-plane and they coexist with a common speed.

When $t\to-\infty$, the lump wave nearly disappears and the soliton dominates the waves, the two branches of which are of the common speed $v_s=-1$ along the x-axis and the speed along the y-axis is zero. As time goes on, the lump wave gradually appears from the left branch of the soliton with the speed $v_l = -1$ along the y-axis and the speed along the x-axis is zero. With the soliton waves moving from the positive direction of the x-axis to the negative direction, the lump leaves gradually from the left branch of the soliton to the right branch. Then the lump will interact with the soliton and go together with the right branch. When $t \to +\infty$, the lump wave nearly disappears again and the soliton waves play the dominant role.

As can be seen in both Figs. 3 and 4, when $t = -5$ in Fig. 3(a) and $t = -8$ in Fig. 4(a), the wave consists of two parts, including a lump wave and a two-soliton wave. The lump appears from one branch of the two-soliton wave and begins to move to the other one. The amplitude of the lump changes with the variable t, and especially when $t = 0$, the lump is located in the middle of the two branches of the soliton waves. Then the lump continues moving, until it attaches to the other branch of the two-soliton wave. The process of interaction changes the amplitudes, shapes and velocities of both waves. This type of interaction solutions provides a method to forecast the appearance of rogue waves, such as financial rogue wave, optical rogue wave and plasma rogue wave, through analyzing the relations between the lump wave part and the soliton wave part[29].

4 Concluding remarks

Besides of finding lump solutions to the (2+1)-dimensional nonlinear evolution equations, we have focused on the interaction behavior associated with a generalized (2+1)-dimensional Hirota bilinear equation. With symbolic computation, two types of interaction solutions including lump-kink and lump-soliton ones have been derived to Eq. (1.4), and the relations among all the parameters and the coefficients have been obtained. We have analyzed the asymptotic properties of the interaction solutions and shown the interaction process via four sets of figures under some selections of parameters. The phenomena are useful in understanding the propagation of nonlinear waves.

We have obtained four cases of interaction solutions between lump and stripe by considering mixing two positive quadratic functions with an exponential function. The interaction process shows that the lump moves closely to the stripe and begins to interact with the stripe. They collide over a period and the amplitudes, shapes and velocities of both waves change. When $t \to +\infty$, the lump is swallowed or drowned by the stripe, which presents the completely non-elastic interaction between the two different types of waves. This kind of interaction solution can be used in the fields of shallow water so as to reveal the propagation and interaction between stripe wave and lump wave.

We have also obtained two cases of interaction solutions between lump and two-soliton by considering mixing two positive quadratic functions with a hyperbolic cosine function. It is noted that the cases we discussed are under the situation $k_2 = 0$. The interaction process is as follows: The lump moves from one branch to the other branch of the two-soliton. When $t = 0$, the lump is located in between of the two branches of the soliton wave. Both the lump and the soliton change their amplitudes, shapes and velocities when they interact. The interaction solution of this type is valuable in forecasting the appearance of rogue waves by analyzing the relations between soliton wave and lump wave.

Acknowledgements This work is supported by the Fundamental Research Funds for the Central Universities of China under grant number 2018RC031.

References

[1] Tang Y N, Tao S Q, Guan Q. Lump solitons and the interaction phenomena of them for two classes of nonlinear evolution equations [J]. Comput. Math. Appl., 2016, 72(9): 2334–2342.
[2] Aspe H, Depassier M C. Evolution equation of surface waves in a convecting fluid [J]. Phys. Rev. A, 1990, 41(6): 3125-3128.
[3] Lü X, Wang J P, Lin F H, Zhou X W. Lump dynamics of a generalized two-dimensional Boussinesq equation in shallow water [J]. Nonlinear Dyn., 2018, 91(2): 1249–1259.
[4] Sun B N, Wazwaz A M. Interaction of lumps and dark solitons in the Mel'nikov equation [J]. Nonlinear Dyn., 2018, 92(4): 2049–2059.
[5] Zhao Z L, Chen Y, Han B. Lump soliton, mixed lump stripe and periodic lump solutions of a (2+1)-dimensional asymmetrical Nizhnik-Novikov-Veselov equation [J]. Mod. Phys. Lett. B, 2017,31(14): 1750157.
[6] Moslem W M. Langmuir rogue waves in electron-positron plasmas [J]. Phys. Plasmas, 2011, 18(5): 032301.
[7] Bailung H, Sharma S K, Nakamura Y. Observation of peregrine solitons in a multicomponent plasma with negative ions [J]. Phys. Rev. Lett., 2011, 107(25): 255005.
[8] Montina A, Bortolozzo U, Residori S, Arecchi F T. Non-Gaussian statistics and extreme waves in a nonlinear optical cavity [J]. Phys. Rev. Lett., 2009, 103(17): 173901.

[9] Bludov Y V, Konotop V V, Akhmediev N. Vector rogue waves in binary mixtures of Bose-Einstein condensates [J]. Eur. Phys. J. Spec. Top., 2010, 185(1): 169–180.

[10] Yin Y H, Ma W X, Liu J G, Lü X. Diversity of exact solutions to a (3+1)-dimensional nonlinear evolution equation and its reduction [J]. Comput. Math. Appl., 2018, 76(6): 1275–1283.

[11] Gao L N, Zi Y Y, Yin Y H, Ma W X, Lü X. Bäcklund transformation, multiple wave solutions and lump solutions to a (3 + 1)-dimensional nonlinear evolution equation [J]. Nonlinear Dyn., 2017, 89(3): 2233–2240.

[12] Hirota R. The Direct Method in Soliton Theory [M]. Cambridge University Press, 2004.

[13] Dehghan M, Heris J M, Saadatmandi A. Application of the Exp-function method for solving a partial differential equation arising in biology and population genetics [J]. Int. J. Numer. Methods Heat Fluid Flow, 2011, 21(6): 736–753.

[14] Jalil M. On the complex structures of the Biswas-Milovic equation for power, parabolic and dual parabolic law nonlinearities [J]. Eur. Phys. J. Pluss, 2015, 130: 1–20.

[15] He J H. Application of homotopy perturbation method to nonlinear wave equations [J]. Chaos Solitons Fractals, 2005, 26(3): 695–700.

[16] Dehghan M, Heris J M, Saadatmandi A. Application of semi-analytic methods for the Fitzhugh-Nagumo equation, which models the transmission of nerve impulses [J]. Math. Methods Appl. Sci., 2010, 33(11): 1384–1398.

[17] Seyedi S H, Saray B N, Nobari M R H. Using interpolation scaling functions based on Galerkin method for solving non-newtonian fluid flow between two vertical flat plates [J]. Appl. Math. Comput., 2015, 269: 488–496.

[18] Ali R. On the multiscale simulation of squeezing nanofluid flow by a high precision scheme [J]. Powder Technol., 2018, 340: 264–273.

[19] Lou S Y, Tang X Y. Nonlinear Mathematical Physics Method [M]. Academic Press, 2006.

[20] Ma W X. Lump solutions to the Kadomtsev-Petviashvili equation [J]. Phys. Lett. A, 2015, 379(36): 1975–1978.

[21] Ma H C, Deng A P. Lump solution of (2+1)-dimensional Boussinesq equation [J]. Commun. Theor. Phys., 2016, 65(5): 546–552.

[22] Manakov S V, Zakharov V E, Bordag L A, Its A R, Matveev V B. Two-dimensional solitons of the Kadomtsev-Petviashvili equation and their interaction [J]. Phys. Lett. A, 1977, 63(3): 205–206.

[23] Lu Z, Tian E M, Grimshaw R. Interaction of two lump solitons described by the Kadomtsev-Petviashvili i equation [J]. Wave Motion, 2004, 40(2): 123–135.

[24] Gilson C R, Nimmo J J C. Lump solutions of the BKP equation [J]. Phys. Lett. A, 1990, 147(8-9): 472–476.

[25] Lü Z, Chen Y. Construction of rogue wave and lump solutions for nonlinear evolution equations [J]. Eur. Phys. J. B, 2015, 88(7): 1–5.

[26] Peng W Q, Tian S F, Zou L, Zhang T T. Characteristics of the solitary waves and lump waves with interaction phenomena in a (2+1)-dimensional generalized Caudrey-Dodd-Gibbon-Kotera-Sawada equation [J]. Nonlinear Dyn., 2018, 93(4):1841–1851.

[27] Foroutan M, Manafian J, Ranjbaran A. Lump solution and its interaction to (3+1)-d potential-YTSF equation [J]. Nonlinear Dyn., 2018, 92(4): 2077–2092.

[28] Jia M, Lou S Y. A novel type of rogue waves with predictability in nonlinear physics [J]. Nonlinear Sci., 2017: 171006604.

[29] Zhang X E, Chen Y, Tang X Y. Rogue wave and a pair of resonance stripe solitons to KP equation [J]. Comput. Math. Appl., 2018, 76(8): 1938–1949.

[30] Lü X, Ma W X. Study of lump dynamics based on a dimensionally reduced Hirota bilinear equation [J]. Nonlinear Dyn., 2016, 85(2): 1217–1222.

[31] Kofane T C, Fokou M, Mohamadou A, Yomba E. Lump solutions and interaction phenomenon to the third-order nonlinear evolution equation [J]. Eur. Phys. J. Plus, 2017, 132(11): 465–472.

The Riemann–Hilbert Problem to Coupled Nonlinear Schrödinger Equation: Long-Time Dynamics on the Half-Line*

Guo Boling (郭柏灵), Liu Nan(刘男)

Abstract We derive the long-time asymptotics for the solution of initial-boundary value problem of coupled nonlinear Schrödinger equation whose Lax pair involves 3×3 matrix in present paper. Based on a nonlinear steepest descent analysis of an associated 3×3 matrix Riemann–Hilbert problem, we can give the precise asymptotic formulas for the solution of the coupled nonlinear Schrödinger equation on the half-line.

Keywords coupled nonlinear Schrödinger equation; Riemann–Hilbert problem; initial-boundary value problem; long-time asymptotics

1 Introduction

The well-known nonlinear steepest descent method first introduced by Deift and Zhou in [14] provides a powerful technique for determining asymptotics of solutions of nonlinear integrable evolution equations. This approach since then has been successfully applied in analyzing the long-time asymptotics of initial-value problems (IVPs) for a number of nonlinear integrable partial differential equations (PDEs) associated with 2×2 matrix spectral problems including the modified Korteweg-de Vries (mKdV) equation [14], the defocusing nonlinear Schrödinger (NLS) equation [15], the KdV equation [19], the derivative NLS equation [25], the Fokas–Lenells equation [28], the short pulse equation [10] and the Kundu–Eckhaus equation [27]. Moreover, by combining the ideas of [14] with the so-called "g-function mechanism" [13], it is also possible to study asymptotics

---
* J. Nonlinear Math. Phys., 2019, 26:3 483–508. DOI:10.1080/14029251.2019.1613055.

of solutions of the IVPs with shock-type oscillating initial data [11], nondecaying step-like initial data [6, 22], and nonzero boundary conditions at infinity [4] for various integrable equations. There also exists some meaningful papers [8, 9, 17] about the study of long-time asymptotics for the IVPs of integrable nonlinear evolution equations associated with 3×3 matrix spectral problems. For the large-time asymptotic analysis of the initial-boundary value problems (IBVPs) of integrable nonlinear PDEs, Lenells et al. derived some interesting asymptotic formulas for the solutions of mKdV equation [23] and derivative NLS equation [3] by using the steepest descent method. Furthermore, the long-time asymptotics for the focusing NLS equation with t-periodic boundary condition on the half-line is analyzed in [5]. We also have done some work about determining the long-time asymptotics for integrable equations on the half-line, see [20, 21]. However, there is only a little of literature [7] to consider the asymptotic behaviors for integrable nonlinear PDEs with Lax pairs involving 3×3 matrices on the half-line. Thus, it is necessary and important to consider the large-time asymptotic behaviors for the IBVPs of integrable equations with 3×3 Lax pairs on the half-line.

In particular, the purpose of this paper is aim to consider the long-time asymptotics for the IBVP of the coupled nonlinear Schrödinger (CNLS) equation

$$\begin{cases} iu_t + u_{xx} + 2(|u|^2 + |v|^2)u = 0, \\ iv_t + v_{xx} + 2(|u|^2 + |v|^2)v = 0, \end{cases} \tag{1.1}$$

posed in the quarter-plane domain

$$\Omega = \{0 \leqslant x < \infty,\ 0 \leqslant t < \infty\}.$$

We will denote the initial data, Dirichlet and Neumann boundary values of (1.1) as follows:

$$\begin{aligned} u(x,0) &= u_0(x), \quad v(x,0) = v_0(x), \quad x \geqslant 0, \\ u(0,t) &= g_0(t), \quad u_x(0,t) = g_1(t), \quad v(0,t) = h_0(t), \quad v_x(0,t) = h_1(t), \quad t \geqslant 0. \end{aligned} \tag{1.2}$$

We also suppose that $\{u_0(x), v_0(x)\}$ and $\{g_j(t), h_j(t)\}_0^1$ belong to the Schwartz class $S(\mathbf{R}_+)$. Equation (1.1) was also called Manakov model, which can be used to describe the propagation of an optical pulse in a birefringent optical fiber [26]. Subsequently, this system also arises in the context of multi-component Bose-Einstein condensates [12]. Due to its physical interest, equation (1.1) has been widely studied. It is noted that the IBVP for (1.1) on the half-line has been investigated via the Fokas method [16], where it was shown that the solution $\{u(x,t), v(x,t)\}$ can be expressed in terms of the unique solution of a 3×3 matrix Riemann–Hilbert (RH) problem formulated in the complex k-plane (see [18]). Meanwhile, the leading-order long-time asymptotics of the Cauchy problem of equation (1.1) was obtained in [17].

Our goal here is to derive the long-time asymptotics of the solution of (1.1) on the half-line by performing a nonlinear steepest descent analysis of the associated RH problem. Compared with other integrable equations, the asymptotic analysis of (1.1) presents some distinctive features: (1) Since the RH problem associated with (1.1) involves 3×3 jump matrix $J(x,t,k)$, we first introduce two 2×1 vector-valued spectral functions $r_1(k)$, $h(k)$ and a 1×2 vector-valued spectral function $r(k)$, and then rewrite our main RH problem as a 2×2 block one. This procedure is more convenient for the following long-time asymptotic analysis. (2) As we all know, the important step of the steepest descent method is to split the jump matrix $J(x,t,k)$ into an appropriate upper/lower triangular form. This immediately leads to construct a $\delta(k)$ function to remove the middle matrix term, however, the function δ satisfies a 2×2 matrix RH problem in our present problem. The unsolvability of the 2×2 matrix function $\delta(k)$ is a challenge when we perform the scaling transformation to reduce the RH problem to a model RH problem. Fortunately, we can follow the idea introduced in [18] to use the available function $\det \delta(k)$ which can be explicitly solved by the Plemelj formula to approximate the function $\delta(k)$ by error control. (3) The relevant RH problem for the Cauchy problem (1.1) considered in [17] only has a jump across $\mathbf{R}$, whereas the RH problem for the IBVP also has a jump across $i\mathbf{R}$, and the jump across this line involves the spectral function $h(k)$. Moreover, during the asymptotic analysis, one should find an analytic approximation $h_a(t,k)$ of $h(k)$. (4) Recalling the meaningful work about analyzing the long-time asymptotics for the Degasperis–Procesi equation on the half-line, the analysis presented in [7] shows that the structure of the jump matrix, which is 3×3, is essentially 2×2 (under an appropriate change of basis), whereas the analysis given in our present paper is more general.

The main result of this paper is stated as the following theorem.

Theorem 1.1 *Assume the assumption 2.1 be valid. Then, for any positive constant N, as $t \to \infty$, the solution $\begin{pmatrix} u(x,t) & v(x,t) \end{pmatrix}$ of the IBVP for the CNLS* equation (1.1) on the half-line satisfies the following asymptotic formula

$$\begin{pmatrix} u(x,t) & v(x,t) \end{pmatrix} = \frac{\begin{pmatrix} u_a(x,t) & v_a(x,t) \end{pmatrix}}{\sqrt{t}} + O\left(\frac{\ln t}{t}\right), \quad t \to \infty, \ 0 \leqslant x \leqslant Nt, \tag{1.3}$$

where the error term is uniform with respect to x in the given range, and the leading-order coefficient $\begin{pmatrix} u_a(x,t) & v_a(x,t) \end{pmatrix}$ is defined by

$$\begin{pmatrix} u_a(x,t) & v_a(x,t) \end{pmatrix} = \frac{e^{-\frac{\pi\nu}{2}}\nu\Gamma(-i\nu)r(k_0)}{2\sqrt{\pi}} e^{i\alpha(\xi,t)}, \tag{1.4}$$

where the 1×2 vector-valued spectral function $r(k)$ is defined by (3.3), which is determined by all the initial and boundary values,

$$\xi = \frac{x}{t}, \quad \nu = \frac{1}{2\pi}\ln(1 + r(k_0)r^\dagger(k_0)), \quad k_0 = -\frac{\xi}{4},$$

and

$$\alpha(\xi,t) = -\frac{\pi}{4} + \nu\ln(8t) + 4k_0^2 t + \frac{1}{\pi}\int_{-\infty}^{k_0} \ln(k_0-s)d\ln(1+r(s)r^\dagger(s)).$$

Remark 1.1 *The asymptotic formula for the half-line problem obtained in Theorem 1.1 has the exact same functional form for the pure initial value problem. The only difference is that the definition of the spectral function $r(k)$ for the half-line problem, which enters the asymptotic formula, involves not only the initial data but also the boundary values. In other words, the only effect of the boundary is to modify $r(k)$. We can understand this as follows: since x grows faster than t in the given region, the distance to the boundary eventually gets so big that what happens at the boundary has a small effect on the solution (recall that we assume that the boundary values decay as $t \to \infty$). Therefore, the boundary values and the initial data play similar roles and it is natural that they enter the asymptotic formula in similar ways.*

The outline of the paper is following. In Section 2, we recall how the solution of CNLS equation (1.1) on the half-line can be expressed in terms of the solution of a 3×3 matrix RH problem. In section 3, we present the detailed derivation of the long-time asymptotics for the solution of CNLS equation, that is, we prove Theorem 1.1.

2 Preliminaries

In this section, we give a short review of the RH problem for (1.1) on the half-line, see [18] for further details. The Lax pair of equation (1.1) is

$$\begin{cases} \mu_x(x,t,k) - ik[\sigma, \mu(x,t,k)] = U(x,t)\mu(x,t,k), \\ \mu_t(x,t,k) - 2ik^2[\sigma, \mu(x,t,k)] = V(x,t,k)\mu(x,t,k), \end{cases} \quad (2.1)$$

where $\mu(x,t,k)$ is a 3×3 matrix-valued eigenfunction, $k \in \mathbf{C}$ is the spectral parameter, and

$$\sigma = \begin{pmatrix} -1 & 0 & 0 \\ 0 & 1 & 0 \\ 0 & 0 & 1 \end{pmatrix}, \quad U(x,t) = i \begin{pmatrix} 0 & u(x,t) & v(x,t) \\ \bar{u}(x,t) & 0 & 0 \\ \bar{v}(x,t) & 0 & 0 \end{pmatrix}, \quad (2.2)$$

$$V(x,t,k) = 2kU(x,t) + iU_x(x,t)\sigma + iU^2(x,t)\sigma. \quad (2.3)$$

Let $\{\gamma_j\}_1^3$ denote contours in the (x,t)-plane connecting (x_j, t_j) with (x,t), and $(x_1, t_1) = (0, \infty)$, $(x_2, t_2) = (0, 0)$, $(x_3, t_3) = (\infty, t)$. The contours can be chosen to consist of straight line segments parallel to the x- or t-axis. We define three eigenfunctions $\{\mu_j\}_1^3$ of the Lax pair (2.1) by the solutions of the following integral equations

$$\mu_j(x,t,k) = I + \int_{\gamma_j} e^{i(kx+2k^2t)\hat{\sigma}} w_j(x',t',k), \quad j = 1,2,3, \quad (2.4)$$

where the closed one-form $w_j(x,t,k)$ is defined by

$$w_j(x,t,k) = e^{-i(kx+2k^2t)\hat{\sigma}}[U(x,t)dx + V(x,t,k)dt]\mu_j(x,t,k), \quad (2.5)$$

and $\hat{\sigma}$ denote the operators which act on a 3×3 matrix X by $\hat{\sigma}X = [\sigma, X]$, then $e^{\hat{\sigma}}X = e^{\sigma}Xe^{-\sigma}$. It then can be shown that the functions $\{\mu_j\}_1^3$ are bounded and analytical for $k \in \mathbf{C}$ while k belongs to

$$\mu_1 : (D_2, D_3, D_3), \quad \mu_2 : (D_1, D_4, D_4), \quad \mu_3 : (\mathbf{C}_-, \mathbf{C}_+, \mathbf{C}_+), \quad (2.6)$$

where D_n denotes nth quadrant, $1 \leqslant n \leqslant 4$, $\mathbf{C}_+$ and $\mathbf{C}_-$ denote the upper and lower half complex k-plane, respectively.

We define the matrix-valued spectral functions $s(k)$ and $S(k)$ by the relations

$$\mu_3(x,t,k) = \mu_2(x,t,k)e^{i(kx+2k^2t)\hat{\sigma}}s(k),$$
$$\mu_1(x,t,k) = \mu_2(x,t,k)e^{i(kx+2k^2t)\hat{\sigma}}S(k). \tag{2.7}$$

Evaluation of (2.7) at $(x,t) = (0,0)$ gives the following expressions

$$s(k) = \mu_3(0,0,k), \quad S(k) = \mu_1(0,0,k). \tag{2.8}$$

Hence, the functions $s(k)$ and $S(k)$ can be obtained respectively from the evaluations at $x = 0$ and at $t = 0$ of the functions $\mu_3(x,0,k)$ and $\mu_1(0,t,k)$.

On the other hand, we can deduce from the properties of μ_j that $s(k)$ and $S(k)$ have the following properties:

(i) $s(k)$ is bounded and analytic for $k \in (D_3 \cup D_4, D_1 \cup D_2, D_1 \cup D_2)$, $S(k)$ is bounded and analytic for $k \in (D_2 \cup D_4, D_1 \cup D_3, D_1 \cup D_3)$;
(ii) $\det s(k) = 1$ for $k \in \mathbf{R}$, $\det S(k) = 1$ for $k \in \mathbf{R} \cup i\mathbf{R}$;
(iii) $s(k) = I + O(k^{-1})$ and $S(k) = I + O(k^{-1})$ uniformly as $k \to \infty$;
(iv)
$$\bar{s}(k) = (s^{-1})^{\mathrm{T}}(k), \quad \bar{S}(k) = (S^{-1})^{\mathrm{T}}(k), \tag{2.9}$$

where $\bar{s}(k) = \overline{s(\bar{k})}$ and $\bar{S}(k) = \overline{S(\bar{k})}$ denote the Schwartz conjugates.

The initial and boundary values of a solution of the CNLS equation (1.1) are not independent. It turns out that the spectral functions $s(k)$ and $S(k)$ must satisfy a surprisingly simple relation

$$S^{-1}(k)s(k) = 0, \quad k \in (\bar{D}_4, \bar{D}_1, \bar{D}_1), \tag{2.10}$$

which called the global relation.

For each $n = 1, \cdots, 4$, we define solution $M_n(x,t,k)$ of (2.1) by the solution of the following Fredholm integral equation:

$$(M_n)_{jl}(x,t,k) = \delta_{jl} + \int_{\gamma_{jl}^n} \left(e^{i(kx+2k^2t)\hat{\sigma}}w_n(x',t',k)\right)_{jl}, \quad k \in D_n, \ j,l = 1,2,3, \tag{2.11}$$

where w_n is given by (2.5), and the contours γ_{jl}^n, $n = 1,2,3,4$, $j,l = 1,2,3$ are defined by

$$\gamma_{jl}^n = \begin{cases} \gamma_1, & \text{if } \mathrm{Re}y_j < \mathrm{Re}y_l \text{ and } \mathrm{Re}z_j \geqslant \mathrm{Re}z_l, \\ \gamma_2, & \text{if } \mathrm{Re}y_j < \mathrm{Re}y_l \text{ and } \mathrm{Re}z_j < \mathrm{Re}z_l, \\ \gamma_3, & \text{if } \mathrm{Re}y_j \geqslant \mathrm{Re}y_l, \end{cases} \tag{2.12}$$

and we denote $ik\sigma = \mathrm{diag}(y_1, y_2, y_3)$, $2ik^2\sigma = \mathrm{diag}(z_1, z_2, z_3)$. For each $n = 1, \cdots, 4$, we define spectral functions $S_n(k)$ by

$$M_n(x,t,k) = \mu_2(x,t,k) e^{i(kx+2k^2t)\hat{\sigma}} S_n(k), \quad k \in D_n. \tag{2.13}$$

According to (2.11), the $\{S_n(k)\}_1^4$ can be computed from initial and boundary values alone as well as the spectral functions $s(k)$ and $S(k)$ (the calculation results can see [18]).

Then, equation (2.13) can be rewritten in the form of a 3×3 RH problem as follows:

$$M_+(x,t,k) = M_-(x,t,k) J(x,t,k), \quad k \in \mathbf{R} \cup i\mathbf{R}, \tag{2.14}$$

where the matrices $M_+(x,t,k)$, $M_-(x,t,k)$, and $J(x,t,k)$ are defined by

$$M(x,t,k) = \begin{cases} M_+(x,t,k), & k \in D_1 \cup D_3, \\ M_-(x,t,k), & k \in D_2 \cup D_4, \end{cases}$$

and

$$J_1 = \begin{pmatrix} 1 & 0 & 0 \\ -\dfrac{s_{33}S_{21} - s_{23}S_{31}}{\bar{s}_{11}\bar{W}_{11}} e^{2i\theta} & 1 & 0 \\ -\dfrac{s_{22}S_{31} - s_{32}S_{21}}{\bar{s}_{11}\bar{W}_{11}} e^{2i\theta} & 0 & 1 \end{pmatrix},$$

$$J_2 = J_1 J_4^{-1} J_3,$$

$$J_3 = \begin{pmatrix} 1 & -\dfrac{\bar{s}_{33}\bar{S}_{21} - \bar{s}_{23}\bar{S}_{31}}{s_{11}W_{11}} e^{-2i\theta} & -\dfrac{\bar{s}_{22}\bar{S}_{31} - \bar{s}_{32}\bar{S}_{21}}{s_{11}W_{11}} e^{-2i\theta} \\ 0 & 1 & 0 \\ 0 & 0 & 1 \end{pmatrix}, \tag{2.15}$$

$$J_4 = \begin{pmatrix} 1 + \dfrac{|s_{12}|^2}{|s_{11}|^2} + \dfrac{|s_{13}|^2}{|s_{11}|^2} & \dfrac{s_{12}}{s_{11}} e^{-2i\theta} & \dfrac{s_{13}}{s_{11}} e^{-2i\theta} \\ \dfrac{\bar{s}_{12}}{\bar{s}_{11}} e^{2i\theta} & 1 & 0 \\ \dfrac{\bar{s}_{13}}{\bar{s}_{11}} e^{2i\theta} & 0 & 1 \end{pmatrix},$$

$$W_{11} = \bar{S}_{11}s_{11} + \bar{S}_{21}s_{21} + \bar{S}_{31}s_{31}, \quad \theta(k) = kx + 2k^2 t. \tag{2.16}$$

The contour for this RH problem is depicted in Fig. 1.

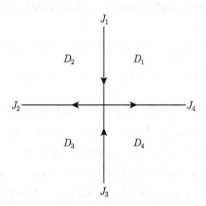

Fig. 1   The contour for the RH problem

In what follows, we will make the following simple assumptions.

Assumption 2.1   *We assume that the following conditions hold*:
- *The initial and boundary values lie in the Schwartz class.*
- *The spectral functions $s(k), S(k)$ defined in (2.7) satisfy the global relation (2.10).*
- *$s_{11}(k)$ and $W_{11}(k)$ have no zeros in $\bar{D}_3 \cup \bar{D}_4$ and $\bar{D}_3$, respectively.*
- *All the initial and boundary values are compatible with equation (1.1) to all orders at $x = t = 0$.*

We have the following representation theorem (the proof can be found in [18]).

Theorem 2.1   *Let $u_0(x), g_0(t), g_1(t), v_0(x), h_0(t), h_1(t)$ be functions in the Schwartz class $\mathcal{S}([0,\infty))$, and the assumption 2.1 is satisfied. Then the RH problem (2.14) with the jump matrices given by (2.15) and the following asymptotics*:

$$M(x,t,k) = I + O\left(\frac{1}{k}\right), \quad k \to \infty \tag{2.17}$$

admits a unique solution $M(x,t,k)$ for each $(x,t) \in \Omega$. Define $\{u(x,t), v(x,t)\}$ in terms of $M(x,t,k)$ by

$$\begin{aligned} u(x,t) &= 2\lim_{k\to\infty}(kM(x,t,k))_{12}, \\ v(x,t) &= 2\lim_{k\to\infty}(kM(x,t,k))_{13}. \end{aligned} \tag{2.18}$$

Then $\{u(x,t), v(x,t)\}$ satisfies the CNLS equation (1.1). Furthermore, $\{u(x,t), v(x,t)\}$ satisfies the initial and boundary value conditions

$$\begin{aligned} u(x,0) = u_0(x), \quad u(0,t) = g_0(t), \quad u_x(0,t) = g_1(t), \\ v(x,0) = v_0(x), \quad v(0,t) = h_0(t), \quad v_x(0,t) = h_1(t). \end{aligned} \quad (2.19)$$

Remark 2.1 *In the case when $s_{11}(k)$ and $W_{11}(k)$ have no zeros, the unique solvability of the above RH problem (2.14) is a consequence of the 'vanishing' lemma (the proof can be found in the appendix of* [18]).

Remark 2.2 *For a well-posed problem, only a subset of the initial and boundary values can be independently prescribed. However, all boundary values are needed for the definition of $S(k)$, and hence for the formulation of the main RH problem as well as the spectral function $r(k)$ defined in (3.3). In general, the computation of the unknown boundary values, namely, the construction of the generalized Dirichlet-to-Neumann map, involves the solution of a nonlinear Volterra integral equation. We do not consider this part in present paper since the detailed analysis has been studied in the paper* [18], *(see Sections 5 and 6 in* [18]). *Our main concern in present paper is the derivation of the long-time asymptotics for the solution of the IBVP of the CNLS equation (1.1).*

3 Long-time asymptotic analysis

Before we proceed to the following analysis, we will follow the ideas used in [2,17] to rewrite our main 3×3 matrix RH problem as a 2×2 block one. This procedure is more convenient for the following long-time asymptotic analysis. More precisely, we rewrite a 3×3 matrix A as a block form

$$A = \begin{pmatrix} A_{11} & A_{12} \\ A_{21} & A_{22} \end{pmatrix},$$

where A_{11} is scalar. Define the vector-valued spectral functions $r_1(k)$ and $h(k)$ by

$$r_1(k) = \begin{pmatrix} \frac{\bar{s}_{12}}{\bar{s}_{11}} \\ \frac{\bar{s}_{13}}{\bar{s}_{11}} \end{pmatrix}, \quad k \in \mathbf{R}, \quad (3.1)$$

$$h(k) = \begin{pmatrix} \dfrac{s_{33}S_{21} - s_{23}S_{31}}{\bar{s}_{11}\bar{W}_{11}} \\ \dfrac{s_{22}S_{31} - s_{32}S_{21}}{\bar{s}_{11}\bar{W}_{11}} \end{pmatrix}, \quad k \in \bar{D}_2, \qquad (3.2)$$

and let $r(k)$ denote the sum given by

$$r(k) = r_1^\dagger(\bar{k}) + h^\dagger(\bar{k}), \quad k \in \mathbf{R}_-. \qquad (3.3)$$

Then we can rewrite the RH problem (2.14) as

$$\begin{cases} M_+(x,t,k) = M_-(x,t,k) J(x,t,k), & k \in \Sigma = \mathbf{R} \cup i\mathbf{R}, \\ M(x,t,k) \to I, & k \to \infty, \end{cases} \qquad (3.4)$$

with the jump matrix $J(x,t,k)$ is given by

$$J(x,t,k) = \begin{cases} \begin{pmatrix} 1 & 0 \\ -h(k)e^{t\Phi(\xi,k)} & I \end{pmatrix}, & k \in i\mathbf{R}_+, \\[1em] \begin{pmatrix} 1 & -r(k)e^{-t\Phi(\xi,k)} \\ -r^\dagger(\bar{k})e^{t\Phi(\xi,k)} & I + r^\dagger(\bar{k})r(k) \end{pmatrix}, & k \in \mathbf{R}_-, \\[1em] \begin{pmatrix} 1 & -h^\dagger(\bar{k})e^{-t\Phi(\xi,k)} \\ 0 & I \end{pmatrix}, & k \in i\mathbf{R}_-, \\[1em] \begin{pmatrix} 1 + r_1^\dagger(\bar{k}) r_1(k) & r_1^\dagger(\bar{k}) e^{-t\Phi(\xi,k)} \\ r_1(k) e^{t\Phi(\xi,k)} & I \end{pmatrix}, & k \in \mathbf{R}_+, \end{cases} \qquad (3.5)$$

where

$$\Phi(\xi,k) = 4ik^2 + 2i\xi k, \quad \xi = \frac{x}{t}, \qquad (3.6)$$

and $\mathbf{R}_+ = [0,\infty)$ and $\mathbf{R}_- = (-\infty, 0]$ denote the positive and negative halves of the real axis. In view of (2.18), we have

$$\bigl(u(x,t) \quad v(x,t)\bigr) = 2\lim_{k\to\infty}(kM(x,t,k))_{12}. \qquad (3.7)$$

Moreover, we can deduce from the properties of $s(k)$ and $S(k)$ that the functions $r_1(k)$, $h(k)$, $r(k)$ defined by (3.1), (3.2) and (3.3) possess the following properties:
- $r_1(k)$ is smooth and bounded on $\mathbf{R}$;

- $h(k)$ is smooth and bounded on $\bar{D}_2$ and analytic in D_2;
- $r(k)$ is smooth and bounded on $\mathbf{R}_-$;
- There exist complex constants $\{r_{1,j}\}_{j=1}^{\infty}$ and $\{h_j\}_{j=1}^{\infty}$ such that, for any $N \geqslant 1$,

$$r_1(k) = \sum_{j=1}^{N} \frac{r_{1,j}}{k^j} + O\left(\frac{1}{k^{N+1}}\right), \quad |k| \to \infty, \ k \in \mathbf{R}, \tag{3.8}$$

$$h(k) = \sum_{j=1}^{N} \frac{h_j}{k^j} + O\left(\frac{1}{k^{N+1}}\right), \quad k \to \infty, \ k \in \bar{D}_2. \tag{3.9}$$

3.1 Transformations of the RH problem

Let $N > 1$ be given, and let $\mathcal{I}$ denote the interval $\mathcal{I} = (0, N]$. The jump matrix J defined in (3.5) involves the exponentials $e^{\pm t\Phi}$. It follows that there is a single stationary point located at the point where $\dfrac{\partial \Phi}{\partial k} = 0$, i.e., at $k = k_0 = -\dfrac{\xi}{4}$. By performing a number of transformations in the following, we can bring the RH problem (3.4) to a form suitable for determining the long-time asymptotics.

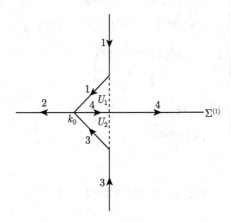

Fig. 2   The contour $\Sigma^{(1)}$ in the complex k-plane

The first transformation is to deform the vertical part of Σ so that it passes through the critical point k_0. Letting U_1 and U_2 denote the triangular domains shown in Fig. 2. Then the first transform is as follows:

$$M^{(1)}(x,t,k) = M(x,t,k) \times \begin{cases} \begin{pmatrix} 1 & 0 \\ -h(k)e^{t\Phi(\xi,k)} & I \end{pmatrix}, & k \in U_1, \\ \begin{pmatrix} 1 & h^\dagger(\bar{k})e^{-t\Phi(\xi,k)} \\ 0 & I \end{pmatrix}, & k \in U_2, \\ I, & \text{elsewhere.} \end{cases} \qquad (3.10)$$

Then we obtain the RH problem

$$\begin{cases} M_+^{(1)}(x,t,k) = M_-^{(1)}(x,t,k) J^{(1)}(x,t,k), & k \in \Sigma^{(1)}, \\ M^{(1)}(x,t,k) \to I, & k \to \infty. \end{cases} \qquad (3.11)$$

The jump matrix $J^{(1)}(x,t,k)$ is given by

$$J_1^{(1)} = \begin{pmatrix} 1 & 0 \\ -he^{t\Phi} & I \end{pmatrix}, \quad J_2^{(1)} = \begin{pmatrix} 1 & -re^{-t\Phi} \\ -r^\dagger e^{t\Phi} & I + r^\dagger r \end{pmatrix},$$

$$J_3^{(1)} = \begin{pmatrix} 1 & -h^\dagger e^{-t\Phi} \\ 0 & I \end{pmatrix}, \quad J_4^{(1)} = \begin{pmatrix} 1 + r_1^\dagger r_1 & r_1^\dagger e^{-t\Phi} \\ r_1 e^{t\Phi} & I \end{pmatrix}, \qquad (3.12)$$

where $J_i^{(1)}$ denotes the restriction of $J^{(1)}$ to the contour labeled by i in Fig. 2. The next transformation is:

$$M^{(2)}(x,t,k) = M^{(1)}(x,t,k)\Delta(k), \qquad (3.13)$$

where

$$\Delta(k) = \begin{pmatrix} \dfrac{1}{\det \delta(k)} & 0 \\ 0 & \delta(k) \end{pmatrix}, \qquad (3.14)$$

and $\delta(k)$ satisfies the following 2×2 matrix RH problem across $(-\infty, k_0)$ oriented in Fig. 2:

$$\begin{cases} \delta_-(k) = (I + r^\dagger(k) r(k)) \delta_+(k), & k < k_0, \\ \delta(k) \to I, & k \to \infty. \end{cases} \qquad (3.15)$$

Furthermore,

$$\begin{cases} \det \delta_-(k) = (1 + r(k) r^\dagger(k)) \det \delta_+(k), & k < k_0, \\ \det \delta(k) \to 1, & k \to \infty. \end{cases} \qquad (3.16)$$

Remark 3.1 *It is noted that we introduce a 2×2 matrix-valued function $\delta(k)$ to remove the middle matrix term while we split the jump matrix $J_2^{(1)}(x,t,k)$ into an appropriate upper/lower triangular form. This may be a main difference compared with the long-time asymptotic analysis of integrable nonlinear evolution equations associated with 2×2 matrix spectral problems on the half-line [3,20,21].*

Since the jump matrix $I + r^\dagger(k) r(k)$ is positive definite, the vanishing lemma [1] yields the existence and uniqueness of the function $\delta(k)$. By Plemelj formula, $\det \delta(k)$ can be solved by

$$\det \delta(k) = \exp\left\{ \frac{1}{2\pi i} \int_{-\infty}^{k_0} \frac{\ln(1 + r(s) r^\dagger(s))}{s - k} ds \right\}$$
$$= (k - k_0)^{-i\nu} e^{\chi(k)}, \quad k \in \mathbf{C} \setminus (-\infty, k_0], \tag{3.17}$$

where

$$\nu = \frac{1}{2\pi} \ln(1 + r(k_0) r^\dagger(k_0)) > 0, \tag{3.18}$$

$$\chi(k) = -\frac{1}{2\pi i} \int_{-\infty}^{k_0} \ln(k - s) d \ln(1 + r(s) r^\dagger(s)). \tag{3.19}$$

On the other hand, a direct calculation as in [17] shows that

$$|\delta(k)| \leqslant \mathrm{const} < \infty, \quad |\det \delta(k)| \leqslant \mathrm{const} < \infty, \tag{3.20}$$

for all k, where we define

$$|A| = (\mathrm{tr} A^\dagger A)^{\frac{1}{2}}, \quad \text{for any matrix } A,$$
$$\|B(\cdot)\|_{L^p} = \||B(\cdot)|\|_{L^p}, \quad \text{for any matrix function } B(\cdot). \tag{3.21}$$

Then we find that $M^{(2)}(x,t,k)$ satisfies the following RH problem

$$\begin{cases} M_+^{(2)}(x,t,k) = M_-^{(2)}(x,t,k) J^{(2)}(x,t,k), & k \in \Sigma^{(2)}, \\ M^{(2)}(x,t,k) \to I, & k \to \infty. \end{cases} \tag{3.22}$$

and the contour $\Sigma^{(2)} = \Sigma^{(1)}$, the jump matrix $J^{(2)} = \Delta_-^{-1} J^{(1)} \Delta_+$, namely,

$$J_1^{(2)} = \begin{pmatrix} 1 & 0 \\ -\delta^{-1} h (\det \delta)^{-1} e^{t\Phi} & I \end{pmatrix},$$

$$J_2^{(2)} = \begin{pmatrix} 1 & -r_2\delta_- \det \delta_- e^{-t\Phi} \\ 0 & I \end{pmatrix} \begin{pmatrix} 1 & 0 \\ -\delta_+^{-1} r_2^\dagger (\det \delta_+)^{-1} e^{t\Phi} & I \end{pmatrix},$$

$$J_3^{(2)} = \begin{pmatrix} 1 & -h^\dagger \delta \det \delta e^{-t\Phi} \\ 0 & I \end{pmatrix},$$

$$J_4^{(2)} = \begin{pmatrix} 1 & r_1^\dagger \delta \det \delta e^{-t\Phi} \\ 0 & I \end{pmatrix} \begin{pmatrix} 1 & 0 \\ \delta^{-1} r_1 (\det \delta)^{-1} e^{t\Phi} & I \end{pmatrix},$$

where we define $r_2(k)$ by

$$r_2(k) = \frac{r(k)}{1 + r(k) r^\dagger(\bar{k})}. \tag{3.23}$$

Before processing the next deformation, we will follow the idea of [3, 23, 24] and decompose each of the functions h, r_1, r_2 into an analytic part and a small remainder because the spectral functions have limited domains of analyticity. The analytic part of the jump matrix will be deformed, whereas the small remainder will be left on the original contour.

Lemma 3.1 *There exists a decomposition*

$$h(k) = h_a(t,k) + h_r(t,k), \quad t > 0, \quad k \in i\mathbf{R}_+,$$

where the functions h_a and h_r have the following properties:

(i) *For each $t > 0$, $h_a(t,k)$ is defined and continuous for $k \in \bar{D}_1$ and analytic for $k \in D_1$.*

(ii) *For each $\xi \in \mathcal{I}$ and each $t > 0$, the function $h_a(t,k)$ satisfies*

$$|h_a(t,k) - h(0)| \leqslant C e^{\frac{t}{4}|\text{Re}\Phi(\xi,k)|},$$

$$|h_a(t,k)| \leqslant \frac{C}{1 + |k|^2} e^{\frac{t}{4}|\text{Re}\Phi(\xi,k)|}, \tag{3.24}$$

for $k \in \bar{D}_1$, where the constant C is independent of ξ, k, t.

(iii) *The L^1, L^2 and L^∞ norms of the function $h_r(t, \cdot)$ on $i\mathbf{R}_+$ are $O(t^{-3/2})$ as $t \to \infty$.*

Proof Since $h(k) \in C^5(i\mathbf{R}_+)$, we find that

$$h^{(n)}(k) = \frac{d^n}{dk^n}\left(\sum_{j=0}^{4} \frac{h^{(j)}(0)}{j!} k^j \right) + O(k^{5-n}), \quad k \to 0,\ k \in i\mathbf{R}_+,\ n = 0,1,2. \tag{3.25}$$

On the other hand, we have

$$h^{(n)}(k) = \frac{d^n}{dk^n}\left(\sum_{j=1}^{3} h_j k^{-j}\right) + O(k^{-4-n}), \quad k \to \infty, \ k \in i\mathbf{R}_+, \ n = 0, 1, 2. \quad (3.26)$$

Let

$$f_0(k) = \sum_{j=2}^{9} \frac{a_j}{(k+i)^j}, \quad (3.27)$$

where $\{a_j\}_2^9$ are complex such that

$$f_0(k) = \begin{cases} \sum_{j=0}^{4} \dfrac{h^{(j)}(0)}{j!} k^j + O(k^5), & k \to 0, \\ \sum_{j=1}^{3} h_j k^{-j} + O(k^{-4}), & k \to \infty. \end{cases} \quad (3.28)$$

It is easy to verify that (3.28) imposes eight linearly independent conditions on the a_j, hence the coefficients a_j exist and are unique. Letting $f = h - f_0$, it follows that

(1) $f_0(k)$ is a rational function of $k \in \mathbf{C}$ with no poles in $\bar{D}_1$;

(2) $f_0(k)$ coincides with $h(k)$ to four order at 0 and to third order at ∞, more precisely,

$$\frac{d^n}{dk^n} f(k) = \begin{cases} O(k^{5-n}), & k \to 0, \\ O(k^{-4-n}), & k \to \infty, \end{cases} \quad k \in i\mathbf{R}_+, \ n = 0, 1, 2. \quad (3.29)$$

The decomposition of $h(k)$ can be derived as follows. The map $k \mapsto \psi = \psi(k)$ defined by $\psi(k) = 4k^2$ is a bijection $[0, i\infty) \mapsto (-\infty, 0]$, so we may define a function $F : \mathbf{R} \to \mathbf{C}$ by

$$F(\psi) = \begin{cases} (k+i)^2 f(k), & \psi \leqslant 0, \\ 0, & \psi > 0. \end{cases} \quad (3.30)$$

$F(\psi)$ is C^5 for $\psi \neq 0$ and

$$F^{(n)}(\psi) = \left(\frac{1}{8k}\frac{\partial}{\partial k}\right)^n \left((k+i)^2 f(k)\right), \quad \psi \leqslant 0.$$

By (3.29), $F \in C^2(\mathbf{R})$ and $F^{(n)}(\psi) = O(|\psi|^{-1-n})$ as $|\psi| \to \infty$ for $n = 0, 1, 2$. In particular,
$$\left\|\frac{d^n F}{d\psi^n}\right\|_{L^2(\mathbf{R})} < \infty, \quad n = 0, 1, 2, \tag{3.31}$$

that is, F belongs to $H^2(\mathbf{R})$. By the Fourier transform $\hat{F}(s)$ defined by
$$\hat{F}(s) = \frac{1}{2\pi} \int_{\mathbf{R}} F(\psi) e^{-i\psi s} d\psi, \tag{3.32}$$

where
$$F(\psi) = \int_{\mathbf{R}} \hat{F}(s) e^{i\psi s} ds, \tag{3.33}$$

it follows from Plancherel theorem that $\|s^2 \hat{F}(s)\|_{L^2(\mathbf{R})} < \infty$. Equations (3.30) and (3.33) imply

$$f(k) = \frac{1}{(k+i)^2} \int_{\mathbf{R}} \hat{F}(s) e^{i\psi s} ds, \quad k \in i\mathbf{R}_+. \tag{3.34}$$

Writing
$$f(k) = f_a(t, k) + f_r(t, k), \quad t > 0, \ k \in i\mathbf{R}_+,$$

where the functions f_a and f_r are defined by

$$f_a(t, k) = \frac{1}{(k+i)^2} \int_{-\frac{t}{4}}^{\infty} \hat{F}(s) e^{4ik^2 s} ds, \quad t > 0, \ k \in \bar{D}_1, \tag{3.35}$$

$$f_r(t, k) = \frac{1}{(k+i)^2} \int_{-\infty}^{-\frac{t}{4}} \hat{F}(s) e^{4ik^2 s} ds, \quad t > 0, \ k \in i\mathbf{R}_+, \tag{3.36}$$

we infer that $f_a(t, \cdot)$ is continuous in $\bar{D}_1$ and analytic in D_1. Moreover, since $|\mathrm{Re}\, 4ik^2| \leq |\mathrm{Re}\,\Phi(\xi, k)|$ for $k \in \bar{D}_1$ and $\xi \in \mathcal{I}$, we can get

$$|f_a(t, k)| \leq \frac{1}{|k+i|^2} \|\hat{F}(s)\|_{L^1(\mathbf{R})} \sup_{s \geq -\frac{t}{4}} e^{s \mathrm{Re}\, 4ik^2}$$

$$\leq \frac{C}{1+|k|^2} e^{\frac{t}{4}|\mathrm{Re}\,\Phi(\xi, k)|}, \quad t > 0, \ k \in \bar{D}_1, \ \xi \in \mathcal{I}. \tag{3.37}$$

Furthermore, we have
$$|f_r(t, k)| \leq \frac{1}{|k+i|^2} \int_{-\infty}^{-\frac{t}{4}} s^2 |\hat{F}(s)| s^{-2} ds$$

$$\leqslant \frac{C}{1+|k|^2}\|s^2\hat{F}(s)\|_{L^2(\mathbf{R})}\sqrt{\int_{-\infty}^{-\frac{t}{4}} s^{-4}ds},$$

$$\leqslant \frac{C}{1+|k|^2}t^{-3/2}, \quad t>0,\ k\in i\mathbf{R}_+,\ \xi\in\mathcal{I}. \tag{3.38}$$

Hence, the L^1, L^2 and L^∞ norms of f_r on $i\mathbf{R}_+$ are $O(t^{-3/2})$. Letting

$$\begin{aligned} h_a(t,k) &= f_0(k) + f_a(t,k), & t>0,\ k\in\bar{D}_1, \\ h_r(t,k) &= f_r(t,k), & t>0,\ k\in i\mathbf{R}_+, \end{aligned} \tag{3.39}$$

we find a decomposition of h with the properties listed in the statement of the lemma.

We next introduce the open subsets $\{\Omega_j\}_1^8$, as displayed in Fig. 3. The following lemma describes how to decompose r_j, $j=1,2$ into an analytic part $r_{j,a}$ and a small remainder $r_{j,r}$.

Lemma 3.2 *There exist decompositions*

$$\begin{aligned} r_1(k) &= r_{1,a}(x,t,k) + r_{1,r}(x,t,k), & k>k_0, \\ r_2(k) &= r_{2,a}(x,t,k) + r_{2,r}(x,t,k), & k<k_0, \end{aligned} \tag{3.40}$$

where the functions $\{r_{j,a}, r_{j,r}\}_1^2$ have the following properties:

(1) *For each $\xi\in\mathcal{I}$ and each $t>0$, $r_{j,a}(x,t,k)$ is defined and continuous for $k\in\bar{\Omega}_j$ and analytic for Ω_j, $j=1,2$.*

(2) *The functions $r_{1,a}$ and $r_{2,a}$ satisfy, for $\xi\in\mathcal{I}$, $t>0$, $j=1,2$,*

$$\begin{aligned} |r_{j,a}(x,t,k) - r_j(k_0)| &\leqslant C|k-k_0|e^{\frac{t}{4}|Re\Phi(\xi,k)|}, \\ |r_{j,a}(x,t,k)| &\leqslant \frac{C}{1+|k-k_0|^2}e^{\frac{t}{4}|Re\Phi(\xi,k)|}, \end{aligned} \quad k\in\bar{\Omega}_j, \tag{3.41}$$

where the constant C is independent of ξ, k, t.

(3) *The L^1, L^2 and L^∞ norms of the function $r_{1,r}(x,t,\cdot)$ on (k_0,∞) are $O(t^{-3/2})$ as $t\to\infty$ uniformly with respect to $\xi\in\mathcal{I}$.*

(4) *The L^1, L^2 and L^∞ norms of the function $r_{2,r}(x,t,\cdot)$ on $(-\infty,k_0)$ are $O(t^{-3/2})$ as $t\to\infty$ uniformly with respect to $\xi\in\mathcal{I}$.*

Proof Analogous to the proof of Lemma 3.1. One can also see [17, 24].

The purpose of the next transformation is to deform the contour so that the jump matrix involves the exponential factor $e^{-t\Phi}$ on the parts of the contour where ReΦ is positive and the factor $e^{t\Phi}$ on the parts where ReΦ is negative according to the signature table for ReΦ. More precisely, we put

$$M^{(3)}(x,t,k) = M^{(2)}(x,t,k)G(k), \qquad (3.42)$$

where

$$G(k) = \begin{cases} \begin{pmatrix} 1 & 0 \\ -\delta^{-1}r_{1,a}(\det\delta)^{-1}e^{t\Phi} & I \end{pmatrix}, & k \in \Omega_1, \\ \begin{pmatrix} 1 & -r_{2,a}\delta\det\delta e^{-t\Phi} \\ 0 & I \end{pmatrix}, & k \in \Omega_2, \\ \begin{pmatrix} 1 & 0 \\ \delta^{-1}r_{2,a}^\dagger(\det\delta)^{-1}e^{t\Phi} & I \end{pmatrix}, & k \in \Omega_3, \\ \begin{pmatrix} 1 & r_{1,a}^\dagger\delta\det\delta e^{-t\Phi} \\ 0 & I \end{pmatrix}, & k \in \Omega_4, \\ \begin{pmatrix} 1 & -h_a^\dagger\delta\det\delta e^{-t\Phi} \\ 0 & I \end{pmatrix}, & k \in \Omega_5, \\ \begin{pmatrix} 1 & 0 \\ \delta^{-1}h_a(\det\delta)^{-1}e^{t\Phi} & I \end{pmatrix}, & k \in \Omega_6, \\ I, & k \in \Omega_7 \cup \Omega_8. \end{cases} \qquad (3.43)$$

Then the matrix $M^{(3)}(x,t,k)$ satisfies the following RH problem

$$\begin{cases} M_+^{(3)}(x,t,k) = M_-^{(3)}(x,t,k)J^{(3)}(x,t,k), & k \in \Sigma^{(3)}, \\ M^{(3)}(x,t,k) \to I, & k \to \infty, \end{cases} \qquad (3.44)$$

with the jump matrix $J^{(3)} = G_-^{-1}(k)J^{(2)}G_+(k)$ given by

$$J_1^{(3)} = \begin{pmatrix} 1 & 0 \\ \delta^{-1}(r_{1,a}+h)(\det\delta)^{-1}e^{t\Phi} & I \end{pmatrix}, \quad J_2^{(3)} = \begin{pmatrix} 1 & -r_{2,a}\delta\det\delta e^{-t\Phi} \\ 0 & I \end{pmatrix},$$

$$J_3^{(3)} = \begin{pmatrix} 1 & 0 \\ -\delta^{-1}r_{2,a}^\dagger(\det\delta)^{-1}e^{t\Phi} & I \end{pmatrix}, \quad J_4^{(3)} = \begin{pmatrix} 1 & (r_{1,a}^\dagger+h^\dagger)\delta\det\delta e^{-t\Phi} \\ 0 & I \end{pmatrix},$$

$$J_5^{(3)} = \begin{pmatrix} 1 & (r_{1,a}^\dagger+h_a^\dagger)\delta\det\delta e^{-t\Phi} \\ 0 & I \end{pmatrix}, \quad J_6^{(3)} = \begin{pmatrix} 1 & 0 \\ \delta^{-1}(r_{1,a}+h_a)(\det\delta)^{-1}e^{t\Phi} & I \end{pmatrix},$$

$$J_7^{(3)} = \begin{pmatrix} 1 & r_{1,r}^\dagger \delta \det \delta e^{-t\Phi} \\ 0 & I \end{pmatrix} \begin{pmatrix} 1 & 0 \\ \delta^{-1} r_{1,r} (\det \delta)^{-1} e^{t\Phi} & I \end{pmatrix}, \qquad (3.45)$$

$$J_8^{(3)} = \begin{pmatrix} 1 & -r_{2,r} \delta_- \det \delta_- e^{-t\Phi} \\ 0 & I \end{pmatrix} \begin{pmatrix} 1 & 0 \\ -\delta_+^{-1} r_{2,r}^\dagger (\det \delta_+)^{-1} e^{t\Phi} & I \end{pmatrix},$$

$$J_9^{(3)} = \begin{pmatrix} 1 & 0 \\ -\delta^{-1} h_r (\det \delta)^{-1} e^{t\Phi} & I \end{pmatrix}, \qquad J_{10}^{(3)} = \begin{pmatrix} 1 & -h_r^\dagger \delta \det \delta e^{-t\Phi} \\ 0 & I \end{pmatrix},$$

with $J_i^{(3)}$ denoting the restriction of $J^{(3)}$ to the contour labeled by i in Fig. 3.

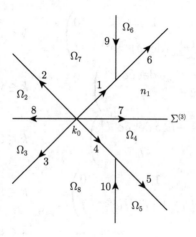

Fig. 3   The contour $\Sigma^{(3)}$ and the open sets $\{\Omega_j\}_1^8$ in the complex k-plane

Remark 3.2   *Recalling the meaningful work [3], begin performing the asymptotic analysis, the authors introduced a new spectral variable which led to a new phase function $\Phi(\zeta, \lambda) = 4i\lambda^2 + 2i\zeta\lambda$ only possed a single critical point, which has the same form as our phase function (3.6). Thus, it is not surprising that the figures of deformation contours are similar to [3]. However, our jump matrices involved are all 3×3 (that is, the spectral functions $r_1(k), r_2(k), h(k)$ are vectors) and the function $\delta(k)$ is a 2×2 matrix, which have the essential difference compared with [3]. As a result, the analysis in the next section will present some new skills, see Lemma 3.3.*

3.2 Local model near k_0

It is easy to check that the jump matrix $J^{(3)}$ decays to identity matrix I as $t \to \infty$ everywhere except near k_0. Thus, the main contribution to the long-time asymptotics should come from a neighborhood of the stationary phase point k_0. To focus on k_0, we make a scaling transformation by

$$N_{k_0} : k \mapsto \frac{z}{\sqrt{8t}} + k_0. \tag{3.46}$$

Let $D_\varepsilon(k_0)$ denote the open disk of radius ε centered at k_0 for a small $\varepsilon > 0$. Then, the map $k \mapsto z$ is a bijection from $D_\varepsilon(k_0)$ to the open disk of radius $\sqrt{8t}\varepsilon$ centered at the origin. Since the function $\delta(k)$ satisfying a 2×2 matrix RH problem (3.15) can not be solved explicitly, to proceed the next step, we will follow the idea developed in [17] to use the available function $\det \delta(k)$ to approximate $\delta(k)$ by error estimate. More precisely, we rewrite the (12) entry of $J_2^{(3)}$ as

$$\left(r_{2,a} \delta \det \delta e^{-t\Phi}\right)(k) = \left(r_{2,a}(\det \delta)^2 e^{-t\Phi}\right)(k)$$

$$+ \left(r_{2,a}(\delta - \det \delta I) \det \delta e^{-t\Phi}\right)(k). \tag{3.47}$$

For the first part on the right-hand side of (3.47), we have

$$N_{k_0}\left(r_{2,a}(\det \delta)^2 e^{-t\Phi}\right)(z) = \eta^2 \varrho^2 r_{2,a}\left(\frac{z}{\sqrt{8t}} + k_0\right), \tag{3.48}$$

where

$$\eta = (8t)^{\frac{i\nu}{2}} e^{2ik_0^2 t + \chi(k_0)}, \quad \varrho = z^{-i\nu} e^{-\frac{iz^2}{4}} e^{\chi(\frac{z}{\sqrt{8t}} + k_0) - \chi(k_0)}. \tag{3.49}$$

Let $\tilde{\delta}(k) = r_{2,a}(k)(\delta(k) - \det \delta(k) I) e^{-t\Phi(\xi, k)}$, then we have the following estimate.

Lemma 3.3 *For $z \in \hat{L} = \{z = \alpha e^{\frac{3i\pi}{4}} : -\infty < \alpha < +\infty\}$, as $t \to \infty$, the following estimate for $\tilde{\delta}(k)$ hold:*

$$|(N_{k_0}\tilde{\delta})(z)| \leqslant C t^{-1/2}, \tag{3.50}$$

where the constant $C > 0$ is independent of z, t.

Proof The idea of the proof comes from [17]. It follows from (3.15) and (3.16) that $\tilde{\delta}$ satisfies the following RH problem across $(-\infty, k_0)$ oriented in Fig.2:

$$\begin{cases} \tilde{\delta}_-(k) = (1 + r(k)r^\dagger(k))\tilde{\delta}_+(k) + e^{-t\Phi(\xi, k)} f(k), & k < k_0, \\ \tilde{\delta}(k) \to 0, & k \to \infty, \end{cases} \tag{3.51}$$

where $f(k) = [r_{2,a}(r^\dagger r - rr^\dagger I)\delta_+](k)$. Then the function $\tilde{\delta}(k)$ can be expressed by

$$\tilde{\delta}(k) = X(k) \int_{-\infty}^{k_0} \frac{e^{-t\Phi(\xi,s)} f(s)}{X_-(s)(s-k)} ds,$$

$$X(k) = \exp\left\{\frac{1}{2\pi i} \int_{-\infty}^{k_0} \frac{\ln(1+r(s)r^\dagger(s))}{s-k} ds\right\}. \qquad (3.52)$$

It follows from $r_{2,a}(r^\dagger r - rr^\dagger I) = r_{2,r}(rr^\dagger I - r^\dagger r)$ that $f(k) = O(t^{-3/2})$. Similar to Lemma 3.2, $f(k)$ can be decomposed into two parts: $f_a(x,t,k)$ which has an analytic continuation to Ω_2, and $f_r(x,t,k)$, which is a small remainder. In particular,

$$|f_a(x,t,k)| \leqslant \frac{C}{1+\left|k-k_0+\frac{1}{t}\right|^2} e^{\frac{t}{4}|\text{Re}\Phi(\xi,k)|}, \quad k \in L_t,$$

$$|f_r(x,t,k)| \leqslant \frac{C}{1+\left|k-k_0+\frac{1}{t}\right|^2} t^{-3/2}, \quad k \in (-\infty, k_0), \qquad (3.53)$$

where $L_t = \left\{k = k_0 - \frac{1}{t} + \alpha e^{\frac{3i\pi}{4}} : 0 \leqslant \alpha < \infty\right\}$. Therefore, for $z \in \hat{L}$, we can find

$$(N_{k_0}\tilde{\delta})(z) = X\left(k_0 + \frac{z}{\sqrt{8t}}\right) \int_{k_0-\frac{1}{t}}^{k_0} \frac{e^{-t\Phi(\xi,s)} f(s)}{X_-(s)\left(s-k_0-\frac{z}{\sqrt{8t}}\right)} ds$$

$$+ X\left(k_0 + \frac{z}{\sqrt{8t}}\right) \int_{-\infty}^{k_0-\frac{1}{t}} \frac{e^{-t\Phi(\xi,s)} f_a(x,t,s)}{X_-(s)\left(s-k_0-\frac{z}{\sqrt{8t}}\right)} ds$$

$$+ X\left(k_0 + \frac{z}{\sqrt{8t}}\right) \int_{-\infty}^{k_0-\frac{1}{t}} \frac{e^{-t\Phi(\xi,s)} f_r(x,t,s)}{X_-(s)\left(s-k_0-\frac{z}{\sqrt{8t}}\right)} ds$$

$$= I_1 + I_2 + I_3.$$

Then,

$$|I_1| \leqslant \int_{k_0-\frac{1}{t}}^{k_0} \frac{|f(s)|}{\left|s-k_0-\frac{z}{\sqrt{8t}}\right|} ds \leqslant Ct^{-3/2} \left|\ln\left|1-\frac{2\sqrt{2}}{zt^{1/2}}\right|\right| \leqslant Ct^{-2},$$

$$|I_3| \leqslant \int_{-\infty}^{k_0 - \frac{1}{t}} \frac{|f_r(x,t,s)|}{\left|s - k_0 - \dfrac{z}{\sqrt{8t}}\right|} ds \leqslant Ct^{-1/2}.$$

By Cauchy's theorem, we can evaluate I_2 along the contour L_t instead of the interval $\left(-\infty, k_0 - \dfrac{1}{t}\right)$. Using the fact $\mathrm{Re}\Phi(\xi,k) > 0$ in Ω_2, we can obtain $|I_2| \leqslant Ce^{-ct}$. This completes the proof of the lemma.

Remark 3.3 *The estimate* (3.50) *also holds if* $r_{2,a}$ *is replaced with* $r^\dagger_{1,a} + h^\dagger$ *or* $r^\dagger_{1,a} + h^\dagger_a$. *There is a similar estimate*

$$|(N_{k_0}\hat{\delta})(z)| \leqslant Ct^{-1/2}, \quad t \to \infty, \tag{3.54}$$

for $z \in \bar{\hat{L}}$, *where* $\hat{\delta}(k) = [\delta^{-1}(k) - (\det \delta)^{-1}I]\rho(k)e^{t\Phi(\xi,k)}$, $\rho = r^\dagger_{2,a}, r_{1,a} + h$ *or* $r_{1,a} + h_a$.

Remark 3.4 *As mentioned above, here we use function* $\det \delta(k)$ *which can be explicitly written down to approximate the unsolvable function* $\delta(k)$ *by error control. This procedure has never appeared in the long-time asymptotic analysis of the integrable nonlinear evolution equations associated with* 2×2 *matrix spectral problems.*

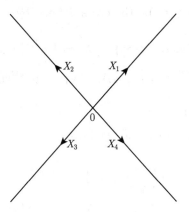

Fig. 4   The contour $X = X_1 \cup X_2 \cup X_3 \cup X_4$

In other words, we have the following important relation:

$$(N_{k_0} J^{(3)}_i)(x,t,z) = \tilde{J}(x,t,z) + O(t^{-1/2}), \quad i = 1,\cdots,6, \tag{3.55}$$

where $\tilde{J}(x,t,z)$ is given by

$$\tilde{J}(x,t,z) = \begin{cases} \begin{pmatrix} 1 & 0 \\ (r_{1,a}+h_a)\eta^{-2}\varrho^{-2} & I \end{pmatrix}, & k \in \mathcal{X}_1 \cap D_1, \\ \begin{pmatrix} 1 & 0 \\ (r_{1,a}+h)\eta^{-2}\varrho^{-2} & I \end{pmatrix}, & k \in \mathcal{X}_1 \cap D_2, \\ \begin{pmatrix} 1 & -r_{2,a}\eta^2\varrho^2 \\ 0 & I \end{pmatrix}, & k \in \mathcal{X}_2, \\ \begin{pmatrix} 1 & 0 \\ -r_{2,a}^\dagger \eta^{-2}\varrho^{-2} & I \end{pmatrix}, & k \in \mathcal{X}_3, \\ \begin{pmatrix} 1 & (r_{1,a}^\dagger+h^\dagger)\eta^2\varrho^2 \\ 0 & I \end{pmatrix}, & k \in \mathcal{X}_4 \cap D_3, \\ \begin{pmatrix} 1 & (r_{1,a}^\dagger+h_a^\dagger)\eta^2\varrho^2 \\ 0 & I \end{pmatrix}, & k \in \mathcal{X}_4 \cap D_4, \end{cases} \quad (3.56)$$

where $\mathcal{X} = X + k_0$ denotes the cross X defined by (3.57) centered at k_0, and $X = X_1 \cup X_2 \cup X_3 \cup X_4 \subset \mathbf{C}$ be the cross defined by

$$\begin{aligned} X_1 &= \{le^{\frac{i\pi}{4}} | 0 \leqslant l < \infty\}, \quad X_2 = \{le^{\frac{3i\pi}{4}} | 0 \leqslant l < \infty\}, \\ X_3 &= \{le^{-\frac{3i\pi}{4}} | 0 \leqslant l < \infty\}, \quad X_4 = \{le^{-\frac{i\pi}{4}} | 0 \leqslant l < \infty\}, \end{aligned} \quad (3.57)$$

and oriented as in Fig. 4.

For any fixed $z \in X$ and $\xi \in \mathcal{I}$, we have

$$\begin{aligned} (r_{1,a}+h)\left(\frac{z}{\sqrt{8t}}+k_0\right) &\to r^\dagger(k_0), \\ r_{2,a}\left(\frac{z}{\sqrt{8t}}+k_0\right) &\to \frac{r(k_0)}{1+r(k_0)r^\dagger(k_0)}, \\ \varrho^2 &\to e^{-\frac{iz^2}{2}} z^{-2i\nu}, \end{aligned} \quad (3.58)$$

as $t \to \infty$. This implies that the jump matrix $\tilde{J}$ tends to the matrix $J^{(k_0)}$ for large t, where

$$J^{(k_0)}(x,t,z) = \begin{cases} \begin{pmatrix} 1 & 0 \\ \eta^{-2}e^{\frac{iz^2}{2}}z^{2i\nu}r^\dagger(k_0) & I \end{pmatrix}, & z \in X_1, \\[2mm] \begin{pmatrix} 1 & -\eta^2 e^{-\frac{iz^2}{2}}z^{-2i\nu}\dfrac{r(k_0)}{1+r(k_0)r^\dagger(k_0)} \\ 0 & I \end{pmatrix}, & z \in X_2, \\[4mm] \begin{pmatrix} 1 & 0 \\ -\eta^{-2}e^{\frac{iz^2}{2}}z^{2i\nu}\dfrac{r^\dagger(k_0)}{1+r(k_0)r^\dagger(k_0)} & I \end{pmatrix}, & z \in X_3, \\[4mm] \begin{pmatrix} 1 & \eta^2 e^{-\frac{iz^2}{2}}z^{-2i\nu}r(k_0) \\ 0 & I \end{pmatrix}, & z \in X_4. \end{cases} \quad (3.59)$$

Theorem 3.1 *The following RH problem:*

$$\begin{cases} M_+^X(x,t,z) = M_-^X(x,t,z)J^X(x,t,z), & z \in X, \\ M^X(x,t,z) \to I, & z \to \infty, \end{cases} \quad (3.60)$$

with the jump matrix $J^X(x,t,z) = \eta^{\hat{\sigma}} J^{(k_0)}(x,t,z)$ *has a unique solution* $M^X(x,t,z)$. *This solution satisfies*

$$M^X(x,t,z) = I - \frac{i}{z}\begin{pmatrix} 0 & \beta^X \\ (\beta^X)^\dagger & 0 \end{pmatrix} + O\left(\frac{1}{z^2}\right), \quad z \to \infty, \quad (3.61)$$

where the error term is uniform with respect to $\arg z \in [0, 2\pi]$ *and the function* β^X *is given by*

$$\beta^X = \frac{e^{\frac{i\pi}{4} - \frac{\pi\nu}{2}}\nu\Gamma(-i\nu)}{\sqrt{2\pi}}r(k_0), \quad (3.62)$$

where $\Gamma(\cdot)$ *denotes the standard Gamma function. Moreover, for each compact subset* $\mathcal{D}$ *of* $\mathbf{C}$,

$$\sup_{r(k_0)\in\mathcal{D}}\sup_{z\in\mathbf{C}\setminus X} |M^X(x,t,z)| < \infty. \quad (3.63)$$

Proof The proof relies on deriving an explicit formula for the solution M^X in terms of parabolic cylinder functions. One can see [17].

Let $D_\varepsilon(k_0)$ denote the open disk of radius ε centered at k_0 for a small $\varepsilon > 0$ and $\mathcal{X}^\varepsilon = \mathcal{X} \cap D_\varepsilon(k_0)$. We can approximate $M^{(3)}$ in the neighborhood $D_\varepsilon(k_0)$ of k_0 by

$$M^{(k_0)}(x,t,k) = \eta^{-\hat{\sigma}}M^X(x,t,z). \quad (3.64)$$

Lemma 3.4 For each $\xi \in \mathcal{I}$ and $t > 0$, the function $M^{(k_0)}(x, t, k)$ defined in (3.64) is an analytic function of $k \in D_\varepsilon(k_0) \setminus \mathcal{X}^\varepsilon$. Furthermore,

$$|M^{(k_0)}(x, t, k) - I| \leqslant C, \quad t > 3, \ \xi \in \mathcal{I}, \ k \in \overline{D_\varepsilon(k_0)} \setminus \mathcal{X}^\varepsilon. \tag{3.65}$$

Across $\mathcal{X}^\varepsilon$, $M^{(k_0)}$ satisfied the jump condition $M_+^{(k_0)} = M_-^{(k_0)} J^{(k_0)}$, where the jump matrix $J^{(k_0)}$ satisfies the following estimates for $1 \leqslant p \leqslant \infty$:

$$\|J^{(3)} - J^{(k_0)}\|_{L^p(\mathcal{X}^\varepsilon)} \leqslant C t^{-\frac{1}{2} - \frac{1}{2p}} \ln t, \quad t > 3, \ \xi \in \mathcal{I}, \tag{3.66}$$

where $C > 0$ is a constant independent of t, ξ, z. Moreover, as $t \to \infty$,

$$\|(M^{(k_0)})^{-1}(x, t, k) - I\|_{L^\infty(\partial D_\varepsilon(k_0))} = O(t^{-1/2}), \tag{3.67}$$

and

$$-\frac{1}{2\pi i} \int_{\partial D_\varepsilon(k_0)} ((M^{(k_0)})^{-1}(x, t, k) - I) dk = \frac{\eta^{-\hat{\sigma}} M_1^X}{\sqrt{8t}} + O(t^{-1}), \tag{3.68}$$

where M_1^X is defined by

$$M_1^X = -i \begin{pmatrix} 0 & \beta^X \\ (\beta^X)^\dagger & 0 \end{pmatrix}. \tag{3.69}$$

Proof The analyticity of $M^{(k_0)}$ is obvious. Since $|\eta| = 1$, the estimate (3.65) follows from the definition of $M^{(k_0)}$ in (3.64) and the estimate (3.63).

On the other hand, we have $J^{(3)} - J^{(k_0)} = J^{(3)} - \tilde{J} + \tilde{J} - J^{(k_0)}$. However, according to the Lemma 89 in [15], we conclude that

$$\|\tilde{J} - J^{(k_0)}\|_{L^\infty(\mathcal{X}_1^\varepsilon)} \leqslant C|e^{\frac{i\gamma}{2} z^2}| t^{-1/2} \ln t, \quad 0 < \gamma < \frac{1}{2}, \ t > 3, \ \xi \in \mathcal{I},$$

for $k \in \mathcal{X}_1^\varepsilon$, that is, $z = \sqrt{8t} u e^{\frac{i\pi}{4}}$, $0 \leqslant u \leqslant \varepsilon$. Thus, for ε small enough, it follows from (3.55) that

$$\|J^{(3)} - J^{(k_0)}\|_{L^\infty(\mathcal{X}_1^\varepsilon)} \leqslant C|e^{\frac{i\gamma}{2} z^2}| t^{-1/2} \ln t. \tag{3.70}$$

Then we have

$$\|J^{(3)} - J^{(k_0)}\|_{L^1(\mathcal{X}_1^\varepsilon)} \leqslant C t^{-1} \ln t, \quad t > 3, \ \xi \in \mathcal{I}. \tag{3.71}$$

By the general inequality $\|f\|_{L^p} \leqslant \|f\|_{L^\infty}^{1-1/p}\|f\|_{L^1}^{1/p}$, we find

$$\|J^{(2)} - J^{(k_0)}\|_{L^p(\mathcal{X}_1^\varepsilon)} \leqslant Ct^{-1/2-1/2p}\ln t, \quad t > 3, \ \xi \in \mathcal{I}. \tag{3.72}$$

The norms on $\mathcal{X}_j^\varepsilon$, $j = 2, 3, 4$, are estimated in a similar way. Therefore, (3.66) follows.

If $k \in \partial D_\varepsilon(k_0)$, the variable $z = \sqrt{8t}(k - k_0)$ tends to infinity as $t \to \infty$. It follows from (3.61) that

$$M^X(x,t,z) = I + \frac{M_1^X}{\sqrt{8t}(k-k_0)} + O\left(\frac{1}{t}\right), \quad t \to \infty, \ k \in \partial D_\varepsilon(k_0). \tag{3.73}$$

Since
$$M^{(k_0)}(x,t,k) = \eta^{-\hat\sigma} M^X(x,t,z),$$

thus we have

$$(M^{(k_0)})^{-1}(x,t,k) - I = -\frac{\eta^{-\hat\sigma} M_1^X}{\sqrt{8t}(k-k_0)} + O\left(\frac{1}{t}\right), \quad t \to \infty, \ k \in \partial D_\varepsilon(k_0). \tag{3.74}$$

The estimate (3.67) immediately follows from (3.74) and $|M_1^X| \leqslant C$. By Cauchy's formula and (3.74), we derive (3.68).

3.3 Derivation of the asymptotic formula

Define the approximate solution $M^{(app)}(x,t,k)$ by

$$M^{(app)} = \begin{cases} M^{(k_0)}, & k \in D_\varepsilon(k_0), \\ I, & \text{elsewhere}. \end{cases} \tag{3.75}$$

Let $\hat{M}(x,t,k)$ be
$$\hat{M} = M^{(3)}(M^{(app)})^{-1}, \tag{3.76}$$

then $\hat{M}(x,t,k)$ satisfies the following RH problem

$$\begin{cases} \hat{M}_+(x,t,k) = \hat{M}_-(x,t,k)\hat{J}(x,t,k), & k \in \hat{\Sigma}, \\ \hat{M}(x,t,k) \to I, & k \to \infty, \end{cases} \tag{3.77}$$

where the jump contour $\hat{\Sigma} = \Sigma^{(3)} \cup \partial D_\varepsilon(k_0)$ is depicted in Fig. 5, and the jump matrix $\hat{J}(x,t,k)$ is given by

$$\hat{J} = \begin{cases} M_-^{(k_0)} J^{(3)} (M_+^{(k_0)})^{-1}, & k \in \hat{\Sigma} \cap D_\varepsilon(k_0), \\ (M^{(k_0)})^{-1}, & k \in \partial D_\varepsilon(k_0), \\ J^{(3)}, & k \in \hat{\Sigma} \setminus \overline{D_\varepsilon(k_0)}. \end{cases} \tag{3.78}$$

Let $\hat{W} = \hat{J} - I$, and we rewrite $\hat{\Sigma}$ as follows:
$$\hat{\Sigma} = \partial D_\varepsilon(k_0) \cup \mathcal{X}^\varepsilon \cup \hat{\Sigma}_1 \cup \hat{\Sigma}_2,$$
where
$$\hat{\Sigma}_1 = \bigcup_1^6 \Sigma_j^{(3)} \setminus D_\varepsilon(k_0), \quad \hat{\Sigma}_2 = \bigcup_7^{10} \Sigma_j^{(3)},$$
and $\{\Sigma_j^{(3)}\}_1^{10}$ denoting the restriction of $\Sigma^{(3)}$ to the contour labeled by j in Fig. 3. Then the following inequalities are valid.

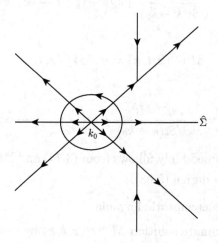

Fig. 5   The contour $\hat{\Sigma}$

Lemma 3.5   For $1 \leqslant p \leqslant \infty$, the following estimates hold for $t > 3$ and $\xi \in \mathcal{I}$,

$$\|\hat{W}\|_{L^p(\partial D_\varepsilon(k_0))} \leqslant Ct^{-\frac{1}{2}}, \tag{3.79}$$

$$\|\hat{W}\|_{L^p(\mathcal{X}^\varepsilon)} \leqslant Ct^{-\frac{1}{2}-\frac{1}{2p}} \ln t, \tag{3.80}$$

$$\|\hat{W}\|_{L^p(\hat{\Sigma}_1)} \leqslant Ce^{-ct}, \tag{3.81}$$

$$\|\hat{W}\|_{L^p(\hat{\Sigma}_2)} \leqslant Ct^{-\frac{3}{2}}. \tag{3.82}$$

Proof   The inequality (3.79) is a consequence of (3.67) and (3.78). For $k \in \mathcal{X}^\varepsilon$, we find
$$\hat{W} = M_-^{(k_0)}(J^{(3)} - J^{(k_0)})(M_+^{(k_0)})^{-1}.$$
Therefore, it follows from (3.65) and (3.66) that the estimate (3.80) holds. For $k \in D_2 \cap (\mathcal{X}_1 \setminus \overline{D_\varepsilon(k_0)})$, $\hat{W}$ only has a nonzero $\delta^{-1}(r_{1,a} + h)(\det \delta)^{-1} e^{t\Phi}$ in (3.81)

entry. Hence, for $t \geqslant 1$, by (3.20), we get

$$|\hat{W}_{21}| \leqslant C|r_{1,a} + h_a|e^{t\operatorname{Re}\Phi}$$

$$\leqslant \frac{C}{1+|k|^2}e^{-3|k-k_0|^2 t} \leqslant Ce^{-3\varepsilon^2 t}.$$

In a similar way, the estimates on $\mathcal{X}_j \setminus \overline{D_\varepsilon(k_0)}$, $j = 2, 3, 4$ hold. This proves (3.81). Since the matrix $\hat{W}$ on $\hat{\Sigma}_2$ only involves the small remainders h_r, $r_{1,r}$ and $r_{2,r}$, thus, by Lemmas 3.1 and 3.2, the estimate (3.82) follows.

The results in Lemma 3.5 imply that:

$$\|\hat{W}\|_{L^\infty(\hat{\Sigma})} \leqslant Ct^{-1/2}\ln t,$$
$$\|\hat{W}\|_{L^1 \cap L^2(\hat{\Sigma})} \leqslant Ct^{-1/2}, \qquad t > 3, \ \xi \in \mathcal{I}, \tag{3.83}$$

This uniformly vanishing bound on $\hat{W}$ shows that the RH problem (3.77) is a small-norm RH problem, for which there is a well-known existence and uniqueness theorem. In fact, we may write

$$\hat{M}(x,t,k) = I + \frac{1}{2\pi i}\int_{\hat{\Sigma}} \frac{(\hat{\mu}\hat{W})(x,t,\zeta)}{\zeta - k}d\zeta, \quad k \in \mathbf{C} \setminus \hat{\Sigma}, \tag{3.84}$$

where the 3×3 matrix-valued function $\hat{\mu}(x,t,k)$ is the unique solution of

$$\hat{\mu} = I + \hat{C}_{\hat{W}}\hat{\mu}, \tag{3.85}$$

and

$$\|\hat{\mu}(x,t,\cdot) - I\|_{L^2(\hat{\Sigma})} = O(t^{-1/2}), \quad t \to \infty, \ \xi \in \mathcal{I}. \tag{3.86}$$

The singular integral operator $\hat{C}_{\hat{W}}: L^2(\hat{\Sigma}) + L^\infty(\hat{\Sigma}) \to L^2(\hat{\Sigma})$ is defined by

$$\hat{C}_{\hat{W}}f = \hat{C}_-(f\hat{W}),$$

$$(\hat{C}_-f)(k) = \lim_{k \to \hat{\Sigma}_-}\int_{\hat{\Sigma}}\frac{f(\zeta)}{\zeta - k}\frac{d\zeta}{2\pi i},$$

where $\hat{C}_-$ is the well-known Cauchy operator. Moreover, by (3.83), we find

$$\|\hat{C}_{\hat{W}}\|_{B(L^2(\hat{\Sigma}))} \leqslant C\|\hat{W}\|_{L^\infty(\hat{\Sigma})} \leqslant Ct^{-1/2}\ln t, \tag{3.87}$$

where $B(L^2(\hat{\Sigma}))$ denotes the Banach space of bounded linear operators $L^2(\hat{\Sigma}) \to L^2(\hat{\Sigma})$. Therefore, the resolvent operator $(I - \hat{C}_{\hat{W}})^{-1}$ is existent and thus of both $\hat{\mu}$ and $\hat{M}$ for large t.

It follows from (3.84) that

$$\lim_{k\to\infty} k(\hat{M}(x,t,k) - I) = -\frac{1}{2\pi i}\int_{\hat{\Sigma}}(\hat{\mu}\hat{W})(x,t,k)dk. \tag{3.88}$$

Using (3.80) and (3.86), we have

$$\int_{\mathcal{X}^\varepsilon}(\hat{\mu}\hat{W})(x,t,k)dk = \int_{\mathcal{X}^\varepsilon}\hat{W}(x,t,k)dk + \int_{\mathcal{X}^\varepsilon}(\hat{\mu}(x,t,k) - I)\hat{W}(x,t,k)dk$$

$$\leqslant \|\hat{W}\|_{L^1(\mathcal{X}^\varepsilon)} + \|\hat{\mu} - I\|_{L^2(\mathcal{X}^\varepsilon)}\|\hat{W}\|_{L^2(\mathcal{X}^\varepsilon)}$$

$$\leqslant Ct^{-1}\ln t, \quad t \to \infty.$$

Similarly, by (3.81) and (3.86), the contribution from $\hat{\Sigma}_1$ to the right-hand side of (3.88) is

$$O(\|\hat{W}\|_{L^1(\hat{\Sigma}_1)} + \|\hat{\mu} - I\|_{L^2(\hat{\Sigma}_1)}\|\hat{W}\|_{L^2(\hat{\Sigma}_1)}) = O(e^{-ct}), \quad t \to \infty.$$

By (3.82) and (3.86), the contribution from $\hat{\Sigma}_2$ to the right-hand side of (3.88) is

$$O(\|\hat{W}\|_{L^1(\hat{\Sigma}_2)} + \|\hat{\mu} - I\|_{L^2(\hat{\Sigma}_2)}\|\hat{W}\|_{L^2(\hat{\Sigma}_2)}) = O(t^{-3/2}), \quad t \to \infty.$$

Finally, by (3.68), (3.79) and (3.86), we can get

$$-\frac{1}{2\pi i}\int_{\partial D_\varepsilon(k_0)}(\hat{\mu}\hat{W})(x,t,k)dk$$

$$= -\frac{1}{2\pi i}\int_{\partial D_\varepsilon(k_0)}\hat{W}(x,t,k)dk - \frac{1}{2\pi i}\int_{\partial D_\varepsilon(k_0)}(\hat{\mu}(x,t,k) - I)\hat{W}(x,t,k)dk$$

$$= -\frac{1}{2\pi i}\int_{\partial D_\varepsilon(k_0)}\left((M^{(k_0)})^{-1}(x,t,k) - I\right)dk + O(\|\hat{\mu} - I\|_{L^2(\partial D_\varepsilon(k_0))}\|\hat{W}\|_{L^2(\partial D_\varepsilon(k_0))})$$

$$= \frac{\eta^{-\hat{\sigma}}M_1^X}{\sqrt{8t}} + O(t^{-1}), \quad t \to \infty.$$

Thus, we obtain the following important relation

$$\lim_{k\to\infty} k(\hat{M}(x,t,k) - I) = \frac{\eta^{-\hat{\sigma}}M_1^X}{\sqrt{8t}} + O(t^{-1}\ln t), \quad t \to \infty. \tag{3.89}$$

Taking into account that (3.7), (3.10), (3.13), (3.42), (3.76) and (3.89), for sufficient large $k \in \mathbf{C} \setminus \hat{\Sigma}$, we get

$$\begin{pmatrix}u(x,t) & v(x,t)\end{pmatrix} = 2\lim_{k\to\infty}(kM(x,t,k))_{12}$$

$$= 2\lim_{k\to\infty} k(\hat{M}(x,t,k) - I)_{12}$$

$$= \frac{(\eta^{-\hat{\sigma}} M_1^X)_{12}}{\sqrt{2t}} + O\left(\frac{\ln t}{t}\right)$$

$$= \frac{-i\beta^X \eta^2}{\sqrt{2t}} + O\left(\frac{\ln t}{t}\right), \quad t \to \infty. \tag{3.90}$$

In view of (3.49) and (3.62), we obtain our main results stated as Theorem 1.1.

References

[1] Ablowitz MJ, Fokas AS. Complex Variables: Introduction and Applications [M]. 2nd ed. Cambridge: Cambridge University Press, 2003.

[2] Ablowitz MJ, Prinari P, Trubatch AD. Discrete and Continuous Nonlinear Schrodinger Systems [M]. Cambridge University Press, Cambridge, 2004.

[3] Arruda LK, Lenells J. Long-time asymptotics for the derivative nonlinear Schrödinger equation on the half-line [J]. Nonlinearity, 2017, 30: 4141–4172.

[4] Biondini G, Mantzavinos D. Long-time asymptotics for the focusing nonlinear Schrödinger equation with nonzero boundary conditions at infinity and asymptotic stage of modulational instability [J]. Comm. Pure Appl. Math., 2017, 70: 2300–2365.

[5] Boutet de Monvel A, Its A, Kotlyarov V. Long-time asymptotics for the focusing NLS equation with time-periodic boundary condition on the half-line [J]. Comm. Math. Phys., 2009, 290: 479–522.

[6] Boutet de Monvel A, Kotlyarov VP, Shepelsky D. Focusing NLS equation: long-time dynamics of step-like initial data [J]. Int. Math. Res. Not., 2011, 2011: 1613–1653.

[7] Boutet de Monvel A, Lenells J, Shepelsky D. Long-time asymptotics for the Degasperis-Procesi equation on the half-line [J]. Ann. Inst. Fourier, 2019, 69(1): 171–230.

[8] Boutet de Monvel A, Shepelsky D. The Ostrovsky–Vakhnenko equation by a Riemann–Hilbert approach [J]. J. Phys. A Math. Theor., 2015, 48: 035204.

[9] Boutet de Monvel A, Shepelsky D. A Riemann–Hilbert approach for the Degasperis-Procesi equation [J]. Nonlinearity, 2013, 26(7): 2081–2107.

[10] Boutet de Monvel A, Shepelsky D, Zielinski L. The short pulse equation by a Riemann–Hilbert approach [J]. Lett. Math. Phys., 2017, 107(7): 1345–1373.

[11] Buckingham R, Venakides S. Long-time asymptotics of the nonlinear Schrödinger equation shock problem [J]. Comm. Pure. Appl. Math., 2007, 60: 1349–1414.

[12] Busch TH, Anglin JR. Dark-bright solitons in inhomogeneous Bose-Einstein condensates [J]. Phys. Rev. Lett., 2001, 87: 010401.

[13] Deift P, Kriecherbauer T, McLaughlin KT-R, Venakides S, Zhou X. Uniform asymptotics for polynomials orthogonal with respect to varying exponential weights

and applications to universality questions in random matrix theory [J]. Comm. Pure Appl. Math., 1999, 52: 1335–1425.

[14] Deift P, Zhou X. A steepest descent method for oscillatory Riemann–Hilbert problems. Asymptotics for the MKdV equation [J]. Ann Math., 1993, 137: 295–368.

[15] Deift P, Zhou X. Long-time behavior of the non-focusing nonlinear Schrödinger equation-a case study [J]. Lectures in Mathematical Sciences, University of Tokyo, 5 1995.

[16] Fokas AS. A unified transform method for solving linear and certain nonlinear PDEs [J]. Proc. R. Soc. Lond. A, 1997, 453: 1411–1443.

[17] Geng X, Liu H. The nonlinear steepest descent method to long-time asymptotics of the coupled nonlinear Schrödinger equation [J]. J. Nonlinear Sci., 2018, 28: 739–763.

[18] Geng X, Liu H, Zhu J. Initial-boundary value problems for the coupled nonlinear Schrödinger equation on the half-line [J]. Stud. Appl. Math., 2015, 135: 310–346.

[19] Grunert K, Teschl G. Long-time asymptotics for the Korteweg–de Vries equation via nonlinear steepest descent [J]. Math. Phys. Anal. Geom., 2009, 12: 287–324.

[20] Guo B, Liu N. Long-time asymptotics for the Kundu–Eckhaus equation on the half-line [J]. J. Math. Phys., 2018, 59: 61505.

[21] Guo B, Liu N, Wang Y. Long-time asymptotics for the Hirota equation on the half-line [J]. Nonlinear Anal. TMA, 2018, 174: 118–140.

[22] Kotlyarov V, Minakov A. Riemann–Hilbert problem to the modified Korteveg–de Vries equation: long-time dynamics of the steplike initial data [J]. J. Math. Phys., 2010, 51: 093506.

[23] Lenells J. The nonlinear steepest descent method: asymptotics for initial-boundary value problems [J]. SIAM J. Math. Anal., 2016, 48(3): 2076–2118.

[24] Lenells J. The nonlinear steepest descent method for Riemann–Hilbert problems of low regularity [J]. Indiana Univ. Math. J., 2017, 66(4): 1287–1332.

[25] Liu J, Perry PA, Sulem C. Long-time behavior of solutions to the derivative nonlinear Schrödinger equation for soliton-free initial data [J]. Ann. Inst. H. Poincaré Anal. Non Linéaire, 2018, 35: 217–265.

[26] Manakov SV. On the theory of two-dimensional stationary self-focusing of electromagenic waves [J]. Sov. Phys. JETP, 1974, 38: 248–253.

[27] Wang DS, Wang XL. Long-time asymptotics and the bright N-soliton solutions of the Kundu–Eckhaus equation via the Riemann–Hilbert approach [J]. Nonlinear Anal. Real World Appl., 2018, 41: 334–361.

[28] Xu J, Fan EG. Long-time asymptotics for the Fokas–Lenells equation with decaying initial value problem: Without solitons [J]. J. Differential Equations, 2015, 259(3): 1098–1148.

郭柏灵院士

郭柏灵论文集

第十七卷（下）

Selected Papers of Guo Boling

Volume 17 (II)

郭柏灵 著

科学出版社

北京

内 容 简 介

《郭柏灵论文集第十七卷》收集的是郭柏灵先生发表于 2019 年度的主要科研论文，涉及的方程范围宽广，有确定性偏微分方程和随机偏微分方程，研究的问题包括适定性、爆破性、渐近性、孤立波等。

这些论文具有很高的学术价值，对偏微分方程、数学物理、非线性分析、计算数学等方向的科研工作者和研究生，是极好地参考著作。

图书在版编目（CIP）数据

郭柏灵论文集. 第 17 卷 / 郭柏灵著. -- 北京：科学出版社，2024.7.
-- ISBN 978-7-03-078966-2
Ⅰ. O175-53
中国国家版本馆 CIP 数据核字第 20242RU899 号

责任编辑：李 欣／责任校对：彭珍珍
责任印制：张 伟／封面设计：陈 敬

科学出版社 出版
北京东黄城根北街 16 号
邮政编码：100717
http://www.sciencep.com
北京厚诚则铭印刷科技有限公司印刷
科学出版社发行 各地新华书店经销
*
2024 年 7 月第 一 版 开本：720×1000 1/16
2024 年 7 月第一次印刷 印张：31 1/4 插页：1
字数：630 000
定价：398.00 元（全 2 册）
（如有印装质量问题，我社负责调换）

目 录

2019 年 (下)

Nonuniform Dependence and Well-Posedness for the Generalized Camassa-Holm
　　Equation ··· 523
General Propagation Lattice Boltzmann Model for a Variable-Coefficient Compound
　　KdV-Burgers Equation ··· 564
Decay Rates for the Viscous Incompressible MHD Equations with and without Surface
　　Tension ·· 598
The Cauchy Problem for a Generalized Camassa-Holm Equation ····················· 640
A Non-Homogeneous Boundary-Value Problem for the KdV-Burgers Equation Posed on
　　a Finite Domain ·· 675
Low Mach Number Limit of Compressible Nematic Liquid Crystal Flows with
　　Well-prepared Initial Data in a 3D Bounded Domain ······························· 681
Some Mathematical Models and Mathematical Analysis about Rayleigh-Taylor
　　Instability ·· 697
Stochastic 2D Primitive Equations: Central Limit Theorem and Moderate Deviation
　　Principle ··· 757
Long-Time Asymptotics of the Focusing Kundu-Eckhaus Equation with Nonzero
　　Boundary Conditions ··· 788
Existence and Blow-Up of the Solutions to the Viscous Quantum Magnetohy Drodynamic
　　Nematic Liquid Cryasal Model ·· 842
The Well-Posedness for the Cauchy Problem of the Double-Diffusive Convection
　　System ·· 904
Long-Time Asymptotics for the Sasa-Satsuma Equation via Nonlinear Steepest
　　Descent Method ··· 926
Long-Time Behavior of Solutions for the Compressible Quantum
　　Magnetohydrodynamic in R^3 ·· 966
Global Solution to the 3-D Inhomogeneous Incompressible MHD System with
　　Discontinuous Density ··· 988

Nonuniform Dependence and Well-Posedness for the Generalized Camassa-Holm Equation*

Mi Yongsheng(米永生), Wang Linsong(王林松),

Guo Boling(郭柏灵), and Mu Chunlai(穆春来)

Abstract In this paper, we are concerned with the Cauchy problem of the generalized Camassa-Holm equation. Using a Galerkin-type approximation scheme, it is shown that this equation is well posed in Sobolev spaces $H^s, s > 5/2$ for both the periodic and the nonperiodic case in the sense of Hadamard. That is, the data-to-solution map is continuous. Furthermore, it is proved that this dependence is sharp by showing that the solution map is not uniformly continuous. The nonuniform dependence is proved using the method of approximate solutions and well-posedness estimates.

Keywords Besov spaces; Camassa-Holm type equation; local well–posedness

1 Introduction

In this paper, we consider the Cauchy problem for the following generalized Camassa–Holm equation

$$\begin{cases} (1-\partial_x^2)u_t = \partial_x(2+\partial_x)\left[(2-\partial_x)u\right]^2, & t>0, x \in \mathbf{R}, \\ u(x,0) = u_0(x), & x \in \mathbf{R}. \end{cases} \quad (1.1)$$

Equation (1.1) was proposed by Novikov[1]. Recently, the Cauchy problem (1.1) was studied by Tu and Yin in [2-5], they obtained the well-posedness, blow–up phenomenathe, global existence, analyticity, persistence properties, the existence and uniqueness of global weak solutions. In [1], Novikov showed that Equation.

---

* Appl. Anal., 2018, 98(8): 1520-1548. Doi: 10.1080/00036811.2018.1430785.

(1.1) is integrable using as definition of integrability the existence of an infinite hierarchy of quasi–local higher symmetries and it belongs to the following class

$$(1 - \partial_x^2)u_t = F(u, u_x, u_{xx}, u_{xxx}), \tag{1.2}$$

which has attracted much interest, particularly in the possible integrable members of (1.3).

The most celebrated integrable member of (1.2) is the well-known Camassa–Holm equation

$$\begin{cases} (1 - \partial_x^2)u_t = -3uu_x 2u_x u_{xx} + uu_{xxx}, & t > 0, x \in \mathbf{R}, \\ u(x,0) = u_0(x), & x \in \mathbf{R}, \end{cases} \tag{1.3}$$

modeling the unidirectional propagation of shallow water waves over a flat bottom, $u(t,x)$ stands for the fluid velocity at time t in the spatial direction x. It is a well-known integrable equation describing the velocity dynamics of shallow water waves, in-depth discussion in [6]. This equation spontaneously exhibits the emergence of singular solutions from smooth initial conditions. It has a bi–Hamilton structure [7] and is completely integrable [8, 9]. In particular, it possesses an infinity of conservation laws and is solvable by its corresponding inverse scattering transform. After the birth of the Camassa–Holm equation, many works have been carried out to probe its dynamic properties. Such as, Equation (1.2) has traveling wave solutions of the form $ce^{-|x-ct|}$, called peakons, which describes an essential feature of the traveling waves of the largest amplitude (see [10-14]). It is shown in [15-17] that the inverse spectral or scattering approach is a powerful tool to handle the Camassa–Holm equation and analyze its dynamics. It is worthwhile to mention that Equation (1.2) gives rise to the geodesic flow of a certain invariant metric on the Bott–Virasoro group [18, 19], and this geometric illustration leads to a proof that the least action principle holds. It is shown in [20] that the blow-up occurs in the form of breaking waves, namely the solution remains bounded but its slope becomes unbounded in finite time. Moreover, the Camassa–Holm equation has global conservative solutions [21, 22] and dissipative solutions [23, 24]. Another aspect of the Camassa–Holm equation that attracted attention is related to the relevance of integrable equations to the modeling of tsunami waves, cf. the discussions in [25-28], The Cauchy problem of the generalized Camassa–Holm equation was investigated in [29, 30].

The second celebrated integrable member of (1.4) is the famous Degasperis–Procesi (DP) equation [31]

$$\begin{cases} (1-\partial_x^2)u_t = -4uu_x + 2u_x u_{xx} + uu_{xxx}, & x \in \mathbf{R}, t > 0, \\ u(x,0) = u_0(x), & x \in \mathbf{R}. \end{cases} \quad (1.4)$$

Degasperis, Holm and Hone [32] proved the formal integrability of Equation (1.4) by constructing a Lax pair. They also showed that it has bi–Hamiltonian structure and an infinite sequence of conserved quantities and admits exact peakon solutions. The direct and inverse scattering approach to pursue it can be seen in [33]. Moreover, in [34], they also presented that the DP equation has a bi–Hamiltonian structure and an infinite number of conservation laws, and admits exact peakon solutions which are analogous to the Camassa–Holm peakons. It is worth pointing out that solutions of this type are not mere abstractizations: the peakons replicate a feature that is characteristic for the waves of great height–waves of the largest amplitude that are exact solutions of the governing equations for irrotational water waves cf. the papers [10, 11, 35]. The DP equation is a model for nonlinear shallow water dynamics cf. the discussion in [32]. The numerical stability of solitons and peakons, the multisoliton solutions and their peakon limits, together with an inverse scattering method to compute n–peakon solutions to Degasperis–Procesi equation have been investigated, respectively in [36-38]. Furthermore, the traveling wave solutions and the classification of all weak traveling wave solutions to Degasperis–Procesi equation were presented in [39, 40].

The third well–known integrable member of (1.2) is the Novikov equation

$$\begin{cases} (1-\partial_x^2)u_t = -4u^2 u_x + 3uu_x u_{xx} + u^2 u_{xxx}, & x \in \mathbf{R}, t > 0, \\ u(x,0) = u_0(x), & x \in \mathbf{R}, \end{cases} \quad (1.5)$$

which has been recently discovered by Vladimir Novikov in a symmetry classification of nonlocal PDEs with quadratic or cubic nonlinearity [1]. The perturbative symmetry approach yields necessary conditions for a PDE to admit infinitely many symmetries. Using this approach, Novikov was able to isolate Equation (1.4) and find its first few symmetries, and he subsequently found a scalar Lax pair for it, then proved that the equation is integrable, which can be thought as a generalization of the Camassa–Holm equation. In [41], it is shown that the Novikov equation admits peakon solutions like the Camassa–Holm. Also, it has a Lax pair in matrix form and a bi–Hamiltonian structure. Furthermore, it has infinitely many conserved quantities. Like Camassa–Holm, the most important quantity conserved by a solution u to Novikov equation is its H^1-norm $\|u\|_{H^1}^2 = \int_R (u^2 + u_x^2)$, which plays an important role in the study of Equation (1.1). In [42-45], the authors study well-posedness and dependence on initial data for the Cauchy problem for Novikov equation. Recently, in [46], a global existence result and conditions on the initial data were considered. The Novikov equation with dissipative term was considered in [47]. Multipeakon solutions were studied in [6, 41, 48].

Nonuniform dependence of the Camassa–Holm type equation has been studied by several authors. In [49], Himonas and Misiołek provides the first nonuniform dependence result for CH on the circle for $s \geqslant 2$ using explicitly constructed travelling wave solutions. This result was sharpened to $s > 3/2$ in [50] on the line and [51] on the circle. Both of these more recent works utilize the method of approximate solutions in conjunction with delicate commutator and multiplier estimates. For the DP equation, the first result for nonuniform dependence can be found in Christov and Hakkaev [52]. In the periodic case with $s \geqslant 2$, nonuniform dependence of the data–to– solution map for DP is established following the method of travelling wave solutions developed in [49] for the CH equation. This result has been recently sharpened in [53], where nonuniform dependence on the initial data for DP is proven on both the circle and line for $s > 3/2$ using the method of approximate solutions in tandem with a twisted L^2-norm that is conserved by the DP equation. This method of approximate solutions has also been adapted to a homogeneous setting in Holliman [54] to prove nonuniform dependence of the flow map for the Hunter–Saxton equation on the circle with $s > 3/2$, and to the higher dimensional setting in [55] to prove nonuniform dependence for the Euler equations.

Next we state the first result of this work, which refers to the well-posedness of the Cauchy problem (1.1) in Sobolev spaces in the sense of Hadamard. This result specifically reads as follows

Theorem 1.1 *If* $s > \dfrac{5}{2}$ *and* $u_0 \in H^s$, *then there exists* $T > 0$ *and a unique solution* $u \in C([0,T]; H^s)$ *of the Cauchy problem*(1.1), *which depends*

continuously on the initial data u_0. Furthermore, we have the estimate

$$\|u(t)\|_{H^s} \leqslant 2\|u_0\|_{H^s}, \quad for \ 0 \leqslant T \leqslant \frac{1}{2c_s\|u_0\|_{H^s}}, \tag{1.6}$$

where $c_s > 0$ is a constant depending on s.

Our proof of Theorem 1.1 is based on a Galerkin–type approximation method, which for quasi–linear symmetric hyperbolic systems can be found in Taylor [56, 57].

The second result of this work demonstrates the fact that the continuous dependence of the solution map on initial data, as established in Theorem 1.1, is sharp. More precisely, it reads as follows.

Theorem 1.2 *If $s > \frac{5}{2}$ then the data–to–solution map for both the periodic and the nonperiodic Cauchy problem (1.1) is not uniformly continuous from any bounded subset in H^s into $C([0,T];H^s)$.*

The proof of Theorem 1.2 is based on the method of approximate solutions and well-posedness estimates for the solution and its lifespan, which is motivated by the works of the [29, 30, 50, 53-55].

The paper is structured as follows. In Section 2, we transform the generalized equation into a system of equations in $H^{s-1} \times H^{s-1}$ for which we show existence and uniqueness of solution on the circle for $s > 5/2$. We then determine the relation between this solution and the solution of (1.1), and hence we obtain the lifespan and the size estimate stated in Theorem 1.1. Section 3 completes the proof of Hadamard well-posedness for the system by establishing that the dependence of the solution on the initial data is continuous, and hence Theorem 1.1 is proved. Section 4 provides an outline of proof of Hadamard well-posedness for the Equation (1.1) on the line. In Section 5 we prove Theorem 1.2, namely we show that the continuous dependence established in Section 3 is not uniform. The proof of this result in the case of the line is briefly discussed in Section 6.

2   Well-posedness on the circle

For the sake of brevity, we shall rewrite the Cauchy problem (1.1) as follows:

$$\begin{cases} \partial_t u = 4u\partial_x u - (\partial_x u)^2 + \Phi(u), & t > 0, x \in \mathbf{R}, \\ u(x,0) = u_0(x), & x \in \mathbf{R}, \end{cases} \tag{2.1}$$

where
$$\Phi(u) = D^{-2}(\partial_x u)^2 + D^{-2}\partial_x[2(\partial_x u)^2 + 6u^2]. \tag{2.2}$$

For any $s \in \mathbf{R}$, we take the operator $D^s = (1 - \partial_x^2)^{\frac{s}{2}}$ to be defined by
$$\widehat{D^s f}(k) \triangleq (1+k^2)^{s/2}\widehat{f}(k),$$
where $\widehat{f}(k)$ is the Fourier transform
$$\widehat{f}(k) \triangleq \int_{\mathbf{T}} e^{-ixk} f(x)\, dx.$$

Therefore, the inverse relation is given by the Fourier series
$$f(x) = \frac{1}{2\pi}\sum_{k\in\mathbf{Z}} \widehat{f}(k)e^{-ixk}.$$

Then, for $f \in H^s(\mathbf{T})$ we have
$$\|f\|_{H^s(\mathbf{T})}^2 = \frac{1}{2\pi}\sum_{k\in\mathbf{Z}}(1+k^2)^s|\widehat{f}(k)|^2 = \|D^s f\|_{L^2(\mathbf{T})}^2.$$

We will prove that the Cauchy problem (2.1) on the circle with initial data $u_0 \in H^s$ is well posed in the sense of Hadamard for $s > \frac{5}{2}$. In the current section, we will show that there exists a solution to this problem which is unique. The continuity of the data–to–solution map will be established in Section 3.

Existence. The Equation(2.1) cannot be treated as an ODE in H^s due to the terms $(u_x)^2$ and uu_x. What is more, although the issue caused by the latter term can be resolved via mollification, this is not the case concerning the term $(u_x)^2$. Thus, we follow a different route to bypass this problem. Differentiating the Equation (2.1) with respect to x and simplifying the resulting expression, we obtain

$$\partial_t(\partial_x u) = 2(\partial_x u)^2 + 4u\partial_x^2 u - 2\partial_x u \partial_x^2 u - 6u^2$$
$$+ D^{-2}\partial_x((\partial_x u)^2) + D^{-2}(2(\partial_x u)^2 + 6u^2). \tag{2.3}$$

Then, letting $w = \partial_x u$ in Equations (2.1) and (2.3) motivates the study of the system

$$\begin{cases} \partial_t u = 4uw - w^2 + F(u,w), \\ \partial_t w = 2w^2 + 4u\partial_x w - 2w\partial_x w - 6u^2 + G(u,w), \\ u(x,0) = u_0(x), w(x,0) = \partial_x u_0(x) = w_0(x), \end{cases} \tag{2.4}$$

where the nonlocal terms F, G are defined by

$$F(u,w) = D^{-2}w^2 + D^{-2}\partial_x(2w^2 + 6u^2),$$
$$G(u,w) = D^{-2}\partial_x(w^2) + D^{-2}(2w^2 + 6u^2). \quad (2.5)$$

Next, we introduce the Friedrichs mollifier $j_\varepsilon, \varepsilon \in (0,1]$, as the operator

$$J_\varepsilon f \triangleq j_\varepsilon * f, \quad (2.6)$$

where the function J_ε is defined by the formula

$$j_\varepsilon(x) = \frac{1}{2\pi} \sum_{k \in \mathbb{Z}} \widehat{j}(\varepsilon k) e^{ikx},$$

so that $\widehat{j_\varepsilon}(k) = \widehat{j}(\varepsilon k)$. Note that in the nonperiodic case, the function j_ε is defined by first fixing a function $j \in \mathcal{S}(\mathbb{R})$ such that $\widehat{j}(\xi) = 0$ on $[-1,1]$ and finally defining $j_\varepsilon(x) = \frac{1}{\varepsilon} j\left(\frac{x}{\varepsilon}\right)$.

Applying J_ε to system (2.4) yields the following mollified system

$$\begin{cases} \partial_t u_\varepsilon = 4J_\varepsilon(J_\varepsilon u_\varepsilon w_\varepsilon)^3 - J_\varepsilon(J_\varepsilon w_\varepsilon)^2 + F(u_\varepsilon, w_\varepsilon), \\ \partial_t w_\varepsilon = 2J_\varepsilon(J_\varepsilon w_\varepsilon) + 4J_\varepsilon[J_\varepsilon u_\varepsilon(J_\varepsilon w_\varepsilon)] - 2J_\varepsilon[J_\varepsilon w_\varepsilon(J_\varepsilon w_\varepsilon)] \\ \qquad -6u_\varepsilon^2 + G(u_\varepsilon, w_\varepsilon)), \\ u_\varepsilon(x,0) = u_0(x), w_\varepsilon(x,0) = \partial_x u_0(x) = w_0(x). \end{cases} \quad (2.7)$$

Consider the vector $U_\varepsilon = (u_\varepsilon, w_\varepsilon)$ with Sobolev norm defined by

$$\|U_\varepsilon\|_{H^s} = \|u_\varepsilon\|_{H^s} + \|w_\varepsilon\|_{H^s}. \quad (2.8)$$

In particular, at time $t = 0$ we have that $U_\varepsilon(0) = (u_\varepsilon(0), w_\varepsilon(0)) = (u_0, \partial_x u_0)$.

For $u_0 \in H^s$, the mollified system (2.7) is a system of ODEs in $H^{s-1} \times H^{s-1}$. Thus, by the Fundamental Theorem for ODEs in Banach spaces (see [58], 10.4.5, chapter 10, p. 283), there exists a unique solution U_ε for this system in the space $C([0, T_\varepsilon]; H^{s-1} \times H^{s-1})$.

Proof of Theorem 1.1 We will prove the Theorem via energy estimates. We will make extensive use of the inequalities

$$\|J_\varepsilon f\|_{H^s} \leqslant \|f\|_{H^s}, \quad \|\partial_x f\|_{H^{s-1}} \leqslant \|f\|_{H^s}, \quad \|D^{-2}f\|_{H^{s-1}} \leqslant \|f\|_{H^{s-2}}. \quad (2.9)$$

Note that the estimate
$$\|fg\|_{H^s} \leqslant c_s(\|f\|_{H^s}\|h\|_{L^\infty} + \|g\|_{H^s}\|f\|_{L^\infty}), \quad s > 0, c_s > 0,$$
together with the Sobolev lemma for $s > \frac{1}{2}$ yields the algebra property
$$\|fg\|_{H^s} \leqslant c_s\|f\|_{H^s}\|g\|_{H^s}, \quad c_s > 0. \tag{2.10}$$

Finally, we will often employ the following lemma.

Lemma 2.1 (Kato–Ponce) *If $s > 0$ then there is $c_s > 0$ such that*
$$\|[D^s, f]g\|_{L^2} \leqslant c_s(\|D^s f\|_{L^2}\|g\|_{L^\infty} + \|\partial_x f\|_{L^\infty}\|D^{s-1}g\|_{L^2}).$$

Energy estimates for w_ε and u_ε Applying the operator D^{s-1} on the left of the second equation of system (2.7), multiplying by $D^{s-1}w_\varepsilon$ on the right, commuting the operators J_ε and D^{s-1} twice and integrating over $\mathbf{T}$, we arrive at the equation

$$\begin{aligned}\frac{1}{2}\|w_\varepsilon\|_{H^{s-1}}^2 = {}& 2\int_{\mathbf{T}} D^{s-1}(J_\varepsilon w_\varepsilon)^2 \cdot D^{s-1} J_\varepsilon w_\varepsilon \, dx \\ & + 4\int_{\mathbf{T}} D^{s-1}[J_\varepsilon u_\varepsilon \partial_x(J_\varepsilon w_\varepsilon)] \cdot D^{s-1} J_\varepsilon w_\varepsilon \, dx \\ & - 2\int_{\mathbf{T}} D^{s-1}[(J_\varepsilon w_\varepsilon)\partial_x(J_\varepsilon w_\varepsilon)] \cdot D^{s-1} J_\varepsilon w_\varepsilon \, dx \\ & - 6\int_{\mathbf{T}} D^{s-1} u_\varepsilon^2 \cdot D^{s-1} w_\varepsilon \, dx \\ & + \int_{\mathbf{T}} D^{s-3}\partial_x(w_\varepsilon^2) \cdot D^{s-1} w_\varepsilon \, dx \\ & + \int_{\mathbf{T}} D^{s-3}(2w^2 + 6u_\varepsilon^2) \cdot D^{s-1} w_\varepsilon \, dx, \\ = {}& A_1 + A_2 + A_3 + A_4 + A_5 + A_6.\end{aligned} \tag{2.11}$$

Using the Cauchy–Schwarz inequality and the algebra property, we find
$$A_1 \lesssim \|(J_\varepsilon w_\varepsilon)^2\|_{H^{s-1}}\|J_\varepsilon w_\varepsilon\|_{H^{s-1}} \lesssim \|w_\varepsilon\|_{H^{s-1}}^3. \tag{2.12}$$

Note that
$$\int_{\mathbf{T}} D^{s-1}[J_\varepsilon u_\varepsilon \partial_x(J_\varepsilon w_\varepsilon)] D^{s-1} J_\varepsilon w_\varepsilon \, dx = \int_{\mathbf{T}} [D^{s-1}, J_\varepsilon u_\varepsilon] \partial_x J_\varepsilon w_\varepsilon \cdot D^{s-1} J_\varepsilon w_\varepsilon \, dx$$

$$+ \int_{\mathbb{T}} J_\varepsilon u_\varepsilon D^{s-1} \partial_x J_\varepsilon w_\varepsilon \cdot D^{s-1} J_\varepsilon w_\varepsilon \, dx,$$

hence, by the Cauchy–Schwarz inequality and the Kato–Ponce Lemma 2.1,

$$A_2 \lesssim \left\| [D^{s-1}, J_\varepsilon u_\varepsilon] \partial_x J_\varepsilon w_\varepsilon \right\|_{L^2} \left\| D^{s-1} J_\varepsilon w_\varepsilon \right\|_{L^2}$$

$$+ \left| \int_{\mathbb{T}} J_\varepsilon u_\varepsilon D^{s-1} \partial_x J_\varepsilon w_\varepsilon \cdot D^{s-1} J_\varepsilon w_\varepsilon \, dx \right|$$

$$\lesssim \left(\|J_\varepsilon u_\varepsilon\|_{H^{s-1}} \|\partial_x J_\varepsilon w_\varepsilon\|_{L^\infty} + \|\partial_x J_\varepsilon u_\varepsilon\|_{L^\infty} \|\partial_x J_\varepsilon w_\varepsilon\|_{H^{s-2}} \right) \|J_\varepsilon w_\varepsilon\|_{H^{s-1}}$$

$$+ \left| \int_{\mathbb{T}} J_\varepsilon u_\varepsilon \partial_x (D^{s-1} J_\varepsilon w_\varepsilon)^2 \, dx \right|.$$

Thus, by the algebra property for $s > \frac{3}{2}$ and integration by parts we find

$$A_2 \lesssim \|J_\varepsilon u_\varepsilon\|_{H^{s-1}} \|J_\varepsilon w_\varepsilon\|_{C^1} \|J_\varepsilon w_\varepsilon\|_{H^{s-1}} + \|J_\varepsilon u_\varepsilon\|_{C^1} \|J_\varepsilon w_\varepsilon\|_{H^{s-1}} \|J_\varepsilon w_\varepsilon\|_{H^{s-1}}$$

$$+ \|J_\varepsilon u_\varepsilon\|_{H^{s-1}} \|J_\varepsilon w_\varepsilon\|_{C^1}^2$$

$$\lesssim \|u_\varepsilon\|_{H^{s-1}} \|w_\varepsilon\|_{H^{s-1}}^2, \tag{2.13}$$

with the last inequality following by the Sobolev lemma for $s > \frac{5}{2}$. The third term of Equation (2.11) is estimated in a similar fashion to yield

$$A_3 \lesssim \|w_\varepsilon\|_{H^{s-1}}^3, \tag{2.14}$$

while regarding the nonlocal terms, we have

$$A_4 \lesssim \|u_\varepsilon\|_{H^{s-1}}^2 \|w_\varepsilon\|_{H^{s-1}}, \quad A_5 \lesssim \|w_\varepsilon\|_{H^{s-1}}^3, \tag{2.15}$$

and

$$A_6 \lesssim \|u_\varepsilon\|_{H^{s-1}}^2 \|w_\varepsilon\|_{H^{s-1}} + \|w_\varepsilon\|_{H^{s-1}}^3. \tag{2.16}$$

Combining (2.11)–(2.16) we deduce that for $s > \frac{5}{2}$,

$$\frac{1}{2} \frac{d}{dt} \|w_\varepsilon\|_{H^{s-1}}^2 \lesssim \|w_\varepsilon\|_{H^{s-1}} (\|u_\varepsilon\|_{H^{s-1}} + \|w_\varepsilon\|_{H^{s-1}})^2$$

which yields the inequality

$$\frac{d}{dt} \|w_\varepsilon\|_{H^{s-1}}^2 \lesssim (\|u_\varepsilon\|_{H^{s-1}} + \|w_\varepsilon\|_{H^{s-1}})^2. \tag{2.17}$$

Similar computations on the first equation of system (2.7) lead to the inequality

$$\frac{d}{dt}\|u_\varepsilon\|^2_{H^{s-1}} \lesssim (\|u_\varepsilon\|_{H^{s-1}} + \|w_\varepsilon\|_{H^{s-1}})^2. \tag{2.18}$$

Adding inequalities (2.17) and (2.18), while recalling definition (2.8), yields

$$\frac{d}{dt}\|U_\varepsilon(t)\|^2_{H^{s-1}} \leqslant c_s \|U_\varepsilon(t)\|^2_{H^{s-1}}$$

for some constant $c_s > 0$ which depends only on s. Solving this differential inequality, we find

$$\|U_\varepsilon(t)\|^2_{H^{s-1}} \leqslant \frac{\|U_\varepsilon(0)\|_{H^{s-1}}}{1 - c_s t \|U_\varepsilon(0)\|_{H^{s-1}}}. \tag{2.19}$$

Thus, for lifespan T_ε equal to

$$T_\varepsilon = \frac{1}{2c_s \|U_\varepsilon(0)\|_{H^{s-1}}}, \tag{2.20}$$

the Fundamental Theorem for ODEs in Banach spaces [58] implies that there exists a unique solution u_ε to system (2.7) for $0 \leqslant t \leqslant T$ satisfying the estimate

$$\|U_\varepsilon(t)\|_{H^{s-1}} \leqslant 2\|U_\varepsilon(0)\|_{H^{s-1}}, \quad 0 \leqslant t \leqslant T_\varepsilon. \tag{2.21}$$

Note, however, that according to the prescribed initial condition

$$\|U_\varepsilon(0)\|_{H^{s-1}} = \|u_0\|_{H^{s-1}} + \|\partial_x u_0\|_{H^{s-1}} \leqslant 2\|u_0\|_{H^s}.$$

Note, however, that according to the prescribed initial conditions with the last inequality following by the second property in (2.9) and the fact that $u_0 \in H^s$. Therefore, for the common lifespan T given by

$$T = \frac{1}{4c_s \|u_0\|_{H^{s-1}}}, \quad 0 \leqslant t \leqslant T, \tag{2.22}$$

the estimate (2.20) implies the size estimate

$$\|U_\varepsilon(0)\|_{H^{s-1}} \leqslant 4\|u_0\|_{H^s}, \quad 0 \leqslant t \leqslant T.$$

Furthermore, returning to system (2.7) we have

$$\|\partial_t u_\varepsilon\|_{H^{s-2}} \lesssim \|J_\varepsilon(J_\varepsilon u_\varepsilon J_\varepsilon w_\varepsilon)\|_{H^{s-2}} + \|J_\varepsilon(J_\varepsilon w_\varepsilon)^2\|_{H^{s-2}} + \|F(u_\varepsilon, w_\varepsilon)\|_{H^{s-2}},$$

$$\|\partial_t w_\varepsilon\|_{H^{s-2}} \lesssim \|J_\varepsilon (J_\varepsilon w_\varepsilon)^2\|_{H^{s-2}} + \|J_\varepsilon[J_\varepsilon u_\varepsilon \partial_x (J_\varepsilon w_\varepsilon)]\|_{H^{s-2}}$$
$$+ \|J_\varepsilon[(J_\varepsilon w_\varepsilon)\partial_x(J_\varepsilon w_\varepsilon)]\|_{H^{s-2}} + |u_\varepsilon^2|_{H^{s-2}} + \|G(u_\varepsilon, w_\varepsilon))\|_{H^{s-2}}.$$

Hence, using the size estimate (2.22) it is straightforward to deduce the following estimate for the derivative of U_ε with respect to t:

$$\|\partial_t U_\varepsilon(t)\|_{H^{s-2}} \leqslant \|u_0\|_{H^s}^2, \quad 0 \leqslant t \leqslant T. \tag{2.23}$$

Return to the original system (2.4). So far, we have established the existence of a unique solution $U_\varepsilon \in C([0,T]; H^{s-1} \times H^{s-1})$ to the mollified ivp (2.7) for $s > \frac{5}{2}$, with lifespan T defined by (2.21) and such that it satisfies the size estimates (2.22) and (2.23). We now need to show that for $s > \frac{5}{2}$, we have $U_\varepsilon \to U = (u,w)$ in $C([0,T]; H^{s-1} \times H^{s-1}))$ where U is the solution of (2.4) satisfying the size estimate

$$\|U(t)\|_{H^{s-1}} \leqslant 4\|u_0\|_{H^s}, \quad 0 \leqslant t \leqslant T, \tag{2.24}$$

and the derivative size estimate

$$\|\partial_t U(t)\|_{H^{s-2}} \leqslant \|u_0\|_{H^s}^2, \quad 0 \leqslant t \leqslant T, \tag{2.25}$$

with the lifespan T defined by Equation (2.21).

Proving the size estimate (2.24). For simplicity of notation, we hereafter let $I = [0,T]$ and denote the space $C([0,T]; H^{s-1} \times H^{s-1})$ by $C(I; H^{s-1})$. We know that $\{U_\varepsilon\}_{\varepsilon \in (0,1]} \subset C(I; H^{s-1}) \subset L^\infty(I; H^{s-1})$. Since $L^\infty(I; H^{s-1})$ is the dual of

$$L^1(I; H^{s-1}),$$

Alaoglu's theorem implies that the closed unit ball in $L^\infty(I; H^{s-1})$ is compact in the weak* sense. What is more, the size estimate (2.22) implies that $\{U_\varepsilon\}_{\varepsilon \in (0,1]} \subset \overline{B}(0, 4\|u_0\|_{H^s})$ and therefore, there exists a subsequence $\{U_{\varepsilon_j}\}_{\varepsilon_j \in (0,1]}$ that converges to some $U \in L^\infty(I; H^{s-1})$ in the weak* sense, $\langle U_{\varepsilon_j}, \varphi \rangle \to \langle U, \varphi \rangle$ for all $\varphi \in L^1(I; H^{s-1})$ with U satisfying the estimate (2.24).

Convergence in $C(I; H^{s-1-\sigma}), \sigma \in (0,1)$. We will prove convergence in this space via Ascoli's theorem. For simplicity of notation, we drop the index j and denote U_{ε_j} by U_ε. We have that $\{U_\varepsilon\}_{\varepsilon \in (0,1]} \subset C(I; H^{s-1})$ and since T is compact, the inclusion mapping $H^{s-1}(\mathbb{T}) \to H^{s-1-\sigma}(\mathbb{T})$ is compact. Rellich's theorem then implies that $\{U_\varepsilon\}_{\varepsilon \in (0,1]}$ is precompact in $H^{s-1-\sigma}(\mathbb{T})$. Thus, the first condition of Ascoli's theorem is met. It remains to show that the second condition is also met, i.e. that U_ε is equicontinuous on I:

$$\|U_\varepsilon(t_1) - U_\varepsilon(t_2)\|_{H^{s-1-\sigma}} \leqslant |t_1 - t_2|^\sigma, \quad \forall t_1, t_2 \in I. \tag{2.26}$$

Using the fact that $a^\sigma < 1 + a, \sigma \in (0,1)$ for $a = \dfrac{1}{(1+k^2)|t_1 - t_2|}$, the triangle inequality and the size estimate (2.22), we find

$$\sup_{t_1 \neq t_2} \frac{\|U_\varepsilon(t_1) - U_\varepsilon(t_2)\|_{H^{s-1-\sigma}}}{|t_1 - t_2|^\sigma} \leqslant 8\|u_0\|_{H^s} + \sup_{t_1 \neq t_2} \frac{\|U_\varepsilon(t_1) - U_\varepsilon(t_2)\|_{H^{s-2}}}{|t_1 - t_2|}.$$

Moreover, by the Mean Value Theorem for $U_\varepsilon(t)$ on $[t_1, t_2]$ we have

$$\|U_\varepsilon(t_1) - U_\varepsilon(t_2)\|_{H^r} \leqslant \sup_{t \in I} \|\partial_t U_\varepsilon(t)\|_{H^r} |t_1 - t_2|, \quad r > 0, \tag{2.27}$$

hence, using (2.27) together with the derivative size estimate (2.23), we infer

$$\|U_\varepsilon(t_1) - U_\varepsilon(t_2)\|_{H^{s-2}} \leqslant \sup_{t \in I} \|\partial_t U_\varepsilon(t)\|_{H^{s-2}} |t_1 - t_2| \lesssim \|u_0\|_{H^s}^2 |t_1 - t_2|.$$

Therefore, we conclude that

$$\sup_{t_1 \neq t_2} \frac{\|U_\varepsilon(t_1) - U_\varepsilon(t_2)\|_{H^{s-1-\sigma}}}{|t_1 - t_2|^\sigma} \lesssim \|u_0\|_{H^s} + \|u_0\|_{H^s}^2$$

and equicontinuity has been established. Thus, by Ascoli's theorem we can extract a subsequence of $\{U_\varepsilon\}_{\varepsilon \in (0,1]}$ which converges uniformly to U in $C(I; H^{s-1-\sigma}), \sigma \in (0,1)$.

Convergence in $C(I; C^1)$. Since U_ε converges to U in $C(I; H^{s-1-\sigma})$ as $\varepsilon \to 0$, it follows that $\lim_{\varepsilon \to 0} \sup_{t \in I} \|U_\varepsilon - U\|_{H^{s-1-\sigma}} = 0$. For $\sigma \in (0,1)$ such that $\sigma < s - \dfrac{5}{2}$ or, equivalently, $s - 1 - \sigma > \dfrac{3}{2}$, the Sobolev lemma yields $\lim_{\varepsilon \to 0} \|U_\varepsilon - U\|_{C(I;C^1)} = 0$. Therefore, convergence in $C(I; C^1)$ has been established.

Proving that U is a solution to the ivp (2.4). We wish to make use of the generalization to Sobolev spaces of the following theorem from real analysis. Suppose that $f_n : I \to \mathbf{R}$ is continuous. If $f_n(t_0) \to f(t_0)$ for some $t_0 \in I$ as $n \to \infty$ and if f_n' is uniformly convergent on I, then f_n converges uniformly to f on I, with $f' = \lim_{n \to \infty} f_n'(t)$.

We will apply this theorem to the sequence $\{U_\varepsilon\}_{\varepsilon \in (0,1]}$. For this sequence, we have shown convergence to $U(t)$ in $C(I; C^1)$, hence the first condition of the theorem is met. It remains to show that $\partial_t U_\varepsilon$ is uniformly convergent.

By the definition of U_ε we have that $\partial_t U_\varepsilon = (\partial_t u_\varepsilon, \partial_t w_\varepsilon)$. Recall the mollified system (2.7)

$$\partial_t u_\varepsilon = 4J_\varepsilon(J_\varepsilon u_\varepsilon J_\varepsilon w_\varepsilon) - J_\varepsilon[(J_\varepsilon w_\varepsilon)^2 + F(u_\varepsilon, w_\varepsilon), \tag{2.28}$$

$$\partial_t w_\varepsilon = 2J_\varepsilon((J_\varepsilon w_\varepsilon)^3)^2 + 4J_\varepsilon[J_\varepsilon u_\varepsilon \partial_x(J_\varepsilon w_\varepsilon)]$$
$$- 2J_\varepsilon[J_\varepsilon w_\varepsilon \partial_x(J_\varepsilon w_\varepsilon)] - 6u_\varepsilon^2 - G(u_\varepsilon, w_\varepsilon)). \tag{2.29}$$

Since $U_\varepsilon \to U$ in $C(I; C^1)$, the term u_ε^2 on the rhs of (2.29) converges uniformly to u^2 in $C(I; C^1)$ and thus in $C(I; C)$. Also, by the continuity of the operators $D^{-2}, D^{-2}\partial_x$ involved in the nonlocal terms of (2.28) and (2.29) we have that $F(u_\varepsilon, w_\varepsilon) \to F(u, w)$ and $G(u_\varepsilon, w_\varepsilon) \to G(u, w)$ uniformly in $C(I; C^1)$ and hence in $C(I; C)$. Regarding the second term on the rhs of (2.28), we note that

$$\|(J_\varepsilon w_\varepsilon)^2 - w^2\|_{C(I;C^1)} \leqslant \|(J_\varepsilon w_\varepsilon)^2 - w_\varepsilon^2\|_{C(I;C^1)} + \|w_\varepsilon^2 - w^2\|_{C(I;C^1)}$$
$$\leqslant \|(J_\varepsilon w_\varepsilon)^2 + w_\varepsilon\|_{C(I;C^1)} \|J_\varepsilon w_\varepsilon - w_\varepsilon\|_{C(I;C^1)}$$
$$+ \|w_\varepsilon^2 - w^2\|_{C(I;C^1)}. \tag{2.30}$$

Employing the Sobolev lemma for $\frac{3}{2} < r < s - 1$ and $s > \frac{5}{2}$ and also the fact that

$$\|I - J_\varepsilon\|_{\mathcal{L}(H^s;H^r)} = O(\varepsilon^{s-r}),$$

we find

$$\|J_\varepsilon w_\varepsilon(t) - w_\varepsilon(t)\|_{C^1} = \|(I - J_\varepsilon)w_\varepsilon(t)\|_{C^1} \lesssim \|(I - J_\varepsilon)w_\varepsilon(t)\|_{H^r}$$
$$\lesssim \|I - J_\varepsilon\|_{\mathcal{L}(H^{s-1};H^r)} \|w_\varepsilon(t)\|_{H^{s-1}}$$
$$\lesssim O(\varepsilon^{s-r-1}) \|u_0\|_{H^s} \to 0, \quad \varepsilon \to 0.$$

Thus,

$$\|J_\varepsilon w_\varepsilon - w_\varepsilon\|_{C(I;C^1)} \to 0, \quad \varepsilon \to 0.$$

Similarly,

$$\|J_\varepsilon u_\varepsilon - u_\varepsilon\|_{C(I;C^1)} \to 0, \quad \varepsilon \to 0.$$

Since $\|(J_\varepsilon w_\varepsilon) + w_\varepsilon\|_{C(I;C^1)}$ is is bounded, (2.30) yields

$$\|(J_\varepsilon w_\varepsilon)^2 - w^2\|_{C(I;C^1)} \lesssim \|w_\varepsilon^2 - w^2\|_{C(I;C^1)} \to 0, \quad \varepsilon \to 0.$$

The first term in (2.28), the second term and the third term in (2.29) can be treated similarly. Therefore, we conclude that $\partial_t U_\varepsilon$ is uniformly convergent in $C(I;C)$ and hence, we deduce that

$$\partial_t U = (\partial_t u, \partial_t w) = \Big(4uw - w^2 + F(u,w),$$
$$2w^2 + 4u\partial_x w - 2w\partial_x w - 6u^2 + G(u,w)\Big),$$

that is, U solves the ivp (2.4). What is more, due to the uniform convergence of $\partial_t U_\varepsilon$ to $\partial_t U$ in $C(I;C)$ the derivative size estimate (2.23) yields the estimate (2.25) for $\partial_t U$. Then, the inequality

$$\|U(t_1 - U(t_2)\|_{H^{s-2}} \leqslant \sup_{t\in I}\|\partial_t U\|_{H^{s-2}}|t_1 - t_2|, \quad \forall t_1, t_2 \in I$$

implies

$$\|U(t_1 - U(t_2)\|_{H^{s-2}} \leqslant \|u_0\|_{H^s}^2|t_1 - t_2|, \quad \forall t_1, t_2 \in I, \tag{2.31}$$

i.e. that $U \in \mathrm{Lip}(I; H^{s-2})$.

Improving the regularity of U up to $C(I; H^{s-1})$. We have shown that

$$U \in L^\infty(I; H^{s-1}) \cap \mathrm{Lip}(I; H^{s-2}).$$

We now want to show that $U \in C(I; H^{s-1})$, i.e., that if $\{t_n\}_{n\in N} \subset I$ converge to $t \in I$ then $\lim n \to \infty \|U(t_n) - U(t)\|_{H^{s-1}} = 0$. By the definition of the H^{s-1} norm, this is equivalent to showing that

$$\lim_{n\to\infty} (\|U(t_n)\|_{H^{s-1}}^2 - \langle U(t_n), U(t)\rangle_{H^{s-1}}$$
$$- \langle U(t), U(t_n)\rangle_{H^{s-1}} + \|U(t)\|_{H^{s-1}}^2) = 0. \tag{2.32}$$

Lemma 2.2 *The solution $U \in L^\infty(I; H^{s-1}) \cap \mathrm{Lip}(I; H^{s-2})$ is continuous on I with respect to the weak topology in H^{s-1}, i.e.*

$$\lim_{n\to\infty} \langle U(t_n) - U(t), \varphi\rangle_{H^{s-1}} = 0, \quad \forall \varphi \in H^{s-1}.$$

Proof Let $\varphi \in H^{s-1}(\mathbf{T})$ (note that φ is actually a vector in $H^{s-1} \times H^{s-1}$). Given $\varepsilon > 0$, choose $\psi \in \mathfrak{L}(\mathbf{T})$ such that Let $\varphi \in H^{s-1}(\mathbf{T})$ (note that φ is actually a vector in $H^{s-1} \times H^{s-1}$). Given $\varepsilon > 0$, choose $\psi \in \mathfrak{L}(\mathbf{T})$ such that

$$\|\varphi - \psi\|_{H^{s-1}} < \frac{\varepsilon}{16\|u_0\|_{H^s}}. \tag{2.33}$$

Applying the triangle inequality and the Cauchy-Schwarz inequality, we find

$$\langle U(t_n) - U(t), \varphi \rangle_{H^{s-1}}$$
$$\leqslant \|U(t_n) - U(t)\|_{H^{s-1}} \|\varphi - \psi\|_{H^{s-1}} + \|U(t_n) - U(t)\|_{H^{s-2}} \|\psi\|_{H^s}.$$

But from the size estimate (2.24) we have

$$\|U(t_n) - U(t)\|_{H^{s-1}} \leqslant \|U(t_n)\|_{H^s} + \|U\|_{H^{s-1}} \leqslant 8\|u_0\|_{H^s}.$$

Hence, using the Lipschitz estimate (2.31) together with inequality (2.33), we find

$$\langle U(t_n) - U(t), \varphi \rangle_{H^{s-1}} \leqslant \frac{\varepsilon}{2} + \|u_0\|_{H^s}^2 |t_n - t| \|\psi\|_{H^s}.$$

Since $\|\psi\|_{H^s}$ is bounded, there is an N such that $|t_n - t| < \dfrac{\varepsilon}{2\|u_0\|_{H^s}^2 \|\psi\|_{H^s}}$ for all $n > N$. Thus, we conclude that $|\langle U(t_n) - U(t), \varphi \rangle_{H^{s-1}}| \leqslant \varepsilon$ for all $n > N$, which completes the proof.

Employing Lemma 2.2, we see that (2.32) reduces to proving that

$$\lim_{n \to \infty} \|U(t_n)\|_{H^{s-1}} = \|U(t)\|_{H^{s-1}},$$

i.e., that the map $t \mapsto \|U(t)\|_{H^{s-1}}$ continuous. Let $Y_\varepsilon(t) = \|J_\varepsilon U(t)\|_{H^{s-1}}$ and $Y(t) = \|JU(t)\|_{H^{s-1}}$. We know that $Y_\varepsilon(t)$ converges to Y pointwise in t as $\varepsilon \to 0$. Thus, if Y_ε is Lipschitz on I, that is if

$$|Y_\varepsilon(t_1) - Y_\varepsilon(t_2)| \leqslant |t_1 - t_2|, \quad \forall t_1, t_2 \in I, \tag{2.34}$$

then by pointwise convergence $|Y(t_1) - Y(t_2)| \leqslant |t_1 - t_2|$ for all $t_1, t_2 \in I$ i.e. Y will also be Lipschitz on I and hence, Y will be continuous on I. To prove (2.34), we apply the Friedrichs mollifier to the first equation of system (2.4) to obtain the equation

$$\partial_t J_\varepsilon u = -J_\varepsilon w^2 + 4J_\varepsilon(uw) + J_\varepsilon F(u, w).$$

This expression implies

$$\int_\mathbb{T} D^{s-1}(\partial_t J_\varepsilon u) D^{s-1}(J_\varepsilon u)\, dx = -\int_\mathbb{T} D^{s-1}(J_\varepsilon w^2) D^{s-1}(J_\varepsilon u)\, dx$$
$$+ 4\int_\mathbb{T} D^{s-1} J_\varepsilon(uw) D^{s-1}(J_\varepsilon u)\, dx$$

$$+ \int_{\mathbf{T}} D^{s-1} J_\varepsilon F(u,w) D^{s-1}(J_\varepsilon u) \, dx.$$

Consequently,

$$\frac{1}{2}\frac{d}{dt}\|J_\varepsilon u\|_{H^{s-1}}^2 \lesssim \left|\int_{\mathbf{T}} D^{s-1}(J_\varepsilon w^2) D^{s-1}(J_\varepsilon u) \, dx\right| + \left|\int_{\mathbf{T}} D^{s-1} J_\varepsilon(uw) D^{s-1}(J_\varepsilon u) \, dx\right|$$

$$+ \left|\int_{\mathbf{T}} D^{s-1} J_\varepsilon F(u,w) D^{s-1}(J_\varepsilon u) \, dx\right|.$$

Estimating the first term, we find

$$\left|\int_{\mathbf{T}} D^{s-1}(J_\varepsilon w^2) D^{s-1}(J_\varepsilon u) \, dx\right| \leqslant \|J_\varepsilon w^2\|_{H^{s-1}} \|J_\varepsilon u\|_{H^{s-1}} \leqslant \|w\|_{H^{s-1}}^2 \|u\|_{H^{s-1}} \tag{2.35}$$

while the second term yields

$$\left|\int_{\mathbf{T}} D^{s-1} J_\varepsilon(uw) D^{s-1}(J_\varepsilon u) \, dx\right| \leqslant \|w\|_{H^{s-1}} \|u\|_{H^{s-1}}^2 \tag{2.36}$$

and for the third term, we have

$$\left|\int_{\mathbf{T}} D^{s-1} J_\varepsilon F(u,w) D^{s-1}(J_\varepsilon u) \, dx\right| \lesssim \|w\|_{H^{s-1}}^2 \|u\|_{H^{s-1}} + \|u\|_{H^{s-1}}^3. \tag{2.37}$$

Combining (2.35)-(2.37) and also using the size estimate (2.24), we find

$$\frac{d}{dt}\|J_\varepsilon u(t)\|_{H^{s-1}} \lesssim (\|u(t)\|_{H^{s-1}} + \|w(t)\|_{H^{s-1}})^2 \lesssim \|u_0\|_{H^s}^3. \tag{2.38}$$

The corresponding equation satisfied by $J_\varepsilon w$ reads

$$\partial_t J_\varepsilon w = 2 J_\varepsilon w^2 + 4 J_\varepsilon(u \partial_x w) - 2 J_\varepsilon(w \partial_x w) - 6 J_\varepsilon u^2 + J_\varepsilon G(u,w),$$

thus

$$\frac{1}{2}\frac{d}{dt}\|J_\varepsilon w\|_{H^{s-1}}^2 \lesssim \left|\int_{\mathbf{T}} D^{s-1}[J_\varepsilon(4u-2w)\partial_x w] D^{s-1}(J_\varepsilon w) \, dx\right|$$

$$+ \left|\int_{\mathbf{T}} D^{s-1} J_\varepsilon w^2 D^{s-1}(J_\varepsilon w) \, dx\right|$$

$$+ \left|\int_{\mathbf{T}} D^{s-1} J_\varepsilon u^2 D^{s-1}(J_\varepsilon w) \, dx\right|$$

$$+ \left| \int_{\mathbb{T}} D^{s-1} J_\varepsilon G(u,w) D^{s-1}(J_\varepsilon w) \, dx \right|$$
$$= B_1 + B_2 + B_3 + B_4. \tag{2.39}$$

We note that B_1 above can be written in the form

$$B_1 = \left| \int_{\mathbb{T}} [D^{s-1}, J_\varepsilon(4u - 2w)] \partial_x w D^{s-1}(J_\varepsilon w) \, dx \right.$$
$$\left. + \int_{\mathbb{T}} [J_\varepsilon(4u - 2w)] D^{s-1} \partial_t w \cdot D^{s-1}(J_\varepsilon w) \, dx \right|.$$

Then, using the Cauchy-Schwarz inequality in the first integral and commuting the mollifier in the second integral, we have

$$B_1 = \|[D^{s-1}, J_\varepsilon(4u - 2w)] \partial_x w\|_{L^2} \|D^{s-1}(J_\varepsilon w)\|_{L^2}$$
$$+ \left| \int_{\mathbb{T}} J_\varepsilon [J_\varepsilon(4u - 2w)] D^{s-1} \partial_t w \cdot D^{s-1} w \, dx \right|.$$

By the Kato-Ponce Lemma

$$\|[D^{s-1}, J_\varepsilon(4u - 2w)] \partial_x w\|_{L^2}$$
$$\lesssim \|4u - 2w\|_{H^{s-1}} \|\partial_x w\|_{L^\infty} + \|w\|_{H^{s-1}} \|\partial_x (4u - 2w)\|_{L^\infty}$$

and then, by the algebra property and the Sobolev lemma

$$\|[D^{s-1}, J_\varepsilon(4u - 2w)] \partial_x w\|_{L^2} \lesssim \|w\|_{H^{s-1}}^2 + \|w\|_{H^{s-1}} \|u\|_{H^{s-1}}.$$

In addition, integrating by parts yields

$$\int_{\mathbb{T}} J_\varepsilon [J_\varepsilon(4u - 2w)] D^{s-1} \partial_t w \cdot D^{s-1} w \, dx$$
$$\lesssim \|\partial_x J_\varepsilon [J_\varepsilon(4u - 2w)]\|_{L^\infty} \|w\|_{H^{s-1}}^2.$$

Hence,

$$B_1 \lesssim \|w\|_{H^{s-1}}^3 + \|w\|_{H^{s-1}}^2 \|u\|_{H^{s-1}}. \tag{2.40}$$

Next, using the algebra property we find

$$B_2 \lesssim \|w\|_{H^{s-1}}^3, \quad B_3 \lesssim \|u\|_{H^{s-1}}^2 \|w\|_{H^{s-1}}. \tag{2.41}$$

Finally, for the nonlocal term we use the Cauchy-Schwarz inequality and the algebra property to obtain

$$B_4 \lesssim \|w\|_{H^{s-1}}^3 + \|u\|_{H^{s-1}}^2 \|w\|_{H^{s-1}}. \tag{2.42}$$

Combining the estimates (2.40)-(2.42) we conclude via the size estimate (2.24) that

$$\frac{d}{dt}\|J_\varepsilon w\|_{H^{s-1}} \lesssim (\|w\|_{H^{s-1}} + \|u\|_{H^{s-1}})^2 \lesssim \|u_0\|_{H^s}^2. \tag{2.43}$$

The estimates (2.38) and (2.43) imply $\frac{d}{dt}Y_\varepsilon(t) \lesssim \|u_0\|_{H^s}^2$ therefore

$$|Y_\varepsilon(t_1) - Y_\varepsilon(t_2)| \lesssim \|u_0\|_{H^s}^2 |t_1 - t_2|,$$

i.e., $Y_\varepsilon(t)$ is Lipschitz which implies that $Y(t)$ is also Lipschitz and hence, $U \in C(I; H^{s-1})$.

Uniqueness. We have shown that there exists a solution $U = (u, w)$ to the ivp (2.4) in $C([0, T]; H^{s-1})$ which satisfies the estimates (2.24) and (2.25) with lifespan T given by (2.21). We now wish to show that this solution is unique. Suppose there exists some $W = (v, z) \neq U$ in $C([0, T]; H^{s-1})$ satisfying the ivp (2.4). Letting $\varphi = u - v$ and $\psi = w - z$, we find that system (2.4) yields the following ivp for (ϕ, ψ)

$$\partial_t \varphi = (4u - z - w)\psi - 4z\varphi + F(u, w) - F(v, z)$$
$$\partial_t \psi = (4u - w)\partial_x \psi + (w + z)\psi - (6(u + v) - 2\partial_x z)\varphi + G(u, w) - G(v, z)$$
$$\varphi(x, 0) = \psi(x, 0) = 0, \tag{2.44}$$

where F, G are defined by Equations (2.5). Consider $\sigma \in \left(\frac{1}{2}, s-2\right)$. Applying D^σ on the left of the first and second equation of system (2.44) and multiplying on the right by $D^\sigma \varphi$ and $D^\sigma \psi$, respectively, we find

$$\frac{1}{2}\frac{d}{dt}\|\varphi\|_{H^\sigma}^2 = \int_\mathbb{T} D^\sigma (4u - z - w)\psi D^\sigma \varphi \, dx + 4\int_\mathbb{T} D^\sigma z\varphi D^\sigma \varphi \, dx$$
$$+ \int_\mathbb{T} D^\sigma[F(u, w) - F(v, z)] D^\sigma \varphi \, dx \tag{2.45}$$

and

$$\frac{1}{2}\frac{d}{dt}\|\psi\|_{H^\sigma}^2 = \int_\mathbb{T} D^\sigma (4u - 2w)\partial_x \psi D^\sigma \psi \, dx$$

$$+ \int_{\mathbb{T}} D^\sigma(w+z)\psi D^\sigma \psi \, dx$$
$$- \int_{\mathbb{T}} D^\sigma(6(u+v) - 2\partial_x z)\varphi D^\sigma \psi \, dx$$
$$+ \int_{\mathbb{T}} D^\sigma [G(u,w) - G(v,z)] D^\sigma \psi \, dx \, dx.$$
$$= C_1 + C_2 + C_3 + C_4. \tag{2.46}$$

We first consider Equation (2.45). By the algebra property and the size estimate (2.24),

$$\int_{\mathbb{T}} D^\sigma (4u - z - w) \psi D^\sigma \varphi \, dx \lesssim (\|u\|_{H^\sigma} + \|z\|_{H^\sigma} + \|w\|_{H^\sigma}) \|\varphi\|_{H^\sigma} \|\psi\|_{H^\sigma}$$
$$\lesssim \|u_0\|_{H^s} \|\varphi\|_{H^\sigma} \|\psi\|_{H^\sigma}.$$

Similarly,

$$\int_{\mathbb{T}} D^\sigma z \varphi D^\sigma \varphi \, dx \lesssim \|u_0\|_{H^s} \|\varphi\|_{H^\sigma}^2.$$

Also, noting that the difference of the nonlocal terms can be expressed as

$$F(u,w) - F(v,z) = D^{-2}[(w+z)\psi] + D^{-2}\partial_x(2[(w+z)\psi] + 6[(u+v)\varphi]),$$

we obtain the estimate

$$\int_{\mathbb{T}} D^\psi [F(u,w) - F(v,z)] D^\sigma \varphi \, dx \lesssim \|u_0\|_{H^s}(\|\varphi\|_{H^\sigma}^2 + \|\varphi\|_{H^\sigma}\|\psi\|_{H^\sigma}).$$

Thus, overall we deduce that $\frac{1}{2}\frac{d}{dt}\|\varphi\|_{H^\sigma}^2 \lesssim \|u_0\|_{H^s}(\|\varphi\|_{H^\sigma}^2 + \|\varphi\|_{H^\sigma}\|\psi\|_{H^\sigma})$, which yields the estimate

$$\frac{d}{dt}\|\varphi\|_{H^\sigma} \lesssim \|u_0\|_{H^s}(\|\varphi\|_{H^\sigma} + \|\psi\|_{H^\sigma}). \tag{2.47}$$

Next, we consider Equation (2.46). By commuting ∂_x with $4u - 2w$, we have

$$C_1 = \int_{\mathbb{T}} [D^\sigma \partial_x, (4u - 2w)]\psi \cdot D^\sigma \psi \, dx + \int_{\mathbb{T}} (4u - 2w) D^\sigma \partial_x \psi \cdot D^\sigma \psi \, dx$$
$$- \int_{\mathbb{T}} D^\sigma \partial_x (4u - 2w)]\psi \cdot D^\sigma \psi \, dx. \tag{2.48}$$

The first integral above can be estimated with the help of the following lemma.

Lemma 2.3 (Calderon-Coifman-Meyer) If $\sigma + 1 \geqslant 0$ then for $f \in H^\rho$ and $g \in H^\sigma$,
$$\|[D^\sigma \partial_x, f]g\|_{L^2} \lesssim \|f\|_{H^\rho}\|g\|_{H^\sigma}$$
provided that $\rho > \frac{3}{2}$ and $\sigma + 1 \leqslant \rho$.

Indeed, Lemma 2.3 for $\rho = s - 1$ together with the size estimate (2.24) imply
$$\int_{\mathbf{T}} [D^\sigma \partial_x, (4u - 2w)]\psi \cdot D^\sigma \psi \, dx \lesssim \|4u - 2w\|_{H^{s-1}}\|\psi\|_{H^\sigma}^2 \lesssim \|\psi\|_{H^\sigma}^2 \|u_0\|_{H^s}.$$

For the second integral of (2.48), integration by parts, the algebra property, the Sobolev lemma for $s > \frac{5}{2}$ and the size estimate (2.24) yield
$$\int_{\mathbf{T}} (4u - 2w) D^\sigma \partial_x \psi \cdot D^\sigma \psi \, dx \lesssim \|\partial_x(4u - 2w)\|_{L^\infty}\|\psi\|_{H^\sigma}^2 \lesssim \|\psi\|_{H^\sigma}^2 \|u_0\|_{H^s}.$$

The third integral can be estimated by recalling that $\sigma < s - 2$:
$$\int_{\mathbf{T}} D^\sigma \partial_x (4u - 2w)]\psi \cdot D^\sigma \psi \, dx \lesssim \|\partial_x(4u - 2w)\psi\|_{H^\sigma}\|\psi\|_{H^\sigma}$$
$$\lesssim \|4u - 2w\|_{H^{s-1}}\|\psi\|_{H^\sigma}^2$$
$$\lesssim \|\psi\|_{H^\sigma}^2 \|u_0\|_{H^s}.$$

Overall, we obtain
$$C_1 \lesssim \|\psi\|_{H^\sigma}^2 \|u_0\|_{H^s}. \tag{2.49}$$

For C_3, we have
$$C_3 \lesssim (\|6(u+v)\|_{H^{s-2}} + \|2\partial_x z\|_{H^{s-2}})\|\varphi\|_{H^\sigma}\|\psi\|_{H^\sigma} \lesssim \|u_0\|_{H^s}\|\varphi\|_{H^\sigma}\|\psi\|_{H^\sigma}. \tag{2.50}$$

The C_2 and C_4 can be estimated similarly. In particular,
$$C_2 \lesssim \|\psi\|_{H^\sigma}^2 \|u_0\|_{H^s} \tag{2.51}$$

and noting that
$$G(u, w) - G(v, z) = D^{-2}\partial_x[(w+z)\psi] + D^{-2}(2[(w+z)\psi] + 6[(u+v)\varphi]),$$

we have
$$C_4 \lesssim \|u_0\|_{H^s}\|\varphi\|_{H^\sigma}\|\psi\|_{H^\sigma}. \qquad (2.52)$$

Combining (2.49)-(2.52) we conclude that for $\sigma \in \left(\frac{1}{2}, s-2\right)$,
$$\frac{1}{2}\frac{d}{dt}\|\psi\|_{H^\sigma}^2 \lesssim \|u_0\|_{H^s}(\|\psi\|_{H^\sigma}^2 + \|\psi\|_{H^\sigma}\|\varphi\|_{H^\sigma}),$$
which yields the estimate
$$\frac{d}{dt}\|\psi\|_{H^\sigma} \lesssim \|u_0\|_{H^s}(\|\psi\|_{H^\sigma} + \|\varphi\|_{H^\sigma}). \qquad (2.53)$$

Adding inequalities (2.47) and (2.53) we deduce
$$\frac{d}{dt}(\|\psi\|_{H^\sigma} + \|\varphi\|_{H^\sigma}) \lesssim \|u_0\|_{H^s}(\|\psi\|_{H^\sigma} + \|\psi\|_{H^\sigma}\|\varphi\|_{H^\sigma}). \qquad (2.54)$$

Solving this inequality, we obtain
$$\|\varphi\|_{H^\sigma} + \|\psi\|_{H^\sigma} \lesssim (\|\varphi(0)\|_{H^\sigma} + \|\psi(0)\|_{H^\sigma})e^{\|u_0\|_{H^s}t}. \qquad (2.55)$$

Since $\varphi(0) = \psi(0) = 0$, we conclude that $\varphi = \psi = 0$ and therefore, the solution U to the ivp (2.4) is unique in $C([0,T]; H^{s-1})$.

Back to Equation (1.1). We now wish to relate the solution U obtained for the ivp (2.4) to the solution u of Equation (1.1).

Lemma 2.4 *The ivp (2.4) implies that $w = \partial_x u$.*

Proof Differentiating the first equation of (2.4) and then rearranging, we have
$$\begin{cases} \partial_t(\partial_x u) = 2w\partial_x u + 4u\partial_x w - 2w\partial_x w - 6u^2 + D^{-2}\partial_x(w^2) + D^{-2}(2w^2 + 6u^2), \\ \partial_t w = 2w^2 + 4u\partial_x w - 2w\partial_x w - 6u^2 + D^{-2}\partial_x(w^2) + D^{-2}(2w^2 + 6u^2), \\ \partial_x u(x,0) = w(x,0) = \partial_x u_0(x). \end{cases}$$

Subtracting the two PDEs above and letting $y = \partial_x u - w$, yields the ivp
$$\partial_t y = 2wy, \quad y(0) = 0.$$

Solving this equation gives $y(x,t) = f(x)e^{2\int_0^t w dt}$ and, furthermore, the initial condition implies that $f \equiv 0$. Thus, we deduce that $y \equiv 0$ or, equivalently, $w = \partial_x u$. Since $w \in C([0,T]; H^{s-1})$, Lemma 2.3 implies that $u \in C([0,T]H^s)$. Moreover, returning to inequality (2.17) and using the same lemma, we obtain the size estimate and lifespan for u given in Theorem 1.1.

3 Continuity of the data-to-solution map

In the current section, We complete the proof of the Hadamard well-posedness for the ivp (1.1) on the circle by showing that the data-to-solution map $u_0(x) \in H^s \mapsto u(x,t) \in C([0,T]; H^s)$ is continuous. From the sequential point of view, this amounts to showing that for the solutions $u_n(x,t)$ and $u(x,t)$ corresponding to the initial data $u_{0,n}(x)$ and $u_0(x)$ respectively, if $u_{0,n}(x) \to u_0(x)$ in H^s then $u_n(x,t) \to u(x,t)$ in $C([0,T]; H^s)$.

We estimate the distance between u_n and u indirectly. First, note that u_n, u solve

$$\begin{cases} \partial_t u = -(\partial_x u)^2 + 4u\partial_x u + \Phi(u), \\ u(x,0) = u_0(x), \end{cases} \tag{3.1}$$

and

$$\begin{cases} \partial_t u_n = -(\partial_x u_n)^2 + 4u_n \partial_x u_n + \Phi(u_n), \\ u_n(x,0) = u_{0,n}(x), \end{cases} \tag{3.2}$$

where Φ is defined by (2.2). Also, we introduce u_n^ε and u^ε as the solutions of the following ivps with mollified initial conditions:

$$\begin{cases} \partial_t u^\varepsilon = -(\partial_x u^\varepsilon)^2 + 4u^\varepsilon \partial_x u^\varepsilon + \Phi(u^\varepsilon), \\ u^\varepsilon(x,0) = J_\varepsilon u_0(x) = j_\varepsilon * u_0(x), \end{cases} \tag{3.3}$$

and

$$\begin{cases} \partial_t u_n^\varepsilon = -(\partial_x u_n^\varepsilon)^2 + 4u_n^\varepsilon \partial_x u_n^\varepsilon + \Phi(u_n^\varepsilon), \\ u_n^\varepsilon(x,0) = J_\varepsilon u_{0,n}(x) = j_\varepsilon * u_{0,n}(x). \end{cases} \tag{3.4}$$

Note that by the triangle inequality

$$\|u_n - u\|_{H^{s-1}} \leqslant \|u_n - u^\varepsilon\|_{H^{s-1}} + \|u^\varepsilon - u_n^\varepsilon\|_{H^{s-1}} + \|u^\varepsilon - u\|_{H^{s-1}} \tag{3.5}$$

and similarly for w, w_n and their mollified counterparts. Hence, instead of estimating the difference $u_n - u$ we will estimate the relevant differences on the rhs of (3.5), and similarly for the difference $w_n - w$.

Due to the presence of the terms $(\partial_x u)^2$, $(\partial_x u^\varepsilon)^2$ and $(\partial_x u_n^\varepsilon)^2$, we employ the approach of transforming the ivps into systems as in Section 2. The systems that originate from (3.1) and (3.2) are

$$\begin{cases} \partial_t u = 4uw - w^2 + F(u,w), \\ \partial_t w = 2w^2 + 4u\partial_x w - 2w\partial_x w - 6u^2 + G(u,w)), \\ u(x,0) = u_0(x), \quad w(x,0) = \partial_x u_0(x) = w_0(x), \end{cases} \tag{3.6}$$

and
$$\begin{cases} \partial_t u_n = 4u_n w_n - u_n^2 + F(u_n, w_n), \\ \partial_t w_n = 2w_n^2 + 4u_n \partial_x w_n - 2w_n \partial_x w - 6u_n^2 - G(u_n, w_n)), \\ u_n(x,0) = u(0,n)(x), \quad w_n(x,0) = \partial_x u_{0,n}(x) = w_{0,n}(x), \end{cases} \quad (3.7)$$

while the ivps (3.3) and (3.4) correspond to the systems

$$\begin{cases} \partial_t u^\varepsilon = 4u^\varepsilon w^\varepsilon - (u^\varepsilon)^2 - F(u^\varepsilon, w^\varepsilon), \\ \partial_t w^\varepsilon = 2(u^\varepsilon)^2 + 4u^\varepsilon \partial_x w^\varepsilon - 2w^\varepsilon \partial_x w^\varepsilon - 6(u^\varepsilon)^2 + G(u^\varepsilon, w^\varepsilon), \\ u^\varepsilon(x,0) = J_\varepsilon u_0(x), \quad w^\varepsilon(x,0) = \partial_x u_0^\varepsilon(x) = w_0^\varepsilon(x), \end{cases} \quad (3.8)$$

and

$$\begin{cases} \partial_t u_n^\varepsilon = 4u_n^\varepsilon w^\varepsilon - (u_n^\varepsilon)^2 - F(u_n^\varepsilon, w_n^\varepsilon), \\ \partial_t w_n^\varepsilon = 2(u_n^\varepsilon)^2 + 4u_n^\varepsilon \partial_x w_n^\varepsilon - 2w_n^\varepsilon \partial_x w_n^\varepsilon - 6(u^\varepsilon)^2 + G(u_n^\varepsilon, w_n^\varepsilon), \\ u_n^\varepsilon(x,0) = J_\varepsilon u_{0,n}(x), \quad w_n^\varepsilon(x,0) = \partial_x u_{0,n}^\varepsilon(x) = w_{0,n}^\varepsilon(x), \end{cases} \quad (3.9)$$

with F, G defined by Equations (2.5). Let

$$v = u^\varepsilon - u_n^\varepsilon, \qquad z = w^\varepsilon - w_n^\varepsilon. \quad (3.10)$$

Subtracting systems (3.8) and (3.9), we obtain the system

$$\begin{cases} \partial_t v = (4u^\varepsilon - w_n^\varepsilon - w^\varepsilon)z + 4w_n^\varepsilon v + F(u^\varepsilon, w^\varepsilon) - F(u_n^\varepsilon, w_n^\varepsilon) \\ \partial_t z = (4u^\varepsilon - 2w^\varepsilon)\partial_x z + (w^\varepsilon + w_n^\varepsilon)z - (6(u^\varepsilon + u_n^\varepsilon) - 2\partial_x w_n^\varepsilon)v \\ \quad + G(u^\varepsilon, w^\varepsilon) - G(u_n^\varepsilon, w_n^\varepsilon) \\ v(x,0) = J_\varepsilon u_0(x) - J_\varepsilon u_{0,n}(x), \quad z(x,0) = J_\varepsilon \partial_x u_0(x) - J_\varepsilon \partial_x u_{0,n}(x). \end{cases} \quad (3.11)$$

Energy estimate for v in (3.10). We consider the first equation of (3.11). Applying the operator D^{s-1} on the left, multiplying by $D^{s-1}v$ on the right and integrating over $\mathbf{T}$ yields

$$\frac{1}{2}\frac{d}{dt}\|v\|_{H^{s-1}}^2 = \int_\mathbf{T} D^{s-1}(4u^\varepsilon - w_n^\varepsilon - w^\varepsilon)zD^{s-1}v\,dx$$
$$+ 4\int_\mathbf{T} D^{s-1}w_n^\varepsilon v D^{s-1}v\,dx$$
$$+ \int_\mathbf{T} D^{s-1}[F(u^\varepsilon, w^\varepsilon) - F(u_n^\varepsilon, w_n^\varepsilon)]D^{s-1}v\,dx. \quad (3.12)$$

For the first term, we use the algebra property and the size estimate (2.24) to obtain

$$\int_{\mathbf{T}} D^{s-1}\left(4u^{\varepsilon} - w_n^{\varepsilon} - w^{\varepsilon}\right) z D^{s-1} v \, dx$$

$$\lesssim (\|u^{\varepsilon}\|_{H^s} + \|w_n^{\varepsilon}\|_{H^s} + \|w^{\varepsilon}\|_{H^s})\|z\|_{H^{s-1}}\|v\|_{H^{s-1}}\|z\|_{H^{s-1}}$$

$$\lesssim (\|u_0^{\varepsilon}\|_{H^s} + \|u_{0,n}^{\varepsilon}\|_{H^s})\|v\|_{H^{s-1}}\|z\|_{H^{s-1}}$$

$$\lesssim \|v\|_{H^{s-1}}\|z\|_{H^{s-1}}. \tag{3.13}$$

Similarly,

$$\int_{\mathbf{T}} D^{s-1} w_n^{\varepsilon} v D^{s-1} v \, dx \lesssim \|u_{0,n}^{\varepsilon}\|_{H^s}\|v\|_{H^{s-1}}^2 \lesssim \|v\|_{H^{s-1}}^2.$$

Also, noting that the difference of the nonlocal term of (3.12) can be expressed as

$$F(u^{\varepsilon}, w^{\varepsilon}) - F(u_n^{\varepsilon}, w_n^{\varepsilon})$$
$$= D^{-2}[(w^{\varepsilon} + w_n^{\varepsilon})z] + D^{-2}\partial_x(2[(w^{\varepsilon} + w_n^{\varepsilon})z] + 6[(u^{\varepsilon} + w_n^{\varepsilon})v]),$$

hence, by the algebra property and the size estimate (2.24) we obtain the estimate

$$\int_{\mathbf{T}} D^{s-1}[F(u^{\varepsilon}, w^{\varepsilon}) - F(u_n^{\varepsilon}, w_n^{\varepsilon})] D^{s-1} v \, dx \lesssim \|v\|_{H^{s-1}}^2 + \|v\|_{H^{s-1}}\|z\|_{H^{s-1}}.$$

Overall, we deduce that

$$\frac{1}{2}\frac{d}{dt}\|v\|_{H^{s-1}}^2 \lesssim \|v\|_{H^{s-1}}^2 + \|v\|_{H^{s-1}}\|z\|_{H^{s-1}},$$

which yields the estimate

$$\frac{d}{dt}\|v\|_{H^{s-1}} \lesssim \|v\|_{H^{s-1}} + \|z\|_{H^{s-1}}. \tag{3.14}$$

Energy estimate for z in (3.10). Applying the operator D^{s-1} on the left of the second equation of (3.11), multiplying by $D^{s-1}v$ on the right and integrating over $\mathbf{T}$ yields

$$\frac{1}{2}\frac{d}{dt}\|z\|_{H^{s-1}}^2 = \int_{\mathbf{T}} D^{s-1}\left(4u^{\varepsilon} - 2w^{\varepsilon}\right) \partial_x z D^{s-1} z \, dx$$

$$+ \int_{\mathbb{T}} D^{s-1}(w^\varepsilon + w_n^\varepsilon)zD^{s-1}z \, dx$$
$$- \int_{\mathbb{T}} D^{s-1}(6(u^\varepsilon + u_n^\varepsilon) - 2\partial_x w_n^\varepsilon)vD^{s-1}z \, dx$$
$$+ \int_{\mathbb{T}} D^{s-1}[G(u^\varepsilon, w^\varepsilon) - G(u_n^\varepsilon, w_n^\varepsilon)]D^{s-1}z \, dx.$$
$$= D_1 + D_2 + D_3 + D_4. \qquad (3.15)$$

Commuting D^{s-1} with $4u^\varepsilon - 2w^\varepsilon$ in in the first term, we have

$$D_1 = \int_{\mathbb{T}} [D^{s-1}\partial_x, (4u^\varepsilon - 2w^\varepsilon)]z \cdot D^{s-1}z \, dx$$
$$+ \int_{\mathbb{T}} (4u^\varepsilon - 2w^\varepsilon)D^{s-1}\partial_x z \cdot D^{s-1}z \, dx$$
$$- \int_{\mathbb{T}} D^{s-1}\partial_x(4u^\varepsilon - 2w^\varepsilon)]z \cdot D^{s-1}z \, dx. \qquad (3.16)$$

Employing the Kato-Ponce Lemma 2.1 in the commutator, integrating by parts in the integral and using the algebra property, the Sobolev lemma and the size estimate (2.24), we obtain

$$E_1 \lesssim \|z\|_{H^{s-1}}^2. \qquad (3.17)$$

Regarding D_2 we have

$$D_2 \lesssim \|w^\varepsilon + w_n^\varepsilon\|_{H^{s-1}}\|z\|_{H^{s-1}}^2$$
$$\lesssim (\|u_0\|_{H^{s-1}} + \|u_{0,n}\|_{H^{s-1}})\|z\|_{H^{s-1}}^2$$
$$\lesssim \|z\|_{H^{s-1}}^2. \qquad (3.18)$$

Regarding D_3 we have

$$D_3 \lesssim \|6(u^\varepsilon + u_n^\varepsilon) - 2\partial_x w_n^\varepsilon\|_{H^{s-1}}\|z\|_{H^{s-1}}^2$$
$$\lesssim (\|u_0^\varepsilon\|_{H^{s-1}} + \|u_n^\varepsilon\|_{H^{s-1}} + \|\partial_x w_n^\varepsilon\|_{H^{s-1}})\|z\|_{H^{s-1}}^2.$$

Using the size estimate (2.24) and the fact that $\|\partial_x J_\varepsilon f\|_{H^s} \lesssim \frac{1}{\varepsilon}\|f\|_{H^s}$, we find

$$D_3 \lesssim \frac{1}{\varepsilon}\|z\|_{H^{s-1}}^2. \qquad (3.19)$$

Noting that the difference of the nonlocal term of (3.12) can be expressed as

$$G(u^\varepsilon, w^\varepsilon) - G(u_n^\varepsilon, w_n^\varepsilon)$$
$$= D^{-2}[(w^\varepsilon + w_n^\varepsilon)z] + D^{-2}(2[(w^\varepsilon + w_n^\varepsilon)z] + 6[(u^\varepsilon + w_n^\varepsilon)v]),$$

hence, by the algebra property and the size estimate (2.24), we obtain the estimate

$$D_4 = \int_{\mathbf{T}} D^{s-1}[G(u^\varepsilon, w^\varepsilon) - G(u_n^\varepsilon, w_n^\varepsilon)] D^{s-1}v \, dx$$
$$\lesssim \|z\|_{H^{s-1}}^2 + \|v\|_{H^{s-1}} \|z\|_{H^{s-1}}. \tag{3.20}$$

Inequalities (3.17)-(3.20) along with Equation (3.15) yield the inequality

$$\frac{1}{2}\frac{d}{dt}\|z\|_{H^{s-1}}^2 \lesssim \frac{1}{\varepsilon}(\|z\|_{H^{s-1}}^2 + \|z\|_{H^{s-1}} \|v\|_{H^{s-1}}),$$

from which we deduce that

$$\frac{d}{dt}\|z\|_{H^{s-1}} \lesssim \frac{1}{\varepsilon}(\|z\|_{H^{s-1}} + \|v\|_{H^{s-1}}). \tag{3.21}$$

Adding inequalities (3.14) and (3.21) we deduce

$$\frac{d}{dt}(\|v\|_{H^{s-1}} + \|z\|_{H^{s-1}}) \lesssim \frac{c_s}{\varepsilon}(\|z\|_{H^{s-1}} + \|v\|_{H^{s-1}}). \tag{3.22}$$

Solving this differential inequality for all $t \in [0, T]$ with T defined by (2.21) gives

$$\|v(t)\|_{H^{s-1}} + \|z(t)\|_{H^{s-1}} \leqslant (\|v(0)\|_{H^\sigma} + \|z(0)\|_{H^\sigma}) e^{\frac{c_s}{\varepsilon} t}$$
$$\leqslant (\|v(0)\|_{H^\sigma} + \|z(0)\|_{H^\sigma}) e^{\frac{c_s}{\varepsilon} T}. \tag{3.23}$$

Hence, under the definition (3.10) of v, z we conclude that

$$\|u^\varepsilon(t) - u_n^\varepsilon(t)\|_{H^{s-1}} + \|w^\varepsilon(t) - w_n^\varepsilon(t)\|_{H^{s-1}}$$
$$\leqslant (\|u^\varepsilon(t) - u_n^\varepsilon(t)\|_{H^{s-1}} + \|w^\varepsilon(0) - w_n^\varepsilon(0)\|_{H^{s-1}}) e^{\frac{c_s}{\varepsilon} T}. \tag{3.24}$$

Estimating the remaining differences. Now let v and z denote the differences of the components of the exact solution (u, w) from either $(u^\varepsilon, w^\varepsilon)$ or $(u_n^\varepsilon, w_n^\varepsilon)$, i.e., $v = u^\varepsilon - w, v = u_n^\varepsilon - w_n$.

Subtracting systems (3.8) and (3.9), and using the above notation, we respectively obtain the following systems:

$$\begin{cases} \partial_t v = (4u^\varepsilon - w - w^\varepsilon)z + 4wv + F(u^\varepsilon, w^\varepsilon) - F(u, w), \\ \partial_t z = (4u^\varepsilon - 2w^\varepsilon)\partial_x z + (w^\varepsilon + w)z - (6(u^\varepsilon + u) - 2\partial_x w)v \\ \qquad + G(u^\varepsilon, w^\varepsilon) - G(u, w), \\ v(x, 0) = J_\varepsilon u_0(x) - u_0(x), z(x, 0) = J_\varepsilon \partial_x u_0(x) - \partial_x u_0(x), \end{cases} \quad (3.25)$$

and

$$\begin{cases} \partial_t v = (4u_n^\varepsilon - w_n - w_n^\varepsilon)z + 4w_n v + F(u_n^\varepsilon, w_n^\varepsilon) - F(u_n, w_n), \\ \partial_t z = (4u_n^\varepsilon - 2w_n^\varepsilon)\partial_x z + (w_n^\varepsilon + w_n)z - (6(u_n^\varepsilon + u_n) - 2\partial_x w_n)v \\ \qquad + G(u_n^\varepsilon, w_n^\varepsilon) - G(u_n, w_n), \\ v(x, 0) = J_\varepsilon u_{0,n}(x) - u_{0,n}(x), z(x, 0) = J_\varepsilon \partial_x u_{0,n}(x) - \partial_x u_{0,n}(x). \end{cases} \quad (3.26)$$

Systems (3.25) and (3.26) are similar to each other, hence we will only provide the details for the energy estimates of the first system. Concerning the energy estimate for v, following the same approach as in the case of system (3.11) we arrive at

$$\frac{d}{dt}\|z\|_{H^{s-1}} \lesssim \|z\|_{H^{s-1}} + \|v\|_{H^{s-1}}. \quad (3.27)$$

Regarding z, however, we will obtain a finer estimate than (2.76). Applying the operator D^{s-1} on the left of the second equation in (2.80), multiplying by $D^{s-1}z$ on the right and integrating over $\mathbf{T}$ gives

$$\frac{1}{2}\frac{d}{dt}\|z\|_{H^{s-1}}^2 = \int_{\mathbf{T}} D^{s-1}(4u^\varepsilon - 2w^\varepsilon)\partial_x z D^{s-1}z\,dx$$

$$+ \int_{\mathbf{T}} D^{s-1}(w^\varepsilon + w^\varepsilon)z D^{s-1}z\,dx$$

$$- 6\int_{\mathbf{T}} D^{s-1}v(u^\varepsilon + u)D^{s-1}z\,dx + 2\int_{\mathbf{T}} D^{s-1}v\partial_x w D^{s-1}z\,dx$$

$$+ \int_{\mathbf{T}} D^{s-1}[G(u^\varepsilon, w) - G(u^\varepsilon, w^\varepsilon)]D^{s-1}z\,dx.$$

$$= E_1 + E_2 + E_3 + E_4 + E_5. \quad (3.28)$$

Commuting D^{s-1} with $4u\varepsilon - 2w\varepsilon$ in the first term, we have

$$D_1 = \int_{\mathbf{T}} [D^{s-1}\partial_x, (4u^\varepsilon - 2w^\varepsilon)]z \cdot D^{s-1}z\,dx$$

$$+ \int_{\mathbf{T}} (4u^\varepsilon - 2w^\varepsilon) D^{s-1} \partial_x z \cdot D^{s-1} z \, dx$$

$$- \int_{\mathbf{T}} D^{s-1} \partial_x (4u^\varepsilon - 2w^\varepsilon)] z \cdot D^{s-1} z \, dx. \tag{3.29}$$

Employing the Kato-Ponce Lemma 2.1 in the commutator, integrating by parts in the integral and using the algebra property, the Sobolev lemma and the size estimate (2.24), we obtain

$$E_1 \lesssim \|z\|_{H^{s-1}}^2. \tag{3.30}$$

Similarly, we find

$$E_2 \lesssim \|z\|_{H^{s-1}}^2, \quad E_3 \lesssim \|v\|_{H^{s-1}} \|z\|_{H^{s-1}}, \quad E_5 \lesssim \|z\|_{H^{s-1}}^2 + \|v\|_{H^{s-1}} \|z\|_{H^{s-1}}. \tag{3.31}$$

Regarding E_3, we first commute z with D^{s-1} and then proceed as in the case of E_1 to obtain

$$E_4 \lesssim \|z\|_{H^{s-1}}^2 + \frac{1}{\varepsilon} \|z\|_{H^{s-2}} \|z\|_{H^{s-1}}. \tag{3.32}$$

Combining (3.30)-(3.32) with Equation (2.83), we arrive at the inequality

$$\frac{1}{2} \frac{d}{dt} \|z\|_{H^{s-1}}^2 \lesssim \|z\|_{H^{s-1}}^2 + \|v\|_{H^{s-1}} \|z\|_{H^{s-1}} + \frac{1}{\varepsilon} \|z\|_{H^{s-2}} \|z\|_{H^{s-1}}. \tag{3.33}$$

Adding (3.27) and (3.33), we obtain

$$\frac{d}{dt}(\|v\|_{H^{s-1}} + \|z\|_{H^{s-1}}) \lesssim (\|v\|_{H^{s-1}} + \|z\|_{H^{s-1}}) + \frac{1}{\varepsilon}(\|v\|_{H^{s-2}} + \|z\|_{H^{s-2}}). \tag{3.34}$$

At this point we note that, via energy estimates similar to the ones employed in the proof of uniqueness, it can be shown that

$$\|v\|_{H^\sigma} = o(\varepsilon^{s-1-\sigma}). \tag{3.35}$$

Lemma 3.1 (interpolation) *Suppose $r_1 < r < r_2$ and $f \in H^r$. Then,*

$$\|f\|_{H^r} \lesssim \|f\|_{H^{r_1}}^{\frac{r_2-r}{r_2-r_1}} \|f\|_{H^{r_2}}^{\frac{r-r_1}{r_2-r_1}}.$$

Employing this lemma for $r = s-2, r_1 = \sigma$ and $r_2 = s-1$ yields

$$\|v\|_{H^{s-2}} + \|z\|_{H^{s-2}} \lesssim \left(\|v\|_{H^\sigma}^{\frac{1}{s-1-\sigma}} + \|z\|_{H^\sigma}^{\frac{1}{s-1-\sigma}} \right) \left(\|v\|_{H^{s-1}}^{\frac{s-2-\sigma}{s-1-\sigma}} + \|z\|_{H^{s-1}}^{\frac{s-2-\sigma}{s-1-\sigma}} \right).$$

Then (3.35) yields that $\|v\|_{H^{s-2}} + \|z\|_{H^{s-2}} \lesssim o(\varepsilon)$ and hence, (3.34) implies

$$\frac{d}{dt}(\|v\|_{H^{s-1}} + \|z\|_{H^{s-1}}) \lesssim (\|v\|_{H^{s-1}} + \|z\|_{H^{s-1}}) + \delta_1$$

with $\delta_1 = 0(\varepsilon)/\varepsilon \to 0$ as $\varepsilon \to 0$. Solving this inequality, we obtain

$$\|v\|_{H^{s-1}} + \|z\|_{H^{s-1}} \lesssim (\|v_0\|_{H^{s-1}} + \|z_0\|_{H^{s-1}})e^t - \delta_1. \tag{3.36}$$

A similar derivation produces the analogous estimate from system (3.26):

$$\|v\|_{H^{s-1}} + \|z\|_{H^{s-1}} \lesssim (\|v_0\|_{H^{s-1}} + \|z_0\|_{H^{s-1}})e^t - \delta_2 \tag{3.37}$$

with $\delta_2 = o(\varepsilon)/\varepsilon \to 0$ as $\varepsilon \to 0$. Recalling that v, z denote the distances between the exact solution u, w and the approximate solutions $u^\varepsilon, w^\varepsilon$ and $u_n^\varepsilon, w_n^\varepsilon$ the estimates (3.36) and (3.37) become

$$\|u^\varepsilon(t) - u(t)\|_{H^{s-1}} + \|w^\varepsilon(t) - w(t)\|_{H^{s-1}}$$
$$\lesssim (\|u^\varepsilon(0) - u(0)\|_{H^{s-1}} + \|w^\varepsilon(0) - w(0)\|_{H^{s-1}})e^t - \delta_1 \tag{3.38}$$

and

$$\|u_n^\varepsilon(t) - u_n(t)\|_{H^{s-1}} + \|w_n^\varepsilon(t) - w_n(t)\|_{H^{s-1}}$$
$$\lesssim (\|u_n^\varepsilon(0) - u_n(0)\|_{H^{s-1}} + \|w_n^\varepsilon(0) - w_n(0)\|_{H^{s-1}})e^t - \delta_2. \tag{3.39}$$

Recalling further that $u^\varepsilon(0) = J_\varepsilon u_0(x)$, $u(0) = u_0(x)$, $u_n^\varepsilon(0) = J_\varepsilon u_{0,n}(x)$, $u(0) = u_{0,n}(x)$ and similarly for the w-terms, we can choose ε sufficiently small so that given $\eta > 0$,

$$\|u^\varepsilon(t) - u(t)\|_{H^{s-1}} + \|w^\varepsilon(t) - w(t)\|_{H^{s-1}} < \frac{\eta}{3} \tag{3.40}$$

and

$$\|u_n^\varepsilon(t) - u_n(t)\|_{H^{s-1}} + \|w_n^\varepsilon(t) - w_n(t)\|_{H^{s-1}} < \frac{\eta}{3}. \tag{3.41}$$

Moreover, returning to (3.24) we can choose N sufficiently large so that for ε as specified earlier

$$\|u^\varepsilon(0) - u_n^\varepsilon(0)\|_{H^{s-1}} + \|w^\varepsilon(0) - w_n(0)\|_{H^{s-1}} < \frac{\eta}{3}e^{-\frac{c_s}{\varepsilon}}, \quad \forall n > N, \tag{3.42}$$

and

$$\|u^\varepsilon(t) - u_n^\varepsilon(t)\|_{H^{s-1}} + \|w^\varepsilon(t) - w_n(t)\|_{H^{s-1}} < \frac{\eta}{3}e^{-\frac{c_s}{\varepsilon}}, \quad \forall n > N. \tag{3.43}$$

Then, the relations (3.40)-(3.43) together with inequality (3.5) imply

$$\|u_n(t) - u(t)\|_{H^{s-1}} + \|w_n(t) - w(t)\|_{H^{s-1}} < \eta, \quad \forall n > N, \qquad (3.44)$$

therefore, the continuity of the data-to-solution map for system (2.4) has been established.

4 Continuity of the data-to-solution map on the line

The Equation (2.1) on the line with initial data $u_0 H_s(\mathbf{R}), s > \frac{5}{2}$, can be treated similarly to the periodic problem. First, one has to establish well-posedness of the mollified system (2.7) and then, infer well-posedness for system (2.4) which was shown at the end of Section 2 to be equivalent to Equation (1.1).

The main modification required when $x \in \mathbf{R}$ resides in the transition from the mollified problem to the original one. While obtaining the size estimate (2.24) for $(u, w) \in L^\infty(I; H^{s-1}(\mathbf{R}))$ is straight-forward via Alaoglu's theorem, showing the convergence of the family $\{u_\varepsilon, w_\varepsilon\}_{\varepsilon \in (0,1]}$ to (u, w) in $C(I; H^{s-1-\sigma}(\mathbf{R}))$ for $\sigma \in (0, 1)$ is more subtle. In particular, one now needs to introduce a function $\varphi \in S(\mathbf{R})$ such that $\varphi(x) > 0$ for all $x \in \mathbf{R}$ and consider the sequence $\{\varphi u_\varepsilon, \varphi w_\varepsilon\}$ as opposed to just $\{(u_\varepsilon, w_\varepsilon)\}$.

The presence of the Schwartz function φ ensures that $(\varphi u_\varepsilon, \varphi w_\varepsilon)$ has compact support and hence, by Rellich's theorem the set $\{\varphi u_\varepsilon(t), \varphi w_\varepsilon(t)\} \subset H^{s-1}(\mathbf{R})$ is precompact in $H^{s-1-\sigma}(\mathbf{R})$. Recall that precompactness is the first condition needed for Ascoli's theorem, which in turn allows us to deduce the desired convergence. Subsequently, it is straightforward to obtain convergence of $(\varphi u_\varepsilon, \varphi w_\varepsilon)$ to $(u_\varepsilon, w_\varepsilon)$ in $C(I; C(\mathbf{R}))$. However, note that, similarly to the periodic case, one now has to prove that (u, w) obtained through pointwise convergence using the convergence of $(\varphi u_\varepsilon, \varphi w_\varepsilon)$ to $(u_\varepsilon, w_\varepsilon)$ in $C(I; H^{s-1-\sigma}(\mathbf{R})) \cap C(I; C(\mathbf{R}))$, is actually a solution to Equation (2.1). The first step toward this direction is to show the following convergence in $C(I; H^{s-\sigma-2}(\mathbf{R}))$:

$$\varphi \partial_t u_\varepsilon \to \varphi 4uw - \varphi w^2 + \varphi F(u, w), \quad \varepsilon \to 0,$$

$$\varphi \partial_t w_\varepsilon \to 2\varphi w^2 + 4\varphi u \partial_x w - 2\varphi w \partial_x w - 6u^2 + \varphi G(u, w)), \quad \varepsilon \to 0.$$

This requires reinvoking Ascoli's theorem, which is the second variation of the ivp on the line compared to the periodic case.

5 Nonuniform dependence on the circle

We aim to disprove uniform continuity of the data-to-solution map for the ivp (1.1). It therefore suffices to construct two sequences of solutions, denoted by $\{u_n\}$, $\{v_n\}$ in $C([0,T]:H^s)$, to (1.1)-(1.2) that share a common lifespan and satisfy

$$\|u_n(t)\|_{H^s} \lesssim 1, \|v_n(t)\|_{H^s} \lesssim 1,$$

$$\lim_{n\to\infty} \|u_n(0) - v_n(0)\|_{H^s} = 0, \tag{5.1}$$

and

$$\liminf_{n\to\infty} \|u_{1,n}(t) - v_{-1,n}(t)\|_{H^s} \gtrsim |\sin t| \text{ for } t \in (0,T]. \tag{5.2}$$

Approximate solutions. For any n, a positive integer, we define the approximate solution $u^{\omega,n} = u^{\omega,n}(x,t)$ as

$$u^{\omega,n}(x,t) \triangleq n^{-1} + n^{-s}\cos(nx - \omega t), \tag{5.3}$$

where $\omega = \{-1, 1\}$.

Now, we compute the error of the approximate solutions (5.3). Note that

$$\partial_t u^{\omega,n} + (\partial_x u^{\omega,n})^2 - 4u^{\omega,n}\partial_x u^{\omega,n}$$
$$= \omega n^{-s}\sin(nx - \omega t) + n^{2-2s}\omega^2\sin^2(nx - \omega t)$$
$$+ 4n^{1-s}[\omega n^{-1} + n^{-s}\cos(nx - \omega t)]\sin(nx - \omega t)$$
$$\triangleq F_1,$$

$$D^{-2}(\partial_x u)^2$$
$$= D^{-2}\left[n^{2-2s}\omega^2\sin^2(nx - \omega t)\right]$$
$$= D^{-2}\left[\frac{1}{2}n^{2-2s}\omega^2(1 - \cos^2(2nx - 2\omega t))\right]$$
$$\triangleq F_2 \tag{5.4}$$

and

$$D^{-2}\partial_x[2(\partial_x u^{\omega,n})^2] = D^{-2}\partial_x[n^{2-2s}[1 - \cos(2nx - 2\omega t)]]$$
$$= D^{-2}\partial_x[n^{3-2s}\sin(2nx - 2\omega t)]$$

$$\triangleq F_3, \tag{5.5}$$

$$D^{-2}\partial_x[(u^{\omega,n})^2] = D^{-2}\partial_x\left[\omega^2 n^{-2} + n^{-2s}\left[\frac{1}{2} + \frac{1}{2}\cos(2nx - 2\omega t)\right]\right.$$
$$\left. + 2\omega n^{-1-s}\cos(nx - \omega t)\right]$$
$$= -n^{-2s+1}D^{-2}[\sin(2nx - 2\omega t)] - 2\omega n^{-s}D^{-2}[\sin(nx - \omega t)]$$
$$\triangleq F_4. \tag{5.6}$$

Thus, the error F of the approximate solutions (5.3) is

$$F(t) = F_1(t) + F_2(t) + F_3(t) + F_4(t). \tag{5.7}$$

We will now proceed to estimate the size of the error E in H^σ.

Lemma 5.1 Let $\sigma > 2, s > \dfrac{5}{2}$, then the H^σ-norm of E is bounded by

$$\|F(t)\|_{H^\sigma} \lesssim n^{-s+\sigma}. \tag{5.8}$$

Proof Applying the triangle inequality, we have

$$\|F(t)\|_{H^\sigma} \leqslant \|F_1(t)\|_{H^\sigma} + \|F_2(t)\|_{H^\sigma} + \|F_3(t)\|_{H^\sigma} + \|F_4(t)\|_{H^\sigma}. \tag{5.9}$$

Using the formulas

$$\|\cos(nx - \alpha)\|_{H^\sigma} \approx n^\sigma \quad \text{and} \quad \|\sin(nx - \alpha)\|_{H^\sigma} \approx n^\sigma, \tag{5.10}$$

the Algebra Property inequality, $n \gg 1$ we have

$$\|F_1\|_{H^\sigma} = \left\|\omega n^{-s}\sin(nx - \omega t) + n^{2-2s}\omega^2\sin^2(nx - \omega t)\right.$$
$$\left. + 4n^{1-s}[\omega n^{-1} + n^{-s}\cos(nx - \omega t)]\sin(nx - \omega t)\right\|_{H^\sigma}$$
$$\lesssim n^{-s+\sigma} + n^{2-2s} + n^{2-2s+\sigma} + n^{1-2s+\sigma}$$
$$\lesssim n^{-s+\sigma} + n^{2-2s+\sigma} + n^{1-2s+\sigma}, \tag{5.11}$$

$$\|F_2\|_{H^\sigma} = \left\|D^{-2}\left[\frac{1}{2}n^{2-2s}\omega^2(1 - \cos^2(2nx - 2\omega t))\right]\right\|_{H^\sigma}$$
$$\lesssim n^{2-2s} + n^{-2s+\sigma}$$
$$\lesssim n^{-2s+\sigma} \tag{5.12}$$

and

$$\|F_3\|_{H^\sigma} = \|D^{-2}\partial_x[n^{3-2s}\sin(2nx - 2\omega t)]\|_{H^\sigma}$$
$$\lesssim n^{1-2s+\sigma},$$

$$\|F_4\|_{H^\sigma} = \|-n^{-2s+1}D^{-2}[\sin(2nx - 2\omega t)] - 2\omega n^{-s}D^{-2}[\sin(nx - \omega t)]\|_{H^{\sigma-1}}$$
$$\lesssim n^{-s-2+\sigma} + n^{-2s-1+\sigma}. \tag{5.13}$$

The proof is completed by combining the above estimates.

Estimating the difference between approximate and actual solutions. Now that we have these initial estimates for the approximate solutions, we will proceed to construct our family of actual solutions. Let $u_{\omega,n}(x,t)$ be the actual solutions to the i.v.p. (1.1) with initial data $u_{\omega,n}(x,0) = u^{\omega,n}(x,0)$. More precisely, $u_{\omega,n}(x,t)$ solves

$$\partial_t u_{\omega,n} + (\partial_x u_{\omega,n})^2 - 4u_{\omega,n}\partial_x u_{\omega,n}$$
$$= -D^{-2}(\partial_x u_{\omega,n})^2 - D^{-2}\partial_x[2(\partial_x u_{\omega,n})^2 + (u_{\omega,n})^2], \tag{5.14}$$

$$u_{\omega,n}(x,0) = \omega n^{-1} + n^{-s}\cos nx, \quad \omega \in \{0,1\}. \tag{5.15}$$

Notice the initial data $u^{\omega,n}(x,0) \in H^s$ for all $s \geq 0$, since

$$\|u_{\omega,n}(\cdot,0)\|_{H^s} = \omega n^{-1} + n^{-s}\|\cos nx\|_{H^s} \approx 1, \tag{5.16}$$

for n sufficiently large. Hence, by Theorem 1.1, there is a $T > 0$ such that for $n > 1$, the Cauchy problem (5.14)-(5.15) has a unique solution in $C([0,T]: H^s)$ with lifespan $T > 0$ such that $u_{\omega,n}$ satisfies (1.6) for $t \in [0,T]$.

To estimate the difference between the actual solutions and the approximate solutions, we define $v = u^{\omega,n} - u_{\omega,n}$ which satisfies the following i.v.p.

$$\partial_t v = F - g\partial_x v + 2\partial_x(fv) - D^{-2}(g\partial_x v) - \partial_x D^{-2}(2g\partial_x v + fv),$$
$$v(x,0) = 0, \tag{5.17}$$

where $f = u^{\omega,n} + u_{\omega,n}, g = \partial_x u^{\omega,n} + \partial_x u_{\omega,n}$, E satisfies the estimate in (5.8).

We will now show that the H^s norm of the difference v decays to zero as n goes to infinity.

Lemma 5.2 *Let $\sigma + 1 > 0, s > \max\left\{\dfrac{5}{2}, \sigma + 1\right\}$, then the H^σ norm of the difference v can be estimated by*

$$\|v(t)\|_{H^\sigma} \lesssim n^{-s+\sigma}. \tag{5.18}$$

Proof Apply the operator D^σ to both sides of (5.17), multiply by $D^\sigma v$, and integrate over the torus to get

$$\frac{1}{2}\frac{d}{dt}\|v(t)\|_{H^s}^2 = \int_{\mathbb{T}} D^\sigma F D^\sigma v \, dx + 2\int_{\mathbb{T}} D^\sigma g \partial_x(fv) D^\sigma v \, dx$$
$$- \int_{\mathbb{T}} D^\sigma g \partial_x v D^\sigma v \, dx - \int_{\mathbb{T}} D^\sigma D^{-2} \partial_x(g\partial_x v) D^\sigma v \, dx$$
$$+ \int_{\mathbb{T}} D^\sigma \partial_x D^{-2} \left[2g\partial_x v + fv\right] D^\sigma v \, dx. \tag{5.19}$$

By the definition of g and the algebra property it follows that

$$\|g\|_{H^\sigma} \lesssim \|u^{\omega,n}\|_{H^{\sigma+1}} + \|u_{\omega,n}\|_{H^{\sigma+1}},$$

thus, restricting $\sigma < s - 1$ we find

$$\|g\|_{H^\sigma} \lesssim \|u^{\omega,n}\|_{H^\sigma} + \|u_{\omega,n}\|_{H^\sigma}.$$

Moreover, the well-posedness size estimate (2.22) implies

$$\|u^{\omega,n}(t)\|_{H^\sigma} \lesssim \|u^{\omega,n}(0)\|_{H^\sigma} \lesssim 1$$

and

$$\|u_{\omega,n}(t)\|_{H^\sigma} \lesssim \|u_{\omega,n}(0)\|_{H^\sigma} \lesssim 1.$$

Inequality (5.17) then implies that $\|g\|_{H^\sigma} \lesssim 1$. Similarly, we have $\|f\|_{H^\sigma} \lesssim 1$.

Estimate of the first integral. Applying the Cauchy-Schwarz inequality, we have

$$\left|\int_{\mathbb{T}} D^\sigma F D^\sigma v \, dx\right| \leqslant \|F\|_{H^\sigma}\|v\|_{H^\sigma} \lesssim n^{-s+\sigma}\|v\|_{H^\sigma}. \tag{5.20}$$

Estimate of the second integral. We begin by rewriting this term by commuting v with $D^\sigma \partial_x$ to arrive at

$$\int_{\mathbb{T}} D^\sigma \partial_x(gv) D^\sigma v \, dx = \int_{\mathbb{T}} [D^\sigma \partial_x, f] v D^\sigma v \, dx + \int_{\mathbb{T}} f D^\sigma \partial_x v D^\sigma v \, dx. \tag{5.21}$$

The first integral of (5.21) can be handled by Lemma 2.3,

$$\left|\int_{\mathbb{T}} [D^\sigma \partial_x, f] v D^\sigma v \, dx\right| \lesssim \|f\|_{H^s}\|v\|_{H^\sigma}^2. \tag{5.22}$$

Integrating by parts and using the Sobolev Theorem, we have

$$\left|\int_{\mathbb{T}} f D^\sigma \partial_x v D^\sigma v \, dx\right| = \frac{1}{2}\left|\int_{\mathbb{T}} (D^\sigma v)^2 \partial_x g \, dx\right|$$

$$\lesssim \|\partial_x f\|_{L^\infty} \|v\|_{H^\sigma}^2$$

$$\lesssim \|f\|_{H^s} \|v\|_{H^\sigma}^2. \tag{5.23}$$

We obtain

$$\int_{\mathbb{T}} D^\sigma \partial_x (gv) D^\sigma v \, dx \lesssim \|f\|_{H^\sigma} \|v\|_{H^\sigma}^2 \lesssim \|v\|_{H^\sigma}^2. \tag{5.24}$$

Estimate of the third integral. using Cauchy–Schwarz and Kato-Ponce we find

$$\int_{\mathbb{T}} D^\sigma g \partial_x v \cdot D^\sigma v \, dx$$

$$= \int_{\mathbb{T}} [D^\sigma, g] \partial_x \cdot D^\sigma v \, dx + \int_{\mathbb{T}} g D^\sigma \partial_x v \cdot D^\sigma v \, dx$$

$$\lesssim (\|g\|_{H^\sigma} \|\partial_x v\|_{L^\infty} + \|\partial_x g\|_{L^\infty} \|\partial_x v\|_{H^{\sigma-1}}) \|v\|_{H^\sigma}$$

$$+ \|\partial_x g\|_{L^\infty} \|v\|_{H^\sigma}^2. \tag{5.25}$$

Hence, restricting $\sigma > \frac{3}{2}$ it follows by the Sobolev lemma that

$$\int_{\mathbb{T}} D^\sigma g \partial_x v \cdot D^\sigma v \, dx \lesssim \|g\|_{H^\sigma} \|v\|_{H^\sigma}^2 \lesssim \|v\|_{H^\sigma}^2.$$

Then, for the nonlocal terms involved in (5.19), the algebra property, the size estimate (2.24) and the fact that $\sigma < s - 1 < s$, we have

$$\int_{\mathbb{T}} D^\sigma D^{-2}(g\partial_x v) D^\sigma v \, dx + \int_{\mathbb{T}} D^\sigma \partial_x D^{-2}[2g\partial_x v + fv] D^\sigma v \, dx \lesssim \|v\|_{H^\sigma}^2. \tag{5.26}$$

Combining (5.19)-(5.20) and (5.24)-(5.26), we obtain an energy estimate for v

$$\frac{1}{2}\frac{d}{dt}\|v(t)\|_{H^s} \lesssim n^{-s+\sigma} \|v\|_{H^\sigma} + \|v\|_{H^\sigma}^2. \tag{5.27}$$

Solving (5.27). We arrive at the desired estimate (5.18)

Proof of Theorem 1.2 in the periodic case The basis of our proof rests upon finding two sequences of solutions to the i.v.p. (1.1)-(1.2) that share a common lifespan and satisfy three conditions. We take the sequence of solutions with $\omega = \{0,1\}$. The three conditions they satisfy are as follows

(1)
$$\|w_{\omega,n}(t)\|_{H^s} \lesssim 1 \text{ for } t \in [0,T],$$

(2)
$$\|u_{0,n}(0) - u_{1,n}(0)\|_{H^s} \to 0 \text{ as } n \to \infty,$$

(3)
$$\liminf_{n \to \infty} \|u_{0,n}(t) - u_{1,n}(t)\|_{H^s} \gtrsim |\sin t| \text{ for } t \in (0,T].$$

Property (1), by the solution size estimate (1.6), we have
$$\|u_{\omega,n}(t)\|_{H^s} \lesssim \|u_{\omega,n}(0)\|_{H^s} \lesssim 1. \tag{5.28}$$

Property (2), follows from the definition of our approximate solutions (2.12). We have
$$\|u_{0,n}(0) - u_{1,n}(0)\|_{H^s} = \|u^{0,n}(0) - v^{1,n}(0)\|_{H^s}$$
$$= \|n^{-s}\cos nx + n^{-\frac{1}{2}} - n^{-s}\cos nx\|_{H^s}$$
$$= 4\pi n^{-\frac{1}{2}} \to 0 \text{ as } n \to \infty.$$

For property (3), Using the reverse triangle inequality we get
$$\|u_{1,n}(t) - u_{0,n}(t)\|_{H^s} \geq \|u^{1,n}(t) - u^{0,n}(t)\|_{H^s} - \|u^{1,n}(t) - u_{1,n}(t)\|_{H^s} - \|u^{0,n}(t) - u_{0,n}(t)\|_{H^s}$$
$$\gtrsim \|u^{1,n}(t) - u^{0,n}(t)\|_{H^s} - n^{-\varepsilon},$$

from which we obtain that
$$\liminf_{n \to \infty} \|u_{1,n}(t) - u_{0,n}(t)\|_{H^s} \gtrsim \liminf_{n \to \infty} \|u^{1,n}(t) - u^{0,n}(t)\|_{H^s}. \tag{5.29}$$

Since, by the trigonometric identity $\cos(nx-t) - \cos(nx) = 2\sin\left(nx - \frac{t}{2}\right)\sin\left(\frac{t}{2}\right)$, we have
$$\|u^{1,n}(t) - u^{0,n}(t)\|_{H^s} = \left\|n^{-\frac{1}{2}} - 2n^{-s}\sin\left(nx - \frac{t}{2}\right)\sin\left(\frac{t}{2}\right)\right\|_{H^s}$$

$$\gtrsim \|n^{-s}\sin\left(nx-\frac{t}{2}\right)\sin\left(\frac{t}{2}\right)\|_{H^s} - \|n^{-s}\|_{H^s}$$

$$\gtrsim |\sin\left(\frac{t}{2}\right)| - n^{-\frac{1}{2}} \to |\sin\left(\frac{t}{2}\right)|, \quad n\to\infty.$$

we have

$$\liminf_{n\to\infty} \|u_{1,n}(t) - u_{0,n}(t)\|_{H^s} \gtrsim |\sin\frac{t}{2}|, \qquad (5.30)$$

which completes the proof of property (3). We complete the proof of Theorem 1.2 in the periodic case.

6 Nonuniform dependence on the line

For each $\omega \in \{0,1\}$, we consider the following sequence of high frequency approximate solutions

$$u^h = u^{h,\omega,n}(x,t) = n^{-\frac{\delta}{2}-s}\varphi\left(\frac{x}{n^\delta}\right)\cos(nx-\omega t), \qquad (6.1)$$

where φ is a C^∞ function such that

$$\varphi = \begin{cases} 1, & \text{if } |x| < 1, \\ 0, & \text{if } |x| \geqslant 2. \end{cases} \qquad (6.2)$$

We then modify u^h by adding to it the solution $u_\ell = u_{\ell,\omega,n}(x,t)$ of the following Cauchy problem with the low-frequency initial data $\omega n^{-1}\varphi\left(\frac{x}{n^\delta}\right)$,

$$\partial_t u_\ell + (\partial_x u_\ell)^2 - 4u_\ell \partial_x u_\ell = -D^2(\partial_x u_\ell)^2 - D^2\partial_x[2(\partial_x u_\ell)^2 + (u_\ell)^2], \qquad (6.3)$$

$$u_\ell(x,0) = \omega n^{-1}\tilde\varphi\left(\frac{x}{n^\delta}\right), \qquad (6.4)$$

where $\tilde\varphi$ is a $C_0^\infty(\mathbf{R})$ function such that

$$\tilde\varphi(x) = 1, \quad \text{if } x \in \mathrm{supp}\varphi. \qquad (6.5)$$

Thus, the two sequences of approximate solutions that will be used here to prove nonuniform dependence for generalized Camassa-Holm equation (1.1) are

$$u^{\omega,n} = u_\ell + u^h, \quad \omega = 0 \text{ or } 1. \qquad (6.6)$$

Note that for ω we have $u_\ell(x,0) = 0$. Therefore, in this case $u_\ell = 0$ which gives

$$u^{0,n}(x,t) = n^{-\frac{\delta}{2}-s}\varphi\left(\frac{x}{n^\delta}\right)\cos(nx), \tag{6.7}$$

that is this sequence has no low–frequency term.

Proof of Theorem 1.2 in the nonperiodic case The proof is quite similar to [29, 30, 50, 53-55], so it is omitted to make the paper concise. This completes the proof of Theorem 1.2.

Acknowledgements This work was supported by NSF of China [grant numbers 11671055, 11771062].

References

[1] Novikov V. Generalization of the Camassa–Holm equation [J]. J. Phys. A., 2009, 342002: 1–14.

[2] Tu X, Yin Z. Local well-posedness and blow–up phenomena for a generalized Camassa–Holm equation with peakon solutions [J]. Discrete Contin. Dyn. Syst. A., 2016, 128: 1–19.

[3] Tu X, Yin Z. Blow–up phenomena and local well-posedness for a generalized Camassa–Holm equation in the critical Besov space [J]. Nonlinear Anal. TMA., 2015, 128: 1–19.

[4] Tu X, Yin Z. Global weak solutions for a generalized Camassa–Holm equation. arXiv:1511.02848v2 [math.AP]. J. Phys. A., 2016.

[5] Tu X, Yin Z. Analyticity of the Cauchy problem and persistence properties for a generalized Camassa–Holm equation Generalization of the Camassa–Holm equation. arXiv:1511.02316v1 [math.AP], 2015.

[6] Geng X, Xue B. An extension of integrable peakon equations with cubic nonlinearity [J]. Nonlinearity, 2009, 22: 1847–1856.

[7] Fokas A, Fuchssteiner B. Symplectic structures, their Backlund transformation and hereditary symmetries [J]. Physica D., 1981, 4: 47–66.

[8] Camassa R, Holm D. An integrable shallow water equation with peaked solitons [J]. Phys. Rev. Lett., 1993, 71: 1661–1664.

[9] Constantin A. On the scattering problem for the Camassa–Holm equation [J]. Proc. R. Soc. Lond. A., 2001, 7457: 953–970.

[10] Constantin A. The trajectories of particles in Stokes waves [J]. Invent. Math., 2006, 166: 523–535.

[11] Constantin A, Escher J. Particle trajectories in solitary water waves [J]. Bull. Amer. Math., 2007, 44: 423–431.

[12] Constantin A, Strauss W. Stability of peakons [J]. Commun Pure Appl. Math., 2000, 53: 603–610.

[13] Constantin A, Escher J. Analyticity of periodic traveling free surface water waves with vorticity [J]. Ann. Math., 2011, 173: 559–568.

[14] Amick C.J, Fraenkel L.E, and Toland J.F. On the Stokes conjecture for the wave of extreme form [J]. Acta Math., 1982, 173: 93–214.

[15] Constantin A, McKean H.P. A shallow water equation on the circle [J]. Commun Pure Appl. Math., 1999, 52: 949–982.

[16] Constantin A. On the inverse spectral problem for the Camassa–Holm equation [J]. J. Funct. Anal., 1998, 155: 352–363.

[17] Constantin A, Gerdjikov V, and Ivanov R.I. Inverse scattering transform for the Camassa–Holm equation [J]. Inverse Probl., 2006, 22: 2197–2207.

[18] Constantin A, Kappeler T, Kolev B, and Topalov T. On Geodesic exponential maps of the Virasoro group [J]. Ann. Global Anal. Geom., 2007, 31: 155–180.

[19] Misiołek G.A. Shallow water equation as a geodesic flow on the Bott–Virasoro group [J]. J. Geom. Phys., 1998, 24: 203–208.

[20] Constantin A, Escher J. Wave breaking for nonlinear nonlocal shallow water equations [J]. Acta Math., 1998, 24: 229–243.

[21] Bressan A, Constantin A. Global conservative solutions of the Camassa–Holm equation [J]. Arch. Ration. Mech. Anal., 2007, 183: 215–239.

[22] Holden H, Raynaud X. Global conservative solutions of the Camassa–Holm equations— a Lagrangianpoiny of view [J]. Commun Partial Differ Equ., 2007, 32: 1511–1549.

[23] Bressan A, Constantin A. Global dissipative solutions of the Camassa–Holm equation [J]. Anal. Appl., 2007, 5: 1–27.

[24] Holden H, Raynaud X. Dissipative solutions for the Camassa–Holm equation [J]. Discrete Contin. Dyn. Syst., 2009, 24: 1047–1112.

[25] Lakshmanan M. Integrable nonlinear wave equations and possible connections to tsunami dynamics [A]. In: Lakshmanan M, editor. Tsunami and nonlinear waves [C], Berlin: Springer, 2007, 31–49.

[26] Segur H. Integrable models of waves in shallow water, in: Probability, geometry and integrable systems [A]. Probability, geometry and integrable systems. Vol. 55, MSRI publications [C], Cambridge: Cambridge University Press, 2008, 345–371.

[27] Segur H. Waves in shallow water with emphasis on the tsunami of 2004 [A]. In: Segur H, editor. Tsunami and nonlinear waves [C], Berlin: Springer, 2007, 3–29.

[28] Constantin A, Johnson R.S. Propagation of very long water waves with vorticity, over variable depth, with applications to tsunamis [J]. Fluid Dynam. Res., 2008, 40: 175–211.

[29] Himonas A, Mantzavinos D. The Cauchy problem for the Fokas–Olver–Rosenau–Qiao equation [J]. Nonlinear Appl., 2014, 95: 499–529.

[30] Himonas A, Mantzavinos D. The Cauchy problem for a 4–parameter family of equations with peakon traveling waves [J]. Nonlinear Appl., 2016, 133: 161–199.

[31] Degasperis A, Procesi M. Asymptotic integrability [J]. Symmetry and Perturbation

Theory, 1999, 1: 23–37.
- [32] Degasperis A, Holm D.D, and Hone A.N.W. A new integrable equation with peakon solutions [J]. Theor. Math. Phys., 2002, 133: 1461–1472.
- [33] Constantin A, Ivanov R, and Lenells J. Inverse scattering transform for the Degasperis–Procesi equation [J]. Nonlinearity, 2010, 23: 2559–2575.
- [34] Degasperis A, Procesi M. Asymptotic integrability [A]. In: Degasperis A, Gaeta G, editors. Symmetry and perturbation theory [C]. Singapore: World Scientific, 1999, 23–37.
- [35] Toland J.F. Stokes waves [J]. Topol. Methods Nonlinear Anal., 1996, 7: 1–48.
- [36] Holm D.D, Staley M.F. Wave structure and nonlinear balances in a family of evolutionary PDEs [J]. SIAM J. Appl. Dyn. Syst., 2003, 2: 323–380.
- [37] Lundmark H, Szmigielski J. Multi–peakon solutions of the Degasperis–Procesi equation, Inverse Problems [J]. Inverse Probl., 2005, 21: 1553–1570.
- [38] Matsuno Y. Multisoliton solutions of the Degasperis–Procesi equation and their peakon limit [J]. Inverse Probl., 2003, 19: 1241–1245.
- [39] Lenells J. Traveling wave solutions of the Degasperis–Procesi equation [J]. J. Math. Anal. Appl., 2005, 306: 72–82.
- [40] Vakhnenko V.O, Parkes E.J. Periodic and solitary–wave solutions of the Degasperis–Procesi equation [J]. Chaos Solitons Fractals, 2004, 20: 1059–1073.
- [41] Home A.N.W, Wang J.P. Integrable peakon equations with cubic nonlinearity [J]. J. Phys. A., 2008, 41: 1–11.
- [42] Wu X, Yin Z. Well–posedness and global existence for the Novikov equation [J]. Annali della Scuola Normale Superiore di Pisa Classe di Sci., 2012, 11: 707–727.
- [43] Ni L, Zhou Y. Well–posedness and persistence properties for the Novikov equationn [J]. J Differ Equ., 2011, 250: 3002–3021.
- [44] Wu X, Yin Z. A note on the Cauchy problem of the Novikov equation [J]. Appl. Anal., 2013, 92: 1116–1137.
- [45] Himonas A, Holliman C. The Cauchy problem for the Novikov equation [J]. Nonlinearity, 2012, 25: 449–479.
- [46] Jiang Z, Ni L. Blow–up phenomenon for the integrable Novikov equation [J]. J. Math. Anal. Appl., 2012, 385: 551–558.
- [47] Yan W, Li Y, and Zhang Y. Global existence and blow–up phenomena for the weakly dissipative Novikov equation [J]. Nonlinear Anal., 2012, 75: 2464–2473.
- [48] Hone W, Lundmark H, and Szmigielski J. Explicit multipeakon solutions of Novikov cubically nonlinear integrable Camassa–Holm type equation [J]. Dyn Partial Differ Equ., 2009, 6: 253–289.
- [49] Himonas A, Misiołek G. High–frequency smooth solutions and well-posedness of the Camassa–Holm equation [J]. Int. Math. Res. Not., 2005, 51: 3135–3151.
- [50] Himonas A, Kenig C. Non–uniform dependence on initial data for the CH equation on the line [J]. Differ Integral Equ., 2009, 22: 201–224.

[51] Himonas A, Kenig C, and Misiołek G. Non-uniform dependence for the periodic CH equation, Comm [J]. Commun Partial Differ Equ., 2010, 35: 1145–1162.

[52] Christov O, Hakkaev S. On the Cauchy problem for the periodic b-family of equations and of the non–uniform continuity of Degasperis–Procesi equation [J]. J. Math. Anal. Appl., 2009, 360: 47–56.

[53] Himonas A, Holliman C. On well-posedness of the Degasperis–Procesi equation [J]. Discrete Contin. Dyn. Syst., 2011, 31: 469–488.

[54] Holliman C. On well-posedness of the Degasperis–Procesi equation [J]. Differ. Int. Equ., 2010, 23: 1150–1194.

[55] Himonas A, Misiołek G. Non–uniform dependence on initial data of solutions to the Euler equations of hydrodynamics [J]. Commun Math Phys., 2009, 296: 285–301.

[56] Taylor M. Pseudodifferential Operators and Nonlinear PDE [M]. Boston: Birkhäuser, 1991.

[57] Taylor M. Partial Differential Equations [M]. III: Nonlinear Equations, New York: Springer, 1996.

[58] Dieudonné J. oundations of Modern Analysis [M]. New York: Academic Press, 1969.

General Propagation Lattice Boltzmann Model for a Variable-Coefficient Compound KdV-Burgers Equation[*]

Lan Zhongzhou(兰中周), Hu Wenqiang(胡文强),
and Guo Boling(郭柏灵)

Abstract In this paper, a general propagation lattice Boltzmann model for a variable-coefficient compound Korteweg-de Vries-Burgers (vc-cKdVB) equation is investigated through selecting an equilibrium distribution function and adding a compensation function, which can provide some more realistic models than their constant-coefficient counterparts in fluids or plasmas. Chapman-Enskog analysis shows that the vc-gKdVB equation can be recovered correctly from the present model. Numerical simulations in different situations of this equation are conducted, including the propagation and interaction of the bell-type, kink-type and periodic-depression solitons and the evolution of the shock-wave solutions. It is found that the numerical results match well with the analytical solutions, which demonstrates that the current lattice Boltzmann model is a satisfactory and efficient algorithm. In addition, it is also shown the present model could be more stable and more accurate than the standard lattice Bhatnagar-Gross-Krook model by adjusting the two free parameters introduced in the propagation step.

Keywords general propagation lattice Boltzmann model; variable-coefficient compound KdV-Burgers equation; numerical simulations; soliton solutions

---

[*] Appl. Math. Model., 2019, 73: 695-714. Doi: 10.1016/j.apm.2019.04.013.

1 Introduction

Lattice Boltzmann method (LBM) has been used in simulating some fluid flows [1-3], and extended to simulate some nonlinear evolution equations (NLEEs), such as the anisotropic dispersion equation [4], convection-diffusion equation [5,6], nonlinear advection-diffusion equations [7], Burgers equation [8], Korteweg-de Vries (KdV) equation [9], modified KdV (mKdV) equation [10], KdV-Burgers equation [12], generalized Boussinesq equation [13], the generalized nonlinear wave equations [14,15], and generalized Gardner equation with time-dependent variable coefficients [16]. Unlike traditional numerical methods which discretize the governing equations in time and space, LBM is based on kinetic theory which tracks the dynamics of ensembles of microscopic particles [3]. Through the particle distribution function and equilibrium distribution function, the macroscopic variables are educed and the macroscopic equations are restored exactly [15].

Most of these previous LBM works solving NLEEs focus on the standard lattice Bhatnagar-Gross-Krook (SLBGK) model, which is popular and numerically efficient. However, it may suffer from some problems. For example, the SLBGK model may turn out to be unstable for the convection-dominated problem when solving the nonlinear advection-diffusion equations [6], where the diffusion coefficient is very small, resulting in a relaxation parameter closed to 0.5. Based on the Lax-Wendroff (LW) scheme [17] and fractional propagation (FP) scheme [18], Ref. [19] proposes a general propagation lattice Boltzmann (GPLB) scheme for fluid dynamics in the framework of the time-splitting method. It is shown that this scheme is more general, and both the LW and FP schemes can be considered as its special cases. The GPLB scheme can also be used to improve the numerical stability by adjusting the two free parameters introduced into the propagation step [19]. Ref. [7] develops a GPLB model for the nonlinear advection-diffusion equations with variable coefficients, and discusses the advantages of the GPLB model than the SLBGK model.

Besides, the numerical studies for the NLEEs based on the LBM generally consider the constant-coefficient NLEEs [14, 15]. In physical situations, on the other hand, when the inhomogeneities of media and nonuniformities of boundaries are taken into account, the variable-coefficient NLEEs can provide some more realistic models than their constant-coefficient counterparts [20-24].

For this reason, in this paper, based on the previous works, we develop a GPLB model for a variable-coefficient compound Korteweg-de Vries-Burgers (vc-cKdVB) equation [25, 26], which is of the following forms,

$$u_t + l(t)uu_x + m(t)u^{2p}u_x + n(t)u_{xx} - r(t)u_{xxx} + s(t)u_x = 0. \tag{1.1}$$

Eq. (1.1) can represent the various physical models widely used in such fields as solid-state materials, plasmas and fluids, where u is the wave-amplitude function of the scaled space coordinate x and time coordinate t, the parameter p is non-negative constant, the coefficients $l(t)$, $m(t)$, $n(t)$, $r(t)$ and $s(t)$ are all analytic functions of time t, and the subscripts represent the partial derivatives. In plasmas and fluids, special cases of Eq. (1.1) have been seen as follows:

(i) When $l(t) = \alpha$, $r(t) = -\beta$, $m(t) = n(t) = s(t) = 0$, Eq. (1.1) can be transformed into the KdV equation,

$$u_t + \alpha u u_x + \beta u_{xxx} = 0, \tag{1.2}$$

which has been seen to describe the wave propagation and the stratified internal wave in fluids, and the ion-acoustic wave in plasma [23, 27].

(ii) When $l(t) = 2\alpha$, $m(t) = -3\beta$, $r(t) = -1$, $p = 1$, $n(t) = s(t) = 0$, Eq. (1.1) becomes a constant-coefficient combined KdV-modified KdV equation or the Gardner equation, for the wave propagation of the sound wave and thermal pulse [28, 29],

$$u_t + 2\alpha u u_x - 3\beta u^2 u_x + u_{xxx} = 0. \tag{1.3}$$

(iii) When $l(t) = 1$, $n(t) = -\theta$, $r(t) = -\delta$, $m(t) = s(t) = 0$, Eq. (1.1) can be reduced into the constant-coefficient KdV-Burgers equation,

$$u_t + uu_x - \theta u_{xx} + \delta u_{xxx} = 0, \tag{1.4}$$

which plays an important role in studying liquid with bubbles inside, the flow of liquid in elastic tubes, and the problems of turbulence [12, 30].

(iv) When $l(t) = k(t)$, $m(t) = f(t)$, $r(t) = -g(t)$, $s(t) = h(t)$, $n(t) = 0$, $p = 1$ and $f(t)g(t) \neq 0$, Eq. (1.1) is transformed into a variable-coefficient KdV-modified KdV equation,

$$u_t + k(t)uu_x + f(t)u^2 u_x + g(t)u_{xxx} + h(t)u_x = 0, \tag{1.5}$$

which has been investigated for its exhaustive classification by using the groups of equivalence transformations [31].

The outline of our paper is given by: In Sect. 2, the general propagation lattice Boltzmann model for Eq. (1.1) will be derived. In Sect. 3, effects of the free parameters induced in our lattice Boltzmann model will be discussed. In Sect. 4, detailed numerical simulations on preceding cases will be performed in order to examine the accuracy and stability of our model. Conclusions will be summarized In Sect. 5.

2 General propagation lattice Boltzmann model

For Eq. (1.1), the evolution law of the particle distribution function can be the corresponding discrete velocity Boltzmann equation [32-35],

$$\frac{\partial f_i}{\partial t} + \xi_i \frac{\partial f_i}{\partial x} = \Omega_i, \qquad (2.1)$$

where $f_i(x,t)$ is a scalar function describing the particle distribution at position x and time t, $\{\xi_i, i = 0, 1, \cdots, n-1\}$ is the set of discrete velocities in the d-dimensional space with n different velocity directions ($DdQn$) lattice model, while Ω_i is the collision term describing the binary collision between particles, which is a complex integro-differential term. For simplicity, we adopt the Bhatnagar-Gross-Krook (BGK) collision operator [3],

$$\Omega_i = -\frac{1}{\tau_0}\left[f_i - f_i^{(eq)}\right], \qquad (2.2)$$

where $f_i^{(eq)}$ is the local equilibrium distribution function, and τ_0 is the single relaxation time.

Besides, the $D1Q5$ velocity model is utilized in this paper, where the discrete velocities can be defined as,

$$\vec{\xi} = c\vec{e}, \quad \vec{e} = \{e_0, e_1, e_2, e_3, e_4\} = \{0, 1, -1, 2, -2\}, \qquad (2.3)$$

where c is a scale factor constant, which represents the lattice velocity and defined as

$$c = a\frac{\Delta x}{\Delta t}, \qquad (2.4)$$

with Δx and Δt as the lattice space and time step, respectively, while a is a free parameter to adjust the propagation step of our model.

Applying the time-splitting method, Eq. (2.1) can be decomposed into collision and propagation steps [19],

$$\frac{\partial f_i}{\partial t} = -\frac{1}{\tau_0}\left[f_i - f_i^{(eq)}\right], \tag{2.5a}$$

$$\frac{\partial f_i}{\partial t} + e_i \frac{\partial f_i}{\partial x} = 0, \tag{2.5b}$$

which are solved sequentially at each time step. The advantage of this replacement is that the collision and propagation steps can be treated with different numerical schemes, respectively.

For Eq. (2.5a), no spatial derivative is involved. Hence, the explicit Euler scheme is used to discretize Eq. (2.5a) into the following form,

$$f_i^+(x,t) = \left(1 - \frac{1}{\tau}\right) f_i(x,t) + \frac{1}{\tau} f_i^{(eq)}(x,t) + \Delta t G_i(x,t), \tag{2.6}$$

where $\tau = \tau_0/\Delta t$ is the dimensionless relaxation time, and the correction term $G_i(x,t)$ is introduced into the collision step to eliminate the effect of the additional term [7]. It should be noted that the collision process is the same as that in the standard LBGK models.

For Eq. (2.5b), we adopt the explicit two-level and three-point scheme [19] to discretize it,

$$f_i(x, t+\Delta t) = p_0 f_i^+(x,t) + p_{-1} f_i^+(x-L_i, t) + p_1 f_i^+(x+L_i, t), \quad L_i = \Delta x \cdot e_i, \tag{2.7}$$

where p_0, p_{-1} and p_1 are free parameters satisfying

$$p_0 + p_{-1} + p_1 = 1, \quad p_{-1} - p_1 = a = \Delta t \cdot \frac{\xi_i}{L_i}. \tag{2.8}$$

One solution of Eq. (2.8) can be expressed as follows:

$$p_0 = 1 - q, \quad p_{-1} = \frac{q+a}{2}, \quad p_1 = \frac{q-a}{2}, \tag{2.9}$$

where q is also one introduced free parameter. Clearly, the propagation process of our general model is different with that in the standard LBGK models, which is just a special case, i.e., $a = q = 1$. The discretized Eq. (2.7) can be rewritten into

$$f_i(x, t+\Delta t) = f_i^+(x,t) + \frac{q}{2}\left[f_i^+(x+L_i,t) - 2f_i^+(x,t) + f_i^+(x-L_i,t)\right]$$
$$- \frac{a}{2}\left[f_i^+(x+L_i,t) - f_i^+(x-L_i,t)\right]. \quad (2.10)$$

Based on the stability analysis in the numerical stability condition, these two parameters should satisfy $a^2 \leqslant q \leqslant 1$ ([7]).

From the above, the combination of the collision scheme given by Eq. (2.6) and the propagation scheme given by Eq. (2.10) constructs the general propagation lattice Boltzmann model.

In the following, we will apply the multi-scale Chapman-Enskog [36] and Taylor expansions to obtain the specific expressions of the local equilibrium distribution function f_i^{eq} and the correction term G_i, completing our general propagation lattice Boltzmann model for Eq. (1.1).

Firstly, applying the Taylor expansion to $f_i^+(x+L_i,t)$ and $f_i^+(x-L_i,t)$, and retaining the terms up to $O(\Delta t^4)$, we have

$$f_i^+(x+L_i,t) = f_i^+(x,t) + L_i \cdot \partial_x f_i^+(x,t) + \frac{(L_i \cdot \partial_x)^2}{2} f_i^+(x,t) + \frac{(L_i \cdot \partial_x)^3}{6} f_i^+(x,t)$$
$$+ O(L_i^4)$$
$$= f_i^+(x,t) + \frac{\Delta t}{a}(\xi_i \cdot \partial_x) f_i^+(x,t) + \frac{\Delta t^2}{2a^2}(\xi_i \cdot \partial_x)^2 f_i^+(x,t)$$
$$+ \frac{\Delta t^3}{6a^3}(\xi_i \cdot \partial_x)^3 f_i^+(x,t) + O(\Delta t^4), \quad (2.11a)$$

$$f_i^+(x-L_i,t) = f_i^+(x,t) - \frac{\Delta t}{a}(\xi_i \cdot \partial_x) f_i^+(x,t) + \frac{\Delta t^2}{2a^2}(\xi_i \cdot \partial_x)^2 f_i^+(x,t)$$
$$- \frac{\Delta t^3}{6a^3}(\xi_i \cdot \partial_x)^3 f_i^+(x,t) + O(\Delta t^4). \quad (2.11b)$$

Substituting Eqs. (2.11) into Eq. (2.10), we obtain

$$f_i(x, t+\Delta t) = f_i^+(x,t) - \Delta t (\xi_i \cdot \partial_x) f_i^+(x,t) + \frac{\Delta t^2 q}{2a^2}(\xi_i \cdot \partial_x)^2 f_i^+(x,t)$$
$$- \frac{\Delta t^3}{6a^2}(\xi_i \cdot \partial_x)^3 f_i^+(x,t) + O(\Delta t^4). \quad (2.12)$$

Besides, using the Taylor expansion to the left side of the above equation to $O(\Delta t^4)$,

$$f_i(x, t + \Delta t) = f_i(x,t) + \Delta t \partial_t f_i(x,t) + \frac{\Delta t^2}{2} \partial_t^2 f_i(x,t) + \frac{\Delta t^3}{6} \partial_t^3 f_i(x,t) + O(\Delta t^4), \quad (2.13)$$

we have

$$f_i(x,t) + \Delta t \partial_t f_i(x,t) + \frac{\Delta t^2}{2} \partial_t^2 f_i(x,t) + \frac{\Delta t^3}{6} \partial_t^3 f_i(x,t)$$

$$= f_i^+(x,t) - \Delta t \left(\xi_i \cdot \partial_x\right) f_i^+(x,t) \quad (2.14)$$

$$+ \frac{\Delta t^2 q}{2a^2} \left(\xi_i \cdot \partial_x\right)^2 f_i^+(x,t) - \frac{\Delta t^3}{6a^2} \left(\xi_i \cdot \partial_x\right)^3 f_i^+(x,t) + O(\Delta t^4).$$

Secondly, applying multi-scale Chapman-Enskog expansion up to the third-order in time t, the first-order in space x and the local particle distribution function f_i and the correction term G_i, we have

$$\partial_t = \epsilon \partial_{t_1} + \epsilon^2 \partial_{t_2} + \epsilon^3 \partial_{t_3}, \quad (2.15a)$$

$$\partial_x = \epsilon \partial_{x_1}, \quad (2.15b)$$

$$f_i = \sum_{n=0}^{\infty} \epsilon^n f_i^{(n)} = f_i^{(0)} + \epsilon f_i^{(1)} + \epsilon^2 f_i^{(2)} + \epsilon^3 f_i^{(3)} + \cdots, \quad (2.15c)$$

$$G_i = \epsilon G_i^{(1)}, \quad (2.15d)$$

where ϵ is a small expansion parameter.

Substituting Eqs. (2.15a)-(2.15d) into Eq. (2.14) and coupling with Eq. (2.6), we can obtain a series of differential equations for the first three orders of ϵ,

$$O(\epsilon^0): \quad f_i^{(0)} = \left(1 - \frac{1}{\tau}\right) f_i^{(0)} + \frac{1}{\tau} f_i^{(eq)}, \quad \text{i.e.,} \quad f_i^{(0)} = f_i^{(eq)}, \quad (2.16a)$$

$$O(\epsilon^1): \quad f_i^{(1)} + \Delta t \, \partial_{t_1} f_i^{(0)} = \left(1 - \frac{1}{\tau}\right) f_i^{(1)} + \Delta t G_i^{(1)} - \Delta t \left(\xi_i \cdot \partial_{x_1}\right) f_i^{(0)}, \quad (2.16b)$$

$$O(\epsilon^2): \quad f_i^{(2)} + \Delta t \left[\partial_{t_1} f_i^{(1)} + \partial_{t_2} f_i^{(0)}\right] + \frac{\Delta t^2}{2} \partial_{t_1}^2 f_i^{(0)}$$

$$= \left(1 - \frac{1}{\tau}\right) f_i^{(2)} - \Delta t \left(\xi_i \cdot \partial_{x_1}\right) \left[\left(1 - \frac{1}{\tau}\right) f_i^{(1)} + \Delta t G_i^{(1)}\right] \quad (2.16c)$$

$$+ \frac{\Delta t^2 q}{2a^2} \left(\xi_i \cdot \partial_{x_1}\right)^2 f_i^{(0)},$$

$$O(\epsilon^3): f_i^{(3)} + \Delta t \left[\partial_{t_1} f_i^{(2)} + \partial_{t_2} f_i^{(1)} + \partial_{t_3} f_i^{(0)} \right] + \frac{\Delta t^2}{2} \left[\partial_{t_1}^2 f_i^{(1)} + 2\partial_{t_1,t_2}^2 f_i^{(0)} \right]$$

$$+ \frac{\Delta t^3}{6} \partial_{t_1}^3 f_i^{(0)} = \left(1 - \frac{1}{\tau}\right) f_i^{(3)} - \Delta t \left(\xi_i \cdot \partial_{x_1}\right) \left(1 - \frac{1}{\tau}\right) f_i^{(2)}$$

$$- \frac{\Delta t^3}{6a^2} (\xi_i \cdot \partial_{x_1})^3 f_i^{(0)}$$

$$+ \frac{\Delta t^2 q}{2a^2} (\xi_i \cdot \partial_{x_1})^2 \left[\left(1 - \frac{1}{\tau}\right) f_i^{(1)} + \Delta t G_i^{(1)} \right].$$

(2.16d)

Simplifying Eq. (2.16) can obtain

$$-\frac{f_i^{(1)}}{\tau \Delta t} = (\partial_{t_1} + \xi_i \cdot \partial_{x_1}) f_i^{(0)} - G_i^{(1)},$$

(2.17a)

$$-\frac{f_i^{(2)}}{\tau \Delta t} = \left[\frac{\Delta t}{2} \partial_{t_1}^2 + \partial_{t_2} - \frac{q\Delta t}{2a^2} (\xi_i \cdot \partial_{x_1})^2 \right] f_i^{(0)} + \left[\partial_{t_1} + \left(1 - \frac{1}{\tau}\right) \xi_i \cdot \partial_{x_1} \right] f_i^{(1)}$$

$$+ \Delta t (\xi_i \cdot \partial_{x_1}) G_i^{(1)},$$

(2.17b)

$$-\frac{f_i^{(3)}}{\tau \Delta t} = -\frac{q\Delta t^2}{2a^2} (\xi_i \cdot \partial_{x_1})^2 G_i^{(1)} + \left[\frac{\Delta t^2}{6} \partial_{t_1}^3 + \Delta t \partial_{t_1,t_2}^2 + \partial_{t_3} \right.$$

$$+ \left. \frac{\Delta t^2}{6a^2} \left(1 - \frac{1}{\tau}\right) (\xi_i \cdot \partial_{x_1})^3 \right] f_i^{(0)}$$

$$+ \left[\frac{\Delta t}{2} \partial_{t_1}^2 + \partial_{t_2} - \frac{q\Delta t}{2a^2} (\xi_i \cdot \partial_{x_1})^2 \right] f_i^{(1)} + \left[\partial_{t_1} + \left(1 - \frac{1}{\tau}\right) \xi_i \cdot \partial_{x_1} \right] f_i^{(2)}.$$

(2.17c)

From Eq. (2.17a), one can obtain

$$f_i^{(1)} = \tau \Delta t \left[G_i^{(1)} - (\partial_{t_1} + \xi_i \cdot \partial_{x_1}) f_i^{(0)} \right].$$

(2.18)

Substituting Eq. (2.18) into Eq. (2.17b), we have

$$f_i^{(2)} = -\tau \Delta t^2 \left\{ \left[\left(\frac{1}{2} - \tau\right) \partial_{t_1}^2 + \frac{1}{\Delta t} \partial_{t_2} + (1 - 2\tau) \partial_{t_1} (\xi_i \cdot \partial_{x_1}) \right. \right.$$

$$- \left. \left(\tau - 1 + \frac{q}{2a^2}\right) (\xi_i \cdot \partial_{x_1})^2 \right] f_i^{(0)}$$

$$+ \tau (\partial_{t_1} + \xi_i \cdot \partial_{x_1}) G_i^{(1)} \Big\}.$$

(2.19)

Similarly, by substituting Eq. (2.18) and (2.19) into Eq. (2.17c), we have

$$-\frac{f_i^{(3)}}{\tau \Delta t^3} = \left\{ \left[\left(\tau^2 - \tau + \frac{1}{6}\right) \partial_{t_1}^3 + \frac{1}{\Delta t}(1-2\tau) \partial_{t_1,t_2}^2 + \frac{1}{\Delta t^2} \partial_{t_3} \right] \right.$$
$$+ \left[3\left(\tau^2 - \tau + \frac{1}{6}\right) \partial_{t_1}^2 + \frac{1}{\Delta t}(1-2\tau) \partial_{t_2} \right] (\xi_i \cdot \partial_{x_1})$$
$$+ \left[3\tau^2 + \left(\frac{q}{a^2} - 4\right)\tau + 1 - \frac{q}{2a^2} \right] \partial_{t_1} (\xi_i \cdot \partial_{x_1})^2$$
$$+ \left. \left[\tau^2 + \left(\frac{q}{a^2} - 2\right)\tau + 1 - \frac{q}{a^2} + \frac{1}{6a^2} \right] (\xi_i \cdot \partial_{x_1})^3 \right\} f_i^{(0)}$$
$$+ \left\{ \left[-\tau\left(\tau - \frac{1}{2}\right) \partial_{t_1}^2 + \frac{\tau}{\Delta t} \partial_{t_2} \right] \right.$$
$$\left. + (1-2\tau)\tau \partial_{t_1}(\xi_i \cdot \partial_{x_1}) + \left[(1 - \frac{q}{2a^2})\tau - \tau^2 \right] (\xi_i \cdot \partial_{x_1})^2 \right\} G_i^{(1)}. \quad (2.20)$$

Similar to general LBM, we define the macroscopic physical quantity u as a distribution function

$$u(x,t) = \sum_i f_i(x,t). \quad (2.21)$$

Due to the conservation law of local mass, f_i and $f_i^{(eq)}$ satisfy

$$u(x,t) = \sum_i f_i(x,t) = \sum_i f_i^{(eq)}(x,t). \quad (2.22)$$

From Eq. (2.16a), we have

$$\sum_i f_i^{(0)}(x,t) = u(x,t), \quad \sum_i f_i^{(n)}(x,t) = 0, \quad n > 0. \quad (2.23)$$

In order to recover Eq. (1.1), some constraints on the equilibrium distribution function and compensation function are imposed as follows:

$$\sum_i \xi_i f_i^{(0)} = 0, \quad (2.24a)$$

$$\sum_i \xi_i^2 f_i^{(0)} = \frac{n(t)u}{\left(1 - \frac{q}{2a^2} - \tau\right)\Delta t}, \quad (2.24b)$$

$$\sum_i \xi_i^3 f_i^{(0)} = -\frac{r(t)u}{\left[\tau^2 - \left(2 - \frac{q}{2a^2}\right)\tau + 1 + \frac{1}{6a^2} - \frac{q}{a^2}\right]\Delta t^2}, \qquad (2.24c)$$

$$\sum_i G_i = 0, \qquad (2.24d)$$

$$\sum_i \xi_i G_i = \frac{1}{\tau \Delta t}\left[s(t)u + \frac{l(t)}{2}u^2 + \frac{m(t)}{2p+1}u^{2p+1}\right], \qquad (2.24e)$$

$$\sum_i \xi_i^2 G_i = 0. \qquad (2.24f)$$

Summing Eq. (2.18) over i and using Eqs. (2.23) and (2.24), we have
$$\partial_{t_1} u = 0. \qquad (2.25)$$

Similarly, by summing Eq. (2.19) over i and recalling the definition (2.15d), we have
$$\partial_{t_2} u + \frac{1}{\epsilon^2}\left[n(t)u_{xx} + s(t)u_x + l(t)uu_x + m(t)u^{2p}u_x\right] = \Delta t\left(\tau - \frac{1}{2}\right)\partial_{t_1}^2 u = 0. \qquad (2.26)$$

Analogously, summing Eq. (2.20) over i and recalling the definition (2.15d), we have
$$\partial_{t_3} u - \frac{1}{\epsilon^3}r(t)u_{xxx} = \mathscr{R}, \qquad (2.27)$$

with

$$\mathscr{R} = -\Delta t^2\left(\tau^2 - \tau + \frac{1}{6}\right)\partial_{t_1}^3 u + \Delta t(2\tau - 1)\partial_{t_2}\partial_{t_1} u - \frac{\Delta t}{\epsilon^2}\Bigg\{\left[3\tau^2 + \left(\frac{q}{a^2} - 4\right)\tau + 1\right.$$
$$\left. - \frac{q}{2a^2}\right]\partial_{t_1}\left[\frac{n(t)u_{xx}}{(1-\frac{q}{2a^2}-\tau)}\right] + (1-2\tau)\tau\partial_{t_1}\left[\frac{s(t)u_x + l(t)uu_x + m(t)u^{2p}u_x}{\tau}\right]\Bigg\}$$

$$= -\frac{\Delta t}{\epsilon^2}\Bigg\{\frac{6a^2\tau^2 + \tau(2q - 8a^2) + 2a^2 - q}{2a^2\tau - 2a^2 + q}\partial_{t_1}[n(t)u_{xx}]$$
$$+ (1-2\tau)\partial_{t_1}\left[s(t)u_x + l(t)uu_x + m(t)u^{2p}u_x\right]\Bigg\}$$

$$= -\frac{\Delta t}{\epsilon^2}\Bigg\{\frac{6a^2\tau^2 + \tau(2q - 8a^2) + 2a^2 - q}{2a^2\tau - 2a^2 + q}u_{xx}\partial_{t_1}n(t)$$
$$+ (1-2\tau)\left[u_x\partial_{t_1}s(t) + uu_x\partial_{t_1}l(t) + u^{2p}u_x\partial_{t_1}m(t)\right]\Bigg\}.$$

It should be noted that Eq. (2.25) has been used repeatedly in the process of simplification for the residual $\mathscr{R}$.

Taking Eq. (2.25)$\times\epsilon$+Eq. (2.26)$\times\epsilon^2$+Eq. (2.27)$\times\epsilon^3$ and assuming $\epsilon = \Delta t$, we have

$$u_t + l(t)uu_x + m(t)u^{2p}u_x + n(t)u_{xx} - r(t)u_{xxx} + s(t)u_x = \mathcal{R}, \qquad (2.28)$$

with

$$\mathcal{R} = \epsilon^3 \mathscr{R} = -\epsilon^2 \left\{ \frac{6a^2\tau^2 + \tau(2q - 8a^2) + 2a^2 - q}{2a^2\tau - 2a^2 + q} u_{xx}\partial_{t_1} n(t) + (1 - 2\tau)\left[u_x \partial_{t_1} s(t) \right.\right.$$
$$\left.\left. + uu_x \partial_{t_1} l(t) + u^{2p} u_x \partial_{t_1} m(t)\right]\right\} = O(\epsilon^2).$$

This means that Eq. (2.28) is the recovered approximate calculation formula of the macroscopic equation, i.e., Eq. (1.1). It should be noted that our model can recover the generalized Korteweg-de Vries-Burgers equation exactly if the variable coefficients are set as constants.

In order to derive the specific expressions of the local equilibrium distribution function, we should introduce one new constraint other than Eqs. (2.22)-(2.24c). Considering the moments of f are all polynomial functions of u, we assume that the equilibrium distribution function satisfies the following constraint,

$$f_4^{(0)} = \sigma u, \qquad (2.29)$$

where σ is introduced to adjust the equilibrium distribution function.

Coupling Eq. (2.29) and Eqs. (2.22)-(2.24c), we can derive the equilibrium distribution function,

$$f_i^{(0)} = \begin{cases} \left[1 + 6\sigma - \dfrac{n(t)}{c^2 \Delta t \tau_2} - \dfrac{r(t)}{2c^3 \Delta t^2 \tau_3}\right] u, & i=0, \\[2mm] \left[-4\sigma + \dfrac{n(t)}{2c^2 \Delta t \tau_2} + \dfrac{r(t)}{2c^3 \Delta t^2 \tau_3}\right] u, & i=1, \\[2mm] \left[-4\sigma + \dfrac{n(t)}{2c^2 \Delta t \tau_2} + \dfrac{r(t)}{6c^3 \Delta t^2 \tau_3}\right] u, & i=2, \\[2mm] \left[\sigma - \dfrac{r(t)}{6c^3 \Delta t^2 \tau_3}\right] u, & i=3, \\[2mm] \sigma u, & i=4, \end{cases} \qquad (2.30)$$

with
$$\tau_2 = \tau + \frac{q}{2a^2} - 1, \quad \tau_3 = \tau^2 - \left(2 - \frac{q}{a^2}\right)\tau + 1 - \frac{q}{a^2} + \frac{1}{6a^2}.$$

Coupling Eqs. (2.24d)-(2.24f) and assuming the collisionless function in the directions of velocity-1 and velocity-2 are the same, we can derive the compensation function,

$$G_i = \begin{cases} \frac{1}{2}\mathscr{G}, & i=0, \\ -\frac{1}{3}\mathscr{G}, & i=1, \\ -\frac{1}{3}\mathscr{G}, & i=2, \\ \frac{1}{3}\mathscr{G}, & i=3, \\ -\frac{1}{6}\mathscr{G}, & i=4, \end{cases} \quad (2.31)$$

with
$$\mathscr{G} = \frac{1}{\tau \Delta t c}\left[s(t)u + \frac{l(t)}{2}u^2 + \frac{m(t)}{2p+1}u^{2p+1}\right].$$

Considering the time-dependent variable coefficients in Eq. (1.1), we think the single relaxation time τ should not be a constant with fixed value in the specific numerical simulations. Informed by Ref. [16] and the nonzero value of the variable coefficient $r(t)$, we can simply set the parameter κ as a given constant with fixed value,

$$\kappa = -\frac{r(t)}{c^3 \Delta t^2 \tau_3}. \quad (2.32)$$

Recalling the definitions of c and τ_3 and due to the stability of the equation which requires $\tau > 0.5$ ([37]), we can obtain the following relation,

$$\tau(t) = 1 - \frac{q}{2a^2} + \sqrt{\frac{q^2}{4a^4} - \frac{1}{6a^2} - \frac{\Delta t}{a^3 \Delta x^3 \kappa}r(t)}. \quad (2.33)$$

3 Discussions for the introduced free parameters

As presented in Section 2, the constant σ is introduced to provide compensatory constraint of the equilibrium distribution functions to adjust them into appropriate relations, and the parameter κ is used to adjust and obtain the single relaxation time τ. Hence, the parameters σ and κ are all adjustment factors, which are important for the numerical simulations. In addition, the free parameters a and q are introduced to adjust the propagation step of our model, which is important in our simulations.

In this section, we will give the numerical simulation procedures firstly, and then discuss the effects of κ, σ, a and q, in our current general propagation lattice Boltzmann model, which can guide us to choose suitable values of the free parameters and obtain nice numerical simulation results.

3.1 Numerical simulation procedures

Denote $f_{i,j}^n = f_i(x_j, t_n), u_j^n = u(x_j, t_n), G_{i,j}^n = G_i(x_j, t_n), x_j = x_{min} + j\Delta x, t_n = n\Delta t$, where $(i = 0, 1, 2, 3, 4; j = 1, 2, \cdots, N)$, n denotes the n-th layer time, j is a spatial grid, and x_{min} is the left border of calculation section.

Then, we can rewrite Eq. (2.6) and Eq. (2.10) by classical finite difference notation:

$$g_{i,j}^n = \left(1 - \frac{1}{\tau}\right) f_{i,j}^n + \frac{1}{\tau} f_{i,j}^{n,(0)} + \Delta t\, G_{i,j}^n, \tag{3.1a}$$

$$f_{i,j}^{n+1} = g_{i,j}^n + \frac{q}{2}(g_{i,j+\mathscr{E}_i}^n - 2g_{i,j}^n + g_{i,j-\mathscr{E}_i}^n) - \frac{a}{2}(g_{i,j+\mathscr{E}_i}^n - g_{i,j-\mathscr{E}_i}^n), \tag{3.1b}$$

with $\mathscr{E} = \{0, 1, -1, 2, -2\}$. And the non-equilibrium extrapolation scheme proposed in Ref. [38] is used for boundary conditions.

Flowchart of our numerical simulation procedures is shown in Fig. 1. We think one can conduct the numerical simulation based on our LBM via this flowchart easily. It should be noted that all parameters should be recalculated at the beginning of every time step, due to the time-dependent variable coefficients.

Besides, the validation of our work is confirmed by comparing the analytical solutions and the numerical results. The root-mean-square error E_2, the max error E_∞, and the global relative error **GRE** are used to measure the accuracy of our model [39]:

$$E_2 = \frac{1}{N}\sqrt{\sum_{j=1}^{N}[u(x_j, t) - u^*(x_j, t)]^2}, \tag{3.2a}$$

$$E_\infty = \max_{j=1,2,\cdots,N} |u(x_j, t) - u^*(x_j, t)|, \tag{3.2b}$$

$$\mathbf{GRE} = \frac{\sum_{j=1}^{N}|u(x_j, t) - u^*(x_j, t)|}{\sum_{j=1}^{N}|u^*(x_j, t)|}, \tag{3.2c}$$

where $u(x_j, t)$ and $u^*(x_j, t)$ are the numerical result and the analytical solution, respectively, and N is the number of grid points.

In addition, a practical method to calculate the rate of convergence for a discretization method is to implement the following formula [39]:

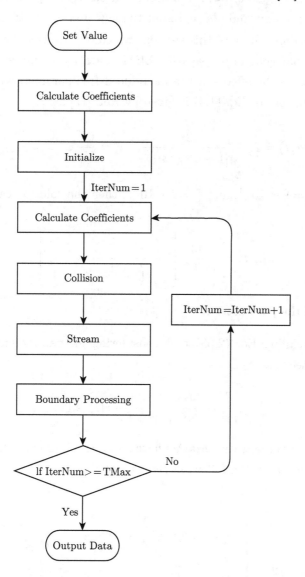

Fig. 1 The flowchart of our numerical simulation procedures

$$p \approx \left| \frac{\log E_{\text{new}} - \log E_{\text{old}}}{\log H_{\text{new}} - \log H_{\text{old}}} \right|, \tag{3.3}$$

where p is the convergent order, E represents the error, H represents the sequence, and the subscripts *new* and *old* represent the new and old statuses, respectively.

In the remaining part of this section, we consider one-soliton solutions for Eq. (1.1) using our general propagation LBM model as the numerical simulation example to discuss the effects of the introduced free parameters. An analytical one-solitary solution for Eq. (1.4) is given by Ref. [12] as,

$$u(x,t) = 2\xi - \frac{2\xi}{[1 + e^{2\nu(x-\xi t)}]^2}, \quad \xi = \frac{6\theta^2}{25\delta}, \quad \nu = -\frac{\theta}{10\delta}. \tag{3.4}$$

Here, we set $\theta = 9 \times 10^{-4}$, and $\delta = 2 \times 10^{-5}$, then the solution can be rewritten as,

$$u(x,t) = \frac{243}{12500} \left[1 - \frac{1}{\left(e^{\frac{2187}{25000}t - 9x} + 1 \right)^2} \right].$$

3.2 Effects of the parameter κ

Firstly, recalling Eq. (2.33) and considering all parameters are real, one should guarantee that

$$\frac{q^2}{4a^4} - \frac{1}{6a^2} - \frac{\Delta t}{a^3 \Delta x^3 \kappa} r(t) \geqslant 0. \tag{3.5}$$

With the help of the software **Mathematica** and coupling with the definitions of a and q, we can obtain the following eight cases:

Case 1

$$r(t) < 0, \quad 0 < a < \sqrt{\frac{2}{3}}, \quad \begin{cases} \text{(I)} \ a^2 \leqslant q < \sqrt{\frac{2}{3}}a, & 0 < \kappa \leqslant \frac{12a\,\Delta t}{\Delta x^3 (3q^2 - 2a^2)} r(t), \\ \text{(II)} \ q = \sqrt{\frac{2}{3}}a, & \kappa > 0, \\ \text{(III)} \ \sqrt{\frac{2}{3}}a < q \leqslant 1, & \kappa \leqslant \frac{12a\Delta t}{\Delta x^3 (3q^2 - 2a^2)} r(t), \text{ or } \kappa > 0; \end{cases}$$

Case 2

$$r(t) < 0, \quad a = \sqrt{\frac{2}{3}}, \quad \begin{cases} \text{(I)} \ \frac{2}{3} < q \leqslant 1, & \kappa < \dfrac{12\sqrt{6}\Delta t}{\Delta x^3 (9q^2 - 4)} r(t) \\ \text{(II)} \ \frac{2}{3} \leqslant q \leqslant 1, & \kappa > 0; \end{cases}$$

Case 3

$$r(t) < 0, \ \sqrt{\frac{2}{3}} < a < 1, \ a^2 \leqslant q \leqslant 1, \ \kappa \leqslant \frac{12a\Delta t}{\Delta x^3 (3q^2 - 2a^2)} r(t), \ \text{or} \ \kappa > 0;$$

Case 4

$$r(t) < 0, \ a = q = 1, \ \kappa \leqslant \frac{12\Delta t}{\Delta x^3} r(t), \ \text{or} \ \kappa > 0;$$

Case 5

$$r(t) > 0, \ 0 < a < \sqrt{\frac{2}{3}}, \ \begin{cases} \text{(I)} \ a^2 \leqslant q < \sqrt{\frac{2}{3}}a, & \dfrac{12a\Delta t}{\Delta x^3 (3q^2 - 2a^2)} r(t) \leqslant \kappa < 0, \\ \text{(II)} \ q = \sqrt{\frac{2}{3}}a, & \kappa < 0, \\ \text{(III)} \ \sqrt{\frac{2}{3}}a < q \leqslant 1, & \kappa \geqslant \dfrac{12a\Delta t}{\Delta x^3 (3q^2 - 2a^2)} r(t), \ \text{or} \ \kappa < 0; \end{cases}$$

Case 6

$$r(t) > 0, \quad a = \sqrt{\frac{2}{3}}, \quad \begin{cases} \text{(I)} \ \frac{2}{3} < q \leqslant 1, & \kappa \geqslant \dfrac{12\sqrt{6}\Delta t}{\Delta x^3 (9q^2 - 4)} r(t), \\ \text{(II)} \ \frac{2}{3} \leqslant q \leqslant 1, & \kappa < 0; \end{cases}$$

Case 7

$$r(t) > 0, \ \sqrt{\frac{2}{3}} < a < 1, \ a^2 \leqslant q \leqslant 1, \ \kappa \geqslant \frac{12a\Delta t}{\Delta x^3 (3q^2 - 2a^2)} r(t), \ \text{or} \ \kappa < 0;$$

Case 8

$$r(t) > 0, \ a = q = 1, \ \kappa \geqslant \frac{12\Delta t}{\Delta x^3} r(t), \ \text{or} \ \kappa < 0.$$

From the above results, it can be observed that the value range of the introduced parameter κ is basically opposite to the variable-coefficient $r(t)$, where a more narrow value range comes up with the special groups of a and q, e.g., $0 < a < \sqrt{2/3}$ and $a^2 \leqslant q < \sqrt{2/3}a$. Hence, following the above analyses, the approximate value range of κ can be given. In the remaining part of this paper, we set $r(t) < 0$ and then $\kappa > 0$ basically.

Then, we can investigate the effects of the parameter κ in our simulations. Let $\sigma = -0.08$, $\Delta x = 0.0025$, $c = 1$ and $a = q = 1$, we select the cases of $\kappa = 0.2, 0.7, 1.0, 1.3, 1.31$ for comparisons. The E_2, E_∞ and **GRE** of numerical result and one-soliton solutions for Eq. (1.4) at different times with different values of κ, which is presented in Tabs. 1-3.

Table. 1 The root-mean-square error E_2 of numerical result and one-soliton solution for Eq. (1.4) at different times with different values of κ ($\star$ represents the divergency)

E_2	$t = 5.0$	$t = 10.0$	$t = 15.0$	$t = 20.0$
$\kappa = 0.2$	1.9767 e-009	4.2772 e-009	5.9545 e-009	7.1027 e-009
$\kappa = 0.7$	2.8139 e-011	6.4884 e-011	9.4743 e-011	1.1732 e-010
$\kappa = 1.0$	6.6146 e-012	1.5899 e-011	2.4328 e-011	3.1325 e-011
$\kappa = 1.3$	3.1164 e-012	7.5778 e-012	1.1834 e-011	1.5492 e-011
$\kappa = 1.31$	3.0829 e-012	1.1345 e-008	3.2693 e-008	$\star$

Table. 2 The max error E_∞ of numerical result and one-soliton solution for Eq. (1.4) at different times with different values of κ ($\star$ represents the divergency)

E_∞	$t = 5.0$	$t = 10.0$	$t = 15.0$	$t = 20.0$
$\kappa = 0.2$	1.6821 e-004	2.4053 e-004	2.8045 e-004	3.0451 e-004
$\kappa = 0.7$	2.2291 e-005	3.2698 e-005	3.8828 e-005	4.2735 e-005
$\kappa = 1.0$	1.0433 e-005	1.5610 e-005	1.8924 e-005	2.1189 e-005
$\kappa = 1.3$	6.8812 e-006	1.0428 e-005	1.2821 e-005	1.4518 e-005
$\kappa = 1.31$	6.8401 e-006	9.7022 e-004	1.3187 e-003	$\star$

Table. 3 The global relative error GRE of numerical result and one-soliton solution for Eq. (1.4) at different times with different values of κ ($\star$ represents the divergency)

GRE	$t = 5.0$	$t = 10.0$	$t = 15.0$	$t = 20.0$
$\kappa = 0.2$	2.4101 e-003	3.5154 e-003	4.0668 e-003	4.3334 e-003
$\kappa = 0.7$	2.6138 e-004	3.9522 e-004	4.6758 e-004	5.0667 e-004
$\kappa = 1.0$	1.2617 e-004	1.9867 e-004	2.4242 e-004	2.6911 e-004
$\kappa = 1.3$	8.9780 e-005	1.4236 e-004	1.7577 e-004	1.9691 e-004
$\kappa = 1.31$	8.2301 e-005	2.2305 e-003	4.3800 e-003	$\star$

It can be found that the E_2, E_∞ and **GRE** of $\kappa = 1.31$ are at the minimum values among these cases at $t = 5$. While with the evolution of the macroscopic equation continuing, the numerical results with $\kappa = 1.31$ diverge from the analytical solutions after a period of time (at $t = 20$), while the numerical results with $\kappa = 0.2, 0.7, 1.0, 1.3$ can be coupled with the analytical solutions for a longer time. Besides, it can be observed that the distinction between our numerical simulation and analytical solution is decreased following with the increase of κ in the convergence interval.

Hence, we can make a conjecture and regard the free parameter $1/\kappa$ as the damping. When the damping κ is selected as a large value, i.e. $1/\kappa$ as a small value, the numerical perturbations, e.g., truncation error, in the numerical simulation can not be restrained and get amplification as the evolution continues, which lead to the divergency results. When the damping κ is selected as a small value, i.e. $1/\kappa$ as a large value, the inhibiting effects are strong to make the numerical results less than the analytical solutions.

3.3 Effects of the constant σ

Secondly, we consider the constant σ induced in Eq. (2.30). Let $\kappa = 1.0$, $\Delta x = 0.0025$, $c = 1$ and $a = q = 1$, we select the cases of $\sigma = -0.059$, $\sigma = -0.08$ and $\sigma = -0.15$ for comparisons. The E_2, E_∞ and **GRE** of numerical result and one-soliton solutions for Eq. (1.4) at different times with different values of σ, which is presented in Tabs. 4-6.

Table. 4 The root-mean-square error E_2 of numerical result and one-soliton solution for Eq. (1.4) at different times with different values of σ

E_2	$t = 5.0$	$t = 10.0$	$t = 15.0$	$t = 20.0$
$\sigma = -0.059$	5.1811 e-012	1.2714 e-011	1.9909 e-011	2.6097 e-011
$\sigma = -0.08$	6.6146 e-012	1.5899 e-011	2.4328 e-011	3.1325 e-011
$\sigma = -0.15$	3.4512 e-011	7.6847 e-011	1.0851 e-010	1.3057 e-010

Table. 5 The max error E_∞ of numerical result and one-soliton solution for Eq. (1.4) at different times with different values of σ

E_∞	$t = 5.0$	$t = 10.0$	$t = 15.0$	$t = 20.0$
$\sigma = -0.059$	8.8791 e-006	1.3517 e-005	1.6646 e-005	1.8869 e-005
$\sigma = -0.08$	1.0433 e-005	1.5610 e-005	1.8924 e-005	2.1189 e-005
$\sigma = -0.15$	2.4437 e-005	3.5387 e-005	4.1496 e-005	4.5188 e-005

From the above results, it can be observed that the value range of the introduced parameter κ is basically opposite to the variable-coefficient $r(t)$, where more narrow value range comes up with the special groups of a and q, e.g., $0 < a < \sqrt{2/3}$ and $a^2 \leqslant q < \sqrt{2/3}a$. Hence, following the above analyses, the approximate value range of κ can be given. In the remaining part of this paper, we set $r(t) < 0$ and then $\kappa > 0$ basically.

Table. 6 The global relative error GRE of numerical result and one-soliton solution for Eq. (1.4) at different times with different values of σ

GRE	$t = 5.0$	$t = 10.0$	$t = 15.0$	$t = 20.0$
$\sigma = -0.059$	1.1658 e-004	1.8405 e-004	2.2649 e-004	2.5324 e-004
$\sigma = -0.08$	1.2617 e-004	1.9867 e-004	2.4242 e-004	2.6911 e-004
$\sigma = -0.15$	2.9954 e-004	4.4189 e-004	5.1304 e-004	5.4758 e-004

3.4 Effects of the lattice space step Δx and the lattice velocity c

Finally, we consider the effects of the lattice space step Δx and the lattice velocity c. Other parameters are set as $\kappa = 1.0$, $a = 0.8$, $q = 0.64$ and the final evolution time is $t = 20$.

The errors and numerical order for different groups of Δx and c are listed in Tabs. 7-9. From the space-time evolution graphs of the numerical results, which are omitted here, we find that the numerical results of all these above cases are in Tabs. 7-9 match the analytical solutions well. Meanwhile, from Tabss. 7-9, It can be found that the convergence rate order of the approach of increasing the lattice velocity c is about 2.0503 (based on the root-mean-square errors E_2), while the approach of decreasing the lattice space step Δx can hardly aggrandize the accuracy of the numerical results. Nevertheless, the stability of the numerical simulations may be reduced, which means that we should give new values of other free parameters to obtain nice numerical simulation results.

Table. 7 The E_2 of numerical result and one-soliton solution (3.4) and the convergence rates at $t = 20$

$t = 20$	$\Delta x = 0.005$		$\Delta x = 0.0025$		$\Delta x = 0.00125$	
	E_2	Order	E_2	Order	E_2	Order
$c = 1$	3.1385e-011	\	3.1255e-011	\	3.1659e-011	\
$c = 2$	8.1581e-012	1.9437	7.6385e-012	2.0327	7.6686e-012	2.0456
$c = 5$	1.6515e-012	1.8296	1.2082e-012	2.0212	1.1679e-012	2.0503

Table. 8 The E_∞ of numerical result and one-soliton solution (3.4) and the convergence rates at $t = 20$

$t = 20$	$\Delta x = 0.005$		$\Delta x = 0.0025$		$\Delta x = 0.00125$	
	E_∞	Order	E_∞	Order	E_∞	Order
$c = 1$	2.1174e-005	\	2.1160e-005	\	2.1337e-005	\
$c = 2$	1.0822e-005	0.9683	1.0415e-005	1.0227	1.0448e-005	1.0301
$c = 5$	4.9722e-006	0.9003	4.1246e-006	1.0160	4.0397e-006	1.0341

Table. 9 The GRE of numerical result and one-soliton solution (3.4) and the convergence rates at $t = 20$

$t = 20$	$\Delta x = 0.005$		$\Delta x = 0.0025$		$\Delta x = 0.00125$	
	GRE	Order	GRE	Order	GRE	Order
$c = 1$	2.7082e-004	\	2.6868e-004	\	2.6941e-004	\
$c = 2$	1.4007e-004	0.9512	1.3550e-004	0.9876	1.3534e-004	0.9932
$c = 5$	6.2999e-005	0.9061	5.5110e-005	0.9843	5.4226e-005	0.9960

3.5 Comparisons between GPLB and SLBGK models

At the end, we compare the GPLB and SLBGK models by taking different values of the introduced parameters a and q. We set $\kappa = 1.0$, $\sigma = -0.08$, $\Delta x = 0.0025$, $c = 1$ and the final evolution time is $t = 20$. In this part, we only consider the following special cases: (I) $a = q = 1$, i.e., the standard lattice BGK model (SLBM); (II) $q = a^2$, i.e., the Lax-Wendroff (LW) scheme; (III) $a^2 < q < a$, here we choose $q = (a + a^2)/2$; (IV) $q = a$, i.e., the fractional propagation (FP) scheme; (V) $a < q < 1$, here we choose $q = a + 0.1$.

The distinction errors between our numerical simulations and the analytical solutions (3.4) with different groups of a and q are listed in Tab. 10. As shown in the table, one can find that the errors by LW scheme are less than those by SLBM, FP scheme and other cases, which demonstrate that our present model could be more accurate than the SLBM model by adjusting the parameters a and q properly. Besides, we can also find that there exist the unstable situations, or divergency, in the numerical evolution when $a = 0.2$ and $q = 0.3$. The errors increase following with the diminution of a. Hence, it is suitable to choose the value of a close to 1, to obtain a nice numerical simulation result.

Table. 10 Errors between our numerical simulations and the analytical solutions (3.4) with different groups of a and q at $t = 20$ ($\star$ represents the divergency)

Different models		E_2	E_∞	GRE
$a = 1$	$q = 1$	1.5492 e-011	1.4518 e-005	1.9691 e-004
$a = 0.8$	$q = 0.64$	1.4506 e-011	1.3547 e-005	1.8448 e-004
	$q = 0.72$	1.5069 e-011	1.4080 e-005	1.9014 e-004
	$q = 0.8$	1.5546 e-011	1.4512 e-005	1.9703 e-004
	$q = 0.9$	1.2715 e-010	6.7760 e-005	4.6543 e-003
$a = 0.5$	$q = 0.25$	1.5090 e-011	1.4112 e-005	1.8966 e-004
	$q = 0.375$	1.5629 e-011	1.4584 e-005	1.9502 e-004
	$q = 0.5$	1.5882 e-011	1.4756 e-005	1.9973 e-004
	$q = 0.6$	8.9522 e-008	1.9157 e-003	7.7793 e-003
$a = 0.2$	$q = 0.04$	1.5640 e-011	1.4609 e-005	1.9513 e-004
	$q = 0.12$	1.5919 e-011	1.4802 e-005	1.9715 e-004
	$q = 0.2$	1.6123 e-011	1.4900 e-005	2.0145 e-004
	$q = 0.3$	$\star$	$\star$	$\star$

3.6　Brief summary

From the above discussion, we can summarize as follows: (1) The parameter κ is an adjustment factor of the single relaxation time τ, whose reciprocal effects the evolution of the NLEEs as the damping, which should be a positive value; (2) The constant σ is an adjustment factor of the equilibrium functions, which can adjust simulation results in the process of the evolution and should be in a suitable range. Generally, σ is always a negative tiny value; (3) The lattice space step Δx and the lattice velocity c can impact the accuracy and the stability of the numerical simulation results; (4) Our present model could be more accurate than the SLBM model by adjusting the parameters a and q properly. It is suitable to choose a value of a close to 1.

Furthermore, via plentiful computational examples for the preceding cases based on our current general propagation LBM, we can give the approximate value range of the parameters κ and σ,

$$0.0 < \kappa < 1.5, \quad -\frac{1}{12} < \sigma < -0.01,$$

which may be a guidance to choose the appropriate free parameters. Setting the initial value of the free parameters as arbitrary values in the above scopes, adjusting the value of the free parameters, and comparing with the analytical solutions in the meantime continually, we can obtain the appropriate free parameters for Eq. (1.1) at this group of $l(t), m(t), n(t), r(t), a$ and q. After that, we can use our present general propagation LBM to figure out the evolutions and search for some new solutions of Eq. (1.1) at this set with some other specific initial conditions, which have no analytical solution at present.

4 Numerical simulations

In this section, following the guidance of the discussions in Section 3, we will select appropriate free parameters and present some numerical simulations for preceding cases, which implies that our current lattice Boltzmann model is a satisfactory and efficient algorithm.

Example 4.1 By setting $\alpha = 6$ and $\beta = 1$ in Eq. (1.2), the two-soliton solution of the constant-coefficient KdV equation can be obtained with the help of the Hirota method, which is of the following form,

$$u(x,t) = \frac{\partial^2}{\partial x^2}\big\{2\log[F(x,t)]\big\}, \qquad (4.1)$$

with

$$F(x,t) = 1 + a_1 e^{\kappa_1 x - \kappa_1^3 t} + a_2 e^{\kappa_2 x - \kappa_2^3 t} + \frac{a_1 a_2 (\kappa_1 - \kappa_2)^2 e^{(\kappa_1+\kappa_2)x - (\kappa_1^3+\kappa_2^3)t}}{(\kappa_1 + \kappa_2)^2},$$

where a_1, a_2, κ_1 and κ_2 are constant parameters.

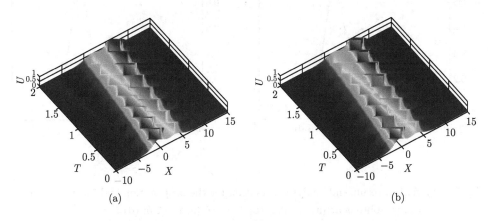

Fig. 2 Numerical result (a) and analytical solution(b) for the propagation and interaction of the two-solitons from $t = 0$ to $t = 2$

In this simulation, we set $a_1 = 2$, $a_2 = 1$, $\kappa_1 = 1.5$, $\kappa_2 = 1$. Some other parameters are $a = q = 1$, $\kappa = 0.11$, $\sigma = -0.0095$, $\Delta x = 0.05$, $c = 500$ (i.e., the corresponding time step is $\Delta t = 0.0001$) and the computation domain is fixed on $I = [-10, 15]$. We present the comparison between detailed numerical results and analytical solutons. Fig. 2 shows the two-dimensional visual comparisons at some different times. The space-time evolution graph of the numerical results is shown in Fig. 3. In addition, the E_2, E_∞ and **GRE** of the numerical results at some different times are presented in Tab. 11. All of them clearly show that the numerical results agree with the analytical solutions well.

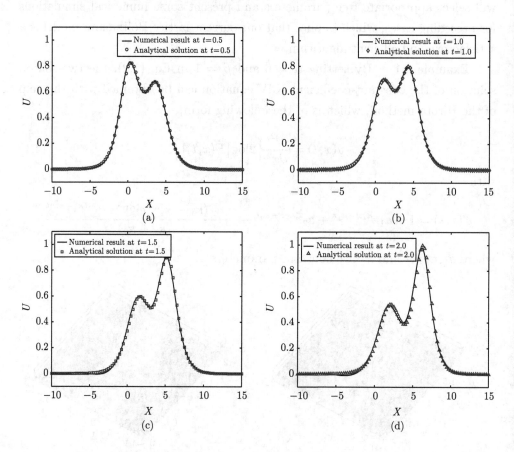

Fig. 3   Numerical result and analytical solution for the propagation and interaction of the two-solitons at (a) $t = 0.5$; (b) $t = 1.0$; (c) $t = 1.5$; (d) $t = 2.0$

Table. 11 Comparison of numerical result and soliton solution for the constant-coefficient KdV equation at different times

	$t=0.5$	$t=1.0$	$t=1.5$	$t=2.0$
E_2	5.2517e-005	9.0765e-005	1.8428e-004	4.8216e-004
E_∞	2.8818e-003	3.3485e-003	3.8751e-003	6.8101e-003
GRE	1.8440e-003	2.8667e-003	4.4121e-003	6.7999e-003

Example 4.2 The soliton solution of the constant-coefficient combined KdV-modified KdV equation in Ref. [28] is given by

$$u(x,t) = \frac{6\gamma}{2\alpha + \sqrt{4\alpha^2 - 18\beta\gamma}\cosh\left[\sqrt{\gamma}(x-\gamma t)\right]}, \qquad (4.2)$$

where $l(t) = 2\alpha$, $m(t) = -3\beta$, $r(t) = 1$, $n(t) = 0$ and $p = 1$. In this simulation, we set $\alpha = 5, \beta = 0.5, \gamma = 2$. Hence, Eq. (1.3) has the form as follows,

$$u_t + 10uu_x - 1.5u^2 u_x + u_{xxx} = 0. \qquad (4.3)$$

Some other parameters are $a = 0.9$, $q = 0.81$, $\kappa = 1.1135$, $\sigma = -0.0958$, $\Delta x = 0.02$, $c = 90$ (i.e., the corresponding time step is $\Delta t = 0.0002$) and the computation domain is fixed on $I = [-7, 13]$. We present the comparison between detailed numerical results and analytical solutions. Fig. 4 shows the two-dimensional visual comparisons at some different times. The space-time evolution graph of the numerical results is shown in Fig. 5. Tab. 12 gives the E_2, E_∞ and GRE of the numerical results at some different times. All of them clearly show that the numerical results agree with the analytical solutions well.

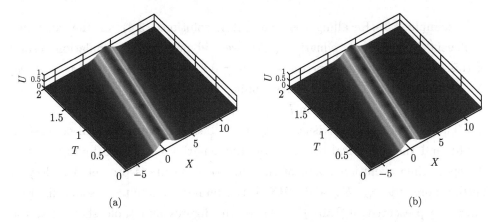

Fig. 4 Numerical result(a) and analytical solution(b) for the propagation of the soliton from $t = 0$ to $t = 2$

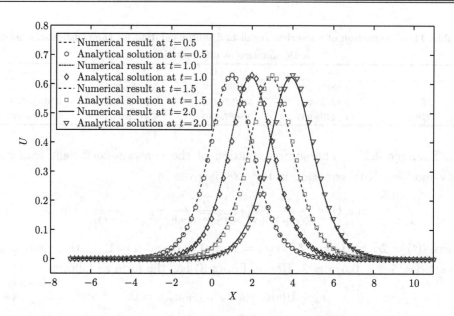

Fig. 5 Numerical result and analytical solution at $t = 0.5, 1.0, 1.5, 2.0$

Table. 12 Comparison of numerical result and soliton solution for the constant-coefficient combined KdV-modified KdV equation at different times

	$t = 0.5$	$t = 1.0$	$t = 1.5$	$t = 2.0$
E_2	1.7380e-005	5.2115e-005	9.1861e-005	1.3514e-004
E_∞	1.3321e-003	2.1879e-003	2.8300e-003	3.3899e-003
GRE	2.3984e-003	4.4335e-003	6.0706e-003	7.4274e-003

Example 4.3 Recalling the one-soliton solutions (3.4) for the constant-coefficient KdV-Burgers equation (1.4), we will present the simulation results of this case. The parameters are set as $a = 0.8$, $q = 0.64$, $\kappa = 1.0$, $\sigma = -0.08$, $\Delta x = 0.005$, $c = 5$ (i.e., the corresponding time step is $\Delta t = 0.0008$) and the computation domain is fixed on $I = [-2, 4]$.

The two-dimensional visual comparison between the numerical simulation results and the analytical solutions at different times is shown in Fig. 6, and the space-time evolution graph of the numerical results is presented in Fig. 7. Furthermore, the E_2, E_∞ and **GRE** of the numerical results at some different times are presented in Tab. 13. All of the figures and table show that the numerical results agree with the analytical solutions well.

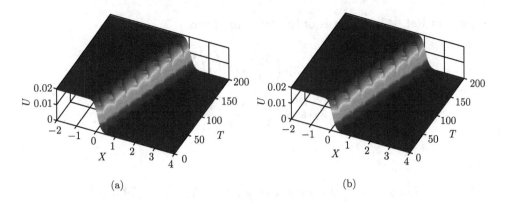

Fig. 6 Numerical result(a) and analytical solution(b) for the propagation of the soliton from $t = 0$ to $t = 200$

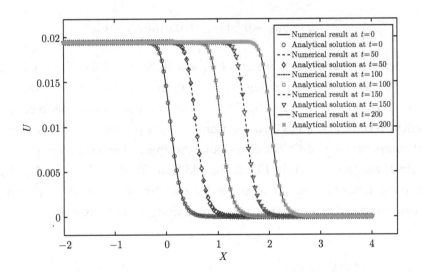

Fig. 7 Numerical result and analytical solution at $t = 0, 50, 100, 150, 200$

Table. 13 Comparison of numerical result and soliton solution for the constant-coefficient KdV-Burgers equation equation at different times

	$t = 50$	$t = 100$	$t = 150$	$t = 200$
E_2	1.0884e-012	1.2032e-012	1.2164e-012	1.2181e-012
E_∞	6.0964e-006	6.3494e-006	6.3706e-006	6.3718e-006
GRE	4.0381e-005	3.5983e-005	3.1266e-005	2.7542e-005

Example 4.4 One shock-wave solution of the KdV-Burgers equation (1.4)

is given in Ref. [40], which is of the following form,

$$u(x,t) = \begin{cases} \lambda_1 + \dfrac{1}{2}(\lambda_1 - \lambda_2)\exp\left(\dfrac{\theta\xi}{2\delta}\right)\cos\left(\sqrt{\dfrac{\lambda_1-\lambda_2}{2\delta} - \dfrac{\theta^2}{4\delta^2}}\cdot\xi\right), & \xi < 0, \\ \lambda_2 + \dfrac{3}{2}(\lambda_1-\lambda_2)\operatorname{sech}^2\left(\sqrt{\dfrac{\lambda_1-\lambda_2}{8\delta}}\cdot\xi\right), & \xi \geqslant 0, \end{cases}$$

(4.4)

with

$$\xi = x - ct, \quad \lambda_1 = c + \sqrt{c^2 + 2A}, \quad \lambda_2 = c - \sqrt{c^2 + 2A},$$

where c and A are all constants. In this simulation, we select $\theta = 0.005$ and $\delta = 0.015$. Hence, Eq. (1.4) can be rewritten as,

$$u_t + uu_x - 0.005 u_{xx} + 0.015 u_{xxx} = 0.$$

Some other parameters are $a = 0.5$, $q = 0.25$, $\kappa = 1.05$, $\sigma = -0.09$, $\Delta x = 0.01$, $c = 2$ (i.e., the corresponding time step is $\Delta t = 0.0025$) and the computation domain is fixed on $I = [-20, 10]$. We present the comparison between detailed numerical results and analytical solutions. Fig. 8 presents the two-dimensional visual comparisons at different times, and the space-time evolution graph of the numerical results is shown in Fig. 9. In addition, Tab. 14 gives the E_2, E_∞ and **GRE** of the numerical results at some different times. All of them clearly show that the numerical results agree with the analytical solutions well.

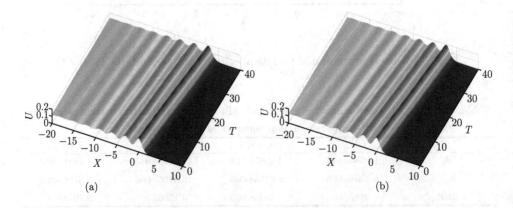

Fig. 8   Numerical result(a) and analytical solution(b) for the propagation of the one shock-wave solution from $t = 0$ to $t = 200$

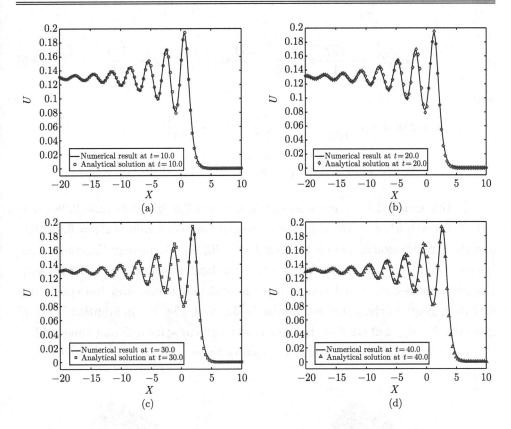

Fig. 9 Numerical result and analytical solution for the evolution of the shock-wave solutions at (a) $t = 10.0$; (b) $t = 20.0$; (c) $t = 30.0$; (d) $t = 40.0$

Table. 14 Comparison of numerical result and analytical solution for the KdV-Burgers equation at different times

	$t = 10$	$t = 20$	$t = 30$	$t = 40$
E_2	4.5096e-006	1.3597e-005	2.5193e-005	3.8816e-005
E_∞	1.0344e-003	1.6786e-003	2.0226e-003	2.2295e-003
GRE	1.2865e-003	2.2217e-003	3.0224e-003	3.7839e-003

Example 4.5 Let $l(t) = -6$, $m(t) = 0.06$, $n(t) = 0$, $r(t) = -0.01$, $s(t) = -15\cos(\pi t/2)$ and $p = 1$. One periodic depression soliton solution of the generalized variable-coefficient KdV-mKdV equation (1.5) can be obtained of the following form by the solution (16) in Ref. [41],

$$u(x,t) = -\frac{202}{\Phi} \exp\left[\frac{t}{100} + \frac{30}{\pi}\sin\left(\frac{\pi t}{2}\right) + x + \frac{1}{100}\right], \quad (4.5)$$

with

$$\Phi = 20200 \exp\left[\frac{t}{100} + \frac{30}{\pi}\sin\left(\frac{\pi t}{2}\right) + x + \frac{1}{100}\right]$$
$$+ 10001 \exp\left[\frac{60}{\pi}\sin\left(\frac{\pi t}{2}\right) + 2x + \frac{1}{50}\right] + 10201 \exp\left(\frac{t}{50}\right).$$

In this simulations, parameters are set as $a = 0.8$, $q = 0.64$, $\kappa = 0.985$, $\sigma = -0.19$, $\Delta x = 0.01$, $c = 320$ (i.e., the corresponding time step is $\Delta t = 0.000025$) and the computation domain is fixed on $I = [-20, 10]$. We present the comparison between detailed numerical results and analytical solutions. Fig. 10 presents the two-dimensional visual comparisons at different times, and the space-time evolution graph of the numerical results is shown in Fig. 11. In addition, Tab. 15 gives the E_2, E_∞ and **GRE** of the numerical results at some different times. All of them clearly show that the numerical results agree with the analytical solutions well.

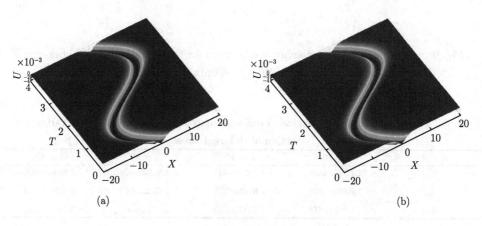

Fig. 10   Numerical result(a) and analytical solution(b) for the propagation of the soliton from $t = 0.0$ to $t = 4.0$

From the front five examples, it can be seen clearly that the numerical results have a commendable accuracy compared with the analytical solutions when we take the appropriate κ and σ, which implies that the present lattice Boltzmann model is a satisfactory and efficient algorithm.

Therefore, it is expected that the lattice Boltzmann model can be used to figure out the evolutions and search for some new solutions of these NLEEs which can be represented by Eq. (1.1) with some specific initial conditions.

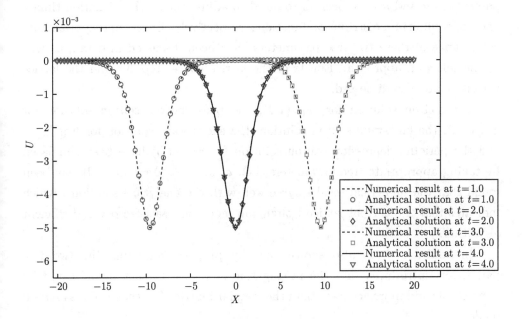

Fig. 11 Numerical result and analytical solution at $t = 1.0, 2.0, 3.0, 4.0$

Table. 15 Comparison of numerical result and analytical solution for the generalized variable-coefficient KdV-mKdV equation at different times

	$t = 1.0$	$t = 2.0$	$t = 3.0$	$t = 4.0$
E_2	9.4246e-013	2.1935e-012	5.0952e-012	8.5617e-012
E_∞	5.1096e-006	7.3044e-006	1.1358e-005	1.4313e-005
GRE	7.4281e-004	1.2207e-003	1.8416e-003	2.4443e-003

5 Conclusions

In this paper, a general propagation lattice Boltzmann model for a variable-coefficient compound Korteweg-de Vries-Burgers equation, i.e., Eq. (1.1), has been proposed through selecting an equilibrium distribution function and adding a compensation function, appropriately. $D1Q5$ velocity model has been used in numerical simulations with different forms of Eq. (1.1). Through the Chapman-Enskog analysis, it has been proven that Eq. (1.1) can be recovered correctly from our present general propagation lattice Boltzmann model. Numerical simulation procedures have been given in detail. The positive parameters a and q have been introduced to adjust the propagation step of our model, and another two free parameters σ and κ have been introduced to adjust the single relaxation time τ and the equilibrium distribution functions, respectively. Effects and approximate value range of these two free parameters have been discussed in detail, as well as the selection approach. Effects of the lattice space step Δx and the lattice velocity c are also discussed.

Two-soliton solution for Eq. (1.2), the bell-type one-soliton solution for Eq. (1.3), the kink-type soliton solution, the shock-wave solution for Eq. (1.4), and the periodic depression soliton solution for Eq. (1.5) have been simulated by taking appropriate free parameters, i.e., a, q, σ, κ, Δx and c. It has been found that the numerical results agree well with the analytical solutions, which indicates that the current lattice Boltzmann model is a satisfactory and efficient algorithm.

In addition, it is also shown that by properly adjusting the two free parameters introduced into the propagation step, the present model could be more stable and more accurate than the standard lattice Bhatnagar-Gross-Krook model.

Acknowledgements This work has been supported by the Science Research Project of Higher Education in Inner Mongolia Autonomous Region under Grant No. NJZZ18117, by the Natural Science Foundation of Inner Mongolia Autonomous Region under Grant No. 2018BS01004, by the Program for Young Talents of Science and Technology in Universities of Inner Mongolia Autonomous Region under Grant No. NJYT-19-B21, by the China Postdoctoral Science Foundation under Grant No. 2018M640094, and by the National Natural Science Foundation of China under Grant No. 11272023.

References

[1] Chen S Y, Doolen G D. Lattice Boltzmann method for fluid flows [J], Annu. Rev. Fluid Mech., 1972, 30: 329-64.

[2] Qian Y H, Succi S, Orszag S A. Recent advances in lattice Boltzmann computing [J], Annu. Rev. Comput. Phys., 1995, 3: 195-242.

[3] Benzi R, Succi S, Vergassola M. The lattice Boltzmann equation: theory and applications [J], Phys. Rep., 1992, 222: 145-197.

[4] Ginzburg I. Equilibrium-type and link-type lattice Boltzmann models for generic advection and anisotropic-dispersion equation [J], Adv. Water Resour., 2005, 28: 1171-1195.

[5] Shi B C, Guo Z L. Lattice Boltzmann model for nonlinear convection-diffusion equations [J], Phys. Rev. E, 2009, 79: 016701-13.

[6] Chai Z H, Shi B C, Guo Z L. A multiple-relaxation-time lattice Boltzmann model for general nonlinear anisotropic convection-diffusion equations [J], J. Sci. Comput., 2016, 69: 355-390.

[7] Guo X Y, Shi B C, Chai Z H. General propagation lattice Boltzmann model for nonlinear advection-diffusion equations [J], Phys. Rev. E, 2018, 97: 043310-11.

[8] Zhang J Y, Yan G W. Lattice Boltzmann method for one and two-dimensional Burgers equation [J], Phys. A, 2008, 387: 4771-4786.

[9] Yan G W. A lattice Boltzmann equation for waves [J], J. Comput. Phys., 2000, 161: 61-69.

[10] Li H B, Huang P H, Liu M R, Kong L J. Simulation of the MKdV equation with lattice Boltzmann method [J], Acta. Phys. Sini., 2001, 50: 837-840.

[11] Ma C F. A lattice BGK model for simulating solitary waves of the combined KdV-MKdV equation [J], Int. J. Mod. Phys. B, 2011, 25: 589-597.

[12] Zhang C Y, Tan H L, Liu M R, Kong L J. A lattice Boltzmann model and simulation of KdV-Burgers equation [J], Commun. Theor. Phys., 2004, 42: 281-284.

[13] Liu F, Shi W P, Wu F F. A lattice Boltzmann model for the generalized Boussinesq equation [J], Appl. Math. Comp., 2016, 274: 331-342.

[14] Chai Z H, He N Z, Guo Z L, Shi B C. Lattice Boltzmann model for high-order nonlinear partial differential equations [J], Phys. Rev. E, 2018, 97: 013304-13.

[15] Chai Z H, Shi B C, Zheng L. A unified lattice Boltzmann model for some nonlinear partial differential equations [J], Chaos, Soliton. Fract., 2008 36: 874-882.

[16] Hu W Q, Gao Y T, Lan Z Z, Su C Q, Feng Y J. Lattice Boltzmann model for a generalized Gardner equation with time-dependent variable coefficients [J], Appl. Math. Model., 2017, 46: 126-140.

[17] McNamara G R, Garcia A L, Alder B J. Stabilization of thermal lattice Boltzmann models [J], J. Stat. Phys., 1995, 81: 395-408.

[18] Qian Y H. Fractional propagation and the elimination of staggered invariants in lattice-BGK models [J]. Int. J. Mod. Phys. C, 1997, 8: 753-761.

[19] Guo Z L, Zheng C G, Zhao T S. A Lattice BGK Scheme with General Propagation [J], J. Sci. Comput., 2003, 16: 569-585.

[20] Nakamura Y, Tsukabayashi I. Observation of modified Korteweg-de Vries solitons in a multicomponent plasma with negative ions [J], Phys. Rev. Lett., 1984 52: 2356-2361.

[21] de la Rosa R, Gandarias M L, Bruzón M S. Equivalence transformations and conservation laws for a generalized variable-coefficient Gardner equation [J], Results in Physics, 2016, 40: 71-79.

[22] Kumar V, Kaur L, Kumar A, Koksal M E. Lie symmetry based-analytical and numerical approach for modified Burgers-KdV equation [J], Results Phys., 2018, 8: 1136-1142.

[23] de la Rosa R, Bruzón M S. Differential invariants of a generalized variable-coefficient Gardner equation [J], Discrete Cont. Dyn. S, 2018, 11: 747-757.

[24] Bruzón M S, de la Rosa R, Tracinà R. Exact solutions via equivalence transformations of variable-coefficient fifth-order KdV equations [J], Appl. Math. Comput., 2018, 325: 239-245.

[25] Younis M, Ali S, Mahmood S A. Solitons for compound KdV-Burgers equation with variable coefficients and power law nonlinearity [J], Nonlinear Dyn., 2015, 81: 1191-1196.

[26] Hassan M M. Exact solitary wave solutions for a generalized KdV-Burgers equation [J], Chaos Soliton. Fract. 2004, 19: 1201-1206.

[27] Brugarino T. Painlevé property, auto-Bäcklund transformation, Lax pairs, and reduction to the standard form for the Korteweg-de Vries equation with nonuniformities [J], J. Math. Phys., 1989, 30: 1013-1015.

[28] Wazwaz A M. New solitons and kink solutions for the Gardner equation [J], Commun. Non. Sci. Numer. Simul., 2007, 12: 1395-1404.

[29] Chow K W, Grimshaw R H J, Ding E. Interactions of breathers and solitons in the extended Korteweg-de Vries equation [J], Wave Motion, 2005, 43: 158-166.

[30] Guan K Y, Gao G. Qualitative analysis about traveling wave solution of burgers-KdV mixed type equation [J], Science China Ser. A, 1987, 1: 64-73.

[31] Vaneeva O, Kuriksha O, Sophocleous C. Enhanced group classification of Gardner equations with time-dependent coefficients [J], Commun. Nonlinear Sci. Numer. Simul., 2015, 22: 1243-1251.

[32] Li Z H, Zhang H X. Study on gas kinetic unified algorithm for flows from rarefied transition to continuum [J], J. Comput. Phys., 2004, 193: 708-738.

[33] Li Z H, Peng A P, Zhang H X, Yang J Y. Rarefied gas flow simulations using high-order gas-kinetic unified algorithms for Boltzmann model equations [J], Prog. Aerosp. Sci., 2015, 74: 81-113.

[34] Hu W Q, Li Z H. Investigation on different discrete velocity quadrature rules in gas-kinetic unified algorithm solving Boltzmann model equation [J], Comput. Math. Appl., 2018, 75: 4179-4200.

[35] Li Z H, Zhang H X. Gas-kinetic numerical studies of three-dimensional complex flows on spacecraft re-entry [J], J. Comput. Phys., 2009, 228: 1116-1138.

[36] Chapman S, Cowling T G. The mathematical theory of non-uniform gases $3^{rd}ed$ [M], Cambridge University Press (1970).

[37] Sterling J D, Chen S Y. Stability analysis of lattice Boltzmann methods [J], J. Comp. Phys., 1996, 123: 196-206.

[38] Guo Z L, Zheng C G, Shi B C. Non-equilibrium extrapolation method for velocity and

pressure boundary conditions in the lattice Boltzmann method [J], Chin.Phys., 2002, 11: 366-09.

[39] Burden R L, Faires J D. Numerical Analysis $7^{th}ed$ [M], Brooks/Cole (2001).

[40] Liu S K, Liu S D. Nonlinear Equations in Physics [M], Peking University Press, Beijing (2001).

[41] Wang P, Tian B, Liu W J, Jiang Y, Xue Y S. Interactions of breathers and solitons of a generalized variable-coefficient Korteweg-de Vries-modified Korteweg-de Vries equation with symbolic computation [J], Eur. Phys. J. D, 2012, 66: 233-10.

Decay Rates for the Viscous Incompressible MHD Equations with and without Surface Tension*

Guo Boling (郭柏灵),   Zeng Lan (曾兰) , and Ni Guoxi (倪国喜)

Abstract   In this paper, we consider a layer of a viscous incompressible electrically conducting fluid interacting with the magnetic filed in a horizontally periodic setting. The upper boundary is bounded by a free boundary and the below is bounded by a flat rigid interface. We prove the global well-posedness of the problem for both the case with and without surface tension. Moreover, we show that the global solution decays to the equilibrium exponentially in the case with surface tension, however the global solution decays to the equilibrium at an almost exponential rate in the case without surface tension.

Keywords   MHD equations; global well-posedness;decay rates; with and without surface tension

1   Introduction

1.1   Formulation in Eulerian coordinates

We consider the motion of a viscous incompressible electrically conducting fluid interacting with the magnetic field in a 3D moving domain

$$\Omega(t) = \{y \in \Sigma \times R | -1 < y_3 < \eta(y_1, y_2, t)\}. \qquad (1.1)$$

We assume $\Omega(t)$ is horizontally periodic by setting $\Sigma = (L_1 \mathbb{T}) \times (L_2 \mathbb{T})$ for $\mathbb{T} = \mathbf{R}/\mathbf{Z}$ the 1-torus and $L_1, L_2 > 0$ periodicity lengths. The upper boundary $\{y_3 = \eta(y_1, y_2, t)\}$ is a free surface that is the graph of the unknown function $\eta : \Sigma \times \mathbf{R}^+ \to \mathbf{R}$. The dynamics of the fluid is described by the velocity, the pressure

---

* Comput. Math. Appl. 2019 , 77(12): 3224–3249. DOI:10.1016/j.camwa.2019.02.008.

and the magnetic field, which are given for each $t \geqslant 0$ by $\tilde{u}(t,\cdot) : \Omega(t) \to \mathbf{R}^3$, $\tilde{p}(t,\cdot) : \Omega(t) \to \mathbf{R}$ and $\tilde{B}(t,\cdot) : \Omega(t) \to \mathbf{R}^3$, respectively. For each $t > 0$, $(\tilde{u}, \tilde{p}, \tilde{B}, \eta)$ is required to satisfy the following free boundary problem for the incompressible viscous and resistive magnetohydrodynamic (MHD) equations:

$$\begin{cases} \partial_t \tilde{u} + \tilde{u} \cdot \nabla \tilde{u} - \mu \Delta \tilde{u} + \nabla \tilde{p} = \tilde{B} \cdot \nabla \tilde{B}, & \text{in } \Omega(t), \\ \operatorname{div} \tilde{u} = 0, & \text{in } \Omega(t), \\ \partial_t \tilde{B} + \tilde{u} \cdot \nabla \tilde{B} - \kappa \Delta \tilde{B} = \tilde{B} \cdot \nabla u, & \text{in } \Omega(t), \\ \operatorname{div} B = 0 & \text{in } \Omega(t), \\ \partial_t \eta = u_3 - u_1 \partial_{y_1} \eta - u_2 \partial_{y_2} \eta & \text{on}\{y_3 = \eta(t, y_1, y_2)\}, \\ (\tilde{p}I - \mu \mathbb{D}(\tilde{u}))\nu = g\eta\nu + \sigma M\nu, \ \tilde{B} = \bar{B}, & \text{on}\{y_3 = \eta(t, y_1, y_2)\}, \\ \tilde{u} = 0, \ \tilde{B} = \bar{B}, & \text{on}\{y_3 = -1\}. \end{cases} \quad (1.2)$$

Here ν is the outward-pointing unit normal on $\{y_3 = \eta\}$, $\bar{B}$ is the constant magnetic field in the outside of the fluid. $\mu > 0$, $\kappa > 0$ are the kinematic viscosity and magnetic diffusion coefficient, respectively. The first four equations in (1.2) are the usual viscous incompressible MHD equations. The fifth equation implies that the free surface is advected with the fluid. The sixth equation is the balance of the stress on the free surface, where I is the 3×3 identity matrix, and $(\mathbb{D}\tilde{u})_{ij} = \partial_i \tilde{u}_j + \partial_j \tilde{u}_i$ is the symmetric gradient of $\tilde{u}$. The tensor $(\tilde{p}I - \mu\mathbb{D}(\tilde{u}))$ is known as the viscous stress tensor, g is the strength of gravity. M is the mean curvature of the free surface and is given by $M = \partial_i(\partial_i \eta/\sqrt{1 + |D\eta|^2})$. Note that, in (1.2), we have shifted the gravitational forcing to the free boundary and eliminated the constant atmospheric pressure, P_{atm}, the magnetic pressure $|\tilde{B}|^2/2$ and the constant outside magnetic pressure $|\bar{B}|^2/2$, in the usual way by adjusting the actual pressure $\bar{p}$ according to

$$\tilde{p} = \bar{p} + gy_3 - P_{atm} + |\tilde{B}|^2/2 - |\bar{B}|^2/2. \quad (1.3)$$

To complete the statement of the problem, we assume the problem satisfies the following initial conditions.

$$\eta(0) = \eta_0, \quad \tilde{u}(0) = u_0, \quad \tilde{B}(0) = B_0, \quad (1.4)$$

furthermore, we will assume $\eta_0 > -1$, which means at the initial time the boundaries do not intersect with each other.

In the global well-posedness theory of the problem (1.2), we suppose that the initial surface function satisfies the following "zero average" condition

$$\frac{1}{L_1 L_2}\int_\Sigma \eta_0 = 0. \tag{1.5}$$

Notice that for sufficiently regular solutions to the periodic problem, the condition (1.5) persists in time, indeed, according to $\partial_t \eta = \tilde{u}\cdot \nu\sqrt{1+(\partial_{y_1}\eta)^2+(\partial_{y_2}\eta)^2}$,

$$\frac{d}{dt}\int_\Sigma \eta = \int_\Sigma \partial_t \eta = \int_{\{y_3=\eta(t,y_1,y_2)\}} \tilde{u}\cdot\nu = \int_{\Omega(t)} \mathrm{div}\tilde{u} = 0, \tag{1.6}$$

which allows us to apply the Poincaré inequality on Σ for η for all $t \geqslant 0$.

1.2 Formulation in flattening coordinates

The moving free boundary and the subsequent change of the domain generate plentiful mathematical difficulties. To overcome these, as usual, we will use a coordinate transformation to flatten the free surface. Here we will not use a Lagrangian coordinate transformation, but rather a flatting transformation introduced by Beale in [2]. To this end, we consider the fixed equilibrium domain

$$\Omega := \{x \in \Sigma \times \mathbf{R} \mid -1 < x_3 < 0\}, \tag{1.7}$$

for which we will write the coordinates as $x \in \Omega$. We will think of Σ as the upper boundary of Ω, and we will write $\Sigma_{-1} := \{x_3 = -1\}$ for the lower boundary. We continue to view η as a function on $\Sigma \times R^+$. We then define

$$\bar{\eta} := \mathcal{P}\eta = \text{ harmonic extension of } \eta \text{ into the lower half space},$$

where $\mathcal{P}\eta$ is defined by (6.1). The harmonic extension $\bar{\eta}$ allows us to flatten the coordinate domain via the mapping

$$\Omega \ni x \mapsto (x_1, x_2, x_3 + \bar{\eta}(x,t)(1+x_3)) := \Phi(x,t) = (y_1, y_2, y_3) \in \Omega(t). \tag{1.8}$$

Note that $\Phi(\Sigma, t) = \{y_3 = \eta(y_1, y_2, t)\}$ and $\Phi(\cdot, t)|_{\Sigma_{-1}} = Id_{\Sigma_{-1}}$, i.e., Φ maps Σ to the free surface and keeps the lower surface fixed. We have

$$\nabla\Phi = \begin{pmatrix} 1 & 0 & 0 \\ 0 & 1 & 0 \\ A & B & J \end{pmatrix} \text{ and } \mathcal{A} := (\nabla\Phi^{-1})^{\mathrm{T}} = \begin{pmatrix} 1 & 0 & -AK \\ 0 & 1 & -BK \\ 0 & 0 & K \end{pmatrix} \tag{1.9}$$

for

$$A = \partial_1 \bar{\eta}\tilde{b}, \quad B = \partial_2 \bar{\eta}\tilde{b}, \quad \tilde{b} = (1+x_3), \tag{1.10}$$

$$J = 1 + \bar{\eta} + \partial_3 \bar{\eta}\tilde{b}, \quad K = J^{-1}. \tag{1.11}$$

Here $J = \det(\nabla\Phi)$ is the Jacobian of the coordinate transformation. If η is sufficiently small in an appropriate Sobolev space, then the mapping is a diffeomorphism. It allows us to transform the problem to one on the fixed spatial domain. Note that the following useful relation will be frequently used throughout this paper:

$$\partial_k(J\mathcal{A}_{jk}) = 0. \tag{1.12}$$

Without loss of generality, we will assume that $\mu = g = \kappa = 1$. Indeed, a standard scaling argument allows us to scale so that $\mu = g = \kappa = 1$. Furthermore, we define the transformed quantities as

$$u(t,x) := \tilde{u}(t,\Phi(t,x)), \quad p(t,x) := \tilde{p}(t,\Phi(t,x)), \quad b(t,x) := \tilde{B}(t,\Phi(t,x)) - \bar{B}.$$

In the new coordinates, (1.2) can be written as

$$\begin{cases} \partial_t u - \partial_t \bar{\eta}\bar{b}K\partial_3 u + u\cdot\nabla_\mathcal{A} u - \Delta_\mathcal{A} u + \nabla_\mathcal{A} p = (b+\bar{B})\cdot\nabla_\mathcal{A} b, & \text{in } \Omega, \\ \text{div}_\mathcal{A} u = 0, & \text{in } \Omega, \\ \partial_t b - \partial_t \bar{\eta}\bar{b}K\partial_3 b + u\cdot\nabla_\mathcal{A} b - \Delta_\mathcal{A} b = (b+\bar{B})\cdot\nabla_\mathcal{A} u, & \text{in } \Omega, \\ \text{div}_\mathcal{A} b = 0, & \text{in } \Omega, \\ (pI - \mathbb{D}_\mathcal{A} u)\mathcal{N} = \eta\mathcal{N} + \sigma M\mathcal{N}, b = 0 & \text{on } \Sigma, \\ \partial_t \eta + u_1\partial_1\eta + u_2\partial_2\eta = u_3, & \text{on } \Sigma, \\ u = 0, \ b = 0, & \text{on } \Sigma_{-1}, \\ u(x,0) = u_0(x), \ b(x,0) = b_0(x), \eta(x_1,x_2,0) = \eta_0(x_1,x_2). \end{cases} \tag{1.13}$$

Here we have written the differential operators $\nabla_\mathcal{A}$, $\text{div}_\mathcal{A}$, and $\Delta_\mathcal{A}$ with their actions given by $(\nabla_\mathcal{A} f)_i := \mathcal{A}_{ij}\partial_j f$, $\text{div}_\mathcal{A} X = \mathcal{A}_{ij}\partial_j X_i$, and $\Delta_\mathcal{A} f = \text{div}_\mathcal{A}\nabla_\mathcal{A} f$ for approximate f and X; for $u\cdot\nabla_\mathcal{A} u$ we mean $(u\cdot\nabla_\mathcal{A} u)_i := u_j\mathcal{A}_{jk}\partial_k u_i$. We have also written $(\mathbb{D}_\mathcal{A} u)_{ij} = \mathcal{A}_{ik}\partial_k u_j + \mathcal{A}_{jk}\partial_k u_i$. Also, $\mathcal{N} := -\partial_1\eta e_1 - \partial_2\eta e_2 + e_3$ denotes the non-unit normal on Σ.

1.3 Related works

The problem of the free boundary in fluid mechanics has been deeply studied in the field of mathematics, and there are a huge number of impressive results. Here, we only introduce briefly some works related to our problem.

When $B = 0$ in model (1.2), it reduces to the well-known viscous surface wave problem. The reduced problem without surface tension was studied firstly by Beale in [2], in which the local well-posedness in the Sobolev spaces had been proved. And Sylvester studie the global well-posedness by using Beale's method in [16]. For the periodic case, Hataya in [9] proved the global existence of small solutions with an algebraic decay rate. In [6-8], Guo and Tice used a new two-tier energy method to prove the local well-posedness, the global solution decays to the equilibrium at an algebraic decay rate in the non-periodic case and the global solution decays to equilibrium at an almost exponential rate in the periodic case. For the case with surface tension, the global well-posedness was proved in the Sobolev spaces by Beale in [3], and Bae in [1] proved the global solvability in Sobolev spaces via the energy method. Beale et al. in [4] and Nishida et al. in [13] proved that the global solution obtained in [3] decays at an optimal algebraic rate in the non-periodic case and decays at an exponential decay rate in the periodic case, respectively. Tani in [18] and Tani et al. in [19] considered the solvability of the problem with or without surface tension under the Beale-Solonnikov's function framework. Furthermore, in [17], Tan and Wang proved the zero surface tension limit within a local time interval and the global one under the small initial data. Furthermore, in [10, 20, 21], Tice et al. researched the effect of the more general surface tension on the decay rate for the viscous surface waves problem.

Correspondingly, for the case $B \neq 0$, namely, the free boundary problem for the viscous MHD equations, there are only a few results. The local-well posedness for the viscous MHD equations in a bounded variable domain with surface tension was proved by Padula et al. in [14], and the small initial data global solvability for the same model was obtained by Solonnikov et al. in [15]. In [11], Lee used the method developed by Masmoudi [12] to derive the vanishing viscosity limit with surface tension under the initial magnetic field is zero on the free boundary and in a vacuum. Recently, for the model (1.2), Wang and Xin in [22] studied the 2D case with $\mu = 0$ and $\sigma > 0$, and they proved the global solution decays to the equilibrium at an almost exponential decay rate, in which they use the structure of the equations sufficiently to find a damping structure for the fluid vorticity which plays an important role in closing the energies estimates.

Motivated by these articles mentioned above, in this paper, we focus on the global well-posedness for the MHD system (1.2) in $\Omega(t)$ both the case with and without surface tension, namely $\sigma > 0$ and $\sigma = 0$. Moreover, we will show that the two cases in different rates decay to the equilibrium state. When $\sigma > 0$, because of the term σM, we can use the classical energy method to obtain enough dissipation for η to close the energy estimates. However, for the case $\sigma = 0$, the regularity for η is not enough, the methods used in the former case can not work. Then, we will apply the method mentioned in [20] to overcome the lack of regularity for η. In [20], the authors used the structure $\text{div}_\mathcal{A} u = 0$ to write $\partial_3 u_3 = -(\partial_2 u_1 + \partial_2 u_2) + G^2$ to improve the full dissipation estimates of u, where G^2 are some quadratic nonlinearities. It is worth noting that, in this paper, we will apply a much more simple method used in [17] to obtain the full dissipation estimates for u and p, in which they had a crucial observation that they can get higher regularity estimates of u on the boundary Σ only from the horizontal dissipation estimates. Moreover, in this paper, we consider the MHD system which is a more complicated system and the nonlinear estimates will be much more complicated.

1.4 Some definitions and notations

Now, we state some definitions and notations that will be used throughout this paper. The Einstein convention of summing over repeated indices for vector and tensor operations. In this paper, $C > 0$ will denote a generic constant that can depend on N and Ω, but does not depend on the initial data and time. We refer to such constants as "universal", which are allowed to change from line to line. We use the notation $A \lesssim B$ to mean that $A \leqslant CB$ where $C > 0$ is a universal constant. We will use $\mathbf{N}^{1+m} = \{\alpha = (\alpha_0, \alpha_1, \cdots, \alpha_m)\}$ to emphasize that the 0-index term is related to temporal derivatives. For $\alpha \in \mathbf{N}^{1+m}$ we write $\partial^\alpha = \partial_t^{\alpha_0} \partial_1^{\alpha_1} \cdots \partial_m^{\alpha_m}$. For just spatial derivatives we write $\mathbf{N}^m$, namely $\alpha_0 = 0$. We define the parabolic counting of such multi-indices by writing $|\alpha| = 2\alpha_0 + \alpha_1 + \cdots + \alpha_m$. We will write Df for the horizontal gradient of f, that is, $Df = \partial_1 f e_1 + \partial_2 f e_2$, while ∇f will denote the usual full gradient.

We write $H^k(\Omega)$ with $k \geqslant 0$ and $H^s(\Sigma)$ with $s \in \mathbf{R}$ for the usual Sobolev spaces, and we will denote $H^0 = L^2$. In this paper, for simplicity, we will avoid writing $H^k(\Omega)$ or $H^s(\Sigma)$ and write only $\|\cdot\|_k$. When we write $\|\partial_t^j u\|_k$, it means that the space is $H^k(\Omega)$ and when we write $\|\partial_t^j \eta\|_k$, it will means that the space is $H^k(\Sigma)$.

For a given norm $\|\cdot\|$ and integers $k, m \geqslant 0$, we introduce the following notation for sums of spatial derivatives:

$$\|D_m^k f\|^2 := \sum_{\alpha \in \mathbf{N}^2, m \leqslant |\alpha| \leqslant k} \|\partial^\alpha f\|^2 \quad \text{and} \quad \|\nabla_m^k f\|^2 := \sum_{\alpha \in \mathbf{N}^3, m \leqslant |\alpha| \leqslant k} \|\partial^\alpha f\|^2. \quad (1.14)$$

The convention we adopt in this notation is that D refers to only horizontal spatial derivatives, while ∇ refers to full spatial derivatives. For space-time derivatives we add bars to our notation:

$$\|\bar{D}_m^k f\|^2 := \sum_{\alpha \in \mathbf{N}^{1+2}, m \leqslant |\alpha| \leqslant k} \|\partial^\alpha f\|^2 \quad \text{and} \quad \|\bar{\nabla}_m^k f\|^2 := \sum_{\alpha \in \mathbf{N}^{1+3}, m \leqslant |\alpha| \leqslant k} \|\partial^\alpha f\|^2. \tag{1.15}$$

When $k = m \geqslant 0$, we denote

$$\|D^k f\|^2 = \|D_k^k f\|^2, \quad \|\nabla^k f\|^2 = \|\nabla_k^k f\|^2,$$
$$\|\bar{D}^k f\|^2 = \|\bar{D}_k^k f\|^2, \quad \|\bar{\nabla}^k f\|^2 = \|\bar{\nabla}_k^k f\|^2. \tag{1.16}$$

The rest of this paper unfolds as follows. In section 2, we first define the energies and dissipations, and then state our main results. In section 3, we prove some preliminary lemmas that we will use in our a priori estimates. In section 4, we complete the a priori estimates for the case $\sigma > 0$. In section 5, we close the a priori estimates for the case $\sigma = 0$.

2 Main results

We first state the result for (1.13) in the case $\sigma > 0$. Firstly, we define some energy functions in this case. We define the energy as

$$\begin{aligned}\mathcal{E} := &\|u\|_2^2 + \|\partial_t u\|_0^2 + \|b\|_2^2 + \|\partial_t b\|_0^2 + \|p\|_1^2 + \|\eta\|_3^2 \\ &+ \|\partial_t \eta\|_{3/2}^2 + \|\partial_t^2 \eta\|_{-1/2}^2,\end{aligned} \tag{2.1}$$

and define the dissipation as

$$\begin{aligned}\mathcal{D} := &\|u\|_3^2 + \|\partial_t u\|_1^2 + \|b\|_3^2 + \|\partial_t b\|_1^2 + \|p\|_2^2 + \|\eta\|_{7/2}^2 \\ &+ \|\partial_t \eta\|_{5/2}^2 + \|\partial_t^2 \eta\|_{1/2}^2.\end{aligned} \tag{2.2}$$

In the case $\sigma > 0$, the global well-posedness result is stated as follows.

Theorem 2.1 *For $\sigma > 0$, we assume that the initial datum $u_0 \in H^2(\Omega)$, $\eta_0 \in H^3(\Sigma)$, $b_0 \in H^2(\Omega)$ and satisfy some appropriate compatibility conditions as well as the zero-average condition (1.5). Then there exists a universal constant $\kappa > 0$ such that, if*

$$\|u_0\|_2^2 + \|\eta_0\|_3^2 + \|b_0\|_2^2 \leqslant \kappa,$$

then, for all $t \geqslant 0$, there exists a unique strong solution (u, p, η, b) to (1.13) satisfying the estimate

$$e^{\lambda t} \mathcal{E}(t) + \int_0^t \mathcal{D}(s) ds \lesssim \mathcal{E}(0). \tag{2.3}$$

Remark 2.1 Since η is such that the mapping $\Phi(\cdot, t)$, defined by (1.8), is a diffeomorphism for each $t \geqslant 0$, one may change coordinate to $y \in \Omega(t)$ to produce a global-in-time decaying solution to (1.2).

Remark 2.2 Theorem 2.1 implies that $\mathcal{E}(t) \lesssim e^{-\lambda t}$, which means that for $\sigma > 0$ the solution returns to the stable state at an exponential decay rate.

We then state our results for (1.13) in the case $\sigma = 0$. And we first define some energy functionals corresponding to this case. For a generic integer n, we define the energy as

$$\mathcal{E}_n := \sum_{j=0}^{n} \left(\left\| \partial_t^j u \right\|_{2n-2j}^2 + \left\| \partial_t^j b \right\|_{2n-2j}^2 + \left\| \partial_t^j \eta \right\|_{2n-2j}^2 \right) + \sum_{j=0}^{n-1} \left\| \partial_t^j p \right\|_{2n-2j-1}^2, \quad (2.4)$$

and define the corresponding dissipation as

$$\mathcal{D}_n := \sum_{j=0}^{n} \left(\left\| \partial_t^j u \right\|_{2n-2j+1}^2 + \left\| \partial_t^j b \right\|_{2n-2j+1}^2 \right) + \sum_{j=0}^{n-1} \left\| \partial_t^j p \right\|_{2n-2j}^2$$

$$+ \|\eta\|_{2n-1/2}^2 + \|\partial_t \eta\|_{2n-1/2}^2 + \sum_{j=2}^{n+1} \left\| \partial_t^j \eta \right\|_{2n-2j+5/2}^2. \quad (2.5)$$

We write the high-order spatial derivatives of η as

$$\mathcal{F}_{2N} := \|\eta\|_{4N+1/2}^2. \quad (2.6)$$

Finally, we define the total energy as

$$\mathcal{G}_{2N}(t) := \sup_{0 \leqslant r \leqslant t} \mathcal{E}_{2N}(r) + \int_0^t \mathcal{D}_{2N}(r) dr + \sup_{0 \leqslant r \leqslant t} (1+r)^{4N-8} \mathcal{E}_{N+2}(r)$$

$$+ \sup_{0 \leqslant r \leqslant t} \frac{\mathcal{F}_{2N}(r)}{(1+r)}. \quad (2.7)$$

Our main results state as follows.

Theorem 2.2 For $\sigma = 0$, we assume that the initial data $u_0 \in H^{4N}(\Omega)$, $b_0 \in H^{4N}(\Omega)$ and $\eta_0 \in H^{4N+1/2}(\Sigma)$ satisfy some appropriate compatibility conditions as well as the zero-average condition (1.5), where $N \geqslant 3$. There exists a constant $\varepsilon_0 > 0$ such that if

$$\mathcal{E}_{2N}(0) + \mathcal{F}_{2N}(0) \leqslant \varepsilon_0,$$

then, for all $t \geqslant 0$, there exists a global unique solution (u, p, b, η) to (1.13) satisfying the estimate

$$\mathcal{G}_{2N}(t) \lesssim \mathcal{E}_{2N}(0) + \mathcal{F}_{2N}(0). \quad (2.8)$$

Remark 2.3 Theorem 2.2 implies that $\mathcal{E}_{N+2}(t) \lesssim (1+t)^{-4N-8}$, which is integrable in time for $N \geqslant 3$. Since N may be taken to be arbitrarily large, this decay results can be regarded as an "almost exponential" decay rate. Comparing the two different cases for $\sigma > 0$ and $\sigma = 0$, reveal that the surface tension can enhance the decay rate.

Remark 2.4 We refer to [3, 6] for the local well-posedness of the system (1.13) for both the case $\sigma > 0$ and $\sigma = 0$. Then, by a continuity argument, to prove Theorem 2.1 and Theorem 2.2 it suffices to derive the a priori estimates Theorem 4.1 and Theorem 5.12, respectively.

3 Preliminaries for a priori estimates

In this section, we will present some preliminary results and give the proofs respectively. We state two forms of equations to (1.13) and describe the corresponding energy evolution structures.

3.1 Geometric form

We now give a linear formation of the problem (1.13) in its geometric form. Assume that u, η, b are known and that $\mathcal{A}$, $\mathcal{N}$, J, etc., are given in terms of η as usual. We then consider the linear equations for (v, H, q, h) given by

$$\begin{cases} \partial_t v - \partial_t \bar{\eta} \tilde{b} K \partial_3 v + u \cdot \nabla_\mathcal{A} v + \mathrm{div}_\mathcal{A}(qI - \mathbb{D}_\mathcal{A} v) = (b + \bar{B}) \cdot \nabla_\mathcal{A} H + F^1, & \text{in } \Omega, \\ \mathrm{div}_\mathcal{A} v = F^2, & \text{in } \Omega, \\ \partial_t H - \partial_t \bar{\eta} \tilde{b} K \partial_3 H + u \cdot \nabla_\mathcal{A} H - \Delta_\mathcal{A} H = (b + \bar{B}) \cdot \nabla_\mathcal{A} v + F^3, & \text{in } \Omega, \\ (qI - \mathbb{D}_\mathcal{A} v)\mathcal{N} = (h - \sigma \Delta_\star h)\mathcal{N} + F^4, \quad H = 0, & \text{on } \Sigma, \\ \partial_t h - v \cdot \mathcal{N} = F^5, & \text{on } \Sigma, \\ v = 0, \quad H = 0, & \text{on } \Sigma_{-1}, \end{cases}$$
(3.1)

where $\Delta_\star = \partial_{x_1}^2 + \partial_{x_2}^2$.

Lemma 3.1 Let u and η be given and solve (1.13). If (v, H, q, h) solve (3.1) then

$$\frac{d}{dt}\left(\int_\Omega \frac{|v|^2}{2}J + \int_\Omega \frac{|H|^2}{2}J + \int_\Sigma \frac{|h|^2}{2} + \sigma \int_\Sigma \frac{|Dh|^2}{2}\right) + \int_\Omega \frac{|\mathbb{D}_\mathcal{A} v|^2}{2}J + \int_\Omega |\nabla_\mathcal{A} H|^2 J$$
$$= \int_\Omega (v \cdot F^1 + qF^2 + v \cdot F^3)J - \int_\Sigma v \cdot F^4 + \int_\Sigma (h - \sigma\Delta_\star h)F^5.$$
(3.2)

Proof We take the inner product of the first equation in (3.1) with Jv and the third equation with JH, then integrate over Ω to find that

$$I_1 + I_2 + I_3 + I_4 = I_5,$$

where

$$I_1 = \int_\Omega (\partial_t v_i J v_i - \partial_t \bar{\eta} \tilde{b} \partial_3 v_i v_i + u_j \mathcal{A}_{jk} \partial_k v_i J v_i),$$

$$I_2 = \int_\Omega \mathcal{A}_{ik} \partial_k q J v_i - \int_\Omega \mathcal{A}_{jk} \partial_k (\mathcal{A}_{jl} \partial_l v_i + \mathcal{A}_{il} \partial_l v_j) J v_i,$$

$$I_3 = \int_\Omega (\partial_t H_i J H_i - \partial_t \bar{\eta} \tilde{b} \partial_3 H_i H_i + u_j \mathcal{A}_{jk} \partial_k H_i J H_i)$$
$$- \int_\Omega \mathcal{A}_{jk} \partial_k (\mathcal{A}_{jl} \partial_l H_i) J H_i,$$

$$I_4 = \int_\Omega (\bar{B}_j + b_j) \mathcal{A}_{jk} \partial_k H_i J v_i + \int_\Omega (\bar{B}_j + b_j) \mathcal{A}_{jk} \partial_k v_i J H_i,$$

$$I_5 = \int_\Omega F_i^1 J v_i + \int_\Omega F_i^3 J H_i.$$

Integrating by parts and using (1.12), one has

$$I_1 = \partial_t \int_\Omega \frac{|v|^2 J}{2} - \int_\Omega \frac{|v|^2 \partial_t J}{2} - \int_\Omega \partial_t \bar{\eta} \tilde{b} \partial_3 \frac{|v|^2}{2} + \int_\Omega u_j \partial_k (J\mathcal{A}_{jk} \frac{|v|^2}{2})$$

$$= \partial_t \int_\Omega \frac{|v|^2 J}{2} - \int_\Omega \frac{|v|^2 \partial_t J}{2} + \int_\Omega \frac{|v|^2}{2} (\partial_t \bar{\eta} + \tilde{b} \partial_t \partial_3 \bar{\eta})$$
$$- \int_\Omega J\mathcal{A}_{jk} \partial_k u_j \frac{|v|^2}{2} - \frac{1}{2} \int_\Sigma (\partial_t \eta |v|^2 - u_j J \mathcal{A}_{jk} e_3 \cdot e_k |v|^2)$$

$$= \partial_t \int_\Omega \frac{|v|^2 J}{2},$$

where according to (1.11), we know that $\partial_t J = \partial_t \bar{\eta} + \tilde{b} \partial_t \partial_3 \bar{\eta}$ and $J\mathcal{A}_{jk} e_3 \cdot e_k = \mathcal{N}_j$ on Σ, then use the condition $\partial \eta = u \cdot \mathcal{N}$. Similarly, an integration by parts reveals that

$$I_2 = -\int_\Omega \mathcal{A}_{jk}(qI - \mathbb{D}_\mathcal{A} v)_{ij} J \partial_k v_i + \int_\Sigma J\mathcal{A}_{j3}(qI - \mathbb{D}_\mathcal{A} v)_{ij} v_i$$

$$= \int_\Omega (-q \mathcal{A}_{ik} \partial_k v_i J + J \frac{|\mathbb{D}_\mathcal{A} v|^2}{2}) + \int_\Sigma (qI - \mathbb{D}_\mathcal{A} v)_{ij} \mathcal{N}_j v_i$$

$$= \int_\Omega (-qJF^2 + J \frac{|\mathbb{D}_\mathcal{A} v|^2}{2}) + \int_\Sigma (h - \sigma \Delta_* h) \mathcal{N} \cdot v + F^4 \cdot v$$

$$= \int_\Omega (-qJF^2 + J \frac{|\mathbb{D}_\mathcal{A} v|^2}{2}) + \int_\Sigma (h - \sigma \Delta_* h)(\partial_t h - F^5) + F^4 \cdot v$$

$$= \int_\Omega (-qJF^2 + J\frac{|\mathbb{D}_A v|^2}{2}) + \partial_t \int_\Sigma (\frac{|h|^2}{2} + \sigma|Dh|^2)$$
$$+ \int_\Sigma v \cdot F^4 - \int_\Sigma (h - \sigma\Delta_* h) \cdot F^5.$$

By using $H = 0$ on $\partial\Omega$, $\text{div}_A u = 0$ and (1.12), one has

$$I_3 = \partial_t \int_\Omega \frac{|H|^2 J}{2} - \int_\Omega \frac{|H|^2 \partial_t J}{2} + \int_\Omega (\partial_t \bar{\eta} + \partial_3 \partial_t \tilde{\eta} b)\frac{|H|^2}{2}$$
$$- \int_\Omega J\mathcal{A}_{jk}\partial_k u_j + \int_\Omega J|\nabla_A H|^2$$
$$= \partial_t \int_\Omega \frac{|H|^2 J}{2} + \int_\Omega J|\nabla_A H|^2,$$

and, similarly, by using $H = 0$ on $\partial\Omega$, $\text{div}_A b = 0$ and (1.12), we deduce

$$I_4 = \int_\Omega (\bar{B}_j + b_j)J\mathcal{A}_{jk}\partial_k(H_i v_i) = -\int_\Omega J\mathcal{A}_{jk}\partial_k b_j H_i v_i = 0. \tag{3.3}$$

Then, (3.2) follows from the estimates of I_1, I_2, I_3 and I_4.

3.2 Perturbed linear form

In many parts of this paper, we will apply the PDE in a different formulation, which looks like a perturbation of the linearized problem. The utility of this form of the equations lies in the fact that the linear operator has constant coefficients. The equations in this form are

$$\begin{cases} \partial_t u + \nabla p - \Delta u = G^1, & \text{in } \Omega, \\ \text{div} u = G^2, & \text{in } \Omega, \\ \partial_t b - \Delta b = G^3, & \text{in } \Omega, \\ (pI - \mathbb{D}u)e_3 = (\eta - \sigma\Delta_*\eta)e_3 + G^4, \ b = 0, & \text{on } \Sigma, \\ \partial_t \eta - u_3 = G^5, & \text{on } \Sigma, \\ u = 0, \ b = 0, & \text{on } \Sigma_{-1}. \end{cases} \tag{3.4}$$

Here we have written the nonlinear terms G^i for $i = 1, \cdots, 5$ as follows. We write $G^1 := \Sigma_{l=1}^5 G^{1,l}$, for

$$G_i^{1,1} := (\delta_{ij} - \mathcal{A}_{ij})\partial_j p, \ G_i^{1,2} := \partial_t \tilde{\eta} b K \partial_3 u_i,$$
$$G_i^{1,3} := -u_j \mathcal{A}_{jk}\partial_k u_i + (b_j + \bar{B}_j)\mathcal{A}_{jk}\partial_k b_i,$$
$$G_i^{1,4} := [K^2(1 + A^2 + B^2) - 1]\partial_{33}u_i - 2AK\partial_{13}u_i - 2BK\partial_{23}u_i, \tag{3.5}$$
$$G_i^{1,5} := [-K^3(1 + A^2 + B^2)\partial_3 J + AK^2(\partial_1 J + \partial_3 A)]\partial_3 u_i$$
$$+ [BK^2(\partial_2 J + \partial_3 B) - K(\partial_1 A + \partial_2 B)]\partial_3 u_i.$$

G^2 is the function
$$G^2 := AK\partial_3 u_1 + BK\partial_3 u_2 + (1-K)\partial_3 u_3, \quad (3.6)$$

and $G^3 = G^{3,1} + G^{3,2} + G^{3,3} + G^{3,4}$, for

$$\begin{aligned} G_i^{3,1} &:= \partial_t \tilde{\eta} \bar{b} K \partial_3 b_i \\ G_i^{3,2} &:= -u_j \mathcal{A}_{jk}\partial_k b_i + (b_j + \bar{B}_j)\mathcal{A}_{jk}\partial_k u_i, \\ G_i^{3,3} &:= [K^2(1+A^2+B^2)-1]\partial_{33}b_i - 2AK\partial_{13}b_i - 2BK\partial_{23}b_i, \quad (3.7) \\ G_i^{3,4} &:= [-K^3(1+A^2+B^2)\partial_3 J + AK^2(\partial_1 J + \partial_3 A)]\partial_3 b_i \\ &\quad + [BK^2(\partial_2 J + \partial_3 B) - K(\partial_1 A + \partial_2 B)]\partial_3 u_1, \end{aligned}$$

$$G^4 := \partial_1 \eta N_1 + \partial_2 \eta N_2 + N_3 + \sigma(H - \Delta_\star \eta)\mathcal{N} + \sigma\Delta_\star(\mathcal{N} - e_3), \quad (3.8)$$

where

$$N_1 = \begin{pmatrix} p - \eta - 2(\partial_1 u_1 - AK\partial_3 u_1) \\ -\partial_2 u_1 - \partial_1 u_2 + BK\partial_3 u_1 + AK\partial_3 u_2 \\ -\partial_1 u_3 - K\partial_3 u_1 + AK\partial_3 u_3 \end{pmatrix},$$

$$N_2 = \begin{pmatrix} -\partial_2 u_1 - \partial_1 u_2 + BK\partial_3 u_1 + AK\partial_3 u_2 \\ p - \eta - 2(\partial_2 u_2 - BK\partial_3 u_2) \\ -\partial_2 u_3 - K\partial_3 u_2 + BK\partial_3 u_3 \end{pmatrix},$$

and

$$N_3 = \begin{pmatrix} (K-1)\partial_3 u_1 + AK\partial_3 u_3 \\ (K-1)\partial_3 u_2 + BK\partial_3 u_3 \\ 2(K-1)\partial_3 u_3 \end{pmatrix}.$$

$$G^5 = -D\eta \cdot u. \quad (3.9)$$

Lemma 3.2 *Suppose* (v, H, q, h) *solve*

$$\begin{cases} \partial_t v + \nabla q - \Delta v = \Phi^1, & \text{in } \Omega, \\ \operatorname{div} v = \Phi^2, & \text{in } \Omega, \\ \partial_t H - \Delta H = \Phi^3, & \text{in } \Omega, \\ (qI - \mathbb{D}v)e_3 = (h - \sigma\Delta_\star h)e_3 + \Phi^4, \ H = 0, & \text{on } \Sigma, \\ \partial_t h - v_3 = \Phi^5, & \text{on } \Sigma, \\ v = H = 0, & \text{on } \Sigma_{-1}. \end{cases} \quad (3.10)$$

Then
$$\partial_t\left(\int_\Omega \frac{|v|^2}{2} + \int_\Omega \frac{|H|^2}{2} + \int_\Sigma \frac{|h|^2}{2} + \sigma\int_\Sigma \frac{|Dh|^2}{2}\right) + \int_\Omega \frac{|\mathbb{D}v|^2}{2} + \int_\Omega |\nabla H|^2$$
$$= \int_\Omega v\cdot(\Phi^1 - \nabla\Phi^2) + \int_\Omega(q\Phi^2 + H\cdot\Phi^3) - \int_\Sigma v\cdot\Phi^4 + \int_\Sigma (h-\sigma\Delta_\star h)\Phi^5. \quad (3.11)$$

Proof From the first and second equation in (3.10), we can rewrite the first one as
$$\partial_t v + \text{div}(qI - \mathbb{D}v) = \Phi^1 - \nabla\Phi^2. \quad (3.12)$$

Taking the inner product of the (3.12) with v and the third equation in (3.10) with H, integrating by parts over Ω and then adding the resulting equations together, one has
$$\partial_t\left(\int_\Omega \frac{|v|^2}{2} + \int_\Omega \frac{|H|^2}{2} + \int_\Sigma \frac{|h|^2}{2} + \sigma\int_\Sigma \frac{|Dh|^2}{2}\right) - \int_\Omega q\,\text{div}\,v + \int_\Omega \frac{|\mathbb{D}v|^2}{2}$$
$$+ \int_\Sigma (qI - \mathbb{D}v)e_3\cdot v + \int_\Omega |\nabla H|^2 = \int_\Omega (\Phi^1 - \nabla\Phi^2)\cdot v + \int_\Omega \Phi^3\cdot H.$$

Furthermore, we bring $\text{div}\,v = \Phi^2$, $(qI - \mathbb{D}v)e_3 = (h - \sigma\Delta_\star h)e_3 + \Phi^4$ and $v_3 = \partial_t h - \Phi^5$ into the above equation, then (3.11) follows.

3.3 Some useful estimates

Before having a priori estimates on the nonlinear terms, we give the useful L^∞ estimates for removing the appearance of J factors.

Lemma 3.3 *There exists a universal $0 < \delta < 1$ so that if $\|\eta\|_{5/2}^2 \leqslant \delta$, then we have the estimate*
$$\|J-1\|_{L^\infty}^2 + \|A\|_{L^\infty}^2 + \|B\|_{L^\infty}^2 \leqslant \frac{1}{2}, \quad \text{and} \quad \|K\|_{L^\infty}^2 + \|\mathcal{A}\|_{L^\infty}^2 \lesssim 1. \quad (3.13)$$

Proof According to the definitions of A, B, J given in (1.10)-(1.11) and Lemma 6.1, we have that
$$\|J-1\|_{L^\infty}^2 + \|A\|_{L^\infty}^2 + \|B\|_{L^\infty}^2 \lesssim \|\bar\eta\|_3^2 \lesssim \|\eta\|_{5/2}^2. \quad (3.14)$$

Then if δ is sufficiently small, (3.13) follows.

Furthermore, we provide an estimate for $\partial_t^n \mathcal{A}$.

Lemma 3.4 *For $n = 2N$ or $n = N+2$, we have*
$$\|\partial_t^{n+1}J\|_0^2 + \|\partial_t^{n+1}\mathcal{A}\|_0^2 \lesssim \mathcal{D}_n. \quad (3.15)$$

Proof Since temporal derivatives commute with the Poisson integral, applying Lemma 6.1, we have

$$\|\partial_t^{m+1}\bar{\eta}\|_1^2 = \|\partial_t^{m+1}\bar{\eta}\|_0^2 + \|\nabla\partial_t^{m+1}\bar{\eta}\|_0^2 \lesssim \|\partial_t^{m+1}\eta\|_{1/2}^2, \quad \text{for } m \geq 0.$$

From the definition of $\mathcal{D}_n$, we have

$$\|\partial_t^{n+1}\eta\|_{1/2}^2 \lesssim \mathcal{D}_n, \quad \text{for } n = 2N \text{ or } n = N+2. \tag{3.16}$$

Then, according to the definition of J, A, B and K, we have

$$\|\partial_t^{n+1}J\|_0^2 + \|\partial_t^{n+1}A\|_0^2 + \|\partial_t^{n+1}B\|_0^2 + \|\partial_t^{n+1}K\| \lesssim \mathcal{D}_n, \quad \text{for } n = 2N \text{ or } n = N+2.$$

Using the Sobolev embeddings we complete the proof of $\mathcal{A}$ since the components of $\mathcal{A}$ are either unity, K, AK or BK.

4 For the case $\sigma > 0$

4.1 Nonlinear estimates

We will employ the form (3.1) to study the temporal derivative of solutions to (1.13). That is, we employ ∂_t to (1.13) and set $(v, H, q, h) = (\partial_t u, \partial_t b, \partial_t p, \partial_t \eta)$ satisfying (3.1) for certain terms F^i. Below we record the form of these forcing terms $F^i, i = 1, 2, 3, 4, 5$.

$$F_i^1 = \partial_t(\partial_t\bar{\eta}\tilde{b}K)\partial_3 u - \partial_t(u_j\mathcal{A}_{jk})\partial_k u_i - \partial_t\mathcal{A}_{ik}\partial_k p + \partial_t\mathcal{A}_{jk}\partial_k(\mathcal{A}_{im}\partial_m u_j + \mathcal{A}_{jm}\partial_m u_i)$$
$$+ \mathcal{A}_{jk}\partial_k(\partial_t\mathcal{A}_{im}\partial_m u_j + \partial_t\mathcal{A}_{jm}\partial_m u_i) + \partial_t((b_j + \bar{B}_j)\mathcal{A}_{jk})\partial_k b_i, \tag{4.1}$$

$$F_i^2 = -\partial_t\mathcal{A}_{ij}\partial_j u_i, \tag{4.2}$$

$$F^3 = \partial_t(\partial_t\bar{\eta}\tilde{b}K)\partial_3 b - \partial_t(u_j\mathcal{A}_{jk})\partial_k b + \partial_t\mathcal{A}_{il}\partial_l\mathcal{A}_{im}\partial_m b$$
$$+ \mathcal{A}_{il}\partial_l\partial_t\mathcal{A}_{im}\partial_m b + \partial_t((b_j + \bar{B}_j)\mathcal{A}_{jk})\partial_k b, \tag{4.3}$$

$$F_i^4 = (\mathcal{A}_{ik}\partial_k u_j + \mathcal{A}_{jk}\partial_k u_i)\partial_t\mathcal{N}_j + (\partial_t\mathcal{A}_{ik}\partial_k u_j + \partial_t\mathcal{A}_{jk}\partial_k u_i)\mathcal{N}_j$$
$$+ (\eta - p)\partial_t\mathcal{N}_i - (\sigma\partial_t M - \sigma\partial_t\Delta_\star\eta)\mathcal{N}_i - \sigma M\partial_t\mathcal{N}_i, \tag{4.4}$$

$$F^5 = \partial_t D\eta \cdot u. \tag{4.5}$$

Next, we will estimate the nonlinear terms F^i for $i = 1, \cdots, 6$, which will be used principally to estimate the interaction terms on the right side of (3.2).

Lemma 4.1 Let $F^1, \cdots, F^5$ be defined by (4.1)-(4.5), and let $\mathcal{E}$ and $\mathcal{D}$ be defined by (2.1) and (2.2). Suppose that $\mathcal{E} \leqslant \delta$, where $\delta \in (0,1)$ is the universal constant given in Lemma 3.3. Then,

$$\|F^1\|_0 + \|F^2\|_0 + \|F^3\|_0 + \|F^4\|_0 + \|F^5\|_0 \lesssim \sqrt{\mathcal{E}\mathcal{D}}, \tag{4.6}$$

$$\left|\int_\Omega p\partial_t(F^2 J)\right| \lesssim \sqrt{\mathcal{E}\mathcal{D}}, \quad \text{and} \quad \left|\int_\Omega pF^2 J\right| \lesssim \mathcal{E}^{\frac{3}{2}}. \tag{4.7}$$

Proof Throughout this lemma, we will employ Holder's inequality, Sobolev embeddings, trace theory, Lemma 3.3 and Lemma 6.1. Firstly, we give the estimates for F^1.

$$\|\partial_t(\partial\bar{\eta}\tilde{b}K)\partial_3 u\|_0 \lesssim \|\partial_t^2\bar{\eta}\|_0\|K\|_{L^\infty}\|\partial_3 u\|_{L^\infty} + \|\partial_t\bar{\eta}\|_{L^\infty}\|\partial_t\bar{\eta}\|_1\|\partial_3 u\|_{L^\infty}$$
$$\lesssim \|\partial_t^2\eta\|_{-1/2}\|u\|_3 + \|\partial_t\eta\|_{3/2}\|\partial_t\eta\|_{1/2}\|u\|_3$$
$$\lesssim (\sqrt{\mathcal{E}} + \mathcal{E})\sqrt{\mathcal{D}} \lesssim \sqrt{\mathcal{E}\mathcal{D}},$$

and the other terms of F^1 can be bounded in a similar way. Next, we control the term F^2 as follows,

$$\|F^2\|_0 \lesssim \|\partial_t\nabla\bar{\eta}\|_0\|\nabla u\|_{L^\infty} \lesssim \sqrt{\mathcal{E}\mathcal{D}}.$$

Similar to the estimates of F^1, F^2, we can obtain

$$\|F^3\|_0 + \|F^4\|_0 + \|F^5\|_0 \lesssim \sqrt{\mathcal{E}\mathcal{D}}.$$

For the terms involved in (4.7), we have

$$\left|\int_\Omega p(\partial_t J F^2 + J\partial_t F^2)\right|$$
$$\lesssim \|p\|_{L^\infty}\|\partial_t\bar{\eta}\|_1\|F^2\|_0 + \|p\|_{L^\infty}\|J\|_{L^\infty}(\|u\|_1\|\partial_t^2\bar{\eta}\|_1 + \|\partial_t\bar{\eta}\|_1\|\partial_t u\|_1)$$
$$\lesssim \|p\|_2\|\partial_t\eta\|_{1/2}\|F^2\|_0 + \|p\|_2(\|u\|_1\|\partial_t^2\eta\|_{1/2} + \|\partial_t\eta\|_{1/2}\|\partial_t u\|_1)$$
$$\lesssim (\mathcal{E}\mathcal{D} + \sqrt{\mathcal{E}\mathcal{D}}) \lesssim \sqrt{\mathcal{E}\mathcal{D}},$$

and

$$\left|\int_\Omega pJF^2\right| \lesssim \|p\|_{L^6}\|\partial_t\nabla\bar{\eta}\|_{L^2}\|\nabla u\|_{L^3}\|J\|_{L^\infty} \lesssim \|p\|_1\|u\|_2\|\partial_t\eta\|_{1/2} \lesssim \mathcal{E}^{3/2}.$$

Then we complete the proof of this lemma.

Then, we turn our attention to the nonlinear terms G^i for $i = 1, \cdots, 5$ defined by (3.5)-(3.12).

Lemma 4.2 Let $G^1, \cdots, G^5$ be defined by (3.5)-(3.12) and let $\mathcal{E}$ and $\mathcal{D}$ be given by (2.1) and (2.2). Suppose that $\mathcal{E} \leqslant \delta$, where $\delta \in (0,1)$ is the universal constant given in Lemma 3.3, and that $D < \infty$. Then,

$$\|G^1\|_1 + \|G^2\|_2 + \|G^3\|_1 + \|G^4\|_{3/2} + \|G^5\|_{5/2} + \|\partial_t G^5\|_{1/2} \lesssim \sqrt{\mathcal{E}\mathcal{D}}, \quad (4.8)$$

and

$$\|G^1\|_0 + \|G^2\|_1 + \|G^2\|_{-1} + \|G^3\|_0 + \|G^4\|_{1/2} + \|G^5\|_{3/2} + \|G^5\|_{-1/2} \lesssim \mathcal{E}. \quad (4.9)$$

Proof Here the estimates of $G^1, \cdots, G^5$ are similar to [[10], Theorem 4.3], so we omit it.

4.2 A priori estimates

In this section we combine energy-dissipation estimates with various elliptic estimates to deduce a system of a priori estimates.

4.2.1 Energy-dissipation estimates

In order to state our energy-dissipation estimates, we must introduce some notations firstly. Recall that for a multi-index $\alpha = (\alpha_0, \alpha_1, \alpha_2) \in \mathbf{N}^{1+2}$, we write $|\alpha| = 2\alpha_0 + \alpha_1 + \alpha_2$ and $\partial^\alpha = \partial_t^{\alpha_0} \partial_1^{\alpha_1} \partial_2^{\alpha_2}$. For $\alpha \in \mathbf{N}^{1+2}$, we set

$$\overline{\mathcal{E}}_\alpha := \int_\Omega \frac{1}{2}|\partial^\alpha u|^2 + \int_\Sigma (\frac{1}{2}|\partial^\alpha \eta|^2 + \frac{\sigma}{2}|D\partial^\alpha \eta|^2) + \int_\Omega \frac{1}{2}|\partial^\alpha b|^2,$$

$$\overline{\mathcal{D}}_\alpha := \int_\Omega \frac{1}{2}|\mathbb{D}\partial^\alpha u|^2 + \int_\Omega |\nabla \partial^\alpha b|^2. \quad (4.10)$$

We then define

$$\overline{\mathcal{E}} := \sum_{|\alpha|\leqslant 2} \overline{\mathcal{E}}_\alpha \quad \text{and} \quad \overline{\mathcal{D}} := \sum_{|\alpha|\leqslant 2} \overline{\mathcal{D}}_\alpha. \quad (4.11)$$

We will also need to use the functional

$$\mathcal{F} := \int_\Omega p F^2 J. \quad (4.12)$$

Our next result encodes the energy-dissipation inequality associated with $\overline{\mathcal{E}}$ and $\overline{\mathcal{D}}$.

Lemma 4.3 Suppose that (u, b, p, η) solves (1.13). Let $\mathcal{E}$ and $\mathcal{D}$ be defined by (2.1) and (2.2). Assume that $\mathcal{E} \leqslant \delta$, where $\delta \in (0,1)$ is the universal constant given in Lemma 3.3. Let $\overline{\mathcal{E}}$ and $\overline{\mathcal{D}}$ be given by (4.11) and $\mathcal{F}$ be given by (4.12). Then

$$\frac{d}{dt}(\overline{\mathcal{E}} - \mathcal{F}) + \overline{\mathcal{D}} \lesssim \sqrt{\mathcal{E}\mathcal{D}}, \quad (4.13)$$

for all $t \in [0, T]$.

Proof Let $\alpha \in \mathbf{N}^{1+2}$ with $|\alpha| \leq 2$. We apply ∂^α to (1.13) to derive an equation for $(\partial^\alpha u, \partial^\alpha b, \partial^\alpha \eta, \partial^\alpha p)$.

Suppose that $\alpha = (1,0,0)$, namely, $\partial^\alpha = \partial_t$. Then $v = \partial_t u$, $q = \partial_t p$, $H = \partial_t b$ and $h = \partial_t \eta$ satisfy (3.1) with $F^1, \cdots, F^5$ defined by (4.1)-(4.5). Then according to Lemma 3.1 and Lemma 4.1, we deduce

$$\frac{d}{dt}\left(\int_\Omega \frac{|\partial_t u|^2}{2}J + \int_\Omega \frac{|\partial_t b|^2}{2}J + \int_\Sigma \frac{|\partial_t \eta|^2}{2} + \int_\Sigma \sigma \frac{|D\partial_t \eta|^2}{2}\right)$$
$$+ \mu \int_\Omega \frac{|\mathbb{D}_\mathcal{A}\partial_t u|^2}{2}J + \kappa \int_\Omega |\nabla_\mathcal{A}\partial_t b|^2 J$$
$$= \int_\Omega (\partial_t u \cdot F^1 + \partial_t p F^2 + \partial_t b \cdot F^3)J - \int_\Sigma \partial_t u \cdot F^4 + \int_\Sigma (\partial_t \eta - \sigma \Delta_\star \partial_t \eta)F^5$$
$$\lesssim (\|\partial_t u\|_0 \|F^1\|_0 + \|\partial_t b\|_0 \|F^3\|_0)\|J\|_{L^\infty} + \| + \|\partial_t u\|_{1/2}\|F^4\|_{-1/2}$$
$$+ (\|\partial_t \eta\|_{1/2} + \sigma\|\partial_t \eta\|_{5/2})\|F^5\|_{-1/2} + \int_\Omega \partial_t p F^2 J$$
$$\lesssim \sqrt{\mathcal{E}}\mathcal{D} + \int_\Omega \partial_t p F^2 J.$$

Since there is no temporal derivative on p in $\mathcal{D}$, for the term involving $\partial_t p$, we have to handle it as follow.

$$\int_\Omega \partial_t p F^2 J = \frac{d}{dt}\int_\Omega p F^2 J - \int_\Omega p \partial_t(F^2 J).$$

Then, it follows (4.7) that

$$\frac{d}{dt}(\overline{\mathcal{E}}_{(1,0,0)} - \mathcal{F}) + \overline{\mathcal{D}}_{(1,0,0)} \lesssim \sqrt{\mathcal{E}}\mathcal{D}, \qquad (4.14)$$

where $\overline{\mathcal{E}}_{(1,0,0)}$ and $\overline{\mathcal{D}}_{(1,0,0)}$ are as defined in (4.10).

Next, we consider $\alpha \in \mathbf{N}^{1+2}$ with $|\alpha| \leq 2$ and $\alpha_0 = 0$, that is, no temporal derivatives. In this case, we view (u,b,p,η) in terms of (3.4), which then means that $(v,H,q,h) = (\partial^\alpha u, \partial^\alpha b, \partial^\alpha p, \partial^\alpha \eta)$ satisfy (3.10) with $\Phi^i = \partial^\alpha G^i$ for $i = 1,\cdots,5$, where the nonlinearities G^i are defined by (3.5)-(3.9). we may then apply Lemma 3.2 to see that for $|\alpha| \leq 2$ and $\alpha_0 = 0$ we have the identity

$$\frac{d}{dt}\overline{\mathcal{E}_\alpha} + \overline{\mathcal{D}_\alpha} = \int_\Omega (\partial^\alpha u \cdot \partial^\alpha(G^1 - \nabla G^2) + \partial^\alpha P \partial^\alpha G^2 + \partial^\alpha b \cdot \partial^\alpha G^3)$$
$$- \int_\Sigma \partial^\alpha u \cdot \partial^\alpha G^4 + \int_\Sigma \partial^\alpha \eta \partial^\alpha G^5 - \sigma \int_\Sigma \partial^\alpha G^5 \Delta_\star \partial^\alpha \eta. \qquad (4.15)$$

When $|\alpha| = 2$ and $\alpha_0 = 0$, we denote $\partial^\alpha = \partial^{\beta+\omega}$ for $|\beta| = |\omega| = 1$. Then, we integrate by parts for the terms involve G^1, G^5 in (4.15) to obtain

RHS of (4.15)

$$= \int_\Omega (-\partial^{\alpha+\beta} u \cdot \partial^\omega (G^1 - \nabla G^2) + \partial^\alpha p \partial^\alpha G^2 + \partial^{\alpha+\beta} b \cdot \partial^\omega G^3)$$

$$- \int_\Sigma \partial^\alpha u \cdot \partial^\alpha G^4 - \int_\Sigma \partial^\omega \eta \partial^{\alpha+\beta} G^5 + \sigma \int_\Sigma \partial^{\alpha+\beta} G^5 \Delta_\star \partial^\omega \eta,$$

$$\lesssim \|u\|_3(\|G^1\|_1 + \|G^2\|_2) + \|p\|_2\|G^2\|_2 + \|b\|_3\|G^3\|_1 + \|D^2 u\|_{1/2}\|D^2 G^4\|_{1/2}$$
$$+ \|D^3 G^5\|_{-1/2}(\|D\eta\|_{1/2} + \|D^3 \eta\|_{1/2})$$

$$\lesssim \sqrt{\mathcal{D}}(\|G^1\|_1 + \|G^2\|_2 + \|G^3\|_1 + \|G^4\|_{3/2} + \|G^5\|_{5/2}).$$

According to (4.8), we deduce

$$RHS \ of \ (4.15) \lesssim \sqrt{\mathcal{E}\mathcal{D}},$$

and so we have the inequality

$$\frac{d}{dt} \sum_{|\alpha|=2, \alpha_0=0} \overline{\mathcal{E}}_\alpha + \sum_{|\alpha|=2, \alpha_0=0} \overline{\mathcal{D}}_\alpha \lesssim \sqrt{\mathcal{E}\mathcal{D}}. \tag{4.16}$$

On the other hand, if $|\alpha| < 2$, then we must have that $\alpha_0 = 0$, and we can directly apply Lemma 4.2 to see that

$$\frac{d}{dt} \sum_{|\alpha|<2, \alpha_0=0} \overline{\mathcal{E}}_\alpha + \sum_{|\alpha|<2, \alpha_0=0} \overline{\mathcal{D}}_\alpha \lesssim \sqrt{\mathcal{E}\mathcal{D}}. \tag{4.17}$$

Now, to deduce (4.13) we only need to sum (4.14), (4.15), and (4.17).

4.2.2 Enhanced energy estimates

From the energy-dissipative estimates of Lemma 4.3 we have controlled $\overline{\mathcal{E}}$ and $\overline{\mathcal{D}}$. Our goal now is to show that these can be used to control $\mathcal{E}$ and $\mathcal{D}$ up to some error terms which we will be able to guarantee are small. Here we firstly focus on estimating the energy $\mathcal{E}$.

Lemma 4.4 *Let $\mathcal{E}$ be as defined by (2.1). Suppose that $\mathcal{E} \leqslant \delta$, where $\delta \in (0, 1)$ is the universal constant given in Lemma 3.3. Then, we obtain*

$$\mathcal{E} \lesssim \overline{\mathcal{E}} + \mathcal{E}^2. \tag{4.18}$$

Proof According to the definitions of $\overline{\mathcal{E}}$ and $\mathcal{E}$, in order to prove (4.18), it suffices to prove that

$$\|u\|_2^2 + \|p\|_1^2 + \|b\|_2^2 + \|\partial_t \eta\|_{3/2}^2 + \|\partial_t^2 \eta\|_{-1/2}^2 \lesssim \overline{\mathcal{E}} + \mathcal{E}^2. \tag{4.19}$$

To estimate u and p we will apply the standard Stokes estimates. Now, according to (3.4), we have that

$$\begin{cases} -\Delta u + \nabla p = -\partial_t u + G^1, & \text{in } \Omega, \\ \operatorname{div} v = G^2, & \text{in } \Omega, \\ (pI - \mathbb{D}u)e_3 = (\eta I + \sigma \Delta_\star \eta)e_3 + G^4, & \text{on } \Sigma, \\ u = 0, & \text{on } \Sigma_{-1}, \end{cases} \quad (4.20)$$

and hence we may apply Lemma 6.3 and (4.9) to see that

$$\begin{aligned} \|u\|_2 + \|P\|_1 &\lesssim \|\partial_t u\|_0 + \|G^1\|_0 + \|G^2\|_1 + \|(\eta I + \sigma \Delta_\star \eta)e_3\|_{1/2} + \|G^4\|_{1/2}, \\ &\lesssim \sqrt{\mathcal{E}} + \|G^1\|_0 + \|G^2\|_1 + \|G^4\|_{1/2}, \\ &\lesssim \sqrt{\mathcal{E}} + \mathcal{E}. \end{aligned} \quad (4.21)$$

From this we deduce that the estimates for u, p in (4.19) hold.

Similarly, for estimating b, we have

$$\begin{cases} -\Delta b = -\partial_t b + G^3, & \text{in } \Omega, \\ b = 0, & \text{on } \Sigma, \\ b = 0, & \text{on } \Sigma_{-1}. \end{cases} \quad (4.22)$$

It follows from Lemma 6.2 that

$$\|b\|_2 \lesssim \|\partial_t b\|_0 + \|G^3\|_0 \lesssim \sqrt{\mathcal{E}} + \mathcal{E}.$$

To estimate the $\partial_t \eta$ term in (4.19), we use the fifth equation of (3.4) in conjunction with (4.9) and the usual trace estimates to see that

$$\|\partial_t \eta\|_{3/2} \lesssim \|u_3\|_{3/2} + \|G^5\|_{3/2} \lesssim \|u\|_2 + \mathcal{E} \lesssim \sqrt{\mathcal{E}} + \mathcal{E}.$$

From this we deduce that the estimate for $\partial_t \eta$ in (4.19) holds.

It remains to estimate the $\partial_t^2 \eta$ term in (4.19). We apply a temporal derivative to the fifth equation of (3.4) and integrate against a function $\phi \in H^{\frac{1}{2}}(\Sigma)$ to see that

$$\int_\Sigma \partial_t^2 \eta \phi = \int_\Sigma \partial_t u_3 \phi + \int_\Sigma \partial_t G^5 \phi.$$

Choose an extension $E\phi \in H^1(\Omega)$ with $E\phi|_\Sigma = \phi$, $E\phi|_{\Sigma_{-1}} = \phi$, and $\|E\phi\|_1 \lesssim \|\phi\|_{1/2}$. Then

$$\int_\Sigma \partial_t u_3 \phi = \int_\Omega \partial_t u \cdot \nabla_x E\phi + \int_\Omega \partial_t G^2 E\phi \leqslant (\|\partial_t u\|_0 + \|\partial_t G^2\|_{-1}) \|\phi\|_{1/2},$$

and Lemma 4.2 implies that

$$\|\partial_t^2 \eta\|_{-1/2} \lesssim \|\partial_t u\|_0 + \|\partial_t G^2\|_{-1} + \|\partial_t G^5\|_{-1/2} \lesssim \sqrt{\mathcal{E}} + \mathcal{E}.$$

Then, we complete the estimates in (4.19).

4.2.3 Enhanced dissipate estimates

We now show a corresponding result for the dissipation.

Lemma 4.5 *Let $\mathcal{E}$ and $\mathcal{D}$ be defined by (2.1) and (2.2). Suppose that $\mathcal{E} \leqslant \delta$, where $\delta \in (0, 1)$ is the universal constant given in Lemma 3.3. Then, we deduce*

$$\mathcal{D} \lesssim \overline{\mathcal{D}} + \mathcal{E}\mathcal{D}. \tag{4.23}$$

Proof For the dissipation estimates of u, we apply the Lemma 6.4 to (4.20) with $r = 3$ and $\phi = -\partial_t u + G^1$, $\psi = G^2$, $f_1 = u|_\Sigma$, and $f_2 = 0$ and deduce

$$\|u\|_3 + \|\nabla p\|_1 \lesssim \|-\partial_t u + G^1\|_1 + \|G^2\|_2 + \|u\|_{5/2}. \tag{4.24}$$

We know that
$$\|u\|_1 + \|Du\|_1 + \|D^2 u\|_1 \lesssim \sqrt{\overline{\mathcal{D}}},$$
and so trace theory provides us an estimate
$$\|u\|_{5/2} \lesssim \sqrt{\overline{\mathcal{D}}}.$$

We also have that $\|\partial_t u\|_1 \lesssim \sqrt{\overline{\mathcal{D}}}$, and Lemma 4.2 tells that
$$\|G^1\|_1 + \|G^2\|_2 \lesssim \sqrt{\mathcal{E}\mathcal{D}}.$$

Then, we bring the above estimates into (4.24) to complete the dissipation estimates of u, that is,

$$\|u\|_3 + \|\nabla p\|_1 \lesssim \sqrt{\overline{\mathcal{D}}} + \sqrt{\mathcal{E}\mathcal{D}}. \tag{4.25}$$

For the b dissipative estimate, we directly apply the elliptic estimates to (4.22) to know

$$\|b\|_3 \lesssim \|\partial_t b\|_1 + \|G^3\|_1 \lesssim \sqrt{\overline{\mathcal{D}}} + \|G^3\|_1 \lesssim \sqrt{\overline{\mathcal{D}}} + \sqrt{\mathcal{E}\mathcal{D}}. \tag{4.26}$$

We now turn to the η estimates. For $\alpha \in \mathbf{N}^2$ and $|\alpha| = 1$, we apply ∂^α to the fourth equation for (3.4) to obtain

$$(1 - \sigma\Delta_*)\partial^\alpha \eta = \partial^\alpha p - \partial_3 \partial^\alpha u_3 - \partial^\alpha G_3^4. \tag{4.27}$$

Then the elliptic estimates, the trace estimates and (4.25) imply that

$$\|D\eta\|_{5/2} = \sum_{|\alpha|=1} \|\partial^\alpha \eta\|_{5/2} \lesssim \sum_{|\alpha|=1} \|\partial^\alpha p - \partial_3 \partial^\alpha u_3 - \partial^\alpha G_3^4\|$$

$$\lesssim \|\nabla p\|_1 + \|u\|_3 + \|G^4\|_{3/2} \lesssim \sqrt{\overline{\mathcal{D}}} + \sqrt{\mathcal{ED}}. \quad (4.28)$$

According to the zero average condition for η and by using the Poincaré inequality, we deduce

$$\|\eta\|_0 \leqslant \|D\eta\|_0, \quad (4.29)$$

then, (4.28) and (4.29) reveal that

$$\|\eta\|_{7/2} \lesssim \|\eta\|_0 + \|D\eta\|_{5/2} \lesssim \|D\eta\|_{5/2} \lesssim \sqrt{\overline{\mathcal{D}}} + \sqrt{\mathcal{ED}}. \quad (4.30)$$

For the $\partial_t \eta$ estimates, we use the fifth equation of (3.4), (4.8) and (4.25) to know

$$\|\partial_t \eta\|_{5/2} \lesssim \|u_3\|_{5/2} + \|G^5\|_{5/2} \lesssim \|u\|_3 + \|G^5\|_{5/2} \lesssim \sqrt{\overline{\mathcal{D}}} + \sqrt{\mathcal{ED}}, \quad (4.31)$$

and

$$\|\partial_t^2 \eta\|_{1/2} \lesssim \|\partial_t u_3\|_{1/2} + \|\partial_t G^5\|_{1/2} \lesssim \|\partial_t u\|_1 + \|\partial_t G^5\|_{1/2} \lesssim \sqrt{\overline{\mathcal{D}}} + \sqrt{\mathcal{ED}}. \quad (4.32)$$

Now we complete the estimate of the pressure by obtaining a bound for $\|P\|_0$. To this end we combine the estimates (4.24) and (4.30) with the Stokes estimate Lemma 6.3 with $\phi = -\partial_t u + G^1, \psi = G^2$, and $\alpha = (\eta I - \sigma \Delta_\star \eta)e_3 + G^3 e_3$ to bound

$$\|u\|_3 + \|P\|_2 \lesssim \|-\partial_t u + G^1\|_1 + \|G^2\|_2 + \|(\eta I - \sigma \delta_\star \eta)e_3\|_{3/2},$$
$$\lesssim \|\partial_t u\|_1 + \|G^1\|_1 + \|G^2\|_2 + \|\eta\|_{7/2},$$
$$\lesssim \sqrt{\overline{\mathcal{D}}} + \sqrt{\mathcal{ED}}.$$

Thus,

$$\|P\|_2 \lesssim \sqrt{\overline{\mathcal{D}}} + \sqrt{\mathcal{ED}}. \quad (4.33)$$

Finally, (4.23) follows from (4.25),(4.26),(4.30)∼(4.33).

4.3 Proof of Theorem 2.1

We now combine the estimates of the previous section in order to deduce our primary a priori estimates for solutions. It shows that under a smallness condition on the energy, the energy decays exponentially and the dissipation integral is bounded by the initial data.

Theorem 4.1 *Suppose that (u, b, p, η) solves (1.13) on the temporal interval $[0, T]$. Let $\mathcal{E}$ and $\mathcal{D}$ be as defined in (2.1) and (2.2). Then there exists universal constant $0 < \delta_\star < \delta$, where $\delta \in (0, 1)$ is the universal constant given in Lemma 3.3, such that if*

$$\sup_{0 \leqslant t \leqslant T} \mathcal{E}(t) \leqslant \delta_\star,$$

then
$$\sup_{0\leqslant t\leqslant T} e^{\lambda t}\mathcal{E}(t)+\int_0^T \mathcal{D}(t)dt \lesssim \mathcal{E}(0), \tag{4.34}$$

for all $t \in [0,T]$, where $\lambda > 0$ is a universal constant.

Proof According to the definition of $\bar{\mathcal{E}}$ and $\bar{\mathcal{D}}$, Lamma 4.4 and Lamma 4.5, we find
$$\bar{\mathcal{E}} \leqslant \mathcal{E} \lesssim \bar{\mathcal{E}}, \quad \text{and} \bar{\mathcal{D}} \leqslant \mathcal{D} \lesssim \bar{\mathcal{D}}, \tag{4.35}$$

as δ_* small enough.

Furthermore, by substituting (4.35) into Lamma 4.3, one has
$$\frac{d}{dt}(\mathcal{E}-\mathcal{F})+\mathcal{D} \leqslant 0, \tag{4.36}$$

as δ_* small enough. Moreover, the estimates in (4.7) tell us that $|\mathcal{F}| \leqslant \mathcal{E}^{3/2} \leqslant \sqrt{\mathcal{E}}\mathcal{E}$, hence
$$\frac{d}{dt}\mathcal{E}+\mathcal{D} \leqslant 0. \tag{4.37}$$

On the one hand, we integrate (4.37) in time over $(0,T)$ to obtain that
$$C\int_0^T \mathcal{D}(t)dt \leqslant \mathcal{E}(T)+C\int_0^T \mathcal{D}(t)dt \leqslant \mathcal{E}(0). \tag{4.38}$$

On the other hand, obviously, we have the bound $\mathcal{E} \leqslant \mathcal{D}$, then we obtain
$$\frac{d}{dt}\mathcal{E}+\mathcal{E} \leqslant 0. \tag{4.39}$$

Then, by using Gronwall's inequality, we complete the proof of (4.34).

5 For the case $\sigma = 0$

5.1 Nonlinear estimates

We will employ the form (3.1) to study the temporal derivatives of solutions for (1.13). That is, we apply ∂^α to (1.13) to deduce that the new functions
$$(v,\ H,\ q,\ h) = (\partial^\alpha u, \partial^\alpha b, \partial^\alpha p, \partial^\alpha \eta),$$

satisfy (3.1) for certain terms F^i for $\partial^\alpha = \partial_t^{\alpha_0}$ with $\alpha_0 \leqslant 2N$. Below we record the form of these forcing terms $F^i, i = 1,2,3,4,5$ for this particular problem, where $F^1 = \sum_{l=1}^7 F^{1,l}$, for
$$F_i^{1,1} := \sum_{0<\beta<\alpha} C_{\alpha,\beta}\partial^\beta(\partial_t\bar{\eta}\tilde{b}K)\partial^{\alpha-\beta}\partial_3 u_i + \sum_{0<\beta\leqslant\alpha} C_{\alpha,\beta}\partial^{\alpha-\beta}\partial_t\bar{\eta}\partial^\beta(\tilde{b}K)\partial_3 u_i,$$

$$F_i^{1,2} := - \sum_{0<\beta\leqslant\alpha} C_{\alpha,\beta}(\partial^\beta(u_j\mathcal{A}_{jk})\partial^{\alpha-\beta}\partial_k u_i + \partial^\beta \mathcal{A}_{ik}\partial^{\alpha-\beta}\partial_k p),$$

$$F_i^{1,3} := \sum_{0<\beta\leqslant\alpha} C_{\alpha,\beta}\partial^\beta \mathcal{A}_{jl}\partial^{\alpha-\beta}\partial_l(\mathcal{A}_{im}\partial_m u_j + \mathcal{A}_{jm}\partial_m u_i),$$

$$F_i^{1,4} := \sum_{0<\beta\leqslant\alpha} C_{\alpha,\beta}\mathcal{A}_{jk}\partial_k(\partial_\beta \mathcal{A}_{il}\partial^{\alpha-\beta}\partial_l u_j + \partial^\beta \mathcal{A}_{jl}\partial^{\alpha-\beta}\partial_l u_i), \quad (5.1)$$

$$F_i^{1,5} := \partial^\alpha \partial_t \bar{\eta}\tilde{b}K\partial_3 u_i \quad \text{and} \quad F_i^{1,6} := \mathcal{A}_{jk}\partial_k(\partial^\alpha \mathcal{A}_{il}\partial_l u_j + \partial^\alpha \mathcal{A}_{jl}\partial_l u_i),$$

$$F_i^{1,7} := \sum_{0<\beta\leqslant\alpha} C_{\alpha,\beta}\partial^\beta[(b_j + \bar{B}_j)\mathcal{A}_{jk}]\partial^{\alpha-\beta}(\partial_k b_i).$$

$$F^{2,1} := - \sum_{0<\beta<\alpha} C_{\alpha,\beta}\partial^\beta \mathcal{A}_{ij}\partial^{\alpha-\beta}\partial_j u_i, \quad \text{and} \quad F^{2,2} = -\partial^\alpha \mathcal{A}_{ij}\partial_j u_i. \quad (5.2)$$

$$F_i^{3,1} := \sum_{0<\beta<\alpha} C_{\alpha,\beta}\partial^\beta(\partial_t\bar{\eta}\tilde{b}K)\partial^{\alpha-\beta}\partial_3 b_i + \sum_{0<\beta\leqslant\alpha} C_{\alpha,\beta}\partial^{\alpha-\beta}\partial_t\bar{\eta}\partial^\beta(\tilde{b}K)\partial_3 b_i,$$

$$F_i^{3,2} := - \sum_{0<\beta\leqslant\alpha} C_{\alpha,\beta}\partial^\beta(u_j\mathcal{A}_{jk})\partial^{\alpha-\beta}\partial_k b_i,$$

$$F_i^{3,3} := \sum_{0<\beta\leqslant\alpha} C_{\alpha,\beta}\partial^\beta \mathcal{A}_{jl}\partial^{\alpha-\beta}(\mathcal{A}_{jm}\partial_m b_i), \quad (5.3)$$

$$F_i^{3,4} := - \sum_{0<\beta\leqslant\alpha} C_{\alpha,\beta}\partial^\beta(u_j\mathcal{A}_{jk})\partial^{\alpha-\beta}\partial_k b_i,$$

$$F_i^{3,5} := \partial^\alpha \partial_t\eta\tilde{b}K\partial_3 b_i \quad \text{and} \quad F_i^{3,6} := \mathcal{A}_{jk}\partial^\alpha \mathcal{A}_{jl}\partial_l b_i,$$

$$F_i^{3,7} := \sum_{0<\beta\leqslant\alpha} C_{\alpha,\beta}\partial^\beta[(b_j + \bar{B}_j)\mathcal{A}_{jk}]\partial^{\alpha-\beta}(\partial_k u_i).$$

$F_i^4 = F_i^{4,1} + F_i^{4,2}$, we have

$$F_i^{4,1} := -(\sum_{0<\beta\leqslant\alpha} C_{\alpha,\beta}\partial^\beta D\eta(\partial^{\alpha-\beta}\eta - \partial^{\alpha-\beta}p),$$

$$F_i^{4,2} := \sum_{0<\beta\leqslant\alpha} C_{\alpha,\beta}(\partial^\beta(\mathcal{N}_j\mathcal{A}_{im})\partial^{\alpha-\beta}\partial_m u_j + \partial^\beta(\mathcal{N}_j\mathcal{A}_{jm})\partial^{\alpha-\beta}\partial_m u_i), (5.4)$$

$$F^5 := - \sum_{0<\beta\leqslant\alpha} C_{\alpha,\beta}\partial^\beta D\eta \cdot \partial^{\alpha-\beta}u. \quad (5.5)$$

Now we present the estimates for F^i ($i = 1,\cdots,5$), when $\partial^\alpha = \partial_t^{\alpha_0}$ for $\alpha_0 \leqslant n$.

Lemma 5.1 Let F^i ($i = 1,\cdots,5$) be defined by (5.1)-(5.5), and let $\partial^\alpha = \partial_t^{\alpha_0}$ with $\alpha_0 \leqslant n$ for $n = 2N$ or $n = N+2$. Then, we have

$$\|F^1\|_0^2 + \|F^2\|_0^2 + \|\partial_t(JF^2)\|_0^2 + \|F^3\|_0^2 + \|F^4\|_0^2 + \|F^5\|_0^2 \lesssim \mathcal{E}_{2N}\mathcal{D}_n, \quad (5.6)$$

and
$$\|F^2\|_0^2 \lesssim \mathcal{E}_{2N}\mathcal{E}_n. \tag{5.7}$$

Proof Firstly, we consider the estimates for F^1. Note that each term in F^1 is at least quadratic, and each such term can be written in the form XY, where X involves fewer derivative counts than Y. We may apply the usual Sobolev embeddings Lemmas along with the definitions of $\mathcal{E}_{2N}$ and $\mathcal{D}_n$ to estimate $\|X\|_{L^\infty}^2 \lesssim \mathcal{E}_{2N}$ and $\|Y\|_0^2 \lesssim \mathcal{D}_n$. Hence $\|XY\|_0^2 \leq \|X\|_{L^\infty}^2 \|Y\|_0^2 \lesssim \mathcal{E}_{2N}\mathcal{D}_n$. The estimates of F^2, F^3 and (5.7) are similar. A similar argument also employing trace estimates obtain the estimates of F^4 and F^5. The same argument also works for $\partial_t(JF^{2,1})$. To bound $\partial_t(JF^{2,2})$ for $\alpha_0 = n$ we have to estimate $\|\partial_t^{n+1}\mathcal{A}\|_0^2 \lesssim \mathcal{D}_n$, but this is possible due to Lemma 3.4. Then a similar splitting into L^∞ and H^0 estimates shows that $\|\partial_t(JF^{2,2})\| \lesssim \mathcal{E}_{2N}\mathcal{D}_n$, and then we complete the proof of (5.6).

Now, for the case $\sigma = 0$, we first estimate the G^i terms defined in (3.5)-(3.11) at the $2N$ level.

Lemma 5.2 Let $G^1, \cdots, G^5$ be defined in (3.5)-(3.11). There exists a $\theta > 0$ such that,

$$\left\|\bar{\nabla}_0^{4N-2}G^1\right\|_0^2 + \left\|\bar{\nabla}_0^{4N-2}G^2\right\|_1^2 + \left\|\bar{\nabla}_0^{4N-2}G^3\right\|_0^2$$
$$+ \left\|\bar{D}_0^{4N-2}G^4\right\|_{1/2}^2 \lesssim \mathcal{E}_{2N}^{1+\theta}, \tag{5.8}$$

$$\left\|\bar{\nabla}_0^{4N-2}G^1\right\|_0^2 + \left\|\bar{\nabla}_0^{4N-2}G^2\right\|_1^2 + \left\|\bar{\nabla}_0^{4N-2}G^3\right\|_0^2 + \left\|\bar{D}_0^{4N-2}G^4\right\|_{1/2}^2$$
$$+ \left\|\bar{D}_0^{4N-2}G^5\right\|_{1/2}^2 + \left\|\bar{\nabla}^{4N-3}\partial_t G^1\right\|_0^2 + \left\|\bar{\nabla}^{4N-3}\partial_t G^2\right\|_1^2 + \left\|\bar{\nabla}^{4N-3}\partial_t G^3\right\|_0^2$$
$$+ \left\|\bar{D}^{4N-3}\partial_t G^4\right\|_{1/2}^2 + \left\|\bar{D}^{4N-2}\partial_t G^5\right\|_{1/2}^2 \lesssim \mathcal{E}_{2N}^\theta \mathcal{D}_{2N}, \tag{5.9}$$

and

$$\|\nabla^{4N-1}G^1\|_0^2 + \|\nabla^{4N-1}G^2\|_1^2 + \|\nabla^{4N-1}G^3\|_0^2 + \|D^{4N-1}G^4\|_{1/2}^2$$
$$+ \|D^{4N-1}G^5\|_{1/2}^2 \lesssim \mathcal{E}_{2N}^\theta \mathcal{D}_{2N} + \mathcal{E}_{N+2}\mathcal{F}_{2N}. \tag{5.10}$$

Proof These estimates can be proved similar as [[8], Theorem 3.3].

Similarly, we can obtain the estimates of G^i defined by (3.5)-(3.11) at the $N+2$ level as $\sigma = 0$.

Lemma 5.3 Let $G^1, \cdots, G^5$ be defined by (3.5)-(3.11). There exists a $\theta > 0$ such that,

$$\left\|\bar{\nabla}_0^{2(N+2)-2} G^1\right\|_0^2 + \left\|\bar{\nabla}_0^{2(N+2)-2} G^2\right\|_1^2 + \left\|\bar{\nabla}_0^{2(N+2)-2} G^3\right\|_0^2$$
$$+ \left\|\bar{D}_0^{2(N+2)-2} G^4\right\|_{1/2}^2 \lesssim \mathcal{E}_{2N}^\theta \mathcal{E}_{N+2}, \tag{5.11}$$

and

$$\left\|\bar{\nabla}_0^{2(N+2)-1} G^1\right\|_0^2 + \left\|\bar{\nabla}_0^{2(N+2)-1} G^2\right\|_1^2 + \left\|\bar{\nabla}_0^{2(N+2)-1} G^3\right\|_0^2 + \left\|\bar{D}^{2(N+2)-1} G^4\right\|_{1/2}^2$$
$$+ \left\|\bar{D}_0^{2(N+2)-1} G^5\right\|_{1/2}^2 + \left\|\bar{D}^{2(N+2)-2} \partial_t G^5\right\|_{1/2}^2 \lesssim \mathcal{E}_{2N}^\theta \mathcal{D}_{N+2}. \tag{5.12}$$

5.2 Energy evolution

We define the temporal energy and dissipation, respectively, as

$$\bar{\mathcal{E}}_n^0 := \sum_{j=0}^n (\left\|\sqrt{J}\partial_t^j u\right\|_0^2 + \left\|\sqrt{J}\partial_t^j b\right\|_0^2 + \left\|\partial_t^j \eta\right\|_0^2), \tag{5.13}$$

$$\bar{\mathcal{D}}_n^0 := \sum_{j=0}^n (\left\|\mathbb{D}\partial_t^j u\right\|_0^2 + \left\|\nabla \partial_t^j b\right\|_0^2). \tag{5.14}$$

Then, we define the horizontal energies and dissipation, respectively, as

$$\bar{\mathcal{E}}_n := \left\|\bar{D}_0^{2n-1} u\right\|_0^2 + \left\|D\bar{D}^{2n-1} u\right\|_0^2 + \left\|\bar{D}_0^{2n-1} b\right\|_0^2$$
$$+ \left\|D\bar{D}^{2n-1} b\right\|_0^2 + \left\|\bar{D}^{2n-1} \eta\right\|_0^2 + \left\|D\bar{D}^{2n-1} \eta\right\|_0^2, \tag{5.15}$$

and

$$\bar{\mathcal{D}}_n := \left\|\bar{D}_0^{2n-1}\mathbb{D}(u)\right\|_0^2 + \left\|D\bar{D}^{2n-1}\mathbb{D}(u)\right\|_0^2 + \left\|\bar{D}_0^{2n-1}\nabla b\right\|_0^2 + \left\|D\bar{D}^{2n-1}\nabla b\right\|_0^2. \tag{5.16}$$

5.2.1 Energy Evolution of temporal derivatives

First, we present the temporal derivatives estimates at $2N$ level.

Lemma 5.4 There exist a $\theta > 0$ so that

$$\bar{\mathcal{E}}_{2N}^0(t) + \int_0^t \bar{\mathcal{D}}_{2N}^0 \lesssim \mathcal{E}_{2N}(0) + (\mathcal{E}_{2N}(t))^{3/2} + \int_0^t (\mathcal{E}_{2N})^\theta \mathcal{D}_{2N}. \tag{5.17}$$

Proof We apply $\partial^\alpha = \partial_t^{\alpha_0}$ with $0 \leqslant |\alpha| \leqslant 2N$ to (1.13) and let $v = \partial_t^{\alpha_0} u$, $q = \partial_t^{\alpha_0} p$, $H = \partial_t^{\alpha_0} b$, $h = \partial_t^{\alpha_0} \eta$ satisfy (3.1). Then, according to Lemma 3.1 and integrating in time from 0 to t, we deduce

$$\int_\Omega \left(\frac{|\partial_t^{\alpha_0} u|^2}{2} + \frac{|\partial_t^{\alpha_0} b|^2}{2}\right) J + \int_\Sigma \frac{|\partial_t^{\alpha_0} \eta|^2}{2} + \int_0^t \int_\Omega \left(\frac{|\mathbb{D}_\mathcal{A} \partial_t^{\alpha_0} u|^2}{2} + |\nabla_\mathcal{A} \partial_t^{\alpha_0} b|^2\right) J$$
$$= \int_\Omega \left(\frac{|\partial_t^{\alpha_0} u(0)|^2}{2} + \frac{|\partial_t^{\alpha_0} b(0)|^2}{2}\right) J + \int_\Sigma \frac{|\partial_t^{\alpha_0} \eta(0)|^2}{2} + \int_0^t \int_\Sigma (-\partial_t^{\alpha_0} u \cdot F^4 + \partial_t^{\alpha_0} \eta F^5)$$
$$+ \int_0^t \int_\Omega (\partial_t^{\alpha_0} u \cdot F^1 + \partial_t^{\alpha_0} p F^2 + \partial_t^{\alpha_0} b \cdot F^3) J.$$

Next, we will estimate the right hand side terms involving F^i of the above equation. For the F^1 term, according to Lemma 3.3 and Lemma 5.1, one has

$$\int_0^t \int_\Omega \partial_t^{\alpha_0} u \cdot F^1 J \lesssim \int_0^t \|\partial_t^{\alpha_0} u\|_0 \|J\|_{L^\infty} \|F^1\|_0$$
$$\lesssim \int_0^t \sqrt{\mathcal{D}_{2N}} \sqrt{\mathcal{E}_{2N} \mathcal{D}_{2N}} = \int_0^t \sqrt{\mathcal{E}_{2N}} \mathcal{D}_{2N}.$$

Similarly,

$$\int_0^t \int_\Omega \partial_t^{\alpha_0} u \cdot F^3 \lesssim \int_0^t \sqrt{\mathcal{E}_{2N}} \mathcal{D}_{2N}, \tag{5.18}$$

and

$$\int_\Sigma (-\partial_t^{\alpha_0} u \cdot F^4 + \partial_t^{\alpha_0} \eta F^5) \lesssim \int_0^t (\|\partial_t^{\alpha_0} u\|_{H^0(\Sigma)} \|F^4\|_0 + \|\partial_t^{\alpha_0} \eta\|_0 \|F^5\|_0)$$
$$\lesssim \int_0^t \sqrt{\mathcal{D}_{2N}} \sqrt{\mathcal{E}_{2N} \mathcal{D}_{2N}} = \int_0^t \sqrt{\mathcal{E}_{2N}} \mathcal{D}_{2N}. \tag{5.19}$$

For the term $\partial_t^{\alpha_0} p F^2$, when $\alpha_0 = 2N$, there is one more time derivative on p than can be controlled by $\mathcal{D}_{2N}$. Hence, we have to consider the cases $\alpha_0 < 2N$ and $\alpha_0 = 2N$ separately. In the case $\alpha_0 = 2N$, we have

$$\int_0^t \int_\Omega \partial_t^{2N} p F^2 = -\int_0^t \int_\Omega \partial_t^{2N-1} p \partial_t(JF^2) + \int_\Omega (\partial_t^{2N-1} p J F^2)(t)$$
$$- \int_\Omega (\partial_t^{2N-1} p J F^2)(0). \tag{5.20}$$

According to Lemma 5.1, one has

$$-\int_0^t \int_\Omega \partial_t^{2N-1} p \partial_t(JF^2) \lesssim \int_0^t \|\partial_t^{2N-1} p\|_0 \|\partial_t(JF^2)\|_0$$
$$\lesssim \int_0^t \sqrt{\mathcal{D}_{2N}} \sqrt{\mathcal{E}_{2N} \mathcal{D}_{2N}} = \int_0^t \sqrt{\mathcal{E}_{2N}} \mathcal{D}_{2N}. \tag{5.21}$$

Then, it follows from (5.7) and Lemma 3.3 that

$$\int_\Omega (\partial_t^{2N-1} pJF^2)(t) \lesssim \|\partial_t^{2N} p\|_0 \|F^2\|_0 \|J\|_{L^\infty} \lesssim (\mathcal{E}_{2N})^{3/2}. \tag{5.22}$$

Combining the estimates (5.21) and (5.22), we obtain

$$\int_0^t \int_\Omega \partial_t^{2N} pF^2 J \lesssim \mathcal{E}_{2N}(0) + (\mathcal{E}_{2N})^{3/2} + \int_0^t \sqrt{\mathcal{E}_{2N}} \mathcal{D}_{2N}. \tag{5.23}$$

In the other case $0 \leqslant \alpha_0 < 2N$, by using (5.6), we directly have

$$\int_0^t \int_\Omega \partial_t^{\alpha_0} pF^2 J \lesssim \int_0^t \|\partial_t^{\alpha_0} p\| \|F^2\|_0 \lesssim \int_0^t \sqrt{\mathcal{D}_{2N}} \sqrt{\mathcal{E}_{2N} \mathcal{D}_{2N}} = \int_0^t \sqrt{\mathcal{E}_{2N}} \mathcal{D}_{2N}. \tag{5.24}$$

Furthermore, according to Lemma 3.3, we can easily deduce

$$\int_0^t \int_\Omega \frac{|\mathbb{D} \partial_t^{\alpha_0} u|^2}{2} J \lesssim \int_0^t \int_\Omega \frac{|\mathbb{D}_\mathcal{A} \partial_t^{\alpha_0} u|^2}{2} J + \int_0^t \sqrt{\mathcal{E}_{2N}} \mathcal{D}_{2N}, \tag{5.25}$$

and

$$\int_0^t \int_\Omega |\nabla \partial_t^{\alpha_0} b|^2 J \lesssim \int_0^t \int_\Omega |\nabla_\mathcal{A} \partial_t^{\alpha_0} b|^2 J + \int_0^t \sqrt{\mathcal{E}_{2N}} \mathcal{D}_{2N}. \tag{5.26}$$

Therefore, we complete the proof of Lemma 5.4.

Now, we present the corresponding estimates at the $N+2$ level.

Lemma 5.5 *In the case $0 \leqslant \alpha_0 \leqslant N+2$, we have*

$$\partial_t(\bar{\mathcal{E}}_{N+2}^0 - 2\int_\Omega \partial_t^{N+1} pF^2 J) + \bar{\mathcal{D}}_{N+2}^0 \lesssim \sqrt{\mathcal{E}_{2N}} \mathcal{D}_{N+2}. \tag{5.27}$$

Proof The proof of Lemma 5.5 is similar to Lemma 5.4. Here, for brevity, we omit the proof.

5.2.2 Energy evolution of horizontal derivatives

In this subsection, we will show how the horizontal energies evolve at the $2N$ and $N+2$ level, respectively.

Lemma 5.6 *Let $\alpha \in \mathbf{N}^{1+2}$, $0 \leqslant \alpha_0 \leqslant 2N-1$ and $|\alpha| \leqslant 4N$. Then, there exists a $\theta > 0$ so that*

$$\bar{\mathcal{E}}_{2N}(t) + \int_0^t \bar{\mathcal{D}}_{2N} \lesssim \mathcal{E}_{2N}(0) + \int_0^t (\mathcal{E}_{2N})^\theta \mathcal{D}_{2N} + \int_0^t \sqrt{\mathcal{D}_{2N} \mathcal{F}_{2N} \mathcal{E}_{N+2}}. \tag{5.28}$$

Proof We apply ∂^α ($\alpha \in \mathbf{N}^{1+2}, 0 \leqslant \alpha_0 \leqslant 2N-1$ and $|\alpha| \leqslant 4N$) to (3.4) and let $v = \partial_t^{\alpha_0} u$, $q = \partial_t^{\alpha_0} p$, $H = \partial_t^{\alpha_0} b$, $h = \partial_t^{\alpha_0} \eta$ satisfy (3.10) with $\Phi^i = \partial^\alpha G_i$ ($i =$

$1, \cdots, 5$). Then, according to Lemma 3.2 and integrating in time from 0 to t, we obtain

$$\partial_t \int_\Omega \left(\frac{|\partial^\alpha u|^2}{2} + \frac{|\partial^\alpha \eta|^2}{2} + \frac{|\partial^\alpha b|^2}{2} \right) + \int_\Omega \left(\frac{|\mathbb{D}\partial^\alpha u|^2}{2} + |\nabla \partial^\alpha b|^2 \right)$$
$$= \int_\Omega \partial^\alpha u \cdot (\partial^\alpha G^1 - \nabla \partial^\alpha G^2) + \int_\Omega \partial^\alpha p \partial^\alpha G^2 + \int_\Omega \partial^\alpha b \cdot \partial^\alpha G^3$$
$$+ \int_\Sigma (-\partial^\alpha u \cdot \partial^\alpha G^4 + \partial^\alpha \eta \cdot \partial^\alpha G^5). \qquad (5.29)$$

We first consider the case $0 \leqslant \alpha_0 \leqslant 2N-1$, $|\alpha| \leqslant 4N-1$. According to (5.8), one has

$$\int_\Omega \partial^\alpha u \cdot (\partial^\alpha G^1 - \nabla \partial^\alpha G^2) + \int_\Omega \partial^\alpha p \partial^\alpha G^2 + \int_\Omega \partial^\alpha b \cdot \partial^\alpha G^3$$
$$\lesssim \|\partial^\alpha u\|_0 (\|\partial^\alpha G^1\|_0 + \|\partial^\alpha G^2\|_1) + \|\partial^\alpha b\|_0 \|\partial^\alpha G^3\|_0 + \|\partial^\alpha p\|_0 \|\partial^\alpha G^2\|_0$$
$$\lesssim \sqrt{\mathcal{D}_{2N}} \sqrt{\mathcal{E}_{2N}^\theta \mathcal{D}_{2N} + \mathcal{E}_{N+2} \mathcal{F}_{2N}} \lesssim \mathcal{E}_{2N}^{\theta/2} \mathcal{D}_{2N} + \sqrt{\mathcal{D}_{2N} \mathcal{F}_{2N} \mathcal{E}_{N+2}}. \qquad (5.30)$$

It follows from the trace estimate $\|\partial^\alpha u\|_{H^0(\Sigma)} \lesssim \|\partial^\alpha u\|_1 \lesssim \sqrt{\mathcal{D}_{2N}}$ that

$$\int_\Sigma (-\partial^\alpha u \cdot \partial^\alpha G^4 + \partial^\alpha \eta \cdot \partial^\alpha G^5) \lesssim \|\partial^\alpha u\|_{H^0(\Sigma)} \|\partial^\alpha G^4\|_0 + \|\partial^\alpha \eta\|_0 \|\partial^\alpha G^5\|_0$$
$$\lesssim \sqrt{\mathcal{D}_{2N}} \sqrt{\mathcal{E}_{2N}^\theta \mathcal{D}_{2N} + \mathcal{E}_{N+2} \mathcal{F}_{2N}} \lesssim \mathcal{E}_{2N}^{\theta/2} \mathcal{D}_{2N} + \sqrt{\mathcal{D}_{2N} \mathcal{F}_{2N} \mathcal{E}_{N+2}}. \qquad (5.31)$$

For the other case $|\alpha| = 4N$, since $\alpha_0 \leqslant 2N-1$, we can write $\alpha = \beta + (\alpha - \beta)$, where $\eta \in \mathbf{N}^2$ satisfies $|\beta| = 1$. Hence, ∂^α involves at least one spatial derivative. Since $|\alpha - \beta| = 4N-1$, we integrate by parts and use (5.8) to find that

$$\left| \int_\Omega \partial^\alpha u \cdot (\partial^\alpha G^1 - \nabla \partial^\alpha G^2) + \int_\Omega \partial^\alpha b \cdot \partial^\alpha G^3 \right| + \left| \int_\Omega \partial^\alpha p \partial^\alpha G^2 \right|$$
$$\lesssim \left| \int_\Omega \partial^{\alpha+\beta} u \cdot \partial^{\alpha-\beta} G^1 + \partial^{\alpha+\beta} b \cdot \partial^{\alpha-\beta} G^3 - \partial^{\alpha+\beta} u \cdot \nabla \partial^{\alpha-\beta} G^2 \right|$$
$$+ \left| \int_\Omega \partial^\alpha p \partial^{\alpha-\beta+\beta} G^2 \right|$$
$$\lesssim \|\partial^{\alpha+\beta} u\|_0 (\|\partial^{\alpha-\beta} G^1\|_0 + \|\partial^{\alpha-\beta} G^2\|_1) + \|\partial^{\alpha+\beta} b\|_0 \|\partial^{\alpha-\beta} G^3\|_0$$
$$+ \|\partial^\alpha p\|_0 \|\bar{\nabla}^{4N-1} G^2\|_1$$
$$\lesssim \mathcal{E}_{2N}^{\theta/2} \mathcal{D}_{2N} + \sqrt{\mathcal{D}_{2N} \mathcal{F}_{2N} \mathcal{E}_{N+2}}. \qquad (5.32)$$

Integrating by parts and using the trace theorem to find that

$$\int_\Sigma \partial^\alpha u \partial^\alpha G^4 = \left| \int_\Sigma \partial^{\alpha+\beta} u \partial^{\alpha-\beta} G^4 \right| \lesssim \|\partial^{\alpha+\beta} u\|_{H^{-1/2}} \|\partial^{\alpha-\beta} G^4\|_{1/2}$$

$$\lesssim \|\partial^\alpha u\|_{H^{1/2}(\Sigma)} \|\bar{D}^{4N-1}G^4\|_{1/2} \lesssim \|\partial^\alpha u\|_1 \|\bar{D}^{4N-1}G^4\|_{1/2}$$
$$\lesssim \mathcal{E}_{2N}^{\theta/2}\mathcal{D}_{2N} + \sqrt{\mathcal{D}_{2N}\mathcal{F}_{2N}\mathcal{E}_{N+2}}. \tag{5.33}$$

For the term involves η, we need to apart it into two cases $\alpha_0 \geqslant 1$ and $\alpha_0 = 0$. In the former case, there is at least one temporal derivative in ∂^α. Thus, we have
$$\|\partial^\alpha \eta\|_{1/2} \lesssim \|\partial^{\alpha-2}\partial_t \eta\|_{1/2} \lesssim \sqrt{\mathcal{D}_{2N}}.$$

Then, integrating by parts, one has
$$\int_\Sigma \partial^\alpha \eta \partial^\alpha G^5 \lesssim \left|\int_\Sigma \partial^{\alpha+\beta}\eta \partial^{\alpha-\beta}G^5\right| \lesssim \|\partial^{\alpha+\beta}\eta\|_{-1/2}\|\partial^{\alpha-\beta}G^5\|_{1/2}$$
$$\lesssim \|\partial^\alpha \eta\|_{1/2}\|\partial^{\alpha-\beta}G^5\|_{1/2} \lesssim \mathcal{E}_{2N}^{\theta/2}\mathcal{D}_{2N} + \sqrt{\mathcal{D}_{2N}\mathcal{F}_{2N}\mathcal{E}_{N+2}}. \tag{5.34}$$

In the other case, $\alpha_0 = 0$, ∂^α involves only spatial derivatives, we need to analyze in detail. Firstly, we denote
$$-\partial^\alpha G^5 = \partial^\alpha(D\eta \cdot u) = D\partial^\alpha \cdot u + \sum_{0<\beta\leqslant\alpha,|\beta|=1} C_{\alpha,\beta}D\partial^{\alpha-\beta}\eta \cdot \partial^\beta u$$
$$+ \sum_{0<\beta\leqslant\alpha,|\beta|\geqslant 2} C_{\alpha,\beta}D\partial^{\alpha-\beta}\eta \cdot \partial^\beta u$$
$$= I_1 + I_2 + I_3.$$

By integrating by parts, one has
$$\left|\int_\Sigma \partial^\alpha \eta I_1\right| = \frac{1}{2}\left|\int_\Sigma D|\partial^\alpha \eta|^2 \cdot u\right| = \frac{1}{2}\left|\int_\Sigma \partial^\alpha \eta \partial^\alpha \eta(\partial_1 u_1 + \partial_2 u_2)\right|$$
$$\lesssim \|\partial^\alpha \eta\|_{1/2}\|\partial^\alpha \eta\|_{-1/2}\|\partial_1 u_1 + \partial_2 u_2\|_{L^\infty}$$
$$\lesssim \|\eta\|_{4N+1/2}\|D\eta\|_{4N-3/2}\mathcal{E}_{N+2}$$
$$\lesssim \sqrt{\mathcal{D}_{2N}\mathcal{F}_{2N}\mathcal{E}_{N+2}}.$$

Similarly, for the estimate of I_2, we have
$$\left|\int_\Sigma \partial^\alpha \eta I_2\right| \lesssim \sqrt{\mathcal{D}_{2N}\mathcal{F}_{2N}\mathcal{E}_{N+2}}.$$

Finally, for I_3, we find that
$$\left|\int_\Sigma \partial^\alpha \eta I_3\right| \lesssim \|\partial^\alpha \eta\|_{-1/2}\|D\partial^{\alpha-\beta}\eta \cdot \partial^\beta u\|_{H^{1/2}(\Sigma)}$$
$$\lesssim \sqrt{\mathcal{D}_{2N}}\sqrt{\mathcal{E}_{2N}\mathcal{D}_{2N}} = \sqrt{\mathcal{E}_{2N}}\mathcal{D}_{2N}.$$

Thus, we deduce

$$\left|\int_\Sigma \partial^\alpha \eta \partial^\alpha G^5\right| \lesssim \sqrt{\mathcal{E}_{2N}}\mathcal{D}_{2N} + \sqrt{\mathcal{D}_{2N}\mathcal{F}_{2N}\mathcal{E}_{N+2}}. \tag{5.35}$$

Bring the estimates (5.30)-(5.35) into (5.29), we conclude

$$\partial_t \int_\Omega \left(\frac{|\partial^\alpha u|^2}{2} + \frac{|\partial^\alpha \eta|^2}{2} + \frac{|\partial^\alpha b|^2}{2}\right) + \int_\Omega \left(\frac{|\mathbb{D}\partial^\alpha u|^2}{2} + |\nabla\partial^\alpha b|^2\right)$$
$$\lesssim (\mathcal{E}_{2N})^{\theta/2}\mathcal{D}_{2N} + \sqrt{\mathcal{D}_{2N}\mathcal{F}_{2N}\mathcal{E}_{N+2}}, \tag{5.36}$$

and then (5.28) follows from (5.36).

Similar to the estimates in Lemma 5.6, by using (5.10), we can obtain the horizontal energies estimates corresponding estimate at the $N+2$ level.

Lemma 5.7 *Let $\alpha \in \mathbb{N}^{1+2}$ satisfy $\alpha_0 \leqslant N+1$ and $|\alpha| \leqslant 2(N+2)$. Then,*

$$\partial_t(\|\partial^\alpha u\|_0^2 + \|\partial^\alpha b\|_0^2 + \|\partial^\alpha \eta\|_0^2) + \|\mathbb{D}\partial^\alpha u\|_0^2 + \|\nabla\partial^\alpha b\|_0^2 \lesssim (\mathcal{E}_{2N})^{\theta/2}\mathcal{D}_{N+2}. \tag{5.37}$$

Furthermore, we deduce

$$\partial_t \bar{\mathcal{E}}_{N+2} + \bar{\mathcal{D}}_{N+2} \lesssim (\mathcal{E}_{2N})^{\theta/2}\mathcal{D}_{N+2}. \tag{5.38}$$

5.2.3 Energy improvement

In this subsection, we will show that, up to some error terms, the total energy $\mathcal{E}_n$ can be controlled by $\bar{\mathcal{E}}_n + \bar{\mathcal{E}}_n^0$ and the total dissipation $\mathcal{D}_{2N}$ can be bounded by $\bar{\mathcal{D}}_n + \bar{\mathcal{D}}_n^0$.

Lemma 5.8 *There exists a $\theta > 0$ so that*

$$\mathcal{E}_{2N} \lesssim \bar{\mathcal{E}}_{2N} + \bar{\mathcal{E}}_{2N}^0 + (\mathcal{E}_{2N})^{\theta+1}, \tag{5.39}$$

and

$$\mathcal{E}_{N+2} \lesssim \bar{\mathcal{E}}_{N+2} + \bar{\mathcal{E}}_{N+2}^0 + (\mathcal{E}_{2N})^\theta \mathcal{E}_{N+2}. \tag{5.40}$$

Proof Let n denote either $2N$ or $N+2$ throughout the proof, and define

$$W_n = \sum_{j=0}^{n-1}\left(\left\|\partial_t^j G^1\right\|_{2n-2j-2}^2 + \left\|\partial_t^j G^2\right\|_{2n-2j-1}^2 + \left\|\partial_t^j G^3\right\|_{2n-2j-2}^2 \right.$$
$$\left. + \left\|\partial_t^j G^4\right\|_{2n-2j-3/2}^2\right).$$

According to the definitions of $\bar{\mathcal{E}}_n^0$ and $\bar{\mathcal{E}}_n$, we know

$$\|\partial_t^n u\|_0^2 + \|\partial_t^n b\|_0^2 + \sum_{j=0}^{n} \left\|\partial_t^j \eta\right\|_{2n-2j}^2 \lesssim \bar{\mathcal{E}}_n^0 + \bar{\mathcal{E}}_n. \qquad (5.41)$$

To control u and p, we will apply the standard Stokes estimates. According to (3.4), we have the form

$$\begin{cases} -\Delta u + \nabla p = -\partial_t u + G^1, & \text{in } \Omega, \\ \text{div} u = G^2, & \text{in } \Omega, \\ (pI - \mathbb{D}u)e_3 = \eta e_3 + G^4, & \text{on } \Sigma, \\ u = 0, & \text{on } \Sigma_{-1}. \end{cases} \qquad (5.42)$$

Then apply ∂_t^j ($j = 0, 1, \cdots, n-1$) to (5.42) and use Lemma 6.3 to find that

$$\left\|\partial_t^j u\right\|_{2n-2j}^2 + \left\|\partial_t^j p\right\|_{2n-2j-1}^2$$
$$\lesssim \left\|\partial_t^{j+1} u\right\|_{2n-2j-2}^2 + \left\|\partial_t^j G^1\right\|_{2n-2j-2}^2 + \left\|\partial_t^j G^2\right\|_{2n-2j-1}^2$$
$$+ \left\|\partial_t^j \eta\right\|_{2n-2j-3/2}^2 + \left\|\partial_t^j G^4\right\|_{2n-2j-3/2}^2$$
$$\lesssim \left\|\partial_t^{j+1} u\right\|_{2n-2(j+1)}^2 + \bar{\mathcal{E}}_n + \bar{\mathcal{E}}_n^0 + \mathcal{W}_n. \qquad (5.43)$$

To control b, we will apply the standard elliptic estimate, and according to (3.4) b satisfies

$$\begin{cases} -\Delta b = \partial_t b + G^3, & \text{in } \Omega, \\ b = 0, & \text{on } \Sigma \cup \Sigma_{-1}. \end{cases} \qquad (5.44)$$

Then, applying ∂_t^j to (5.44) and using Lemma 6.2, one has

$$\left\|\partial_t^j b\right\|_{2n-2j}^2 \lesssim \left\|\partial_t^{j+1} b\right\|_{2n-2j-2}^2 + \left\|\partial_t^j G^3\right\|_{2n-2j-2}^2$$
$$\lesssim \left\|\partial_t^{j+1} b\right\|_{2n-2(j+1)}^2 + \mathcal{W}_n. \qquad (5.45)$$

Combining (5.43) and (5.45), we use the estimates obtained in (5.8) and (5.11) to obtain

$$\left\|\partial_t^j u\right\|_{2n-2j}^2 + \left\|\partial_t^j p\right\|_{2n-2j-1}^2 + \left\|\partial_t^j b\right\|_{2n-2j}^2$$
$$\lesssim \left\|\partial_t^{j+1} u\right\|_{2n-2(j+1)}^2 + \left\|\partial_t^{j+1} b\right\|_{2n-2(j+1)}^2 + \bar{\mathcal{E}}_n + \bar{\mathcal{E}}_n^0 + \mathcal{W}_n. \qquad (5.46)$$

After a simple induction on (5.46), we yield that

$$\sum_{j=0}^{n-1} \left(\left\|\partial_t^j u\right\|_{2n-2j}^2 + \left\|\partial_t^j p\right\|_{2n-2j}^2 + \left\|\partial_t^j b\right\|_{2n-2j}^2 \right)$$
$$\lesssim \bar{\mathcal{E}}_n + \bar{\mathcal{E}}_n^0 + \mathcal{W}_n + \|\partial_t^n u\|_0^2 + \|\partial_t^n b\|_0^2$$
$$\lesssim \bar{\mathcal{E}}_n + \bar{\mathcal{E}}_n^0 + \mathcal{W}_n. \qquad (5.47)$$

Thus, it follows from (5.41) and (5.47) that

$$\mathcal{E}_n \lesssim \bar{\mathcal{E}}_n + \bar{\mathcal{E}}_n^0 + \mathcal{W}_n. \qquad (5.48)$$

Finally, for $n = 2N$, we employ (5.8) to bound $\mathcal{W}_{2N} \lesssim \mathcal{E}_{2N}^{1+\theta}$. Thus, the estimate and (5.48) imply (5.39). Similarly, for $n = N + 2$, we employ (5.11) to bound $\mathcal{W}_{N+2} \lesssim \mathcal{E}_{2N}^\theta \mathcal{E}_{N+2}$. Hence, the estimate and (5.48) imply (5.40).

5.2.4 Dissipation improvement

Lemma 5.9 *There exists a $\theta > 0$, so that*

$$\mathcal{D}_{2N} \lesssim \bar{\mathcal{D}}_{2N} + \bar{\mathcal{D}}_{2N}^0 + \mathcal{F}_{2N}\mathcal{E}_{N+2} + \mathcal{E}_{2N}^\theta \mathcal{D}_{2N}, \qquad (5.49)$$

and

$$\mathcal{D}_{N+2} \lesssim \bar{\mathcal{D}}_{N+2} + \bar{\mathcal{D}}_{N+2}^0 + \mathcal{E}_{2N}^\theta \mathcal{D}_{N+2}. \qquad (5.50)$$

Proof Let n denote $2N$ or $N+2$, and define

$$\mathcal{Y}_n = \left\|\bar{\nabla}_0^{2n-1} G^1\right\|_0^2 + \left\|\bar{\nabla}_0^{2n-1} G^2\right\|_1^2 + \left\|\bar{\nabla}_0^{2n-1} G^3\right\|_0^2$$
$$+ \left\|D^{2n-1} G^4\right\|_{1/2}^2 + \left\|D^{2n-1} G^5\right\|_{1/2}^2 + \left\|\bar{D}_0^{2n-2} \partial_t G^5\right\|_{1/2}^2.$$

Firstly, by the definitions of $\bar{\mathcal{D}}_n^0$, $\bar{\mathcal{D}}_n$ and Korn's inequality, we deduce

$$\|\bar{D}_0^{2n-1} u\|_0^2 + \|D\bar{D}^{2n-1} u\|_1^2 \lesssim \bar{\mathcal{D}}_n,$$

and

$$\sum_{j=0}^{n} \left\|\partial_t^j u\right\|_1^2 \lesssim \bar{\mathcal{D}}_n^0. \qquad (5.51)$$

Summing up the the above inequalities, one has

$$\|\bar{D}_0^{2n} u\|_1^2 \lesssim \bar{\mathcal{D}}_n + \bar{\mathcal{D}}_n^0. \qquad (5.52)$$

Now, we show the estimates of p and u. Since we do not have an estimate of η in terms of dissipation, we can not use the boundary condition on Σ as in (5.42). Fortunately, we can obtain higher regularity estimates of u on Σ, then we have the form

$$\begin{cases} -\Delta u + \nabla p = -\partial_t u + G^1, & \text{in } \Omega, \\ \text{div} u = G^2, & \text{in } \Omega, \\ u = u, & \text{on } \Sigma \cup \Sigma_{-1}. \end{cases} \quad (5.53)$$

Apply ∂_t^j $(j = 0, 1, \cdots, n-1)$ to (5.53) and employ Lemma 6.3 to deduce

$$\left\|\partial_t^j u\right\|_{2n-2j+1}^2 + \left\|\nabla \partial_t^j p\right\|_{2n-2j-1}^2$$
$$\lesssim \left\|\partial_t^{j+1} u\right\|_{2n-2j-1}^2 + \left\|\partial_t^j G^1\right\|_{2n-2j-1}^2 + \left\|\partial_t^j G^2\right\|_{2n-2j}^2 + \left\|\partial_t^j u\right\|_{H^{2n-2j+1/2}(\Sigma)}^2$$
$$\lesssim \left\|\partial_t^{j+1} u\right\|_{2n-2j-1}^2 + \left\|\partial_t^j u\right\|_{H^{2n-2j+1/2}(\Sigma)}^2 + \mathcal{Y}_n + \bar{\mathcal{D}}_n + \bar{\mathcal{D}}_n^0. \quad (5.54)$$

Since Σ and Σ_{-1} are flat, by the definition of Sobolev norm on T^2 and the trace theorem, for $j = 0, 1, \cdots, n-1$, we have

$$\left\|\partial_t^j u\right\|_{H^{2n-2j+1/2}(\Sigma)}^2 \lesssim \|\partial_t^j u\|_{H^{1/2}(\Sigma)}^2 + \|D^{2n-2j} \partial_t^j u\|_{H^{1/2}(\Sigma)}^2$$
$$\lesssim \|\partial_t^j u\|_1^2 + \|D^{2n-2j} \partial_t^j u\|_1$$
$$\lesssim \bar{\mathcal{D}}_n + \bar{\mathcal{D}}_n^0, \quad (5.55)$$

where we have used the result obtained in (5.52). Hence, we deduce

$$\left\|\partial_t^j u\right\|_{2n-2j+1}^2 + \left\|\nabla \partial_t^j p\right\|_{2n-2j-1}^2 \lesssim \left\|\partial_t^{j+1} u\right\|_{2n-2j-1}^2 + \mathcal{Y}_n + \bar{\mathcal{D}}_n + \bar{\mathcal{D}}_n^0. \quad (5.56)$$

For the total dissipative estimate of b, similar in Lemma 5.8, using the standard elliptic estimates to (5.44), we obtain

$$\left\|\partial_t^j b\right\|_{2n-2j+1}^2 \lesssim \left\|\partial_t^{j+1} b\right\|_{2n-2j-1}^2 + \left\|\partial_t^j G^3\right\|_{2n-2j-1}^2$$
$$\lesssim \left\|\partial_t^{j+1} b\right\|_{2n-2j-1}^2 + \mathcal{Y}_n. \quad (5.57)$$

Combining the estimates in (5.55) and (5.57), one has, for $j = 0, 1, \cdots, n-1$,

$$\left\|\partial_t^j u\right\|_{2n-2j+1}^2 + \left\|\nabla \partial_t^j p\right\|_{2n-2j-1}^2 + \left\|\partial_t^j b\right\|_{2n-2j+1}^2$$
$$\lesssim \left\|\partial_t^{j+1} u\right\|_{2n-2j-1}^2 + \left\|\partial_t^{j+1} b\right\|_{2n-2j-1}^2 + \mathcal{Y}_n + \bar{\mathcal{D}}_n + \bar{\mathcal{D}}_n^0. \quad (5.58)$$

After a simple induction on (5.58), the definitions of $\bar{\mathcal{D}}_n$ and $\bar{\mathcal{D}}_n^0$, one has

$$\sum_{j=0}^{n} \left\|\partial_t^j u\right\|_{2n-2j+1}^2 + \sum_{j=0}^{n-1} \left\|\partial_t^j p\right\|_{2n-2j}^2 + \sum_{j=0}^{n} \left\|\partial_t^j b\right\|_{2n-2j+1}^2 \lesssim \mathcal{Y}_n + \bar{\mathcal{D}}_n + \bar{\mathcal{D}}_n^0, \quad (5.59)$$

where for $j = n$, we have used the result in (5.51).

Note that the dissipation estimates in $\bar{\mathcal{D}}_n$ and $\bar{\mathcal{D}}_n^0$ only contains u and b, then we have to recover certain dissipation estimates of η. We may derive some estimates of $\partial_t^j \eta$ for $j = 0, 1, \cdots, n+1$ on Σ by employing the boundary conditions of (3.4):

$$\eta = p - 2\partial_3 u_3 - G^4, \quad (5.60)$$

and

$$\partial_t \eta = u_3 + G^5. \quad (5.61)$$

For $j = 0$, we use the boundary condition (5.60). Note that we do not have any bound on p on the boundary Σ, but we have bounded ∇p in Ω. Thus, we differentiate (5.60) and employ (5.59) to find that

$$\|D\eta\|_{2n-3/2}^2 \lesssim \|Dp\|_{H^{2n-3/2}(\Sigma)}^2 + \|D\partial_3 u_3\|_{H^{2n-3/2}(\Sigma)}^2 + \|DG^4\|_{H^{2n-3/2}(\Sigma)}^2$$
$$\lesssim \|\nabla p\|_{2n-1}^2 + \|u_3\|_{2n+1}^2 + \|G^4\|_{2n-1/2}^2$$
$$\lesssim \mathcal{Y}_n + \bar{\mathcal{D}}_n + \bar{\mathcal{D}}_n^0.$$

Thanks to the critical zero average condition

$$\int_{T^2} \eta = 0,$$

allow us to use Poincaré inequality on Σ to know

$$\|\eta\|_{2n-1/2}^2 \lesssim \|\eta\|_0^2 + \|D\eta\|_{2n-3/2}^2 \lesssim \|D\eta\|_{2n-3/2}^2 \lesssim \mathcal{Y}_n + \bar{\mathcal{D}}_n + \bar{\mathcal{D}}_n^0. \quad (5.62)$$

For $j = 1$, we use (5.61), the definition of $\mathcal{Y}_n$ and (5.59) to see

$$\|\partial_t \eta\|_{2n-1/2}^2 \lesssim \|u_3\|_{H^{2n-1/2}(\Sigma)}^2 + \|G^5\|_{H^{2n-1/2}(\Sigma)}^2$$
$$\lesssim \|u_3\|_{2n}^2 + \|G^5\|_{2n-1/2}^2$$
$$\lesssim \mathcal{Y}_n + \bar{\mathcal{D}}_n + \bar{\mathcal{D}}_n^0. \quad (5.63)$$

Finally, for $j = 2, \cdots, n+1$, we apply ∂_t^{j-1} to (5.61) and use trace estimate to see that

$$\left\|\partial_t^j \eta\right\|_{2n-2j+5/2}^2 \lesssim \left\|\partial_t^{j-1} u_3\right\|_{H^{2n-2j+5/2}(\Sigma)}^2 + \left\|\partial_t^{j-1} G^5\right\|_{H^{2n-2j+5/2}(\Sigma)}^2$$

$$\lesssim \left\|\partial_t^{j-1} u_3\right\|^2_{2n-2(j-1)+1} + \left\|\partial_t^{j-1} G^5\right\|^2_{2n-2(j-1)+1/2}$$
$$\lesssim \mathcal{Y}_n + \bar{\mathcal{D}}_n + \bar{\mathcal{D}}^0. \tag{5.64}$$

Summing (5.62), (5.63) and (5.64), we complete the estimate for η, namely,

$$\|\eta\|^2_{2n-1/2} + \|\partial_t \eta\|^2_{2n-1/2} + \sum_{j=2}^{n+1} \left\|\partial_t^j \eta\right\|^2_{2n-2j+5/2} \lesssim \mathcal{Y}_n + \bar{\mathcal{D}}_n + \bar{\mathcal{D}}_n^0. \tag{5.65}$$

It follows from (5.59) and (5.65) that, for $n = 2N$ or $n = N + 2$, we have

$$\mathcal{D}_n \lesssim \bar{\mathcal{D}}_n + \bar{\mathcal{D}}_n^0 + \mathcal{Y}_n. \tag{5.66}$$

Setting $n = 2N$ in (5.66) and using the estimates (5.9)-(5.10) in Lemma 5.2 to estimate $\mathcal{Y}_{2N} \lesssim (\mathcal{E}_{2N})^\theta \mathcal{D}_{2N} + \mathcal{E}_{N+2} \mathcal{F}_{2N}$. On the other hand, we set $n = N + 2$ and apply the estimate (5.12) in Lemma 5.3 to bound $\mathcal{Y}_{N+2} \lesssim (\mathcal{E}_{2N})^\theta \mathcal{D}_{N+2}$.

5.3 Global energy estimates

We first need to control $\mathcal{F}_{2N}$. This is achieved by the following proposition.

Proposition 5.1 *There exists a universal constant $0 < \delta < 1$ so that if $\mathcal{G}_{2N}(T) \leqslant \delta$, then*

$$\sup_{0 \leqslant r \leqslant t} \mathcal{F}_{2N}(r) \lesssim \mathcal{F}_{2N}(0) + t \int_0^t \mathcal{D}_{2N}, \quad \text{for all } 0 \leqslant t \leqslant T. \tag{5.67}$$

Proof Based on the transport estimate on the kinematic boundary condition, we may show as in Lemma 7.1 of [21] that

$$\sup_{0 \leqslant r \leqslant t} \mathcal{F}_{2N}(r) \lesssim \exp(C \int_0^t \sqrt{\mathcal{E}_{N+2}(r)} dr)$$
$$\times \left[\mathcal{F}_{2N}(0) + t \int_0^t (1 + \mathcal{E}_{2N}(r)) \mathcal{D}_{2N}(r) dr + \left(\int_0^t \sqrt{\mathcal{E}_{N+2}(r) \mathcal{F}_{2N}(r)}\right)^2\right]. \tag{5.68}$$

According to $\mathcal{G}_{2N} \leqslant \delta$, we know

$$\int_0^t \sqrt{\mathcal{E}_{N+2}(r)} dr \lesssim \sqrt{\delta} \int_0^t \frac{1}{(1+r)^{2N-4}} dr \lesssim \sqrt{\delta}. \tag{5.69}$$

Since $\delta \leqslant 1$, this implies that for any constant $C > 0$,

$$\exp\left(C \int_0^t \sqrt{\mathcal{E}_{N+2}(r)} dr\right) \lesssim 1. \tag{5.70}$$

Then by (5.69) and (5.70), we deduce from (5.68) that

$$\sup_{0 \leqslant r \leqslant t} \mathcal{F}_{2N}(r) \lesssim \mathcal{F}_{2N}(0) + t \int_0^t \mathcal{D}_{2N}(r)dr + \sup_{0 \leqslant r \leqslant t} \mathcal{F}_{2N}(r) \left(\int_0^t \sqrt{\mathcal{E}_{N+2}(r)}dr \right)^2$$

$$\lesssim \mathcal{F}_{2N}(0) + t \int_0^t \mathcal{D}_{2N}(r)dr + \delta \sup_{0 \leqslant r \leqslant t} \mathcal{F}_{2N}(r). \tag{5.71}$$

By taking δ small enough, (5.67) follows.

This bound on $\mathcal{F}_{2N}$ allows us to estimate the integral of $\mathcal{E}_{N+2}\mathcal{F}_{2N}$ and the integral of $\sqrt{\mathcal{D}_{2N}\mathcal{E}_{N+2}\mathcal{F}_{2N}}$ as in Corollary 7.3 of [21].

Corollary 5.10 *There exists $0 < \delta < 1$ so that if $\mathcal{G}_{2N}(T) \leqslant \delta$, then*

$$\int_0^t \mathcal{E}_{N+2}\mathcal{F}_{2N} \lesssim \delta \mathcal{F}_{2N}(0) + \delta \int_0^t \mathcal{D}_{2N}(r)dr, \tag{5.72}$$

and

$$\int_0^t \sqrt{\mathcal{D}_{2N}\mathcal{E}_{N+2}\mathcal{F}_{2N}} \lesssim \mathcal{F}_{2N}(0) + \sqrt{\delta} \int_0^t \mathcal{D}_{2N}(r)dr, \tag{5.73}$$

for all $0 \leqslant t \leqslant T$.

Now we show the boundness of the high-order terms.

Proposition 5.2 *There exists $0 < \delta < 1$ so that if $\mathcal{G}_{2N}(T) \leqslant \delta$, then*

$$\sup_{0 \leqslant r \leqslant t} \mathcal{E}_{2N}(r) + \int_0^t \mathcal{D}_{2N} + \sup_{0 \leqslant r \leqslant t} \frac{\mathcal{F}_{2N}(r)}{(1+r)} \lesssim \mathcal{E}_{2N}(0) + \mathcal{F}_{2N}(0), \tag{5.74}$$

for all $0 \leqslant t \leqslant T$.

Proof Fix $0 \leqslant t \leqslant T$. We sum up the results of Lemma 5.4 and Lemma 5.6 to know

$$\bar{\mathcal{E}}_{2N}^0(t) + \bar{\mathcal{E}}_{2N}(t) + \int_0^t (\bar{\mathcal{D}}_{2N}^0 + \bar{\mathcal{D}}_{2N})$$

$$\leqslant C_1 \mathcal{E}_{2N}(0) + C_1 (\mathcal{E}_{2N}(t))^{3/2} + C_1 \int_0^t (\mathcal{E}_{2N}^\theta \mathcal{D}_{2N} + \sqrt{\mathcal{D}_{2N}\mathcal{E}_{N+2}\mathcal{F}_{2N}}). \tag{5.75}$$

Then, combining with Lemma 5.8 and Lemma 5.9, we deduce

$$\mathcal{E}_{2N}(t) + \int_0^t \mathcal{D}_{2N} \leqslant C_2 \left(\bar{\mathcal{E}}_{2N}^0 + \bar{\mathcal{E}}_{2N} + \int_0^t (\bar{\mathcal{D}}_{2N}^0 + \bar{\mathcal{D}}_{2N}) \right) + C_2 (\mathcal{E}_{2N}(t))^{1+\theta}$$

$$+ C_2 \int_0^t (\mathcal{E}_{2N}^\theta \mathcal{D}_{2N} + \sqrt{\mathcal{D}_{2N}\mathcal{E}_{N+2}\mathcal{F}_{2N}} + \mathcal{E}_{N+2}\mathcal{F}_{2N})$$

$$\leqslant C_3 (\mathcal{E}_{2N}(0) + \mathcal{E}_{2N}(t))^{1+\theta} + \mathcal{E}_{2N}(t))^{3/2})$$

$$+ C_3 \int_0^t (\mathcal{E}_{2N}^\theta \mathcal{D}_{2N} + \sqrt{\mathcal{D}_{2N}\mathcal{E}_{N+2}\mathcal{F}_{2N}} + \mathcal{E}_{N+2}\mathcal{F}_{2N}). \tag{5.76}$$

Let us assume that $\delta \in (0,1)$ is as small as in Corollary 5.10. Thus we conclude

$$\sup_{0 \leqslant r \leqslant t} \mathcal{E}_{2N}(t) + \int_0^t \mathcal{D}_{2N} \lesssim \mathcal{E}_{2N}(0) + \mathcal{F}_{2N}(0). \tag{5.77}$$

It remains to show the decay estimates of $\mathcal{E}_{N+2}$. Before that, we show that the pressure term involved in Lemma 5.5 can be absorbed into $\bar{\mathcal{E}}^0_{N+2} + \bar{\mathcal{E}}_{N+2}$.

Lemma 5.11 *Let F^2 be defined in (5.2) with $\partial^\alpha = \partial_t^{N+2}$, then there exists a constant $\delta \in (0,1)$ so that if $\mathcal{G}_{2N} \leqslant \delta$, then*

$$\frac{1}{2}(\bar{\mathcal{E}}^0_{N+2} + \bar{\mathcal{E}}_{N+2}) \leqslant \bar{\mathcal{E}}^0_{N+2} + \bar{\mathcal{E}}_{N+2} - 2\int_\Omega \partial_t^{N+1} p F^2 J$$

$$\leqslant \frac{3}{2}(\bar{\mathcal{E}}^0_{N+2} + \bar{\mathcal{E}}_{N+2}). \tag{5.78}$$

Proof Let us assume that $\delta \in (0,1)$ is as small as in Corollary 5.10. According to Proposition 5.2, one has

$$\mathcal{E}_{N+2} \lesssim \bar{\mathcal{E}}^0_{N+2} + \bar{\mathcal{E}}_{N+2} + \mathcal{E}^\theta_{2N} \mathcal{E}_{N+2} \leqslant C(\bar{\mathcal{E}}^0_{N+2} + \bar{\mathcal{E}}_{N+2}) + C\delta \mathcal{E}_{N+2}.$$

Hence, we deduce

$$\mathcal{E}_{N+2} \lesssim \bar{\mathcal{E}}^0_{N+2} + \bar{\mathcal{E}}_{N+2}. \tag{5.79}$$

Combining the estimates obtained in (3.13) and (5.7), we know that

$$\left| 2\int_\Omega \partial_t^{N+1} p F^2 J \right| \leqslant 2 \left\| \partial_t^{N+1} p \right\|_0 \| F^2 \|_0 \| J \|_{L^\infty}$$

$$\leqslant C \sqrt{\mathcal{E}_{N+2}} \sqrt{\mathcal{E}^\theta_{2N} \mathcal{E}_{N+2}}$$

$$= \mathcal{E}^{\theta/2}_{2N} \mathcal{E}_{N+2} \leqslant C \mathcal{E}^{\theta/2}_{2N} (\bar{\mathcal{E}}^0_{N+2} + \bar{\mathcal{E}}_{N+2})$$

$$\leqslant C \delta^{\theta/2} (\bar{\mathcal{E}}^0_{N+2} + \bar{\mathcal{E}}_{N+2}).$$

If δ is small enough, (5.78) follows.

Proposition 5.3 *There exists $0 < \delta < 1$ so that if $\mathcal{G}_{2N}(T) \leqslant \delta$, then*

$$(1 + t^{4N-8}) \mathcal{E}_{N+2}(t) \lesssim \mathcal{E}_{2N}(0) + \mathcal{F}_{2N}(0) \quad \text{for all } 0 \leqslant t \leqslant T. \tag{5.80}$$

Proof Fix $0 \leqslant t \leqslant T$. According to Lemma 5.5, Lemma 5.7 and Lemma 5.11, we know

$$\partial_t \left(\bar{\mathcal{E}}^0_{N+2}(t) + \bar{\mathcal{E}}_{N+2}(t) \right) + \bar{\mathcal{D}}^0_{N+2} + \bar{\mathcal{D}}_{N+2} \leqslant \mathcal{E}^{\theta/2}_{2N} \mathcal{D}_{N+2} + \sqrt{\mathcal{E}_{2N}} \mathcal{D}_{N+2}. \tag{5.81}$$

Let us assume that $\delta \in (0,1)$ is as small as in Corollary 5.10, thus we have $\mathcal{E}_{2N}(t) \leqslant \mathcal{G}_{2N}(T) \leqslant \delta$. Similar in (5.79), we can obtain

$$\mathcal{D}_{N+2} \lesssim \bar{\mathcal{D}}^0_{N+2} + \bar{\mathcal{D}}_{N+2}. \tag{5.82}$$

Thus, combining (5.79), (5.82) and (5.81) we deduce

$$\partial_t \mathcal{E}_{N+2} + \mathcal{D}_{N+2} \lesssim \mathcal{E}_{2N}^{\theta/2} \mathcal{D}_{N+2} + \sqrt{\mathcal{E}_{2N}} \mathcal{D}_{N+2} \lesssim \delta^{\theta/2} \mathcal{D}_{N+2} + \sqrt{\delta} \mathcal{D}_{N+2}. \tag{5.83}$$

Hence, if δ is small enough, we obtain

$$\partial_t \mathcal{E}_{N+2} + \mathcal{D}_{N+2} \leqslant 0.$$

On the other hand, based on the Sobolev interpolation inequality we can prove

$$\mathcal{E}_{N+2} \lesssim \mathcal{D}_{N+2}^\theta \mathcal{E}_{2N}^{1-\theta}, \quad \text{where } \theta = \frac{4N-8}{4N-7}. \tag{5.84}$$

Now since we know that the boundness of high energy estimate Proposition 5.2, we get

$$\sup_{0 \leqslant r \leqslant t} \mathcal{E}_{2N}(r) \lesssim \mathcal{E}_{2N}(0) + \mathcal{F}_{2N}(0) := \mathcal{M}_0. \tag{5.85}$$

We obtain from (5.84) that

$$\mathcal{E}_{N+2} \lesssim \mathcal{M}_0^{1-\theta} \mathcal{D}_{N+2}^\theta, \quad \text{where } \theta = \frac{4N-8}{4N-7}. \tag{5.86}$$

Hence by (5.85) and (5.83), there exists some constant $C_1 > 0$ such that

$$\frac{d}{dt} \mathcal{E}_{N+2} + \frac{C_1}{\mathcal{M}_0^s} \mathcal{E}_{N+2}^{1+s} \lesssim 0, \quad \text{where } s = \frac{1}{\theta} - 1 = \frac{1}{4N-8}. \tag{5.87}$$

Solving this differential inequality directly, we obtain

$$\mathcal{E}_{N+2}(t) \lesssim \frac{\mathcal{M}_0}{(\mathcal{M}_0^s + sC_1(\mathcal{E}_{N+2}(0))^s t)^{1/s}} \mathcal{E}_{N+2}(0). \tag{5.88}$$

Using that $\mathcal{E}_{N+2}(0) \lesssim \mathcal{M}_0$ and the fact $1/s = 4n-8 > 1$, we obtain that

$$\mathcal{E}_{N+2}(t) \lesssim \frac{\mathcal{M}_0}{1+sC_1 t)^{1/s}} \lesssim \frac{\mathcal{M}_0}{(1+t)^{1/s}} \lesssim \frac{\mathcal{M}_0}{(1+t)^{4N-8}}. \tag{5.89}$$

This implies (5.80).

Now we combine propositions to arrive at our ultimate energy estimates for $\mathcal{G}_{2N}$.

Theorem 5.12 There exists a universal $0 < \delta < 1$ so that if $\mathcal{G}_{2N}(T) \leqslant \delta$, then
$$\mathcal{G}_{2N}(t) \lesssim \mathcal{E}_{2N}(0) + \mathcal{F}_{2N}(0) \quad \text{for all } 0 \leqslant t \leqslant T. \tag{5.90}$$

Proof The conclusion follows directly from the definition of $\mathcal{G}_{2N}$ and Proposition 5.1-Proposition 5.3.

6 Appendix A. Analytic tools

A.1 Harmonic extension

We define the appropriate Poisson integral in $\mathbb{T} \times (-\infty, 0)$ by
$$\mathcal{P}\eta(x) = \sum_{n \in (L_1^{-1}\mathbb{Z}) \times (L_2^{-1}\mathbb{Z})} e^{2\pi i n \cdot x'} e^{2\pi |n| x_3} \hat{\eta}(n), \tag{6.1}$$

where we have written
$$\hat{\eta}(n) = \int_\Sigma \eta(x') \frac{e^{-2\pi i n \cdot x'}}{L_1 L_2} dx'.$$

It is well known that $\mathcal{P} : H^s(\Sigma) \to H^{s+1/2}(\mathbb{T} \times (-\infty, 0))$ is a bounded linear operator for $s > 0$. However, if restricted to the domain Ω, one has the following result.

Lemma 6.1 It holds that for all $s \in \mathbf{R}$,
$$\|\mathcal{P}f\|_s \lesssim |f|_{s-1/2}. \tag{6.2}$$

Proof See [6].

A.2 Elliptic estimates

Lemma 6.2 Suppose $u \in H^r(\Omega)$ solve
$$\begin{cases} -\mu \Delta u = f \in H^{r-2}(\Omega), \\ u|_{\Sigma \cup \Sigma_{-1}} = 0, \end{cases} \tag{6.3}$$

then for $r \geqslant 2$, one has
$$\|u\|_r \lesssim \|f\|_{r-2}. \tag{6.4}$$

Proof See [17].

Lemma 6.3 *Suppose* (u, p) *solve*

$$\begin{cases} -\mu \Delta u + \nabla p = \phi \in H^{r-2}(\Omega), \\ \operatorname{div} u = \psi \in H^{r-1}(\Omega), \\ (pI - \mathbb{D}(u))e_3 = \alpha \in H^{r-3/2}(\Sigma), \quad u|_{\Sigma_{-1}} = 0. \end{cases} \quad (6.5)$$

Then for $r \geq 2$, *one has*

$$\|u\|_{H^r}^2 + \|p\|_{H^{r-1}}^2 \lesssim \|\phi\|_{H^{r-2}}^2 + \|\psi\|_{H^{r-1}}^2 + \|\alpha\|_{H^{r-3/2}}^2.$$

Proof See [6].

Lemma 6.4 *Suppose* $r \geq 2$ *and let* $\phi \in H^{r-2}(\Omega)$, $\psi \in H^{r-1}(\Omega)$, $f_1 \in H^{r-1/2}(\Sigma)$, $f_2 \in H^{r-1/2}(\Sigma_{-1})$ *be given such that*

$$\int_\Omega \psi = \int_\Sigma f_1 \cdot \nu + \int_{\Sigma_{-1}} f_2 \cdot \nu.$$

Then there exists unique $u \in H^r(\Omega)$, $p \in H^{r-1}(\Omega)$ *solving*

$$\begin{cases} -\mu \Delta u + \nabla p = \phi, & \text{in } \Omega, \\ \operatorname{div} u = \psi, & \text{in } \Omega, \\ u = f_1, & \text{on } \Sigma, \\ u = f_2, & \text{on } \Sigma_{-1}. \end{cases} \quad (6.6)$$

Moreover,

$$\|u\|_{H^r(\Omega)}^2 + \|\nabla p\|_{H^{r-2}(\Omega)}^2 \lesssim \|\phi\|_{H^{r-2}(\Omega)}^2 + \|\psi\|_{H^{r-1}(\Omega)}^2 + \|f_1\|_{H^{r-1/2}(\Sigma)}^2 + \|f_2\|_{H^{r-1/2}(\Sigma_{-1})}^2.$$

Proof See [17].

Acknowledgements Guo is supported by the NSFC of China under a grant number 11731014, and Ni is supported by the Science Challenge Project under grand number TZ2016002.

References

[1] Bae H. Solvability of the free boundary value problem of the Navier-Stokes equations [J]. Discret. Contin. Dyn. Syst., 2011, 29(3): 769–801.

[2] Beale J. The initial value problem for the Navier-Stokes equations equations with a free surface [J]. Comm. Pure Appl. Math., 1981, 34(3): 359–392.

[3] Beale J. Large-time regularity of viscous surface waves [J]. Arch. Ration. Mech. Anal., 1983, 84(4): 307–352.

[4] Beale J, Nishida T. Large-time behavior of viscous surface waves [J]. North-Holland Math. Stud., 1985, 8: 1–14.

[5] Drazin P, Reid W. Hydrodynamic stability, 2nd [M]. Cambridge: Cambridge University Press, 2004.

[6] Guo Y, Tice I. Local well-posedness of the viscous surface wave problem without surface tension [J]. Anal PDE., 2013, 6: 287–369.

[7] Guo Y, Tice I. Decay of viscous surface waves without surface tension in horizontally infinite domains [J]. Anal PDE., 2013, 6: 1429–1533.

[8] Guo Y, Tice I. Almost exponential decay of periodic viscous surface waves without surface tension [J]. Arch. Rational Mech. Anal., 2013, 207: 459–531.

[9] Hataya Y. Decaying solution of a Navier-Stokes flow without surface tension [J]. J. Math. Kyoto Univ., 2009, 49(4): 691–717.

[10] Kim C, Tice I. Dynamics and stability of Surfactant-driven surface waves [J]. SIAM J. Math. Anal., 2017, 49: 1295–1332.

[11] Lee D. Uniform estimate of viscous free-boundary magnetohydrodynamics with zero vacuum magnetic field [J]. SIAM J. Math. Anal., 2017, 49(4): 2710–2789.

[12] Masmoudi N, Rousset F. Uniform regularity and vanishing viscosity limit for the free surface Navier-Stokes equations [J]. Arch. Ration. Mech. Anal., 2017, 223: 301–417.

[13] Nishida T, Teramoto Y, Yoshihara H. Global in time behavior of viscous surface waves: horizontally periodic motion [J]. J. Math. Kyoto Univ., 2004, 44(2): 271–323.

[14] Padula M, Solonnikov V. On the free-boundary problem of magnetohydrodynamics [J]. J. Math. Sci. (N.Y.), 2011, 178: 313–344.

[15] Solonnikov V, Frolova E. Solvability of a free boundary problem of magneto-hydrodynamics in an in an infinite time interval [J]. J. Math. Sci., 2013, 195: 76–97.

[16] Sylvester D. Large time existence of small viscous surfacewaveswithout surface tension [J]. Commun. Part. Differ. Equ., 1990, 15(6): 823–903.

[17] Tan Z, Wang Y. Zero surface tension limit of viscous surface waves [J]. Commun. Math. Phys., 2014, 328: 733–807.

[18] Tani A. Small-time existence for the three-dimensional Navier-Stokes equations for an incompressible fluid with a free surface [J]. Arch. Ration. Mech. Anal., 1996, 133(4): 299–331.

[19] Tani A, Tanaka N. Large-time existence of surface waves in incompressible viscous fluids with or without surface tension [J]. Arch. Ration. Mech. Anal., 1995, 130(4): 303–314.

[20] Tice I, Zbarsky S. Decay of solutions to the linearized free surface Navier-Stokes equations with fractional boundary operators [J]. J. Math. Fluid Mech., 2020, 22(4): Paper No. 48, 28 pp.

[21] Remond-Tiedrez I, Tice I. The viscous surface wave problem with generalized surface energies [J]. SIAM J. Math. Anal., 2019, 51(6): 4894--4952.

[22] Y. Wang, Z. Xin, Incompressible inviscid resistive MHD surface waves in 2D. http://arXiv:1801.04694v1 [math.AP] 15 Jan 2018.

The Cauchy Problem for a Generalized Camassa-Holm Equation*

Mi Yongsheng (米永生), Liu Yue (刘跃),

Guo Boling (郭柏灵), and Luo Ting (罗婷)

Abstract In this paper, we are concerned with the Cauchy problem of the generalized Camassa-Holm equation. Using a Galerkin-type approximation scheme, it is shown that this equation is well-posed in Sobolev spaces $H^s, s > 3/2$ for both the periodic and the nonperiodic case in the sense of Hadamard. That is, the data-to-solution map is continuous. Furthermore, it is proved that this dependence is sharp by showing that the solution map is not uniformly continuous. The nonuniform dependence is proved using the method of approximate solutions and well-posedness estimates. Moreover, it is shown that the solution map for the generalized Camassa-Holm equation is Hölder continuous in H^r-topology. Finally, with initial analytic data, we show that its solutions are analytical in both variables, globally in space and locally in time.

Keywords Besov spaces; Camassa-Holm type equation; local well-posedness

1 Introduction

In this paper, we consider the Cauchy problem of the generalized Camassa-Holm equation in both the periodic and the nonperiodic cases, respectively.

$$u_t - u_{txx} = \partial_x(2 - \partial_x)(1 + \partial_x)u^2, \tag{1.1}$$

$$u(0, x) = u_0(x). \tag{1.2}$$

Recently, Novikov in [40] proposed the following integrable quasi-linear scaler

---

* J. Differential Equations, 2019, 266(10): 6739–6770. DOI:10.1016/j.jde.2018.11.019.

evolution equation of order 2,

$$(1 - \epsilon^2 \partial_x^2)u_t = \partial_x(2 - \epsilon\partial_x)(1 + \epsilon\partial_x)u^2, \tag{1.3}$$

where $\epsilon \neq 0$ is a real constant. It was shown in [40] that Eq. (1.3) possesses a hierarchy of local higher symmetries and the first non-trivial one is $u_\tau = \partial_x[(1 - \epsilon\partial_x u)]^{-1}$. Letting $v(t,x) = u(\epsilon t, \epsilon x)$, then one can transform Eq. (1.1) into the equivalent Eq. (1.1).

Eq. (1.1) belongs to the following class [40]

$$(1 - \partial_x^2)u_t = F(u, u_x, u_{xx}, u_{xxx}) \tag{1.4}$$

which has attracted much attention on the Cauchy problem of the possible integrable members of (1.4).

The first well-known integrable member of (1.4) is the Camassa-Holm equation [15]

$$u_t - u_{txx} + 3uu_x = 2u_x u_{xx} + uu_{xxx} \tag{1.5}$$

modeling the unidirectional propagation of shallow water waves over a flat bottom, $u(t,x)$ stands for the fluid velocity at time t in the spatial direction x. It is a well-known integrable equation describing the velocity dynamics of shallow water waves. This equation spontaneously exhibits the emergence of singular solutions from smooth initial conditions. It has a bi-Hamilton structure [19] and is completely integrable [5,15]. In particular, it possesses an infinity of conservation laws and is solvable by its corresponding inverse scattering transform. After the birth of the Camassa-Holm equation, many works have been carried out to probe its dynamic properties. Such as, Eq. (1.2) has traveling wave solutions of the form $ce^{-|x-ct|}$, called peakons, which describes an essential feature of the travelling waves of the largest amplitude (see [6,9,10]). It is shown in [7,11,14] that the inverse spectral or scattering approach is a powerful tool to handle the Camassa-Holm equation and analyze its dynamics. It is worthwhile to mention that Eq. (1.2) gives rise to geodesic flow of a certain invariant metric on the Bott-Virasoro group [12,38], and this geometric illustration leads to a proof that the Least Action Principle holds. It is shown in [8] that the blow-up occurs in the form of breaking waves, namely, the solution remains bounded but its slope becomes unbounded in finite time. Moreover, the Camassa-Holm equation has global conservative solutions [3,27] and dissipative solutions [2,26].

The second well-known integrable member of (1.4) is the Degasperis-Procesi equation [17]

$$u_t - u_{txx} + 4uu_x = 3u_x u_{xx} + uu_{xxx}. \qquad (1.6)$$

Degasperis, Holm and Hone [17] proved the formal integrability of Eq. (1.2) by constructing a Lax pair. They also showed that it has a bi-Hamiltonian structure and an infinite sequence of conserved quantities and admits exact peakon solutions. The direct and inverse scattering approaches to pursue it can be seen in [16]. Moreover, in [18], they also presented that the Degasperis-Procesi equation has a bi-Hamiltonian structure and an infinite number of conservation laws, and admits exact peakon solutions which are analogous to the Camassa-Holm peakons. It is worth pointing out that solutions of this type are not mere abstractizations: the peakons replicate a feature that is characteristic for the waves of great height-waves of the largest amplitude that are exact solutions of the governing equations for irrotational water waves cf. the papers [6, 10, 43]. The Degasperis-Procesi equation is a model for nonlinear shallow water dynamics cf. the discussion in [13]. The numerical stability of solitons and peakons, the multisoliton solutions and their peakon limits, together with an inverse scattering method to compute N-peakon solutions to Degasperis-Procesi equation have been investigated respectively in [29, 36, 37]. Furthermore, the traveling wave solutions and the classification of all weak traveling wave solutions to Degasperis-Procesi equation were presented in [34, 44].

The third well-known integrable member of (1.4) is the Novikov equation [40]

$$u_t - u_{txx} + 4u^2 u_x = 3uu_x u_{xxx} + u^2 u_{xxx} \qquad (1.7)$$

which has been recently discovered by Vladimir Novikov in a symmetry classification of nonlocal PDEs with quadratic or cubic nonlinearity [40]. The perturbative symmetry approach yields necessary conditions for a PDE to admit infinitely many symmetries. Using this approach, Novikov was able to isolate Eq. (1.3) and find its first few symmetries, and he subsequently found a scalar Lax pair for it, then proved that the equation is integrable, which can be thought as a generalization of the Camassa-Holm equation. In [31], it is shown that the Novikov equation admits peakon solutions like the Camassa-Holm. Also, it has a Lax pair in matrix form and a bi-Hamiltonian structure. Furthermore, it has infinitely many conserved quantities. Like Camassa-Holm. the most important quantity conserved by a solution u to Novikov equation is its H^1-norm $\|u\|_{H^1}^2 = \int_{\mathbb{R}} (u^2 + u_x^2)$, which plays an important role in the study of Eq. (1.1). In [20, 39, 45], the authors study well-posedness and dependence on initial data for the Cauchy problem for Novikov equation. Recently, in [32], a global existence result and conditions on the initial data were considered. Existence and uniqueness of the global weak solution to Novikov equation with initial data under some conditions were proved in [46]. The Novikov equation with the dissipative term was considered in [47]. Multipeakon solutions were studied in [30, 31]. The Cauchy problem of the Novikov equation on the circle was investigated in [42].

Non-uniform dependence of the Camassa-Holm type equation has been studied by several authors. In [24], Himonas and Misiołek provide the first nonuniform dependence result for CH on the circle for $s \geqslant 2$ using explicitly constructed traveling wave solutions. Both of these more recent works utilize the method of approximate solutions in conjunction with delicate commutator and multiplier estimates. For the DP equation, the first result for nonuniform dependence can be found in Christov and Hakkaev [4]. In the periodic case with $s \geqslant 2$, nonuniform dependence of the data-to- solution map for DP is established following the method of traveling wave solutions developed in [24] for the CH equation. This result has been recently sharpened in [21], where nonuniform dependence on the initial data for DP is proven on both the circle and line for $s > 3/2$ using the method of approximate solutions in tandem with a twisted L_2-norm that is conserved by the DP equation. This method of approximate solutions has also been adapted to a homogeneous setting in Holliman [28] to prove nonuniform dependence of the flow map for the Hunter-Saxton equation on the circle with $s > 3/2$, and to the higher dimensional setting in [25] to prove nonuniform dependence for the Euler equations.

In [35], Yin and Li first establish the local existence and uniqueness of strong solutions for the Cauchy problem (1.1)–(1.2). In this paper, we mainly study the Cauchy problem (1.1)–(1.2) with initial data in H^s where $s > 3/2$, in both the periodic and the nonperiodic case, respectively. Our main results are stated as follows.

Theorem 1.1 *If $s > 3/2$ and $u_0 \in H^s$ then there exists $T > 0$ and a unique solution $u \in C([0,T]; H^s)$ of the i.v.p. (1.1)–(1.2), which depends continuously on the initial data u_0. Furthermore, we have the estimate*

$$\|u(t)\|_{H^s} \leqslant 2\|u_0\|_{H^s}, \quad \text{for } 0 \leqslant T \leqslant \frac{1}{c_s\|u_0\|_{H^s}^2}, \tag{1.8}$$

where $c_s > 0$ is a constant depending on s.

Using this result of Theorem 1.1 and the method of approximate solutions we prove the following nonuniform dependence result.

Theorem 1.2 *If $s > 3/2$, then the data-to-solution map for both the periodic and the nonperiodic the generalized Camassa-Holm equation defined by the Cauchy problem (1.1)–(1.2) is not uniformly continuous from any bounded subset in H^s into $C([0,T]; H^s)$.*

Theorems 1.1 and 1.2 show that the generalized Camassa-Holm equation is well-posed in Sobolev spaces H^s on both the line and the circle for $s > 3/2$ and its data-to-solution map is continuous but not uniformly continuous. Here, we show that the solution map for the generalized Camassa-Holm equation is Hölder continuous in H^r-topology for all $0 \leqslant r < s$. More precisely, we prove the following result.

Theorem 1.3 *If $s > 3/2$ and $0 \leqslant r < s$, then the data-to-solution map for the Cauchy problem (1.1)–(1.2), on both the line and the circle, is Hölder continuous on the space Hs equipped with the H^r norm. More precisely, for initial data $u(0), w(0)$ in a ball $B(0,\rho) = \{\varphi \in H^s : \|\varphi\|_{H^s} \leqslant \rho\}$ of H^s, the corresponding solutions $u(t), w(t)$ satisfy the inequality*

$$\|u(t) - w(t)\|_{C([0,T];H^r)} \leqslant c\|u(0) - w(0)\|_{H^r}^\alpha, \tag{1.9}$$

where the exponent α is given by

$$\alpha = \begin{cases} 1, & \text{if } (s,r) \in A_1, \\ \dfrac{2(s-1)}{s-r}, & \text{if } (s,r) \in A_1, \\ s-r & \text{if } (s,r) \in A_1 \end{cases}$$

and the regions A_1, A_2 and A_3 in the sr-plane are defined by

$$A_1 = \{(s,r) : s > \frac{3}{2}, 0 \leqslant r \leqslant s-1, r+s \geqslant 2\},$$

$$A_1 = \{(s,r) : 2 > s > \frac{3}{2}, 0 \leqslant r \leqslant 2-s\},$$

$$A_1 = \{(s,r) : s > \frac{3}{2}, s-1 \leqslant r \leqslant s\}.$$

The lifespan T and the constant c depend on s, r and ρ.

Theorem 1.4 *If the initial data u_0 is analytic on $\mathbf{R}$ then there exists an $\varepsilon > 0$ and a unique solution $u(t,x)$ to the Cauchy problem (1.1)–(1.2) that is analytic both in x and t on $\mathbf{R}$ for all t in $(-\varepsilon, \varepsilon)$.*

The rest of this paper is organized as follows. In Section 2, using the method of approximate solutions, we show that the solution map is not uniformly continuous for the periodic and the nonperiodic cases, respectively. Section 3 is devoted to the study of the local wellposedness result for $s > 3/2$ with an accompanying solution size estimate. Moreover, in Section 4, it is shown that the solution map for the generalized Camassa-Holm equation is Hölder continuous in H^r-topology. Finally, Section 5 is devoted to the study of the analyticity of the Cauchy problem (1.1)–(1.2) based on a contraction type argument in a suitably chosen scale of the Banach spaces.

2 Non-uniform dependence

In this section, it is shown that the solution map for the generalized Camassa-Holm equation (1.1) is not uniformly continuous on bounded sets of Sobolev spaces with exponent greater than 3/2 in both the periodic and the non-periodic case.

2.1 Nonuniform dependence in the periodic case

For the sake of brevity, we shall rewrite the Cauchy problem (1.1)–(1.2) as follows

$$u_t - 2uu_x = \partial_x D^{-2}\left[\partial_x u^2\right] + \partial_x D^{-2}\left[u^2\right], \tag{2.1}$$

$$u(x, 0) = u_0(x). \tag{2.2}$$

For any $s \in \mathbf{R}$, we take the operator $D^s = (1 - \partial_x^2)^{\frac{s}{2}}$ to be defined by

$$\widehat{D^s f}(k) \doteq (1 + k^2)^{s/2} \widehat{f}(k), \tag{2.3}$$

where $\widehat{f}(k)$ is the Fourier transform

$$\widehat{f}(k) \doteq \int_{\mathbf{T}} e^{-ixk} f(x)\, dx. \tag{2.4}$$

Therefore, the inverse relation is given by the Fourier series

$$f(x) = \frac{1}{2\pi} \sum_{k \in \mathbf{Z}} \widehat{f}(k) e^{-ixk}. \tag{2.5}$$

Then, for $f \in H^s(\mathbb{T})$, we have

$$\|f\|_{H^s(\mathbb{T})}^2 = \frac{1}{2\pi}\sum_{k\in\mathbb{Z}}(1+k^2)^s|\widehat{f}(k)|^2 = \|D^s f\|_{L^2(\mathbb{T})}^2. \tag{2.6}$$

In particular, D^{-2} is of order -2 satisfying the estimate

$$\|D^{-2}f\|_{H^r(\mathbb{T})} \lesssim \|f\|_{H^{r-2}(\mathbb{T})}. \tag{2.7}$$

Also, we shall use the companion estimate

$$\|\partial_x D^{-2}f\|_{H^r(\mathbb{T})} \lesssim \|f\|_{H^{r-1}(\mathbb{T})}. \tag{2.8}$$

We define the relation

$$x \lesssim y \iff \text{there is a constant } c > 0 \text{ s.t. } x \leqslant cy$$

and

$$x \approx y \iff x \lesssim y \text{ and } y \lesssim x. \tag{2.9}$$

We will often use the notation $H^s = H^s(\mathbb{T})$ in this section.

Next, we will prove Theorem 1.2 for Sobolev exponents $s > 3/2$. The basis of our proof rests upon finding two sequences of solutions, $\{u_n\}$, $\{v_n\}$ in $C([0,T] : H^s)$, to (1.1)–(1.2) that share a common lifespan and satisfy

$$\|u_n(t)\|_{H^s} + \|v_n(t)\|_{H^s} \lesssim 1,$$
$$\lim_{n\to\infty}\|u_n(0) - v_n(0)\|_{H^s} = 0, \tag{2.10}$$

and

$$\liminf_{n\to\infty}\|u_{1,n}(t) - v_{-1,n}(t)\|_{H^s} \gtrsim |\sin t| \quad \text{for } t \in (0,T]. \tag{2.11}$$

2.1.1 Approximate solutions

For any n, a positive integer, we define the approximate solution $u^{\omega,n} = u^{\omega,n}(x,t)$ as

$$u^{\omega,n}(x,t) \doteq \omega n^{-1} + n^{-s}\cos(nx - \omega t), \tag{2.12}$$

where

$$\omega = \{-1, 1\}. \tag{2.13}$$

Now we compute the error of the approximate solutions (2.12). Note that

$$\partial_t u^{\omega,n} - 2u^{\omega,n}\partial_x u^{\omega,n} = \omega n^{-s}\sin(nx-\omega t)$$
$$+ 2n^{1-s}[\omega n^{-1} + n^{-s}\cos(nx-\omega t)]\sin(nx-\omega t)$$
$$= 3\omega n^{-s}\sin(nx-\omega t) + \frac{1}{2}n^{-2s+1}\sin 2(nx-\omega t)$$
$$\triangleq E_1, \tag{2.14}$$

$$D^{-2}\partial_x^2[(u^{\omega,n})^2] = D^{-2}\partial_x^2\left[\omega^2 n^{-2} + n^{-2s}\left[\frac{1}{2} + \frac{1}{2}\cos(2nx-2\omega t)\right]\right.$$
$$\left. + 2\omega n^{-1-s}\cos(nx-\omega t)\right]$$
$$= -n^{-2s+1}D^{-2}\partial_x[\sin(2nx-2\omega t)] - 2\omega n^{-s}D^{-2}\partial_x[\sin(nx-\omega t)]$$
$$= -2n^{-2s+2}D^{-2}[\cos(2nx-2\omega t)] - 2\omega n^{-s+1}D^{-2}[\cos(nx-\omega t)]$$
$$\triangleq E_2, \tag{2.15}$$

and

$$D^{-2}\partial_x[(u^{\omega,n})^2] = D^{-2}\partial_x\left[\omega^2 n^{-2} + n^{-2s}\left[\frac{1}{2} + \frac{1}{2}\cos(2nx-2\omega t)\right]\right.$$
$$\left. + 2\omega n^{-1-s}\cos(nx-\omega t)\right]$$
$$= -n^{-2s+1}D^{-2}[\sin(2nx-2\omega t)] - 2\omega n^{-s}D^{-2}[\sin(nx-\omega t)]$$
$$\triangleq E_3. \tag{2.16}$$

Thus, the error E of the approximate solutions (2.12) is

$$E(t) = E_1(t) + E_2(t) + E_3(t), \tag{2.17}$$

where E_j $(j=1,2,3)$ are defined in formulas (2.14), (2.15), and (2.16).

Next, we shall estimate the H^σ-norm of the error E in (2.17) by estimating separately the H^σ-norm of each term E_j.

Lemma 2.1 Let $\sigma \in \mathbf{Z}^+$ and $n \gg 1$, then

$$\|\cos(nx-\alpha)\|_{H^\sigma} \approx n^\sigma \text{ and } \|\sin(nx-\alpha)\|_{H^\sigma} \approx n^\sigma, \quad \alpha \in \mathbf{R}. \tag{2.18}$$

Estimating the H^σ-Norm of E_1 Applying the triangle inequality, we have

$$\|E_1\|_{H^\sigma} = \|3\omega n^{-s}\sin(nx-\omega t) + \frac{1}{2}n^{-2s+1}\sin 2(nx-\omega t)\|_{H^\sigma}$$
$$\leqslant 3\omega n^{-s}\|\sin(nx-\omega t)\|_{H^\sigma} + \frac{1}{2}n^{-2s+1}\|\sin 2(nx-\omega t)\|_{H^\sigma}, \tag{2.19}$$

which, by Lemma 2.1, gives

$$\|E_1\|_{H^\sigma} \lesssim n^{-s+\sigma} + n^{-2s+1+\sigma}, \quad n \gg 1. \tag{2.20}$$

Estimating the H^σ-Norm of E_2   Also, we have

$$\|E_2\|_{H^\sigma} = \| - 2n^{-2s+2} D^{-2}[\cos(2nx - 2\omega t)] - 2\omega n^{-s+1} D^{-2}[\cos(nx - \omega t)]\|_{H^\sigma}$$

$$\leqslant 2n^{-2s+2} \|D^{-2}[\cos(2nx - 2\omega t)]\|_{H^\sigma} + 2\omega n^{-s+1} \|D^{-2}[\cos(nx - \omega t)]\|_{H^\sigma}$$

$$\leqslant 2n^{-2s+2} \|\cos(2nx - 2\omega t)\|_{H^{\sigma-2}} + 2\omega n^{-s+1} \|\cos(nx - \omega t)\|_{H^{\sigma-2}}. \tag{2.21}$$

Thus, by Lemma 2.1, we obtain

$$\|E_2\|_{H^\sigma} \lesssim n^{-2s+\sigma} + n^{-s+\sigma-1}, \quad n \gg 1. \tag{2.22}$$

Estimating the H^σ-Norm of E_3   Finally, we have

$$\|E_3\|_{H^\sigma} = \| - n^{-2s+1} D^{-2}[\sin(2nx - 2\omega t)] - 2\omega n^{-s} D^{-2}[\sin(nx - \omega t)]\|_{H^\sigma}$$

$$\leqslant \|n^{-2s+1} D^{-2}[\sin(2nx - 2\omega t)]\|_{H^\sigma} + \|2\omega n^{-s} D^{-2}[\sin(nx - \omega t)]\|_{H^\sigma}$$

$$\leqslant n^{-2s+1} \|\sin(2nx - 2\omega t)\|_{H^{\sigma-2}} + 2\omega n^{-s} \|\sin(nx - \omega t)\|_{H^{\sigma-2}}$$

which by Lemma 2.1 gives

$$\|E_3\|_{H^\sigma} \lesssim n^{-2s+\sigma-1} + n^{-s+\sigma-2}, \quad n \gg 1. \tag{2.23}$$

Lemma 2.2   *If ω is bounded then for $n \gg 1$, we have*

$$\|E(t)\|_{H^\sigma} \lesssim n^{-2s+\sigma-1} + n^{-s+\sigma-2} + n^{-2s+\sigma} + n^{-s+\sigma-1} + n^{-s+\sigma} + n^{-2s+1+\sigma},$$
$$n \gg 1. \tag{2.24}$$

In particular, if $s > (1+\sigma)/2$, then

$$\|E(t)\|_{H^\sigma} \lesssim n^{-r_s}, \quad n \gg 1,$$

where $r_s > 0$ and

$$r_s = \begin{cases} 2s - 1 - \sigma, & \text{if } (1+\sigma)/2 < s \leqslant 1, \\ s - \sigma, & \text{if } s \geqslant 1. \end{cases} \tag{2.25}$$

2.1.2 Estimating the difference between approximate and actual solutions

Now that we have these initial estimates for the approximate solutions, we will proceed to construct our family of actual solutions. Let $u_{\omega,n}(x,t)$ be the actual solutions to the i.v.p. (1.1)–(1.2) with initial data $u_{\omega,n}(x,0) = u^{\omega,n}(x,0)$. More precisely, $u_{\omega,n}(x,t)$ solves

$$\partial_t u_{\omega,n} - 2u_{\omega,n}\partial_x u_{\omega,n} = \partial_x D^{-2}\left[\partial_x u_{\omega,n}^2 + u_{\omega,n}^2\right], \tag{2.26}$$

$$u_{\omega,n}(x,0) = \omega n^{-1} + n^{-s}\cos nx. \tag{2.27}$$

Notice the initial data $u^{\omega,n}(x,0) \in H^s$ for all $s \geq 0$, since

$$\|u_{\omega,n}(\cdot,0)\|_{H^s} = \omega n^{-1} + n^{-s}\|\cos nx\|_{H^s} \approx 1, \tag{2.28}$$

for n sufficiently large. Hence, by Theorem 1.1, there is a $T > 0$ such that for $n > 1$, the Cauchy problem (2.26)–(2.27) has a unique solution in $C([0,T] : H^s)$ with lifespan $T > 0$ such that $u_{\omega,n}$ satisfies (1.8) for $t \in [0,T]$.

To estimate the difference between the actual solutions and the approximate solutions, we define $v = u^{\omega,n} - u_{\omega,n}$ which satisfies the following i.v.p.,

$$\partial_t v = E + \partial_x(fv) + \partial_x D^{-2}[\partial_x(fv) + fv], \tag{2.29}$$

$$v(x,0) = 0,$$

where $f = u^{\omega,n} + u_{\omega,n}$, E satisfies the estimate in (2.17).

We will now show that the H^s norm of the difference v decays to zero as n goes to infinity.

Lemma 2.3 *Let $s > 1 + \sigma$, then the H^σ-norm of E is bounded by*

$$\|v(t)\|_{H^\sigma} \lesssim n^{-s+\sigma}. \tag{2.30}$$

Proof Apply the operator D^σ to both sides of (2.29), multiply by $D^\sigma v$, and integrate over the torus to get

$$\frac{1}{2}\frac{d}{dt}\|v(t)\|_{H^s}^2 = \int_\mathbb{T} D^\sigma E D^\sigma v\, dx + \int_\mathbb{T} D^\sigma \partial_x(fv)D^\sigma v\, dx$$
$$+ \int_\mathbb{T} D^\sigma \partial_x D^{-2}\left[\partial_x(fv) + fv\right]D^\sigma v\, dx. \tag{2.31}$$

By the definition of f and the algebra property it follows that

$$\|f\|_{H^\sigma} \lesssim \|u^{\omega,n}\|_{H^{\sigma+1}} + \|u_{\omega,n}\|_{H^{\sigma+1}},$$

thus, restricting $\sigma < s - 1$, we find

$$\|f\|_{H^\sigma} \lesssim \|u^{\omega,n}\|_{H^\sigma} + \|u_{\omega,n}\|_{H^\sigma}.$$

Moreover, the well-posedness size estimate (2.20) implies

$$\|u^{\omega,n}(t)\|_{H^\sigma} \lesssim \|u^{\omega,n}(0)\|_{H^\sigma} \lesssim 1$$

and

$$\|u_{\omega,n}(t)\|_{H^\sigma} \lesssim \|u_{\omega,n}(0)\|_{H^\sigma} \lesssim 1,$$

then implies that $\|g\|_{H^\sigma} \lesssim 1$. Similarly, we have $\|f\|_{H^\sigma} \lesssim 1$.

We now estimate the integrals on the right-hand side of (2.31).

Estimate of the first integral Applying the Cauchy-Schwarz inequality, we have

$$\left| \int_{\mathbb{T}} D^\sigma E D^\sigma v \, dx \right| \leq \|E\|_{H^\sigma} \|v\|_{H^\sigma} \lesssim n^{-s+\sigma} \|v\|_{H^\sigma}. \tag{2.32}$$

Estimate of the second integral We begin by rewriting this term by commuting v with $D^\sigma \partial_x$ to arrive at

$$\int_{\mathbb{T}} D^\sigma \partial_x (fv) D^\sigma v \, dx = \int_{\mathbb{T}} [D^\sigma \partial_x, f] v D^\sigma v \, dx + \int_{\mathbb{T}} f D^\sigma \partial_x v D^\sigma v \, dx. \tag{2.33}$$

The first integral of (2.33) can be handled by the following Calderon-Coifman-Meyer type commutator estimate that can be found in [22].

Lemma 2.4 *If $\sigma + 1 \geq 0$, then*

$$\|[D^\sigma \partial_x, w] v\|_{L^2} \leq C \|w\|_{H^\rho} \|v\|_{H^\sigma} \tag{2.34}$$

provided that $\rho > \dfrac{3}{2}$ and $\sigma + 1 \leq \rho$.

Applying Lemma 2.4 with $\sigma + 1 \geq 0, \rho = s > \dfrac{3}{2}$ and $\sigma + 1 \leq s$ tells us

$$\left| \int_{\mathbb{T}} [D^\sigma \partial_x, f] v D^\sigma v \, dx \right| \lesssim \|f\|_{H^s} \|v\|_{H^\sigma}^2. \tag{2.35}$$

Integrating by parts and using the Sobolev Theorem, we have

$$\left| \int_{\mathbb{T}} f D^\sigma \partial_x v D^\sigma v \, dx \right| = \frac{1}{2} \left| \int_{\mathbb{T}} (D^\sigma v)^2 \partial_x f \, dx \right| \lesssim \|\partial_x f\|_{L^\infty} \|v\|_{H^\sigma}^2 \lesssim \|f\|_{H^s} \|v\|_{H^\sigma}^2. \tag{2.36}$$

We obtain

$$\int_\mathbb{T} D^\sigma \partial_x(fv) D^\sigma v \, dx \lesssim \|f\|_{H^s} \|v\|_{H^\sigma}^2 \lesssim \|v\|_{H^\sigma}^2. \tag{2.37}$$

Estimate of the third integral For the third term, we observe that after applying the Cauchy-Schwarz inequality, the quantity $fv + \partial_x fv$ will be in the $H^{\sigma-1}$ space, which precludes the use of the algebra property. To overcome this obstacle, we will apply the following multiplier estimate, also found in [22].

Lemma 2.5 *Let $\sigma \in \left(\frac{1}{2}, 1\right)$, then*

$$\|fg\|_{H^{\sigma-1}} \lesssim \|f\|_{H^{\sigma-1}} \|g\|_{H^\sigma}.$$

Applying Lemma 2.5, we obtain

$$\int_\mathbb{T} D^\sigma \partial_x D^{-2}(fv + \partial_x(fv)) D^\sigma v \, dx \leqslant \|\partial_x D^{-2}(fv + \partial_x(fv))\|_{H^\sigma} \|v\|_{H^\sigma}$$
$$\leqslant \|\partial_x D^{-2}(fv)\|_{H^\sigma} \|v\|_{H^\sigma}$$
$$\quad + \|\partial_x D^{-2}(\partial_x(gv))\|_{H^\sigma} \|v\|_{H^\sigma}$$
$$\lesssim \|fv\|_{H^{\sigma-1}} \|v\|_{H^\sigma} + \|\partial_x(fv)\|_{H^{\sigma-1}} \|v\|_{H^\sigma}$$
$$\lesssim \|f\|_{H^{\sigma-1}} \|v\|_{H^\sigma}^2 + \|f\|_{H^\sigma} \|v\|_{H^\sigma}^2$$
$$\lesssim \|v\|_{H^\sigma}^2. \tag{2.38}$$

Finally, combining (2.32), (2.33) and (2.38), we obtain an energy estimate for v,

$$\frac{1}{2}\frac{d}{dt}\|v(t)\|_{H^s}^2 \lesssim n^{-s+\sigma} \|v\|_{H^\sigma} + \|v\|_{H^\sigma}^2. \tag{2.39}$$

Solving (2.39) and using the error estimate (2.24), we arrive at the desired estimate (2.30).

Proof of Theorem 1.2 in the periodic case The basis of our proof rests upon finding two sequences of solutions to the i.v.p. (1.1)–(1.2) that share a common lifespan and satisfy three conditions. We take the sequence of solutions with $\omega = \pm 1$. The three conditions they satisfy are as follows

(1)
$$\|w_{\omega,n}(t)\|_{H^s} \lesssim 1 \text{ for } t \in [0,T],$$

(2)
$$\|u_{1,n}(0) - u_{-1,n}(0)\|_{H^s} \to 0 \text{ as } n \to \infty,$$

(3)
$$\liminf_{n\to\infty} \|u_{1,n}(t) - u_{-1,n}(t)\|_{H^s} \gtrsim |\sin t| \quad \text{for } t \in (0, T].$$

Property (1), by the solution size estimate (1.8), we have
$$\|u_{\omega,n}(t)\|_{H^s} \lesssim \|u_{\omega,n}(0)\|_{H^s} \lesssim 1. \tag{2.40}$$

Property (2), follows from the definition of our approximate solutions (2.12), we have
$$\|u_{1,n}(0) - u_{-1,n}(0)\|_{H^s} = \|u^{1,n}(0) - v^{-1,n}(0)\|_{H^s}$$
$$= \|n^{-1} + n^{-s}\cos nx + n^{-1} - n^{-s}\cos nx\|_{H^s}$$
$$= 4\pi n^{-1} \to 0 \quad as \quad n \to \infty.$$

For property (3), using the reverse triangle inequality, we get
$$\|u_{1,n}(t) - u_{-1,n}(t)\|_{H^s} \geqslant \|u^{1,n}(t) - u^{-1,n}(t)\|_{H^s} - \|u^{1,n}(t) - u_{1,n}(t)\|_{H^s}$$
$$- \|u^{-1,n}(t) - u_{-1,n}(t)\|_{H^s}$$
$$\gtrsim \|u^{1,n}(t) - u^{-1,n}(t)\|_{H^s} - n^{-\varepsilon},$$

from which we obtain that
$$\liminf_{n\to\infty} \|u_{1,n}(t) - u_{-1,n}(t)\|_{H^s} \gtrsim \liminf_{n\to\infty} \|u^{1,n}(t) - u^{-1,n}(t)\|_{H^s}. \tag{2.41}$$

Since, by the trigonometric identity $\cos(\alpha - \beta) - \cos(\alpha + \beta) = 2\sin\alpha\sin\beta$, we have
$$u^{1,n}(t) - u^{-1,n}(t) = 2n^{-1} + 2n^{-s}\sin nx \sin t.$$

Then, inequality (2.41) gives
$$\liminf_{n\to\infty} \|u_{1,n}(t) - u_{-1,n}(t)\|_{H^s} \gtrsim \liminf_{n\to\infty}(|\sin t| - n^{-1}) \gtrsim |\sin t|, \tag{2.42}$$

which completes the proof of property (3). We complete the proof of Theorem 1.2 in the periodic case.

2.2 Nonuniform dependence in the Nonperiodic case

For each $\omega \in \{0, 1\}$, we consider the following sequence of high frequency approximate solutions
$$u^h = u^{h,\omega,n}(x, t) = n^{-\frac{\delta}{2}-s}\varphi\left(\frac{x}{n^\delta}\right)\cos(nx - \omega t), \tag{2.43}$$

where φ is a C^∞ function such that

$$\varphi = \begin{cases} 1, & \text{if } |x| < 1, \\ 0, & \text{if } |x| \geq 2. \end{cases} \quad (2.44)$$

We then modify u^h by adding to it the solution $u_\ell = u_{\ell,\omega,n}(x,t)$ of the following Cauchy problem with the low-frequency initial data $\omega n^{-1} \varphi\left(\dfrac{x}{n^\delta}\right)$,

$$\partial_t u_\ell - 2u_\ell \partial_x u_\ell - \partial_x D^{-2} \left[\partial_x u_\ell^2 + u_\ell^2\right] = 0, \quad x \in \mathbf{R}, t \in \mathbf{R} \quad (2.45)$$

$$u_\ell(x,0) = \omega n^{-1} \varphi\left(\dfrac{x}{n^\delta}\right), \quad x \in \mathbf{R}, \quad (2.46)$$

where $\tilde{\varphi}$ is a $C_0^\infty(\mathbf{R})$ function such that

$$\tilde{\varphi}(x) = 1, \quad \text{if } x \in \operatorname{supp}\varphi. \quad (2.47)$$

Thus, the two sequences of approximate solutions that will be used here for proving nonuniform dependence for Eq. (1.1) are

$$u^{\omega,n} = u_\ell + u^h, \quad \omega = 0 \text{ or } 1. \quad (2.48)$$

Note that for ω, we have $u_\ell(x,0) = 0$. Therefore, in this case $u_\ell = 0$ which gives

$$u^{0,n}(x,t) = n^{-\frac{\delta}{2}-s} \varphi\left(\dfrac{x}{n^\delta}\right) \cos(nx), \quad (2.49)$$

that is this sequence has no low-frequency term.

Proof of Theorem 1.2 in the Nonperiodic case The proof is quite similar to that for the CH equation [22] and Novikov equation [20], so it is omitted to make the paper concise. This completes the proof of Theorem 1.2.

3 Well-posedness

We will now prove well-posedness for the generalized Camassa-Holm equation (1.1)–(1.2) for the periodic and the nonperiodic case, respectively.

3.1 Well-posedness on the circle

We observe that for $u \in H^s$ the product $u\partial_x u$ is in H^{s-1}, thus the equation (1.1) can not be thought as an ODE on the space H^s. To deal with this problem we replace the MHS i.v.p. (1.1)–(1.2) by a mollified version

$$\partial_t u = J_\epsilon [J_\epsilon 2u \partial_x J_\epsilon u] + \partial_x D^{-2}\left[\partial_x u^2\right] + \partial_x D^{-2}\left[u^2\right] \doteq F_\epsilon, \quad (3.1)$$

$$u(0,x) = u_0(x), \quad u_0(x) \in H^s, \qquad (3.2)$$

where for each $\epsilon \in (0,1]$, the operator J_ϵ is the Friedrichs mollifier. We fix a Schwartz function $j \in \mathcal{S}(\mathbf{R})$ that satisfies $0 \leqslant \widehat{j}(\xi) \leqslant 1$ for every $\xi \in \mathbf{R}$ and $\widehat{j}(\xi)$ for $\xi \in [-1.1]$. This allows us to define the periodic functions j_ϵ as

$$j_\epsilon(x) = \frac{1}{2\pi} \sum_{n \in \mathbf{Z}} \widehat{j}(\epsilon n) e^{inx}. \qquad (3.3)$$

Then J_ϵ is defined by

$$J_\epsilon f(x) = j_\epsilon f(x). \qquad (3.4)$$

This construction of j_ϵ results in two lemmas that will prove repeatedly useful throughout the paper.

Lemma 3.1 ([28]) *Let $s > 0$ and J_ϵ be defined as in (3.4). Then for any $f \in H^s$, we have $J_\epsilon f \to f$ in H^s.*

Lemma 3.2 ([28]) *Let $r < s$, the map $I - J_\epsilon : H^s \to H^r$ and J_ϵ be defined as in (3.4). Then for any $f \in H^s$, we have $J_\epsilon f \to f$ in H^s satisfies the operator norm estimate*

$$\|I - J_\epsilon\|_{\mathcal{L}(H^s, H^r)} = o(\epsilon^{s-r}). \qquad (3.5)$$

Hence, for each $\epsilon \in (0,1]$, (3.1)–(3.2) has a unique solution u_ϵ with lifespan $T_\epsilon > 0$.

Our strategy is now to demonstrate that the Cauchy problem (3.1)–(3.1) satisfies the hypotheses of the Fundamental ODE Theorem. We will therefore obtain a unique solution $u_\epsilon(\cdot, t) \in H^s, |t| < T_\epsilon$, for some $T_\epsilon > 0$. This idea is summarized in the following lemma.

Lemma 3.3 *For all $\epsilon \in (0,1]$ the mollified i.v.p. (3.1)–(3.2) has a unique solution $u_\epsilon \in C([0,T]; H^s)$ with lifespan $T_\epsilon > 0$.*

Proof The map $F_\epsilon : H^s \to H^s$ is well-defined. The only remaining hypothesis of the Fundamental ODE Theorem, we need to show is satisfied is that F_ϵ is a continuously differentiable map. In this case, derivative of F_ϵ can be explicitly calculated at each $u_0 \in H^s$ as

$$F'_\epsilon(u_0)u = J_\epsilon(J_\epsilon u \partial_x J_\epsilon u_0 + J_\epsilon u_0 \partial_x J_\epsilon u) + 2\partial_x D^{-2}(u_0 u + u \partial_x u_0 + u_0 \partial_x u).$$

Hence, for each $\epsilon \in (0,1]$, (3.1)–(3.2) has a unique solution u_ϵ with lifespan $T_\epsilon > 0$.

3.1.1 Energy estimate and lifespan of solution u_ϵ

For each ϵ, there is a solution u_ϵ to the mollified generalized Camassa-Holm equation (3.1)–(3.2). The lifespan of each of these solutions has a lower bound T_ϵ. In this subsection, we shall demonstrate that there is actually a lower bound $T > 0$ that does not depend upon ϵ. This estimate is crucial in our proofs. To show the existence of T, we shall derive an energy estimate for the u_ϵ. Applying the operator D^s to both sides of (3.1), multiplying by $D^s u_\epsilon$, and integrating over the torus yields the H^s-energy of u_ϵ

$$\frac{1}{2}\frac{d}{dt}\|u_\epsilon(t)\|_{H^s}^2 = 2\int_{\mathbb{T}} D^s [J_\epsilon(J_\epsilon u_\epsilon \partial_x J_\epsilon u_\epsilon)] D^s J_\epsilon u_\epsilon \, dx + \int_{\mathbb{T}} D^s \partial_x D^{-2} [\partial_x(u_\epsilon^2)] D^s u_\epsilon \, dx$$
$$+ \int_{\mathbb{T}} D^s \partial_x D^{-2} [u_\epsilon^2] D^s u_\epsilon \, dx. \tag{3.6}$$

To bound the energy, we will need the following Kato-Ponce [33] commutator estimate.

Lemma 3.4 (Kato-Ponce) *If $s > 0$, then there is $c_s > 0$ such that*

$$\|[D^s, f]g\|_{L^2} \lesssim c_s(\|D^s f\|_{L^2}\|g\|_{L^\infty} + \|\partial_x\|_{L^\infty}\|D^{s-1}f\|_{L^2}). \tag{3.7}$$

We now rewrite the first term of (3.6) by first commuting the exterior J_ϵ and then commuting the operator D^s with $J_\epsilon u_\epsilon$ arriving at

$$\int_{\mathbb{T}} D^s[J_\epsilon u_\epsilon \partial_x J_\epsilon u_\epsilon] D^s J_\epsilon u_\epsilon \, dx = \int_{\mathbb{T}} [D^s, J_\epsilon u_\epsilon] \partial_x J_\epsilon u_\epsilon D^s J_\epsilon u_\epsilon \, dx$$
$$+ \int_{\mathbb{T}} J_\epsilon u_\epsilon \partial_x D^s J_\epsilon u_\epsilon D^s J_\epsilon u_\epsilon \, dx. \tag{3.8}$$

Setting $v = J_\epsilon u_\epsilon$, we can bound the first term of (3.8) by first using the Cauchy-Schwarz inequality and then applying lemma 3.4 to arrive at

$$\int_{\mathbb{T}} [D^s, v]\partial_x v D^s v \, dx \lesssim (\|v\|_{H^s}\|\partial_x v\|_{L^\infty} + \|\partial_x v\|_{H^{s-1}}\|\partial_x v\|_{L^\infty})\|v\|_{H^s}.$$
$$\lesssim \|v\|_{C^1}\|v\|_{H^s}^2. \tag{3.9}$$

For the second term of (3.8), we have

$$\int_{\mathbb{T}} v\partial_x D^s v D^s v \, dx = -\frac{1}{2}\int_{\mathbb{T}} \partial_x v (D^s v)^2 \, dx$$
$$\lesssim \sup|\partial_x v| \int_{\mathbb{T}} (D^s v)^2 \, dx$$

$$\lesssim \|v\|_{C^1}\|v\|_{H^s}. \tag{3.10}$$

The second term on the right-hand side of (3.6) is bounded by the first applying the Cauchy-Schwarz inequality and then using of the estimate (2.8) and the algebra property of H^s. Here we have

$$\int_{\mathbb{T}} D^s \partial_x D^{-2}\left(\partial_x u_\epsilon^2\right) \cdot D^s u_\epsilon \, dx \leqslant \|\partial_x D^{-2}\partial_x u_\epsilon^2\|_{H^s}\|u_\epsilon\|_{H^s}$$
$$\leqslant \|\partial_x u_\epsilon^2\|_{H^{s-1}}\|u_\epsilon\|_{H^s}$$
$$\lesssim \|u_\epsilon\|_{C^1}^2\|u_\epsilon\|_{H^s} \tag{3.11}$$

and

$$\int_{\mathbb{T}} D^s \partial_x D^{-2} u_\epsilon^2 \cdot D^s u_\epsilon \, dx \leqslant \|\partial_x D^{-2} u_\epsilon^2\|_{H^s}\|u_\epsilon\|_{H^s}$$
$$\leqslant \|u_\epsilon^2\|_{H^{s-1}}\|u_\epsilon\|_{H^s}$$
$$\lesssim \|u_\epsilon\|_{C^1}^2\|u_\epsilon\|_{H^s}. \tag{3.12}$$

Using the fact that

$$\|J_\epsilon u_\epsilon\|_{H^s} \leqslant \|u_\epsilon\|_{H^s} \tag{3.13}$$

and combining these results we conclude that u_ϵ satisfies the differential inequality

$$\frac{1}{2}\frac{d}{dt}\|u_\epsilon(t)\|_{H^s}^2 \lesssim \|u_\epsilon(t)\|_{C^1}^2\|u_\epsilon(t)\|_{H^s}^2. \tag{3.14}$$

This together with Sobolev's lemma gives

$$\frac{1}{2}\frac{d}{dt}\|u_\epsilon(t)\|_{H^s}^2 \leqslant c_s\|u_\epsilon(t)\|_{H^s}^4. \tag{3.15}$$

Set $y = \|u_\epsilon(t)\|_{H^s}^2$, the differential inequality (3.15)

$$\frac{1}{2}y^{-2}\frac{dy}{dt} \leqslant c_s, \quad y(0) = \|u_0\|_{H^s}^2. \tag{3.16}$$

Integrating (3.16) from 0 to t gives

$$\frac{1}{u(0)} - \frac{1}{u(t)} \leqslant c_s 2t. \tag{3.17}$$

Solving for $y(t)$ and substituting back in $y = \|u_\epsilon(t)\|_{H^s}^2$ gives us the inequality

$$\|u_\epsilon(t)\|_{H^s} \leqslant \frac{\|u_0\|_{H^s}}{(1 - 2c_s\|u_0\|_{H^s}^2 t)^{\frac{1}{2}}}. \tag{3.18}$$

Setting $T = \dfrac{1}{4c_s\|u_0\|_{H^s}^2}$, we see from (3.18) that the solution u_ϵ exists for $0 \leqslant t \leqslant T$ and satisfies a solution size bound

$$\|u_\epsilon(t)\|_{H^s} \leqslant 2\|u_0\|_{H^s}, \quad 0 \leqslant t \leqslant T. \tag{3.19}$$

Therefore we see that T is a lower bound for the lifespan of u_ϵ independent of ϵ.

3.1.2 Existence

Proposition 3.1 *There exists a solution u to the i.v.p. (1.1)–(1.2) in the space $L^\infty([0,1]; H^s) \cap \text{Lip}([0,1]; H^{s-1})$. Furthermore, the H^s norm of u satisfies (1.6).*

Proof Our proof revolves around refining the convergence of the family $\{u_\epsilon\}$ several times by extracting subsequences $\{u_{\epsilon_\nu}\}$. After each such extraction, it is assumed that the resulting subsequence is relabelled as $\{u_\epsilon\}$.

Weak* convergence in $L^\infty([0,T]; H^s)$ The family $\{u_\epsilon\}_{\epsilon\in(0,1]}$ is bounded in the space $C([0,T]; H^s) \subset L^\infty([0,T]; H^s)$. By the inequality (3.19), we have

$$\|u_\epsilon\|_{L^\infty([0,1],H^s)} = \sup_{t\in[0,T]} \|u_\epsilon(t)\|_{H^s} \leqslant 2\|u_0\|_{H^s}$$

$$\Rightarrow \{u_\epsilon\}_{\epsilon\in(0,1]} \subset \overline{B}(0, 2\|u_0\|_{H^s}) \subset L^\infty([0,T]; H^s). \tag{3.20}$$

The Riesz representation theorem implies that $L^\infty([0,T]; H^s) = (L^1([0,T]; H^s))^*$, where we define the duality relation for $\varphi \in L^1([0,T]; H^s)$ by

$$\langle u_\epsilon \varphi \rangle = \int_0^T \sum_{k\in\mathbf{Z}} (1+k^2)^s \widehat{u}_\epsilon(t,k) \widehat{\varphi}(t,k)\, dt. \tag{3.21}$$

We may thus apply Alaoglu's theorem to deduce that $\{u_\epsilon\}$ is precompact in $\overline{B}(0, 2\|u_0\|_{H^s}) \subset L^\infty([0,T]; H^s)$ with respect to the weak* topology. Therefore we may extract a subsequence $\{u_{\epsilon_\nu}\}$ that converges to an element $u \in \overline{B}(0, 2\|u_0\|_{H^s})$ weakly*. Given this construction, u will satisfy our solution size estimate (1.8).

Strong convergence in $C([0,T]; H^{s-1})$ We will prove that the family $\{u_\epsilon\}_{\epsilon\in(0,1]}$ satisfies the hypotheses of Ascoli's Theorem. We begin with the equicontinuity condition. For $t_1, t_2 \in [0,T]$, we have

$$\|u_\epsilon(t_1) - u_\epsilon(t_1)\|_{H^{s-1}} \leqslant \sup_{t\in[0,T]} \|\partial_t u_\epsilon(t)\|_{H^{s-1}} |t_1 - t_2|. \tag{3.22}$$

Using the fact that u_ϵ satisfies (3.1)–(3.2), we see that the H^{s-1} norm of $\partial_t u_\epsilon$ can be bounded independently of ϵ as follows

$$\sup_{t\in[0,T]} \|\partial_t u_\epsilon(t)\|_{H^{s-1}} \tag{3.23}$$

$$= \sup_{t\in[0,T]} \|2J_\epsilon[J_\epsilon u \partial_x J_\epsilon u] + \partial_x D^{-2}\partial_x u^2 + \partial_x D^{-2} u^2\|_{H^{s-1}}$$

$$\lesssim \sup_{t\in[0,T]} (\|J_\epsilon[J_\epsilon u \partial_x J_\epsilon u]\|_{H^{s-1}} + \|\partial_x D^{-2}\partial_x u^2\|_{H^{s-1}} + \|\partial_x D^{-2} u^2\|_{H^{s-1}})$$

$$\lesssim \sup_{t\in[0,T]} (\|J_\epsilon[J_\epsilon u \partial_x J_\epsilon u]\|_{H^{s-1}} + \|\partial_x u^2]\|_{H^{s-1}} + \|u^2\|_{H^{s-2}})$$

$$\lesssim \|u_0\|_{H^s}^2. \tag{3.24}$$

Next, we observe that for each $t \in [0,T]$ the set $U(t) = \{u_\epsilon\}_{\epsilon\in(0,1]}$ is bounded in H^s. Since T is a compact manifold, the inclusion mapping $i : H^s \to H^{s-1}$ is a compact operator, and therefore we may deduce that $U(t)$ is a precompact set in H^{s-1}. As the two hypotheses of Ascoli's Theorem have been satisfied, we have a subsequence $\{u_{\epsilon_v}\}$ that converges in $([0,T]; H^{s-1})$. By uniqueness of limits, this subsequence must converge to u.

Strong convergence in $C([0,T]; H^{s-\sigma})$ for $\sigma \in (0,1]$ As in the previous case, we will prove that the family $\{u_\epsilon\}$ satisfies the hypotheses of Ascoli's Theorem. To establish the equicontinuity condition, we see that for $t_1, t_2 \in [0,T]$ that we have

$$\|u_\epsilon(t_1) - u_\epsilon(t_2)\|_{H^{s-\sigma}} \leqslant \|u_\epsilon\|_{C^\sigma([0,T]; H^{s-\sigma})} |t_1 - t_2|^\sigma. \tag{3.25}$$

Our objective therefore is to bound $\|u_\epsilon\|_{C^\sigma([0,T]; H^{s-\sigma})}$ independently of ϵ. We begin with the definition of this norm as

$$\|u_\epsilon\|_{C^\sigma([0,T]; H^{s-\sigma})} \doteq \sup_{t\in[0,T]} \|u_\epsilon(t)\|_{H^{s-\sigma}} + \sup_{t\neq t'} \frac{\|u_\epsilon(t) - u_\epsilon(t')\|_{H^{s-\sigma}}}{|t_1 - t_2|^\sigma}. \tag{3.26}$$

The first term on the right-hand side of (3.26) is bounded by the application of (3.20), giving us

$$\sup_{t\in[0,T]} \|u_\epsilon(t)\|_{H^{s-\sigma}} \leqslant \sup_{t\in[0,T]} \|u_\epsilon(t)\|_{H^s} \leqslant 2\|u_0\|_{H^s}. \tag{3.27}$$

For the second term, we begin with two elementary inequalities. First, as $\sigma \in (0,1)$, we have

$$\frac{1}{(1+k^2)\sigma|t-t'|^{2\sigma}} \leqslant \left(1 + \frac{1}{(1+k^2)|t-t'|^2}\right)^\sigma \leqslant 1 + \frac{1}{(1+k^2)|t-t'|^2}. \tag{3.28}$$

Using this inequality, we may further deduce that

$$\frac{(1+k^2)^{s-\sigma}}{|t-t'|^{2\sigma}} \leqslant (1+k^2)^s + \frac{(1+k^2)^{s-\sigma}}{|t-t'|^{2\sigma}}. \tag{3.29}$$

We therefore can bound this term by

$$\sup_{t \neq t'} \frac{\|u_\epsilon(t) - u_\epsilon(t')\|_{H^{s-\sigma}}}{|t_1 - t_2|^{2\sigma}} = \sup_{t \neq t'} \sum_{k \in \mathbb{Z}} \frac{(1+k^2)^{s-\sigma}}{|t-t'|^{2\sigma}} |\widehat{u}_\epsilon(k,t) - \widehat{u}_\epsilon(k,t')|^2$$

$$\leqslant \sup_{t \neq t'} \sum_{k \in \mathbb{Z}} 1 + k^2)^s |\widehat{u}_\epsilon(k,t) - \widehat{u}_\epsilon(k,t')|^2$$

$$+ \sup_{t \neq t'} \sum_{k \in \mathbb{Z}} \frac{(1+k^2)^{s-\sigma}}{|t-t'|^{2\sigma}} |\widehat{u}_\epsilon(k,t) - \widehat{u}_\epsilon(k,t')|^2$$

$$\lesssim \sup_{t \in [0,T]} \|u_\epsilon(t)\|_{H^s}^2 + \sup_{t \in [0,T]} \|\partial_x u_\epsilon(t)\|_{H^{s-1}}^2$$

$$\lesssim \|u_0\|_{H^s}^2. \qquad (3.30)$$

Combining (3.27) and (3.30), we have the ϵ independent bound

$$\|u_\epsilon\|_{C^\sigma([0,T];H^{s-\sigma})} \lesssim \|u_0\|_{H^s} + \|u_0\|_{H^s}^2. \qquad (3.31)$$

The precompactness condition is established in exactly the same fashion as the previous case as the inclusion mapping of H^s into $H^{s-\sigma}$ is a compact operator. As the two hypotheses of Ascoli have been satisfied, we may extract a subsequence that converges to u in $C([0,T]; H^{s-\sigma})$.

Strong convergence in $C([0,T]; C^1(\mathbb{T}))$ We will make $\sigma \in (0,1)$ so that $s - \sigma > \frac{3}{2}$. The Sobolev lemma then tells us that $H^{s-\sigma}$ embeds continuously into $C^1(\mathbb{T})$, which therefore implies that $u_\epsilon \to u$ in $C([0,T]; C^1(\mathbb{T}))$. We will next prove that $\partial_\epsilon u_\epsilon \to -u^k \partial_x u + \frac{1}{2} \partial_x^{-1} \left(\partial_x(u^k) \partial_x u \right)$ in $C([0,T]; C(\mathbb{T}))$.

Strong convergence of $\partial_t u_\epsilon$ in $C([0,T]; C^1(\mathbb{T}))$ From (3.1) we have

$$\partial_t u_\epsilon = 2J_\epsilon [J_\epsilon u_\epsilon \partial_x J_\epsilon u_\epsilon] + \partial_x D^{-2} \partial_x u_\epsilon^2 + \partial_x D^{-2} u_\epsilon^2. \qquad (3.32)$$

As we have already established that $u_\epsilon \to u$ in $C([0,T]; C^1(\mathbb{T}))$, it follows that $\partial_x u_\epsilon \to \partial_x u$. Using the fact that this space is an algebra, then the continuity of $\partial_x D^s$ implies the convergence of the nonlocal term

$$\partial_x D^{-2} \partial_x u_\epsilon^2 \to \partial_x D^{-2} \partial_x u^2, \quad \partial_x D^{-2} u_\epsilon^2 \to \partial_x D^{-2} u^2, \quad \text{in } C([0,T]; C(\mathbb{T})). \qquad (3.33)$$

Next, we observe that $J_\epsilon u_\epsilon \to u$ in $C([0,T]; C(\mathbb{T}))$ as

$$\|J_\epsilon u_\epsilon - u\|_{C([0,T]; C(\mathbb{T}))} = \|J_\epsilon u_\epsilon \pm u_\epsilon - u\|_{C([0,T]; C(\mathbb{T}))}$$

$$= \|J_\epsilon u_\epsilon - u_\epsilon\|_{C([0,T]; C(\mathbb{T}))} + \|u_\epsilon - u\|_{C([0,T]; C(\mathbb{T}))}. \qquad (3.34)$$

For the first term of this sum, choose r with $\frac{1}{2} < r < s$. Then applying Lemma 3.2, we see that for $t \in [0,T]$,

$$\|J_\epsilon u_\epsilon(t) - u(t)\|_{C(\mathbb{T})} \lesssim \|J_\epsilon u_\epsilon(t) \pm u_\epsilon(t) - u(t)\|_{H^s}$$
$$\lesssim \|I - J_\epsilon\|_{\mathcal{L}(H^s, H^r)}$$
$$= o(\epsilon^{s-r})\|u_t\|_{H^s}. \tag{3.35}$$

Estimating the second term immediately follows from the fact that we have established $u_\epsilon \to u$ in $C([0,T]; C(\mathbb{T}))$. We then examine ∂_x as above, and similarly conclude that $J_\epsilon \partial_x u_\epsilon \to \partial_x u$ in $C([0,T]; C(\mathbb{T}))$. Finally, proceeding via additive and multiplicative properties of limits we may deduce that $\partial_t u_\epsilon \to 2u\partial_x u + \partial_x D^{-2}\partial_x u^2 + \partial_x D^{-2} u^2$ in $C([0,T]; C(\mathbb{T}))$.

We are now ready to complete the proof of Proposition 3.1. We have established that in the space $C([0,T]; C(\mathbb{T}))$ we have $\partial_t u_\epsilon \to 2u^2 \partial_x u + \partial_x D^{-2}\partial_x(u^2 + \partial_x D^{-2} u^2$ in $C([0,T]; C(\mathbb{T}))$. Therefore we have that $u \mapsto u(t)$ is differentiable map from as that $I \to C([0,T]; C(\mathbb{T}))$ with $\partial_t u = 2u^2\partial_x u + \partial_x D^{-2}\partial_x u^2 + \partial_x D^{-2} u^2$ in $C([0,T]; C(\mathbb{T}))$. Thus $u \in L^\infty([0,T]; H^s)$ is a solution to (1.1)–(1.2). With this information, we can now establish that $u \in L^\infty([0,1]; H^s) \cap \text{Lip}([0,1]; H^{s-1})$ as we have

$$\|u_{t_1} - u(t_2)\|^2_{H^{s-1}} \sup_t \|\partial_t\|^2_{H^{s-1}} |t_1 - t_2| \lesssim \|u_0\|^2_{H^s} |t_1 - t_2|, \tag{3.36}$$

which implies that the Proposition 3.1 holds.

Now that we have established the existence of a solution u we will improve the conclusions we have established about its regularity.

Proposition 3.2 *The solution u to the i.v.p (3.1)–(3.2). constructed in Proposition 3.2 is an element of the space $C([0,T]; H^s)$.*

Proof Fix $t \in [0,T]$ and take a sequence $t_n \to t$. For the solution u to be continuous at t, we must have $\|u_{t_n} - u(t)\|_{H^s} \to 0$. This property is equivalent to $\|u_{t_n} - u(t)\|^2_{H^s} \to 0$. Using the definition of the norm, we have

$$\|u_{t_n} - u(t)\|^2_{H^s} = \langle u_{t_n} - u(t), u_{t_n} - u(t)\rangle_{H^s}$$
$$= \|u_{t_n}\|^2_{H^s} + \|u(t)\|^2_{H^s} - \langle u_{t_n}, u(t)\rangle_{H^s} - \langle u(t), u_{t_n}\rangle_{H^s}. \tag{3.37}$$

It suffices to show that u is continuous at t with respect to the weak topology on H^s, or

$$\lim_{n\to\infty} \langle u_{t_n}, u(t)\rangle_{H^s} = \lim_{n\to\infty} \langle u(t), u_{t_n}\rangle_{H^s} = \|u(t)\|^2_{H^s} \tag{3.38}$$

and that the map $t \mapsto \|u(t)\|_{H^s}^2$ is continuous.

Verifying u is continuous given the weak topology on H^s Let $\varphi \in H^s$. We must demonstrate that given a sequence $t_n \to t$ that we have

$$\lim_{n \to \infty} \langle u(t_n) - u(t), \varphi \rangle_{H^s} = 0. \tag{3.39}$$

Let $\epsilon > 0$. We will choose a $\psi \in C^\infty(\mathbb{T})$ with $\|\varphi - \psi\|_{H^s} < \dfrac{\epsilon}{8\|u_0\|_{H^s}}$. We then see that

$$|\langle u(t_n) - u(t), \varphi - \psi \rangle_{H^s}| \leq |\langle u(t_n) - u(t), \varphi - \psi \rangle_{H^s}| + |\langle u(t_n) - u(t), \psi \rangle_{H^s}|. \tag{3.40}$$

For the first term, we have

$$|\langle u(t_n) - u(t), \varphi - \psi \rangle_{H^s}| \leq \|u(t_n) - u(t)\|_{H^s} \|\varphi - \psi\|_{H^s} < \frac{\epsilon}{2}. \tag{3.41}$$

For the second term, using the Lipschitz property of u in H^{s-1}, we have

$$|\langle u(t_n) - u(t), \psi \rangle_{H^s}| \leq \|u(t_n) - u(t)\|_{H^{s-1}} \|\psi\|_{H^{s+1}} \tag{3.42}$$

$$\lesssim \|u_0\|_{H^s}^3 \|\psi\|_{H^{s+1}} |t_n - t|, \tag{3.43}$$

which is bounded by $\dfrac{\epsilon}{2}$ for sufficiently large n.

Verifying $t \mapsto \|u(t)\|_{H^s}^2$ is continuous. We begin by defining the functions

$$G(t) = \|u(t)\|_{H^s}^2, \quad G_\epsilon(t) = \|J_\epsilon u(t)\|_{H^s}^2. \tag{3.44}$$

Lemma 3.1 implies that $G_\epsilon \to G$ pointwise in t as $\epsilon \to 0$. Thus it suffices to show that each G_ϵ is Lipschitz, and that the Lipschitz constants for this family of functions are bounded. Applying the operator J_ϵ to the equation (1.1) we obtain the following H^s energy inequality for G_ϵ,

$$\frac{1}{2}|G'_\epsilon(t)| \leq \left|2\int_\mathbb{T} D^s J_\epsilon(u\partial_x u) D^s J_\epsilon u\, dx\right| + \left|\int_\mathbb{T} D^s J_\epsilon(\partial_x D^{-2}\partial_x u^2) D^s J_\epsilon u\, dx\right|$$
$$+ \left|\int_\mathbb{T} D^s J_\epsilon(\partial_x D^{-2} u^2) D^s J_\epsilon u\, dx\right|. \tag{3.45}$$

To bound the first term on the right-hand side of (3.45), we first commute the operator D^s with u^k to obtain

$$\left|\int_\mathbb{T} D^s J_\epsilon(u\partial_x u) D^s J_\epsilon u\, dx\right| \leq \left|\int_\mathbb{T} [D^s, u]\partial_x u \cdot D^s J_\epsilon^2 u\, dx\right|$$

$$+ \left| \int_{\mathbb{T}} J_\epsilon u D^s \partial_x u \cdot D^s J_\epsilon u \, dx \right|. \tag{3.46}$$

Using the Cauchy-Schwarz inequality and the lemma 3.4 (Kato-Ponce), we have

$$\left| \int_{\mathbb{T}} [D^s, u] \partial_x u \cdot D^s J_\epsilon^2 u \, dx \right| \lesssim \|[D^s, u] \partial_x u\|_{L^2} \|D^s J_\epsilon^2 u\|_{L^2}$$

$$\lesssim \|u\|_{H^s} (\|u\|_{H^s} \|\partial_x u\|_{L^\infty} + \|\partial_x u\|_{H^{s-1}} \|\partial_x u\|_{L^\infty})$$

$$\lesssim \|u_0\|_{H^s}^3, \tag{3.47}$$

where in the last inequality we used the solution size estimate $\|u(t)\|_{H^s} \lesssim \|u_0\|_{H^s}$.

For the second term, we commute J_ϵ with u and make an integration by parts to obtain

$$\left| \int_{\mathbb{T}} J_\epsilon u D^s \partial_x u \cdot D^s J_\epsilon u \, dx \right| \lesssim \left| \int_{\mathbb{T}} [J_\epsilon, u] D^s \partial_x u \cdot D^s J_\epsilon u \, dx \right| + \left| \int_{\mathbb{T}} \partial_x u \cdot (D^s J_\epsilon u)^2 \, dx \right|$$

$$\lesssim \|u_0\|_{H^s}^3, \tag{3.48}$$

where for the estimation of the first integral, we used the following lemma applied with $w = u$ and $f = D^s u$, whose proof can be found in [41].

Lemma 3.5 *Let w be such that $\|\partial_x w\|_{L^\infty}$. Then, there is a constant $c > 0$ such that for any $f \in L^2$, we have*

$$\|[J_\epsilon, w] \partial_x f\|_{L^2} \leqslant c \|f\|_{L^2} \|\partial_x w\|_{L^2}. \tag{3.49}$$

To estimate the third term of (3.45), we apply the Cauchy-Schwarz inequality, the inequality (2.8) and the well-posedness estimate (1.6) to obtain

$$\left| \int_{\mathbb{T}} D^s J_\epsilon (\partial_x D^{-2} \partial_x u^2) D^s J_\epsilon u \, dx \right| \lesssim \|u_0\|_{H^s}^3$$

$$\left| \int_{\mathbb{T}} D^s J_\epsilon (\partial_x D^{-2} u^2) D^s J_\epsilon u \, dx \right| \lesssim \|u_0\|_{H^s}^3. \tag{3.50}$$

Putting these results together, we have the Lipschitz constants for the family G_ϵ bounded by $c_s \|u_0\|_{H^s}^3$ for some constant c_s. We may therefore conclude that G is Lipschitz and that the solution is continuous, which completes the proof of Proposition 3.2.

3.1.3 Uniqueness

Having established the existence of a solution u in $C([0, T]; H^s)$ to the Cauchy problem (1.1)–(1.2), in this section, we shall show that this solution is unique.

Proposition 3.3 (uniqueness) *For initial data $u_0 \in H^s$, $s > \frac{3}{2}$, the Cauchy problem (1.1)–(1.2) has a unique solution in the space $C([0,T]; H^s)$.*

Proof Let $u_0 \in H^s$ and let u and w be two solutions to the Cauchy problem (1.1)–(1.2) with $u(x,0) = u_0(x) = w(x,0)$. Then the difference $v = u - w$ satisfies the following Cauchy problem

$$\partial_t v = \partial_x(wv) + \partial_x D^{-2}\partial_x(wv) + \partial_x D^{-2}(wv),$$
$$v(x,0) = 0, \qquad (3.51)$$

where

$$w = u + v.$$

Let $\sigma \in [0, s-1]$. The H^σ-energy estimate is then given by

$$\frac{1}{2}\frac{d}{dt}\|v(t)\|_{H^\sigma}^2 = \int_{\mathbb{T}} D^\sigma \partial_x(wv) D^\sigma v \, dx + \int_{\mathbb{T}} D^\sigma \partial_x D^{-2}\partial_x(wv) D^\sigma v \, dx$$
$$+ \int_{\mathbb{T}} D^\sigma \partial_x D^{-2}(wv) D^\sigma v \, dx. \qquad (3.52)$$

To bound (3.52), we commute $D^\sigma \partial_x$ and v, which results in two integrals. The commutator integral is estimated by applying the Cauchy-Schwarz inequality followed by Lemma 2.4 and the solution size estimate (1.6). The second integral is bounded using integration by parts, the Sobolev Embedding Theorem, and the solution size bound (1.8). The non-local term is estimated by the Cauchy-Schwarz inequality and the inequality (2.8). The resulting energy estimate

$$\frac{1}{2}\frac{d}{dt}\|v(t)\|_{H^\sigma}^2 \lesssim \|v(t)\|_{H^\sigma}^2, \qquad (3.53)$$

which we solve to find the inequality

$$\|v(t)\|_{H^\sigma}^2 \leqslant \|v(0)\|_{H^\sigma}^2 e^{2c_s T}. \qquad (3.54)$$

We recall that $v = u - w$ where u and w are both solutions to the i.v.p. (1.1)–(1.2). This means we have

$$\|v(t)\|_{H^\sigma} \leqslant \|v(0)\|_{H^\sigma} e^{c_s T} \leqslant \|u_0 - u_0\|_{H^\sigma} e^{c_s T} = 0. \qquad (3.55)$$

Thus, we have uniqueness.

3.1.4 Continuity of the data-to-solution map

Proposition 3.4 *The data-to-solution map for the i.v.p. (1.1)–(1.2) from H^s to $C([0,T];H^s)$ given by $u_0 \to u$ is continuous.*

Proof Fix $u_0 \in H^s$ and let $\{u_{0,n}\} \in H^s$ be a sequence such that

$$\lim_{n\to\infty} u_{0,n} = u_0 \quad \text{in} \quad H^s. \tag{3.56}$$

Then if u is the solution to the equation (1.1) with initial data u_0 and if u_n is the solution to the eqution (1.1) with initial data $u_{0,n}$, we will prove that

$$\lim_{n\to\infty} u_n = u \quad \text{in} \quad C([0,T];H^s). \tag{3.57}$$

Our approach is to use energy estimates. To avoid some of the difficulties of estimating the term involving $u^k \partial_x u$, we will use the J_ϵ convolution operator to smooth out the initial data. Let $\epsilon \in (0,1]$. We take u^ϵ to be the solution to the the equation (1.1) with smoothed initial data $J_\epsilon u_0 = j_\epsilon * u_0$. Similarly, let u_n^ϵ be the solution with initial data $J_\epsilon u_{0,n}$. Applying the triangle inequality, we arrive at

$$\|u - u_n\|_{C([0,T];H^s)} \leqslant \|u - u^\epsilon\|_{C([0,T];H^s)} + \|u^\epsilon - u_n^\epsilon\|_{C([0,T];H^s)}$$
$$+ \|u_n^\epsilon - u_n\|_{C([0,T];H^s)}. \tag{3.58}$$

We will prove that for any $\eta > 0$, there exists an N such that for all $n > N$, each of these terms can be bounded by $\dfrac{\eta}{3}$ for suitable choices of ϵ and N. We note that the choice of a sufficiently small ϵ will be independent of N and will only depend on η; whereas, the choice of N will depend on both η and ϵ. However, after ϵ has been chosen, N can be chosen so as to force each of the three terms to be small.

Estimation of $\|u^\epsilon - u_n^\epsilon\|_{C([0,T];H^s)}$ We can bound this term directly using an H^s-energy estimate. Let $v = u^\epsilon - u_n^\epsilon$. Then v satisfies the following Cauchy problem

$$\partial_t v = \partial_x [wv] + \partial_x D^{-2} \partial_x [wv] + \partial_x D^{-2} [wv]$$
$$v(x,0) = u^\epsilon(0) - u_n^\epsilon(0) = J_\epsilon u_0 - J_\epsilon u_{0,n}, \tag{3.59}$$

where

$$\widetilde{w} = u + v.$$

Apply the operator D^s to both sides of (3.59), multiply by $D^s v$ and integrate over the torus to obtain the H^s-energy

$$\frac{1}{2}\frac{d}{dt}\|v(t)\|_{H^s}^2 = \int_{\mathbb{T}} D^s\partial_x(\widetilde{w}v)D^s v\, dx + \int_{\mathbb{T}} D^s\partial_x D^{-2}\partial_x(\widetilde{w}v)D^s v\, dx$$
$$+ \int_{\mathbb{T}} D^s\partial_x D^{-2}(\widetilde{w}v)D^s v\, dx. \qquad (3.60)$$

To estimate the first integral of (3.60), we commute the operator $D^s\partial_x$ with the function $\widetilde{w}$ and apply Lemma 2.4 and the Sobolev Embedding theorem to get

$$\left|\int_{\mathbb{T}} D^s\partial_x(\widetilde{w}v)D^s v\, dx\right| \lesssim \|\widetilde{w}\|_{H^{s+1}}\|v\|_{H^s}^2. \qquad (3.61)$$

We shall consider $\|\widetilde{w}\|_{H^{s+1}}$. From our construction of J_ϵ, we can see that our initial data $J_\epsilon u_0, J_\epsilon u_{0,n} \in C^\infty$. Therefore, we may apply our solution size estimate (1.8), in conjunction with the Algebra Property for $s+1 > \frac{1}{2}$, to the definition of $\|\widetilde{w}\|_{H^{s+1}}$ to find

$$\|\widetilde{w}\|_{H^{s+1}} \lesssim \sum_{j=0}^{k} \|J_\epsilon u_0\|_{H^{s+1}}^j \|J_\epsilon u_{0,n}\|_{H^{s+1}}^{k-j}.$$

In examining $\|J_\epsilon u_0\|_{H^{s+1}}^j$, we shall use that if $f \in H^{s+1}$, then

$$\|f\|_{H^{s+1}} \leqslant \|f\|_{H^s} + \|\partial_x f\|_{H^s}.$$

Using this inequality and $\|\partial_x J_\epsilon f\|_{H^s} \leqslant \frac{\alpha}{\epsilon}\|f\|_{H^s}$, we can write

$$\|J_\epsilon u_0\|_{H^{s+1}} \leqslant \|J_\epsilon u_0\|_{H^s} + \|\partial_x J_\epsilon u_0\|_{H^s} \lesssim \frac{1}{\epsilon}.$$

Similarly, we have

$$\|J_\epsilon u_{0,n}\|_{H^{s+1}} \lesssim \frac{1}{\epsilon}\|u_0\|_{H^s} \lesssim \frac{1}{\epsilon}.$$

Therefore, we have

$$\left|\int_{\mathbb{T}} D^s\partial_x(\widetilde{w}v)D^s v\, dx\right| \lesssim \frac{1}{\epsilon}\|v\|_{H^s}. \qquad (3.62)$$

For the second and third integral of (3.60), applying the Cauchy-Schwarz inequality and the inequality (2.8) allows us to obtain

$$\left|\int_{\mathbb{T}} D^s\partial_x D^{-2}\partial_x(\widetilde{w}v)D^s v\, dx\epsilon\right| \lesssim \|v\|_{H^s}^2, \quad \left|\int_{\mathbb{T}} D^s\partial_x D^{-2}(\widetilde{w}v)D^s v\, dx\epsilon\right| \lesssim \|v\|_{H^s}^2.$$
$$(3.63)$$

Combining (3.61) and (3.63), we have the following energy estimate

$$\frac{1}{2}\frac{d}{dt}\|v(t)\|_{H^s}^2 \lesssim \frac{1}{\epsilon}\|v\|_{H^s},$$

which implies

$$\|v(t)\|_{H^s}^2 \leqslant \|v(0)\|_{H^s}^2 e^{\frac{2c_s t}{\epsilon}} = \|u_0^\epsilon - u_{0,n}^\epsilon\|_{H^s}^2 e^{\frac{2c_s t}{\epsilon}}.$$

Recalling that the solutions are mollified, we write

$$\|u_0^\epsilon - u_{0,n}^\epsilon\|_{H^s} e^{\frac{2c_s t}{\epsilon}} \leqslant \|J_\epsilon(u_0^\epsilon - u_{0,n}^\epsilon)\|_{H^s} e^{\frac{2c_s t}{\epsilon}} \leqslant \|u_0 - u_{0,n}\|_{H^s} e^{\frac{2c_s t}{\epsilon}}. \quad (3.64)$$

When we bound the first and third terms of (3.58), we will force ϵ to be small. After ϵ (independent of N) is chosen, we can bound $\|u^\epsilon - u_n^\epsilon\|_{C([0,T];H^s)}$ by taking N large enough that

$$\|u_0 - u_{0,n}\|_{H^s} < \frac{\eta}{3}e^{-\frac{c_\epsilon T}{\epsilon}}.$$

Then we have

$$\|u^\epsilon - u_n^\epsilon\|_{C([0,T];H^s)} \leqslant \frac{\eta}{3}.$$

Estimation of $\|u^\epsilon - u\|_{C([0,T];H^s)}$ **and** $\|u_n^\epsilon - u_n\|_{C([0,T];H^s)}$ Since the arguments will be largely the same for both terms, we will omit the subscript n until such a time as differences in their handling emerge. For convenience, let $v = u^\epsilon - u$, a direct calculation verifies that v solves the Cauchy problem

$$\partial_t v = \partial_x v(u^\epsilon + u) + \partial_x D^{-2}\partial_x(v(u^\epsilon + u)) + \partial_x D^{-2}(v(u^\epsilon + u))$$

with initial condition $v_0 = J_\epsilon u_0 - u_0$. We begin by calculating the H^s energy of v. We have

$$\frac{1}{2}\frac{d}{dt}\|v(t)\|_{H^s}^2 = \int_{\mathbb{T}} D^s \partial_x v(u^\epsilon + u) D^s v \, dx + \int_{\mathbb{T}} D^s \partial_x D^{-2}\partial_x v(u^\epsilon + u) D^s v \, dx$$
$$+ \int_{\mathbb{T}} D^s \partial_x D^{-2} v(u^\epsilon + u) D^s v \, dx. \quad (3.65)$$

We rewrite each term as a commutator and then apply the Cauchy-Schwarz inequality before using Lemma 3.4, the Sobolev Embedding Theorem, and the Algebra Property. For the non-local terms, we employ the Cauchy-Schwarz inequality and the inequality (2.8). The resulting energy estimate is

$$\frac{1}{2}\frac{d}{dt}\|v\|_{H^s}^2 \lesssim \|v\|_{H^s}^2(\|\tilde{u}\|_{H^s}^2 + \|u\|_{H^s}^2). \quad (3.66)$$

By interpolating between 0 and t, we have

$$\|v\|_{H^{s-1}} \lesssim \|v\|_{H^0}^{\frac{1}{s}}\|v\|_{H^s}^{1-\frac{1}{s}} \lesssim \|v\|_{L^2}^{\frac{1}{s}}. \qquad (3.67)$$

Note that the solution size estimate (1.8) implies that $\|v(t)\|_{H^s} \lesssim 1$. By utilizing an L^2-energy estimate, it can be shown that $\|v\|_{L^2} = 0(\epsilon^s)$. This is used to reduce the energy estimate to the differential inequality

$$\frac{dy}{dt} \lesssim y + \delta, \qquad (3.68)$$

where $y = y(t) = \|v(t)\|_{H^s}$ and $\delta = \delta(\epsilon) \to 0$ as $\epsilon \to 0$. Solving (3.68) gives

$$y(t) \lesssim y(0) + \delta. \qquad (3.69)$$

From here, we will treat the cases $y = \|u^\epsilon - u\|_{H^s}$ and $y = \|u_n^\epsilon - u_n\|_{H^s}$ separately.

Case of $y = \|u^\epsilon - u\|_{H^s}$. Since $y(0) = \|J_\epsilon u_0 - u_0\|_{H^s} \to 0$ as $\epsilon \to 0$ and $\delta = \delta(\epsilon) \to 0$ as $\epsilon \to 0$, we see that for sufficiently small ϵ we can bound the first term of (3.64) by $\dfrac{\eta}{3}$.

Case of $y = \|u_n^\epsilon - u_n\|_{H^s}$. Since

$$y(0) = \|J_\epsilon u_{0,n} - u_{0,n}\|_{H^s} \leqslant 2\|u_{0,n} - u_0\|_{H^s} + 2\|J_\epsilon u_0 - u_0\|_{H^s},$$

we may now further refine the choice of ϵ and N so that $y(t) < \dfrac{\eta}{3}$, completing this case. Collecting our results completes the proof of continuous dependence.

3.2 Well-posedness in the line

In the proof of existence on the line, since the inclusion $H^s(\mathbf{R}) \subset H^{s-\sigma}(\mathbf{R})$ is not compact for $\sigma > 0$ (contrast this with the situation on the circle). However, by Rellich's Theorem, the map $f \mapsto \varphi f$ is a compact operator from $H^s(\mathbf{R})$ to $H^{s-\sigma}(\mathbf{R})$ for any $\varphi \in \mathcal{S}(\mathbf{R})$. Hence, considering the family φu_ϵ instead. The proof for uniqueness does not rely on this fact and will not require any other adjustments.

For the proof of continuous dependence, the method mirrors that of the periodic case. However, we must choose a different mollifier J_ϵ. Define $J_\epsilon f(x) = j_\epsilon * f(x)$, $\epsilon > 0$, where $j_\epsilon(x) = \dfrac{1}{\epsilon} j\left(\dfrac{1}{\epsilon}\right)$. Here $j(x) \in \mathcal{S}(\mathbf{R})$ such that $0 \leqslant \hat{j}(x\xi) \leqslant 1$ and $\hat{j}(x\xi) = 1$ if $\xi \leqslant 1$. Hence, how we construct the mollifier J_ϵ plays a critical role in the proofs of well-posedness for the i.v.p (1.1)–(1.2), in both the periodic and non-periodic cases.

4 Hölder continuity

Proof of Theorem 1.3 Lipschitz Continuity in A_1. $u_0, w_0 \in H^s$ and let u and w be two solutions to the Cauchy problem (1.1)–(1.2) with $u(x,0) = u_0(x)$ and $w(x,0) = w_0$. Then the difference $v = u - w$ satisfies the following Cauchy problem

$$\partial_t v = [zv] + \partial_x D^{-2}\partial_x (zv) + \partial_x D^{-2}[zv],$$
$$v(x,0) = u_0 - w_0, \tag{4.1}$$

where

$$z = w + u.$$

Let $0 \leqslant r \leqslant s - 1$ and $r + s > 2$. The H^r-energy estimate is then given by

$$\frac{1}{2}\frac{d}{dt}\|v(t)\|_{H^r}^2 = \int_{\mathbb{T}} D^r \partial_x(zv) D^r v\, dx + \int_{\mathbb{T}} D^r \partial_x D^{-2}\partial_x(zv) D^r v\, dx$$
$$+ \int_{\mathbb{T}} D^r \partial_x D^{-2}(zv) D^r v\, dx. \tag{4.2}$$

To bound (4.2), we commute $D^r \partial_x$ and v, which results in two integrals. The commutator integral is estimated by applying the Cauchy-Schwarz inequality followed by Lemma 2.3 and the solution size estimate (1.8). The second integral is bounded using integration by parts, the Sobolev Embedding Theorem and the solution size bound (1.8). The non-local term is estimated by the Cauchy-Schwarz inequality and the continuity of ∂_x^{-1}. The resulting energy estimate

$$\frac{1}{2}\frac{d}{dt}\|v(t)\|_{H^r}^2 \lesssim c\|v(t)\|_{H^r}^2, \tag{4.3}$$

which we solve to find the inequality

$$\|v(t)\|_{H^r}^2 \leqslant \|v(0)\|_{H^r}^2 e^{2c_s T} \tag{4.4}$$

or equivalently

$$\|v(t)\|_{H^r} \leqslant \|v(0)\|_{H^r} e^{c_s T} \leqslant \|u_0 - w_0\|_{H^r} e^{c_s T} \tag{4.5}$$

which is the desired Lipschitz continuity in A_1.

Hölder Continuity in A_2. By the Lipschitz continuity in A_1 and the condition $r < 2 - s$, we have

$$\|u(t) - w(t)\|_{H^r} \leqslant \|u(t) - w(t)\|_{H^{2-s}} \leqslant \|u_0 - w_0\|_{H^{2-s}} e^{c_s T}. \tag{4.6}$$

By interpolating between the H^r and the H^s- norms, we have

$$\|v(0)\|_{H^{2-s}} \leqslant \|v(0)\|_{H^r}^{\frac{2(s-1)}{s-r}} \|v(0)\|_{H^s}^{\frac{2-s-r}{s-r}} \leqslant \|v(0)\|_{H^r}^{\frac{2(s-1)}{s-r}}, \tag{4.7}$$

which shows the Hölder Continuity in A_2.

Hölder Continuity in A_3. Since $s - 1 \leqslant r \leqslant s$ by interpolating between H^{s-1} and Hs norms, we get

$$\|v(t)\|_{H^s} \leqslant \|v(t)\|_{H^{s-1}}^{s-r} \|v(t)\|_{H^s}^{r-s+1}. \tag{4.8}$$

Furthermore, from the well-posedness size estimate (1.8), we have

$$\|v(t)\|_{H^s} \lesssim \|u_0\|_{H^s} + \|w_0\|_{H^s} \lesssim 1 \tag{4.9}$$

and therefore

$$\|v(t)\|_{H^r} \leqslant c \|v(t)\|_{H^{s-1}}^{s-r}. \tag{4.10}$$

By the Lipschitz continuity in A_1 and the condition $s - 1 \leqslant r$, we get

$$\|v(t)\|_{H^r} \leqslant c \|v(0)\|_{H^{s-1}}^{s-r} \leqslant c \|v(0)\|_{H^r}^{s-r}, \tag{4.11}$$

which is the desired Hölder continuity.

5 Analyticity of solutions

In this section, we look for real analytic solutions of the Cauchy problem (1.1)–(1.2) and give the proof of Theorem 1.4. We will need a suitable scale of Banach spaces are as follows. For any $s > 0$, we set

$$E_s = \{u \in C^\infty(\mathbf{R}) : |||u|||_s = \sup_{k \in B_0} \frac{s^k \|\partial^k u\|_{H^2}}{k!/(k+1)^2} < \infty\},$$

where $H^2(\mathbf{R})$ is the Sobolev space of order two on the real line and N_0 is the set of nonnegative integers. One can easily verify that E_s is equipped with the norm $||| \cdot |||_s$ is a Banach space and that, for any $0 < s' < s$, E_s is continuously embedded in $E_{s'}$ with

$$|||u|||_{s'} \leqslant |||u|||_s.$$

Another simple consequence of the definition is that any u in E_s is a real analytic function on $\mathbf{R}$. Crucial for our purposes is the fact that each E_s forms an algebra under pointwise multiplication of functions.

Lemma 5.1 ([23]) *Let $0 < s < 1$. There is a constant $c > 0$, independent of s, such that for any u and v in E_s we have*

$$|||uv|||_s \leqslant c|||u|||_s|||v|||_s.$$

Lemma 5.2 ([23]) *There is a constant $c > 0$ such that for any $0 < s' < s < 1$, we have $|||\partial_x u|||_{s'} \leqslant \frac{c}{s-s'}|||u|||_s$, and $|||(1-\partial_x^2)^{-1} u|||_{s'} \leqslant |||u|||_s, |||(1-\partial_x^2)^{-1}\partial_x u|||_{s'} \leqslant |||u|||_s$.*

Theorem 5.1 *Let $\{X_s\}_{0<s<1}$ be a scale of decreasing Banach spaces, namely for any $s' < s$ we have $X_s \subset X_{s'}$ and $||| \cdot |||_{s'} \leqslant ||| \cdot |||_s$. Consider the Cauchy problem*

$$\begin{cases} \dfrac{du}{dt} = F(t, u(t)), \\ u(0) = 0. \end{cases} \quad (5.1)$$

Let T, R and C be positive constants and assume that F satisfies the following conditions

1) *If for $0 < s' < s < 1$ the function $t \mapsto u(t)$ is holomorphic in $|t| < T$ and continuous on $|t| \leqslant T$ with values in X_s and*

$$\sup_{|t| \leqslant T} |||u(t)|||_s < R,$$

then $t \mapsto F(t, u(t))$ is a holomorphic function on $|t| < T$ with values in $X_{s'}$.

2) *For any $0 < s' < s < 1$ and any $u, v \in X_s$ with $|||u|||_s < R, |||v|||_s < R$,*

$$\sup_{|t| \leqslant T} |||F(t, u) - F(t, v)|||_{s'} \leqslant \frac{C}{s-s'}|||u - v|||_s.$$

3) *There exists $M > 0$ such that for any $0 < s < 1$,*

$$\sup_{|t| \leqslant T} |||F(t, 0)|||_s \leqslant \frac{M}{1-s}.$$

Then there exists a $T_0 \in (0, T)$ and a unique function $u(t)$, which for every $s \in (0, 1)$ is holomorphic in $|t| < (1-s)T_0$ with values in X_s, and is a solution to the Cauchy problem (5.1).

We restate the Cauchy problem (1.1)–(1.2) in a more convenient form. Let $v = u_x$. Then the problem (1.1)–(1.2) can be written as a system for u and

$$\begin{cases} u_t = 2uv - (1-\partial_x^2)^{-1}(u^2 + v^2) = F(u, v), \\ v_t = 2v^2 + 2uv_x - \partial_x(1-\partial_x^2)^{-1}(u^2 + v^2) = G(u, v), \\ u(x, 0) = u_0(x), v(x, 0) = u_0'(x). \end{cases} \quad (5.2)$$

Now we have all the tools we need to prove the theorem.

Proof of Theorem 1.4 Note that it is sufficient to verify the conditions 1) and 2) in the statement of the abstract Cauchy-Kowalevski theorem above for both $F(u,v)$ and $G(u,v)$ in the system (5.2) since neither F nor G depends on t explicitly. We observe that, for $0 < s' < s < 1$, the estimates in Lemma 5.1–5.2 imply the following bounds

$$|||F(u,v)|||_{s'} \leqslant c |||u|||_s |||v|||_s + c|||u|||_s^2 + c|||v|||_s^2$$

and

$$|||F(u,v)|||_{s'} \leqslant \frac{c}{s-s'} |||u|||_s |||v|||_s + c|||u|||_s^2 + c|||v|||_s^2$$

hence condition 1) holds.

Note that to verify the second condition, it suffices to estimate

$$|||F(u_1,v) - F(u_2,v)|||_{s'}, \quad |||F(u,v_1) - F(u,v_2)|||_{s'},$$

and

$$|||G(u_1,v) - G(u_2,v)|||_{s'}, \quad |||G(u,v_1) - G(u,v_2)|||_{s'}.$$

Since

$$|||F(u_1,v_1) - F(u_2,v_2)|||_{s'} \leqslant |||F(u_1,v_1) - F(u_1,v_2)|||_{s'} + |||F(u_1,v_2) - F(u_2,v_2)|||_{s'},$$

and

$$|||G(u_1,v_1) - G(u_2,v_2)|||_{s'} \leqslant |||G(u_1,v_1) - G(u_1,v_2)|||_{s'} + |||G(u_1,v_2) - G(u_2,v_2)|||_{s'}.$$

Using this together with Lemma 5.1–5.2, we get the following estimates

$$|||F(u_1,v) - F(u_2,v)|||_{s'} \leqslant c|||u_1 - u_2|||_s |||v|||_s + |||u_1 - u_2|||_s |||u_1 + u_2|||$$
$$\leqslant C_R \left(|||u_1 - u_2|||_s + \frac{1}{s-s'} |||u_1 - u_2|||_s \right),$$

$$|||F(u,v_1) - F(u,v_2)|||_{s'} \leqslant c|||v_1 - v_2|||_s |||u|||_s + |||v_1 - v_2|||_s |||v_1 - v_2|||_s$$
$$\leqslant C_R \left(|||v_1 - v_2|||_s + \frac{1}{s-s'} |||v_1 - v_2|||_s \right), \quad (5.3)$$

$$|||G(u_1,v) - F(u_2,v)|||_{s'} \frac{1}{s-s'} |||u_1 - u_2|||_s |||v|||_s + |||u_1 - u_2|||_s |||u_1 + u_2|||$$
$$\leqslant C_R \left(|||u_1 - u_2|||_s + \frac{1}{s-s'} |||u_1 - u_2|||_s \right),$$

$$|||G(u,v_1) - F(u,v_2)|||_{s'} \frac{1}{s-s'} |||v_1-v_2|||_s |||u|||_s + |||v_1-v_2|||_s |||v_1-v_2|||_s$$
$$\leqslant C_{\mathbf{R}} \left(|||v_1-v_2|||_s + \frac{1}{s-s'} |||v_1-v_2|||_s \right),$$

where the constant $C_{\mathbf{R}}$ depends only on $\mathbf{R}$. The conditions (1) through (3) above are now easily verified. The proof of Theorem 1.4 is complete.

Acknowledgements This work is supported by NSF of China under grant number 11671055, by NSF of Chongqing under grant number cstc2018jcyjAX0273, by Key project of science and technology research program of Chongqing Education Commission under grant number KJZD-K20180140, by Simons Foundation under grant number grant-499875 and by China Postdoctoral Science Foundation under grant number 2018M641271.

References

[1] Baouendi M, Goulaouic C. Sharp estimates for analytic pseudodifferential operators and application to the Cauchy problems [J]. J. Differential Equations, 1983, 48: 241–268.

[2] Bressan A, Constantin A. Global dissipative solutions of the Camassa-Holm equation [J]. Anal. Appl., 2007, 5: 1–27.

[3] Bressan A, Constantin A. Global conservative solutions of the Camassa-Holm equation [J]. Arch. Ration. Mech. Anal., 2007, 183: 215–239.

[4] Christov Q, Hakkaev S. On the Cauchy problem for the periodic b-family of equations and of the non-uniform continuity of Degasperis-Procesi equation J [J]. Math. Anal. Appl., 2009, 360: 47–56

[5] Constantin A. On the scattering problem for the Camassa-Holm equation [J]. Proc. R. Soc. Lond. A, 2001, 457: 953–970.

[6] Constantin A. The trajectories of particles in Stokes waves [J]. Invent. Math., 2006, 166: 23–535.

[7] Constantin A. On the inverse spectral problem for the Camassa-Holm equation [J]. J. Funct. Anal., 1998, 155: 352–363.

[8] Constantin A, Escher J. Wave breaking for nonlinear nonlocal shallow water equations [J]. Acta Math., 1998, 181: 229–243.

[9] Constantin A, Escher J. Analyticity of periodic traveling free surface water waves with vorticity [J]. Ann. of Math., 2011, 173: 559–568.

[10] Constantin A, Escher J. Particle trajectores in solitary water waves [J]. Bull. Amer. Math. Soc., 2007, 44: 423–431.

[11] Constantin A, Gerdjikov V, Ivanov R I. Inverse scattering transform for the Camassa-Holm equation [J]. Inverse Probl, 2006, 22: 2197–2207.

[12] Constantin A, Kappeler T, Kolev B, Topalov T. On Geodesic exponential maps of the Virasoro group [J]. Ann. Global Anal. Geom., 2007, 31: 155–180.

[13] Constantin A, Lannes D. The hydro-dynamical relevance of the Camassa-Holm and Degasperis-Procesi equations [J]. Arch. Ration. Mech. Anal., 2009, 193: 165–186.

[14] Constantin A, McKean H P. A shallow water equation on the circle [J]. Comm. Pure Appl. Math., 1999, 52: 949–982.

[15] Camassa R, Holm D. An integrable shallow water equation with peaked solitons [J]. Phys. Rev. Lett., 1993, 71: 1661–1664.

[16] Constantin A, Ivanov R, Lenells J. Inverse scattering transform for the Degasperis-Procesi equation [J]. Nonlinearity, 2010, 23: 2559–2575.

[17] Degasperis A, Holm D D, Hone A N W. A new integrable equation with peakon solutions [J]. Theor. Math. Phys., 2002, 133: 1475–1489.

[18] Degasperis A, Procesi M. Asymptotic integrability, in: Symmetry and Perturbation Theory [M]. World Scientific, 1999. 23–37.

[19] Fokas A, Fuchssteiner B. Symplectic structures, their Backlund transformation and hereditary symmetries [J]. Phys D, 1981, 4: 47–66.

[20] Himonas A, Holliman C. The Cauchy problem for the Novikov equation [J]. Nonlinearity, 2012, 25: 449–479.

[21] Himonas A, Holliman C. On well-posedness of the Degasperis-Procesi equation [J]. Discrete Contin. Dyn. Syst., 2011, 31: 469–488.

[22] Himonas A, Kenig C, Misiołek G. Non-uniform dependence for the periodic CH equation [J]. Comm. Partial Differential Equations, 2010, 35: 1145–1162.

[23] Himonas A, Misiołek G. Analyticity of the Cauchy problem for an integrable evolution equation [J]. Math. Ann., 2003, 327: 575–584.

[24] Himonas A, Misiołek G. High-frequency smooth solutions and well-posedness of the Camassa-Holm equation [J]. Int. Math. Res. Not., 2005, 51: 3135–3151.

[25] Himonas A, Misiołek G. Non-uniform dependence on initial data of solutions to the Euler equations of hydrodynamics [J]. Comm. Math. Phys., 2009, 296; 285–301.

[26] Holden H, Raynaud X. Dissipative solutions for the Camassa-Holm equation [J]. Discrete Contin. Dyn. Syst., 2009, 24: 1047–1112.

[27] Holden H, Raynaud X. Global conservative solutions of the Camassa-Holm equations-a Lagrangianpoiny of view [J]. Comm. Partial Differential Equations, 2007, 32: 1511–1549.

[28] Holliman C. Non-uniform dependence and well-posedness for the periodic Hunter-Saxton equation [J]. Differ. Integral. Equ., 2010, 23: 1150–1194.

[29] Holm D D, Staley M F. Wave structure and nonlinear balances in a family of evolutionary PDEs [J]. SIAM J. Appl. Dyn. Syst., 2003, 2: 323–380.

[30] Hone W, Lundmark H, Szmigielski J. Explicit multipeakon solutions of Novikov cubically nonlinear integrable Camassa-Holm type equation [J]. Dyn. Partial Differ. Equ., 2009, 6: 253–289.

[31] Home A N W, Wang J P. Integrable peakon equations with cubic nonlinearity [J]. J. Phys. A, 2008, 41: 372002.

[32] Jiang Z, Ni L. Blow-up phenomenon for the integrable Novikov equation [J]. J. Math. Anal. Appl., 2012, 385: 551–558.

[33] Kato T, Ponce G. Commutator estimates and the Euler and Navier-Stokes equations [J]. Comm. Pure Appl. Math., 1988, 41: 891–907.

[34] Lenells J. Traveling wave solutions of the Degasperis-Procesi equation [J]. J. Math. Anal. Appl., 2005, 306: 72–82.

[35] Li J L, Yin Z Y. Well-posedness and global existence for a generalized Degasperis-Procesi equation [J]. Nonlinear Anal. Real World Appl., 2016, 28: 72–90.

[36] Lundmark H, Szmigielski J. Multi-peakon solutions of the Degasperis-Procesi equation [J]. Inverse Probl., 2005, 21: 1553–1570.

[37] Matsuno Y. Multisoliton solutions of the Degasperis-Procesi equation and their peakon limit [J]. Inverse Probl., 2003, 19: 1241–1245.

[38] Misiolek G A. Shallow water equation as a geodesic flow on the Bott-Virasoro group [J]. J. Geom. Phys., 1998, 24: 203–208.

[39] Ni L, Zhou Y. Well-posedness and persistence properties for the Novikov equation [J]. J. Differential Equations, 2011, 250: 3002–3021.

[40] Novikov V. Generalization of the Camassa-Holm equation [J]. J. Phys. A, 2009, 42: 342002.

[41] Taylor M. Partial differential equations III, nonlinear equations [M]. Springer, 1996. ISBN: 0-387-946528-X.

[42] Tiglay F. The periodic Cauchy problem for Novikov equation [J]. Int. Math. Res. Not., 2011, 2011: 4633–4648.

[43] Toland J F. Stokes waves [J]. Topol. Methods Nonlinear Anal., 1996, 7: 1–48.

[44] Vakhnenko V.O, Parkes E.J. Periodic and solitary-wave solutions of the Degasperis-Procesi equation [J]. Chaos Solitons Fractals, 2004, 20: 1059–1073.

[45] Wu X, Yin Z. Well-posedness and global existence for the Novikov equation [J]. Annali Sc. Norm. Sup. Pisa. XI, 2012, 707–727.

[46] Wu X, Yin Z. Global weak solutions for the Novikov equation [J]. J. Phys. A, 2011, 44: 055202.

[47] Yan W, Li Y, Zhang Y. Global existence and blow-up phenomena for the weakly dissipative Novikov equation [J]. Nonlinear Anal., 2012, 75: 2464–2473.

A Non-Homogeneous Boundary-Value Problem for the KdV-Burgers Equation Posed on a Finite Domain*

You Shujun (游淑军), Guo Boling(郭柏灵)

Abstract   We consider the initial-boundary value problem for the KdV-Burgers equation posed on a bounded interval $[a, b]$. This problem features non-homogeneous boundary conditions applied at $x = a$ and $x = b$ and is known to have a unique global smooth solution.

Keywords   KdV-Burgers equation; Initial-boundary value problem; Global smooth solution

1 Introduction

KdV-Burgers equation mainly arises from nonlinear wave models of fluid in an elastic tube, liquid with small bubbles and turbulence. It attracts the attention of many mathematicians. There are many articles on the pure initial-value problem set on the whole line and the initial-boundary value problem posed on the half line [1-3,5-15]. When the equation is used in practical situations, one inevitably encounters a finite domain where lateral boundary conditions must be imposed. Hence, the theory for such boundary-value problems, while more complicated and less elegant than the theory on the whole line, is important [4].

In this paper, we are interested in the study of the initial-boundary value problem for the KdV-Burgers equation

$$u_t + (f(u))_x = \alpha u_{xx} + \beta u_{xxx}, \quad \alpha > 0, \ \beta > 0, \tag{1.1}$$

posed on a finite interval $[a, b]$, which without loss of generality is taken to be

---

* Appl. Math. Lett., 2019, 94: 155–159. DOI:10.1016/j.aml.2019.03.002.

[0, 1], subject to an initial condition

$$u|_{t=0} = u_0(x), \quad x \in [0,1], \tag{1.2}$$

and the non-homogeneous boundary conditions

$$u|_{x=0} = g_1(t), \ u|_{x=1} = g_2(t), \ u_x|_{x=0} = g_3(t), \quad t \geqslant 0, \tag{1.3}$$

where $f(\xi)$, the initial value u_0 and the boundary data g_1, g_2, g_3 are given functions.

In order to study the problem (1.1)-(1.3), we consider the following transformation. Set $v = u - h(x,t)$, where

$$h(x,t) = (1-x)g_1(t) + xg_2(t) + x(1-x)(g_1(t) - g_2(t) + g_3(t)).$$

A direct calculation shows that

$$v_t + (f(v+h))_x = \alpha v_{xx} + \beta v_{xxx} + F(x,t), \quad \alpha > 0, \ \beta > 0, \tag{1.4}$$

with initial-boundary conditions

$$v|_{x=0} = v|_{x=1} = v_x|_{x=0} = 0, \quad t \geqslant 0, \tag{1.5}$$

$$v|_{t=0} = u_0(x) - h(x,0) := v_0(x), \quad x \in [0,1], \tag{1.6}$$

where $F(x,t) = \alpha h_{xx} - h_t$.

Our main results are as follows:

Theorem 1.1 *Suppose that $u_0(x) \in H^6[0,1]$, $g_1(t), g_2(t), g_3(t) \in C^2[0, +\infty)$, $f(\xi) \in C^2(\mathbf{R})$ and $|f'(\xi)| \leqslant A|\xi|^p + B$, $\xi \in \mathbf{R}$, $A, B > 0$, $0 \leqslant p \leqslant 1$. Then there exists a unique global smooth solution of the initial-boundary value problem (1.1)-(1.3),*

$$u \in L^\infty(0,T;H^6), \quad u_t \in L^\infty(0,T;H^3), \quad u_{tt} \in L^\infty(0,T;L^2).$$

We use C to represent various positive constants that can depend on initial-boundary data.

2 A priori estimates

Lemma 2.1 *Suppose that $v_0(x) \in L^2[0,1]$, $g_1(t), g_2(t), g_3(t) \in C[0, +\infty)$, $f(\xi) \in C^1(\mathbf{R})$ and $|f'(\xi)| \leqslant A|\xi|^p + B$, $\xi \in \mathbf{R}$, $A, B > 0$, $0 \leqslant p \leqslant 1$. Then for the solution of Problem (1.4)-(1.6) we have*

$$\sup_{0\leqslant t\leqslant T}\|v(\cdot,t)\|_{L^2}^2 + \alpha\int_0^T \|v_x\|_{L^2}^2\, dt + \beta\int_0^T (v_x(1,t))^2\, dt \leqslant C.$$

Proof Taking the inner product of (1.4) and v, it follows that

$$\frac{1}{2}\frac{d}{dt}\|v\|_{L^2}^2 + \alpha\|v_x\|_{L^2}^2 + \frac{\beta}{2}(v_x(1,t))^2$$
$$= \int_{g_1(t)}^{g_2(t)} f(\xi)\, d\xi - \int_0^1 f(v+h)h_x\, dx + \int_0^1 F(x,t)v\, dx. \tag{2.1}$$

By the Hölder's inequality, Young's inequality, and applying Gronwall's inequality provides Lemma 2.1.

Lemma 2.2 *Suppose that $v_0(x) \in H^3[0,1]$, $g_1(t), g_2(t), g_3(t) \in C^1[0,+\infty)$, $f(\xi) \in C^1(\mathbf{R})$ and $|f'(\xi)| \leqslant A|\xi|^p + B$, $\xi \in \mathbf{R}$, $A, B > 0$, $0 \leqslant p \leqslant 1$. Then for the solution of Problem (1.4)-(1.6) we have*

$$\sup_{0\leqslant t\leqslant T}\left(\|v_t\|_{L^2}^2 + \|v\|_{H^3}^2\right) + \alpha\int_0^T \|v_{xt}\|_{L^2}^2\, dt + \beta\int_0^T (v_{xt}(1,t))^2\, dt \leqslant C.$$

Proof Differentiating (1.4) with respect to t, then taking the inner products of the resulting equation and v_t. It follows that

$$\frac{1}{2}\frac{d}{dt}\|v_t\|_{L^2}^2 + \alpha\|v_{xt}\|_{L^2}^2 + \frac{\beta}{2}(v_{xt}(1,t))^2$$
$$= \int_0^1 f'(v+h)(v+h)_t v_{xt}\, dx + \int_0^1 F_t(x,t)v_t\, dx. \tag{2.2}$$

Note that $|f'(\xi)| \leqslant A|\xi|^p + B$, by the Hölder's inequality and the Young's inequality we have

$$\left|\int_0^1 f'(v+h)(v+h)_t v_{xt}\, dx\right| \leqslant \frac{\alpha}{2}\|v_{xt}\|_{L^2}^2 + C(\|v_x\|_{L^2}^p + 1)\|v_t\|_{L^2}^2 + C.$$

Note that $\int_0^T \|v_x\|_{L^2}^2\, dt \leqslant C$ in Lemma 2.1, from (2.2), we get

$$\sup_{0\leqslant t\leqslant T}\|v_t(\cdot,t)\|_{L^2}^2 \leqslant C.$$

Note equality (1.4), one can show that

$$\sup_{0\leqslant t\leqslant T}\|v\|_{H^3}^2 \leqslant C.$$

Lemma 2.3 *Suppose that $v_0(x) \in H^6[0,1]$, $g_1(t), g_2(t), g_3(t) \in C^2[0,+\infty)$, $f(\xi) \in C^2(\mathbf{R})$ and $|f'(\xi)| \leqslant A|\xi|^p + B$, $\xi \in \mathbf{R}$, $A, B > 0$, $0 \leqslant p \leqslant 1$. Then for the solution of Problem (1.4)-(1.6) we have*

$$\sup_{0\leqslant t\leqslant T}\left(\|v_{tt}\|_{L^2}^2+\|v_t\|_{H^3}^2+\|v\|_{H^6}^2\right)\leqslant C,$$

$$\alpha\int_0^T\|v_{xtt}(x,t)\|_{L^2}^2\,dt+\beta\int_0^T(v_{xtt}(1,t))^2\,dt\leqslant C.$$

Proof Differentiating (1.4) two times with respect to t, then taking the inner products of the resulting equation and v_{tt}. It follows that

$$\frac{1}{2}\frac{d}{dt}\|v_{tt}\|_{L^2}^2+\alpha\|v_{xtt}\|_{L^2}^2+\frac{\beta}{2}(v_{xtt}(1,t))^2$$
$$=((f(v+h))_{tt},\,v_{xtt})+\int_0^1 F_{tt}(x,t)v_{tt}\,dx. \qquad (2.3)$$

By the Gagliardo-Nirenberg inequality and the Young inequality we have

$$|((f(v+h))_{tt},\,v_{xtt})|\leqslant\frac{\alpha}{2}\|v_{xtt}\|_{L^2}^2+C\left(\|v_{tt}\|_{L^2}^2+\|v_{xt}\|_{L^2}^2+1\right).$$

From (2.3) we obtain

$$\frac{d}{dt}\|v_{tt}\|_{L^2}^2+\alpha\|v_{xtt}\|_{L^2}^2+\beta(v_{xtt}(1,t))^2\leqslant C\left(\|v_{tt}\|_{L^2}^2+\|v_{xt}\|_{L^2}^2+1\right). \qquad (2.4)$$

Multiplying the inequality (2.4) by e^{-Ct}, integrating over $[0,\,t]$, leads to the formula

$$e^{-Ct}\|v_{tt}\|_{L^2}^2+\alpha\int_0^t e^{-C\tau}\|v_{xtt}(x,\tau)\|_{L^2}^2\,d\tau+\beta\int_0^t e^{-C\tau}(v_{xtt}(1,\tau))^2\,d\tau$$
$$\leqslant\|v_{tt}(x,0)\|_{L^2}^2+C\int_0^t e^{-C\tau}\left(\|v_{xt}(x,\tau)\|_{L^2}^2+1\right)d\tau.$$

Applying Lemma 2.2 yields that

$$\sup_{0\leqslant t\leqslant T}\|v_{tt}\|_{L^2}^2+\alpha\int_0^T\|v_{xtt}(x,t)\|_{L^2}^2\,dt+\beta\int_0^T(v_{xtt}(1,t))^2\,dt\leqslant C.$$

Note equality (1.4), one can show that

$$\sup_{0\leqslant t\leqslant T}\left(\|v_t\|_{H^3}^2+\|v\|_{H^6}^2\right)\leqslant C.$$

3 The existence and uniqueness of solution

In this section, we study the well-posedness of the initial-boundary value problem (1.1)-(1.3).

Theorem 3.1 *Suppose that $v_0(x) \in H^6[0,1]$, $g_1(t), g_2(t), g_3(t) \in C^2[0,+\infty)$, $f(\xi) \in C^2(\mathbf{R})$ and $|f'(\xi)| \leqslant A|\xi|^p + B$, $\xi \in \mathbf{R}$, $A, B > 0$, $0 \leqslant p \leqslant 1$. Then there exists a unique global smooth solution of the initial-boundary value problem (1.4)-(1.6),*

$$v \in L^\infty(0,T;H^6), \quad v_t \in L^\infty(0,T;H^3), \quad v_{tt} \in L^\infty(0,T;L^2).$$

Proof First of all, by Galerkin method, compact argument and Sobolev imbedding theorem, we can get the existence of global smooth solution for problem (1.4)-(1.6).

Next, we prove the uniqueness of the solution. Suppose that there are two solutions v_1 and v_2. Let $w = v_1 - v_2$, From eq. (1.4)-(1.6) we can get the following equality

$$\frac{1}{2}\frac{d}{dt}\|w\|_{L^2}^2 + \alpha\|w_x\|_{L^2}^2 + \frac{\beta}{2}(w_x(1,t))^2 = \int_0^1 (f(v_1+h) - f(v_2+h))w_x\,dx. \quad (3.1)$$

Using the differential mean value theorem, we arrive at

$$f(v_1+h) - f(v_2+h) = f'(v_2 + h + \theta(v_1-v_2))w, \quad \theta \in (0,1).$$

Then, we obtain the following estimate

$$\left|\int_0^1 (f(v_1+h) - f(v_2+h))w_x\,dx\right| \leqslant C\|w\|_{L^2}^2 + \frac{\alpha}{2}\|w_x\|_{L^2}^2.$$

From (3.1), employing Gronwall's inequality and initial data, we conclude that $w \equiv 0$. This completes the proof of Theorem 3.1. Furthermore, The proof of Theorem 1.1 is complete.

References

[1] Molinet L, Ribaud F. On the low regularity of the Korteweg-de Vries-Burgers equation [J]. Int. Math. Res. Notices, 2002, 37: 1979–2005.

[2] Cavalcanti M M, Domingos Cavalcanti V N, Komornik V, et al. Global well-posedness and exponential decay rates for a KdV-Burgers equation with indefinite damping [J]. Ann. I. H. Poincaré - AN, 2014, 31: 1079–1100.

[3] Bona J L, Sun S, Zhang B. Non-homogeneous boundary value problems for the Korteweg-de Vries and the Korteweg-de Vries-Burgers equations in a quarter plane [J]. Ann. I. H. Poincaré–AN, 2008, 25: 1145–1185.

[4] Bona J L, Sun S, Zhang B. A non-homogeneous boundary-value problem for the Korteweg-de Vries equation posed on a finite domain II [J]. Journal of Differential Equations, 2009, 247: 2558–2596.

[5] Dlotko T. The generalized Korteweg-de Vries-Burgers equation in $H^2(R)$ [J]. Nonlinear Anal., 2011, 74: 721–732.

[6] Guo Z, Wang B. Global well-posedness and inviscid limit for the Korteweg-de Vries-Burgers equation [J]. J. Differential Equations, 2009, 246: 3864–3901.

[7] Chugainova A P, Shargatov V A. Stability of nonstationary solutions of the generalized KdV-Burgers equation [J]. Comput. Math. and Math. Phys., 2015, 55(2): 251–263.

[8] Zhang B Y. Analyticity of solutions for the generalized Korteweg-de Vries equation with respect to their initial datum [J]. SIAM J. Math. Anal., 1995, 26: 1488–1513.

[9] Kenig C E, Ponce G, Vega L. Well-posedness and scattering results for the generalized Korteweg-de Vries equation via the contraction principle [J]. Comm. Pure Appl. Math., 1993, 46: 527–620.

[10] Colliander J E, Kenig C E. The generalized KdV equation on the half-line [J]. Comm. Partial Differential Equations, 2002, 27: 2187–2266.

[11] Bona J L, Sun S, Zhang B Y. Forced oscillations of a damped KdV equation in a quarter plane [J]. Commun. Contemp. Math., 2003, 5: 369–400.

[12] Faminskii A V. An initial boundary-value problem in a half-strip for the Korteweg-de Vries equation in fractional-order Sobolev spaces [J]. Comm. Partial Differential Equations, 2004, 29: 1653–1695.

[13] Faminskii A V, Larkin N A. Odd-order quasilinear evolution equations posed on a bounded interval [J]. Bol. Soc. Parana. Mat., 2010, 28: 67–77.

[14] Larkin N A. The 2D Zakharov-Kuznetsov-Burgers equation on a strip [J]. Bol. Soc. Parana. Mat., 2016, 34: 151–172.

[15] Gallego F A, Pazoto A F. On the well-posedness and asymptotic behavior of the generalized KdV-Burgers equation. Proceedings of the Royal Society of Edinburgh Section A: Mathematics [DB/OL]. https://doi.org/10.1017/S0308210518000240.

Low Mach Number Limit of Compressible Nematic Liquid Crystal Flows with Well-prepared Initial Data in a 3D Bounded Domain*

Guo Boling (郭柏灵), Zeng Lan (曾兰), and Ni Guoxi (倪国喜)

Abstract In this paper, we consider the low Mach number limit of the compressible nematic liquid crystal flows in a 3D bounded domain. We establish the uniform estimates with respect to the Mach number for the strong solutions in a short time interval. Consequently, we obtain the convergence of the compressible nematic liquid crystal system to the incompressible nematic liquid crystals system as the Mach number tends to zero.

Keywords low Mach number limit; compressible nematic liquid crystal flows; bounded domain.

1 Introduction

In this paper, we establish the uniform estimates of strong solutions with respect to the Mach number in a bounded domain $\Omega \subset \mathbf{R}^3$ to the compressible nematic liquid crystal flows.

$$\rho_t + \mathrm{div}(\rho u) = 0, \tag{1.1}$$

$$(\rho u)_t + \mathrm{div}(\rho u \otimes u) + \frac{1}{\epsilon^2}\nabla P(\rho) - \mu\Delta u - (\lambda+\mu)\nabla\mathrm{div} u = -\nabla d \cdot \Delta d, \tag{1.2}$$

$$d_t + u \cdot \nabla d = \Delta d + |\nabla d|^2 d, \quad |d| = 1, \tag{1.3}$$

where the unknowns ρ, u and d stand for the density, velocity, and the

---
*Electron. J. Differential Equations, 2019, 2019(14): 1–13.

macroscopic of the nematic liquid crystal orientation field, respectively. The pressure $P(\rho)$ is a C^1 function satisfying $P'(\cdot) > 0$ and $P'(0) = 0$, such as the well-known γ-law $P(\rho) = a\rho^\gamma (\gamma > 1)$ which satisfies the assumptions. The parameter $\epsilon > 0$ is the scaled Mach number. The physical constants μ and λ denote the shear viscosity and bulk viscosity of the flow and satisfy

$$\mu > 0, \quad 2\mu + 3\lambda \geqslant 0.$$

In fluid mechanics, the Mach number is an important physical quantity to determine whether the fluid is compressible or incompressible. If the Mach number is small, the fluid should behave asymptotically like an incompressible one, provided velocity and viscosity are small. As a result, the low Mach number limit problem has attracted much attention in recent years. When d is a constant vector field, the system (1.1)-(1.3) becomes the compressible Navier-Stokes system, of which the low Mach number limit problem has obtained a great number of results in the past decades. The readers may refer to [5, 10-12, 14], for instance, and the references therein for details.

Furthermore, a lot of progress have been made on the low Mach number limit for the compressible nematic liquid crystal equations. In [4], the authors are concerned with the low Mach number limit of system (1.1)-(1.3) with periodic boundary conditions. In [2], Bie et al. obtained global existence and the low Mach number limit for compressible flow of liquid crystals in critical spaces. Particularly, for the bounded domain case, the low Mach number limit of weak solutions to the compressible flow of liquid crystals was proved in [13], and Yang in [15] firstly studied the low Mach number limit of the strong solution to the system (1.1)-(1.3) provided the initial data small enough. Motivated by the articles mentioned above, in this paper, we intend to establish the low Mach number limit of the strong solution for the system (1.1)-(1.3) with the larger initial data in a short time interval. The main difficulty comparing to the periodic case in [4] and the whole space case in [2] is the uniform high-norm estimates with respect to the Mach number and a time interval independent of the Mach number. In a bounded domain, after integrating by parts for the high-order derivatives, we have to estimate the boundary term which we will skillfully apply the slip conditions to control.

The low Mach number fluid can be regarded as a perturbation near the background isentropic fluid, where the density is usually set to be constant.

Hence, we introduce the density variation by σ^ϵ as follows:
$$\rho^\epsilon = 1 + \epsilon\sigma^\epsilon,$$
and we will take $P'(1) = 1$. Then, the non-dimensional system (1.1)-(1.3) can be rewritten as the following form:

$$\sigma_t^\epsilon + \text{div}(\sigma^\epsilon u^\epsilon) + \frac{1}{\epsilon}\text{div}u^\epsilon = 0, \tag{1.4}$$

$$\rho^\epsilon(u_t^\epsilon + u^\epsilon \cdot \nabla u^\epsilon) + \frac{1}{\epsilon}P'(1+\epsilon\sigma^\epsilon)\nabla\sigma^\epsilon - \mu\Delta u^\epsilon - (\lambda+\mu)\nabla\text{div}u^\epsilon = -\nabla d^\epsilon \cdot \Delta d^\epsilon, \tag{1.5}$$

$$d_t^\epsilon + u^\epsilon \cdot \nabla d^\epsilon = (\Delta d^\epsilon + |\nabla d^\epsilon|^2 d^\epsilon), \quad |d^\epsilon| = 1. \tag{1.6}$$

The system (1.4)-(1.6) is supplemented with the following initial and boundary value conditions:

$$(\sigma^\epsilon, u^\epsilon, d^\epsilon)(\cdot, 0) = (\sigma_0^\epsilon, u_0^\epsilon, d_0^\epsilon)(\cdot) \quad \text{in} \quad \Omega, \tag{1.7}$$

$$u^\epsilon \cdot n = 0, \ \text{curl}u^\epsilon \times n = 0, \ \frac{\partial d^\epsilon}{\partial n} = 0, \quad \text{on} \quad \partial\Omega, \tag{1.8}$$

where n is the unit outer normal vector to the smooth boundary $\partial\Omega$.

Firstly, the local existence results for the problem (1.4)-(1.8) can be established in a similar way as in [8].

Proposition 1.1 (local existence) *Let $\Omega \subset \mathbf{R}^3$ be a bounded, simply connected domain with smooth boundary $\partial\Omega$. Assume the initial data $(\sigma_0^\epsilon, u_0^\epsilon, d_0^\epsilon)$ satisfy the following conditions,*

$$\begin{cases} (\partial_t^k \sigma^\epsilon(0), \partial_t^k u^\epsilon(0)) \in H^{2-k}(\Omega), \quad \partial_t^k d^\epsilon(0) \in H^{3-k}(\Omega), \quad k = 0, 1, 2, \\ \int_\Omega \sigma_0 dx = 0, \quad 1 + \epsilon\sigma_0^\epsilon \geqslant m \ \text{for some constant} \ m > 0. \end{cases} \tag{1.9}$$

Moreover, assume the following compatibility conditions are satisfied:

$$\begin{cases} \partial_t^k u^\epsilon(0) \cdot n = 0, \quad n \times \text{curl}u_0^\epsilon = n \times \text{curl}u_t^\epsilon(0) = 0, \quad \text{on} \ \partial\Omega, \ k = 0, 1, \\ \partial_t^k \frac{\partial d^\epsilon(0)}{\partial n} = 0 \ \text{on} \ \partial\Omega, \quad k = 0, 1. \end{cases} \tag{1.10}$$

Then, there exists a constant $T^\epsilon > 0$ such that the initial boundary value problem (1.4)-(1.8) has a unique solution $(\sigma^\epsilon, u^\epsilon, d^\epsilon)$ satisfying

$$\begin{cases} 1 + \epsilon\sigma^\epsilon > 0 \ in \ \Omega \times (0, T^\epsilon), \\ \partial_t^k \sigma^\epsilon \in C([0, T^\epsilon], H^{2-k}), \\ \partial_t^k u^\epsilon \in C([0, T^\epsilon], H^{2-k}) \cap L^2(0, T^\epsilon; H^{3-k}), \\ \partial_t^k d^\epsilon \in C([0, T^\epsilon], H^{3-k}) \cap L^2(0, T^\epsilon; H^{4-k}), \quad k = 0, 1, 2. \end{cases}$$

Remark 1.1 To simplify the statement, we have used $\sigma_t^\epsilon(0)$ to denote the quantity $\sigma_t^\epsilon|_{t=0}$ which can be obtained from (1.4), and the other quantities are defined in a similar way.

For simplicity, we denote

$$M^\epsilon(t) = \sup_{0 \leqslant s \leqslant t} \left\{ \|(\sigma^\epsilon, u^\epsilon, \nabla d^\epsilon)(\cdot, s)\|_{H^2} + \|(\sigma_s^\epsilon, u_s^\epsilon, \nabla d_s^\epsilon)(\cdot, s)\|_{H^1} \right.$$
$$+ \left\| \frac{1}{1 + \epsilon \sigma(\cdot, s)} \right\|_{L^\infty} + \epsilon \|(\sigma_{ss}^\epsilon, u_{ss}^\epsilon, \nabla d_{ss}^\epsilon)(\cdot, s)\|_{L^2} \bigg\}$$
$$+ \left\{ \int_0^t \left(\|u^\epsilon\|_{H^3}^2 + \|u_s^\epsilon\|_{H^2} + \|\epsilon(\sigma_{ss}^\epsilon, u_{ss}^\epsilon, \nabla d_{ss}^\epsilon)\|_{H^1} \right) ds \right\}^{\frac{1}{2}}.$$

Then, we state the main results in this paper as follows.

Theorem 1.1 Assume that $(\sigma^\epsilon, u^\epsilon, d^\epsilon)$ is the solution obtained in Proposition 1.1, and the initial datum $(\sigma_0^\epsilon, u_0^\epsilon, d_0^\epsilon)$ further satisfies

$$\|(\sigma_0^\epsilon, u_0^\epsilon, \nabla d_0^\epsilon)\|_{H^2} + \|(\sigma_t^\epsilon, u_t^\epsilon, \nabla d_t^\epsilon)(0)\|_{H^1} + \epsilon \|(\sigma_{tt}^\epsilon, u_{tt}^\epsilon, \nabla d_{tt}^\epsilon)(0)\|_{L^2} \leqslant D_0.$$

Then there exist two positive constants T_0 and D such that $(\sigma^\epsilon, u^\epsilon, d^\epsilon)$ satisfies the uniform estimates:

$$M^\epsilon(T_0) \leqslant D, \tag{1.11}$$

where D_0, T_0 and D are constants independent of $\epsilon \in (0,1)$.

Based on the above uniform estimates, by applying the Arzelá-Ascolis theorem, we can prove the following convergence result in a standard way.

Theorem 1.2 Let $(\sigma^\epsilon, u^\epsilon, d^\epsilon)$ be the solution obtained in Theorem 1.1, and the initial data $(\sigma_0^\epsilon, u_0^\epsilon, d_0^\epsilon)$ further satisfies that

$$\begin{cases} (u_0^\epsilon, \nabla d_0^\epsilon) \to (u_0, \nabla d_0) \text{ strongly in } H^s \text{ for any } 0 \leqslant s < 2 \text{ as } \epsilon \to 0, \\ \epsilon \sigma_0^\epsilon \to 0 \text{ strongly in } H^s \text{ for any } 0 \leqslant s < 1 \text{ as } \epsilon \to 0. \end{cases} \tag{1.12}$$

Then $(\rho^\epsilon, u^\epsilon, \nabla d^\epsilon) \to (1, u, \nabla d)$ strongly in $C([0,T_0]; H^1)$ as the Mach number $\epsilon \to 0$, and there exists a function $\pi(x,t)$ such that (u, π, d) satisfies the following classical incompressible nematic crystal equations

$$\begin{cases} u_t + u \cdot \nabla u + \nabla \pi - \mu \Delta u = -\nabla d \cdot \Delta d, \\ \operatorname{div} u = 0, \\ d_t + u \cdot \nabla d = \Delta d + |\nabla d|^2 d, \quad |d| = 1, \end{cases} \tag{1.13}$$

with the following initial and boundary conditions

$$\begin{cases} (u,d)|_{t=0} = (u_0, d_0), & \text{in } \Omega, \\ u \cdot n = 0, \ \mathrm{curl}\, u \times n = 0, \ \dfrac{\partial d}{\partial n} = 0, & \text{on } \partial\Omega. \end{cases} \tag{1.14}$$

The remainder of this paper is devoted to the proof of Theorem 1.1.

2 Proof of Theorem 1.1

In this paper, we will use the methods applied in [5-7] to prove Theorem 1.1. According to the similar arguments in [5,6], we know that to prove Theorem 1.1 it is suffices to prove

$$M^\epsilon(T_0) \leqslant C_0(M_0^\epsilon) \exp(t^{1/4} C(M^\epsilon(t))), \tag{2.1}$$

for all $t \in [0, T^\epsilon]$ and for some given nondecreasing continuous functions $C_0(\cdot)$ and $C(\cdot)$.

For the sake of simplicity, we will drop the superscript ϵ of σ^ϵ, u^ϵ, d^ϵ and so on. Moreover, in the following, we will write $M^\epsilon(t)$ and M_0^ϵ as M and M_0, respectively. The symbol C denotes a generic constant and its value may change from line to line.

Firstly, we list some lemmas which will be frequently used throughout this paper.

Lemma 2.1 ([9]) Let Ω be a bounded domain in $\mathbf{R}^N$ with a smooth boundary $\partial\Omega$ and outward normal n. For any $u \in H^1(\Omega)$ with $u \cdot n = 0$ or $u \times n = 0$ on $\partial\Omega$, there exists a positive constant C independent of u such that

$$\|u\|_{L^2(\Omega)} \leqslant C(\|\mathrm{div}\, u\|_{L^2(\Omega)} + \|\mathrm{curl}\, u\|_{L^2(\Omega)}), \tag{2.2}$$

where the vorticity $\mathrm{curl}\, u = (\partial_2 u_3 - \partial_3 u_2, \partial_3 u_1 - \partial_1 u_3, \partial_1 u_2 - \partial_2 u_1)^\mathrm{T}$.

Lemma 2.2 ([14]) Let Ω be a bounded domain in $\mathbf{R}^N$ with smooth boundary $\partial\Omega$ and outward normal n. Then, for any $u \in H^1(\Omega)$, $s \geqslant 1$, there exists a constant $C > 0$ independent of u, such that

$$\|u\|_{H^s(\Omega)} \leqslant C(\|\mathrm{div}\, u\|_{H^{s-1}(\Omega)} + \|\mathrm{curl}\, u\|_{H^{s-1}(\Omega)}$$
$$+ \|u \times n\|_{H^{s-\frac{1}{2}}(\partial\Omega)} + \|u\|_{H^{s-1}(\Omega)}). \tag{2.3}$$

Lemma 2.3 ([3]) *Let Ω be a bounded domain in $\mathbf{R}^N$ with smooth boundary $\partial\Omega$ and outward normal n. Then, for any $u \in H^1(\Omega)$, $s \geqslant 1$, there exists a constant $C > 0$ independent of u, such that*

$$\|u\|_{H^s(\Omega)} \leqslant C(\|\mathrm{div} u\|_{H^{s-1}(\Omega)} + \|\mathrm{curl} u\|_{H^{s-1}(\Omega)} + \|u \cdot n\|_{H^{s-\frac{1}{2}}(\partial\Omega)} + \|u\|_{H^{s-1}(\Omega)}). \tag{2.4}$$

According to Lemma 2.1, Lemma 2.2 and Lemma 2.3, we derive

$$\|\mathrm{curl} u\|_{H^2} \leqslant C(\|\Delta \mathrm{curl} u\|_{L^2} + \|u\|_{H^2}), \tag{2.5}$$

for $u \cdot n = 0$ and $\mathrm{curl} u \times n = 0$ on $\partial\Omega$. In fact, the latter one gives (see [1,7])

$$\mathrm{curl}\mathrm{curl} u \cdot n = 0 \quad \text{on } \partial\Omega.$$

Firstly, we know that ρ and its derivatives always appear as a coefficient of u and its derivatives. Thus, for simplicity, we use the standard energy method in [5,12] to obtain

$$\|\rho(\cdot,t)\|_{H^2} + \|\rho_t(\cdot,t)\|_{H^1} + \|\rho_{tt}(\cdot,t)\|_{L^2} + \left\|\frac{1}{\rho}(\cdot,t)\right\|_{L^\infty} \leqslant C_0(M_0)(\sqrt{t}C(M)). \tag{2.6}$$

Now, we use the method in [5,6,15] to prove a priori estimates on σ, u and d. Multiplying (1.4)-(1.5) by σ and u, respectively, and integrating over $\Omega \times (0,t)$, we obtain

$$\frac{1}{2}\|(\sigma, \sqrt{\rho}u)\|_{L^2}^2 + \int_0^t \|(\sqrt{\mu}\mathrm{curl} u, \sqrt{\lambda + 2\mu}\mathrm{div} u)\|_{L^2}^2 ds$$

$$= -\frac{1}{2}\int_0^t \int_\Omega \sigma^2 \mathrm{div} u \, dx ds + \int_0^t \int_\Omega \frac{P'(1) - P'(1+\epsilon\sigma)}{\epsilon} u \nabla \sigma \, dx ds$$

$$- \int_0^t \int_\Omega (u \cdot \nabla)d \cdot \Delta d \, dx ds + \frac{1}{2}\|(\sigma_0, \sqrt{\rho_0}u_0)\|_{L^2}^2$$

$$\leqslant C_0(M_0) + C\int_0^t \|\nabla \sigma\|_{L^2}^2 \|\nabla u\|^2 ds + C\int_0^t \|u\|_{L^6}\|\sigma\|_{L^3}\|\nabla \sigma\|_{L^2} ds$$

$$+ C\int_0^t \|u\|_{L^6}\|\nabla d\|_{L^3}\|\Delta d\|_{L^2} ds$$

$$\leqslant C_0(M_0)\exp(tC(M)), \tag{2.7}$$

where we have used

$$-\Delta u = -\nabla \mathrm{div} u + \mathrm{curl}\mathrm{curl} u. \tag{2.8}$$

Multiplying (1.5) by $\nabla \mathrm{div} u$ and integrating the result over $\Omega \times (0, t)$, we find

$$(\lambda + 2\mu) \int_0^t \|\nabla \mathrm{div} u\|_{L^2}^2 ds - \frac{1}{\epsilon} \int_0^t \int_\Omega \nabla \mathrm{div} u \cdot \nabla \sigma dx ds$$

$$= \int_0^t \int_\Omega (\rho u_t + \rho u \cdot \nabla u + \nabla d \cdot \Delta d) \nabla \mathrm{div} u dx ds$$

$$+ \int_0^t \int_\Omega \frac{P'(1+\epsilon\sigma) - P'(1)}{\epsilon} \nabla \sigma \cdot \nabla \mathrm{div} u dx dx$$

$$= -\frac{1}{2} \int_\Omega \rho (\mathrm{div} u)^2 dx + \frac{1}{2} \int_\Omega \rho (\mathrm{div} u_0)^2 dx + \frac{1}{2} \int_0^t \int_\Omega \rho_t (\mathrm{div} u)^2 dx ds$$

$$- \int_0^t \int_\Omega \nabla \rho \cdot u_t \mathrm{div} u dx ds + \int_0^t \int_\Omega (\rho u \cdot \nabla u + \nabla \cdot \Delta d) \nabla \mathrm{div} u dx ds$$

$$+ \int_0^t \int_\Omega \frac{P'(1+\epsilon\sigma) - P'(1)}{\epsilon} \nabla \sigma \cdot \nabla \mathrm{div} u dx ds.$$

Then we obtain

$$\int_\Omega \rho (\mathrm{div} u)^2 dx + \int_0^t \|\nabla \mathrm{div} u\|_{L^2}^2 ds - \frac{1}{\epsilon} \int_0^t \int_\Omega \nabla \mathrm{div} u \cdot \nabla \sigma dx ds$$

$$\leqslant C_0(M_0) + \int_0^t (\|\nabla u\|_{L^2} \|u\|_{H^2} + \|\nabla d\|_{H^2} \|\Delta d\|_{L^2}) \|\nabla \mathrm{div} u\|_{L^2} ds$$

$$+ \int_0^t \|\sigma\|_{H^2} \|\nabla \sigma\|_{L^2} \|\nabla \mathrm{div} u\|_{L^2} ds$$

$$\leqslant C_0(M_0) \exp(tC(M)). \tag{2.9}$$

To eliminate the singular term in (2.9), we take ∇ to (1.4) and multiply the result by $\nabla \sigma$ to find

$$\frac{1}{2} \int_\Omega |\nabla \sigma|^2 dx + \frac{1}{\epsilon} \int_0^t \int_\Omega \nabla \sigma \cdot \nabla \mathrm{div} u dx ds$$

$$= \frac{1}{2} \int_\Omega |\nabla \sigma_0|^2 dx - \int_0^t \int_\Omega \nabla \mathrm{div}(\sigma u) \cdot \nabla \sigma dx$$

$$\leqslant C_0(M_0) \exp(tC(M)). \tag{2.10}$$

Summing up (2.9) and (2.10), we derive

$$\|(\mathrm{div} u, \nabla \sigma)\|_{L^2}^2 + \int_0^t \|\nabla \mathrm{div} u\|_{L^2}^2 ds \leqslant C_0(M_0) \exp(tC(M)). \tag{2.11}$$

Denote $\omega = \mathrm{curl} u$. Taking curl to (1.4), we have

$$\rho \partial_t \omega + \rho u \cdot \nabla \omega - \mu \Delta \omega = f, \tag{2.12}$$

where $f = \nabla\rho \times \partial_t u + \nabla(\rho u_i) \times \partial_i u - \nabla\Delta d_j \times \nabla d_j$.

Multiplying (2.12) by ω, we derive

$$\|\text{curl} u\|_{L^2}^2 + \int_0^t \int_\Omega |\text{curlcurl} u|^2 dx ds \leqslant C_0(M_0)\exp(\sqrt{t}C(M)). \tag{2.13}$$

According to the Lemma 2.3 and the boundary condition $\dfrac{\partial d}{\partial n} = 0$ on $\partial\Omega$, we know that

$$\|\nabla d\|_{H^1} \leqslant C(\|\text{div}\nabla d\|_{L^2} + \|\text{curl}\nabla d\|_{L^2}) = C\|\Delta d\|_{L^2}. \tag{2.14}$$

Taking ∇ to (1.6), we derive

$$\nabla d_t - \nabla\Delta d = \nabla(|\nabla d|^2 d) - \nabla(u\cdot\nabla d). \tag{2.15}$$

Multiplying (2.15) by ∇d_t and integrating over $\Omega \times (0,t)$, we get

$$\frac{1}{2}\int_\Omega |\Delta d|^2 dx + \int_0^t\int_\Omega |\nabla d_t|^2 dxds$$

$$= \frac{1}{2}\int_\Omega |\Delta d_0|^2 dx + \int_0^t\int_\Omega (\nabla(|\nabla d|^2 d) - \nabla(u\cdot\nabla d))\cdot\nabla d_t dxds$$

$$\leqslant \int_0^t\int_\Omega (|\nabla d|^3 + |\nabla d||\nabla^2 d| + |\nabla u||\nabla d| + |u||\nabla^2 d|)\nabla d_t dxds$$

$$\leqslant \int_0^t \|d\|_{H^3}^2(\|\nabla d\|_{L^2} + \|\nabla^2 d\|_{L^2} + \|\nabla u\|_{L^2})\|\nabla d_t\|_{L^2} ds$$

$$+ \int_0^t \|u\|_{H^2}\|\nabla^2 d\|_{L^2}\|\nabla d_t\|_{L^2} ds$$

$$\leqslant C_0(M_0)\exp(tC(M)). \tag{2.16}$$

Combining (2.14) with (2.16), we obtain

$$\|\nabla d\|_{H^1}^2 + \int_0^t\int_\Omega |\nabla d_t|^2 dxds \leqslant C_0(M_0)\exp(tC(M)). \tag{2.17}$$

Multiplying (2.12) by $\partial_t\omega - \Delta\omega$, we obtain

$$\frac{\mu}{2}\frac{d}{dt}\int_\Omega |\text{curlcurl} u|^2 dx + \int_\Omega (\mu|\Delta\omega|^2 + \rho|\omega_t|^2)dx$$

$$= \int_\Omega \rho\omega_t \Delta\omega dx - \int_\Omega \rho(u\cdot\nabla)\omega(\omega_t - \Delta\omega)dx + \int_\Omega f(\omega_t - \Delta\omega)dx$$

$$= I_1 + I_2 + I_3, \tag{2.18}$$

where by using (2.8), we have

$$-\mu \int_\Omega \Delta\omega \cdot \omega_t dx = \mu \int_\Omega \text{curlcurl}\omega \cdot \omega_t dx$$
$$= \mu \int_\Omega \text{curl}\omega \cdot \text{curl}\omega_t dx + \int_{\partial\Omega}(\omega_t \times n)\text{curl}\omega dS$$
$$= \frac{\mu}{2}\frac{d}{dt}\int_\Omega |\text{curl}\omega|^2 dx.$$

Then, we estimate I_1, I_2 and I_3 as follows.

$$I_1 = -\int_\Omega \rho\omega_t \text{curlcurl}\omega dx$$
$$= -\int_\Omega \rho\text{curl}\omega\text{curl}\omega_t dx - \int_\Omega \text{curl}\omega \cdot (\nabla\rho \times \omega_t)dx$$
$$= -\frac{1}{2}\frac{d}{dt}\int_\Omega \rho|\text{curl}\omega|^2 dx + C\|\rho\|_{L^\infty}\|\text{curl}\omega\|_{L^2}\|\nabla\text{curl}\omega\|_{L^2}$$
$$+ \|\nabla\rho\|_{L^6}\|\text{curl}\omega\|_{L^3}\|\omega_t\|_{L^2}$$
$$\leqslant -\frac{1}{2}\frac{d}{dt}\int_\Omega \rho|\text{curl}\omega|^2 dx + C\|\rho\|_{H^2}\|u\|_{H^2}^2\|u\|_{H^3}$$
$$+ \|\rho\|_{H^2}\|u_t\|_{H^1}\|u\|_{H^2}^{\frac{1}{2}}\|u\|_{H^3}^{\frac{1}{2}},$$

$$|I_2| \leqslant C\|\rho\|_{L^\infty}\|\nabla\omega\|_{L^2}(\|\omega_t\|_{L^2} + \|\Delta\omega\|_{L^2})$$
$$\leqslant C\|\rho\|_{H^2}\|u\|_{H^2}(\|u\|_t\|_{H^1} + \|u\|_{H^3}),$$

and

$$|I_3| \leqslant C\|f\|_{L^2}(\|u_t\|_{H^1} + \|u\|_{H^3}),$$

where

$$\|f\| \leqslant C\|(|\nabla\rho||\partial_t u|, |\nabla(\rho u)||\nabla u|, |\nabla^3 d||\nabla d|)\|_{L^2}$$
$$\leqslant C\|\rho\|_{H^2}\|\partial_t u\|_{L^2}^{\frac{1}{2}}\|\partial_t u\|_{H^1}^{\frac{1}{2}} + C\|\rho\|_{H^2}\|u\|_{H^1}^{\frac{1}{2}}\|u\|_{H^2}^{\frac{3}{2}} + C\|d\|_{H^3}^2$$
$$\leqslant C(M).$$

Substituting the above estimates into (2.18) and integrating over $(0,t)$, we obtain

$$\|\text{curlcurl}u\|_{L^2}^2 + \int_0^t \int_\Omega (|\Delta\text{curl}u|^2 + |\text{curl}u_t|^2)dxds$$
$$\leqslant C_0(M_0)\exp(\sqrt{t}C(M)). \tag{2.19}$$

Applying ∂_t to (1.4) and (1.5), respectively, we obtain

$$\sigma_{tt} + \frac{1}{\epsilon}\text{div}u_t = -\text{div}(\sigma u)_t, \tag{2.20}$$

$$\rho u_{tt} + \rho u \cdot \nabla u_t - \mu \Delta u_t - (\lambda + \mu)\nabla \mathrm{div} u_t = -\rho_t u_t - (\rho u)_t \cdot \nabla u$$
$$- \frac{1}{\epsilon}(P'(1+\epsilon\sigma)\nabla\sigma)_t - \nabla d_t \cdot \Delta d - \nabla d \cdot \Delta d_t. \tag{2.21}$$

Multiplying (2.21) by $-\nabla \mathrm{div} u$, we derive

$$\frac{\lambda+2\mu}{2}\int_\Omega |\nabla \mathrm{div} u|^2 dx - \frac{P'(1)}{\epsilon}\int_0^t \int_\Omega \nabla\sigma_t \nabla \mathrm{div} u\, dxds$$
$$= \frac{\lambda+2\mu}{2}\int_\Omega |\nabla \mathrm{div} u_0|^2 dx + \int_0^t \int_\Omega \left(\frac{P'(1+\epsilon\sigma)-P'(1)}{\epsilon}\nabla\sigma\right)_t \nabla \mathrm{div} u\, dxds$$
$$+ \int_0^t \int_\Omega (\rho u_{tt} + \rho u \cdot \nabla u_t + \rho_t u_t - (\rho u)_t \cdot \nabla u)\nabla \mathrm{div} u\, dxds$$
$$+ \int_0^t \int_\Omega (\nabla d_t \cdot \Delta d + \nabla d \cdot \Delta d_t)\nabla \mathrm{div} u\, dxds$$
$$= \frac{\lambda+2\mu}{2}\int_\Omega |\nabla \mathrm{div} u_0|^2 dx + I_4 + I_5 + I_6. \tag{2.22}$$

We estimate I_4, I_5 and I_6 as follows.

$$|I_4| \leqslant C \int_0^t \|\sigma\|_{H^2}\|\nabla \mathrm{div} u\|_{L^2}(\|\sigma_t\|_{L^3} + \|\nabla\sigma_t\|_{L^2})ds \leqslant tC(M),$$

$$|I_5| \leqslant C \int_0^t \|\rho\|_{H^2}\|u_{tt}\|_{L^2}\|u\|_{H^2}ds + tC(M) \leqslant tC(M),$$

$$|I_6| \leqslant \int_0^t \|d\|_{H^2}\|\nabla \mathrm{div} u\|_{L^2}(\|\nabla d_t\|_{L^3} + \|\Delta d_t\|_{L^2})ds \leqslant tC(M).$$

To eliminate the singular term, we take ∇ to (1.4) and multiply the result by $\nabla\sigma_t$ to obtain

$$\int_0^t \|\nabla\sigma_t\|_{L^2}^2 ds + \frac{1}{\epsilon}\int_0^t \int_\Omega \nabla\sigma_t \nabla \mathrm{div} u\, dxds$$
$$= -\int_0^t \int_\Omega \nabla\mathrm{div}(\sigma u)\cdot \nabla\sigma_t\, dxds \leqslant tC(M). \tag{2.23}$$

Summing up (2.22) and (2.23), we derive

$$\int_\Omega |\nabla \mathrm{div} u|^2 dx + \int_0^t \|\nabla\sigma_t\|_{L^2}^2 ds \leqslant C_0(M_0)\exp(tC(M)). \tag{2.24}$$

Taking ∂_i to (1.5) and multiplying the result by $\partial_i \nabla \mathrm{div} u$, we derive

$$\int_0^t \int_\Omega |\partial_i \nabla \mathrm{div} u|^2 dxds - \frac{1}{\epsilon}\int_0^t \int_\Omega \partial_i\nabla\sigma \cdot \partial_i\nabla\mathrm{div} u\, dxds$$

$$\leqslant \int_0^t \int_\Omega (|\nabla(\rho u_t + \rho u \cdot \nabla u)|^2 + |\nabla(\sigma \nabla \sigma)|^2 + |\nabla(\nabla d \cdot \Delta d)|^2) dx ds$$
$$\leqslant tC(M). \tag{2.25}$$

To eliminate the singular term, taking $\partial_i \nabla$ to (1.4) and multiplying the result by $\partial_i \nabla \sigma$, we obtain

$$\frac{1}{2} \int_\Omega |\partial_i \nabla \sigma|^2 dx + \frac{1}{\epsilon} \int_0^t \int_\Omega \partial_i \nabla \mathrm{div} u \cdot \partial_i \nabla \sigma dx ds$$
$$= \frac{1}{2} \int_\Omega |\partial_i \nabla \sigma_0|^2 dx + \int_0^t \int_\Omega \partial_i \nabla (\sigma \mathrm{div} u + \nabla \sigma \cdot u) \partial_i \nabla \sigma dx ds$$
$$\leqslant C_0(M_0) + \int_0^t \|u\|_{H^3} \|\sigma\|_{H^2}^2 ds$$
$$\leqslant C_0(M_0) \exp(\sqrt{t}C(M)). \tag{2.26}$$

Summing up (2.25) with (2.26), we obtain

$$\|\nabla^2 \sigma\|_{L^2}^2 + \int_0^t \int_\Omega |\nabla^2 \mathrm{div} u|^2 dx ds \leqslant C_0(M_0) \exp(\sqrt{t}C(M)). \tag{2.27}$$

To obtain a priori estimate on $\|d\|_{L_t^\infty(H^3)}$, by using the elliptic regularity theory, (2.17), (2.19) and (2.24), we obtain

$$\|\nabla d\|_{H^2} \leqslant C \|\nabla d_t\|_{L^2} + C \|\nabla u \cdot \nabla d\|_{L^2} + C \|u \cdot \nabla^2 d\|_{L^2}$$
$$+ C \||\nabla d|^3\|_{L^2} + \||\nabla d| |\nabla^2 d|\|_{L^2}$$
$$\leqslant C \|\nabla d_t\|_{L^2} + C \|\nabla u\|_{L^3} \|\nabla d\|_{L^6} + C \|u\|_{L^6} \|\nabla^2 d\|_{L^3} + C \|\nabla d\|_{L^6} \|\nabla^2 d\|_{L^3}$$
$$\leqslant C \|\nabla d_t\|_{L^2} + \frac{1}{2} \|\nabla d\|_{H^2} + C_0(M_0) \exp(tC(M)),$$

where we have used Nirenberg's interpolation inequality and Young inequality frequently. Then, we conclude

$$\|\nabla d\|_{H^2} \leqslant C \|\nabla d_t\|_{L^2} + C_0(M_0) \exp(tC(M)). \tag{2.28}$$

Hence, to obtain the estimate on $\|\nabla d\|_{L_t^\infty(H^2)}$, we have to estimate $\|\nabla d_t\|_{L_t^\infty(L^2)}$. Taking ∂_t to (1.6), we obtain

$$d_{tt} + (u \cdot \nabla d)_t = \Delta d_t + (|\nabla d|^2 d)_t. \tag{2.29}$$

Multiplying (2.29) by $-\Delta d_t$ and integrating over $\Omega \times (0, t)$, we derive

$$\frac{1}{2} \int_\Omega |\nabla d_t|^2 dx + \int_0^t \int_\Omega |\Delta d_t|^2 dx dt$$

$$= \frac{1}{2} \int_\Omega |\nabla d_t(0)|^2 dx + \int_0^t \int_\Omega (u_t \cdot \nabla d + u \cdot \nabla d_t) \Delta d_t dx ds$$

$$- \int_0^t \int_\Omega (|\nabla d|^2 d_t + d\partial_t |\nabla d|^2) \Delta d_t dx ds$$

$$\leqslant C_0(M_0) + C \int_0^t (\|\nabla d\|_{L^\infty} \|u_t\|_{L^2} + \|u\|_{L^\infty} \|\nabla d_t\|_{L^2}) \|\Delta d_t\|_{L^2} ds$$

$$+ C \int_0^t (\|\nabla d\|_{L^\infty}^2 \|d_t\|_{L^2} + \|\nabla d\|_{L^\infty} \|\nabla d_t\|_{L^2}) \|\Delta d_t\|_{L^2} ds$$

$$\leqslant C_0(M_0) \exp(tC(M)). \tag{2.30}$$

Substituting (2.30) into (2.28), we obtain

$$\|\nabla d\|_{H^2} \leqslant C_0(M_0) \exp(tC(M)). \tag{2.31}$$

Then, by using similar calculations in [5], we can obtain the basic a priori estimates on σ_t, u_t. Multiplying (2.20), and (2.21) by σ_t and u_t, respectively, and integrating over $\Omega \times (0, t)$, we conclude

$$(\|\sigma_t\|_{L^2}^2 + \|u_t\|_{L^2}^2) + \int_0^t \|(\operatorname{curl} u_t, \operatorname{div} u_t)\|_{L^2}^2 ds \leqslant C_0(M_0) \exp(tC(M)). \tag{2.32}$$

Multiplying (2.20), (2.21) by $-\Delta \sigma_t$ and $-\nabla \operatorname{div} u_t$, respectively, we obtain

$$\frac{1}{2} \int_\Omega |\nabla \sigma_t|^2 dx + \frac{1}{\epsilon} \int_0^t \int_\Omega \nabla \operatorname{div} u_t \cdot \nabla \sigma_t dx ds$$

$$= \frac{1}{2} \int_\Omega |\nabla \sigma_t(0)|^2 dx + \int_0^t \int_\Omega \operatorname{div}(\sigma_t u + \sigma u_t) \Delta \sigma_t dx ds$$

$$= \frac{1}{2} \int_\Omega |\nabla \sigma_t(0)|^2 dx + I_7, \tag{2.33}$$

where

$$I_7 = \int_0^t \int_\Omega u \cdot \nabla \sigma_t \Delta \sigma_t dx ds - \int_0^t \int_\Omega \nabla(\sigma_t \operatorname{div} u + u_t \nabla \sigma + \sigma \operatorname{div} u_t) dx ds$$

$$= - \int_0^t \int_\Omega \partial_j u_i \partial_i \sigma_t \partial_j \sigma_t dx ds + \frac{1}{2} \int_0^t \int_\Omega \operatorname{div} u |\nabla \sigma_t|^2 dx ds$$

$$- \int_0^t \int_\Omega \nabla(\sigma_t \operatorname{div} u + u_t \nabla \sigma + \sigma \operatorname{div} u_t) dx ds$$

$$\leqslant tC(M) + C(M) \int_0^t \|u\|_{H^3} ds + C(M) \int_0^t \|u_t\|_{H^2} ds$$

$$\leqslant \sqrt{t} C(M), \tag{2.34}$$

and

$$\frac{1}{2}\int_\Omega \rho(\mathrm{div}u_t)^2 dx + (\lambda+2\mu)\int_0^t\int_\Omega |\nabla \mathrm{div}u_t|^2 dx$$
$$-\frac{P'(1)}{\epsilon}\int_0^t\int_\Omega \nabla\sigma_t\cdot\nabla\mathrm{div}u_t dxds$$
$$=\frac{1}{2}\int_\Omega \rho_0(\mathrm{div}u_t(0))^2 dx + \int_0^t\int_\Omega \left(\frac{P'(1+\epsilon\sigma)-P'(1)}{\epsilon}\nabla\sigma\right)_t\nabla\mathrm{div}u_t dxds$$
$$+\int_0^t\int_\Omega (\frac{\epsilon}{2}\sigma_t(\mathrm{div}u_t)^2 - \epsilon u_{tt}\cdot\nabla\sigma\mathrm{div}u_t)dxds$$
$$-\int_0^t\int_\Omega (\rho_t u_t + (\rho u\cdot\nabla u)_t)\nabla\mathrm{div}u_t dxds$$
$$+\int_0^t\int_\Omega (\nabla d_t\cdot\Delta d + \nabla d\cdot\Delta d_t)\nabla\mathrm{div}u_t dxds$$
$$\leqslant C_0(M_0) + \sqrt{t}C(M). \tag{2.35}$$

Summing up (2.33), (2.34) and (2.35), we derive

$$\int_\Omega (|\nabla\sigma_t|^2 + (\mathrm{div}u_t)^2)dx + \int_0^t\int_\Omega |\nabla\mathrm{div}u_t|^2 dxds \leqslant C_0(M_0)\exp(\sqrt{t}C(M)). \tag{2.36}$$

To complete the estimate of $\|u_t\|_{L_t^\infty(H^1)}$, we apply ∂_t to (2.12) to know

$$\rho_t\omega_t + \rho\omega_{tt} + (\rho u)_t\cdot\nabla\omega + \rho u\cdot\nabla\omega_t - \mu\Delta\omega_t = \nabla\rho_t\times u_t + \nabla\rho\times u_{tt}$$
$$+\nabla\Delta(d_j)_t\times\nabla d_j + \nabla\Delta d_j\times\nabla(d_j)_t + \nabla(\rho u_i)_t\times\partial_i u + \nabla(\rho u_i)\times\partial_i u_t. \tag{2.37}$$

Multiplying (2.37) by ω_t in $L^2(\Omega\times(0,t))$, we deduce

$$\frac{1}{2}\int_\Omega \rho|\omega_t|^2 dx + \mu\int_0^t\int_\Omega |\mathrm{curl}\omega_t|^2 dxds$$
$$=\frac{1}{2}\int_\Omega \rho|\omega_t(0)|^2 dx + \int_0^t\int_\Omega \left(\frac{\epsilon}{2}\sigma_t|\omega_t|^2 - \epsilon\sigma_t\omega_t - (\rho u)_t\cdot\nabla\omega - \rho u\cdot\nabla\omega_t\right)\omega_t dxds$$
$$+\epsilon\int_0^t\int_\Omega (\nabla\sigma_t\times u_t + \nabla\sigma\times u_{tt})\omega_t dxds + \int_0^t\int_\Omega \nabla\Delta(d_j)_t\times\nabla d_k\omega_t dxds$$
$$+\int_0^t\int_\Omega (\nabla\Delta d_j\times\nabla(d_j)_t + \nabla(\rho u_i)_t\times\partial_i u + \nabla(\rho u_i)\times\partial_i u_t)\omega_t dxds$$
$$=C_0(M_0) + I_8 + I_9 + I_{10} + I_{11}, \tag{2.38}$$

where, by using (2.32) and the integrating by parts formula, we have

$$-\mu \int_0^t \int_\Omega \Delta\omega \cdot \omega_t dxds = \mu \int_0^t \int_\Omega \text{curlcurl}\omega_t \cdot \omega_t dxds$$

$$= \mu \int_0^t \int_\Omega |\text{curl}\omega_t|^2 dxds + \mu \int_0^t \int_{\partial\Omega} \text{curl}\omega_t \cdot (\omega_t \times n) dS$$

$$= \mu \int_0^t \int_\Omega |\text{curl}\omega_t|^2 dxds.$$

We estimate $I_i (i = 8, 9, 10, 11)$ as follows.

$$I_{10} = \int_0^t \int_\Omega \nabla\Delta(d_j)_t \cdot (\nabla d_j \times \omega_t) dxds$$

$$= -\int_0^t \int_\Omega \Delta(d_j)_t \text{div}(\nabla d_j \times \omega_t) dxds + \int_0^t \int_{\partial\Omega} \Delta(d_j)_t \nabla(d_j)_t \cdot (\omega_t \times n) dS$$

$$= \int_0^t \int_\Omega \Delta(d_j)_t \nabla d_j \cdot \text{curl}\omega_t dxds \leqslant \sqrt{t} C(M).$$

With similar calculations in [5], we have

$$|J_8| + |J_9| + |J_{11}| \leqslant \sqrt{t} C(M).$$

Substituting the above estimates into (2.38), we obtain

$$\int_\Omega \rho |\text{curl}u_t|^2 dx + \int_0^t \int_\Omega |\text{curlcurl}u_t|^2 dxds \leqslant C_0(M_0) \exp C(\sqrt{t} C(M)). \quad (2.39)$$

Now, we have a priori estimate on $\|\nabla d_t\|_{L_t^\infty(H^1)}$. Multiplying (2.29) by $-\Delta d_{tt}$ and integrating over $\Omega \times (0, t)$, we derive

$$\frac{1}{2} \int_\Omega |\Delta d_t|^2 dx + \int_0^t \int_\Omega |\nabla d_{tt}|^2 dxds$$

$$= \frac{1}{2} \int_\Omega |\Delta d_t(0)|^2 dx + \int_0^t \int_\Omega [(u \cdot \nabla d)_t - (|\nabla d|^2 d)_t] \Delta d_{tt} dxds$$

$$\leqslant C_0(M_0) + C \int_0^t (\|\nabla d\|_{H^2} \|u_t\|_{L^2} + \|u\|_{H^2} \|\nabla d_t\|_{L^2}$$

$$+ \|\nabla d\|_{H^2} \|\nabla d_t\|_{L^2}) \|\Delta d_{tt}\|_{L^2} ds$$

$$\leqslant C_0(M_0) \exp(\sqrt{t} C(M)). \quad (2.40)$$

For the same reason as (2.14), we conclude that

$$\|\nabla d_t\|_{H^1}^2 + \int_0^t \int_\Omega |\nabla d_{tt}|^2 dxds \leqslant C_0(M_0) \exp(\sqrt{t} C(M)). \quad (2.41)$$

Finally, we only need to estimate $\epsilon\sigma_{tt}, \epsilon u_{tt}, \epsilon\nabla d_{tt}$ to close the energy estimates. Multiplying $\partial_{tt}(1.4)$, $\partial_{tt}(1.5)$, $\partial_{tt}(1.6)$ by $\epsilon^2\sigma_{tt}$, $\epsilon^2 u_{tt}$ and $\epsilon^2\Delta d_{tt}$, respectively, and integrating over $\Omega \times (0,t)$, we can derive that

$$\epsilon\|(\sigma_{tt}, u_{tt}, \nabla d_{tt})\|_{L^2}^2 + \epsilon\int_0^t \|(u_{tt}, \nabla d_{tt})\|_{H^1}^2 ds \leqslant C_0(M_0)\exp(t^{1/4}C(M)). \tag{2.42}$$

Collecting the estimates obtained in (2.7), (2.11), (2.13), (2.17), (2.19), (2.24), (2.27), (2.31), (2.32), (2.36), (2.39), (2.41), and (2.42), we derive

$$\|(\sigma, u)\|_{L^2} + \|(\mathrm{div}u, \mathrm{curl}u, \mathrm{curlcurl}u, \nabla\mathrm{div}u)\|_{L^2} + \|(\nabla\sigma, \nabla d)\|_{H^1} + \|\nabla d\|_{H^2}$$
$$+ \|(\sigma_t, u_t)\|_{L^2} + \|(\nabla\sigma_t, \mathrm{div}u_t, \mathrm{curl}u_t)\|_{L^2} + \|\nabla d_t\|_{H^1} + \epsilon\|(\sigma_{tt}, u_{tt}, \nabla d_{tt})\|_{L^2}$$
$$+ \|(\mathrm{div}u, \mathrm{curl}u, \mathrm{curlcurl}u)\|_{L_t^2(L^2)} + \|(\nabla^2\mathrm{div}u, \Delta\mathrm{curl}u)\|_{L_t^2(L^2)}$$
$$+ \|(\mathrm{div}u_t, \mathrm{curl}u_t, \mathrm{curlcurl}u_t, \nabla\mathrm{div}u_t)\|_{L_t^2(L^2)} + \epsilon\|(\sigma_{tt}, u_{tt}, \nabla d_{tt})\|_{L_t^2(H^1)}$$
$$\leqslant C_0(M_0)\exp(t^{1/4}C(M)). \tag{2.43}$$

Thus, (2.1) holds true. Then, we complete the proof of Theorem 1.1.

Acknowledgements Guo is supported by NSFC under grant numbers 11731014, 11571254, and Ni is supported by NSFC under grant number 11771055 and Science Challenge Project under grand number TZ2016002.

References

[1] Bendali A, Dminguez J, Gallic S. A variational approach for the vector potential formulation of the Stokes and Navier-Stokes problems in three dimensional domains [J]. J. Math. Anal. Appl., 1985, 107: 537–560.

[2] Bie Q, Bo H, Wang Q, Yao Z. Global existence and incompressible limit in critical spaces for compressible flow of liquid crystals [J]. Z. Angew. Math. Phys., 2017, 68(5): 113.

[3] Bourguignon J, Brezis H. Remarks on the Euler equation [J]. J. Funct. Anal., 1974, 15: 341–363.

[4] Ding S, Huang J, Wen H, Zi R. Incompressible limit of the compressible hydrodynamic flow of liquid crystals [J]. J. Funct. Anal., 2013, 264: 1711–1756.

[5] Dou C, Jiang S, Ou Y. Low Mach number limit of full Navier-Stokes equations in a 3D bounded domain [J]. J. Differential Equations, 2015, 258: 379–398.

[6] Fan J, Li F, Nakamura G. Convergence of the full compressible Navier-Stokes-Maxwell system to the incompressible magnetohydrodynamic equations in a bounded domain [J]. Kinet. Relat. Models, 2016, 9: 443–453.

[7] Fan J, Li F, Nakamura G. Uniform well-posedness and singular limits of the isentropic Navier-Stokes-Maxwell system in a bounded domain [J]. Z. Angew. Math. Phys., 2015, 66: 1581–1593.

[8] Huang T, Wang C, Wen H. Strong solutions of the compressible nematic liquid crystal flow [J]. J. Differential Equations, 2012, 252(3): 2222–2265.

[9] Lions P. Mathematical Topics in Fluid Mechanics, Compressible Models, vol. 2 [M]. New York: Oxford University Press, 1998.

[10] Lions P, Masmoudi N. Incompressible limit for a viscous compressible fluid [J]. J. Math. Pures Appl., 1998, 77: 585–627.

[11] Masmoudi N. Incompressible, inviscid limit of the compressible Navier-Stokes system [J]. Ann. Inst. H. Poincaré Anal. Non Linéaire, 2001, 18: 199–224.

[12] Ou Y. Low Mach number limit of viscous polytropic fluid flows [J]. J. Differential Equations, 2011, 251: 2037–2065.

[13] Wang D, Yu C. Incompressible limit for the compressible flow of liquid crystals [J]. J. Math. Fluid Mech., 2014, 16: 771–786.

[14] Xiao Y, Xin Z. On the vanishing viscosity limit for the 3D Navier-Stokes equations with a slip boundary condition [J]. Commun. Pure Appl. Math., 2007, 60: 1027–1055.

[15] Yang X. Uniform well-posedness and low Mach number limit to the compressible Nematic liquid crystal flows in a bounded domain [J]. Nonlinear Analysis. 2015, 120: 118–120.

[16] Yang Y, Dou C, Ju Q. Weak-strong uniqueness property for the compressible flow of liquid crystals [J]. J. Differential Equations, 2013, 255(6): 1233–1253.

Some Mathematical Models and Mathematical Analysis about Rayleigh-Taylor Instability *

Guo Boling (郭柏灵), and Xie Binqiang (解斌强)

Abstract In this paper we will review some mathematical models and mathematical analysis about Rayleigh-Taylor instability.

Keywords mathematical models; mathematical analysis; Rayleigh-Taylor instability.

1 Introduction

The instability of the interface between two different densities of fluid under the action of gravity or inertial force, as early as 1950, G. L. Taylor clearly pointed out this instability, and often named after him, actually earlier than him, in 1900 L. Rayleigh and S.H. Lamb in 1932 also talked about this problem in some sense, people sometimes called Rayleigh-Taylor or Rayleigh-Lamb-Taylor instability. This interfacial instability phenomenon can be found not only in astrophysics, but also in laster fusion and high-speed collision. It is very important even for hydraulic machinery and various engines. The linear development stage of interface instability is relatively clear. However, there are still many problems in nonlinear development that need to be recognized. The relevant research has very important practical and theoretical value.

In the mathematical analysis theory, since 2003, there have been some breakthrough works on the RT instability of compressible fluids [1], [2], free boundary problems [3] and MHD fluids [4]. Of course, these theoretical results are still far away from the actual physical mechanics. There is still a big gap in the problem.

---

* Annal. Appl. Math., 2019, 1: 1-46. Doi: CNKI:SUN:WFFN.0.2019-01-001.

2 Some RT instability mathematical models

2.1 Double infinite fluid Taylor instability

We consider the two-layer infinite fluid shown in Figure 1, which has densities of ρ_1 and ρ_2, each occupying a half plane of $y > 0$ and $y < 0$, and gravity $\vec{g}$ parallel to y axis, pointing to its negative direction, i.e.

$$\vec{g} = -g\vec{j}.$$

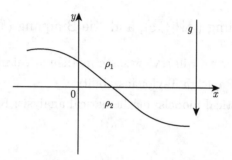

Fig. 1

The deviation of the initial position of the interface from $y = 0$ is a small amount, i.e,

$$y = \varepsilon \cos kx.$$

Assumed that the two-layer fluid is in a static state at the initial moment, and thus is initially non-rotating. For an ideal incompressible fluid, the fluid remains free-curl in the subsequent movement. Thus, the velocity potential $\Phi_i (i = 1, 2)$ can be introduced to correspond to the upper layer and lower layer of fluid. Then under two-dimensional conditions, $\Phi_i (i = 1, 2)$ satisfies the *Laplace* equation

$$\Phi_{ixx} + \Phi_{iyy} = 0, \quad i = 1, 2. \tag{2.1}$$

If we suppose that the interface position during the movement

$$y = \eta(x, t).$$

And its derivative is the first-order small amount of ϵ, then

$$F(r, t) = y - \eta(x, t) = 0. \tag{2.2}$$

Therefore,

$$\frac{\partial F}{\partial t} = -\eta_t \sim \varepsilon,$$
$$\frac{\partial F}{\partial \eta} = 1, \quad \frac{\partial F}{\partial x} = -\eta_x \sim \varepsilon, \qquad (2.3)$$
$$\nabla F = -\eta_x \vec{i} + \vec{j}.$$

The normal vector of the interface

$$\vec{n} = \frac{\nabla F}{|\nabla F|}, \qquad (2.4)$$

and the normal velocity of the interface

$$D_n = -\frac{\partial F/\partial t}{|\nabla F|}. \qquad (2.5)$$

The following subscripts $i = 1$ and 2 respectively denote the physical quantities in the flow on both sides of the interface, and the conditions of equal pressure and normal velocities on both sides of the fluid interface are continuous.

$$p_1 = p_2, \quad D_n = q_1 \cdot n = q_2 \cdot n, \qquad (2.6)$$

where p is pressure and q is velocity.

On the interface $F(r,t) = 0$,

$$\rho_1 \left\{ \Phi_{1t} + \frac{1}{2}(\nabla \Phi_1)^2 + U(r) + f_1(t) \right\}$$
$$= \rho_2 \left\{ \Phi_{2t} + \frac{1}{2}(\nabla \Phi_2)^2 + U(r) + f_2(t) \right\} \qquad (2.7)$$
$$-\frac{\partial F}{\partial t} = \nabla \Phi_1 \cdot \nabla F = \nabla \Phi_2 \cdot \nabla F.$$

In the case of potential, the momentum equation can be integrated.

$$\Phi_t + \frac{1}{2}(\nabla \Phi)^2 + \frac{P}{\rho} + U(r) = f'(t), \qquad (2.8)$$

where $U(r)$ is the potential of mass power, $f'(t)$ is an arbitrary function. In (2.8), we can choose $\rho_1 f_1(t) = \rho_2 f_2(t)$, Then, under the condition of retaining a first-order small amount, on $y = \eta(x,t)$,

$$\eta_t = \Phi_{1y} = \Phi_{2y},$$
$$\rho_1(\Phi_{1t} + g\eta(x,t)) = \rho_2(\Phi_{2t} + g\eta(x,t)). \qquad (2.9)$$

Since η is a first-order small quantity, Φ_t is also a first-order small quantity, and the relevant quantity on $y = \eta$ can be performed Taylor expands, i.e.,

$$\Phi_i|_{y=\eta} = \Phi_i|_{y=0} + \eta \frac{\partial \Phi_i}{\partial y} = \Phi_i|_{y=0} + O(\varepsilon^2). \tag{2.10}$$

That is, the relevant derivative of (2.9) can be expanded by the value of $y = 0$, that is, (2.9) still holds on the unknown active boundary $y = \eta$, but it is also fixed on the fixed boundary $y = 0$ after linearization. This greatly simplifies the solution of the problem. Φ_i obviously meets infinity disappearing condition, i.e.

$$\begin{aligned} x = \infty, & \quad \Phi_1 = 0, \\ y = -\infty, & \quad \Phi_2 = 0, \end{aligned} \tag{2.11}$$

For the two-material interface, initial conditions should also be given. We only discuss harmonics with zero initial velocity of k:

$$\begin{aligned} \eta(x, 0) &= \varepsilon \cos kx, \\ \eta_t(x, 0) &= 0. \end{aligned} \tag{2.12}$$

In general, for a variety of k harmonics, we can use the Fourier series expansion method, which can be solved by the linear superposition method.

Since the equation is linear, the variable x appears only in the initial condition as $\cos kx$, so it is desirable to choose

$$\begin{aligned} \Phi_i(x, y, t) &= \varphi_i(x, t) \cos kx, \\ \eta(x, t) &= A(t) \cos kx. \end{aligned} \tag{2.13}$$

We substitute into (2.12), we get

$$A(0) = \varepsilon, \quad A'(0) = 0. \tag{2.14}$$

Also we insert them into the Laplace equation, we have

$$\varphi_{iyy} - k^2 \varphi_i = 0. \tag{2.15}$$

Therefore, the solution is the following form:

$$\varphi_i = e^{\pm ky} A_i(t). \tag{2.16}$$

By the infinity condition (2.11), we know

$$\varphi_1 = A_1(t) e^{-ky}, \quad \varphi_2 = A_2(t) e^{ky}. \tag{2.17}$$

Substituting these results into the boundary condition $y = 0$ in (2.9), we have

$$\rho_1(\ddot{A}_1(t) + gA_1(t)) = \rho_2(\ddot{A}_2(t) + gA_2(t)),$$

$$\ddot{A}(t) = -k_1 A_1 = kA_2.$$

Differentiating the second formula of the above formula with respect to t, then substitute into the previous formula of the above formula, and eliminating $\ddot{A}_1(t)$ and $\ddot{A}_2(t)$. Finally, the second-order equation of $A(t)$ is obtained.

$$\ddot{A}(t) - n_0^2 A(t) = 0, \tag{2.18}$$

$$n_0 = \sqrt{\frac{\rho_1 - \rho_2}{\rho_1 + \rho_2} \cdot kg}. \tag{2.19}$$

Now, we will discuss it in two situations.

(1) $\rho_1 > \rho_2$ that is, the heavy fluid is located above the light fluid. At this time n_0 is purely real, $A(t) = ae^{-n_0 t} + be^{n_0 t}$. Due to the existence of the $e^{n_0 t}$ formal solution, $A(t)$ grows exponentially over time, that is, the disturbance is unstable. This phenomenon is explained in Fig. 2. When a fluid with a density of ρ_1 invades into a fluid with a density of ρ_2, the gravity of the portion of the fluid is

$$G \backsim V\rho_1 g.$$

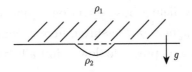

Fig. 2

And the buoyancy it receives is the weight of the invading fluid that is discharged in the same volume, that is, the buoyancy is

$$F \sim V\rho_2 g.$$

When $\rho_1 > \rho_2$, Gravity is greater than buoyancy, and this part of the invading fluid continues to sink, causing the disturbance on the interface to develop further, causing instability.

Taking into account initial conditions (2.12), we have

$$y = \eta(x,t) = \varepsilon \cosh n_0 t + \cos kx. \tag{2.20}$$

Here $\cosh n_0 t = \dfrac{e^{n_0 t} + e^{-n_0 t}}{2}$ is the hyperbolic cosine. If the initial condition (2.12) is changed to

$$\eta(x,0) = 0,$$

$$\eta_t(x,0) = \varepsilon \cos kx.$$

The form of the interface is

$$y = \frac{\varepsilon}{n_0} \sinh nt \cos kx.$$

In a more general case, it should be a linear superposition of two solutions.

After finding the expression of the interface shape $\eta(x,t)$, it is not difficult to get the form of $\varphi_i(t)$

(2) $\rho_1 < \rho_2$, $n_0 = im_0$ n_0 the pure imaginary number case. At this time

$$y = \eta(x,t) = \varepsilon \cos m_0 t + \cos kx. \tag{2.21}$$

Then the disturbance on the interface no longer develops, but is disturbed with time. The interface is stable, corresponding to the gravity less than buoyancy in Fig. 2, and the density ρ_1 of the fluid invading into the density ρ_2 of the fluid is retracted again, making the interface disturbances no longer considered.

Now we discuss the incompressible fluid with a density of ρ, a thickness of h under the gravitational field whose direction is down. In order for the fluid to be suspended and stationary, a pressure difference must be given to the lower surface. Since the ideal incompressible flow is initially at rest, it is non-curl.

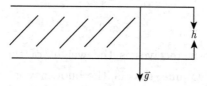

We introduce the velocity potential. Let Φ^0 be the case corresponding to the undisturbed static, and φ is the additional velocity potential formed by the interface disturbance, the velocity potential is

$$\Phi = \Phi^{(0)} + \varphi.$$

In the case of ignoring high order quantities, the pressure can be expressed as
$$p = -\rho(\Phi_t^0 + \varphi_t + gyf(t)).$$
Firstly, we look at the zero-order situation.
  On the upper surface $y = 0$, $p_1 = -\rho(\Phi_t^0 + f(t))$.
  On the lower surface $y = -1$, $p_2 = -\rho(\Phi_t^{(0)} - gh + f(t))$.
  Obviously, Φ_t^0 is only a function of t, subtracting two equations
$$p_2 - p_1 = \rho gh.$$

Now, let's discuss the disturbance on the interface, on the upper surface, we have
$$y = \eta_1(x, t).$$
On the lower surface
$$y = -h + \eta_2(x, t).$$
Since the pressures of the upper and lower surfaces, p_1 and p_2, are satisfied by the zero-order quantity, the disturbance pressure is
$$y = 0, \quad \varphi_t + g\eta_1 = 0,$$
$$y = -h, \quad \varphi_t + g\eta_2 = 0.$$

2.2 Finite thickness double interface fluid

In practice, we often encounter complex situations of fluids with limited thickness.

In addition, the second condition of (2.7) can be written as linearized
$$y = 0, \quad \eta_{1,t} = \varphi_y,$$
$$y = -h, \quad \eta_{2,t} = \varphi_y.$$

The perturbation of the Fourier component with $\cos kx$ is still discussed, so that φ satisfies the Laplace equation, by choosing
$$\varphi = (Ae^{ky} + Be^{-ky})e^{nt}\cos kx,$$
and
$$\eta_1 = a_1 e^{nt} \cos kx,$$
$$\eta_2 = a_2 e^{nt} \cos kx.$$

Substituting into the boundary condition of $y = 0$, we get
$$n(A + B) + ga_1 = 0,$$
$$na_1 = (A - B)k.$$

Substituting into the condition of $y = -h$, we have
$$n(Ae^{-kh} + Be^{kh}) + ga_2 = 0,$$
$$na_2 = k(Ae^{-kh} - Be^{kh}).$$

Eliminating a_1 and a_2 from these four equations, thus we get
$$A(n^2 + kg) + B(n^2 - kg) = 0,$$
$$A(n^2 + kg)e^{-kh} + B(n^2 - kg)e^{kh} = 0.$$

Thus, the conditions for the non-trivial solutions of A and B are
$$n^2 = \mp kg.$$

For the sake of simplicity, we take
$$m_0 = \sqrt{kg}.$$

Then the two real roots of n are
$$n = \pm m_0.$$

Obviously corresponding to $A = 0$ i.e. $\varphi = Be^{-ky}e^{\pm m_0 t}\cos kx$ is the solution of the lower surface problem. Corresponding to $B = 0$ the case, $\varphi = Ae^{ky}e^{\pm im_0 t}\cos kx$ is the solution of the upper surface problem. Considering that $t = 0$, the fluid should be at rest, so the solution in the general form can be written as
$$\varphi = [\bar{A}\sin(m_0 t)e^{ky} + B sh(m_0 t)e^{-ky}]\cos kx.$$

Obviously, the first term in square brackets represents a solution that vibrates over time, while the second term represents an unstable solution. Correspondingly, the two surface disturbances should also have the following form of solution:
$$\eta_1 = [A_1\cos(m_0 t) + B_1 ch m_0 t]\cos kx,$$
$$\eta_2 = [A_2\cos(m_0 t) + B_2 ch m_0 t]\cos kx.$$

Substituting into the kinematic conditions of $y = 0, y = -h$, we can get

$$m_0[-A_1 \sin(m_o t) + B_1 sh(m_0 t)] = k[\bar{A}\sin(m_o t) - \bar{B} sh(m_0 t)],$$
$$m_0[-A_2 \sin(m_o t) + B_2 sh(m_0 t)] = k[\bar{A}e^{-kh}\sin(m_o t) - \bar{B}e^{kh} sh(m_0 t)].$$

Therefore,

$$-\bar{A} = \frac{m_0 A_1}{k} = \frac{m_0 A_2 e^{kh}}{k},$$
$$-\bar{B} = \frac{m_0 B_1}{k} = \frac{m_0 B_2 e^{-kh}}{k},$$

where

$$A_2 = A_1 e^{-kh},$$
$$B_2 = B_1 e^{kh}.$$

Assuming that the upper and lower surface disturbances at the initial moment are

$$t = 0, \quad \begin{aligned} \eta_1 &= \epsilon_1 \cos kx, \\ \eta_2 &= \epsilon_2 \cos kx. \end{aligned}$$

Obviously, we have

$$A_1 + B_1 = \epsilon_1,$$
$$A_2 + B_2 = \epsilon_2.$$

The latter formula can be written as

$$A_1 e^{-kh} + B_1 e^{kh} = \varepsilon_2.$$

Thus, we have

$$A_1 = \frac{\varepsilon_1 - \varepsilon_2 e^{-kh}}{1 - e^{-2kh}},$$
$$B_1 = \frac{\varepsilon_2 - \varepsilon_1}{1 - e^{-2kh}} e^{-kh},$$
$$A_2 = \frac{\varepsilon_1 - \varepsilon_2 e^{-kh}}{1 - e^{-2kh}} e^{-kh},$$
$$B_2 = \frac{\varepsilon_2 - \varepsilon_1 e^{-kh}}{1 - e^{-2kh}},$$

and
$$\begin{cases} \eta_1(x,t) = \dfrac{1}{1-e^{-2k\eta}}[(\varepsilon_1 - \varepsilon_2 e^{-kh})\cos(m_0 t) \\ \qquad\qquad + (\varepsilon_2 - \varepsilon_1 e^{-kh})e^{-kh}\cosh(m_0 t)]\cos kx, \\ \eta_2(x,t) = \dfrac{1}{1-e^{-2kh}}[(\varepsilon_1 - \varepsilon_2 e^{-kh})e^{-kh}\cos(m_0 t) \\ \qquad\qquad + (\varepsilon_2 - \varepsilon_1 e^{-kh})\cosh(m_0 t)]\cos kx. \end{cases} \quad (2.22)$$

For $\varepsilon_2 = 0$ and $\varepsilon_1 = 0$, the expression of (2.22) is very interesting, so it can be seen as the correlation and perturbation between the upper and lower surfaces.

2.3 Fluid interface movement of finite thickness two-sided vacuum

As shown in Fig. 3, it is now considered to have a finite thickness of d, which is an incompressible fluid with a density of ρ, a vacuum outside the horizontal infinite upper and lower surfaces, and under a gravity field force $\overrightarrow{g} = g$. Overrightarrow$_y$.

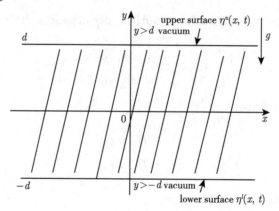

Fig. 3

Set
$$F^u(x,y,t) = y - \eta^u(x,t), \quad (2.23)$$
$$F^l(x,y,t) = y + d - \eta^l(x,t). \quad (2.24)$$

They are the equations of motion for the upper and lower interfaces, respectively, where $\eta^u(x,t), \eta^l(x,t)$ respectively are perturbation of the upper and lower interfaces. Let $\Phi^0(t)$ be the initial undisturbed velocity potential. The velocity potential of the disturbing fluid are
$$\Phi(x,y,t) = \Phi^0(t) + \varphi(x,y,t). \quad (2.25)$$

Here φ is the additional velocity potential caused by the disturbance interface, $\nabla^2 \varphi = 0$. According to the Bernoulli equation, the pressure on the upper and lower interfaces are

$$p^u = -\rho \left[\frac{\partial \Phi}{\partial t} + \frac{1}{2}(\nabla \Phi)^2 + g\eta^u + f(t) \right], \qquad (2.26)$$

$$p^l = -\rho \left[\frac{\partial \Phi}{\partial t} + \frac{1}{2}(\nabla \Phi)^2 - gd + g\eta^l + f(t) \right], \qquad (2.27)$$

where $f(t)$ is a function of any t. For undisturbed situations,

$$p^{u(0)} = -\rho \left[\frac{\partial \Phi^0}{\partial t} + f(t) \right], \qquad (2.28)$$

$$p^{l(0)} = -\rho \left[\frac{\partial \Phi^0}{\partial t} + f(t) \right], \qquad (2.29)$$

which satisfies the pressure balance condition $p^{l(0)} - p^{u(0)} = \rho g d$. Therefore the disturbance pressure is zero on both interfaces, i.e.

$$\rho \left[\frac{\partial \varphi}{\partial t} + \frac{1}{2}(\nabla \varphi)^2 + g\eta^u \right] = 0, \ y = \eta^u, \qquad (2.30)$$

$$\rho \left[\frac{\partial \varphi}{\partial t} + \frac{1}{2}(\nabla \varphi)^2 + g\eta^l \right] = 0, \ y = \eta^l. \qquad (2.31)$$

As $-\frac{\partial F}{\partial t}/|\nabla F|$ is known as interface normal velocity, $\vec{n} = \nabla F/|\nabla F|$ is the normal vector, we have

$$-\frac{\frac{\partial F^u}{\partial t}}{|\nabla F^u|} = \nabla \Phi \cdot \frac{\nabla F^u}{|\nabla F^u|}, \quad y = \eta^u, \qquad (2.32)$$

$$-\frac{\frac{\partial F^l}{\partial t}}{|\nabla F^l|} = \nabla \Phi \cdot \frac{\nabla F^l}{|\nabla F^l|}, \quad y = \eta^l. \qquad (2.33)$$

Substituting into the equation (2.23), (2.24), we get

$$\frac{\partial \eta^u}{\partial t} = -\frac{\partial \varphi}{\partial x}\frac{\partial \eta^u}{\partial x} + \frac{\partial \varphi}{\partial y}, \quad y = \eta^u, \qquad (2.34)$$

$$\frac{\partial \eta^l}{\partial t} = -\frac{\partial \varphi}{\partial x}\frac{\partial \eta^l}{\partial x} + \frac{\partial \varphi}{\partial y}, \quad y = \eta^l. \qquad (2.35)$$

According to (2.30), (2.31), (2.34), (2.35), we expand η^u, η^l, φ as the power of the small parameter ε

$$\eta^u(x,t) = \varepsilon \eta^u_{1,1}(t) \cos kx + \varepsilon^2 [\eta^u_{2,0}(t) + \eta^u_{2,2}(t) \cos(2kx)] + \cdots, \qquad (2.36)$$

$$\eta^l(x,t) = \varepsilon\eta_{1,1}^l(t)\cos kx + \varepsilon^2[\eta_{2,0}^l(t) + \eta_{2,2}^l(t)\cos(2kx)] + \cdots, \tag{2.37}$$

$$\varphi(x,y,t) = \varepsilon[e^{ky}A_{1,1}(t) + e^{-ky}B_{1,1}(t)]\cos kx$$
$$+ \varepsilon^2\{A_{2,0}(t) + [e^{2ky}A_{2,2}(t) + e^{-2ky}B_{2,2}(t)]\cos 2kx\} + \cdots. \tag{2.38}$$

As a first-order approximation $o(\varepsilon)$, a second-order coupled ordinary differential equation system is obtained.

$$\frac{d^2\eta_{1,1}^u}{dt^2} - \frac{2e^{-\xi}}{1+e^{-2\xi}}\frac{d^2\eta_{1,1}^l}{dt^2} + \gamma^2\frac{1-e^{-2\xi}}{1+e^{2\xi}}\eta_{1,1}^u = 0, \tag{2.39}$$

$$\frac{d^2\eta_{1,1}^l}{dt^2} - \frac{2e^{-\xi}}{1+e^{-2\xi}}\frac{d^2\eta_{1,1}^u}{dt^2} - \gamma^2\frac{1-e^{-2\xi}}{1+e^{2\xi}}\eta_{1,1}^l = 0, \tag{2.40}$$

where $\gamma^2 = gk$ is the square of the linear growth rate of classical $RT1(A_T = 1)$, $\xi = kd$.

The initial conditions are

$$\eta_{1,1}^u|_{t=0} = a, \quad \frac{d\eta_{1,1}^u}{dt}|_{t=0} = 0,$$
$$\eta_{1,1}^l|_{t=0} = b, \quad \frac{d\eta_{1,1}^l}{dt}|_{t=0} = 0. \tag{2.41}$$

From this, we get

$$\eta_{1,1}^u(t) = \frac{1}{1-e^{-2\xi}}[\bar{a}\cos\gamma t + e^{-\xi}\bar{b}\cosh\gamma t], \tag{2.42}$$

$$\eta_{1,1}^l(t) = \frac{1}{1-e^{-2\xi}}[e^{-\xi}\bar{a}\cos\gamma t + \bar{b}\cosh\gamma t], \tag{2.43}$$

where $\bar{a} = a - be^{-\xi}, \bar{b} = b - ae^{-\xi}$.

In the second order of $o(\varepsilon^2)$, the coupled ODE is the following

$$\frac{d^2\eta_{2,2}^u}{dt^2} - \frac{2e^{-2\xi}}{1+e^{-4\xi}}\frac{d^2\eta_{2,2}^l}{dt^2} + 2\gamma^2\frac{1-e^{-4\xi}}{1+e^{-4\xi}}\eta_{2,2}^u$$
$$+\gamma^2 k\left[\frac{1}{2}\bar{a}^2 c_1\cosh(2T) - \frac{1}{2}e^{-2\xi}c_1\bar{b}^2\cos(2T)\right.$$
$$-\frac{2}{e^{2\xi}+e^{-2\xi}}c_{ab}\cos(T)\cosh(T)$$
$$\left.+2e^{-2\xi}c_1 c_{ab}\sin(T)\sinh(T) + \frac{e^{\xi}+e^{2\xi}}{1+e^{4\xi}}\bar{a}\bar{b}\right] = 0, \tag{2.44}$$

$$\frac{d^2\eta_{2,2}^l}{dt^2} - \frac{2e^{-2\xi}}{1+e^{-4\xi}}\frac{d^2\eta_{2,2}^u}{dt^2} - 2\gamma^2\frac{1-e^{-4\xi}}{1+e^{4\xi}}\eta_{2,2}^l$$
$$+\gamma^2 k\left[\frac{1}{2}\bar{b}^2 c_1 \cosh(2T) - \frac{1}{2}e^{-2\xi}c_1\bar{a}^2\cos(2T)\right.$$
$$-\frac{2}{e^{2\xi}+e^{-2\xi}}c_{ab}\cos(T)\cosh(T)$$
$$\left.-2e^{-2\xi}c_1 c_{ab}\sin(T)\sinh(T) + \frac{e^{\xi}+e^{2\xi}}{1+e^{4\xi}}b\bar{a}\right] = 0, \qquad (2.45)$$

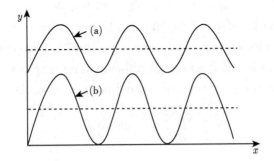

Fig. 4 (a) $kd = 0.2$, $\gamma t = 4.2, \cdots$, (b) $kd = 0.8$, $\gamma t = 4.9$

where $\bar{a} = a - b(e^{-\xi}+e^{\xi})/2$, $\bar{b} = b - a(e^{-\xi}+e^{\xi})/2$, $c_1 = (1+e^{-2\xi})/(1-e^{-2\xi}+e^{-4\xi}-e^{-6\xi})$, $c_{ab} = a^2+b^2-ab(e^{-\xi}+e^{\xi})$. $T = \gamma t$, the initial conditions are

$$\eta_{2,2}^u(t=0) = 0, \quad \frac{d\eta_{2,2}^u}{dt}(t=0) = 0,$$
$$\eta_{2,2}^l(t=0) = 0, \quad \frac{d\eta_{2,2}^l}{dt}(t=0) = 0. \qquad (2.46)$$

From these, we deduce

$$\eta_{2,2}^u(t) = \frac{k}{2}\left(\frac{1}{1-e^{-2\xi}}\right)^2\left[-e^{-2\xi}\bar{b}^2\cosh^2(T) + \bar{a}^2\cos^2(T)\right.$$
$$-\frac{e^{-2\xi}}{1+e^{-2\xi}}D_1\bar{b}\cosh(\sqrt{2}T) - \frac{1}{1+e^{-2\xi}}D_2\bar{a}\cos(\sqrt{2}T) \qquad (2.47)$$
$$\left.-2e^{-\xi}\frac{1-e^{-2\xi}}{1+e^{-2\xi}}\bar{a}\bar{b}\cos(T)\cosh(T)\right],$$

$$\eta_{2,2}^l(t) = \frac{k}{2}\left(\frac{1}{1-e^{-2\xi}}\right)^2\left[-\bar{b}^2\cosh^2(T) + e^{-2\xi}\bar{a}^2\cos^2(T)\right.$$

$$-\frac{1}{1+e^{-2\xi}}D_1\bar{b}\cosh(\sqrt{2}T) - \frac{e^{-2\xi}}{1+e^{-2\xi}}D_2\bar{a}\cos(\sqrt{2}T) \qquad (2.48)$$
$$+2e^{-\xi}\frac{1-e^{-2\xi}}{1+e^{-2\xi}}\bar{a}\bar{b}\cos(T)\cosh(T)\bigg].$$

From the above upper and lower interface expressions, it can be graphically represented as

2.4 Variable density fluid layer

Here we don't discuss a two-layer fluid with a density discontinuity, but a single-layer fluid with a density distribution in the y direction. This distribution is also unstable under certain conditions. The fluid is heated from the bottom, and the fluid has different densities at different temperatures. Then it develops into convection. Without disturbance, the fluid is at rest. At this time,

$$\frac{dp^{(0)}}{dy} = -\rho^{(0)}(y)g.$$

Therefore we get

$$p^{(0)} = -\int \rho^{(0)}(y)g dy + f(t).$$

In case of disturbance

$$p = p^{(0)} + p',$$
$$\rho = \rho^{(0)} + \rho'.$$

And u, v are first-order small quantities. The continuity equation can be written as

$$\frac{\partial \rho'}{\partial t} + \rho^{(0)}\left(\frac{\partial u}{\partial x} + \frac{\partial v}{\partial y}\right) + v\frac{d\rho^{(0)}}{dy} = 0.$$

We take

$$\frac{\partial u}{\partial x} + \frac{\partial v}{\partial y} = 0.$$

Then,

$$\frac{\partial \rho'}{\partial t} = -v\frac{d\rho^{(0)}}{dy}.$$

In the momentum equation, for the disturbance term, we have

$$\rho^{(0)}\frac{\partial u}{\partial t} + \frac{\partial p'}{\partial x} = 0,$$

$$\rho^{(0)} \frac{\partial v}{\partial t} + \frac{\partial p'}{\partial y} = -\rho' g.$$

Assume that the disturbance has the form e^{ikx+nt}, i.e.,

$$p' = \bar{p}(y) e^{nt+ikx},$$
$$\rho' = \bar{\rho}(y) e^{nt+ikx},$$
$$u' = \bar{u}(y) e^{nt+ikx},$$
$$v' = \bar{v}(y) e^{nt+ikx}.$$

Substituting into the above equation, we get

$$ik\bar{u} = -\frac{d\bar{v}}{dy},$$
$$n\bar{\rho} = -\bar{v} \frac{d\rho^{(0)}}{dy},$$
$$\rho^{(0)} n\bar{u} + ik\bar{p} = 0,$$
$$\rho^{(0)} n\bar{v} + \frac{d\bar{p}}{dy} = -g\bar{\rho}.$$

Eliminating $\bar{u}$ from the first and third of the four equations, we deduce

$$\rho^{(0)} n \frac{d\bar{v}}{dy} + k^2 \bar{p} = 0.$$

And by the second, and fourth equations, $\bar{\rho}$ is eliminated,

$$\rho^{(0)} n\bar{v} + \frac{d\bar{p}}{dy} = \frac{\bar{v}}{n} g \frac{d\rho^{(0)}}{dy}.$$

Eliminating $\bar{p}$ from the last two equations, we have

$$\frac{d}{dy}\left(\rho^{(0)} \frac{d\bar{v}}{dy}\right) - k^2 \rho^0 \bar{v} + \frac{k^2 g}{n^2} \frac{d\rho^{(0)}}{dy} \bar{v} = 0. \tag{2.49}$$

If you do not consider the density distribution in the y direction, $\frac{d\rho^{(0)}}{dy} = 0$, then

$$\frac{d^2 \bar{v}(y)}{dy^2} - k^2 \bar{v}(y) = 0.$$

That is, $\bar{v} = a e^{\pm ky}$. Now we discuss the case of $\rho^{(0)} = \rho^{(0)}(y)$. For the sake of simplicity, we only study the case:

$$\rho^{(0)} = \rho_0 e^{\beta y}, \quad \beta = const.$$

Substituting into (2.49), we have

$$\frac{d^2\bar{v}}{dy^2} + \beta\frac{d\bar{v}}{dy} - k^2\left(1 - \frac{g\beta}{n^2}\right)\bar{v} = 0. \tag{2.50}$$

Taking the form solution of $\bar{v} \sim e^{qy}$, there is

$$q^2 + \beta q - k^2\left(1 - \frac{g\beta}{n^2}\right) = 0.$$

Therefore, we have

$$\bar{v}(y) = A_1 e^{q_1 y} + A_2 e^{q_2 y},$$

Here q_1, q_2 is the two roots above for the q equation, i.e.,

$$\begin{cases} q_1 = \dfrac{1}{2}\left\{-\beta + \sqrt{\beta^2 + 4k^2\left(1 - \dfrac{g\beta}{n^2}\right)}\right\}, \\ q_2 = \dfrac{1}{2}\left\{-\beta - \sqrt{\beta^2 + 4k^2\left(1 - \dfrac{g\beta}{n^2}\right)}\right\}. \end{cases} \tag{2.51}$$

Assuming that the layer of fluid is between the two solid walls of $y = -h$ and $y = 0$, by

$$y = 0, \quad \bar{v}(y) = 0.$$

We get

$$A_2 - A_1 = A,$$

and

$$\bar{v}(y) = A(e^{q_1 y} - e^{q_2 y}).$$

Due to

$$y = -h, \quad \bar{v}(y) = 0,$$

we have

$$e^{(q_1 - q_2)h} = 1.$$

To meet this condition, only when

$$q_1 - q_2 = \frac{2m\pi i}{h}. \tag{2.52}$$

Thus,

$$\bar{v}(y) = Ae^{\frac{1}{2}(q_1+q_2)y}\left\{e^{\frac{1}{2}(q_1-q_2)y} - e^{-\frac{1}{2}(q_1-q_2)y}\right\}.$$

In accordance with (2.51), (2.52), we have

$$\bar{v} = Be^{-\frac{1}{2}\beta y} \sin\frac{m\pi y}{h}. \tag{2.53}$$

and

$$n = \frac{\sqrt{g\beta}}{\sqrt{1 + \left[\frac{\beta^2}{4} + \frac{m^2\pi^2}{h^2}\right]k^{-2}}}. \tag{2.54}$$

It can be seen from the above equation that $\beta > 0$, that is, As the density gradient of the fluid is the same direction as g, the above density distribution of the fluid layer is unstable. For short waves,

$$n \sim \sqrt{g\beta}.$$

That is, the index n is proportional to $\beta^{\frac{1}{2}}$, the greater the density gradient, the more unstable the instability.

2.5 The Taylor instability of viscous fluids

The two-layer semi-infinite fluids occupy the $y > 0$ and $y < 0$ planes, respectively, with densities of ρ_1 and ρ_2. The gravitational acceleration is perpendicular to the interface, along the negative direction of y. In the case of incompressibility, the continuous equation can be written as

$$u_x + v_y = 0.$$

The momentum equation is NS equation

$$\frac{\partial u}{\partial t} + (u \cdot \nabla)u + \frac{1}{\rho}\nabla p = \frac{\mu}{\rho}\Delta u + g.$$

On the material interface $F(\vec{r}, t) = 0$, the interface normal velocity is continuous

$$-\frac{\partial F}{\partial t} = (u_1 \cdot \nabla)F = (u_2 \cdot \nabla)F.$$

And the tangential velocity is continuous, the stress equation has

$$n_j\sigma_{1jk} = n_j\sigma_{2jk},$$

where $\sigma_{ijk}(i = 1, 2)$ is the viscous stress tensor of the two fluids.

$$\sigma_{ijk} = -p_i\delta_{jk} + \mu_i\left(\frac{\partial u_{ij}}{\partial x_i} + \frac{\partial u_{ik}}{\partial x_j}\right), \quad i = 1, 2,$$

p_i is the fluid pressure, $i = 1, 2$ means different fluids.

Setting the form of the interface is

$$F(\vec{r}, t) = y - \eta(x, t) = 0,$$

where $\vec{\eta}(x, t)$ is a small amount. Then

$$\vec{n} = -\eta_x \vec{i} + \vec{j}.$$

And the corresponding u, p are first-order small quantities, then the momentum equation can be linearized to

$$\begin{cases} u_{it} + \dfrac{p_{ix}}{\rho_i} = \dfrac{\mu_i}{\rho_i}(u_{ixx} + u_{iyy}), \\ v_{it} + \dfrac{p_{iy}}{\rho_i} = \dfrac{\mu_i}{\rho_i}(v_{ixx} + v_{iyy}) - g, \end{cases} \quad i = 1, 2.$$

For viscous fluids, the flow is not no-curl. One is that the potential function is insufficient to satisfy the momentum equation. We choose

$$\begin{cases} u_i = \Phi_{ix} + \Psi_{iy}, \\ v_i = \Phi_{iy} - \Psi_{ix}, \end{cases} \quad i = 1, 2.$$

Substituting into the momentum equation, we have

$$\left(\Phi_{it} + \dfrac{P_i}{\rho_i}\right)_x = \dfrac{\mu_i}{\rho_i}(\Psi_{ixx} + \Psi_{iyy})_y - \Psi_{ity},$$

$$\left(\Phi_{it} + \dfrac{P_i}{\rho_i}\right)_y = \dfrac{\mu_i}{\rho_i}(\Psi_{ixx} + \Psi_{iyy})_x - \Psi_{itx} - g.$$

If we set $\Psi_{it} = \dfrac{\mu_i}{\rho_i}(\Psi_{ixx} + \Psi_{iyy})$, then

$$\Phi_{it} + \dfrac{P_i}{\rho_i} + gy = f(t), \quad p_i = -\rho_i(\Phi_{it} + gy).$$

According to the existing vector expression of $\vec{n}$, the equation of motion and stress conditions on the interface can be written as

$$y = 0, \quad u_1 = u_2,$$

$$\eta_t = v_1 = v_2,$$

$$\sigma_{1xy} = \sigma_{2xy},$$

$$\sigma_{1xx} - \sigma_{2yy} = \Sigma\left(\dfrac{1}{R_1} + \dfrac{1}{R_2}\right) = -\Sigma\eta_{xx}.$$

Here, the surface tension is considered in the normal pressure difference, Σ is the surface tension coefficient, and the corresponding stress is

$$\sigma_{ixy} = \mu_i \left[\frac{\partial u_i}{\partial y} + \frac{\partial v_i}{\partial x} \right],$$

$$\sigma_{iyy} = -p_i + 2\mu \frac{\partial v_i}{\partial \eta}.$$

Representing u_i, v_i as a function expression for Φ_i and Ψ_i, substituting into the above expression and converting the condition on the interface $y = 0$ to

$$\begin{cases} \Phi_{1x} + \Psi_{1y} = \Phi_{2x} + \Psi_{2y}, \\ \eta_t = \Phi_{1y} - \Psi_{1x} = \Phi_{2y} - \Psi_{2x}, \end{cases} \quad y = 0, \qquad (2.55)$$

$$\mu_1[2\Phi_{1xy} + \Psi_{1yy} - \Psi_{1xx}] = \mu_2[2\Phi_{2xy} + \Psi_{2yy} - \Psi_{2xx}],$$

$$\rho_1(\Phi_{1t} + g\eta) + 2\mu_1(\Phi_{1yy} - \Psi_{1xy}) + \Sigma \eta_{xx} \qquad (2.56)$$

$$= \rho_2(\Phi_{2t} + y\eta) + 2\mu_2(\Phi_{2yy} - \Psi_{2xy}).$$

Assume that the initial perturbation of the interface is

$$\eta(x, 0) = \varepsilon \cos kx.$$

In order for Φ_i to satisfy the conditions of the Laplace equation and $x = \pm\infty$, we take

$$\eta(x, t) = ae^{nt} \cos kx,$$
$$\Phi_1 = Ae^{ht-ky} \cos kx,$$
$$\Phi_2 = Ce^{nt+ky} \cos kx.$$

Setting

$$\Psi_1 = Be^{-m_1 y + nt} \sin kx,$$
$$\Psi_2 = De^{m_2 y + nt} \sin kx.$$

Substituting into (2.54) and we consider the disappearance condition at $y = \pm\infty$, thus we deduce

$$m_i = \sqrt{k^2 + \frac{\rho_i n}{\mu_i}}.$$

Substituting the above Φ_i and Ψ_i expressions into the inner boundary conditions (2.55) and (2.56), eliminating a, we get

$$A + B + C + D = 0,$$

$$kA + m_1B - kC + m_2D = 0,$$
$$2\mu_1 k^2 A + \mu_1(m^2 + k^2)B + 2\mu_2 k^2 C - \mu_2(m_2^2 + k^2)D = 0,$$

$$\left(\frac{a}{h} - n\rho_1 - 2\mu_1 k^2\right)A + \left(\frac{a}{n} - 2\mu_1 m_1 k\right)B$$
$$+ (\rho_2 n + 2\mu_2 k^2)C - 2\mu_2 m_2 k D = 0.$$

Here
$$a = (\rho_1 - \rho_2)kg - k^3\Sigma.$$

The non-trivial solution of the above homogeneous equations of A, B, C and D exists on the condition that the corresponding coefficient determinant is zero, i.e.,

$$\begin{vmatrix} 1 & 1 & 1 & -1 \\ k & m_1 & -k & m_2 \\ 2\mu_1 & \mu_1(k^2 + m_1^2) & 2\mu_2 k^2 & -\mu_2(k^2 + m_2^2) \\ \dfrac{a}{n} - n\rho_1 - 2\mu_1 k^2 & \dfrac{a}{n} - 2\mu_1 m_1 k & n\rho_2 + 2\mu_2 k^2 & -2\mu_2 m_2 k \end{vmatrix} = 0.$$

Expanding this determinant and substituting m into this, we get

$$[-a + (\rho_1 + \rho_2)n^2]\left\{\frac{1}{\mu_1 k + \sqrt{\mu_2^2 + \mu_2\rho_2 n}} + \frac{1}{\mu_2 k + \sqrt{\mu_1^2 + \mu_1\rho_1 n}}\right\}$$
$$+ 4nk = 0. \qquad (2.57)$$

When the viscosity μ_i is large, n becomes smaller, we ignore the term with n in the root formula in the formula (2.57), i.e., we take

$$\mu_i k^2 \gg \rho_i n.$$

Or approximately, we choose

$$m_i \simeq k.$$

Thus, we get
$$[-a + (\rho_1 + \rho_2)n^2]\frac{2}{k(\mu_1 + \mu_2)} + 4nk = 0.$$

Meanwhile we choose
$$\nu = \frac{\mu_1 + \mu_2}{\rho_1 + \rho_2}.$$

Therefore we deduce

$$n = \nu k^2 \pm \sqrt{\frac{\rho_1 - \rho_2}{\rho_1 + \rho_2} kg + \nu^2 k^4 - \frac{k^4 \Sigma}{\rho_1 \rho_2}}. \qquad (2.58)$$

We focus on the unstable root that is positive before the root number, so that the invicid n is n_0,

$$n_0 = \sqrt{\frac{\rho_1 - \rho_2}{\rho_1 + \rho_2} kg - \frac{k^3 \Sigma}{\rho_1 \rho_2}}. \qquad (2.59)$$

We discuss two situations:

(1) $n_0 \gg \nu k^2$, That is, when k or viscosity is small, by (2.58) we obtain

$$n = n_0 \left[1 - \frac{\nu k^3}{n_0} + \frac{1}{2} \left(\frac{\nu k^2}{n_0} \right)^2 - \cdots \right]. \qquad (2.60)$$

From these we can see viscosity reduces RT instability. The larger the k, the more significant.

(2) $n_0 \ll \nu k^2$, by (2.58), we get

$$n = n_0 \frac{n_0}{2\nu k^2}. \qquad (2.61)$$

That is, n quickly decreases with increasing k, especially at $k \to k_c$, by (2.59) we can obtain

$$n_0 = \sqrt{\frac{\rho_1 - \rho_2}{\rho_1 + \rho_2} kg \left(1 - \frac{k^2}{k_c^2} \right)}.$$

This is the case where n tends to zero near k_c. When $\Sigma = 0$, regardless of surface tension,

$$n \sim \frac{\rho_1 - \rho_2}{\rho_1 + \rho_2} \frac{g}{2\nu k}.$$

That is, when k is large, n asymptotically approaches the k axis in the form of a curve.

2.6 Ablation instability

We consider a simplify model, discussing the ideal incompressible fluid, and thinking that the effect of heat conduction is mainly the motion that drives the phase change. It has a certain velocity or acceleration, which is ignored in the equation. Let the section be at an average speed of U_0 moves forward in the direction of y with an acceleration of g, the density before the cross section is ρ_1, and the density after the cross section is ρ_2, taking the coordinate system moving along with the section, here the speed of the "1" zone in the coordinate system

$$V_1 = -U_0, \qquad (2.62)$$

and along the opposite direction of y.

Conservation of mass momentum in a coordinate system with a constant cross-section can be written as

$$\begin{aligned} \rho_1 q_1 \cdot n &= \rho_2 q_2 \cdot n, \\ p_1 + \rho_2 (q \cdot n)^2 &= p_2 + (q_2 \cdot n)^2, \\ q_1 \cdot \tau &= q_2 \cdot \tau. \end{aligned} \qquad (2.63)$$

Here n and τ represent the normal and tangential directions of the section, respectively. $\rho_i, p_i (i=1,2)$ are the density and pressure on both sides, $q_i (i=1,2)$ denotes the velocity of the section on both sides that are moving in motion. Assume that the section is the following form:

$$y = \eta(x, t),$$

where η is a small amount, then

$$\begin{aligned} \vec{n} &= -\eta_x \vec{i} + \vec{j}, \\ \tau &= \vec{i} + \eta_x \vec{j}, \\ D_n &= \eta_t. \end{aligned}$$

Here D_n is the normal velocity of the section. In this way, the pressure and speed can be expanded, i.e.

$$\begin{aligned} p_i &= p_i^{(0)} + p_i^{(1)} + \cdots, \\ u_i &= V_i \vec{j} + u_i^{(1)} + \cdots. \end{aligned} \qquad (2.64)$$

Here $p_i^{(0)}, V_i \vec{j}$ is the zero-order quantity corresponding to the undisturbed case, and the second term is the first-order small quantity. The previously selected coordinate system is just the coordinate system that does not move the no section between the no disturbances, rather than the coordinate system that corresponds to the point in the section between the disturbances. In order to apply the conservation condition on the section, the following local coordinate transformation must be taken,

$$q_i = u_i - D_n n = u_i - \eta_t j. \qquad (2.65)$$

Substituting the relevant quantities into (2.63), respectively, we can obtain the relationship between the following zero-order and first-order quantities on both sides of the discontinuity:

Zero order quantity

$$\rho_1 V_1 = \rho_2 V_2 = m,$$
$$p_1^{(0)} + \rho_1 V_1^2 = p_2^{(1)} + \rho_2 V_2^2. \tag{2.66}$$

First order quantity

$$\rho_1(v_1^{(1)} - \eta_t) = \rho_2(v_2^{(1)} - \eta_t),$$
$$p_1^{(1)} + 2m(v_1^{(1)} - \eta_t) = p_2^{(1)} + 2m(v_2^{(1)} - \eta_t), \tag{2.67}$$
$$u_1^{(1)} + V_1 \eta_x = u_2^{(1)} + V_2 \eta_x.$$

Below, we solve the first-order quantity. The front of the ablation section is originally static and therefore non-curl. However, the rear of the section, that is, the "2" area may be non-rotating, but it may also be rotated. We discuss the two cases below.

Case 1 There is no swirling flow after the cross section, since it is non-rotating in both areas, it is desirable to choose

$$u_i^{(1)} = \nabla \Phi_i, \quad i = 1, 2. \tag{2.68}$$

Under incompressible conditions, we have

$$\Delta \Phi_i = 0, \quad i = 1, 2. \tag{2.69}$$

Integrating the momentum equation, taking the first-order small quantity, the following pressure disturbance expression can be obtained:

$$p_i^{(1)} = -\rho_i \{\Phi_{it} + V_i \Phi_{iy} + gy + f_i(t)\}. \tag{2.70}$$

Substituting into (2.67), we can get the condition of the section between $y = \eta(x,t)$. Since the zero-order quantity is uniform, The value of the first-order quantity on $y = \eta(x,t)$ can be Taylor expanded, ignoring the high order small quantity, we deduce the condition on the $y = 0$ fixed boundary,

$$\rho_1(\Phi_{1y} - \eta_t) = \rho_2(\Phi_{2y} - \eta_t),$$
$$\rho_1\{\Phi_{1t} - V_1\Phi_{1y} + g\eta\} = \rho_2\{\Phi_{2t} - V_2\Phi_{2y} + g\eta\}, \tag{2.71}$$
$$\Phi_{1x} + V_1\eta_x = \Phi_{2x} + V_2\eta_x.$$

For the sake of simplicity, we discuss the case where both layers are semi-infinite fluids, assuming that a fluid with a density of ρ_1 occupies a semi-infinite plane of $y > 0$, and a fluid with a density of ρ_2 occupies $y < 0$ semi-infinite plane.

Assume that the initial moment of disturbance is in the form of $\cos kx$, in order for Φ_i to satisfy the Laplace equation and the corresponding infinity disappearance condition, we choose

$$\eta = ae^{nt} \cos kx,$$
$$\Phi_1 = Ae^{nt-ky} \cos kx, \qquad (2.72)$$
$$\Phi_2 = Be^{nt+ky} \cos kx.$$

Substituting this formula into (2.71), we get

$$\rho_1(-kA - na) = \rho_2(kB - na), \qquad (2.73)$$

$$\rho_1(nA + V_1 kA + ga) = \rho_2(nB - V_2 kB + ga),$$
$$-kA - V_1 ka = -kB - V_2 ka. \qquad (2.74)$$

This is a homogeneous linear algebraic equation for a, A and B. The condition for this non-trivial solution is that the corresponding coefficient determinant is zero, thus solving the following dispersion relation:

$$n = \pm\sqrt{\frac{\rho_1 + \rho_2}{\rho_1 - \rho_2}kg - \frac{\rho V_1 (V_2 - V_1) k^2}{\rho_1 - \rho_2}}. \qquad (2.75)$$

Considering that (2.66), we have

$$n = \pm\sqrt{\frac{\rho_1 + \rho_2}{\rho_1 - \rho_2}kg - \frac{\Delta p^{(0)} k^2}{\rho_1 - \rho_2}}. \qquad (2.76)$$

Here $\Delta p^{(0)} = p_1^{(1)} - p_2^{(0)} = \rho_1 V_1 (V_2 - V_1)$ is the ablation pressure. We have previously assumed the area "1" is an ablated medium, so $\rho_1 > \rho_2$. By (2.66) $|v_2| > |v_1|$, and V_i are negative, so the ablation pressure is positive. When $g > 0$, i.e., the accelerating ablation, instability may occur, but when $g < 0$, n is pure, i.e., the deceleration ablation is stable.

Below, we discuss a few special cases

(1) $\rho_1 \gg \rho_2$

(2.71) can be written as

$$n = \pm\sqrt{kg - \frac{\Delta p^{(0)}}{\rho_1} k^2}. \qquad (2.77)$$

$g > 0$ corresponds to the RT instability case, but slows the development of instability due to the presence of ablation pressure. When $\Delta p^{(0)} = \rho_1 g/k$, instability can be truncated.

(2) $\rho_1 \sim \rho_2$

At this time (2.76) can be written as

$$n = \pm\sqrt{\frac{\rho_1 + \rho_2}{\rho_1 - \rho_2} kg - V_1^2 k^2}.$$

When $g > 0$, the instability develops rapidly. When $g < 0$, the section is quickly oscillated.

Now consider that the ablation product area is a swirling flow, and in the case of a spin in the "2" area, we choose

$$u_2^{(1)} = \Phi_{2x} + \Psi_{2x},$$
$$v_2^{(1)} = \Phi_{2y} - \Psi_{2x}.$$

The "1" area is still non-rotating. There are four physical quantities to be determined, which are η, Φ_1, Φ_2, and Ψ_2. But the boundary condition (2.67) contains only three equations. The problem is uncertain. Landau assumed zero normal velocity when discussing combustion.

$$\eta_t = v_1^{(1)} = v_2^{(1)}.$$

There is a supplementary condition from (2.67),

$$p_1^{(1)} = p_2^{(1)}.$$

To get the expression of (2.70) for the perturbation pressure $p_2^{(1)}$ in the "2" area, we must take

$$\Phi_{2t} + V_2 \Psi_{2y} = 0.$$

Thus the momentum equation can be satisfied. If we take $\Psi_2 \sim e^{kt+m_2 y} \sin kx$, then

$$m_2 = -n/V_2.$$

As mentioned earlier, $V_2 < 0$, so $m_2 > 0$. Therefore Φ_2 satisfies the disappearing condition at $y \to -\infty$.

Setting

$$\eta = ae^{nt}\cos kx,$$
$$\Phi_1 = Ae^{nt-ky}\cos kx,$$
$$\Phi_2 = Be^{nt+ky}\cos kx,$$
$$\Psi_2 = ce^{nt+m_2 y}\sin kx.$$

Substituting into the boundary condition of $y = \eta(x,t)$ and expanding on y, we can get four constants that need to be determined relative to a, A, B and c. These four constants satisfy homogeneous linear algebraic equations, the existence condition is that the corresponding coefficient determinant is zero.

Thus,

$$n = \frac{\rho_\gamma k V_1}{\rho_1 + \rho_2} \pm \sqrt{\frac{\rho_1 - \rho_2}{\rho_1 + \rho_2}kg + \frac{\rho_1 V_1 k^2}{\rho_1 + \rho_2}\left(V_1 - \frac{\rho_1 V_1}{\rho_1 + \rho_2}\right)}. \qquad (2.78)$$

Since $V_1 < 0$, $V_2 < 0$, when $\rho_1 \gg \rho_2$, the second item in (2.78) square brackets can be omitted, we get

$$n = -\frac{\rho_1 k |V_1|}{\rho_1 + \rho_2} \pm \sqrt{\frac{\rho_1 - \rho_2}{\rho_1 + \rho_2}kg + \frac{\rho_1 V_1 V_2 k^2}{\rho_1 + \rho_2}}. \qquad (2.79)$$

$g = 0$ is the combustion instability case discussed by Landau, $\rho_1 \gg \rho_2$,

$$n = -k|V_1| \pm \sqrt{kg + V_1 V_2 k^2}.$$

When the burning speed is $v_\gamma = -V_1$, there is

$$n \sim \sqrt{kg + V_1 V_2 k^2}.$$

At this time, the instability is more serious. When k is a small amount, that is, for long waves, the squared term of k in the root number is ignored, there is

$$n = -k|V_1| + \sqrt{kg}.$$

Storm used a similar formula and multiplied a correction factor to estimate the instability of the laser ablation interface.

2.7 RT instability of magnetic field in MHD

The ideal magnetohydrodynamics equations can be expressed as:

$$\frac{\partial \rho}{\partial t} + \rho(\nabla \cdot u) = 0, \qquad (2.80)$$

$$\rho \frac{du}{dt} + \nabla p = \rho g + \frac{1}{\mu_0}(\nabla \times B) \times B, \tag{2.81}$$

$$\nabla \times E = -\frac{\partial B}{\partial t}, \tag{2.82}$$

$$E = -u \times B. \tag{2.83}$$

Linearized (2.80)-(2.83) and expressed as the "0" order and the "1" order small perturbation scale.

$$\rho = \rho^{(0)} + \rho^{(1)},$$
$$u = u^{(0)} + u^{(1)},$$
$$p = p^{(0)} + p^{(1)},$$
$$B = B^{(0)} + B^{(1)}.$$

Substituting into (2.80)-(2.83), we obtain the linearized equations

$$\frac{\partial \rho^{(1)}}{\partial t} + u^{(0)} \cdot \nabla \rho^{(1)} + u^{(1)} \cdot \nabla \rho^{(0)} = 0, \tag{2.84}$$

$$\rho^0 (\frac{\partial u^{(1)}}{\partial t} + u^0 \cdot \nabla u^{(1)} + u^{(1)} \cdot \nabla u^{(0)}) + \rho^{(1)}(\frac{\partial u^{(0)}}{\partial t} + u^{(0)} \cdot \nabla u^{(0)}) = \rho^{(1)} g$$
$$- \nabla (p^{(1)} + \frac{B^{(0)} \cdot B^{(1)}}{\mu_0}) + \frac{1}{\mu_0}(B^{(0)} \cdot \nabla B^{(1)} + B^{(1)} \cdot \nabla B^{(0)}), \tag{2.85}$$

$$\frac{\partial B^{(1)}}{\partial t} = B^{(0)} \cdot \nabla u^{(1)} + B^{(1)} \cdot \nabla u^{(0)} - u^{(0)} \cdot \nabla B^{(1)} - u^{(1)} \cdot \nabla B^{(0)}. \tag{2.86}$$

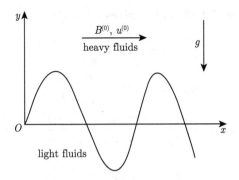

Fig. 5

In addition, using incompressible conditions

$$\nabla \cdot u^{(1)} = 0. \tag{2.87}$$

Use the following initial flow field distribution: $u^{(0)} = 0$, $B^0 = B^0(y)e_x$, $g^{(0)} = -ge_y$, and set $u^{(1)} = u^{(1)}e_x + v^{(1)}e_y$, $B^{(1)} = B^{(1)}e_x + C^{(1)}e_y$. Substituting the above initial fluid distribution and small perturbation into the equation (2.84)-(2.86), there is

$$0 = \frac{\partial \rho^{(1)}}{\partial t} + v^{(1)}\frac{d\rho^{(0)}}{dy} + u^{(0)}\frac{\partial \rho^{(1)}}{\partial x}, \tag{2.88}$$

$$\rho^{(0)}\left(\frac{\partial u^{(1)}}{\partial t} + u^{(0)}\frac{\partial u^{(1)}}{\partial x} + v^{(1)}\frac{du^{(0)}}{dy}\right) = -\frac{\partial p^{(1)}}{\partial x} + \frac{C^{(1)}}{\mu_0}\frac{dB^{(0)}}{dy}, \tag{2.89}$$

$$\rho^{(0)}\left(\frac{\partial v^{(1)}}{\partial t} + u^{(0)}\frac{\partial v^{(1)}}{\partial x}\right) = -\rho^{(1)}g - \frac{\partial p^{(1)}}{\partial y} - \frac{B^{(0)}}{\mu_0}\frac{\partial B^{(1)}}{\partial y}$$

$$-\frac{B^{(1)}}{\mu_0}\frac{dB^{(0)}}{dy} + \frac{B^{(0)}}{\mu_0}\frac{\partial C^{(1)}}{\partial x}, \tag{2.90}$$

$$\frac{\partial B^{(1)}}{\partial t} = B_0\frac{\partial u^{(1)}}{\partial x} + C^{(1)}\frac{du^{(0)}}{dy} - v^{(1)}\frac{dB^{(0)}}{dy} - u^{(0)}\frac{\partial B^{(1)}}{\partial x}, \tag{2.91}$$

$$\frac{\partial C^{(1)}}{\partial t} = B^{(0)}\frac{\partial v^{(1)}}{\partial x} - u^{(0)}\frac{\partial C^{(1)}}{\partial x}, \tag{2.92}$$

$$0 = \frac{\partial u^{(1)}}{\partial x} + \frac{\partial v^{(1)}}{\partial y}. \tag{2.93}$$

Setting

$$\left(\rho^{(1)}, u^{(1)}, v^{(1)}, p^{(1)}, B^{(1)}, C^{(1)}\right)$$
$$= \left(\bar{\rho}(y), \bar{u}(y), \bar{v}(y), \bar{p}(y), \bar{B}(y), \bar{C}(y)\right)e^{ikx+nt}. \tag{2.94}$$

Substituting (2.94) into (2.88)-(2.93) gives $u^{(0)} = 0$

$$0 = n\bar{\rho} + \bar{v}\frac{d\rho^{(0)}}{dy}, \tag{2.95}$$

$$n\rho^{(0)}\bar{u} = -ik\bar{p} + \frac{\bar{C}}{\mu_0}\frac{dB^{(0)}}{dy}, \tag{2.96}$$

$$n\rho^{(0)}\bar{v} = -\bar{\rho}g - \frac{d\bar{p}}{dy} - \frac{B^{(0)}}{\mu_0}\frac{d\bar{B}}{dy} - \frac{\bar{B}}{\mu_0}\frac{dB^{(0)}}{dy} + ik\frac{B^{(0)}}{\mu_0}\bar{C}, \tag{2.97}$$

$$n\bar{B} = ikB^{(0)}\bar{u} - \bar{v}\frac{dB^{(0)}}{dy}, \tag{2.98}$$

$$n\bar{C} = ikB^{(0)}\bar{v}, \tag{2.99}$$

$$0 = ik\bar{u} + \frac{d\bar{v}}{dy}. \tag{2.100}$$

Where $n = \gamma - iw$, γ represents the linear growth rate, and $w = w(k)$ is the disturbance frequency.

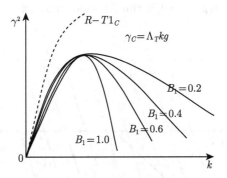

Fig. 6

The second-order eigenvalue equation for the perturbation velocity $\bar{v}$ can be obtained by simplifying the above algebraic equations.

$$n^2 \rho^{(0)} \frac{d^2\bar{v}}{dy^2} + n^2 \frac{d\rho^{(0)}}{dy} \frac{d\bar{v}}{dy} - k^2 n^2 \rho^{(0)} \bar{v} + k^2 g \frac{d\rho^{(0)}}{dy} \bar{v}$$
$$= \frac{1}{\mu_0} \left[-k^2 (B^{(0)})^2 \frac{d^2\bar{v}}{dy^2} - 2k^2 B^{(0)} \frac{dB^{(0)}}{dy} \frac{d\bar{v}}{dy} + k^4 (B^{(0)})^2 \bar{v} \right]. \quad (2.101)$$

Integrating (2.101) from $-\infty$ to ∞, we can obtain the following simple form:

$$-n^2 k \int_{-\infty}^{\infty} \rho^{(0)} \bar{v} dy + k^2 g \int_{-\infty}^{\infty} \frac{d\rho^{(0)}}{dy} \bar{v} dy = \frac{k^4}{\mu_0} \int_{-\infty}^{\infty} (B^{(0)})^2 \bar{v} dy. \quad (2.102)$$

Setting

$$\bar{v} = \bar{v}_{\max} e^{-k|y|} = \begin{cases} \bar{v}_{\max} e^{-ky}, & y > 0, \\ \bar{v}_{\max} e^{ky}, & y < 0, \end{cases} \quad k = \frac{2\pi}{\lambda}. \quad (2.103)$$

$$\rho^{(0)} = \begin{cases} \rho_1 - \dfrac{\rho_1 - \rho_2}{e^{-\frac{y}{2L_p}}}, & y > 0, \\ \rho_2 + \dfrac{\rho_1 - \rho_2}{e^{\frac{y}{2L_p}}}, & y < 0, \end{cases} \quad (2.104)$$

$$B^{(0)} = \begin{cases} B_1 - \dfrac{B_1 - B_2}{2} e^{-\frac{y}{2L_B}}, & y > 0, \\ B_2 + \dfrac{B_1 - B_2}{2} e^{\frac{y}{2L_B}}, & y < 0. \end{cases} \quad (2.105)$$

Where L_p and L_B are the density filter layer width and the magnetic field filter layer width, respectively.

$$n^2 = \frac{\Lambda_T k g}{k L_\rho + 1} - \frac{k^2}{\mu_0(\rho_1+\rho_2)}\left[B_1^2 + B_2^2 - \frac{(B_1-B_2)^2}{1+1/kL_B} + \frac{(B_1-B_2)^2}{2+4/kK_B}\right], \quad (2.106)$$

Where $\Lambda_T = (\rho_1 - \rho_2)/(\rho_1 + \rho_2)$ is the Atwood number.

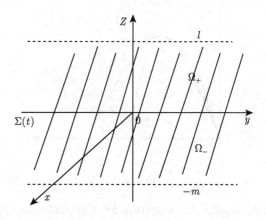

Fig. 7

3 Some mathematical analysis theory results

In this section, we mainly introduce some theoretical analysis results of Guo Yan and his collaborators.

3.1 Linear instability of compressible viscous fluids

We consider a horizontal infinite flat area $\Omega = \mathbf{R}^2 \times (-m, l) \subset \mathbf{R}^3$.
Compressible NS equation:

$$\begin{cases} \partial_t \rho_\pm + \mathrm{div}(\rho_\pm u_\pm) = 0, \\ \rho_\pm(\partial_t u_\pm + u_\pm \cdot \nabla u_\pm) + \mathrm{div} S_\pm = -g\rho_\pm \vec{e}_3, \end{cases} \quad (3.1)$$

Where

$$S_\pm = P_\pm(\rho_\pm I) - \varepsilon_\pm(\rho_\pm)\left(Du_\pm + Du_\pm^{\mathrm{T}} - \frac{2}{3}\mathrm{div} u_\pm I\right) - \delta_\pm(\rho_\pm)\mathrm{div} u_\pm I.$$

The jump condition on the interface

$$\begin{cases} (u_+)|_{\Sigma(t)} - (u_-)|_{\Sigma(t)} = 0, \\ (S_+ \cdot \nu)|_{\Sigma(t)} - (S_- \cdot \nu)|_{\Sigma(t)} = \sigma H \cdot \nu, \end{cases} \quad (3.2)$$

$\sigma \geqslant 0$, H is mean curvature of the surface.

The boundary

$$u_-(x_1, x_2, -m, t) = u_+(x_1, x_2, l, t) = 0, \quad \forall (x_1, x_2) \in \mathbf{R}^2, \quad t > 0. \tag{3.3}$$

Setting the density, the speed is periodic on the space $\Omega_\pm(t)$

$$\begin{cases} \rho_\pm(x + 2\pi L k_1 e_1 + 2\pi L k_2 e_2, t) = \rho_\pm(x, t), & x \in \Omega_\pm(t), \\ u_\pm(x + 2\pi L k_1 e_1 + 2\pi L k_2 e_2, t) = u_\pm(x, t), & x \in \Omega_\pm(t). \end{cases} \tag{3.4}$$

In the Lagrange coordinates, we can rewrite the equations in another form. Thus, we can solve this equation in the fixed domain.

Setting $\Omega_- = \mathbf{R}^2 \times (-m, 0)$, $\Omega_+ = \mathbf{R}^2 \times (0, l)$.

There exists a mapping:

$$\eta_\pm : \Omega_\pm \to \Omega_\pm(0). \tag{3.5}$$

Such that $\Sigma_0 = \eta_\pm^0 \{x_3 = 0\}$, $\eta_+^0 = \{x_3 = l\}$, $\eta_-^0 = \{x_3 = -m\}$.

Defining the flow mapping $\eta_\pm$:

$$\begin{cases} \partial_t \eta_\pm(x, t) = u_\pm(\eta_\pm(x, t), t), \\ \eta(x, 0) = \eta_\pm^0(x), \end{cases} \tag{3.6}$$

$\Omega_\pm(t) = \eta_\pm(\Omega_\pm, t)$, $\Sigma(t) = \eta_\pm\{\{x_3 = 0\}, t\}$.

Defining Lagrange unknowns

$$\begin{cases} v_\pm(x, t) = u_\pm(\eta_\pm(x, t), t), \\ q_\pm(x, t) = \rho_\pm(\eta_\pm(x, t), t). \end{cases} \tag{3.7}$$

Introducing $\eta_\pm, v_\pm, q_\pm$ equation, A, $A^\mathrm{T} = (D\eta)^{-1}$. In Lagrange coordinates, the evolution equation for v, q, η is

$$\begin{cases} \partial_t \eta_i = v_i, \\ \partial_t q + q A_{ij} \partial_j v_i = 0, \\ q \partial_t v_i + A_{jk} \partial_k T_{ij} = -gq A_{ij} \partial_j \eta_3. \end{cases} \tag{3.8}$$

Where the viscous stress tensor T is

$$\begin{aligned} T_{ij} = & p(q) I_{ij} - \varepsilon(q)(A_{jk}\partial_k v_i + A_{ik}\partial_k v_j - \frac{2}{3}(A_{lk}(\partial_k v_l) I_{ij})) \\ & - \delta(q)(A_{lk}\partial_k v_l) I_{ij}. \end{aligned} \tag{3.9}$$

I_{ij} is an element of 3×3 matrix I.

To write the jump conditions, for a quantity $f = f_\pm$, we define the interfacial jump as
$$[[f]] := f_+|_{x_3=0} - f_-|_{x_3=0}. \qquad (3.10)$$
The jump conditions across the interface are
$$[[v]] = 0, \qquad (3.11)$$
$$[[Tn]] = \sigma H n, \qquad (3.12)$$
where we have written
$$n = \frac{\partial_1 \eta \times \partial_2 \eta}{|\partial_1 \eta \times \partial_2 \eta|_{x_3=0}}, \qquad (3.13)$$
for the unit normal to the surface $\Sigma(t) = \eta(x_3 = 0, t)$ and H for twice the mean curvature of $\Sigma(t)$.
$$H = \left(\frac{|\partial_1 \eta|^2 \partial_2^2 \eta - 2(\partial_1 \eta \cdot \partial_2 \eta)\partial_1\partial_2 \eta + |\partial_2 \eta|^2 \partial_1^2 \eta}{|\partial_1 \eta|^2 |\partial_2 \eta|^2 - |\partial_1 \eta \cdot \partial_2 \eta|^2} \right) \cdot n. \qquad (3.14)$$
Finally, we require the no-slip boundary condition
$$v_-(x_1, x_2, -m, t) = v_+(x_1, x_2, l, t) = 0. \qquad (3.15)$$
In the following we seek a steady-state solution, by taking
$$v = 0, \quad \eta = Id, \quad q(x,t) = \rho_0(x_3),$$
$$\eta(\{x_3 = 0\}) = \{x_3 = 0\}, \quad H = 0, \quad n = e_3, \quad A = I.$$
We deduce the following ODE
$$\frac{dp(\rho_0)}{dx_3} = -g\rho_0, \qquad (3.16)$$
subject to the jump condition
$$[[p(\rho_0)]] = 0. \qquad (3.17)$$
To solve this we introduce the enthalpy function defined by
$$h_\pm(z) = \int_1^z \frac{P'_\pm(r)}{r} dr. \qquad (3.18)$$
Since we are interested in Rayleigh–Taylor instability, we want the fluid to be denser above the interface, i.e. $\rho_0^+ > \rho_0^-$, then
$$[[\rho_0]] = \rho_0^+ - \rho_0^- > 0. \qquad (3.19)$$

The solution is then given by $P_\pm(\rho) = K_\pm \rho^\gamma$, $K_\pm > 0$, $\gamma_\pm \geq 1$,

$$\rho_0(x_3) = \begin{cases} \left((\rho_0^-)^{\gamma_--1} - \dfrac{g(\gamma_- - 1)}{K_-\gamma_-} x_3\right)^{\frac{1}{\gamma_--1}}, & x_3 < 0, \\ \left((\rho_0^+)^{\gamma_+-1} - \dfrac{g(\gamma_+ - 1)}{K_+\gamma_+} x_3\right)^{\frac{1}{\gamma_+-1}}, & 0 < x_3 < \dfrac{K_+\gamma_+}{g(\gamma_+ - 1)}(\rho_0^+)^{\gamma_+-1}, \\ 0, & x_3 > \dfrac{K_+\gamma_+}{g(\gamma_+ - 1)}(\rho_0^+)^{\gamma_+-1}. \end{cases}$$
(3.20)

For a polytropic gas law, the condition that $\rho_0^+ > \rho_0^-$ is equivalent to the conditions

$$\left(\frac{K_-}{K_+}\right)^{\frac{1}{\gamma_+}} (\rho_0^-)^{\frac{\gamma_-}{\gamma_+}} > \rho_0^-, \Leftrightarrow (\rho_0^-)^{\gamma_--\gamma_+} > \frac{K_+}{K_-}. \tag{3.21}$$

If $\gamma_+ = \gamma_-$, this requires $K_- > K_+$, and any choice of $\rho_0^- > 0$, If $\gamma_+ \neq \gamma_-$, then $K_-, K_+ > 0$ can be arbitrary, but we must require that $\rho_0^- > 0$ satisfies

$$\begin{cases} \rho_0^- > \left(\dfrac{K_+}{K_-}\right)^{\frac{1}{(\gamma_--\gamma_+)}}, & \gamma_- > \gamma_+, \\ \rho_0^- < \left(\dfrac{K_-}{K_+}\right)^{\frac{1}{(\gamma_+-\gamma_-)}}, & \gamma_+ > \gamma_-. \end{cases}$$

We now linearize the nonlinear problem (3.8) around the steady-state solution $v = 0, \eta = \text{Id}, q = \rho_0$. The resulting linearized equations are, writing η, v, q for the unknowns,

$$\begin{cases} \partial_t \eta = v, \\ \partial_t q + \rho_0 \text{div} v = 0, \end{cases} \tag{3.22}$$

and

$$\rho_0 \partial_t v + \nabla(P'(\rho_0)q) + g\rho e_3 + g\rho_0 \nabla \eta_3 \\ = \text{div}(\varepsilon_0(Dv + Dv^T - \frac{2}{3}(\text{div} v)I) + \delta_0(\text{div} v)I), \tag{3.23}$$

where $\varepsilon_0 = \varepsilon_0(\rho_0)$, $\delta_0 = \delta_0(\rho_0)$.

The jump conditions linearize to

$$\begin{cases} [[v]] = 0, \\ [[P'(\rho_0)qI - \varepsilon_0(Dv + Dv^T) - (\delta_0 - \dfrac{2\varepsilon_0}{3})(\text{div} v)I]]e_3 = \sigma \Delta_{x_1,x_2} \eta_3 e_3. \end{cases} \tag{3.24}$$

We assume that initial data are provided as $\eta(0) = \eta_0, v(0) = v_0, q(0) = q_0$, that satisfy the jump and boundary conditions in addition to the assumption that $[[\eta_0]] = 0$, which implies that $\eta(t)$ is continuous across $x_3 = 0$, $\forall t \geq 0$.

3.2 Growing mode ansatz

We will look for a growing normal mode solution to (3.22), (3.23) by first assuming an ansatz

$$v(x,t) = w(x)e^{\lambda t}, \quad q(x,t) = \bar{q}(x)e^{\lambda t}, \quad \eta(x,t) = \bar{\eta}(x)e^{\lambda t}, \quad (3.25)$$

for some $\lambda > 0$. Plugging the ansatz into (3.22)-(3.23), we may solve the first and second equations for $\bar{q}, \bar{\eta}$ in terms of v. By doing so and eliminating them from the third equation, we arrive at the time-invariant equation

$$\lambda^2 \rho_0 w - \nabla(P'(\rho_0)\rho_0 \mathrm{div} w) + g\rho_0 \nabla w_3 - g\rho_0 \mathrm{div} w e_3$$
$$= \mathrm{div}(\lambda \varepsilon_0 (Dw + Dw^T - \frac{2}{3}(\mathrm{div} w)I) + \lambda \delta_0 (\mathrm{div} w)I), \quad (3.26)$$

This is coupled to the jump conditions $[[w]] = 0$ and

$$[[\lambda \delta_0 - 2\lambda \varepsilon_0/3 + P'(\rho_0)\mathrm{div} w I + \lambda \varepsilon_0 (Dw + Dw^T)]] = -\sigma \Delta_{x_1,x_2} w_3 e_3, \quad (3.27)$$

and the boundary conditions $w_-(x_1, x_2, -m) = w_+(x_1, x_2, l) = 0$, Notice that the first jump condition implies that the assumptions on

$$\eta(0) = \bar{\eta}(0) = w(0)/\lambda.$$

Since the coefficients of the linear problem (3.26) depend only on the x_3 variable. We define the new unknowns:

$$w_1(x) = -i\varphi(x_3)e^{ix'\cdot\xi}, \quad w_2(x) = -i\theta(x_3)e^{ix'\cdot\xi}, \quad w_3 = \psi(x_3)e^{ix'\cdot\xi}, \quad (3.28)$$

$$\mathrm{div} w = (\xi_1 \varphi + \xi_2 \theta + w_3')e^{ix'\cdot\xi}, \quad (3.29)$$

$$Dw + Dw^T = \begin{pmatrix} 2\xi_1\varphi & \xi_1\theta + \xi_2\varphi & i(\xi_1\psi - \varphi') \\ \xi_1\theta + \xi_2\varphi & 2\xi_2\theta & i(\xi_2\psi - \varphi') \\ i(\xi_1\psi - \varphi') & i(\xi_2\psi - \theta') & -2\psi' \end{pmatrix} e^{ix'\cdot\xi}. \quad (3.30)$$

For each fixed ξ, we arrive at the following system of ODEs for $\varphi(x_3), \theta(x_3), \psi(x_3), \lambda$.

$$-(\lambda\varepsilon_0\varphi')' + [\lambda^2\rho_0 + \lambda\varepsilon_0|\xi|^2 + \xi_1^2(\lambda_0\delta_0 + \lambda\varepsilon_0/3 + P'(\rho_0)\rho_0)]\varphi$$
$$= -\xi_1[(\lambda\delta_0 + \lambda\varepsilon_0/3 + P'(\rho_0)\rho_0)\psi' + (\lambda\varepsilon_0' - g\rho_0)\psi]$$
$$-\xi_1\xi_2[\lambda\delta_0 + \lambda\varepsilon_0/3 + P'(\rho_0)\rho_0]\theta, \quad (3.31)$$

$$-(\lambda\varepsilon_0\theta')' + [\lambda^2\rho_0 + \lambda\varepsilon_0|\xi|^2 + \xi_2^2(\lambda\delta_0 + \lambda\varepsilon_0/3 + P'(\rho_0)\rho_0)]\theta$$
$$= -\xi_2[(\lambda\delta_0 + \lambda\varepsilon_0/3 + P'(\rho_0)\rho_0)\psi' + (\lambda\varepsilon_0' - g\rho_0)\psi]$$
$$-\xi_1\xi_2[\lambda\delta_0 + \lambda\varepsilon_0/3 + P'(\rho_0)\rho_0]\varphi, \tag{3.32}$$

$$-\left[\left(\frac{4\lambda\varepsilon_0}{3} + \lambda\delta_0 + P'(\rho_0)\rho_0\right)\psi'\right]' + (\lambda^2\rho_0 + \lambda\varepsilon_0|\xi|^2)\psi$$
$$= [\lambda\delta_0 + \lambda\varepsilon_0/3 + P'(\rho_0)(\xi_1\varphi + \xi_2\theta)]' + (g\rho_0 - \lambda\varepsilon_0')(\xi_1\varphi + \xi_2\theta). \tag{3.33}$$

The first jump condition yields jump conditions for the new unknowns

$$[[\varphi]] = [[\theta]] = [[\psi]] = 0. \tag{3.34}$$

The second jump condition becomes

$$\left[\left[(\lambda\delta_0 - 2\lambda\varepsilon_0/3 + P'(\rho_0))(\psi' + \xi_1\varphi + \xi_2\theta)e_3 + \lambda\varepsilon_0\begin{pmatrix} i(\xi_1\psi - \varphi') \\ i(\xi_2\psi - \theta') \\ 2\psi' \end{pmatrix}\right]\right] = \sigma|\xi|^2\psi e_3, \tag{3.35}$$

which implies that

$$[[\lambda\varepsilon_0(\varphi' - \xi_1\psi)]] = [[\lambda\varepsilon_0(\theta' - \xi_2\psi)]] = 0, \tag{3.36}$$

and that

$$[[(\lambda\delta_0 + \lambda\varepsilon_0/3 + P'(\rho_0)\rho_0)(\psi' + \xi_1\varphi + \xi_2\theta)]] + [[\lambda\varepsilon_0(\psi' - \xi_1\varphi - \xi_2\theta)]] = \sigma|\xi|^2\psi. \tag{3.37}$$

The boundary conditions

$$\varphi(-m) = \varphi(l) = \theta(-m) = \theta(l) = \psi(-m) = \psi(l) = 0, \tag{3.38}$$

must also hold. if (φ, θ, ψ) solve (3.31)-(3.33) for $\xi \in R^2$ and λ, then for any rotation operator $R \in SO(2), (\tilde{\psi}, \tilde{\theta}) = R(\psi, \theta)$ solve the same equations for $\tilde{\xi} = R\xi$ with ψ, λ unchanged. So, by choosing an appropriate rotation, we may assume without loss of generality that $\xi_2 = 0$, and $\xi_1 = |\xi|$. In this setting θ solves

$$\begin{cases} -(\lambda\varepsilon_0\theta')' + (\lambda^2\rho_0 + \varepsilon_0|\xi|^2)\theta = 0, \\ \theta(-m) = \theta(l) = 0, \\ [[\theta]] = [[\lambda\varepsilon\theta']] = 0. \end{cases} \tag{3.39}$$

Multiplying this equation by θ, integrating over $(-m, l)$, integrating by parts, and using the jump conditions then yields

$$\int_{-l}^{m} \lambda\varepsilon_0|\theta'|^2 + (\lambda^2\rho_0 + \lambda\varepsilon_0|\xi|^2)\theta^2 = 0, \tag{3.40}$$

which implies that $\theta = 0$, since we assume $\lambda > 0$. This reduces to the pair of equations for φ, ψ:

$$-\lambda^2\rho_0\varphi = -(\lambda\varepsilon_0\varphi')' + |\xi|^2(4\lambda\varepsilon_0/3 + \lambda\delta_0/3 + P'(\rho_0)\rho_0)\varphi$$
$$+ |\xi|[(\lambda\delta_0 + \lambda\varepsilon_0/3 + P'(\rho_0)\rho_0)\psi' + (\lambda\varepsilon_0' - g\rho_0)\psi], \tag{3.41}$$

$$-\lambda^2\rho_0\psi = -[(\frac{4\lambda\varepsilon_0}{3} + \lambda\delta_0 + P'(\rho_0)\rho_0)\psi']' + \lambda\varepsilon_0|\xi|^2\psi$$
$$- |\xi|[(\lambda\delta_0 + \lambda\varepsilon_0/3 + P'(\rho_0))\varphi' + (g\rho_0 - \lambda\varepsilon_0')\varphi], \tag{3.42}$$

along with the jump conditions

$$[[\varphi]] = [[\psi]] = [[\lambda\varepsilon_0(\varphi' - |\xi|\psi)]] = 0, \tag{3.43}$$

$$[[(\lambda\delta_0 + \lambda\varepsilon_0/3 + P'(\rho_0)\rho_0)(\psi' + |\xi|\varphi)]] + [[\lambda\varepsilon_0(\varphi') - |\xi|\psi]] = \sigma|\xi|^2\psi, \tag{3.44}$$

and the boundary conditions

$$\varphi(-m) = \varphi(l) = \psi(-m) = \psi(l) = 0. \tag{3.45}$$

3.3 Statement of main results

In the absence of viscosity, $\varepsilon = \delta = 0$ and for a fixed spatial frequency $\xi \neq 0$, (3.41), (3.42) can be viewed as an eigenvalue problem with eigenvalue $-\lambda^2$. Such a problem has a natural variational structure that allows for the construction of solutions via the direct methods and for a variational characterization of the eigenvalue according to

$$-\lambda^2 = \inf \frac{E(\varphi, \psi)}{J(\varphi, \psi)}, \tag{3.46}$$

where

$$E(\varphi, \psi) = \frac{1}{2}\int_{-m}^{l} P'(\rho_0)\rho_0(\psi' + |\xi|\varphi)^2 - 2g\rho_0|\xi|\psi\varphi, \tag{3.47}$$

and

$$J(\varphi, \psi) = \frac{1}{2}\int_{-m}^{l} \rho_0(\varphi^2 + \psi^2). \tag{3.48}$$

Unfortunately, when viscosity is present, the natural variational structure breaks down since λ^2 appears quadratically as a multiplier of ρ_0 and linearly as a multiplier of ε_0 and δ_0. Since the equations imply a quadratic relationship between λ and various integrals of the solution, which can be solved for λ to determine the sign of $\operatorname{Re}\lambda$. On the other hand, the appearance of λ both quadratically and linearly eliminates the capacity to use constrained minimization techniques to produce solutions to the equations.

In order to circumvent this problem and restore the ability to use variational methods, we artificially remove the linear dependence on λ. To this end, we define the modified viscosities $\tilde{\varepsilon} = s\varepsilon_0, \tilde{\delta} = s\delta$, where $s > 0$ is an arbitrary parameter. We then introduce a family of modified problems given by

$$-\lambda^2 \rho_0 \varphi = -(\tilde{\varepsilon}\varphi')' + |\xi|^2(4\lambda\tilde{\varepsilon}/3 + \tilde{\delta}P'(\rho_0)\rho_0)]\varphi \\ + |\xi|[(\tilde{\delta} + \tilde{\varepsilon}/3 + P'(\rho_0)\rho_0)\psi' + (\tilde{\varepsilon}' - g\rho_0)\psi], \qquad (3.49)$$

$$-\lambda^2 \rho_0 \psi = -\left[\left(\frac{4\tilde{\varepsilon}}{3} + \tilde{\delta} + P'(\rho_0)\rho_0\right)\psi'\right]' + \tilde{\varepsilon}|\xi|^2\psi \\ - |\xi|[((\tilde{\delta} + \tilde{\varepsilon}/3 + P'(\rho_0)\rho_0)\varphi)' + (g\rho_0 - \tilde{\varepsilon}')\varphi], \qquad (3.50)$$

along with the jump conditions

$$[\![\varphi]\!] = [\![\psi]\!] = [\![\lambda\varepsilon_0(\varphi' - |\xi|\psi)]\!] = 0, \qquad (3.51)$$

$$[\![(\tilde{\delta} + \tilde{\varepsilon}/3 + P'(\rho_0)\rho_0)(\psi' + |\xi|\varphi)]\!] + [\![\tilde{\varepsilon}(\varphi') - |\xi|\psi]\!] = \sigma|\xi|^2\psi, \qquad (3.52)$$

and the boundary conditions

$$\varphi(-m) = \varphi(l) = \psi(-m) = \psi(l) = 0. \qquad (3.53)$$

A solution to the modified problem with $\lambda = s$ corresponds to a solution to the original problem.

Modifying the problem in this way restores the variational structure and allows us to apply a constrained minimization to the viscous analogue of the energy E defined above to find a solution to (3.49)-(3.50) with $\lambda = \lambda(|s|, s) > 0$, when $s > 0$ is sufficiently small and precisely when

$$0 < |\xi| \leqslant |\xi|_c = \sqrt{\frac{g[\![\rho_0]\!]}{\sigma}}. \qquad (3.54)$$

We then further exploit the variational structure to show that λ a continuous function and is strictly increasing in s. Using this, we show in Theorem 3.8 that the parameter s can be uniquely chosen so that

$$s = \lambda(|s|, s), \qquad (3.55)$$

which implies that we have found a solution to the original problem (3.41)-(3.42). This choice of s allows us to think of $\lambda = \lambda(|\xi|)$, it gives rise to a solution to the system of equations (3.32)-(3.33) as well.

Theorem 3.1 *For $\xi \in \mathbb{R}^2$, so that $0 < |\xi|^2 < g[[\rho_0]]/\sigma$, there exists a solution $\varphi = \varphi(\xi, x_3), \theta = \theta(\xi, x_3), \psi = \psi(\xi, x_3), \lambda = \lambda(\xi) > 0$ to (3.32)-(3.33) satisfying the appropriate jump and boundary conditions so that $\psi = \psi(\xi, 0) \neq 0$. The solutions are smooth when restricted to $(-m, 0)$ or $(0, l)$, and they are equivariant in ξ in the sense that if $R \in SO(2)$, is a rotation operator, then*

$$\begin{pmatrix} \varphi = \rho(R\xi, x_3) \\ \theta = \theta(R\xi, x_3) \\ \psi = \psi(R\xi, x_3) \end{pmatrix} = \begin{pmatrix} R_{11} & R_{12} & 0 \\ R_{21} & R_{22} & 0 \\ 0 & 0 & 1 \end{pmatrix} \begin{pmatrix} \varphi(\xi, x_3) \\ \theta(\xi, x_3) \\ \psi(\xi, x_3) \end{pmatrix}. \qquad (3.56)$$

Without surface tension $((\sigma = 0))$, it is possible to construct a solution to (3.55) with $\lambda > 0, \forall \xi \neq 0$, but with surface tension ($\sigma > 0$) there is a critical frequency $|\xi|_c = \sqrt{g[[\rho_0]]/\sigma}$ for which no solution with $\lambda > 0$ is available if $|\xi| \geqslant |\xi|_c$. In the nonperiodic case, we capture a continuum $|\xi| \in (0, |\xi|_c)$ of growing mode solutions, but in the $2\pi L$ periodic case, we find only finitely many. Indeed, if

$$\sqrt{\frac{\sigma}{g[[\rho_0]]}} < L, \qquad (3.57)$$

then a positive but finite number of spatial frequencies $\xi \in (L^{-1}Z)^2$, satisfy $|\xi| < |\xi|_c$. On the other hand, if

$$L \leqslant \sqrt{\frac{\sigma}{g[[\rho_0]]}}, \qquad (3.58)$$

then our method fails to construct any growing mode solutions at all. It is important to know the behavior of $\lambda(|\xi|)$ as $|\xi|$ varies within $0 < |\xi| < |\xi|_c$, We show in Proposition 3.9 that $\lambda(|\xi|)$ is continuous and satisfies

$$\lim_{|\xi| \to 0} \lambda(|\xi|) = \lim_{|\xi| \to |\xi|_c} \lambda(|\xi|) = 0. \qquad (3.59)$$

In the nonperiodic case, this implies that there is a largest growth rate

$$0 < \Lambda = \max_{0 \leqslant |\xi| \leqslant |\xi|_c} \lambda(|\xi|). \tag{3.60}$$

In the periodic case for L satisfying (3.57), the largest rate is always achieved and is given by

$$0 < \Lambda_L = \sup\{\lambda(|\xi|) | \xi \in (L^{-1}Z)^2, |\xi| \in (0, |\xi|_c)\}. \tag{3.61}$$

Note that in general $\Lambda_L < \Lambda$.

The stabilizing effects of viscosity and surface tension are evident in these results. Without viscosity or surface tension, there is $\lambda(\xi) \to \infty, |\xi| \to \infty$. With viscosity but no surface tension, all spatial frequencies remain unstable, but the growth rate $\lambda(\xi)$ bounded. With viscosity and surface tension, only a critical interval of spatial frequencies is unstable, and $\lambda(|\xi|)$ remains bounded. Finally, with viscosity and surface tension and the periodicity L satisfying (3.58), no growing modes exist.

In the periodic case when L satisfies (3.57), the solutions to (3.32)-(3.33) constructed in Theorem 3.1 immediately give rise to growing mode solutions to (3.22)-(3.23).

Theorem 3.2 Suppose that L satisfies (3.57), and let $\xi_1, \xi_2 \in (L^{-1}E)^2$ be lattice points such that $\xi_1 = -\xi_2, \lambda(\xi_i) = \Lambda_L$, where Λ_L is defined by (3.67). Define

$$\hat{w}(\xi, x_3) = -i\varphi(\xi, x_3)e_1 - i\theta(\xi, x_3)e_2 + \psi(\xi, x_3)e_3, \tag{3.62}$$

where φ, θ, ψ are the solutions provided by Theorem 3.1. Writing $x' = x_1e_1 + x_2e_2$, we define

$$\eta(x, t) = e^{\Lambda_L t} \sum_{j=1}^{3} \hat{w}(\xi_j, x_3) e^{ix'\xi_j}, \tag{3.63}$$

$$v(x, t) = \Lambda_L e^{\Lambda_L t} \sum_{j=1}^{2} \hat{w}(\xi_j, x_3) e^{ix'\xi_j}, \tag{3.64}$$

and

$$q(x, t) = e^{-\Lambda_L t} \rho_0(x_3) \sum_{j=1}^{2} (e_i \xi_j \varphi(\xi_j, x_3)) + e_i \xi_j \theta(\xi_j, x_3) + \partial_3 \psi(\xi_j, x_3)) e^{ix'\xi}. \tag{3.65}$$

Then η, v, q are real solutions to (3.22)-(3.23) and the corresponding jump and boundary conditions. For every $t > 0$, we have $\eta(t), v(t), q(t) \in H^k(\Omega)$, and

$$\begin{cases} \|\eta(t)\|_{H^k} = e^{t\Lambda_L}\|\eta(0)\|_{H^k}, \\ \|v(t)\|_{H^k} = e^{t\Lambda_L}\|v(0)\|_{H^k}, \\ \|q(t)\|_{H^k} = e^{t\Lambda_L}\|q(0)\|_{H^k}. \end{cases} \qquad (3.66)$$

Remark 3.3 *In this theorem, the space $H^k(\Omega)$ is not the usual Sobolev space of order k, but what we call the piecewise Sobolev space of order k. In the nonperiodic case, although $\Lambda = \lambda(|\xi|)$ may be achieved for some $\xi \in R^2, |\xi| \in (0, |\xi|_c)$, no $L^2(\Omega)$ solution to (3.22)-(3.23) may be constructed.*

Theorem 3.4 *Let $f \in C_c^\infty(0, |\xi|_c)$ be a real-valued function. For $\xi \in R^2$ with $|\xi| \in (0, |\xi|_c)$, define*

$$\hat{w}(\xi, x_3) = -i\varphi(\xi, x_3)e_1 - i\theta(\xi, x_3)e_2 + \psi(\xi, x_3)e_3, \qquad (3.67)$$

where φ, θ, ψ are the solutions provided by Theorem 2.1. Writing $x' = x_1 e_1 + x_2 e_2$, we define

$$\eta(x,t) = \frac{1}{4\pi^2} \int_{R^2} f(|\xi|)\hat{w}(\xi, x_3)e^{\lambda(|\xi|)t}e^{ix'\cdot\xi}d\xi, \qquad (3.68)$$

$$v(x,t) = \frac{1}{4\pi^2} \int_{R^2} \lambda(|\xi|)f(|\xi|)\hat{w}(\xi, x_3)e^{\lambda(|\xi|)t}e^{ix'\cdot\xi}d\xi, \qquad (3.69)$$

$$q(x,t) = \frac{1}{4\pi^2} \int_{R^2} f(|\xi|)(\xi_1\varphi(\xi, x_3) + \xi_2\theta(\xi, x_3) + \partial_{x_3}\psi(\xi, x_3))e^{\lambda(|\xi|)t}e^{ix'\cdot\xi}d\xi. \qquad (3.70)$$

Then η, v, q are real-valued solutions to the linearized equations (3.22)-(3.23) along with the corresponding jump and boundary conditions. The solutions are equivariant in the sense that if $R \in SO(3)$ is a rotation that keeps the vector e_3 fixed, then

$$\eta(Rx,t) = R\eta(x,t), \quad v(Rx,t) = Rv(x,t), \quad q(Rx,t) = q(x,t). \qquad (3.71)$$

$\forall k \in N$, *we have the estimate*

$$\|\eta(0)\|_{H^k} + \|v(0)\|_{H^k} + \|q(0)\|_{H^k} \leqslant \bar{C}_k \left(\int_{R^2} (1+|\xi|^2)^{k+1}|f(|\xi|)|^2 d\xi \right)^{\frac{1}{2}} < \infty, \qquad (3.72)$$

for a constant $\bar{C}$ depending on the parameters $\rho_0^\pm, p_\pm, g, \sigma, m, l$; moreover, $\forall t > 0$, we have $\eta(t), v(t), q(t) \in H^k$ and

$$\begin{cases} e^{t\lambda_0(f)}\|\eta(0)\|_{H^k} \leqslant \|\eta(t)\|_{H^k} \leqslant e^{t\Lambda}\|\eta(0)\|_{H^k}, \\ e^{t\lambda_0(f)}\|v(0)\|_{H^k} \leqslant \|v(t)\|_{H^k} \leqslant e^{t\Lambda}\|v(0)\|_{H^k}, \\ e^{t\lambda_0(f)}\|q(0)\|_{H^k} \leqslant \|q(t)\|_{H^k} \leqslant e^{t\Lambda}\|q(0)\|_{H^k}, \end{cases} \qquad (3.73)$$

where
$$\lambda_0(f) = \inf_{|\xi|\in supp(f)} \lambda(\xi) > 0, \tag{3.74}$$

and Λ is given by (3.60).

To state the result, we first define the weighted L^2 norm and the viscosity seminorm by

$$\|v\|_1^2 = \int_\Omega \rho_0|v|^2, \quad \|v\|_2^2 = \int_\Omega \frac{\epsilon_0}{2}|Dv + Dv^T - \frac{2}{3}(\text{div}v)I|^2 + \delta_0|\text{div}v|^2. \tag{3.75}$$

Theorem 3.5 *Let v, η, q be a solution to (3.22)-(3.23) along with the corresponding jump and boundary conditions. Then, in the nonperiodic case,*

$$\|v(t)\|_1^2 + \|v\|_2^2 + \|\partial_t v(t)\|_2^2$$
$$\leqslant Ce^{2\Lambda t}(\|\partial_t v(0)\|_1^2 + \|v(0)\|_1^2 + \|v(0)\|_2^2 + \sigma \int_{R^2} |\nabla_{x_1 x_2} v_3(0)|^2)$$

for a constant $C = C(\rho_0^\pm, p_\pm, \Lambda, \epsilon, \delta, \sigma, g, m, l) > 0$. In the periodic case with L satisfying (3.57), the same inequality holds with Λ replaced with Λ_L.

Theorem 3.6 *In the periodic case let L satisfy (3.58). For $j \geqslant 1$, define the constants $K_j \geqslant 0$ in terms of the initial data via*

$$K_j = \int_\Omega \rho_0 \frac{|\partial_t^j v(0)|^2}{2} + \int_\Omega \frac{P'(\rho_0)\rho_0}{2}\left|\text{div}\partial_t^{j-1}v(0) - \frac{g}{p'(\rho_0)}\partial_t^{j-1}v_3(0)\right|^2$$
$$+ \int_{(2\pi LT)^2} \frac{\sigma}{2}|\nabla_{x_1,x_2}\partial_t^{j-1}v_3(0)|^2. \tag{3.76}$$

Then the solutions to (3.22)-(3.23) satisfy

$$\|\eta(t)\|_1 + \|\eta(t)\|_2 \leqslant \|\eta(0)\|_1 + \|\eta(0)\|_2 + t(\|v(0)\|_1 + \|v(0)\|_2)$$
$$+2t^{\frac{3}{2}}\sqrt{K_1}, \tag{3.77}$$
$$\|v(t)\|_1 + \|v(t)\|_2 \leqslant \|v(0)\|_1 + \|v(0)\|_2 + 3\sqrt{t}\sqrt{K_1}, \tag{3.78}$$

and for $j \geqslant 1$,

$$\sup_{t\geqslant 0} \frac{1}{2}\|\partial_t^j v(t)\|_1^2 + \int_0^\infty \|\partial_t^j v(t)\|_2^2 dt \leqslant 2K_j, \tag{3.79}$$

$$\sup_{t\geqslant 0} \|\partial_t^j v(t)\|_2^2 \leqslant \|\partial_t^j v(0)\|_2^2 + 2\sqrt{K_j}\sqrt{K_{j+1}}. \tag{3.80}$$

3.4 A family of modified variational problems.

In order to understand λ in a variational framework, we consider the two energies

$$E(\varphi,\psi) = \frac{\sigma}{2}|\xi|^2(\psi(0))^2 + \frac{1}{2}\int_{-m}^{l}(\tilde{\delta}+p'(\rho_0)\rho_0)(\psi'+|\xi|\varphi)^2 - 2g\rho_0|\xi|\varphi\psi$$
$$+\frac{1}{2}\int_{-m}^{l}\tilde{\varepsilon}\Big((\varphi'-|\xi|\psi)^2 + (\psi'-|\xi|\varphi)^2 + \frac{1}{3}(\psi'+|\xi|\varphi)^2\Big), \qquad (3.81)$$

and

$$J(\varphi,\psi) = \frac{1}{2}\int_{-m}^{l}\rho_0(\varphi^2+\psi^2), \qquad (3.82)$$

which are both well defined on the space $H_0^1((-m,l)) \times H_0^1((-m,l))$. Consider the set

$$\mathcal{A} = \{(\varphi,\psi) \in H_0^1((-m,l)) \times H_0^1((-m,l)) | J(\varphi,\psi) = 1\}. \qquad (3.83)$$

We want to show that the infimum $\inf_{\varphi,\psi\in\mathcal{A}} E(\varphi,\psi)$ solves (3.49)-(3.50) along with the corresponding jump and boundary conditions. Also notice that by employing the identity $-2ab = (a-b)^2 - (a^2+b^2)$, and the constraint on $J(\varphi,\psi)$, we may rewrite

$$\begin{aligned}E(\varphi,\psi) &= -g|\xi| + \frac{\sigma|\xi|^2}{2}(\psi(0))^2 + \frac{1}{2}\int_{-m}^{l}(\tilde{\delta}+p'(\rho_0)\rho_0)(\psi'+|\xi|\varphi)^2 \\ &+g|\xi|\rho_0(\varphi-\psi)^2 + \frac{1}{2}\int_{-m}^{l}\tilde{\varepsilon}\Big((\varphi'-|\xi|\psi)^2 + (\psi'-|\xi|\varphi)^2 \\ &+\frac{1}{3}(\psi'+|\xi|\varphi)^2\Big) \geqslant -g|\xi|,\end{aligned} \qquad (3.84)$$

$\tilde{\varepsilon} = s\varepsilon(\rho_0), s \in (0,\infty)$.

$$E(\varphi,\psi) = E(\varphi,\psi;s), \qquad (3.85)$$

$$\mu(s) = \inf_{(\varphi,\psi)} E(\varphi,\psi;s). \qquad (3.86)$$

Proposition 3.7 *E achieves its infimum on $\mathcal{A}$.*

Proof First note that (3.84) shows that E is bounded below on $\mathcal{A}$. Let $(\varphi_n,\psi_n) \in \mathcal{A}$ be a minimizing sequence. Then φ_n,ψ_n are bounded in $H_0^1((-m,l))$ and $\psi_n(0)$ is bounded in R. so up to the extraction of a subsequence $(\varphi_n,\psi_n) \rightharpoonup (\varphi,\psi)$ weakly in $H_0^1 \times H_0^1$, and $(\varphi_n,\psi_n) \to (\varphi,\psi)$ strongly in $L^2 \times L^2$. The compact embedding $H_0^1 \subset\subset H^{\frac{2}{3}} \hookrightarrow C^0$ implies that $\psi_n(0) \to \psi(0)$ as well. Because of the quadratic structure of all the terms in the integrals defining E, weak lower semicontinuity and strong L^2 convergence imply that

$$E(\varphi, \psi) \leqslant \liminf_{n \to \infty} E(\varphi_n, \psi_n) = \inf_{\mathcal{A}} E. \tag{3.87}$$

That $(\varphi, \psi) \in \mathcal{A}$ follows from the strong L^2 convergence.

We now show that the minimizer constructed in the previous result satisfies Euler-Langrange equations equivalent to (3.49)-(3.50).

Proposition 3.8 *Let $(\varphi, \psi) \in \mathcal{A}$ be the minimizers of E, let $\mu = E(\varphi, \psi)$, then (φ, ψ) are smooth when restricted to $(-m, 0)$ or $(0, l)$ and satisfy*

$$\mu \rho_0 \varphi = -(\bar{\varepsilon}\varphi')' + |\xi|^2(4\bar{\varepsilon}/3 + \bar{\delta} + p'(\rho_0)\rho_0)\varphi$$
$$+ |\xi|[(\bar{\delta} + \bar{\varepsilon}/3 + p'(\rho_0)\rho_0)\psi' + (\bar{\varepsilon}' - g\rho_0)\psi], \tag{3.88}$$
$$\mu \rho_0 \psi = -[(4\bar{\varepsilon}/3 + \bar{\delta} + p'(\rho_0)\rho_0)\psi']' + \bar{\varepsilon}|\xi|^2\psi$$
$$- |\xi|[((\bar{\delta} + \bar{\varepsilon}/3 + p'(\rho_0)\rho_0)\varphi)' + (g\rho_0 - \bar{\varepsilon}')\varphi], \tag{3.89}$$

along with the jump conditions

$$[[\varphi]] = [[\psi]] = [[\xi(\varphi' - |\xi|\psi)]] = 0, \tag{3.90}$$
$$[[(\bar{\delta} + \bar{\varepsilon}/3 + p'(\rho_0)\rho_0)(\psi' + |\xi|\varphi)]] + [[\bar{\varepsilon}(\psi' - |\xi|\varphi)]]$$
$$= \sigma|\xi|^2\psi(0), \tag{3.91}$$

and the boundary conditions $\varphi(-m) = \varphi(l) = \psi(-m) = \psi(l) = 0$.

Proof Fix $(\varphi_0, \psi_0) \in H_0^1((-m, l)) \times H_0^1((-m, l))$. Define

$$j(t, \tau) = J(\varphi + t\varphi_0 + \tau\varphi, \psi + t\psi_0 + \tau\psi), \tag{3.92}$$

and note that $j(0, 0) = 1$. Moreover, j is smooth,

$$\frac{\partial j}{\partial t}(0,0) = \int_{-m}^{l} \rho_0(\varphi_0\varphi + \psi_0\psi), \quad \frac{\partial j}{\partial \tau}(0,0) = \int_{-m}^{l} \rho_0(\varphi^2 + \psi^2) = 2. \tag{3.93}$$

So, by the inverse function theorem, we can solve for $\tau = \tau(t)$ in a neighborhood of 0 as a C^1 function of t so that $\tau(0) = 0, j(t, z(t)) = 1$. We may differentiate the last equation to find

$$\frac{\partial j}{\partial t}(0,0) + \frac{\partial j}{\partial \tau}(0,0)\tau'(0) = 0, \tag{3.94}$$

and hence that
$$\tau'(0) = -\frac{1}{2}\frac{\partial j}{\partial t}(0,0) = -\frac{1}{2}\int_{-m}^{l}\rho_0(\varphi_0\varphi + \psi_0\psi). \tag{3.95}$$

Since (φ, ψ) are minimizers over $\mathcal{A}$, we may make variations with respect to (φ_0, ψ_0) to find that
$$0 = \frac{d}{dt}\Big|_{t=0} E(\varphi + t\varphi_0 + \tau(t)\varphi, \psi + t\psi_0 + \tau(t)\psi), \tag{3.96}$$
which implies that
$$0 = \sigma|\xi|^2\psi(0)(\psi_0(0) + \tau'(0)\psi(0))$$
$$+ \int_{-m}^{l}(\bar{\delta} + \bar{\varepsilon}/3 + p'(\rho_0)\rho_0)(\psi' + |\xi|\varphi)(\psi_0' + \tau'(0)\psi' + |\xi|\varphi_0 + |\xi|\tau'(0)\varphi)$$
$$- \int_{-m}^{l}g|\xi|\rho_0(\psi(\rho_0 + \tau'(0)\varphi) + \varphi(\psi_0 + \tau'(0)\psi))$$
$$+ \int_{-m}^{l}\bar{\varepsilon}(\psi' - |\xi|\varphi)(\psi_0' + \tau'(0)\psi' - |\xi|\varphi_0 - |\xi|\tau'(0)\varphi)$$
$$+ \int_{-m}^{l}\bar{\varepsilon}(\psi' - |\xi|\varphi)(\varphi_0' + \tau'(0)\varphi' - |\xi|\psi_0 - |\xi|\tau'(0)\psi). \tag{3.97}$$

Rearranging and plugging in the value of $\tau'(0)$, we may rewrite this equation as
$$\sigma|\xi|^2\psi(0)\psi_0(0) + \int_{-m}^{l}(\bar{\delta} + \bar{\varepsilon}/3 + p'(\rho_0)\rho_0)(\psi' + |\xi|\varphi)(\psi_0' - |\xi|\varphi_0)$$
$$- g|\xi|\rho_0(\psi\rho_0 + \varphi\psi_0)$$
$$+ \int_{-m}^{l}\bar{\varepsilon}((\psi' - |\xi|\varphi)(\psi_0' - |\xi|\varphi_0) + (\varphi' - |\xi|\psi)(\varphi_0' + |\xi|\psi_0))$$
$$= \mu\int_{-m}^{l}\rho_0(\varphi_0\varphi + \psi_0\psi), \tag{3.98}$$

where the Lagrange multiplier is $\mu = E(\varphi, \psi)$. Since φ_0, ψ_0 are independent, this gives rise to the pair of equations
$$\int_{-m}^{l}\bar{\varepsilon}\varphi'\varphi_0' + (\bar{\delta} + 4\bar{\varepsilon}/3 + p'(\rho_0)\rho_0)|\xi|^2\varphi\varphi_0 - \int_{-m}^{l}(\bar{\varepsilon}|\xi|\psi\rho_0)'$$
$$+|\xi|\int_{-m}^{l}[(\bar{\delta} + \bar{\varepsilon}/3 + p'(\rho_0)\rho_0)\psi' + (\bar{\varepsilon}' - g\rho_0)\psi]\varphi_0$$
$$= \mu\int_{-m}^{l}\rho_0\varphi\varphi_0, \tag{3.99}$$

and

$$\sigma|\xi|^2\psi(0)\psi_0(0) + \int_{-m}^{l} [(\bar{\delta} + 4\bar{\epsilon}/3 + p'(\rho_0)\rho_0)\psi'$$
$$+ (\bar{\delta} + \bar{\epsilon}/3 + p'(\rho_0)\rho_0)|\xi|\varphi]\psi' - \int_{-m}^{l} (\bar{\epsilon}|\xi|\varphi\psi_0)'$$
$$+ \int_{-m}^{l} [\bar{\epsilon}|\xi|^2\psi + (\bar{\epsilon} - g\rho_0)|\xi|\varphi]\psi_0$$
$$= \mu \int_{-m}^{l} \rho_0 \psi \psi_0. \qquad (3.100)$$

By making variations with φ_0, ψ_0 compactly supported in either $(-m, 0)$ or $(0, l)$, we find that φ, ψ satisfy (3.88)-(3.89) in a weak sense in $(-m, 0)$ and $(0, l)$. Standard bootstrapping arguments then show that (φ, ψ) are in $H^k(-m, 0)(H^k(0, l))$, $\forall k \geqslant 0$, and hence the functions are smooth when restricted to either interval. To show that the jump conditions are satisfied. Integrating the terms in (3.99), we find that

$$[[\bar{\epsilon}(\varphi' - |\xi|\psi)]]\varphi(0) = 0. \qquad (3.101)$$

Performing a similar integration by parts in (3.100) yields the jump condition

$$0 = \sigma|\xi|^2\psi(0) - [[(\bar{\delta} + \bar{\epsilon}/3 + p'(\rho_0)\rho_0) + \psi' + |\xi|\varphi]] - [[\bar{\epsilon}(\psi' - |\xi|\varphi)]]. \qquad (3.102)$$

The conditions $[[\varphi]] = [[\psi]] = 0$, $\varphi(-m) = \varphi(l) = \psi(-m) = \psi(l) = 0$ are satisfied trivially since

$$\varphi, \psi \in H_0^1(-m, l) \hookrightarrow C_0^{0, \frac{1}{2}}((-m, l)).$$

We now show that for s sufficiently small, the infimum inf E over $\mathcal{A}$ is in fact negative.

Proposition 3.9 *Suppose that $0 < |\xi|^2 < g[[\rho_0]]/\sigma$. Then there exists $s_0 > 0$ depending on the quantities $\rho_0^\pm$, $p_\pm$, g, $\epsilon_\pm$, σ, $m, l, |\xi|$ so that for $s \leqslant s_0$, it holds that $\mu(s) < 0$.*

Proof Since both E and J are homogeneous of degree 2, it suffices to show that

$$\inf_{(\varphi,\psi) \in H_0^1 \times H_0^1} \frac{E(\varphi, \psi)}{J(\varphi, \psi)} < 0, \qquad (3.103)$$

but since J is positive definite, we may reduce to constructing any pair $(\varphi, \psi) \in H_0^1 \times H_0^1$ such that $E(\varphi, \psi) < 0$. We will assume that $\varphi = -\psi'/|\xi|$ so that the first integrand term in $E(\varphi,\psi)$ vanished. We must then construct $\psi \in H_0^2$, so that

$$\bar{E}(\psi) = E(-\psi'/|\xi|, \psi) = \frac{\sigma|\xi|^2}{2}(\psi(0))^2$$
$$+ \int_{-m}^{l} g\rho_0 \psi\psi' + \frac{\bar{\epsilon}}{2}\left(\left(\frac{\psi''}{|\xi|} + |\xi|\psi\right)^2 + 4(\psi')^2\right) < 0.$$

We employ the identity $\psi\psi' = ((\psi)^2)'/2$, an integration by parts, and the fact that ρ_0 solves

$$\frac{dp(\rho_0)}{dx_3} = -g\rho_0$$

to write

$$\int_{-m}^{l} g\rho_0 \psi\psi' = \left[\frac{g\rho_0 \psi^2}{2}\right]_0^l - \frac{1}{2}\int_0^l g\rho_0' \psi^2$$
$$+ \left[\frac{g\rho_0 \psi^2}{2}\right]_{-m}^0 - \frac{1}{2}\int_{-m}^0 g\rho_0' \psi^2$$
$$= -\frac{g[\psi(0)]^2}{2}[[\rho_0]] + \frac{g^2}{2}\int_{-m}^{l} \frac{\rho_0}{p'(\rho_0)}\psi^2. \quad (3.104)$$

Note that $[[\rho_0]] = \rho^+ - \rho^- > 0$ so that the right-hand side is not positive definite.

For $\alpha > 5$, we define the test function $\psi_\alpha \in H_0^2(-m, l)$ according to

$$\psi_\alpha(x_3) = \begin{cases} \left(1 - \dfrac{x_3^2}{l^2}\right)^{\alpha/2}, & x_3 \in (0, l), \\ \left(1 + \dfrac{x_3^2}{m^2}\right)^{\alpha/2}, & x_3 \in (-m, 0). \end{cases} \quad (3.105)$$

Simple calculations then show that

$$\int_{-m}^{l} (\psi_\alpha)^2 = \frac{\sqrt{\pi}(m+l)\Gamma(\alpha+1)}{2\Gamma\left(\alpha + \dfrac{3}{2}\right)} = o_\alpha(1), \quad (3.106)$$

where $o_\alpha(1)$ is a quantity that vanishes as $\alpha \to \infty$, and that

$$\int_{-m}^{l}\left(\left(\frac{\psi''}{|\xi|} + |\xi|\psi\right)^2 + 4(\psi')^2\right) \leqslant C, \quad (3.107)$$

for a constant C depending on $\alpha, m, l, |\xi|$. Combining these, we find that

$$\tilde{E}(\psi_\alpha) \leqslant \frac{\sigma|\xi|^2 - g[[\rho_0]]}{2} + o_\alpha(1) + C, \quad (3.108)$$

for a constant C depending on $\alpha, \rho_0^\pm, p_\pm, g, \epsilon_\pm, m, l, |\xi|$. Since $\sigma|\xi|^2 < g[[\rho_0]]$ we may then fix α sufficiently large so that the first two terms sum to something strictly negative. Then there exists $s_0 > 0$ depending on the various parameters so that for $s \leqslant s_0$, it holds that $\tilde{E}(\psi_\alpha) < 0$, thereby proving the result..

Remark 3.10 We can get $\mu(s) \leqslant -C_0 + sC_1$, from
$$\frac{\sigma|\xi|^2 - g[[\rho_0]]}{2}(\psi(0))^2 < 0,$$
$\psi(0) \neq 0, |\xi|^2 < g[[\rho_0]]/\sigma$, so we get $E(\varphi, \psi) < 0$.

Lemma 3.11 Suppose that $(\varphi, \psi) \in A$ satisfy $E(\varphi, \psi) < 0$. Then $\psi(0) \neq 0, |\xi|^2 < g[[\rho_0]]/\sigma$.

Proof A completion of the square allows us to write
$$p'(\rho_0)\rho_0(\psi' + |\xi|\varphi)^2 - 2g\rho_0|\xi|\psi\varphi$$
$$= \left(\sqrt{p'(\rho_0)\rho_0}(\psi' + |\xi|\varphi) - \frac{g\sqrt{\rho_0}}{\sqrt{p'(\rho_0)}}\psi\right)^2 + 2g\rho_0\psi\psi' - \frac{g^2\varphi_0}{p'(\rho_0)}\psi^2. \quad (3.109)$$

Integrating by parts as in (3.104), we know that
$$\int_{-m}^{l} 2g\rho_0\psi\psi' - \frac{g^2\varphi_0}{p'(\rho_0)}\psi^2 = -g[[\rho_0]](\psi(0))^2. \quad (3.110)$$

Combining these equalities, we can rewrite $E(\varphi, \psi)$ as
$$E(\varphi, \psi) = \frac{1}{2}\int_{-m}^{l}(\bar{\delta} + \bar{\epsilon}/3)(\psi' + |\xi|\varphi)^2 + p'(\rho_0)\rho_0\left((\psi' + |\xi|\varphi) - \frac{g}{p'(\rho_0)}\psi\right)^2$$
$$+ \frac{1}{2}\int_{-m}^{l}\bar{\epsilon}((\varphi' - |\xi|\psi)^2 + (\psi' - |\xi|\varphi)^2) + \frac{\sigma|\xi|^2 - g[[\rho_0]]}{2}\psi^2(0).$$

From the nonnegativity of the integrals, we deduce that if $E(\varphi, \psi) < 0$, then $\psi(0) \neq 0$, $|\xi|^2 < g[[\rho_0]]/\sigma$

Proposition 3.12 Let $\mu : (0, \infty) \to R$ be given by (3.86), Then the following hold:

1) $\mu \in C^{0,1}_{loc}((0, \infty))$, and in particular $\mu \in C^0((0, \infty))$.
2) There exists a positive constant $C_2 = C_2(\rho_0^\pm, \rho_\pm, g, \epsilon_\pm, \sigma, m, l)$ so that
$$\mu(s) \geqslant -g|\xi| + sC_2. \quad (3.111)$$
3) $\mu(s)$ is strictly increasing.

Proof Fix a compact interval $Q = [a,b] \subset C(0,\infty)$ and fix any pair $(\varphi_0, \psi_0) \in A$. We may decompose E according to

$$E(\varphi, \psi, s) = E_0(\varphi, \psi) + sE_1(\varphi, \psi), \tag{3.112}$$

for

$$E_0(\varphi, \psi) = \frac{\sigma|\xi|^2}{2}(\psi(0))^2 + \frac{1}{2}\int_{-m}^{l} p'(\rho_0)\rho_0(\psi' + |\xi|\varphi)^2 - 2g\rho_0|\xi|\psi\varphi \tag{3.113}$$

and

$$E_1(\varphi, \psi) = \frac{1}{2}\int_{-m}^{l} (\delta_0 + \epsilon_0/3)(\psi' + |\xi|\varphi)^2 + \epsilon_0((\psi' - |\xi|\varphi)^2 \\ + (\psi' + |\xi|\varphi)^2) \geq 0. \tag{3.114}$$

The nonnegativity of E_1 implies that E is nondecreasing in s with $(\varphi, \psi) \in A$ kept fixed.

Now, by Proposition 3.1, $\forall s \in (0, +\infty)$ we can find a pair $(\varphi, \psi) \in A$ so that

$$E(\varphi_s, \psi_s; s) = \inf_{(\varphi,\psi)\in A} E(\varphi, \psi; s) = \mu(s). \tag{3.115}$$

We deduce from the nonnegativity of E_1, the minimality of (φ_s, ψ_s), and the equality (3.84) that

$$E(\varphi_0, \psi_0; b) \geq E(\varphi_0, \psi_0; s) \geq sE_1(\varphi_s, \psi_s) - g|\xi|, \tag{3.116}$$

for all $s \in Q$. This implies that there exists a constant $0 < K = K(a, b, \varphi_0, \psi_0, g, |\xi|) < \infty$ so that

$$\sup_{s \in Q} E_1(\varphi_s, \psi_s) < K. \tag{3.117}$$

Let $s \in Q, i = 1, 2$. Using the minimality of $(\varphi_{s_1}, \psi_{s_1})$ compared to $(\varphi_{s_2}, \psi_{s_2})$ we know that

$$\mu(s_1) = E(\varphi_{s_1}, \psi_{s_1}; s_1) \leq E(\varphi_{s_2}, \psi_{s_2}; s_1), \tag{3.118}$$

but from our decomposition (3.112), we may bound

$$E(\varphi_{s_2}, \psi_{s_2}; s_1) \leq E(\varphi_{s_2}, \psi_{s_2}; s_2) + |s_1 - s_2|E_1(\varphi_{s_2}, \psi_{s_2}) \\ = \mu(s_2) + |s_1 - s_2|E_1(\varphi_{s_2}, \psi_{s_0}). \tag{3.119}$$

Chaining these two inequalities together and employing (3.117), we find that

$$\mu(s_1) \leq K|s_1 - s_2|. \tag{3.120}$$

Reversing the role of the indices 1, 2 in the derivation of this inequality gives the same bound with the indices switched. We deduce that

$$|\mu(s_1) - \mu(s_2)| \leq K|s_1 - s_2|, \tag{3.121}$$

which proves the first assertion.

To prove (3.111), we note that equality (3.84) and the nonnegativity of E_1 imply that

$$\mu(s) \geq -g|\xi| + s \inf_{(\varphi,\psi)\in\mathcal{A}} E_1(\varphi,\psi). \tag{3.122}$$

It is a simple matter to see that this infimum, which we call the constant C_2, is positive.

Finally, to prove the third assertion, note that if $0 < s_1 < s_2 < \infty$, then the decomposition (3.112) implies that

$$\mu(s_1) = E(\varphi_{s_1}, \psi_{s_1}; s_1) \leq E(\varphi_{s_2}, \psi_{s_2}; s_1) \leq E(\varphi_{s_2}, \psi_{s_2}; s_2) = \mu(s_2). \tag{3.123}$$

This shows that μ is nondecreasing in s. Now suppose by way of contradiction that $\mu(s_1) = \mu(s_2)$. Then the previous inequality implies that

$$s_1 E(\varphi_{s_2}, \psi_{s_2}) = s_2 E(\varphi_{s_2}, \psi_{s_2}), \tag{3.124}$$

which means that $E_1(\varphi_{s_2}, \psi_{s_2}) = 0$. This in turn forces $\varphi_{s_2} = \psi_{s_2} = 0$, which contradicts the fact that $(\varphi_{s_2}, \psi_{s_2}) \in \mathcal{A}$. Hence equality cannot be achieved, and μ is strictly increasing in s.

Now we know that when $0 < |\xi|^2 < g[[\rho_0]]/\gamma$, the eigenvalue $\mu(s)$ is a continuous function. We can then define the open set

$$S = \mu^{-1}((-\infty, 0)) \subset (0, \infty), \tag{3.125}$$

on which we can calculate $\lambda = \sqrt{-\mu}$. Note that S is nonempty by Proposition 3.9.

We can now state a result giving the existence of solutions to (3.49)-(3.50) for these values of $|\xi|, s$. To emphasize the dependence on the parameters, we write

$$\varphi = \varphi_s(|\xi|, x_3), \quad \psi = \psi_s(|\xi|, x_3), \quad \lambda = \lambda(|\xi|, s). \tag{3.126}$$

Proposition 3.13 *For each $s \in S$, $0 < |\xi|^2 < g[[\rho_0]]/\sigma$ there exists a solution $\varphi_s(|\xi|, x_3)$, $\psi_s(|\xi|, x_3)$ with $\lambda = \lambda(|\xi|, s) > 0$ to the problem (3.49)-(3.50)* along with the corresponding jump and boundary conditions. For these solutions $\psi_s(|\xi|, 0) \neq 0$, and the solutions are smooth when restricted to either $(-m, 0)$ or $(0, l)$.

Proof Let $\varphi_s(|\xi|, \cdot), \psi_s(|\xi|, \cdot) \in \mathcal{A}$ be the solutions to (3.88)-(3.89). Since $s \in S$, we may write $\mu = -\lambda^2$, $\lambda > 0$, which means that the pair $\varphi_s(|\xi|, s), \psi_s(|\xi|, s)$ solve the problem (3.49)-(3.50). The fact that $\psi_s(|\xi|, 0) \neq 0$ follows from Lemma 3.11.

In order for these solutions to give rise to solutions to the original problem, we must be able to find $s \in S$, so that $s = \lambda(|\xi|, s)$. It turns out that the set S is sufficiently large to accomplish this.

Theorem 3.14 There exists a unique $s \in S$, so that $\lambda(|\xi|, s) = \sqrt{-\mu(s)} > 0$ and

$$s = \lambda(|\xi|, s). \tag{3.127}$$

Proof According to Remark 3.4, we know that $\mu(s) \leqslant -C_0 + sC_1$. Moreover, the lower bound (3.111) implies that $\mu(s) \to +\infty$ as $s \to \infty$. Since μ is continuous and strictly increasing, there exists $s \in (0, \infty)$ so that

$$S = \mu^{-1}((-\infty, 0)) = (0, s_*). \tag{3.128}$$

Since $\mu > 0$ on S, we may define $\lambda = \sqrt{-\mu}$. Now define the function $\Phi : (0, s_*) \to (0, \infty)$ according to

$$\Phi(s) = s/\lambda(|\xi|, s). \tag{3.129}$$

It is a simple matter to check that Φ is continuous and strictly increasing in s. Also, $\lim_{s \to 0} \Phi(s) = 0$, $\lim_{s \to s_*} \Phi(s) = +\infty$. Then by the intermediate value theorem, there exists $s \in (0, s_*)$, so that $\Phi(s) = 1$, i.e., $s = \lambda(|\xi|, s)$. This s is unique since Φ is strictly increasing.

We may now use Theorem 3.14 to think of $s = s(|\xi|)$. Since for each fixed $0 < |\xi|^2 < g[[\rho_0]]/\sigma$, we can uniquely find $s \in S$ so that (3.127) holds. As such, we may also write $\lambda = \lambda(|\xi|)$ from now on. Using this new notation and the solutions to the equations given by Proposition 3.13, we can construct solutions to the system (3.49)-(3.50) as well.

Proof of Theorem 3.1 We may find a rotation operator $R \in SO(2)$, so that $R\xi = (|\xi|, 0)$. For $s = s(\xi)$ given by Theorem 3.14, define $(\varphi(\xi, x_3), \theta(\xi, x_3)) = R^{-1}(\varphi_s(|\xi|, x_3), 0)$ and $\psi(\xi, x_3) = \psi_s(|\xi|, x_3)$, where the functions $\varphi_s(|\xi|, x_3)$,

$\psi_s(|\xi|, x_3)$ are the solutions from Proposition 3.13, This gives a solution to (3.31)-(3.33). The equivariance in follows from the definition.

3.5 Solutions to (3.22)-(3.23)

In this section, we will construct growing solutions to (3.22)-(3.23) by using the solutions to (3.31)-(3.33) constructed in Theorem 3.1. In the periodic case, this can only be done when L satisfies (2.23), but the construction is essentially trivial since normal mode solutions are in $L^2(\Omega)$. In the nonperiodic case, we must resort to a Fourier synthesis of the normal modes in order to produce L^2 solutions.

We begin by defining some terms. For a function $f \in L^2(\Omega)$, we define the horizontal Fourier transform in the nonperiodic case via

$$\hat{f}(\xi_1, \xi_2, x_3) = \int_{R^2} f(x_1, x_2, x_3) e^{-i(x_1\xi_1 + x_2\xi_2)} dx_1 dx_2, \qquad (3.130)$$

for $\xi \in R^2$. In the periodic case the integral over R^2 must be replaced with an integral over $(2\pi L_T)^2$ for $\xi \in (L^{-1}Z)^2$. In the nonperiodic case, by the *Fubini* and *Parseval* theorems, we have that $\hat{f} \in L^2(\Omega)$ and

$$\int_\Omega |f(x)|^2 dx = \frac{1}{4\pi^2} \int_\Omega |\hat{f}(\xi, x_3)|^2 d\xi dx_3. \qquad (3.131)$$

The periodic case replaces $4\pi^2$ with $4\pi L^2$ and the integral with a sum over $(L^{-1}Z)^2$ on the right-hand side.

We now define the piecewise Sobolev spaces. For a function f defined on Ω we write f_+ for the restriction to Ω_+ and f_- for the restriction to Ω_-. For $k \in N$, define the piecewise Sobolev space of order k by

$$H^k(\Omega) = \{f | f_+ \in H^k(\Omega_+), f_- \in H^k(\Omega_-)\}, \qquad (3.132)$$

endowed with the norm $\|f\|_{H^k}^2 = \|f\|_{H^k(\Omega_+)}^2 + \|f\|_{H^k(\Omega_-)}^2$. Writing $I_- = (-m, 0)$, $I_+ = (0, l)$, we can take the norms to be given as

$$\|f\|_{H^k}^2 = \sum_{j=0}^{k} \int_{R^2} (1+|\xi|^2)^{k-j} \|\partial_{x_3}^j \hat{f}_\pm(\xi, \cdot)\|_{L^2(I_\pm)}^2 d\xi. \qquad (3.133)$$

Proof of Theorem 3.2 It is clear that η, v, q defined in this way are solutions to (3.22)-(3.23). That they are real-valued follows from the equivariance in ξ stated in Theorem 3.1 The solutions are in $H^k(\Omega)$ at $t = 0$ because of Lemma 3.11. The growth in time stated in (3.66) follows from the definition of η, v, q.

Proof of Theorem 3.4 For each fixed $\xi \in \mathbb{R}^2$ so that $|\xi| \in (0, |\xi|_c)$.

$$\eta(x,t) = f(|\xi|)\hat{w}(\xi, x_3)e^{\lambda(|\xi|)t}e^{ix'\cdot\xi}, \tag{3.134}$$

$$v(x,t) = \lambda(|\xi|)f(|\xi|)\hat{w}(\xi, x_3)e^{\lambda(|\xi|)t}e^{ix'\cdot\xi}, \tag{3.135}$$

$$q(x,t) = -\rho_0(x_3)f(|\xi|)(\xi_1\varphi(\xi, x_3) + \xi_2\theta(\xi, x_3) + \partial_3\psi(\xi, x_3))e^{ix'\cdot\xi}, \tag{3.136}$$

constitute a solution to (3.22)-(3.23).

Since $\operatorname{supp}(f) \subset (0, |\xi|_c)$, Lemma 3.11 implies that

$$\sup_{\xi \in \operatorname{supp}(f)} \|\partial_{x_3}^k \hat{w}(\xi, \cdot)\|_{L^\infty} \leqslant \infty, \quad \forall k \in N. \tag{3.137}$$

These bounds, the definition of Λ, and the dominated convergence theorem imply that the Fourier synthesis of these solutions given by (3.68)-(3.70) is also a solution that is smooth when restricted to $\Omega_\pm$. The Fourier synthesis is real-valued because $f(|\xi|)$ is real-valued and radial and because of the equivariance in ξ given in Theorem 3.1. This equivariance in ξ also implies the equivariance of η, v, q written in (3.71).

The bound (3.72) follows by applying Lemma 3.11 with arbitrary $k \geqslant 0$ and utilizing the fact that f is compactly supported. The compact support of f also implies that $\lambda_0(f) > 0$, so that $\lambda_0(f) \leqslant \lambda(|\xi|) \leqslant \Lambda$, $|\xi| \in supp(f)$. This then yields the bounds (3.73).

3.6 Growth of solutions to the linearized problem

In this section, we will prove estimates for the growth in time of arbitrary solutions to (3.22)-(3.23) in terms of the largest growing mode: Λ in the nonperiodic case and Λ_L in the periodic case, defined by (3.60) and (3.61) respectively. To this end, we suppose that η, v, q are real-valued solutions to (3.22)-(3.23) along with the corresponding jump and boundary conditions.

It will be convenient to work with a second-order formulation of the equations. To arrive at this, we differentiate the third equation in time and eliminate the q, η terms using the other equations. This yields the equation

$$\rho_0 \partial_{tt} v - \nabla(P'(\rho_0)\rho_0 \operatorname{div} v) + g\rho_0 \nabla v_3 - g\rho_0 \operatorname{div} v e_3$$
$$= \operatorname{div}(\varepsilon_0(D\partial_t v + D\partial_t v^T - \frac{2}{3}(\operatorname{div}\partial_t v)I + \delta_0(\operatorname{div}\partial_t v)I), \tag{3.138}$$

coupled to the jump conditions

$$[[\partial_t v]] = 0, \tag{3.139}$$

and

$$[[(p'(\rho_0)\rho_0 \operatorname{div} v)I + \varepsilon_0(D\partial_t v + D\partial_t v^T) + (\delta_0 - 2\varepsilon_0/3)\operatorname{div}\partial_t v I]]e_3$$
$$= -\sigma \Delta_{x_1,x_2} v_3 e_3. \tag{3.140}$$

The functions $\partial_t v$ also satisfies $\partial_t v(x_1, x_2, -m, t) = \partial_t v(x_1, x_2, l, t) = 0$ at the upper and lower boundaries. The initial data for $\partial_t v(x_1, x_2, -m, t) = \partial_t v(x_1, x_2, l, t) = 0$ are given in terms of the initial data $q(0), v(0)$ and $\eta(0)$ via the third linear equation; i.e., $\partial_t v(0)$ satisfies

$$\rho_0 \partial_t v(0) = -g\rho(0)e_3 - g\rho_0 \nabla \eta_3(0)$$
$$+ \operatorname{div}(\varepsilon_0(Dv(0) + Dv(0)^T - \frac{2}{3}(\operatorname{div}(v(0))I)) + \delta_0(\operatorname{div}(v(0)))I). \tag{3.141}$$

Our first result gives an energy and its evolution equation for solutions to the second-order problem.

Lemma 3.15 Let v solve (3.138) and the corresponding jump and boundary conditions. Then in the nonperiodic case,

$$\partial_t \int_\Omega \rho_0 \frac{|\partial_t v|^2}{2} + \frac{p'(\rho_0)\rho_0}{2}\left|\operatorname{div} v - \frac{g}{p'(\rho_0)}v_3\right|^2$$
$$+ \int_\Omega \frac{\varepsilon_0}{2}|D\partial_t v + D\partial_t v^T - \frac{2}{3}(\operatorname{div}\partial_t v)I|^2$$
$$+ \int_\Omega \delta_0|\operatorname{div}\partial_t v|^2$$
$$= \partial_t \int_\Omega \frac{g[[\rho_0]]}{2}|v_3|^2 - \frac{\sigma}{2}|\nabla_{x_1,x_2}v_3|^2. \tag{3.142}$$

In the periodic case, the same equation holds with the integral over R^2 replaced with an integral over $(2\pi L\mathbb{T})^2$.

Proof We will prove the result in the nonperiodic case. The periodic case follows similarly. Take the dot product of (3.138) with v_t and integrate over Ω_t. After integrating by parts, we get

$$\int_{\Omega_+} \rho_0 \partial_t v \cdot \partial_{tt} v + p'(\rho_0)\rho_0(\operatorname{div} v)(\operatorname{div}\partial_t v)$$
$$- g\rho_0(v_3 \operatorname{div}\partial_t v + \partial_t v_3 \operatorname{div} v) + \frac{g^3 \rho_0}{p'(\rho_0)}v_3 \partial_t v_3$$
$$+ \int_{\Omega_+} \frac{\varepsilon_0}{2}|D\partial_t v + D\partial_t v^T - \frac{2}{3}(\operatorname{div}\partial_t v)I|^2 + \int_{\Omega_+} \delta_0|\operatorname{div}\partial_t v|^2$$

$$= \int_{\mathbf{R}^2} g\rho_0^+ v_3 \partial_t v_3 - \int_{\mathbf{R}^2} p'_+(\rho_0^+)\rho_0^+ \mathrm{div}v \partial_t v_3 - \int T e_3 \cdot \partial_t v, \qquad (3.143)$$

where we have written

$$T = (p'(\rho_0)\rho_0 \mathrm{div} v)I + \varepsilon_0 \left(D\partial_t v + D\partial_t v^T - \frac{2}{3}(\mathrm{div}\partial_t v)I \right) + \delta_0 \mathrm{div}\partial_t v I. \qquad (3.144)$$

We may pull time derivatives out of the first integrals on each side of the equation to arrive at the equality

$$\frac{\partial}{\partial t} \int_{\Omega_+} \rho_0 \frac{|\partial_t v|^2}{2} + \frac{p'(\rho_0)\rho_0}{2} \left| \mathrm{div} v - \frac{g}{p'(\rho_0)} v_3 \right|^2$$

$$+ \int_{\Omega_+} \frac{\varepsilon_0}{2} \left| D\partial_t v + D\partial_t v^T - \frac{2}{3}(\mathrm{div}\partial_t v)I \right|^2$$

$$+ \int_{\Omega_+} \delta_0 |\mathrm{div}\partial_t v|^2$$

$$= \frac{\partial}{\partial t} \int_{\mathbf{R}^2} g\rho_0^+ \frac{|v_3|^2}{2} - \int_{\mathbf{R}^2} T e_3 \cdot \partial_t v. \qquad (3.145)$$

A similar result holds on $\Omega_- = \mathbf{R}^2 \times (-m, 0)$ with the opposite sign on the right-hand side. Adding the two together yields

$$\frac{\partial}{\partial t} \int_\Omega \rho_0 \frac{|\partial_t v|^2}{2} + \frac{p'(\rho_0)\rho_0}{2} \left| \mathrm{div} v - \frac{g}{p'(\rho_0)} v_3 \right|^2$$

$$+ \int_\Omega \frac{\varepsilon_0}{2} \left| D\partial_t v + D\partial_t v^T - \frac{2}{3}(\mathrm{div}\partial_t v)I \right|^2$$

$$+ \int_\Omega \delta_0 |\mathrm{div}\partial_t v|^2$$

$$= \partial_t \int_{\mathbf{R}^2} \frac{g[[\rho_0]]}{2} |v_3|^2 - \int_{\mathbf{R}^2} [[T e_3 \cdot \partial_t v]].$$

Using the jump conditions, we find that

$$-\int_{\mathbf{R}^2} [[T e_3 \cdot \partial_t v]] = \int_{\mathbf{R}^2} \sigma \triangle_{x_1,x_2} v_3 \partial_t v_3$$

$$= -\sigma \int_{\mathbf{R}^2} \nabla_{x_1,x_2} v_3 \nabla_{x_1,x_2} \partial_t v_3 = -\partial_t \int_{\mathbf{R}^2} \frac{\sigma}{2} |\nabla_{x_1,x_2} v_3|^2.$$

The result follows by plugging this in above.

The next result allows us to estimate the energy in terms of Λ.

Lemma 3.16 Let $v \in H^1(\Omega)$ be so that $v(x_1, x_2, -m) = v(x_1, x_2, l) = 0$. In the nonperiodic case we have the inequality

$$\int_{\mathbf{R}^2} \frac{g[[\rho_0]]}{2} |v_3|^2 - \frac{\sigma}{2} |\nabla_{x_1,x_2} v_3|^2 - \int_\Omega \frac{p'(\rho_0)\rho_0}{2} \left| \mathrm{div} v - \frac{g}{p'(\rho_0)} v_3 \right|^2$$

$$\leqslant \frac{\Lambda^2}{2}\int_\Omega \rho_0|v|^2 + \frac{\Lambda^2}{2}\int_\Omega \frac{\varepsilon_0}{2}\left|Dv + Dv^T - \frac{2}{3}(\operatorname{div} v)I\right|^2 + \delta_0|\operatorname{div} v|^2. \quad (3.146)$$

In the periodic case, if $\sqrt{\sigma/g[[\rho_0]]} < L$, then the same inequality holds with the $\mathbb{R}^2$ integral replaced with an integral over $(2\pi L\mathbb{T})^2$, and Λ replaced with Λ_L.

Proof We will again prove only the nonperiodic version. Take the horizontal Fourier transform and apply (3.131) to see that

$$4\pi^2 \int_{\mathbb{R}^2} \frac{g[[\rho_0]]}{2}|v_3|^2 - \frac{\sigma}{2}|\nabla_{x_1,x_2}v_3|^2 - 4\pi^2 \int_\Omega \frac{p'(\rho_0)\rho_0}{2}\left|\operatorname{div} v - \frac{g}{p'(\rho_0)}v_3\right|^2$$

$$= \int_{\mathbb{R}^2} \frac{g[[\rho_0]] - \sigma|\xi|^2}{2}|\hat{v}_3|^2$$

$$\quad - \int_\Omega \frac{p'(\rho_0)}{2}\rho_0 \left|i\xi_1\hat{v}_1 + i\xi_2\hat{v}_2 + \partial_3\hat{v}_3 - \frac{g}{p'(\rho_0)}\hat{v}_3\right|^2 d\xi dx_3$$

$$= \int_{\mathbb{R}^2} \Big(\frac{g[[\rho_0]] - \sigma|\xi|^2}{2}|\hat{v}_3|^2$$

$$\quad - \int_{-m}^l \frac{p'(\rho_0)}{2}\rho_0\left|i\xi_1\hat{v}_1 + i\xi_2\hat{v}_2 + \partial_3\hat{v}_3 - \frac{g}{p'(\rho_0)}\hat{v}_3\right|^2 dx_3\Big)d\xi. \quad (3.147)$$

Consider the last integrand for fixed $\xi \neq 0$, writing $\varphi(x_3) = i\hat{v}_1(\xi, x_3), Q(x_3) = i\hat{v}_2(\xi, x_3), \psi(x_3) = \hat{v}_3(\xi, x_3)$. That is, define

$$Z(\varphi, \theta, \psi; \xi)$$
$$= \frac{g[[\rho_0]] - \sigma|\xi|^2}{2}|\psi|^2 - \int_{-m}^l \frac{p'(\rho_0)\rho_0}{2}\left|\xi_1\varphi + \xi_2\theta - \frac{g}{p'(\rho_0)}\psi\right|^2 dx_3. \quad (3.148)$$

By splitting

$$Z(\varphi, \theta, \psi; \xi) = Z(\Re\varphi, \Re\theta, \Re\psi; \Re\xi) + Z(\Im\varphi, \Im\theta, \Im\psi; \Im\xi), \quad (3.149)$$

we may reduce to bounding Z when φ, θ, ψ are real-valued functions and then apply the bound to the real and imaginary parts of φ, θ, ψ.

The expression for Z is invariant under simultaneous rotations of ξ and (φ, θ), so without loss of generality we may assume that $\xi = (|\xi|, 0), |\xi| > 0$ and $\theta = 0$. If $\sigma > 0$, then we assume for now that $|\xi| < |\xi|_c$ as well. Then, using (3.81) with $\tilde{\epsilon} = \lambda(|\xi|)\epsilon_0, \tilde{\delta} = \lambda(|\xi|)\delta_0$, we may rewrite Z as

$$Z(\varphi, \theta, \psi; \xi) = -E(\varphi, \psi; \lambda(|\xi|)) + \frac{\lambda(|\xi|)}{2}\int_{-m}^l \delta_0|\psi'| + |\xi|\varphi|^2$$

$$+\frac{\lambda(|\xi|)}{2}\int_{-m}^{l}\epsilon_0\left(|\varphi'-|\xi|\varphi|^2+|\psi'-|\xi|\varphi|^2+\frac{1}{3}|\psi'+|\xi|\varphi|^2\right) \tag{3.150}$$

and hence

$$Z(\varphi,\theta,\psi;\xi)\leqslant\frac{\Lambda^2}{2}\int_{-m}^{l}\rho_0(|\varphi|^2+|\psi|^2)+\frac{\Lambda}{2}\int_{m}^{l}\delta|\psi'+|\xi|\varphi|^2$$
$$+\frac{\Lambda}{2}\int_{-m}^{l}\epsilon_0(|\varphi'-|\xi|\psi|^2+|\psi'-|\xi|\varphi|^2+\frac{1}{3}|\psi'+|\xi|\varphi|^2). \tag{3.151}$$

For $|\xi|\geqslant\xi_c$, the expression for Z is nonpositive, so the previous inequality holds trivially, and so we deduce that it holds for all $|\xi|>0$.

Translating the inequality back to the original notation for fixed ξ, we find that

$$\frac{g[[\rho_0]]-\sigma|\xi|^2}{2}|\hat{v}_3|^2$$
$$-\int_{-m}^{l}\frac{p'(\rho_0)\rho_0}{2}\left|i\xi_1\hat{v}_1+i\xi_2\hat{v}_2+\partial_3\hat{v}_3-\frac{g}{p'(\rho_0)}\hat{v}_3\right|^2 dx_3$$
$$\leqslant\frac{\Lambda^2}{2}\int_{-m}^{l}\rho|\hat{v}|^2+\frac{\Lambda}{2}\int_{-m}^{l}|i\xi_1\hat{v}_1+i\xi_2\hat{v}_2+\partial_3\hat{v}_3|^2+\frac{\epsilon_0}{2}|\hat{B}|^2, \tag{3.152}$$

where

$$B=Dv+Dv^{\mathrm{T}}-\frac{2}{3}(\mathrm{div}V)I. \tag{3.153}$$

Integrating each side of this inequality over all $\xi\in R^2$, and using (3.81) then proves the result.

When $\sigma>0$ and L is sufficiently small, a better result is available in the periodic case.

Lemma 3.17 Let $v\in H^1(\Omega)$ be so that $v(x_1,x_2,-m)=v(x_1,x_2,l)=0$ and suppose in the periodic case that L satisfies (3.58). Then

$$\int_{(2\pi(T))^2}\frac{g[[\rho_0]]}{2}|v_3|^2$$
$$-\frac{\sigma}{2}|\nabla_{x_1x_2}v_3|^2-\int_{\Omega}\frac{p'(\rho_0)\rho_0}{2}\left|\mathrm{div}v-\frac{g}{p'(\rho_0)}v_3\right|^2\leqslant 0. \tag{3.154}$$

Proof Apply the horizontal Fourier transform to see that

$$4\pi^2 L^2\int_{(2\pi(T))^2}\frac{g[[\rho_0]]}{2}|v_3|^2-\frac{\sigma}{2}|\nabla_{x_1x_2}v_3|^2$$

$$-4\pi^2 L^2 \int_\Omega \frac{p'(\rho_0)\rho_0}{2}\left|\operatorname{div} v - \frac{g}{p'(\rho_0)}v_3\right|^2$$
$$= \sum_{\xi\in(L^{-1}Z)^2} \frac{g[[\rho_0]] - \sigma|\xi|^2}{2}|\hat v_3|^2$$
$$- \sum_{\xi\in(L^{-1}Z)^2}\int_{-m}^{l}\frac{p'(\rho_0)\rho_0}{2}\left|i\xi_1\hat v_1 + i\xi_2\hat v_2 + \partial_3\hat v_3 - \frac{g}{p'(\rho_0)}\hat v_3\right|^2 dx_3. \tag{3.155}$$

Because of (3.58), the only $\xi \in (L^{-1}Z)^2$, for which $g[[\rho_0]] - g|\xi|^2 \geq 0$, is $\xi = 0$. Since all but the $\xi = 0$ term on the right side of the last equation are nonpositive, we reduce to showing that

$$\frac{g[[\rho_0]]}{2}|\hat v_3|^2 - \int_{-m}^{l}\frac{p'(\rho_0)\rho_0}{2}\left|\partial_3\hat v_3 - \frac{g}{p'(\xi)}\hat v_3\right|^2 dx_3 \leq 0. \tag{3.156}$$

For this, we expand the term in the integral and integrate it by parts to get

$$\frac{g[[\rho_0]]}{2}|\hat v_3|^2 - \int_{-m}^{l}\frac{p'(\rho_0)\rho_0}{2}\left|\partial_3\hat v_3 - \frac{g}{p'(\xi)}\hat v_3\right|^2 dx_3$$
$$= -\frac{1}{2}\int_{-m}^{l} p'(\rho_0)\rho_0|\partial_3\hat v_3|^2, \tag{3.157}$$

which yields the desired inequality.

3.7 Proof of Theorem 3.5 and 3.6

Proof of Theorem 3.5 Again, we will only prove the nonperiodic case. Integrate the result of Lemma 4.1 in time from 0 to t to find that

$$\int_\Omega \rho_0\frac{|\partial_t v(t)|^2}{2} + \int_0^t\int_\Omega \frac{\varepsilon_0}{2}|D\partial_t v(s) + D\partial_t v(s)^{\mathrm T} - \frac{2}{3}(\operatorname{div}\partial_t v(s))|^2$$
$$+ \delta_0|\operatorname{div}\partial_t v(s)|ds$$
$$\leq K_0 + \int_{R^2}\frac{g[[\rho_0]]}{2}|v_3(t)|^2 - \frac{\sigma}{2}|\nabla_{x_1 x_2}v_3(t)|^2$$
$$- \int_\Omega \frac{p'(\rho_0)\rho_0}{2}\left|\operatorname{div} v(t) - \frac{g}{p'(\rho_0)}v_3(t)\right|^2, \tag{3.158}$$

where

$$K_0 = \int_\Omega \rho_0\frac{|\partial_t v(0)|^2}{2} + \int_\Omega \frac{p'(\rho_0)\rho_0}{2}\left|\operatorname{div} v(0) - \frac{g}{p'(\rho_0)}v_3(0)\right|^2 + \frac{\sigma}{2}|\nabla_{x_1 x_2}v_3(0)|^2. \tag{3.159}$$

We may then apply Lemma 4.2 to get the inequality

$$\int_\Omega \rho_0\frac{|\partial_t v(t)|^2}{2} + \int_0^t\int_\Omega \frac{\varepsilon}{2}(Dv_t(s) + D\partial_t v(s)^{\mathrm T} - \frac{2}{3}(\operatorname{div}\partial_t v(s))I)$$

$$+\delta_0|\mathrm{div}\partial_t v(s)|^2 ds$$
$$\leqslant K_0 + \frac{\Lambda^2}{2}\int_\Omega \rho_0|v(t)|^2$$
$$+\frac{\Lambda}{2}\int_\Omega \frac{\varepsilon_0}{2}(D\partial_t v(s)+D\partial_t v(s)^{\mathrm{T}}-\frac{2}{3}(\mathrm{div}\partial_t v(s))I)$$
$$+\delta_0|\mathrm{div}\partial_t v(t)|^2. \tag{3.160}$$

Using the definitions of the norms $\|\cdot\|_1, \|\cdot\|_2$, we may compactly rewrite the previous inequality as

$$\frac{1}{2}\|\partial_t v(t)\|_1^2 + \int_0^t \|\partial_t v(s)\|_2^2 ds \leqslant K_0 + \frac{\Lambda^2}{2}\|v(t)\|_1^2 + \frac{\Lambda}{2}\|v(t)\|_2^2. \tag{3.161}$$

Integrating in time and using Cauchy's inequality, we may bound

$$\Lambda\|v(t)\|_2^2 = \Lambda\|v(0)\|_2^2 + \Lambda\int_0^t 2(v(s),\partial_t v(s))ds$$
$$\leqslant \Lambda\|v(0)\|_2^2 + \int_0^t \|\partial_t v(s)\|_2^2 ds + \Lambda^2\int_0^t \|v(s)\|_1^2 ds. \tag{3.162}$$

On the other hand

$$\Lambda\partial_t\|v(s)\|_1^2 = 2\Lambda(\partial_t v(t),v(t))_1 \leqslant \Lambda^2\|v(s)\|_1^2 + \|\partial_t v(s)\|_1^2, \tag{3.163}$$

we may combine these two inequalities with (3.161) to derive the differential inequality

$$\partial_t\|v(t)\|_1^2 + \|v(t)\|_2^2 \leqslant K_1 + 2\Lambda\|v(t)\|_1^2 + 2\Lambda\int_0^t \|v(s)\|_2^2 ds \tag{3.164}$$

for $K_1 = 2K_0/\Lambda + 2\|v(0)\|_2^2$. An application of Gronwall's theorem then shows that

$$\|v(t)\|_1^2 + \int_0^t \|v(s)\|_2^2 ds \leqslant e^{2\Lambda t}\|v(0)\|_1^2 + \frac{K_1}{2\Lambda}\left(e^{2\Lambda t}-1\right), \tag{3.165}$$

To derive the corresponding bound for $\|v(t)\|_2^2, \|\partial_t v(t)\|_1^2$, we return to (3.161), and plug in (3.162) and (3.165) to see that

$$\frac{1}{\Lambda}\|\partial_t v(t)\|_1^2 + \|v(t)\|_2^2 \leqslant K_1 + \Lambda\|v(t)\|_1^2 + 2\Lambda\int_0^t \|v(s)\|_2^2 ds$$
$$\leqslant e^{2\Lambda t}(2\Lambda\|v(0)\|_1^2 + K_1), \tag{3.166}$$
$$K_0 \leqslant C(\|\partial_t v(0)\|_1^2 + \|v(0)\|_1^2 + \|v(0)\|_2^2) + \sigma\int_{\mathbf{R}^2}|\nabla_{x_1 x_2}v_3(0)|^2, \tag{3.167}$$

for a constant $C > 0$ depending on $\rho_0^\pm, p_\pm, \Lambda, \varepsilon_\pm, \delta_\pm, \sigma, g, m, l$.

Proof of Theorem 3.6 We again integrate the result of Lemma 4.1 in time from 0 to t to find that

$$\int_\Omega \rho_0 \frac{|\partial_t v(t)|^2}{2} + \int_0^t \int_\Omega \frac{\varepsilon_0}{2} |Dv_t(s) + D\partial_t v(s)^T - \frac{2}{3} \text{div} \partial_t v(s) I|^2$$
$$+ \delta_0 |\text{div} \partial_t v(s)|^2 ds$$
$$\leqslant K_1 + \int_{(2\pi LT)^2} \frac{g[[\rho_0]]}{2} |v_3(t)|^2 - \frac{\sigma}{2} |\nabla_{x_1 x_2} v_3(t)|^2$$
$$- \int_\Omega \frac{p'(\rho_0)\rho_0}{2} |\text{div} v(t) - \frac{g}{p'(\rho_0)} v_3(t)|^2. \quad (3.168)$$

We may apply Lemma 3.17 to see that the sum of all of the integrals on the right side of the previous inequality is nonpositive, and hence

$$\frac{1}{2} \|\partial_t v(t)\|_1^2 + \int_0^t \|\partial_t v(s)\|_2^2 ds \leqslant K_1. \quad (3.169)$$

From this we deduce that

$$\|v(t)\|_1 + \|v(t)\|_2 \leqslant \|v(0)\|_1 + \|v(0)\|_2 + 3\sqrt{t}\sqrt{K_1}. \quad (3.170)$$

Then, using that $\partial_t \eta = v$, we get

$$\|\eta(t)\|_1 + \|\eta(t)\|_2 \leqslant \|\eta(0)\|_1 + \|\eta(0)\|_2 + t(\|v(0)\|_1 + \|v(0)\|_2) + 2t^{\frac{3}{2}}\sqrt{K_1}. \quad (3.171)$$

To derive the estimates for $\partial_t^j v (j \geqslant 2)$ we apply ∂^j to (3.138). Then $w = \partial^j v$ satisfies the same equation and boundary conditions as v, which allows us to argue as above to derive the inequality

$$\frac{1}{2} \|\partial_t^j v(t)\|_1^2 + \int_0^t \|\partial_t^j v(s)\|_2^2 ds \leqslant K_3, \quad \forall j \geqslant 1.$$

This trivially implies (3.79). To get (3.80), we bound

$$\|\partial_t^j v(t)\|_2^2 \leqslant \|\partial_t^j v(0)\|_2^2 + 2\int_0^t \|\partial_t^j v(s)\|_2 \|\partial_t^{j+1} v(s)\|_2 ds$$
$$\leqslant \|\partial_t^j v(0)\|_2^2 + 2\left(\int_0^t \|\partial_t^j v(s)\|_2^2 ds\right)^{\frac{1}{2}} \left(\int_0^t \|\partial_t^{j+1} v(s)\|_2^2\right)^{\frac{1}{2}}$$
$$\leqslant \|\partial_t^j v(0)\|_2^2 + 2\sqrt{K_j}\sqrt{K_{j+1}}. \quad (3.172)$$

References

[1] Yan. Guo, Ian. Tice, Linear Rayleigh-Taylor instability for viscous, compressible fluids, SIAM J. MATH. ANAL. Vol. 42, No. 4, pp. 1688-1720.

[2] Yan. Guo, Ian. Tice, Compressible, inviscid Rayleigh-Taylor instability. Indiana Univ. Math. J. 60 (2011), no. 2, 677-711.

[3] Yanjin. Wang, Ian. Tice, The viscous surface-internal wave problem: nonlinear Rayleigh-Taylor instability. Comm. Partial Differential Equations 37 (2012), no. 11, 1967-2028.

[4] Fei. Jiang, Song. Jiang, Yanjin. Wang, On the Rayleigh-Taylor instability for the incompressible viscous magnetohydrodynamic equations. Comm. Partial Differential Equations 39 (2014), no. 3, 399-438.

Stochastic 2D Primitive Equations: Central Limit Theorem and Moderate Deviation Principle *

Zhang Rangrang (张让让), Zhou Guoli (周国立),
and Guo Boling (郭柏灵)

Abstract In this paper, we establish a central limit theorem and a moderate deviation for two-dimensional stochastic primitive equations driven by multiplicative noise. This is the first result about the limit theorem and the moderate deviations for stochastic primitive equations. The proof of the results relies on the weak convergence method and some delicate and careful *a priori* estimates.

Keywords primitive equations; central limit theorem; moderate deviations

1 Introduction

As a fundamental model in meteorology, the primitive equations were derived from the Navier-Stokes equations, with rotation, coupled with thermodynamics and salinity diffusion-transport equations, by assuming two important simplifications: the Boussinesq approximation and the hydrostatic balance (see [21, 22, 27] and the references therein). This model, in the deterministic case, has been intensively investigated because of the interests stemmed from physics and mathematics. For example, the mathematical study of the primitive equations started in a series of articles by Lions, Temam, and Wang in the early 1990s (see [21-24]), where they set up the mathematical framework and showed the global existence of weak solutions. Cao and Titi [4] developed a beautiful approach to dealing with the L^6-norm of the fluctuation of horizontal velocity

---

* Computers and Math. with Appl., 2019, 77, 928-946.

and obtained the global well-posedness of the 3D viscous primitive equations.

Due to the existence of some uncertainties, it is natural and reasonable to consider the primitive equations in random cases (for reasons, see [14, 18, 25, 28] and the references therein). In the past two decades, there have been numerous works about the stochastic primitive equations both in two-dimensional (2D) case and in three-dimensional (3D) case. Glatt-Holtz and Ziane [16] studied the 2D primitive equations with multiplicative noises. Using an approximation method, the authors proved the existence and uniqueness of strong solutions to 2D stochastic primitive equations. In [15], Gao and Sun proved the Freidlin-Wentzell's large deviation principles (LDP) holds for 2D stochastic primitive equations. Debussche et al. [5] established the global well-posedness of the strong solution of 3D primitive equations driven by multiplicative random noises. The ergodic theory of 3D stochastic primitive equations driven by regular multiplicative noise was studied in [8], where the authors established that all weak solutions, which are limitations of spectral Galerkin approximations share the same invariant measure. Using a new method, Zhou [30] proved the existence of random attractors for 3D stochastic primitive equations driven by fractional Brownian motions. Dong et al. [9] proved the LDP holds for the strong solution of 3D stochastic primitive equations.

The purpose of this paper is to investigate deviations from the strong solution Y^ε of 2D stochastic primitive equations (see equation (3.7)) from the solution Y^0 of the deterministic equation (see (3.8)), as ε decreases to 0. That is, we seek the asymptotic behavior of the trajectory,

$$Z^\varepsilon(t) = \frac{1}{\sqrt{\varepsilon}\lambda(\varepsilon)}(Y^\varepsilon - Y^0)(t), \quad t \in [0, T],$$

where $\lambda(\varepsilon)$ is some deviation scale which strongly influences the asymptotic behavior of Z^ε. Concretely, three cases are involved:

(1) The case $\lambda(\varepsilon) = \frac{1}{\sqrt{\varepsilon}}$ provides LDP, which has been proved by [15].

(2) The case $\lambda(\varepsilon) = 1$ provides the central limit theorem (CLT). We will show that Z^ε converges to a solution of a stochastic equation, as ε decreases to 0 in Section 4.

(3) To fill in the gap between the CLT's scale ($\lambda(\varepsilon) = 1$) and the large deviations' scale $\left(\lambda(\varepsilon) = \frac{1}{\sqrt{\varepsilon}}\right)$, we will study the so-called moderate deviation

principle (MDP) in Section 5. Here, the deviations' scale satisfies

$$\lambda(\varepsilon) \to +\infty, \quad \sqrt{\varepsilon}\lambda(\varepsilon) \to 0, \quad \text{as} \quad \varepsilon \to 0. \tag{1.1}$$

Similar to LDP, MDP arises in the theory of statistical inference naturally, which can provide us with the rate of convergence and a useful method for constructing asymptotic confidence intervals (see, e.g. [11, 19, 20] and references therein). The proof of moderate deviations is mainly based on the weak convergence approach, which was developed by Dupuis and Ellis in [10]. The key idea is to prove some variational representation formula about the Laplace transform of bounded continuous functionals, which will lead to proving an equivalence between the Laplace principle and LDP. In particular, for Brownian functionals, an elegant variational representation formula has been established by Boué, Dupuis [2] and Budhiraja, Dupuis [3].

Up to now, there are some results about the moderate deviations for fluid dynamics models and other processes. For example, Wang et al. [26] established the CLT and MDP for 2D Navier-Stokes equations driven by multiplicative Gaussian noise in $C([0,T];H) \cap L^2([0,T];V)$. Further, Dong et al. [7] considered the MDP for 2D Navier-Stokes equations driven by multiplicative Lévy noises in $D([0,T];H) \cap L^2([0,T];V)$. In view of the characterization of the super-Brownian motion (SBM) and the Fleming-Viot process (FVP), Fatheddin and Xiong [12] obtained MDP for those processes.

In this paper, as stated above, we consider two kinds of asymptotic behaviors of Z^ε: the CLT and MDP in $C([0,T];H) \cap L^2([0,T];V)$, which provide the exponential decay of small probabilities associated with the corresponding stochastic dynamical systems with small noise. We divide the proof into two parts. For the CLT, we show that Z^ε converges to a solution of a stochastic equation, as ε decrease to 0. For the MDP, it can be changed to prove that Z^ε satisfies a large deviation principle in $C([0,T];H) \cap L^2([0,T];V)$ with $\lambda(\varepsilon)$ satisfying (1.1). The proof will be based on the weak convergence approach introduced by Boué and Dupuis [2], Budhiraja and Dupuis [3]. Moreover, some delicate and careful *a priori* estimates and tightness arguments are involved in the proof.

This paper is organized as follows. The mathematical framework of 2D primitive equations are in Section 2. The formulation of 2D primitive equations is presented in Section 3. The central limit theorem is proved in Section 4. The moderate deviation principle is established in Section 5.

2 Preliminaries

The 2D primitive equations can be formally derived from the full three dimensional system under the assumption of invariance with respect to the second horizontal variable y as in [16]. The 2D primitive equations driven by a stochastic forcing in a Cartesian system can be written as

$$\frac{\partial v}{\partial t} - \mu_1 \Delta v + v\partial_x v + \theta\partial_z v + \partial_x P = \psi_1(t, v, T)\frac{dW_1}{dt}, \tag{2.1}$$

$$\partial_z P + T = 0, \tag{2.2}$$

$$\partial_x v + \partial_z \theta = 0, \tag{2.3}$$

$$\frac{\partial T}{\partial t} - \mu_2 \Delta T + v\partial_x T + \theta\partial_z T = \psi_2(t, v, T)\frac{dW_2}{dt}, \tag{2.4}$$

where the velocity $v = v(t, x, z) \in \mathbf{R}$, the vertical velocity θ, the temperature T and the pressure P are all unknown functionals. $(x, z) \in \mathcal{M} = (0, L) \times (-h, 0)$, $t > 0$, W_1 and W_2 are two independent cylindrical Wiener processes, which will be given in Section 3. $\Delta = \partial_x^2 + \partial_z^2$ is the Laplacian operator. Without loss of generality, we assume that

$$\mu_1 = \mu_2 = 1.$$

We impose the following boundary conditions on (2.1)-(2.4):

$$\partial_z v = 0, \ \theta = 0, \ \partial_z T = 0, \quad \text{on } \Gamma_u := (0, L) \times \{0\}, \tag{2.5}$$

$$\partial_z v = 0, \ \theta = 0, \ \partial_z T = 0, \quad \text{on } \Gamma_b := (0, L) \times \{-h\}, \tag{2.6}$$

$$v = 0, \ \partial_x T = 0, \quad \text{on } \Gamma_l := \{0, L\} \times (-h, 0). \tag{2.7}$$

Integrating (2.3) from $-h$ to z and using (2.5)-(2.6), we have

$$\theta(t, x, z) := \Phi(v)(t, x, z) = -\int_{-h}^{z} \partial_x v(t, x, z')dz'. \tag{2.8}$$

Moreover, in view of (2.3) and (2.5)-(2.7), we deduce that

$$\int_{-h}^{0} v\,dz = 0.$$

Integrating (2.2) from $-h$ to z, set p_b be a certain unknown function at Γ_b satisfying

$$P(x, z, t) = p_b(x, t) - \int_{-h}^{z} T(x, z', t)dz'.$$

Then, (2.2)-(2.4) can be rewritten as

$$\frac{\partial v}{\partial t} - \Delta v + v\partial_x v + \Phi(v)\partial_z v + \partial_x p_b - \int_{-h}^{z} \partial_x T dz' = \psi_1(t,v,T)\frac{dW_1}{dt}, \quad (2.9)$$

$$\frac{\partial T}{\partial t} - \Delta T + v\partial_x T + \Phi(v)\partial_z T = \psi_2(t,v,T)\frac{dW_2}{dt}, \quad (2.10)$$

$$\int_{-h}^{0} v dz = 0. \quad (2.11)$$

The boundary value conditions for (2.9)-(2.11) are given by

$$\partial_z v = 0, \ \partial_z T = 0, \quad \text{on } \Gamma_u, \quad (2.12)$$

$$\partial_z v = 0, \ \partial_z T = 0, \quad \text{on } \Gamma_b, \quad (2.13)$$

$$v = 0, \ \partial_x T = 0, \quad \text{on } \Gamma_l. \quad (2.14)$$

Denote $Y = (v, T)$ and the initial condition

$$Y(0) = Y_0 := (v_0, T_0). \quad (2.15)$$

3 Formulation of primitive equations

Let $\mathcal{L}(K_1; K_2)$ (resp. $\mathcal{L}_2(K_1; K_2)$) be the space of bounded (resp. Hilbert-Schmidt) linear operators from the Hilbert space K_1 to K_2, whose norm is denoted by $\|\cdot\|_{\mathcal{L}(K_1;K_2)}$ ($\|\cdot\|_{\mathcal{L}_2(K_1;K_2)}$). For $p \in \mathbb{Z}^+$, set

$$|\phi|_p = \begin{cases} \left(\int_{\mathcal{M}} |\phi(x,z)|^p dxdz\right)^{\frac{1}{p}}, & \phi \in L^p(\mathcal{M}), \\ \left(\int_0^L |\phi(x)|^p dx\right)^{\frac{1}{p}}, & \phi \in L^p((0,L)). \end{cases}$$

In particular, $|\cdot|$ and $(\cdot,\cdot)$ represent norm and inner product of $L^2(\mathcal{M})$ (or $L^2((0,L))$), respectively. For $m \in \mathbb{N}^+$, $(W^{m,p}(\mathcal{M}), \|\cdot\|_{m,p})$ stands for the classical Sobolev space, see [1]. When $p = 2$, denote by $H^m(\mathcal{M}) = W^{m,2}(\mathcal{M})$,

$$\begin{cases} H^m(\mathcal{M}) = \{Y \big| \partial_\alpha Y \in (L^2(\mathcal{M}))^2 \text{ for } |\alpha| \leq m\}, \\ |Y|^2_{H^m(\mathcal{M})} = \sum_{0 \leq |\alpha| \leq m} |\partial_\alpha Y|^2. \end{cases}$$

It is well-known that $(H^m(\mathcal{M}), |\cdot|_{H^m(\mathcal{M})})$ is a Hilbert space. Let $|\cdot|_{H^p((0,L))}$ stand for the norm of $H^p((0,L))$ for $p \in \mathbb{Z}^+$.

Define working spaces for (2.9)-(2.15):

$$\mathcal{V} := \{(v,T) \in (C^\infty(\mathcal{M}))^2 : (v,T) \text{ satisfies boundary conditions (2.12)-(2.14)}\}.$$

Let H and V be the closure of $\mathcal{V}$ under the topology of $L^2(\mathcal{M})$ and $H^1(\mathcal{M})$ respectively. We equip H with an inner product

$$(Y,\tilde{Y}) := (v,\tilde{v}) + (T,\tilde{T}) = \int_\mathcal{M} (v\tilde{v} + T\tilde{T})dxdz, \quad |Y| := (Y,Y)^{\frac{1}{2}}.$$

Taking into account the boundary conditions (2.12)-(2.14), the inner product and norm on V can be given by

$$((Y,\tilde{Y})) = ((v,\tilde{v}))_1 + ((T,\tilde{T}))_2,$$
$$((v,\tilde{v}))_1 = \int_\mathcal{M} (\partial_x v \partial_x \tilde{v} + \partial_z v \partial_z \tilde{v})dxdz,$$
$$((T,\tilde{T}))_2 = \int_\mathcal{M} (\partial_x T \partial_x \tilde{T} + \partial_z T \partial_z \tilde{T})dxdz,$$

and take $\|\cdot\| = \sqrt{((\cdot,\cdot))}$, where $Y = (v,T), \tilde{Y} = (\tilde{v},\tilde{T}) \in V$. Note that under the above definition, a Poincaré inequality $|Y| \leqslant C\|Y\|$ holds for all $Y \in V$. Let $\check{V}$ be the closure of $V \cap (C^\infty(\mathcal{M}))^2$ in $(H^2(\mathcal{M}))^2$ and equip this space with the norm and inner product of $H^2(\mathcal{M})$.

Define an intermediate space

$$\mathcal{H} = \{Y = (v,T) \in H, \; \partial_z Y = (\partial_z v, \partial_z T) \in H\}.$$

Let V' be the dual space of V. Then, the dense and continuous embeddings

$$V \hookrightarrow H = H' \hookrightarrow V',$$

hold and denote by $\langle x,y \rangle$ the duality between $x \in V$ and $y \in V'$.

3.1 Some functionals

Define P_H is the Leray-type projection operator from $(L^2(\mathcal{M}))^2$ onto H. The principle linear portion of the equation is defined by

$$AY := P_H \begin{pmatrix} -\Delta v \\ -\Delta T \end{pmatrix}, \quad \text{for } Y = (v,T) \in D(A),$$

where

$D(A)$ is the closure of $\mathcal{V}$ with respect to the topology of $H^2(\mathcal{M})$.

It's well-known that A is a self-adjoint and positive definite operator. Due to the regularity results of the Stokes problem of geophysical fluid dynamics, we have $|AY|$ is equivalent to $|Y|_{H^2(\mathcal{O})}$ (see [31]).

For $Y = (v, T)$, $\tilde{Y} = (\tilde{v}, \tilde{T}) \in D(A)$, define $B(Y, \tilde{Y}) = B_1(v, \tilde{Y}) + B_2(v, \tilde{Y})$, where

$$B_1(v, \tilde{Y}) := P_H \begin{pmatrix} v\partial_x \tilde{v} \\ v\partial_x \tilde{T} \end{pmatrix}, \quad B_2(v, \tilde{Y}) := P_H \begin{pmatrix} \Phi(v)\partial_z \tilde{v} \\ \Phi(v)\partial_z \tilde{T} \end{pmatrix}.$$

By interpolation inequalities (see [15, 16], etc.), we have

Lemma 3.1 *For any $Y = (v, T)$, $\tilde{Y}, \hat{Y} \in V$, there exists a constant C such that*

$$\langle v\partial_x v + \Phi(v)\partial_z v, \partial_{zz} v \rangle = 0, \tag{3.1}$$

$$\langle B(Y, \tilde{Y}), \hat{Y} \rangle = -\langle B(Y, \hat{Y}), \tilde{Y} \rangle, \quad \langle B(Y, \tilde{Y}), \tilde{Y} \rangle = 0, \tag{3.2}$$

$$|\langle B(Y, \tilde{Y}), \hat{Y} \rangle| \leq C\|\tilde{Y}\|\|Y\|^{\frac{1}{2}}\|Y\|^{\frac{1}{2}}|\hat{Y}|^{\frac{1}{2}}\|\hat{Y}\|^{\frac{1}{2}} + C|\partial_z \tilde{Y}|\|Y\|\|\hat{Y}|^{\frac{1}{2}}\|\hat{Y}\|^{\frac{1}{2}}. \tag{3.3}$$

For the pressure term in (2.9), define

$$G(Y) := P_H \begin{pmatrix} -\int_{-h}^{z} \partial_x T dz' \\ 0 \end{pmatrix}, \quad \text{for } Y = (v, T) \in V.$$

Using the above functionals, we obtain

$$\begin{cases} dY(t) + AY(t)dt + B(Y(t), Y(t))dt + G(Y(t))dt = \psi(t, Y(t))dW(t), \\ Y(0) = Y_0, \end{cases} \tag{3.4}$$

where

$$W = \begin{pmatrix} W_1 \\ W_2 \end{pmatrix}, \quad \psi(t, Y(t)) = \begin{pmatrix} \psi_1(t, Y(t)) & 0 \\ 0 & \psi_2(t, Y(t)) \end{pmatrix}.$$

3.2 Hypotheses

For the strong solution of (3.4), we shall fix a single stochastic basis $\mathcal{T} := (\Omega, \mathcal{F}, \{\mathcal{F}_t\}_{t \geq 0}, P, W)$ with the expectation $\mathbb{E}$. Here,

$$W = \begin{pmatrix} W_1 \\ W_2 \end{pmatrix}$$

is a cylindrical Brownian motion with the form $W(t,\omega) = \sum_{i\geqslant 1} r_i w_i(t,\omega)$, where $\{r_i\}_{i\geqslant 1}$ is a complete orthonormal basis of a Hilbert space $U = U_1 \times U_2$, U_1 and U_2 are separable Hilbert spaces, $\{w_i\}_{i\geqslant 1}$ is a sequence of independent one-dimensional standard Brownian motions on $(\Omega, \mathcal{F}, \{\mathcal{F}_t\}_{t\geqslant 0}, P)$.

In order to obtain the global well-posedness and moderate deviations of equation (3.4), we introduce the following Hypotheses:

Hypothesis A $\psi : [0,T] \times H \to \mathcal{L}_2(U;H)$ satisfies that there exists a constant K such that

$$\|\psi(t,\phi)\|^2_{\mathcal{L}_2(U;H)} \leqslant K(1+|\phi|^2), \quad \forall t \in [0,T], \phi \in H;$$

$$\|\psi(t,\phi_1) - \psi(t,\phi_2)\|^2_{\mathcal{L}_2(U;H)} \leqslant K|\phi_1-\phi_2|^2, \quad \forall t \in [0,T], \phi_1, \phi_2 \in H.$$

Hypothesis B For the same constant K, we suppose that

$$\|\partial_z \psi(t,\phi)\|^2_{\mathcal{L}_2(U;H)} \leqslant K(1+|\partial_z \phi|^2), \quad \forall t \in [0,T], \partial_z \phi \in H.$$

Now, we state the definition of a solution to (3.4) introduced by [15, 16].

Definition 3.1 Let $\mathcal{T} = (\Omega, \mathcal{F}, \{\mathcal{F}_t\}_{t\geqslant 0}, P, W)$ be a fixed stochastic basis and the initial condition $Y_0 \in \mathcal{H}$. An $\mathcal{F}_t$-predictable stochastic process $Y(t,\omega)$ is called a strong solution of (3.4) on $[0,T]$ with the initial value $Y_0 \in \mathcal{H}$ if for P-a.s. $\omega \in \Omega$,

$$Y(\cdot) \in C([0,T]; H) \bigcap L^2([0,T]; V), \quad \forall T > 0,$$

and for every $t \in [0,T]$,

$$(Y(t), \phi) - (Y_0, \phi) + \int_0^t \left[\langle Y(s), A\phi \rangle + \langle B(Y,Y), \phi \rangle + (G(Y), \phi) \right] ds$$
$$= \int_0^t (\psi(s, Y(s))dW(s), \phi), \quad P\text{-}a.s.,$$

for all $\phi \in D(A)$.

Remark 3.1 The solution $Y(t, \cdot)$ is a strong solution in the probabilistic sense.

To study the long-time behavior of the system (3.4), some norm estimates are necessary. We recall the following result, which was obtained by [15, 16].

Theorem 3.1 Let the initial value $Y_0 \in \mathcal{H}$. Assume Hypothesis A and Hypothesis B hold, then there exists a unique global solution Y of (3.4). Furthermore, there exists a constant $C(K,T)$ such that

$$\mathbb{E}\left(\sup_{0\leqslant s\leqslant T} |Y(s)|^p + \int_0^T |Y(s)|^{p-2} \|Y(s)\|^2 ds \right) \leqslant C(K,T), \tag{3.5}$$

and

$$\mathbb{E}\left(\sup_{0\leqslant s\leqslant T}|\partial_z Y(s)|^4 + \int_0^T |\partial_z Y(s)|^2 \|\partial_z Y(s)\|^2 ds\right) \leqslant C(K,T). \tag{3.6}$$

In this paper, we consider the following equations with small multiplicative noise

$$\begin{cases} dY^\varepsilon(t) + AY^\varepsilon(t)dt + B(Y^\varepsilon(t), Y^\varepsilon(t))dt + G(Y^\varepsilon(t))dt = \sqrt{\varepsilon}\psi(t, Y^\varepsilon(t))dW(t), \\ Y^\varepsilon(0) = Y_0 \in \mathcal{H}. \end{cases} \tag{3.7}$$

As the parameter ε tends to 0, the solution Y^ε of (3.7) will tend to the solution of the following SPDE,

$$\begin{cases} dY^0(t) + AY^0(t)dt + B(Y^0(t), Y^0(t))dt + G(Y^0(t))dt = 0, \\ Y^0(0) = Y_0 \in \mathcal{H}. \end{cases} \tag{3.8}$$

As stated in the introduction, we will investigate the asymptotic behavior of the trajectory,

$$Z^\varepsilon(t) = \frac{1}{\sqrt{\varepsilon}\lambda(\varepsilon)}(Y^\varepsilon - Y^0)(t), \quad t \in [0,T], \tag{3.9}$$

where $\lambda(\varepsilon)$ is equal to 1 or satisfies (1.1). We shall divide their proof into two sections: Section 4 and Section 5.

4 Central limit theorem

In this part, we devote to establishing the central limit theorem, i.e., taking $\lambda(\varepsilon) = 1$ in (3.9). Let Y^ε be the unique solution of (3.7) in $L^2(\Omega; C([0,T];H)) \cap L^2(\Omega; L^2([0,T];V))$ and Y^0 be the unique solution of (3.8). The following norm estimates were obtained by [15, 16].

Lemma 4.1 *Let $Y_0 \in \mathcal{H}$. Assume Hypotheses A-B are in force, then there exists $0 < \varepsilon_0^1 < 1$ such that*

$$\sup_{\varepsilon\in[0,\varepsilon_0^1]} \mathbb{E}\left(\sup_{0\leqslant s\leqslant T}|Y^\varepsilon(s)|^p + \int_0^T |Y^\varepsilon(s)|^{p-2}\|Y^\varepsilon(s)\|^2 ds\right) \leqslant C(K,T), \tag{4.1}$$

and

$$\sup_{\varepsilon\in[0,\varepsilon_0^1]} \mathbb{E}\left(\sup_{0\leqslant s\leqslant T}|\partial_z Y^\varepsilon(s)|^4 + \int_0^T |\partial_z Y^\varepsilon(s)|^2\|\partial_z Y^\varepsilon(s)\|^2 ds\right) \leqslant C(K,T). \tag{4.2}$$

In particular, it holds that

$$\sup_{0\leqslant s\leqslant T}|Y^0(s)|^p + \int_0^T |Y^0(s)|^{p-2}\|Y^0(s)\|^2 ds \leqslant C(K,T), \qquad (4.3)$$

and

$$\sup_{0\leqslant s\leqslant T}|\partial_z Y^0(s)|^4 + \int_0^T |\partial_z Y^0(s)|^2 \|\partial_z Y^0(s)\|^2 ds \leqslant C(K,T), \qquad (4.4)$$

where $C(K,T)$ is independent of ε_0^1.

Now, we explore the convergence of Y^ε as $\varepsilon \to 0$.

Proposition 4.1 Let $Y_0 \in \mathcal{H}$. Assume Hypotheses A-B hold, then for any $0 \leqslant \varepsilon \leqslant \varepsilon_0^1$,

$$\mathbb{E}\left(\sup_{0\leqslant t\leqslant T}|Y^\varepsilon(t) - Y^0(t)|^2 + \int_0^T \|Y^\varepsilon(s) - Y^0(s)\|^2 ds\right) \leqslant \varepsilon C(K,T), \qquad (4.5)$$

where ε_0^1 is determined by Lemma 4.1.

Proof Let $0 \leqslant \varepsilon \leqslant \varepsilon_0^1$ and $N > 1$. Define $\tau_N = \inf\left\{t : |Y^\varepsilon(t)|^2 + \int_0^t \|Y^\varepsilon(s)\|^2 ds > N\right\}$. Set $X^\varepsilon = Y^\varepsilon - Y^0$, we deduce from (3.7) and (3.8) that

$$\begin{cases} dX^\varepsilon(t) + AX^\varepsilon(t)dt + \Big(B(Y^\varepsilon(t), Y^\varepsilon(t)) - B(Y^0, Y^0)\Big)dt + G(X^\varepsilon(t))dt \\ \qquad = \sqrt{\varepsilon}\psi(t, Y^\varepsilon(t))dW(t), \\ X^\varepsilon(0) = 0. \end{cases} \qquad (4.6)$$

Applying Itô formula to $|X^\varepsilon|^2$ and by (3.2), we have

$$|X^\varepsilon(t \wedge \tau_N)|^2 + 2\int_0^{t\wedge\tau_N} \|X^\varepsilon(s)\|^2 ds$$

$$= -2\int_0^{t\wedge\tau_N} \langle B(Y^\varepsilon, X^\varepsilon) + B(X^\varepsilon, Y^0), X^\varepsilon\rangle ds - 2\int_0^{t\wedge\tau_N} (G(X^\varepsilon), X^\varepsilon) ds$$

$$+ 2\sqrt{\varepsilon}\int_0^{t\wedge\tau_N} (\psi(s, Y^\varepsilon)dW(s), X^\varepsilon) + \varepsilon\int_0^{t\wedge\tau_N} \|\psi(s, Y^\varepsilon)\|_{\mathcal{L}_2(U;H)}^2 ds$$

$$\leqslant 2\int_0^{t\wedge\tau_N} |\langle B(X^\varepsilon, Y^0), X^\varepsilon\rangle| ds + 2\int_0^{t\wedge\tau_N} |(G(X^\varepsilon), X^\varepsilon)| ds$$

$$+ 2\sqrt{\varepsilon}|\int_0^{t\wedge\tau_N} (\psi(s, Y^\varepsilon)dW(s), X^\varepsilon)| + \varepsilon\int_0^{t\wedge\tau_N} \|\psi(s, Y^\varepsilon)\|_{\mathcal{L}_2(U;H)}^2 ds$$

$$:= I_1^\varepsilon(t) + I_2^\varepsilon(t) + I_3^\varepsilon(t) + I_4^\varepsilon(t). \qquad (4.7)$$

Taking the supremum up to time $t \wedge \tau_N$ in (4.7), and then taking the expectation, we have

$$\mathbb{E}\left(\sup_{0 \leqslant s \leqslant t \wedge \tau_N} \left(|X^\varepsilon(s)|^2 + 2\int_0^s \|X^\varepsilon(l)\|^2 dl\right)\right)$$

$$\leqslant \mathbb{E}\left(I_1^\varepsilon(t) + I_2^\varepsilon(t) + \sup_{s \in [0, t \wedge \tau_N]} I_3^\varepsilon(s) + I_4^\varepsilon(t)\right).$$

Using (3.3) and the Young inequality, we deduce that

$$\mathbb{E} I_1^\varepsilon(t) \leqslant C \mathbb{E} \int_0^{t \wedge \tau_N} (\|Y^0\| \|X^\varepsilon\| \|X^\varepsilon\| + |\partial_z Y^0| \|X^\varepsilon\| |X^\varepsilon|^{\frac{1}{2}} \|X^\varepsilon\|^{\frac{1}{2}}) ds$$

$$\leqslant \frac{1}{4} \mathbb{E} \int_0^{t \wedge \tau_N} \|X^\varepsilon\|^2 ds + C \mathbb{E} \int_0^{t \wedge \tau_N} (\|Y^0\|^2 + |\partial_z Y^0|^4) |X^\varepsilon|^2 ds.$$

Utilizing the Cauchy-Schwarz inequality and the Young inequality, it follows that

$$\mathbb{E} I_2^\varepsilon(t) \leqslant C \mathbb{E} \int_0^{t \wedge \tau_N} |X^\varepsilon| \|X^\varepsilon\| ds$$

$$\leqslant \frac{1}{4} \mathbb{E} \int_0^{t \wedge \tau_N} \|X^\varepsilon\|^2 ds + C \mathbb{E} \int_0^{t \wedge \tau_N} |X^\varepsilon|^2 ds.$$

Moreover, by Hypothesis A, we deduce that

$$\mathbb{E} I_4^\varepsilon(t) \leqslant \varepsilon K \mathbb{E} \int_0^{t \wedge \tau_N} (1 + |Y^\varepsilon|^2) ds.$$

Applying the Burkholder-Davis-Gundy inequality and Hypothesis A, it yields

$$\mathbb{E} \sup_{s \in [0, t \wedge \tau_N]} I_3^\varepsilon(s) \leqslant 2\sqrt{\varepsilon} \mathbb{E} \left(\int_0^{t \wedge \tau_N} |X^\varepsilon|^2 \|\psi(s, Y^\varepsilon)\|_{\mathcal{L}_2(U;H)}^2 ds\right)^{\frac{1}{2}}$$

$$\leqslant 4\sqrt{\varepsilon K} \mathbb{E} \left(\int_0^{t \wedge \tau_N} |X^\varepsilon|^2 (1 + |Y^\varepsilon|^2) ds\right)^{\frac{1}{2}}$$

$$\leqslant \frac{1}{2} \mathbb{E} \sup_{0 \leqslant s \leqslant t \wedge \tau_N} |X^\varepsilon|^2 + 8\varepsilon K \mathbb{E} \int_0^{t \wedge \tau_N} (1 + |Y^\varepsilon|^2) ds.$$

Combing the above estimates and using (4.1) in Lemma 4.1, we get

$$\mathbb{E}\left(\sup_{0 \leqslant s \leqslant t \wedge \tau_N} \left(|X^\varepsilon(s)|^2 + \int_0^s \|X^\varepsilon(l)\|^2 dl\right)\right)$$

$$\leqslant C \mathbb{E} \int_0^{t \wedge \tau_N} (\|Y^0\|^2 + |\partial_z Y^0|^4 + 1) |X^\varepsilon|^2 ds + C\varepsilon K \mathbb{E} \int_0^{t \wedge \tau_N} (1 + |Y^\varepsilon|^2) ds$$

$$\leqslant C \mathbb{E} \int_0^{t \wedge \tau_N} (\|Y^0\|^2 + |\partial_z Y^0|^4 + 1) |X^\varepsilon|^2 ds + \varepsilon C(K, T). \tag{4.8}$$

By Gronwall inequality and taking into account (4.3) and (4.4), we have

$$\mathbb{E}\left(\sup_{0\leqslant s\leqslant t\wedge\tau_N}\left(|X^\varepsilon(s)|^2 + \int_0^s \|X^\varepsilon(l)\|^2 dl\right)\right)$$
$$\leqslant \varepsilon C(K,T)\cdot\exp\left\{\int_0^{t\wedge\tau_N}(\|Y^0\|^2 + |\partial_z Y^0|^4 + 1)ds\right\}$$
$$\leqslant \varepsilon C(K,T). \tag{4.9}$$

Letting $N\to\infty$, we further obtain

$$\mathbb{E}\left(\sup_{0\leqslant t\leqslant T}|X^\varepsilon(t)|^2 + \int_0^T \|X^\varepsilon(t)\|^2 dt\right) \leqslant \varepsilon C(K,T).$$

$\square$

Let $\tilde{V}^0 = (\tilde{v}^0, \tilde{T}^0)$ be the solution of the following SPDE:

$$d\tilde{V}^0(t) + A\tilde{V}^0(t)dt + \Big(B(Y^0, \tilde{V}^0) + B(\tilde{V}^0, Y^0)\Big)dt + G(\tilde{V}^0(t))dt$$
$$= \psi(t, Y^0(t))dW(t) \tag{4.10}$$

with $\tilde{V}^0(0) = 0$. Assume Hypothesis A and Hypothesis B hold, by the same arguments as the proof of Lemma 4.1, there exists a unique strong solution to (4.10). Meanwhile, we can also obtain

$$\mathbb{E}\left(\sup_{0\leqslant t\leqslant T}|\tilde{V}^0|^p + \int_0^T |\tilde{V}^0|^{p-2}\|\tilde{V}^0(s)\|^2 ds\right) \leqslant C(K,T), \tag{4.11}$$

and

$$\mathbb{E}\left(\sup_{0\leqslant t\leqslant T}|\partial_z \tilde{V}^0|^4 + \int_0^T |\partial_z \tilde{V}^0|^2 \|\partial_z \tilde{V}^0(s)\|^2 ds\right) \leqslant C(K,T). \tag{4.12}$$

Define

$$\tilde{V}^\varepsilon(t) := \frac{Y^\varepsilon - Y^0}{\sqrt{\varepsilon}} = \frac{X^\varepsilon}{\sqrt{\varepsilon}} = (\tilde{v}^\varepsilon(t), \tilde{T}^\varepsilon(t)),$$

then $\tilde{V}^\varepsilon$ is the solution of the following SPDE,

$$d\tilde{V}^\varepsilon(t) + A\tilde{V}^\varepsilon(t)dt + \Big(B(\tilde{V}^\varepsilon(t), Y^\varepsilon) + B(Y^0, \tilde{V}^\varepsilon(t))\Big)dt + G(\tilde{V}^\varepsilon(t))dt$$
$$= \psi(t, Y^\varepsilon)dW(t),$$

with $\tilde{V}^\varepsilon(0) = 0$. We claim that

Lemma 4.2 Let $Y_0 \in \mathcal{H}$. Assume Hypotheses A-B hold, then

$$\sup_{0<\varepsilon\leqslant\varepsilon_0^1} \mathbb{E} \int_0^T |\tilde{V}^\varepsilon|^2 \|\tilde{V}^\varepsilon\|^2 ds \leqslant C(K,T),$$

where ε_0^1 is the same as the one appeared in Lemma 4.1 or Proposition 4.1.

Proof Applying Itô formula to $|\tilde{V}^\varepsilon|^2$ and by (3.2), we have

$$d|\tilde{V}^\varepsilon(t)|^2 = -2\|\tilde{V}^\varepsilon\|^2 dt - 2\langle B(\tilde{V}^\varepsilon(t), Y^\varepsilon) + B(Y^0, \tilde{V}^\varepsilon(t)), \tilde{V}^\varepsilon \rangle dt$$
$$- 2\left(G(\tilde{V}^\varepsilon(t)), \tilde{V}^\varepsilon\right) dt + 2(\psi(t, Y^\varepsilon) dW(t), \tilde{V}^\varepsilon) + \|\psi(t, Y^\varepsilon)\|^2_{\mathcal{L}_2(U;H)} dt,$$

then

$$\langle |\tilde{V}^\varepsilon(\cdot)|^2 \rangle_t = 4 \int_0^t \|\psi(s, Y^\varepsilon)\|^2_{\mathcal{L}_2(U;H)} |\tilde{V}^\varepsilon|^2 ds.$$

Applying Itô formula to $|\tilde{V}^\varepsilon|^4$, we deduce that

$$d|\tilde{V}^\varepsilon(t)|^4 = 2|\tilde{V}^\varepsilon|^2 d|\tilde{V}^\varepsilon|^2 + d\langle |\tilde{V}^\varepsilon(\cdot)|^2 \rangle_t$$
$$= 2|\tilde{V}^\varepsilon|^2 \Big[-2\|\tilde{V}^\varepsilon\|^2 dt - 2\langle B(\tilde{V}^\varepsilon(t), Y^\varepsilon), \tilde{V}^\varepsilon \rangle dt - 2\left(G(\tilde{V}^\varepsilon(t)), \tilde{V}^\varepsilon\right) dt$$
$$+ 2(\psi(t, Y^\varepsilon) dW(t), \tilde{V}^\varepsilon) + \|\psi(t, Y^\varepsilon)\|^2_{\mathcal{L}_2(U;H)} dt \Big]$$
$$+ 4\|\psi(t, Y^\varepsilon)\|^2_{\mathcal{L}_2(U;H)} |\tilde{V}^\varepsilon(t)|^2 dt.$$

Define $\tau_N := \inf\left\{t : |\tilde{V}^\varepsilon(t)|^4 + \int_0^t \|\tilde{V}^\varepsilon\|^2 ds > N\right\}$, then

$$|\tilde{V}^\varepsilon(t \wedge \tau_N)|^4 + 4\int_0^{t\wedge\tau_N} |\tilde{V}^\varepsilon|^2 \|\tilde{V}^\varepsilon\|^2 ds$$
$$\leqslant 4\int_0^{t\wedge\tau_N} |\tilde{V}^\varepsilon|^2 |\langle B(\tilde{V}^\varepsilon, Y^\varepsilon), \tilde{V}^\varepsilon \rangle| ds + 4\int_0^{t\wedge\tau_N} |\tilde{V}^\varepsilon|^2 \left|\left(G(\tilde{V}^\varepsilon), \tilde{V}^\varepsilon\right)\right| ds$$
$$+ 6\int_0^{t\wedge\tau_N} |\tilde{V}^\varepsilon|^2 \|\psi(s, Y^\varepsilon)\|^2_{\mathcal{L}_2(U;H)} ds + 4\left|\int_0^{t\wedge\tau_N} |\tilde{V}^\varepsilon|^2 (\psi(s, Y^\varepsilon) dW(s), \tilde{V}^\varepsilon)\right|$$
$$:= K_1^\varepsilon(t) + K_2^\varepsilon(t) + K_3^\varepsilon(t) + K_4^\varepsilon(t).$$

Taking the expectation, we obtain

$$\mathbb{E}\left(\sup_{0\leqslant s\leqslant t\wedge\tau_N} \left(|\tilde{V}^\varepsilon(s)|^4 + 4\int_0^{s\wedge\tau_N} |\tilde{V}^\varepsilon|^2 \|\tilde{V}^\varepsilon\|^2 dl\right)\right)$$
$$\leqslant \mathbb{E}K_1^\varepsilon(t) + \mathbb{E}K_2^\varepsilon(t) + \mathbb{E}K_3^\varepsilon(t) + \mathbb{E}\sup_{0\leqslant s\leqslant t\wedge\tau_N} |K_4^\varepsilon(s)|. \qquad (4.13)$$

Notice that

$$Y^\varepsilon = Y^0 + \sqrt{\varepsilon}\tilde{V}^\varepsilon, \qquad (4.14)$$

then by (3.2), we get

$$\mathbb{E}K_1^\varepsilon(t) = 4\mathbb{E}\int_0^{t\wedge\tau_N} |\tilde{V}^\varepsilon|^2 |\langle B(\tilde{V}^\varepsilon, Y^0 + \sqrt{\varepsilon}\tilde{V}^\varepsilon), \tilde{V}^\varepsilon\rangle|ds$$
$$= 4\mathbb{E}\int_0^{t\wedge\tau_N} |\tilde{V}^\varepsilon|^2 |\langle B(\tilde{V}^\varepsilon, Y^0), \tilde{V}^\varepsilon\rangle|ds.$$

By (3.3) and the Young inequality, we deduce that

$$\mathbb{E}K_1^\varepsilon(t) \leqslant 4\mathbb{E}\int_0^{t\wedge\tau_N} |\tilde{V}^\varepsilon|^2 (\|\tilde{V}^\varepsilon\|\|\tilde{V}^\varepsilon\|\|Y^0\| + |\partial_z Y^0|\|\tilde{V}^\varepsilon\|^{\frac{3}{2}}|\tilde{V}^\varepsilon|^{\frac{1}{2}})ds$$
$$\leqslant 2\mathbb{E}\int_0^{t\wedge\tau_N} |\tilde{V}^\varepsilon|^2 \|\tilde{V}^\varepsilon\|^2 ds + C\mathbb{E}\int_0^{t\wedge\tau_N} (\|Y^0\|^2 + |\partial_z Y^0|^4)|\tilde{V}^\varepsilon|^4 ds.$$

Using the Cauchy-Schwarz inequality and the Young inequality, we obtain

$$\mathbb{E}K_2^\varepsilon(t) \leqslant 4\mathbb{E}\int_0^{t\wedge\tau_N} |\tilde{V}^\varepsilon|^3 \|\tilde{V}^\varepsilon\| ds$$
$$\leqslant \frac{1}{4}\mathbb{E}\int_0^{t\wedge\tau_N} |\tilde{V}^\varepsilon|^2 \|\tilde{V}^\varepsilon\|^2 ds + C\mathbb{E}\int_0^{t\wedge\tau_N} |\tilde{V}^\varepsilon|^4 ds.$$

By Hypothesis A, (4.1), (4.3) and (4.14), for any $0 < \varepsilon \leqslant \varepsilon_0^1$, we deduce that

$$\mathbb{E}K_3^\varepsilon(t) \leqslant 6K\mathbb{E}\int_0^{t\wedge\tau_N} (1+|Y^\varepsilon|^2)|\tilde{V}^\varepsilon|^2 ds$$
$$\leqslant 6K\mathbb{E}\int_0^{t\wedge\tau_N} |\tilde{V}^\varepsilon|^2 ds + 12K\mathbb{E}\int_0^{t\wedge\tau_N} (|Y^0|^2 + \varepsilon|\tilde{V}^\varepsilon|^2)|\tilde{V}^\varepsilon|^2 ds$$
$$\leqslant 12K\varepsilon\mathbb{E}\int_0^{t\wedge\tau_N} |\tilde{V}^\varepsilon|^4 ds + C(K,T). \tag{4.15}$$

Applying the Burkholder-Davis-Gundy inequality, the Young inequality, (4.1), (4.3) and (4.14), for any $0 < \varepsilon \leqslant \varepsilon_0^1$, it follows that

$$\mathbb{E}\sup_{0\leqslant s\leqslant t\wedge\tau_N}|K_4^\varepsilon(s)| \leqslant C\mathbb{E}\left(\int_0^{t\wedge\tau_N} |\tilde{V}^\varepsilon|^4 \|\psi(s,Y^\varepsilon)\|_{\mathcal{L}_2(U;H)}^2 |\tilde{V}^\varepsilon|^2 ds\right)^{\frac{1}{2}}$$
$$\leqslant C\mathbb{E}\left[\sup_{0\leqslant s\leqslant t\wedge\tau_N}|\tilde{V}^\varepsilon|^2 \cdot \left(\int_0^{t\wedge\tau_N} \|\psi(s,Y^\varepsilon)\|_{\mathcal{L}_2(U;V)}^2 |\tilde{V}^\varepsilon|^2 ds\right)^{\frac{1}{2}}\right]$$
$$\leqslant \frac{1}{4}\mathbb{E}\sup_{0\leqslant s\leqslant t\wedge\tau_N}|\tilde{V}^\varepsilon(s)|^4 + CK\mathbb{E}\int_0^{t\wedge\tau_N}(1+|Y^\varepsilon|^2)|\tilde{V}^\varepsilon|^2 ds$$
$$\leqslant \frac{1}{4}\mathbb{E}\sup_{0\leqslant s\leqslant t\wedge\tau_N}|\tilde{V}^\varepsilon(s)|^4 + CK\mathbb{E}\int_0^{t\wedge\tau_N}(1+|Y^0|^2+\varepsilon|\tilde{V}^\varepsilon|^2)|\tilde{V}^\varepsilon|^2 ds$$
$$\leqslant \frac{1}{4}\mathbb{E}\sup_{0\leqslant s\leqslant t\wedge\tau_N}|\tilde{V}^\varepsilon(s)|^4 + \varepsilon CK\mathbb{E}\int_0^{t\wedge\tau_N}|\tilde{V}^\varepsilon|^4 ds + C(K,T). \tag{4.16}$$

As a result of (4.13)-(4.16), we have

$$\mathbb{E}\left[\sup_{0\leqslant s\leqslant t\wedge\tau_N}\left(\frac{1}{2}|\tilde{V}^\varepsilon(s)|^4 + \int_0^{s\wedge\tau_N}|\tilde{V}^\varepsilon|^2\|\tilde{V}^\varepsilon\|^2 dl\right)\right]$$
$$\leqslant C\mathbb{E}\int_0^{t\wedge\tau_N}(\|Y^0\|^2 + |\partial_z Y^0|^4 + \varepsilon CK + 1)|\tilde{V}^\varepsilon|^4 ds + C(K,T).$$

By Gronwall inequality, (4.3) and (4.4), we have

$$\mathbb{E}\left[\sup_{0\leqslant s\leqslant t\wedge\tau_N}\left(|\tilde{V}^\varepsilon(s)|^4 + 2\int_0^{s\wedge\tau_N}|\tilde{V}^\varepsilon|^2\|\tilde{V}^\varepsilon\|^2 dl\right)\right]$$
$$\leqslant C(K,T)\cdot\exp\left\{\int_0^T(\|Y^0\|^2 + |\partial_z Y^0|^4 + \varepsilon CK + 1)ds\right\}$$
$$\leqslant C(K,T)(1+\varepsilon).$$

Letting $N \to \infty$, we get

$$\sup_{0<\varepsilon\leqslant\varepsilon_0^1}\mathbb{E}\left[\sup_{0\leqslant s\leqslant T}|\tilde{V}^\varepsilon(s)|^4 + 2\int_0^T|\tilde{V}^\varepsilon|^2\|\tilde{V}^\varepsilon\|^2 ds\right] \leqslant C(K,T),$$

which implies the result. □

Now, we are able to prove the following central limit theorem.

Theorem 4.2 (central limit theorem) *Let $Y_0 \in \mathcal{H}$. Assume Hypotheses A-B hold, then $\dfrac{Y^\varepsilon - Y^0}{\sqrt{\varepsilon}}$ converges to $\tilde{V}^0$ in the space $L^2(\Omega; C([0,T];H)) \cap L^2(\Omega; L^2([0,T];V))$, that is*

$$\lim_{\varepsilon\to 0}\mathbb{E}\left\{\sup_{0\leqslant t\leqslant T}\left|\frac{Y^\varepsilon - Y^0}{\sqrt{\varepsilon}} - \tilde{V}^0\right|^2 + \int_0^T\left\|\frac{Y^\varepsilon - Y^0}{\sqrt{\varepsilon}} - \tilde{V}^0(s)\right\|^2 ds\right\} = 0. \quad (4.17)$$

Proof Let $0 < \varepsilon \leqslant \varepsilon_0^1$, where ε_0^1 is the constant determined by Lemma 4.1. Recall that $\tilde{V}^\varepsilon = \dfrac{Y^\varepsilon - Y^0}{\sqrt{\varepsilon}}$ satisfies

$$\begin{cases} d\tilde{V}^\varepsilon(t) + A\tilde{V}^\varepsilon(t)dt + \Big(B(\tilde{V}^\varepsilon(t), Y^\varepsilon) + B(Y^0, \tilde{V}^\varepsilon(t))\Big)dt + G(\tilde{V}^\varepsilon(t))dt \\ \quad = \psi(t, Y^\varepsilon)dW(t), \\ \tilde{V}^\varepsilon(0) = 0. \end{cases} \quad (4.18)$$

Define

$$\tilde{\rho}^\varepsilon(t) = \tilde{V}^\varepsilon(t) - \tilde{V}^0(t),$$

where $\tilde{V}^0$ is the solution of (4.10). Then, we have

$$d\tilde{\rho}^\varepsilon(t) + A\tilde{\rho}^\varepsilon(t)dt + \Big(B(\tilde{V}^\varepsilon(t), Y^\varepsilon) - B(\tilde{V}^0(t), Y^0)\Big)dt$$
$$+ B(Y^0, \tilde{\rho}^\varepsilon(t))dt + G(\tilde{\rho}^\varepsilon(t))dt$$
$$= (\psi(t, Y^\varepsilon) - \psi(t, Y^0))dW(t),$$

Define
$$\tau_N := \inf\left\{t : |\tilde{\rho}^\varepsilon(t)|^2 + \int_0^t \|\tilde{\rho}^\varepsilon(s)\|^2 ds > N\right\}.$$

Applying Itô formula to $|\tilde{\rho}^\varepsilon(t)|^2$ and by (3.2), we have

$$|\tilde{\rho}^\varepsilon(t \wedge \tau_N)|^2 + 2\int_0^{t \wedge \tau_N} \|\tilde{\rho}^\varepsilon(s)\|^2 ds$$
$$= -2\int_0^{t \wedge \tau_N} \langle B(\tilde{V}^\varepsilon, Y^\varepsilon - Y^0) + B(\tilde{\rho}^\varepsilon, Y^0), \tilde{\rho}^\varepsilon\rangle ds - 2\int_0^{t \wedge \tau_N} (G(\tilde{\rho}^\varepsilon), \tilde{\rho}^\varepsilon)ds$$
$$+ 2\int_0^{t \wedge \tau_N} ((\psi(s, Y^\varepsilon) - \psi(s, Y^0))dW(s), \tilde{\rho}^\varepsilon)$$
$$+ \int_0^{t \wedge \tau_N} \|\psi(s, Y^\varepsilon) - \psi(s, Y^0)\|^2_{\mathcal{L}_2(U;H)} ds$$
$$\leqslant 2\int_0^{t \wedge \tau_N} |\langle B(\tilde{V}^\varepsilon, Y^\varepsilon - Y^0), \tilde{\rho}^\varepsilon\rangle|ds + 2\int_0^{t \wedge \tau_N} |\langle B(\tilde{\rho}^\varepsilon, Y^0), \tilde{\rho}^\varepsilon\rangle|ds$$
$$+ 2\int_0^{t \wedge \tau_N} |(G(\tilde{\rho}^\varepsilon), \tilde{\rho}^\varepsilon)|ds + 2|\int_0^{t \wedge \tau_N} ((\psi(s, Y^\varepsilon) - \psi(s, Y^0))dW(s), \tilde{\rho}^\varepsilon)|$$
$$+ \int_0^{t \wedge \tau_N} \|\psi(s, Y^\varepsilon) - \psi(s, Y^0)\|^2_{\mathcal{L}_2(U;H)} ds$$
$$:= J_1^\varepsilon(t) + J_2^\varepsilon(t) + J_3^\varepsilon(t) + J_4^\varepsilon(t) + J_5^\varepsilon(t).$$

Taking the supremum up to time $t \wedge \tau_N$ in the above equation, and taking the expectation, we obtain

$$\mathbb{E}\left(\sup_{0 \leqslant s \leqslant t \wedge \tau_N}\left(|\tilde{\rho}^\varepsilon(s)|^2 + 2\int_0^s \|\tilde{\rho}^\varepsilon(l)\|^2 dl\right)\right)$$
$$\leqslant \mathbb{E}\left(J_1^\varepsilon(t) + J_2^\varepsilon(t) + J_3^\varepsilon(t) + \sup_{0 \leqslant s \leqslant t \wedge \tau_N} J_4^\varepsilon(s) + J_5^\varepsilon(t)\right).$$

By definition, we have $Y^\varepsilon - Y^0 = \sqrt{\varepsilon}\tilde{V}^\varepsilon$. With the aid of (3.2) and (3.3), using integration by parts formula and the Young inequality, we have

$$\mathbb{E}J_1^\varepsilon(t) = 2\sqrt{\varepsilon}\mathbb{E}\int_0^{t \wedge \tau_N} |\langle B(\tilde{V}^\varepsilon, \tilde{V}^\varepsilon), \tilde{\rho}^\varepsilon\rangle|ds$$
$$= 2\sqrt{\varepsilon}\mathbb{E}\int_0^{t \wedge \tau_N} |\langle B(\tilde{V}^\varepsilon, \tilde{V}^0), \tilde{V}^\varepsilon\rangle|ds$$

$$\leqslant 2\sqrt{\varepsilon}\mathbb{E}\int_0^{t\wedge\tau_N}(\|\tilde{V}^0\|\|\tilde{V}^\varepsilon\|\|\tilde{V}^\varepsilon\|+|\partial_z\tilde{V}^0|\|\tilde{V}^\varepsilon\|\|\tilde{V}^\varepsilon\|^{\frac{1}{2}}\|\tilde{V}^\varepsilon\|^{\frac{1}{2}})ds$$

$$\leqslant C\sqrt{\varepsilon}\mathbb{E}\int_0^{t\wedge\tau_N}|\tilde{V}^\varepsilon|^2\|\tilde{V}^\varepsilon\|^2 ds + C\sqrt{\varepsilon}\mathbb{E}\int_0^{t\wedge\tau_N}(\|\tilde{V}^0\|^2+\|\tilde{V}^\varepsilon\|^2+|\partial_z\tilde{V}^0|^4)ds$$

$$\leqslant C\sqrt{\varepsilon}\mathbb{E}\int_0^{t\wedge\tau_N}|\tilde{V}^\varepsilon|^2\|\tilde{V}^\varepsilon\|^2 ds + \sqrt{\varepsilon}C(K,T).$$

Using (3.3), we deduce that

$$\mathbb{E}J_2^\varepsilon(t) \leqslant 2C\mathbb{E}\int_0^{t\wedge\tau_N}(\|Y^0\|\|\tilde{\rho}^\varepsilon\|\|\tilde{\rho}^\varepsilon\|+|\partial_z Y^0|\|\tilde{\rho}^\varepsilon\|\|\tilde{\rho}^\varepsilon\|^{\frac{1}{2}}\|\tilde{\rho}^\varepsilon\|^{\frac{1}{2}})ds$$

$$\leqslant \frac{1}{4}\mathbb{E}\int_0^{t\wedge\tau_N}\|\tilde{\rho}^\varepsilon\|^2 ds + C\mathbb{E}\int_0^{t\wedge\tau_N}(\|Y^0\|^2+|\partial_z Y^0|^4)|\tilde{\rho}^\varepsilon|^2 ds.$$

By Cauchy-Schwarz inequality and the Young inequality, we get

$$\mathbb{E}J_3^\varepsilon(t) \leqslant \frac{1}{4}\mathbb{E}\int_0^{t\wedge\tau_N}\|\tilde{\rho}^\varepsilon\|^2 ds + C\mathbb{E}\int_0^{t\wedge\tau_N}|\tilde{\rho}^\varepsilon|^2 ds.$$

Applying the Burkholder-Davis-Gundy inequality and by Hypothesis A, we obtain

$$\mathbb{E}\sup_{0\leqslant s\leqslant t\wedge\tau_N}J_4^\varepsilon(s) \leqslant 4C\mathbb{E}\left(\int_0^{t\wedge\tau_N}|\tilde{\rho}^\varepsilon(s)|^2\|\psi(s,Y^\varepsilon)-\psi(s,Y^0)\|_{\mathcal{L}_2(U;H)}^2 ds\right)^{\frac{1}{2}}$$

$$\leqslant 4C\sqrt{K}\mathbb{E}(\int_0^{t\wedge\tau_N}|\tilde{\rho}^\varepsilon(s)|^2|X^\varepsilon|^2 ds)^{\frac{1}{2}}$$

$$\leqslant \frac{1}{2}\mathbb{E}\sup_{0\leqslant s\leqslant t\wedge\tau_N}|\tilde{\rho}^\varepsilon(s)|^2 + C(K)\mathbb{E}\int_0^{t\wedge\tau_N}|X^\varepsilon|^2 ds.$$

Using Hypothesis A, we have

$$\mathbb{E}J_5^\varepsilon(t) \leqslant K\mathbb{E}\int_0^{t\wedge\tau_N}|X^\varepsilon|^2 ds.$$

Based on the above estimates and utilizing (4.1), (4.3), we have

$$\mathbb{E}\left(\sup_{0\leqslant s\leqslant T\wedge\tau_N}\left(|\tilde{\rho}^\varepsilon(s)|^2+\int_0^s\|\tilde{\rho}^\varepsilon(l)\|^2 dl\right)\right)$$

$$\leqslant C\sqrt{\varepsilon}\mathbb{E}\int_0^{T\wedge\tau_N}|\tilde{V}^\varepsilon|^2\|\tilde{V}^\varepsilon\|^2 ds + C\sqrt{\varepsilon}\mathbb{E}\int_0^{T\wedge\tau_N}(1+\|Y^0\|^2+|\partial_z Y^0|^4)|\tilde{\rho}^\varepsilon|^2 ds$$

$$+\sqrt{\varepsilon}C(K,T).$$

Applying Lemma 4.2, it follows that

$$\mathbb{E}\left(\sup_{0\leqslant s\leqslant T\wedge\tau_N}\left(|\tilde{\rho}^\varepsilon(s)|^2+\int_0^s\|\tilde{\rho}^\varepsilon(l)\|^2 dl\right)\right)$$
$$\leqslant C\sqrt{\varepsilon}\mathbb{E}\int_0^{T\wedge\tau_N}(1+\|Y^0\|^2+|\partial_z Y^0|^4)|\tilde{\rho}^\varepsilon|^2 ds+\sqrt{\varepsilon}C(K,T). \tag{4.19}$$

Applying Gronwall inequality to (4.19) and using (4.3), (4.4), we deduce that

$$\mathbb{E}\left(\sup_{0\leqslant s\leqslant T\wedge\tau_N}\left(|\tilde{\rho}^\varepsilon(s)|^2+\int_0^s\|\tilde{\rho}^\varepsilon(l)\|^2 dl\right)\right)$$
$$\leqslant \varepsilon C(K,T)\cdot\exp\left\{\int_0^T(1+\|Y^0\|^2+|\partial_z Y^0|^4)ds\right\}$$
$$\leqslant \varepsilon C(K,T).$$

Letting $N\to\infty$ and $\varepsilon\to 0$, we conclude the result. □

5 Moderate deviation principle

In this part, we are concerned with the moderate deviation principle of Y^ε. As stated in the introduction, we need to prove $\dfrac{Y^\varepsilon-Y^0}{\sqrt{\varepsilon}\lambda(\varepsilon)}$ satisfies a large deviation principle on $C([0,T];H)\cap L^2([0,T];V)$ with $\lambda(\varepsilon)$ satisfying (1.1). From now on, we assume (1.1) holds.

5.1 The weak convergence approach

Let $Z^\varepsilon=\dfrac{Y^\varepsilon-Y^0}{\sqrt{\varepsilon}\lambda(\varepsilon)}$, we will use the weak convergence approach introduced by Budhiraja and Dupuis in [3] to verify Z^ε satisfies a large deviation principle. Firstly, we recall some standard definitions and results from the large deviation theory (see [6]).

Let $\{Z^\varepsilon\}$ be a family of random variables defined on a probability space $(\Omega,\mathcal{F},P)$ taking values in some Polish space $\mathcal{E}$.

Definition 5.1 (Rate function) *A function $I:\mathcal{E}\to[0,\infty]$ is called a rate function if I is lower semicontinuous. A rate function I is called a good rate function if the level set $\{x\in\mathcal{E}:I(x)\leqslant M\}$ is compact for each $M<\infty$.*

Definition 5.2 (LDP) *The sequence $\{Z^\varepsilon\}$ is said to satisfy the large deviation principle with rate function I if for each Borel subset A of $\mathcal{E}$,*

$$-\inf_{x\in A^\circ}I(x)\leqslant\liminf_{\varepsilon\to 0}\varepsilon\log P(Z^\varepsilon\in A)\leqslant\limsup_{\varepsilon\to 0}\varepsilon\log P(Z^\varepsilon\in A)\leqslant-\inf_{x\in\bar{A}}I(x).$$

Suppose $W(t)$ is a cylindrical Wiener process on a Hilbert space U defined on a probability space $(\Omega, \mathcal{F}, \{\mathcal{F}_t\}_{t\in[0,T]}, P)$ (the paths of W take values in $C([0,T]; \mathcal{U})$, where $\mathcal{U}$ is another Hilbert space such that the embedding $U \subset \mathcal{U}$ is Hilbert-Schmidt). Now we define

$$\mathcal{A} = \{\phi : \phi \text{ is a } U\text{- valued } \{\mathcal{F}_t\}$$
$$- \text{ predictable process s.t. } \int_0^T |\phi(s)|_U^2 ds < \infty \ P\text{-a.s.}\};$$
$$T_M = \{h \in L^2([0,T]; U) : \int_0^T |h(s)|_U^2 ds \leqslant M\};$$
$$\mathcal{A}_M = \{\phi \in \mathcal{A} : \phi(\omega) \in T_M, \ P\text{-a.s.}\}.$$

Here and in the sequel in this paper, we will always refer to the weak topology on the set T_M.

Suppose $\mathcal{G}^\varepsilon : C([0,T]; \mathcal{U}) \to \mathcal{E}$ is a measurable mapping and $Z^\varepsilon = \mathcal{G}^\varepsilon(W)$. Now, we list below sufficient conditions for the large deviation principle of the sequence Z^ε as $\varepsilon \to 0$.

Hypothesis H There exists a measurable map $\mathcal{G}^0 : C([0,T]; \mathcal{U}) \to \mathcal{E}$ satisfying

(i) For every $M < \infty$, let $\{h^\varepsilon : \varepsilon > 0\} \subset \mathcal{A}_M$. If h^ε converges to h as T_M-valued random elements in distribution, then $\mathcal{G}^\varepsilon(W(\cdot) + \lambda(\varepsilon) \int_0^\cdot h^\varepsilon(s) ds)$ converges in distribution to $\mathcal{G}^0(\int_0^\cdot h(s) ds)$.

(ii) For every $M < \infty$, $K_M = \{\mathcal{G}^0(\int_0^\cdot h(s) ds) : h \in T_M\}$ is a compact subset of $\mathcal{E}$.

The following result is due to Budhiraja et al. in [3].

Theorem 5.1 *If $\mathcal{G}^0$ satisfies Hypothesis H, then Z^ε satisfies a large deviation principle on $\mathcal{E}$ with the good rate function I given by*

$$I(f) = \inf_{\{h \in L^2([0,T]; U) : f = \mathcal{G}^0(\int_0^\cdot h(s) ds)\}} \left\{\frac{1}{2} \int_0^T |h(s)|_U^2 ds\right\}, \quad \forall f \in \mathcal{E}. \tag{5.1}$$

By convention, $I(\emptyset) = \infty$.

In this part, we are concerned with the following SPDE driven by small multiplicative noise

$$\begin{cases} dZ^\varepsilon(t) + AZ^\varepsilon(t) dt + B(Z^\varepsilon(t), Y^0 + \sqrt{\varepsilon}\lambda(\varepsilon) Z^\varepsilon) dt + B(Y^0, Z^\varepsilon(t)) dt \\ \qquad + G(Z^\varepsilon(t)) dt = \lambda^{-1}(\varepsilon) \psi(t, Y^0 + \sqrt{\varepsilon}\lambda(\varepsilon) Z^\varepsilon) dW(t), \quad (5.2) \\ Z^\varepsilon(0) = 0. \end{cases}$$

Under Hypotheses A-B, by Theorem 3.1, there exists a unique strong solution $Z^\varepsilon \in C([0,T];H) \cap L^2([0,T];V)$. Therefore, there exists a Borel-measurable function

$$\mathcal{G}^\varepsilon : C([0,T];\mathcal{U}) \to C([0,T];H) \cap L^2([0,T];V), \tag{5.3}$$

such that $Z^\varepsilon(\cdot) = \mathcal{G}^\varepsilon(W(\cdot))$.

Let $h \in L^2([0,T];U)$, we consider the following skeleton equation

$$\begin{cases} dR^h(t) + AR^h(t)dt + B(R^h, Y^0)dt + B(Y^0, R^h(t))dt + G(R^h(t))dt \\ \quad = \psi(t, Y^0(t))h(t)dt, \\ R^h(0) = 0. \end{cases} \tag{5.4}$$

The solution R^h, whose existence will be proved in next subsection, defines a measurable mapping $\mathcal{G}^0 : C([0,T];\mathcal{U}) \to C([0,T];H) \cap L^2([0,T];V)$ such that $\mathcal{G}^0(\int_0^\cdot h(s)ds) := R^h(\cdot)$.

The main result in this part reads as

Theorem 5.2 *Let $Y_0 \in \mathcal{H}$. Assume Hypotheses A-B hold. Then Z^ε satisfies a large deviation principle on $C([0,T];H) \cap L^2([0,T];V)$ with the good rate function I defined by (5.1) with respect to the uniform convergence.*

From now on, we always assume the initial data $Y_0 \in \mathcal{H}$ and Hypotheses A-B hold.

5.2 The skeleton equations

Theorem 5.3 *Let $h \in T_M \subset L^2([0,T];U)$. The equation (5.4) has a unique strong solution $R^h \in C([0,T];V) \cap L^2([0,T];D(A))$. Moreover, there exists a constant $C(M,K,T)$ such that*

$$\sup_{h \in T_M} \left\{ \sup_{0 \leqslant t \leqslant T} \|R^h(t)\|^2 + \int_0^T \|R^h(t)\|_{D(A)}^2 dt \right\} \leqslant C(M,K,T). \tag{5.5}$$

Proof Since the equation (5.4) is a linear equation, the proof of global well-posedness is standard, hence, we omit it here. In the following, we prove (5.5). Referring to [4], we have

$$\sup_{0 \leqslant t \leqslant T} \|Y^0(t)\|^2 + \int_0^T \|Y^0(t)\|_{D(A)}^2 d\ell \leqslant C(T). \tag{5.6}$$

Similar to the proof of the global well-posedness of the skeleton equation in [8], we deduce that (5.5) holds.

$\square$

5.2.1 Compactness of the Skeleton equations

In order to prove the compactness of R^h, as in [13], we introduce the following spaces. Let K be a separable Hilbert space. Given $p > 1, \alpha \in (0,1)$, let $W^{\alpha,p}([0,T];K)$ be the Sobolev space of all $u \in L^p([0,T];K)$ such that

$$\int_0^T \int_0^T \frac{\|u(t)-u(s)\|_K^p}{|t-s|^{1+\alpha p}} dt ds < \infty,$$

endowed with the norm

$$\|u\|_{W^{\alpha,p}([0,T];K)}^p = \int_0^T \|u(t)\|_K^p dt + \int_0^T \int_0^T \frac{\|u(t)-u(s)\|_K^p}{|t-s|^{1+\alpha p}} dt ds.$$

The following result can be found in [13].

Lemma 5.1 *Let $B_0 \subset B \subset B_1$ be Banach spaces, B_0 and B_1 reflexive, with compact embedding of B_0 in B. Let $p \in (1,\infty)$ and $\alpha \in (0,1)$ be given. Let Λ be the space*

$$\Lambda = L^p([0,T];B_0) \cap W^{\alpha,p}([0,T];B_1),$$

endowed with the natural norm. Then the embedding of Λ in $L^p([0,T];B)$ is compact.

The following is the result we obtained in this part.

Proposition 5.4 *The solution R^h of the skeleton equations (5.4) is compact in $L^2([0,T];V)$.*

Proof Define

$$F(R^h, Y^0) = AR^h + B(R^h, Y^0) + B(Y^0, R^h) + G(R^h).$$

From (5.4), we have

$$R^h(t) = -\int_0^t F(R^h(s), Y^0(s))ds + \int_0^t \psi(s, Y^0)h(s)ds$$
$$:= I_1(t) + I_2(t).$$

With the aid of the Hölder inequality and the Cauchy-Schwarz inequality, we have

$$\|F(R^h, Y^0)\|_{V'} \leqslant C\|R^h\| + C\|R^h\|\|Y^0\| + C\|R^h\|\|\partial_z Y^0\| + C\|Y^0\|\|\partial_z R^h\| + C|R^h|.$$

Applying the Hölder inequality, we deduce that

$$\|I_1(t) - I_1(s)\|_{V'}^2 = \|\int_s^t F(R^h, Y^0)dl\|_{V'}^2$$

$$\leqslant C(t-s)\int_s^t \|F(R^h,Y^0)\|_{V'}^2 dl$$
$$\leqslant C(t-s)^2\Big[\sup_{t\in[0,T]}\|R^h(t)\|^2 + \sup_{t\in[0,T]}\|R^h(t)\|^2 \sup_{t\in[0,T]}\|Y^0(t)\|^2$$
$$+ \sup_{t\in[0,T]}|\partial_z Y^0|^2 \sup_{t\in[0,T]}\|R^h(t)\|^2$$
$$+ \sup_{t\in[0,T]}|\partial_z R^h|^2 \sup_{t\in[0,T]}\|Y^0(t)\|^2 + \sup_{t\in[0,T]}|R^h(t)|^2\Big].$$

From the definition of $W^{r,2}([0,T];V')$, for $r\in(0,\frac{1}{2})$, we have

$$\|I_1\|_{W^{r,2}([0,T];V')}^2 = \int_0^T \|I_1(t)\|_{V'}^2 dt + \int_0^T\int_0^T \frac{\|I_1(t)-I_1(s)\|_{V'}^2}{|t-s|^{1+2r}}dsdt$$
$$\leqslant C(r,T)\Big[\sup_{t\in[0,T]}\|R^h(t)\|^2 + \sup_{t\in[0,T]}\|R^h(t)\|^2 \sup_{t\in[0,T]}\|Y^0(t)\|^2$$
$$+ \sup_{t\in[0,T]}|\partial_z Y^0|^2 \sup_{t\in[0,T]}\|R^h(t)\|^2$$
$$+ \sup_{t\in[0,T]}|\partial_z R^h|^2 \sup_{t\in[0,T]}\|Y^0(t)\|^2 + \sup_{t\in[0,T]}|R^h(t)|^2\Big].$$

Using (5.5) and (5.6), it follows that

$$\mathbb{E}\|I_1\|_{W^{r,2}([0,T];V')} \leqslant C_1(r,M,K,T).$$

With the aid of the Hölder inequality and Hypothesis A, for $r\in(0,\frac{1}{2})$,

$$\|I_2\|_{W^{r,2}([0,T];H)}^2 \leqslant C(r,T)\sup_{t\in[0,T]}(1+|Y^0|^2)\int_0^T |h(s)|_U^2 ds.$$

Then

$$\mathbb{E}\|I_2\|_{W^{r,2}([0,T];H)} \leqslant C_2(r,M,K,T).$$

Collecting all the previous inequalities, for $r\in\left(0,\frac{1}{2}\right)$, we obtain

$$\mathbb{E}\|R^h\|_{W^{r,2}([0,T];V')} \leqslant C_3(r,M,K,T).$$

In view of Theorem 5.3, R^h are uniformly bounded in the space

$$\Lambda := L^2([0,T];D(A))\cap W^{r,2}([0,T];V').$$

By Lemma 5.1, we conclude that R^h is compact in $L^2([0,T];V)$. □

5.2.2 Tightness of primitive equations with small perturbations

For any $h^\varepsilon \in \mathcal{A}_M$, consider the following SPDE

$$d\bar{Z}^\varepsilon(t) + A\bar{Z}^\varepsilon(t) + B(\bar{Z}^\varepsilon(t), Y^0 + \sqrt{\varepsilon}\lambda(\varepsilon)\bar{Z}^\varepsilon)dt + B(Y^0, \bar{Z}^\varepsilon(t))dt + G(\bar{Z}^\varepsilon(t))dt$$
$$= \lambda^{-1}(\varepsilon)\psi(t, Y^0 + \sqrt{\varepsilon}\lambda(\varepsilon)\bar{Z}^\varepsilon)dW(t) + \psi(t, Y^0 + \sqrt{\varepsilon}\lambda(\varepsilon)\bar{Z}^\varepsilon)h^\varepsilon(t)dt, \qquad (5.7)$$

with $\bar{Z}^\varepsilon(0) = 0$, then $\mathcal{G}^\varepsilon(W(\cdot) + \lambda(\varepsilon)\int_0^\cdot h^\varepsilon(s)ds) = \bar{Z}^\varepsilon$. In view of the properties of Z^ε and R^h, by the same method as the proof of Theorem 5.3, we obtain

Lemma 5.2 *For any family $\{h^\varepsilon; \varepsilon > 0\} \subset \mathcal{A}_M$, there exists a constant $0 < \varepsilon_0^2 < 1$ such that*

$$\sup_{0\leqslant\varepsilon\leqslant\varepsilon_0^2} \mathbb{E}\left(\sup_{0\leqslant t\leqslant T} |\bar{Z}^\varepsilon|^2 + \int_0^T \|\bar{Z}^\varepsilon\|^2 dt\right) \leqslant C(M, K, T), \qquad (5.8)$$

$$\sup_{0\leqslant\varepsilon\leqslant\varepsilon_0^2} \mathbb{E}\left(\sup_{0\leqslant t\leqslant T} |\partial_z \bar{Z}^\varepsilon|^2 + \int_0^T \|\partial_z \bar{Z}^\varepsilon\|^2 dt\right) \leqslant C(M, K, T), \qquad (5.9)$$

where the constant $C(M, K, T)$ is independent of ε_0^2.

Let $\mathcal{D}(Z)$ be the distribution of a random variable Z.

Proposition 5.5 $\mathcal{D}(\bar{Z}^\varepsilon)_{\varepsilon \in [0,\varepsilon_0^2]}$ *is tight in $L^2([0,T]; H)$.*

Proof Define

$$F(\bar{Z}^\varepsilon, Y^0) := A\bar{Z}^\varepsilon + B(\bar{Z}^\varepsilon, Y^0 + \sqrt{\varepsilon}\lambda(\varepsilon)\bar{Z}^\varepsilon) + B(Y^0, \bar{Z}^\varepsilon) + G(\bar{Z}^\varepsilon).$$

From (5.7), we have

$$\bar{Z}^\varepsilon(t) = -\int_0^t F(\bar{Z}^\varepsilon(s), Y^0(s))ds$$
$$+ \lambda^{-1}(\varepsilon)\int_0^t \psi(s, Y^0 + \sqrt{\varepsilon}\lambda(\varepsilon)\bar{Z}^\varepsilon)dW(s) + \int_0^t \psi(s, Y^0 + \sqrt{\varepsilon}\lambda(\varepsilon)\bar{Z}^\varepsilon)h^\varepsilon(s)ds$$
$$:= J_1^\varepsilon(t) + J_2^\varepsilon(t) + J_3^\varepsilon(t).$$

Let $\alpha > 1$ be fixed, we deduce from the interpolation inequality that $D(A^{\frac{\alpha}{2}}) \subset L^\infty(\mathcal{M})$. Then, with the aid of the Hölder inequality and the Cauchy-Schwarz inequality, we have

$$\|F(\bar{Z}^\varepsilon, Y^0)\|_{D(A^{-\frac{\alpha}{2}})} \leqslant C\|\bar{Z}^\varepsilon\| + C|\bar{Z}^\varepsilon|\|Y^0\| + C\|\bar{Z}^\varepsilon\||\partial_z Y^0|$$
$$+ C\sqrt{\varepsilon}\lambda(\varepsilon)(|\bar{Z}^\varepsilon| + |\partial_z \bar{Z}^\varepsilon|)\|\bar{Z}^\varepsilon\|$$
$$+ C|Y^0|\|\bar{Z}^\varepsilon\| + C|\partial_z \bar{Z}^\varepsilon|\|Y^0\| + C|\bar{Z}^\varepsilon|.$$

Using the Hölder inequality, we obtain

$$\|J_1^\varepsilon(t) - J_1^\varepsilon(s)\|^2_{D(A^{-\frac{\alpha}{2}})}$$
$$\leqslant C(t-s)\int_s^t \|F(\bar{Z}^\varepsilon, Y^0)\|^2_{D(A^{-\frac{\alpha}{2}})} dl$$
$$\leqslant C(t-s)\Big[\int_0^T \|\bar{Z}^\varepsilon(t)\|^2 dt + C \sup_{t\in[0,T]} |\bar{Z}^\varepsilon|^2 \int_0^T \|Y^0(t)\|^2 dt$$
$$+ C \sup_{t\in[0,T]} |\partial_z Y^0|^2 \int_0^T \|\bar{Z}^\varepsilon\|^2 dt + C\varepsilon\lambda^2(\varepsilon) \sup_{t\in[0,T]} (|\bar{Z}^\varepsilon|^2 + |\partial_z \bar{Z}^\varepsilon|^2) \int_0^T \|\bar{Z}^\varepsilon\|^2 dt$$
$$+ C \sup_{t\in[0,T]} |Y^0(t)|^2 \int_0^T \|\bar{Z}^\varepsilon(t)\|^2 dt$$
$$+ C \sup_{t\in[0,T]} |\partial_z \bar{Z}^\varepsilon(t)|^2 \int_0^T \|Y^0(t)\|^2 dt + C(t-s) \sup_{t\in[0,T]} |\bar{Z}^\varepsilon(t)|\Big].$$

From the definition of $W^{r,2}([0,T]; D(A^{-\frac{\alpha}{2}}))$, for $r \in \left(0, \frac{1}{2}\right)$,

$$\|J_1^\varepsilon\|^2_{W^{r,2}([0,T];D(A^{-\frac{\alpha}{2}}))}$$
$$= \int_0^T \|J_1^\varepsilon(t)\|^2_{D(A^{-\frac{\alpha}{2}})} dt + \int_0^T \int_0^T \frac{\|J_1^\varepsilon(t) - J_1^\varepsilon(s)\|^2_{D(A^{-\frac{\alpha}{2}})}}{|t-s|^{1+2r}} ds dt$$
$$\leqslant C(r,T)\Big[\int_0^T \|\bar{Z}^\varepsilon(t)\|^2 dt + C \sup_{t\in[0,T]} |\bar{Z}^\varepsilon|^2 \int_0^T \|Y^0(t)\|^2 dt$$
$$+ C \sup_{t\in[0,T]} |\partial_z Y^0|^2 \int_0^T \|\bar{Z}^\varepsilon\|^2 dt + C\varepsilon\lambda^2(\varepsilon) \sup_{t\in[0,T]} (|\bar{Z}^\varepsilon|^2 + |\partial_z \bar{Z}^\varepsilon|^2) \int_0^T \|\bar{Z}^\varepsilon\|^2 dt$$
$$+ C \sup_{t\in[0,T]} |Y^0(t)|^2 \int_0^T \|\bar{Z}^\varepsilon(t)\|^2 dt$$
$$+ C \sup_{t\in[0,T]} |\partial_z \bar{Z}^\varepsilon(t)|^2 \int_0^T \|Y^0(t)\|^2 dt + C(t-s) \sup_{t\in[0,T]} |\bar{Z}^\varepsilon(t)|\Big].$$

Using (4.3)- (4.4), we deduce from Lemma 5.2 that, for any $0 \leqslant \varepsilon \leqslant \varepsilon_0^2$,

$$\mathbb{E}\|J_1^\varepsilon\|_{W^{r,2}([0,T];D(A^{-\frac{\alpha}{2}}))} \leqslant C_1(r,M,K,T).$$

Referring to (4.3), (5.8) and by Hypothesis A, for any $0 \leqslant \varepsilon \leqslant \varepsilon_0^2$, $r \in \left(0, \frac{1}{2}\right)$ and $p \geqslant 2$, we have

$$\mathbb{E}\|J_2^\varepsilon\|_{W^{r,2}([0,T];H)} \leqslant C(r)\mathbb{E}\int_0^T \|\psi(s, Y^0 + \sqrt{\varepsilon}\lambda(\varepsilon)\bar{Z}^\varepsilon)\|^2_{\mathcal{L}_2(U;H)} ds$$

$$\leqslant C(r,K,T)\mathbb{E}\sup_{t\in[0,T]}(1+|Y^0|^2+\varepsilon\lambda^2(\varepsilon)|\bar{Z}^\varepsilon|^2)$$
$$\leqslant C_2(r,K,T).$$

By Hypothesis A, for $r \in \left(0, \frac{1}{2}\right)$,

$$\|J_3^\varepsilon\|^2_{W^{r,2}([0,T];H)} \leqslant C(r,T)\sup_{t\in[0,T]}(1+|Y^0|^2+\varepsilon\lambda^2(\varepsilon)|\bar{Z}^\varepsilon|^2)\int_0^T |h^\varepsilon(s)|^2_U ds.$$

Then, for any $0 \leqslant \varepsilon \leqslant \varepsilon_0^2$, using (4.3) and (5.8), we deduce that

$$\mathbb{E}\|J_3^\varepsilon\|_{W^{r,2}([0,T];H)} \leqslant C_3(r,M,K,T).$$

Collecting all the previous inequalities, we conclude that for $r \in \left(0, \frac{1}{2}\right)$ and $\alpha > 1$,

$$\mathbb{E}\|\bar{Z}^\varepsilon\|_{W^{r,2}([0,T];D(A^{-\frac{\alpha}{2}}))} \leqslant C_4(r,M,K,T). \tag{5.10}$$

In view of Lemma 5.2, $\bar{Z}^\varepsilon$ are bounded uniformly in ε in the space

$$\Lambda := L^2([0,T];V) \cap W^{r,2}([0,T];D(A^{-\frac{\alpha}{2}})).$$

By Lemma 5.1, we know Λ is compactly imbedded in $L^2([0,T];H)$. Denote

$$|\cdot|_\Lambda := |\cdot|_{L^2([0,T];V)} + \|\cdot\|_{W^{r,2}([0,T];D(A^{-\frac{\alpha}{2}}))},$$

then for any $L > 0$,

$$K_L = \{Y \in L^2([0,T];H), |Y|_\Lambda \leqslant L\}$$

is compact in $L^2([0,T];H)$. Finally, notice that

$$P(\bar{Z}^\varepsilon \in K_L^c) \leqslant P(|\bar{Z}^\varepsilon|_\Lambda > L) \leqslant \frac{E(|\bar{Z}^\varepsilon|_\Lambda)}{L} \leqslant \frac{C}{L},$$

choosing a constant L sufficiently large, we deduce that $\mathcal{D}(\bar{Z}^\varepsilon)_{\varepsilon\in[0,\varepsilon_0^2]}$ is tight in $L^2([0,T];H)$. $\square$

Since the imbedding $C([0,T];H) \cap W^{\alpha,2}([0,T];D(A^{-\frac{\alpha}{2}})) \subset C([0,T];D(A^{-\frac{\alpha}{2}}))$ is compact, as a consequence of (5.10), we also have

Corollary 5.6 $\mathcal{D}(\bar{Z}^\varepsilon)_{\varepsilon\in[0,\varepsilon_0^2]}$ *is tight in* $C([0,T];D(A^{-\frac{\alpha}{2}}))$, *for some* $\alpha > 1$.

5.3 The proof of MDP

According to Theorem 5.1, the proof of MDP (Theorem 5.2) will be completed if the following Theorem 5.7 and Theorem 5.8 are established.

Theorem 5.7 *The family*

$$K_M = \left\{ \mathcal{G}^0\left(\int_0^\cdot h(s)ds\right) : h \in T_M \right\}$$

is a compact subset of $C([0,T];H) \cap L^2([0,T];V)$.

Proof Let $\{R^{h^n} = \mathcal{G}^0(\int_0^\cdot h^n(s)ds); n \geqslant 1\}$ be a sequence of elements in K_M. With the aid of Theorem 5.3, we assert that there exist a subsequence still denoted by $\{n\}$ and an element $h \in T_M$ such that

(i) $h^n \to h$ in T_M weakly as $n \to \infty$.
(ii) $R^{h^n} \to R^h$ in $L^2([0,T];D(A))$ weakly.
(iii) $R^{h^n} \to R^h$ in $L^\infty([0,T];V)$ weak-star.
(iv) $R^{h^n} \to R^h$ in $L^2([0,T];V)$ strongly.

Using similar method as Section 5.3 in [9], we have $R^h = \mathcal{G}^0\left(\int_0^\cdot h(s)ds\right)$ and $R^h \in C([0,T];V) \cap L^2([0,T];D(A))$. In the following, we only need to prove that $R^{h^n} \to R^h$ strongly in $C([0,T];H) \cap L^2([0,T];V)$.

Denote $\rho^n(t) = R^{h^n} - R^h$, from (5.4), we get

$$d\rho^n(t) + A\rho^n(t)dt + B(\rho^n(t), Y^0)dt + B(Y^0, \rho^n(t))dt + G(\rho^n(t))dt$$
$$= \psi(t, Y^0)(h^n - h)dt. \tag{5.11}$$

Applying the chain rule to (5.11), it follows that

$$|\rho^n(t)|^2 + 2\int_0^t \|\rho^n(s)\|^2 ds = -2\int_0^t \langle B(\rho^n(s), Y^0(s)), \rho^n(s)\rangle ds$$
$$- 2\int_0^t \langle B(Y^0(s), \rho^n(s)), \rho^n(s)\rangle ds$$
$$- 2\int_0^t (G(\rho^n(s)), \rho^n(s))ds$$
$$+ 2\int_0^t (\psi(s, Y^0(s))(h^n - h), \rho^n(s))ds$$
$$\leqslant 2\int_0^t |\langle B(\rho^n(s), Y^0(s)), \rho^n(s)\rangle| ds$$
$$+ 2\int_0^t |(G(\rho^n(s)), \rho^n(s))| ds$$

$$+2\int_0^t |(\psi(s,Y^0(s))(h^n-h), \rho^n(s))|ds.$$

Using (3.3) and the Hölder inequality, we get

$$\sup_{t\in[0,T]} |\rho^n(t)|^2 + 2\int_0^T \|\rho^n(t)\|^2 dt$$

$$\leq C\int_0^T (\|Y^0\|\|\rho^n\|\|\rho^n\| + |\partial_z Y^0|\|\rho^n\||\rho^n|^{\frac{1}{2}}\|\rho^n\|^{\frac{1}{2}})ds + C\int_0^T |\rho^n(t)|\|\rho^n(t)\|dt$$

$$+C\int_0^T \|\psi(s,Y^0(s))\|_{\mathcal{L}_2(U;H)}|h^n-h|_U|\rho^n(s))|ds$$

$$\leq \int_0^T \|\rho^n(t)\|^2 dt + C\int_0^T (\|Y^0\|^2 + |\partial_z Y^0|^4 + 1)|\rho^n(s)|^2 ds$$

$$+CM^{\frac{1}{2}}(1+\sup_{t\in[0,T]}|Y^0(t)|)\left(\int_0^T |\rho^n(s)|^2 ds\right)^{\frac{1}{2}},$$

where

$$\int_0^T |h^n(t)-h(t)|_U^2 dt \leq 2\int_0^T |h^n(t)|^2 dt + 2\int_0^T |h(t)|^2 dt \leq 4M$$

is used. Then

$$\sup_{t\in[0,T]} |\rho^n(t)|^2 + \int_0^T \|\rho^n(t)\|^2 dt$$

$$\leq C\int_0^T (\|Y^0\|^2 + |\partial_z Y^0|^4 + 1)|\rho^n(s)|^2 ds$$

$$+CM^{\frac{1}{2}}(1+\sup_{t\in[0,T]}|Y^0(t)|)\left(\int_0^T |\rho^n(s)|^2 ds\right)^{\frac{1}{2}}. \quad (5.12)$$

Applying Gronwall inequality to (5.12), we deduce that

$$\sup_{t\in[0,T]} |\rho^n(t)|^2 + \int_0^T \|\rho^n(t)\|^2 dt$$

$$\leq CM^{\frac{1}{2}}(1+\sup_{t\in[0,T]}|Y^0(t)|)\left(\int_0^T |\rho^n(s)|^2 ds\right)^{\frac{1}{2}}$$

$$\times \exp\left\{\int_0^T (\|Y^0(t)\|^2 + |\partial_z Y^0(t)|^4 + 1)dt\right\}.$$

With the aid of (iv), (4.3) and (4.4), we have

$$\sup_{t\in[0,T]} |\rho^n(t)|^2 + \int_0^T \|\rho^n(t)\|^2 dt \to 0.$$

We complete the proof.

Theorem 5.8 Let $\{h^\varepsilon; \varepsilon > 0\} \subset \mathcal{A}_M$ be a sequence that converges in distribution to h as $\varepsilon \to 0$. Then

$$\mathcal{G}^\varepsilon\left(W(\cdot) + \lambda(\varepsilon)\int_0^\cdot h^\varepsilon(s)ds\right) \text{ converges in distribution to } \mathcal{G}^0\left(\int_0^\cdot h(s)ds\right),$$

in $C([0,T]; H) \cap L^2([0,T]; V)$.

Proof Suppose that $\{h^\varepsilon; \varepsilon > 0\} \subset \mathcal{A}_M$ and h^ε converges to h as T_M-valued random elements in distribution. By Girsanov's Theorem, we obtain $\bar{Z}^\varepsilon = \mathcal{G}^\varepsilon\left(W(\cdot) + \lambda(\varepsilon)\int_0^\cdot h^\varepsilon(s)ds\right)$. Consider

$$dX^\varepsilon(t) + AX^\varepsilon(t)dt = \lambda^{-1}(\varepsilon)\psi(t, Y^0(t) + \sqrt{\varepsilon}\lambda(\varepsilon)X^\varepsilon(t))dW(t), \tag{5.13}$$

with initial value $X^\varepsilon(0) = 0$. Applying Itô formula to (5.13), using Hypothesis A and (4.3), we get

$$\lim_{\varepsilon \to 0}\left[\mathbb{E}\sup_{t\in[0,T]}\|X^\varepsilon(t)\|^2 + \mathbb{E}\int_0^T \|X^\varepsilon(t)\|^2_{D(A)}dt\right] = 0. \tag{5.14}$$

Using the same method as the proof of Proposition 5.4, we deduce that there exists a constant $0 < \varepsilon_0^3 < 1$ such that $(X^\varepsilon)_{0<\varepsilon<\varepsilon_0^3}$ is tight in $C([0,T]; H) \cap L^2([0,T]; V)$. Let $\varepsilon_0 := \varepsilon_0^2 \wedge \varepsilon_0^3$. Set

$$\prod = \left(L^2([0,T]; H) \cap C([0,T]; D(A^{-\frac{\alpha}{2}})), T_M, C([0,T]; H) \cap L^2([0,T]; V)\right).$$

By Proposition 5.5 and Proposition 5.6, we know that the family $\{(\bar{Z}^\varepsilon, h^\varepsilon, X^\varepsilon), \varepsilon \in (0, \varepsilon_0)\}$ is tight in $\prod$. Let $(Z, h, 0)$ be any limit point of $\{(\bar{Z}^\varepsilon, h^\varepsilon, X^\varepsilon), \varepsilon \in (0, \varepsilon_0)\}$. We will show that Z has the same law as $\mathcal{G}^0(\int_0^\cdot h(s)ds)$, and in fact $\bar{Z}^\varepsilon$ converges in distribution to Z in $C([0,T]; H) \cap L^2([0,T]; V)$ as $\varepsilon \to 0$, which implies Theorem 5.8.

By the Skorokhod representative theorem, there exists a stochastic basis $(\Omega^1, \mathcal{F}^1, \{\mathcal{F}_t^1\}_{t\in[0,T]}, P^1)$ and, on this basis, $\prod$-valued random variables $(\tilde{U}^\varepsilon, \tilde{h}^\varepsilon, \tilde{X}^\varepsilon), (\tilde{U}, \tilde{h}, 0)$ such that $(\tilde{U}^\varepsilon, \tilde{h}^\varepsilon, \tilde{X}^\varepsilon)$ (respectively $(\tilde{U}, \tilde{h}, 0)$) has the same law as $(\bar{Z}^\varepsilon, h^\varepsilon, X^\varepsilon)$ (respectively $(Z, h, 0)$), and $(\tilde{U}^\varepsilon, \tilde{h}^\varepsilon, \tilde{X}^\varepsilon) \to (\tilde{U}, \tilde{h}, 0)$, P^1-a.s. in $\prod$. From the equation satisfied by $(\bar{Z}^\varepsilon, h^\varepsilon, X^\varepsilon)$, we see that $(\tilde{U}^\varepsilon, \tilde{h}^\varepsilon, \tilde{X}^\varepsilon)$ satisfies the following equation in the distribution sense as follows:

$$d(\tilde{U}^\varepsilon(t) - \tilde{X}^\varepsilon) + A(\tilde{U}^\varepsilon(t) - \tilde{X}^\varepsilon)dt + (B(\tilde{U}^\varepsilon(t), Y^0(t)$$
$$+\sqrt{\varepsilon}\lambda(\varepsilon)\tilde{X}^\varepsilon(t)) + B(Y^0(t), \tilde{U}^\varepsilon(t)))dt$$

$$= \psi(t, Y^0(t) + \sqrt{\varepsilon}\lambda(\varepsilon)\tilde{U}^\varepsilon)\tilde{h}^\varepsilon(t), \tag{5.15}$$

and

$$P^1(\tilde{U}^\varepsilon - \tilde{X}^\varepsilon \in C([0,T]; H) \cap L^2([0,T]; V))$$
$$= P(\bar{Z}^\varepsilon - X^\varepsilon \in C([0,T]; H) \cap L^2([0,T]; V))$$
$$= 1.$$

Let Ω_0^1 be the subset of Ω^1 such that $(\tilde{U}^\varepsilon, \tilde{h}^\varepsilon, \tilde{X}^\varepsilon) \to (\tilde{U}, \tilde{h}, 0)$ in $\prod$, we have $P^1(\Omega_0^1) = 1$. For any $\tilde{\omega} \in \Omega_0^1$, we have

$$\sup_{t\in[0,T]} |\tilde{U}^\varepsilon(\tilde{\omega},t) - \tilde{U}(\tilde{\omega},t)|^2 + \int_0^T \|\tilde{U}^\varepsilon(\tilde{\omega},t) - \tilde{U}(\tilde{\omega},t)\|^2 dt \to 0, \quad \text{as } \varepsilon \to 0. \tag{5.16}$$

Set $\tilde{\eta}^\varepsilon = \tilde{U}^\varepsilon - \tilde{X}^\varepsilon$, then $\tilde{\eta}^\varepsilon(\tilde{\omega}) \in C([0,T];H) \cap L^2([0,T];V)$, and $\tilde{\eta}^\varepsilon(\tilde{\omega})$ satisfies

$$d\tilde{\eta}^\varepsilon(t) + A\tilde{\eta}^\varepsilon(t)dt + \Big[B\big(\tilde{\eta}^\varepsilon(t) + \tilde{X}^\varepsilon(t), Y^0(t) + \sqrt{\varepsilon}\lambda(\varepsilon)(\tilde{\eta}^\varepsilon(t) + \tilde{X}^\varepsilon(t))\big)$$
$$+ B\big(Y^0(t), \tilde{\eta}^\varepsilon(t) + \tilde{X}^\varepsilon(t)\big)\Big]dt$$
$$= \psi\Big(t, Y^0(t) + \sqrt{\varepsilon}\lambda(\varepsilon)(\tilde{\eta}^\varepsilon(t) + \tilde{X}^\varepsilon(t))\Big)\tilde{h}^\varepsilon(t)dt, \tag{5.17}$$

with initial value $\tilde{\eta}^\varepsilon(0) = 0$.

Since

$$\lim_{\varepsilon \to 0} \Big[\sup_{t\in[0,T]} |\tilde{X}^\varepsilon(\tilde{\omega},t)|^2 + \int_0^T \|\tilde{X}^\varepsilon(\tilde{\omega},t)\|^2 dt\Big] = 0, \quad \tilde{U}^\varepsilon = \tilde{\eta}^\varepsilon + \tilde{X}^\varepsilon,$$

using (5.17) and by the same method as Theorem 5.7, we obtain

$$\lim_{\varepsilon \to 0} \Big[\sup_{t\in[0,T]} |\tilde{U}^\varepsilon(\tilde{\omega},t) - \hat{U}(\tilde{\omega},t)|^2 + \int_0^T \|\tilde{U}^\varepsilon(\tilde{\omega},t) - \hat{U}(\tilde{\omega},t)\|^2 dt\Big] = 0, \tag{5.18}$$

where $\hat{U}(t)$ satisfies the following equations

$$\hat{U}(t) = -\int_0^t A\hat{U}(s)ds - \int_0^t (B(\hat{U}(s), Y^0(s)) + B(Y^0(s), \hat{U}(s)))ds$$
$$+ \int_0^t \psi(s, Y^0(s))\tilde{h}(s)ds. \tag{5.19}$$

Hence, by (5.16) and (5.18), we deduce that $\tilde{U} = \hat{U} = \mathcal{G}^0(\int_0^\cdot \tilde{h}(s)ds)$, and $\tilde{U}$ has the same law as $\mathcal{G}^0(\int_0^\cdot h(s)ds)$. As $\bar{Z}^\varepsilon$ and $\tilde{U}^\varepsilon$ have the same law on $C([0,T];H) \cap$

$L^2([0,T];V)$ and by (5.18), we deduce that $\mathcal{G}^\varepsilon\left(W(\cdot)+\lambda(\varepsilon)\int_0^\cdot h^\varepsilon(s)ds\right)$ converges in distribution to $\mathcal{G}^0\left(\int_0^\cdot h(s)ds\right)$ as $\varepsilon\to 0$. We complete the proof. □

Acknowledgements We are grateful to the anonymous referees for their many valuable comments and suggestions. This work was also partially supported by NNSF of China (Grant No.11801032, 11401057), Natural Science Foundation Project of CQ (Grant No. cstc2016jcyjA0326), Fundamental Research Funds for the Central Universities(Grant No. 2018CDXYST0024, 106112015CDJXY100005) and China Scholarship Council (Grant No.201506055003).

References

[1] R.A. Adams: *Sobolev Space*. New York: Academic Press, 1975.

[2] M. Boué, P. Dupuis: *A variational representation for certain functionals of Brownian motion*. Ann. Probab. 26, 1641-1659 (1998).

[3] A. Budhiraja, P. Dupuis: *A variational representation for positive functionals of infinite dimensional Brownian motion*. Probab. Math. Statist. 20, 39-61 (2000).

[4] C. Cao, E.S. Titi: *Global well-posedness of the three-dimensional viscous primitive equations of large-scale ocean and atmosphere dynamics*. Ann. of Math. 166, 245-267 (2007).

[5] A. Debussche, N. Glatt-Holtz, R. Temam, M. Ziane: *Global existence and regularity for the 3D stochastic primitive equations of the ocean and atmosphere with multiplicative white noise*. Nonlinearity 25, 2093-2118 (2012).

[6] A. Dembo, O. Zeitouni: *Large deviations techniques and applications*. 2nd ed., New York: Springer, 1998.

[7] Z. Dong, J. Xiong, J. Zhai, T. Zhang: *A moderate deviation principle for 2D stochastic Navier-Stokes equations driven by multiplicative Lévy noises*. J. Funct. Anal. 1, 227-254 (2017).

[8] Z. Dong, J. Zhai, R. Zhang: *Exponential mixing for 3D stochastic primitive equations of the large scale ocean*. arXiv:1506.08514.

[9] Z. Dong, J. Zhai, R. Zhang: *Large deviation principles for 3D stochastic primitive equations*. J. Differ. Equ. 5, 3110-3146 (2017).

[10] P. Dupuis, R.S. Ellis: *A weak convergence approach to the theory of large deviations*. New York: Wiley, 1997.

[11] M.S. Ermakov: *The sharp lower bound of asymptotic efficiency of estimators in the zone of moderate deviation probabilities*. Electron. J. Stat. 6, 2150-2184 (2012).

[12] P. Fatheddin, J. Xiong: *Moderate deviation principle for a class of stochastic partial differential equations*. J. Appl. Probab. 1, 279-292 (2016).

[13] F. Flandoli, D. Gatarek: *Martingale and stationary solutions for stochastic Navier-Stokes equations*. Probab. Theory Related Fields 3, 367-391 (1995).

[14] C. Frankignoul, K. Hasselmann: *Stochastic climate models, Part II: Application to sea-surface temperature anomalies and thermocline variability*. Tellus 29, 289-305 (1977).

[15] H. Gao, C. Sun: *Well-posedness and large deviations for the stochastic primitive equations in two space dimensions*. Commun. Math. Sci. 2, 575-593 (2012).

[16] N. Glatt-Holtz, M. Ziane: *The stochastic primitive equations in two space dimensions with multiplicative noise*. Discrete and Contin. Dyn. Syst. Series B, 10, 801-822 (2008).

[17] N. Glatt-Holtz, M. Ziane: *Strong pathwise solutions of the stochastic Navier-Stokes system*. Adv. Diff. Eqns. 14, 567-600 (2009).

[18] B. Guo, D. Huang: *3D stochastic primitive equations of the large-scale ocean: global well-posedness and attractors*. Comm. Math. Phys. 2, 697-723 (2009).

[19] I.A. Ibragimov, R.Z. Khasminskii: *Asymptotically normal families of distributions and efficient estimation*. Ann. Statist. 19, 1681-1721 (1991).

[20] W.C.M. Kallenberg: *Intermediate efficiency, theory, and examples*. Ann. Statist, 11, 170-182 (1983).

[21] J.L. Lions, R. Temam, S. Wang: *New formulations of the primitive equations of atmosphere and applications*. Nonlinearity 5, 237-288 (1992).

[22] J.L. Lions, R. Temam, S. Wang: *On the equations of the large scale ocean*. Nonlinearity 5, 1007-1053 (1992).

[23] J.L. Lions, R. Temam, S. Wang: *Models of the coupled atmosphere and ocean*. Computational Mechanics Advance 1, 1-54 (1993).

[24] J.L. Lions, R. Temam, S. Wang: *Mathematical theory for the coupled atmosphere-ocean models*. J. Math. Pures Appl. 74, 105-163 (1995).

[25] U. Mikolajewicz, E. Maier-Reimer: *Internal secular variability in an OGCM*. Climate Dyn. 4, 145-156 (1990).

[26] R. Wang, J. Zhai, T. Zhang: *A moderate deviation principle for 2-D stochastic Navier-Stokes equations*. J. Differ. Equ. 10, 3363-3390 (2015).

[27] J. Pedlosky: *Geophysical Fluid Dynamics*. New York: Springer, 1987.

[28] O.M. Phillips: *On the generation of waves by turbulent winds*. J. Fluid Mech. 2, 417-445 (1957).

[29] R. Samelson, R. Temam, S. Wang: *Some mathematical properties of the planetary geostrophic equations for large-scale ocean circulation*. Appl. Anal. 2, 147-173 (1998).

[30] G. Zhou: *Random attractor of the 3D viscous primitive equations driven by fractional noises*. arXiv:1604.05376.

[31] M. Ziane: *Regularity results of Stokes type system*. App. Anal. 4, 263-292 (1995).

Long-Time Asymptotics of the Focusing Kundu-Eckhaus Equation with Nonzero Boundary Conditions*

Wang Dengshan (王灯山), Guo Boling (郭柏灵),
and Wang Xiaoli (王晓丽)

Abstract The long-time asymptotics of the focusing Kundu-Eckhaus equation with nonzero boundary conditions at infinity is investigated by the nonlinear steepest descent method of Deift and Zhou. Three asymptotic sec-tors in space-time plane are found: the plane wave sector I, plane wave sector II and an intermediate sector with a modulated one-phase elliptic wave. The asymptotic solutions of the three sectors are proposed by successively deforming the corresponding Riemann-Hilbert problems to solvable model problems. Moreover, a time-dependent g-function mechanism is introduced to remove the exponential growths of the jump matrices in the modulated one-phase elliptic wave sector. Finally, the modulational instability is studied to reveal the criterion for the existence of modulated elliptic waves in the central region.

Keywords long-time asymptotics; Kundu-Eckhaus equation; Riemann-Hilbert problem; nonlinear steepest descent method

1 Introduction

The focusing Kundu-Eckhaus (KE) equation reads

$$iq_t + \frac{1}{2}q_{xx} + |q|^2 q + 2\beta^2|q|^4 q - 2i\beta(|q|^2)_x q = 0, \tag{1.1}$$

where β is a constant, which was proposed by Kundu [1] when considering the gauge connections among some generalized Landau-Lifshitz equations [2] and

---
* J. Differential Equations 2019, 266: 5209-5253. Doi: 10.1016/j.jde.2018.10.053.

derivative nonlinear Schrödinger (NLS) type equations [3, 4]. The focusing KE equation is a completely integrable system having Lax pair [5]-[7], soliton solutions [5] and rogue wave solutions [6, 7] by the Darboux transformation and generalized Darboux transformation [8, 9].

In the past years, some effective methods [10]-[17] were developed to investigate the nonlinear problems, among which the inverse scattering transform [18] had been successfully applied to many interesting integrable systems. Particularly, Zakharov et al. [19] developed a Riemann-Hilbert formulation [20,21] which is a modern version of inverse scattering transform. In 1993, P. Deift and X. Zhou [22] proposed the nonlinear steepest descent method for solving the oscillatory matrix Riemann-Hilbert problems, which opens a way to study the long-time asymptotics of nonlinear integrable systems [23]-[28]. Recent years, much work has been done to investigate the asymptotic solutions of different nonlinear integrable models with initial values [29]-[34] or initial-boundary values [35]-[37]. Another development of inverse scattering transform was the so-called Fokas method [38], which was proposed to study the initial-boundary value problems of linear and nonlinear integrable partial different equations [39]-[43].

More recent years, there are some studies on the long-time asymptotics of Kundu-Eckhaus equation with zero boundary conditions [44, 45]. However, no work has been done to explore the asymptotic solutoins of Kundu-Eckhaus equation with nonzero boundary conditions. It is worth referring that Biondini et al. [31]-[34] have investigated the NLS equations with nonzero boundary conditions by formulating the Riemann-Hilbert problems. In particular, they studied the long-time asymptotics and modulational instability of the focusing NLS equation with nonzero boundary conditions [31]-[33].

Thus in this paper, we consider the long-time asymptotics of the focusing KE equation (1.1) with initial condition $q(x, 0)$ on line. Especially, we mainly study the nonzero boundary conditions at infinity, i.e.,

$$\lim_{x \to \pm\infty} q(x, 0) = A e^{i(\mu x + \theta_\pm)}, \tag{1.2}$$

where $A > 0$ and $\theta_\pm$ are real constants. Our calculations are based on the Lax pair of the focusing KE equation (1.1) as follows

$$\begin{aligned} \phi_x + iz\sigma_3\phi &= (U - i\beta U^2 \sigma_3)\phi, \\ \phi_t + iz^2\sigma_3\phi &= \frac{1}{2}(V + 4i\beta^2 U^4 \sigma_3 - \beta(UU_x - U_xU))\phi, \end{aligned} \tag{1.3}$$

with
$$\sigma_3 = \begin{pmatrix} 1 & 0 \\ 0 & -1 \end{pmatrix}, \quad U = \begin{pmatrix} 0 & q \\ -q^* & 0 \end{pmatrix}, \quad (1.4)$$

and $V = 2zU - 2\beta U^3 - i(U^2 + U_x)\sigma_3$, where z is the spectral parameter. It is found that there are three asymptotic sectors in the x-t plane, i.e., the plane wave sector I, plane wave sector II and an intermediate sector with a modulated one-phase elliptic wave. The analysis of modulational instability reveals the criterion for the existence of modulated elliptic waves in the central region.

Remark 1.1 *It is known that the Lax pair (1.3) with (1.4) of the focusing KE equation (1.1) can be converted to the Lax pair of the standard focusing NLS equation as follows:*

Take the transformation
$$\varphi(x,t,z) = e^{-i\beta \int |q|^2 dx \sigma_3} \phi(x,t,z), \quad (1.5)$$

then it is found that the matrix function $\varphi(x,t,z)$ satisfies the following Lax pair of the standard focusing NLS equation
$$\varphi_x + iz\sigma_3 \varphi = \widehat{U}\varphi,$$
$$\varphi_t + iz^2\sigma_3 \varphi = \frac{1}{2}\widehat{V}\varphi, \quad (1.6)$$

with matrix functions $\widehat{U}$ and $\widehat{V}$ satisfying
$$\widehat{U} = e^{-i\beta \int |q|^2 dx \, \mathrm{ad}\sigma_3} U = \begin{pmatrix} 0 & \widehat{q} \\ -\widehat{q}^* & 0 \end{pmatrix},$$
$$\widehat{V} = e^{-i\beta \int |q|^2 dx \, \mathrm{ad}\sigma_3} V = 2z\widehat{U} - i(\widehat{U}^2 + \widehat{U}_x)\sigma_3,$$

where
$$\widehat{q} = q e^{-2i\beta \int |q|^2 dx},$$

from which we have
$$q = \widehat{q} e^{2i\beta \int |\widehat{q}|^2 dx}. \quad (1.7)$$

It seems that one can only study the Lax pair (1.6) to derive the potential function $\widehat{q}(x,t)$ and then get the potential function $q(x,t)$ from Eq. (1.7). However, this is not feasible because it is hard to integrate Eq.(1.7) in the case of the modulated one-phase elliptic wave solution of function $\widehat{q}(x,t)$. So, in this paper, different from Refs. [44, 45] we investigate the long-time asymptotics of the focusing KE equation (1.1) starting with the Lax pair (1.3) with (1.4) instead of Lax pair (1.6), which is one of the innovation points of our work. In all the procedures, one doesn't need to calculate the complicated integral in Eq. (1.7).

This paper is organized as follows. In Section 2, the original Riemann-Hilbert problem for the focusing KE equation (1.1) with nonzero boundary conditions in Eq. (1.2) is proposed based on inverse scattering transform. The main results of the present paper are listed in Section 3. Sections 4-6 prove the three main theorems of this paper, respectively. Finally, the modulational instability of the focusing KE equation with initial plane wave is studied in Section 7.

2 The original Riemann-Hilbert problem

In this section, we study the inverse scattering transform and original Riemann-Hilbert problem of the focusing KE equation (1.1) with initial value $q(x, 0)$ satisfying nonzero boundary conditions in Eq. (1.2).

2.1 Scattering problem

The Lax pair (1.3) of the focusing KE equation (1.1) can be rewritten as

$$\begin{cases} \phi_x = F\phi, & F = -iz\sigma_3 + U - i\beta U^2 \sigma_3, \\ \phi_t = G\phi, & G = -iz^2\sigma_3 + \frac{1}{2}(V + 4i\beta^2 U^4 \sigma_3 - \beta(UU_x - U_x U)), \end{cases} \quad (2.8)$$

where $V = 2zU - 2\beta U^3 - i(U^2 + U_x)\sigma_3$.

Consider the nonzero boundary conditions (1.2) of the initial value $q(x, 0)$. For any time t, potential function $q(x, t)$ should behave asymptotically like plane waves as $x \to \pm\infty$, which is $q(x, t) \sim Ae^{i[\mu x + (A^2 + 2\beta^2 A^4 - \frac{\mu^2}{2})t + \theta_-]}$ as $x \to -\infty$ and $q(x, t) \sim Ae^{i[\mu x + (A^2 + 2\beta^2 A^4 - \frac{\mu^2}{2})t + \theta_+]}$ as $x \to \infty$. Take the gauge transformation

$$\phi(x, t) = \begin{pmatrix} e^{i[\mu x + (A^2 + 2\beta^2 A^4 - \frac{\mu^2}{2})t]/2} & 0 \\ 0 & e^{-i[\mu x + (A^2 + 2\beta^2 A^4 - \frac{\mu^2}{2})t]/2} \end{pmatrix} \psi(x, t), \quad (2.9)$$

then the Lax pair (2.8) becomes

$$\psi_x = \left[-i\left(z + \frac{\mu}{2}\right)\sigma_3 + Q\right]\psi, \quad (2.10)$$

$$\psi_t = \left[-i\left(z^2 + \frac{1}{2}\left(A^2 + 2\beta^2 A^4 - \frac{\mu^2}{2}\right)\right)\sigma_3 + \widehat{Q}\right]\psi, \quad (2.11)$$

where

$$Q = \begin{pmatrix} i\beta |q|^2 & qe^{-i[\mu x+(A^2+2\beta^2 A^4-\frac{\mu^2}{2})t]} \\ -q^* e^{i[\mu x+(A^2+2\beta^2 A^4-\frac{\mu^2}{2})t]} & -i\beta |q|^2 \end{pmatrix}.$$

and

$$\widehat{Q} = \begin{pmatrix} \frac{i}{2}|q|^2 + 2i\beta^2|q|^4 + \frac{\beta}{2}(qq_x^* - q_x q^*) & \Omega_1 e^{-i[\mu x+(A^2+2\beta^2 A^4-\frac{\mu^2}{2})t]} \\ \Omega_2 e^{i[\mu x+(A^2+2\beta^2 A^4-\frac{\mu^2}{2})t]} & -\frac{i}{2}|q|^2 - 2i\beta^2|q|^4 - \frac{\beta}{2}(qq_x^* - q_x q^*) \end{pmatrix},$$

where $\Omega_1 = \frac{i}{2}q_x + zq + \beta |q|^2 q$ and $\Omega_2 = \frac{i}{2}q_x^* - zq^* - \beta |q|^2 q^*$.

For $x \to \pm\infty$, denote $Q_\pm = \lim_{x\to\pm\infty} Q$, $\widehat{Q}_\pm = \lim_{x\to\pm\infty} \widehat{Q}$ and remember the nonzero boundary conditions (1.2) of the initial value $q(x,0)$, then the Lax pair in Eqs. (2.10)-(2.11) simplifies to

$$\psi_{\pm x} = \left[-i\left(z+\frac{\mu}{2}\right)\sigma_3 + Q_\pm\right]\psi_\pm = \begin{pmatrix} i\beta A^2 - i\left(z+\frac{\mu}{2}\right) & Ae^{i\theta_\pm} \\ -Ae^{-i\theta_\pm} & -i\beta A^2 + i\left(z+\frac{\mu}{2}\right) \end{pmatrix}\psi_\pm, \quad (2.12)$$

$$\psi_{\pm t} = \begin{pmatrix} i\beta^2 A^4 - i\mu\beta A^2 + \frac{i\mu^2}{4} - iz^2 & A\left(z+\beta A^2 - \frac{\mu}{2}\right)e^{i\theta_\pm} \\ A\left(\frac{\mu}{2} - z - \beta A^2\right)e^{-i\theta_\pm} & -\left(i\beta^2 A^4 - i\mu\beta A^2 + \frac{i\mu^2}{4} - iz^2\right) \end{pmatrix}\psi_\pm, \quad (2.13)$$

which have explicitly exact solution for real z

$$\psi_\pm(x,t,z) = \begin{pmatrix} 1 & \frac{i\left(z+\frac{\mu}{2}-\lambda-\beta A^2\right)e^{i\theta_\pm}}{A} \\ \frac{i\left(z+\frac{\mu}{2}-\lambda-\beta A^2\right)e^{-i\theta_\pm}}{A} & 1 \end{pmatrix} e^{-i\lambda[x+(z-\frac{\mu}{2}+\beta A^2)t]\sigma_3}, \quad (2.14)$$

where λ is the complex square root defined by $\lambda = \sqrt{\left(z+\frac{\mu}{2}-\beta A^2\right)^2 + A^2}$. The branch cut of λ is $\gamma = \gamma^+ \cup \gamma^-$ with $\gamma^+ = [B, E]$ and $\gamma^- = [\bar{E}, B]$, where $B = \beta A^2 - \frac{\mu}{2}$, $E = \beta A^2 - \frac{\mu}{2} + iA$ and $\bar{E} = \beta A^2 - \frac{\mu}{2} - iA$ as shown in Fig. 1, where the branch cut γ is oriented upward.

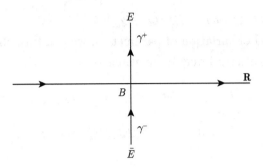

Fig. 1   The contour $\Sigma^{(0)}$ of the original RH problem $P^{(0)}$

Define two simultaneous solutions $\Psi_\pm(x,t,z)$ of the Lax pair in Eq. (2.10)-(2.11) so that they satisfy the asymptotic conditions $\Psi_\pm(x,t,z) = \psi_\pm(x,t,z) + o(1)$ as $x \to \infty$. It is obvious that $\Psi_\pm(x,t,z)$ are bounded for $z \in \Sigma^{(0)} = \mathbf{R} \cup \gamma$ as $x \to \pm\infty$, where contour $\Sigma^{(0)}$ is displayed in Fig. (1). Further define two matrix-valued functions $\mu_\pm$ by

$$\mu_\pm(x,t,z) = \Psi_\pm(x,t,z) e^{i\lambda[x+(z-\frac{\mu}{2}+\beta A^2)t]\hat{\sigma}_3}, \qquad (2.15)$$

then the asymptotic conditions of $\Psi_\pm(x,t,z)$ and Eq. (2.15) shows that

$$\mu_\pm(x,t,z) = J_\pm + o(1), \quad x \to \pm\infty, \quad k \in \Sigma^{(0)}, \qquad (2.16)$$

where

$$J_\pm = \begin{pmatrix} 1 & \dfrac{i\left(z+\frac{\mu}{2}-\lambda-\beta A^2\right)e^{i\theta_\pm}}{A} \\ \dfrac{i\left(z+\frac{\mu}{2}-\lambda-\beta A^2\right)e^{-i\theta_\pm}}{A} & 1 \end{pmatrix}.$$

Making use of the definitions of matrix-valued functions $\Psi_\pm(x,t,z)$ and $\mu_\pm(x,t,z)$ along with Eqs. (2.15)-(2.16), the Lax pair in Eq. (2.10)-(2.11) becomes

$$(J_\pm^{-1}\mu_\pm)_x = -i\lambda[\sigma_3, J_\pm^{-1}\mu_\pm] + J_\pm^{-1}(Q-Q_\pm)\mu_\pm,$$

$$(J_\pm^{-1}\mu_\pm)_t = -i\lambda\left(z-\frac{\mu}{2}+\beta A^2\right)[\sigma_3, J_\pm^{-1}\mu_\pm] + J_\pm^{-1}(\hat{Q}-\hat{Q}_\pm)\mu_\pm,$$

which can be written in differential form as

$$d(e^{i\lambda[x+(z-\frac{\mu}{2}+\beta A^2)t]\hat{\sigma}_3} J_\pm^{-1}\mu_\pm) = e^{i\lambda[x+(z-\frac{\mu}{2}+\beta A^2)t]\hat{\sigma}_3} W\mu_\pm, \qquad (2.17)$$

where $W = J_\pm^{-1}(Q - Q_\pm)dx + J_\pm^{-1}(\hat{Q} - \hat{Q}_\pm)dt$.

By the method of variation of parameters, we can turn the differential form (2.17) above into Volterra integral equations of $\mu_\pm$

$$\mu_-(x,t,z) = J_- + \int_{-\infty}^{x} J_- e^{-i\lambda(x-s)\sigma_3} J_-^{-1}(Q - Q_-)\mu_-(x,t,z) e^{i\lambda(x-s)\sigma_3} ds,$$

$$\mu_+(x,t,z) = J_+ - \int_{x}^{+\infty} J_+ e^{-i\lambda(x-s)\sigma_3} J_+^{-1}(Q - Q_+)\mu_+(x,t,z) e^{i\lambda(x-s)\sigma_3} ds.$$
(2.18)

Suppose $(q(x,t) - Ae^{i(\mu x + \theta_\pm)}) \in L^1(\mathbf{R}^\pm)$ then $\mu_\pm(x,t,z)$ can analytical continue off the contour $\Sigma^{(0)}$ provided that the right hand sides of Volterra integrals converge. Denote $\mu_\pm = [\mu_{\pm 1}, \mu_{\pm 2}]$. It is easy to see that the integral equation for the second column of μ_- contains only the exponential factor $e^{-i\lambda(x-s)}$ which decays when z is in the lower half plane except γ^-, i.e. $z \in \mathbb{C}_-\backslash\gamma^-$, and the integral equation for the first column of μ_+ contain only the exponential factor $e^{-i\lambda(s-x)}$ which also decays for $z \in \mathbb{C}_-\backslash\gamma^-$. Thus μ_{-2} and μ_{+1} can be analytically continued to $\mathbb{C}_-\backslash\gamma^-$. Similarly, it is found that μ_{-1} and μ_{+2} can be analytically continued to upper half plane except γ^+, i.e. $\mathbb{C}_+\backslash\gamma^+$.

Further, by Abel's theorem, we have

$$D(z) = \det\Psi_\pm(x,t,z) = \lim_{x \to \pm\infty} \det\Psi_\pm(x,t,z) = \det J_\pm = \frac{2\lambda}{\lambda + z + \frac{\mu}{2} - \beta A^2},$$
(2.19)

which is nonzero and non-singular for $\Sigma^{(0)*} = \Sigma^{(0)}\backslash\{E, \bar{E}\}$. It is known that the matrices $\Psi_\pm(x,t,z)$ are fundamental solutions of the Lax pair (2.10)-(2.11) for $z \in \Sigma^{(0)*}$. Thus we can define the scattering matrix $S(z)$ s.t.

$$\Psi_-(x,t,z) = \Psi_+(x,t,z) S(z), \quad z \in \Sigma^{(0)*}.$$
(2.20)

The symmetry $\overline{F(q,\bar{q},\bar{z})} = -\sigma F(q,\bar{q},z)\sigma$ with $\sigma = \begin{pmatrix} 0 & 1 \\ -1 & 0 \end{pmatrix}$ indicates that

$$\overline{\Psi_\pm(x,t,\bar{z})} = -\sigma \Psi_\pm(x,t,z)\sigma, \quad \overline{S(\bar{z})} = -\sigma S(z)\sigma, \quad z \in \Sigma^{(0)*},$$

which implies that the scattering matrix $S(z)$ is

$$S(z) = \begin{pmatrix} a(z) & -\bar{b}(z) \\ b(z) & \bar{a}(z) \end{pmatrix}, \quad a(z)\bar{a}(z) + b(z)\bar{b}(z) = 1,$$

where $\bar{a}(z) = \overline{a(\bar{z})}$ and $\bar{b}(z) = \overline{b(\bar{z})}$. Then we have

$$\Psi_{-1}(x,t,z) = a(z)\Psi_{+1}(x,t,z) + b(z)\Psi_{+2}(x,t,z),$$

$$\Psi_{-2}(x,t,z) = \bar{a}(z)\Psi_{+2}(x,t,z) - \bar{b}(z)\Psi_{+1}(x,t,z).$$

The jump discontinuity of λ across the branch cut γ leads to the jump of scattering data on γ. As usual, define

$$a^-(z) = \lim_{\epsilon \to 0^+} a(z+\epsilon) = a(z) = -a^+(z), \quad b^-(z) = \lim_{\epsilon \to 0^+} b(z+\epsilon) = a(z) = -b^+(z).$$

The the following lemma holds:

Lemma 2.1 *Assume $e^{\pm \epsilon x}(q(x,t) - Ae^{i(\mu x + \theta_\pm)}) \in L^1(\mathbf{R}^\pm)$ then the columns of fundamental solutions $\Psi_\pm$ and scattering data $a(z), b(z)$ satisfy the following jump conditions across the branch cut γ:*

$$\Psi_{-1}^+ = \frac{ie^{-i\theta_-}}{A}\left(z + \frac{\mu}{2} + \lambda - \beta A^2\right)\Psi_{-2}, \quad \Psi_{+2}^+ = \frac{ie^{i\theta_+}}{A}\left(z + \frac{\mu}{2} + \lambda - \beta A^2\right)\Psi_{+1},$$

$$\Psi_{+1}^+ = \frac{ie^{-i\theta_+}}{A}\left(z + \frac{\mu}{2} + \lambda - \beta A^2\right)\Psi_{+2}, \quad \Psi_{-2}^+ = \frac{ie^{i\theta_-}}{A}\left(z + \frac{\mu}{2} + \lambda - \beta A^2\right)\Psi_{-1},$$

$$a^+(z) = e^{i(\theta_+ - \theta_-)}\bar{a}(z), \quad \bar{a}^+(z) = e^{i(\theta_- - \theta_+)}a(z), \quad z \in \gamma,$$

$$b^+(z) = -e^{-i(\theta_+ + \theta_-)}\bar{b}(z), \quad \bar{b}^+(z) = -e^{i(\theta_+ + \theta_-)}b(z), \quad z \in \gamma,$$

and

$$\rho^+(z) = -e^{-2i\theta_+}\bar{\rho}(z), \quad \bar{\rho}^+(z) = -e^{2i\theta_+}\rho(z), \quad z \in \gamma,$$

where $\rho(z) = b(z)/a(z)$ is the reflection coefficient.

Proof Because $\Psi_\pm$ are the fundamental solutions of the Lax pair (2.10)-(2.11) for $z \in \Sigma^{(0)*}$, every column of each solution can be expressed by the either pair. Thus as z approaches the branch cut γ from the left, the limit of Ψ_{-1}, i.e. Ψ_{-1}^+ is also a column solution of the Lax pair (2.10)-(2.11), which implies that

$$\Psi_{-1}^+ = k_1 \Psi_{-1} + k_2 \Psi_{-2}, \quad z \in \gamma,$$

where k_1 and k_2 are only dependent of t and k. Letting $x \to -\infty$ on both sides of this equation, we obtain that $k_1 = 0$ and $k_2 = \frac{ie^{-i\theta_-}}{A}\left(z + \frac{\mu}{2} + \lambda - \beta A^2\right)$. Thus we have

$$\Psi_{-1}^+ = \frac{ie^{-i\theta_-}}{A}\left(z + \frac{\mu}{2} + \lambda - \beta A^2\right)\Psi_{-2}.$$

Applying Cramer's rule to the algebraic equation (2.20) we find that for $z \in \Sigma^{(0)*}$

$$a(z) = \mathrm{Wr}(\Psi_{-1}(x,t,z), \Psi_{+2}(x,t,z))/D(z),$$

$$\bar{a}(z) = \mathrm{Wr}(\Psi_{+1}(x,t,z), \Psi_{-2}(x,t,z))/D(z),$$

where Wr indicates the Wronskian determinant. So we have

$$\begin{aligned}a^+(z) &= \mathrm{Wr}(\Psi_{-1}^+, \Psi_{+2}^+)/D^+(z) \\ &= \frac{ie^{-i\theta_-}}{A}\left(z+\frac{\mu}{2}+\lambda-\beta A^2\right)\frac{ie^{i\theta_+}}{A}\left(z+\frac{\mu}{2}+\lambda-\beta A^2\right)\mathrm{Wr}(\Psi_{-2},\Psi_{+1})/D^+(z) \\ &= e^{i(\theta_+-\theta_-)}\mathrm{Wr}(\Psi_{+1},\Psi_{-2})/D(z) \\ &= e^{i(\theta_+-\theta_-)}\bar{a}(z).\end{aligned}$$

In the same way, the other jump conditions can also be proved. □

2.2 Inverse problem

Define the fundamental matrix-value function to be

$$M^{(0)}(x,t;z) = \begin{cases} e^{i\beta\int_x^{+\infty}(|q(y,t)|^2-A^2)dy\sigma_3}\left[\dfrac{\Psi_{-1}(x,t,z)}{a(z)D(z)}, \Psi_{+2}(x,t,z)\right] \\ e^{i\lambda[x+(z-\frac{\mu}{2}+\beta A^2)t]\sigma_3}, \quad z \in \mathbb{C}_+\backslash\gamma^+, \\ e^{i\beta\int_x^{+\infty}(|q(y,t)|^2-A^2)dy\sigma_3}\left[\Psi_{+1}(x,t,z), \dfrac{\Psi_{-2}(x,t,z)}{\bar{a}(z)D(z)}\right] \\ e^{i\lambda[x+(z-\frac{\mu}{2}+\beta A^2)t]\sigma_3}, \quad z \in \mathbb{C}_-\backslash\gamma^-, \end{cases} \quad (2.21)$$

where $\det M^{(0)}(x,t;z) = 1$ for $z \in \mathbb{C}\backslash\gamma$.

To apply the nonlinear steepest descent method of Deift and Zhou [22], we first formulate the RH problem of the focusing KE equation (1.1) with initial value $q(x,0)$ in Eq. (1.2). To do so, define $M_{\pm}^{(0)}(x,t;z)$ to be the limiting values of $M^{(0)}(x,t;z)$ for $\mathrm{Im}(z) \to 0^{\pm}$. Then the jump condition of the matrix-value function $M^{(0)}(x,t;z)$ across $\mathbf{R}$ is

$$M_+^{(0)}(x,t;z) = M_-^{(0)}(x,t;z)\begin{pmatrix} \dfrac{1}{D(z)}(1+\rho(z)\bar{\rho}(z)) & \bar{\rho}(z)e^{-ift} \\ \rho(z)e^{ift} & D(z) \end{pmatrix}, \quad (2.22)$$

where $f = 2\lambda\left(\xi+z-\dfrac{\mu}{2}+\beta A^2\right)$ with $\xi = x/t$.

Furthermore, define $M_+^{(0)}(x,t;z)$ to be the limiting value of $M^{(0)}(x,t;z)$ approaching from the left side of the oriented contour γ^+, i.e. $\mathrm{Re}(z) = \beta A^2 - \frac{\mu}{2} - \epsilon$ as $\epsilon \to 0$. Similarly, define $M_-^{(0)}(x,t;z)$ to be the limiting values from the right side, that is $\mathrm{Re}(z) = \beta A^2 - \frac{\mu}{2} + \epsilon$ as $\epsilon \to 0$. According to the jump conditions in Lemma (2.1), the jump condition of the matrix-value function $M^{(0)}(x,t;z)$ across γ^+ is

$$M_+^{(0)}(x,t;z)$$
$$= M_-^{(0)}(x,t;z) \begin{pmatrix} \dfrac{\lambda - z - \frac{\mu}{2} + \beta A^2}{iAe^{i\theta_+}} \bar{\rho}(z) e^{-ift} & \dfrac{2i\lambda}{Ae^{-i\theta_+}} \\ \dfrac{iA}{2\lambda e^{i\theta_+}} (1+|\rho(z)|^2) & \dfrac{\lambda + z + \frac{\mu}{2} - \beta A^2}{iAe^{-i\theta_+}} \rho(z) e^{ift} \end{pmatrix},$$
(2.23)

where $|\rho(z)|^2 = \rho(z)\bar{\rho}(z)$.

In the same way, from Lemma (2.1) and after straightforward calculations, the jump condition of the matrix-value function $M^{(0)}(x,t;z)$ across γ^- can also be obtained as

$$M_+^{(0)}(x,t;z)$$
$$= M_-^{(0)}(x,t;z) \begin{pmatrix} \dfrac{i(\lambda + z + \frac{\mu}{2} - \beta A^2)}{Ae^{i\theta_+}} \bar{\rho}(z) e^{-ift} & \dfrac{iA}{2\lambda e^{-i\theta_+}}(1+|\rho(z)|^2) \\ \dfrac{2i\lambda}{Ae^{i\theta_+}} & \dfrac{i(\lambda - z - \frac{\mu}{2} + \beta A^2)}{Ae^{-i\theta_+}} \rho(z) e^{ift} \end{pmatrix}.$$
(2.24)

Now the matrix-value function $M^{(0)} = M^{(0)}(x,t;z)$ is the solution to the original Riemann-Hilbert problem:

$$\begin{cases} \text{(i). } M^{(0)} \text{ is analytic off } \Sigma^{(0)} = \mathbf{R} \cup \gamma^+ \cup \gamma^-; \\ \text{(ii). } M_+^{(0)} = M_-^{(0)} V^{(0)} \text{ on } \Sigma^{(0)}; \\ \text{(iii). } M^{(0)} \sim I, \quad \text{as } z \to \infty, \end{cases} \quad (2.25)$$

where the contour $\Sigma^{(0)}$ is shown in Fig. 1 and the jump matrices $V^{(0)} = \{V_j^{(0)}\}_{j=1}^3$

are

$$V_1^{(0)} = \begin{pmatrix} \dfrac{1}{D(z)}(1+\rho(z)\bar\rho(z)) & \bar\rho(z)e^{-ift} \\ \rho(z)e^{ift} & D(z) \end{pmatrix},$$

$$V_2^{(0)} = \begin{pmatrix} \dfrac{\lambda - z - \dfrac{\mu}{2} + \beta A^2}{iAe^{i\theta_+}}\bar\rho(z)e^{-ift} & \dfrac{2i\lambda}{Ae^{-i\theta_+}} \\ \dfrac{iA}{2\lambda e^{i\theta_+}}(1+\rho(z)\bar\rho(z)) & \dfrac{\lambda + z + \dfrac{\mu}{2} - \beta A^2}{iAe^{-i\theta_+}}\rho(z)e^{ift} \end{pmatrix},$$

$$V_3^{(0)} = \begin{pmatrix} \dfrac{i\left(\lambda + z + \dfrac{\mu}{2} - \beta A^2\right)}{Ae^{i\theta_+}}\bar\rho(z)e^{-ift} & \dfrac{iA}{2\lambda e^{-i\theta_+}}(1+\rho(z)\bar\rho(z)) \\ \dfrac{2i\lambda}{Ae^{i\theta_+}} & \dfrac{i\left(\lambda - z - \dfrac{\mu}{2} + \beta A^2\right)}{Ae^{-i\theta_+}}\rho(z)e^{ift} \end{pmatrix},$$

where $\rho(z) = b(z)/a(z)$, $\lambda = \sqrt{\left(z + \dfrac{\mu}{2} - \beta A^2\right)^2 + A^2}$ and

$$f = f(z,\xi) = 2\lambda\left(\xi + z - \dfrac{\mu}{2} + \beta A^2\right), \quad D(z) = \dfrac{2\lambda}{\lambda + z + \dfrac{\mu}{2} - \beta A^2}. \tag{2.26}$$

Thus the original RH problem is written as

$$P^{(0)} = \{M^{(0)}, \Sigma^{(0)}, V^{(0)}, I \text{ as } z \to \infty\}. \tag{2.27}$$

2.3 Reconstructing the potential function $q(x,t)$

Now we reconstruct the potential function $q(x,t)$ from the matrix-value function $M^{(0)}(x,t;z)$ of the original RH problem $P^{(0)}$. The conditions of the RH problem in Eq. (2.25) denotes that function $M^{(0)}(x,t;z)$ admits the large-z asymptotic expansion

$$M^{(0)}(x,t;z) = I + \dfrac{M_1^{(0)}(x,t;z)}{z} + \dfrac{M_2^{(0)}(x,t;z)}{z^2} + O\left(\dfrac{1}{z^3}\right), \quad z \to \infty. \tag{2.28}$$

Further, from the definition of $M^{(0)}(x,t;z)$ in Eq. (2.21) it is observed that function $\psi(x,t) = M^{(0)}(x,t;z)e^{-i\lambda x\sigma_3}$ satisfies the fist equation of the Lax pair in Eq. (2.10). So combining Eqs. (2.10) and (2.28) we have

$$\begin{pmatrix} 0 & qe^{-i[\mu x+(A^2+2\beta^2 A^4-\frac{\mu^2}{2})t]} \\ & e^{i\beta\int_x^{+\infty}(|q(y,t)|^2-A^2)dy} \\ -q^*e^{i[\mu x+(A^2+2\beta^2 A^4-\frac{\mu^2}{2})t]} & 0 \\ e^{-i\beta\int_x^{+\infty}(|q(y,t)|^2-A^2)dy} & \end{pmatrix}$$

$$= -ie^{-i\beta\int_x^{+\infty}(|q(y,t)|^2-A^2)dy\sigma_3}[M_1^{(0)}(x,t;z),\sigma_3],$$

which reconstruct the solution of the focusing Kundu-Eckhaus equation (1.1) in the form

$$q(x,t) = 2i(M_1^{(0)}(x,t;z))_{12}e^{i[\mu x+(A^2+2\beta^2 A^4-\frac{\mu^2}{2})t]}e^{-2i\beta\int_x^{+\infty}(|q(y,t)|^2-A^2)dy}, \quad (2.29)$$

where $(M_1^{(0)}(x,t;z))_{12}$ is the 12-entry of matrix $M_1^{(0)}(x,t;z)$.

Reconstructing the potential function $q(x,t)$.

2.4 Behavior of Im(f)

From the jump matrices $V^{(0)}$ it is seen that the sign signature of Im(f) is very important for the analysis of the long-time asymptotics of the RH problem $P^{(0)}$. So in this subsection, we analyze the behavior of Im(f) on the complex plane. Note that

$$f = f(z,\xi) = 2\sqrt{\left(z+\frac{\mu}{2}-\beta A^2\right)^2 + A^2}\left(z+\xi-\frac{\mu}{2}+\beta A^2\right). \quad (2.30)$$

Taking $z = \eta + i\nu$, then we have

$$f(z,\xi) = 2\sqrt{\left(\eta+\frac{\mu}{2}+i\nu-\beta A^2\right)^2 + A^2}\left(\eta+\xi-\frac{\mu}{2}+i\nu+\beta A^2\right). \quad (2.31)$$

For $0 < \nu < 1$, the square root $\sqrt{\left(\eta+\frac{\mu}{2}+i\nu-\beta A^2\right)^2 + A^2}$ in function $f(z,\xi)$ can be written as

$$\begin{cases} -\sqrt{A^2+\left(\eta+\frac{\mu}{2}-\beta A^2\right)^2} - \dfrac{\nu\left(\eta+\frac{\mu}{2}-\beta A^2\right)}{\sqrt{A^2+\left(\eta+\frac{\mu}{2}-\beta A^2\right)^2}}i, & \eta < \beta A^2 - \dfrac{\mu}{2}, \\ \sqrt{A^2+\left(\eta+\frac{\mu}{2}-\beta A^2\right)^2} + \dfrac{\nu\left(\eta+\frac{\mu}{2}-\beta A^2\right)}{\sqrt{A^2+\left(\eta+\frac{\mu}{2}-\beta A^2\right)^2}}i, & \eta \geqslant \beta A^2 - \dfrac{\mu}{2}. \end{cases}$$

Thus when $\eta \geqslant \beta A^2 - \dfrac{\mu}{2}$, we have

$$\mathrm{Im}(f) \sim 2\nu\sqrt{A^2 + \left(\eta + \dfrac{\mu}{2} - \beta A^2\right)^2}$$
$$+ \left(\eta + \xi - \dfrac{\mu}{2} + \beta A^2\right) \dfrac{2\nu\left(\eta + \dfrac{\mu}{2} - \beta A^2\right)}{\sqrt{A^2 + \left(\eta + \dfrac{\mu}{2} - \beta A^2\right)^2}},$$

so we have $\mathrm{Im}(f) > 0$ for $\eta \geqslant \max\left\{\dfrac{\mu}{2} - \beta A^2, \beta A^2 - \dfrac{\mu}{2}\right\}$.

When $\eta < \beta A^2 - \dfrac{\mu}{2}$ we have

$$\mathrm{Im}(f) \sim -2\nu\sqrt{A^2 + \left(\eta + \dfrac{\mu}{2} - \beta A^2\right)^2}$$
$$+ \left(\eta + \xi - \dfrac{\mu}{2} + \beta A^2\right) \dfrac{-2\nu\left(\eta + \dfrac{\mu}{2} - \beta A^2\right)}{\sqrt{A^2 + \left(\eta + \dfrac{\mu}{2} - \beta A^2\right)^2}}.$$

Then we have $\mathrm{Im}(f) = 0$ on the real axis if

$$\eta = \eta_\pm = \dfrac{2\beta A^2 - \mu - \xi \pm \sqrt{(\xi + \mu - 2\beta A^2)^2 - 4[2A^2 + \xi(\mu - 2\beta A^2)]}}{4}. \quad (2.32)$$

For $(\xi+\mu-2\beta A^2)^2 - 4[2A^2+\xi(\mu-2\beta A^2)] = 0$, i.e., $\xi = \xi_1 = \mu - 2\beta A^2 - 2\sqrt{2}A$ or $\xi = \xi_2 = \mu - 2\beta A^2 + 2\sqrt{2}A$, there is only one real value $\eta = \eta_+ = \eta_-$. For $(\xi + \mu - 2\beta A^2)^2 - 4[2A^2 + \xi(\mu - 2\beta A^2)] < 0$, i.e. $\xi_1 < \xi < \xi_2$, there are no real values of $\eta_\pm$. In contrast, there are two real values $\eta_\pm$ for $\xi > \xi_2$ or $\xi < \xi_1$, and we have $\mathrm{Im}(f) > 0$ above the real axis along the segment (η_-, η_+).

Because there are several parameters in the focusing KE equation (1.1) and the boundary conditions of the initial value $q(x, 0)$ in Eq. (1.2), for simplicity, in this paper we constrain ourself within the case of $\xi_2 > 0$, i.e. $\mu > 2\beta A^2 - 2\sqrt{2}A$. Furthermore, in what follows we only consider the case of $\xi_1 < 0$, i.e. $\mu < 2\beta A^2 + 2\sqrt{2}A$. The case of $\mu \geqslant 2\beta A^2 + 2\sqrt{2}A$, i.e. $\xi_1 > 0$, can also be studied in the same way. The sign signature of $\mathrm{Im}(f)$ for $2\beta A^2 + 2\sqrt{2}A > \mu > 2\beta A^2 - 2\sqrt{2}A$ are shown in Fig. 2, where we have choose parameters $A = 1, \beta = \frac{1}{2}\mu$ and $\mu = 3$. It is observed that the curves $\mathrm{Im}(f) = 0$ intersect with the real axis at two real points $\eta_\pm$ for large $|\xi|$, but the curves $\mathrm{Im}(f) = 0$ will not intersect with the real axis for small $|\xi|$.

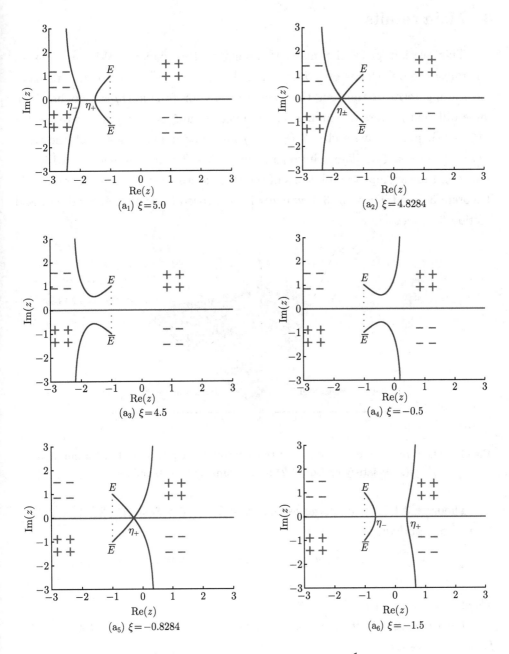

Fig. 2 Sign signatures of $\mathrm{Im}(f)$ for $A=1, \beta=\dfrac{1}{2}$ and $\mu=3$

3 Main results

This section gives the main theorems of this paper. Without loss of generality, we only focus on $2\beta A^2 + 2\sqrt{2}A > \mu > 2\beta A^2 - 2\sqrt{2}A$. In this case, the leading term of the long-time asymptotics for function $q(x,t)$ behaves in three different sectors depending on the magnitude $\xi = x/t$: plane wave sector I ($\xi > \xi_2$), plane wave sector II ($\xi < \xi_1$) and modulated genus 1 elliptic wave sector ($\xi_1 < \xi < \xi_2$). The solution regions in $x - t$ plane are shown in Fig. 3. The long-time asymptotic solutions of the three sectors are given in Theorem 3.1, Theorem 3.2 and Theorem 3.3, which will be proved in Section 4, Section 5 and Section 6, respectively.

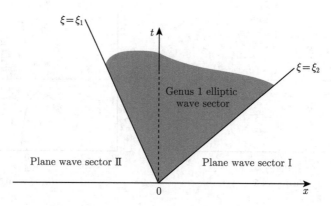

Fig. 3   The three asymptotic sectors of the focusing KE equation (1.1) with initial value $q(x,0)$ satisfying Eq. (1.2) in x-t plane for $\xi_1 < 0$ and $\xi_2 > 0$

Theorem 3.1   *In the plane wave sector I, i.e.* $\xi > \xi_2 = \mu - 2\beta A^2 + 2\sqrt{2}A$, *for* $t \to \infty$, *we have*

$$q(x,t)e^{2i\beta \int_x^{+\infty}(|q(y,t)|^2 - A^2)dy} = Ae^{i[\mu x + (A^2 + 2\beta^2 A^4 - \frac{\mu^2}{2})t + 2g(\infty) + \theta_+]} + O(t^{-1/2}), \tag{3.33}$$

where $g(\infty)$ *is given in Eq.* (4.53).

Theorem 3.2   *In the plane wave sector II, i.e.* $\xi < \xi_1 = \mu - 2\beta A^2 - 2\sqrt{2}A$, *for* $t \to \infty$, *we have*

$$q(x,t)e^{2i\beta \int_x^{+\infty}(|q(y,t)|^2 - A^2)dy} = Ae^{i[\mu x + (A^2 + 2\beta^2 A^4 - \frac{\mu^2}{2})t + 2g(\infty) + \theta_-]} + O(t^{-1/2}), \tag{3.34}$$

where $g(\infty)$ *is given in Eq.* (5.94).

Theorem 3.3 *In the modulated genus 1 elliptic wave sector, i.e., $0 < \xi < \xi_2 = \mu - 2\beta A^2 + 2\sqrt{2}A$, for $t \to \infty$, we have*

$$q(x,t)e^{2i\beta \int_x^{+\infty}(|q(y,t)|^2 - A^2)dy}$$

$$= (A + \text{Im}(\alpha)) \frac{\Theta\left(\frac{\Omega t}{2\pi} + \frac{\omega}{2\pi} + \Xi - u_\infty + c\right)\Theta(u_\infty + c)}{\Theta\left(\frac{\Omega t}{2\pi} + \frac{\omega}{2\pi} + \Xi + u_\infty + c\right)\Theta(-u_\infty + c)}$$

$$\cdot e^{i[\mu x + (A^2 + 2\beta^2 A^4 - \frac{\mu^2}{2} + 2g(\infty))t + 2\widehat{G}(\infty) + e^{i\theta} +]} + O(t^{-1/2}), \qquad (3.35)$$

where $\text{Im}(\alpha) = b, \Omega, \hat{g}(\infty), \omega, \widehat{G}(\infty), c, u_\infty$ *are given by Eqs. (6.115), (6.128), (6.129), (6.134), (6.135), (6.153), (6.156), respectively.*

Remark 3.1 *For $\xi_1 < \xi < 0$ the asymptotic solution $q(x,t)$ is also in the modulated genus 1 elliptic wave sector, which can also be derived by the same procedure in Section 6. We don't give the details here.*

Remark 3.2 *If $\xi_1 > 0$, i.e., $\mu \geqslant 2\beta A^2 + 2\sqrt{2}A$, we have $\xi_2 > 0$ because of $A > 0$. We can also get the long-time asymptotics of solution function $q(x,t)$ by repeating the processes in Section 4-6. In this case, the solution regions in $x - t$ plane are shown in Fig. 4. We also don't give the details in the present paper.*

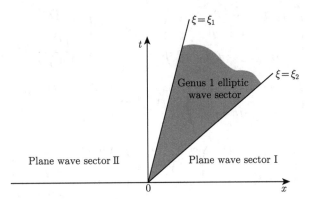

Fig. 4   The three asymptotic sectors of the focusing KE equation (1.1) with initial value $q(x,0)$ satisfying Eq. (1.2) in x-t plane for $\xi_1 > 0$

4 Plane wave region I

Assume $\xi > \xi_2$ then $\mathrm{Im}(f) = 0$ has two real roots $\eta_\pm$ on the real axis, which are on the left side of the γ cut, see Fig. 2(a_1). Starting from the original RHP $P^{(0)}$ and deforming the contour $\Sigma^{(0)}$ we explore the long-time asymptotics of solution for the focusing KE equation (1.1) with boundary conditions of the initial value $q(x,0)$ in Eq. (1.2). In this case, the point η_- is a critical point which is of great importance for the deformation of the original contour $\Sigma^{(0)}$.

4.1 $P^{(0)} \to P^{(1)}$: A transformation between $M^{(0)}$ and $M^{(1)}$

Note that the jump matrices $V_1^{(0)}, V_2^{(0)}$ and $V_3^{(0)}$ are factored as

$$V_1^{(0)} = V_2^{(1)} V_6^{(1)} V_1^{(1)}, \quad \text{on } (-\infty, \eta_-),$$
$$V_1^{(0)} = V_4^{(1)} V_3^{(1)}, \quad \text{on } (\eta_-, \infty),$$
$$V_2^{(0)} = (V_{3-}^{(1)})^{-1} V_5^{(1)} V_{3+}^{(1)}, \quad \text{on the } \gamma^+ \text{ cut},$$
$$V_3^{(0)} = V_{4-}^{(1)} V_5^{(1)} (V_{4+}^{(1)})^{-1}, \quad \text{on the } \gamma^- \text{ cut},$$

where $V_j^{(1)}, j = 1, 2, \cdots, 6$ are defined by

$$V_1^{(1)} = \begin{pmatrix} D^{-\frac{1}{2}} & \dfrac{D^{\frac{1}{2}} \bar{\rho}}{1 + \rho\bar{\rho}} e^{-ift} \\ 0 & D^{\frac{1}{2}} \end{pmatrix}, \quad V_2^{(1)} = \begin{pmatrix} D^{-\frac{1}{2}} & 0 \\ \dfrac{D^{\frac{1}{2}} \rho}{1 + \rho\bar{\rho}} e^{ift} & D^{\frac{1}{2}} \end{pmatrix},$$

$$V_3^{(1)} = \begin{pmatrix} D^{-\frac{1}{2}} & 0 \\ D^{-\frac{1}{2}} \rho e^{ift} & D^{\frac{1}{2}} \end{pmatrix}, \quad V_4^{(1)} = \begin{pmatrix} D^{-\frac{1}{2}} & D^{-\frac{1}{2}} \bar{\rho} e^{-ift} \\ 0 & D^{\frac{1}{2}} \end{pmatrix},$$

$$V_5^{(1)} = i \begin{pmatrix} 0 & e^{i\theta_+} \\ e^{-i\theta_+} & 0 \end{pmatrix}, \quad V_6^{(1)} = \begin{pmatrix} 1 + \rho^2 & 0 \\ 0 & \dfrac{1}{1 + \rho^2} \end{pmatrix}. \quad (4.36)$$

Following the procedure in Refs. [29-31] and making use of the factorizations above, deform the contour $\Sigma^{(0)}$ as in Fig. 5 then we obtain a new contour $\Sigma^{(1)}$ in Fig. 6.

Define

$$M^{(1)} = M^{(0)} G(z), \quad (4.37)$$

where

$$G(z) = \begin{cases} (V_1^{(1)})^{-1}, & \text{on } z \in \text{I}, \\ V_2^{(1)}, & \text{on } z \in \text{II}, \\ (V_3^{(1)})^{-1}, & \text{on } z \in \text{III} \cup \text{V}, \\ V_4^{(1)}, & \text{on } z \in \text{IV} \cup \text{VI}, \\ I, & \text{on } z \in \text{other regions}. \end{cases} \tag{4.38}$$

Then this new matrix-value function $M^{(1)}$ solves the equivalent RH problem

$$P^{(1)} = \{M^{(1)}, \Sigma^{(1)}, V^{(1)}, I \text{ as } z \to \infty\}, \tag{4.39}$$

where $V^{(1)}$ is given in Eq. (4.36) and the contour $\Sigma^{(1)}$ is shown in Fig. 6.

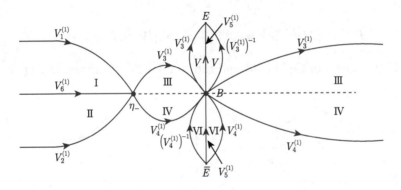

Fig. 5   The initial deformation of the contour $\Sigma^{(0)}$

4.2 $P^{(1)} \to P^{(2)}$: Removing the jump across the cut $(-\infty, \eta_-)$

In this section, we remove the jump across the cut $(-\infty, \eta_-)$ by introducing the function $\delta(z)$, which solves the scalar RH problem

$$P^{(\delta)} = \{\delta(z), (-\infty, \eta_-), V^{(\delta)} = 1 + \rho^2(z), \text{ as } z \to \infty\}, \tag{4.40}$$

with jump condition $\delta_+ = \delta_-(1 + \rho^2(z))$. The solution $\delta(z)$ of this scalar RH problem is

$$\delta(z) = e^{\frac{1}{2\pi i} \int_{-\infty}^{\eta_-} \frac{\ln(1+\rho^2(s))}{s-z} ds}. \tag{4.41}$$

Define

$$M^{(2)} = M^{(1)} \delta^{-\sigma_3}, \tag{4.42}$$

then we have $V^{(2)} = \delta_-^{\sigma_3} V^{(1)} \delta_+^{-\sigma_3}$ and

$$V_1^{(2)} = \begin{pmatrix} D^{-\frac{1}{2}} & \dfrac{\delta^2 D^{\frac{1}{2}} \bar{\rho}}{1+\rho\bar{\rho}} e^{-ift} \\ 0 & D^{\frac{1}{2}} \end{pmatrix}, \quad V_2^{(2)} = \begin{pmatrix} D^{-\frac{1}{2}} & 0 \\ \dfrac{\delta^{-2} D^{\frac{1}{2}} \rho}{1+\rho\bar{\rho}} e^{ift} & D^{\frac{1}{2}} \end{pmatrix},$$

$$V_3^{(2)} = \begin{pmatrix} D^{-\frac{1}{2}} & 0 \\ \delta^{-2} D^{-\frac{1}{2}} \rho e^{ift} & D^{\frac{1}{2}} \end{pmatrix},$$

$$V_4^{(2)} = \begin{pmatrix} D^{-\frac{1}{2}} & \delta^2 D^{-\frac{1}{2}} \bar{\rho} e^{-ift} \\ 0 & D^{\frac{1}{2}} \end{pmatrix}, \quad V_5^{(2)} = i \begin{pmatrix} 0 & \delta^2 e^{i\theta_+} \\ \delta^{-2} e^{-i\theta_+} & 0 \end{pmatrix}. \quad (4.43)$$

Since $\delta \sim 1$ as $z \to \infty$, we have $M^{(2)} \sim I$ as $z \to \infty$. Then $M^{(2)}$ solves the RH problem

$$P^{(2)} = \{M^{(2)}, \Sigma^{(2)} = \Sigma^{(1)} \setminus (-\infty, \eta_-), V^{(2)}, I \text{ as } z \to \infty\}, \quad (4.44)$$

where $V^{(2)}$ is given in Eq. (4.43) and the contour $\Sigma^{(2)}$ is shown in Fig. 7.

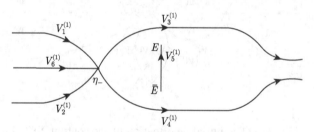

Fig. 6 The contour $\Sigma^{(1)}$ of the RH problem $P^{(1)}$ for the plane wave sector I

4.3 $P^{(2)} \to P^{(3)}$: Removing the term $D(z)$ in the jump matrices $V^{(2)}$

Now we cancel the term $D(z)$ in the jump matrices $V^{(2)}$ such that the diagonal elements in matrices $V^{(2)}$ are identity elements. In doing so, taking the following transformation on complex plane with contour $\Sigma^{(2)}$,

$$M^{(3)} = M^{(2)} \widetilde{G}(z), \quad (4.45)$$

where

$$\widetilde{G}(z) = \begin{cases} D^{\sigma_3/2}, & \text{on } z \in \text{I}, \\ D^{-\sigma_3/2}, & \text{on } z \in \text{II}, \\ I, & \text{on } z \in \text{III} \cup \text{IV}. \end{cases} \quad (4.46)$$

Because $D(z) \sim 1$ as $z \to \infty$, we have $M^{(3)} \sim I$ as $z \to \infty$. Then the new matrix function $M^{(3)}$ solves the equivalent RH problem

$$P^{(3)} = \{M^{(3)}, \Sigma^{(3)} = \Sigma^{(2)}, V^{(3)}, I \text{ as } z \to \infty\}, \qquad (4.47)$$

where the contour $\Sigma^{(3)}$ is shown in Fig. 7 and the jump matrices $V^{(3)}$ are given below

$$V_1^{(3)} = \begin{pmatrix} 1 & \dfrac{\delta^2 \bar{\rho}}{1+\rho\bar{\rho}} e^{-ift} \\ 0 & 1 \end{pmatrix}, \quad V_2^{(3)} = \begin{pmatrix} 1 & 0 \\ \dfrac{\delta^{-2}\rho}{1+\rho\bar{\rho}} e^{ift} & 1 \end{pmatrix},$$

$$V_3^{(3)} = \begin{pmatrix} 1 & 0 \\ \delta^{-2}\rho e^{ift} & 1 \end{pmatrix}, \quad V_4^{(3)} = \begin{pmatrix} 1 & \delta^2 \bar{\rho} e^{-ift} \\ 0 & 1 \end{pmatrix},$$

$$V_5^{(3)} = i \begin{pmatrix} 0 & \delta^2 e^{i\theta_+} \\ \delta^{-2} e^{-i\theta_+} & 0 \end{pmatrix}. \qquad (4.48)$$

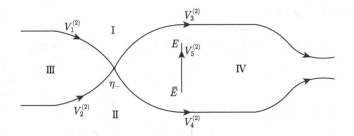

Fig. 7 The contour $\Sigma^{(2)}$ of the RH problem $P^{(2)}$ for the plane wave sector I

4.4 $P^{(3)} \to P^{(4)}$: g function and model problem $P^{(\mathrm{mod})}$

Introduce a g-function and define

$$M^{(4)} = M^{(3)} e^{ig(z)\sigma_3}, \qquad (4.49)$$

where $g(z)$ is analytic off the γ branch cut. On γ cut the jump matrix $V_5^{(3)}$ becomes

$$V_5^{(4)} = e^{-ig_-(z)\sigma_3} V_5^{(3)} e^{ig_+(z)\sigma_3} = \begin{pmatrix} 0 & i\delta^2 e^{i\theta_+} e^{-i(g_+ + g_-)} \\ i\delta^{-2} e^{-i\theta_+} e^{i(g_+ + g_-)} & 0 \end{pmatrix}. \qquad (4.50)$$

If imposing
$$g_+ + g_- + \frac{1}{\pi}\int_{-\infty}^{\eta_-}\frac{\ln(1+\rho(\zeta)^2)}{\zeta-z}d\zeta = 0, \tag{4.51}$$
the jump matrix $V_5^{(4)}$ is a constant matrix
$$V_5^{(4)} = i\begin{pmatrix} 0 & e^{i\theta_+} \\ e^{-i\theta_+} & 0 \end{pmatrix}.$$

Noting $\lambda_+ = -\lambda_-$ on the λ cut, by the Plemelj's formula [46], we have
$$g(z) = \frac{\lambda(z)}{2\pi^2 i}\int_{\gamma\,cut}\frac{1}{(s-z)\lambda(s)}\int_{-\infty}^{\eta_-}\frac{\ln(1+\rho(s)^2)}{\zeta-s}d\zeta ds. \tag{4.52}$$

When $z \to \infty$, we have $g(z) \sim g(\infty)$ with
$$g(\infty) = -\frac{1}{2\pi^2 i}\int_{\gamma\,cut}\frac{1}{\lambda(s)}\int_{-\infty}^{\eta_-}\frac{\ln(1+\rho(s)^2)}{\zeta-s}d\zeta ds, \tag{4.53}$$

which is dependent on parameters ξ, A, β and μ, but is independent of z. Thus $M^{(4)}$ solve the RH problem
$$P^{(4)} = \{M^{(4)}, \Sigma^{(4)} = \Sigma^{(3)}, V^{(4)}, e^{ig(\infty)\sigma_3} \text{ as } z \to \infty\}, \tag{4.54}$$

with
$$V_1^{(4)} = \begin{pmatrix} 1 & \frac{\delta^2\bar{\rho}}{1+\rho\bar{\rho}}e^{-i(ft+2g)} \\ 0 & 1 \end{pmatrix}, \quad V_2^{(4)} = \begin{pmatrix} 1 & 0 \\ \frac{\delta^{-2}\rho}{1+\rho\bar{\rho}}e^{i(ft+2g)} & 1 \end{pmatrix},$$
$$V_3^{(4)} = \begin{pmatrix} 1 & 0 \\ \delta^{-2}\rho e^{i(ft+2g)} & 1 \end{pmatrix}, \quad V_4^{(4)} = \begin{pmatrix} 1 & \delta^2\bar{\rho}e^{-i(ft+2g)} \\ 0 & 1 \end{pmatrix},$$
$$V_5^{(4)} = i\begin{pmatrix} 0 & e^{i\theta_+} \\ e^{-i\theta_+} & 0 \end{pmatrix}. \tag{4.55}$$

The jump matrices $V_j^{(4)}$ ($j = 1, 2, 3, 4$) decay exponentially to the identity away from the point η_- as $t \to \infty$. The model problem will provide the leading term of the asymptotic solution $q(x, t)$. Let $M^{(\text{mod})}$ solve the RH problem
$$P^{(\text{mod})} = \{M^{(\text{mod})}, \Sigma^{(\text{mod})} = \gamma\,cut, V^{(\text{mod})}, e^{ig(\infty)\sigma_3} \text{ as } z \to \infty\}, \tag{4.56}$$

with
$$V^{(mod)} = V_5^{(4)} = i \begin{pmatrix} 0 & e^{i\theta_+} \\ e^{-i\theta_+} & 0 \end{pmatrix},$$

which has an exact solution
$$M^{(mod)} = e^{ig(\infty)\sigma_3} \begin{pmatrix} \frac{1}{2}(K + K^{-1}) & \frac{e^{i\theta_+}}{2}(K - K^{-1}) \\ -\frac{e^{-i\theta_+}}{2}(K - K^{-1}) & \frac{1}{2}(K + K^{-1}) \end{pmatrix}, \quad (4.57)$$

where
$$K = \left(\frac{z - \beta A^2 + \frac{\mu}{2} - iA}{z - \beta A^2 + \frac{\mu}{2} + iA} \right)^{1/4}. \quad (4.58)$$

4.5 Proof of Theorem 3.1

The matrices $\{V_j^{(3)}\}_{j=1}^4$ decay exponentially to identity away from the stationary phase point η_- as $t \to \infty$. However, the decay on the contour $\Sigma^{(3)}$ is not uniform in z near η_-, so a condition is necessary for the error estimate. Let $r_{\eta_-}^\epsilon$ be the circle of radius ϵ centered at $z = \eta_-$. Here ϵ is a sufficiently small real number. Deform the contour $\Sigma^{(3)}$ to the contours $\Sigma^{(app)}$ and $\Sigma^{(err)}$, as shown in Fig. 8. Thus we can rewrite $M^{(3)}$ as
$$M^{(3)} = M^{(app)} M^{(err)}, \quad (4.59)$$

where $M^{(app)}$ includes the jump across the γ branch cut and near η_-, i.e. $\Sigma^{(app)}$ in Fig. 8, and the higher-order contribution from the contour $\Sigma^{(3)}$ is factored out into the error term $M^{(err)}$, see $\Sigma^{(err)}$ in Fig. 8.

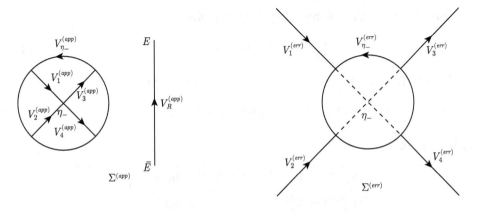

Fig. 8 The contours $\Sigma^{(app)}$ and $\Sigma^{(err)}$ from deformation of contour $\Sigma^{(3)}$

Define
$$M^{(app)} = \begin{cases} \text{parametrix of } M^{(3)}, & \text{inside } r_{\eta_-}^\epsilon, \\ M^{(\text{mod})} e^{-ig(\infty)\sigma_3}, & \text{outside } r_{\eta_-}^\epsilon, \end{cases} \quad (4.60)$$

which means that the function $M^{(app)}$ satisfies the same jump conditions as $M^{(3)}$ inside $r_{\eta_-}^\epsilon$, and it will have a jump $V_{\eta_-}^{(app)}$ across the circle $r_{\eta_-}^\epsilon$.

Take the Laurent series for matrices $M^{(3)}$, $M^{(app)}$ and $M^{(err)}$ as

$$\begin{aligned} M^{(3)} &= I + M_1^{(3)} z^{-1} + M_2^{(3)} z^{-2} + \cdots, & z \to \infty, \\ M^{(app)} &= I + M_1^{(app)} z^{-1} + M_2^{(app)} z^{-2} + \cdots, & z \to \infty, \\ M^{(err)} &= I + M_1^{(err)} z^{-1} + M_2^{(err)} z^{-2} + \cdots, & z \to \infty, \end{aligned} \quad (4.61)$$

where the $M_i^{(3)}$, $M_i^{(app)}$ and $M_i^{(err)}$ are independent of z, then from the factorization (4.59), comparing the coefficients of z^{-1}, we have

$$M_1^{(3)} = M_1^{(app)} + M_1^{(err)}, \quad (4.62)$$

which indicates that

$$(M_1^{(3)})_{12} = (M_1^{(app)} + M_1^{(err)})_{12}. \quad (4.63)$$

The following Lemmas aims to formulate the expressions of $(M_1^{(app)})_{12}$ and $(M_1^{(err)})_{12}$, which are useful to the proof of Theorem 3.1.

Lemma 4.1 *For large z, the term $(M_1^{(app)})_{12}$ is of the following form*

$$(M_1^{(app)})_{12} = -\frac{iA}{2} e^{i(2g(\infty)+\theta_+)}. \quad (4.64)$$

Proof According to Ref. [29], it is seen that

$$V_{\eta_-}^{(app)} = I + O(t^{-\frac{1}{2}}), \quad \text{on } r_{\eta_-}^\epsilon. \quad (4.65)$$

So the matrix-value function $M^{(app)}$ satisfies the RH problem

$$P^{(app)} = \{M^{(app)}, \Sigma^{(app)}, V^{(app)}, I \text{ as } z \to \infty\}, \quad (4.66)$$

with jump matrices

$$\begin{aligned} V_i^{(app)} &= V_i^{(3)} \ (i = 1, 2, 3, 4), & \text{inside } r_{\eta_-}^\epsilon, \\ V_R^{(app)} &= V_5^{(3)}, & \text{on } \gamma, \\ V_{\eta_-}^{(app)} &= I + O(t^{-\frac{1}{2}}), & \text{on } r_{\eta_-}^\epsilon. \end{aligned}$$

Outside $r^{\epsilon}_{\eta_-}$, function $M^{(app)} = M^{(mod)}e^{-ig(\infty)\sigma_3}$. So take the Laurent series of $M^{(mod)}$ at infinity, i.e.

$$M^{(mod)} = e^{ig(\infty)\sigma_3} + M_1^{(mod)}z^{-1} + M_2^{(mod)}z^{-2} + \cdots, \quad z \to \infty,$$

where the $M_i^{(mod)}$ are independent of z. By using Eq. (4.61), we have $M_1^{(app)} = M_1^{(mod)}e^{-ig(\infty)\sigma_3}$, where $M_1^{(mod)}$ is given by the explicit solution $M^{(mod)}$ in Eq. :

$$M_1^{(mod)} = \lim_{z \to \infty} z[M^{(mod)} - e^{ig(\infty)\sigma_3}]$$

$$= e^{ig(\infty)\sigma_3} \lim_{z \to \infty} z \begin{pmatrix} \frac{1}{2}(K + K^{-1}) - 1 & \frac{e^{i\theta_+}}{2}(K - K^{-1}) \\ -\frac{e^{-i\theta_+}}{2}(K - K^{-1}) & \frac{1}{2}(K + K^{-1}) - 1 \end{pmatrix}.$$

Note from the expression of K in Eq. (4.58) that

$$K - K^{-1} = -iAz^{-1} + O(z^{-1}), \quad \text{as } z \to \infty,$$

so the term $(M_1^{(app)})_{12}$ can be calculated as

$$(M_1^{(app)})_{12} = e^{2ig(\infty)} \lim_{z \to \infty} z \frac{e^{i\theta_+}}{2}(K - K^{-1}) = -\frac{iA}{2}e^{i(2g(\infty)+\theta_+)}.$$

□

Lemma 4.2 *Matrix-value function $M^{(err)}$ satisfies the RH problem*

$$P^{(err)} = \{M^{(err)}, \Sigma^{(err)}, V^{(err)}, I \text{ as } z \to \infty\}, \tag{4.67}$$

with the jump matrices

$$V_i^{(err)} = M^{(app)}V_i^{(3)}(M^{(app)})^{-1} = I + O(e^{-4\epsilon^2 t}) \ (i = 1, 2, 3, 4), \quad \text{outside } r^{\epsilon}_{\eta_-},$$
$$V_{\eta_-}^{(err)} = M_-^{(app)}(V_{\eta_-}^{(app)})^{-1}(M_-^{(app)})^{-1} = I + O(t^{-\frac{1}{2}}), \quad \text{on } r^{\epsilon}_{\eta_-}.$$

Proof By the factorization (4.59), we have

$$\begin{aligned}V^{(err)} &= (M_-^{(err)})^{-1}M_+^{(err)} \\ &= M_-^{(app)}(M_-^{(3)})^{-1}M_+^{(3)}(M_+^{(app)})^{-1} \\ &= M_-^{(app)}V^{(3)}(M_-^{(app)}V^{(app)})^{-1} \\ &= M_-^{(app)}V^{(3)}(V^{(app)})^{-1}(M_-^{(app)})^{-1}.\end{aligned}$$

Because $M_-^{(app)} = M^{(app)}$ and as $z \to \infty$ we have $V_i^{(app)} = I$ ($i = 1, 2, 3, 4$) outside $r_{\eta_-}^\epsilon$, and $V^{(3)} = I$ on $r_{\eta_-}^\epsilon$, it follows

$$V_i^{(err)} = M^{(app)} V_i^{(3)} (M^{(app)})^{-1} \ (i = 1, 2, 3, 4), \text{ outside } r_{\eta_-}^\epsilon,$$
$$V_{\eta_-}^{(err)} = M_-^{(app)} (V_{\eta_-}^{(app)})^{-1} (M_-^{(app)})^{-1}, \text{ on } r_{\eta_-}^\epsilon.$$

Next, it is necessary to investigate the convergence rate of $V_i^{(err)}$ ($i = 1, 2, 3, 4$) outside $r_{\eta_-}^\epsilon$. Here we only give the details for $V_1^{(err)}$ outside $r_{\eta_-}^\epsilon$.

On the jump contour $z = \eta_- + u e^{\frac{3\pi}{4} i}$ ($u \geqslant \epsilon$), where ϵ is large enough, the jump matrix $V_1^{(err)}$ is

$$V_1^{(err)} = V_1^{(3)} = \begin{pmatrix} 1 & \dfrac{\delta^{-2} \bar\rho}{1 + \rho \bar\rho} e^{-ift} \\ 0 & 1 \end{pmatrix}.$$

For $z \to \infty$, we have

$$f = 2\sqrt{\left(z + \frac{\mu}{2} - \beta A^2\right)^2 + A^2} \left(\xi + z - \frac{\mu}{2} + \beta A^2\right)$$
$$\sim 2\left(z + \frac{\mu}{2} - \beta A^2\right)\left(\xi + z - \frac{\mu}{2} + \beta A^2\right),$$

which denotes that

$$\text{Im}(f) \sim 2u\left(-u + \sqrt{2}\eta_- + \frac{\sqrt{2}}{2}\xi\right) \sim -2u^2,$$

so it is observed that

$$|e^{-ift}| = |e^{\text{Im}(f)t}| \sim e^{-2tu^2} \leqslant e^{-2\epsilon^2 t}.$$

Thus it follows

$$V_1^{(err)} = M^{(app)} V_1^{(3)} (M^{(app)})^{-1} = I + O(e^{-2\epsilon^2 t}).$$

Similarly, we can also get

$$V_i^{(err)} = I + O(e^{-2\epsilon^2 t}), \quad i = 2, 3, 4.$$

By using Eq. (4.65), it is easy to see that

$$V_{\eta_-}^{(err)} = M_-^{(app)} (V_{\eta_-}^{(app)})^{-1} (M_-^{(app)})^{-1} = I + O(t^{-\frac{1}{2}}), \quad \text{on } r_{\eta_-}^\epsilon.$$

This completes the proof of this Lemma. □

Lemma 4.3 For large z, the term $M_1^{(err)}(x,t)$ admits the estimate as follow

$$|M_1^{(err)}(x,t)| = O(t^{-1/2}), \quad \text{as } z \to \infty. \tag{4.68}$$

Proof Given an oriented contour Σ, define the Cauchy transform C_Σ as

$$(C_\Sigma(f))(z) = \frac{1}{2\pi i} \int_\Sigma \frac{f(\zeta)}{\zeta - z} \, d\zeta.$$

Further define $C_\Sigma^+(f)$ and $C_\Sigma^-(f)$ to be the nontangential limits of $C_\Sigma(f)$ approaching Σ from left and right, respectively. For any matrix V defined on Σ take

$$C_V^- f = C_\Sigma^-(f(V-I)).$$

It is easy to know that both operators C_Σ and C_V^- are bounded and linear.

From the definition of the RH problem $P^{(err)}$ in Lemma 4.2, we have

$$M_+^{(err)} = M_-^{(err)} V^{(err)},$$

where $V^{(err)}$ is given in Lemma 4.2. Thus we obtain

$$(M_+^{(err)} - I) - (M_-^{(err)} - I) = M_+^{(err)} - M_-^{(err)} = M_-^{(err)}(V^{(err)} - I),$$

which is solved by Plemelj's formula [46] as

$$M^{(err)} - I = C_{\Sigma^{(err)}} M_-^{(err)}(V^{(err)} - I)$$
$$= \frac{1}{2\pi i} \int_{\Sigma^{(err)}} \frac{M_-^{(err)}(\zeta)(V^{(err)}(\zeta) - I)}{\zeta - z} d\zeta$$
$$= -\frac{1}{2\pi i z} \int_{\Sigma^{(err)}} M_-^{(err)}(\zeta)(V^{(err)}(\zeta) - I) d\zeta + O(z^{-2}). \tag{4.69}$$

Thus we have

$$M_1^{(err)} = \lim_{z \to \infty} z(M^{(err)} - I)$$
$$= -\frac{1}{2\pi i} \int_{\Sigma^{(err)}} M_-^{(err)}(\zeta)(V^{(err)}(\zeta) - I) d\zeta. \tag{4.70}$$

Since

$$M_-^{(err)}(\zeta)(V^{(err)}(\zeta) - I) = (M_-^{(err)}(\zeta) - I)(V^{(err)}(\zeta) - I) + (V^{(err)}(\zeta) - I),$$

by Holder's inequality, we have

$$|M_1^{(err)}| \leqslant C_1 \|M_-^{(err)} - I\|_{L^2} \|V^{(err)} - I\|_{L^2} + C_2 \|V^{(err)} - I\|_{L^1} \tag{4.71}$$

for certain positive constants C_1 and C_2. By Eq. (4.69), we have

$$\begin{aligned}M_-^{(err)} - I &= C_{\Sigma(err)}^- M_-^{(err)}(V^{(err)} - I) \\ &= C_{V(err)}^- M_-^{(err)} \\ &= C_{V(err)}^-(M_-^{(err)} - I) + C_{V(err)}^- I.\end{aligned}$$

So the following identity holds

$$(I - C_{V(err)}^-)(M_-^{(err)} - I) = C_{V(err)}^- I,$$

i.e.,

$$M_-^{(err)} - I = (I - C_{V(err)}^-)^{-1} C_{V(err)}^- I.$$

Thus it follows

$$\begin{aligned}\|M_-^{(err)} - I\|_{L^2} &\leqslant \|(I - C_{V(err)}^-)^{-1}\| \|C_{V(err)}^- I\|_{L^2} \\ &= \|(I - C_{V(err)}^-)^{-1}\| \|C_{\Sigma(err)}^- (V^{(err)} - I)\|_{L^2} \\ &\leqslant \|(I - C_{V(err)}^-)^{-1}\| \|C_{\Sigma(err)}^-\| \|V^{(err)} - I\|_{L^2}.\end{aligned}$$

Since the operators $(I - C_{V(err)}^-)^{-1}$ and $C_{\Sigma(err)}^-$ are bounded, we have

$$\|M_-^{(err)} - I\|_{L^2} \leqslant C_3 \|V^{(err)} - I\|_{L^2}, \tag{4.72}$$

for certain positive constant C_3. Substituting Eq. (4.72) into Eq. (4.71) follows

$$|M_1^{(err)}| \leqslant C_1 C_3 \|V^{(err)} - I\|_{L^2} + C_2 \|V^{(err)} - I\|_{L^1}.$$

Furthermore, Lemma 4.2 shows

$$\|V^{(err)} - I\|_{L^p} = O(t^{-\frac{1}{2}}), \quad p = 1, 2,$$

so we finally have

$$|M_1^{(err)}| = O(t^{-\frac{1}{2}}).$$

$\square$

Proof of Theorem 3.1 Note that

$$\begin{aligned}M^{(1)} &= M^{(0)} G(z), \quad G(z) \sim I, \quad z \to \infty, \\ M^{(2)} &= M^{(1)} \delta^{-\sigma_3}, \quad \delta \sim 1, \quad z \to \infty,\end{aligned}$$

$$M^{(3)} = M^{(2)}\tilde{G}(z), \quad \tilde{G}(z) \sim I, \quad z \to \infty.$$

From the expression of $q(x,t)$ in Eq. (2.29) and applying the successive deformation of the original RH problem, we get

$$q(x,t) = 2i(M_1^{(3)})_{12} e^{i[\mu x + (A^2 + 2\beta^2 A^4 - \frac{\mu^2}{2})t]} e^{-2i\beta \int_x^{+\infty}(|q(y,t)|^2 - A^2)dy}. \quad (4.73)$$

Further making use of Eq. (4.63), Lemma 4.1, and Lemma 4.3, we finally obtain the asymptotic solution of $q(x,t)$ in the plane wave sector I as

$$q(x,t) e^{2i\beta \int_x^{+\infty}(|q(y,t)|^2 - A^2)dy}$$
$$= 2i(M_1^{(app)} + M_1^{(err)})_{12} e^{i[\mu x + (A^2 + 2\beta^2 A^4 - \frac{\mu^2}{2})t]}$$
$$= A e^{i[\mu x + (A^2 + 2\beta^2 A^4 - \frac{\mu^2}{2})t + 2g(\infty) + \theta_+]} + O(t^{-\frac{1}{2}}), \quad z \to \infty, \quad (4.74)$$

where $g(\infty)$ is given by Eq. (4.53), which is only dependent of A, μ, β and ξ with $\xi > \xi_2$. So, Theorem 3.1 is proved completely.

5 Plane wave region II

When $\xi < \xi_1 = \mu - 2\beta A^2 - 2\sqrt{2}A < 0$, we will go to another plane wave sector, where $\text{Im}(f) = 0$ has two real roots $\eta_\pm$ on the real axis and now the point η_+ is more important than the point η_-, see Fig. 2(a_6). Starting from the original RHP $P^{(0)}$ and deforming the contour $\Sigma^{(0)}$ we continue to study the long-time asymptotics of solution for the focusing KE equation (1.1) with boundary conditions of the initial value $q(x,0)$ in Eq. (1.2).

5.1 $P^{(0)} \to P^{(1)}$: A preliminary transformation between $M^{(0)}$ and $M^{(1)}$

We take the following transformation to rescale the original RH problem $P^{(0)}$ in (2.27)

$$M^{(1)} = M^{(0)} G(z), \quad (5.75)$$

where

$$G(z) = \begin{cases} \begin{pmatrix} a(z) & 0 \\ 0 & a^{-1}(z) \end{pmatrix}, & \text{on } z \in \mathbb{C}_+ \backslash \gamma^+, \\ \begin{pmatrix} \bar{a}^{-1}(z) & 0 \\ 0 & \bar{a}(z) \end{pmatrix}, & \text{on } z \in \mathbb{C}_- \backslash \gamma^-. \end{cases} \quad (5.76)$$

Because $a(z) \sim 1$ as $z \to \infty$, it follows $M^{(1)} \sim I$ as $z \to \infty$. Then $M^{(1)}$ solves the RH problem

$$P^{(1)} = \{M^{(1)}, \quad \widehat{\Sigma}^{(1)} = \Sigma^{(0)}, \quad V^{(1)}, \quad I \text{ as } z \to \infty\}, \tag{5.77}$$

where $V^{(1)} = \{V_j^{(1)}\}_{j=1}^3$ is given below

$$V_1^{(1)} = \begin{pmatrix} D^{-1} & \bar{r}(z)e^{-ift} \\ r(z)e^{ift} & D(1+r\bar{r}) \end{pmatrix},$$

$$V_2^{(1)} = \begin{pmatrix} \dfrac{\lambda - z - \dfrac{\mu}{2} + \beta A^2}{iAe^{i\theta_-}}\bar{r}(z)e^{-ift} & \dfrac{2i\lambda}{Ae^{-i\theta_-}}(1+r(z)\bar{r}(z)) \\ \dfrac{iA}{2\lambda e^{i\theta_-}} & \dfrac{\lambda + z + \dfrac{\mu}{2} - \beta A^2}{iAe^{-i\theta_-}}r(z)e^{ift} \end{pmatrix}, \tag{5.78}$$

$$V_3^{(1)} = \begin{pmatrix} \dfrac{i(\lambda + z + \dfrac{\mu}{2} - \beta A^2)}{Ae^{i\theta_-}}\bar{r}(z)e^{-ift} & \dfrac{iA}{2\lambda e^{-i\theta_-}} \\ \dfrac{2i\lambda}{Ae^{i\theta_-}}(1+r(z)\bar{r}(z)) & \dfrac{i\left(\lambda - z - \dfrac{\mu}{2} + \beta A^2\right)}{Ae^{-i\theta_-}}r(z)e^{ift} \end{pmatrix},$$

where $r(z) = b(z)/\bar{a}(z)$.

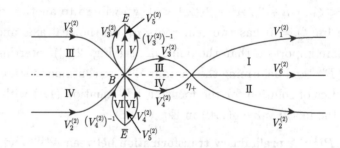

Fig. 9   The initial deformation of the contour $\widehat{\Sigma}^{(1)} = \Sigma^{(0)}$

5.2 $P^{(1)} \to P^{(2)}$: A transformation between $M^{(1)}$ and $M^{(2)}$

Note that the jump matrices $V_1^{(1)}, V_2^{(1)}$ and $V_3^{(1)}$ are factored as

$$V_1^{(1)} = V_2^{(2)} V_6^{(2)} V_1^{(2)}, \qquad \text{on } (\eta_+, \infty),$$

$$V_1^{(1)} = V_4^{(2)} V_3^{(2)}, \qquad \text{on } (-\infty, \eta_+),$$

$$V_2^{(1)} = (V_{3-}^{(2)})^{-1} V_5^{(2)} V_{3+}^{(2)}, \qquad \text{on the } \gamma^+ \text{ cut},$$

$$V_3^{(1)} = V_{4-}^{(2)} V_5^{(2)} (V_{4+}^{(2)})^{-1}, \qquad \text{on the } \gamma^- \text{ cut},$$

where $V_j^{(2)}, j = 1, 2, \cdots, 6$ are defined by

$$V_1^{(2)} = \begin{pmatrix} D^{-\frac{1}{2}} & 0 \\ \dfrac{D^{-\frac{1}{2}}r}{1+r\bar{r}}e^{ift} & D^{\frac{1}{2}} \end{pmatrix}, \quad V_2^{(2)} = \begin{pmatrix} D^{-\frac{1}{2}} & \dfrac{D^{-\frac{1}{2}}\bar{r}}{1+r\bar{r}}e^{-ift} \\ 0 & D^{\frac{1}{2}} \end{pmatrix},$$

$$V_3^{(2)} = \begin{pmatrix} D^{-\frac{1}{2}} & D^{\frac{1}{2}}\bar{r}e^{-ift} \\ 0 & D^{\frac{1}{2}} \end{pmatrix}, \quad V_4^{(2)} = \begin{pmatrix} D^{-\frac{1}{2}} & 0 \\ D^{\frac{1}{2}}re^{ift} & D^{\frac{1}{2}} \end{pmatrix},$$

$$V_5^{(2)} = i\begin{pmatrix} 0 & e^{i\theta_-} \\ e^{-i\theta_-} & 0 \end{pmatrix}, \quad V_6^{(2)} = \begin{pmatrix} \dfrac{1}{1+r\bar{r}} & 0 \\ 0 & 1+r\bar{r} \end{pmatrix}. \quad (5.79)$$

Deform the contour $\widehat{\Sigma}^{(1)}$ as in Fig. 9, then we arrive at a new contour $\widehat{\Sigma}^{(2)}$ in Fig. 10. Further, define

$$M^{(2)} = M^{(1)} \begin{cases} (V_1^{(2)})^{-1}, & \text{on } z \in \mathrm{I}, \\ V_2^{(2)}, & \text{on } z \in \mathrm{II}, \\ (V_3^{(2)})^{-1}, & \text{on } z \in \mathrm{III} \cup \mathrm{V}, \\ V_4^{(2)}, & \text{on } z \in \mathrm{IV} \cup \mathrm{VI}, \\ I, & \text{on } z \in \text{other regions}. \end{cases} \quad (5.80)$$

Then this new function $M^{(2)}$ solves the equivalent RH problem

$$P^{(2)} = \{M^{(2)}, \ \widehat{\Sigma}^{(2)}, \ V^{(2)}, \ I \text{ as } z \to \infty\}, \quad (5.81)$$

where $V^{(2)}$ is given in Eq. (5.79), and the contour $\widehat{\Sigma}^{(2)}$ is shown in Fig. 10.

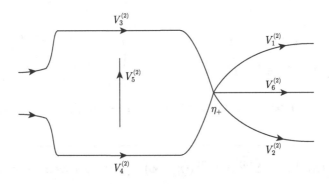

Fig. 10 The contour $\widehat{\Sigma}^{(2)}$ of the RH problem $P^{(2)}$ for the plane wave sector II

5.3 $P^{(2)} \to P^{(3)}$: Removing the jump across the cut (η_+, ∞)

In this subsection, similar to Sec. 4.2, we remove the jump across the cut (η_+, ∞) by introducing the function $\delta(z)$, which solves the scalar RH problem

$$P^{(\delta)} = \left\{ \delta(z), \quad (\eta_+, \infty), \quad V^{(\delta)} = \frac{1}{1 + r(z)\bar{r}(z)}, \quad 1 \text{ as } z \to \infty \right\}, \qquad (5.82)$$

with jump condition $\delta_+ = \delta_- \dfrac{1}{1 + r(z)\bar{r}(z)}$. The solution $\delta(z)$ is

$$\delta(z) = e^{\frac{1}{2\pi i} \int_{\eta_+}^{\infty} \frac{\ln\left(\frac{1}{1+r(s)\bar{r}(s)}\right)}{s - z} ds}. \qquad (5.83)$$

Further define

$$M^{(3)} = M^{(2)} \delta^{-\sigma_3}, \qquad (5.84)$$

we have $V^{(3)} = \delta_-^{\sigma_3} V^{(2)} \delta_+^{-\sigma_3}$ and

$$V_1^{(3)} = \begin{pmatrix} D^{-\frac{1}{2}} & 0 \\ \dfrac{\delta^{-2} D^{-\frac{1}{2}} r}{1 + r\bar{r}} e^{ift} & D^{\frac{1}{2}} \end{pmatrix}, \quad V_2^{(3)} = \begin{pmatrix} D^{-\frac{1}{2}} & \dfrac{\delta^2 D^{-\frac{1}{2}} \bar{r}}{1 + r\bar{r}} e^{-ift} \\ 0 & D^{\frac{1}{2}} \end{pmatrix},$$

$$V_3^{(3)} = \begin{pmatrix} D^{-\frac{1}{2}} & \delta^2 D^{\frac{1}{2}} \bar{r} e^{-ift} \\ 0 & D^{\frac{1}{2}} \end{pmatrix},$$

$$V_4^{(3)} = \begin{pmatrix} D^{-\frac{1}{2}} & 0 \\ \delta^{-2} D^{\frac{1}{2}} r e^{ift} & D^{\frac{1}{2}} \end{pmatrix}, \quad V_5^{(3)} = i \begin{pmatrix} 0 & \delta^2 e^{i\theta_-} \\ \delta^{-2} e^{-i\theta_-} & 0 \end{pmatrix}. \qquad (5.85)$$

Since $\delta \sim 1$ as $z \to \infty$, we have $M^{(3)} \sim I$ as $z \to \infty$. Then $M^{(3)}$ solves the RH problem

$$P^{(3)} = \{ M^{(3)}, \quad \widehat{\Sigma}^{(3)} = \widehat{\Sigma}^{(2)} \backslash (\eta_+, \infty), \quad V^{(3)}, \quad I \text{ as } z \to \infty \}, \qquad (5.86)$$

where $V^{(3)}$ is given in Eq. (5.85) and the contour $\widehat{\Sigma}^{(3)}$ is shown in Fig. 11.

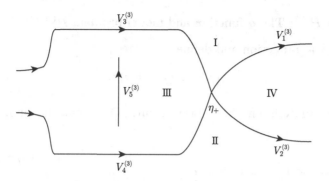

Fig. 11 The contour $\widehat{\Sigma}^{(3)}$ of the RH problem $P^{(3)}$ for the plane wave sector II

5.4 $P^{(3)} \to P^{(4)}$: Removing the term $D(z)$ in the jump matrices $V^{(3)}$

Now we cancel the term $D(z)$ in the jump matrices $V^{(3)}$ such that the diagonal elements in matrices $V^{(3)}$ are 1. In doing so, taking the following transformation on complex plane with contour $\widehat{\Sigma}^{(3)}$

$$M^{(4)} = M^{(3)}\tilde{G}(z), \qquad (5.87)$$

where matrix $\tilde{G}(z)$ is given in Eq. (4.46) and the regions I-IV are shown in Fig. 9. Because $D(z) \sim 1$ as $z \to \infty$, we have $M^{(4)} \sim I$ as $z \to \infty$. Then the new matrix function $M^{(4)}$ solves the equivalent RH problem

$$P^{(4)} = \{M^{(4)}, \quad \widehat{\Sigma}^{(4)} = \widehat{\Sigma}^{(3)}, \quad V^{(4)}, \quad I \text{ as } z \to \infty\}, \qquad (5.88)$$

where the contour $\widehat{\Sigma}^{(4)}$ is shown in Fig. 11 and the jump matrices $V^{(4)}$ are given below

$$V_1^{(4)} = \begin{pmatrix} 1 & 0 \\ \frac{\delta^{-2}r}{1+r\bar{r}}e^{ift} & 1 \end{pmatrix}, \quad V_2^{(4)} = \begin{pmatrix} 1 & \frac{\delta^2 \bar{r}}{1+r\bar{r}}e^{-ift} \\ 0 & 1 \end{pmatrix},$$

$$V_3^{(4)} = \begin{pmatrix} 1 & \delta^2 \bar{r} e^{-ift} \\ 0 & 1 \end{pmatrix}, \quad V_4^{(4)} = \begin{pmatrix} 1 & 0 \\ \delta^{-2}r e^{ift} & 1 \end{pmatrix},$$

$$V_5^{(4)} = i \begin{pmatrix} 0 & \delta^2 e^{i\theta_-} \\ \delta^{-2} e^{-i\theta_-} & 0 \end{pmatrix}. \qquad (5.89)$$

5.5 $P^{(4)} \to P^{(5)}$: The g function and model problem $P^{(\mathrm{mod})}$

Introduce a g-function and define

$$M^{(5)} = M^{(4)} e^{ig(z)\sigma_3}, \tag{5.90}$$

where $g(z)$ is analytic off the γ branch cut. On γ cut the jump matrix $V_5^{(4)}$ becomes

$$V_5^{(5)} = e^{-ig_-(z)\sigma_3} V_5^{(4)} e^{ig_+(z)\sigma_3} = \begin{pmatrix} 0 & i\delta^2 e^{i\theta_-} e^{-i(g_++g_-)} \\ i\delta^{-2} e^{-i\theta_-} e^{i(g_++g_-)} & 0 \end{pmatrix}. \tag{5.91}$$

If imposing

$$g_+ + g_- + \frac{1}{\pi} \int_{-\infty}^{\eta_-} \frac{\ln\left(\frac{1}{1+r(s)\bar{r}(s)}\right)}{\zeta - z} d\zeta = 0, \tag{5.92}$$

the jump matrix $V_5^{(5)}$ will be a constant matrix

$$V_5^{(4)} = \begin{pmatrix} 0 & ie^{i\theta_-} \\ ie^{-i\theta_-} & 0 \end{pmatrix}.$$

Noting $\lambda_+ = -\lambda_-$ on the λ cut, by the Plemelj's formula [46], we have

$$g(z) = \frac{\lambda(z)}{2\pi^2 i} \int_{\gamma \text{ cut}} \frac{1}{(s-z)\lambda(s)} \int_{\eta_+}^{\infty} \frac{\ln\left(\frac{1}{1+r(s)\bar{r}(s)}\right)}{\zeta - s} d\zeta ds. \tag{5.93}$$

When $z \to \infty$, we have $g(z) \sim g(\infty)$ with

$$g(\infty) = -\frac{1}{2\pi^2 i} \int_{\gamma \text{ cut}} \frac{1}{\lambda(s)} \int_{\eta_+}^{\infty} \frac{\ln\left(\frac{1}{1+r(s)\bar{r}(s)}\right)}{\zeta - s} d\zeta ds, \tag{5.94}$$

which is dependent on parameters ξ, A, β and μ, but is independent of z. Thus $M^{(5)}$ solve the RH problem

$$P^{(5)} = \{M^{(5)}, \widehat{\Sigma}^{(5)} = \widehat{\Sigma}^{(4)}, V^{(5)}, e^{ig(\infty)\sigma_3} \text{ as } z \to \infty\}, \tag{5.95}$$

with

$$V_1^{(5)} = \begin{pmatrix} 1 & 0 \\ \dfrac{\delta^{-2} r}{1+r\bar{r}} e^{i(ft+2g)} & 1 \end{pmatrix}, \quad V_2^{(5)} = \begin{pmatrix} 1 & \dfrac{\delta^2 \bar{r}}{1+r\bar{r}} e^{-i(ft+2g)} \\ 0 & 1 \end{pmatrix}, \tag{5.96}$$

$$V_3^{(5)} = \begin{pmatrix} 1 & \delta^2 \bar{r} e^{-i(ft+2g)} \\ 0 & 1 \end{pmatrix}, \quad V_4^{(5)} = \begin{pmatrix} 1 & 0 \\ \delta^{-2} r e^{i(ft+2g)} & 1 \end{pmatrix},$$

$$V_5^{(5)} = i \begin{pmatrix} 0 & e^{i\theta_-} \\ e^{-i\theta_-} & 0 \end{pmatrix}.$$

The jump matrices $V_j^{(5)}$ ($j = 1, 2, 3, 4$) decay exponentially to the identity away from the point η_+ as $t \to \infty$. As Sec. 4.4 and Sec. 4.5, the model problem will provide the leading terms for the solution. Let $M^{(\mathrm{mod})}$ solve the the RH problem

$$P^{(\mathrm{mod})} = \{M^{(\mathrm{mod})}, \quad \Sigma^{(\mathrm{mod})} = \gamma \text{ cut}, \quad V^{(\mathrm{mod})}, \quad e^{ig(\infty)\sigma_3} \text{ as } z \to \infty\}, \quad (5.97)$$

with

$$V^{(\mathrm{mod})} = V_5^{(5)} = \begin{pmatrix} 0 & ie^{i\theta_-} \\ ie^{-i\theta_-} & 0 \end{pmatrix},$$

which has solution

$$M^{(\mathrm{mod})} = e^{ig(\infty)\sigma_3} \begin{pmatrix} \frac{1}{2}(K + K^{-1}) & \frac{e^{i\theta_-}}{2}(K - K^{-1}) \\ -\frac{e^{-i\theta_-}}{2}(K - K^{-1}) & \frac{1}{2}(K + K^{-1}) \end{pmatrix}, \quad (5.98)$$

where

$$K = \left(\frac{z - \beta A^2 + \frac{\mu}{2} - iA}{z - \beta A^2 + \frac{\mu}{2} + iA} \right)^{1/4}.$$

Following the same procedure as Sec. 4.5 and replacing the parameters η_-, θ_+ in Sec. 4.5 with η_+, θ_-, respectively, the asymptotic solution of $q(x,t)$ in the plane wave sector II is obtained as

$$q(x,t) e^{2i\beta \int_x^{+\infty} (|q(y,t)|^2 - A^2) dy} = A e^{i[\mu x + (A^2 + 2\beta^2 A^4 - \frac{\mu^2}{2})t + 2g(\infty) + \theta_-]} + O(t^{-\frac{1}{2}}), \quad z \to \infty, \tag{5.99}$$

where $g(\infty)$ is given by Eq. (5.94), which is only dependent of A, μ, β and ξ with $\xi < \xi_1 < 0$.

6 Modulated elliptic wave sector

When $0 < \xi < \xi_2$, there are no real intersection points of $\mathrm{Im} f(z) = 0$ with the real axis, where the signed signature of $\mathrm{Im} f(z)$ is shown in Fig. 2(a_3). Then the asymptotic solution of $q(x,t)$ will go to the modulated elliptic wave region. Based on Biondini's way in Ref. [31], which extended the methods of Refs. [29, 30], we study the long-time asymptotics of $q(x,t)$ in this region. In this case, the time-dependent g-function should be introduced. Taking the first, second and third deformations as done in the plane-wave sector I in Sec. 4, we arrive at the RH problem $P^{(3)}$ with jump matrices in Eq. (4.48), where the contour $\Sigma^{(3)} = \Sigma^{(2)}$ in Fig. 7 and matrix-value function $M^{(3)}(z)$ has the normalization condition

$$M^{(3)}(z) = I + O\left(\frac{1}{z}\right), \text{ as } z \to \infty. \tag{6.100}$$

It is worth pointing out that in the current case the point η_- in contour $\Sigma^{(3)}$ is replaced with the point η_0 to be determined below, so we rename contour $\Sigma^{(3)}$ as $\widetilde{\Sigma}^{(3)}$. In what follows, we continue to deform the contour $\Sigma^{(3)}$ to derive the solvable model problem.

6.1 $P^{(3)} \to P^{(4)}$: Removal of the exponential growth in matrices $V_3^{(3)}$ and $V_3^{(4)}$

When $0 < \xi < \xi_2$, the sign signature of $\mathrm{Im} f(z)$ is shown in Fig. 2(a_3), which indicates that the jump matrices $V_3^{(3)}$ and $V_3^{(4)}$ in Eq. (4.48) can be exponential growth, or exponential decreasing in certain segments of the branch cuts. To be specific, as t goes to infinity the matrix $V_3^{(3)}$ increases exponentially in segment $[\eta_0, \alpha]$ and matrix $V_3^{(4)}$ increases exponentially in segment $[\eta_0, \bar{\alpha}]$, see the red parts of Fig. 12. In order to remove the parts of exponential growth, we take the factorizations of matrices $V_3^{(3)}$ and $V_4^{(3)}$ (see also Fig. 12) as follows

$$V_3^{(3)} = V_6^{(3)} V_7^{(3)} V_6^{(3)}, \tag{6.101}$$

$$V_4^{(3)} = V_8^{(3)} V_9^{(3)} V_8^{(3)}, \tag{6.102}$$

where $V_j^{(3)}$ ($j = 6, 7, 8, 9$) are defined by

$$V_6^{(3)} = \begin{pmatrix} 1 & \rho^{-1}\delta^2 e^{-ift} \\ 0 & 1 \end{pmatrix}, \quad V_7^{(3)} = \begin{pmatrix} 0 & -\rho^{-1}\delta^2 e^{-ift} \\ \rho\delta^{-2} e^{ift} & 0 \end{pmatrix},$$

$$V_8^{(3)} = \begin{pmatrix} 1 & 0 \\ \bar{\rho}^{-1}\delta^{-2}e^{ift} & 1 \end{pmatrix}, \quad V_9^{(3)} = \begin{pmatrix} 0 & \bar{\rho}\delta^2 e^{-ift} \\ -\bar{\rho}^{-1}\delta^{-2}e^{ift} & 0 \end{pmatrix}.$$

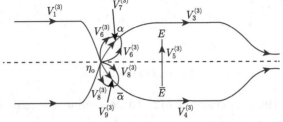

Fig. 12 The contour $\widetilde{\Sigma}^{(3)}$ of the RH problem $P^{(3)}$ for the modulated genus 1 elliptic wave sector

Further introduce a time-dependent g-function mechanism by taking the transformation

$$M^{(4)} = M^{(3)} e^{i\hat{g}(z)t\sigma_3}, \tag{6.103}$$

where function $\hat{g}(z)$ is analytic off the cuts $\gamma \cup \varsigma$, where $\varsigma = \varsigma^+ \cup \varsigma^-$ with $\varsigma^+ = [\eta_0, \alpha]$ and $\varsigma^- = [\eta_0, \bar{\alpha}]$ (see Fig. 12 or Fig. 13). Then the new jump matrices $V^{(4)}$ with branch cuts in Fig. 13 are of the following forms

$$V_1^{(4)} = \begin{pmatrix} 1 & \dfrac{\delta^2 \bar{\rho}}{1+\rho\bar{\rho}} e^{-i(f+2\hat{g})t} \\ 0 & 1 \end{pmatrix}, \quad V_2^{(4)} = \begin{pmatrix} 1 & 0 \\ \dfrac{\delta^{-2}\rho}{1+\rho\bar{\rho}} e^{i(f+2\hat{g})t} & 1 \end{pmatrix},$$

$$V_3^{(4)} = \begin{pmatrix} 1 & 0 \\ \delta^{-2}\rho e^{i(f+2\hat{g})t} & 1 \end{pmatrix}, \quad V_4^{(4)} = \begin{pmatrix} 1 & \delta^2 \bar{\rho} e^{-i(f+2\hat{g})t} \\ 0 & 1 \end{pmatrix},$$

$$V_5^{(4)} = i \begin{pmatrix} 0 & \delta^2 e^{i\theta_+} e^{-i(\hat{g}_+ + \hat{g}_-)t} \\ \delta^{-2} e^{-i\theta_+} e^{i(\hat{g}_+ + \hat{g}_-)t} & 0 \end{pmatrix}, \tag{6.104}$$

$$V_6^{(4)} = \begin{pmatrix} 1 & \rho^{-1}\delta^2 e^{-i(f+2\hat{g})t} \\ 0 & 1 \end{pmatrix},$$

$$V_7^{(4)} = \begin{pmatrix} 0 & -\rho^{-1}\delta^2 e^{-i(f+\hat{g}_+ + \hat{g}_-)t} \\ \rho\delta^{-2} e^{i(f+\hat{g}_+ + \hat{g}_-)t} & 0 \end{pmatrix},$$

$$V_8^{(4)} = \begin{pmatrix} 1 & 0 \\ \bar{\rho}^{-1}\delta^{-2}e^{i(f+2\hat{g})t} & 1 \end{pmatrix},$$

$$V_9^{(4)} = \begin{pmatrix} 0 & \bar{\rho}\delta^2 e^{-i(f+\hat{g}_++\hat{g}_-)t} \\ -\bar{\rho}^{-1}\delta^{-2}e^{i(f+\hat{g}_++\hat{g}_-)t} & 0 \end{pmatrix}.$$

For simplicity, set

$$h(z) = f(z) + 2\hat{g}(z). \tag{6.105}$$

We now analyze the properties of $h(z)$ to determine the real parameter η_0 and complex parameter α. Define a function $\hat{\lambda}(z)$ as

$$\hat{\lambda}(z) = \sqrt{\left[\left(z + \frac{\mu}{2} - \beta A^2\right)^2 + A^2\right](z-\alpha)(z-\bar{\alpha})}, \tag{6.106}$$

which has branch cuts $\gamma \cup \varsigma$ and as usual we take $\hat{\lambda}(z) = \hat{\lambda}_-(z) = -\hat{\lambda}_+(z)$.

Following the similar procedure as in [30, 31], assume the function $h(z)$ has the form of Abelian integral

$$h(z) = \frac{1}{2}\left(\int_E^z + \int_{\bar{E}}^z\right) dh(s), \tag{6.107}$$

where we have defined the Abelian differential dh as

$$dh(z) = \frac{4(z-\eta_0)(z-\alpha)(z-\bar{\alpha})}{\hat{\lambda}(z)} dz. \tag{6.108}$$

In the following, we determine the real parameter η_0 and complex parameter α by removing the possible exponential growths in jump matrices $V_6^{(4)}, V_7^{(4)}, V_8^{(4)}$ and $V_9^{(4)}$. At the same time, we should make sure there are not new exponential growth jump matrices to be introduced.

Sign signature of $\mathrm{Im}(h)(z)$

Let $\alpha = a + ib$ $(b > 0)$ then we have

$$(z-\eta_0)(z-\alpha)(z-\bar{\alpha}) = z^3 - (\eta_0+2a)z^2 + (a^2+2a\eta_0+b^2)z - \eta_0(a^2+b^2). \tag{6.109}$$

Setting $c_0 = -\eta_0(a^2+b^2)$, $c_1 = a^2 + 2a\eta_0 + b^2$ and $c_2 = -(\eta_0+2a)$ then we have

$$h(z) = 2\left(\int_E^z + \int_{\bar{E}}^z\right) \frac{s^3 + c_2 s^2 + c_1 s + c_0}{\hat{\lambda}(s)} ds. \tag{6.110}$$

In order to remove the exponential growths in jump matrices $V_6^{(4)}, V_7^{(4)}, V_8^{(4)}$ and $V_9^{(4)}$, we suppose that the sign signatures of $\mathrm{Im}(h)(z)$ and $\mathrm{Im}(f)(z)$ are the same in complex plane for large z, i.e.

$$\mathrm{Im}(h)(z) = \mathrm{Im}(f)(z) + O\left(\frac{1}{z}\right), \quad \text{as } z \to \infty. \tag{6.111}$$

From the expression of $\hat{\lambda}(z)$ we have for $z \to \infty$

$$\hat{\lambda}(z) = z\left(z + \frac{\mu}{2} - \beta A^2\right)$$
$$\left(1 - \frac{\alpha + \bar{\alpha}}{2z} - \frac{(\alpha - \bar{\alpha})^2}{8z^2} + \frac{A^2}{2\left(z + \frac{\mu}{2} - \beta A^2\right)^2} + O\left(\frac{1}{z^3}\right)\right),$$

which denotes that for $z \to \infty$,

$$\frac{dh}{dz} = 4\left(a + \beta A^2 - \frac{\mu}{2} + c_2 + z + \frac{\chi}{z} + O\left(\frac{1}{z^2}\right)\right), \tag{6.112}$$

where $\chi = (c_2 + a)\left(\beta A^2 + a - \frac{\mu}{2}\right) + c_1 - \frac{A^2}{2} - \frac{b^2}{2} + \left(\beta A^2 - \frac{\mu}{2}\right)^2$. Integrating Eq. (6.112) we have

$$h(z) = 2z^2 + 4\left(a + \beta A^2 - \frac{\mu}{2} + c_2\right)z + 4\chi\ln(z) + h_0 + O\left(\frac{1}{z}\right), \quad z \to \infty, \tag{6.113}$$

where h_0 is a constant to be determined below.

The definition of $f(z)$ in Eq. (2.30) indicates that

$$f(z) = 2z^2 + 2\xi z + \xi(\mu - 2\beta A^2) - 2\left(\frac{\mu}{2} - \beta A^2\right)^2 + A^2 + O\left(\frac{1}{z}\right), \quad z \to \infty. \tag{6.114}$$

In order to make the equality in Eq. (6.111) to be held, from Eqs. (6.112) and (6.114) we find $\chi = 0$ and c_1, c_2 are

$$c_1 = \frac{A^2}{2} + \frac{b^2}{2} - \left(\beta A^2 - \frac{\mu}{2}\right)^2 - \left[\frac{1}{2}(\xi + \mu) - \beta A^2\right]\left(\beta A^2 + a - \frac{\mu}{2}\right),$$

$$c_2 = \frac{1}{2}(\xi + \mu) - a - \beta A^2,$$

which denotes that

$$a = \beta A^2 - \eta_0 - \frac{1}{2}(\xi + \mu),$$

$$b = \left[2\eta_0^2 + (\xi + \mu - 2\beta A^2)\eta_0 + A^2 + \frac{\xi\mu}{2} - \xi\beta A^2\right]^{1/2}, \qquad (6.115)$$

where parameter η_0 is still underdetermined below. Note that

$$2\left(\int_E^z + \int_{\bar{E}}^z\right)\left(s + \frac{\xi}{2}\right)ds = 2z^2 + 2\xi z - 2\left[\left(\beta A - \frac{\mu}{2}\right)^2 - A^2\right] - \xi(2\beta A - \mu). \qquad (6.116)$$

Then the expression of $h(z)$ in Eq. (6.110) can be written as

$$h(z) = 2\left(\int_E^z + \int_{\bar{E}}^z\right)\left[\frac{s^3 + c_2 s^2 + c_1 s + c_0}{\widehat{\lambda}(s)} - \left(s + \frac{\xi}{2}\right)\right]ds + 2z^2 + 2\xi z$$
$$- 2\left[\left(\beta A - \frac{\mu}{2}\right)^2 - A^2\right] - \xi(2\beta A - \mu). \qquad (6.117)$$

Noting $\chi = 0$ and taking $z \to \infty$ to Eqs. (6.113) and (6.117) we have

$$h_0 = 2\left(\int_E^\infty + \int_{\bar{E}}^\infty\right)\left[\frac{s^3 + c_2 s^2 + c_1 s + c_0}{\widehat{\lambda}(s)} - \left(s + \frac{\xi}{2}\right)\right]ds$$
$$- 2\left[\left(\beta A - \frac{\mu}{2}\right)^2 - A^2\right] - \xi(2\beta A - \mu), \qquad (6.118)$$

which is well-defined because of the following asymptotics of the integrand in Eq. (6.118)

$$\frac{s^3 + c_2 s^2 + c_1 s + c_0}{\widehat{\lambda}(s)} - \left(s + \frac{\xi}{2}\right) = O\left(\frac{1}{z^2}\right), \quad z \to \infty.$$

Until now the key point is to determine the parameter η_0 on the real line. In the following, we analyze the sign signature of $\mathrm{Im}h(z)$ near point $\alpha = a + ib$ to keep the jump matrices $V_3^{(4)}, V_6^{(4)}$ and $V_7^{(4)}$ bounded. The asymptotic behaviors of $\lambda(z)$ and $[(z-\alpha)(z-\bar{\alpha})]^{1/2}$ near point α are as follows

$$\lambda = \left[\left(\alpha + \frac{\mu}{2} - \beta A^2\right)^2 + A^2\right]^{1/2}$$
$$\left[1 + \frac{\left(\alpha + \frac{\mu}{2} - \beta A^2\right)(z - \alpha)}{\left(\alpha + \frac{\mu}{2} - \beta A^2\right)^2 + A^2} + O((z-\alpha)^2)\right], \quad z \to \alpha, \qquad (6.119)$$

$$\sqrt{(z-\alpha)(z-\bar{\alpha})} = (\alpha - \bar{\alpha})^{1/2}\left[(z-\alpha)^{1/2} + \frac{(z-\alpha)^{3/2}}{2(\alpha - \bar{\alpha})} + O((z-\alpha)^{\frac{5}{2}})\right], \quad z \to \alpha. \qquad (6.120)$$

Then near point α, Eq. (6.108) can be written as

$$dh(z) = \frac{4(\alpha-\eta_0)(\alpha-\bar{\alpha})^{1/2}}{\left[\left(\alpha+\frac{\mu}{2}-\beta A^2\right)^2+A^2\right]^{1/2}} \left[(z-\alpha)^{1/2} + \left[\frac{1}{2(\alpha-\bar{\alpha})} + \frac{1}{\alpha-\eta_0}\right] \right.$$

$$\left. - \frac{\alpha+\frac{\mu}{2}-\beta A^2}{\left(\alpha+\frac{\mu}{2}-\beta A^2\right)^2+A^2} \left[(z-\alpha)^{3/2} + (z-\alpha)^{5/2}\right]\right], \quad z \to \alpha,$$

which denotes that

$$h(z) = h(\alpha) + \frac{4(\alpha-\eta_0)(\alpha-\bar{\alpha})^{1/2}}{\left[\left(\alpha+\frac{\mu}{2}-\beta A^2\right)^2+A^2\right]^{1/2}} \left[\frac{2}{3}(z-\alpha)^{3/2} + \frac{2}{5}\left[\frac{1}{2(\alpha-\bar{\alpha})} + \frac{1}{\alpha-\eta_0}\right]\right.$$

$$\left. - \frac{\alpha+\frac{\mu}{2}-\beta A^2}{\left(\alpha+\frac{\mu}{2}-\beta A^2\right)^2+A^2} \left[(z-\alpha)^{5/2} + (z-\alpha)^{7/2}\right]\right], \quad z \to \alpha, \quad (6.121)$$

where

$$h(\alpha) = 2\left(\int_E^\alpha + \int_{\bar{E}}^\alpha\right) \frac{s^3+c_2s^2+c_1s+c_0}{\widehat{\lambda}(s)} ds. \quad (6.122)$$

According to the statement in Ref. [29] the sign signature of $\mathrm{Im}\,h(z)$ needs three branches of $\mathrm{Im}(h(z)) = 0$ emanating from point α. Function $h(\alpha)$ is written as $h(\alpha) = h_\mathrm{I}(\alpha) + h_\mathrm{II}(\alpha)$ with

$$h_\mathrm{I}(\alpha) = 2\left(\int_E^\alpha + \int_{\bar{E}}^{\bar{\alpha}}\right) \frac{s^3+c_2s^2+c_1s+c_0}{\widehat{\lambda}(s)} ds, \quad (6.123)$$

$$h_\mathrm{II}(\alpha) = 2\int_{\bar{\alpha}}^\alpha \frac{s^3+c_2s^2+c_1s+c_0}{\widehat{\lambda}(s)} ds. \quad (6.124)$$

It is observed that $h_\mathrm{I}(\alpha) = \overline{h_\mathrm{I}(\alpha)}$ and $h_\mathrm{II}(\alpha) = -\overline{h_\mathrm{II}(\alpha)}$, and thus $h_\mathrm{I}(\alpha)$ is real and $h_\mathrm{II}(\alpha)$ is purely imaginary. So $\mathrm{Im}(h(z)) = 0$ is just $h_\mathrm{II}(\alpha) = 0$, which is equivalent to

$$\int_{\bar{E}}^E \frac{s^3+c_2s^2+c_1s+c_0}{\widehat{\lambda}(s)} ds = \int_{\bar{E}}^E \left(\frac{b^2+(s-a)^2}{\left(s+\frac{\mu}{2}-\beta A^2\right)^2+A^2}\right)^{1/2} (s-\eta_0) ds = 0,$$
$$(6.125)$$

which determines the point η_0 uniquely (reminding the expressions a and b in Eq. (6.115)) and hence the point α in term of ξ, μ, A and β. In the meantime, the sign signature of Im(h)(z) in Fig. 13 denotes that the jump matrices $V_3^{(4)}, V_6^{(4)}$ and $V_7^{(4)}$ bounded in the corresponding branch cuts.

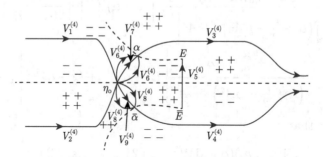

Fig. 13 The contour $\widetilde{\Sigma}^{(4)}$ of the RH problem $P^{(4)}$ for the modulated genus 1 elliptic wave sector and the sign signature of Im(h)(z)

The jump conditions of function $h(z)$

Considering the bounded condition of matrices $V_5^{(4)}, V_7^{(4)}$ and $V_9^{(4)}$, we set the following jump conditions

$$\hat{g}_+(z) + \hat{g}_-(z) = 0, \quad \text{on } \gamma,$$
$$f(z) + \hat{g}_+(z) + \hat{g}_-(z) = \Omega, \quad \text{on } \varsigma.$$

which indicates the jump conditions of function $h(z)$ as

$$h_+(z) + h_-(z) = 0, \quad \text{on } \gamma, \tag{6.126}$$
$$h_+(z) + h_-(z) = 2\Omega, \quad \text{on } \varsigma, \tag{6.127}$$

where Ω is a real constant defined by

$$\Omega = 2\left(\int_E^\alpha + \int_{\bar{E}}^{\bar{\alpha}}\right) \frac{(z-\eta_0)(z-\alpha)(z-\bar{\alpha})}{\widehat{\lambda}(z)} dz. \tag{6.128}$$

Remind that for large z function $h(z)$ satisfies the asymptotic condition

$$h(z) = 2z^2 + 2\xi z + h_0 + O\left(\frac{1}{z}\right), \quad z \to \infty,$$

where h_0 is a constant given by Eq. (6.118).

In addition, the asymptotic condition of function $f(z)$ for large z is

$$f(z) = 2z^2 + 2\xi z + \xi(\mu - 2\beta A^2) - 2\left(\frac{\mu}{2} - \beta A^2\right)^2 + A^2 + O\left(\frac{1}{z}\right), \quad z \to \infty.$$

So from the definition of function $h(z)$ in Eq. (6.105), we know that

$$\hat{g}(\infty) = \frac{1}{2}h_0 - \frac{\xi}{2}(\mu - 2\beta A^2) + \left(\frac{\mu}{2} - \beta A^2\right)^2 - \frac{1}{2}A^2. \tag{6.129}$$

Then the matrix-values function $M^{(4)}$ satisfies the RH problem

$$P^{(4)} = \{M^{(4)}, \ \widetilde{\Sigma}^{(4)}, \ V^{(4)}, \ e^{i\hat{g}(\infty)t\sigma_3} \text{ as } z \to \infty\}, \tag{6.130}$$

where $\hat{g}(\infty)$ is given by Eq. (6.129), the contour $\widetilde{\Sigma}^{(4)}$ is shown in Fig. 13 and the jump matrices $V^{(4)}$ are obtained from Eq. (6.104) as

$$V_1^{(4)} = \begin{pmatrix} 1 & \frac{\delta^2 \bar{\rho}}{1+\rho\bar{\rho}} e^{-iht} \\ 0 & 1 \end{pmatrix}, \quad V_2^{(4)} = \begin{pmatrix} 1 & 0 \\ \frac{\delta^{-2}\rho}{1+\rho\bar{\rho}} e^{iht} & 1 \end{pmatrix},$$

$$V_3^{(4)} = \begin{pmatrix} 1 & 0 \\ \delta^{-2}\rho e^{iht} & 1 \end{pmatrix}, \quad V_4^{(4)} = \begin{pmatrix} 1 & \delta^2 \bar{\rho} e^{-iht} \\ 0 & 1 \end{pmatrix},$$

$$V_5^{(4)} = i\begin{pmatrix} 0 & \delta^2 e^{i\theta_+} \\ \delta^{-2} e^{-i\theta_+} & 0 \end{pmatrix}, \tag{6.131}$$

$$V_6^{(4)} = \begin{pmatrix} 1 & \rho^{-1}\delta^2 e^{-iht} \\ 0 & 1 \end{pmatrix}, \quad V_7^{(4)} = \begin{pmatrix} 0 & -\rho^{-1}\delta^2 e^{-i\Omega t} \\ \rho\delta^{-2} e^{i\Omega t} & 0 \end{pmatrix},$$

$$V_8^{(4)} = \begin{pmatrix} 1 & 0 \\ \bar{\rho}^{-1}\delta^{-2} e^{iht} & 1 \end{pmatrix}, \quad V_9^{(4)} = \begin{pmatrix} 0 & \bar{\rho}\delta^2 e^{-i\Omega t} \\ -\bar{\rho}^{-1}\delta^{-2} e^{i\Omega t} & 0 \end{pmatrix}.$$

6.2 $P^{(4)} \to P^{(5)}$: A transformation between $M^{(4)}$ and $M^{(5)}$

In order to remove the variable z from the jump matrices $V_5^{(4)}$, $V_7^{(4)}$ and $V_9^{(4)}$, take

$$M^{(5)} = M^{(4)} e^{i\widehat{G}(z)\sigma_3}, \tag{6.132}$$

where $\widehat{G}(z)$ is chosen to satisfy the RH problem
(i) $\widehat{G}(z)$ is analytic off the $\Sigma^{(\omega)} = \gamma \cup \varsigma^+ \cup \varsigma^-$ branch cuts.
(ii) On the branch cuts $\gamma \cup \varsigma^+ \cup \varsigma^-$ we have

$$\widehat{G}_+ + \widehat{G}_- = \widehat{\mathbb{G}}(z,\omega) = \begin{cases} \omega - i\ln(\delta^2 \rho^{-1}), & \text{on the } \varsigma^+ \text{ branch cut,} \\ \omega - i\ln(\delta^2 \bar{\rho}), & \text{on the } \varsigma^- \text{ branch cut,} \\ -i\ln(\delta^2), & \text{on the } \gamma \text{ branch cut.} \end{cases}$$

(iii) $\widehat{G}(z) \sim \widehat{G}(\infty)$, as $z \to \infty$,

where ω and $\widehat{G}(\infty)$ are real constants. Solve this RH problem by the Plemelj's formula [46], we have

$$\widehat{G}(z) = -\frac{\widehat{\lambda}(z)}{2\pi i} \int_{\Sigma^{(\omega)}} \frac{\mathbb{G}(z,\omega)}{\widehat{\lambda}_-(\zeta)(\zeta-z)} d\zeta. \tag{6.133}$$

Recalling that $\widehat{\lambda}(z) = z^2 + \left(\frac{\mu}{2} - \beta A^2 - \frac{\alpha+\bar{\alpha}}{2}\right)z + O(1)$, we further have as $z \to \infty$

$$\widehat{G}(z) = \frac{\widehat{\lambda}(z)}{2\pi i z} \int_{\Sigma^{(\omega)}} \frac{\mathbb{G}(z,\omega)}{\widehat{\lambda}_-(\zeta)} \left(1 + \frac{\zeta}{z} + O\left(\frac{1}{z^2}\right)\right) d\zeta$$
$$= \frac{1}{2\pi i} \left(z + \frac{\mu}{2} - \beta A^2 - \frac{\alpha+\bar{\alpha}}{2}\right) \int_{\Sigma^{(\omega)}} \frac{\mathbb{G}(z,\omega)}{\widehat{\lambda}_-(\zeta)} d\zeta$$
$$+ \frac{1}{2\pi i} \int_{\Sigma^{(\omega)}} \frac{\mathbb{G}(z,\omega)}{\widehat{\lambda}_-(\zeta)} \zeta d\zeta + O\left(\frac{1}{z}\right).$$

Applying the condition $\widehat{G}(z) \sim \widehat{G}(\infty)$, as $z \to \infty$ follows

$$\int_{\Sigma^{(\omega)}} \frac{\mathbb{G}(z,\omega)}{\widehat{\lambda}_-(\zeta)} d\zeta = 0, \tag{6.134}$$

which determines the constants ω. And the constant $\widehat{G}(\infty)$ is given by

$$\widehat{G}(\infty) = \frac{1}{2\pi i} \int_{\Sigma^{(\omega)}} \frac{\mathbb{G}(z,\omega)}{\widehat{\lambda}_-(\zeta)} \zeta d\zeta. \tag{6.135}$$

Combining all these properties of $\widehat{G}(z)$ with the transformation in Eq. (6.132), the RH problem for $M^{(5)}$ is formulated as

$$P^{(5)} = \{M^{(5)}, \quad \widetilde{\Sigma}^{(5)} = \widetilde{\Sigma}^{(4)}, \quad V^{(5)} = e^{-i\widehat{G}_-\sigma_3} V^{(4)} e^{i\widehat{G}_+\sigma_3}, \quad e^{i(g(\infty)t+\widehat{G}(\infty))\sigma_3} \text{ as } z \to \infty\}, \tag{6.136}$$

where $V^{(5)}$ is expressed by

$$V_1^{(5)} = \begin{pmatrix} 1 & \frac{\delta^2 \bar{\rho}}{1+\rho\bar{\rho}} e^{-i(ht+2\widehat{G})} \\ 0 & 1 \end{pmatrix}, \quad V_2^{(5)} = \begin{pmatrix} 1 & 0 \\ \frac{\delta^{-2}\rho}{1+\rho\bar{\rho}} e^{i(ht+2\widehat{G})} & 1 \end{pmatrix},$$

$$V_3^{(5)} = \begin{pmatrix} 1 & 0 \\ \delta^{-2}\rho e^{i(ht+2\widehat{G})} & 1 \end{pmatrix}, \quad V_4^{(5)} = \begin{pmatrix} 1 & \delta^2 \bar{\rho} e^{-i(ht+2\widehat{G})} \\ 0 & 1 \end{pmatrix},$$

$$V_5^{(5)} = i\begin{pmatrix} 0 & e^{i\theta_+} \\ e^{-i\theta_+} & 0 \end{pmatrix}, \quad V_6^{(5)} = \begin{pmatrix} 1 & \rho^{-1}\delta^2 e^{-i(ht+2\widehat{G})} \\ 0 & 1 \end{pmatrix},$$

$$V_7^{(5)} = \begin{pmatrix} 0 & -e^{-i(\Omega t+\omega)} \\ e^{i(\Omega t+\omega)} & 0 \end{pmatrix},$$

$$V_8^{(5)} = \begin{pmatrix} 1 & 0 \\ \rho^{-1}\delta^{-2}e^{i(ht+2\widehat{G})} & 1 \end{pmatrix}, \quad V_9^{(5)} = \begin{pmatrix} 0 & e^{-i(\Omega t+\omega)} \\ -e^{i(\Omega t+\omega)} & 0 \end{pmatrix}. \tag{6.137}$$

6.3 Model problem $P^{(\mathrm{mod})}$ and the results

The jump matrices $V_j^{(5)}$ ($j = 1,2,3,4,6,8$) decay exponentially to the identity away from the points η_0, α and $\bar{\alpha}$ as $t \to \infty$, see Fig. 13. The model problem will provide the leading term of the solution. Let $M^{(\mathrm{mod})}$ solve the the RH problem

$$P^{(\mathrm{mod})} = \{M^{(\mathrm{mod})}, \ \Sigma^{(\mathrm{mod})} = \gamma \cup \varsigma^+ \cup (-\varsigma^-), \ V^{(\mathrm{mod})}, \ e^{i(g(\infty)t + \widehat{G}(\infty))\sigma_3} \text{ as } z \to \infty\}, \tag{6.138}$$

with

$$V_{\varsigma^+\cup(-\varsigma^-)}^{(\mathrm{mod})} = \begin{pmatrix} 0 & -e^{-i(\Omega t+\omega)} \\ e^{i(\Omega t+\omega)} & 0 \end{pmatrix}, \quad V_\gamma^{(\mathrm{mod})} = V_5^{(5)} = i\begin{pmatrix} 0 & e^{i\theta_+} \\ e^{-i\theta_+} & 0 \end{pmatrix}, \tag{6.139}$$

where by $-\varsigma^-$ we means the opposite direction of cut ς^-.

For large z, introduce the factorization $M^{(5)} = M^{(\mathrm{err})} M^{(\mathrm{mod})}$, where the higher-order contribution from the contours besides the branch cuts $\gamma \cup \varsigma^+ \cup (-\varsigma^-)$ have been factored out into an error term. Then by using the Eq (2.29) we have the solution $q(x,t)$ of the focusing KE equation (1.1) in the form

$$q(x,t)e^{2i\beta \int_x^{+\infty}(|q(y,t)|^2 - A^2)dy}$$
$$= 2i(M_1^{(\mathrm{mod})} + M_1^{(\mathrm{err})})_{12} e^{i[\mu x + (A^2 + 2\beta^2 A^4 - \frac{\mu^2}{2} + g(\infty))t + \widehat{G}(\infty)]}. \tag{6.140}$$

Following the similar procedure as Sec. 4 and using the idea in Ref. [29], we can show that

$$|M_1^{(\mathrm{err})}| = O(t^{-1/2}). \tag{6.141}$$

In what follows, we solve the model problem $P^{(\mathrm{mod})}$ explicitly in terms of theta functions as done in Refs. [29]-[31].

Let Γ_1 and Γ_2 be two cycles described in Fig. 14. Before defining the theta function $\Theta(z)$, we first introduce the Abelian differential form as Refs. [29]-[31],

$$dw = \frac{w_0}{\widehat{\lambda}(z)} dz, \tag{6.142}$$

where w_0 is a real constant to be determined by $w_0 = 1 \Big/ \oint_{\Gamma_2} \frac{dz}{\widehat{\lambda}(z)}$. Normalize the basis by taking

$$\oint_{\Gamma_2} dw = 1. \tag{6.143}$$

Further assume the Abelian differential has Riemann period of the form

$$\tau = \oint_{\Gamma_1} dw. \tag{6.144}$$

Farkas and Kra [47] have shown that τ is purely imaginary and $i\tau < 0$. So we can define the genus-1 theta function $\Theta(z)$ as

$$\Theta(z) = \sum_{\iota} e^{2\pi i \iota z + \pi i \iota^2 \tau}, \tag{6.145}$$

where $\Theta(z)$ has the following properties

$$\Theta(z) = \Theta(-z), \quad \Theta(z+n) = \Theta(z),$$
$$\Theta(z+n\tau) = e^{-2\pi i n z - \pi i n^2 \tau} \Theta(z), \quad n \in \mathbb{Z}.$$

Further consider the Abelian map

$$u(z) = \int_{\beta A^2 - \frac{\mu}{2} - iA}^{z} dw, \tag{6.146}$$

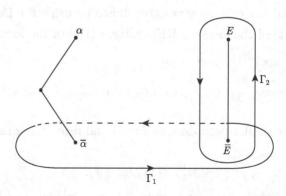

Fig. 14 The canonical homology cycle of $\{\Gamma_1, \Gamma_2\}$ of the genus 1 Riemann surface

which is analytic off $\gamma \cup \bar{\gamma} \cup \varsigma^+ \cup (-\varsigma^-)$. Then we can define the vector-valued function M as

$$M = (M_1, M_2)$$

$$= \left(\frac{\Theta\left(\frac{\Omega t}{2\pi} + \frac{\omega}{2\pi} + \Xi + u(z) + c\right)}{\sqrt{e^{i\theta_+}/i\Theta(u(z) + c)}}, \frac{\Theta\left(\frac{\Omega t}{2\pi} + \frac{\omega}{2\pi} + \Xi - u(z) + c\right)}{\sqrt{ie^{-i\theta_+}\Theta(-u(z) + c)}} \right), \quad (6.147)$$

where $\Xi = -\ln(ie^{-i\theta_+})/(2i\pi)$, Ω and ω are defined before and c is a complex constant to be determined. The Γ_1 and Γ_2 cycles of the genus 1 Riemann surface given by $\hat{\lambda}(z)$ imply

$$u_+(z) + u_-(z) = \begin{cases} 0, & z \in \gamma, \\ n - \tau, & z \in \varsigma^+ \cup (-\varsigma^-), n \in \mathbb{Z}, \end{cases} \quad (6.148)$$

which can be used to calculate the jump matrices of M on branch cuts. On the γ cut we have

$$M_+ = \left(\frac{\Theta\left(\frac{\Omega t}{2\pi} + \frac{\omega}{2\pi} + \Xi + u_+ + c\right)}{\sqrt{e^{i\theta_+}/i\Theta(u_+ + c)}}, \frac{\Theta\left(\frac{\Omega t}{2\pi} + \frac{\omega}{2\pi} + \Xi - u_+ + c\right)}{\sqrt{ie^{-i\theta_+}\Theta(-u_+ + c)}} \right)$$

$$= \left(\frac{\Theta\left(\frac{\Omega t}{2\pi} + \frac{\omega}{2\pi} + \Xi - u_- + c\right)}{\sqrt{e^{i\theta_+}/i\Theta(-u_- + c)}}, \frac{\Theta\left(\frac{\Omega t}{2\pi} + \frac{\omega}{2\pi} + \Xi + u_- + c\right)}{\sqrt{ie^{-i\theta_+}\Theta(u_- + c)}} \right),$$

$$= \left(ie^{-i\theta_+} M_{2-}, -ie^{i\theta_+} M_{1-} \right),$$

$$= M_- \begin{pmatrix} 0 & -ie^{i\theta_+} \\ ie^{-i\theta_+} & 0 \end{pmatrix}.$$

On the $\varsigma^+ \cup (-\varsigma^-)$ cut we have

$$M_+ = \left(\frac{\Theta\left(\frac{\Omega t}{2\pi} + \frac{\omega}{2\pi} + \Xi + u_+ + c\right)}{\sqrt{e^{i\theta_+}/i\Theta(u_+ + c)}}, \frac{\Theta\left(\frac{\Omega t}{2\pi} + \frac{\omega}{2\pi} + \Xi - u_+ + c\right)}{\sqrt{ie^{-i\theta_+}\Theta(-u_+ + c)}} \right)$$

$$= \left(\frac{\Theta\left(\frac{\Omega t}{2\pi} + \frac{\omega}{2\pi} + \Xi + n - \tau - u_- + c\right)}{\sqrt{e^{i\theta_+}/i\Theta(n - \tau - u_- + c)}}, \right.$$

$$\left. \frac{\Theta\left(\frac{\Omega t}{2\pi} + \frac{\omega}{2\pi} + \Xi + u_- - n + \tau + c\right)}{\sqrt{ie^{-i\theta_+}}\Theta(u_- - n + \tau + c)} \right),$$

$$= \left[\frac{\Theta\left(\frac{\Omega t}{2\pi} + \frac{\omega}{2\pi} + \Xi - u_- + c\right)e^{-2i\pi(-\frac{\Omega t}{2\pi} - \frac{\omega}{2\pi} + \frac{\ln(ie^{-i\theta_+})}{2i\pi} + u_- - c) - i\pi\tau}}{\sqrt{e^{i\theta_+}/i}\Theta(-u_- + c)e^{-2i\pi(u_- - c) - i\pi\tau}}, \right.$$

$$\left. \frac{\Theta\left(\frac{\Omega t}{2\pi} + \frac{\omega}{2\pi} + \Xi + u_- + c\right)e^{-2i\pi(\frac{\Omega t}{2\pi} + \frac{\omega}{2\pi} - \frac{\ln(ie^{-i\theta_+})}{2i\pi} + u_- + c) - i\pi\tau}}{\sqrt{ie^{-i\theta_+}}\Theta(u_- + c)e^{-2i\pi(u_- + c) - i\pi\tau}} \right],$$

$$= \left(\frac{\Theta\left(\frac{\Omega t}{2\pi} + \frac{\omega}{2\pi} + \Xi - u_- + c\right)e^{i(\Omega t + \omega)}}{\sqrt{ie^{-i\theta_+}}\Theta(-u_- + c)}, \right.$$

$$\left. \frac{\Theta\left(\frac{\Omega t}{2\pi} + \frac{\omega}{2\pi} + \Xi + u_- + c\right)e^{-i(\Omega t + \omega)}}{\sqrt{e^{i\theta_+}/i}\Theta(u_- + c)} \right)$$

$$= \left(e^{i(\Omega t + \omega)}\mathbb{M}_{2-}, e^{-i(\Omega t + \omega)}\mathbb{M}_{1-} \right)$$

$$= \mathbb{M}_- \begin{pmatrix} 0 & e^{-i(\Omega t + \omega)} \\ e^{i(\Omega t + \omega)} & 0 \end{pmatrix}.$$

It is obvious that the vector function $\mathbb{M}$ in Eq. (6.147) is not the solution of the model problem $P^{(\mathrm{mod})}$. To modify it we introduce the following matrix-values function $\mathbb{N}$ as

$$\mathbb{N} = \frac{1}{2} \begin{pmatrix} (\nu(z) + \nu^{-1}(z))\mathbb{M}_1(z, c) & i(\nu(z) - \nu^{-1}(z))\mathbb{M}_2(z, c) \\ -i(\nu(z) - \nu^{-1}(z))\mathbb{M}_1(z, -c) & (\nu(z) + \nu^{-1}(z))\mathbb{M}_2(z, -c) \end{pmatrix}, \quad (6.149)$$

where $\mathbb{M}_1(z, c)$ and $\mathbb{M}_2(z, c)$ are the first and second elements of the vector function $\mathbb{M}$ in Eq. (6.147) and $\nu(z)$ is given by

$$\nu(z) = \left(\frac{(z - \alpha)\left[z - \left(\beta A^2 - \frac{\mu}{2} + iA\right)\right]}{(z - \bar{\alpha})\left[z - \left(\beta A^2 - \frac{\mu}{2} - iA\right)\right]} \right)^{1/4}, \quad (6.150)$$

which has asymptotic properties $\nu(z) \to 1$ and $\nu - \nu^{-1} \to 0$ as $z \to \infty$ and $\nu(z)$ has branch cut along the γ and $\varsigma^+ \cup (-\varsigma^-)$ cuts with $\nu_+(z) = i\nu_-(z)$. Then $\mathbb{N}$

satisfies the correct jump conditions as follows

$$\mathbb{M}_+ = \mathbb{M}_- \begin{pmatrix} 0 & ie^{i\theta_+} \\ ie^{-i\theta_+} & 0 \end{pmatrix}, \quad z \in \gamma, \tag{6.151}$$

$$\mathbb{M}_+ = \mathbb{M}_- \begin{pmatrix} 0 & -e^{-i(\Omega t + \omega)} \\ e^{i(\Omega t + \omega)} & 0 \end{pmatrix}, \quad z \in \varsigma^+ \cup (-\varsigma^-). \tag{6.152}$$

Let z_1 be the unique zero of $\nu - \nu^{-1}$, and let $X_1(z_1)$ and $X_2(z_1)$ be the preimages of z_1 on the first and second sheet of the elliptic surface, respectively. Then we have

$$c = -\int_{\beta A^2 - \frac{\mu}{2} - iA}^{X_2(z_1)} dw. \tag{6.153}$$

Denote $\mathbb{N}(\infty, c) = \lim_{z \to \pm\infty} \mathbb{N}(z, c), \mathbb{M}_1(\infty, c) = \lim_{z \to \pm\infty} \mathbb{M}_1(z, c)$ and $\mathbb{M}_2(\infty, -c) = \lim_{z \to \pm\infty} \mathbb{M}_2(z, -c)$. Then due to the definition of $\mathbb{M}, \mathbb{N}$ and asymptotic properties of $\nu(z)$, i.e.

$$\nu(z) + \nu^{-1}(z) = 2 + O\left(\frac{1}{z^2}\right), \quad z \to \infty,$$

$$\nu(z) + \nu^{-1}(z) = -i(A + \mathrm{Im}(\alpha)) + O\left(\frac{1}{z^2}\right), \quad z \to \infty,$$

we have

$$\mathbb{N}(\infty, c) = \begin{pmatrix} \mathbb{M}_1(\infty, c) & 0 \\ 0 & \mathbb{M}_2(\infty, -c) \end{pmatrix}, \tag{6.154}$$

where $\mathbb{M}(\infty, c) = (\mathbb{M}_1(\infty, c), \mathbb{M}_2(\infty, c))$ and

$$\mathbb{M}(\infty, c) = \left(\frac{\Theta\left(\frac{\Omega t}{2\pi} + \frac{\omega}{2\pi} + \Xi + u_\infty + c\right)}{\sqrt{e^{i\theta_+} / i\Theta(u_\infty + c)}}, \frac{\Theta\left(\frac{\Omega t}{2\pi} + \frac{\omega}{2\pi} + \Xi - u_\infty + c\right)}{\sqrt{ie^{-i\theta_+}\Theta(-u_\infty + c)}} \right), \tag{6.155}$$

where

$$u(\infty) = \int_{\beta A^2 - \frac{\mu}{2} - iA}^{\infty} dw. \tag{6.156}$$

Then we obtain the unique solution of the RH problem $P^{(\mathrm{mod})}$, which is

$$M^{(\mathrm{mod})} = e^{i(g(\infty)t + \widehat{G}(\infty))\sigma_3} \mathbb{N}^{-1}(\infty, c)\mathbb{N}(z, c), \tag{6.157}$$

which denotes that
$$(M^{(\mathrm{mod})})_{12} = \frac{1}{2}(A + \mathrm{Im}(\alpha))\frac{\mathrm{M}_2(\infty,c)}{\mathrm{M}_1(\infty,c)}e^{i(g(\infty)t+\widehat{G}(\infty))}. \tag{6.158}$$

Combining Eqs. (6.141), (6.155), (6.158) along with (6.140), we finally derive the asymptotic solution of the KE equation (1.1) with the initial value $q(x,0)$ satisfying boundary conditions in Eq. (1.2) as

$$q(x,t)e^{2i\beta \int_x^{+\infty}(|q(y,t)|^2 - A^2)dy}$$
$$= (A + \mathrm{Im}(\alpha))\frac{\Theta\left(\frac{\Omega t}{2\pi} + \frac{\omega}{2\pi} + \Xi - u_\infty + c\right)\Theta(u_\infty + c)}{\Theta\left(\frac{\Omega t}{2\pi} + \frac{\omega}{2\pi} + \Xi + u_\infty + c\right)\Theta(-u_\infty + c)}$$
$$\cdot e^{i[\mu x + (A^2 + 2\beta^2 A^4 - \frac{\mu^2}{2} + 2g(\infty))t + 2\widehat{G}(\infty) + e^{i\theta}+]} + O(t^{-1/2}), \tag{6.159}$$

where $\mathrm{Im}(\alpha) = b, \Omega, \hat{g}(\infty), \omega, \widehat{G}(\infty), c, u_\infty$ are given by Eqs. (6.115), (6.128), (6.129), (6.134), (6.135), (6.153), (6.156), respectively.

This finishes the proof of Theorem 3.3.

7 Modulational instability

In order to clarify the physical formation mechanism of the modulated elliptic waves in the Genus 1 elliptic wave sector in Fig. 3, we investigate the modulational instability [48] of the initial plane wave and find the criterion for the existence of modulated elliptic waves. The modulational instability is also known as the Benjamin-Feir instability [49, 50], who discovered modulation instability for nonlinear Stokes waves on the water surface.

Now we check whether the following plane wave solution of the focusing KE equation (1.1)
$$q(x,t) = Ae^{i[\mu x + (A^2 + 2\beta^2 A^4 - \frac{\mu^2}{2})t + \theta]}, \tag{7.160}$$
where A, μ and θ are constants, is stable against small perturbations or not. To answer this question, by using the linear stability analysis we perturb the plane wave solution (7.160) slightly in a way such that
$$q(x,t) = (A + \epsilon p(x,t))e^{i[\mu x + (A^2 + 2\beta^2 A^4 - \frac{\mu^2}{2})t + \theta]}, \tag{7.161}$$
where $p(x,t) \ll 1$ is a weak perturbation and ϵ is a positive parameter. Substituting the perturbation solution (7.161) in Eq. (1.1) yields a linearized

equation (only considering the linear term of ϵ) of p and p^* as

$$ip_t + \frac{1}{2}p_{xx} - i(2A^2\beta - \mu)p_x - 2i\beta A^2 p_x^* + A^2(1 + 4A^2\beta^2)(p + p^*) = 0, \quad (7.162)$$

which can be solved easily in the frequency domain. Assume a general solution for Eq. (7.162) of the form

$$p(x,t) = \sigma e^{i(kx+\Omega t)} + v e^{-i(kx+\Omega t)}, \quad (7.163)$$

where k is the wave number (Fourier variable) and Ω is the frequency of perturbation, respectively. Inserting Eq. (7.163) along with it conjugate into the linear equation (7.162) we derive two homogeneous equations for σ and v as

$$\left(2\beta A^2 k - \frac{1}{2}k^2 - \Omega + A^2 - \mu k + 4\beta^2 A^4\right)\sigma + (4\beta^2 A^4 + 2\beta A^2 k + A^2)v = 0,$$

$$(A^2 + 4\beta^2 A^4 - 2\beta A^2 k)\sigma + \left(\mu k + 4\beta^2 A^4 + \Omega + A^2 - 2\beta A^2 k - \frac{1}{2}k^2\right)v = 0,$$

which have nontrivial exact solutions only if the determinant of their coefficient matrix vanishes. Thus we have the relationship between k and Ω as

$$\Omega^2 - (4\beta A^2 k - 2\mu k +)\Omega + k^2\left(4\beta^2 A^4 + \mu^2 + A^2 - 4\beta A^2 \mu - \frac{1}{4}k^2\right) = 0,$$

which results in an explicit dispersion relation

$$\Omega = (2\beta A^2 - \mu)k \pm \frac{1}{2}\sqrt{k^2(k^2 - 4A^2)}, \quad (7.164)$$

which clearly displays the stability regions of the plane wave solution (7.160). The frequency of perturbation Ω is real for $k^2 \geqslant 4A^2$ but becomes imaginary for $|k| < 2A$, where the small perturbation will grow exponentially and instability occurs. We call this instability as the modulational instability since it leads to a spontaneous modulation of the plane wave. The gain spectrum of this instability is obtained from Eq. (7.164) for $|k| < 2A$

$$g(k) = \frac{1}{2}|k|\sqrt{4A^2 - k^2}. \quad (7.165)$$

Fig. 15 shows the gain spectra of modulation instability for three different power levels. It is observed that the gain spectrum is symmetric with respect to wave number $k = 0$ and has maximum at two values of k, i.e. $k_{\max} = \pm\sqrt{2}A$, where the peak value of the gain spectra is $g_{\max} = A^2$.

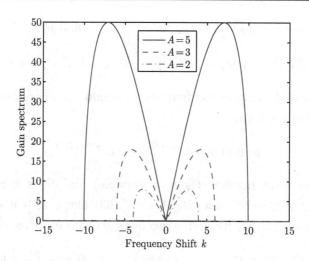

Fig. 15 Gain spectra of modulation instability for three different power levels $A = 2$, $A = 3$ and 5

It is remarked that unstable Fourier modes and gain spectrum of the focusing KE equation (1.1) with plane wave background are similar to the case of focusing nonlinear Schrödinger equation with constant background which is studied in Ref. [33]. In this paper, we have investigated the long-time asymptotic behaviors and modulational instability of the focusing KE equation (1.1) with initial value $q(x,0)$ satisfying boundary conditions in Eq. (1.2). It is shown that as time goes to infinity the unstable modes yield two asymptotic plane wave solutions for $x, t \to \infty$, which are separated by a modulated one-phase elliptic wave solution, see Fig. 3.

Acknowledgements D. Wang and X. Wang are grateful to Prof. Stephanos Venakides for his guidance on the nonlinear steepest descent method. D. Wang thanks J. Lenells, J. Xu, G. Q. Zhang and L. Huang for many insightful conversations on Riemann – Hilbert problems. This work is supported by Beijing Natural Science Foundation under Grant No. 1182009, the Beijing Great Wall Talents Cultivation Program under Grant No. CIT&TCD20180325, Qin Xin Talents Cultivation Program (Nos. QXTCP A201702 and QXTCP B201704) of Beijing Information Science and Technology University and the Beijing Finance Funds of Natural Science Program for Excellent Talents under Grant No. 2014000026833ZK19. X. Wang is supported by the Natural Science Foundation of China under Grant No. 11801292.

References

[1] Kundu A. Landau-Lifshitz and higher-order nonlinear systems gauge generated from nonlinear Schrödinger-type equations [J], J. Math. Phys., 1984, 25: 3433.

[2] Orfanidis S J. SU(n) Heisenberg spin chain [J], Phys. Lett. A, 1980, 75: 304-306.

[3] Kaup D J, Newell A C. An exact solution for a derivative nonlinear Schrödinger equation [J], J. Math. Phys., 1978, 19: 798.

[4] Chen H H, Lee Y C, Liu C S. Integrability of nonlinear Hamiltonian system by inverse scattering method [J], Phys. Scr., 1979, 20: 490.

[5] Geng X G, Tam H W. Darboux transformation and soliton solutions for generalized nonlinear Schrödinger equations [J], J. Phys. Soc. Jpn., 1999, 68: 1508.

[6] Zhao L C, Liu C, Yang Z Y. The rogue waves with quintic nonlinearity and nonlinear dispersion effects in nonlinear optical fibers [J], Commu. Nonl. Sci. Numer. Simul., 2015, 20: 9-13.

[7] Wang X, Yang B, Chen Y, Yang Y Q. Higher-order rogue wave solutions of the Kundu-Eckhaus equation [J], Phys. Scr., 2014, 89: 095210.

[8] Ma W X. Darboux transformations for a Lax integrable system in 2n-dimensions [J], Lett. in Math. Phys., 1997, 39: 33-49.

[9] Zeng Y B, Ma W X, Shao Y J. Two binary Darboux transformations for the KdV hierarchy with self-consistent sources [J], J. Math. Phys., 2001(4): 2113-2128.

[10] Xu X X, Sun Y P. An integrable coupling hierarchy of Dirac integrable hierarchy, its Liouville integrability and Darboux transformation [J], J. Nonl. Sci. Appl., 2017, 10: 3328-3343.

[11] Li X Y, Zhao Q L. A new integrable symplectic map by the binary nonlinearization to the super AKNS system [J], J. Geom. Phys., 2017, 121: 123-137.

[12] Guo M, Zhang Y, Wang M, Chen Y D, Yang H W. A new ZK-ILW equation for algebraic gravity solitary waves in finite depth stratified atmosphere and the research of squall lines formation mechanism [J], Comput. Math. Appl., 2018, 75: 3589-3603.

[13] Zhu S D, Song J F. Residual symmetries, nth Bäcklund transformation and interaction solutions for (2+1)-dimensional generalized Broer-Kaup equations [J], Appl. Math. Lett., 2018, 83: 33-39.

[14] Ma W X, Zhou Y. Lump solutions to nonlinear partial differential equations via Hirota bilinear forms [J], J. Diff. Equ., 2018, 264(4): 2633-2659.

[15] Ma W X. Abundant lumps and their interaction solutions of (3+1)-dimensional linear PDEs [J], J. Geom. Phys., 2018, 133: 10-16.

[16] Zhang J B, Ma W X. Mixed lump-kink solutions to the BKP equation [J], Compu. Math. Appl., 2017, 74: 591-596.

[17] Ma W X, Yong X L, Zhang H Q. Diversity of interaction solutions to the (2+1)-dimensional Ito equation [J], Comput. Math. Appl., 2018, 75: 289-295.

[18] Gardner C S, Greene J M, Kruskal M D, Miura R M. Method for solving the Korteweg-deVries equation [J], Phys. Rev. Lett., 1967, 19: 1095.

[19] Zakharov V E, Manakov S V, Novikov S P, Pitaevskii L P. The Theory of Solitons: The Inverse Scattering Method [M], Consultants Bureau, New York (1984).

[20] Ma W X. Riemann-Hilbert problems and N-soliton solutions for a coupled mKdV system [J], J. Geom. Phys., 2018, 132: 45-54.

[21] Hu B B, Xia T C, Ma W X. Riemann-Hilbert approach for an initial-boundary value problem of the two-component modified Korteweg-de Vries equation on the half-line [J], Appl. Math. Comput., 2018, 332: 148-159.

[22] Deift P, Zhou X. A steepest descent method for oscillatory Riemann-Hilbert problems [J], Ann. Math., 1993, (2)137: 295-368.

[23] Tovbis A, Venakides S, Zhou X. On semiclassical (zero dispersion limit) solutions of the focusing nonlinear Schrödinger equation [J], Commum. Pure Appl. Math., 2004, 57: 877-985.

[24] Xu J, Fan E, Chen Y. Long-time Asymptotic for the Derivative Nonlinear Schrödinger Equation with Step-like Initial Value [J], Math. Phys. Anal. Geom., 2013, 16: 253-288.

[25] Huang L, Xu J, Fan E G. Long-time asymptotic for the Hirota equation via nonlinear steepest descent method [J], Nonlinear Anal. Real World Appl., 2015, 26: 229-262.

[26] Jenkins R, McLaughlin K D. The semiclassical limit of focusing NLS for a family of square barrier initial data [J], Commun. Pure Appl. Math., 2014, 67: 246-320.

[27] Yamane H. Long-time asymptotics for the defocusing integrable discrete nonlinear Schrödinger equation [J], J. Math. Soc. Japan, 2014, 6: 765-803.

[28] Tovbis A, El G A. Semiclassical limit of the focusing NLS: Whitham equations and the Riemann-Hilbert Problem approach [J], Physica D, 2016, 333: 171-184.

[29] Buckingham R, Venakides S. Long-Time Asymptotics of the Nonlinear Schrödinger Equation Shock Problem [J], Commum. Pure Appl. Math., 2007, Vol. LX: 1349-1414.

[30] de Monvel A B, Kotlyarov V P, Shepelsky D. Focusing NLS equation: long-time dynamics of step-like initial data [J], Int. Math. Res. Not. IMRN, 2011, 7: 1613-1653.

[31] Biondini G, Mantzavinos D. Long-Time Asymptotics for the Focusing Nonlinear Schrödinger Equation with Nonzero Boundary Conditions at Infinity and Asymptotic Stage of Modulational Instability [J], Commum. Pure Appl. Math., 2017, Vol. LXX: 2300-2365.

[32] Biondini G, Fagerstrom E. The integrable nature of modulational instability [J], SIAM J. Appl. Math., 2015, 75: 136-163.

[33] Biondini G, Mantzavinos D. Universal nature of the nonlinear stage of modulational instability [J], Phys. Rev. Lett., 2016, 116: 043902.

[34] Prinari B, Biondini G, Vitale F. Dark-bright soliton solutions with nontrivial polarization interactions for the three-component defocusing nonlinear Schrödinger equation with nonzero boundary conditions [J], J. Math. Phys., 2015, 56: 071505.

[35] de Monvel A B, Its A, Kotlyarov V. Long-Time Asymptotics for the Focusing NLS Equation with Time-Periodic Boundary Condition on the Half-Line [J], Commun. Math. Phys., 2009, 290: 479-522.

[36] Lenells J. The Nonlinear Steepest Descent Method: Asymptotics for Initial-Boundary Value Problems [J], SIAM J. Math. Anal., 2016, 48: 2076-2118.
[37] de Monvel A B, Lenells J, Shepelsky D. Long-time asymptotics for the Degasperis-Procesi equation on the half-line, Ann. Inst. Fourier, 2019, 69(1): 171-230.
[38] Fokas A S. A unified transform method for solving linear and certain nonlinear PDEs [J], Proc. R. Soc. Lond. Ser. A, 1997, 453: 1411-1443.
[39] Fokas A S. Integrable Nonlinear Evolution Equations on the Half-Line [J], Commun. Math. Phys., 2002, 230: 1-39.
[40] Pelloni B. Advances in the study of boundary value problems for nonlinear integrable PDEs [J], Nonlinearity, 2015, 28: R1-R38.
[41] Fokas A S, Lenells J, Pelloni B. Boundary value problems for the elliptic sine-Gordon equation in a semi-strip [J], J. Nonlinear Sci., 2013, 23: 241-282.
[42] Miller P D, Qin Z. Initial-boundary value problems for the defocussing nonlinear Schrödinger equation in the semiclassical limit [J], Stud. Appl. Math., 2015, 134: 276.
[43] Tian S F. Initial-boundary value problems for the general coupled nonlinear Schrödinger equation on the interval via the Fokas method [J], J. Differ. Equ., 2017, 262: 506-558.
[44] Zhu Q Z, Xu J, Fan E G. The Riemann-Hilbert problem and long-time asymptotics for the Kundu-Eckhaus equation with decaying initial value [J], Appl. Math. Lett., 2018, 76: 81-89.
[45] Wang D S, Wang X L. Long-time asymptotics and the bright N-soliton solutions of the Kundu-Eckhaus equation via the Riemann-Hilbert approach [J], Nonlinear Anal. Real World Appl., 2018, 41: 334-361.
[46] Deift P. Orthogonal Polynomials and Random Matrices: A Riemann-Hilbert Approach [M], New York University-Courant Institute of Mathematical Sciences, New York, NY (1999).
[47] Farkas H, Kra I. Riemann surfaces [M], 2nd ed. Graduate Texts in Mathematics, 71. Springer, New York (1992).
[48] Zakharov V E, Ostrovsky L A. Modulation instability: The beginning [J], Physica D, 2009, 238: 540-548.
[49] Benjamin T B, Feir J E. The disintegration of wave trains on deep water. Part 1. Theory [J], J. Fluid Mech., 1967, 27: 417-430.
[50] Benjamin T B. Instability of periodic wavetrains in nonlinear dispersive systems [J], Proc. Roy. Soc. A., 1967, 299: 59-75.

Existence and Blow-Up of the Solutions to the Viscous Quantum Magnetohy Drodynamic Nematic Liquid Crystal Model*

Wang Guangwu (王光武), Guo Boling (郭柏灵)

Abstract In this paper, we investigate the coupled viscous quantum magnetohydrodynamic equations and nematic liquid crystal equations which describe the motion of the nematic liquid crystals under the magnetic field and the quantum effects in two-dimensional case. We prove the existence of the global finite energy weak solutions by use of a singular pressure close to vacuum. Then we obtain the local-in-time existence of the smooth solution. In the final, the blow-up of the smooth solutions is studied. The main techniques are Faedo-Galerkin method, compactness theory, Arzela-Ascoli Theorem and construction of the functional differential inequality.

Keywords viscous quantum magnetohydrodynamic equations; nematic liquid crystal; global weak solution; singular pressure; smooth solution; blow-up

1 Introduction

In this paper, we investigate the viscous quantum magnetohydrodynamic nematic liquid crystal model (QMHD-NLC) in $(0, T) \times \Omega$:

$$\partial_t \rho + \mathrm{div}(\rho u) = \nu \Delta \rho, \tag{1.1}$$

$$\partial_t(\rho u) + \mathrm{div}(\rho u \otimes u) + \nabla P = \nu \Delta(\rho u) - \lambda \nabla \cdot \left(\nabla d \odot \nabla d - \frac{|\nabla d|^2}{2} I \right)$$
$$+ \frac{\hbar^2}{2} \rho \nabla \left(\frac{\Delta \sqrt{\rho}}{\sqrt{\rho}} \right) + \kappa (\nabla \times B) \times B, \tag{1.2}$$

---

* Sci. China Math., 2019, 62(3): 469–508. Doi: https://doi.org/10.1007/s11425-017-9165-4.

$$d_t + u \cdot \nabla d = \Delta d + |\nabla d|^2 d, \quad |d| = 1, \tag{1.3}$$
$$\partial_t B - \nabla \times (u \times B) = -\delta \nabla \times (\nabla \times B), \tag{1.4}$$
$$\text{div} B = 0. \tag{1.5}$$

Here we denote by $\Omega \subset \mathbf{R}^2$ the two-dimensional periodic domain, i.e. $\Omega = \{x = (x_1, x_2) | |x_i| < D; (i = 1, 2)\}$. ρ and u represent the density and the velocity field of the flow respectively. P is the pressure function. We only consider the isentropic case i.e. $P = A\rho^\alpha (A > 0$ is a constant, $\alpha > 1)$. $d(x,t): \Omega \to S^1$, the unit sphere in $\mathbf{R}^2$, is a unit-vector field that represents the macroscopic molecular orientation of the liquid crystal material. B is the magnetic field. $\hbar$ is the scaled Planck constant, and ν, κ, δ are positive constants. $\nabla \cdot$ denotes the divergence operator, and $\nabla d \odot \nabla d$ denotes the 2×2 matrix whose (i,j)-the entry is given by $\nabla_i d \cdot \nabla_j d$ for $1 \leqslant i, j \leqslant 2$. The expression $\dfrac{\Delta\sqrt{\rho}}{\sqrt{\rho}}$ can be interpreted as the quantum Bohm potential.

The system (1.1)-(1.5) is coupled with the viscous quantum magnetohydrodynamic equations and Ericsen-Leslie equations. This model can be used to describe the motion of the crystal body in the quantum Hall devices [13], fermionic superfluidity [56], Fermi-Dirac or Bose-Einstein statistics [45], resonant atomic gases [47], metallic systems and itinerant Fermi systems with dipole-dipole interactions [14].

If $\nu = B = 0$ and d is a constant vector, the equations (1.1)-(1.5) become the usual quantum hydrodynamic(QHD) model. The finite energy weak solution of the QHD model coupled with the Poisson equation in two dimensions was investigated by Antonelli and Marcati in [1]. For the multi-dimensional case, Gasser and Markowich [15] obtained the existence of the global finite energy weak solution to the QHD equations coupled with an electric potential term. We [19] established the local-in-time existence and blow-up of the smooth solutions to the QHD model. The blow-up of the smooth solution of the QHD model in half space was proved in [52].

If $\nu = B = 0$ and d is a constant vector, and adding a viscous term $\mu\text{div}\left(\rho\dfrac{\nabla u + \nabla^T u}{2}\right)$ or $\mu\text{div}\left(h(\rho)\dfrac{\nabla u + \nabla^T u}{2}\right) + \lambda \nabla(g(\rho)\text{div}u)$ to the momentum equation (1.2), we call this system as quantum Navier-Stokes equation with density-dependent viscosity (DQNS). For $\alpha > 3$(3-dimension) or $\alpha > 1$(2-dimension), Jüngel [31] proved the global existence of DQNS when the scaled

Planck constant was bigger than the viscosity constant ($\hbar > \nu$). Dong [9] extended that result when the viscosity constant was equal to the scaled Plank constant ($\hbar = \nu$), and Jiang [32] showed that the global existence still hold when the viscosity constant was bigger than the scaled Planck constant ($\hbar < \nu$). Notice that the definition of global weak solutions used in [9, 31, 32] followed the idea introduced in [5] by testing the momentum equation by $\rho\phi$ with a test function ϕ. In 2015, Gisclon and Violet [16] investigated the existence of weak solution to the DQNS equations with singular pressure, where the authors adopt some arguments from [58] to make use of the cold pressure for compactness in the case of $\hbar < \nu$, $\alpha \geqslant 1$, $1 \leqslant d \leqslant 3$($d$ is the dimensional number). Recently, Antonelli and Spirito [2, 3] obtained the existence of a global weak solution to the DQNS when $d = 2$, $\hbar < \nu$, $\alpha > 1$ or $d = 3$, $\hbar^2 < \nu^2 < \frac{9}{8}\hbar^2$, $1 < \alpha < 3$. Furthermore, when $\hbar = 0$, the system is called the compressible Navier-Stokes equation with degenerate viscosity(DCNS). The stability of barotropic DCNS without any additional drag term was studied by Mellet, Vasseur in [42]. Recently, Vasseur and Yu [51] got the global weak solution to DCNS by constructing some smooth multipliers allowing to derive the Mellet-Vasseur type inequality for the weak solutions, almost at the same time, Li and Xin gave another approach in [38].

If $\nu = B = 0$ and d is a constant vector, and adding a viscous term $\mu\mathrm{div}\left(\dfrac{\nabla u + \nabla^T u}{2}\right)$ or $\mu\mathrm{div}\left(\dfrac{\nabla u + \nabla^T u}{2}\right) + \lambda\nabla(\mathrm{div} u)$ to the momentum equation (1.2), this system become the quantum Navier-Stokes equation with constant viscosity(CQNS). There are quite a few results about this model. Only the blow-up of the smooth solution to CQNS was investigated by us in [20]. Furthermore, if $\hbar = 0$, the CQNS model becomes the compressible Navier-Stokes equations with constant viscosity (CNS). The one-dimensional problem has been studied extensively [23, 35]. For the multi-dimensional case, the global classical solutions with the density strictly away from the vacuum were first obtained by Matsumura and Nishida [41] for initial data close to a non-vacuum equilibrium. Recently Huang, Li and Xin [28] obtained the global classical solutions to CNS with the density containing vacuum if the initial energy was suitably small. The weak solutions to CNS for discontinuous initial data was studied by Hoff [22, 24, 25]. When the initial total energy was finite(Which means that the initial density may vanish), Lions [40] obtained the global existence of weak solutions provided the exponent $\alpha > \frac{9}{5}$, and then Feireisl, Novotny and Petzeltová [11] and Feireisl [12] extended the existence results to $\alpha > \frac{3}{2}$, and even to Navier-Stokes-Fourier system.

If $\hbar = \nu = B = 0$, and adding a viscous term $\mu \text{div}\left(\frac{\nabla u + \nabla^T u}{2}\right)$ or $\mu \text{div}\left(\frac{\nabla u + \nabla^T u}{2}\right) + \lambda \nabla (\text{div} u)$ to the momentum equation (1.2), this system becomes the compressible nematic liquid crystal equation (CNLC) or compressible Ericsen-Leslie (CEL) equations. For the 3D compressible case, the existence of the strong solutions has been investigated extensively. The local existence of strong solutions and blow-up criterion were obtained in [29, 30], while the existence and uniqueness of global strong solutions to the Cauchy problem in critical Besov spaces were proved in [27] provided that the initial data were close to an equilibrium state, and the global existence of classical solutions to the Cauchy problem was shown in [39] with smooth initial data which had small energy but possibly large oscillations with possible vacuum and constant state as far-field condition. Recently, great progress has also been achieved about the existence of weak solution to multi-dimensional problem (1.1)-(1.5). Jiang et al. [33, 34] investigated the existence of global weak solution to the two dimensional problem in a bounded domain under a restriction imposed on the initial energy including the case of small energy and multi-dimensional with large initial energy. Moreover they also obtained the global existence of finite energy weak solutions to the two-dimensional Cauchy problem, provided that the second component of initial data of direction field satisfied some geometric angle condition. Meanwhile, Wu and Tan [53] established the global existence of weak solutions to Cauchy problem (1.1)-(1.5) by using Suen and Hoff's method [50]. The global existence of large weak solutions to (1.1)-(1.5) was shown in the 1D case in [8]. The local well-posedness and blow-up criteria of strong solutions to the Ericksen-Leslie system in $\mathbf{R}^3$ for the well-known Oseen-Frank model have been achieved by Hong et al. in [26]. Cavaterra et al. [6] proved the global existence of the weak solution and a Beale-Kato-Majda(BKM) type criterion. We [18] have investigated the blow-up of the smooth solution to the CEL model.

If $\nu = \hbar = 0$, and d is a constant vector, the system is called as compressible magnetohydrodynamic model (MHD). There are amount of papers about the MHD including the existence, the asymptotic behavior and so on, here we omit it.

If d is a constant vector, we call the system as the viscous quantum magnetohydrodynamic model (vQMHD) [21, 55].

If $B = 0$, and d is a constant vector, the system becomes the viscous quantum hydrodynamic model (vQHD). The existence of the finite energy weak solution to vQHD was obtained by Jüngel in [31].

The initial conditions are:

$$(\rho_0, u_0, d, B)|_{t=0} = (\rho_0, u_0, d_0, B_0), \quad |d_0| = 1, \quad \mathrm{div} B_0 = 0, \quad x \in \Omega. \tag{1.6}$$

Here ρ_0, u_0 and d_0 are 2D-periodic functions. Moreover, the initial data should satisfies the following properties:

$$|d_0(x)| = 1, \quad \inf_x d_0^2 > 0, \tag{1.7}$$

where d_0^2 is the second component of the vector d_0.

Since it is very difficult to deal with the velocity individually, we replace the concept of global weak solutions obtained by multiplying the momentum equation by ϕ by a more standard formulation which required the addition of an extra cold pressure as introduced in [4], [16], [43] and [57]. Therefore we prove the global existence of a weak solution to the following system

$$\partial_t \rho + \mathrm{div}(\rho u) = \nu \Delta \rho, \tag{1.8}$$

$$\partial_t(\rho u) + \mathrm{div}(\rho u \otimes u) + \nabla(P + P_c) = \nu \Delta(\rho u) - \lambda \nabla \cdot \left(\nabla d \odot \nabla d - \frac{|\nabla d|^2}{2} I \right)$$
$$+ \frac{\hbar^2}{2} \rho \nabla \left(\frac{\Delta \sqrt{\rho}}{\sqrt{\rho}} \right) + \kappa (\nabla \times B) \times B, \tag{1.9}$$

$$d_t + u \cdot \nabla d = \Delta d + |\nabla d|^2 d, \quad |d| = 1, \tag{1.10}$$

$$\partial_t B - \nabla \times (u \times B) = -\delta \nabla \times (\nabla \times B), \tag{1.11}$$

$$\mathrm{div} B = 0. \tag{1.12}$$

The extra cold pressure as introduced in [4], [43] and [57]. P_c is a suitable increasing function satisfying

$$\lim_{\rho \to 0} P_c(\rho) = +\infty.$$

In this paper, we assume that

$$P'_c(\rho) = \begin{cases} A_1 \rho^{-4k-1}, & n \leqslant 1, \ k > 1, \\ \rho^{\alpha-1}, & n > 1, \ \alpha > 1, \end{cases} \quad (1.13)$$

for some constant $A_1 > 0$.

Remark 1.1 *As mentioned in [4], the singular pressure may be viewed as mathematical assumptions designed to preserve stability. By this way, we are able to get the compactness of the velocity. The addition of this extra cold pressure is fully developed for chemical reactive flows in [43] and [57], also the quantum hydrodynamic [16].*

The main results of this paper are as follows. Firstly, we state the existence of global weak solutions.

Theorem 1.1 (global existence of weak solutions) *Let $T > 0$, $P(\rho) = A\rho^\alpha (\alpha \geqslant 1)$, $\Omega \subset \mathbf{R}^2$. (1.7) holds. And ρ_0, u_0, d_0, B_0 are 2D-periodic functions. Furthermore*

$$E(\rho_0, u_0, d_0, E_0, B_0) = \int_\Omega H(\rho_0) + H_c(\rho_0) + \frac{\hbar^2}{2}|\nabla\sqrt{\rho_0}|^2 + \frac{1}{2}\rho_0|u_0|^2$$
$$+ \lambda\frac{1}{2}|\nabla d_0|^2 + +\frac{\kappa}{2}|B_0|^2 dx < \infty, \quad (1.14)$$

where $H(\rho_0) = \dfrac{A\rho_0^\alpha}{\alpha - 1}$ and $H''_c(\rho_0) = \dfrac{P'_c(\rho_0)}{\rho_0}$. There exists a weak solution (ρ, u, d, B) to (1.8)-(1.12), (1.6) with the regularity

$$\sqrt{\rho} \in L^\infty([0,T]; H^1(\Omega)) \cap L^2([0,T]; H^2(\Omega)), \quad \rho \geqslant 0, \quad (1.15)$$
$$\rho \in H^1([0,T]; L^2(\Omega)) \cap L^\infty([0,T]; L^\alpha(\Omega)) \cap L^2([0,T]; W^{1,3}(\Omega)), \quad (1.16)$$
$$\sqrt{\rho}u \in L^\infty([0,T]; L^2(\Omega)), \quad \rho u \in L^2([0,T]; W^{1,3/2}(\Omega)), \quad (1.17)$$
$$\rho^{\frac{1}{2}}\nabla u \in L^2([0,T]; L^2(\Omega)), \quad (1.18)$$
$$P(\rho) \in L^\infty([0,T]; L^1(\Omega)), \quad P_c(\rho) \in L^\infty([0,T]; L^1(\Omega)), \quad (1.19)$$
$$d \in L^2([0,T]; H^2(\Omega)) \cap L^\infty([0,T]; L^2(\Omega)), \nabla d \in L^4([0,T]; L^4(\Omega)), \quad (1.20)$$
$$B \in L^\infty([0,T]; L^2(\Omega)) \cap L^2([0,T]; H^1(\Omega)), \quad (1.21)$$

satisfying (1.1) pointwise and for all smooth test functions satisfying $\phi(\cdot, T) = 0$,

$$-\int_\Omega \rho_0 u_0 \phi = \int_0^T \int_\Omega \Big(\rho u \cdot \partial_t \phi + \rho u \otimes u : \nabla \phi + (P(\rho) + P_c(\rho))\mathrm{div}\phi$$

$$+ \kappa(\nabla \times B) \times B \cdot \rho\phi$$

$$-\frac{\hbar^2}{2}\left(2\Delta\sqrt{\rho}\nabla\rho \cdot \phi + \nabla\sqrt{\rho} \otimes \nabla\sqrt{\rho} : \nabla\phi + \rho Du : \nabla\phi\right)$$

$$+ \lambda\left(\nabla d \odot \nabla d - \frac{|\nabla d|^2}{2}I\right) : \nabla\phi\bigg)dxdt, \qquad (1.22)$$

$$-\int_\Omega d_0\phi(x,0)dx = \int_0^T\int_\Omega \left(d\phi_t + u\cdot\nabla d\cdot\phi + \nabla d\cdot\nabla\phi + |\nabla d|^2 d\cdot\phi\right)dxdt, \quad (1.23)$$

$$-\int_\Omega B_0\cdot\phi(x,0)dx = \int_0^T\int_\Omega (B\cdot\phi_t - \nabla\times(u\times B)\cdot\phi - \delta\nabla B\cdot\nabla\phi dxdt. \quad (1.24)$$

Then we will give the local-in-time existence of the smooth solution to the equations (1.1)-(1.5).

Theorem 1.2 *Set*

$$0 < \rho_* \leqslant \rho_0 \leqslant \rho^*, \quad |d_0| = 1, \quad \mathrm{div}\, B_0 = 0, \qquad (1.25)$$

and

$$(\rho_0, u_0, d_0, B_0) \in H^6(\Omega) \times H^5(\Omega) \times H^5(\Omega) \times H^5(\Omega). \qquad (1.26)$$

Here $\Omega \subset \mathbf{R}^2$. Then there exist a positive time T_* and a smooth solution (ρ, u, d, B) to the equations (1.1)-(1.5) for any $t \in [0, T_*]$. And the solution also satisfies

$$\rho \in C^j([0,T_*]; H^{6-2j}(\Omega)), \quad j = 0,1,2,$$
$$u \in C^j([0,T_*]; H^{5-2j}(\Omega)), \quad j = 0,1,2,$$
$$d \in C^j([0,T_*]; H^{5-2j}(\Omega)), \quad j = 0,1,2,$$
$$B \in C^j([0,T_*]; H^{5-2j}(\Omega)), \quad j = 0,1,2.$$

Moreover by the regularity of the parabolic equation we have

$$\rho \in C^{4+\alpha, 2+\alpha/2}((0,T_*)\times\Omega),$$
$$u \in C^{2+\alpha, 1+\alpha/2}((0,T_*)\times\Omega),$$
$$d \in C^{2+\alpha, 1+\alpha/2}((0,T_*)\times\Omega),$$
$$B \in C^{2+\alpha, 1+\alpha/2}((0,T_*)\times\Omega).$$

In the last we will give the blow-up of the smooth solution.

Theorem 1.3 Set (ρ, u, d, B) is the smooth solution of (1.1)-(1.5) with the initial data satisfying for $x \in \mathbf{R}^2$,

$$\begin{aligned} \rho(x,0) &= \rho_0 > 0, & \rho_0 &= \bar{\rho} \ (|x| > R), \\ u(x,0) &= u_0, & u_0 &= 0 \ (|x| > R), \\ d(x,0) &= d_0, & d_0 \text{ is a constant vector } (|x| > R), \ |d_0| &= 1, \\ B(x,0) &= B_0, & B_0 &= 0 \ (|x| > R), \quad \operatorname{div} B_0 = 0. \end{aligned} \tag{1.27}$$

Moreover the initial data also satisfies the following conditions:

$$M(0) > 0, \quad F(0) > 0, \tag{1.28}$$

and

(1) $\alpha > 2$, $\Delta_1 = (F(0)+2\nu M(0))^2 - 4(\alpha-1)\mathcal{E}(0)G(0)$, $-\lambda_1 = -\dfrac{b-\sqrt{b^2-4ac}}{2a}$,

$-\lambda_2 = -\dfrac{b+\sqrt{b^2-4ac}}{2a}$, $a = (\alpha-1)\mathcal{E}(0)$, $b = (F(0)+2\nu M(0))$, $c = G(0)$,

(a) $\Delta_1 > 0$, $\lambda_2 - \lambda_1 \exp\left(\dfrac{a(\lambda_2 - \lambda_1)}{F(0)}\right) > 0$,

(b) $\Delta_1 = 0$, $F(0) \geqslant \dfrac{8\nu M(0)}{3}$;

(c) $\Delta_1 < 0$ without other condition.

(2) $1 < \alpha \leqslant 2$, $\Delta_1 = (F(0)+2\nu M(0))^2 - 4\mathcal{E}(0)G(0)$, $-\lambda_1 = -\dfrac{b-\sqrt{b^2-4ac}}{2a}$,

$-\lambda_2 = -\dfrac{b+\sqrt{b^2-4ac}}{2a}$, $a = \mathcal{E}(0)$, $b = (F(0)+2\nu M(0))$, $c = G(0)$,

(a) $\Delta_1 > 0$, $\lambda_2 - \lambda_1 \exp\left(\dfrac{a(\lambda_2 - \lambda_1)}{F(0)}\right) > 0$,

(b) $\Delta_1 = 0$, $F(0) \geqslant \dfrac{8\nu M(0)}{3}$;

(c) $\Delta_1 < 0$, without other condition.

Here

$$\mathcal{E}(t) = \int_\Omega \frac{1}{2}\rho|u|^2 dx + \int_\Omega (H(\rho) - H(\bar{\rho})) dx + \int_\Omega \frac{\hbar^2}{2}|\nabla\sqrt{\rho}|^2 dx + \int_\Omega \frac{\lambda}{2}|\nabla d|^2 dx + \int_\Omega \frac{\kappa}{2}|B|^2 dx.$$

Then the life span T_{**} of the smooth solution is finite, which satisfies:

(1) $\alpha > 2$,

(a) $\Delta_1 > 0$

$$T_{**} \leqslant \frac{\left(\exp\left(\frac{a(\lambda_2 - \lambda_1)}{F(0)}\right) - 1\right)\lambda_1\lambda_2}{\left(\lambda_2 - \lambda_1 \exp\left(\frac{a(\lambda_2 - \lambda_1)}{F(0)}\right)\right)},$$

(b) $\Delta_1 = 0$

$$T_{**} \leqslant \frac{a\lambda_1^2}{F(0) - a\lambda_1},$$

(c) $\Delta_1 < 0$

$$T_{**} \leqslant \frac{4ac - b^2}{4a^2} \tan\left\{\frac{4ac - b^2}{4F(0)a^2} + \arctan\left\{\frac{2ab}{4ac - b^2}\right\}\right\} - \frac{b}{2a}.$$

(2) $1 < \alpha \leqslant 2$,

(a) $\Delta_1 > 0$

$$T_{**} \leqslant \frac{\left(\exp\left(\frac{a(\lambda_2 - \lambda_1)}{F(0)}\right) - 1\right)\lambda_1\lambda_2}{\left(\lambda_2 - \lambda_1 \exp\left(\frac{a(\lambda_2 - \lambda_1)}{F(0)}\right)\right)},$$

(b) $\Delta_1 = 0$

$$T_{**} \leqslant \frac{a\lambda_1^2}{F(0) - a\lambda_1},$$

(c) $\Delta_1 < 0$

$$T_{**} \leqslant \frac{4ac - b^2}{4a^2} \tan\left\{\frac{4ac - b^2}{4F(0)a^2} + \arctan\left\{\frac{2ab}{4ac - b^2}\right\}\right\} - \frac{b}{2a}.$$

Remark 1.2 *From the proof of Theorem 1.3, it is not difficult to find that if ν is small enough, it always holds that $\Delta_1 < 0$ only with the condition $M(0), F(0) > 0$, thus the blow-up of the smooth solution to the equations (1.1)-(1.5) is naturally true.*

Remark 1.3 *We will emphasize that whether the global finite energy weak solutions of the equations (1.1)-(1.5) exist or not, there is non contradiction between the global existence of the weak solution and the blow-up of the smooth solution, because we can not derive the existence of global smooth solution from the global existence of the weak solution.*

It is not difficult to find that (1.8)-(1.12) lacks compactness which we need to get the estimate of L^2 or H^1-norm for u. To overcome this difficult, we first add the right hand of (1.2) by a viscosity term $\gamma\Delta u - \gamma u$:

$$\partial_t \rho + \text{div}(\rho u) = \nu\Delta\rho, \tag{1.29}$$

$$\partial_t(\rho u) + \mathrm{div}(\rho u \otimes u) + \nabla(P + P_c)$$
$$= \nu \Delta(\rho u) - \lambda \nabla \cdot \left(\nabla d \odot \nabla d - \frac{|\nabla d|^2}{2}I\right)$$
$$+ \frac{\hbar^2}{2}\rho\nabla\left(\frac{\Delta\sqrt{\rho}}{\sqrt{\rho}}\right) + \kappa(\nabla \times B) \times B + \gamma\Delta u - \gamma u, \qquad (1.30)$$

$$d_t + u \cdot \nabla d = \Delta d + |\nabla d|^2 d, \quad |d| = 1, \qquad (1.31)$$
$$\partial_t B - \nabla \times (u \times B) = -\delta \nabla \times (\nabla \times B), \qquad (1.32)$$
$$\mathrm{div} B = 0. \qquad (1.33)$$

The organization of this paper are the following. In Section 2, we give some notations, lemmas and theorems which will be used in the rest sections. In Section 3, we prove the local-in-time existence of weak solutions. Then we prove the global-in-time existence of finite energy weak solutions to the equations (1.8)-(1.12) in Section 4. The main techniques are the Faedo-Galerkin method and compactness theory. In Section 5, we use the Arzela-Ascoli Theorem to prove the local-in-time existence of the smooth solutions to the equations (1.1)-(1.5). In the last section, we prove the blow-up of the smooth solution to the equations (1.1)-(1.5).

2 Preliminaries

In this section we first give some notations, lemmas and theorems which will be used in the following section. In this paper we denote that C is the constant independent on N and γ. $L^p([0,T]; L^q(\Omega))(p,q \geqslant 1)$ is a space whose element is the p-integrable respect to time variable and q-integrable respect to space variable function. $W^{k,p}$ and H^s are the Sobolev spaces. $(H^s)^*$ is the dual space of H^s. Ω is a periodic domain in $\mathbf{R}^2$.

Lemma 2.1 (Gagliardo-Nirenberg inequality [44]) *Let $\Omega \subset \mathbf{R}^d (d \geqslant 1)$ be a bounded open set with $\partial\Omega \in C^{0,1}$, $m \in \mathbf{N}$, $1 \leqslant p,q,r \leqslant \infty$. Then there exists a constant $C > 0$ such that for all $u \in W^{m,p}(\Omega) \cap L^q(\Omega)$,*

$$\|D^\alpha u\|_{L^r(\Omega)} \leqslant C\|u\|_{W^{m,p}(\Omega)}^\theta \|u\|_{L^q(\Omega)}^{1-\theta},$$

where $0 \leqslant |\alpha| \leqslant m-1$, $\theta = |\alpha|/m$, and $|\alpha| - d/r = \theta(m - d/p) - (1-\theta)d/q$. If $m - |\alpha| - d/p \notin \mathbf{N}_0$, then $\theta \in [|\alpha|/m, 1]$ is allowed.

Lemma 2.2 (Aubin-Lions Lemma [49])   Assume $X \subset Y \subset Z$ are Banach spaces and $X \hookrightarrow\hookrightarrow Y$. Then the following imbedding are compact:

$$L^q([0,T];X) \cap \{\varphi : \partial_t \varphi \in L^1([0,T];Z)\} \hookrightarrow\hookrightarrow L^q([0,T];Y), \quad \forall 1 \leqslant q \leqslant \infty, \tag{2.34}$$

$$L^\infty([0,T];X) \cap \{\varphi : \partial_t \varphi \in L^r([0,T];Z)\} \hookrightarrow\hookrightarrow C([0,T];Y), \quad \forall 1 \leqslant r \leqslant \infty. \tag{2.35}$$

Lemma 2.3 (Egoroffs theorem [46])   Let $f_n \to f$ a.e. in Ω a bounded measurable set in $\mathbf{R}^n$, with f finite a.e.. Then for any $\varepsilon > 0$ there exists a measurable subset $\Omega_\varepsilon \subset \Omega$ such that $|\Omega \backslash \Omega_\varepsilon| < \varepsilon$ and $f_n \to f$ uniformly in Ω_ε, moreover, if

$$f_n \to f \quad a.e. \ in \ \Omega,$$

$f_n \in L^p(\Omega)$ and uniformly bounded, for any $1 < p \leqslant +\infty$,

then, we have

$$f_n \to f \quad strongly \ in \ L^s(\Omega), \quad for \ any \ s \in [1,p).$$

Lemma 2.4 (Sobolev imbedding theorem [10])   Let U be a bounded open subset of $\mathbf{R}^n$, with a C^1 boundary. Assume $u \in W^{k,p}(U)$.

(i) If $kp < n$, then $u \in L^q(U)$, where $\dfrac{1}{q} = \dfrac{1}{p} - \dfrac{k}{n}$. We have in addition to the estimate $\|u\|_{L^q(U)} \leqslant C\|u\|_{W^{k,p}(U)}$, and $W^{k,p}(\Omega) \hookrightarrow\hookrightarrow L^{q^*}(\Omega)(1 \leqslant q^* < q)$.

(ii) If $kp = n$, then $u \in L^q(U)$, where $\forall 1 \leqslant q < \infty$, and $W^{k,p} \hookrightarrow\hookrightarrow L^q(\Omega)$.

(iii) If $kp > n$, then $u \in C^{k-[\frac{n}{p}]-1,\sigma}(\bar{U})$, where

$$\sigma = \begin{cases} \left[\dfrac{n}{p}\right] + 1 - \dfrac{n}{p}, & if \ \dfrac{n}{p} \ is \ not \ an \ integer, \\ any \ positive \ number < 1, & if \ \dfrac{n}{p} \ is \ an \ integer. \end{cases}$$

We have, in addition, the estimate

$$\|u\|_{C^{k-[\frac{n}{p}]-1,\sigma}(\bar{U})} \leqslant C\|u\|_{W^{k,p}(U)},$$

and $W^{k,p} \hookrightarrow\hookrightarrow C^{k-[\frac{n}{p}]-1,\sigma}(\bar{U})$.

Here the constat C depending only on k,p,n and U.

Theorem 2.1 ([10]) Let $T > 0$, denote the letter L as a second-order uniformly elliptic operator in the spatial variable x. (i) Assume

$$g \in H_0^1(\Omega), \quad f \in L^2(0,T;L^2(\Omega)). \tag{2.36}$$

Suppose also $u \in L^2(0,T;H_0^1(\Omega))$, with $u' \in L^2(0,T;H^{-1}(\Omega))$, is the weak solution of

$$\begin{cases} u_t + Lu = f, & \text{in } \Omega \times (0,T), \\ u = g, & \text{in } \Omega \times \{t=0\}. \end{cases} \tag{2.37}$$

Then

$$u \in L^2(0,T;H^2(\Omega)) \cap L^\infty(0,T;H_0^1(\Omega)), \quad u' \in L^2(0,T;L^2(\Omega)), \tag{2.38}$$

and we have the estimate

$$\operatorname*{ess\,sup}_{0 \leqslant t \leqslant T} \|u(t)\|_{H_0^1(\Omega)} + \|u\|_{L^2(0,T;H_0^1(\Omega))} + \|u'\|_{L^2(0,T;L^2(\Omega))}$$
$$\leqslant C(\|f\|_{L^2(0,T;L^2(\Omega))} + \|g\|_{H_0^1(\Omega)}).$$

(ii) If, in addition,

$$g \in H^2(\Omega), \quad f' \in L^2(0,T;L^2(\Omega)),$$

then

$$\begin{aligned} u \in L^\infty(0,T;H^2(\Omega)), \quad u' \in L^\infty(0,T;L^2(\Omega)) \cap L^2(0,T;H_0^1(\Omega)), \\ u'' \in L^2(0,T;H^{-1}(\Omega)), \end{aligned} \tag{2.39}$$

with the estimate

$$\operatorname*{ess\,sup}_{0 \leqslant t \leqslant T} \|u(t)\|_{H_0^1(\Omega)} + \|u\|_{L^2(0,T;H_0^1(\Omega))} + \|u'\|_{L^2(0,T;L^2(\Omega))}$$
$$+ \|u''\|_{L^2(0,T;H^{-1}(\Omega))} \leqslant C(\|f\|_{L^2(0,T;L^2(\Omega))} + \|g\|_{H_0^1(\Omega)}).$$

(iii) If

$$g \in H^{2m+1}(\Omega), \quad \frac{d^k f}{dt^k} \in L^2(0,T;H^{2m-2k}(\Omega)), \quad k=0,1,\cdots,m.$$

Suppose also that the following m^{th}-order compatibility conditions hold:

$$\begin{cases} g_0 = g \in H_0^1(\Omega), \\ g_1 = f(0) + \Delta g_0 \in H_0^1(\Omega), \\ \vdots \\ g_m = \dfrac{d^{m-1}f}{dt^{m-1}}(0) + \Delta g_{m-1} \in H_0^1(\Omega). \end{cases}$$

Then
$$\frac{d^k u}{dt^k} \in L^2(0, T; H^{2m+2-2k}(\Omega)), \quad k = 0, \cdots, m+1;$$

and we have the estimate

$$\sum_{k=0}^{m+1} \|\frac{d^k u}{dt^k}\|_{L^2(0,T;H^{2m+2-2k}(\Omega))} \leqslant C \left(\sum_{k=0}^{m} \|\frac{d^k f}{dt^k}\|_{L^2(0,T;H^{2m-2k}(\Omega))} + \|g\|_{H^{2m+1}(\Omega)} \right).$$

The constant C depends only on Ω, T, and the coefficient of L.

Finally, let us consider the abstract initial value problem in the periodic Hilbert space $L^2(\Omega)$ [37],

$$u_{tt} - 2\nu \Delta u_t + \left(\nu + \frac{\hbar^2}{4}\right) \Delta^2 u - \nu \beta \Delta u = F, \tag{2.40}$$

$$u(0) = u_0, \quad u'(0) = u_1. \tag{2.41}$$

Here $\nu, \hbar$ are positive constants, Δ is the Laplacian operator.

By applying the Faedo-Galerkin method, we can obtain the existence of solutions to (2.40)-(2.41) in a standard way. We omit the details.

Theorem 2.2 Let $T > 0$, and assume that $F \in H^1(0, T; L^2(\Omega))$, Then, if $u_0 \in H^4(\Omega)$ and $u_1 \in H^2(\Omega)$, the solution to (2.40)-(2.41) exists and satisfies

$$u \in C^j(0, T; H^{4-2j}(\Omega)), \quad j = 0, 1, \quad u_{tt} \in L^\infty(0, T; L^2(\Omega)). \tag{2.42}$$

Moreover, assume that

$$F_t, F \in L^2(0, T; H^2(\Omega)), \tag{2.43}$$

then, if $u_0 \in H^6$ and $u_1 \in H^4$, it follows

$$u \in C^j(0, T; H^{6-2j}(\Omega)), \quad j = 0, 1, 2, \quad u_{ttt} \in L^\infty(0, T; L^2(\Omega)). \tag{2.44}$$

Next, we will give some notations about the Hölder space.

Denote $Q_T = \Omega \times (0,T)$. For any points $P(x,t), Q(y,s) \in Q_T$, define the parabolic distance between them as [54]

$$d(P,Q) = \sqrt{|x-y|^2 + |t-s|}.$$

Let $u(x,t)$ be a function on Q_T. For $0 < \alpha < 1$, define

$$[u]_{\sigma,\sigma/2;Q_T} = \sup_{P,Q \in Q_T, P \neq Q} \frac{|u(P) - u(Q)|}{d^\sigma(P,Q)},$$

which is a semi-norm, and denote by $C^{\sigma,\sigma/2}(\bar{Q}_T)$ the set of all functions on Q_T such that $[u]_{\sigma,\sigma/2;Q_T} < +\infty$, endowed with the norm

$$|u|_{\sigma,\sigma/2;Q_T} = |u|_{0;Q_T} + [u]_{\sigma,\sigma/2;Q_T},$$

where $|u|_{0;Q_T}$ is the maximum norm of $u(x,t)$ on Q_T, i.e.,

$$|u|_{0;Q_T} = \sup_{(x,t) \in Q_T} |u(x,t)|.$$

Furthermore, for any nonnegative integer k, denote

$$C^{2k+\sigma, k+\sigma/2}(\bar{Q}_T) = \{u; D^\beta D_t^r u \in C^{\sigma,\sigma/2}(\bar{Q}_T), \text{ for any } \beta, r \text{ such that } |\beta| + 2r \leqslant 2k\},$$

and define the semi-norm

$$[u]_{2k+\sigma, k+\sigma/2; Q_T} = \sum_{|\beta|+2r=2k} [D^\beta D_t^r u]_{\sigma,\sigma/2;Q_T},$$

$$[u]_{2k,k;Q_T} = \sum_{|\beta|+2r=2k} |D^\beta D_t^r u|_{0;Q_T},$$

and the norm

$$|u|_{2k+\sigma, k+\sigma/2; Q_T} = \sum_{|\beta|+2r \leqslant 2k} |D^\beta D_t^r u|_{\sigma,\sigma/2;Q_T},$$

$$|u|_{2k,k;Q_T} = \sum_{|\beta|+2r \leqslant 2k} |D^\beta D_t^r u|_{0;Q_T}.$$

$C^{2k+\sigma, k+\sigma/2}(\bar{Q}_T)$ is a Banach space.

$$|u|_{2,1;Q_T} = |u|_{0;Q_T} + |Du|_{0;Q_T} + |D^2 u|_{0;Q_T} + |u_t|_{0;Q_T},$$
$$\|u\|_{2+\sigma, 1+\sigma/2; Q_T} = |u|_{2+\sigma, 1+\sigma/2; Q_T} + |Du|_{\sigma,\sigma/2;Q_T} + |D^2 u|_{\sigma,\sigma/2;Q_T} + |u_t|_{\sigma,\sigma/2;Q_T},$$

where $|D^2 u|_{0,Q_T}$, $|D^2 u|_{\sigma,\sigma/2;(Q_T)}$ denote the sums of the corresponded norms of the second order derivatives of u with respect to x.

3 Local existence of weak solutions

In this section we will show the local existence of weak solutions to the viscosity system (1.29)-(1.33) by Faedo-Galerkin method. Let $T^* > 0$, and let $\{\omega_j\}$ be an orthogonal basis of $L^2(\Omega)$ which is also an orthogonal basis of $H^1(\Omega)$. Introduce the finite-dimensional space $X_N = span\{\omega_1, \cdots, \omega_N\}$, $n \in N$. Denote the approximate solutions of the problem (1.29)-(1.33) by $u_N^\gamma, d_N^\gamma, E_N^\gamma, B_N^\gamma$ in the following form

$$u_N^\gamma(x,t) = \sum_{j=1}^{N} \lambda_j(t) \omega_j(x).$$

For some functions $\lambda_i(t)$, the norm of v in $C^0([0,T^*]; X_N)$ can be formulated as

$$\|v\|_{C^0([0,T^*]; X_N)} = \max_{t \in [0,T^*]} \sum_{i=1}^{N} |\lambda_i(t)|.$$

As a consequence, v can be bounded in $C^0([0,T^*]; C^k(\Omega))$ for any $k \in N$, and there exists a constant $C > 0$ depending on k such that

$$\|u_N^\gamma\|_{C^0([0,T^*]; C^k(\Omega))} \leqslant C \|u_N^\gamma\|_{C^0([0,T^*]; L^2(\Omega))}. \tag{3.45}$$

3.1 Solvability of the density equation

The approximate system is defined as follows. Let $\rho \in C^1([0,T^*]; C^3(\Omega))$ be the classical solution to

$$\rho_t + \mathrm{div}(\rho u) = \nu \Delta \rho, \quad \rho|_{t=0} = \rho_0(x). \tag{3.46}$$

The maximum principle provides the lower and upper bounds ([12], chapter 7.3)

$$\inf_{x \in \Omega} \rho_0(x) \exp\left(-\int_0^t \|\mathrm{div} u\|_{L^\infty(\Omega)} ds\right) \leqslant \rho(x) \leqslant \sup_{x \in \Omega} \rho_0(x) \exp\left(\int_0^t \|\mathrm{div} u\|_{L^\infty} ds\right).$$

Since we assumed that $\rho_0(x) > \bar{\rho} > 0$, $\rho(x)$ is strictly positive. In view of (3.45), for $\|u\|_{C^0([0,T^*]; L^2(\Omega))} \leqslant C$, there exists constant $\underline{\rho}(C)$ and $\bar{\rho}(C)$ such that

$$0 < \underline{\rho}(C) \leqslant \rho(x,t) \leqslant \bar{\rho}(C).$$

We introduce the operator $S_1 : C^0([0,T^*]; X_N) \to C^0([0,T^*]; C^3(\Omega))$ by $S_1(u) = \rho$. Since the equation for ρ is linear, S_1 is Lipshitz continuous in the following sense:

$$\|S_1(u_1) - S_1(u_2)\|_{C^0([0,T^*]; C^k(\Omega))} \leqslant C(N,k) \|u_1 - u_2\|_{C^0([0,T^*]; L^2(\Omega))}. \tag{3.47}$$

3.2 Solvability of the direction vector equation

Now we turn to show the local solvability of the direction vector. More precisely, we will show that for

$$u \in V := \{v | v \in C^0((0,T^*); H^1(\Omega) \cap L^\infty(\Omega)), \partial_t v \in L^2((0,T^*); H^1(\Omega))\},$$

there exists a $T_d^* \in (0, T^*]$, such that the equation

$$\partial_t d + u \cdot \nabla d = \Delta d + |\nabla d|^2 d \tag{3.48}$$

has a unique strong solution $d(x,t)$ on $[0, T_d^*)$ satisfying the initial condition

$$d(x, 0) = d_0(x) \in H^3(\Omega). \tag{3.49}$$

The local existence can be shown by modifying the arguments in [8] and it is the same as [33]. Here we also give a detailed proof. Moreover, we shall show that the solution d continuously depends on u, which will also be used in the proof of the local existence of solutions to the third approximate problem.

3.2.1 Linearized problems

Denote

$$\mathcal{D}^* := \{b \in C^0([0, T_d^*]; H^2(\Omega)) | \partial_t b \in C^0([0, T_d^*]; L^2(\Omega)) \cap L^2((0, T_d^*); H_0^1(\Omega))\},$$
$$V_k := \{v \in V | \|v\|_V \leqslant K\},$$
$$\mathcal{D}_{\kappa_d}^* := \{b \in \mathcal{D}^* | b(x, 0) = d_0(x), \|b\|_{\mathcal{D}^*} \leqslant \kappa_d\},$$

where K and κ_d are positive constants

$$\begin{cases} \|v\|_V := \left(\|v\|_{C^0([0,T];H^1(\Omega))}^2 + \|v\|_{C^0([0,T];L^\infty(\Omega))}^2 + \|\partial_t v\|_{L^2((0,T);H^1(\Omega))}^2 \right)^{\frac{1}{2}}, \\ \|b\|_{\mathcal{D}^*} := \left(\|\partial_t b\|_{L^2((0,T_d^*);H^1(\Omega))}^2 + \|\partial_t b\|_{L^\infty((0,T_d^*);L^2(\Omega))}^2 \right. \\ \left. + \|b\|_{L^\infty((0,T_d^*);H^2(\Omega))}^2 + \|b\|_{L^2((0,T_d^*);H^3(\Omega))}^2 \right)^{\frac{1}{2}}. \end{cases}$$
$$\tag{3.50}$$

Without loss of generality, we assume $\kappa_d \geqslant 1$.

To show the local existence of solutions, we will construct a sequence of approximate solutions to (3.48)-(3.49) and use the technique of iteration based on the set $\mathcal{D}_{\kappa_d}^*$. First, we linearize the original system (3.48) for $b \in \mathcal{D}_{\kappa_d}^*$ as follows:

$$\partial_t d - \Delta d = |\nabla b|^2 b - u \cdot \nabla b, \tag{3.51}$$

with a given function $v \in V_k$, and initial condition:

$$d|_{t=0} = d_0, \quad x \in \Omega. \tag{3.52}$$

Since

$$F(b) := |\nabla b|^2 b - u \cdot \nabla b \in L^2((0, T_d^*); H^1(\Omega))$$

and $F(b) \in L^2((0, T_d^*); (H^1(\Omega))^*)$, we can apply a standard method, such as a semi-discrete Galerkin method in [10], to prove that the initial value problem (3.51)-(3.52) has a unique solution

$$d \in C^0([0, T_d^*]; H^2(\Omega)) \cap L^2((0, T_d^*); H^3(\Omega)), \quad \partial_{tt} d \in L^2((0, T_d^*); (H^1(\Omega))^*),$$

$$\partial_t d \in C^0([0, T_d^*]; L^2(\Omega)) \cap L^2((0, T_d^*); H_0^1(\Omega)).$$

Here we have used $(H^1(\Omega))^*$ to denote the dual space of $H^1(\Omega)$. Therefore, we can get a solution d^1 to (3.51) with b replaced by some given $d^0 \in \mathcal{D}_{\kappa_d}^*$. Assuming $d^{k-1} \in \mathcal{D}_{k_d}^*$ for $k \geqslant 1$, we can construct an approximate solution d^k satisfying

$$\partial_t d^k - \Delta d^k = |\nabla d^{k-1}|^2 d^{k-1} - u \cdot \nabla d^{k-1}, \tag{3.53}$$

with the initial and boundary conditions:

$$d^k|_{t=0} = d_0, \quad x \in \Omega.$$

3.2.2 Uniform estimates

Next we derive the uniform estimates:

$$\|d^k\|_{\mathcal{D}^*} \leqslant \kappa_d \tag{3.54}$$

for some constant κ_d depending on d_0 and the given constant K. We mention that in the following estimates, the letter $c \geqslant 1$ will denote various positive constants independent of κ_d, K, v, d_0 and Ω; the letter $C_1(\Omega)$ will denote various positive constants depending on Ω, and the letter $C(\cdots)$ various positive constants depending on its variable(it may depend on Ω sometimes, however we omit them for simplicity), and is nondecreasing in its variables. Of course, they may also depend on the fixed value θ.

First, from (3.53) and integrating by parts, we can deduce that

$$\frac{1}{2}\frac{d}{dt}\int_\Omega |d^k|^2 dx + \int_\Omega |\nabla d^k|^2 dx$$
$$= \int_\Omega |\nabla d^{k-1}|^2 d^{k-1} \cdot d^k dx - \int_\Omega u \cdot \nabla d^{k-1} \cdot d^k dx,$$

and

$$\frac{1}{2}\frac{d}{dt}\int_\Omega |\nabla d^k|^2 dx + \int_\Omega |\Delta d^k|^2 dx$$
$$= -\int_\Omega |\nabla d^{k-1}|^2 d^{k-1} \cdot \Delta d^k dx + \int_\Omega u \cdot \nabla d^{k-1} \cdot \Delta d^k dx,$$

which, together with Cauchy-Schwartz's and Hölder's inequalities, imply that

$$\frac{1}{2}\frac{d}{dt}\|d^k\|_{H^1(\Omega)}^2 \leqslant \Big(\|\,|\nabla d^{k-1}|\,\|_{L^4(\Omega)}^4 \|d^{k-1}\|_{L^\infty(\Omega)}^2 + \|u\|_{L^4(\Omega)}^2 \|\,|\nabla d^{k-1}|\,\|_{L^4(\Omega)}^2\Big)$$
$$+ \|d^k\|_{L^2(\Omega)}^2 + \|\nabla d\|_{L^2(\partial\Omega)} \|d_0\|_{L^2(\partial\Omega)}.$$

Hence, using the embedding theorem $H^2(\Omega) \hookrightarrow L^\infty(\Omega)$ for d, $H^1(\Omega) \hookrightarrow L^4(\Omega)$ for $(\nabla d, v)$, and trace theorem, we have

$$\frac{d}{dt}\|d^k\|_{H^1(\Omega)}^2 \leqslant C(K, \|d_0\|_{H^2(\Omega)})(\kappa_d^6 + \kappa_d^2 K^2) + 2\|d^k\|_{H^1(\Omega)}^2, \tag{3.55}$$

which, together with Gronwall's inequality, implies

$$\|d^k\|_{H^1(\Omega)}^2 \leqslant \Big(\|d_0\|_{H^1(\Omega)}^2 + C(K, \|d_0\|_{H^2(\Omega)})(\kappa_d^6 + \kappa_d^2 K^2)t\Big)e^{2t}$$

for any $t \in (0, T_d^*)$. Taking $T_1^* = \kappa_d^{-6} \leqslant 1$ and letting $T_d^* \leqslant T_1^*$, we conclude

$$\|d^k\|_{H^1(\Omega)}^2 \leqslant C(K, \|d_0\|_{H^2(\Omega)}). \tag{3.56}$$

Next we derive bounds on $\partial_t d$. It follows from (3.53) and integration by parts that

$$\frac{1}{2}\frac{d}{dt}\int_\Omega |\partial_t d^k|^2 dx + \int_\Omega |\nabla \partial_t d^k|^2 dx$$
$$= -2\Big(\int_\Omega (\Delta d^{k-1} \cdot \partial d^{k-1}) d^{k-1} \cdot \partial_t d^k dx + \int_\Omega \nabla d_i^{k-1} \cdot \nabla d_j^{k-1} \partial_t d_i^{k-1} \partial_t d_j^k dx$$
$$+ \int_\Omega \partial_t d_i^{k-1} d_j^{k-1} \nabla d_i^{k-1} \cdot \nabla \partial_t d_j^k dx\Big) + \int_\Omega |\nabla d^{k-1}|^2 \partial_t d^{k-1} \cdot \partial_t d^k dx$$
$$- \int_\Omega u \cdot \nabla \partial_t d^{k-1} \cdot \partial_t d^k dx - \int_\Omega \partial_t u \cdot \nabla d^{k-1} \cdot \partial_t d^k dx. \tag{3.57}$$

Applying Cauchy-Schwartz's inequality and Hölder's inequality, and Gagliardo-Nirenberg inequality, the right-hand side of (3.57) can be bounded from above by

$$c\Big(\|d^{k-1}\|^2_{L^\infty}\|\nabla d^{k-1}\|^2_{L^4(\Omega)}\|\partial_t d^{k-1}\|_{L^2(\Omega)}\|\nabla\partial_t d^{k-1}\|_{L^2(\Omega)}$$
$$+ (\|\nabla\partial_t d^{k-1}\|_{L^2(\Omega)} + \|\partial_t u\|_{L^2(\Omega)}\|\nabla\partial_t u\|_{L^2(\Omega)})\|\partial_t d^k\|^2_{L^2(\Omega)}$$
$$+ \|\partial_t d^{k-1}\|_{L^2(\Omega)}\|\nabla\partial_t d^{k-1}\|^{\frac{1}{2}}_{L^2(\Omega)}(\|\Delta d^{k-1}\|^2_{L^2(\Omega)}\|d^{k-1}\|^2_{L^\infty(\Omega)} + \|\nabla d^{k-1}\|^4_{L^4(\Omega)})$$
$$+ \|v\|^2_{L^\infty(\Omega)}\|\nabla\partial_t d^{k-1}\|^2_{L^2(\Omega)} + \|\nabla d^{k-1}\|^2_{L^4(2)}\Big) + \frac{1}{2}\|\nabla\partial_t d^k\|^2_{L^2(\Omega)}.$$

Using $H^2(\Omega) \hookrightarrow L^\infty(\Omega)$ and $H^1(\Omega) \hookrightarrow L^4(\Omega)$ for d and ∇d respectively, the above term can be further bounded from above by

$$C_1(\Omega)\Big((\kappa_d^5 + K^2)(\|\nabla\partial_t d^{k-1}\|_{L^2(\Omega)} + \|\nabla\partial_t d^{k-1}\|^{\frac{1}{2}}_{L^2(\Omega)}) + \kappa_d^2$$
$$+ (\|\nabla\partial_t d^{k-1}\|_{L^2(\Omega)} + \|\partial_t u\|_{L^2(\Omega)}\|\nabla\partial_t u\|_{L^2(\Omega)})\|\partial_t d^k\|^2_{L^2(\Omega)}\Big) + \frac{1}{2}\|\nabla\partial_t d^k\|^2_{L^2(\Omega)}.$$
(3.58)

Hence, we have

$$\frac{d}{dt}\|\partial_t d^k\|^2_{L^2(\Omega)} + \|\nabla\partial_t d^k\|^2_{L^2(\Omega)}$$
$$\leqslant C_1(\Omega)\Big((\kappa_d^5 + K^2)(\|\nabla\partial_t d^{k-1}\|_{L^2(\Omega)} + \|\nabla\partial_t d^{k-1}\|^{\frac{1}{2}}_{L^2(\Omega)}) + \kappa_d^2$$
$$+ (\|\nabla\partial_t d^{k-1}\|_{L^2(\Omega)} + \|\partial_t u\|_{L^2(\Omega)}\|\nabla\partial_t u\|_{L^2(\Omega)})\|\partial_t d^k\|^2_{L^2(\Omega)}\Big). \quad (3.59)$$

Consequently, using Gronwall's and Hölder's inequalities, we get

$$\|\partial_t d^k\|^2_{L^2(\Omega)} \leqslant \Big(\|\partial_t d_0\|^2_{L^2(\Omega)} + C_1(\Omega)(\kappa_d^5 + K^2)(t + \kappa_d\sqrt{t} + \sqrt{\kappa_d}t^{\frac{3}{4}})\Big)e^{C_1(\Omega)(\kappa_d\sqrt{t}+K^2)}$$

for any $t \in (0, T_d^*)$. Now, taking $T_2^* = \kappa_d^{-12} \leqslant 1$, let $T_d^* \leqslant T_2^*$, and noting that

$$\|\partial_t d(0)\|_{L^2(\Omega)} = \|(\Delta d_0 + |\nabla d_0|^2 d_0) - u_0\cdot\nabla d_0\|_{L^2(\Omega)},$$

we conclude

$$\sup_{t\in(0,T_d^*)}\|\partial_t d^k\|_{L^2(\Omega)} \leqslant C(K, \|d_0\|_{H^2(\Omega)}). \quad (3.60)$$

Moreover, using (3.53), (3.59),(3.60), Cauchy's inequality, and the elliptic theory, we obtain the following estimates,

$$\|\partial_t d^k\|_{L^2((0,T_d^*);H^1(\Omega))} \leqslant C(K, \|d_0\|_{H^2(\Omega)}), \quad (3.61)$$

$$\|\nabla^2 d^k\|_{L^2(\Omega)}^2 \leq c\Big(\|\nabla^2 d_0\|_{L^2(\Omega)}^2 + \|\partial_t d^k\|_{L^2(\Omega)}^2 + \|u \cdot \nabla d^{k-1}\|_{L^2(\Omega)}^2$$
$$+ \||\nabla d^{k-1}|^2 d^{k-1}\|_{L^2(\Omega)}^2\Big)$$
$$\leq C(K, \|d_0\|_{H^2(\Omega)}) + c\|u \cdot \nabla d^{k-1}\|_{L^2(\Omega)}^2 + c\||\nabla d^{k-1}|^2 d^{k-1}\|_{L^2(\Omega)}^2, \tag{3.62}$$

$$\|\nabla^3 d^k\|_{L^2((0,T_d^*);L^2(\Omega))}^2 \leq C(K, \|d_0\|_{H^2(\Omega)}, \|\nabla^3 d_0\|_{L^2(\Omega)})$$
$$+ c\|\nabla(u \cdot \nabla d^{k-1})\|_{L^2((0,T_d^*);L^2(\Omega))}^2$$
$$+ c\|\nabla(|\nabla d^{k-1}|^2 d^{k-1})\|_{L^2((0,T_d^*);L^2(\Omega))}^2, \tag{3.63}$$

where

$$\|\nabla^3 d^k\|_{L^2((0,T_d^*);L^2(\Omega))}^2 := \sum_{1 \leq i,j,l,m \leq 2} \|\partial_i \partial_j \partial_m d_l^k\|_{L^2((0,T_d^*);L^2(\Omega))}^2.$$

To complete the derivation of (3.54), it suffices to deduce the uniform estimates of the last two terms in (3.62) and (3.63) as follows.

Since $d \in \mathcal{D}^*$, in view of [46](lemmas 1.65, 1.66), we have

$$d(t) - d_0 = \int_0^t \partial_t d(\tau) d\tau \quad \text{for any } t \in (0, T_d^*),$$

which implies that

$$\|d(t) - d_0\|_{H^1(\Omega)} \leq \sqrt{t} \|\partial_t d\|_{L^2((0,T_d^*);H^1(\Omega))}. \tag{3.64}$$

Using (3.64), we obtain

$$\|u \cdot \nabla d^{k-1}\|_{L^2(\Omega)}^2 \leq c\|v\|_{L^\infty(\Omega)}^2 \Big(\|\nabla d^{k-1} - \nabla d_0^{k-1}\|_{L^2(\Omega)}^2 + \|\nabla d_0^{k-1}\|_{L^2(\Omega)}\Big)$$
$$\leq cK^2(t\kappa_d^2 + \|d_0\|_{H^1(\Omega)}^2).$$

Recalling that $0 \leq t \leq T_d^* \leq \kappa_d^{-12}$, we have

$$\|u \cdot \nabla d^{k-1}\|_{L^2(\Omega)}^2 \leq C(K, \|d_0\|_{H^1(\Omega)}). \tag{3.65}$$

From Lemma 2.1, we can get that

$$\|u\|_{L^\infty(\Omega)}^2 \leq c_1(\Omega)\|u\|_{H^2(\Omega)}\|u\|_{L^2(\Omega)}, \tag{3.66}$$
$$\|u\|_{L^4(\Omega)}^2 \leq c_1(\Omega)\|u\|_{H^1(\Omega)}\|u\|_{L^2(\Omega)} \tag{3.67}$$

for some constants $c_1(\Omega)$ depending on Ω.

Making use of (3.64), (3.66) and (3.67), it is easy to see that

$$\||\nabla d^{k-1}|^2 d^{k-1}(t)\|_{L^2(\Omega)}^2$$
$$\leqslant c\|\nabla d^{k-1}\|_{L^4(\Omega)}^4 \|d^{k-1}\|_{L^\infty(\Omega)}^2$$
$$\leqslant c\Big(\|\nabla d^{k-1}\|_{L^4(\Omega)}^4 \|d^k - d_0^{k-1}\|_{L^\infty(\Omega)}^2 + \|\nabla d_0^{k-1}\|_{L^4(\Omega)}^4 \|d_0^{k-1}\|_{L^\infty(\Omega)}^2$$
$$+ \|\nabla d^{k-1} - \nabla d_0^{k-1}\|_{L^4(\Omega)}^4 \|d_0^{k-1}\|_{L^\infty(\Omega)}^2\Big)$$
$$\leqslant C_1(\Omega)\Big(\kappa_d^5 \|d^{k-1} - d_0\|_{H^1(\Omega)} + \|d_0\|_{H^2(\Omega)}^6\Big)$$
$$\leqslant C_1(\Omega)\Big(\kappa_d^5 \|\partial_t d\|L^2((0,T_d^*); L^2(\Omega))\sqrt{t} + \|d_0\|_{H^2(\Omega)}^6\Big),$$

for any $t \in (0, T_d^*)$. Since $T_d^* \leqslant \kappa_d^{-12}$, we have

$$\||\nabla d^{k-1}|^2 d^{k-1}\|_{L^2(\Omega)} \leqslant C_1(\Omega)(1 + \|d_0\|_{H^2(\Omega)}^6). \tag{3.68}$$

Using Hölder's inequality, (3.66) and (3.67), we infer that

$$\|\nabla(u \cdot \nabla d^{k-1})\|_{L^2((0,T_d^*);L^2(\Omega))}^2 + c\|\nabla(|\nabla d^{k-1}|^2 d^{k-1})\|_{L^2((0,T_d^*);L^2(\Omega))}^2$$
$$\leqslant c\bigg(\|\nabla u\|_{L^\infty((0,T_d^*);L^2(\Omega))}^2 \|\nabla d^{k-1}\|_{L^2((0,T_d^*);L^\infty(\Omega))}^2$$
$$+ \|u\|_{L^\infty(Q_T)}^2 \|\nabla^2 d^{k-1}\|_{L^2((0,T_d^*));L^2(\Omega)}^2$$
$$+ \|d^{k-1}\|_{L^\infty(Q_T)}^2 \||\nabla d^{k-1}|\nabla^2 d^{k-1}\|_{L^2((0,T_d^*);L^2(\Omega))}^2$$
$$+ \int_0^{T_d^*} \|\nabla d^{k-1}\|_{L^\infty(\Omega)}^2 \|\nabla d^{k-1}\|_{L^4(\Omega)}^4 ds\bigg)$$
$$\leqslant C_1(\Omega)\Big(K^2 \|\nabla d^{k-1}\|_{L^\infty((0,T_d^*);L^2(\Omega))} \|d^{k-1}\|_{L^1((0,T_d^*);H^3(\Omega))}$$
$$+ K^2 \|\nabla d^{k-1}\|_{L^\infty((0,T_d^*);L^2(\Omega))}^2 T_d^*$$
$$+ \|d^{k-1}\|_{L^\infty((0,T_d^*);H^2(\Omega))}^5 \|d^{k-1}\|_{L^1((0,T_d^*);H^3(\Omega))}\Big)$$
$$\leqslant C_1(\Omega)\Big(K^2 \kappa_d^2 \sqrt{T_d^*} + K^2 \kappa_d^2 T_d^* + \kappa_d^6 \sqrt{T_d^*}\Big),$$

which, together with the fact $T_d^* \leqslant \kappa_d^{-12} \leqslant 1$, yields

$$\|\nabla(u \cdot \nabla d^{k-1})\|_{L^2((0,T_d^*);L^2(\Omega))}^2 + c\|\nabla(|\nabla d^{k-1}|^2 d^{k-1})\|_{L^2((0,T_d^*);L^2(\Omega))}^2 \leqslant C_1(\Omega)(1+K^2). \tag{3.69}$$

Finally, putting (3.56), (3.60)-(3.63), (3.68)-(3.69) together, we then obtain that
$$\|d^k\|_{\mathcal{D}^*} \leqslant C(K, \|d_0\|_{H^2(\Omega)}, \|\nabla^3 d_0\|_{L^2(\Omega)}).$$

Choosing $\kappa_d \geqslant \max\{C(K, \|d_0\|_{H^3(\Omega)}, \|\nabla^3 d_0\|_{L^2(\Omega)}), \|d_0\|_{H^3(\Omega)}, 1\}$, we get (3.54).

3.2.3 Taking limits

In order to take limits in (3.53), we shall further show that $\{d^k\}_{k=1}^\infty$ is a Cauchy sequence. To this end, we define
$$\bar{d}^{k+1} = d^{k+1} - d^k,$$

which satisfies
$$\partial_t \bar{d}^{k+1} - \Delta \bar{d}^{k+1} = [\nabla \bar{d}^k : (\nabla d^k + \nabla d^{k-1})] d^k + |\nabla d^{k-1}|^2 \bar{d}^k - u \cdot \nabla \bar{d}^k, \quad (3.70)$$

with initial data
$$\bar{d}^{k+1}|_{t=0} = 0, \quad x \in \Omega.$$

Using Cauchy-Schwarz's and Hölder's inequalities, (3.66) and (3.67) again, it follows from (3.70) that

$$\frac{d}{dt}\|\bar{d}^{k+1}\|_{H^1(\Omega)} + 2(\|\nabla \bar{d}^{k+1}\|_{L^2(\Omega)}^2 + \|\Delta \bar{d}^{k+1}\|_{L^2(\Omega)}^2)$$
$$\leqslant \|\bar{d}^{k+1}\|_{H^1(\Omega)}^2 + c\Big(\|\nabla \bar{d}^k\|_{L^2(\Omega)}^2 \|\nabla(d^k + d^{k-1})\|_{L^\infty(\Omega)}^2 \|d^k\|_{L^\infty(\Omega)}^2$$
$$+ \|\bar{d}^k\|_{L^4(\Omega)}^2 \|\nabla d^{k-1}\|_{L^\infty(\Omega)}^2 \|\nabla d^k\|_{L^4(\Omega)}^2 + \|u^k\|_{L^\infty(\Omega)}^2 \|\nabla \bar{d}^k\|_{L^2(\Omega)}^2\Big)$$
$$\leqslant C_1(\Omega) \|\bar{d}^k\|_{H^1(\Omega)}^2 \Big(K^2 + \kappa_d^3(\|d^k\|_{H^3(\Omega)} + \|d^{k-1}\|_{H^3(\Omega)})\Big) + \|\bar{d}^{k+1}\|_{H^1(\Omega)}^2. \quad (3.71)$$

Then, using Gronwall's and Hölder inequalities, we have

$$\|\bar{d}^{k+1}\|_{H^1(\Omega)}^2 \leqslant C_1(\Omega) \Big(K^2 t + \kappa_d^3 \int_0^t (\|d^k\|_{H^3(\Omega)} + \|d^{k-1}\|_{H^3(\Omega)}) ds\Big) e^t \sup_{0 \leqslant s \leqslant t} \|\bar{d}^k\|_{H^1(\Omega)}^2$$
$$\leqslant C(K, \kappa_d)(t + \sqrt{t}) e^t \sup_{0 \leqslant s \leqslant t} \|\bar{d}^k\|_{H^1(\Omega)}^2.$$

Taking
$$T_3^* = \min\{T_2^*, C^{-2}(K, \kappa_d)/(64e^2), 1/2\}$$

and letting $T_d^* \leqslant T_3^*$, we have

$$\sup_{t \in (0, T_d^*)} \|\bar{d}^{k+1}\|_{H^1(\Omega)}^2 \leqslant \frac{1}{4} \sup_{t \in (0, T_d^*)} \|\bar{d}^k\|_{H^1(\Omega)}^2. \quad (3.72)$$

Furthermore, from (3.71), (3.72) and the elliptic estimates we get

$$\|\bar{d}^{k+1}\|^2_{L^2((0,T_d^*);H^2(\Omega))} \leq \frac{1}{4} \sup_{t\in(0,T_d^*)} \|\bar{d}^k\|^2_{H^1(\Omega)}.$$

By iteration, we have

$$\sup_{t\in(0,T_d^*)} \|\bar{d}^{k+1}\|^2_{H^1(\Omega)} + \|\bar{d}^{k+1}\|^2_{L^2((0,T_d^*);H^2(\Omega))} \leq \frac{1}{2^{k-1}} \sup_{t\in(0,T_d^*)} \|\bar{d}^k\|^2_{H^1(\Omega)},$$

for any $k \geq 2$. In particular,

$$\sum_{k=2}^{\infty} \left(\|\bar{d}^k\|_{L^\infty((0,T_d^*);H^1(\Omega))} + \|\bar{d}^k\|_{L^2(0,T_d^*);H^2(\Omega))} \right) < \infty,$$

thus $\{d^k\}_{k=1}^\infty$ is a Cauchy sequence in $L^\infty((0,T_d^*);H^1(\Omega)) \cap L^2((0,T_d^*);H^2(\Omega))$. Therefore,

$$d^k \to d \text{ strongly in } L^\infty((0,T_d^*);H^1(\Omega)) \cap L^2((0,T_d^*);H^2(\Omega)). \tag{3.73}$$

As a consequence, d is a solution to the problem (3.48)-(3.49) satisfying the regularity $\|d\|_{\mathcal{D}^*} \leq \kappa_d$, by virtue of (3.53) and (3.73), the lower semi-continuity, the elliptic estimate and the compactness theorem with time ([8],Lemma2.5).

3.2.4 Continuous dependence

Finally, we show that the solution d continuously dependents on u. Let $T_d^* \in (0, T_3^*]$, $u_1, u_2 \in V_k$, d_1 and d_2 be two solutions of the initial value problem (3.48)-(3.49), corresponding to $u = u_1$ and $u = u_2$ respectively. Moreover, two solutions satisfy

$$\|d_1\|_{\mathcal{D}^*} + \|d_2\|_{\mathcal{D}^*} < \infty.$$

Multiplying the difference of the equations for $d_1 - d_2$ by $d_1 - d_2$ and integrating by parts, we obtain

$$\frac{d}{dt} \int_\Omega |d_1 - d_2|^2 dx + 2 \int_\Omega |\nabla(d_1 - d_2)|^2 dx$$
$$= 2 \int_\Omega |\nabla d_1|^2 |d_1 - d_2|^2 dx + 2 \int_\Omega (\nabla(d_1 - d_2) : \nabla(d_1 + d_2)) d_2 \cdot (d_1 - d_2) dx$$
$$- 2 \int_\Omega (u_1 - u_2) \cdot \nabla d_1 \cdot (d_1 - d_2) dx - 2 \int_\Omega u_2 \cdot \nabla(d_1 - d_2) \cdot (d_1 - d_2) dx.$$

Similarly to the derivation of (3.71), we use (3.66) and (3.67) to see that the right hand side of the above identity can be bounded from above by

$$\|\nabla(d_1-d_2)\|^2_{L^2(\Omega)} + C(\|d_1\|_{\mathcal{D}^*}, \|d_2\|_{\mathcal{D}^*}, K)(1+\|d_1\|_{H^3(\Omega)}$$
$$+ \|d_2\|_{H^3(\Omega)})\Big(\|d_1-d_2\|^2_{L^2(\Omega)} + \|u_1-u_2\|_{L^2(\Omega)}\|d_1-d_2\|_{L^2(\Omega)}\Big),$$

Hence,

$$\frac{d}{dt}\|d_1-d_2\|^2_{L^2(\Omega)} + \|\nabla(d_1-d_2)\|^2_{L^2(\Omega)}$$
$$\leqslant C(\|d\|_{\mathcal{D}^*}, \|d_2\|_{\mathcal{D}^*}, K)(1+\|d_1\|_{H^3(\Omega)}+\|d_2\|_{H^3(\Omega)})\Big(\|d_1-d_2\|^2_{L^2(\Omega)}$$
$$+ \|u_1-u_2\|_{L^2(\Omega)}\|d_1-d_2\|_{L^2(\Omega)}\Big),$$

in particular,

$$\frac{d}{dt}\|d_1-d_2\|^2_{L^2(\Omega)}$$
$$\leqslant C(\|d\|_{\mathcal{D}^*}, \|d_2\|_{\mathcal{D}^*}, K)(1+\|d_1\|_{H^3(\Omega)}+\|d_2\|_{H^3(\Omega)})\Big(\|d_1-d_2\|^2_{L^2(\Omega)}$$
$$+ \|u_1-u_2\|_{L^2(\Omega)}\|d_1-d_2\|_{L^2(\Omega)}\Big).$$

Thus, applying Gronwall inequality and Cauchy-Schwartz inequality, we conclude

$$\|(d_1-d_2)(t)\|_{L^2(\Omega)} \leqslant (t+\sqrt{t})C(\|d_1\|_{\mathcal{D}^*}, \|d_2\|_{\mathcal{D}^*}, K, T_d^*)\|u_1-u_2\|_{L^\infty((0,t);L^2(\Omega))} \tag{3.74}$$

for any $t \in (0, T_d^*)$. Moreover,

$$\|\nabla(d_1-d_2)\|_{L^2((0,t);L^2(\Omega))} \leqslant (t+\sqrt{t})C(\|d_1\|_{\mathcal{D}^*}, \|d_2\|_{\mathcal{D}^*}, K, T_d^*)\|u_1-u_2\|_{L^\infty((0,t);L^2(\Omega))}.$$

Obviously, the uniqueness of local solutions in the function class $\mathcal{D}^*$ follows from (3.74) immediately. In particular, $\|d_1\|_{\mathcal{D}^*}, \|d_2\|_{\mathcal{D}^*} \leqslant \kappa_d$. To conclude this subsection, we summarize the above results on the local existence of d.

Proposition 3.1 *Let $K > 0$, $0 < \alpha < 1$, $u \in V_k$, and $d_0 \in H^3(\Omega)$ Then there exists a finite time*

$$T_d^K := h_1(K, \|d_0\|_{H^2(\Omega)}, \|\nabla^3 d_0\|_{L^2(\Omega)}) \in (0, \min\{0, T\}),$$

and a corresponding unique mapping

$$S_2 : V_k(\text{ with } T_d^k \text{ in place of } T) \to C^0([0, T_d^K]; H^2(\Omega)),$$

where h_1 is non-increasing in its first two variables and $Q_{T_d^K} := \Omega \times (0, T_d^K)$, such that

(1) S_2 belongs to the following function class

$$\mathcal{R}_{T_d^K} := \{d | d \in L^2((0,T_d^K); H^3(\Omega)), \partial_{tt} d \in L^2((0,T_d^K); (H^1(\Omega))^*)$$

$$\partial_t d \in C^0([0, T_d^k]; L^2(\Omega)) \cap L^2((0, T_d^K); H^1(\Omega))\}. \tag{3.75}$$

(2) $d = S_2(u)$ satisfies (3.48) almost everywhere in $Q_{T_d^K}$, (3.49) in Ω. Moreover, $d|_{t=0} = d_0$.

(3) $S_2(u)$ enjoys the following estimate:

$$\|S_2(u)\|_{\mathcal{D}^*} \leqslant C(K, \|d_0\|_{H^2(\Omega)}, \|\nabla^3 d_0\|_{L^2(\Omega)}), \tag{3.76}$$

where C is nondecreasing in its variables, and T_d^* in the definition of $\mathcal{D}^*$ should be replaced by T_d^K. Moreover,

$$|d(t)| = 1, \quad \text{for any } t, \quad \text{if } |d_0| = 1.$$

(4) $S_2(u)$ continuously depends on u in the following sense:

$$\|[S_2(u_1) - S_2(u_2)](t)\|_{L^2(\Omega)} + \|\nabla(S_2(u_1) - S_2(u_2))\|_{L^2(Q_t)}$$
$$\leqslant \sqrt{t} C(K, \|d_0\|_{H^2(\Omega)}, \|\nabla^3 d_0\|_{L^2(\Omega)}) \|u_1 - u_2\|_{L^\infty((0,t); L^2(\Omega))} \tag{3.77}$$

for any $t \in [0, T_d^k]$, where C is non-decreasing in its variables.

3.3 Solvability of the magnetic equation

Firstly, we can rewrite the magnetic equation (1.32) as follows

$$\partial_t B - B \cdot \nabla u + u \cdot \nabla B + (\nabla \cdot u) B = \delta \Delta B, \quad \text{div} B = 0. \tag{3.78}$$

Similar to Section 3.2, we can use the operator $S_3 : C^0([0, T_B^*]; X_N) \to C^0([0, T]; H^2(\Omega))(T_B^* < T^*)$ by $S_3(u) = B$. S_3 is Lipshitz continuous in the following sense:

$$\|S_3(u_1) - S_3(u_2)\|_{C^0([0, T_B^*]; C^k(\Omega))} \leqslant C(N, k) \|u_1 - u_2\|_{C^0([0, T_B^*]; L^2(\Omega))}. \tag{3.79}$$

3.4 Solvability of the velocity equation

Next we wish to solve (1.30) on the space X_N. To this end, for given $\rho = S_1(u)$, $d = S_2(u)$, $B = S(u)$, we are looking for functions $u_N^\gamma \in (C^0([0, T^*]; X_N))$ such that

$$-\int_\Omega \rho_0 u_0 \cdot \phi(\cdot, 0) dx = \int_0^T \int_\Omega \left(\rho u_N^\gamma \cdot \phi_t + \rho(u_N^\gamma \otimes u_N^\gamma) : \nabla \phi + (P(\rho) + P_c(\rho)) \text{div} \phi \right.$$

$$-\frac{\hbar^2}{2}\frac{\Delta\sqrt{\rho}}{\sqrt{\rho}}\operatorname{div}(\rho\phi) - \nu\nabla(\rho u_N^\gamma):\nabla\phi$$
$$+\lambda\left(\nabla d\odot\nabla d - \frac{|\nabla d|^2}{2}I\right)\nabla\phi$$
$$+\kappa(\nabla\times B)\times B\cdot\phi - \gamma(\nabla u_N^\gamma:\nabla\phi + u_N^\gamma\cdot\phi)\Big)dxdt,$$
(3.80)

for all $\phi\in C^1([0,T^*];X_N)$ such that $\phi(\cdot,T^*)=0$. Notice that we have added the regularization term $\gamma(\Delta u_N^\gamma - u_N^\gamma)$. The reason is that we will apply Banach fixed point theorem to prove the local-in-time existence of solutions. The regularization yields the H^1 regularity of u_N^γ needed to conclude the global existence of solutions.

To solve (3.80), we follow ([40], Chapter 7.3.3) and introduce the following family of operators, given a function $\varrho\in L^1(\Omega)$ with $\varrho\geqslant\underline{\varrho}>0$:

$$\mathcal{M}[\varrho]:X_N\to X_N^*, \langle\mathcal{M}[\varrho]u,w\rangle = \int_\Omega \varrho u\cdot w, \quad u,w\in X_N.$$

These operators are symmetric and positive definite with the smallest eigenvalue

$$\int_{\|u\|_{L^2(\Omega)}=1}\langle\mathcal{M}[\varrho]u,u\rangle = \int_{\|u\|_{L^2(\Omega)}=1}\int_\Omega \varrho|u|^2 dx \geqslant \inf_{x\in\Omega}\varrho(x)\geqslant\underline{\varrho}.$$

Hence, since X_N is finite-dimensional, the operators are invertible with

$$\|\mathcal{M}^{-1}[\varrho]\|_{L(X_N^*,X_N)} \leqslant \underline{\varrho}^{-1},$$

where $L(X_N^*,X_N)$ is the set of bounded linear mappings from X_N^* to X_N. Moreover (see [40] Chapter 7.3.3), M^{-1} is Lipschitz continuous in the sense

$$\|\mathcal{M}^{-1}[\varrho_1] - \mathcal{M}^{-1}[\varrho_2]\|_{L(X_N^*,X_N)} \leqslant C(N,\underline{\varrho})\|\varrho_1 - \varrho_2\|_{L^1(\Omega)} \quad (3.81)$$

for all $\varrho_1,\varrho_2\in L^1(\Omega)$ such that $\varrho_1,\varrho_2\geqslant\underline{\varrho}>0$.

Now the integral equation (3.80) can be rephrased as an ordinary differential equation on the finite-dimensional space X_N:

$$\frac{d}{dt}(\mathcal{M}[\rho(t)]w_N) = \mathcal{N}[\rho,u,d,B,w_N], \quad t>0, \quad \mathcal{M}[\rho_0]w_N(0) = \mathcal{M}[\rho_0]w_0, \quad (3.82)$$

where $\rho = S_1(u)$, $d = S_2(u)$, $B = S_3(u)$ and

$$\langle\mathcal{N}[\rho,u,d,B],\phi\rangle = \int_\Omega\Big(\rho(u\otimes w_N):\nabla\phi + (P(\rho)+P_c(\rho))\operatorname{div}\phi - \frac{\hbar^2}{2}\frac{\Delta\sqrt{\rho}}{\sqrt{\rho}}\operatorname{div}(\rho\phi)$$
$$-\nu\nabla(\rho w_N):\nabla\phi + \kappa(\nabla\times B)\times B\cdot\phi$$
$$+\lambda\left(\nabla d\odot\nabla d - \frac{|\nabla d|^2}{2}I\right)\nabla\phi - \delta(\nabla w_N:\nabla\phi + w_N\cdot\phi)\Big)dx,$$

$\forall \phi \in X_N$. For operator $\mathcal{N}[\rho, u, d, B, \cdot]$, defined for every $t \in [0, T]$ as an operator from X_N to X_N^*, is continuous in time. Standard theory for systems of ordinary differential equations provides the existence of a unique classical solution to (3.82), i.e., for a given $u \in C^0([0,T]; X_N)$, there exists a unique solution $u_N^\gamma \in C^1([0, T^*]; X_N)$ to (3.80).

Integrating (3.82) over $(0, t)$ yields the following nonlinear equation:

$$u_N^\gamma(t) = \mathcal{M}^{-1} S_1(u_N^\gamma)(t) \Big(\mathcal{M}[\rho_0]u_0 + \int_0^t \mathcal{N}[S_1(u_N^\gamma), u_N^\gamma, S_2(u_N^\gamma), S_3(u_N^\gamma), u_N^\gamma(s)] ds \Big),$$
$$\text{in } X_N. \qquad (3.83)$$

Since the operators S_1, S_2, S_3 and M are Lipschitz type, (3.83) can be solved by evoking the fixed point theorem of Banach on a short time interval $[0, T']$, where $T' \leqslant T^*$, in the space $C^0([0, T]; X_N)$. In fact, we have even $u_N^\gamma \in C^1([0, T']; X_N)$. Then we can solve the equation (3.80). Thus, there exists a unique local-in-time solution $(\rho_N^\gamma, u_N^\gamma, d_N^\gamma, B_N^\gamma)$ to (1.29)-(1.33).

4 Global existence of the weak solutions

In this section, we will prove the existence of the global finite weak solution. Set $(\rho_N^\gamma, u_N^\gamma, d_N^\gamma, B_N^\gamma)$ are solutions constructed in the above section. Then we will prove the limit of $(\rho_N^\gamma, u_N^\gamma, d_N^\gamma, B_N^\gamma)$ is the weak solution to (1.29)-(1.33) as $N \to \infty$ and $\gamma \to 0$.

4.1 A-priori estimate

Firstly, we give some a-priori estimates. For simplicity, we omit the subscript N and superscript γ in this subsection.

Theorem 4.1 *For $0 < T < \infty$, Suppose (ρ, u, d, B) is the weak solution of (1.29)-(1.33). Then, for any $0 < t \leqslant T$, there holds*

$$\frac{d}{dt} E(\rho, u, d, B) + \int_\Omega \nu H''(\rho)|\nabla \rho|^2 + \frac{\nu \hbar^2}{4} \rho |\nabla^2 \log \rho|^2 + \nu \rho |\nabla u|^2$$
$$+ \lambda |\Delta d + |\nabla d|^2 d|^2 + \kappa \delta |\nabla B|^2 + \gamma |\nabla u|^2 + \gamma |u|^2 dx \leqslant 0, \qquad (4.84)$$

where

$$E(\rho, u, d, E, B) = \int_\Omega H(\rho) + \frac{\hbar^2}{2} |\nabla \sqrt{\rho}|^2 + \frac{1}{2} \rho |u|^2 + \lambda \frac{1}{2} |\nabla d|^2 + + \frac{\kappa}{2} |B|^2 dx, \quad (4.85)$$

here, $H(\rho) = \dfrac{A \rho^\alpha}{\alpha - 1}$ and $H_c''(\rho) = \dfrac{P_c'(\rho)}{\rho}$.

Proof Multiplying (1.30) by u, and integrating by parts respect in Ω, we can obtain

$$\frac{d}{dt}\int_\Omega \frac{1}{2}\rho|u|^2 + H(\rho) + H_c(\rho) + \frac{\hbar^2}{2}|\nabla\sqrt{\rho}|^2 + \frac{\lambda}{2}|\nabla d|^2 + \frac{\kappa}{2}|B|^2 dx$$
$$+ \int_\Omega \nu H''(\rho)|\nabla\rho|^2 dx + \int_\Omega \nu H_c''(\rho)|\nabla\rho|^2 dx + \int_\Omega \nu\rho|\nabla u|^2 dx + \frac{\nu\hbar^2}{4}\int_\Omega \rho|\nabla^2\log\rho|^2 dx$$
$$+ \lambda\int_\Omega \nabla\cdot\left(\nabla d\odot\nabla d - \frac{|\nabla d|^2}{2}I\right)\cdot u dx - \kappa\int_\Omega (\nabla\times B)\times B\cdot u dx \leq 0. \quad (4.86)$$

Here we use the following facts:

$$\int_\Omega (\rho u)_t \cdot u + \mathrm{div}(\rho u\otimes u)\cdot u dx$$
$$= \int_\Omega \rho_t|u|^2 + \rho\partial_t\left(\frac{|u|^2}{2}\right) + \mathrm{div}(\rho u)|u|^2 + \rho u\cdot\nabla\left(\frac{|u|^2}{2}\right) dx$$
$$= \int_\Omega \rho_t|u|^2 + \rho\partial_t\left(\frac{|u|^2}{2}\right) + \frac{1}{2}\mathrm{div}(\rho u)|u|^2 dx$$
$$= \partial_t\int_\Omega \frac{1}{2}\rho|u|^2 dx + \frac{1}{2}\int_\Omega \nu\Delta\rho|u|^2 dx,$$

$$\int_\Omega \nabla\rho^\alpha\cdot u dx = \int_\Omega \frac{\alpha}{\alpha-1}\nabla\rho^{\alpha-1}\cdot(\rho u) dx$$
$$= \int_\Omega \frac{\alpha}{\alpha-1}\rho^{\alpha-1}\mathrm{div}(\rho u) dx = \frac{\alpha}{\alpha-1}\int_\Omega \rho^{\alpha-1}(\rho_t - \nu\Delta\rho) dx$$
$$= \frac{1}{\alpha-1}\partial_t\int_\Omega \rho^\alpha dx + \nu\int_\Omega \alpha\rho^{\alpha-2}|\nabla\rho|^2 dx$$
$$= \partial_t\int_\Omega H(\rho) dx + \nu\int_\Omega H''(\rho)|\nabla\rho|^2 dx.$$

From the equation (1.13), we know that

$$P_c(\rho) = \begin{cases} A_1\dfrac{1}{-4k}\rho^{-4k}, & \rho\leq 1,\ k\geq 1, \\ \dfrac{1}{\alpha}\rho^\alpha, & \rho > 1,\ \alpha > 1. \end{cases}$$

Thus for $\rho > 1$, we can get

$$\int_\Omega \frac{1}{\alpha}\nabla\rho^\alpha\cdot u dx = \frac{1}{\alpha(\alpha-1)}\partial_t\int_\Omega \rho^\alpha dx + \nu\int_\Omega \rho^{\alpha-2}|\nabla\rho|^2 dx$$
$$= \partial_t\int_\Omega H_c(\rho) dx + \nu\int_\Omega H_c''(\rho)|\nabla\rho|^2 dx;$$

for $\rho \leqslant 1$, we can get

$$\int_\Omega \frac{1}{-4k}\nabla\rho^{-4k} = \int_\Omega \rho^{-4k-2}\nabla\rho\cdot(\rho u)dx$$

$$= -\int_\Omega \frac{1}{-4k-1}\rho^{-4k-1}\mathrm{div}(\rho u)dx = \int_\Omega \frac{1}{-4k-1}\rho^{-4k-1}(\rho_t - \nu\Delta\rho)dx$$

$$= \frac{1}{4k(4k+1)}\partial_t\int_\Omega \rho^{-4k}dx + \nu\int_\Omega \rho^{-4k-2}|\nabla\rho|^2 dx$$

$$= \partial_t\int_\Omega H_c(\rho)dx + \nu\int_\Omega H_c''(\rho)|\nabla\rho|^2 dx,$$

$$\int_\Omega \rho\nabla\left(\frac{\Delta\sqrt{\rho}}{\sqrt{\rho}}\right)\cdot u\,dx = -\int_\Omega \left(\frac{\Delta\sqrt{\rho}}{\sqrt{\rho}}\right)\mathrm{div}(\rho u)dx$$

$$= \int_\Omega \Delta\sqrt{\rho}\,2\partial_t(\sqrt{\rho})dx - \nu\int_\Omega \frac{\Delta\sqrt{\rho}}{\sqrt{\rho}}\Delta\rho\,dx$$

$$= -\partial_t\int_\Omega |\nabla\sqrt{\rho}|^2 dx + \nu\int_\Omega \rho\nabla\left(\frac{\Delta\sqrt{\rho}}{\sqrt{\rho}}\right)\cdot\nabla\log\rho\,dx$$

$$= -\partial_t\int_\Omega |\nabla\sqrt{\rho}|^2 dx + \frac{1}{2}\nu\int_\Omega \mathrm{div}(\rho\nabla^2\log\rho)\cdot\nabla\log\rho\,dx$$

$$= -\partial_t\int_\Omega |\nabla\sqrt{\rho}|^2 dx - \frac{1}{2}\nu\int_\Omega \rho|\nabla^2\log\rho|^2 dx,$$

$$\nu\int_\Omega \Delta(\rho u)\cdot u\,dx = -\int_\Omega \nu\nabla(\rho u):\nabla u\,dx$$

$$= -\nu\int_\Omega \nabla\rho:\nabla u:u\,dx - \nu\int_\Omega \rho|\nabla u|^2 dx$$

$$= \frac{1}{2}\nu\int_\Omega \Delta\rho|u|^2 dx - \nu\int_\Omega \rho|\nabla u|^2 dx.$$

Multiplying (1.32) by $\Delta d + |\nabla d|^2 d$, and integrating by parts respect in Ω, we get

$$-\frac{d}{dt}\int_\Omega \frac{1}{2}|\nabla d|^2 dx - \int_\Omega \nabla d\odot\nabla d:\nabla u\,dx + \int_\Omega \frac{|\nabla d|^2}{2}\nabla u\,dx - \int_\Omega |\Delta d + |\nabla d|^2 d|^2 dx = 0. \tag{4.87}$$

here we use the following computation:

$$\int_\Omega d_t\cdot\Delta d\,dx = -\frac{d}{dt}\int_\Omega \frac{1}{2}|\nabla d|^2 dx,$$

$$\int_\Omega (u\cdot\nabla d)\cdot\Delta d\,dx = \int_\Omega d_{jj}u^i d_i dx$$

$$= \int_\Omega [(d_j u^i d_i)_j - d_i\cdot d_j u^i_j - u^i\left(\frac{|\nabla d|^2}{2}\right)_i]dx$$

$$= -\int_\Omega d_i \cdot d_j u_j^i + u_i^i \left(\frac{|\nabla d|^2}{2}\right) dx$$

$$= -\int_\Omega \nabla d \odot \nabla d : \nabla u dx + \int_\Omega \frac{|\nabla d|^2}{2} \nabla u dx,$$

$$\int_\Omega (d_t - u \cdot \nabla d) \cdot |\nabla d|^2 d dx = 0.$$

Multiplying (1.33) by B, and integrating by parts respect in Ω, we have

$$\frac{1}{2}\frac{d}{dt}\int_\Omega |B|^2 dx - \int_\Omega (\nabla \times B) \times B \cdot u dx = -\delta \int_\Omega |\nabla B|^2 dx. \qquad (4.88)$$

Here we use the following facts

$$\int_\Omega \nabla \times (u \times B) \cdot B dx = \int_\Omega (\nabla \times B) \times B \cdot u dx$$

$$- \delta \int_\Omega \nabla \times (\nabla \times B) \cdot B dx$$

$$= \delta \int_\Omega \Delta B \cdot B dx = -\delta \int_\Omega |\nabla B|^2 dx.$$

Combining (4.86)-(4.88), we can get

$$\frac{d}{dt}\int_\Omega H(\rho) + H_c(\rho) + \frac{\hbar^2}{2}|\nabla\sqrt{\rho}|^2 + \frac{1}{2}\rho|u|^2 + \lambda\frac{1}{2}|\nabla d|^2 + \frac{\kappa}{2}|B|^2 dx$$

$$+ \int_\Omega \nu H''(\rho)|\nabla\rho|^2 + H_c''(\rho)|\nabla\rho|^2 dx + \frac{\nu\hbar^2}{4}\rho|\nabla^2\log\rho|^2 + \nu\rho|\nabla u|^2 + \kappa\delta|\nabla B|^2$$

$$+ \lambda|\Delta d + |\nabla d|^2 d|^2 + \gamma|\nabla u|^2 + \gamma|u|^2 dx \leqslant 0. \qquad (4.89)$$

$\square$

4.2 The limit of the solutions as N tends to infinity

Setting $(\rho_N^\gamma, u_N^\gamma, d_N^\gamma, B_N^\gamma)$ are the solutions to (1.29)-(1.33), which also satisfy Theorem 4.1. In this subsection, we will omit the superscript for simplicity. From Theorem 4.1, by Gronwall inequality, we can easily get the following estimates:

Corollary 4.1 *Set Theorem 4.1 holds, it yields:*

$$\|\sqrt{\rho_N}\|_{L^\infty([0,T];H^1(\Omega))} \leqslant C, \qquad (4.90)$$

$$\|\rho_N\|_{L^\infty([0,T];L^\alpha(\Omega))} + \|\nabla\rho_N^{\frac{\alpha}{2}}\|_{L^2([0,T];L^2(\Omega))} \leqslant C, \qquad (4.91)$$

$$\|\rho_N^{-4k}\|_{L^\infty([0,T];L^1(\Omega))} + \|\nabla\rho_N^{-2k}\|_{L^2([0,T];L^2(\Omega))} \leqslant C, \quad \rho_N \leqslant 1, \qquad (4.92)$$

$$\|\sqrt{\rho_N}u_N\|_{L^\infty(0,T);L^2(\Omega)} + \|\sqrt{\rho_N}\nabla u_N\|_{L^2([0,T];L^2(\Omega))} \leqslant C, \qquad (4.93)$$

$$\|B_N\|_{L^\infty([0,T];L^2(\Omega))\cap L^2([0,T];H^1(\Omega))} \leqslant C, \qquad (4.94)$$

$$\|\nabla d_N\|_{L^\infty([0,T];L^2(\Omega))} \leqslant C, \qquad (4.95)$$

$$\|\Delta d_N + |\nabla d_N|^2 d_N\|_{L^2([0,T];L^2(\Omega))} \leqslant C, \qquad (4.96)$$

$$\|\sqrt{\rho_N}\nabla^2 \log\rho_N\|_{L^2([0,T];L^2(\Omega))} \leqslant C, \qquad (4.97)$$

$$\gamma\|u_N\|_{L^2([0,T];H^1(\Omega))} \leqslant C. \qquad (4.98)$$

Proof Here we only prove the equations (4.91) and (4.92). Since

$$P(\rho_N) = A\rho_N^\alpha, \quad P_c(\rho_N) = \begin{cases} A_1 \dfrac{1}{-4k}\rho_N^{-4k}, & \rho_N \leqslant 1,\ k \geqslant 1, \\ \dfrac{1}{\alpha}\rho_N^\alpha, & \rho_N > 1,\ \alpha > 1, \end{cases}$$

we have

$$H(\rho_N) = \frac{A}{\alpha-1}\rho_N^\alpha, \quad H_c(\rho_N) = \begin{cases} A_1 \dfrac{1}{4k(4k+1)}\rho_N^{-4k}, & \rho_N \leqslant 1,\ k \geqslant 1, \\ \dfrac{1}{\alpha(\alpha-1)}\rho_N^\alpha, & \rho_N > 1,\ \alpha > 1. \end{cases}$$

From the equation (4.84), we can easily obtain

$$\int_\Omega \rho_N^\alpha dx + \int_0^T \int_\Omega |\nabla \rho_N^{\frac{\alpha}{2}}|^2 dx \leqslant C,$$

and for $\rho_N \leqslant 1$,

$$\int_\Omega \rho_N^{-4k} dx + \int_0^T \int_\Omega |\nabla \rho_N^{-2k}|^2 dx \leqslant C.$$

Thus we can get (4.91) and (4.92). □

The energy inequality (4.84) and Corollary 4.1 allow us to conclude some estimates.

Lemma 4.1 *The following uniform estimate holds for some constant $C > 0$,*

$$\|\sqrt{\rho_N}\|_{L^2([0,T];H^2(\Omega))} + \|\sqrt[4]{\rho_N}\|_{L^4([0,T];W^{1,4}(\Omega))} \leqslant C. \qquad (4.99)$$

Proof This lemma follows from the energy estimate in Theorem 4.1, the inequality

$$\int_\Omega \rho_N |\nabla^2 \log\rho_N|^2 dx \geqslant \kappa_2 \int_\Omega |\nabla^2 \sqrt{\rho_N}|^2 dx, \qquad (4.100)$$

with κ_2, which is shown in [31], the inequality

$$\int_\Omega \rho_N |\nabla^2 \log \rho_N|^2 dx \geqslant \kappa \int_\Omega |\nabla \sqrt[4]{\rho_N}|^4 dx, \quad \kappa > 0,$$

which is proved in [31]. □

We are able to deduce more regularity from the H^2 bound for $\sqrt{\rho_N}$.

Lemma 4.2 (space regularity for ρ_N and $\rho_N u_N$) *The following uniform estimates hold for some constant $C > 0$ independent on N and γ:*

$$\|\rho_N u_N\|_{L^2([0,T];W^{1,q}(\Omega))} \leqslant C, \qquad (4.101)$$

$$\|\rho_N\|_{L^2([0,T];W^{2,q}(\Omega))} \leqslant C, \qquad (4.102)$$

$$\|\rho_N\|_{L^{4\alpha/3+1}([0,T];L^{4\alpha/3+1}(\Omega))} \leqslant C, \qquad (4.103)$$

where $1 < q < 2$.

Proof Since the space $H^2(\Omega)$ embeds continuously into $L^\infty(\Omega)$, it shows that

$$\sqrt{\rho_N} \text{ is bounded in } L^2([0,T]; L^\infty(\Omega)).$$

Thus, in view of (4.92)-(4.93), $\rho_N u_N = \sqrt{\rho_N}\sqrt{\rho_N}u_N$ is uniformly bounded in $L^2([0,T]; L^2(\Omega))$. By (4.90) and (4.99), $\nabla\sqrt{\rho_N}$ is bounded in $L^2([0,T]; L^p(\Omega))(2 < p < \infty)$ and $\sqrt{\rho_N}$ is bounded in $L^\infty([0,T]; L^p(\Omega))(2 < p < \infty)$. This, together with (4.90), implies that

$$\nabla(\rho_N u_N) = 2\nabla\sqrt{\rho_N} \otimes (\sqrt{\rho_N}u_N) + \sqrt{\rho_N}\nabla u_N\sqrt{\rho_N}$$

is uniformly bounded in $L^2([0,T]; L^q(\Omega))\left(q = \dfrac{2p}{p+2}\right)$.

Since

$$\nabla^2\rho_N = 2(\sqrt{\rho_N}\nabla^2\sqrt{\rho_N} + \nabla\sqrt{\rho_N} \otimes \nabla\sqrt{\rho_N})$$

and $H^1(\Omega) \hookrightarrow\hookrightarrow L^{p'}(1 \leqslant p' < \infty)$, we have $\nabla^2\rho_N \in L^2([0,T]; L^q(\Omega))$.

Then, the Gagliardo-Nirenberg inequality, with $\theta = 3/(4\alpha+3)$ and $q_1 = 2(4\alpha+3)/3$,

$$\|\sqrt{\rho_N}\|^{q_1}_{L^{q_1}([0,T];L^{q_1}(\Omega))} \leqslant C\int_0^T \|\sqrt{\rho_N}\|^{q_1\theta}_{H^2(\Omega)}\|\sqrt{\rho_N}\|^{q_1(1-\theta)}_{L^{2\alpha}(\Omega)} dt$$

$$\leqslant C\|\rho_N\|^{q_1(1-\theta)}_{L^\infty([0,T];L^\alpha(\Omega))}\int_0^T \|\sqrt{\rho_N}\|^2_{H^2(\Omega)} dt \leqslant C,$$

shows that ρ_N is bounded in $L^{q_1/2}([0,T]; L^{q_1/2}(\Omega))$.

This completes the proof. □

Lemma 4.3 *Under the assumption of Theorem 1.1, there yields that*

$$\|\rho_N^\alpha\|_{L^{\frac{5}{3}}([0,T];L^{\frac{5}{3}}(\Omega))} \leqslant C, \qquad (4.104)$$

$$\|P_c(\rho_N)\|_{L^{\frac{5}{3}}([0,T];L^{\frac{5}{3}}(\Omega))} \leqslant C. \qquad (4.105)$$

Proof Noticing that $\nabla \rho_N^{\frac{\alpha}{2}}$ is bounded in $L^2([0,T];L^2(\Omega))$, using the Sobolev embedding theorem, we can get that ρ_N^α is bounded in $L^1([0,T];L^3(\Omega))$, then we apply Hölder inequality to have

$$\|\rho_N^\alpha\|_{L^{\frac{5}{3}}([0,T];L^{\frac{5}{3}}(\Omega))} \leqslant \|\rho_N^\alpha\|_{L^\infty([0,T];L^1(\Omega))}^{\frac{2}{5}} \|\rho_N^\alpha\|_{L^1([0,T];L^3(\Omega))}^{\frac{3}{5}} \leqslant C.$$

Similarly, we can obtain (4.105). □

From (4.97) we can get the estimate about $\|\Delta d_N\|_{L^2([0,T];H^2(\Omega))}$.

Lemma 4.4

$$\|\Delta d_N\|_{L^2([0,T];L^2(\Omega))} \leqslant C. \qquad (4.106)$$

Proof Since

$$|d_N|^2 = 1,$$

and

$$d_N \cdot \nabla d_N = 0, \quad |\nabla d_N|^2 + d_N \Delta d_N = 0,$$

we have

$$|\Delta d_N + |\nabla d_N|^2 d_N|^2 = |\Delta d_N|^2 + 2|\nabla d_N|^2 d \Delta d_N + |\nabla d_N|^4$$
$$= |\Delta d_N|^2 - |\nabla d_N|^4.$$

Then, it yields that

$$\|\Delta d_N\|_{L^2((0,T);L^2(\Omega))} \leqslant \|\Delta d_N + |\nabla d_N|^2 d_N\|_{L^2((0,T);L^2(\Omega))} + \|\nabla d_N\|_{L^4((0,T);L^4(\Omega))}.$$

From [36], we know that if $d: \mathbf{R}^2 \to S^1$, $\nabla d \in H^1(\mathbf{R}^2)$, $\|\nabla d\|_{L^2} \leqslant C$, $d_2 \geqslant \varepsilon_0$, then $\|\nabla d\|_{L^4}^4 \leqslant (1-\delta_0)\|\Delta d\|_{L^2}^2 (\delta_0 > 0)$.

Therefore, if $d_0^2 > \varepsilon_0$, we can get (4.106). □

Lemma 4.5 (time regularity for ρ_N and ρu_N) *The following uniform estimates hold for $s > d/2 + 1$:*

$$\|\partial_t \rho_N\|_{L^2([0,T];L^{3/2}(\Omega))} \leqslant C, \qquad (4.107)$$

$$\|\partial_t (\rho_N u_N)\|_{L^{4/3}([0,T];(H^s(\Omega))^*)} \leqslant C. \qquad (4.108)$$

Proof By (4.101)-(4.102), we find that $\partial_t \rho_N = -\text{div}(\rho_N u_N) - \nu \Delta \rho_N$ is uniformly bounded in $L^2([0,T]; L^{3/2}(\Omega))$.

The sequence $\rho_N u_N \otimes u_N$ is bounded in $L^\infty([0,T]; L^1(\Omega))$; hence, $\text{div}(\rho_N u_N \otimes u_N)$ is bounded in $L^\infty([0,T]; (W^{1,\infty}(\Omega))^*)$, because of the continuous embedding of $H^s(\Omega)$ into $W^{1,\infty}(\Omega)$ for $s > d/2 + 1$, and also in $L^\infty([0,T]; (H^s(\Omega))^*)$. The estimate

$$\int_0^T \int_\Omega \rho_N \nabla \left(\frac{\Delta \sqrt{\rho_N}}{\sqrt{\rho_N}}\right) \cdot \phi\, dx dt = -\int_0^T \int_\Omega \Delta \sqrt{\rho_N} (2\nabla \sqrt{\rho_N} \cdot \phi + \sqrt{\rho_N} \text{div} \phi) dx dt$$

$$\leq \|\Delta \sqrt{\rho_N}\|_{L^2([0,T]; L^2(\Omega))} (2\|\sqrt{\rho_N}\|_{L^\infty([0,T]; L^2(\Omega))} \|\phi\|_{L^4([0,T]; L^6(\Omega))})$$
$$+ \|\sqrt{\rho_N}\|_{L^\infty([0,T]; L^4(\Omega))} \|\phi\|_{L^2([0,T]; W^{1,4}(\Omega))}$$

$$\leq C \|\phi\|_{L^2([0,T]; W^{1,4}(\Omega))}$$

for all $\phi \in L^2([0,T]; W^{1,4}(\Omega))$ proves that $\rho \Delta \sqrt{\rho_N}/\sqrt{\rho_N}$ is uniformly bounded in $L^2([0,T]; (W^{1,4}(\Omega))^*) \hookrightarrow L^2([0,T]; (H^s(\Omega))^*)$. In view of (4.103), ρ_N^γ is bounded in $L^{4/3}([0,T]; L^{4/3}(\Omega)) \hookrightarrow L^{4/3}([0,T]; (H^s(\Omega))^*)$. Furthermore, by (4.101), $\Delta(\rho_N u_N)$ is uniformly bounded in $L^2([0,T]; (W^{1,3}(\Omega))^*)$, and by (4.98), $\delta \Delta u_N$ is bounded in $L^2([0,T]; (H^1(\Omega))^*)$. Therefore, using Corollaries 4.1 and 4.4, we get that

$$(\rho_N u_N)_t = -\text{div}(\rho_N u_N \otimes u_N) - \nabla P(\rho_N) + \nu \Delta(\rho_N u_N)$$
$$- \lambda \nabla \cdot \left(\nabla d_N \odot \nabla d_N - \frac{|\nabla d_N|^2}{2} I\right)$$
$$+ \frac{\hbar^2}{2} \rho_N \nabla \left(\frac{\Delta \sqrt{\rho_N}}{\sqrt{\rho_N}}\right) + \kappa(\nabla \times B_N) \times B_N$$

is uniformly bounded in $L^2([0,T]; (H^s(\Omega))^*)$. □

The bound $\sqrt[4]{\rho_N} \in L^4([0,T]; W^{1,4}(\Omega)$ in (4.99) provides a uniform estimate for $\partial_t \sqrt{\rho_N}$.

Lemma 4.6 (time regularity for $\sqrt{\rho_N}$) *The following estimate holds:*

$$\|\partial_t \sqrt{\rho_N}\|_{L^2([0,T]; (H^1(\Omega))^*)} \leq C. \tag{4.109}$$

Proof Dividing the mass equation (1.29) by $2\sqrt{\rho}$, we have

$$\partial_t \sqrt{\rho_N} = -\nabla \sqrt{\rho_N} \cdot u_N - \frac{1}{2} \sqrt{\rho_N} \text{div} u_N + \nu(\Delta \sqrt{\rho_N} + 4|\nabla \sqrt[4]{\rho_N}|^2)$$
$$= -\text{div}(\sqrt{\rho_N} u_N) + \frac{1}{2} \sqrt{\rho_N} \text{div} u_N + \nu(\Delta \sqrt{\rho_N} + 4|\nabla \sqrt[4]{\rho_N}|^2).$$

The first term on the right-hand side is bounded in $L^2([0,T];(H^1(\Omega))^*)$ by (4.92) and (4.93). The remaining terms are uniformly bounded in $L^2([0,T];L^2(\Omega))$; see (4.92), (4.93), (4.98). □

Lemma 4.7 *There holds that*

$$\|\partial_t d_N\|_{L^2([0,T];L^2(\Omega))} \leqslant C. \tag{4.110}$$

Proof Multiplying (1.31) by $\partial_t d_N$ and integrating by parts respect to x in Ω, we have

$$\int_\Omega |\partial_t d_N|^2 dx \leqslant \int_\Omega |u_N \cdot \nabla d_N \partial_t d_N| dx + \int_\Omega \Delta d_N \partial_t d_N dx + \int_\Omega ||\nabla d_N|^2 d_N \partial_t d_N| dx$$
$$\leqslant \|u_N\|_{L^4(\Omega)} \|\nabla d_N\|_{L^4(\Omega)} \|\partial_t d_N\|_{L^2(\Omega)} - \frac{1}{2} \|\nabla d_N\|_{L^2(\Omega)}^2.$$

Then we have

$$\|\partial_t d_N\|_{L^2(\Omega)} \leqslant \|u_N\|_{L^4(\Omega)} \|\nabla d_N\|_{L^4(\Omega)} + \|\nabla d_N\|_{L^4(\Omega)},$$

and

$$\|\partial_t d_N\|_{L^2((0,T);L^2(\Omega))} \leqslant C.$$

Thus we complete the proof of this lemma. □

Lemma 4.8 *There holds*

$$\|\partial_t B_N\|_{L^2([0,T];H^{-1}(\Omega))} \leqslant C. \tag{4.111}$$

Proof For any $\varphi \in L^2([0,T];H_0^1(\Omega))$, it yields that

$$\left|\int_0^T \int_\Omega \partial_t B_N \cdot \varphi dx dt\right|$$
$$= \left|\int_0^T \int_\Omega \nabla \times (u_N \times B_N)\varphi dx dt - \int_0^T \int_\Omega \nabla B_N : \nabla \varphi dx dt\right|$$
$$\leqslant \|u_N\|_{L^2([0,T];H^1(\Omega))} \|B_N\|_{L^\infty([0,T];L^2(\Omega))} \|\varphi\|_{L^2([0,T];H_0^1(\Omega))}$$
$$\quad + \|\nabla B_N\|_{L^2([0,T];L^2(\Omega))} \|\varphi\|_{L^2([0,T];H_0^1(\Omega))}$$
$$\leqslant C\|\varphi\|_{L^2([0,T];H_0^1(\Omega))}.$$

□

Lemma 4.9 *Under the assumption of Theorem 1.1, there holds that*

$$\left\|\nabla\left(\frac{1}{\sqrt{\rho_N}}\right)\right\|_{L^2([0,T];L^2(\Omega))} \leq C. \tag{4.112}$$

Proof Since $\nabla\sqrt{\rho_N} \in L^\infty([0,T]; L^2(\Omega))$, and

$$\nabla\left(\frac{1}{\sqrt{\rho_N}}\right) = -\frac{1}{2}\frac{\nabla\rho_N}{\rho_N^{\frac{3}{2}}} = -\frac{\nabla\sqrt{\rho_N}}{\rho_N},$$

we have, if $\rho_N > 1$, $\nabla(1/\sqrt{\rho_N}) \in L^\infty([0,T]; L^2(\Omega))$.

Now we take into account of the case $\rho_N \leq 1$. From $\int_0^T \int_\Omega H_c''(\rho_N)|\nabla\rho_N|^2 dxdt \leq C$, we get

$$\int_0^T \int_\Omega \frac{A_1}{4k^2}|\nabla\rho_N^{-2k}|^2 dxdt \leq C.$$

Make the connection between $\nabla(\rho_N^{-2k})$ and $\nabla\rho_N^{-1/2}$:

$$\nabla\left(\frac{1}{\sqrt{\rho_N}}\right) = \nabla\left(\frac{1}{\rho_N^{2k}}\rho_N^{2k-1/2}\right)$$
$$= \rho_N^{2k-1/2}\nabla\left(\frac{1}{\rho_N^{2k}}\right) + \frac{1}{\rho_N^{2k}}\nabla\rho_N^{2k-1/2}$$
$$= \rho_N^{2k-1/2}\nabla\left(\frac{1}{\rho_N^{2k}}\right) + (2k-1/2)\rho_N^{-2k}\nabla\rho_N\rho_N^{2k-3/2}$$
$$= \rho_N^{2k-1/2}\nabla\left(\frac{1}{\rho_N^{2k}}\right) + (2k-1/2)\nabla\rho_N\rho_N^{-3/2}$$
$$= \rho_N^{2k-1/2}\nabla\left(\frac{1}{\rho_N^{2k}}\right) - 2(2k-1/2)\nabla\left(\frac{1}{\sqrt{\rho_N}}\right),$$

we have

$$(1+4k-1)\nabla\left(\frac{1}{\sqrt{\rho_N}}\right) = \rho_N^{2k-1/2}\nabla\left(\frac{1}{\rho^{2k}}\right),$$

and

$$\left|\nabla\left(\frac{1}{\sqrt{\rho_N}}\right)\right|^2 = \frac{1}{16k^2}\rho^{4k-1}\left|\nabla\left(\frac{1}{\rho_N^{2k}}\right)\right|^2$$
$$= \frac{1}{16k^2}\rho_N^{4k-1}4k^2\frac{1}{A_1}H_c''(\rho_N)|\nabla\rho_N|^2 = \frac{1}{4A_1}\rho^{4k-1}H_c''(\rho_N)|\nabla\rho_N|^2.$$

Such as $\rho_N \leq 1$ and $\int_0^T \int_\Omega H_c''(\rho_N)|\nabla\rho_N|^2 dxdt < +\infty$, we have $\nabla\left(\frac{1}{\sqrt{\rho_N}}\right) \in L^2([0,T]; L^2(\Omega))$. □

Next we will show the limit of the Fadeo-Galerkin approximated solution. We perform first the limit $N \to \infty$, $\gamma > 0$ being fixed. The limit $\gamma \to 0$ is carried out in the last part of the next subsection. We consider both limits separately since the weak formulation (1.22) for the vQMHD-NLC model is different from its approximation (3.80).

We conclude from the Aubin-Lions lemma, taking into account the regularity (4.102) and (4.107) for ρ_N, the regularity (4.99) and (4.109) for $\sqrt{\rho_N}$, and the regularity (4.101) and (4.108) for $\rho_N u_N$, that there exists a subsequence of ρ_N, $\sqrt{\rho_N}$, and $\rho_N u_N$, which are not related, such that, for some function ρ and J, as $N \to \infty$,

$$\rho_N \to \rho \text{ strongly in } L^2([0,T]; L^\infty(\Omega)),$$
$$\sqrt{\rho_n} \to \sqrt{\rho} \text{ weakly in } L^2([0,T]; H^2(\Omega)),$$
$$\sqrt{\rho_n} \to \sqrt{\rho} \text{ strongly in } L^2([0,T]; H^1(\Omega)),$$
$$\rho_n u_N \to J \text{ strongly in } L^2([0,T]; L^2(\Omega)).$$

Here we have used that the embedding $W^{2,p}(\Omega) \hookrightarrow L^\infty(\Omega)(p > 3/2)$, $H^2(\Omega) \hookrightarrow H^1(\Omega)$, and the fact that $W^{1,3/2}(\Omega) \hookrightarrow L^2(\Omega)$ are compact. The estimate (4.98) on u_N provides further the existence of a subsequence (not relabeled) such that, as $N \to \infty$,

$$u_N \rightharpoonup u \text{ weakly in } L^2([0,T]; H^1(\Omega)).$$

Then, since $\rho_n u_N$ converges weakly to ρu in $L^1([0,T]; L^6(\Omega))$, we infer that $J = \rho u$.

Now let $N \to \infty$ in the approximate system (3.45) and (3.48)-(3.51) with $\rho = \rho_N$, $u = u_N$, $d = d_N$, $E = E_N$ and $B = B_N$. Clearly, the limit $N \to \infty$ shows immediately that n solves

$$\rho_t + \text{div}(\rho u) = \nu \Delta \rho.$$

Next, we consider the weak formulation (3.80) term by term. The strong convergence of $\rho_N u_N$ in $L^2([0,T]; L^2(\Omega))$ and the weak convergence of u_N in $L^2([0,T]; H^1(\Omega))$ lead to

$$\rho_N u_N \otimes u_N \rightharpoonup \rho u \otimes u \text{ weakly in } L^1([0,T]; L^{3/2}(\Omega)).$$

Furthermore, in view of (4.102) (up to a subsequence),

$$\nabla(\rho_n u_N) \rightharpoonup \nabla(\rho u) \text{ weakly in } L^2([0,T]; L^{3/2}(\Omega)).$$

From the $\sqrt{\rho_N} \to \sqrt{\rho}$ strongly in $L^2([0,T]; H^1(\Omega))$, we know $\rho_N \to \rho$ a.e., then we have $\rho_N^\alpha \to \rho^\alpha$ a.e.. Thus combing the Egoroffs theorem and $\rho_N^\alpha \in L^{\frac{5}{3}}([0,T]; L^{\frac{5}{3}}(\Omega))$, we can get

$$\rho_N^\alpha \to \rho^\alpha, \quad \text{strongly in } L^1([0,T]; L^1(\Omega)).$$

Similarly, we can also get the convergence of ρ_N^{-4k} for $\rho_N \leqslant 1$:

$$\rho_N^{-4k} \to \rho^{-4k}, \quad \text{strongly in } L^1([0,T]; L^1(\Omega)).$$

Finally, the above convergence results show that the limit $N \to \infty$ of

$$\int_\Omega \frac{\Delta\sqrt{\rho_N}}{\sqrt{\rho_N}} \mathrm{div}(\rho_N \phi) dx = \int_\Omega \Delta\sqrt{\rho_N}(2\nabla\sqrt{\rho_N} \cdot \phi + \sqrt{\rho_N}\mathrm{div}\phi) dx$$

equals, for sufficiently smooth test functions,

$$\int_\Omega \Delta\sqrt{\rho}(2\nabla\sqrt{\rho} \cdot \phi + \sqrt{\rho}\mathrm{div}\phi) dx.$$

Using Corollary 4.1, Lemma 4.4, Lemma 4.7 and Lemma 4.8, we can get the convergence of d_N and B_N:

$$d_N \to d, \quad \text{strongly in } L^2([0,T]; H^1(\Omega)) \cap C([0,T]; L^2(\Omega)),$$
$$B_N \to B, \quad \text{strongly in } L^2([0,T]; L^2(\Omega)),$$
$$d_N \rightharpoonup d, \quad \text{weakly in } L^2([0,T]; H^2(\Omega)),$$
$$B_N \rightharpoonup B, \quad \text{weakly in } L^2([0,T]; H^1(\Omega)),$$
$$d_N \rightharpoonup d, \quad \text{weakly } * \text{ in } L^\infty([0,T]; H^1(\Omega)),$$
$$B_N \rightharpoonup B, \quad \text{weakly } * \text{ in } L^\infty([0,T]; L^2(\Omega)).$$

Therefore, we prove that the limit of (ρ_N, u_N, d_N, B_N) is the weak solution to the equations (1.29)-(1.33). Moreover, the limit function (ρ, u, d, B) satisfies the follow equality, for all test function $\forall \phi(x,T) = 0 \in \mathcal{D}(\Omega)$:

$$-\int_\Omega \rho_0(x) u_0(x) \cdot \phi(x,0) dx = \int_0^T \int_\Omega \Big(\rho u \phi_t + \rho(u \otimes u_N) : \nabla\phi + P(\rho)\mathrm{div}\phi$$
$$- \frac{\hbar^2}{2} \frac{\Delta\sqrt{\rho}}{\sqrt{\rho}} \mathrm{div}(\rho\phi) - \nu \nabla(\rho u) : \nabla\phi$$
$$+ \kappa(\nabla \times B) \times B \cdot \phi + \lambda \Big(\nabla d \odot \nabla d - \frac{|\nabla d|^2}{2} I \Big) \nabla\phi$$

$$-\gamma(\nabla u : \nabla\phi + u \cdot \phi)\bigg)dxdt. \tag{4.113}$$

$$-\int_\Omega d_0(x)\varphi(x,0)dx - \int_0^T \int_\Omega d\varphi_t dxdt + \int_0^T \int_\Omega u \cdot \nabla d\varphi dxdt$$
$$= -\int_0^T \int_\Omega \nabla d \cdot \nabla \varphi dxdt + \int_0^T |\nabla d|^2 d\varphi dxdt,$$

$$-\int_\Omega B_0(x)\varphi(x,0)dx = \int_0^T \int_\Omega B\phi_t dxdt + \int_\Omega (\nabla \times (u \times B))\varphi dxdt$$
$$+ \int_0^T \int_\Omega \nabla B \cdot \nabla \varphi dxdt.$$

4.3 The limit of solutions as γ vanishing

Assume that $(\rho^\gamma, u^\gamma, d^\gamma, B^\gamma)$ are the weak solutions constructed in the above subsection. We will show that the limit of that solution $(\rho^\gamma, u^\gamma, d^\gamma, B^\gamma)$ satisfies the (1.8)-(1.12). From the above subsection, we know that the Theorem 4.1, Corollary 4.1, and Lemmas 4.1-4.6, 4.9 still hold. We only need to estimate Lemma 4.7 and 4.8. To prove these lemmas, we need the estimates for u^γ and ∇u^γ. Therefore, we will estimate the u^γ and ∇u^γ firstly.

Lemma 4.10 *Under the assumption of Theorem 1.1, there holds*

$$\|\nabla u^\gamma\|_{L^{p_2}([0,T];L^{q_2})} \leqslant C, \tag{4.114}$$

with $p_2 = \dfrac{8k}{4k+1}$ and $q_2 = \dfrac{24k}{12k+1}$.

Proof Let us write

$$\nabla u^\gamma = \frac{1}{\sqrt{\rho^\gamma}}\sqrt{n^\gamma}\nabla u^\gamma.$$

Using (4.93), we have an estimate in $L^2([0,T]; L^2(\Omega))$ for $\sqrt{\rho^\gamma}u^\gamma$. Then it remains to obtain an estimate for $\dfrac{1}{\sqrt{\rho}}$ in the appropriate space. Using (4.92),

$$\|(\rho^\gamma)^{-2k}\|_{L^2([0,T\rho];L^6(\Omega))}^2 = \int_0^T \left(\int_\Omega \left(\frac{1}{\sqrt{\rho^\gamma}}\right)^{24k} dx\right)^{8k/24k} dt,$$

we obtain

$$\left\|\frac{1}{\sqrt{\rho^\gamma}}\right\|_{L^{8k}([0,T];L^{24k}(\Omega))} \leqslant C. \tag{4.115}$$

As previously said, since $\nabla u^\gamma = \dfrac{1}{\sqrt{\rho^\gamma}}\sqrt{\rho^\gamma}\nabla u^\gamma$, using (4.93) and (4.115), we have the result. □

Using the Sobolev embedding, a direct consequence of Lemma 4.10 is the following one.

Lemma 4.11 *Under the assumption of Theorem 1.1, there holds*

$$\|u^\gamma\|_{L^{p_2}([0,T];L^{q_3}(\Omega))} \leqslant C, \qquad (4.116)$$

with $p_2 = \dfrac{8k}{4k+1}$ and $q_3 = \dfrac{24k}{4k+1}$.

Using the previous lemmas, we are now able to establish the following proposition.

Lemma 4.12 *Under the assumption of Theorem 1.1, it holds that*

$$\|\sqrt{\rho^\gamma}u^\gamma\|_{L^{p_4}([0,T];L^{p_4}(\Omega))} \leqslant C, \qquad (4.117)$$

with $p_4, q_4 > 2$.

Proof Let $1/2 > r > 0$ which will be chosen later on. We write

$$\sqrt{\rho^\gamma}u^\gamma = (\sqrt{\rho^\gamma}u^\gamma)^{2r}(u^\gamma)^{1-2r}(\rho^\gamma)^{1/2-r}.$$

Using (4.89), and the Sobolev imbedding theorem from $H^1(\Omega) \hookrightarrow L^{\tilde{q}_4}(1 \leqslant \tilde{q}_4 < \infty)$, we know $\|\rho^\gamma\|_{L^\infty([0,T];L^3(\Omega))} \leqslant C$, which gives

$$\|(\rho^\gamma)^{1/2-r}\|_{L^\infty([0,T];L^{\frac{3}{1/2-r}}(\Omega))} \leqslant C.$$

Then we can choose $1/2 > r > 1/10$ in a way that

$$\dfrac{1}{p_4} = \dfrac{1-2r}{p_2}, \quad \text{and} \quad \dfrac{1}{q_4} = \dfrac{2r}{2} + \dfrac{1-2r}{q_3} + \dfrac{1/2-r}{3},$$

satisfy $p_4 > 2$ and $q_4 > 2$. Indeed, it is easy to see that the condition $q_4 > 2$ is true for all $0 < r < 1/2$. The condition p_4 leads to $r > \dfrac{1}{8k+2}$. Since $k > 1$, $\dfrac{1}{8k+2} < \dfrac{1}{10}$ and then $p_4 > 2$ is true as soon as $r > 1/10$. □

Combining Lemma 4.9-4.11, we can derive that

$$\partial_t\left(\dfrac{1}{\sqrt{\rho^\gamma}}\right) + \nabla\left(\dfrac{u^\gamma}{\sqrt{\rho^\gamma}}\right) + \dfrac{3}{2\sqrt{\rho^\gamma}}\mathrm{div}u^\gamma$$

$$= -\nu \left(\frac{\Delta \sqrt{\rho^\gamma}}{\rho^\gamma} + \frac{|\nabla \sqrt{\rho^\gamma}|^2}{(\rho^\gamma)^{\frac{3}{2}}} \right) \in L^\infty([0,T]; W^{-1,1}(\Omega)).$$

Using the Aubin-Lions Lemma and Egoroffs Theorem we can derive the following convergence.

Lemma 4.13 *Under the assumption of Theorem 1.1, up to a subsequence, the following convergence holds when γ tends to 0:*

$$\sqrt{\rho^\gamma} \to \sqrt{\rho}, \quad \text{strongly in } L^2([0,T]; H^1(\Omega)),$$
$$P(\rho^\gamma) \to P(\rho), \quad \text{strongly in } L^1([0,T]; L^1(\Omega)),$$
$$P_c(\rho^\gamma) \to P_c(\rho), \quad \text{strongly in } L^1([0,T]; L^1(\Omega)),$$
$$\frac{1}{\sqrt{\rho^\gamma}} \to \frac{1}{\sqrt{\rho}}, \quad \text{almost everywhere},$$
$$\sqrt{\rho^\gamma} u^\gamma \to \sqrt{\rho} u, \quad \text{strongly in } L^2([0,T]; L^2(\Omega)),$$
$$\nabla u^\gamma \rightharpoonup \nabla u, \quad \text{weakly in } L^{p_2}([0,T]; L^{q_2}(\Omega)),$$
$$u^\gamma \rightharpoonup u, \quad \text{weakly in } L^{p_2}([0,T]; L^{q_3}(\Omega)),$$
$$d^\gamma \to d, \quad \text{strongly in } L^2([0,T]; H^1(\Omega)) \cap C([0,T]; L^2(\Omega)),$$
$$B^\gamma \to B, \quad \text{strongly in } L^2([0,T]; L^2(\Omega)),$$
$$d^\gamma \rightharpoonup d, \quad \text{weakly in } L^2([0,T]; H^2(\Omega)),$$
$$B^\gamma \rightharpoonup B, \quad \text{weakly in } L^2([0,T]; H^1(\Omega)),$$
$$d^\gamma \rightharpoonup d, \quad \text{weakly} * \text{ in } L^\infty([0,T]; H^1(\Omega)),$$
$$B^\gamma \rightharpoonup B, \quad \text{weakly} * \text{ in } L^\infty([0,T]; L^2(\Omega)).$$

From Lemma 4.12, we can easily get the equations (1.22)-(1.5), thus we complete the proof of Theorem 1.1.

Note that the a-priori estimates in Section 4 are independent of D. By using the diagonal method and letting $D \to \infty$, we can obtain the global existence of a weak solution to the Cauchy problem of system (1.8)-(1.12). For simplicity, we do not state the theorem here.

5 Local existence of smooth solution

In this section, we will prove the local-in-time existence of the smooth solution to the equations (1.1)-(1.5). In this section, we need an assumption

that
$$0 < \rho_* \leqslant \rho_0 \leqslant \rho^*, \qquad (5.118)$$
where $\rho_* = \min_x \rho_0$ and $\rho^* = \max_x \rho_0$.

Since it is very difficult to deal with the third-order term $\rho \nabla \left(\frac{\Delta \sqrt{\rho}}{\sqrt{\rho}} \right)$ directly, we need to formally transform the equation of the density to a forth order hyperbolic equation:

$$\partial_{tt}\rho - 2\nu \Delta \rho_t + \left(\nu^2 + \frac{\hbar^2}{4} \right) \Delta^2 \rho - \Delta P(\rho)$$
$$= \frac{\hbar^2}{2} \frac{|\Delta \rho|^2}{\rho} + \frac{\hbar^2}{2} \frac{\nabla \Delta \rho \nabla \rho}{\rho} + \frac{\hbar^2 |\nabla \rho|^2 \Delta \rho}{\rho^2} - \frac{\hbar^2}{3} \frac{|\nabla \rho|^4}{\rho^3}$$
$$+ u : \nabla^2 \rho : u + 4 \nabla \rho : \nabla u : u + 2\rho \nabla \text{div} u \cdot u + \rho \nabla u : \nabla^T u + \rho |\text{div} u|^2$$
$$+ \lambda \nabla \Delta d \nabla d + \lambda |\Delta d|^2 + \kappa B \cdot \Delta B + \kappa |\nabla \times B|^2. \qquad (5.119)$$

Taking a derivative of the equation (1.1) with respect to time variable t, we have
$$\partial_{tt}\rho + \partial_t(\text{div}(\rho u)) = \nu \Delta \rho_t. \qquad (5.120)$$

From (1.2), we have

$$\partial_t(\text{div}(\rho u)) + \text{div}^2(\rho u \otimes u) + \Delta P(\rho) = \nu \Delta \text{div}(\rho u) + \frac{\hbar^2}{2} \nabla \cdot \left(\rho \nabla \left(\frac{\Delta \sqrt{\rho}}{\sqrt{\rho}} \right) \right)$$
$$- \lambda \text{div}^2 \left(\nabla d \odot \nabla d - \frac{|\nabla d|^2}{2} I \right) + \kappa \nabla \cdot ((\nabla \times B) \times B). \qquad (5.121)$$

Noting the following facts, we can have

$$\nabla \cdot ((\nabla \times B) \times B) = B \cdot \nabla \times (\nabla \times B) - (\nabla \times B) \cdot (\nabla \times B)$$
$$= B \cdot (\nabla(\nabla \cdot B) - \Delta B) - |\nabla \times B|^2 = -B \cdot \Delta B - |\nabla \times B|^2, \quad \nabla \cdot B = 0,$$

$$\nabla \cdot \left(\nabla \cdot \left(\nabla d \odot \nabla d - \frac{|\nabla d|^2}{2} I \right) \right) = \nabla \cdot (\nabla d \Delta d) = \nabla \Delta d \cdot \nabla d + |\Delta d|^2,$$

$$\nabla \cdot \left(\rho \nabla \left(\frac{\Delta \sqrt{\rho}}{\sqrt{\rho}} \right) \right) = \frac{1}{2} \Delta^2 \rho - \frac{|\Delta \rho|^2}{\rho} - \frac{\nabla \Delta \rho \nabla \rho}{\rho} + \frac{2|\nabla \rho|^2 \Delta \rho}{\rho^2} - \frac{2}{3} \frac{|\nabla \rho|^4}{\rho^3}.$$

Therefore we can get the equation (5.119).

We can also get the equations about u, d, B:

$$\partial_t u + u \cdot \nabla u + \frac{\nabla P(\rho)}{\rho} = \frac{\hbar^2}{4} \frac{\nabla \Delta \rho}{\rho} - \frac{\hbar^2}{2} \frac{\nabla \rho \Delta \rho}{\rho^2} + \frac{\hbar^2}{4} \frac{|\nabla \rho|^2 \nabla \rho}{\rho^3} + \nu \frac{\Delta \rho}{\rho} u$$

$$+ \nu \Delta u + \frac{2\nu \nabla \rho \cdot \nabla u}{\rho} - \frac{\lambda}{\rho} \Delta d \nabla d + \frac{\kappa}{\rho} (\nabla \times B) \times B, \tag{5.122}$$

$$d_t + u \cdot \nabla d = \Delta d + |\nabla d|^2 d, \quad |d| = 1, \tag{5.123}$$

$$B_t - \nabla \times (u \times B) = \delta \Delta B, \quad \text{div} B = 0. \tag{5.124}$$

Then we rewrite the equations (5.119), (5.122), (5.123) and (5.124):

$$\rho_{tt} - 2\nu \Delta \rho_t + \left(\nu^2 + \frac{\hbar^2}{4}\right) \Delta^2 \rho = f_1, \tag{5.125}$$

$$u_t - \nu \Delta u = f_2, \tag{5.126}$$

$$d_t - \Delta d = f_3, \tag{5.127}$$

$$B_t - \delta \Delta B = f_4, \tag{5.128}$$

where

$$f_1 = \Delta P(\rho) + \frac{\hbar^2}{2} \frac{|\Delta \rho|^2}{\rho} + \frac{\hbar^2}{2} \frac{\nabla \Delta \rho \nabla \rho}{\rho} + \frac{\hbar^2 |\nabla \rho|^2 \Delta \rho}{\rho^2} - \frac{\hbar^2}{3} \frac{|\nabla \rho|^4}{\rho^3}$$
$$+ u : \nabla^2 \rho : u + 4 \nabla \rho : \nabla u : u + 2\rho \nabla \text{div} u \cdot u + \rho \nabla u : \nabla^T u + \rho |\text{div} u|^2$$
$$+ \lambda \nabla \Delta d \nabla d + \lambda |\Delta d|^2 + \kappa B \cdot \Delta B + \kappa |\nabla \times B|^2,$$

$$f_2 = -u \cdot \nabla u - \frac{\nabla P(\rho)}{\rho} + \frac{\hbar^2}{4} \frac{\nabla \Delta \rho}{\rho} - \frac{\hbar^2}{2} \frac{\nabla \rho \Delta \rho}{\rho^2} + \frac{\hbar^2}{4} \frac{|\nabla \rho|^2 \nabla \rho}{\rho^3} + \nu \frac{\Delta \rho}{\rho} u$$
$$+ \frac{2\nu \nabla \rho \cdot \nabla u}{\rho} - \frac{\lambda}{\rho} \Delta d \nabla d + \frac{\kappa}{\rho} (\nabla \times B) \times B,$$

$$f_3 = -u \cdot \nabla d + |\nabla d|^2,$$

$$f_4 = \nabla \times (u \times B).$$

5.1 Construction of the iteration scheme

We now construct the approximate solutions $\{U^k\}_{k=0}^{\infty}$ with $U^k = (\rho^k, u^k, d^k, B^k)$ based on the equations (5.125)-(5.128). The iteration scheme for the approximate solution $U^{k+1} = (\rho^{k+1}, u^{k+1}, d^{k+1}, B^{k+1})(k \geq 0)$, is defined by solving the following problems:

$$\begin{cases} \rho_{tt}^{k+1} - \nu \Delta \rho_t^{k+1} + \left(\nu^2 + \frac{\hbar^2}{4}\right) \Delta^2 \rho^{k+1} = f_1^k, \\ \rho^{k+1}(x, 0) = \rho_0, \\ \rho_t^{k+1}(x, 0) = \nu \Delta \rho_0 - \nabla \rho_0 \cdot u_0 - \rho_0 \text{div} u_0, \end{cases} \tag{5.129}$$

$$\begin{cases} u_t^{k+1} - \nu\Delta u^{k+1} = f_2^k, \\ u^{k+1}(x,0) = u_0, \end{cases} \tag{5.130}$$

$$\begin{cases} d_t^{k+1} - \Delta d^{k+1} = f_3^k, \\ d^{k+1}(x,0) = d_0, \end{cases} \tag{5.131}$$

$$\begin{cases} B_t^{k+1} - \delta\Delta B^{k+1} = f_4^k, \\ B^{k+1}(x,0) = B_0, \end{cases} \tag{5.132}$$

where

$$f_1^k = \Delta P(\rho^k) + \frac{\hbar^2}{2}\frac{|\Delta\rho^k|^2}{\rho^k} + \frac{\hbar^2}{2}\frac{\nabla\Delta\rho^k\nabla\rho^k}{\rho^k} + \frac{\hbar^2|\nabla\rho^k|^2\Delta\rho^k}{(\rho^k)^2} - \frac{\hbar^2}{3}\frac{|\nabla\rho^k|^4}{(\rho^k)^3}$$
$$+ u^k : \nabla^2\rho^k : u^k + 4\nabla\rho^k : \nabla u^k : u^k$$
$$+ 2\rho^k\nabla\mathrm{div} u^k \cdot u^k + \rho^k \nabla u^k : \nabla^T u^k + \rho^k|\mathrm{div} u^k|^2$$
$$+ \lambda\nabla\Delta d^k \nabla d^k + \lambda|\Delta d^k|^2 + \kappa B^k \cdot \Delta B^k + \kappa|\nabla\times B^k|^2,$$

$$f_2^k = -u^k \cdot \nabla u^k - \frac{\nabla P(\rho^k)}{\rho^k} + \frac{\hbar^2}{4}\frac{\nabla\Delta\rho^k}{\rho^k} - \frac{\hbar^2}{2}\frac{\nabla\rho^k\Delta\rho^k}{(\rho^k)^2} + \frac{\hbar^2}{4}\frac{|\nabla\rho^k|^2\nabla\rho^k}{(\rho^k)^3} + \nu\frac{\Delta\rho^k}{\rho^k}u^k$$
$$+ \frac{2\nu\nabla\rho^k \cdot \nabla u^k}{\rho^k} - \frac{\lambda}{\rho^k}\Delta d^k \nabla d^k + \frac{\kappa}{\rho^k}(\nabla\times B^k)\times B^k,$$

$$f_3^k = -u^k \cdot \nabla d^k + |\nabla d^k|^2,$$

$$f_4^k = \nabla \times (u^k \times B^k).$$

Next, we will prove the convergence of the sequence $\{U^k\}_{k=0}^{\infty}$.

5.2 Convergence of the approximate sequence

Theorem 5.1 *Suppose* $(\rho_0, u_0, d_0, B_0) \in H^6(\Omega) \times H^4(\Omega) \times H^5(\Omega) \times H^4(\Omega)$, *satisfying* (5.118). *Then, there exists a positive time* T^* *and a sequence* $\{U^{k+1}\}_{k=0}^{\infty}$ *of approximate solutions, which solve the system* (5.129)-(5.132) *for any* $t \in [0, T^*]$ *and satisfy*

$$\begin{cases} \rho^{k+1} \in C^j([0,T]; H^{6-2j}(\Omega)) \cap C^3([0,T]; L^2(\Omega)), & j = 0,1,2, \\ u^{k+1} \in C^j([0,T]; H^{5-2j}(\Omega)), & j = 0,1,2, \\ d^{k+1} \in C^j([0,T]; H^{5-2j}(\Omega)), & j = 0,1,2, \\ B^{k+1} \in C^j([0,T]; H^{5-2j}(\Omega)), & j = 0,1,2. \end{cases} \tag{5.133}$$

Moreover, there is a positive constant M_* so that for all $t \in [0, T^*]$, we have

$$\begin{cases} \|u^{k+1}\|^2_{H^5(\Omega)} + \|u^{k+1}_t\|^2_{H^3(\Omega)} + \|u^{k+1}_{tt}\|^2_{H^1(\Omega)} \leqslant M_*, \\ \|d^{k+1}\|^2_{H^5(\Omega)} + \|d^{k+1}_t\|^2_{H^3(\Omega)} + \|d^{k+1}_{tt}\|^2_{H^1(\Omega)} \leqslant M_*, \\ \|B^{k+1}\|^2_{H^5(\Omega)} + \|B^{k+1}_t\|^2_{H^3(\Omega)} + \|B^{k+1}_{tt}\|^2_{H^1(\Omega)} \leqslant M_*, \\ \|\rho^{k+1}\|^2_{H^6(\Omega)} + \|\rho^{k+1}_t\|^2_{H^4(\Omega)} + \|\rho^{k+1}_{tt}\|^2_{H^2(\Omega)} \leqslant M_*. \end{cases} \quad (5.134)$$

Proof Obviously, $U^0 = (\rho^0, u^0, d^0, B^0)$ satisfies (5.133) and (5.134) for time interval $[0, T_1]$ with M_* replaced by some constant $B_0 > 0$.

We start the iterative process with $U^0 = (\rho^0, u^0, d^0, B^0)$; then by some problem (5.129)-(5.132) for $k = 0$, we can prove the (local-in-time) existence of a solution $U^1 = (\rho^1, u^1, d^1, B^1)$ which satisfy (5.129)-(5.132) for a time interval $[0, T_0]$ (which without loss of generality is chosen to be $[0, T_0]$, since we focus on local in-time existence of solutions) and with M_* replaced by another constant $M_1 > 0$. In fact, for $U^0 = (\rho^0, u^0, d^0, E^0)$ the function $f_1^0, f_2^0, f_3^0, f_4^0$ depend only on the initial data ρ_0, u_0, d_0, B_0 and

$$\sum_{k=e,i} \{\|f_1^0\|^2_{H^4(\Omega)} + \|f_2^0\|^2_{H^3(\Omega)} + \|f_3^0\|^2_{H^3(\Omega)} + \|f_4^0\|^2_{H^3(\Omega)}\} \leqslant Na_0 I_0^4 e^{N\|u_0\|^2_{H^5(\Omega)}}.$$

(5.135)

From now on, $N > 0$ denotes a generic constant independent of U^k, $k \geqslant 1$,

$$a_0 = \frac{(1+\rho^*)^m}{\psi_*} \quad \text{for an integer } m \geqslant 10, \quad (5.136)$$

and

$$I_0 = \sum_{k=e,i} \{\|\rho_0\|^2_{H^6(\Omega)} + \|\nabla \rho_0\|^2_{H^5(\Omega)} + \|u_0\|^2_{H^5(\Omega)} + \|d_0\|^2_{H^5(\Omega)} + \|B_0\|^0_{H^5(\Omega)}\}. \quad (5.137)$$

The system (5.129)-(5.132) with $k = 0$ are linear with $U^1 = (\rho^1, u^1, d^1, B^1)$. It can be solved with the help of the estimate (5.135) about the right-hand side terms as follows. Applying Theorem 2.2 to the hyperbolic equation, we have

$$\rho^1 \in C^j([0, T_0]; H^{6-2j}(\Omega)), \quad j = 0, 1, 2.$$

Applying Theorem 2.1 to the parabolic equation, we have

$$u^1 \in C^j([0, T_0]; H^{5-2j}(\Omega)), \quad j = 0, 1, 2,$$
$$d^1 \in C^j([0, T_0]; H^{5-2j}(\Omega)), \quad j = 0, 1, 2,$$

$$B^1 \in C^j([0,T_0]; H^{5-2j}(\Omega)), \quad j=0,1,2.$$

Moreover, based on the estimate (5.135), we conclude that there is a constant M_1 such that U^1 satisfies (5.133) where $k=1$, $M_* = M_1$ and $T_* = T_0$.

Next, let us prove the estimates for $k \geqslant 1$. Assume that $\{U^i\}_{i=1}^k (k \leqslant 1)$ exists uniformly for $t \in [0, T_0]$, solves the system (5.129)-(5.132), and satisfying (5.133)-(5.134) with (the constant M_* replaced by) the upper bound M_k. We shall prove that it still holds for U^{k+1} for $t \in [0, T_1]$. In fact, the system (5.129)-(5.132) are linear with $U^{k+1} = (\rho^{k+1}, u^{k+1}, d^{k+1}, B^{k+1})$ for given U^k. In analogy, the applications of Theorem 2.2 to (5.129) for ρ^{k+1} and Theorem 2.1 to (5.130)-(5.132) for $u^{k+1}, d^{k+1}, B^{k+1}$, shows that $U^{p+1} = (\rho^{k+1}, u^{k+1}, d^{k+1}, B^{k+1})$ exits for $t \in [0, T_0]$ and satisfies

$$\begin{cases} \rho^{p+1} \in C^j([0,T_0]; H^{6-2j}(\Omega)), & j=0,1,2, \\ u^{p+1} \in C^j([0,T_0]; H^{5-2j}(\Omega)), & j=0,1,2, \\ d^{p+1} \in C^j([0,T_0]; H^{5-2j}(\Omega)), & j=0,1,2, \\ B^{p+1} \in C^j([0,T_0]; H^{5-2j}(\Omega)), & j=0,1,2. \end{cases}$$

Now our goal is to deduce the uniform bounds for $U^{j+1} (1 \leqslant j \leqslant p)$, for some time interval. Let first estimate L^2-norms of the initial value $\rho^{k+1}, \rho_t^{k+1}, \rho_{tt}^{k+1}$ where ρ_{tt}^{k+1} is obtained through (5.129) at $t=0$, where ρ^{k+1}, ρ_t^{k+1} are replaced by $\rho(x,0), \rho_t(x,0)$. Hence the initial values will depend only on ρ_0, u_0, d_0, B_0. Obviously, there exists a constant $M_2 > 0$, such that the initial value of $\rho^{k+1}, \rho_t^{k+1}, \rho_{tt}^{k+1}$ for $p \geqslant 1$ are bounded by

$$\max\left\{\|\rho_0\|_{H^6(\Omega)}^2, \|\rho_t(x,0)\|_{H^4(\Omega)}^2, \|\rho_{tt}(x,0)\|_{H^2(\Omega)}^2\right\} \leqslant M_2 I_0.$$

Denote by

$$M_0 = 40 M_2 I_0 \cdot \max\left\{T, \frac{1}{\nu}, \frac{1}{\nu + \frac{\hbar^2}{4}}\right\}, \tag{5.138}$$

$$M_1 = 3 N a_0^2 (1 + I_0 + M_0)^7 \cdot \max\left\{T, \frac{1}{\nu^2}, \frac{1}{\nu + \frac{\hbar^2}{4}}\right\}, \tag{5.139}$$

and choose

$$T_* = \min\left\{T, \frac{\rho_*}{4M_0}, \frac{M_2 I_0}{NM_3}, \frac{\ln 2}{NM_4}, \frac{2M_2 I_0}{NM_5}, \frac{2M_2 I_0}{NM_6}\right\}, \tag{5.140}$$

where

$$M_3 = 5a_0^2(1 + I_0 + M_0 + M_1)^6, \quad M_4 = 2a_0^3(1 + I_0 + M_0 + M_1)^{12},$$
$$M_5 = a_0^2(1 + I_0 + M_0 + M_1)^{16}, \quad M_6 = a_0^5(1 + I_0 + M_0 + M_1)^{24}.$$

As before, $N \geqslant M_2$ denotes a generic constant independent of $U^k(k \geqslant 1)$, and a_0 is defined by (5.136).

We claim that if the solution $\{U^j\}_{j=1}^k (k \geqslant 2)$, to the problem (5.129)-(5.132) satisfies

$$\begin{aligned}
\|\rho^j\|_{H^6(\Omega)}^2 + \|\rho_t^j\|_{H^4(\Omega)}^2 + \|\rho_{tt}^j\|_{H^2(\Omega)}^2 &\leqslant M_0, \\
\|u^j\|_{H^5(\Omega)}^2 + \|u_t^j\|_{H^3(\Omega)}^2 + \|u_{tt}^j\|_{H^1(\Omega)}^2 &\leqslant a_0 M_1, \\
\|d^j\|_{H^5(\Omega)}^2 + \|d_t^j\|_{H^3(\Omega)}^2 + \|d_{tt}^j\|_{H^1(\Omega)}^2 &\leqslant a_0 M_1, \\
\|B^j\|_{H^5(\Omega)}^2 + \|B_t^j\|_{H^3(\Omega)}^2 + \|B_{tt}^j\|_{H^1(\Omega)}^2 &\leqslant a_0 M_1,
\end{aligned} \quad (5.141)$$

for all $1 \leqslant j \leqslant p$ and $t \in [0, T_*]$, then this is also true for U^{k+1}, namely

$$\begin{aligned}
\|\rho^{j+1}\|_{H^6(\Omega)}^2 + \|\rho_t^{j+1}\|_{H^4(\Omega)}^2 + \|\rho_{tt}^{j+1}\|_{H^2(\Omega)}^2 &\leqslant M_0, \\
\|u^{j+1}\|_{H^5(\Omega)}^2 + \|u_t^{j+1}\|_{H^3(\Omega)}^2 + \|u_{tt}^{j+1}\|_{H^1(\Omega)}^2 &\leqslant a_0 M_1, \\
\|d^{j+1}\|_{H^5(\Omega)}^2 + \|d_t^{j+1}\|_{H^3(\Omega)}^2 + \|d_{tt}^{j+1}\|_{H^1(\Omega)}^2 &\leqslant a_0 M_1, \\
\|B^{j+1}\|_{H^5(\Omega)}^2 + \|B_t^{j+1}\|_{H^3(\Omega)}^2 + \|B_{tt}^{j+1}\|_{H^1(\Omega)}^2 &\leqslant a_0 M_1,
\end{aligned} \quad (5.142)$$

for all $t \in [0, T_*]$. Here M_0 and M_1 are given by (5.138) and (5.139).

We firstly obtain the uniform bounds for $U^{j+1}(1 \leqslant j \leqslant k)$ and (5.142), then we can get uniform estimates on the time derivatives of U^{k+1}.

From (5.141), we derive the estimates $\rho^{j+1}(1 \leqslant j \leqslant k)$ by using Theorem 2.2:

$$\begin{aligned}
\|\rho^{j+1}\|_{H^6(\Omega)}^2 &\leqslant N\|f_1^j\|_{H^2(\Omega)}^2 + \|\rho_0^{j+1}\|_{H^6(\Omega)}^2 \\
&\leqslant N(\|\rho^j\|_{H^5(\Omega)}^4 + \|u^j\|_{H^5(\Omega)}^2 + \|d^j\|_{H^5(\Omega)}^2 + \|B^j\|_{H^5(\Omega)}^2) \leqslant NM_0^{12}, \\
\|\rho_t^{j+1}\|_{H^4(\Omega)}^2 &\leqslant N\|\partial_t f_1^j\|_{L^2}^2 + \|\partial_t \rho^{j+1}(x,0)\|_{H^4(\Omega)}^2 \leqslant NM_0^{16}.
\end{aligned}$$

Using (5.129), we can have

$$\|\rho_{tt}^{j+1}\|_{H^2(\Omega)}^2 \leqslant N\|\rho^{j+1}\|_{H^6(\Omega)}^2 + \|\rho_t^{j+1}\|_{H^4(\Omega)}^2 + \|f_1^j\|_{H^2(\Omega)}^2 \leqslant NM_0^{24},$$

Similarly using Theorem 2.1, we can get the bounds of $u^{j+1}, d^{j+1}, B^{j+1}, u_t^{j+1}, d_t^{j+1}, B_t^{j+1}, u_{tt}^{j+1}, d_{tt}^{j+1}, B_{tt}^{j+1}$.

Therefore, by previous estimates about U^{k+1}, we conclude that the approximate solution U^{k+1} is uniformly bounded in the time interval $[0, T_*]$ and it satisfies (5.142) for each $k \geqslant 1$ as long as U^k satisfies (5.141) with M_0, M_1 and T_* defined by (5.138)-(5.140) respectively, which are independent on $U^{k+1}(k \geqslant 1)$. By repeating the procedure used above, we can construct the approximate solution $\{U^i\}_{i=1}^{\infty}$, which solves (5.129)-(5.132) on $[0, T_*]$ with T_* defined in (5.140) and the constant $M_* > 0$ chosen by

$$M_* = \max\{M_0, M_1, Na_0^8(1 + I_0 + M_1 + M_2)^{24}\}.$$

Let us recall that M_0, M_1 and a_0 are defined above, and $N > 0$ is a generic constant independent of $U^{p+1}(p \geqslant 1)$. Therefore, we complete the proof of Theorem 5.1. □

To prove the convergence of the sequence $\{U^j\}_{j=1}^{\infty}$, we only need to prove the uniform equi-continuity of the sequence. So we will estimate the difference $\{Y^{j+1}\}_{j=1}^{\infty}$, where $Y^{j+1} = U^{j+1} - U^j$. Let us denote $Y^{j+1} = (\tilde{\rho}^{j+1}, \tilde{u}^{j+1}, \tilde{d}^{j+1}, \tilde{B}^{j+1})$.

Then we can get the equations about Y^{j+1}:

$$\partial_{tt}\tilde{\rho}^{k+1} - 2\nu\Delta\tilde{\rho}_t^{k+1} + \left(\nu^2 + \frac{\hbar^2}{4}\right)\Delta^2\tilde{\rho}^{k+1} = \tilde{f}_1^k, \tag{5.143}$$

$$\partial_t\tilde{u}^{k+1} - \nu\Delta\tilde{u}^{k+1} = \tilde{f}_2^k, \tag{5.144}$$

$$\partial_t\tilde{d}^{k+1} - \Delta\tilde{d}^{k+1} = \tilde{f}_3^k, \tag{5.145}$$

$$\partial_t\tilde{B}^{k+1} - \delta\Delta\tilde{B}^{k+1} = \tilde{f}_4^k, \tag{5.146}$$

where

$$\tilde{f}_1^k = f_1^k - f_1^{k-1},$$
$$\tilde{f}_2^k = f_2^k - f_2^{k-1},$$
$$\tilde{f}_3^k = f_3^k - f_3^{k-1},$$
$$\tilde{f}_4^k = f_4^k - f_4^{k-1}.$$

Using Theorem 5.1 and Theorem 2.1-2.2, we can obtain the following estimates:

$$\|\tilde{\rho}^{k+1}\|_{H^6(\Omega)} + \|\tilde{\rho}_t^{k+1}\|_{H^4(\Omega)} + \|\tilde{\rho}_{tt}^{k+1}\|_{H^2(\Omega)}$$

$$\leq N_*(\|\tilde{\rho}^k\|_{H^6(\Omega)} + \|\tilde{\rho}^k_t\|_{H^4(\Omega)} + \|\tilde{\rho}^k_{tt}\|_{H^2(\Omega)}), \tag{5.147}$$

$$\|\tilde{u}^{k+1}\|_{H^5(\Omega)} + \|\tilde{u}^{k+1}_t\|_{H^3(\Omega)} + \|\tilde{u}^{k+1}_{tt}\|_{H^1(\Omega)}$$
$$\leq N_*(\|\tilde{u}^k\|_{H^5(\Omega)} + \|\tilde{u}^k_t\|_{H^3(\Omega)} + \|\tilde{u}^k_{tt}\|_{H^1(\Omega)}), \tag{5.148}$$

$$\|\tilde{d}^{k+1}\|_{H^5(\Omega)} + \|\tilde{d}^{k+1}_t\|_{H^3(\Omega)} + \|\tilde{d}^{k+1}_{tt}\|_{H^1(\Omega)}$$
$$\leq N_*(\|\tilde{d}^k\|_{H^5(\Omega)} + \|\tilde{d}^k_t\|_{H^3(\Omega)} + \|\tilde{d}^k_{tt}\|_{H^1(\Omega)}), \tag{5.149}$$

$$\|\tilde{B}^{k+1}\|_{H^5(\Omega)} + \|\tilde{B}^{k+1}_t\|_{H^3(\Omega)} + \|\tilde{B}^{k+1}_{tt}\|_{H^1(\Omega)}$$
$$\leq N_*(\|\tilde{B}^k\|_{H^5(\Omega)} + \|\tilde{B}^k_t\|_{H^3(\Omega)} + \|\tilde{B}^k_{tt}\|_{H^1(\Omega)}), \tag{5.150}$$

where N_* denotes a constant dependent on M_*. By using the previous estimates in Theorem 5.1 and Arzela-Ascoli theorem, we know that there exists a converged subsequence of $\{U^k\}_{k=1}^\infty$ whose limit satisfies (1.1)-(1.5) in $t \in [0, T_*]$, and

$$\rho \in C^j([0, T_*]; H^{6-2j}(\Omega)), \quad j = 0, 1, 2,$$
$$u \in C^j([0, T_*]; H^{5-2j}(\Omega)), \quad j = 0, 1, 2,$$
$$d \in C^j([0, T_*]; H^{5-2j}(\Omega)), \quad j = 0, 1, 2,$$
$$B \in C^j([0, T_*]; H^{5-2j}(\Omega)), \quad j = 0, 1, 2.$$

Thus we get the local-in-time existence of the smooth solution to the equations (1.1)-(1.5).

6 Blow-up of the smooth solution

In this section we will prove the blow-up of the smooth solution to the Cauchy problem of the model (1.1)-(1.5). From the assumption in Theorem 1.3, we know the initial data satisfying:

$$\begin{aligned}
&\rho(x, 0) = \rho_0 > 0, \quad \rho_0 = \bar{\rho}(|x| > R),\\
&u(x, 0) = u_0, \quad u_0 = 0(|x| > R),\\
&d(x, 0) = d_0, \quad d_0 \text{ is a constant vector } (|x| > R), \quad |d_0| = 1,\\
&B(x, 0) = B_0, \quad B_0 = 0(|x| > R), \quad \text{div} B_0 = 0.
\end{aligned} \tag{6.151}$$

Here R is the arbitrary constant.

From the initial condition (6.151), we know that the maximum speed of propagation of the front of a smooth disturbance is governed by the sound speed

$$\sigma = [P(\rho)]^{\frac{1}{2}} = [A\rho^\alpha]^{\frac{1}{2}}, \tag{6.152}$$

since $\bar{u} = 0$. More precisely, letting

$$\mathbf{D}(t) = \{x \in \mathbf{R}^2 : |x| \geqslant R + \sigma t\}, \quad \mathbf{B}(t) = \mathbf{R}^2 \backslash \mathbf{D}(t). \tag{6.153}$$

We have the following proposition:

Proposition 6.1 *If (ρ, u) is a C^1 solution of (1.1)-(1.5) on $\mathbf{D}(t)$, for $0 \leqslant t \leqslant T$, then $(\rho, u) \equiv (\bar{\rho}, 0)$ on $\mathbf{D}(t)$, $0 \leqslant t < T$. This proposition can be easily prove similar the proposition in* [48].

Firstly we need define the following equalities:

$$M(t) = \int_\Omega \rho - \bar{\rho}\, dx, \tag{6.154}$$

$$\mathcal{E}(t) = \int_\Omega \frac{1}{2}\rho|u|^2 + (H(\rho) - H(\bar{\rho})) + \frac{\hbar^2}{2}|\nabla\sqrt{\rho}|^2 + \frac{\lambda}{2}|\nabla d|^2 + \frac{\kappa}{2}|B|^2 dx, \tag{6.155}$$

$$G(t) = \int_\Omega \frac{1}{2}\rho|x|^2 dx \quad \text{(momentum of inertia)}, \tag{6.156}$$

$$F(t) = \int_\Omega \rho u \cdot x\, dx \quad \text{(momentum weight)}. \tag{6.157}$$

Then we have the following properties.

Lemma 6.1 *Set (ρ, u, d, B) is the smooth solution to the (1.1)-(1.5), Then we have*

$$M(t) = M(0), \tag{6.158}$$

$$\mathcal{E}(t) + \int_0^T \int_\Omega \nu H''(\rho)|\nabla\rho|^2 + \frac{\nu\hbar^2}{4}\rho|\nabla^2 \log \rho|^2 + \nu\rho|\nabla u|^2 + \lambda|\Delta d + |\nabla d|^2 d|^2$$
$$+ \kappa\delta|\nabla B|^2 dx = \mathcal{E}(0), \tag{6.159}$$

$$\frac{d}{dt}G(t) = F(t) + 2\nu M(0), \tag{6.160}$$

$$\frac{d}{dt}F(t) = \int_\Omega \rho|u|^2 + 2(\alpha - 1)(H(\rho) - H(\bar{\rho})) + \hbar^2|\nabla\sqrt{\rho}|^2 + \frac{\kappa}{2}|B|^2 dx. \tag{6.161}$$

Proof From (1.1), we have

$$\frac{d}{dt}M(t) = \frac{d}{dt}\int_\Omega \rho - \bar{\rho}\, dx = \int_\Omega \rho_t\, dx = \int_\Omega (-\operatorname{div}(\rho u) + \nu\Delta\rho)\, dx = 0.$$

Multiplying equation (1.1) by $H'(\rho) - \frac{1}{2}|u|^2 - \frac{\hbar^2}{2}\frac{\Delta\sqrt{\rho}}{\sqrt{\rho}}$, equation (1.2) by u, equation (1.3) by $\Delta d + |\nabla d|^2 d$ and equation (1.4) by B, and integrating by parts, using the following facts

$$\int_\Omega \partial_t \rho H'(\rho)\, dx = \frac{d}{dt}\int_\Omega (H(\rho) - H(\bar{\rho}))\, dx,$$

$$-\int_\Omega \partial_t\rho \frac{\hbar^2}{2}\frac{\Delta\sqrt{\rho}}{\sqrt{\rho}}dx = -\hbar^2\int_\Omega \partial_t(\sqrt{\rho})\Delta\sqrt{\rho}dx = \frac{\hbar^2}{2}\partial_t\int_\Omega |\nabla\sqrt{\rho}|^2 dx,$$

$$\int_\Omega \nu\Delta\rho H'(\rho)dx = -\nu\int_\Omega \nabla\rho\nabla H'(\rho)dx = -\nu\int_\Omega H''(\rho)|\nabla\rho|^2 dx,$$

$$-\int_\Omega \nu\Delta\rho \frac{|u|^2}{2}dx = \nu\int_\Omega \nabla\rho : \nabla u : u dx,$$

$$\int_\Omega \frac{\Delta\sqrt{\rho}}{\sqrt{\rho}}\Delta\rho dx = -\int_\Omega \rho\nabla\log\rho\nabla\left(\frac{\Delta\sqrt{\rho}}{\sqrt{\rho}}\right)dx$$
$$= -\frac{1}{2}\int_\Omega \nabla\log\rho\,\mathrm{div}(\rho\nabla^2\log\rho)dx = \frac{1}{2}\int_\Omega \rho|\nabla^2\log\rho|^2 dx,$$

$$\int_\Omega \partial_t(\rho u)\cdot u + \mathrm{div}(\rho u\otimes u)\cdot u dx$$
$$= \int_\Omega \partial_t\rho|u|^2 + \rho\partial_t\left(\frac{|u|^2}{2}\right) + \nabla\cdot(\rho u)|u|^2 + \rho u\cdot\nabla\left(\frac{|u|^2}{2}\right)dx$$
$$= \int_\Omega \partial_t\rho|u|^2 + \rho\partial_t\left(\frac{|u|^2}{2}\right) + \frac{1}{2}\nabla\cdot(\rho u)|u|^2 dx,$$

$$\int_\Omega \nabla P\cdot u dx = \int_\Omega \nabla H'\rho u dx = -\int_\Omega H'(\rho)\mathrm{div}(\rho u)dx,$$

$$\int_\Omega (\nabla\times B)\times B\cdot u dx = -\int_\Omega \nabla\times(u\times B)\cdot B dx,$$

$$\int_\Omega \frac{\hbar^2}{2}\rho\nabla\left(\frac{\Delta\sqrt{\rho}}{\sqrt{\rho}}\right)\cdot u dx = -\int_\Omega \frac{\hbar^2}{2}\left(\frac{\Delta\sqrt{\rho}}{\sqrt{\rho}}\right)\mathrm{div}(\rho u)dx,$$

$$\int_\Omega \nu\Delta(\rho u)\cdot u = -\int_\Omega \nu\nabla(\rho u)\cdot\nabla u = -\int_\Omega \nu\nabla\rho : \nabla u : u dx - \int_\Omega \nu\rho|\nabla u|^2 dx,$$

$$\int_\Omega -\lambda\nabla\cdot\left(\nabla d\odot\nabla d - \frac{|\nabla d|^2}{2}I\right)\cdot u dx = \lambda\int_\Omega \nabla d\odot\nabla d : \nabla u dx - \int_\Omega \frac{|\nabla d|^2}{2}\nabla u dx,$$

$$\int_\Omega d_t\cdot\Delta d dx = -\frac{d}{dt}\int_\Omega \frac{1}{2}|\nabla d|^2 dx,$$

$$\int_\Omega (u\cdot\nabla d)\cdot\Delta d dx = \int_\Omega d_{jj}u^i d_i dx$$
$$= \int_\Omega \left[(d_j u^i d_i)_j - d_i\cdot d_j u^i_j - u^i\left(\frac{|\nabla d|^2}{2}\right)_i\right]dx$$
$$= -\int_\Omega d_i\cdot d_j u^i_j + u^i_i\left(\frac{|\nabla d|^2}{2}\right)dx = -\int_\Omega \nabla d\odot\nabla d : \nabla u dx + \int_\Omega \frac{|\nabla d|^2}{2}\nabla u dx,$$

$$-\int_\Omega \kappa\nabla\times(\nabla\times B)\cdot B dx = \int_\Omega \kappa\Delta B\cdot B dx = -\kappa\int_\Omega |\nabla B|^2 dx,$$

we can get (6.159) holds.

Then we have the following facts

$$G'(t) = \int_\Omega \frac{1}{2}\rho_t|x|^2 dx = \int_\Omega (-\mathrm{div}(\rho u) + \nu\Delta\rho)\frac{1}{2}|x|^2 dx$$

$$= \int_{\mathbf{B}(t)} (-\mathrm{div}(\rho u) + \nu\Delta\rho)\frac{1}{2}|x|^2 dx$$

$$= \int_{\mathbf{B}(t)} \rho u \cdot x\, dx - \int_{\mathbf{B}(t)} \nu\nabla\rho \cdot x\, dx = F(t) + \int_{\mathbf{B}(t)} \nu(\rho - \bar\rho)\nabla \cdot x\, dx$$

$$= F(t) + 2\nu \int_{\mathbf{B}(t)} (\rho - \bar\rho)dx = F(t) + 2\nu M(t) = F(t) + 2\nu M(0),$$

and

$$F'(t) = \int_\Omega (\rho u)_t \cdot x\, dx$$

$$= \int_\Omega \left(-\mathrm{div}(\rho u \otimes u) + \nabla P \cdot x + \nu\Delta(\rho u) - \lambda\nabla \cdot \left(\nabla d \odot \nabla d - \frac{|\nabla d|^2}{2}I \right) \right.$$

$$\left. + \frac{\hbar^2}{2}\rho\nabla\left(\frac{\Delta\sqrt{\rho}}{\sqrt{\rho}}\right) + \kappa(\nabla \times B) \times B \right) \cdot x\, dx$$

$$= F_1 + F_2 + F_3 + F_4 + F_5 + F_6.$$

Next, we will calculate $F_1, F_2, F_3, F_4, F_5, F_6$ respectively.

$$F_1 = -\int_\Omega \mathrm{div}(\rho u \otimes u) \cdot x\, dx = -\int_{\mathbf{B}(t)} \mathrm{div}(\rho u \otimes u) \cdot x\, dx$$

$$= -\sum_{i,j=1}^2 \int_{\mathbf{B}(t)} \partial_i(\rho u_i u_j) \cdot x_j dx = \sum_{i,j=1}^2 \int_{\mathbf{B}(t)} \rho u_i u_j \delta_{ij} dx$$

$$= \int_{\mathbf{B}(t)} \rho|u|^2 dx = \int_\Omega \rho|u|^2 dx,$$

$$F_2 = -\int_\Omega \nabla P \cdot x\, dx = \int_{\mathbf{B}(t)} \nabla(P(\rho) - P(\bar\rho)) \cdot x$$

$$= \sum_{i=1}^2 \int_{\mathbf{B}(t)} \partial_i(P(\rho) - P(\bar\rho))x_i dx$$

$$= \sum_{i=1}^2 \int_{\mathbf{B}(t)} (P(\rho) - P(\bar\rho))\partial_i x_i dx$$

$$= 2\int_\Omega (P(\rho) - P(\bar\rho))dx,$$

$$F_3 = \int_\Omega \nu\Delta(\rho u)\cdot x dx = \int_{\mathbf{B}(t)} \nu\Delta(\rho u)\cdot x dx$$
$$= \sum_{i,j=1}^2 \int_{\mathbf{B}(t)} \partial_i^2(\rho u_j)x_j dx = \sum_{i,j=1}^2 \int_{\mathbf{B}(t)} (\rho u_j)\partial_i^2 x_j dx = 0,$$
$$F_4 = -\lambda \int_\Omega \nabla\cdot\left(\nabla d \odot \nabla d - \frac{|\nabla d|^2}{2}I\right)\cdot x dx$$
$$= -\lambda \int_{\mathbf{B}(t)} \nabla\cdot\left(\nabla d \odot \nabla d - \frac{|\nabla d|^2}{2}I\right)\cdot x dx$$
$$= -\lambda \sum_{i,j,k=1}^2 \int_{\mathbf{B}(t)} \partial_i\left(\partial_i d_k \partial_j d_k - \frac{|\nabla d|^2}{2}\delta_{ij}\right) x_j dx$$
$$= \lambda \sum_{i,j,k=1}^2 \int_{\mathbf{B}(t)} \left(\partial_i d_k \partial_j d_k - \frac{|\nabla d|^2}{2}\delta_i j\right)\delta_{ij} dx$$
$$= \lambda \sum_{i,k=1}^2 \int_{\mathbf{B}(t)} \left(\partial_i d_k \partial_i d_k - \frac{|\nabla d|^2}{2}\right) dx$$
$$= \lambda \int_{\mathbf{B}(t)} |\nabla d|^2 - \frac{2}{2}|\nabla d|^2 dx = 0,$$
$$F_5 = \frac{\hbar^2}{2}\int_\Omega \rho\nabla\left(\frac{\Delta\sqrt\rho}{\sqrt\rho}\right)\cdot x dx = \frac{\hbar^2}{2}\int_{\mathbf{B}(t)} \rho\nabla\left(\frac{\Delta\sqrt\rho}{\sqrt\rho}\right)\cdot x dx$$
$$= -\frac{\hbar^2}{2}\int_{\mathbf{B}(t)} 2\Delta\sqrt\rho\nabla\sqrt\rho\cdot x dx - 2\int_{\mathbf{B}(t)} \frac{\hbar^2}{2}\sqrt\rho\Delta\sqrt\rho dx$$
$$= \int_{\mathbf{B}(t)} \frac{\hbar^2}{2}\nabla(|\nabla\sqrt\rho|^2)\cdot x dx + 2\int_{\mathbf{B}(t)} \frac{\hbar^2}{2}|\nabla\sqrt\rho|^2 dx - \int_{\mathbf{B}(t)} 4\frac{\hbar^2}{2}\sqrt\rho\Delta\sqrt\rho dx$$
$$= -2\int_{\mathbf{B}(t)} \frac{\hbar^2}{2}|\nabla\sqrt\rho|^2 dx + 2\int_{\mathbf{B}(t)} \frac{\hbar^2}{2}|\nabla\sqrt\rho|^2 dx + 2\frac{\hbar^2}{2}\int_{\mathbf{B}(t)} |\nabla\sqrt\rho|^2 dx$$
$$= 2\int_{\mathbf{B}(t)} \frac{\hbar^2}{2}|\nabla\sqrt\rho|^2 dx = \hbar^2 \int_\Omega |\nabla\sqrt\rho|^2 dx,$$
$$F_6 = \kappa\int_\Omega (\nabla\times B)\times B\cdot x dx = \kappa\int_{\mathbf{B}(t)} (\nabla\times B)\times B\cdot x dx$$
$$= \kappa\int_{\mathbf{B}(t)} -b_2(\partial_1 b_2 - \partial_2 b_1)x_1 + b_1(\partial_1 b_2 - \partial_2 b_1)x_2 dx$$
$$= \kappa\int_{\mathbf{B}(t)} -\partial_2 b_2 b_1 + \frac{1}{2}b_2^2 - \partial_1 b_1 b_2 + \frac{1}{2}b_1^2 dx$$
$$= \kappa\int_{\mathbf{B}(t)} \partial_1\left(\frac{b_1^2}{2}\right) + \partial_2\left(\frac{b_2^2}{2}\right) + \frac{1}{2}(b_1^2 + b_2^2) dx$$

$$= \kappa \int_{\mathbf{B}(t)} \nabla |B|^2 dx + \frac{\kappa}{2} \int_{\mathbf{B}(t)} |B|^2 dx = \frac{\kappa}{2} \int_{\mathbf{B}(t)} |B|^2 dx = \frac{\kappa}{2} \int_{\Omega} |B|^2 dx.$$

Thus we complete the proof. □

For simplicity, we denote

$$\mathcal{E}(t) = \int_\Omega \frac{1}{2}\rho|u|^2 dx + \int_\Omega (H(\rho) - H(\bar\rho))dx$$
$$+ \int_\Omega \frac{\hbar^2}{2}|\nabla\sqrt{\rho}|^2 dx + \int_\Omega \frac{\lambda}{2}|\nabla d|^2 dx + \int_\Omega \frac{\kappa}{2}|B|^2 dx$$
$$= E_1 + E_2 + E_3 + E_4 + E_5.$$

Then, we have

$$F'(t) = 2E_1 + 2(\alpha - 1)E_2 + 2E_3 + E_5.$$

The Proposition 1 says that $(\rho, u) = (\bar\rho, 0)$ outside $\mathbf{B}(t)$. By Jensen's inequality, and (6.151) we have

$$\int_{\mathbf{B}(t)} P(\rho)dx = A\int_{\mathbf{B}(t)} \rho^\alpha dx \geqslant A(vol\mathbf{B}(t))^{1-\alpha}\left(\int_{\mathbf{B}(t)} \rho dx\right)^\alpha$$
$$= A(vol\mathbf{B}(t))^{1-\alpha}(m(0) + vol\mathbf{B}(t)\bar\rho)^\alpha \geqslant \bar{p}vol\mathbf{B}(t) = \int_{\mathbf{B}(t)} P(\bar\rho)dx.$$
(6.162)

Thus we have

$$0 < 2E_1(t) \leqslant F'(t) \leqslant \max\{2, 2(\alpha - 1)\}\mathcal{E}(t) \leqslant \max\{2, 2(\alpha - 1)\}\mathcal{E}(0). \quad (6.163)$$

Then we have the following lemma.

Lemma 6.2 *Set (ρ, u, d, B) is the smooth solution of (1.1)-(1.5), we obtain*
(1) $\alpha > 2$

$$G(t) \leqslant (\alpha - 1)\mathcal{E}(0)t^2 + (F(0) + 2\nu M(0))t + G(0); \quad (6.164)$$

(2) $1 < \alpha \leqslant 2$

$$G(t) \leqslant \mathcal{E}(0)t^2 + (F(0) + 2\nu M(0))t + G(0). \quad (6.165)$$

Proof (1) If $\alpha > 2$, from (6.163), we have

$$F'(t) \leqslant 2(\alpha - 1)\mathcal{E}(0),$$

and
$$G'(t) \leqslant 2(\alpha - 1)\mathcal{E}(0)t + F(0) + 2\nu M(0).$$
Therefore,
$$G(t) \leqslant (\alpha - 1)\mathcal{E}(0)t^2 + (F(0) + 2\nu M(0))t + G(0).$$
Similarly, for $1 < \alpha \leqslant 2$, we can get $G(t) \leqslant \mathcal{E}(0)t^2 + (F(0) + 2\nu M(0))t + G(0)$. □

(1) $\alpha > 2$

It holds that
$$\begin{aligned} F^2(t) &\leqslant 4E_1(t)G(t) \leqslant 2F'(t)G(t) \\ &\leqslant 2F'(t)[(\alpha - 1)\mathcal{E}(0)t^2 + (F(0) + 2\nu M(0))t^2 + G(0)]. \end{aligned} \quad (6.166)$$

Set
$$\Delta_1 = (F(0) + 2\nu M(0))^2 - 4(\alpha - 1)\mathcal{E}(0)G(0),$$
$$a = (\alpha - 1)\mathcal{E}(0), \ b = (F(0) + 2\nu M(0)), \ c = G(0).$$

If $\Delta_1 < 0$, we know
$$(\alpha - 1)\mathcal{E}(0)t^2 + (F(0) + 2\nu M(0))t^2 + G(0) > 0, \quad \forall t > 0.$$

Therefore, we have
$$\frac{F'(t)}{F(t)^2} \geqslant \frac{2}{(\alpha - 1)\mathcal{E}(0)t^2 + (F(0) + 2\nu M(0))t + G(0)}. \quad (6.167)$$

From (6.167) we know that the life span T_{**} of the smooth solution to (1.1)-(1.5) can not be infinity. Indeed, from (6.167) we have
$$\begin{aligned}\frac{1}{F(0)} &> \frac{1}{F(0)} - \frac{1}{F(T_{**})} \geqslant \int_0^{T_{**}} \frac{2}{(\alpha - 1)\mathcal{E}(0)t^2 + (F(0) + 2\nu M(0))t + G(0)} dt \\ &= \frac{4a^2}{4ac - b^2} \arctan\left\{\frac{4T_{**}a^2}{4ac - b^2} + \frac{2ab}{4ac - b^2}\right\} - \frac{4a^2}{4ac - b^2} \arctan\left\{\frac{2ab}{4ac - b^2}\right\}.\end{aligned} \quad (6.168)$$

So we can solve out T_{**} from the above equation
$$T_{**} \leqslant \frac{4ac - b^2}{4a^2} \tan\left\{\frac{4ac - b^2}{4F(0)a^2} + \arctan\left\{\frac{2ab}{4ac - b^2}\right\}\right\} - \frac{b}{2a}.$$

If $\Delta_1 > 0$, we know that $(\alpha - 1)\mathcal{E}(0)t^2 + (F(0) + 2\nu M(0))t + G(0) = 0$ has two negative solutions $-\lambda_1 = -\dfrac{b - \sqrt{b^2 - 4ac}}{2a}$, $-\lambda_2 = -\dfrac{b + \sqrt{b^2 - 4ac}}{2a}$. Thus we have

$$\frac{F'(t)}{F^2(t)} \geq \frac{1}{a(t + \lambda_1)(t + \lambda_2)},$$

and

$$\frac{1}{F(0)} \geq \int_0^{T_{**}} \frac{1}{a(t + \lambda_1)(t + \lambda_2)} dt$$

$$\geq \frac{1}{a(\lambda_2 - \lambda_1)} \ln \frac{(t + \lambda_1)}{(t + \lambda_2)} \Big|_0^{T_{**}}$$

$$\geq \frac{1}{a(\lambda_2 - \lambda_1)} \ln \frac{\lambda_2(T_{**} + \lambda_1)}{\lambda_1(T_{**} + \lambda_2)}.$$

Therefore, we have

$$\left(\lambda_2 - \lambda_1 \exp\left(\frac{a(\lambda_2 - \lambda_1)}{F(0)}\right)\right) T_{**} \leq \left(\exp\left(\frac{a(\lambda_2 - \lambda_1)}{F(0)}\right) - 1\right) \lambda_1 \lambda_2.$$

If $\lambda_2 - \lambda_1 \exp\left(\dfrac{a(\lambda_2 - \lambda_1)}{F(0)}\right) > 0$, we have

$$T_{**} \leq \frac{\left(\exp\left(\dfrac{a(\lambda_2 - \lambda_1)}{F(0)}\right) - 1\right) \lambda_1 \lambda_2}{\left(\lambda_2 - \lambda_1 \exp\left(\dfrac{a(\lambda_2 - \lambda_1)}{F(0)}\right)\right)}.$$

If $\Delta_1 = 0$, we know that $(\gamma - 1)\mathcal{E}(0)t^2 + (F(0) + 2\nu M(0))t + G(0) = 0$ has two same solutions $-\lambda_1 = -\lambda_2 = -\dfrac{b}{4a}$. Thus we have

$$\frac{F'(t)}{F^2(t)} \geq \frac{1}{a(t + \lambda_1)^2},$$

and

$$\frac{1}{F(0)} \geq \int_0^{T_{**}} \frac{1}{a(t + \lambda_1)^2} dt$$

$$= \frac{1}{a\lambda_1} - \frac{1}{a(T_{**} + \lambda)}.$$

Therefore, for $F(0) \geq a\lambda_1 = \dfrac{(F(0) + 2\nu M(0))}{4}$, i.e. $F(0) \geq \dfrac{8\nu M(0)}{3}$, we have

$$T_{**} \leq \frac{a\lambda_1^2}{F(0) - a\lambda_1}.$$

Similarly, for the case $1 < \alpha \leqslant 2$, we have

$$F^2(t) \leqslant 4E_1(t)G(t) \leqslant 2F'(t)G(t)$$
$$\leqslant 2F'(t)[\mathcal{E}(0)t^2 + (F(0) + 2\nu M(0))t^2 + G(0)]. \tag{6.169}$$

Set

$$\Delta_1 = (F(0) + 2\nu M(0))^2 - 4\mathcal{E}(0)G(0), \ a = \mathcal{E}(0), \ b = (F(0) + 2\nu M(0)), \ c = G(0).$$

If $\Delta_1 < 0$, we know

$$\mathcal{E}(0)t^2 + (F(0) + 2\nu M(0))t^2 + G(0) > 0, \quad \forall t > 0.$$

Therefore, we have

$$\frac{F'(t)}{F(t)^2} \geqslant \frac{2}{\mathcal{E}(0)t^2 + (F(0) + 2\nu M(0))t + G(0)}. \tag{6.170}$$

From (6.170), we know that the life span T_{**} of the smooth solution to (1.1)-(1.5) can not be infinity. Indeed, from (6.170), we have

$$\frac{1}{F(0)} > \frac{1}{F(0)} - \frac{1}{F(T_{**})} \geqslant \int_0^{T_{**}} \frac{2}{\mathcal{E}(0)t^2 + (F(0) + 2\nu M(0))t + G(0)} dt$$
$$= \frac{4a^2}{4ac - b^2} \arctan\left\{\frac{4T_{**}a^2}{4ac - b^2} + \frac{2ab}{4ac - b^2}\right\} - \frac{4a^2}{4ac - b^2} \arctan\left\{\frac{2ab}{4ac - b^2}\right\}. \tag{6.171}$$

So we can solve out T_{**} from the above equation

$$T_{**} \leqslant \frac{4ac - b^2}{4a^2} \tan\left\{\frac{4ac - b^2}{4F(0)a^2} + \arctan\left\{\frac{2ab}{4ac - b^2}\right\}\right\} - \frac{b}{2a}.$$

Thus we complete the proof of Theorem 1.3.

If $\Delta_1 > 0$, we know that $\mathcal{E}(0)t^2 + (F(0) + 2\nu M(0))t + G(0) = 0$ has two negative solutions $-\lambda_1 = -\dfrac{b - \sqrt{b^2 - 4ac}}{2a}, -\lambda_2 = -\dfrac{b + \sqrt{b^2 - 4ac}}{2a}$. Thus we have

$$\frac{F'(t)}{F^2(t)} \geqslant \frac{1}{a(t + \lambda_1)(t + \lambda_2)},$$

and

$$\frac{1}{F(0)} \geq \int_0^{T_{**}} \frac{1}{a(t+\lambda_1)(t+\lambda_2)} dt$$

$$\geq \frac{1}{a(\lambda_2-\lambda_1)} \ln \frac{(t+\lambda_1)}{(t+\lambda_2)}\bigg|_0^{T_{**}}$$

$$\geq \frac{1}{a(\lambda_2-\lambda_1)} \ln \frac{\lambda_2(T_{**}+\lambda_1)}{\lambda_1(T_{**}+\lambda_2)}.$$

Therefore, we have

$$\left(\lambda_2 - \lambda_1 \exp\left(\frac{a(\lambda_2-\lambda_1)}{F(0)}\right)\right) T_{**} \leq \left(\exp\left(\frac{a(\lambda_2-\lambda_1)}{F(0)}\right) - 1\right) \lambda_1 \lambda_2.$$

If $\lambda_2 - \lambda_1 \exp\left(\frac{a(\lambda_2-\lambda_1)}{F(0)}\right) > 0$, we have

$$T_{**} \leq \frac{\left(\exp\left(\frac{a(\lambda_2-\lambda_1)}{F(0)}\right) - 1\right) \lambda_1 \lambda_2}{\left(\lambda_2 - \lambda_1 \exp\left(\frac{a(\lambda_2-\lambda_1)}{F(0)}\right)\right)}.$$

If $\Delta_1 = 0$, we know that $\mathcal{E}(0)t^2 + (F(0)+2\nu M(0))t + G(0) = 0$ has two same solutions $-\lambda_1 = -\lambda_2 = -\frac{b}{4a}$. Thus we have

$$\frac{F'(t)}{F^2(t)} \geq \frac{1}{a(t+\lambda_1)^2},$$

and

$$\frac{1}{F(0)} \geq \int_0^{T_{**}} \frac{1}{a(t+\lambda_1)^2} dt$$

$$= \frac{1}{a\lambda_1} - \frac{1}{a(T_{**}+\lambda)}.$$

Therefore, for $F(0) \geq a\lambda_1 = \frac{(F(0)+2\nu M(0))}{4}$, i.e. $F(0) \geq \frac{8\nu M(0)}{3}$, we have

$$T_{**} \leq \frac{a\lambda_1^2}{F(0) - a\lambda_1}.$$

Acknowledgements The first author is supported by the Foundation of Guangzhou University No. 2700050357, The second author is supported by National Natural Science Foundation of China (Grant No. 11731014).

References

[1] Antonelli P, Marcati P. The quantum hydrodynamics system in two space in two space dimensions[J]. Archive for Rational Mechanics and Analysis, 2012, 203: 499-527.

[2] Antonelli P, Spirito S. Global existence of finite energy weak solutions of quantum Navier-Stokes equations[J]. Arch. Rational Mech. Anal., 2017, 225: 1161-1199.

[3] Antonelli P, Spirito S. On the compactness of finite energy weak solutions to the quantum Navier-Stokes equations[J]. ArXiv:1512.07496v2, 2015.

[4] Bresch D, Desjardins B. On the existence of global weak solutions to the Navier-Stokes equations for viscous compressible and heat conducting fluids[J]. J. Math. Pures Appl., 2007, 87(1): 57-90.

[5] Bresch D, Desjardins B, Lin C-K. On some compressible fluid models: Korteweg, lubrication, and shallow water systems[J]. Comm. Partial Differential Equations., 2003, 28(3-4): 843-868.

[6] Cavaterra C, Rocca E, Wu H. Global weak solution and blow-up criterion of the general Ericsen-Lesile system for nematic liquid crystal flows[J]. Journal of Differential Equations, 2013, 255: 24-57.

[7] Ding S-J, Huang J-R, Xia F-G, Wen H-Y, Zi R. Z. Incompressible limit of the compressible nematic liquid crystal flow[J]. J. Funct. Anal., 2013, 264(7): 1711-1756.

[8] Ding S-J, Lin J, Wang C-Y, Wen H-Y. Compressible hydrodynamic flow of liquid crystals in 1-D[J]. Discrit. Cont. Dyn. Sys. B., 2011, 15: 357-371.

[9] Dong J-W. A note on barotropic compressible quantum Navier-Stokes equations[J]. Nonlinear Analysis: TMA, 2010, 73: 854-856.

[10] Evens L-C. Partial Differential Equations[M]. American Mathematical Society, Providence, RI, 1998.

[11] Feireisl E, Novotný A, Petzeltová H. On the existence of globally defined weak solutions to the Navier-Stokes equations[J]. J. Math. Fluid Mech., 2001, 3: 358-392.

[12] Feireisl E. Dynamics of viscous compressible fluid[M]. Oxford lecture series in Mathematics and its applications, vol. 26. Oxford Science Publications, The Clarendon Press, Oxford University Press, New York, 2004.

[13] Fradkin E, Kivenlson S-A. Liquid-crystal phases of quantum Hall systems[J]. Phys. Rev. B, 1999, 59(12): 8065-8072.

[14] Fregoso B-M. Quantum liquid crystal phases and unconventional magnetism in electronic and atomic fermi systems[D]. PHD dissertation, University of Illinois at Urbana-Champaign, 2010.

[15] Gasser I, Markowich P. Quantum hydrodynamics, Wigner transforms and the classical limit[J]. Asymptotic Anal., 1997, 14: 97-116.

[16] Gisclon M, Lacroix-Violet I. About the barotropic compressible quantum Navier-Stokes equations[J]. Nonlinear Analysis: TMA, 2015, 128: 106-121.

[17] Guo B-L, Ding S-J. Landau-Lifshitz equations[M]. Word Science: Singapore, 2008.

[18] Guo B-L, Wang G-W. Blow-up of the smooth solution to the compressible nematic liquid crystal system[J], Acta Appl. Math. 2018, 156: 211-224.

[19] Guo B-L, Wang G-W. Blow-up of the smooth solution to quantum hydrodynamic models in $\mathbf{R}^d$[J]. Journal of Differential Equations, 2016, 162(7): 3815-3842.

[20] Guo B-L, Wang G-W. Blow-up of smooth solutions to the quantum Navier-Stokes equations, submitted.

[21] Hass F. A magnetohydrodynamic model for quantum plasmas[J]. Phys. Plasmas, 2005, 12: 062117-062126.

[22] Hoff D. Discontinuous solutions of the Navier-Stokes equations for multidimensional heat-conducting flow[J]. Arch. Rational Mech. Anal., 1997, 139: 303-354.

[23] Hoff D. Global existence for 1D, compressible, isentropic Navier-Stokes equations with large initial data[J]. Trans. Amer. Math. Soc., 1987, 303: 169-181.

[24] Hoff D. Global solutions of the Navier-Stokes equations for multidimensional, compressible flow with discontinuous initial data[J]. J. Diff. Eqns., 2013, 120(2): 215-254.

[25] Hoff D. Strong convergence to global solutions for multidimensional flows of compressible. viscous fluid with polytropic equations of state and discontinuous initial data[J], Arch. Rational Mech. Anal., 1995, 132: 1-14.

[26] Hong M-C, Li J-K, Xin Z-P. Blow-up criteria of strong solutions to the Ericksen-Leslie system in $\mathbf{R}^3$[J]. Comm. in Partial Differential Equations, 2014, 39: 1284-1328.

[27] Hu X-P, Wu H. Global solution to the three-dimensional compressible flow of liquid crystals[J]. SIAM J. Math. Anal., 2013, 45(5): 2678-2699.

[28] Huang X-D, Li J, Xin Z-P. Global well-posedness of classical solutions with large oscillations and vacuum to the three-dimensional isentropic compressible Navier-Stokes equations[J]. Comm. Pure Appl. Math., 2012, 65: 549-585.

[29] Huang X-D, Wang C-Y, Wen H-Y. Blow up criterion for compressible nematic liquid crystal flows in dimension three[J]. Arch. Ration. Mech. Anal., 2012, 204: 285-311.

[30] Huang X-D, Wang C-Y, Wen H-Y. Strong solutions of the compressible nematic liquid crystal flow[J]. J. Diff. Eqns., 2012, 252: 222-2265.

[31] Jüngel A. Global weak solution to compressible Navier-Stokes equations for quantum fluids[J]. SIAM J. Math. Anal., 2010, 42(3): 1025-1045.

[32] Jiang F. A remark on weak solutions to the barotropic compressible quantum Navier-Stokes equations[J]. Nonlinear Analaysis: RWA, 2011, 12: 1733-1735.

[33] Jiang F, Jiang S, Wang D-H. Global weak solutions to the equations of compressible flow of nematic liquid crystals in two dimensions[J]. Arch. Rational Mech. Anal., 2014, 214: 403-451.

[34] Jiang F, Jiang S, Wang D-H. On multi-dimensional compressible flows of nematic liquid crystals with large inital energy in a bounded domain[J]. J. Func. Anal., 2013, 265: 3369-3397.

[35] Kazhikhov A-V, Shelukhin V-V. Unique global solution with respect to time initial-

boundary value problems for one-dimensional equations of a viscous gas[J]. J. Appl. Math. Mach., 1977, 41(2): 273-282.

[36] Lei Z, Li D, Zhang X-Y. Remarks of global wellposedness of liquid crystal flows and heat flow of harmonic maps in two dimensions[J]. Proceedings of American Mathematical Society, 2012, 142(11): 3801-3810.

[37] Li H-L, Marcati P. Existence and asymptotic behavior of multi-dimensional quantum hydrodynamic model for semiconductors[J]. Comm. Math. Phys., 2004, 245: 215-247.

[38] Li J, Xin Z-P. Global existence of weak solutions to the barotropic compressible Navier-Stokes flows with degenerate viscosities[J]. arXiv:1504.06826v2, 2015.

[39] Li J, Xu Z, Zhang J. Global well-posedness with large oscillations and vacuum to the three dimensional equations of compressible nematic liquid crystal flows[J]. arXiv:1204.4966, 2012.

[40] Lions P-L. Mathematical topics in fluid mechanics[M]. Vol. 2. Compressible models. Oxford Lecture Series in Mathematics and its Applications, vol.10. Oxford science publications, the Clarendon Press, Oxford University Press, New York, 1998.

[41] Matsumura A, Nishida T. The initial value problem for the equations of motion of viscous and heat-conductive gases[J]. J. Math. Kyota Univ., 1980, 20(1): 67-104.

[42] Mellet A, Vasseur A. On the barotropic compressible Navier-Stokes equations[J]. Comm. Partial Differential Equations, 2007, 32(1-3): 431-452.

[43] Mucha P-B, Pokorný M, Zatorska E. Chemically reacting mixtures in terms of degenerated parabolic setting[J]. R. Modern Phys., 2013, 69(3): 071501.

[44] Nirenberg L. On elliptic partial differential equations. Ann. Scuola Norm. Sup. Pisa, 1959, 13(3): 115-162.

[45] Nosanow L-H, Parish L-J, Pinski F-J. Zero-temperature properties of matter and the quantum theorem of corresponding states: The liquid-to-crystal phase transition for Fermi and Bose systems[J]. Phys. Rev. B, 1975, 11(1): 191-204.

[46] Novotny A, Straskraba I. Introduction to the mathematical theory of compressible flow[M]. Oxford University Press, Oxford, 2004.

[47] Radzihovsky L. Quantum liquid-crystal order in resonant atomic gases[J]. Physica C, 2012, 481: 189-206.

[48] Sideris T-C. Formation of singularities of solutions to nonlinear hyperbolic equations[J]. Arch. Ration. Mech. Anal., 1984, 86: 369-381.

[49] Simon J. Compact sets in the space $L^p([0,T];B)$[J], Ann. Mat. Pura Appl., 1987, 146(4): 65-96.

[50] Suen A, Hoff D. Global low-energy weak solutions of the equations of three-dimensional compressible magnetohydrodynamics[J]. Arch. Rational Mcch. Anal.,2012, 205: 27-58.

[51] Vasseur A-F, Yu C. Existence of global weak solutions for 3D degenerate compressible Navier-Stokes equations[J]. Invent. Math., 2016, 206(3): 935-974.

[52] Wang G-W, Guo B-L. Blow-up of solutions to quantum hydrodynamic models in half space[J]. J. Math. Phy., 2017, 58: 031505.

[53] Wu G-C, Tan Z. Global low-energy weak solution and large-time behavior for the compressible flow of liquid crystals[J]. arXiv:1210.1269[math.AP], 2012.

[54] Wu Z-Q, Yin J-X, Wang C-P. Elliptic and parabolic equations[M]. World Scientific, Singapore, 2006.

[55] Yang J-W, Ju Q-C. Global existence of the three-dimensional viscous quantum magnetohydrodynamic model[J]. Journal of Mathematical Physics, 2014, 55: 081501.

[56] Yang K. Quantum liquid crystal phases in ferminonic superfluid with pairing between fermion species of enequal densities[J]. Inter. J. Mord. Phys. B, 2013, 27(15): 1362001: 1-7.

[57] Zatorska E. Fundamental problems to equations of compressible chemically reacting flows[D]. PhD Dissertation, 2013.

[58] Zatorska E. On the flow of chemically reacting gaseous mixture[J]. J. Diff. Eqns., 2012, 253(12): 3471-3500.

The Well-Posedness for the Cauchy Problem of the Double-Diffusive Convection System*

Chen Fei (陈菲), Guo Boling (郭柏灵), and Zeng Lan (曾兰)

Abstract In this paper, we consider the well-posedness of the solution for the Cauchy problem of the double-diffusive convection system in $\mathbf{R}^3$. We establish the local existence and uniqueness of the solution for the double-diffusive convection system in $H^1(\mathbf{R}^3)$ with large initial data and global well-posedness under the assumption that the L^2 norm of the initial data is small. Moreover, we also prove that there exists a global unique solution in $H^N(\mathbf{R}^3)$ for any $N \geqslant 2$, without any other smallness conditions of the initial data.

Keywords double-diffusive convection system; Cauchy problem; well-posedness

1 Introduction

In this paper, we study the global unique solution for the Cauchy problem of the nondimensional double-diffusive convection system (see Refs. [7], [10], [26]):

$$\begin{cases} \dfrac{\partial U}{\partial t} + U \cdot \nabla U - \mu \Delta U + \mu \nabla P - \mu(\lambda\theta - \eta S)\mathbf{K} = 0, & \text{in } \mathbf{Q}_T, \\ \dfrac{\partial \theta}{\partial t} + U \cdot \nabla \theta - \Delta \theta - w = 0, & \text{in } \mathbf{Q}_T, \\ \dfrac{\partial S}{\partial t} + U \cdot \nabla S - \tau \Delta S - w = 0, & \text{in } \mathbf{Q}_T, \\ \operatorname{div} U = 0, & \text{in } \mathbf{Q}_T, \\ (U, \theta, S)(x,t)|_{t=0} = (U_0(x), \theta_0(x), S_0(x)), & \text{on } \mathbf{R}^3, \end{cases} \quad (1.1)$$

where $\mathbf{Q}_T \equiv \mathbf{R}^3 \times (0, T)$, $U = (u, v, w)$ stands for the velocity vector of the fluid, θ stands for the temperature of the fluid, and S stands for the salinity.

---

* J. Math. Phys, 2019, 60(1): 011511. Doi: 10.1063/1.5052668.

The nondimensional parameter μ means the Prandtl number, τ means the Lewis number, λ means the thermal Rayleigh number, and η means the solute Rayleigh number. $\mu, \lambda, \eta, \tau > 0$. $\mathbf{K}$ is the vertical unit vector. The first equation of the system (1.1) is the Boussinesq equation, the second one is the diffusion equation of the temperature function, while the third one is that of the salinity function, and the last two equations represent the divergence condition and the initial condition, respectively.

Convective motions occur in a fluid when density variations appear. Double-diffusion convection is named by such convection motions when the density variations are caused by two different components with different rates of diffusion (there are two requirements for the occurrence of the double-diffusive convection: the fluid has two or more components with different molecular diffusivities and those components make opposing contributions to the vertical density gradient). In 1857, Jevons [9] first discovered the phenomenon of the double-diffusion convection, which was forgotten and then rediscovered as an "oceanographic curiosity" a century later (see e.g., the work of Stommel-Arons-Blanchard [23], Veronis [26] and Baines-Gill [2]). Because of the wide use in astrophysics, engineering and geology, etc., a great number of studies for the double-diffusive convection have been carried out (see e.g., Refs. [4], [8], [14], [20] and [21] and the references therein).

There are a lot of literature studies on the study of the double-diffusive convection system not only in physics and engineering but also in mathematics (see Refs. [7], [10], [11], [13], [16-19], [22] and [25] and the references therein). There are some results for the well-posedness of the double-diffusive convection system, especially, for that based on Brinkman-Forchheimer equation (see e.g., Refs. [16-18], [25]). For example, Terasawa-Ôtani [25] and Ôtani-Uchida [16] considered the global solvability with Dirichlet and Neumann boundary conditions, respectively. In Refs. [16] and [25], the global unique strong solution was established for 3-D bounded domains with initial data belonging to $H^1_\sigma \times H^1 \times H^1$ (the definition of H^1_σ will be given in part 2). Furthermore, in 2016, Ôtani-Uchida [18] was concerned with the solvability of the initial boundary value problem with unbounded domains. As for the results of the periodic solutions, see Refs. [16] and [17] and the references therein. We find that, in Refs. [16-18] and [25], they all used a linearized Brinkman-Forchheimer equation (originally, it has a convection term) in their study,

$$\partial_t \mathbf{u} = \nu \Delta \mathbf{u} - a\mathbf{u} - \nabla P + \mathbf{g}T + \mathbf{h}C + f_1, \qquad (1.2)$$

where $\mathbf{u}, T, C, P$ stand for the velocity of the fluid, the temperature of the fluid, the concentration of the solute, and the pressure of fluid, respectively. In (1.2), it ignores the convection term, which, however, will cause a crucial difficulty in the study of the global existence just like that of the Navier-Stokes equations. Here, in order to overcome the difficulties caused by the convection term, we use the method in Refs. [27] and [5]; that is, by the *a priori* assumption $\sup_{0 \leqslant t \leqslant T} \|U(t)\|_{\mathrm{L}^3} \leqslant \varepsilon$ (ε is a small constant), we get the global uniform *a priori* estimates of the solutions, then by the continuous theory, we conclude that the *a priori* assumption holds such that we get the closed global energy estimates.

The rest of the paper unfolds as follows: In Sec. 2, we give some necessary notations and some lemmas which will play an important role in the proof of the theorems; in Sec. 3, we show our main results in this paper; in Sec. 4, first, we prove the local well-posedness part of Theorem 3.1; that is, by the contraction mapping principle, we get the local well-posedness of the strong solution to the double-diffusion convection system (1.1) in $C([0, T_*]; H^1)$, and second, by the continuous theory, we extend the local solution to be a global one with the L^2 norm of the initial data being small. In detail, under the *a priori* assumption $\sup_{0 \leqslant t \leqslant T} \|U(t)\|_{\mathrm{L}^3} \leqslant \varepsilon$, we get the global uniform *a priori* estimates in H^1; next, we prove that the *a priori* assumption holds, which leads to the global uniform *a priori* estimates in H^1 hold; in Sec. 5, we try to obtain the uniform estimates of the derivatives of high order to the solution without any extra smallness condition expect (3.2), which finishes the proof of Theorem 3.2.

2 Preliminaries

Throughout this paper, we use the following notations. For simplicity, we denote

$$\int f = \int_{\mathbf{R}^3} f \, dx, \qquad \iint f = \iint_{\mathbf{Q}_t} f \, dx dt.$$

For $1 \leqslant p \leqslant \infty$, and $m \in \mathbf{N}$, the Sobolev spaces are defined in a standard manner.

$$\mathrm{L}^p \triangleq \mathrm{L}^p(\mathbf{R}^3), \quad \mathrm{H}^m \triangleq W^{m,2}(\mathbf{R}^3), \quad \mathbf{L}^P \triangleq (\mathrm{L}^P)^3, \quad \mathbf{H}^m \triangleq (\mathrm{H}^m)^3.$$

And we define

$$\mathbf{L}_\sigma^2 \triangleq \{f \in \mathbf{L}^2;\ \mathrm{div}\,f = 0\},$$
$$\mathbf{H}_\sigma^1 \triangleq \{f \in \mathbf{H}^1;\ \mathrm{div}\,f = 0\},$$
$$\vdots$$
$$\mathbf{H}_\sigma^N \triangleq \{f \in \mathbf{H}^N;\ \mathrm{div}\,f = 0\},$$

where $N \geqslant 2$ and div f means the divergence of f in the distribution sense. Then we find

$$\mathbf{H}_\sigma^N \hookrightarrow \cdots \hookrightarrow \mathbf{H}_\sigma^1 \hookrightarrow \mathbf{L}_\sigma^2, \quad \mathbf{H}^N \hookrightarrow \cdots \hookrightarrow \mathbf{H}^1 \hookrightarrow \mathbf{L}^2,$$

with continuous and locally dense embedding.

Given a Banach space X, we shall denote by $\|(a,b)\|_X = \|a\|_X + \|b\|_X$, especially,

$$\|(U,\theta,S)\|_{\mathbf{H}^\gamma} = \|U\|_{\mathbf{H}_\sigma^\gamma} + \|\theta\|_{\mathbf{H}^\gamma} + \|S\|_{\mathbf{H}^\gamma},$$

for $\gamma \geqslant 0$.

We also define

$$\partial_i f \triangleq \partial_{x_i} f,\ i=1,2,3,\quad \partial_i^2 f \triangleq \sum_{i=1}^3 \partial_{x_i}^2 f,\quad f^i \partial_i g \triangleq \sum_{i=1}^3 f^i \partial_{x_i} g.$$

For simplicity, we denote by C the positive constant, which may be different from line to line and may depend on time T, the constant μ, τ, λ, η, the regularity index N in $\mathbf{H}^N$, and the norm of the initial data.

In the sequel, we will frequently use the following interpolation inequality which can be found in Refs. [1] and [24]:

Lemma 2.1 *For $2 \leqslant p \leqslant \infty$ and $0 \leqslant \beta, \beta_1 \leqslant \beta_2$; when $p = \infty$, we require further that $\beta_1 \leqslant \beta+1$ and $\beta_2 \geqslant \beta+2$. Then we have that for any $f \in C_0^\infty(\mathbf{R}^3)$,*

$$\|\nabla^\beta f\|_{L^p} \leqslant \|\nabla^{\beta_1} f\|_{L^2}^{1-\gamma} \|\nabla^{\beta_2} f\|_{L^2}^\gamma,$$

where $0 \leqslant \gamma \leqslant 1$ and β satisfy

$$\frac{1}{p} - \frac{\beta}{3} = (1-\gamma)\left(\frac{1}{2} - \frac{\beta_1}{3}\right) + \gamma\left(\frac{1}{2} - \frac{\beta_2}{3}\right).$$

In order to get the continuity of the time of the solution, we introduce the following lemma, which is just Theorem 1.67 in Ref. [15]:

Lemma 2.2 Let H be a Hilbert space and $V \hookrightarrow H$ be dense in H. If $u, v \in L^p((a,b), V)$ with $a, b \in \mathbf{R}$, $a < b$, $1 < p < \infty$, and $u', v' \in L^q((a,b), V^*)$, $\frac{1}{p} + \frac{1}{q} = 1$, then $u, v \in C([a,b], H)$ and

$$(u(t), v(t)) - (u(s), v(s)) = \int_S^t (\langle u'(\tau), v(\tau) \rangle + \langle v'(\tau), u(\tau) \rangle) \, d\tau$$

(here, $\langle \cdot, \cdot \rangle$ is the duality between V and V^*).

Then let us introduce the well-known results on linear evolution equations (see, [3], [6] and [12]).

Lemma 2.3 If $U_0 \in \mathbf{H}_\sigma^1$, $\theta_0 \in H^1$, $S_0 \in H^1$, $G_1 \in L^2(0, T; \mathbf{L}_\sigma^2)$, $G_2 \in L^2(0, T; L^2)$, $G_3 \in L^2(0, T; L^2)$, then for the following problems

$$\begin{cases} \dfrac{\partial U}{\partial t} - \mu \Delta U = G_1, \\ \operatorname{div} U = 0, \\ U(x, 0) = U_0, \end{cases}$$

$$\begin{cases} \dfrac{\partial \theta}{\partial t} - \Delta \theta = G_2, \\ \theta(x, 0) = \theta_0, \end{cases}$$

$$\begin{cases} \dfrac{\partial S}{\partial t} - \tau \Delta S = G_3, \\ S(x, 0) = S_0, \end{cases}$$

there exists a unique solution U, θ and S, respectively, which satisfies

$$\begin{cases} U \in L^2(0, T; \mathbf{H}^2) \cap C([0, T]; \mathbf{H}_\sigma^1), & U_t \in L^2(0, T; \mathbf{L}_\sigma^2), \\ \theta \in L^2(0, T; H^2) \cap C([0, T]; H^1), & \theta_t \in L^2(0, T; L^2), \\ S \in L^2(0, T; H^2) \cap C([0, T]; H^1), & S_t \in L^2(0, T; L^2). \end{cases} \quad (2.1)$$

3   The main results

For any fixed time $T \in (0, \infty)$, one can now state the main results of the paper. The first result in this paper is the local well-posedness of problem (1.1):

Theorem 3.1 Assume the initial data $(U_0, \theta_0, S_0) \in \mathbf{H}_\sigma^1 \times H^1 \times H^1$, and then there exists a time $T_* \in (0, T]$ depending only on $\|U_0\|_{\mathbf{H}_\sigma^1}, \|\theta_0\|_{H^1}, \|S_0\|_{H^1}$ such that the problem (1.1) has a unique local strong solution (U, θ, S, P) in

$\mathbf{R}^3 \times [0, T_*]$ satisfying

$$\begin{cases} U \in L^2(0, T_*; \mathbf{H}^2) \cap C([0, T_*]; \mathbf{H}^1_\sigma), \\ \theta \in L^2(0, T_*; H^2) \cap C([0, T_*]; H^1), \\ S \in L^2(0, T_*; H^2) \cap C([0, T_*]; H^1), \\ \nabla P \in L^2(0, T_*; \mathbf{L}^2). \end{cases} \qquad (3.1)$$

Moreover, if the initial data (U_0, θ_0, S_0) satisfy

$$\|(U_0, \theta_0, S_0)\|_{L^2} \leqslant \varepsilon_1, \qquad (3.2)$$

where the constant ε_1 is small enough, then the unique local strong solution of the problem (1.1) can be extended to be a global one satisfying

$$\|(U, \theta, S)(t)\|_{H^1}^2 + \|(U, \theta, S)\|_{L^2(0,t;H^2)}^2 + \|(U_t, \theta_t, S_t)\|_{L^2(0,t;L^2)}^2$$
$$\leqslant C \|(U_0, \theta_0, S_0)\|_{H^1}^2, \quad \text{for any } t \in [0, T], \qquad (3.3)$$

where the constant C depends on $\mu, \lambda, \eta, \tau, T$.

Remark 3.1 We find from (3.2) that for the initial data $(U_0, \theta_0, S_0) \in H^1_\sigma \times H^1 \times H^1$, it only needs $\|(U_0, \theta_0, S_0)\|_{L^2}$ to be small, while $\|(\nabla U_0, \nabla \theta_0, \nabla S_0)\|_{L^2}$ can be arbitrarily large.

The second result in our paper is the global well-posedness of problem (1.1) in H^N ($N \geqslant 2$):

Theorem 3.2 If the initial data $(U_0, \theta_0, S_0) \in \mathbf{H}^N_\sigma \times H^N \times H^N$ ($N \geqslant 2$) satisfy (3.2), then there exists a unique global solution $(U, \theta, S, \nabla P)$ of the problem (1.1) satisfying

$$\begin{cases} U \in L^2(0, T; \mathbf{H}^{N+1}) \cap C([0, T]; \mathbf{H}^N_\sigma), \\ \theta \in L^2(0, T; H^{N+1}) \cap C([0, T]; H^N), \\ S \in L^2(0, T; H^{N+1}) \cap C([0, T]; H^N), \\ \nabla P \in L^\infty(0, T; \mathbf{H}^{N-2}) \cap L^2(0, T; \mathbf{H}^{N-1}), \end{cases} \qquad (3.4)$$

and

$$\|(U, \theta, S)(t)\|_{H^N}^2 + \|(U, \theta, S)\|_{L^2(0,t;H^{N+1})}^2$$
$$\leqslant C \|(U_0, \theta_0, S_0)\|_{H^N}^2, \quad \text{for any } t \in [0, T], \qquad (3.5)$$

where the constant C depends on $N, \mu, \lambda, \eta, \tau, T, \|(U_0, \theta_0, S_0)\|_{H^1}$.

Remark 3.2 In the result of the global well-posedness of the solution in Theorem 3.2, what we need to emphasize is that the smallness condition only depends on the L^2 norm of the initial data, without any dependence on the norm of the derivatives of the initial data.

Remark 3.3 Our results ameliorate that of Refs. [16-18] and [25] since we consider the convection term in the first equation of (1.1); meanwhile, our results not only consider the case in H^1 but also that in H^N for any $N \geqslant 2$.

4 The proof of Theorem 3.1

4.1 The proof of the local well-posedness part of Theorem 3.1

In this part, we focus on the proof of the local existence and uniqueness of the strong solution to the double-diffusive convection system (1.1). In order to use the results of the linear evolution equations Lemma 2.3 and contraction mapping principle to deal with system (1.1), we need to get rid of the influence of the unknown pressure term. Let us introduce the orthogonal projection $\mathbf{P}$ that projects $\mathbf{L}^2$ into the closed subspace $\mathbf{L}_\sigma^2$, which leads to that for all $f \in \mathbf{L}^2$, and there exists a unique orthogonal decomposition $f = f_1 + \nabla f_2$, where $f_1 \in \mathrm{H} \triangleq \mathbf{L}_\sigma^2$ and $\nabla f_2 \in \mathrm{H}^\perp \triangleq \{\nabla g;\ g \in \mathrm{L}_{\mathrm{loc}}^2, \nabla g \in \mathbf{L}^2\}$, which satisfy

$$\begin{cases} \hat{f}_1 = \hat{f} - |\xi|^{-2}(\xi \cdot \hat{f})\xi, \\ \hat{f}_2 = |\xi|^{-2} \xi \cdot \hat{f}. \end{cases}$$

By acting the orthogonal projection $\mathbf{P}$ on the momentum equation of (1.1), we consider the following system:

$$\begin{cases} \dfrac{\partial U}{\partial t} - \mu \Delta U + \mathbf{P}[U \cdot \nabla U] = \mathbf{P}[\mu(\lambda\theta - \eta S)\mathbf{K}], & \text{in } \mathbf{Q}_\mathrm{T}, \\ \dfrac{\partial \theta}{\partial t} - \Delta \theta + U \cdot \nabla \theta = w, & \text{in } \mathbf{Q}_\mathrm{T}, \\ \dfrac{\partial S}{\partial t} - \tau \Delta S + U \cdot \nabla S = w, & \text{in } \mathbf{Q}_\mathrm{T}, \\ \operatorname{div} U = 0, & \text{in } \mathbf{Q}_\mathrm{T}, \\ (U, \theta, S)(x,t)|_{t=0} = (U_0(x), \theta_0(x), S_0(x)), & \text{on } \mathbf{R}^3. \end{cases} \quad (4.1)$$

Now we have the following proposition:

Proposition 4.1 If the initial data $(U_0, \theta_0, S_0) \in \mathbf{H}_\sigma^1 \times \mathrm{H}^1 \times \mathrm{H}^1$, then there exists a unique local strong solution (U, θ, S) of system (4.1) satisfying (2.1) for some $\mathrm{T}_* \in (0, \mathrm{T}]$.

Proof Now we prove that there exists a unique local solution of the system (4.1) satisfying (2.1).

By Hölder's inequality and Sobolev's inequality, we find that, for any $\bar{U} \in \mathbf{H}^2$, $\bar{\theta} \in H^2, \bar{S} \in H^2$, there hold

$$\|\mathbf{P}[\bar{U} \cdot \nabla \bar{U}]\|_{\mathbf{L}_\sigma^2}^2 \leqslant \|\bar{U} \cdot \nabla \bar{U}\|_{\mathbf{L}^2}^2$$

$$\leqslant \|\bar{U}\|_{\mathbf{L}^6}^2 \|\nabla \bar{U}\|_{\mathbf{L}^3}^2 \leqslant \|\bar{U}\|_{\mathbf{L}^6}^2 \|\nabla \bar{U}\|_{\mathbf{L}^2} \|\nabla \bar{U}\|_{\mathbf{L}^6} \leqslant C\|\bar{U}\|_{\mathbf{H}^1}^3 \|\bar{U}\|_{\mathbf{H}^2},$$

$$\|\bar{U} \cdot \nabla \bar{\theta}\|_{L^2}^2 \leqslant \|\bar{U}\|_{\mathbf{L}^6}^2 \|\nabla \bar{\theta}\|_{L^3}^2$$

$$\leqslant \|\bar{U}\|_{\mathbf{L}^6}^2 \|\nabla \bar{\theta}\|_{L^2} \|\nabla \bar{\theta}\|_{L^6} \leqslant C\|\bar{U}\|_{\mathbf{H}^1}^2 \|\bar{\theta}\|_{H^1} \|\bar{\theta}\|_{H^2},$$

$$\|\bar{U} \cdot \nabla \bar{S}\|_{L^2}^2 \leqslant \|\bar{U}\|_{\mathbf{L}^6}^2 \|\nabla \bar{S}\|_{L^3}^2$$

$$\leqslant \|\bar{U}\|_{\mathbf{L}^6}^2 \|\nabla \bar{S}\|_{L^2} \|\nabla \bar{S}\|_{L^6} \leqslant C\|\bar{U}\|_{\mathbf{H}^1}^2 \|\bar{S}\|_{H^1} \|\bar{S}\|_{H^2}. \tag{4.2}$$

So, by Lemma 2.3, we obtain that for any

$$\bar{U} \in L^\infty(0, T; \mathbf{H}^1) \cap L^2(0, T; \mathbf{H}^2), \quad \bar{\theta}, \bar{S} \in L^\infty(0, T; H^1) \cap L^2(0, T; H^2),$$

the following system

$$\begin{cases} \dfrac{\partial U}{\partial t} - \mu \Delta U = -\mathbf{P}[\bar{U} \cdot \nabla \bar{U}] + \mathbf{P}[\mu(\lambda \bar{\theta} - \eta \bar{S})\mathbf{K}], & \text{in } \mathbf{Q}_T, \\ \dfrac{\partial \theta}{\partial t} - \Delta \theta = -\bar{U} \cdot \nabla \bar{\theta} + \bar{w}, & \text{in } \mathbf{Q}_T, \\ \dfrac{\partial S}{\partial t} - \tau \Delta S = -\bar{U} \cdot \nabla \bar{S} + \bar{w}, & \text{in } \mathbf{Q}_T, \\ \operatorname{div} U = 0, & \text{in } \mathbf{Q}_T, \\ (U, \theta, S)(x, t)|_{t=0} = (U_0(x), \theta_0(x), S_0(x)), & \text{on } \mathbf{R}^3, \end{cases} \tag{4.3}$$

has a unique solution (U, θ, S) satisfying (2.1). Now we can define a mapping J, $J(\bar{U}, \bar{\theta}, \bar{S}) = (U, \theta, S)$, and denote the set $\mathbf{M}$ as

$$\{(\bar{U}, \bar{\theta}, \bar{S}); \ \bar{U} \in C([0, t]; \mathbf{H}_\sigma^1) \cap L^2(0, t; \mathbf{H}^2), \ \bar{\theta}, \bar{S} \in C([0, t]; H^1) \cap L^2(0, t; H^2),$$

$$(\bar{U}, \bar{\theta}, \bar{S})(x, 0) = (\bar{U}_0(x), \bar{\theta}_0(x), \bar{S}_0(x)),$$

$$\|\bar{U}\|_{C([0,t];\mathbf{H}_\sigma^1)}, \|\bar{U}\|_{L^2([0,t];\mathbf{H}^2)}, \|\bar{\theta}\|_{C([0,t];H^1)}, \|\bar{\theta}\|_{L^2(0,t;H^2)},$$

$$\|\bar{S}\|_{C([0,t];H^1)}, \|\bar{S}\|_{L^2(0,t;H^2)} \leqslant B\},$$

where B and t will be determined later.

Next, noting that for any $(\bar{U}, \bar{\theta}, \bar{S}) \in \mathbf{M}$, by (4.2) and Hölder's inequality, one has

$$\|\mathbf{P}[\bar{U} \cdot \nabla \bar{U}]\|_{\mathbf{L}^2(0,t;\mathbf{L}_\sigma^2)}^2 + \|\bar{U} \cdot \nabla \bar{\theta}\|_{L^2(0,t;L^2)}^2 + \|\bar{U} \cdot \nabla \bar{S}\|_{L^2(0,t;L^2)}^2$$

$$\leqslant CB^4 t^{1/2},$$
$$\|\mathbf{P}[\mu(\lambda\bar{\theta}-\eta\bar{S})\mathbf{K}]\|^2_{L^2(0,t;L^2)} + \|\bar{w}\|^2_{L^2(0,t;L^2)}$$
$$\leqslant C\|(\bar{\theta},\bar{S})\|^2_{L^2(0,t;L^2)} + C\|\bar{w}\|^2_{L^2(0,t;L^2)}$$
$$\leqslant CB^2 t, \tag{4.4}$$

where the constant C may depend on μ, λ, η. We will prove that J is a contraction mapping on $\mathbf{M}$, if t is suitable small. Multiplying $(4.3)_1$ by U and then integrating over Q_t, we get by Hölder's inequality and Young's inequality that

$$\frac{1}{2}\int(|U(x,t)|^2 - |U_0(x)|^2) + \mu\iint|\nabla U|^2$$
$$\leqslant \frac{1}{2}\iint\left(\left|\mathbf{P}[\bar{U}\cdot\nabla\bar{U}]\right|^2 + \left|\mathbf{P}[\mu(\lambda\bar{\theta}-\eta\bar{S})\mathbf{K}]\right|^2\right) + \frac{1}{2}\iint|U|^2, \tag{4.5}$$

so, by Gronwall's inequality and noting that $t \in (0,T]$ is finite, we get by (4.4) that

$$\int|U(x,t)|^2 + \mu\iint|\nabla U|^2 \leqslant C\int|U_0(x)|^2 + C\left(B^4 t^{1/2} + B^2 t\right). \tag{4.6}$$

Multiplying $(4.3)_1$ by ΔU, then integrating over Q_t, we obtain

$$\int(|\nabla U(x,t)|^2 - |\nabla U_0(x)|^2) + \mu\iint|\Delta U|^2$$
$$\leqslant C\iint\left(\left|\mathbf{P}[\bar{U}\cdot\nabla\bar{U}]\right|^2 + \left|\mathbf{P}[\mu(\lambda\bar{\theta}-\eta\bar{S})\mathbf{K}]\right|^2\right) + \frac{\mu}{2}\iint|\Delta U|^2, \tag{4.7}$$

and then by (4.4), one has

$$\int|\nabla U(x,t)|^2 + \mu\iint|\Delta U|^2 \leqslant C\int|\nabla U_0(x)|^2 + C\left(B^4 t^{1/2} + B^2 t\right). \tag{4.8}$$

Similarly, we have

$$\int|\theta(x,t)|^2 + \iint|\nabla\theta|^2 \leqslant C\int|\theta_0(x)|^2 + C\left(B^4 t^{1/2} + B^2 t\right), \tag{4.9}$$

$$\int|\nabla\theta(x,t)|^2 + \iint|\Delta\theta|^2 \leqslant \int|\nabla\theta_0(x)|^2 + C\left(B^4 t^{1/2} + B^2 t\right), \tag{4.10}$$

and

$$\int|S(x,t)|^2 + \tau\iint|\nabla S|^2 \leqslant C\int|S_0(x)|^2 + C\left(B^4 t^{1/2} + B^2 t\right), \tag{4.11}$$

$$\int |\nabla S(x,t)|^2 + \tau \iint |\Delta S|^2 \leqslant \int |\nabla S_0(x)|^2 + C\left(B^4 t^{1/2} + B^2 t\right). \quad (4.12)$$

Summing up (4.6)-(4.12), one has

$$\|U\|^2_{C([0,t];\mathbf{H}^1_\sigma)} + \|U\|^2_{L^2(0,t;\mathbf{H}^2)} + \|\theta\|^2_{C([0,t];H^1)}$$
$$+ \|\theta\|^2_{L^2(0,t;H^2)} + \|S\|^2_{C([0,t];H^1)} + \|S\|^2_{L^2(0,t;H^2)}$$
$$\leqslant C_0 \left\{ \|U_0\|^2_{\mathbf{H}^1_\sigma} + \|\theta_0\|^2_{H^1} + \|S_0\|^2_{H^1} + B^4 t^{1/2} + B^2 t \right\}.$$

By choosing

$$B^2 = 3C_0 \left\{ \|U_0\|^2_{\mathbf{H}^1_\sigma} + \|\theta_0\|^2_{H^1} + \|S_0\|^2_{H^1} \right\},$$

and

$$t \leqslant T_1 \triangleq \min\left\{T, \frac{1}{(3C_0 B^2)^2}, \frac{1}{3C_0}\right\},$$

we get

$$J(\mathbf{M}) \subset \mathbf{M}.$$

Let $(\bar{U}_1, \bar{\theta}_1, \bar{S}_1), (\bar{U}_2, \bar{\theta}_2, \bar{S}_2) \in \mathbf{M}$, and the corresponding solutions $(U_1, \theta_1, S_1) = J(\bar{U}_1, \bar{\theta}_1, \bar{S}_1), (U_2, \theta_2, S_2) = J(\bar{U}_2, \bar{\theta}_2, \bar{S}_2)$. Define $(\delta U, \delta \theta, \delta S) = (U_1 - U_2, \theta_1 - \theta_2, S_1 - S_2)$. Then

$$\begin{cases} \dfrac{\partial \delta U}{\partial t} - \mu \Delta \delta U \\ \qquad = -\mathbf{P}[\bar{U}_1 \cdot \nabla \bar{U}_1 - \bar{U}_2 \cdot \nabla \bar{U}_2] + \mu \mathbf{P}[(\lambda(\bar{\theta}_1 - \bar{\theta}_2) - \eta(\bar{S}_1 - \bar{S}_2))\mathbf{K}], \\ \dfrac{\partial \delta \theta}{\partial t} - \Delta \delta \theta = -(\bar{U}_1 \cdot \nabla \bar{\theta}_1 - \bar{U}_2 \cdot \nabla \bar{\theta}_2) + (\bar{w}_1 - \bar{w}_2), \\ \dfrac{\partial \delta S}{\partial t} - \tau \Delta \delta S = -(\bar{U}_1 \cdot \nabla \bar{S}_1 - \bar{U}_2 \cdot \nabla \bar{S}_2) + (\bar{w}_1 - \bar{w}_2), \\ \operatorname{div} \delta U = \operatorname{div} U_1 = \operatorname{div} U_2 = 0, \\ (\delta U, \delta \theta, \delta S)(x,t)|_{t=0} = (0,0,0). \end{cases} \quad (4.13)$$

Just like the inequalities (4.2), we can get

$$\|\mathbf{P}[\bar{U}_1 \cdot \nabla \bar{U}_1 - \bar{U}_2 \cdot \nabla \bar{U}_2]\|^2_{\mathbf{L}^2_\sigma}$$
$$\leqslant \|\mathbf{P}[(\bar{U}_1 - \bar{U}_2) \cdot \nabla \bar{U}_1]\|^2_{\mathbf{L}^2_\sigma} + \|\mathbf{P}[\bar{U}_2 \cdot \nabla(\bar{U}_1 - \bar{U}_2)]\|^2_{\mathbf{L}^2_\sigma}$$
$$\leqslant C\|\bar{U}_1 - \bar{U}_2\|_{\mathbf{H}^1}\|\bar{U}_1\|_{\mathbf{H}^1}\|\bar{U}_1\|_{\mathbf{H}^2} + C\|\bar{U}_2\|^2_{\mathbf{H}^1}\|\bar{U}_1 - \bar{U}_2\|_{\mathbf{H}^1}\|\bar{U}_1 - \bar{U}_2\|_{\mathbf{H}^2},$$
$$\|\bar{U}_1 \cdot \nabla \bar{\theta}_1 - \bar{U}_2 \cdot \nabla \bar{\theta}_2\|^2_{L^2}$$
$$\leqslant \|(\bar{U}_1 - \bar{U}_2) \cdot \nabla \bar{\theta}_1\|^2_{\mathbf{L}^2} + \|\bar{U}_2 \cdot \nabla(\bar{\theta}_1 - \bar{\theta}_2)\|^2_{\mathbf{L}^2}$$

$$\leqslant C\|\bar{U}_1 - \bar{U}_2\|_{\mathbf{H}^1}^2 \|\bar{\theta}_1\|_{H^1}\|\bar{\theta}_1\|_{H^2} + C\|\bar{U}_2\|_{\mathbf{H}^1}^2 \|\bar{\theta}_1 - \bar{\theta}_2\|_{H^1}\|\bar{\theta}_1 - \bar{\theta}_2\|_{H^2},$$
$$\|\bar{U}_1 \cdot \nabla \bar{S}_1 - \bar{U}_2 \cdot \nabla \bar{S}_2\|_{L^2}^2$$
$$\leqslant \|(\bar{U}_1 - \bar{U}_2) \cdot \nabla \bar{S}_1\|_{L^2}^2 + \|\bar{U}_2 \cdot \nabla(\bar{S}_1 - \bar{S}_2)\|_{L^2}^2$$
$$\leqslant C\|\bar{U}_1 - \bar{U}_2\|_{\mathbf{H}^1}^2 \|\bar{S}_1\|_{H^1}\|\bar{S}_1\|_{H^2} + C\|\bar{U}_2\|_{\mathbf{H}^1}^2 \|\bar{S}_1 - \bar{S}_2\|_{H^1}\|\bar{S}_1 - \bar{S}_2\|_{H^2},$$

and Hölder's inequality leads to

$$\|\mathbf{P}[\bar{U}_1 \cdot \nabla \bar{U}_1 - \bar{U}_2 \cdot \nabla \bar{U}_2]\|_{L^2(0,t;\mathbf{L}_\sigma^2)}^2$$
$$\leqslant C\|\bar{U}_1 - \bar{U}_2\|_{C([0,t];\mathbf{H}_\sigma^1)}^2 \|\bar{U}_1\|_{C([0,t];\mathbf{H}_\sigma^1)} \|\bar{U}_1\|_{L^1(0,t;\mathbf{H}^2)}$$
$$+ C\|\bar{U}_2\|_{C([0,t];\mathbf{H}_\sigma^1)}^2 \|\bar{U}_1 - \bar{U}_2\|_{C([0,t];\mathbf{H}_\sigma^1)} \|\bar{U}_1 - \bar{U}_2\|_{L^1(0,t;\mathbf{H}^2)},$$
$$\leqslant CB^2 t^{1/2} \left\{\|\bar{U}_1 - \bar{U}_2\|_{C([0,t];\mathbf{H}_\sigma^1)}^2 + \|\bar{U}_1 - \bar{U}_2\|_{C([0,t];\mathbf{H}_\sigma^1)} \|\bar{U}_1 - \bar{U}_2\|_{L^2(0,t;\mathbf{H}^2)}\right\},$$
$$\|\bar{U}_1 \cdot \nabla \bar{\theta}_1 - \bar{U}_2 \cdot \nabla \bar{\theta}_2\|_{L^2(0,t;L^2)}^2$$
$$\leqslant C\|\bar{U}_1 - \bar{U}_2\|_{C([0,t];\mathbf{H}_\sigma^1)}^2 \|\bar{\theta}_1\|_{C([0,t];H^1)} \|\bar{\theta}_1\|_{L^1(0,t;H^2)}$$
$$+ C\|\bar{U}_2\|_{C([0,t];\mathbf{H}_\sigma^1)}^2 \|\bar{\theta}_1 - \bar{\theta}_2\|_{C([0,t];H^1)} \|\bar{\theta}_1 - \bar{\theta}_2\|_{L^1(0,t;H^2)},$$
$$\leqslant CB^2 t^{1/2} \left\{\|\bar{U}_1 - \bar{U}_2\|_{C([0,t];\mathbf{H}_\sigma^1)}^2 + \|\bar{\theta}_1 - \bar{\theta}_2\|_{C([0,t];H^1)} \|\bar{\theta}_1 - \bar{\theta}_2\|_{L^2(0,t;H^2)}\right\},$$
$$\|\bar{U}_1 \cdot \nabla \bar{S}_1 - \bar{U}_2 \cdot \nabla \bar{S}_2\|_{L^2(0,t;L^2)}^2$$
$$\leqslant C\|\bar{U}_1 - \bar{U}_2\|_{C([0,t];\mathbf{H}_\sigma^1)}^2 \|\bar{S}_1\|_{C([0,t];H^1)} \|\bar{S}_1\|_{L^1(0,t;H^2)}$$
$$+ C\|\bar{U}_2\|_{C([0,t];\mathbf{H}_\sigma^1)}^2 \|\bar{S}_1 - \bar{S}_2\|_{C([0,t];H^1)} \|\bar{S}_1 - \bar{S}_2\|_{L^1(0,t;H^2)},$$
$$\leqslant CB^2 t^{1/2} \left\{\|\bar{U}_1 - \bar{U}_2\|_{C([0,t];\mathbf{H}_\sigma^1)}^2 + \|\bar{S}_1 - \bar{S}_2\|_{C([0,t];H^1)} \|\bar{S}_1 - \bar{S}_2\|_{L^2(0,t;H^2)}\right\}.$$
$$(4.14)$$

And

$$\|\mu\mathbf{P}[(\lambda(\bar{\theta}_1 - \bar{\theta}_2) - \eta(\bar{S}_1 - \bar{S}_2))\mathbf{K}]\|_{L^2(0,t;L^2)}^2 + \|\bar{w}_1 - \bar{w}_2\|_{L^2(0,t;L^2)}^2$$
$$\leqslant Ct \left\{\|\bar{\theta}_1 - \bar{\theta}_2\|_{C([0,t];L^2)}^2 + \|\bar{S}_1 - \bar{S}_2\|_{C([0,t];L^2)}^2 + \|\bar{w}_1 - \bar{w}_2\|_{C([0,t];L^2)}^2\right\}. \quad (4.15)$$

Similar to (4.5), (4.7) and according to (4.13)-(4.15), we have

$$\|\delta U\|_{C([0,t];\mathbf{H}_\sigma^1)}^2 + \|\delta U\|_{L^2(0,t;\mathbf{H}^2)}^2 + \|(\delta\theta, \delta S)\|_{C([0,t];H^1)}^2 + \|(\delta\theta, \delta S)\|_{L^2(0,t;H^2)}^2$$
$$\leqslant \|\mathbf{P}[\bar{U}_1 \cdot \nabla \bar{U}_1 - \bar{U}_2 \cdot \nabla \bar{U}_2]\|_{L^2(0,t;\mathbf{L}_\sigma^2)}^2$$
$$+ \|\bar{U}_1 \cdot \nabla \bar{\theta}_1 - \bar{U}_2 \cdot \nabla \bar{\theta}_2\|_{L^2(0,t;L^2)}^2 + \|\bar{w}_1 - \bar{w}_2\|_{L^2(0,t;L^2)}^2$$
$$+ \|\bar{U}_1 \cdot \nabla \bar{S}_1 - \bar{U}_2 \cdot \nabla \bar{S}_2\|_{L^2(0,t;L^2)}^2$$
$$+ \|\mu\mathbf{P}[(\lambda(\bar{\theta}_1 - \bar{\theta}_2) - \eta(\bar{S}_1 - \bar{S}_2))\mathbf{K}]\|_{L^2(0,t;L^2)}^2$$

$$\leqslant CB^2 t^{1/2} \left\{ \|\bar{U}_1 - \bar{U}_2\|_{C([0,t];\mathbf{H}^1_\sigma)}^2 + \|\bar{U}_1 - \bar{U}_2\|_{L^2(0,t;\mathbf{H}^2)}^2 + \|\bar{\theta}_1 - \bar{\theta}_2\|_{C([0,t];H^1)}^2 \right.$$
$$\left. + \|\bar{\theta}_1 - \bar{\theta}_2\|_{L^2(0,t;H^2)}^2 + \|\bar{S}_1 - \bar{S}_2\|_{C([0,t];H^1)}^2 + \|\bar{S}_1 - \bar{S}_2\|_{L^2(0,t;H^2)}^2 \right\}$$
$$+ Ct \left\{ \|\bar{\theta}_1 - \bar{\theta}_2\|_{C([0,t];L^2)}^2 + \|\bar{S}_1 - \bar{S}_2\|_{C([0,t];L^2)}^2 + \|\bar{w}_1 - \bar{w}_2\|_{C([0,t];L^2)}^2 \right\}$$
$$\leqslant \zeta \left\{ \|\bar{U}_1 - \bar{U}_2\|_{C([0,t];\mathbf{H}^1_\sigma)}^2 + \|\bar{U}_1 - \bar{U}_2\|_{L^2(0,t;\mathbf{H}^2)}^2 + \|\bar{\theta}_1 - \bar{\theta}_2\|_{C([0,t];H^1)}^2 \right.$$
$$\left. + \|\bar{\theta}_1 - \bar{\theta}_2\|_{L^2(0,t;H^2)}^2 + \|\bar{S}_1 - \bar{S}_2\|_{C([0,t];H^1)}^2 + \|\bar{S}_1 - \bar{S}_2\|_{L^2(0,t;H^2)}^2 \right\},$$

where $0 < \zeta < 1$ by choosing $t \leqslant T_* \triangleq \min\{T_1, 1/(4C), 1/(2CB^2)^2\}$. So we conclude that J is a contraction mapping on $\mathbf{M}$. According to the contraction mapping principle, we obtain that there exists a unique local solution of system (4.1) on $[0, T_*]$.

Now we finish the proof of Theorem 3.1. Let (U, θ, S) be the solution of system (4.1). Define

$$\Omega = U_t + U \cdot \nabla U - \mu \Delta U - \mu[\lambda\theta - \eta S]\mathbf{K},$$

then $(4.1)_1$ shows $\mathbf{P}\Omega = 0$, and (2.1) leads to $\Omega \in L^2(0, T_*; L^2)$. So, there exists a unique ∇P satisfying $\Omega = -\nabla P$ such that (1.1) and (3.1) hold.

Here we have finished the proof of the local well-posedness part of Theorem 3.1.

4.2 The proof of the global well-posedness part of Theorem 3.1

Once we obtain the local well-posedness of (1.1) in $\mathbf{H}^1$, the aim in this section is to extend the local strong solution to be a global one. The key ingredient is to get the global uniform *a priori* estimates of the solution on $[0, T]$ for any fixed $T \in (0, \infty)$. That is, we will get the uniform *a priori* estimates of $\|(U, \theta, S)\|_{L^\infty(0,t;L^2)}$, $\|(\nabla U, \nabla\theta, \nabla S)\|_{L^\infty(0,t;L^2)}$, for any $t \in [0, T]$. To do this, first, we suppose that the following *a priori* hypothesis holds:

$$\sup_{0 \leqslant t \leqslant L} \|U(t)\|_{L^3} \leqslant \varepsilon, \tag{4.16}$$

where ε is a constant to be determined later.

Step 1 Taking L^2 inner product of $(1.1)_1$ with U, $(1.1)_2$ with θ, $(1.1)_3$ with S, integrating by parts over $\mathbf{R}^3$, and summing up the resulting equations, we get

$$\frac{d}{dt} \int |(U, \theta, S)|^2 + \int |(\nabla U, \nabla\theta, \nabla S)|^2$$

$$\leqslant C\left\{\left|\int (U\cdot\nabla)U\cdot U + (U\cdot\nabla)\theta\cdot\theta + (U\cdot\nabla)S\cdot S\right.\right.$$
$$\left.\left.+\int(\theta-S)w+\int w\theta+\int wS\right|\right\}. \tag{4.17}$$

Integrating by parts over $\mathbf{R}^3$, noting $\operatorname{div} U = 0$, by (4.17), Hölder's inequality, and Gronwall's inequality, one has

$$\|(U,\theta,S)(t)\|_{L^2}^2 + \|(\nabla U,\nabla\theta,\nabla S)\|_{L^2(0,t;L^2)}^2$$
$$\leqslant C(\mathrm{T})\|(U_0,\theta_0,S_0)\|_{L^2}^2, \quad \text{for any } t\in[0,\mathrm{T}]. \tag{4.18}$$

Step 2 Taking L^2 inner product of $(1.1)_1$ with $U_t - \Delta U$ and integrating by parts over $\mathbf{R}^3$, we get by Hölder's inequality and Young's inequality that

$$\frac{\mathrm{d}}{\mathrm{d}t}\int|\nabla U|^2 + \int(|U_t|^2+|\Delta U|^2)$$
$$\leqslant C\left\{\left|\int(U\cdot\nabla)U\cdot\Delta U + (U\cdot\nabla)U\cdot U_t + \int(\theta-S)\mathbf{K}\cdot(U_t-\Delta U)\right|\right\}$$
$$\leqslant C\{\|U\|_{L^3}\|\nabla U\|_{L^6}(\|U_t\|_{L^2}+\|\Delta U\|_{L^2})$$
$$+(\|\theta\|_{L^2}+\|S\|_{L^2})(\|U_t\|_{L^2}+\|\Delta U\|_{L^2})\}$$
$$\leqslant C\|U\|_{L^3}(\|U_t\|_{L^2}^2+\|\Delta U\|_{L^2}^2)$$
$$+C(\|\theta\|_{L^2}^2+\|S\|_{L^2}^2)+\frac{1}{12}(\|U_t\|_{L^2}^2+\|\Delta U\|_{L^2}^2)$$
$$\leqslant \left(\frac{1}{12}+C\|U\|_{L^3}\right)(\|U_t\|_{L^2}^2+\|\Delta U\|_{L^2}^2)+C(\|\theta\|_{L^2}^2+\|S\|_{L^2}^2). \tag{4.19}$$

In similar way, we get

$$\frac{\mathrm{d}}{\mathrm{d}t}\int|\nabla\theta|^2+\int(|\theta_t|^2+|\Delta\theta|^2)$$
$$\leqslant\left(\frac{1}{12}+C\|U\|_{L^3}\right)(\|\theta_t\|_{L^2}^2+\|\Delta\theta\|_{L^2}^2)+C\|w\|_{L^2}^2, \tag{4.20}$$

$$\frac{\mathrm{d}}{\mathrm{d}t}\int|\nabla S|^2+\int(|S_t|^2+|\Delta S|^2)$$
$$\leqslant\left(\frac{1}{12}+C\|U\|_{L^3}\right)(\|S_t\|_{L^2}^2+\|\Delta S\|_{L^2}^2)+C\|w\|_{L^2}^2. \tag{4.21}$$

Summing up (4.19)-(4.21), we arrive at

$$\frac{\mathrm{d}}{\mathrm{d}t}\|(\nabla U,\nabla\theta,\nabla S)(t)\|_{L^2}^2+\|(U_t,\theta_t,S_t,\Delta U,\Delta\theta,\Delta S)\|_{L^2(0,t;L^2)}^2$$

$$\leqslant \left(\frac{1}{4} + C\|U\|_{L^3}\right) \|(U_t, \theta_t, S_t, \Delta U, \Delta\theta, \Delta S)\|^2_{L^2(0,t;L^2)} + C\|(U, \theta, S)\|^2_{L^2},$$

and then, according to (4.18) and choosing ε in (4.16) satisfying

$$\varepsilon \leqslant \frac{1}{4C}, \qquad (4.22)$$

one has

$$\|(\nabla U, \nabla\theta, \nabla S)(t)\|^2_{L^2} + \|(U_t, \theta_t, S_t, \Delta U, \Delta\theta, \Delta S)\|^2_{L^2(0,t;L^2)}$$
$$\leqslant \|(\nabla U_0, \nabla\theta_0, \nabla S_0)\|^2_{L^2} + C(T)\|(U_0, \theta_0, S_0)\|^2_{L^2}, \quad \text{for any } t \in [0, T]. \qquad (4.23)$$

Step 3 Noting that the deduction of (4.23) is based on the *a priori* assumptions of (4.16), so we need to prove that (4.16) holds, provided the initial data are suitably small in some sense.

Define

$$T_{\max} \triangleq \sup\{t \in [0, T] : \|U(t)\|_{L^3} \leqslant \varepsilon\}, \text{ where } \varepsilon \text{ is determined by (4.22)}.$$

By Lemma 2.1, we get

$$\|U_0\|_{L^3} \leqslant C_1 \|U_0\|^{1/2}_{L^2} \|\nabla U_0\|^{1/2}_{L^2} \leqslant \frac{\varepsilon}{2},$$

provided

$$\|U_0\|^{1/2}_{L^2} \|\nabla U_0\|^{1/2}_{L^2} \leqslant \frac{\varepsilon}{2C_1}, \qquad (4.24)$$

where $C_1 > 0$ is a constant. By continuity, we have $0 < T_{\max} \leqslant T$ and

$$\|U(t)\|_{L^3} \leqslant \varepsilon, \quad 0 \leqslant t < T_{\max}, \quad \|U(T_{\max})\|_{L^3} = \varepsilon. \qquad (4.25)$$

Next, we shall prove that $T_{\max} = T$ under the condition (3.2). Otherwise, if $T_{\max} < T$, considering the initial data $(U_0, \theta_0, S_0) \in H^1_\sigma \times H^1 \times H^1$, then we deduce from (4.18) and (4.23) that

$$\|U(t)\|_{L^3} \leqslant C \|U(t)\|^{1/2}_{L^2} \|\nabla U(t)\|^{1/2}_{L^2}$$
$$\leqslant C\{C(T)\|(U_0, \theta_0, S_0)\|_{L^2}\}^{1/2}$$
$$\times \{C(T)\|(U_0, \theta_0, S_0)\|^{1/2}_{L^2} + \|(\nabla U_0, \nabla\theta_0, \nabla S_0)\|^{1/2}_{L^2}\}$$
$$\leqslant C(T)\{\|(U_0, \theta_0, S_0)\|_{L^2} + \|(U_0, \theta_0, S_0)\|^{1/2}_{L^2} \|(\nabla U_0, \nabla\theta_0, \nabla S_0)\|^{1/2}_{L^2}\}$$
$$\leqslant C(T)\{\|(U_0, \theta_0, S_0)\|_{L^2} + \|(U_0, \theta_0, S_0)\|^{1/2}_{L^2} \cdot C(\|(U_0, \theta_0, S_0)\|_{H^1})^{\frac{1}{2}}\}.$$

Hence
$$\|U(t)\|_{L^3} \leqslant \frac{\varepsilon}{2}, \quad 0 \leqslant t \leqslant T_{\max},$$
provided the initial data satisfy (4.24) and
$$\|(U_0, \theta_0, S_0)\|_{L^2} \leqslant \min\left\{\frac{\varepsilon}{4C(T)}, \frac{\varepsilon^2}{16C(T)^2 C(\|(U_0, \theta_0, S_0)\|_{H^1})}\right\},$$
which obviously contradicts (4.25). So, by choosing
$$\varepsilon_1 = \min\left\{\frac{\varepsilon}{2C_1}, \frac{\varepsilon}{4C(T)}, \frac{\varepsilon^2}{16C(T)^2 C(\|(U_0, \theta_0, S_0)\|_{H^1})}\right\}$$
in (3.2), we get $T_{\max} = T$, which means that (4.16) holds. Then (4.18) and (4.23) all hold on $[0, T]$, which leads to (3.3). Therefore, we have finished the proof of Theorem 3.1.

5 The proof of the global existence part of Theorem 3.2

In this part, we aim at finishing the proof of Theorem 3.2. Once we establish the global solution in H^1, it is enough to get the global solution in H^N by a standard compactness principle, as long as we get the global *a priori* estimates of the k-order derivatives for $2 \leqslant k \leqslant N$. Now we introduce Lemma 5.1 which is a key ingredient in proving Theorem 3.2.

Lemma 5.1 *Let* $k = 2, 3, \cdots, N$ $(N \geqslant 2)$, *and then for any* $t \in [0, T]$, *there exists a constant* C *depending on* $N, \mu, \lambda, \eta, \tau, T, \|(U, \theta, S)\|_{L^\infty(0,T;H^1)}$ *such that*

$$\frac{d}{dt}\|(\nabla^k U, \nabla^k \theta, \nabla^k S)\|_{L^2}^2 + \|(\nabla^{k+1} U, \nabla^{k+1}\theta, \nabla^{k+1} S)\|_{L^2}^2$$
$$\leqslant C\|(\nabla^k U, \nabla^k \theta, \nabla^k S)\|_{L^2}^2. \tag{5.1}$$

Proof Taking k-th spatial derivatives to (1.1), multiplying the resultant equation by $\nabla^k U$ and integrating over $\mathbf{R}^3$ (by parts), we obtain

$$\frac{1}{2}\frac{d}{dt}\int |\nabla^k U|^2 + \mu \int |\nabla^{k+1} U|^2$$
$$= \int \nabla^k (U \cdot \nabla U) \, \nabla^k U + \int \nabla^k [\mu(\lambda\theta - \eta S)\mathbf{K}] \, \nabla^k U,$$

that is,
$$\frac{d}{dt}\int |\nabla^k U|^2 + \int |\nabla^{k+1} U|^2$$

$$\leqslant C(\mu,\lambda,\eta)\left\{\left|\int \nabla^k(U\cdot\nabla U)\cdot\nabla^k U\right| + \int(|\nabla^k\theta| + |\nabla^k S|)\cdot|\nabla^k U|\right\}$$
$$\triangleq C(\mu,\lambda,\eta)\{I_1 + J_1\}. \tag{5.2}$$

Similarly, we get

$$\frac{d}{dt}\int |\nabla^k\theta|^2 + \int |\nabla^{k+1}\theta|^2$$
$$\leqslant C\left\{\left|\int \nabla^k(U\cdot\nabla\theta)\cdot\nabla^k\theta\right| + \int |\nabla^k w||\nabla^k\theta|\right\}$$
$$\triangleq C\{I_2 + J_2\}, \tag{5.3}$$

$$\frac{d}{dt}\int |\nabla^k S|^2 + \int |\nabla^{k+1}S|^2$$
$$\leqslant C(\tau)\left\{\left|\int \nabla^k(U\cdot\nabla S)\cdot\nabla^k S\right| + \int |\nabla^k w||\nabla^k S|\right\}$$
$$\triangleq C(\tau)\{I_3 + J_3\}. \tag{5.4}$$

First of all, we deal with I_1. In fact, applying the Leibnitz formula, integrating by parts and noting $\operatorname{div} U = 0$, and then by Hölder's inequality and Sobolev's inequality, it arrives at

$$I_1 = \left|\int \nabla^k(U\cdot\nabla U)\cdot\nabla^k U\right|$$
$$\leqslant \left|\int \sum_{l=0}^k C_k^l\,(\nabla^l U\cdot\nabla\nabla^{k-l}U)\,\nabla^k U\right|$$
$$\leqslant \left|\int (U\cdot\nabla)\nabla^k U\cdot\nabla^k U\right| + C(k)\sum_{l=1}^k \int |\nabla^l U||\nabla^{k+1-l}U||\nabla^k U|$$
$$= \frac{1}{2}\int |\operatorname{div} U|\,|\nabla^k U|^2 + C(k)\sum_{l=1}^k \int |\nabla^l U||\nabla^{k+1-l}U||\nabla^k U|$$
$$\leqslant C(k)\sum_{l=1}^k \|\nabla^l U\|_{L^3}\|\nabla^{k+1-l}U\|_{L^2}\|\nabla^{k+1}U\|_{L^2}. \tag{5.5}$$

For the case $1 \leqslant l \leqslant [k/2]$, by virtue of the interpolation inequality Lemma 2.1 and Young's inequality, one has

$$\|\nabla^l U\|_{L^3}\|\nabla^{k+1-l}U\|_{L^2}\|\nabla^{k+1}U\|_{L^2}$$
$$\leqslant \|\nabla^\alpha U\|_{L^2}^{1-\frac{l}{k}}\cdot\|\nabla^k U\|_{L^2}^{\frac{l}{k}}\|\nabla U\|_{L^2}^{\frac{l}{k}}\|\nabla^{k+1}U\|_{L^2}^{1-\frac{l}{k}}\|\nabla^{k+1}U\|_{L^2}$$

$$\leqslant \|\nabla^\alpha U\|_{L^2}^{\frac{2k}{l}-2} \|\nabla U\|_{L^2}^2 \|\nabla^k U\|_{L^2}^2 + \frac{1}{12}\|\nabla^{k+1} U\|_{L^2}^2, \tag{5.6}$$

where α is defined by

$$\alpha = \frac{k}{2(k-l)} \in \left(\frac{1}{2},1\right],$$

and for the case $[k/2]+1 \leqslant l \leqslant k$, we have

$$\|\nabla^l U\|_{L^3} \|\nabla^{k+1-l} U\|_{L^2} \|\nabla^{k+1} U\|_{L^2}$$
$$\leqslant \|\nabla U\|_{L^2}^{1-\frac{2l-1}{2k}} \|\nabla^{k+1} U\|_{L^2}^{\frac{2l-1}{2k}} \|\nabla^\alpha U\|_{L^2}^{\frac{2l-1}{2k}} \|\nabla^k U\|_{L^2}^{1-\frac{2l-1}{2k}} \|\nabla^{k+1} U\|_{L^2}$$
$$\leqslant \|\nabla U\|_{L^2}^{1-\frac{2l-1}{2k}} \|\nabla^\alpha U\|_{L^2}^{\frac{2l-1}{2k}} \|\nabla^k U\|_{L^2}^{1-\frac{2l-1}{2k}} \|\nabla^{k+1} U\|_{L^2}^{1+\frac{2l-1}{2k}}$$
$$\leqslant \|\nabla^\alpha U\|_{L^2}^{\frac{2(2l-1)}{2k-2l+1}} \|\nabla U\|_{L^2}^2 \|\nabla^k U\|_{L^2}^2 + \frac{1}{12}\|\nabla^{k+1} U\|_{L^2}^2, \tag{5.7}$$

where α is defined by

$$\alpha = \frac{k}{2l-1} \in \left(\frac{1}{2},1\right],$$

and then we get by substituting (5.6) and (5.7) into (5.5) that

$$I_1 \leqslant C(k)\|U\|_{H^1}^{\varsigma_1} \|\nabla^k U\|_{L^2}^2 + \frac{1}{6}\|\nabla^{k+1} U\|_{L^2}^2, \tag{5.8}$$

where ς_1 is a constant depending on k.

Similarly, we deal with I_2.

$$I_2 = \left|\int \nabla^k (U \cdot \nabla \theta) \cdot \nabla^k \theta\right|$$
$$\leqslant \left|\int \sum_{l=0}^k C_k^l \nabla^l U \nabla^{k+1-l}\theta \nabla^k \theta\right|$$
$$\leqslant \left|\int (U \cdot \nabla)\nabla^k \theta \cdot \nabla^k \theta\right| + C(k)\sum_{l=1}^k \int |\nabla^l U| |\nabla^{k+1-l}\theta| |\nabla^k \theta|$$
$$= \frac{1}{2}\int |\operatorname{div} U| |\nabla^k \theta|^2 + C(k)\sum_{l=1}^k \int |\nabla^l U| |\nabla^{k+1-l}\theta| |\nabla^k \theta|$$
$$\leqslant C(k)\sum_{l=1}^k \|\nabla^l U\|_{L^3} \|\nabla^{k+1-l}\theta\|_{L^2} \|\nabla^{k+1}\theta\|_{L^2}. \tag{5.9}$$

For the case $1 \leqslant l \leqslant [k/2]$, one has

$$\|\nabla^l U\|_{L^3} \|\nabla^{k+1-l}\theta\|_{L^2} \|\nabla^{k+1}\theta\|_{L^2}$$

$$\leqslant \|\nabla^\alpha U\|_{L^2}^{1-\frac{l}{k}} \|\nabla^k U\|_{L^2}^{\frac{l}{k}} \|\nabla\theta\|_{L^2}^{\frac{l}{k}} \|\nabla^{k+1}\theta\|_{L^2}^{1-\frac{l}{k}} \|\nabla^{k+1}\theta\|_{L^2}$$

$$\leqslant \|\nabla^\alpha U\|_{L^2}^{\frac{2k}{l}-2} \|\nabla\theta\|_{L^2}^2 \|\nabla^k U\|_{L^2}^2 + \frac{1}{12}\|\nabla^{k+1}\theta\|_{L^2}^2, \qquad (5.10)$$

where α is defined by

$$\alpha = \frac{k}{2(k-l)} \in \left(\frac{1}{2}, 1\right],$$

and for the case $[k/2]+1 \leqslant l \leqslant k$, we have

$$\|\nabla^l U\|_{L^3} \|\nabla^{k+1-l}\theta\|_{L^2} \|\nabla^{k+1}\theta\|_{L^2}$$

$$\leqslant \|\nabla U\|_{L^2}^{1-\frac{2l-1}{2k}} \|\nabla^{k+1} U\|_{L^2}^{\frac{2l-1}{2k}} \|\nabla^\alpha \theta\|_{L^2}^{\frac{2l-1}{2k}} \|\nabla^k \theta\|_{L^2}^{1-\frac{2l-1}{2k}} \|\nabla^{k+1}\theta\|_{L^2}$$

$$\leqslant \|\nabla U\|_{L^2}^{\frac{2k-2l+1}{k}} \|\nabla^\alpha \theta\|_{L^2}^{\frac{2l-1}{k}} \|\nabla^k \theta\|_{L^2}^{\frac{2k-2l+1}{k}} \|\nabla^{k+1} U\|_{L^2}^{\frac{2l-1}{k}} + \frac{1}{12}\|\nabla^{k+1}\theta\|_{L^2}^2$$

$$\leqslant \|\nabla^\alpha \theta\|_{L^2}^{\frac{2(2l-1)}{2k-2l+1}} \|\nabla U\|_{L^2}^2 \|\nabla^k \theta\|_{L^2}^2 + \frac{1}{12}\|(\nabla^{k+1}U, \nabla^{k+1}\theta)\|_{L^2}^2, \qquad (5.11)$$

where α is defined by

$$\alpha = \frac{k}{2l-1} \in \left(\frac{1}{2}, 1\right],$$

and then we get by substituting (5.10), (5.11) into (5.9) that

$$I_2 \leqslant C(k)\|(U,\theta)\|_{H^1}^{\varsigma_2} \|(\nabla^k U, \nabla^k \theta)\|_{L^2}^2 + \frac{1}{6}\|(\nabla^{k+1}U, \nabla^{k+1}\theta)\|_{L^2}^2, \qquad (5.12)$$

where ς_2 is a constant depending on k.

I_3 satisfies

$$I_3 = \left|\int \nabla^k (U \cdot \nabla S) \cdot \nabla^k S\right|$$

$$\leqslant \left|\int (U \cdot \nabla) \nabla^k S \cdot \nabla^k S\right| + C(k) \sum_{l=1}^{k} \int |\nabla^l U| |\nabla^{k+1-l} S| |\nabla^k S|$$

$$= \frac{1}{2}\int |\operatorname{div} U| |\nabla^k S|^2 + C(k) \sum_{l=1}^{k} \int |\nabla^l U| |\nabla^{k+1-l} S| |\nabla^k S|$$

$$\leqslant C(k) \sum_{l=1}^{k} \|\nabla^l U\|_{L^3} \|\nabla^{k+1-l} S\|_{L^2} \|\nabla^{k+1} S\|_{L^2}. \qquad (5.13)$$

For the case $1 \leqslant l \leqslant [k/2]$, one has

$$\|\nabla^l U\|_{L^3} \|\nabla^{k+1-l} S\|_{L^2} \|\nabla^{k+1} S\|_{L^2}$$

$$\leqslant \|\nabla^\alpha U\|_{L^2}^{1-\frac{l}{k}} \|\nabla^k U\|_{L^2}^{\frac{l}{k}} \|\nabla S\|_{L^2}^{\frac{l}{k}} \|\nabla^{k+1} S\|_{L^2}^{1-\frac{l}{k}} \|\nabla^{k+1} S\|_{L^2}$$
$$\leqslant \|\nabla^\alpha U\|_{L^2}^{\frac{2k}{l}-2} \|\nabla S\|_{L^2}^2 \|\nabla^k U\|_{L^2}^2 + \frac{1}{12} \|\nabla^{k+1} S\|_{L^2}^2, \tag{5.14}$$

where α is defined by

$$\alpha = \frac{k}{2(k-l)} \in \left(\frac{1}{2}, 1\right],$$

and for the case $[k/2] + 1 \leqslant l \leqslant k$, one has

$$\|\nabla^l U\|_{L^3} \|\nabla^{k+1-l} S\|_{L^2} \|\nabla^{k+1} S\|_{L^2}$$
$$\leqslant \|\nabla U\|_{L^2}^{1-\frac{2l-1}{2k}} \|\nabla^{k+1} U\|_{L^2}^{\frac{2l-1}{2k}} \|\nabla^\alpha S\|_{L^2}^{\frac{2l-1}{2k}} \|\nabla^k S\|_{L^2}^{1-\frac{2l-1}{2k}} \|\nabla^{k+1} S\|_{L^2}$$
$$\leqslant \|\nabla^\alpha S\|_{L^2}^{\frac{2(2l-1)}{2k-2l+1}} \|\nabla U\|_{L^2}^2 \|\nabla^k S\|_{L^2}^2 + \frac{1}{12} \|(\nabla^{k+1} U, \nabla^{k+1} S)\|_{L^2}^2, \tag{5.15}$$

where α is defined by

$$\alpha = \frac{k}{2l-1} \in \left(\frac{1}{2}, 1\right],$$

and then we get by substituting (5.14), (5.15) into (5.13) that

$$I_3 \leqslant C(k) \|(U,S)\|_{H^1}^{\varsigma_2} \|(\nabla^k U, \nabla^k S)\|_{L^2}^2 + \frac{1}{6} \|(\nabla^{k+1} U, \nabla^{k+1} S)\|_{L^2}^2, \tag{5.16}$$

where ς_2 is a constant depending on k.

And by Hölder's inequality and Young's inequality, we obtain

$$J_1 + J_2 + J_3 \leqslant C \|(\nabla^k U, \nabla^k \theta, \nabla^k S)\|_{L^2}^2. \tag{5.17}$$

So, summing up (5.2)-(5.4) and then substituting (5.8), (5.12), (5.16), and (5.17) into the resulting inequality and noting $k \leqslant N$, we get

$$\frac{d}{dt} \|(\nabla^k U, \nabla^k \theta, \nabla^k S)\|_{L^2}^2 + \|(\nabla^{k+1} U, \nabla^{k+1} \theta, \nabla^{k+1} S)\|_{L^2}^2$$
$$\leqslant C(N, \mu, \lambda, \eta, \tau) \|(U, \theta, S)\|_{H^1}^\varsigma \|(\nabla^k U, \nabla^k \theta, \nabla^k S)\|_{L^2}^2, \quad \text{for any } t \in [0, T],$$

where constant

$$\varsigma \triangleq \max\{\varsigma_1, \varsigma_2, \varsigma_3\}.$$

Here, we have finished the proof of Lemma 5.1.

Since we have proved that there is a unique local solution with large initial data and a unique global one with the smallness condition (3.2) in

H^1, from (3.3), Lemma 5.1 and Gronwall's inequality, we can deduce that for $k = 2, 3, \cdots, N(N \geqslant 2)$,

$$\|(\nabla^k U, \nabla^k \theta, \nabla^k S)(t)\|_{L^2}^2 + \|(\nabla^{k+1} U, \nabla^{k+1} \theta, \nabla^{k+1} S)\|_{L^2(0,t;L^2)}^2$$
$$\leqslant C(N, \mu, \lambda, \eta, \tau, T, \|(U_0, \theta_0, S_0)\|_{H^1}) \|(\nabla^k U_0, \nabla^k \theta_0, \nabla^k S_0)\|_{L^2}^2, \quad \text{for any } t \in [0, T],$$

from which and (3.3), the proof of (3.5) is finished.

Thus from (3.3), (3.5) and Lemma 2.2, we have proved that if the initial data $U_0, \theta_0, S_0 \in H^N$, for any $N \geqslant 1$, then there exists a unique global solution U, θ, S satisfying

$$U \in C([0, T]; H^N) \cap L^2(0, T; H^{N+1}), \quad \nabla P \in L^2(0, T; L^2),$$
$$\theta, S \in C([0, T]; H^N) \cap L^2(0, T; H^{N+1}).$$

In order to finish the proof of Theorem 3.2, we also need to get the estimate of high-order derivatives of ∇P.

Taking the operator

$$\mathbf{Q} \triangleq -\nabla(-\Delta)^{-1} \text{div}$$

to the Eq. $(1.1)_1$, we know that

$$\mu \nabla P = \nabla(-\Delta)^{-1} \text{div}[U \cdot \nabla U] - \mu \{\nabla(-\Delta)^{-1} \text{div}[(\lambda \theta - \eta S)\mathbf{K}]\}. \quad (5.18)$$

According to the Plancherel theorem, (5.18), and the Sobolev inequality, one has for $N \geqslant 2$,

$$\|\nabla P\|_{H^{N-2}}^2$$
$$\leqslant C \sum_{k=0}^{N-2} \{\|\nabla^k [U \cdot \nabla U]\|_{L^2}^2 + \|(\nabla^k \theta, \nabla^k S)\|_{L^2}^2\}$$
$$\leqslant C \sum_{k=0}^{N-2} \left\{ \sum_{l=0}^{k} \|\nabla^l U\|_{L^3}^2 \|\nabla^{k+2-l} U\|_{L^2}^2 + \|(\nabla^k \theta, \nabla^k S)\|_{L^2}^2 \right\}$$
$$\leqslant C \sum_{k=0}^{N-2} \left\{ \sum_{l=0}^{k} \|\nabla^l U\|_{L^2} \|\nabla^{l+1} U\|_{L^2} \|\nabla^{k+2-l} U\|_{L^2}^2 + \|(\nabla^k \theta, \nabla^k S)\|_{L^2}^2 \right\},$$

where $C = C(\mu, \lambda, \eta)$. Since $0 \leqslant l \leqslant k$, $0 \leqslant k \leqslant N - 2$,

$$l + 1 \leqslant k + 1 \leqslant N - 1 \quad \text{and} \quad k + 2 - l \leqslant N,$$

which lead to

$$\|\nabla P\|_{H^{N-2}}^2 \leqslant C\left\{\|U\|_{H^{N-2}}\|U\|_{H^{N-1}}\|U\|_{H^N}^2 + \|(\theta,S)\|_{H^{N-2}}^2\right\}.$$

So

$$\|\nabla P\|_{L^\infty(0,T;H^{N-2})}^2$$
$$\leqslant C\left\{\|U\|_{L^\infty(0,T;H^{N-2})}\|U\|_{L^\infty(0,T;H^{N-1})}\|U\|_{L^\infty(0,T;H^N)}^2 + \|(\theta,S)\|_{L^\infty(0,T;H^{N-2})}^2\right\}$$
$$\leqslant C(\mu,\lambda,\eta,T,\|(U_0,\theta_0,S_0)\|_{H^N}).$$

Similarly, we can get

$$\|\nabla P\|_{L^2(0,T;H^{N-1})}^2 \leqslant C(\mu,\lambda,\eta,T,\|(U_0,\theta_0,S_0)\|_{H^N}).$$

Thus we have finished the proof of Theorem 3.2.

Acknowledgements This work is supported by the NNSF of China under grant numbers 11731014, 11571254, and 11571118 and by the China Postdoctoral Science Foundation under grant number 2017M620688.

References

[1] Adams R A. Sobolev Spaces, 2nd ed. [M]. Academic Press, New York, NY, USA, 2003.

[2] Baines P G, Gill A. On thermohaline convection with linear gradients [J]. J. Fluid Mech., 1969, 37: 289–306.

[3] Beirão Da Veiga H. On the suitable weak solutions to the Navier-Stokes equations in the whole space [J]. J. Math. pures Appl., 1985, 64: 77–86.

[4] Brandt A, Fernando H J S. Double-Diffusive Convection [M]. American Geophysical Union, Washington, 1995.

[5] Chen F, Li Y S, Zhao Y Y. Global well-posedness for the incompressible MHD equations with variable viscosity and conductivity [J]. J. Math. Anal. Appl., 2017, 447: 1051–1071.

[6] Guo B L, Yuan G W. On the suitable weak solutions for the Cauchy problem of the Boussinesq equations [J]. Nonlinear Anal., 1996, 26: 1367–1385.

[7] Hsia C H, Ma T, and Wang S. Attractor bifurcation of three-dimensional Double-Diffusive Convection [J]. Z. Anal. Anwend., 2008, 27: 233–252.

[8] Huppert H E, Turner J S. Double-diffusive convection [J]. J. Fluid Mech., 1981, 106: 299–329.

[9] Jevons W S. On the cirrous form of cloud [J]. London, Edinburgh, Dublin Philos. Mag. J. Sci., 1857, 14: 22–35.

[10] Joseph D D. Uniqueness criteria for the conduction-diffusion solution of the Boussinesq equations [J]. Arch. Rational. Mech. Anal., 1969, 35: 169–177.

[11] Joseph D D. Global stability of the conduction-diffusion solution [J]. Arch. Rational. Mech. Anal., 1970, 36: 285–292.

[12] Lions J L, Magenes E. Problémes aux Limites non Homogénes et Applications [M]. Vol. I and II, Dunod, Paris, 1969.

[13] Lions J L, Temam R, Wang S H. On the equations of the large-scale ocean [J]. Nonlinearity, 1992, 5: 1007–1053.

[14] Nield D A, Bejan A. Convection in Porous Medium [M]. 3rd ed., Springer, New York, 2006.

[15] Novotny A, Straskraba I. Introduction to the Mathematical Theory of Compressible Flow [M]. Oxford University Press, 2004.

[16] Ôtani M, Uchida S. Global sovability of some double-diffusive convection system coupled with Brinkman-Forchheimer equations [J]. Lib. Math. (N.S.), 2013, 33: 79–107.

[17] Ôtani M, Uchida S. The existence of periodic solutions of some double-diffusive convection system based on Brinkman-Forchheimer equations [J]. Adv. Math. Sci. Appl., 2013, 23: 77–92.

[18] Ôtani M, Uchida S. Global solvability for double-diffusive convection system based on Brinkman-Forchheimer equation in general domains [J]. Osaka J. Math., 2016, 53: 855–872.

[19] Piniewski M. Asymptotic dynamics in double-diffusive convection [J]. Appl. Math., 2008, 35: 223–245.

[20] Radko T, Ball J, Colosi J, Flanagan J. Double-diffusive convection in a stochastic shear [J]. J. Phys. Oceanogr., 2015, 45: 3155–3167.

[21] Radko T, Flanagan J D, Stellmach S, Timmermans M L. Double-fiffusive recipes. part II: Layer-merging events [J]. J. Phys. Oceanogr., 2014, 44: 1285–1305.

[22] Shir C C, Joseph D D. Convective instability in a temperature and concentration field [J]. Arch. Rational. Mech. Anal., 1968, 30: 38–80.

[23] Stommel H, Arons A, Blanchard D. An oceanographical curiosity: The perpetual salt fountain[J]. Deep-Sea Res., 1956, 3: 152–153.

[24] Tan Z, Wang Y. Global solution and large-time behavior of the 3D compressible Euler equations with damping [J]. J. Differ. Equations, 2013, 254: 1686–1704.

[25] Terasawa K, Ôtani M. Global solvability of double-diffusive convection systems based upon Brinkman-Forchheimer equations [J]. GAKUTO Internat. Ser. Math. Sci. Appl., 2010, 32: 505–515.

[26] Veronis G. On finite amplitude instability in the thermohaline convection [J]. J. Marine Res., 1965, 23: 1–17.

[27] Zhang J. Global well-posedness for the incompressible Navier-Stokes equations with density-dependent viscosity coefficient [J]. J. Differ. Equations, 2015, 259: 1722-1742.

Long-Time Asymptotics for the Sasa-Satsuma Equation via Nonlinear Steepest Descent Method*

Liu Nan (刘男), and Guo Boling (郭柏灵)

Abstract We formulate a 3×3 Riemann-Hilbert problem to solve the Cauchy problem for the Sasa-Satsuma equation on the line, which allows us to give a representation for the solution of the Sasa-Satsuma equation. We then apply the method of nonlinear steepest descent to compute the long-time asymptotics of the Sasa-Satsuma equation.

Keywords Sasa-Satsuma equation; Nonlinear steepest descent method; long-time asymptotics

1 Introduction

In the context of inverse scattering, the first work to provide explicit formulas (i.e., depending only on initial conditions) for large-time asymptotics of solutions is due to Zakharov and Manakov [42] in the context of the nonlinear Schrödinger (NLS) equation. In this setting, the inverse scattering map and the reconstruction of the solution (potential) is formulated through an oscillatory Riemann–Hilbert (RH) problem. Then the now well-known nonlinear steepest descent method introduced by Deift and Zhou in [10] provides a detailed rigorous proof to calculate the large-time asymptotic behaviors of the integrable nonlinear evolution equations. This approach has been successfully applied in determining asymptotic formulas for the initial value problems of a number of integrable systems associated with 2×2 matrix sprectral problems, including the mKdV equation [10], the defocusing NLS equation [11], the KdV equation [15], the Hirota

---

* J. Math. Phys., 2019, 60: 011504. DOI:10.1063/1.5061793.

equation [20], the derivative NLS equation [27], the Fokas–Lenells equation [38] and the Kundu–Eckhaus equation [35]. Moreover, by combining the ideas of [10] with the so-called "g-function mechanism" [9], it is also possible to study asymptotics of solutions of the initial value problems with shock-type oscillating initial data [7], nondecaying step-like initial data [4, 23, 40] and the initial-boundary value problems with t-periodic boundary conditions [3, 34] for various integrable equations. Recently, Lenells et al. have derived some interesting asymptotic formulas for the solution of derivative NLS equation on the half-line under the initial and boundary values lie in the Schwartz class [2] by using the steepest descent method. We also have done some work about determining the long-time asymptotics for integrable equations on the half-line, see [16, 18]. However, there is just a little literature [5, 6, 14] about the study of long-time asymptotics for the integrable nonlinear evolution equations associated with 3×3 matrix spectral problems. Therefore, it is necessary and important to consider the large-time asymptotic behavior of the integrable equations with 3×3 Lax pairs.

Our present paper aims to consider the long-time asymptotics for the initial value problem of the Sasa–Satsuma (SS) equation [32],

$$u_t = \varepsilon\{u_{xxx} + 6|u|^2 u_x + 3u(|u|^2)_x\}, \tag{1.1}$$

with the initial datum

$$u(x,0) = u_0(x), \tag{1.2}$$

where $\varepsilon = \pm 1$, $u_0(x)$ belongs to the Schwartz space $S(\mathbf{R})$. We consider the case of $\varepsilon = 1$ in present paper. (After we finish this paper, see [17], we are told that the case $\varepsilon = -1$ is considered in a very recent paper [26]. And we compare this paper with the paper [26], the phase functions in the jump matrices are chosen different form and we find that the analysis of the SS equation by the authors of [26], and the present work was completed independently. In fact, we follow the more recent modifications of the nonlinear steepest descent approach proposed by Lenells, see [25], and we find the asymptotics using Cauchy's formula, but the authors of [26] followed the original paper of Deift and Zhou [10] devoted to the long-time asymptotic behavior of the solution of the SS equation.)

The SS equation derived in [32] is of considerable interest for applications and is widely used in nonlinear optics (see [31] and references therein) because the integrable cases of the so-called higher-order NLS equation [21, 22] describing the propagation of short pulses in optical fibers are related through a gauge transformation either to the SS equation [32] or to the so-called Hirota equation [19]. Due to the important role played in both physics and mathematics, the SS equation has attracted much attention and various works have been presented. For example, by employing 3×3 Lax pair, the inverse scattering transform formalism for the initial value problem of the SS equation has been obtained in [32]. Soliton solution, Bäcklund transformation, and conservation laws for the SS equation were found in the optical fiber communications [28]. Twisted rogue-wave pairs in the SS equation were obtained by the author in [8]. Squared eigenfunctions are derived for the SS equation in [41]. The initial-boundary value problem of the SS equation on the half-line was analyzed in [39] by using the Fokas method. There also exists some interesting results for the so-called coupled SS equations, we refer the readers to see [29, 30, 33, 36, 45].

Our purpose here is to derive the long-time asymptotics of the solution $u(x,t)$ of SS equation (1.1) with a 3×3 Lax pair on the line by performing a nonlinear steepest descent analysis of the associated RH problem. Developing and extending the unified transform approach announced in [12, 24], our primary task is to formulate the main RH problem corresponding to the equation (1.1). As a result, we can give a representation of the solution to the Cauchy problem (1.1) in terms of the solution of the corresponding 3×3 RH problem with the jump matrix $J(x,t,k)$ given in terms of the scattering matrix $s(k)$. By using the perfect symmetry of the jump matrix, we introduce a 1×2 vector-valued spectral function $\rho(k)$ and rewrite our main RH problem as a 2×2 block ones. This procedure is more convenient for the following long-time asymptotic analysis compared with the analysis in [5, 6]. As we all know, the first important step of the steepest descent method is to split the jump matrix $J(x,t,k)$ into an appropriate upper/lower triangular form. This immediately leads to construct a $\delta(k)$ function to remove the middle matrix term, however, the function δ satisfies a 2×2 matrix RH problem in our present problem. The unsolvability of the 2×2 matrix function $\delta(k)$ is a challenge for us when we perform the scaling transformation to reduce the RH problem to a model RH problem. Fortunately, the function $\det \delta(k)$ can be explicitly solved by the Plemelj formula because $\det \delta(k)$ satisfies a scalar RH problem. Therefore, we follow the idea introduced in [14] to use the available function $\det \delta(k)$ to approximate the function $\delta(k)$

by error control. On the other hand, the spectral curve for SS equation (1.1) possesses two stationary points, which are different from the case of coupled NLS system considered in [14] where the phase function has a single critical point. The symmetry of the spectral function $\rho(k)$ plays an important role in studying the solution of the model RH problem near the critical point $-k_0$. Therefore, the study of the long-time asymptotics for the initial value problem of (1.1) on the line is more involved. These are some innovation points of the present paper.

The organization of this paper is as follows. In Section 2, we formulate the main RH problem and show how the solution of the SS equation (1.1) can be expressed in terms of the solution of the 3×3 matrix RH problem. In Section 3, we transform the original RH problem to a suitable form and derive the long-time asymptotic behavior of the solution of the SS equation (1.1).

2 Riemann–Hilbert formalism

In this section, we aim to formulate a RH problem to solve the Cauchy problem of the SS equation (1.1).

2.1 The Lax pair and basic eigenfunctions

The Lax pair of equation (1.1) is

$$\begin{cases} \psi_x(x,t;k) = -ik\sigma\psi(x,t;k) + U(x,t)\psi(x,t;k), \\ \psi_t(x,t;k) = 4ik^3\sigma\psi(x,t;k) + V(x,t;k)\psi(x,t;k), \end{cases} \quad (2.1)$$

where $\psi(x,t;k)$ is a 3×3 matrix-valued function, $k \in \mathbf{C}$ is the spectral parameter, and

$$\sigma = \begin{pmatrix} 1 & 0 & 0 \\ 0 & 1 & 0 \\ 0 & 0 & -1 \end{pmatrix}, \quad U(x,t) = \begin{pmatrix} 0 & 0 & u(x,t) \\ 0 & 0 & \bar{u}(x,t) \\ -\bar{u}(x,t) & -u(x,t) & 0 \end{pmatrix}, \quad (2.2)$$

$$V(x,t;k) = -4k^2 U(x,t) - 2ik V_1(x,t) - 2U^3(x,t) + V_2(x,t), \quad (2.3)$$

with

$$V_1 = \begin{pmatrix} |u|^2 & u^2 & u_x \\ \bar{u}^2 & |u|^2 & \bar{u}_x \\ \bar{u}_x & u_x & -2|u|^2 \end{pmatrix}, \quad V_2 = \begin{pmatrix} u_x\bar{u} - u\bar{u}_x & 0 & u_{xx} \\ 0 & u\bar{u}_x - u_x\bar{u} & \bar{u}_{xx} \\ -\bar{u}_{xx} & -u_{xx} & 0 \end{pmatrix}.$$

Introducing a new eigenfunction $\mu(x,t;k)$ by

$$\psi(x,t;k) = \mu(x,t;k)e^{-i(kx-4k^3t)\sigma},$$

we obtain the equivalent Lax pair

$$\begin{cases} \mu_x(x,t;k) + ik[\sigma,\mu(x,t;k)] = U(x,t)\mu(x,t;k), \\ \mu_t(x,t;k) - 4ik^3[\sigma,\mu(x,t;k)] = V(x,t;k)\mu(x,t;k). \end{cases} \quad (2.4)$$

We define two eigenfunctions $\{\mu_j\}_1^2$ of x-part of equation (2.4) by the following Volterra integral equations

$$\mu_1(x,t;k) = I + \int_{-\infty}^{x} e^{-ik(x-x')\hat{\sigma}}[U(x',t)\mu_1(x',t;k)]dx', \quad (2.5)$$

$$\mu_2(x,t;k) = I - \int_{x}^{+\infty} e^{-ik(x-x')\hat{\sigma}}[U(x',t)\mu_2(x',t;k)]dx', \quad (2.6)$$

where $\hat{\sigma}$ denote the operators which act on a 3×3 matrix X by $\hat{\sigma}X = [\sigma, X]$, then $e^{\hat{\sigma}}X = e^{\sigma}Xe^{-\sigma}$. Thus, it can be shown that the functions $\{\mu_j\}_1^2$ are bounded and analytical for $k \in \mathbf{C}$ while k belongs to

$$\begin{aligned} \mu_1 &: (\mathbf{C}_+, \mathbf{C}_+, \mathbf{C}_-), \\ \mu_2 &: (\mathbf{C}_-, \mathbf{C}_-, \mathbf{C}_+), \end{aligned} \quad (2.7)$$

where $\mathbf{C}_+$ and $\mathbf{C}_-$ denote the upper and lower half complex k-plane, respectively. And if we let $-ik\sigma = \mathrm{diag}(z_1,z_2,z_3)$, then we have

$$\mathbf{C}_+ = \{k \in \mathbf{C} | \mathrm{Re}z_1 = \mathrm{Re}z_2 > \mathrm{Re}z_3\},$$
$$\mathbf{C}_- = \{k \in \mathbf{C} | \mathrm{Re}z_1 = \mathrm{Re}z_2 < \mathrm{Re}z_3\}.$$

The solutions of the system of differential equation (2.4) must be related by a matrix independent of x and t, therefore,

$$\mu_1(x,t;k) = \mu_2(x,t;k)e^{-i(kx-4k^3t)\hat{\sigma}}s(k). \quad (2.8)$$

Evaluation at $x \to +\infty, t = 0$ gives

$$s(k) = \lim_{x \to +\infty} e^{ikx\hat{\sigma}}\mu_1(x,0;k), \quad (2.9)$$

that is,

$$s(k) = I + \int_{-\infty}^{+\infty} e^{ikx\hat{\sigma}}[U(x,0)\mu_1(x,0;k)]dx. \quad (2.10)$$

In fact, the matrix-valued spectral function $s(k)$ can be determined in terms of the initial value $u_0(x)$.

The fact that $U(x,t)$ is traceless together with equations (2.5)-(2.6) implies
$$\det \mu_j(x,t;k) = 1, \quad j = 1, 2.$$

Thus, one get
$$\det s(k) = 1.$$

On the other hand, if we denote $\tilde{U}(x,t;k) = -ik\sigma + U(x,t)$, then we find that
$$\tilde{U}^\dagger(x,t;\bar{k}) = -\tilde{U}(x,t;k), \quad \overline{\tilde{U}(x,t;-\bar{k})} = \Gamma \tilde{U}(x,t;k)\Gamma,$$

where
$$\Gamma = \begin{pmatrix} 0 & 1 & 0 \\ 1 & 0 & 0 \\ 0 & 0 & 1 \end{pmatrix}.$$

Moreover, by (2.1), we have
$$\psi_x^A(x,t;k) = -(-ik\sigma + U(x,t))^T \psi^A(x,t;k),$$

with $\psi^A(x,t;k) = (\psi^{-1}(x,t;k))^T$, where the superscript "T" denotes a matrix transpose. Thus, we get
$$\psi^\dagger(x,t;\bar{k}) = \psi^{-1}(x,t;k), \quad \psi(x,t;k) = \Gamma\overline{\psi(x,t;-\bar{k})}\Gamma.$$

These relations imply that the eigenfunctions $\mu_j(x,t;k)$ satisfy
$$\mu_j^\dagger(x,t;\bar{k}) = \mu_j^{-1}(x,t;k), \quad \mu_j(x,t;k) = \Gamma\overline{\mu_j(x,t;-\bar{k})}\Gamma, \quad j = 1, 2, \qquad (2.11)$$

where "$\dagger$" denotes Hermitian conjugate.

Therefore, the matrix-valued function $s(k)$ obeys the symmetries
$$s^\dagger(\bar{k}) = s^{-1}(k), \quad s(-k) = \Gamma\overline{s(\bar{k})}\Gamma. \qquad (2.12)$$

2.2 The formulation of Riemann–Hilbert problem

In order to formulate a RH problem, we should define the following eigenfunctions. For each $n = 1, 2$, a solution $M_n(x,t;k)$ of x-part of (2.4) is defined by the following system of integral equations:
$$(M_n)_{jl}(x,t;k) = \delta_{jl} + \int_{\infty_{jl}^n}^x \left(e^{-ik(x-x')\hat{\sigma}} [U(x',t)M_n(x',t;k)] \right)_{jl} dx', \qquad (2.13)$$

$$k \in D_n, \quad j,l = 1,2,3,$$

where $D_1 = \mathbf{C}_+, D_2 = \mathbf{C}_-$, and the limits ∞_{jl}^n of integration, $n = 1,2, j,l = 1,2,3$, are defined by

$$\infty_{jl}^n = \begin{cases} +\infty, & \text{if } \text{Re} z_j(k) > \text{Re} z_l(k), \\ -\infty, & \text{if } \text{Re} z_j(k) \leqslant \text{Re} z_l(k). \end{cases} \tag{2.14}$$

According to the definition of the ∞^n, we find that

$$\infty^1 = \begin{pmatrix} -\infty & -\infty & +\infty \\ -\infty & -\infty & +\infty \\ -\infty & -\infty & -\infty \end{pmatrix}, \quad \infty^2 = \begin{pmatrix} -\infty & -\infty & -\infty \\ -\infty & -\infty & -\infty \\ +\infty & +\infty & -\infty \end{pmatrix}. \tag{2.15}$$

It follows from the proof in [24] that we have the following proposition of M_n.

Proposition 2.1 *For each $n = 1,2$, the function $M_n(x,t;k)$ is well defined by equation (2.13) for $k \in \bar{D}_n$. M_n is bounded and analytical as a function of $k \in D_n$ away from a possible discrete set of singularities $\{k_j\}$ at which the Fredholm's determinant vanishes. Moreover, M_n admits a bounded and continuous extension to $\bar{D}_n$ and*

$$M_n(x,t;k) = I + O\left(\frac{1}{k}\right), \quad k \to \infty, \quad k \in D_n. \tag{2.16}$$

The above proposition implies that the $M_n(x,t;k)$ has the properties required for the formulation of RH problem. Hence, to derive the corresponding RH problem, we need to compute the jump matrices of $M_n(x,t;k)$ across the real k-axis.

For each $n = 1,2$, we define spectral functions $S_n(k)$ by

$$S_n(k) = \lim_{x \to +\infty} e^{ikx\hat{\sigma}} M_n(x,0;k), \quad k \in D_n, \quad n = 1,2. \tag{2.17}$$

According to (2.13), the $\{S_n(k)\}_1^2$ can be computed from the spectral function $s(k)$.

Proposition 2.2 *The $S_n(k)$ defined in (2.17) can be expressed in terms of the entries of $s(k)$ as follows:*

$$S_1(k) = \begin{pmatrix} s_{11} & s_{12} & 0 \\ s_{21} & s_{22} & 0 \\ s_{31} & s_{32} & \dfrac{1}{\bar{s}_{33}} \end{pmatrix}, \quad S_2(k) = \begin{pmatrix} \dfrac{\bar{s}_{22}}{s_{33}} & -\dfrac{\bar{s}_{21}}{s_{33}} & s_{13} \\ -\dfrac{\bar{s}_{12}}{s_{33}} & \dfrac{\bar{s}_{11}}{s_{33}} & s_{23} \\ 0 & 0 & s_{33} \end{pmatrix}. \tag{2.18}$$

Proof For a constant $X_0 < 0$, we introduce $\mu_1(x,t;k;X_0)$ as the solution of (2.5) with the lower limit of the integral $-\infty$ replaced by X_0. Similarly, we define $M_n(x,t;k;X_0)$, $n = 1, 2$, as the solutions of (2.13) with the lower limit of the integral $-\infty$ replaced by X_0. We will derive expressions for

$$S_n(k;X_0) := \lim_{x \to +\infty} e^{ikx\hat{\sigma}} M_n(x,0;k;X_0) \qquad (2.19)$$

in terms of

$$s(k;X_0) := \lim_{x \to +\infty} e^{ikx\hat{\sigma}} \mu_1(x,0;k;X_0).$$

Then, expression (2.18) will follow by taking the limit $X_0 \to -\infty$.

We first define $T_n(k;X_0)$, $n = 1, 2$, by

$$T_n(k;X_0) = e^{ikX_0\hat{\sigma}} M_n(X_0, 0; k; X_0), \quad k \in D_n. \qquad (2.20)$$

Then, we have

$$M_n(x,t;k;X_0) = \mu_1(x,t;k;X_0) e^{-i(kx-4k^3t)\hat{\sigma}} T_n(k;X_0), \quad k \in D_n. \qquad (2.21)$$

This relation implies

$$S_n(k;X_0) = s(k;X_0) T_n(k;X_0), \quad k \in D_n. \qquad (2.22)$$

Moreover, the integral equation (2.13) together with the definition (2.15) of ∞_{jl}^n and (2.19) imply that

$$(S_n)_{jl}(k;X_0) = 0, \qquad \text{if } \infty_{jl}^n = +\infty, \qquad (2.23)$$

$$(s^{-1} S_n)_{jl}(k;X_0) = \delta_{jl}, \quad \text{if } \infty_{jl}^n = X_0. \qquad (2.24)$$

Computing the explicit solution of the algebraic system (2.23)-(2.24) and using the relation (2.12), we find that $\{S_n(k;X_0)\}_1^2$ are exactly given by the equation obtained from (2.18) by replacing $s(k)$ with $s(k;X_0)$. Taking $X_0 \to -\infty$, we get (2.18).

For convenience, we assume that s_{33} has no zero in $\mathbf{C}_-$. Then, we have the following main result in this section.

Theorem 2.1 *Let $u_0(x)$ be a function in the Schwartz space $S(\mathbf{R})$ and define the matrix-valued spectral function $s(k)$ via (2.9). Define $M(x,t;k)$ as the solution of the following matrix RH problem:*

- $M(x,t;k) = \begin{cases} M_+(x,t;k) = M_1(x,t;k), & k \in \mathbf{C}_+ = D_1, \\ M_-(x,t;k) = M_2(x,t;k), & k \in \mathbf{C}_- = D_2, \end{cases}$

is a sectionally meromorphic function in $\mathbf{C} \setminus \mathbf{R}$;

- $M(x,t;k)$ satisfies the jump condition

$$M_+(x,t;k) = M_-(x,t;k)J(x,t,k), \quad k \in \mathbf{R}, \qquad (2.25)$$

where the matrix $J(x,t,k)$ is defined by

$$J(x,t,k) = e^{-i(kx-4k^3t)\hat{\sigma}} S_2^{-1}(k) S_1(k)$$

$$= e^{-i(kx-4k^3t)\hat{\sigma}} \begin{pmatrix} 1 & 0 & \dfrac{\bar{s}_{31}}{\bar{s}_{33}} \\ 0 & 1 & \dfrac{\bar{s}_{32}}{\bar{s}_{33}} \\ \dfrac{s_{31}}{s_{33}} & \dfrac{s_{32}}{s_{33}} & 1 + \dfrac{\bar{s}_{31}}{\bar{s}_{33}}\dfrac{s_{31}}{s_{33}} + \dfrac{\bar{s}_{32}}{\bar{s}_{33}}\dfrac{s_{32}}{s_{33}} \end{pmatrix}; \qquad (2.26)$$

- $M(x,t;k)$ has the following asymptotics:

$$M(x,t;k) = I + O\left(\frac{1}{k}\right), \quad k \to \infty. \qquad (2.27)$$

Then $M(x,t;k)$ exists and is unique.

Define $\{u(x,t), \bar{u}(x,t)\}$ in terms of $M(x,t;k)$ by

$$\bar{u}(x,t) = 2i \lim_{k \to \infty} (kM(x,t;k))_{31}, \qquad (2.28)$$

$$u(x,t) = 2i \lim_{k \to \infty} (kM(x,t;k))_{32}. \qquad (2.29)$$

Then $u(x,t)$ solves the Sasa–Satsuma equation (1.1). Furthermore, $u(x,0) = u_0(x)$.

Proof The existence and uniqueness for the solution of above RH problem is a consequence of a 'vanishing lemma' for the associated RH problem with the vanishing condition at infinity $M(k) = O(1/k)$, $k \to \infty$. This fact holds due to the jump matrix $J(x,t,k)$ in (2.26) is positive definite [1]. Substituting the large k asymptotics of $M(x,t;k)$

$$M(x,t;k) = I + \frac{M_1(x,t)}{k} + \frac{M_2(x,t)}{k^2} + \frac{M_3(x,t)}{k^3} + \cdots, \quad k \to \infty, \qquad (2.30)$$

into equation (2.4) and collecting terms with $O(1/k^n)$, we immediately get (2.27). Moreover, based on the symmetry property (2.11), we have $(M_1)_{12} = -(M_1)_{21}^{\dagger}$.

We next will use the standard arguments of the dressing method [43,44] (see also [13]) to prove that $u(x,t)$ defined by (2.29) solves the SS equation (1.1). The main idea of the dressing method is to construct two linear operators $\mathcal{L}$ and $\mathcal{N}$ such that (i) $\mathcal{L}M$ and $\mathcal{N}M$ satisfy the same jump condition as M, and (ii) $\mathcal{L}M$ and $\mathcal{N}M$ are of $O(1/k)$ as $k \to \infty$. Then, it follows that

$$\mathcal{L}M = 0, \quad \mathcal{N}M = 0. \tag{2.31}$$

These equations constitute the Lax pair associated with the above RH problem.

In fact, set

$$\mathcal{L}M = M_x + ik[\sigma, M] - UM, \tag{2.32}$$
$$\mathcal{N}M = M_t - 4ik^3[\sigma, M] - VM, \tag{2.33}$$

where U and V are given by (2.2) and (2.3), respectively. A direct computation yields

$$(\mathcal{L}M)_+ = (\mathcal{L}M)_- J, \quad (\mathcal{N}M)_+ = (\mathcal{N}M)_- J. \tag{2.34}$$

Substituting the asymptotic expansion (2.30) into (2.32), we find that $\mathcal{L}M$ is of $O(1/k)$ as $k \to \infty$. Thus, we get

$$(\mathcal{L}M)_+ = (\mathcal{L}M)_- J, \quad k \in \mathbf{R},$$
$$\mathcal{L}M = O(1/k), \quad k \to \infty,$$

which implies

$$\mathcal{L}M = 0. \tag{2.35}$$

Furthermore, comparing the coefficients of $O(1/k^n)$ in the asymptotic expansion of (2.35), we obtain

$$M_1^{(O)} = -\frac{i}{2}\sigma U, \quad [M_1^{(D)}]_x = \frac{i}{2}\sigma U^2, \quad [M_2]_x + i[\sigma, M_3] = UM_2,$$
$$M_2^{(O)} = \frac{1}{4}U_x - \frac{i}{2}\sigma U M_1^{(D)}, \quad [M_2^{(D)}]_x = UM_2^{(O)},$$

where superscripts '(O)' and '(D)' indicate the off-diagonal and diagonal parts of block matrix, respectively, that is, for a 3×3 matrix A,

$$A^{(O)} = \frac{1}{2}\sigma[\sigma, A], \quad A^{(D)} = A - A^{(O)}.$$

Using these relations and the fact $V_1 = -\sigma U^2 + \sigma U_x$, we can conclude that

$$(\mathcal{N}M)_+ = (\mathcal{N}M)_- J, \quad k \in \mathbf{R},$$

$$\mathcal{N}M = O(1/k), \quad k \to \infty,$$

which means that

$$\mathcal{N}M = 0. \tag{2.36}$$

The compatibility condition of (2.35) and (2.36) yields the SS equation (1.1). This implies that $u(x,t)$ defined by (2.29) solves the SS equation (1.1).

In order to verify the initial condition, one observes that for $t = 0$, the RH problem reduces to that associated with $u_0(x)$, which yields $u(x,0) = u_0(x)$, owing to the uniqueness of the solution of the RH problem.

If we rewrite a 3×3 matrix A as a block form

$$A = \begin{pmatrix} A_{11} & A_{12} \\ A_{21} & A_{22} \end{pmatrix},$$

where A_{22} is scalar. Then, the above RH problem can be rewritten as the following form:

$$\begin{cases} M_+(x,t;k) = M_-(x,t;k) J(x,t,k), & k \in \mathbf{R}, \\ M(x,t;k) \to I, & k \to \infty, \end{cases} \tag{2.37}$$

where

$$J(x,t,k) = \begin{pmatrix} I & \rho^\dagger(\bar{k}) e^{-t\Phi} \\ \rho(k) e^{t\Phi} & 1 + \rho(k) \rho^\dagger(\bar{k}) \end{pmatrix}, \tag{2.38}$$

$$\Phi(\zeta, k) = 2i\zeta k - 8ik^3, \quad \zeta = \frac{x}{t}, \quad \rho(k) = \begin{pmatrix} \dfrac{s_{31}}{s_{33}} & \dfrac{s_{32}}{s_{33}} \end{pmatrix}.$$

According to the symmetry relation (2.12), we can conclude that

$$\rho(k) = \overline{\rho(-\bar{k})} \begin{pmatrix} 0 & 1 \\ 1 & 0 \end{pmatrix}. \tag{2.39}$$

Moreover, we have

$$\begin{pmatrix} \bar{u}(x,t) & u(x,t) \end{pmatrix} = 2i \lim_{k \to \infty} (kM(x,t;k))_{21}. \tag{2.40}$$

In the following long-time asymptotic analysis, we will focus on the new block RH problem (2.37) with the jump matrix defined by (2.38).

3 Long-time asymptotic analysis

An analogue of the classical steepest descent method for RH problems was invented by Deift and Zhou [10], we consider the stationary points of the function Φ, that is, taking

$$\frac{\mathrm{d}\Phi(\zeta,k)}{\mathrm{d}k} = 0,$$

the stationary phase points are obtained for $x > 0$ as

$$\pm k_0 = \pm\sqrt{\frac{\zeta}{12}}, \tag{3.1}$$

and the signature table for $\mathrm{Re}\Phi(\zeta,k)$ is shown in Fig. 1. Let $M > 1$ be a given constant, and let $\mathcal{I}$ denote the interval $\mathcal{I} = (0, M]$. We restrict our attention here to the physically interesting region $\zeta \in \mathcal{I}$.

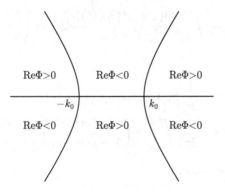

Fig. 1 The signature table for $\mathrm{Re}\Phi(\zeta,k)$ in the complex k-plane

3.1 Transformations of the original RH problem

In this section, we aim to transform the associated original RH problem (2.37) to a solvable RH problem. We note that the jump matrix enjoys two distinct factorizations:

$$J = \begin{cases} \begin{pmatrix} I & 0 \\ \rho e^{t\Phi} & 1 \end{pmatrix} \begin{pmatrix} I & \rho^\dagger e^{-t\Phi} \\ 0 & 1 \end{pmatrix}, \\ \begin{pmatrix} I & \frac{\rho^\dagger e^{-t\Phi}}{1+\rho\rho^\dagger} \\ 0 & 1 \end{pmatrix} \begin{pmatrix} (I+\rho^\dagger\rho)^{-1} & 0 \\ 0 & 1+\rho\rho^\dagger \end{pmatrix} \begin{pmatrix} I & 0 \\ \frac{\rho e^{t\Phi}}{1+\rho\rho^\dagger} & 1 \end{pmatrix}. \end{cases} \tag{3.2}$$

We introduce a function $\delta(k)$ as the solution of the matrix RH problem

$$\begin{cases} \delta_+(k) = (I + \rho^\dagger(k)\rho(k))\delta_-(k), & |k| < k_0, \\ \quad\quad = \delta_-(k), & |k| > k_0, \\ \delta(k) \to I, & k \to \infty. \end{cases} \quad (3.3)$$

Since the jump matrix $I + \rho^\dagger(k)\rho(k)$ is positive definite, the vanishing lemma [1] yields the existence and uniqueness of the function $\delta(k)$. Furthermore, we have

$$\begin{cases} \det \delta_+(k) = (1 + \rho(k)\rho^\dagger(k)) \det \delta_-(k), & |k| < k_0, \\ \quad\quad = \det \delta_-(k), & |k| > k_0, \\ \det \delta(k) \to 1, & k \to \infty. \end{cases} \quad (3.4)$$

By Plemelj formula, we can obtain

$$\det \delta(k) = \exp\left\{ \frac{1}{2\pi i} \int_{-k_0}^{k_0} \frac{\ln(1 + \rho(\xi)\rho^\dagger(\xi))}{\xi - k} d\xi \right\} = \left(\frac{k - k_0}{k + k_0} \right)^{-i\nu} e^{\chi(k)}, \quad (3.5)$$

where

$$\nu = \frac{1}{2\pi} \ln(1 + \rho(k_0)\rho^\dagger(k_0)) > 0, \quad (3.6)$$

$$\chi(k) = \frac{1}{2\pi i} \int_{-k_0}^{k_0} \ln\left(\frac{1 + \rho(\xi)\rho^\dagger(\xi)}{1 + \rho(k_0)\rho^\dagger(k_0)} \right) \frac{d\xi}{\xi - k}. \quad (3.7)$$

Here we have used the relation

$$1 + \rho(k_0)\rho^\dagger(k_0) = 1 + \rho(-k_0)\rho^\dagger(-k_0), \quad (3.8)$$

which follows from the second symmetry relation in (2.12), more precisely,

$$s_{31}(-k) = \bar{s}_{32}(\bar{k}), \quad s_{32}(-k) = \bar{s}_{31}(\bar{k}), \quad s_{33}(-k) = \bar{s}_{33}(\bar{k}).$$

On the other hand, for $|k| < k_0$, it follows from (3.3) that

$$\lim_{\epsilon \to 0^+} \delta(k - i\epsilon) = (I + \rho^\dagger(k)\rho(k))^{-1} \lim_{\epsilon \to 0^+} \delta(k + i\epsilon). \quad (3.9)$$

If let $g(k) = (\delta^\dagger(\bar{k}))^{-1}$, then we get

$$g_+(k) = \lim_{\epsilon \to 0^+} g(k + i\epsilon) = \lim_{\epsilon \to 0^+} (\delta^\dagger(\bar{k} - i\epsilon))^{-1}$$

$$= (I + \rho^\dagger(k)\rho(k)) \lim_{\epsilon \to 0^+} (\delta^\dagger(\bar{k} + i\epsilon))^{-1}$$

$$=(I+\rho^\dagger(k)\rho(k))g_-(k).$$

Therefore, we find
$$(\delta^\dagger(\bar{k}))^{-1}=\delta(k). \tag{3.10}$$

Direct calculation as in [14] and the maximum principle show that
$$|\delta(k)|\leqslant \text{const}<\infty, \quad |\det\delta(k)|\leqslant \text{const}<\infty, \tag{3.11}$$

for all k, where we define $|A|=(\text{tr}A^\dagger A)^{1/2}$ for any matrix A.

Define
$$\Delta(k)=\begin{pmatrix}\delta^{-1}(k) & 0 \\ 0 & \det\delta(k)\end{pmatrix}. \tag{3.12}$$

Introduce
$$M^{(1)}(x,t;k)=M(x,t;k)\Delta^{-1}(k),$$

and reverse the orientation for $|k|<k_0$ as shown in Fig. 2, then $M^{(1)}$ satisfies the following RH problem
$$\begin{cases} M_+^{(1)}(x,t;k)=M_-^{(1)}(x,t;k)J^{(1)}(x,t,k), & k\in\mathbf{R}, \\ M^{(1)}(x,t;k)\to I, & k\to\infty, \end{cases} \tag{3.13}$$

where
$$J^{(1)}=\Delta_- J\Delta_+^{-1}$$
$$=\begin{cases}\begin{pmatrix}I & 0 \\ \rho_1\delta\det\delta e^{t\Phi} & 1\end{pmatrix}\begin{pmatrix}I & \dfrac{\delta^{-1}\rho_4 e^{-t\Phi}}{\det\delta} \\ 0 & 1\end{pmatrix}, & |k|>k_0, \\[2ex] \begin{pmatrix}I & 0 \\ \rho_3\delta_-\det\delta_- e^{t\Phi} & 1\end{pmatrix}\begin{pmatrix}I & \dfrac{\delta_+^{-1}\rho_2 e^{-t\Phi}}{\det\delta_+} \\ 0 & 1\end{pmatrix}, & |k|<k_0, \end{cases} \tag{3.14}$$

and the functions $\{\rho_j(k)\}_1^4$ are defined by
$$\rho_1(k)=\rho(k), \qquad \rho_2(k)=-\dfrac{\rho^\dagger(\bar{k})}{1+\rho(k)\rho^\dagger(\bar{k})},$$
$$\rho_3(k)=-\dfrac{\rho(k)}{1+\rho(k)\rho^\dagger(\bar{k})}, \quad \rho_4(k)=\rho^\dagger(\bar{k}). \tag{3.15}$$

Fig. 2   The oriented contour on **R**

Our next goal is to deform the contour. However, the spectral functions $\{\rho_j(k)\}_1^4$ have limited domains of analyticity, we will follow the idea of [10, 25] and decompose each of the functions $\{\rho_j(k)\}_1^4$ into an analytic part and a small remainder. The analytic part of the jump matrix will be deformed, whereas the small remainder will be left on the original contour. We introduce the open subsets $\{\Omega_j\}_1^4$, as displayed in Fig. 3 such that

$$\Omega_1 \cup \Omega_3 = \{k | \mathrm{Re}\Phi(\zeta, k) < 0\},$$
$$\Omega_2 \cup \Omega_4 = \{k | \mathrm{Re}\Phi(\zeta, k) > 0\}.$$

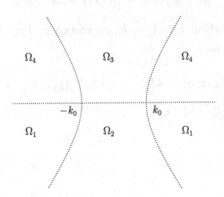

Fig. 3   The open sets $\{\Omega_j\}_1^4$ in the complex k-plane.

In fact, we have the following lemma.

Lemma 3.1   *There exist decompositions*

$$\rho_j(k) = \begin{cases} \rho_{j,a}(x,t,k) + \rho_{j,r}(x,t,k), & |k| > k_0, \ k \in \mathbf{R}, \ j = 1, 4, \\ \rho_{j,a}(x,t,k) + \rho_{j,r}(x,t,k), & |k| < k_0, \ k \in \mathbf{R}, \ j = 2, 3, \end{cases} \qquad (3.16)$$

where the functions $\{\rho_{j,a}, \rho_{j,r}\}_1^4$ have the following properties:

(1) *For each $\zeta \in \mathcal{I}$ and each $t > 0$, $\rho_{j,a}(x,t,k)$ is defined and continuous for $k \in \bar{\Omega}_j$ and analytic for Ω_j, $j = 1, 2$.*

(2) For $K > 0$, the functions $\rho_{j,a}$ for $j = 1, 2, 3, 4$ satisfy

$$|\rho_{j,a}(x,t,k) - \rho_j(k_0)| \leq C_K |k - k_0| e^{\frac{t}{4}|\text{Re}\Phi(\zeta,k)|}, \quad k \in \bar{\Omega}_j, \ |k| < K, \ t > 0, \ \zeta \in \mathcal{I}. \tag{3.17}$$

In particular, the functions $\rho_{1,a}$ and $\rho_{4,a}$ satisfy

$$|\rho_{j,a}(x,t,k)| \leq \frac{C}{1+|k|^2} e^{\frac{t}{4}|\text{Re}\Phi(\zeta,k)|}, \quad t > 0, \ k \in \bar{\Omega}_j, \ \zeta \in \mathcal{I}, \ j = 1, 4, \tag{3.18}$$

where the constants C_K, C are independent of ζ, k, t, but C_K may depend on K.

(3) The L^1, L^2 and L^∞ norms of the functions $\rho_{1,r}(x,t,\cdot)$ and $\rho_{4,r}(x,t,\cdot)$ on $(-\infty, -k_0) \cup (k_0, \infty)$ are $O(t^{-3/2})$ as $t \to \infty$ uniformly with respect to $\zeta \in \mathcal{I}$.

(4) The L^1, L^2 and L^∞ norms of the functions $\rho_{2,r}(x,t,\cdot)$ and $\rho_{3,r}(x,t,\cdot)$ on $(-k_0, k_0)$ are $O(t^{-3/2})$ as $t \to \infty$ uniformly with respect to $\zeta \in \mathcal{I}$.

(5) For $j = 1, 2, 3, 4$, the following symmetries hold:

$$\rho_{j,a}(x,t,k) = \overline{\rho_{j,a}(x,t,-\bar{k})} \begin{pmatrix} 0 & 1 \\ 1 & 0 \end{pmatrix}, \quad \rho_{j,r}(x,t,k) = \overline{\rho_{j,r}(x,t,-\bar{k})} \begin{pmatrix} 0 & 1 \\ 1 & 0 \end{pmatrix}. \tag{3.19}$$

Proof We first consider the decomposition of $\rho_1(k)$. Denote $\Omega_1 = \Omega_1^+ \cup \Omega_1^-$, where Ω_1^+ and Ω_1^- denote the parts of Ω_1 in the right and left half-planes, respectively. We derive a decomposition of $\rho_1(k)$ in Ω_1^+, and then extend it to Ω_1^- by symmetry. Since $\rho_1(k) \in C^\infty(\mathbf{R})$, then for $n = 0, 1, 2$,

$$\rho_1^{(n)}(k) = \frac{d^n}{dk^n} \left(\sum_{j=0}^{6} \frac{\rho_1^{(j)}(k_0)}{j!}(k-k_0)^j \right) + O((k-k_0)^{7-n}), \quad k \to k_0, \tag{3.20}$$

let

$$f_0(k) = \sum_{j=4}^{10} \frac{a_j}{(k-i)^j}, \tag{3.21}$$

where $\{a_j\}_4^{10}$ are complex constants such that

$$f_0(k) = \sum_{j=0}^{6} \frac{\rho_1^{(j)}(k_0)}{j!}(k-k_0)^j + O((k-k_0)^7), \quad k \to k_0. \tag{3.22}$$

It is easy to verify that (3.22) imposes seven linearly independent conditions on the a_j, hence the coefficients a_j exist and are unique. Letting $f = \rho_1 - f_0$, it follows that

(i) $f_0(k)$ is a rational function of $k \in \mathbf{C}$ with no poles in Ω_1^+;

(ii) $f_0(k)$ coincides with $\rho_1(k)$ to six order at k_0, more precisely,

$$\frac{\mathrm{d}^n}{\mathrm{d}k^n}f(k) = \begin{cases} O((k-k_0)^{7-n}), & k \to k_0, \\ O(k^{-4-n}), & k \to \infty, \end{cases} \quad k \in \mathbf{R},\ n = 0,1,2. \tag{3.23}$$

The decomposition of $\rho_1(k)$ can be derived as follows. The map $k \mapsto \phi = \phi(k)$ defined by $\phi(k) = -\mathrm{i}\Phi(\zeta, k)$ is a bijection $[k_0, \infty) \mapsto (-\infty, 16k_0^3]$, so we may define a function F by

$$F(\phi) = \begin{cases} \dfrac{(k-\mathrm{i})^3}{k-k_0}f(k), & \phi < 16k_0^3, \\ 0, & \phi \geqslant 16k_0^3. \end{cases} \tag{3.24}$$

Then,

$$F^{(n)}(\phi) = \left(\frac{1}{24(k_0^2-k^2)}\frac{\partial}{\partial k}\right)^n \left(\frac{(k-\mathrm{i})^3}{k-k_0}f(k)\right), \quad \phi < 16k_0^3.$$

By (3.23), $F \in C^1(\mathbf{R})$ and $F^{(n)}(\phi) = O(|\phi|^{-2/3})$ as $|\phi| \to \infty$ for $n = 0,1,2$. In particular,

$$\left\|\frac{\mathrm{d}^n F}{\mathrm{d}\phi^n}\right\|_{L^2(\mathbf{R})} < \infty, \quad n = 0,1,2, \tag{3.25}$$

that is, F belongs to $H^2(\mathbf{R}) \times H^2(\mathbf{R})$. By the Fourier transform $\hat{F}(s)$ defined by

$$\hat{F}(s) = \frac{1}{2\pi}\int_{\mathbf{R}} F(\phi)\mathrm{e}^{-\mathrm{i}\phi s}\mathrm{d}\phi, \tag{3.26}$$

where

$$F(\phi) = \int_{\mathbf{R}} \hat{F}(s)\mathrm{e}^{\mathrm{i}\phi s}\mathrm{d}s, \tag{3.27}$$

it follows from Plancherel theorem that $\|s^2\hat{F}(s)\|_{L^2(\mathbf{R})} < \infty$. Equations (3.24) and (3.27) imply

$$f(k) = \frac{k-k_0}{(k-\mathrm{i})^3}\int_{\mathbf{R}} \hat{F}(s)\mathrm{e}^{\mathrm{i}\phi s}\mathrm{d}s, \quad k > k_0. \tag{3.28}$$

Writing

$$f(k) = f_a(t,k) + f_r(t,k), \quad t > 0,\ k > k_0,$$

where the functions f_a and f_r are defined by

$$f_a(t,k) = \frac{k-k_0}{(k-\mathrm{i})^3}\int_{-\frac{t}{4}}^{\infty}\hat{F}(s)\mathrm{e}^{s\Phi(\zeta,k)}\mathrm{d}s, \quad t>0,\ k \in \Omega_1^+, \tag{3.29}$$

$$f_r(t,k) = \frac{k-k_0}{(k-i)^3}\int_{-\infty}^{-\frac{t}{4}} \hat{F}(s)e^{s\Phi(\zeta,k)}ds, \quad t>0,\ k>k_0, \tag{3.30}$$

we infer that $f_a(t,\cdot)$ is continuous in $\bar{\Omega}_1^+$ and analytic in Ω_1^+. Moreover, we can get

$$|f_a(t,k)| \leq \frac{|k-k_0|}{|k-i|^3}\|\hat{F}(s)\|_{L^1(\mathbf{R})}\sup_{s\geq -\frac{t}{4}}e^{s\operatorname{Re}\Phi(\zeta,k)}$$

$$\leq \frac{C|k-k_0|}{|k-i|^3}e^{\frac{t}{4}|\operatorname{Re}\Phi(\zeta,k)|}, \quad t>0,\ k\in\bar{\Omega}_1^+,\ \zeta\in\mathcal{I}. \tag{3.31}$$

Furthermore, we have

$$|f_r(t,k)| \leq \frac{|k-k_0|}{|k-i|^3}\int_{-\infty}^{-\frac{t}{4}} s^2|\hat{F}(s)|s^{-2}ds$$

$$\leq \frac{C}{1+|k|^2}\|s^2\hat{F}(s)\|_{L^2(\mathbf{R})}\sqrt{\int_{-\infty}^{-\frac{t}{4}} s^{-4}ds},$$

$$\leq \frac{C}{1+|k|^2}t^{-3/2}, \quad t>0,\ k>k_0,\ \zeta\in\mathcal{I}. \tag{3.32}$$

Hence, the L^1, L^2 and L^∞ norms of f_r on (k_0,∞) are $O(t^{-3/2})$. Letting

$$\rho_{1,a}(t,k) = f_0(k) + f_a(t,k), \quad t>0,\ k\in\bar{\Omega}_1^+, \tag{3.33}$$
$$\rho_{1,r}(t,k) = f_r(t,k), \quad t>0,\ k>k_0. \tag{3.34}$$

For $k<-k_0$, we use the symmetry (3.19) to extend this decomposition. Thus, we find a decomposition of ρ_1 for $|k|>k_0$ with the properties listed in the statement of the lemma.

The decomposition of ρ_3 for the case $|k|<k_0$ is similar, detailed proof can be found in [10, 25], we will be omitted. The decompositions of ρ_2 and ρ_4 can be obtained from ρ_3 and ρ_1 by Schwartz conjugation.

The next transformation is to deform the contour so that the jump matrix involves the exponential factor $e^{-t\Phi}$ on the parts of the contour where $\operatorname{Re}\Phi$ is positive and the factor $e^{t\Phi}$ on the parts where $\operatorname{Re}\Phi$ is negative. More precisely, we put

$$M^{(2)}(x,t;k) = M^{(1)}(x,t;k)G(k), \tag{3.35}$$

where

$$G(k) = \begin{cases} \begin{pmatrix} I & 0 \\ \rho_{1,a}\delta \det\delta e^{t\Phi} & 1 \end{pmatrix}, & k \in U_1, \\ \begin{pmatrix} I & -\dfrac{\delta^{-1}\rho_{2,a}e^{-t\Phi}}{\det\delta} \\ 0 & 1 \end{pmatrix}, & k \in U_2, \\ \begin{pmatrix} I & 0 \\ \rho_{3,a}\delta \det\delta e^{t\Phi} & 1 \end{pmatrix}, & k \in U_3, \\ \begin{pmatrix} I & -\dfrac{\delta^{-1}\rho_{4,a}e^{-t\Phi}}{\det\delta} \\ 0 & 1 \end{pmatrix}, & k \in U_4, \\ I, & k \in U_5 \cup U_6. \end{cases} \qquad (3.36)$$

Then the matrix $M^{(2)}(x,t;k)$ satisfies the following RH problem

$$M_+^{(2)}(x,t;k) = M_-^{(2)}(x,t;k) J^{(2)}(x,t,k), \quad k \in \Sigma, \qquad (3.37)$$

with the jump matrix $J^{(2)} = G_-^{-1}(k) J^{(1)} G_+(k)$ is given by

$$J_1^{(2)} = \begin{pmatrix} I & 0 \\ -\rho_{1,a}\delta \det\delta e^{t\Phi} & 1 \end{pmatrix}, \quad J_2^{(2)} = \begin{pmatrix} I & -\dfrac{\delta^{-1}\rho_{2,a}e^{-t\Phi}}{\det\delta} \\ 0 & 1 \end{pmatrix},$$

$$J_3^{(2)} = \begin{pmatrix} I & 0 \\ -\rho_{3,a}\delta \det\delta e^{t\Phi} & 1 \end{pmatrix}, \quad J_4^{(2)} = \begin{pmatrix} I & -\dfrac{\delta^{-1}\rho_{4,a}e^{-t\Phi}}{\det\delta} \\ 0 & 1 \end{pmatrix},$$

$$J_5^{(2)} = \begin{pmatrix} I & 0 \\ \rho_{1,r}\delta \det\delta e^{t\Phi} & 1 \end{pmatrix} \begin{pmatrix} I & \dfrac{\delta^{-1}\rho_{4,r}e^{-t\Phi}}{\det\delta} \\ 0 & 1 \end{pmatrix},$$

$$J_6^{(2)} = \begin{pmatrix} I & 0 \\ \rho_{3,r}\delta_- \det\delta_- e^{t\Phi} & 1 \end{pmatrix} \begin{pmatrix} I & \dfrac{\delta_+^{-1}\rho_{2,r}e^{-t\Phi}}{\det\delta_+} \\ 0 & 1 \end{pmatrix} \qquad (3.38)$$

with $J_i^{(2)}$ denoting the restriction of $J^{(2)}$ to the contour labeled by i in Fig. 4.

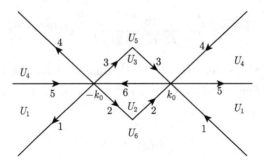

Fig. 4 The jump contour Σ and the open sets $\{U_j\}_1^6$ in the complex k-plane

Obviously, the jump matrix $J^{(2)}$ decays to identity matrix I as $t \to \infty$ everywhere except near the critical points $\pm k_0$. This implies that the main contribution to the long-time asymptotics should come from the neighborhoods of the critical points $\pm k_0$.

3.2 Local models near the stationary phase points $\pm k_0$

To focus on $-k_0$ and k_0, we introduce the following scaling operators

$$S_{-k_0}: k \mapsto \frac{z}{\sqrt{48tk_0}} - k_0,$$
$$S_{k_0}: k \mapsto \frac{z}{\sqrt{48tk_0}} + k_0. \tag{3.39}$$

Let $D_\varepsilon(\pm k_0)$ denote the open disk of radius ε centered at $\pm k_0$ for a small $\varepsilon > 0$. Then, the operator $S_{\pm k_0}$ is a bijection from $D_\varepsilon(\pm k_0)$ to the open disk of radius $\sqrt{48tk_0}\varepsilon$ and centered at the origin for all $\zeta \in \mathcal{I}$.

Since the function $\delta(k)$ satisfying a 2×2 matrix RH problem (3.3) can not be solved explicitly, to proceed to the next step, we will follow the idea developed in [14] to use the available function $\det \delta(k)$ to approximate $\delta(k)$ by error estimate. More precisely, we write

$$\left(\rho_{j,a}\delta \det \delta e^{t\Phi}\right)(k) = \left(\rho_{j,a}(\det \delta)^2 e^{t\Phi}\right)(k) + \left(\rho_{j,a}(\delta - \det \delta I)\det \delta e^{t\Phi}\right)(k), \tag{3.40}$$

for $j = 1, 3$. For the first part in the right-hand side of (3.40), we have

$$S_{k_0}\left(\rho_{j,a}(\det \delta)^2 e^{t\Phi}\right)(z) = \eta^2 \varrho^2 \rho_{j,a}\left(\frac{z}{\sqrt{48tk_0}} + k_0\right), \tag{3.41}$$

where

$$\eta = (192tk_0^3)^{\frac{i\nu}{2}} e^{8itk_0^3 + \chi(k_0)},$$

$$\varrho = z^{-\mathrm{i}\nu} \mathrm{e}^{-\frac{\mathrm{i}z^2}{4}(1+z/\sqrt{432tk_0^3})} \left(\frac{2k_0}{z/\sqrt{48tk_0}+2k_0}\right)^{-\mathrm{i}\nu} \mathrm{e}^{(\chi([z/\sqrt{48tk_0}]+k_0)-\chi(k_0))}. \quad (3.42)$$

Let $\tilde{\delta}(k) = \rho_{j,a}(k)(\delta(k) - \det \delta(k)I)\mathrm{e}^{t\Phi(\zeta,k)}$, then we have the following estimate.

Lemma 3.2 *For $z \in \hat{L} = \{z = \alpha \mathrm{e}^{\frac{3\mathrm{i}\pi}{4}} : -\infty < \alpha < +\infty\}$, as $t \to \infty$, the following estimate for $\tilde{\delta}(k)$ hold:*

$$|(S_{k_0}\tilde{\delta})(z)| \leqslant Ct^{-1/2}, \quad (3.43)$$

where the constant $C > 0$ independent of z, t.

Proof It follows from (3.3) and (3.4) that $\tilde{\delta}$ satisfies the following RH problem

$$\begin{cases} \tilde{\delta}_+(k) = (1+\rho(k)\rho^\dagger(k))\tilde{\delta}_-(k) + \mathrm{e}^{t\Phi(\zeta,k)}f(k), & |k| < k_0, \\ \phantom{\tilde{\delta}_+(k)} = \tilde{\delta}_-(k), & |k| > k_0, \\ \tilde{\delta}(k) \to 0, & k \to \infty, \end{cases} \quad (3.44)$$

where $f(k) = [\rho_{j,a}(\rho^\dagger \rho - \rho \rho^\dagger I)\delta_-](k)$. Then the function $\tilde{\delta}(k)$ can be expressed by

$$\begin{aligned}\tilde{\delta}(k) &= X(k) \int_{-k_0}^{k_0} \frac{\mathrm{e}^{t\Phi(\zeta,\xi)}f(\xi)}{X_+(\xi)(\xi-k)}\mathrm{d}\xi, \\ X(k) &= \exp\left\{\frac{1}{2\pi\mathrm{i}}\int_{-k_0}^{k_0} \frac{\ln(1+\rho(\xi)\rho^\dagger(\xi))}{\xi-k}\mathrm{d}\xi\right\}.\end{aligned} \quad (3.45)$$

On the other hand, we find that $\rho_{j,a}(\rho^\dagger \rho - \rho\rho^\dagger I) = \rho_{j,r}(\rho\rho^\dagger I - \rho^\dagger \rho)$. Thus, we get $f(k) = O(t^{-3/2})$. Similar to Lemma 3.1, $f(k)$ can be decomposed into two parts: $f_a(x,t,k)$ which has an analytic continuation to Ω_3 and $f_r(x,t,k)$ which is a small remainder. In particular,

$$\begin{aligned}|f_a(x,t,k)| &\leqslant \frac{C}{1+|k|^2}\mathrm{e}^{\frac{t}{4}|\mathrm{Re}\Phi(\zeta,k)|}, & k \in L_t, \\ |f_r(x,t,k)| &\leqslant \frac{C}{1+|k|^2}t^{-3/2}, & k \in (-k_0, k_0),\end{aligned} \quad (3.46)$$

where

$$\begin{aligned}L_t = &\left\{k = k_0 - \frac{1}{t} + \alpha \mathrm{e}^{\frac{3\mathrm{i}\pi}{4}} : 0 \leqslant \alpha \leqslant \sqrt{2}\left(k_0 - \frac{1}{2t}\right)\right\} \\ &\cup \left\{k = -k_0 + \alpha \mathrm{e}^{\frac{\mathrm{i}\pi}{4}} : 0 \leqslant \alpha \leqslant \sqrt{2}\left(k_0 - \frac{1}{2t}\right)\right\}.\end{aligned}$$

Therefore, for $z \in \hat{L}$, we can find

$$(S_{k_0}\tilde{\delta})(z) = X\left(k_0 + \frac{z}{\sqrt{48tk_0}}\right) \int_{k_0-\frac{1}{t}}^{k_0} \frac{e^{t\Phi(\zeta,\xi)}f(\xi)}{X_+(\xi)\left(\xi - k_0 - \frac{z}{\sqrt{48tk_0}}\right)} d\xi$$

$$+ X\left(k_0 + \frac{z}{\sqrt{48tk_0}}\right) \int_{-k_0}^{k_0-\frac{1}{t}} \frac{e^{t\Phi(\zeta,\xi)}f_a(x,t,\xi)}{X_+(\xi)\left(\xi - k_0 - \frac{z}{\sqrt{48tk_0}}\right)} d\xi$$

$$+ X\left(k_0 + \frac{z}{\sqrt{48tk_0}}\right) \int_{-k_0}^{k_0-\frac{1}{t}} \frac{e^{t\Phi(\zeta,\xi)}f_r(x,t,\xi)}{X_+(\xi)\left(\xi - k_0 - \frac{z}{\sqrt{48tk_0}}\right)} d\xi$$

$$= I_1 + I_2 + I_3.$$

Then,

$$|I_1| \leqslant \int_{k_0-\frac{1}{t}}^{k_0} \frac{|f(\xi)|}{\left|\xi - k_0 - \frac{z}{\sqrt{48tk_0}}\right|} d\xi \leqslant Ct^{-3/2} \left|\ln\left|1 - \frac{4\sqrt{3}}{zt^{1/2}}\right|\right| \leqslant Ct^{-2},$$

$$|I_3| \leqslant \int_{-k_0}^{k_0-\frac{1}{t}} \frac{|f_r(x,t,\xi)|}{\left|\xi - k_0 - \frac{z}{\sqrt{48tk_0}}\right|} d\xi \leqslant Ct^{-1/2}.$$

By the Cauchy's theorem, we can evaluate I_2 along the contour L_t instead of the interval $\left(-k_0, k_0 - \frac{1}{t}\right)$. Using the fact $\mathrm{Re}\Phi(\zeta,k) < 0$ in Ω_3, we can obtain $|I_2| \leqslant Ce^{-ct}$. This completes the proof of the lemma.

Remark 3.1 *There is a similar estimate*

$$|(S_{k_0}\hat{\delta})(z)| \leqslant Ct^{-1/2}, \quad t \to \infty, \tag{3.47}$$

for $z \in \bar{\hat{L}}$, where $\hat{\delta}(k) = [\delta^{-1}(k) - (\det \delta)^{-1}I]\rho_{j,a}(k)e^{-t\Phi(\zeta,k)}$, $j = 2, 4$.

In other words, we have the following important relation:

$$(S_{k_0}J_i^{(2)})(x,t,z) = \tilde{J}(x,t,z) + O(t^{-1/2}), \quad i = 1,\cdots,4, \tag{3.48}$$

where $\tilde{J}(x,t,z)$ is given by

$$\tilde{J}(x,t,z) = \begin{cases} \begin{pmatrix} I & 0 \\ -\eta^2 \varrho^2(t,z)\rho_{1,a}\left(\dfrac{z}{\sqrt{48tk_0}}+k_0\right) & 1 \end{pmatrix}, & k \in (\mathcal{X}_{k_0}^{\varepsilon})_1, \\[2ex] \begin{pmatrix} I & -\eta^{-2}\varrho^{-2}(t,z)\rho_{2,a}\left(\dfrac{z}{\sqrt{48tk_0}}+k_0\right) \\ 0 & 1 \end{pmatrix}, & k \in (\mathcal{X}_{k_0}^{\varepsilon})_2, \\[2ex] \begin{pmatrix} I & 0 \\ -\eta^2 \varrho^2(t,z)\rho_{3,a}\left(\dfrac{z}{\sqrt{48tk_0}}+k_0\right) & 1 \end{pmatrix}, & k \in (\mathcal{X}_{k_0}^{\varepsilon})_3, \\[2ex] \begin{pmatrix} I & -\eta^{-2}\varrho^{-2}(t,z)\rho_{4,a}\left(\dfrac{z}{\sqrt{48tk_0}}+k_0\right) \\ 0 & 1 \end{pmatrix}, & k \in (\mathcal{X}_{k_0}^{\varepsilon})_4. \end{cases} \qquad (3.49)$$

Here $\mathcal{X}_{k_0} = X + k_0$ denote the cross X defined by (3.50) centered at k_0, $\mathcal{X}_{k_0}^{\varepsilon} = \mathcal{X}_{k_0} \cap D_{\varepsilon}(k_0)$, and $X = X_1 \cup X_2 \cup X_3 \cup X_4 \subset \mathbf{C}$ be the cross defined by

$$\begin{aligned} X_1 &= \{le^{-\frac{i\pi}{4}}|0 \leqslant l < \infty\}, & X_2 &= \{le^{-\frac{3i\pi}{4}}|0 \leqslant l < \infty\}, \\ X_3 &= \{le^{\frac{3i\pi}{4}}|0 \leqslant l < \infty\}, & X_4 &= \{le^{\frac{i\pi}{4}}|0 \leqslant l < \infty\}, \end{aligned} \qquad (3.50)$$

and oriented as in Fig. 5.

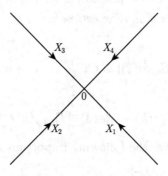

Fig. 5 The contour $X = X_1 \cup X_2 \cup X_3 \cup X_4$

For the jump matrix $\tilde{J}(x,t,z)$, for any fixed $z \in X$ and $\zeta \in \mathcal{I}$, we have

$$\rho_{1,a}\left(\frac{z}{\sqrt{48tk_0}} + k_0\right) \to \rho(k_0),$$

$$\rho_{3,a}\left(\frac{z}{\sqrt{48tk_0}} + k_0\right) \to -\frac{\rho(k_0)}{1+\rho(k_0)\rho^\dagger(k_0)}, \qquad (3.51)$$

$$\varrho^2 \to e^{-\frac{iz^2}{2}} z^{-2i\nu},$$

as $t \to \infty$. This implies that the jump matrix $\tilde{J}$ tend to the matrix $J^{(k_0)}$

$$J^{(k_0)}(x,t,z) = \begin{cases} \begin{pmatrix} I & 0 \\ -\eta^2 e^{-\frac{iz^2}{2}} z^{-2i\nu} \rho(k_0) & 1 \end{pmatrix}, & z \in X_1, \\[2mm] \begin{pmatrix} I & \eta^{-2} e^{\frac{iz^2}{2}} z^{2i\nu} \frac{\rho^\dagger(k_0)}{1+\rho(k_0)\rho^\dagger(k_0)} \\ 0 & 1 \end{pmatrix}, & z \in X_2, \\[4mm] \begin{pmatrix} I & 0 \\ \eta^2 e^{-\frac{iz^2}{2}} z^{-2i\nu} \frac{\rho(k_0)}{1+\rho(k_0)\rho^\dagger(k_0)} & 1 \end{pmatrix}, & z \in X_3, \\[2mm] \begin{pmatrix} I & -\eta^{-2} e^{\frac{iz^2}{2}} z^{2i\nu} \rho^\dagger(k_0) \\ 0 & 1 \end{pmatrix}, & z \in X_4 \end{cases} \qquad (3.52)$$

for large t.

Theorem 3.1 *The following RH problem:*

$$\begin{cases} M_+^X(x,t;z) = M_-^X(x,t;z) J^X(x,t,z), & \text{for almost every } z \in X, \\ M^X(x,t;z) \to I, & \text{as } z \to \infty, \end{cases} \qquad (3.53)$$

with the jump matrix $J^X(x,t,z) = \eta^{\hat\sigma} J^{(k_0)}(x,t,z)$ has a unique solution $M^X(x,t;z)$. If we let

$$M^X(x,t;z) = I + \frac{M_1^X(x,t)}{z} + O\left(\frac{1}{z^2}\right), \quad z \to \infty, \qquad (3.54)$$

then we have the following relation

$$(M_1^X)_{21} = -i\beta^X, \quad \beta^X = \frac{\nu \Gamma(-i\nu) e^{\frac{i\pi}{4} - \frac{\pi\nu}{2}}}{\sqrt{2\pi}} \rho(k_0), \qquad (3.55)$$

where $\Gamma(\cdot)$ denotes the standard Gamma function. Moreover, for each compact subset $\mathcal{D}$ of $\mathbb{C}$,

$$\sup_{\rho(k_0) \in \mathcal{D}} \sup_{z \in \mathbb{C}\setminus X} |M^X(x,t;z)| < \infty. \qquad (3.56)$$

Proof See appendix.

Therefore, we can approximate $M^{(2)}$ in the neighborhood $D_\varepsilon(k_0)$ of k_0 by

$$M^{(k_0)}(x,t;k) = \eta^{-\hat{\sigma}} M^X(x,t;z). \tag{3.57}$$

Lemma 3.3 *For each $\zeta \in \mathcal{I}$ and $t > 0$, the function $M^{(k_0)}(x,t;k)$ defined in (3.57) is an analytic function of $k \in D_\varepsilon(k_0) \setminus \mathcal{X}_{k_0}^\varepsilon$. Across $\mathcal{X}_{k_0}^\varepsilon$, $M^{(k_0)}$ satisfied the jump condition $M_+^{(k_0)} = M_-^{(k_0)} J^{(k_0)}$, where the jump matrix $J^{(k_0)}$ satisfies the following estimates for $1 \leqslant p \leqslant \infty$:*

$$\|J^{(2)} - J^{(k_0)}\|_{L^p(\mathcal{X}_{k_0}^\varepsilon)} \leqslant Ct^{-\frac{1}{2}-\frac{1}{2p}} \ln t, \quad t > 3, \ \zeta \in \mathcal{I}, \tag{3.58}$$

where $C > 0$ is a constant independent of t, ζ, z. Moreover, as $t \to \infty$,

$$\|(M^{(k_0)})^{-1}(x,t;k) - I\|_{L^\infty(\partial D_\varepsilon(k_0))} = O(t^{-1/2}), \tag{3.59}$$

and

$$-\frac{1}{2\pi i} \int_{\partial D_\varepsilon(k_0)} ((M^{(k_0)})^{-1}(x,t;k) - I)dk = \frac{\eta^{-\hat{\sigma}} M_1^X}{\sqrt{48tk_0}} + O(t^{-1}). \tag{3.60}$$

Proof The analyticity of $M^{(k_0)}$ is obvious. On the other hand, we have $J^{(2)} - J^{(k_0)} = J^{(2)} - \tilde{J} + \tilde{J} - J^{(k_0)}$. However, a careful computation as the Lemma 3.35 in [10], we conclude that

$$\|\tilde{J} - J^{(k_0)}\|_{L^\infty((\mathcal{X}_{k_0}^\varepsilon)_4)} \leqslant C|e^{\frac{i\gamma}{2}z^2}|t^{-1/2}\ln t, \quad 0 < \gamma < \frac{1}{2}, \ t > 3, \ \zeta \in \mathcal{I}, \tag{3.61}$$

for $k \in (\mathcal{X}_{k_0}^\varepsilon)_4$, that is, $z = \sqrt{48tk_0}ue^{\frac{i\pi}{4}}$, $0 \leqslant u \leqslant \varepsilon$. Thus, for ε small enough, it follows from (3.48) that

$$\|J^{(2)} - J^{(k_0)}\|_{L^\infty((\mathcal{X}_{k_0}^\varepsilon)_4)} \leqslant C|e^{\frac{i\gamma}{2}z^2}|t^{-1/2}\ln t. \tag{3.62}$$

Then we have

$$\|J^{(2)} - J^{(k_0)}\|_{L^1((\mathcal{X}_{k_0}^\varepsilon)_4)} \leqslant Ct^{-1}\ln t, \quad t > 3, \ \zeta \in \mathcal{I}. \tag{3.63}$$

By the general inequality $\|f\|_{L^p} \leqslant \|f\|_{L^\infty}^{1-1/p} \|f\|_{L^1}^{1/p}$, we find

$$\|J^{(2)} - J^{(k_0)}\|_{L^p((\mathcal{X}_{k_0}^\varepsilon)_4)} \leqslant Ct^{-1/2-1/2p}\ln t, \quad t > 3, \ \zeta \in \mathcal{I}. \tag{3.64}$$

The norms on $(\mathcal{X}_{k_0}^\varepsilon)_j$, $j = 1,2,3$, are estimated in a similar way. Therefore, (3.58) follows.

If $k \in \partial D_\varepsilon(k_0)$, the variable $z = \sqrt{48tk_0}(k - k_0)$ tends to infinity as $t \to \infty$. It follows from (3.54) that

$$M^X(x, t; z) = I + \frac{M_1^X}{\sqrt{48tk_0}(k - k_0)} + O\left(\frac{1}{t}\right), \quad t \to \infty, \ k \in \partial D_\varepsilon(k_0).$$

Since

$$M^{(k_0)}(x, t; k) = \eta^{-\hat{\sigma}} M^X(x, t; z),$$

thus we have

$$(M^{(k_0)})^{-1}(x, t; k) - I = -\frac{\eta^{-\hat{\sigma}} M_1^X}{\sqrt{48tk_0}(k - k_0)} + O\left(\frac{1}{t}\right), \quad t \to \infty, \ k \in \partial D_\varepsilon(k_0). \tag{3.65}$$

The estimate (3.59) immediately follows from (3.65) and $|M_1^X| \leqslant C$. By Cauchy's formula and (3.65), we derive (3.60).

Let

$$\mathcal{X}^\varepsilon_{-k_0} = \cup_{j=1}^4 \Sigma_j \cap D_\varepsilon(-k_0),$$

where $\{\Sigma_j\}_1^6$ denote the restriction of Σ to the contour labeled by j in Fig. 4. Proceeding the analogous steps as above, we can approximate $M^{(2)}$ in the neighborhood $D_\varepsilon(-k_0)$ of $-k_0$ by $M^{(-k_0)}(x, t; k)$, where $M^{(-k_0)}$ satisfies the following RH problem:

$$\begin{aligned} M_+^{(-k_0)}(x, t; k) &= M_-^{(-k_0)}(x, t; k) J^{(-k_0)}(x, t, z), \quad k \in \mathcal{X}^\varepsilon_{-k_0}, \\ M^{(-k_0)}(x, t; k) &\to I, \quad k \to \infty, \end{aligned} \tag{3.66}$$

where the jump matrix $J^{(-k_0)}(x, t, z)$ is given by

$$J^{(-k_0)}(x, t, z) = \hat{\eta}^{-\hat{\sigma}} J^Y(x, t, z), \quad \hat{\eta} = (192tk_0^3)^{-\frac{i\nu}{2}} e^{-8itk_0^3 + \chi(-k_0)}, \tag{3.67}$$

and

$$J^Y(x,t,z) = \begin{cases} \begin{pmatrix} I & -e^{-\frac{iz^2}{2}}(-z)^{-2i\nu}\dfrac{\rho^\dagger(-k_0)}{1+\rho(-k_0)\rho^\dagger(-k_0)} \\ 0 & 1 \end{pmatrix}, & z \in X_1, \\ \begin{pmatrix} I & 0 \\ e^{\frac{iz^2}{2}}(-z)^{2i\nu}\rho(-k_0) & 1 \end{pmatrix}, & z \in X_2, \\ \begin{pmatrix} I & e^{-\frac{iz^2}{2}}(-z)^{-2i\nu}\rho^\dagger(-k_0) \\ 0 & 1 \end{pmatrix}, & z \in X_3, \\ \begin{pmatrix} I & 0 \\ -e^{\frac{iz^2}{2}}(-z)^{2i\nu}\dfrac{\rho(-k_0)}{1+\rho(-k_0)\rho^\dagger(-k_0)} & 1 \end{pmatrix}, & z \in X_4. \end{cases} \quad (3.68)$$

Using the formulae (2.39) and (3.8), one verifies that

$$\begin{pmatrix} 0 & 1 & 0 \\ 1 & 0 & 0 \\ 0 & 0 & 1 \end{pmatrix} \overline{J^Y(x,t,-\bar{z})} \begin{pmatrix} 0 & 1 & 0 \\ 1 & 0 & 0 \\ 0 & 0 & 1 \end{pmatrix} = J^X(x,t,z), \quad (3.69)$$

which in turn implies, by uniqueness, that

$$M^{(-k_0)}(x,t;k) = \hat{\eta}^{-\hat{\sigma}} \begin{pmatrix} 0 & 1 & 0 \\ 1 & 0 & 0 \\ 0 & 0 & 1 \end{pmatrix} \overline{M^X(x,t,-\bar{z})} \begin{pmatrix} 0 & 1 & 0 \\ 1 & 0 & 0 \\ 0 & 0 & 1 \end{pmatrix}. \quad (3.70)$$

Furthermore, we have the similar results about $M^{(-k_0)}(x,t;k)$ as stated in Lemma 3.3. The jump matrix $J^{(-k_0)}$ satisfies the following estimates for $1 \leqslant p \leqslant \infty$:

$$\|J^{(2)} - J^{(-k_0)}\|_{L^p(\mathcal{X}^\varepsilon_{-k_0})} \leqslant Ct^{-\frac{1}{2}-\frac{1}{2p}} \ln t, \quad t > 3, \ \zeta \in \mathcal{I}. \quad (3.71)$$

As $t \to \infty$,

$$\|(M^{(-k_0)})^{-1}(x,t;k) - I\|_{L^\infty(\partial D_\varepsilon(-k_0))} = O(t^{-1/2}), \quad (3.72)$$

and

$$-\frac{1}{2\pi i} \int_{\partial D_\varepsilon(-k_0)} ((M^{(-k_0)})^{-1}(x,t;k) - I) dk = \frac{\hat{\eta}^{-\hat{\sigma}} M_1^Y}{\sqrt{48tk_0}} + O(t^{-1}), \quad (3.73)$$

where M_1^Y is defined by

$$M_1^Y = -\begin{pmatrix} 0 & 1 & 0 \\ 1 & 0 & 0 \\ 0 & 0 & 1 \end{pmatrix} \overline{M_1^X} \begin{pmatrix} 0 & 1 & 0 \\ 1 & 0 & 0 \\ 0 & 0 & 1 \end{pmatrix}, \quad (M_1^Y)_{21} = -i\beta^Y, \quad \beta^Y = \overline{\beta^X} \begin{pmatrix} 0 & 1 \\ 1 & 0 \end{pmatrix}. \tag{3.74}$$

3.3 Find asymptotic formula

Define the approximate solution $M^{(app)}(x,t;k)$ by

$$M^{(app)} = \begin{cases} M^{(k_0)}, & k \in D_\varepsilon(k_0), \\ M^{(-k_0)}, & k \in D_\varepsilon(-k_0), \\ I, & \text{elsewhere.} \end{cases} \tag{3.75}$$

Let $\hat{M}(x,t;k)$ be

$$\hat{M} = M^{(2)}(M^{(app)})^{-1}, \tag{3.76}$$

then $\hat{M}(x,t;k)$ satisfies the following RH problem

$$\begin{aligned} \hat{M}_+(x,t;k) &= \hat{M}_-(x,t;k)\hat{J}(x,t,k), & k \in \hat{\Sigma}, \\ \hat{M}(x,t;k) &\to I, & k \to \infty, \end{aligned} \tag{3.77}$$

where the jump contour $\hat{\Sigma} = \Sigma \cup \partial D_\varepsilon(k_0) \cup \partial D_\varepsilon(-k_0)$ is depicted in Fig. 6, and the jump matrix $\hat{J}(x,t,k)$ is given by

$$\hat{J} = \begin{cases} M_-^{(app)} J^{(2)}(M_+^{(app)})^{-1}, & k \in \hat{\Sigma} \cap (D_\varepsilon(k_0) \cup D_\varepsilon(-k_0)), \\ (M^{(app)})^{-1}, & k \in \partial D_\varepsilon(k_0) \cup \partial D_\varepsilon(-k_0), \\ J^{(2)}, & k \in \hat{\Sigma} \setminus (\overline{D_\varepsilon(k_0)} \cup \overline{D_\varepsilon(-k_0)}). \end{cases} \tag{3.78}$$

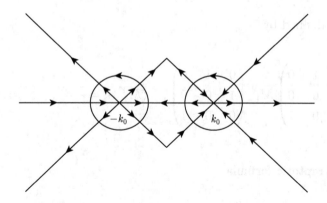

Fig. 6 The contour $\hat{\Sigma}$

For convenience, we rewrite $\hat{\Sigma}$ as follows:

$$\hat{\Sigma} = (\partial D_\varepsilon(-k_0) \cup \partial D_\varepsilon(k_0)) \cup (\mathcal{X}^\varepsilon_{-k_0} \cup \mathcal{X}^\varepsilon_{k_0}) \cup \hat{\Sigma}_1 \cup \hat{\Sigma}_2,$$

where

$$\hat{\Sigma}_1 = \bigcup_1^4 \Sigma_j \setminus (D_\varepsilon(-k_0) \cup D_\varepsilon(k_0)), \quad \hat{\Sigma}_2 = \bigcup_5^6 \Sigma_j.$$

Then we have the following lemma if let $\hat{w} = \hat{J} - I$.

Lemma 3.4 For $1 \leqslant p \leqslant \infty$, the following estimates hold for $t > 3$ and $\zeta \in \mathcal{I}$,

$$\|\hat{w}\|_{L^p(\partial D_\varepsilon(-k_0) \cup \partial D_\varepsilon(k_0))} \leqslant C t^{-1/2}, \tag{3.79}$$

$$\|\hat{w}\|_{L^p(\mathcal{X}^\varepsilon_{-k_0} \cup \mathcal{X}^\varepsilon_{k_0})} \leqslant C t^{-\frac{1}{2}-\frac{1}{2p}} \ln t, \tag{3.80}$$

$$\|\hat{w}\|_{L^p(\hat{\Sigma}_1)} \leqslant C e^{-ct}, \tag{3.81}$$

$$\|\hat{w}\|_{L^p(\hat{\Sigma}_2)} \leqslant C t^{-3/2}. \tag{3.82}$$

Proof The inequality (3.79) is a consequence of (3.59) and (3.78). For $k \in \mathcal{X}^\varepsilon_{k_0}$, we find

$$\hat{w} = M^{(k_0)}_-(J^{(2)} - J^{(k_0)})(M^{(k_0)}_+)^{-1}.$$

Therefore, it follows from (3.58) that the estimate (3.80) holds.

For $k \in \Omega_4 \cap \Sigma_4$, let $k = k_0 + u e^{\frac{i\pi}{4}}$, $\varepsilon < u < \infty$, then

$$\text{Re}\Phi(\zeta, k) = 4u^2(\sqrt{2}u + 6k_0).$$

Since $\hat{w}$ only has a nonzero $-\dfrac{\delta^{-1}\rho_{4,a}e^{-t\Phi}}{\det\delta}$ in (12) entry, hence, for $t \geqslant 1$, by (3.11) and (3.18), we get

$$|\hat{w}_{12}| = \left|-\frac{\delta^{-1}\rho_{4,a}e^{-t\Phi}}{\det\delta}\right| \leqslant C|\rho_{4,a}|e^{-t|\operatorname{Re}\Phi|}$$

$$\leqslant \frac{C}{1+|k|^2}e^{-\frac{3t}{4}|\operatorname{Re}\Phi|} \leqslant Ce^{-c\varepsilon^2 t}.$$

In a similar way, the other estimates on $\hat{\Sigma}_1$ hold. This proves (3.81).

Since the matrix $\hat{w}$ on $\hat{\Sigma}_2$ only involves the small remainders $\{\rho_{j,r}\}_1^4$, thus, by Lemma 3.1, the estimate (3.82) follows.

The estimates in Lemma 3.4 imply that

$$\begin{aligned}\|\hat{w}\|_{(L^1\cap L^2)(\hat{\Sigma})} &\leqslant Ct^{-1/2}, \\ \|\hat{w}\|_{L^\infty(\hat{\Sigma})} &\leqslant Ct^{-1/2}\ln t,\end{aligned} \qquad t > 3,\ \zeta \in \mathcal{I}. \qquad (3.83)$$

Let $\hat{C}$ denote the Cauchy operator associated with $\hat{\Sigma}$:

$$(\hat{C}f)(k) = \int_{\hat{\Sigma}}\frac{f(s)}{s-k}\frac{\mathrm{d}s}{2\pi\mathrm{i}}, \quad k \in \mathbf{C}\setminus\hat{\Sigma},\ f \in L^2(\hat{\Sigma}).$$

We denote the boundary values of $\hat{C}f$ from the left and right sides of $\hat{\Sigma}$ by $\hat{C}_+f$ and $\hat{C}_-f$, respectively. As is well known, the operators $\hat{C}_\pm$ are bounded from $L^2(\hat{\Sigma})$ to $L^2(\hat{\Sigma})$, and $\hat{C}_+ - \hat{C}_- = I$, here I denotes the identity operator.

Define the operator $\hat{C}_{\hat{w}}\colon L^2(\hat{\Sigma}) + L^\infty(\hat{\Sigma}) \to L^2(\hat{\Sigma})$ by $\hat{C}_{\hat{w}}f = \hat{C}_-(f\hat{w})$, that is, $\hat{C}_{\hat{w}}$ is defined by $\hat{C}_{\hat{w}}(f) = \hat{C}_+(f\hat{w}_-) + \hat{C}_-(f\hat{w}_+)$ where we have chosen, for simplicity, $\hat{w}_+ = \hat{w}$ and $\hat{w}_- = 0$. Then, by (3.83), we find

$$\|\hat{C}_{\hat{w}}\|_{B(L^2(\hat{\Sigma}))} \leqslant C\|\hat{w}\|_{L^\infty(\hat{\Sigma})} \leqslant Ct^{-1/2}\ln t, \qquad (3.84)$$

where $B(L^2(\hat{\Sigma}))$ denotes the Banach space of bounded linear operators $L^2(\hat{\Sigma}) \to L^2(\hat{\Sigma})$. Therefore, there exists a $T > 0$ such that $I - \hat{C}_{\hat{w}} \in B(L^2(\hat{\Sigma}))$ is invertible for all $\zeta \in \mathcal{I}$, $t > T$. Following this, we may define the 2×2 matrix-valued function $\hat{\mu}(x,t;k)$ whenever $t > T$ by

$$\hat{\mu} = I + \hat{C}_{\hat{w}}\hat{\mu}. \qquad (3.85)$$

Then

$$\hat{M}(x,t;k) = I + \frac{1}{2\pi\mathrm{i}}\int_{\hat{\Sigma}}\frac{(\hat{\mu}\hat{w})(x,t;s)}{s-k}\mathrm{d}s, \quad k \in \mathbf{C}\setminus\hat{\Sigma} \qquad (3.86)$$

is the unique solution of the RH problem (3.77) for $t > T$.

Moreover, standard estimates using the Neumann series show that the function $\hat{\mu}(x,t;k)$ satisfies

$$\|\hat{\mu}(x,t;\cdot) - I\|_{L^2(\hat{\Sigma})} = O(t^{-1/2}), \quad t \to \infty, \ \zeta \in \mathcal{I}. \tag{3.87}$$

In fact, equation (3.85) is equivalent to $\hat{\mu} = I + (I - \hat{C}_{\hat{w}})^{-1}\hat{C}_{\hat{w}}I$. Using the Neumann series, we get

$$\|(I - \hat{C}_{\hat{w}})^{-1}\|_{B(L^2(\hat{\Sigma}))} \leqslant \frac{1}{1 - \|\hat{C}_{\hat{w}}\|_{B(L^2(\hat{\Sigma}))}},$$

whenever $\|\hat{C}_{\hat{w}}\|_{B(L^2(\hat{\Sigma}))} < 1$. Thus, we find

$$\|\hat{\mu}(x,t;\cdot) - I\|_{L^2(\hat{\Sigma})} = \|(I - \hat{C}_{\hat{w}})^{-1}\hat{C}_{\hat{w}}I\|_{L^2(\hat{\Sigma})}$$
$$\leqslant \|(I - \hat{C}_{\hat{w}})^{-1}\|_{B(L^2(\hat{\Sigma}))} \|\hat{C}_{-}(\hat{w})\|_{L^2(\hat{\Sigma})}$$
$$\leqslant \frac{C\|\hat{w}\|_{L^2(\hat{\Sigma})}}{1 - \|\hat{C}_{\hat{w}}\|_{B(L^2(\hat{\Sigma}))}} \leqslant C\|\hat{w}\|_{L^2(\hat{\Sigma})}$$

for all t large enough and all $\zeta \in \mathcal{I}$. In view of (3.83), this gives (3.87). It follows from (3.86) that

$$\lim_{k \to \infty} k(\hat{M}(x,t;k) - I) = -\frac{1}{2\pi i}\int_{\hat{\Sigma}}(\hat{\mu}\hat{w})(x,t;k)dk. \tag{3.88}$$

Using (3.82) and (3.87), we have

$$\int_{\hat{\Sigma}_2}(\hat{\mu}\hat{w})(x,t;k)dk = \int_{\hat{\Sigma}_2}\hat{w}(x,t;k)dk + \int_{\hat{\Sigma}_2}(\hat{\mu}(x,t;k) - I)\hat{w}(x,t;k)dk$$
$$\leqslant \|\hat{w}\|_{L^1(\hat{\Sigma}_2)} + \|\hat{\mu} - I\|_{L^2(\hat{\Sigma}_2)}\|\hat{w}\|_{L^2(\hat{\Sigma}_2)}$$
$$\leqslant Ct^{-3/2}, \quad t \to \infty.$$

Similarly, by (3.80) and (3.87), the contribution from $\mathcal{X}^{\varepsilon}_{-k_0} \cup \mathcal{X}^{\varepsilon}_{k_0}$ to the right-hand side of (3.88) is

$$O(\|\hat{w}\|_{L^1(\mathcal{X}^{\varepsilon}_{-k_0} \cup \mathcal{X}^{\varepsilon}_{k_0})} + \|\hat{\mu} - I\|_{L^2(\mathcal{X}^{\varepsilon}_{-k_0} \cup \mathcal{X}^{\varepsilon}_{k_0})}\|\hat{w}\|_{L^2(\mathcal{X}^{\varepsilon}_{-k_0} \cup \mathcal{X}^{\varepsilon}_{k_0})}) = O(t^{-1}\ln t), \quad t \to \infty.$$

By (3.81) and (3.87), the contribution from $\hat{\Sigma}_1$ to the right-hand side of (3.88) is

$$O(\|\hat{w}\|_{L^1(\hat{\Sigma}_1)} + \|\hat{\mu} - I\|_{L^2(\hat{\Sigma}_1)}\|\hat{w}\|_{L^2(\hat{\Sigma}_1)}) = O(e^{-ct}), \quad t \to \infty.$$

Finally, by (3.60), (3.73), (3.78), (3.79) and (3.87), we can get

$$-\frac{1}{2\pi i}\int_{\partial D_\varepsilon(-k_0)\cup\partial D_\varepsilon(k_0)} (\hat{\mu}\hat{w})(x,t;k)dk$$

$$= -\frac{1}{2\pi i}\int_{\partial D_\varepsilon(-k_0)\cup\partial D_\varepsilon(k_0)} \hat{w}(x,t;k)dk$$

$$-\frac{1}{2\pi i}\int_{\partial D_\varepsilon(-k_0)\cup\partial D_\varepsilon(k_0)} (\hat{\mu}(x,t;k)-I)\hat{w}(x,t;k)dk$$

$$= -\frac{1}{2\pi i}\int_{\partial D_\varepsilon(-k_0)} \left((M^{(-k_0)})^{-1}(x,t;k)-I\right)dk$$

$$-\frac{1}{2\pi i}\int_{\partial D_\varepsilon(k_0)} \left((M^{(k_0)})^{-1}(x,t;k)-I\right)dk$$

$$+O(\|\hat{\mu}-I\|_{L^2(\partial D_\varepsilon(-k_0)\cup\partial D_\varepsilon(k_0))}\|\hat{w}\|_{L^2(\partial D_\varepsilon(-k_0)\cup\partial D_\varepsilon(k_0))})$$

$$= \frac{\eta^{-\hat{\sigma}}M_1^X}{\sqrt{48tk_0}} + \frac{\hat{\eta}^{-\hat{\sigma}}M_1^Y}{\sqrt{48tk_0}} + O(t^{-1}), \quad t\to\infty.$$

Thus, we obtain the following important relation

$$\lim_{k\to\infty} k(\hat{M}(x,t;k)-I) = \frac{\eta^{-\hat{\sigma}}M_1^X + \hat{\eta}^{-\hat{\sigma}}M_1^Y}{\sqrt{48tk_0}} + O(t^{-1}\ln t), \quad t\to\infty. \quad (3.89)$$

Taking into account that (2.40), (3.3), (3.4), (3.36) and (3.89), for sufficient large $k\in \mathbf{C}\setminus\hat{\Sigma}$, we get

$$\left(\bar{u}(x,t) \quad u(x,t)\right) = 2i\lim_{k\to\infty} (kM(x,t;k))_{21}$$

$$= 2i\lim_{k\to\infty} k(\hat{M}(x,t;k)-I)_{21}$$

$$= 2i\left(\frac{-i\eta^2\beta^X - i\hat{\eta}^2\beta^Y}{\sqrt{48tk_0}}\right) + O\left(\frac{\ln t}{t}\right)$$

$$= \frac{\eta^2\beta^X + \hat{\eta}^2\beta^Y}{\sqrt{12tk_0}} + O\left(\frac{\ln t}{t}\right). \quad (3.90)$$

Collecting the above computations, we obtain our main results stated as the following theorem.

Theorem 3.2 *Let $u_0(x)$ lie in the Schwartz space $S(\mathbf{R})$. Then, for any positive constant $M>1$, as $t\to\infty$, the solution $u(x,t)$ of the initial problem for Sasa-Satsuma equation (1.1) on the line satisfies the following asymptotic formula*

$$u(x,t) = \frac{u_{as}(x,t)}{\sqrt{t}} + O\left(\frac{\ln t}{t}\right), \quad t\to\infty, \zeta\in\mathcal{I}, \quad (3.91)$$

where the error term is uniform with respect to x in the given range, and the leading-order coefficient $u_{as}(x,t)$ is given by

$$u_{as}(x,t) = \frac{\nu e^{-\frac{\pi\nu}{2}}}{\sqrt{24\pi k_0}} \bigg((192tk_0^3)^{i\nu} e^{16itk_0^3 + 2\chi(k_0) + \frac{\pi i}{4}} \Gamma(-i\nu) \frac{s_{32}(k_0)}{s_{33}(k_0)} \\ + (192tk_0^3)^{-i\nu} e^{-16itk_0^3 + 2\chi(-k_0) - \frac{\pi i}{4}} \Gamma(i\nu) \frac{\bar{s}_{31}(k_0)}{\bar{s}_{33}(k_0)} \bigg), \quad (3.92)$$

where k_0, ν and $\chi(k)$ are given by (3.1), (3.6) and (3.7), respectively.

3.4 Discussions

From the point of view of scaling transform, if the spatial variable x is replaced by $-x$, the SS equation (1.1) with $\varepsilon = 1$ considered in present paper is turned into the problem considered in [26], that is, SS equation (1.1) corresponding to $\varepsilon = -1$. Now, we check the coincidence of our asymptotic formula (3.91) with the leading asymptotics obtained in [26].

We first point out a crucial mistake in the subsection 3.7, i.e., the final transformation of [26]. We note that the expression $H(k)$ should be modified by

$$H(k) = (\delta_A)^{-\sigma} M^{A^0}(k) (\delta_A)^{\sigma},$$

where we follow the notation in [26]. This directly leads to an important quantity $(M_1^{A^0})_{12}$ in equation (67) of [26] that should be corrected to

$$(M_1^{A^0})_{12} = i(\delta_A)^2 \beta_{12}.$$

Thus, the formula (65) becomes

$$q(x,t) = (u(x,t), \bar{u}(x,t))^{\mathrm{T}} = \frac{i}{\sqrt{12k_0 t}} \left(M_1^{A^0} - \sigma_1 \overline{M_1^{A^0}} \right)_{12} + O\left(c(k_0) t^{-1} \ln t \right),$$

$$= \frac{i}{\sqrt{12k_0 t}} \left(i(\delta_A)^2 \beta_{12} + i(\delta_A)^{-2} \sigma_1 \bar{\beta}_{12} \right).$$

It then follows that

$$q^\dagger(x,t) = (\bar{u}(x,t), u(x,t)) = \frac{1}{\sqrt{12k_0 t}} \left((\delta_A)^{-2} \beta_{21} + (\delta_A)^2 \bar{\beta}_{21} \sigma_1 \right).$$

Recalling the formula (58) for δ_A, we get $(\delta_A)^{-2} = \eta^2$ since $\chi(-k_0) = -\chi(k_0)$, where η is defined in (3.42).

Therefore, it follows from (3.90) that the coincidence of the asymptotic formulas become to compare the expressions β_{21} and β^X. On the other hand, we know from [26] that

$$\beta_{21} = \frac{\nu \Gamma(i\nu) e^{\frac{\pi \nu}{2} - \frac{3\pi i}{4}}}{\sqrt{2\pi}} \gamma(-k_0). \tag{3.93}$$

In view of the ν in (3.93) corresponding to $-\nu$ in our paper (see (3.6) and (21) in [26]), we should only verify the coincidence of $\rho(k_0)$ and $\gamma(-k_0)$. In fact, we can understand this as follows: Since the signature tables for $\mathrm{Re}\,\Phi$ and the $\mathrm{Re}\,\theta$ are just opposite, that is, the signature table of $\mathrm{Re}\,\Phi$ can be seen as the signature table of $\mathrm{Re}\,\theta$ rotating 180 degrees, from the point of view of the transforming processes of the RH problems or the deformations of the jump contours, we have $\rho(k_0) = \gamma(-k_0)$.

Remark 3.2 *The key difficult point to directly check the relation between spectral functions ρ and γ is the Lax pair representation. If we rewrite our Lax pair (2.1) as*

$$\begin{cases} \phi_x = (ik\sigma - U)\phi, \\ \phi_t = (-4ik^3\sigma - V)\phi, \end{cases}$$

and use it to perform the spectral analysis as well as construct the RH problem, one can easily find the corresponding relation of the two spectral functions ρ and γ.

4 Appendix. Proof of Theorem 3.1

To solve the model RH problem (3.53), it is convenient to introduce the following transformation

$$\Psi(z) = M^X(z) z^{i\nu\sigma} e^{\frac{iz^2}{4}\sigma},$$

which implies that

$$\Psi_+(z) = \Psi_-(z) v(k_0), \quad v(k_0) = z^{-i\nu\hat{\sigma}} e^{-\frac{iz^2}{4}\hat{\sigma}} J^X, \tag{4.1}$$

where we suppress the x and t dependence for clarity. Since the jump matrix is constant along each ray, we have

$$\frac{d\Psi_+(z)}{dz} = \frac{d\Psi_-(z)}{dz} v(k_0),$$

from which it follows that $\frac{\mathrm{d}\Psi(z)}{\mathrm{d}z}\Psi^{-1}(z)$ has no jump discontinuity along any of the rays. Moreover,

$$\frac{\mathrm{d}\Psi(z)}{\mathrm{d}z}\Psi^{-1}(z) = \frac{\mathrm{d}M^X(z)}{\mathrm{d}z}(M^X)^{-1}(z) + \frac{\mathrm{i}}{2}zM^X(z)\sigma(M^X)^{-1}(z)$$
$$+ \frac{\mathrm{i}\nu}{2}M^X(z)\sigma(M^X)^{-1}(z) = \frac{\mathrm{i}z}{2}\sigma - \frac{\mathrm{i}}{2}[\sigma, M_1^X] + O\left(\frac{1}{z}\right).$$

It follows by Liouville's argument that

$$\frac{\mathrm{d}\Psi(z)}{\mathrm{d}z} = \left(\frac{\mathrm{i}z}{2}\sigma + \beta\right)\Psi(z), \tag{4.2}$$

where

$$\beta = -\frac{\mathrm{i}}{2}[\sigma, M_1^X] = \begin{pmatrix} 0 & \beta_{12} \\ \beta_{21} & 0 \end{pmatrix}.$$

In particular,

$$(M_1^X)_{21} = -\mathrm{i}\beta_{21}. \tag{4.3}$$

It is possible to show that the solution of the RH problem (3.53) for $M^X(z)$ is unique since $\det J^X = 1$, and therefore we have $(M^X(\bar{k}))^\dagger = (M^X(k))^{-1}$, which implies that

$$\beta_{12} = -\beta_{21}^\dagger.$$

From equation (4.2) we obtain

$$\begin{aligned}
\frac{\mathrm{d}^2\Psi_{22}(z)}{\mathrm{d}z^2} &= \left(-\frac{\mathrm{i}}{2} - \frac{z^2}{4} + \beta_{21}\beta_{12}\right)\Psi_{22}(z), \\
\beta_{21}\Psi_{12}(z) &= \frac{\mathrm{d}\Psi_{22}(z)}{\mathrm{d}z} + \frac{\mathrm{i}}{2}z\Psi_{22}, \\
\frac{\mathrm{d}^2\beta_{21}\Psi_{11}(z)}{\mathrm{d}z^2} &= \left(\frac{\mathrm{i}}{2} - \frac{z^2}{4} + \beta_{21}\beta_{12}\right)\beta_{21}\Psi_{11}(z), \\
\Psi_{21}(z) &= \frac{1}{\beta_{21}\beta_{12}}\left(\frac{\mathrm{d}\beta_{21}\Psi_{11}(z)}{\mathrm{d}z} - \frac{\mathrm{i}}{2}z\beta_{21}\Psi_{11}\right).
\end{aligned} \tag{4.4}$$

It is well known that the Weber's equation

$$\frac{\mathrm{d}^2 g(s)}{\mathrm{d}s^2} + \left(\frac{1}{2} - \frac{s^2}{4} + a\right)g(s) = 0$$

has the solution

$$g(s) = c_1 D_a(s) + c_2 D_a(-s), \tag{4.5}$$

where $D_a(\cdot)$ denotes the standard parabolic-cylinder function and satisfies

$$\frac{dD_a(s)}{ds} + \frac{s}{2}D_a(s) - aD_{a-1}(s) = 0, \tag{4.6}$$

$$D_{a-1}(s) = \frac{\Gamma(a)}{\sqrt{2\pi}}\left(e^{\frac{i\pi}{2}(a-1)}D_{-a}(is) + e^{-\frac{i\pi}{2}(a-1)}D_{-a}(-is)\right). \tag{4.7}$$

Furthermore, as $s \to \infty$, we know from [37] that

$$D_a(s) = \begin{cases} s^a e^{-\frac{s^2}{4}}(1 + O(s^{-2})), & |\arg s| < \frac{3\pi}{4}, \\ s^a e^{-\frac{s^2}{4}}(1 + O(s^{-2})) - \frac{\sqrt{2\pi}}{\Gamma(-a)} e^{a\pi i + \frac{s^2}{4}} s^{-a-1}(1 + O(s^{-2})), & \frac{\pi}{4} < \arg s < \frac{5\pi}{4}, \\ s^a e^{-\frac{s^2}{4}}(1 + O(s^{-2})) - \frac{\sqrt{2\pi}}{\Gamma(-a)} e^{-a\pi i + \frac{s^2}{4}} s^{-a-1}(1 + O(s^{-2})), & -\frac{5\pi}{4} < \arg s < -\frac{\pi}{4}, \end{cases}$$

where Γ is the Gamma function. Setting $a = i\beta_{21}\beta_{12}$, we have

$$\Psi_{22}(z) = c_1 D_a(e^{-\frac{3i\pi}{4}}z) + c_2 D_a(e^{\frac{\pi i}{4}}z),$$
$$\beta_{21}\Psi_{11}(z) = c_3 D_{-a}(e^{\frac{3i\pi}{4}}z) + c_4 D_{-a}(e^{-\frac{\pi i}{4}}z).$$

Since as $z \to \infty$, we have

$$\Psi_{11} \to z^{i\nu} e^{\frac{iz^2}{4}} I, \quad \Psi_{22} \to z^{-i\nu} e^{-\frac{iz^2}{4}}, \tag{4.8}$$

hence, as $\arg z \in \left(-\frac{3\pi}{4}, -\frac{\pi}{4}\right)$, we immediately arrive that

$$\begin{aligned} \Psi_{22}(z) &= e^{-\frac{\pi\nu}{4}} D_a(e^{\frac{\pi i}{4}}z), \quad a = -i\nu, \\ \beta_{21}\Psi_{11}(z) &= \beta_{21} e^{\frac{3\pi\nu}{4}} D_{-a}(e^{\frac{3\pi i}{4}}z). \end{aligned} \tag{4.9}$$

According to (4.4) and (4.6), we have

$$\begin{aligned} \Psi_{21}(z) &= \beta_{21} e^{\frac{\pi(i+3\nu)}{4}} D_{-a-1}(e^{\frac{3\pi i}{4}}z), \\ \beta_{21}\Psi_{12}(z) &= e^{\frac{\pi(i-\nu)}{4}} a D_{a-1}(e^{\frac{\pi i}{4}}z). \end{aligned} \tag{4.10}$$

Similarly, for $\arg z \in \left(-\frac{\pi}{4}, \frac{\pi}{4}\right)$, we can get

$$\begin{aligned} \Psi_{22}(z) &= e^{-\frac{\pi\nu}{4}} D_a(e^{\frac{\pi i}{4}}z), \quad a = -i\nu, \\ \beta_{21}\Psi_{11}(z) &= \beta_{21} e^{-\frac{\pi\nu}{4}} D_{-a}(e^{-\frac{\pi i}{4}}z), \\ \Psi_{21}(z) &= \beta_{21} e^{-\frac{\pi(3i+\nu)}{4}} D_{-a-1}(e^{-\frac{\pi i}{4}}z), \\ \beta_{21}\Psi_{12}(z) &= e^{\frac{\pi(i-\nu)}{4}} a D_{a-1}(e^{\frac{\pi i}{4}}z). \end{aligned} \tag{4.11}$$

Along the ray $\arg z = -\dfrac{\pi}{4}$, we have

$$\Psi_+(z) = \Psi_-(z) \begin{pmatrix} I & 0 \\ -\rho(k_0) & 1 \end{pmatrix},$$

thus,

$$\beta_{12} e^{\frac{3\pi\nu}{4}} D_{-a}(e^{\frac{3\pi i}{4}} z) = \beta_{21} e^{-\frac{\pi\nu}{4}} D_{-a}(e^{-\frac{\pi i}{4}} z) - e^{\frac{\pi(i-\nu)}{4}} a D_{a-1}(e^{\frac{\pi i}{4}} z)\rho(k_0). \qquad (4.12)$$

On the other hand, it follows from (4.7) that

$$D_{a-1}(e^{\frac{\pi i}{4}} z) = \frac{\Gamma(a)}{\sqrt{2\pi}}\left(e^{\frac{i\pi}{2}(a-1)} D_{-a}(e^{\frac{3\pi i}{4}} z) + e^{-\frac{i\pi}{2}(a-1)} D_{-a}(e^{-\frac{\pi i}{4}} z)\right). \qquad (4.13)$$

Compared the coefficients of (4.12) with (4.13), we can find that

$$\beta_{21} = \frac{\Gamma(-i\nu)}{\sqrt{2\pi}} e^{\frac{\pi i}{4} - \frac{\pi\nu}{2}} \nu \rho(k_0). \qquad (4.14)$$

The estimate (3.56) is an consequence of Lemma A.4 in [25] and the asymptotic formula (3.54).

References

[1] Ablowitz M J, Fokas A S. Complex Variables: Introduction and Applications [M]. 2nd ed. Cambridge: Cambridge University Press, 2003.

[2] Arruda LK, Lenells J. Long-time asymptotics for the derivative nonlinear Schrödinger equation on the half-line [J]. Nonlinearity, 2017, 30: 4141–4172.

[3] Boutet de Monvel A, Its A, Kotlyarov V. Long-time asymptotics for the focusing NLS equation with time-periodic boundary condition on the half-line [J]. Comm. Math. Phys., 2009, 290: 479–522.

[4] Boutet de Monvel A, Kotlyarov VP, Shepelsky D. Focusing NLS equation: long-time dynamics of step-like initial data [J]. Int. Math. Res. Not., 2011, 2011: 1613–1653.

[5] Boutet de Monvel A, Shepelsky D. The Ostrovsky–Vakhnenko equation by a Riemann–Hilbert approach [J]. J. Phys. A Math. Theor., 2015, 48: 035204.

[6] Boutet de Monvel A, Shepelsky D. A Riemann–Hilbert approach for the Degasperis–Procesi equation [J]. Nonlinearity, 2013, 26(7): 2081–2107.

[7] Buckingham R, Venakides S. Long-time asymptotics of the nonlinear Schrödinger equation shock problem [J]. Comm. Pure. Appl. Math., 2007, 60: 1349–1414.

[8] Chen S. Twisted rogue-wave pairs in the Sasa–Satsuma equation [J]. Phys. Rev. E, 2013, 88: 023202.

[9] Deift P, Kriecherbauer T, McLaughlin KT-R, Venakides S, Zhou X. Uniform asymptotics for polynomials orthogonal with respect to varying exponential weights and applications to universality questions in random matrix theory [J]. Comm. Pure Appl. Math., 1999, 52: 1335–1425.

[10] Deift P, Zhou X. A steepest descent method for oscillatory Riemann–Hilbert problems. Asymptotics for the MKdV equation [J]. Ann Math., 1993, 137: 295–368.

[11] Deift P, Zhou X. Long-time asymptotics for solutions of the NLS equation with initial data in a weighted Sobolev space [J]. Comm. Pure Appl. Math., 2003, 56: 1029–1077.

[12] Fokas AS. A unified transform method for solving linear and certain nonlinear PDEs [J]. Proc. R. Soc. Lond. A, 1997, 453: 1411–1443.

[13] Fokas AS. A unified approach to boundary value problems [M]. In: CBMS-NSF Regional Conference Series in Applied Mathematics, SIAM, 2008.

[14] Geng X, Liu H. The nonlinear steepest descent method to long-time asymptotics of the coupled nonlinear Schrödinger equation [J]. J. Nonlinear Sci., 2018, 28: 739–763.

[15] Grunert K, Teschl G. Long-time asymptotics for the Korteweg–de Vries equation via nonlinear steepest descent [J]. Math. Phys. Anal. Geom., 2009, 12: 287–324.

[16] Guo B, Liu N. Long-time asymptotics for the Kundu–Eckhaus equation on the half-line [J]. J. Math. Phys., 2018, 59: 61505.

[17] Guo B, Liu N. Long-time asymptotics for the Sasa–Satsuma equation via nonlinear steepest descent method [J]. e-print arXiv:1801.06420 2018.

[18] Guo B, Liu N, Wang Y. Long-time asymptotics for the Hirota equation on the half-line [J]. Nonlinear Anal. TMA, 2018, 174: 118–140.

[19] Hirota R. Exact envelope-soliton solutions of a nonlinear wave equation [J]. J. Math. Phys., 1973, 14: 805–809.

[20] Huang L, Xu J, Fan E. Long-time asymptotic for the Hirota equation via nonlinear steepest descent method [J]. Nonlinear Anal Real World Appl., 2015, 26: 229–262.

[21] Kodama Y. Optical solitons in a monomode fiber [J]. J. Stat. Phys., 1985, 39: 597–614.

[22] Kodama Y. A. Hasegawa, Nonlinear pulse propagation in a monomode dielectric guide [J]. IEEE J. Quantum Electron., 1987, 23: 510–524.

[23] Kotlyarov V, Minakov A. Riemann–Hilbert problem to the modified Korteveg–de Vries equation: long-time dynamics of the steplike initial data [J]. J. Math. Phys., 2010, 51: 093506.

[24] Lenells J. Initial-boundary value problems for integrable evolution equations with 3×3 Lax pairs [J]. Phys. D, 2012, 241: 857–875.

[25] Lenells J. The nonlinear steepest descent method for Riemann–Hilbert problems of low regularity [J]. Indiana Univ. Math. J., 2017, 66(4): 1287–1332.

[26] Liu H, Geng X, Xue B. The Deift–Zhou steepest descent method to long-time asymptotics for the Sasa–Satsuma equation [J]. J. Differential Equations, 2018, 265: 5984–6008.

[27] Liu J, Perry PA, Sulem C. Long-time behavior of solutions to the derivative nonlinear

Schrödinger equation for soliton-free initial data [J]. Ann. Inst. H. Poincaré Anal. Non Linéaire, 2018, 35: 217–265.

[28] Liu Y, Gao Y, Xu T, Lü X, Sun Z, Meng X, Yu X, Gai X. Soliton solution, Bäcklund transformation, and conservation laws for the Sasa–Satsuma equation in the optical fiber communications [J]. Z. Naturforsch., 2000, 65a: 291–300.

[29] Lü X. Bright-soliton collisions with shape change by intensity redistribution for the coupled Sasa–Satsuma system in the optical fiber communications [J]. Comm. Nonlinear Sci. Numer. Simulat., 2014, 19: 3969–3987.

[30] Nakkeeran K, Porsezian K, Shanmugha Sundaram P, Mahalingam A. Optical solitons in N-coupled higher order nonlinear Schrödinger equations [J]. Phys. Rev. Lett., 1998, 80: 1425–1428.

[31] Porsezian K. Soliton models in resonant and nonresonant optical fibers [J]. Pramana, 2001, 57: 1003–1039.

[32] Sasa N, Satsuma J. New-type of soliton solutions for a higher-order nonlinear Schrödinger equation [J]. J. Phys. Soc. Jpn., 1991, 60: 409–417.

[33] Sergyeyeva A, Demskoi D. Sasa–Satsuma (complex modified Korteweg-de Vries II) and the complex sine-Gordon II equation revisited: Recursion operators, nonlocal symmetries, and more [J]. J. Math. Phys., 2007, 48: 042702.

[34] Tian S, Zhang T. Long-time asymptotic behavior for the Gerdjikov–Ivanov type of derivative nonlinear Schrödinger equation with time-periodic boundary condition [J]. Proc. Amer. Math. Soc., 2018, 146: 1713–1729.

[35] Wang DS, Wang XL. Long-time asymptotics and the bright N-soliton solutions of the Kundu–Eckhaus equation via the Riemann–Hilbert approach [J]. Nonlinear Anal. Real World Appl., 2018, 41: 334–361.

[36] Wu J, Geng X. Inverse scattering transform of the coupled Sasa–Satsuma equation by Riemann–Hilbert approach [J]. Comm. Theor. Phys., 2017, 67: 527–534.

[37] Whittaker ET, Watson GN. A Course of Modern Analysis [M]. 4th ed. Cambridge University Press, Cambridge 1927.

[38] Xu J, Fan EG. Long-time asymptotics for the Fokas–Lenells equation with decaying initial value problem: Without solitons [J]. J. Differential Equations, 2015, 259(3): 1098–1148.

[39] Xu J, Fan EG. The unified transform method for the Sasa–Satsuma equation on the half-line [J]. Proc. R. Soc. A., 2013, 469: 20130068.

[40] Xu J, Fan E, Chen Y. Long-time asymptotic for the derivative nonlinear Schrödinger equation with step-like initial value [J]. Math. Phys. Anal. Geom., 2013, 16: 253–288.

[41] Yang J, Kaup DJ. Squared eigenfunctions for the Sasa–Satsuma equation [J]. J. Math. Phys., 2009, 50: 023504.

[42] Zakharov VE, Manakov SV. Asymptotic behavior of non-linear wave systems integrated by the inverse scattering method [J]. 30 Years of the Landau Institute-Selected Papers 1996, 358–364.

[43] Zakharov VE, Shabat AB. A scheme for integrating the nonlinear equations of numerical physics by the method of the inverse scattering problem I [J]. Funct. Anal. Appl., 1974, 8: 226–235.

[44] Zakharov VE, Shabat AB. A scheme for integrating the nonlinear equations of numerical physics by the method of the inverse scattering problem II [J]. Funct. Anal. Appl., 1979, 13: 166–174.

[45] Zhang H, Wang Y, Ma W. Binary Darboux transformation for the coupled Sasa-Satsuma equations [J]. Chaos, 2017, 27: 073102.

Long-Time Behavior of Solutions for the Compressible Quantum Magnetohydrodynamic in $\mathbf{R}^{3*}$

Xi Xiaoyu (席肖玉), Pu Xueke (蒲学科), and Guo Boling(郭柏灵)

Abstract In this paper, long-time behavior of solutions for the compressible viscous quantum magnetohydrodynamic model in three-dimensional whole space is studied. We establish the optimal time decay rates for higher-order spatial derivatives of density, velocity and magnetic field, which improves the work of Pu and Xu [Z. Angew. Math. Phys. 68:18, (2017)].

Keywords long-time behavior; quantum magnetohydrodynamic model; optimal time decay rates

1 Introduction

The compressible viscous quantum magnetohydrodynamic (vQMHD) model can be written as

$$\begin{cases} (\rho u)_t + \text{div}(\rho u) = 0, \\ (\rho u)_t + \text{div}(\rho u \otimes u) - \mu \Delta u - (\mu + \lambda) \nabla \text{div} u + \nabla P \\ \qquad - \dfrac{\hbar^2}{2} \rho \nabla \left(\dfrac{\Delta \sqrt{\rho}}{\sqrt{\rho}} \right) = (\nabla \times B) \times B, \\ B_t - \nabla \times (u \times B) = -\nabla \times (\nu \nabla \times B), \quad \text{div} B = 0, \\ (\rho, u, B)(x,t)|_{t=0} = (\rho_0(x), u_0(x), B_0(x)), \end{cases} \quad (1.1)$$

where $(x,t) \in \mathbf{R}^3 \times \mathbf{R}^+$ denotes the space-time position. The unknown functions ρ, u and B represent the density, velocity, and magnetic field, respectively.

---

* Z. Angew. Math. Phys. (2019) 70:7. Doi:10.1007/s00033-018-1049-z.

$P = P(\rho)$ is the pressure, $\hbar$ denotes the Planck constant and ν is the magnetic diffusivity. $\mu > 0$ and λ are the coefficients of viscosity and the second coefficient of viscosity, respectively, and satisfy the usual condition $3\lambda + 2\mu \geqslant 0$. The expression $\Delta\sqrt{\rho}/\sqrt{\rho}$ can be interpreted as a quantum potential, i.e. Bohm potential, which satisfies

$$2\rho\nabla\left(\frac{\Delta\sqrt{\rho}}{\sqrt{\rho}}\right) = \mathrm{div}(\rho\nabla^2\ln\rho) = \Delta\nabla\rho + \frac{|\nabla\rho|^2\nabla\rho}{\rho^2} - \frac{\nabla\rho\Delta\rho}{\rho} - \frac{\nabla\rho\cdot\nabla^2\rho}{\rho}.$$

The quantum correction terms, which are closely related to Bohm potential, were initially considered for the thermodynamic equilibrium by Wigner in [34].

For more physical interpretations of the model, one can refer to Hass [11, 12]. Furthermore, as the spatial variable tends to infinity, we assume

$$\lim_{|x|\to\infty}(\rho_0 - 1, u_0, B_0)(x) = 0. \tag{1.2}$$

Many works have been done on the decay of the solutions for the fluid dynamics. For the convergence rates for the compressible Navier-Stokes equations with or without external forces, one can refer to [2]-[5], [10], [13]-[17], [20]-[24]. For higher-order spatial derivatives of solutions, Guo and Wang [10] employed a general energy method to build the time convergence rates by assuming the initial perturbation are bounded in $\dot{H}^{-s}$ for $s \in [0, 3/2)$, instead of L^1.

Motivated by the study of optimal decay rates for Navier-Stokes equations, there have been vast results of time convergence rates of solutions to the MHD equations. First of all, under the H^3-framework, Li and Yu [19] and Chen and Tan [1] not only established the global existence of classical solutions, but also obtained the time decay rates for the three-dimensional compressible MHD equations by assuming the initial data belong to L^1 and L^q for $q \in [1, 6/5)$, respectively. More precisely, Chen and Tan [1] obtained the time decay rates

$$\|\nabla^2(\rho-1, u, B)\|_{H^{3-k}} \leqslant (1+t)^{-\frac{3}{2}(\frac{1}{q}-\frac{1}{2})-\frac{k}{2}}, \tag{1.3}$$

with $k = 0, 1$. Motivated by the work of Guo and Wang [10], Tan and Wang [33] established the optimal time decay rates for the higher-order spatial derivatives of solutions if the initial perturbation belongs to $H^N \cap \dot{H}^{-s}(N \geqslant 3, s \in [0, 3/2))$. Gao, Tao and Yao [6] deduced faster time decay rates for the higher-order spatial derivatives of solutions

$$\|\nabla^2(\rho-1, u)\|_{H^1} \leqslant (1+t)^{-\frac{7}{4}},$$

$$\|\nabla^m B\|_{H^{3-m}} \leqslant (1+t)^{-\frac{3+2m}{4}}, \qquad (1.4)$$

with $m = 2, 3$, which is better than (1.3). For the full compressible MHD equations, Pu and Guo [26] and Gao, Tao and Yao [7] established time decay rates for the classical solutions, and optimal decay rates for higher-order spatial derivatives of classical solutions in three-dimensional whole space, respectively. For equations associated with compressible MHD equation, see [8], [9] and the references therein.

In this paper, we consider the decay rates for the vQMHD system. Using Galerkin approximation, the global existence of weak solutions was deduced by Yang and Ju [36]. The global existence of weak solutions and large time behavior of solutions in $\mathbf{T}^3$ were established by Li, Cheng and Yan [18]. The global existence and decay of classical solutions near the constant steady solution were established by Pu and Xu [28], where the time decay rates are as follows

$$\|\nabla^k(\rho-1)\|_{H^{5-k}} + \|\nabla^k u\|_{H^{4-k}} + \|\nabla^k B\|_{H^{4-k}} \leqslant C(1+t)^{-\frac{3+2k}{4}}, \qquad (1.5)$$

with $k = 0, 1$. For the full compressible quantum model, Pu and Xu [29] established the optimal time decay rates for the higher-order spatial derivatives of solutions if the initial perturbation belongs to $(H^{N+2} \cap \dot{H}^{-s}) \times (H^{N+1} \cap \dot{H}^{-s}) \times (H^N \cap \dot{H}^{-s})$ ($N \geqslant 3, s \in [0, 3/2)$). Recently, Xie, Xi and Guo [35] deduced faster time convergence rates for the higher-order spatial derivatives of solutions when the initial data belong to L^1 rather than some negative Sobolev space, which improves the work of [25]. Inspired by [30, 31], we aim to establish optimal time decay rates for the higher-order spatial derivatives of solutions by using Fourier splitting method (see [32]) and deduce faster time convergence rates for the higher-order spatial derivatives of solutions when the initial data belongs to L^1.

The main result of this paper is the following:

Theorem 1.1 *Assume that the initial data* $(\rho_0 - 1, u_0, B_0) \in (H^5 \cap L^1) \times (H^4 \cap L^1) \times (H^4 \cap L^1)$. *Then there exists a constant* $\delta > 0$ *such that if*

$$\|\rho_0 - 1\|_{H^5} + \|u_0\|_{H^4} + \|B_0\|_{H^4} \leqslant \delta, \qquad (1.6)$$

then the global solution (ρ, u, B) *of problem* (1.1) *has the time decay rates*

$$\|\nabla^k(\rho-1)\|_{H^{5-k}} + \|\nabla^k u\|_{H^{4-k}} + \|\nabla^k B\|_{H^{4-k}} \leqslant C(1+t)^{-\frac{3+2k}{4}},$$

$$\|\nabla^4 B\|_{L^2} \lesssim (1+t)^{-\frac{11}{4}}, \tag{1.7}$$

for any integer $k = 0, 1, 2, 3$.

Remark 1.1 *Obviously, (1.7) provides faster time decay rates for the higher order spatial derivatives of smooth solutions than (1.5).*

Remark 1.2 *If the initial data $(\rho_0 - 1, u_0, B_0) \in (H^{N+2} \cap L^1) \times (H^{N+1} \cap L^1) \times (H^{N+1} \cap L^1)$ for any integer $N \geqslant 3$, then we can obtain the following time decay rates*

$$\|\nabla^k(\rho-1)\|_{H^{N+2-k}} + \|\nabla^k u\|_{H^{N+1-k}} + \|\nabla^k B\|_{H^{N+1-k}} \leqslant C(1+t)^{-\frac{3+2k}{4}},$$

$$\|\nabla^N B\|_{L^2} \lesssim (1+t)^{-\frac{2N+3}{4}},$$

for any integer $k = 0, 1, 2, \cdots, N$.

Remark 1.3 *Under all the assumptions of Theorem 1.1, for the mixed space-time derivatives of solution (ρ, u, B) to problem (1.1), we can deduce the following decay rate*

$$\|\nabla^k \rho_t\|_{H^{3-k}} + \|\nabla^k u_t\|_{L^2} \leqslant C(1+t)^{-\frac{3+2(k+1)}{4}},$$

$$\|\nabla^k B_t\|_{L^2} \leqslant C(1+t)^{-\frac{3+2(k+2)}{4}},$$

for any integer $k = 0, 1, 2$.

Notation 1.1 *Throughout this paper, we use L^p to denote the $L^p(\mathbf{R}^3)$ norm and H^s to denote the $W^{s,2}(\mathbf{R}^3)$ norm. The symbol $\mathcal{F}(f) := \hat{f}$ stands for the usual Fourier transform of the function f. Moreover, the notation $a \lesssim b$ and $a \sim b$ means that $a \leqslant Cb$ and $C_1 b \leqslant a \leqslant C_2 b$. C, C_1, C_2 are generic positive constants independent of t.*

The rest of this paper is organized as follows. In Section 2, some inequalities are given, which will be used throughout this paper. In Section 3, we establish some crucial energy estimates for solutions and their higher-order derivatives. In Section 4, we use the Fourier splitting method to establish faster time decay rates for the higher order spatial derivatives of smooth solutions.

2 Preliminaries

Lemma 2.1 (Gagliardo-Nirenberg inequality) *Let $0 \leqslant m, \alpha \leqslant l$ and the function $f \in C_0^\infty(\mathbf{R}^3)$, then we have*

$$\|\nabla^\alpha f\|_{L^p} \leqslant \|\nabla^m f\|_{L^2}^{1-\theta} \|\nabla^l f\|_{L^2}^\theta,$$

where $0 \leqslant \theta \leqslant 1$ and α satisfies
$$\frac{1}{p} - \frac{\alpha}{3} = \left(\frac{1}{2} - \frac{m}{3}\right)(1-\theta) + \left(\frac{1}{2} - \frac{l}{3}\right)\theta.$$

Lemma 2.2 (see [27]) Let $m \geqslant 1$ be an integer and define the commutator
$$[\nabla^m, f]g = \nabla^m(fg) - f\nabla^m g.$$
Then there exists some constant $C > 0$ such that
$$\|\nabla^m(fg)\|_{L^p} \leqslant C(\|f\|_{L^{p_1}}\|g\|_{\dot{H}^{m,p_2}} + \|g\|_{L^{p_3}}\|f\|_{\dot{H}^{m,p_4}}),$$
$$\|[\nabla^m, f]g\|_{L^p} \leqslant C(\|\nabla f\|_{L^{p_1}}\|g\|_{\dot{H}^{m-1,p_2}} + \|g\|_{L^{p_3}}\|f\|_{\dot{H}^{m,p_4}}),$$
where $f, g \in \mathcal{S}$, and $p_2, p_4 \in (1, \infty)$ such that
$$\frac{1}{p} = \frac{1}{p_1} + \frac{1}{p_2} = \frac{1}{p_3} + \frac{1}{p_4}.$$

3 Energy estimates

Before studying the time decay rates, it's sufficient to give energy estimates for solutions of problem (3.8). We consider solutions near the constant state $(1, 0, 0)$. Denoting $\varrho = \rho - 1$, $u = u/\gamma$, $B = B$ with $\gamma = \sqrt{P'(1)}$, for simplicity, we set $\gamma = 1$, $\nu = 1$, then the system (1.1) can be rewritten in the form
$$\begin{cases} \varrho_t + \mathrm{div}\, u = -\varrho\,\mathrm{div}\, u - u \cdot \nabla \varrho, \\ u_t - \mu\Delta u - (\mu + \lambda)\nabla\mathrm{div}\, u + \nabla\varrho - \dfrac{\hbar^2}{4}\nabla\Delta\varrho = F, \\ B_t - \nu\Delta B = \nabla \times (u \times B). \end{cases} \quad (3.8)$$
Here the nonlinear function F is defined as
$$F = -u \cdot \nabla u - \frac{\varrho}{\varrho+1}[\mu\Delta u + (\mu+\lambda)\nabla\mathrm{div}\, u] - \left(\frac{P'(\varrho+1)}{\varrho+1} - 1\right)\nabla\varrho$$
$$-\frac{\hbar^2}{4}\frac{\varrho}{\varrho+1}\Delta\nabla\varrho + \frac{\hbar^2}{4}\left(\frac{|\nabla\varrho|^2\nabla\varrho}{(\varrho+1)^3} - \frac{\nabla\varrho\Delta\varrho}{(\varrho+1)^2} - \frac{\nabla\varrho \cdot \nabla^2\varrho}{(\varrho+1)^2}\right)$$
$$+\frac{1}{\varrho+1}((\nabla \times B) \times B). \quad (3.9)$$

The associated initial data is given by
$$(\varrho, u, B)|_{t=0} = (\varrho_0, u_0, B_0). \quad (3.10)$$

First, we aim to the estimates for the magnetic field.

Lemma 3.1 *Under the assumption* (1.6), *we have for any integer* $0 \leq k \leq 4$,

$$\frac{1}{2}\frac{d}{dt}\int |\nabla^k B|^2 \, dx + \int |\nabla^{k+1} B|^2 \, dx \lesssim \delta \|\nabla^{k+1} u\|_{L^2}^2. \tag{3.11}$$

Proof Taking k-th spatial derivatives to $(3.8)_3$, multiplying the resultant equation by $\nabla^k B$ and integrating by parts over $\mathbf{R}^3$, we have

$$\frac{1}{2}\frac{d}{dt}\int |\nabla^k B|^2 \, dx + \int |\nabla^{k+1} B|^2 \, dx$$

$$= \int \nabla^k (-u \cdot \nabla B + B \cdot \nabla u - B \operatorname{div} u) \cdot \nabla^k B$$

$$= II_1 + II_2 + II_3. \tag{3.12}$$

Since it's easy to verify (3.11) for the case $k = 0$, we only need to verify the case $k \geq 1$. We deduce from Leibnitz formula and Hölder inequality that

$$II_1 = \int \sum_{l=0}^{k-1} C_{k-1}^l \nabla^l u \nabla^{k-l} B \nabla^{k+1} B \, dx$$

$$\lesssim \sum_{l=0}^{k-1} \|\nabla^l u\|_{L^3} \|\nabla^{k-l} B\|_{L^6} \|\nabla^{k+1} B\|_{L^2}. \tag{3.13}$$

For the case $0 \leq l \leq [(k-1)/2]$, by using Gagliardo-Nirenberg inequality and Young inequality, we have

$$\|\nabla^l u\|_{L^3} \|\nabla^{k-l} B\|_{L^6} \|\nabla^{k+1} B\|_{L^2}$$

$$\lesssim \|\nabla^\alpha u\|_{L^2}^{1-\frac{l}{k+1}} \|\nabla^{k+1} u\|_{L^2}^{\frac{l}{k+1}} \|B\|_{L^2}^{\frac{l}{k+1}} \|\nabla^{k+1} B\|_{L^2}^{1-\frac{l}{k+1}} \|\nabla^{k+1} B\|_{L^2}$$

$$\lesssim \delta(\|\nabla^{k+1} u\|_{L^2}^2 + \|\nabla^{k+1} B\|_{L^2}^2), \tag{3.14}$$

where $\alpha = (k+1)/(2k+2-2l) \in [1/2, 1)$.

For the case $[(k-1)/2] + 1 \leq l \leq k - 1$, we obtain

$$\|\nabla^l u\|_{L^3} \|\nabla^{k-l} B\|_{L^6} \|\nabla^{k+1} B\|_{L^2}$$

$$\lesssim \|u\|_{L^2}^{1-\frac{l+\frac{1}{2}}{k+1}} \|\nabla^{k+1} u\|_{L^2}^{\frac{l+\frac{1}{2}}{k+1}} \|\nabla^\alpha B\|_{L^2}^{\frac{l+\frac{1}{2}}{k+1}} \|\nabla^{k+1} B\|_{L^2}^{1-\frac{l+\frac{1}{2}}{k+1}} \|\nabla^{k+1} B\|_{L^2}$$

$$\lesssim \delta(\|\nabla^{k+1} u\|_{L^2}^2 + \|\nabla^{k+1} B\|_{L^2}^2), \tag{3.15}$$

where $\alpha = (k+1)/(2l+1) \in [1/2, 1]$.

Similarly, for II_2 and II_3, we also have

$$II_2 \lesssim \delta(\|\nabla^{k+1} u\|_{L^2}^2 + \|\nabla^{k+1} B\|_{L^2}^2), \quad II_3 \lesssim \delta(\|\nabla^{k+1} u\|_{L^2}^2 + \|\nabla^{k+1} B\|_{L^2}^2).$$

Then we complete the proof of lemma.

Next, we give the estimates for the density and velocity field.

Lemma 3.2 *Assume that* $0 \leqslant k \leqslant 4$, *then there exist positive constants* $\delta_0 < 1$ *sufficiently small and* $\hbar_0 < 1$ *such that for all* $\delta < \delta_0$ *and* $\hbar < \hbar_0$,

$$\frac{1}{2}\frac{d}{dt}\int |\nabla^k \varrho|^2 + |\nabla^k u|^2 + \frac{\hbar^2}{4}|\nabla^{k+1}\varrho|^2 \, dx + \mu_0 \|\nabla^{k+1} u\|_{L^2}^2$$
$$\leqslant C\delta\|\nabla^{k+1}\varrho\|_{L^2}^2 + C\delta\|\nabla^{k+1} B\|_{L^2}^2 + C\delta\hbar^2\|\nabla^k \triangle \varrho\|_{L^2}^2, \tag{3.16}$$

where C is independent of t.

Proof Since the inequality (3.16) is verified for the case $k = 0$ in [28], now we consider the case $1 \leqslant k \leqslant 4$, Taking k-th spatial derivatives to $(3.8)_1$, $(3.8)_2$, and multiplying the resultant equation by $\nabla^k \varrho$, $\nabla^k u$, respectively, summing up them and integrating by parts over $\mathbf{R}^3$, we have

$$\frac{1}{2}\frac{d}{dt}\int |\nabla^k \varrho|^2 + |\nabla^k u|^2 \, dx + \mu\|\nabla^{k+1} u\|_{L^2}^2 + (\mu+\lambda)\|\nabla^k \operatorname{div} u\|_{L^2}^2$$
$$= \frac{\hbar^2}{4}\int \nabla^k \triangle\nabla\varrho \cdot \nabla^k u \, dx - \int \nabla^k \operatorname{div}(\varrho u)\nabla^k \varrho \, dx$$
$$- \int \nabla^k(u \cdot \nabla u) \cdot \nabla^k u \, dx - \mu \int \nabla^k\left(\frac{\varrho}{\varrho+1}\Delta u\right) \cdot \nabla^k u \, dx$$
$$- (\mu+\lambda)\int \nabla^k\left(\frac{\varrho}{\varrho+1}\nabla \operatorname{div} u\right) \cdot \nabla^k u \, dx$$
$$- \int \nabla^k\left(\left(\frac{P'(\varrho+1)}{\varrho+1} - 1\right)\nabla\varrho\right) \cdot \nabla^k u \, dx$$
$$- \frac{\hbar^2}{4}\int \nabla^k\left(\frac{\varrho}{\varrho+1}\nabla\Delta\varrho\right) \cdot \nabla^k u \, dx$$
$$+ \frac{\hbar^2}{4}\int \nabla^k\left(\frac{|\nabla\varrho|^2 \nabla\varrho}{(\varrho+1)^3} - \frac{\nabla\varrho\Delta\varrho}{(\varrho+1)^2} - \frac{\nabla\varrho \cdot \nabla^2\varrho}{(\varrho+1)^2}\right) \cdot \nabla^k u \, dx$$
$$+ \int \nabla^k\left(\frac{(\nabla \times B) \times B}{\varrho+1}\right) \cdot \nabla^k u \, dx = \sum_{i=1}^{9} I_i.$$

We will estimate each term on the right hand side. First, for the term I_1, by the continuity equation, integration by parts, Hölder's inequality and Lemma 2.2, we have

$$I_1 = \frac{\hbar^2}{4}\int \nabla^k \Delta \varrho \nabla^k(\varrho_t + \nabla\varrho \cdot u + \varrho \operatorname{div} u) \, dx$$
$$\leqslant -\frac{\hbar^2}{8}\frac{d}{dt}\int |\nabla^{k+1}\varrho|^2 \, dx + C\hbar^2(\|\nabla\varrho\|_{L^3}\|u\|_{\dot{H}^{k,6}} + \|u\|_{L^\infty}\|\nabla\varrho\|_{\dot{H}^k})\|\nabla^k \Delta\varrho\|_{L^2}$$

$$+ C\hbar^2(\|\nabla u\|_{L^3}\|\varrho\|_{\dot{H}^{k,6}} + \|\varrho\|_{L^\infty}\|\nabla u\|_{\dot{H}^k})\|\nabla^k \Delta \varrho\|_{L^2}$$

$$\leqslant -\frac{\hbar^2}{8}\frac{d}{dt}\int |\nabla^{k+1}\varrho|^2\,dx + C\hbar^2\delta(\|\nabla^{k+1}u\|_{L^2}^2 + \|\nabla^{k+1}\varrho\|_{L^2}^2 + \|\nabla^k\Delta\varrho\|_{L^2}^2).$$

For the term I_2, by integratoin by parts, Hölder's and Sobolev's inequalities, we have

$$I_2 = \int \nabla^{k-1}\mathrm{div}(\varrho u)\nabla^{k+1}\varrho\,dx$$

$$\leqslant C\sum_{0\leqslant l\leqslant k}\|\nabla^l\varrho\nabla^{k-l}u\|_{L^2}\|\nabla^{k+1}\varrho\|_{L^2}.$$

For $l \leqslant [k/2]$, Hölder's inequality and Gagliardo-Nirenberg inequality imply that

$$\|\nabla^l\varrho\nabla^{k-l}u\|_{L^2} \leqslant C\|\nabla^l\varrho\|_{L^3}\|\nabla^{k-l}u\|_{L^6}$$

$$\leqslant C\|\nabla^\alpha\varrho\|_{L^2}^{1-\frac{l}{k}}\|\nabla^{k+1}\varrho\|_{L^2}^{\frac{l}{k}}\|\nabla u\|_{L^2}^{\frac{l}{k}}\|\nabla^{k+1}u\|_{L^2}^{1-\frac{l}{k}}, \quad (3.17)$$

where α satifies

$$\frac{l}{3} - \frac{1}{3} = \left(\frac{\alpha}{3} - \frac{1}{2}\right)\left(1 - \frac{l}{k}\right) + \left(\frac{k+1}{3} - \frac{1}{2}\right)\frac{l}{k},$$

which implies that

$$\alpha = 2 - \frac{k}{2(k-l)} \in \left[1, \frac{3}{2}\right].$$

When $[k/2] + 1 \leqslant l \leqslant k$, by using Hölder's inequality and Gagliardo-Nirenberg inequality, we have

$$\|\nabla^l\varrho\nabla^{k-l}u\|_{L^2} \leqslant C\|\nabla^l\varrho\|_{L^6}\|\nabla^{k-l}u\|_{L^3}$$

$$\leqslant C\|\varrho\|_{L^2}^{1-\frac{l+1}{k+1}}\|\nabla^{k+1}\varrho\|_{L^2}^{\frac{l+1}{k+1}}\|\nabla^\alpha u\|_{L^2}^{\frac{l+1}{k+1}}\|\nabla^{k+1}u\|_{L^2}^{1-\frac{l+1}{k+1}}, \quad (3.18)$$

where α satisfies

$$\frac{k-l}{3} - \frac{1}{3} = \left(\frac{\alpha}{3} - \frac{1}{2}\right)\frac{l+1}{k+1} + \left(\frac{k+1}{3} - \frac{1}{2}\right)\left(1 - \frac{l+1}{k+1}\right), \quad (3.19)$$

which implies that

$$\alpha = \frac{k+1}{2(1+l)} \in \left[\frac{1}{2}, 1\right).$$

In view of (3.17) and (3.18), we have

$$\|\nabla^l \varrho \nabla^{k-l} u\|_{L^2} \leqslant C\delta(\|\nabla^{k+1}\varrho\|_{L^2} + \|\nabla^{k+1}u\|_{L^2}).$$

Then by Cauchy's inequality, we have

$$I_2 \leqslant C\delta(\|\nabla^{k+1}\varrho\|_{L^2}^2 + \|\nabla^{k+1}u\|_{L^2}^2).$$

For the term I_3, by integration by parts, Hölder's and Sobolev's inequalities, we have

$$I_3 = \int \nabla^{k-1}(u \cdot \nabla u)\nabla^{k+1}u\, dx$$
$$\leqslant C \sum_{0 \leqslant l \leqslant k-1} \|\nabla^l u \nabla^{k-l} u\|_{L^2} \|\nabla^{k+1}u\|_{L^2}.$$

For $l \leqslant [(k-1)/2]$, Hölder's inequality and Gagliardo-Nirenberg inequality imply that

$$\|\nabla^l u \nabla^{k-l} u\|_{L^2} \leqslant C\|\nabla^l u\|_{L^3}\|\nabla^{k-l}u\|_{L^6}$$
$$\leqslant C\|\nabla^\alpha u\|_{L^2}^{1-\frac{l}{k}} \|\nabla^{k+1}u\|_{L^2}^{\frac{l}{k}} \|\nabla u\|_{L^2}^{\frac{l}{k}} \|\nabla^{k+1}u\|_{L^2}^{1-\frac{l}{k}}, \qquad (3.20)$$

where α satisfies

$$\frac{l}{3} - \frac{1}{3} = \left(\frac{\alpha}{3} - \frac{1}{2}\right)\left(1 - \frac{l}{k}\right) + \left(\frac{k+1}{3} - \frac{1}{2}\right)\frac{l}{k}.$$

This implies that

$$\alpha = 2 - \frac{k}{2(k-l)} \in \left(1, \frac{3}{2}\right].$$

When $[(k-1)/2]+1 \leqslant l \leqslant k-1$, by using Hölder's inequality and Gagliardo-Nirenberg inequality, we have

$$\|\nabla^l u \nabla^{k-l} u\|_{L^2} \leqslant C\|\nabla^l u\|_{L^6}\|\nabla^{k-l}u\|_{L^3}$$
$$\leqslant C\|\nabla u\|_{L^2}^{1-\frac{l}{k}} \|\nabla^{k+1}u\|_{L^2}^{\frac{l}{k}} \|\nabla^\alpha u\|_{L^2}^{\frac{l}{k}} \|\nabla^{k+1}u\|_{L^2}^{1-\frac{l}{k}}, \qquad (3.21)$$

where α satisfies

$$\frac{k-l}{3} - \frac{1}{3} = \left(\frac{\alpha}{3} - \frac{1}{2}\right)\frac{l+1}{k+1} + \left(\frac{k+1}{3} - \frac{1}{2}\right)\left(1 - \frac{l+1}{k+1}\right).$$

This implies that
$$\alpha = \frac{k+1}{2(k+l)} \in \left(\frac{1}{2}, 1\right).$$

In view of (3.20) and (3.21), we have
$$\|\nabla^l u \nabla^{k-l} u\|_{L^2} \leqslant C\delta \|\nabla^{k+1} u\|_{L^2}.$$

By Cauchy's inequality, we have
$$I_3 \leqslant C\delta \|\nabla^{k+1} u\|_{L^2}^2.$$

For the term I_4, setting $h(\varrho) = \varrho/(\varrho+1)$, by using integration by parts and Hölder's inequality, we have
$$I_4 = \int \nabla^{k-1}(h(\varrho)\Delta u) \nabla^{k+1} u \, dx$$
$$\leqslant C \sum_{0 \leqslant l \leqslant k-1} \|\nabla^l h(\varrho) \nabla^{k-l} u\|_{L^2} \|\nabla^{k+1} u\|_{L^2}.$$

Now we discuss this term in the following cases:

(a) For $l = 0$, since $|h(\varrho)| \leqslant C|\varrho|$, by using Hölder's and Sobolev's inequalities, we obtain that
$$\|h(\varrho) \nabla^{k+1} u\|_{L^2} \leqslant C\|\varrho\|_{L^\infty} \|\nabla^{k+1} u\|_{L^2} \leqslant C\delta \|\nabla^{k+1} u\|_{L^2}. \quad (3.22)$$

(b) For $l = 1$, since $|h'(\varrho)| \leqslant C$, by using Hölder's and Sobolev's inequalities, we have
$$\|\nabla h(\varrho) \nabla^k u\|_{L^2} \leqslant C\|h'(\varrho) \nabla \varrho\|_{L^3} \|\nabla^k u\|_{L^6} \leqslant C\delta \|\nabla^{k+1} u\|_{L^2}. \quad (3.23)$$

(c) For $2 \leqslant l \leqslant k-1$. If $2 \leqslant l \leqslant [k/2]$, by using Hölder's inequality and Gagliardo-Nirenberg inequality, we have
$$\|\nabla^l h(\varrho) \nabla^{k-l-1} \Delta u\|_{L^2} \leqslant C\|\nabla^l h(\varrho)\|_{L^\infty} \|\nabla^{k-l+1} u\|_{L^2}$$
$$\leqslant C\|\nabla^\alpha \varrho\|_{L^2}^{1-\frac{l}{k}} \|\nabla^{k+1}\varrho\|_{L^2}^{\frac{l}{k}} \|\nabla u\|^{\frac{l}{k}} \|\nabla^{k+1} u\|_{L^2}^{1-\frac{l}{k}}$$
$$\leqslant C\delta (\|\nabla^{k+1} u\|_{L^2} + \|\nabla^{k+1} \varrho\|_{L^2}), \quad (3.24)$$

where α satisfies
$$\frac{l}{3} = \left(\frac{\alpha}{3} - \frac{1}{2}\right)\frac{l}{k} + \left(\frac{k+1}{3} - \frac{1}{2}\right)\left(1 - \frac{l}{k}\right).$$

This implies that
$$\alpha = 1 + \frac{k}{2(k-l)} \in \left(\frac{3}{2}, 2\right].$$

If $[k/2]+1 \leqslant l \leqslant k-1$, by using Hölder's inequality and Gagliardo-Nirenberg inequality, we have

$$\|\nabla^l h(\varrho)\nabla^{k-l-1}\Delta u\|_{L^2} \leqslant C\|\nabla^l h(\varrho)\|_{L^2}\|\nabla^{k-l+1}u\|_{L^\infty}$$
$$\leqslant C\|\nabla\varrho\|_{L^2}^{1-\frac{l-1}{k}}\|\nabla^{k+1}\varrho\|_{L^2}^{\frac{l-1}{k}}\|\nabla^\alpha u\|^{\frac{l-1}{k}}\|\nabla^{k+1}u\|_{L^2}^{1-\frac{l-1}{k}}$$
$$\leqslant C\delta(\|\nabla^{k+1}u\|_{L^2} + \|\nabla^{k+1}\varrho\|_{L^2}), \qquad (3.25)$$

where α satisfies
$$\frac{k-l+1}{3} = \left(\frac{\alpha}{3} - \frac{1}{2}\right)\frac{l-1}{k} + \left(\frac{k+1}{3} - \frac{1}{2}\right)\left(1 - \frac{l-1}{k}\right).$$

This implies that
$$\alpha = 1 + \frac{k}{2(l-1)} \in \left[\frac{3}{2}, 2\right].$$

By using Cauchy's inequality, we have
$$I_4 \leqslant C\delta(\|\nabla^{k+1}u\|_{L^2}^2 + \|\nabla^{k+1}\varrho\|_{L^2}^2).$$

Similarly, for the term I_5 and I_6, we have
$$I_5 \leqslant C\delta(\|\nabla^{k+1}u\|_{L^2}^2 + \|\nabla^{k+1}\varrho\|_{L^2}^2),$$
$$I_6 \leqslant C\delta(\|\nabla^{k+1}u\|_{L^2}^2 + \|\nabla^{k+1}\varrho\|_{L^2}^2).$$

For the term I_7, by using integration by parts and Hölder's inequality, we have
$$I_7 = \frac{\hbar^2}{4}\int \nabla^{k-1}(h(\varrho)\Delta\nabla\varrho)\nabla^{k+1}u\,dx$$
$$\leqslant C\hbar^2 \sum_{0\leqslant l\leqslant k-1} \|\nabla^l h(\varrho)\nabla^{k-l}\Delta\varrho\|_{L^2}\|\nabla^{k+1}u\|_{L^2}.$$

Similar to (3.22)-(3.25), we have
$$\|\nabla^l h(\varrho)\nabla^{k-l}\Delta\varrho\|_{L^2} \leqslant C\delta(\|\nabla^{k+1}\varrho\|_{L^2} + \|\nabla^k \Delta\varrho\|_{L^2}).$$

Hence, Cauchy's inequality implies that

$$I_7 \leqslant C\hbar^2 \delta(\|\nabla^{k+1} u\|_{L^2}^2 + \|\nabla^{k+1}\varrho\|_{L^2}^2 + \|\nabla^k \triangle \varrho\|_{L^2}^2).$$

For the term I_8, setting $g(\varrho) = 1/(\varrho+1)^2$, by integration by parts and Hölder's inequality, we have

$$\frac{\hbar^2}{3}\int \nabla^k\left(\frac{\nabla \varrho \triangle \varrho}{(\varrho+1)^2}\right)\nabla^k u \, dx = -\frac{\hbar^2}{3}\int \nabla^{k-1}(g(\varrho)\nabla\varrho\triangle\varrho)\nabla^{k+1} u \, dx$$

$$\leqslant C\hbar^2 \sum_{0\leqslant l \leqslant k-1} \|\nabla^l(g(\varrho))\nabla^{k-l-1}(\nabla\varrho\triangle\varrho)\|_{L^2}\|\nabla^{k+1} u\|_{L^2}.$$

For $l \leqslant [k/2]$, by using Hölder's inequality and Gagliardo-Nirenberg inequality and Lemma 2.2, we have

$$\|\nabla^l(g(\varrho))\nabla^{k-l-1}(\nabla\varrho\triangle\varrho)\|_{L^2}$$
$$\leqslant \|\nabla^l(g(\varrho))\|_{L^3}\|\nabla^{k-l-1}(\nabla\varrho\triangle\varrho)\|_{L^6}$$
$$\leqslant C\|\nabla^\alpha \varrho\|_{L^2}^{1-\frac{l}{k}}\|\nabla^{k+1}\varrho\|_{L^2}^{\frac{l}{k}}(\|\nabla\varrho\|_{L^\infty}\|\nabla^{k-l-1}\triangle\varrho\|_{L^6} + \|\triangle\varrho\|_{L^\infty}\|\nabla^{k-l+1}\varrho\|_{L^2})$$
$$\leqslant C\delta\|\nabla^\alpha \varrho\|_{L^2}^{1-\frac{l}{k}}\|\nabla^{k+1}\varrho\|_{L^2}^{\frac{l}{k}}(\|\triangle\varrho\|_{L^2}^{\frac{l}{k}}\|\nabla^k\triangle\varrho\|_{L^2}^{1-\frac{l}{k}} + \|\nabla\varrho\|_{L^2}^{\frac{l}{k}}\|\nabla^{k+1}\varrho\|_{L^2}^{1-\frac{l}{k}})$$
$$\leqslant C\delta(\|\nabla^{k+1}\varrho\|_{L^2} + \|\nabla^k\triangle\varrho\|_{L^2}), \tag{3.26}$$

where α satisfies

$$\frac{l}{3} - \frac{1}{3} = \left(\frac{\alpha}{3} - \frac{1}{2}\right)\left(1 - \frac{l}{k}\right) + \left(\frac{k+1}{3} - \frac{1}{2}\right)\frac{l}{k}.$$

This implies that

$$\alpha = 1 - \frac{k}{2(k-l)} \in \left[0, \frac{1}{2}\right].$$

When $[k/2]+1 \leqslant l \leqslant k-1$, by using Hölder's inequality and Gagliardo-Nirenberg inequality, we have

$$\|\nabla^l(g(\varrho))\nabla^{k-l-1}(\nabla\varrho\triangle\varrho)\|_{L^2}$$
$$\leqslant \|\nabla^{l-1}(g'(\varrho)\nabla\varrho)\nabla^{k-l-1}(\nabla\varrho\triangle\varrho)\|_{L^2}$$
$$\leqslant C\sum_{0\leqslant m \leqslant k-1}\|\nabla^m(g'(\varrho))\nabla^{l-m}\varrho\nabla^{k-l-1}(\nabla\varrho\triangle\varrho)\|_{L^2}.$$

For $m = 0$, similar to (3.26), we obtain

$$\|g'(\varrho)\nabla^l \varrho \nabla^{k-l-1}(\nabla\varrho\triangle\varrho)\|_{L^2} \leqslant C\delta(\|\nabla^{k+1}\varrho\|_{L^2} + \|\nabla^k\triangle\varrho\|_{L^2}).$$

For $1 \leqslant m \leqslant l-1$, Hölder's inequality and Gagliardo-Nirenberg inequality and Lemma 2.2 imply that

$$\|\nabla^m(g'(\varrho))\nabla^{l-m}\varrho\nabla^{k-l-1}(\nabla\varrho\triangle\varrho)\|_{L^2}$$
$$\leqslant C\|\nabla^m g'(\varrho)\|_{L^\infty}\|\nabla^{l-m}\varrho\|_{L^\infty}\|\nabla^{k-l-1}(\nabla\varrho\triangle\varrho)\|_{L^2}$$
$$\leqslant C\|\nabla^2\varrho\|_{L^2}^{1-\frac{2m-1}{2(k-1)}}\|\nabla^{k+1}\varrho\|_{L^2}^{\frac{2m-1}{2(k-1)}}\|\nabla^2\varrho\|_{L^2}^{1-\frac{2l-2m-1}{2(k-1)}}\|\nabla^{k+1}\varrho\|_{L^2}^{1-\frac{2l-2m-1}{2(k-1)}}$$
$$\cdot \|\nabla^{k-l-l}(\nabla\varrho\triangle\varrho)\|_{L^2}$$
$$\leqslant C\delta\|\nabla^{k+1}\varrho\|_{L^2}^{\frac{l-1}{k-1}}(\|\nabla^{k-1+1}\varrho\|_{L^2})$$
$$\leqslant C\delta\|\nabla^{k+1}\varrho\|_{L^2}^{\frac{l-1}{k-1}}\|\nabla^{k+1}\varrho\|_{L^2}^{1-\frac{l-1}{k-1}}\|\nabla^\alpha\varrho\|_{L^2}^{\frac{l-1}{k-1}}$$
$$\leqslant C\delta\|\nabla^{k+1}\varrho\|_{L^2}.$$

where α satisfies

$$\frac{k-l+1}{3}-\frac{1}{2}=\left(\frac{\alpha}{3}-\frac{1}{2}\right)\frac{l-1}{k}+\left(\frac{k+1}{3}-\frac{1}{2}\right)\left(1-\frac{l-1}{k-1}\right).$$

This implies that

$$\alpha=2-\frac{k-1}{2(k-l)}\in(0,1).$$

Hence, by Cauchy's inequality, we have

$$\frac{\hbar^2}{3}\int\nabla^k\left(\frac{\nabla\varrho\triangle\varrho}{(\varrho+1)^2}\right)\nabla^k u\,dx\leqslant C\hbar^2\delta(\|\nabla^{k+1}u\|_{L^2}^2+\|\nabla^{k+1}\varrho\|_{L^2}^2+\|\nabla^k\triangle\varrho\|_{L^2}^2).$$

Then

$$I_8\leqslant C\hbar^2\delta(\|\nabla^{k+1}u\|_{L^2}^2+\|\nabla^{k+1}\varrho\|_{L^2}^2+\|\nabla^k\triangle\varrho\|_{L^2}^2).$$

For the term I_9, we have

$$I_9\leqslant C\delta(\|\nabla^{k+1}B\|_{L^2}^2+\|\nabla^{k+1}u\|_{L^2}^2).$$

Owing to the estimates $I_1\sim I_9$ and the smallness of δ, we conclude the proof of the lemma.

Finally, we will derive the dissipation estimate for ϱ.

Lemma 3.3 Assume that $0\leqslant k\leqslant 4$, then there exist some constants $\delta_0<1$ sufficiently small and $\hbar_0<1$ such that

$$\frac{d}{dt}\int\nabla^k u\cdot\nabla^{k+1}\varrho+\frac{2\mu+\lambda}{2}\frac{|\nabla^{k+1}\varrho|^2}{(\varrho+1)^2}\,dx+\|\nabla^{k+1}\varrho\|_{L^2}^2+C\hbar^2\|\nabla^k\triangle\varrho\|_{L^2}^2$$

$$\leqslant C\|\nabla^{k+1}(u,B)\|_{L^2}^2,$$

for all $\delta \leqslant \delta_0$ and $\hbar < \hbar_0$, where C is independent of t.

Proof Appling ∇^k to $(3.8)_2$ and multiplying the resultant identity by $\nabla^{k+1}\varrho$, then integrating over $\mathbf{R}^3$, we have

$$\sum_{i=1}^{8} K_i = \int \nabla^k u_t \cdot \nabla^{k+1}\varrho\,dx - \mu \int \nabla^k\left(\frac{\Delta u}{\varrho+1}\right) \cdot \nabla^{k+1}\varrho\,dx$$

$$- (\mu+\lambda)\int \nabla^k\left(\frac{\nabla \mathrm{div}u}{\varrho+1}\right) \cdot \nabla^{k+1}\varrho\,dx + \int \nabla^k(u \cdot \nabla u) \cdot \nabla^{k+1}\varrho\,dx$$

$$+ \int \nabla^k\left(\left(\frac{P'(\varrho+1)}{\varrho+1}-1\right)\nabla\varrho\right) \cdot \nabla^{k+1}\varrho\,dx$$

$$- \frac{\hbar^2}{4}\int \nabla^k\left(\frac{1}{\varrho+1}\Delta\nabla\varrho\right) \cdot \nabla^{k+1}\varrho\,dx$$

$$+ \frac{\hbar^2}{4}\int \nabla^k\left(\frac{|\nabla\varrho|^2\nabla\varrho}{(\varrho+1)^3} - \frac{\nabla\varrho\Delta\varrho}{(\varrho+1)^2} - \frac{\nabla\varrho\cdot\nabla^2\varrho}{(\varrho+1)^2}\right) \cdot \nabla^{k+1}\varrho\,dx$$

$$+ \int \nabla^k\left(\frac{(\nabla\times B)\times B}{\varrho+1}\right) \cdot \nabla^{k+1}\varrho\,dx = 0.$$

For the term K_1, by using the continuity equation and Lemma 2.2, we have

$$K_1 = \frac{d}{dt}\int \nabla^k u \cdot \nabla^{k+1}\varrho\,dx - \|\nabla^k \mathrm{div}u\|_{L^2}^2 - \int \nabla^k \mathrm{div}u \cdot \nabla^k \mathrm{div}(\varrho u)\,dx$$

$$\geqslant \frac{d}{dt}\int \nabla^k u \cdot \nabla^{k+1}\varrho\,dx - \|\nabla^k \mathrm{div}u\|_{L^2}^2$$

$$- C\|\nabla^{k+1}u\|_{L^2}(\|\varrho\|_{L^\infty}\|\nabla^{k+1}u\|_{L^2} + \|u\|_{L^\infty}\|\nabla^{k+1}\varrho\|_{L^2})$$

$$\geqslant \frac{d}{dt}\int \nabla^k u \cdot \nabla^{k+1}\varrho\,dx - \|\nabla^k \mathrm{div}u\|_{L^2}^2 - C\delta\|\nabla^{k+1}(\varrho,u)\|_{L^2}^2.$$

For the term K_3, we can split it as

$$\frac{K_3}{\mu+\lambda} = -\int \frac{\nabla^{k+1}\mathrm{div}u}{\varrho+1}\nabla^{k+1}\varrho\,dx - \int [\nabla^k, \frac{1}{\varrho+1}]\nabla\mathrm{div}u\nabla^{k+1}\varrho\,dx$$

$$= -\int \frac{\nabla^{k+1}((\varrho+1)\mathrm{div}u)}{(\varrho+1)^2}\nabla^{k+1}\varrho\,dx + \int \frac{[\nabla^{k+1},\varrho+1]\mathrm{div}u}{(\varrho+1)^2}\nabla^{k+1}\varrho\,dx$$

$$- \int [\nabla^k, \frac{1}{\varrho+1}]\nabla\mathrm{div}u\nabla^{k+1}\varrho\,dx$$

$$= K_{31} + K_{32} + K_{33}.$$

For the first term K_{31}, by the continuity equation, integration by parts and

Lemma 2.2, we have

$$K_{31} = \frac{1}{2}\frac{d}{dt}\int \frac{|\nabla^{k+1}\varrho|^2}{(\varrho+1)^2}\,dx + \int \varrho_t \frac{|\nabla^{k+1}\varrho|^2}{(\varrho+1)^3}\,dx$$
$$- \frac{1}{2}\mathrm{div}\left(\frac{u}{(\varrho+1)^2}\right)|\nabla^{k+1}\varrho|^2\,dx + \int \frac{[\nabla^{k+1},u]\nabla\varrho}{(\varrho+1)^2}\nabla^{k+1}\varrho\,dx$$
$$\geq \frac{1}{2}\frac{d}{dt}\int \frac{|\nabla^{k+1}\varrho|^2}{(\varrho+1)^2}\,dx - C\delta\|\nabla^{k+1}(\varrho,u)\|_{L^2}^2.$$

For the last two terms, we deduce from Lemma 2.2 that

$$|K_{32},K_{33}| \leq C\delta\|\nabla^{k+1}(\varrho,u)\|_{L^2}^2.$$

The term K_2 can be treated similarly,

$$K_2 \geq \frac{\mu}{2}\frac{d}{dt}\int \frac{|\nabla^{k+1}\varrho|^2}{(\varrho+1)^2}\,dx - C\delta\|\nabla^{k+1}(\varrho,u)\|_{L^2}^2.$$

For the term K_4, by Lemma 2.2, we have

$$|K_4| = \int |\nabla^k(u\cdot\nabla u)\cdot\nabla^{k+1}\varrho|\,dx$$
$$\leq C\|\nabla^{k+1}\varrho\|_{L^2}(\|\nabla u\|_{L^\infty}\|\nabla u\|_{\dot H^k} + \|\nabla u\|_{L^3}\|u\|_{\dot H^{k,6}})$$
$$\leq C\delta\|\nabla^{k+1}(\varrho,u)\|_{L^2}^2.$$

For the term K_5, we split it as

$$|K_5| = \int |\nabla^{k+1}\varrho|^2\,dx - \int \nabla^k\left(\frac{P'(\varrho+1)}{\varrho+1}\nabla\varrho\right)\cdot\nabla^{k+1}\varrho\,dx$$
$$= K_{51} + K_{52}.$$

For the term K_{52}, by Hölder's inequality and Lemma 2.2, we have

$$|K_{52}| \leq C\|\nabla^k\left(\frac{P'(\varrho+1)}{\varrho+1}\nabla\varrho\right)\|_{L^2}\|\nabla^{k+1}\varrho\|_{L^2}$$
$$\leq C\left(\|\frac{P'(\varrho+1)}{\varrho+1}\|_{L^2}\|\nabla\varrho\|_{\dot H^k} + \|\frac{P'(\varrho+1)}{\varrho+1}\|_{\dot H^{k,6}}\|\nabla\varrho\|_{L^3}\right)\|\nabla^{k+1}\varrho\|_{L^2}$$
$$\leq C\delta\|\nabla^{k+1}\varrho\|_{L^2}^2.$$

Collecting these terms, we get

$$K_5 \geq \|\nabla^{k+1}\varrho\|_{L^2}^2 - C\delta\|\nabla^{k+1}\varrho\|_{L^2}^2.$$

For the term K_6, by integration by parts and Lemma 2.2, we have

$$K_6 = \frac{\hbar^2}{4}\int \frac{1}{\varrho+1}|\nabla^k\Delta\varrho|^2\,dx - \frac{\hbar^2}{12}\int \nabla\left(\frac{1}{\varrho+1}\right)\cdot \nabla^k\Delta\varrho\nabla^{k+1}\varrho\,dx$$
$$- \frac{\hbar^2}{4}\int \left[\nabla^k, \frac{1}{\varrho+1}\right]\nabla\Delta\varrho\nabla^{k+1}\varrho\,dx$$
$$\geqslant \frac{\hbar^2}{4}\int \frac{|\nabla^k\Delta\varrho|^2}{\varrho+1}\,dx - C\delta\hbar^2(\|\nabla^k\Delta\varrho\|_{L^2}^2 + \|\nabla^{k+1}\varrho\|_{L^2}^2).$$

For the term K_7, K_8, we have

$$|K_7| \leqslant \delta_1 \hbar^2 \|\nabla^k\Delta\varrho\|_{L^2}^2 + \frac{C\hbar^2\delta^2}{\delta_1}\|\nabla^{k+1}\varrho\|_{L^2}^2,$$
$$|K_8| \leqslant C\delta\|\nabla^{k+1}(\varrho,B)\|_{L^2}^2.$$

Combing $K_1 \sim K_8$ and taking $\delta_1 = 1/24$, by the smallness of δ, we conclude the lemma.

4 Proof of Theorem 1.1

In this section, we will give the proof for Theorem 1.1. From Lemma 3.1 and Lemma 3.2, choosing δ small enough, we have

$$\frac{d}{dt}\int |\nabla^k\varrho|^2 + |\nabla^k u|^2 + |\nabla^k B|^2 + \hbar^2|\nabla^{k+1}\varrho|^2\,dx$$
$$+ C_1(\|\nabla^{k+1}u\|_{L^2}^2 + \|\nabla^{k+1}B\|_{L^2}^2)$$
$$\leqslant C_2\delta(\|\nabla^{k+1}\varrho\|_{L^2}^2 + \hbar^2\|\nabla^k\Delta\varrho\|_{L^2}^2). \tag{4.27}$$

From Lemma 3.3, we have

$$\frac{d}{dt}\int \nabla^k u \cdot \nabla^{k+1}\varrho + |\nabla^{k+1}\varrho|^2\,dx + C_3(\|\nabla^{k+1}\varrho\|_{L^2}^2 + \hbar^2\|\nabla^k\Delta\varrho\|_{L^2}^2)$$
$$\leqslant C_4\|\nabla^{k+1}(u,B)\|_{L^2}^2. \tag{4.28}$$

Multiplying (4.28) by $2C_2\delta/C_3$, adding with (4.27), choosing $\delta > 0$ small enough, then there exists a constant $C_5 > 0$ such that

$$\frac{d}{dt}\left(\int \frac{2C_2\delta}{C_3}(\nabla^k u \cdot \nabla^{k+1}\varrho + |\nabla^{k+1}\varrho|^2)\,dx + \|\nabla^k(\varrho,u,B,\hbar\nabla\varrho)\|_{L^2}^2\right)$$
$$+ C_5(\|\nabla^{k+1}(\varrho,u,B)\|_{L^2}^2 + \hbar^2\|\nabla^k\Delta\varrho\|_{L^2}^2) \leqslant 0. \tag{4.29}$$

Then we deduce the following inequality for $0 \leqslant l \leqslant m-1$, $m \leqslant 4$,

$$\frac{\mathrm{d}}{\mathrm{d}t}\mathcal{E}_l^m + C(\|\nabla^{l+1}\varrho\|_{H^{m+1-l}}^2 + \|\nabla^{l+1}u\|_{H^{m-l}}^2 + \|\nabla^{l+1}B\|_{H^{m-l}}^2) \leqslant 0, \qquad (4.30)$$

where the function $\mathcal{E}_l^m$ is defined as

$$\mathcal{E}_l^m = \|\nabla^l \varrho\|_{H^{m+1-l}}^2 + \|\nabla^l u\|_{H^{m-l}}^2 + \|\nabla^l B\|_{H^{m-l}}^2 + \delta \sum_{k=l}^{m} \int \nabla^k u \cdot \nabla^{k+1}\varrho \, \mathrm{d}x. \qquad (4.31)$$

By virtue of the fifth term $\hbar^2 \nabla \Delta \varrho / 4$ on the left-hand side of $(3.8)_2$, the inequality (4.30) holds on for $0 \leqslant l \leqslant m \leqslant 4$.

Then, we will improve the time decay rates for the higher-order spatial derivatives of density, velocity, and magnetic field.

Lemma 4.1 *Under all the assumptions of Theorem 1.1, the global solution (ϱ, u, B) of the problem (3.8)-(3.10) has the following decay rates*

$$\|\nabla^k \varrho\|_{H^{5-k}} + \|\nabla^k u\|_{H^{4-k}} + \|\nabla^k B\|_{H^{4-k}} \leqslant C(1+t)^{-\frac{3+2k}{4}}, \qquad (4.32)$$

for any integer $k = 0, 1, 2, 3$.

Proof We will use the induction strategy to prove (4.32). For the case $\alpha = 0, 1$, the inequality (4.32) holds true owing to the inequality (1.5). Now, we assume the inequality (4.32) holds for the case $k = l$, i.e.

$$\|\nabla^l \varrho\|_{H^{5-l}} + \|\nabla^l u\|_{H^{4-l}} + \|\nabla^l B\|_{H^{4-l}} \leqslant C(1+t)^{-\frac{3+2l}{4}}, \qquad (4.33)$$

for any integer $l = 1, 2$. Then, it remains to verify that the inequality (4.32) holds for the case $\alpha = l+1$. Replacing l as $l+1$ in inequality (4.30) implies

$$\frac{\mathrm{d}}{\mathrm{d}t}\mathcal{E}_{l+1}^4 + C(\|\nabla^{l+2}\varrho\|_{H^{4-l}}^2 + \|\nabla^{l+2}u\|_{H^{3-l}}^2 + \|\nabla^{l+2}B\|_{H^{3-l}}^2) \leqslant 0, \qquad (4.34)$$

where the function $\mathcal{E}_{l+1}^4$ is defined as

$$\mathcal{E}_{l+1}^4 = \|\nabla^{l+1}\varrho\|_{H^{4-l}}^2 + \|\nabla^{l+1}u\|_{H^{3-l}}^2 + \|\nabla^{l+1}B\|_{H^{3-l}} + \delta \sum_{i=l+1}^{4} \int \nabla^i u \cdot \nabla^{i+1}\varrho \, \mathrm{d}x. \qquad (4.35)$$

For some constant R defined below, denoting the time sphere S_0 (see [32]) by

$$S_0 : \left\{ \xi \in \mathbf{R}^3 \Big| |\xi| \leqslant \left(\frac{R}{1+t}\right)^{\frac{1}{2}} \right\},$$

then we have

$$\|\nabla^{k+2}u\|_{L^2}^2 \geq \int_{\mathbf{R}^3/S_0} |\xi|^{2(k+2)}|\hat{u}|^2 d\xi$$

$$\geq \frac{R}{1+t}\int_{\mathbf{R}^3/S_0} |\xi|^{2(k+1)}|\hat{u}|^2 d\xi$$

$$\geq \frac{R}{1+t}\int_{\mathbf{R}^3} |\xi|^{2(k+1)}|\hat{u}|^2 d\xi - \frac{R^2}{(1+t)^2}\int_{\mathbf{R}^3} |\xi|^{2k}|\hat{u}|^2 d\xi.$$

Therefore, we have the following inequality

$$\|\nabla^{k+2}u\|_{L^2}^2 \geq \frac{R}{1+t}\|\nabla^{k+1}u\|_{L^2}^2 - \frac{R^2}{(1+t)^2}\|\nabla^k u\|_{L^2}^2. \quad (4.36)$$

Adding (4.36) from $k = l$ to $k = 3$ implies

$$\|\nabla^{l+2}u\|_{H^{3-l}}^2 \geq \frac{R}{1+t}\|\nabla^{l+1}u\|_{H^{3-l}}^2 - \frac{R^2}{(1+t)^2}\|\nabla^l u\|_{H^{3-l}}^2. \quad (4.37)$$

Similarly, we have

$$\|\nabla^{l+2}B\|_{H^{3-l}}^2 \geq \frac{R}{1+t}\|\nabla^{l+1}B\|_{H^{3-l}}^2 - \frac{R^2}{(1+t)^2}\|\nabla^l B\|_{H^{3-l}}^2, \quad (4.38)$$

$$\|\nabla^{l+2}\varrho\|_{H^{4-l}}^2 \geq \frac{R}{1+t}\|\nabla^{l+1}\varrho\|_{H^{4-l}}^2 - \frac{R^2}{(1+t)^2}\|\nabla^l \varrho\|_{H^{4-l}}^2, \quad (4.39)$$

which together with (4.37) yields that

$$\frac{d}{dt}\mathcal{E}_{l+1}^4 + \frac{RC}{2(1+t)}(\|\nabla^{l+1}\varrho\|_{H^{4-l}}^2 + \|\nabla^{l+1}u\|_{H^{3-l}}^2 + \|\nabla^{l+1}B\|_{H^{3-l}}^2)$$

$$\leq \frac{CR^2}{(1+t)^2}(\|\nabla^l \varrho\|_{H^{4-l}}^2 + \|\nabla^l u\|_{H^{3-l}}^2 + \|\nabla^l B\|_{H^{3-l}}^2)$$

$$\leq C(1+t)^{-2}(1+t)^{-\frac{3+2l}{2}} \leq C(1+t)^{-\frac{7+2l}{2}}, \quad (4.40)$$

where we have used (4.33). With the help of Young inequality and the smallness of δ, it arrives at directly

$$\mathcal{E}_{l+1}^4 \sim (\|\nabla^{l+1}\varrho\|_{H^{4-l}}^2 + \|\nabla^{l+1}u\|_{H^{3-l}}^2 + \|\nabla^{l+1}B\|_{H^{3-l}}^2). \quad (4.41)$$

Substituting the equivalent relation (4.41) into the inequality (4.40), we get

$$\frac{d}{dt}\mathcal{E}_{l+1}^4 + \frac{RC}{2(1+t)}\mathcal{E}_{l+1}^4 \leq C(1+t)^{-\frac{7+2l}{2}}. \quad (4.42)$$

Choosing $R = (l+3)/C$ and multiplying (4.42) by $(1+t)^{l+3}$, then we obtain

$$\frac{\mathrm{d}}{\mathrm{d}t}[(1+t)^{l+3}\mathcal{E}_{l+1}^4] \leqslant C(1+t)^{-\frac{1}{2}}. \tag{4.43}$$

The integration of (4.43) over $[0, t]$ implies

$$\mathcal{E}_{l+1}^4 \leqslant C(1+t)^{-\frac{5+2l}{2}},$$

or equivalently

$$\|\nabla^{l+1}\varrho\|_{H^{4-l}}^2 + \|\nabla^{l+1}u\|_{H^{3-l}}^2 + \|\nabla^{l+1}B\|_{H^{3-l}}^2 \leqslant C(1+t)^{-\frac{5+2l}{2}}, \tag{4.44}$$

where we have used (4.41). Therefore, we finish the proof of (4.32) by induction, and thus we complete the proof of Lemma 4.1.

Finally, we aim to improve the time decay rates for the fourth-order spatial derivatives of magnetic field.

Lemma 4.2 *Under the assumption of Theorem 1.1, the magnetic field has the following time decay rate*

$$\|\nabla^4 B\|_{L^2} \lesssim (1+t)^{-\frac{11}{4}}.$$

Proof Taking fourth spatial derivatives to $(3.8)_3$, multiplying the resultant equation by $\nabla^4 B$ and integrating by parts over $\mathbf{R}^3$, we have

$$\frac{1}{2}\frac{\mathrm{d}}{\mathrm{d}t}\int |\nabla^4 B|^2\, \mathrm{d}x + \int |\nabla^5 B|^2\, \mathrm{d}x = \int \nabla^4(-u \cdot \nabla B + B \cdot \nabla u - B\mathrm{div}u) \cdot \nabla^4 B$$
$$= III_1 + III_2 + III_3.$$

By virtue of the Leibnitz formula, Hölder and Young inequality, we have

$$III_1 \lesssim \sum_{k=0}^{3} \|\nabla^k u\|_{L^3}\|\nabla^{4-k} B\|_{L^6}\|\nabla^5 B\|_{L^2}$$
$$\lesssim \sum_{k=1}^{3} \|\nabla^k u\|_{H^1}\|\nabla^{5-k} B\|_{L^2}\|\nabla^5 B\|_{L^2} + \|u\|_{H^1}\|\nabla^5 B\|_{L^2}^2$$
$$\lesssim \sum_{k=1}^{3} \|\nabla^k u\|_{H^1}^2\|\nabla^{5-k} B\|_{L^2}^2 + (\varepsilon + \delta)\|\nabla^5 B\|_{L^2}^2.$$

We can deduce the following inequalities in the similar way as that in Lemma 3.1

$$III_2 \lesssim \delta(\|\nabla^5 B\|_{L^2}^2 + \|\nabla^5 u\|_{L^2}^2), \quad III_3 \lesssim \delta(\|\nabla^5 B\|_{L^2}^2 + \|\nabla^5 u\|_{L^2}^2).$$

Then combing the above estimates, the smallness of δ and ε, and the decay rates (4.32), it arrives that

$$\frac{1}{2}\frac{\mathrm{d}}{\mathrm{d}t}\int |\nabla^4 B|^2\,\mathrm{d}x + \int |\nabla^5 B|^2\,\mathrm{d}x$$
$$\lesssim \sum_{k=1}^{3}\|\nabla^k u\|_{H^1}^2\|\nabla^{5-k}B\|_{L^2}^2$$
$$= \sum_{k=2}^{3}\|\nabla^k u\|_{H^1}^2\|\nabla^{5-k}B\|_{L^2}^2 + \|\nabla u\|_{H^1}^2\|\nabla^4 B\|_{L^2}^2$$
$$\lesssim \sum_{k=2}^{3}(1+t)^{-\frac{3}{2}-k}(1+t)^{-\frac{5}{2}-4+k} + (1+t)^{-\frac{5}{2}}(1+t)^{-\frac{1}{2}-4}$$
$$\lesssim (1+t)^{-8} + (1+t)^{-7}$$
$$\lesssim (1+t)^{-7}$$

Taking the Fourier splitting method as the inequality (4.36) and the decay rates (4.32), we have

$$\frac{\mathrm{d}}{\mathrm{d}t}\int |\nabla^4 B|^2\,\mathrm{d}x + \frac{6}{1+t}\int |\nabla^4 B|^2\,\mathrm{d}x$$
$$\lesssim \left(\frac{6}{1+t}\right)^2 \int |\nabla^3 B|^2\,\mathrm{d}x + (1+t)^{-7}$$
$$\lesssim (1+t)^{-\frac{13}{2}}. \tag{4.45}$$

Multiplying (4.45) by $(1+t)^6$, then integrating over $[0,t]$, we have

$$\|\nabla^4 B\|_{L^2}^2 \lesssim (1+t)^{-\frac{11}{2}},$$

and thus we complete the proof.

Acknowledgements X. Xi was supported by Introduction of talent research start-up fund at Guangzhou University, NSF of Guangdong Province (No. 2018A030310312) and the project for young creative talents of Higher Education of Guangdong Province (No. 2017KQNCX148). X. Pu is supported by NSFC (No. 11471057). B. Guo was partially supported by NSFC (No. 11731014). The authors thank the referees for valuable comments and suggestions.

References

[1] Chen Q, Tan Z. Global existence and convergence rates of smooth solutions for the compressible magnetohydrodynamic equations [J]. Nonlinear Anal., 2010, 72: 4438–4451.

[2] Deckelnick K. Decay estimates for the compressible Navier-Stokes equations in unbounded domains [J]. Math. Z., 1992, 209, 115–130.

[3] Deckelnick K. L^2-decay for the compressible Navier-Stokes equations in unbounded domains [J]. Comm. Partial Differential Equations, 1993, 18: 1445–1476.

[4] Duan R-J, Liu H-X, Ukai S, Yang T. Optimal L^p-L^q convergence rates for the compressible Navier-Stokes equations with potential force [J]. J. Differential Equations, 2007, 238: 220–233.

[5] Duan R-J, Ukai S, Yang T, Zhao H-J. Optimal convergence rates for the compressible Navier-Stokes equations with potential forces [J]. Math. Models Methods Appl. Sci., 2007, 17: 737–758.

[6] Gao J-C, Chen Y-H, Yao Z-A. Long-time behavior of solution to the compressible magnetohydrodynamic equations [J]. Nonlinear Anal., 2015, 128: 122–135.

[7] Gao J-C, Tao Q, Yao Z-A. Optimal decay rates of classical solutions for the full compressible MHD equations [J]. Z. Angew. Math. Phys., 2016, 67(2): 23.

[8] Gao J-C, Tao Q, Yao Z-A. Long-time behavior of solution for the compressible nematic liquid crystal flows in R^3 [J]. J. Differential Equations, 2016, 261(4): 2334–2383.

[9] Gao J-C, Yao Z-A. Global Existence and Optimal Decay Rates of Solutions for Compressible Hall-MHD Equations [J]. Discrete Contin. Dyn. Syst., 2017, 36(6): 3077–3106.

[10] Guo Y, Wang Y-J. Decay of dissipative equations and negative Sobolev spaces [J]. Comm. Partial Differential Equations, 2012, 37: 2165–2208.

[11] Haas F. A magnetohydrodynamic model for quantum plasmas [J]. Phys. Plasmas., 2005, 12(6): 062117.

[12] Haas F. Quantum Plasmas: An Hydrodynamic Approach [M]. Springer, New York, 2011.

[13] Hoff D, Zumbrun K. Multidimensional diffusion waves for the Navier-Stokes equations of compressible flow [J]. Indiana Univ. Math. J., 1995, 44: 604–676.

[14] Hoff D, Zumbrun K. Pointwise decay estimates for multidimensional Navier-Stokes diffusion waves [J]. Z. Angew. Math. Phys., 1997, 48: 597–614.

[15] Kagei Y, Kobayashi T. On large time behavior of solutions to the compressible Navier-Stokes equations in the half space in $\mathbf{R}^3$ [J]. Arch. Ration. Mech. Anal., 2002, 165: 89–159.

[16] Kagei Y, Kobayashi T. Asymptotic behavior of solutions of the compressible Navier-Stokes equations on the half space [J]. Arch. Ration. Mech. Anal., 2005, 177: 231–330.

[17] Kobayashi T, Shibata Y. Decay estimates of solutions for the equations of motion of compressible viscous and heat conductive gases in an exterior domain in $\mathbf{R}^3$ [J]. Commun. Math. Phys., 1999, 200: 621–659.

[18] Li H, Cheng M, Yan W. Global existence and large time behavior of solutions for compressible quantum magnetohydrodynamics flows in $\mathbf{T}^3$ [J]. J. Math. Anal. Appl., 2017, 452: 1209–1228.

[19] Li F-C, Yu H-J. Optimal decay rate of classical solutions to the compressible magnetohydrodynamic equations [J]. Proc. R. Soc. Edinb. Sect. A, 2011, 141: 109–126.

[20] Liu T-P, Wang W-K. The pointwise estimates of diffusion waves for the Navier-Stokes equations in odd multi-dimensions [J]. Commun. Math. Phys., 1998, 196: 145–173.

[21] Matsumura A. An Energy Method for the Equations of Motion of Compressible Viscous and Heat-Condutive Fluids, 1-16 [J]. University of Wisconsin-Madison MRC Technical Summary Report, 2194, 1986.

[22] Matsumura A, Nishida T. The initial value problem for the equations of motion of compressible viscous and heat-conductive fluids [J]. Proc. Jpn. Acad. Ser. A, 1979, 55: 337–342.

[23] Matsumura A, Nishida T. The initial value problems for the equations of motion of viscous and heat-conductive gases [J]. J. Math. Kyoto Univ., 1980 20: 67–104.

[24] Ponce G. Global existence of small solutions to a class of nonlinear evolution equations [J]. Nonlinear Anal., 1985, 9: 339–418.

[25] Pu X-K, Guo B-L. Optimal decay rate of the compressible quantum Navier-Stokes equations [J]. Ann. of Appl. Math., 2016, 32(3): 275–287.

[26] Pu X-K, Guo B-L. Global existence and convergence rates of smooth solutions for the full compressible MHD equations [J]. Z. Angew. Math. Phys., 2013, 64: 519–538.

[27] Pu X-K, Guo B-L. Global existence and semiclassical limit for quantum hydrodynamic equations with viscosity and heat conduction [J]. Kinetic & Related Models, 2015, 9(1): 165–191.

[28] Pu X-K, Xu X-L. Decay rates of the magnetohydrodynamic model for quantum plasmas [J]. Z. Angew. Math. Phys., 2017, 68:18.

[29] Pu X-K, Xu X-L. Asymptotic behaviors of the full quantum hydrodynamic equations [J]. J. Math. Anal. Appl., 2017, 454(1): 219–245.

[30] Schonbek M-E. Large time behaviour of solutions to the Navier-Stokes equations in H^m spaces [J]. Comm. Partial Differential Equations, 1995, 20: 103–117.

[31] Schonbek M-E, Wiegner M. On the decay of higher-order norms of the solutions of Navier-Stokes equations [J]. Proc. R. Soc. Edinb. Sect. A, 1996, 126: 677–685.

[32] Schonbek M-E. L^2 decay for weak solutions of the Navier-Stokes equations [J]. Arch. Ration. Mech. Anal., 1985, 88: 209–222.

[33] Tan Z, Wang H-Q. Optimal decay rates of the compressible magnetohydrodynamic equations [J]. Nonlinear Anal. Real World Appl., 2013, 14: 188–201.

[34] Wigner E. On the quantum correction for thermodynamic equilibrium [J]. Phys. Rev., 1932, 40: 749–759.

[35] Xie B-Q, Xi X-Y, Guo B-L. Long-time behavior of solutions for full compressible quantum model in $\mathbf{R}^3$ [J]. Applied Mathematics Letters, 2018 80: 54–58.

[36] Yang J, Ju Q. Global existence of the three-dimensional viscous quantum magnetohydrodynamic model [J]. J. Math. Phys., 2014, 55(8): 081501.

Global Solution to the 3-D Inhomogeneous Incompressible MHD System with Discontinuous Density*

Chen Fei (陈菲), Guo Boling (郭柏灵), and Zhai Xiaoping (翟小平)

Abstract In this paper, we consider the Cauchy problem of the incompressible MHD system with discontinuous initial density in $\mathbf{R}^3$. We establish the global well-posedness of the MHD system if the initial data satisfies $(\rho_0, u_0, H_0) \in L^\infty(\mathbf{R}^3) \times H^s(\mathbf{R}^3) \times H^s(\mathbf{R}^3)$ with $\frac{1}{2} < s \leqslant 1$ and $0 < \underline{\rho} \leqslant \rho_0 \leqslant \overline{\rho} < +\infty$, $\|(u_0, H_0)\|_{\dot{H}^{\frac{1}{2}}} \leqslant c$, for some small $c > 0$ which only depends on $\underline{\rho}, \overline{\rho}$. As a byproduct, we also get the decay estimate of the solution.

Keywords global well-posedness; incompressible MHD system; Cauchy problem; decay estimate

1 Introduction and the main results

Magnetohydrodynamic (MHD) system describes the interaction between the magnetic field and conductive fluid, which is a nonlinear system that couples Navier-Stokes equations with Maxwell equations. In magnetohydrodynamics, the displacement currents can be neglected in the time dependent Maxwell equations. So it becomes the following system:

$$\begin{cases} \partial_t \rho + \mathrm{div}(\rho u) = 0, \\ \partial_t(\rho u) + \mathrm{div}(\rho u \otimes u) - \Delta u + \nabla P = \mathrm{curl}\, H \times H, \\ \partial_t H - \Delta H = \mathrm{curl}(u \times H), \quad (t,x) \in \mathbf{R}^+ \times \mathbf{R}^3, \\ \mathrm{div}\, u = \mathrm{div}\, H = 0, \\ (\rho, u, H)|_{t=0} = (\rho_0, u_0, H_0), \end{cases} \quad (1.1)$$

where ρ is the density, u is the velocity field, H is the magnetic field and P is the scalar pressure. The body force

$$\operatorname{curl} H \times H = H \cdot \nabla H - \nabla \left(\frac{|H|^2}{2} \right)$$

and

$$\operatorname{curl}(u \times H) = H \cdot \nabla u - u \cdot \nabla H,$$

if $\operatorname{div} u = \operatorname{div} H = 0$.

When $H = 0$, (1.1) turns into the well-known inhomogeneous incompressible Navier-Stokes system, which has been studied by many researchers (see [1], [2], [9], [12], [13], [14], [28], [31], [32], [35], [38]). When the density ρ is a constant, (1.1) reduces to be the classical MHD system which has been studied also by many researchers (see [6], [7], [10], [16], [21], [22], [29], [30], [33]). Duvaut and Lions [16] established the local well-posedness in $H^s(\mathbf{R}^N)$, $s \geqslant N$, and certified global existence of the solution with small initial data. Sermange and Temam [34] proved that the 2-D local strong solution can be extended to global and unique. With mixed partial dissipation and additional magnetic diffusion in the 2-D MHD system, Cao and Wu [7] proved that such a system is globally well-posed for any data in $H^2(\mathbf{R}^2)$. In a recent remarkable paper, Lin, Xu and Zhang [29] proved the global existence of smooth solution of the 2-D MHD system around the non-trivial steady state solution $(x_2, 0)$ (see [30] for 3-D case). In [33], Ren, Wu, Xiang and Zhang got the global existence and the decay estimates of small smooth solution for the 2-D MHD equations without magnetic diffusion. There are also many results on the regularity criteria (see [10], [21], [22]).

When the density ρ is not a constant, (1.1) is the so called inhomogeneous incompressible MHD system. Compared with the Navier-Stokes equations, the dynamic motion of the fluid and the magnetic field interact on each other and both the hydrodynamic and electrodynamic effects in the motion are strongly coupled, the problems of MHD system are considerably more complicated. Even through, in the past several years, there are also many mathematical results related to the incompressible MHD system (see [3], [4], [8], [11], [15], [16], [18], [26], [27], [34], [36], [37], [38]). Gerbeau and Le Bris [18] and also Desjardins and Le Bris [15] researched global existence of weak solution of finite energy in $\mathbf{R}^3$ and in the torus $\mathbf{T}^3$. Chen, Tan and Wang [11] showed the local strong solution in H^2 when the initial data contain vacuum states (i.e. the initial density ρ may vanish in some open set of Ω). Then, Huang and Wang [26] extended the local strong solution to be global in two dimensions (see also Gong and Li [19] for three dimensions).

When the initial density is away from zero and is close enough to a positive constant, Abidi and Hmidi in [3] and Abidi and Paicu in [4] obtained the global solution with small initial data in the critical Besov spaces. By critical, we mean that we want to solve the system (1.1) in functional spaces with invariant norms by the changes of scales which leaves (1.1) invariant. In the case of inhomogeneous incompressible MHD fluids, it is easy to see that the transformations: $(\rho_\lambda, u_\lambda, H_\lambda)(t,x) = (\rho(\lambda^2 t, \lambda x), \lambda u(\lambda^2 t, \lambda x), \lambda H(\lambda^2 t, \lambda x))$ have that property, provided that the pressure term has been changed accordingly. The results in [3], [4] have been extended by Chen, Li and Xu in [8], Zhai, Li and Xu in [36], Zhai, Li and Yan in [37].

When the initial density is away from zero and is not close enough to a positive constant, Gui in [20] considered the global well-posedness of 2-D MHD equations with constant viscosity and variable conductivity for a generic family of the variations of the initial data, and established the global well-posedness of the equations in the critical spaces with constant conductivity. Zhai and Yin in [38] got the global solution in $\mathbf{R}^3$ by applying a new a priori estimate for an elliptic equation with nonconstant coefficients. Hoff [23,24] and Huang, Li and Xin [25] respectively proved the global existence of small energy weak solutions and global well-posedness of small energy classical solutions of the isentropic compressible Navier-Stokes equations, where, [25] is the first for global classical solutions that may have large oscillations and can contain vacuum states. An important idea in [23], [24] and [25] is that by using an appropriate time weight, the energy estimate for space derivatives of the velocity field can be closed although the initial date has low regularity (even only in $L^2(\mathbf{R}^d)$). By using a similar idea, Chen, Zhang and Zhao in [9] obtained the Fujita-Kato solution for the 3-D inhomogeneous Navier-Stokes equations.

In this paper, we will get a similar result to [9] for the 3-D inhomogeneous incompressible MHD system. We establish the global existence and uniqueness of the solution, under the condition that the initial date (u_0, H_0) is small in the critical space $\dot{H}^{\frac{1}{2}}$. We use the idea of time weight in [9], [23], [24] and [25] to deal with less regular initial velocity field and magnetic field, and the Lagrangian idea in [13], [14] to deal with the proof of the uniqueness of the solution. At last, we also get the decay estimate of the solution by the dual method as a generalization of the result in [9].

The main results of this paper are the following theorems:

Theorem 1.1 Let $\frac{1}{2} < s \leqslant 1$. Given $\underline{\rho}, \overline{\rho} \in (0, \infty)$, suppose that

$$0 < \underline{\rho} \leqslant \rho_0 \leqslant \overline{\rho} < +\infty, \quad (u_0, H_0) \in H^s(\mathbf{R}^3) \times H^s(\mathbf{R}^3), \tag{1.2}$$

and there exists a constant c depending only on $\underline{\rho}, \overline{\rho}$ such that

$$\|(u_0, H_0)\|_{H^{\frac{1}{2}}} \leqslant c, \tag{1.3}$$

then the system (1.1) has a unique global solution (ρ, u, H) satisfying

$$\underline{\rho} \leqslant \rho(t, x) \leqslant \overline{\rho} \quad \text{for} \quad (t, x) \in [0, \infty) \times \mathbf{R}^3,$$
$$F_0(t) \leqslant C\|(u_0, H_0)\|_{L^2}^2 \quad \text{for} \quad t \in [0, \infty),$$
$$F_1(t) \leqslant C\|(u_0, H_0)\|_{H^s}^2 \quad \text{for} \quad t \in [0, \infty),$$
$$F_2(t) \leqslant C\|(u_0, H_0)\|_{H^s}^2 \exp\left\{C\|(u_0, H_0)\|_{H^s}^4\right\} \quad \text{for} \quad t \in [0, \infty), \tag{1.4}$$

where the constant C depending only on $\underline{\rho}, \overline{\rho}$ and

$$F_0(t) = \int_{\mathbf{R}^3} |(\sqrt{\rho}u, H)(t,x)|^2 \, dx + \int_0^t \int_{\mathbf{R}^3} |(\nabla u, \nabla H)|^2 \, dx d\tau,$$

$$F_1(t) = \omega(t)^{1-s} \int_{\mathbf{R}^3} |(\nabla u, \nabla H)|^2 \, dx$$
$$+ \int_0^t \int_{\mathbf{R}^3} \omega(\tau)^{1-s} \left(|(\sqrt{\rho}u_t, H_t)|^2 + |(\nabla^2 u, \nabla^2 H)|^2 + |\nabla P|^2\right) dx d\tau,$$

$$F_2(t) = \omega(t)^{2-s} \int_{\mathbf{R}^3} \left(|(\sqrt{\rho}u_t, H_t)|^2 + |(\nabla^2 u, \nabla^2 H)|^2 + |\nabla P|^2\right) dx$$
$$+ \int_0^t \int_{\mathbf{R}^3} \omega(\tau)^{2-s} |(\nabla u_t, \nabla H_t)|^2 \, dx d\tau,$$

in which $\omega(t) \triangleq \min\{1, t\}$.

Remark 1.1 In Theorem 1.1, The powers to the weight $\omega(t)$ in $F_i(t)$, $i = 1, 2$ are inspired by the following characterization of Besov norm (c.f. Theorem 2.34 of [5]):

$$C^{-1}\|u\|_{\dot{B}_{p,q}^{-2r}} \leqslant \left\| \|t^r e^{t\Delta} u\|_{L^p} \right\|_{L^q(\mathbf{R}^+, \frac{dt}{t})} \leqslant C\|u\|_{\dot{B}_{p,q}^{-2r}}, \tag{1.5}$$

$r > 0$, $1 \leqslant p, q \leqslant +\infty$, which means that for $u_0, H_0 \in H^s(\mathbf{R}^3)$ with $s \in (0,1)$, (1.5) and
$$\nabla u_0, \nabla H_0 \in H^{s-1}(\mathbf{R}^3) \hookrightarrow B_{2,\infty}^{s-1}(\mathbf{R}^3), \quad \nabla^2 u_0, \nabla^2 H_0 \in B_{2,\infty}^{s-2}(\mathbf{R}^3),$$
lead to
$$t^{\frac{1-s}{2}} \|(e^{t\Delta}\nabla u_0, e^{t\Delta}\nabla H_0)\|_{L^2} \leqslant C\|(u_0, H_0)\|_{H^s}$$
and
$$t^{\frac{2-s}{2}} \|(e^{t\Delta}\nabla^2 u_0, e^{t\Delta}\nabla^2 H_0)\|_{L^2} \leqslant C\|(u_0, H_0)\|_{H^s}.$$

Remark 1.2 *The aim of the introduction of the time weight $\omega(t)$ in Theorem 1.1 is to close the energy estimates for space derivatives of the velocity field and the magnetic field with low regular initial data $(u_0, H_0) \in H^s(\mathbf{R}^3) \times H^s(\mathbf{R}^3)$ when $\frac{1}{2} < s \leqslant 1$. Here we only consider the low regularity case when $\frac{1}{2} < s \leqslant 1$, since it is much easier to deal with the higher regular cases.*

We also get the following decay estimate of the solution obtained in Theorem 1.1:

Theorem 1.2 *Assume that (u, H) is a solution obtained in Theorem 1.1, and the initial data $(u_0, H_0) \in L^q(\mathbf{R}^3)$, $q \in \left[\frac{6}{5}, 2\right]$. Then the solution satisfies the following inequality*
$$\|(\nabla^k u, \nabla^k H)(t)\|_{L^2} \leqslant C(1+t)^{-\frac{k}{2}-\zeta(q)}, \quad \text{for} \quad t \geqslant 1, \quad k = 0, 1,$$
where $\zeta(q) = \frac{3}{2}\left(\frac{1}{q} - \frac{1}{2}\right)$.

Throughout this paper, we use the following notations. For simplicity, we denote
$$\int f \, dx = \int_{\mathbf{R}^3} f \, dx.$$
For $1 \leqslant p \leqslant \infty$, and $m \in \mathbf{N}$, the Sobolev spaces are defined in a standard manner:
$$L^p \triangleq L^p(\mathbf{R}^3), \quad H^m \triangleq W^{m,2}(\mathbf{R}^3).$$

Given a Banach space X, we shall denote by $\|(a,b)\|_X = \|a\|_X + \|b\|_X$. The rest of the paper unfolds as follows. In Section 2, we prove the global existence part of Theorem 1.1 by using the time weighted energy method; in Section 3, we complete the uniqueness of the solutions by using the Lagrangian coordinate method; in Section 4, we further prove the decay of the solution by applying the dual method.

2 The proof of the existence part of Theorem 1.1

In this section, we concentrate on the proof of the existence part of Theorem 1.1. Let j_σ be the standard Friedrich's mollifier, and define

$$\rho_0^\sigma \triangleq j_\sigma * \rho_0, \quad u_0^\sigma \triangleq j_\sigma * u_0, \quad \text{and} \quad H_0^\sigma \triangleq j_\sigma * H_0.$$

Choose σ small enough such that

$$\underline{\rho}/2 \leqslant \rho_0^\sigma(x) \leqslant 2\overline{\rho}, \quad x \in \mathbf{R}^3.$$

With the initial data

$$(\rho_0^\sigma, u_0^\sigma, H_0^\sigma),$$

there exists a time $T_\sigma > 0$ such that system (1.1) has a unique smooth solution

$$(\rho^\sigma, u^\sigma, H^\sigma)$$

on $[0, T^\sigma]$. In what follows, we'll only present the uniform estimates (1.4) of the smooth approximate solutions

$$(\rho^\sigma, u^\sigma, H^\sigma).$$

Then, the existence part of Theorem 1.1 essentially follows by (1.4) and a standard compactness argument. In the sequel, we omit the superscript σ for simplicity. And we denote by C or C_i $(i = 1, 2)$ the positive constant, which may depend on $\overline{\rho}, \underline{\rho}$ but does not rely on the time T and the superscript σ.

Next, we'll use the bootstrap theory to get the uniform estimate. We suppose that the following *a priori* hypothesis holds:

$$\|(u, H)(t)\|_{L^3} \leqslant c_0, \quad t \in [0, T]. \tag{2.1}$$

For some small enough $c_0 > 0$ determined later. Next, we'll get the estimate (1.4) and deduce the following refined estimate

$$\|(u, H)(t)\|_{L^3} \leqslant \frac{c_0}{2}, \quad t \in [0, T]. \tag{2.2}$$

Step 1 L^2 and H^1 estimates without the time weight.

From the transport equation, we can easily get

$$\|\rho(t)\|_{L^\infty} = \|\rho_0\|_{L^\infty},$$

and with (1.2), we have
$$\underline{\rho} \leqslant \rho(t,x) \leqslant \overline{\rho}, \quad (t,x) \in [0,T] \times \mathbf{R}^3.$$

Taking L^2 inner product of $(1.1)_2$ with u, integrating by parts over $\mathbf{R}^3$, and using the transport equation $(1.1)_1$, we get
$$\frac{1}{2}\frac{\mathrm{d}}{\mathrm{d}t}\int \rho|u|^2\,\mathrm{d}x + \int |\nabla u|^2\,\mathrm{d}x = \int (\mathrm{curl}\, H \times H) \cdot u\,\mathrm{d}x. \tag{2.3}$$

Similarly, taking L^2 inner product of $(1.1)_3$ with H, we obtain
$$\frac{1}{2}\frac{\mathrm{d}}{\mathrm{d}t}\int |H|^2\,\mathrm{d}x + \int |\nabla H|^2\,\mathrm{d}x = \int \mathrm{curl}(u \times H) \cdot H\,\mathrm{d}x$$
$$= \int (u \times H) \cdot \mathrm{curl}\, H\,\mathrm{d}x$$
$$= -\int (\mathrm{curl}\, H \times H) \cdot u\,\mathrm{d}x. \tag{2.4}$$

By combining (2.3) with (2.4) and integrating it over $[0,t]$, $\forall\, t \in [0,T]$, we have
$$\frac{1}{2}\|(\sqrt{\rho}u, H)(t)\|_{L^2}^2 + \int_0^t \|(\nabla u, \nabla H)\|_{L^2}^2\,\mathrm{d}\tau = \frac{1}{2}\|(\sqrt{\rho_0}u_0, H_0)\|_{L^2}^2, \tag{2.5}$$
which means $F_0(t) \leqslant \|(\sqrt{\rho_0}u_0, H_0)\|_{L^2}^2$.

Taking L^2 inner product of $(1.1)_2$ with u_t and integrating by parts over $\mathbf{R}^3$, we obtain
$$\frac{1}{2}\frac{\mathrm{d}}{\mathrm{d}t}\int |\nabla u|^2\,\mathrm{d}x + \int \rho|u_t|^2\,\mathrm{d}x$$
$$= -\int \rho(u \cdot \nabla)u \cdot u_t\,\mathrm{d}x + \int (H \cdot \nabla)H \cdot u_t\,\mathrm{d}x$$
$$= -\int \rho(u \cdot \nabla)u \cdot u_t\,\mathrm{d}x + \int (H \cdot \nabla)H \cdot (\sqrt{\rho}u_t)(\sqrt{\rho})^{-1}$$
$$\leqslant C\int (|u||\nabla u| + |H||\nabla H|)|\sqrt{\rho}u_t|\,\mathrm{d}x$$
$$\leqslant C\|(u,H)\|_{L^3}\|(\nabla u, \nabla H)\|_{L^6}\|\sqrt{\rho}u_t\|_{L^2}$$
$$\leqslant C\|(u,H)\|_{L^3}\left(\|\sqrt{\rho}u_t\|_{L^2}^2 + \|(\nabla^2 u, \nabla^2 H)\|_{L^2}^2\right). \tag{2.6}$$

Similarly, taking L^2 inner product of $(1.1)_3$ with H_t, we achieve
$$\frac{1}{2}\frac{\mathrm{d}}{\mathrm{d}t}\int |\nabla H|^2\,\mathrm{d}x + \int |H_t|^2\,\mathrm{d}x$$
$$\leqslant C\int (|H||\nabla u||H_t| + |u||\nabla H||H_t|)\,\mathrm{d}x$$

$$\leqslant C\|(u,H)\|_{L^3}\left(\|H_t\|_{L^2}^2 + \|(\nabla^2 u, \nabla^2 H)\|_{L^2}^2\right). \tag{2.7}$$

We rewrite $(1.1)_2$ as

$$-\Delta u + \nabla P = -\rho u_t - \rho(u\cdot\nabla)u + \operatorname{curl} H \times H,$$

which along with the classical estimates of the Stokes equation leads to

$$\|\nabla^2 u\|_{L^2} + \|\nabla P\|_{L^2} \leqslant \left(\|\rho u_t\|_{L^2} + \|\rho(u\cdot\nabla)u\|_{L^2} + \|\operatorname{curl} H \times H\|_{L^2}\right)$$
$$\leqslant C\left(\|\sqrt{\rho}u_t\|_{L^2} + \|u\|_{L^3}\|\nabla u\|_{L^6} + \|H\|_{L^3}\|\nabla H\|_{L^6}\right)$$
$$\leqslant C\left(\|\sqrt{\rho}u_t\|_{L^2} + \|(u,H)\|_{L^3}\|(\nabla^2 u,\nabla^2 H)\|_{L^2}\right). \tag{2.8}$$

Similarly,

$$\|\nabla^2 H\|_{L^2} \leqslant C\left(\|H_t\|_{L^2} + \|(H\cdot\nabla)u\|_{L^2} + \|(u\cdot\nabla)H\|_{L^2}\right)$$
$$\leqslant C\left(\|H_t\|_{L^2} + \|(u,H)\|_{L^3}\|(\nabla^2 u,\nabla^2 H)\|_{L^2}\right). \tag{2.9}$$

By combining (2.8) with (2.9), we obtain

$$\|(\nabla^2 u,\nabla^2 H)\|_{L^2} + \|\nabla P\|_{L^2} \leqslant C_1\left(\|(\sqrt{\rho}u_t, H_t)\|_{L^2} + \|(u,H)\|_{L^3}\|(\nabla^2 u,\nabla^2 H)\|_{L^2}\right),$$

and thus

$$\|(\nabla^2 u,\nabla^2 H)\|_{L^2} + \|\nabla P\|_{L^2} \leqslant C\|(\sqrt{\rho}u_t, H_t)\|_{L^2}, \tag{2.10}$$

provided

$$\sup_{0\leqslant t\leqslant T}\|(u,H)(t)\|_{L^3} \leqslant \frac{1}{2C_1}.$$

Summing up (2.6) and (2.7), we get by (2.10) that

$$\frac{d}{dt}\int\left(|\nabla u|^2 + |\nabla H|^2\right)dx + \int\left(\rho|u_t|^2 + |H_t|^2\right)dx \leqslant C_2\|(u,H)\|_{L^3}\|(\sqrt{\rho}u_t,H_t)\|_{L^2}^2,$$

which shows that

$$\frac{d}{dt}\int\left(|\nabla u|^2 + |\nabla H|^2\right)dx + \int\left(\rho|u_t|^2 + |H_t|^2\right)dx \leqslant 0, \tag{2.11}$$

provided the constant c_0 in (2.1) satisfies

$$\sup_{0\leqslant t\leqslant T}\|(u,H)(t)\|_{L^3} \leqslant c_0 \leqslant \min\left\{\frac{1}{2C_1},\frac{1}{2C_2}\right\}.$$

For any $t \in [0, T]$, integrating (2.11) on $[0, t]$, we can get by (2.10) that,

$$\|(\nabla u, \nabla H)(t)\|_{L^2}^2 + \int_0^t \|(\sqrt{\rho}u_t, H_t, \nabla^2 u, \nabla^2 H, \nabla P)\|_{L^2}^2 \, d\tau$$
$$\leqslant C\|(\nabla u_0, \nabla H_0)\|_{L^2}^2. \tag{2.12}$$

Step 2 H^1 estimate with the time weight and interpolation.

Multiplying (2.11) with $\omega(t)$ and integrating it over $[0, t]$, $\forall\, t \in [0, T]$, we get by (2.10) that

$$\omega(t)\|(\nabla u, \nabla H)(t)\|_{L^2}^2 + \int_0^t \omega(\tau)\|\left(\sqrt{\rho}u_t, H_t, \nabla^2 u, \nabla^2 H, \nabla P\right)\|_{L^2}^2 \, d\tau$$
$$\leqslant C \int_0^t \|(\nabla u, \nabla H)(\tau)\|_{L^2}^2 \, d\tau$$
$$\leqslant C\|(\sqrt{\rho_0}u_0, H_0)\|_{L^2}^2$$
$$\leqslant C\|(u_0, H_0)\|_{L^2}^2, \tag{2.13}$$

where we have used (2.5).

Noting that (u, H) is under the assumption (2.1), and by using an interpolation argument, we will get the estimate of $F_1(t)$. Considering the following linear system

$$\begin{cases} \rho(v_t + u \cdot \nabla v) - \Delta v + \nabla \widetilde{P} = H \cdot \nabla B, \\ B_t + u \cdot \nabla B - \Delta B = H \cdot \nabla v, \\ \operatorname{div} v = \operatorname{div} B = 0, \\ (v, B)|_{t=0} = (v_0, B_0), \end{cases}$$

where $\widetilde{P}$ includes the magnetic pressure. From the proof of (2.12) and (2.13), we get

$$\|(\nabla v, \nabla B)(t)\|_{L^2}^2 + \int_0^t \|(\sqrt{\rho}v_t, B_t, \nabla^2 v, \nabla^2 B, \nabla P)\|_{L^2}^2 \, d\tau \leqslant C\|(\nabla v_0, \nabla B_0)\|_{L^2}^2,$$
$$\omega(t)\|(\nabla v, \nabla B)(t)\|_{L^2}^2 + \int_0^t \omega(\tau)\|(\sqrt{\rho}v_t, B_t, \nabla^2 v, \nabla^2 B, \nabla P)\|_{L^2}^2 \, d\tau \leqslant C\|(v_0, B_0)\|_{L^2}^2.$$

Similar to the analysis in [32], by the complex interpolation, we deduce that for any $\theta \in [0, 1]$,

$$\omega(t)^{1-\theta}\|(\nabla v, \nabla B)(t)\|_{L^2}^2 + \int_0^t \omega(\tau)^{1-\theta}\|(\sqrt{\rho}v_t, B_t, \nabla^2 v, \nabla^2 B, \nabla P)\|_{L^2}^2 \, d\tau$$

$$\leqslant C\|(v_0, B_0)\|_{\dot H^s}^2,$$

which implies that for $t \in [0, T]$,

$$F_1(t) \leqslant C\|(u_0, H_0)\|_{\dot H^s}^2, \tag{2.14}$$

$$\|(u, H)(t)\|_{L^3} \leqslant C\|(u, H)(t)\|_{\dot H^{\frac{1}{2}}} \leqslant C\|(u_0, H_0)\|_{\dot H^{\frac{1}{2}}}. \tag{2.15}$$

By choosing $c \leqslant \frac{c_0}{2C}$ in (1.3), we get (2.2). Then we complete the bootstrap arguments.

Step 3 H^2 estimate.

Differentiating $(1.1)_2$ with respect to t, taking the L^2 inner product with u_t, and integrating by parts over $\mathbf{R}^3$, we get by $(1.1)_1$ that

$$\frac{1}{2}\frac{d}{dt}\int \rho|u_t|^2\,dx + \int |\nabla u_t|^2\,dx$$
$$= \int (-2\rho(u\cdot\nabla)u_t\cdot u_t - \rho(u_t\cdot\nabla)u\cdot u_t - \rho u\cdot\nabla(u\cdot\nabla u\cdot u_t))\,dx$$
$$+ \int ((\operatorname{curl} H_t \times H)\cdot u_t + (\operatorname{curl} H \times H_t)\cdot u_t)\,dx. \tag{2.16}$$

By similar arguments as (2.16), from $(1.1)_3$, we obtain

$$\frac{1}{2}\frac{d}{dt}\int |H_t|^2\,dx + \int |\nabla H_t|^2\,dx$$
$$= \int (\operatorname{curl}(u_t \times H)\cdot H_t + \operatorname{curl}(u \times H_t)\cdot H_t)\,dx$$
$$= \int ((u_t \times H)\cdot \operatorname{curl} H_t + (u \times H_t)\cdot \operatorname{curl} H_t)\,dx$$
$$= \int (-(\operatorname{curl} H_t \times H)\cdot u_t + (u \times H_t)\cdot \operatorname{curl} H_t)\,dx. \tag{2.17}$$

By combining (2.16) with (2.17), we have

$$\frac{1}{2}\frac{d}{dt}\int (\rho|u_t|^2 + |H_t|^2)\,dx + \int (|\nabla u_t|^2 + |\nabla H_t|^2)\,dx$$
$$= \int \big(-2\rho(u\cdot\nabla)u_t\cdot u_t - \rho(u_t\cdot\nabla)u\cdot u_t - \rho u\cdot\nabla(u\cdot\nabla u\cdot u_t)$$
$$+(\operatorname{curl} H \times H_t)\cdot u_t + (u \times H_t)\cdot \operatorname{curl} H_t \big)\,dx. \tag{2.18}$$

Define

$$\widetilde{F}_2(t) \triangleq \frac{1}{2}\omega(t)^{2-s}\int (\rho|u_t|^2 + |H_t|^2)\,dx$$

$$+ \int_0^t \omega(\tau)^{2-s} \int \left(|\nabla u_t|^2 + |\nabla H_t|^2\right) \mathrm{d}x\mathrm{d}\tau.$$

Multiplying $\omega(\tau)^{2-s}$ on both sides of (2.18), integrating with respect to τ and then integrating by parts over $\mathbf{R}^3$, we get

$$\begin{aligned}\widetilde{F}_2(t) \leqslant &-2\int_0^t \omega(\tau)^{2-s} \int \rho(u\cdot\nabla)u_t\cdot u_t\,\mathrm{d}x\mathrm{d}\tau \\ &- \int_0^t \omega(\tau)^{2-s} \int \rho(u_t\cdot\nabla)u\cdot u_t\,\mathrm{d}x\mathrm{d}\tau \\ &- \int_0^t \omega(\tau)^{2-s} \int \rho u\cdot\nabla(u\cdot\nabla u\cdot u_t)\,\mathrm{d}x\mathrm{d}\tau \\ &+ \int_0^t \omega(\tau)^{2-s} \int (\operatorname{curl} H \times H_t)\cdot u_t\,\mathrm{d}x\mathrm{d}\tau \\ &+ \int_0^t \omega(\tau)^{2-s} \int (u \times H_t)\cdot \operatorname{curl} H_t\,\mathrm{d}x\mathrm{d}\tau \\ &+ C\left|\int_0^t \omega(\tau)^{1-s} \int (\rho|u_t|^2 + |H_t|^2)\,\mathrm{d}x\mathrm{d}\tau\right| \\ \triangleq &\sum_{i=1}^6 I_i. \end{aligned} \qquad (2.19)$$

From (2.14), Hölder's inequality, Sobolev inequality in [17], and by choosing $c_0 \leqslant \dfrac{1}{16C}$, we get that

$$\begin{aligned} I_1 &\leqslant \int_0^t \omega(\tau)^{2-s}\|\rho\|_{L^\infty}\|u\|_{L^3}\|\nabla u_t\|_{L^2}\|u_t\|_{L^6}\,\mathrm{d}\tau \\ &\leqslant Cc_0 \int_0^t \omega(\tau)^{2-s}\|\nabla u_t\|_{L^2}^2\,\mathrm{d}\tau \\ &\leqslant \frac{1}{16}\widetilde{F}_2(t). \end{aligned} \qquad (2.20)$$

In addition, we deduce by Gagliardo-Nirenberg inequality and Young's inequality that

$$\begin{aligned} I_2 &\leqslant \int_0^t \omega(\tau)^{2-s}\|\rho\|_{L^\infty}\|u_t\|_{L^3}\|\nabla u\|_{L^2}\|u_t\|_{L^6}\,\mathrm{d}\tau \\ &\leqslant C\int_0^t \omega(\tau)^{2-s}\|\nabla u\|_{L^2}\|u_t\|_{L^2}^{\frac{1}{2}}\|\nabla u_t\|_{L^2}^{\frac{3}{2}}\,\mathrm{d}\tau \\ &\leqslant C\int_0^t \left(\omega(\tau)^{\frac{2-s}{4}}\|\nabla u\|_{L^2}\|\sqrt{\rho}u_t\|_{L^2}^{\frac{1}{2}}\right)\left(\omega(\tau)^{\frac{3(2-s)}{4}}\|\nabla u_t\|_{L^2}^{\frac{3}{2}}\right)\mathrm{d}\tau \\ &\leqslant C\int_0^t \omega(\tau)^{2-s}\|\nabla u\|_{L^2}^4\|\sqrt{\rho}u_t\|_{L^2}^2\,\mathrm{d}\tau + \frac{1}{16}\widetilde{F}_2(t) \end{aligned}$$

$$\leqslant C\int_0^t \|\nabla u\|_{L^2}^4 \widetilde{F}_2(\tau)\,d\tau + \frac{1}{16}\widetilde{F}_2(t), \tag{2.21}$$

and

$$I_3 = -\int_0^t \omega(\tau)^{2-s} \int (\rho(u\cdot\nabla)u\cdot\nabla u\cdot u_t + \rho(u\otimes u):\nabla^2 u\cdot u_t$$
$$+ \rho(u\cdot\nabla u)(u\cdot\nabla u_t))\,dx d\tau$$
$$\leqslant \int_0^t \omega(\tau)^{2-s}(\|\rho\|_{L^\infty}\|u\|_{L^6}\|\nabla u\|_{L^2}\|\nabla u\|_{L^6}\|u_t\|_{L^6}$$
$$+\|\rho\|_{L^\infty}\|u\|_{L^6}^2\|\nabla^2 u\|_{L^2}\|u_t\|_{L^6} + \|\rho\|_{L^\infty}\|u\|_{L^6}^2\|\nabla u\|_{L^6}\|\nabla u_t\|_{L^2})\,d\tau$$
$$\leqslant \int_0^t \omega(\tau)^{2-s}\|\nabla u\|_{L^2}^2\|\nabla^2 u\|_{L^2}\|\nabla u_t\|_{L^2}\,d\tau$$
$$\leqslant C\int_0^t \|\nabla u\|_{L^2}^4 \widetilde{F}_2(\tau)\,d\tau + \frac{1}{16}\widetilde{F}_2(t). \tag{2.22}$$

Similarly, we get that

$$I_4 + I_5$$
$$\leqslant C\int_0^t \omega(\tau)^{2-s}\int (|u||H_t||\nabla H_t| + |\nabla H||H_t||u_t|)\,dx d\tau$$
$$\leqslant C\int_0^t \omega(\tau)^{2-s}(\|u\|_{L^6}\|H_t\|_{L^3}\|\nabla H_t\|_{L^2} + \|\nabla H\|_{L^2}\|H_t\|_{L^3}\|u_t\|_{L^6})\,d\tau$$
$$\leqslant C\int_0^t \omega(\tau)^{2-s}\|(\nabla u,\nabla H)\|_{L^2}\|H_t\|_{L^2}^{1/2}\|\nabla H_t\|_{L^2}^{1/2}\|(\nabla u_t,\nabla H_t)\|_{L^2}\,d\tau$$
$$\leqslant \frac{1}{32}\int_0^t \omega(\tau)^{2-s}\|(\nabla u_t,\nabla H_t)\|_{L^2}^2\,d\tau \tag{2.23}$$
$$+ C\int_0^t \omega(\tau)^{2-s}\|(\nabla u,\nabla H)\|_{L^2}^2\|H_t\|_{L^2}\|\nabla H_t\|_{L^2}\,d\tau$$
$$\leqslant \frac{1}{16}\widetilde{F}_2(t) + C\int_0^t \omega(\tau)^{2-s}\|(\nabla u,\nabla H)\|_{L^2}^4\|H_t\|_{L^2}^2\,d\tau$$
$$\leqslant \frac{1}{16}\widetilde{F}_2(t) + C\int_0^t \|(\nabla u,\nabla H)\|_{L^2}^4 \widetilde{F}_2(\tau)\,d\tau, \tag{2.24}$$

and

$$I_6 \leqslant F_1(t) \leqslant C\|(\sqrt{\rho_0}u_0, H_0)\|_{\dot{H}^s}^2. \tag{2.25}$$

Then, by substituting (2.20)–(2.25) into the summation of (2.19), we have

$$\widetilde{F}_2(t) \leqslant C\|(\sqrt{\rho_0}u_0, H_0)\|_{\dot{H}^s}^2 + C\int_0^t \|(\nabla u,\nabla H)\|_{L^2}^4 \widetilde{F}_2(\tau)\,d\tau.$$

So, by Gronwall's inequality, we obtain

$$\widetilde{F}_2(t) \leqslant C\|(\sqrt{\rho_0}u_0, H_0)\|_{\dot{H}^s}^2 \exp\left\{C\int_0^t \|(\nabla u, \nabla H)\|_{L^2}^4 \, d\tau\right\}. \tag{2.26}$$

We get by interpolation, (2.14) and (2.15) that

$$\int_0^{\min(1,t)} \|(\nabla u, \nabla H)\|_{L^2}^4 d\tau$$

$$\leqslant \int_0^{\min(1,t)} \|(u, H)\|_{\dot{H}^{\frac{1}{2}}}^{\frac{8}{3}} \|(\nabla^2 u, \nabla^2 H)\|_{L^2}^{\frac{4}{3}} \, d\tau$$

$$\leqslant C\|(u_0, H_0)\|_{\dot{H}^{\frac{1}{2}}}^{\frac{8}{3}} \int_0^{\min(1,t)} \|(\nabla^2 u, \nabla^2 H)\|_{L^2}^{\frac{4}{3}} \, d\tau$$

$$\leqslant C\|(u_0, H_0)\|_{\dot{H}^{\frac{1}{2}}}^{\frac{8}{3}} \left(\int_0^{\min(1,t)} \tau^{1-s}\|(\nabla^2 u, \nabla^2 H)\|_{L^2}^2 \, d\tau\right)^{\frac{2}{3}} \left(\int_0^1 \tau^{-2(1-s)} \, d\tau\right)^{\frac{1}{3}}$$

$$\leqslant C\|(u_0, H_0)\|_{\dot{H}^{\frac{1}{2}}}^{\frac{8}{3}} \|(u_0, H_0)\|_{\dot{H}^s}^{\frac{4}{3}}$$

$$\leqslant C\|(u_0, H_0)\|_{\dot{H}^s}^4, \tag{2.27}$$

and if $t \geqslant 1$, then

$$\int_1^t \|(\nabla u, \nabla H)\|_{L^2}^4 \, d\tau \leqslant \sup_{1 \leqslant \tau \leqslant t} \|(\nabla u, \nabla H)\|_{L^2}^2 \int_1^t \|(\nabla u, \nabla H)\|_{L^2}^2 \, d\tau$$

$$\leqslant C\|(u_0, H_0)\|_{\dot{H}^s}^2 \|(u_0, H_0)\|_{L^2}^2$$

$$\leqslant C\|(u_0, H_0)\|_{\dot{H}^s}^4, \tag{2.28}$$

so, by substituting (2.27) and (2.28) into the summation of (2.26) and (2.10), we get

$$F_2(t) \leqslant C\|(u_0, H_0)\|_{\dot{H}^s}^2 \exp\{C\|(u_0, H_0)\|_{\dot{H}^s}^4\}.$$

Since we get the uniform energy estimates, the proof of the existence part is completed.

3 The proof of the uniqueness part of Theorem 1.1

3.1 More regularity of the solutions

In this section, the aim is to proof the uniqueness of the solutions in Theorem 1.1. Firstly, we'll focus on some more information on the regularity of the solutions, which will be used in the proof.

Lemma 3.1 If (ρ, u, H) is the solution of system (1.1) obtained in Theorem 1.1, $T \in [0, 1]$ and $\gamma \in [0, 1]$, then

$$\int_0^T t^{1+\gamma-s} \left(\|(\partial_t u, \partial_t H)\|^2_{L^{\frac{6}{3-2\gamma}}} + \|(\nabla^2 u, \nabla^2 H)\|^2_{L^{\frac{6}{3-2\gamma}}} + \|\nabla P\|^2_{L^{\frac{6}{3-2\gamma}}} \right) dt \leq C,$$

$$\int_0^T \|(\nabla u, \nabla H)\|_{L^\infty} dt \leq C T^{\frac{2s-1}{5}},$$

$$\int_0^T t \|(\nabla u, \nabla H)\|^2_{L^\infty} dt \leq C T^{\frac{4s-2}{5}},$$

where the constant C depends on $\|(u_0, H_0)\|_{H^s}$.

Proof (1) We prove the first inequality. By Hölder's inequality, Gagliardo-Nirenberg inequality and (1.4), we have

$$\|t^{\frac{1+\gamma-s}{2}} \|(\partial_t u, \partial_t H)\|_{L^{\frac{6}{3-2\gamma}}} \|_{L^2_T}$$

$$\leq C \|(t^{\frac{1-s}{2}} \|(\partial_t u, \partial_t H)\|_{L^2})^{1-\gamma} (t^{\frac{2-s}{2}} \|(\nabla \partial_t u, \nabla \partial_t H)\|_{L^2})^\gamma \|_{L^2_T}$$

$$\leq C \|t^{\frac{1-s}{2}} \|(\partial_t u, \partial_t H)\|_{L^2} \|_{L^2_T}^{1-\gamma} \|t^{\frac{2-s}{2}} \|(\nabla \partial_t u, \nabla \partial_t H)\|_{L^2} \|_{L^2_T}^\gamma$$

$$\leq C \left(\|(u_0, H_0)\|_{H^s} \right). \tag{3.1}$$

Similarly, we get that

$$\|t^{\frac{1-s}{2} + \frac{\gamma(2-s)}{3}} \|\rho u \cdot \nabla u\|_{L^{\frac{6}{3-2\gamma}}} \|_{L^2_T}$$

$$\leq C \|t^{\frac{1-s}{2} + \frac{\gamma(2-s)}{3}} \|u\|_{L^{\frac{3}{1-\gamma}}} \|\nabla u\|_{L^6} \|_{L^2_T}$$

$$\leq C \sup_{t \in [0,T]} \left(\|u\|_{L^3}^{1-\frac{2\gamma}{3}} (t^{\frac{2-s}{2}} \|\nabla^2 u\|_{L^2})^{\frac{2\gamma}{3}} \right) \|t^{\frac{1-s}{2}} \|\nabla^2 u\|_{L^2} \|_{L^2_T}$$

$$\leq C \left(\|u_0\|_{H^s} \right), \tag{3.2}$$

and

$$\|t^{\frac{1-s}{2} + \frac{\gamma(2-s)}{3}} \|H \cdot \nabla H\|_{L^{\frac{6}{3-2\gamma}}} \|_{L^2_T} \leq C \|t^{\frac{1-s}{2} + \frac{\gamma(2-s)}{3}} \|H\|_{L^{\frac{3}{1-\gamma}}} \|\nabla H\|_{L^6} \|_{L^2_T}$$

$$\leq C \left(\|H_0\|_{H^s} \right). \tag{3.3}$$

Since $-\Delta u + \nabla P = -\rho \partial_t u - \rho u \cdot \nabla u + \mathrm{curl}\, H \times H$, by the $W^{2,p}$ estimate of the Stokes system and noting that $1 - s + \dfrac{2\gamma(2-s)}{3} \leq 1 + \gamma - s$, if $s > \dfrac{1}{2}$, we have

$$\int_0^T t^{1+\gamma-s} \left(\|(\partial_t u, \partial_t H)\|^2_{L^{\frac{6}{3-2\gamma}}} + \|(\nabla^2 u, \nabla^2 H)\|^2_{L^{\frac{6}{3-2\gamma}}} + \|\nabla p\|^2_{L^{\frac{6}{3-2\gamma}}} \right) dt$$

$$\leqslant C\left(\|(u_0, H_0)\|_{H^s}\right).$$

Then by Gagliardo-Nirenberg inequality and (1.4), we obtain that

$$\int_0^T \|(\nabla u, \nabla H)\|_{L^\infty}\, dt$$
$$\leqslant \int_0^T \|(u, H)\|_{\dot{H}^{\frac{1}{2}}}^{\frac{1}{5}} \|(\nabla^2 u, \nabla^2 H)\|_{L^6}^{\frac{4}{5}}\, dt$$
$$\leqslant \sup_{0 \leqslant t \leqslant T} \|(u, H)\|_{\dot{H}^{\frac{1}{2}}}^{\frac{1}{5}} \left(\int_0^T t^{2-s} \|(\nabla^2 u, \nabla^2 H)\|_{L^6}^2\, dt\right)^{\frac{2}{5}} \left(\int_0^T t^{\frac{-2(2-s)}{3}}\, dt\right)^{\frac{3}{5}}$$
$$\leqslant C\left(\|(u_0, H_0)\|_{H^s}\right) T^{\frac{2s-1}{5}}.$$

Similarly, we get that

$$\int_0^T t\|(\nabla u, \nabla H)\|_{L^\infty}^2\, dt \leqslant C\left(\|(u_0, H_0)\|_{H^s}\right) T^{\frac{4s-2}{5}}.$$

Then the proof of Lemma 3.1 is completed.

3.2 The Lagrangian formulation

Next, we show the Lagrangian formulation which will play a vital role in the proof. Define the trajectory of $X(t, y)$ of $u(t, x)$ by

$$\partial_t X(t, y) = u(t, X(t, y)), \quad X(0, y) = y,$$

which shows the following relation between the Eulerian coordinate x and the Lagrangian coordinate y:

$$X(t, y) = y + \int_0^t u(\tau, X(\tau, y))\, d\tau. \tag{3.4}$$

By choosing T small enough, we can get from Lemma 3.1 that

$$\int_0^T \|\nabla u(t)\|_{L^\infty}\, dt \leqslant \frac{1}{2}, \tag{3.5}$$

which ensures that for $t \leqslant T$, $X(t, y)$ is invertible with respect to the variable y, and we denote its inverse mapping $Y(t, x)$.

Define
$$v(t, y) \triangleq u(t, x) = u(t, X(t, y)),$$

and
$$A(t, y) \triangleq (D_y X(t, y))^{-1} = (D_x Y(t, x)),$$

where
$$(D_y X)_{ij} \triangleq \partial_{y_j} X^i.$$

It holds that
$$\nabla_x u(t, x) = A^T(t, y) \nabla_y v(t, y), \quad \text{and} \quad \text{div}_x u(t, x) = \text{div}_y(A(t, y) v(t, y)),$$

which shows
$$\text{div}_y(A \cdot) = A^T : \nabla_y. \tag{3.6}$$

Here and in the sequel, A^T denotes the transpose matrix of A, and $A^T : \nabla_y$ means $Tr(A^T \nabla_y)$. We denote as in [32] that

$$\nabla_u \triangleq A^T \cdot \nabla_y, \quad \text{div}_u \triangleq \text{div}(A \cdot), \quad \Delta_u \triangleq \text{div}_u \nabla_u, \quad \eta(t, y) \triangleq \rho(t, X(t, y)),$$
$$v(t, y) \triangleq u(t, X(t, y)), \quad B(t, y) \triangleq H(t, X(t, y)), \quad \text{and} \quad \Pi(t, y) \triangleq P(t, X(t, y)). \tag{3.7}$$

So the Lagrangian formulation of (1.1) becomes
$$\begin{cases} \partial_t \eta = 0, \\ \eta \partial_t v - \Delta_u v + \nabla_u \Pi = B \cdot \nabla_u B - \nabla_u \dfrac{|B|^2}{2}, \\ \partial_t B - \Delta_u B - B \cdot \nabla_u v = 0, \\ \text{div}_u v = \text{div}_u B = 0, \\ (\eta, v, B)|_{t=0} = (\rho_0, u_0, H_0). \end{cases} \tag{3.8}$$

Then, we transform the regularity information of the solution in the Eulerian coordinates into that in the Lagrangian coordinates.

Lemma 3.2 *We assume that $(\rho, u, H, \nabla P)$ is the solution of system (1.1) obtained in Theorem 1.1 and $(\eta, v, B, \nabla \Pi)$ is given by (3.7), then for any $t \leqslant T$ small enough and $0 \leqslant \gamma \leqslant 1$, one has*

$$\int_0^t \tau^{1+\gamma-s} \left(\|(\partial_t v, \partial_t B)\|^2_{L^{\frac{6}{3-2\gamma}}} + \|\nabla \Pi\|^2_{L^{\frac{6}{3-2\gamma}}} \right) d\tau \leqslant C,$$

$$\int_0^t \|(\nabla v, \nabla B)\|_{L^\infty} d\tau \leqslant C t^{\frac{2s-1}{5}},$$

$$\int_0^t \tau \|(\nabla v, \nabla B)\|^2_{L^\infty} d\tau \leqslant C t^{\frac{4s-2}{5}},$$

and if $\gamma < s$, then we have
$$\int_0^t \tau^{1+\gamma-s}\|(\nabla^2 v, \nabla^2 B)\|_{L^{\frac{6}{3-2\gamma}}}^2 \, d\tau \leqslant C, \quad \text{and} \quad \|\nabla A\|_{L_t^{\infty}(L^{\frac{6}{3-2\gamma}})} \leqslant Ct^{\frac{s-\gamma}{2}}.$$

Here the constant C depends on $\|(u_0, H_0)\|_{H^s}$.

Proof Similar to the proof of Lemma 3.3 in [32], we know by (3.4), (3.5) and Lemma 3.1 that
$$\|\nabla_y X(t,\cdot)\|_{L^{\infty}} \leqslant e^{\left\{\int_0^t \|\nabla_x u(\tau)\|_{L^{\infty}} d\tau\right\}} \leqslant e^{\frac{1}{2}}. \tag{3.9}$$

By (3.9) and Lemma 3.1, we can easily deduce that
$$\int_0^t \|(\nabla v, \nabla B)\|_{L^{\infty}} d\tau \leqslant \|\nabla_y X\|_{L^{\infty}} \int_0^t \|(\nabla_x u, \nabla_x H)\|_{L^{\infty}} d\tau \leqslant Ct^{\frac{2s-1}{5}},$$
$$\int_0^t \tau \|(\nabla v, \nabla B)\|_{L^{\infty}}^2 d\tau \leqslant \|\nabla_y X\|_{L^{\infty}}^2 \int_0^t \tau \|(\nabla_x u, \nabla_x H)\|_{L^{\infty}}^2 d\tau \leqslant Ct^{\frac{4s-2}{5}},$$

and
$$\|\tau^{\frac{1+\gamma-s}{2}} \nabla \Pi\|_{L_t^2(L^{\frac{6}{3-2\gamma}})} \leqslant \|\nabla_y X\|_{L^{\infty}} \|\tau^{\frac{1+\gamma-s}{2}} \nabla_x P\|_{L_t^2(L^{\frac{6}{3-2\gamma}})} \leqslant C.$$

Noting that
$$\partial_t v(t,x) = (\partial_t u + u \cdot \nabla u)(t, X(t,y))$$

and
$$\partial_t B(t,x) = (\partial_t H + u \cdot \nabla H)(t, X(t,y)),$$

then by (3.1)–(3.3), it's easy to get
$$\|\tau^{\frac{1+\gamma-s}{2}}(\partial_t v, \partial_t B)\|_{L_t^2(L^{\frac{6}{3-2\gamma}})} \leqslant \|\tau^{\frac{1+\gamma-s}{2}}(\partial_t u, \partial_t H)\|_{L_t^2(L^{\frac{6}{3-2\gamma}})}$$
$$+ \|\tau^{\frac{1+\gamma-s}{2}}(u \cdot \nabla u, u \cdot \nabla H)\|_{L_t^2(L^{\frac{6}{3-2\gamma}})}$$
$$\leqslant C.$$

From (3.4) and (3.9), we have
$$\|\nabla_y^2 X(t,y)\|_{L^{\frac{6}{3-2\gamma}}} \leqslant C\int_0^t \|\nabla_y^2 X\|_{L^{\frac{6}{3-2\gamma}}} \|\nabla_x u\|_{L^{\infty}} d\tau + C\int_0^t \|\nabla_x^2 u\|_{L^{\frac{6}{3-2\gamma}}} d\tau,$$

in addition, if
$$\gamma < s,$$

then we deduce by Gronwall's inequality and Lemma 3.1 that

$$\|\nabla_y^2 X(t,y)\|_{L^{\frac{6}{3-2\gamma}}} \leqslant C \int_0^t \|\nabla_x^2 u\|_{L^{\frac{6}{3-2\gamma}}} \, d\tau,$$

$$\leqslant C \left(\int_0^t \tau^{1+\gamma-s} \|\nabla_x^2 u\|_{L^{\frac{6}{3-2\gamma}}}^2 \, d\tau\right)^{1/2} \left(\int_0^t \tau^{-(1+\gamma-s)} \, d\tau\right)^{1/2},$$

$$\leqslant C t^{\frac{s-\gamma}{2}}. \tag{3.10}$$

Hence, by (3.9), (3.10) and Lemma 3.1, we get

$$\|\tau^{\frac{1+\gamma-s}{2}}(\nabla^2 v, \nabla^2 B)\|_{L_t^2(L^{\frac{6}{3-2\gamma}})}$$

$$\leqslant \|\tau^{\frac{1+\gamma-s}{2}}(\nabla_x^2 u, \nabla_x^2 H)\|_{L_t^2(L^{\frac{6}{3-2\gamma}})} \|\nabla_y X\|_{L_t^\infty(L^\infty)}^2$$

$$+ \|\tau^{\frac{1}{2}}(\nabla_x u, \nabla_x H)\|_{L_t^2(L^\infty)} \|\tau^{\frac{\gamma-s}{2}} \|\nabla_y^2 X(\tau,\cdot)\|_{L^{\frac{6}{3-2\gamma}}} \|_{L_t^\infty}$$

$$\leqslant C.$$

Finally, we estimate $\|\nabla A\|_{L_t^\infty(L^{\frac{6}{3-2\gamma}})}$. Thanks to (3.5), for any $t \leqslant T$, we have

$$A(t,y) = D_x Y(t,x) = (Id + (D_y X - Id))^{-1}$$

$$= \sum_{k=0}^\infty (-1)^k \left(\int_0^t D_y u(\tau, X(\tau,y)) \, d\tau\right)^k. \tag{3.11}$$

By choosing T in (3.5) small enough, noting $|\nabla u| = |Du|$, we get from (3.9), (3.10) and Lemma 3.1 for any $t \leqslant T$ and $\gamma < s$,

$$\|\nabla A\|_{L_t^\infty(L^{\frac{6}{3-2\gamma}})}$$

$$\leqslant C \|\nabla_x^2 u\|_{L_t^1(L^{\frac{6}{3-2\gamma}})} \|\nabla_y X\|_{L_t^\infty(L^\infty)}^2 + C \|\nabla_x u\|_{L_t^1(L^\infty)} \|\nabla_y^2 X\|_{L_t^\infty(L^{\frac{6}{3-2\gamma}})}$$

$$\leqslant C t^{\frac{s-\gamma}{2}} \|\tau^{\frac{1+\gamma-s}{2}} \nabla_x^2 u\|_{L_t^2(L^{\frac{6}{3-2\gamma}})} + C t^{\frac{s-\gamma}{2}}$$

$$\leqslant C t^{\frac{s-\gamma}{2}}.$$

This complete the proof of Lemma 3.2.

3.3 The proof of the uniqueness

Firstly, we recall the following important lemma from [14], [32].

Lemma 3.3 *Let $\eta \in L^\infty(\mathbf{R}^d)$ be a time independent positive function, and be bounded away from zero. Let R satisfy $R_t \in L^2((0,T) \times \mathbf{R}^d)$ and $\nabla \operatorname{div} R \in$*

$L^2((0,T) \times \mathbf{R}^d)$. Then the following system

$$\begin{cases} \eta \partial_t v - \Delta v + \nabla \Pi = f, \quad (t,x) \in (0,T) \times \mathbf{R}^d, \\ \operatorname{div} v = \operatorname{div} R, \\ v|_{t=0} = v_0, \end{cases}$$

has a unique solution $(v, \nabla \Pi)$ such that

$$\|\nabla v\|_{L_T^\infty(L^2)} + \|(v_t, \nabla^2 v, \nabla \Pi)\|_{L_T^2(L^2)}$$
$$\leqslant C \left(\|\nabla v_0\|_{L^2} + \|(f, R_t)\|_{L_T^2(L^2)} + \|\nabla \operatorname{div} R\|_{L_T^2(L^2)} \right),$$

where the constant C depends on $\inf \eta$ and $\sup \eta$, but independent of T.

Next, we will focus on the proof of the uniqueness of the solutions.

Let $(\rho_i, u_i, H_i, \nabla P_i)$, $i = 1,2$ be two solutions of system (1.1) obtained in Theorem 1.1, and $(\eta_i, v_i, B_i, \nabla \Pi_i)$, $i = 1,2$, be determined by (3.7). Denote $A_i := A(u_i)$, $i = 1,2$, and

$$\delta v = v_2 - v_1, \quad \delta B = B_2 - B_1, \quad \delta \Pi = \Pi_2 - \Pi_1, \quad \delta A = A_2 - A_1.$$

Then by (3.7) and (3.8), we get that

$$\begin{cases} \rho_0 \partial_t \delta v - \Delta \delta v + \nabla \delta \Pi = \delta f_1 + \delta f_2 + \delta f_3, \\ \partial_t \delta B - \Delta \delta B = \delta f_4 + \delta f_5, \\ \operatorname{div} \delta v = \operatorname{div} \delta g_1, \\ \operatorname{div} \delta B = \operatorname{div} \delta g_2, \\ (\delta v, \delta B)|_{t=0} = 0, \end{cases} \quad (3.12)$$

where

$$\begin{cases} \delta f_1 = (Id - A_2^\mathrm{T}) \nabla \delta \Pi - \delta A^\mathrm{T} \nabla \Pi_1, \\ \delta f_2 = \operatorname{div}[-(Id - A_2 A_2^\mathrm{T}) \nabla \delta v + (A_2 A_2^\mathrm{T} - A_1 A_1^\mathrm{T}) \nabla v_1], \\ \delta f_3 = \delta B \cdot A_2^\mathrm{T} \nabla B_2 + B_1 \cdot \delta A^\mathrm{T} \nabla B_2 + B_1 \cdot A_1^\mathrm{T} \nabla \delta B \\ \qquad + A_2^\mathrm{T} \nabla B_2 \cdot \delta B + \delta A^\mathrm{T} \nabla B_2 \cdot B_1 + A_1^\mathrm{T} \nabla \delta B \cdot B_1, \\ \delta f_4 = \operatorname{div}[-(Id - A_2 A_2^\mathrm{T}) \nabla \delta B + (A_2 A_2^\mathrm{T} - A_1 A_1^\mathrm{T}) \nabla B_1], \\ \delta f_5 = \delta B \cdot A_2^\mathrm{T} \nabla v_2 + B_1 \cdot \delta A^\mathrm{T} \nabla v_2 + B_1 \cdot A_1^\mathrm{T} \nabla \delta v, \\ \delta g_1 = (Id - A_2) \delta v - \delta A v_1, \quad \delta g_2 = (Id - A_2) \delta B - \delta A B_1. \end{cases}$$

Denote

$$\delta F(t) \triangleq \|(\nabla \delta v, \nabla \delta B)\|_{L_t^\infty(L^2)} + \|(\partial_t \delta v, \partial_t \delta B)\|_{L_t^2(L^2)}$$

$$+ \|(\nabla^2 \delta v, \nabla^2 \delta B)\|_{L_t^2(L^2)} + \|\nabla \delta \Pi\|_{L_t^2(L^2)}.$$

Lemma 3.4 *The following estimate holds:*

$$\|(\delta f_1, \delta f_2, \delta f_3, \delta f_4, \delta f_5)\|_{L_t^2(L^2)} + \|(\nabla \operatorname{div} \delta g_1, \nabla \operatorname{div} \delta g_2)\|_{L_t^2(L^2)}$$
$$+ \|(\partial_t \delta g_1, \partial_t \delta g_2)\|_{L_t^2(L^2)}$$
$$\leqslant \varepsilon(t) \delta F(t),$$

where $\varepsilon(t)$ tends to zero as t goes to zero.

Now, we assume that Lemma 3.4 holds, then by applying Lemma 3.3 to (3.12), we can deduce from Lemma 3.4 that

$$\delta F(t) \leqslant C(\|(\delta f_1, \delta f_2, \delta f_3, \delta f_4, \delta f_5)\|_{L_t^2(L^2)} + \|(\nabla \operatorname{div} \delta g_1, \nabla \operatorname{div} \delta g_2)\|_{L_t^2(L^2)}$$
$$+ \|(\partial_t \delta g_1, \partial_t \delta g_2)\|_{L_t^2(L^2)})$$
$$\leqslant \varepsilon(t) \delta F(t),$$

which ensures the uniqueness of the solutions obtained in Theorem 1.1 on a sufficiently small time interval $[0, T_1]$. Then the uniqueness on the whole time interval $[0, \infty)$ can be deduced by a bootstrap argument.

3.4 The proof of Lemma 3.4

In the following, we choose t small enough so that

$$\int_0^t \|(\nabla v_i, \nabla B_i)\|_{L^\infty} \, d\tau \leqslant \frac{1}{2}, \quad \text{for} \quad i = 1, 2. \tag{3.13}$$

Denote $\varepsilon(t)$ a function of t tending to zero as $t \to 0$, which may be different in different lines.

Step 1 Estimate of $\|\delta f_1\|_{L_t^2(L^2)}$.

By Lemma 3.2, we have

$$\|(\operatorname{Id} - A_2^{\mathrm{T}}) \nabla \delta \Pi\|_{L_t^2(L^2)} \leqslant \int_0^t \|\nabla_y v_2(\tau)\|_{L^\infty} \, d\tau \, \|\nabla \delta \Pi\|_{L_t^2(L^2)} \leqslant C t^{\frac{2s-1}{5}} \delta F(t). \tag{3.14}$$

We get by (3.11) that

$$\delta A(t) = \left(\int_0^t D\delta v \, d\tau \right) \left(\sum_{k \geqslant 0} \sum_{0 \leqslant j < k} V_1^j V_2^{k-1-j} \right), \tag{3.15}$$

where $V_i(t) \triangleq \int_0^t Dv_i \, d\tau$, $i = 1, 2$. And by (3.13), Sobolev inequality and Hölder's inequality, we obtain

$$\|\delta A(t)\|_{L^6} \leqslant C\| \int_0^\tau |\nabla \delta v(\tau', \cdot)| \, d\tau' \|_{L_t^\infty(L^6)} \leqslant C \int_0^t \|\nabla^2 \delta v\|_{L^2} \, d\tau \leqslant C t^{\frac{1}{2}} \delta F(t). \tag{3.16}$$

So by (3.16) and choosing $\gamma = \dfrac{1}{2}$ in Lemma 3.2, we have

$$\|\delta A^T \nabla \Pi_1\|_{L_t^2(L^2)} \leqslant \|\tau^{-\frac{3}{2}-s}{2} \delta A(\tau)\|_{L_t^\infty(L^6)} \|\tau^{\frac{3}{2}-s}{2} \nabla \Pi_1\|_{L_t^2(L^3)} \leqslant C t^{\frac{2s-1}{4}} \delta F(t). \tag{3.17}$$

According to $s > \dfrac{1}{2}$, (3.14) and (3.17), we get

$$\|\delta f_1\|_{L_t^2(L^2)} \leqslant C\varepsilon(t) \delta F(t).$$

Step 2 Estimate of $\|(\delta f_2, \delta f_4)\|_{L_t^2(L^2)}$.

Noting that

$$\begin{cases} (A_2 A_2^T - A_1 A_1^T)\nabla v_1 = (-\delta A(Id - A_2^T) - (Id - A_1)\delta A^T)\nabla v_1 \\ \qquad\qquad + (\delta A^T + \delta A)\nabla v_1, \\ (A_2 A_2^T - A_1 A_1^T)\nabla B_1 = (-\delta A(Id - A_2^T) - (Id - A_1)\delta A^T)\nabla B_1 \\ \qquad\qquad + (\delta A^T + \delta A)\nabla B_1, \end{cases}$$

we get by (3.15), (3.16), Lemma 3.2 and Hölder's inequality that

$$\|\tau^{-\frac{1}{2}} \nabla \delta A\|_{L_t^\infty(L^2)}$$
$$\leqslant C\|\tau^{-\frac{1}{2}} \int_0^\tau |\nabla^2 \delta v| \, d\tau'\|_{L_t^\infty(L^2)}$$
$$\quad + \|\tau^{-\frac{1}{2}} \int_0^\tau |\nabla \delta v| \, d\tau' \int_0^\tau (|\nabla^2 v_1| + |\nabla^2 v_2|) \, d\tau'\|_{L_t^\infty(L^2)}$$
$$\leqslant C\left(\|\nabla^2 \delta v\|_{L_t^2(L^2)} + \|\tau^{-\frac{1}{2}+\frac{s-\frac{1}{2}}{2}} \int_0^\tau |\nabla \delta v| \, d\tau'\|_{L_t^\infty(L^6)} \|\tau^{\frac{3}{2}-s}{2}(\nabla^2 v_1, \nabla^2 v_2)\|_{L_t^2(L^3)}\right)$$
$$\leqslant C\delta F(t). \tag{3.18}$$

Then by Lemma 3.2, (3.16) and (3.18), we deduce that

$$\|(\mathrm{div}[\delta A(Id - A_2^T)\nabla v_1], \, \mathrm{div}[\delta A(Id - A_2^T)\nabla B_1])\|_{L_t^2(L^2)}$$

$$\leqslant \|\tau^{-\frac{1}{2}}\nabla\delta A\|_{L_t^\infty(L^2)}\|Id - A_2\|_{L_t^\infty(L^\infty)}\|\tau^{\frac{1}{2}}(\nabla v_1, \nabla B_1)\|_{L_t^2(L^\infty)}$$
$$+ \|\tau^{-\frac{1}{2}}\delta A\|_{L_t^\infty(L^6)}\|\nabla A_2\|_{L_t^\infty(L^3)}\|\tau^{\frac{1}{2}}(\nabla v_1, \nabla B_1)\|_{L_t^2(L^\infty)}$$
$$+ \|\tau^{-\frac{1}{2}+\frac{s-\frac{1}{2}}{2}}\delta A\|_{L_t^\infty(L^6)}\|Id - A_2\|_{L_t^\infty(L^\infty)}\|\tau^{\frac{3-s}{2}}(\nabla^2 v_1, \nabla^2 B_1)\|_{L_t^2(L^3)}$$
$$\leqslant \varepsilon(t)\delta F(t).$$

We can get the same estimates for $\mathrm{div}[(Id - A_1)\delta A^T \nabla v_1 + (\delta A^T + \delta A)\nabla v_1]$ and $\mathrm{div}[(Id - A_1)\delta A^T \nabla B_1 + (\delta A^T + \delta A)\nabla B_1]$. So we prove that

$$\|(\mathrm{div}[(A_2 A_2^T - A_1 A_1^T)\nabla v_1], \mathrm{div}[(A_2 A_2^T - A_1 A_1^T)\nabla B_1])\|_{L_t^2(L^2)} \leqslant \varepsilon(t)\delta F(t). \tag{3.19}$$

We write

$$\begin{cases} (Id - A_2 A_2^T)\nabla\delta v = -((Id - A_2)(Id - A_2^T) + (Id - A_2) + (Id - A_2^T))\nabla\delta v, \\ (Id - A_2 A_2^T)\nabla\delta B = -((Id - A_2)(Id - A_2^T) + (Id - A_2) + (Id - A_2^T))\nabla\delta B. \end{cases}$$

Then we infer by Lemma 3.2 that

$$\|(\mathrm{div}[(Id - A_2 A_2^T)\nabla\delta v], \mathrm{div}[(Id - A_2 A_2^T)\nabla\delta B])\|_{L_t^2(L^2)}$$
$$\leqslant C\Big(\|\nabla A_2\|_{L_t^\infty(L^3)}\|(\nabla\delta v, \nabla\delta B)\|_{L_t^2(L^6)} + \|Id - A_2\|_{L_t^\infty(L^\infty)}\|(\nabla^2\delta v, \nabla^2\delta B)\|_{L_t^2(L^2)}\Big)$$
$$\leqslant \varepsilon(t)\delta F(t). \tag{3.20}$$

So, by (3.19) and (3.20), we prove that

$$\|(\delta f_2, \delta f_4)\|_{L_t^2(L^2)} \leqslant \varepsilon(t)\delta F(t).$$

Step 3 Estimate of $\|(\delta f_3, \delta f_5)\|_{L_t^2(L^2)}$.
Noting

$$\delta f_3 + \delta f_5 = \delta B \cdot A_2^T \nabla B_2 + B_1 \cdot \delta A^T \nabla B_2 + B_1 \cdot A_1^T \nabla \delta B$$
$$+ A_2^T \nabla B_2 \cdot \delta B + \delta A^T \nabla B_2 \cdot B_1 + A_1^T \nabla \delta B \cdot B_1$$
$$+ \delta B \cdot A_2^T \nabla v_2 + B_1 \cdot \delta A^T \nabla v_2 + B_1 \cdot A_1^T \nabla \delta v,$$

by (2.15), Gagliardo-Nirenberg inequality, Hölder's inequality and Lemma 3.2, we get

$$\|B_1\|_{L_t^2(L^\infty)} \leqslant \Big(\int_0^t \|B_1\|_{L^3}^{\frac{2}{3}}\|\nabla^2 B_1\|_{L^2}^{\frac{4}{3}}\,d\tau\Big)^{\frac{1}{2}}$$

$$\leqslant \sup_{0\leqslant \tau\leqslant t}\|B_1(\tau)\|_{L^3}^{\frac{1}{3}}\left(\int_0^t \tau^{-2(1-s)}\,\mathrm{d}\tau\right)^{\frac{1}{6}}\left(\int_0^t \tau^{1-s}\|\nabla^2 B_1\|_{L^2}^2\,\mathrm{d}\tau\right)^{\frac{1}{3}}$$
$$\leqslant Ct^{\frac{2s-1}{6}}, \tag{3.21}$$
$$\|\tau^{-\frac{1}{2}}\delta B\|_{L_t^\infty(L^2)} \leqslant \|\tau^{-\frac{1}{2}}\int_0^\tau \partial_t \delta B\,\mathrm{d}\tau'\|_{L_t^\infty(L^2)} \leqslant \|\partial_t\delta B\|_{L_t^2(L^2)}. \tag{3.22}$$

Then, by (2.15), (3.16), (3.21), (3.22), Lemma 3.2 and Hölder's inequality, we find that

$$\|(\delta f_3,\delta f_5)\|_{L_t^2(L^2)} \leqslant C\big(\|\delta B\cdot A_2^{\mathrm{T}}(\nabla B_2,\nabla v_2)\|_{L_t^2(L^2)} + \|B_1\cdot \delta A^{\mathrm{T}}(\nabla B_2,\nabla v_2)\|_{L_t^2(L^2)}$$
$$+ \|B_1\cdot A_1^{\mathrm{T}}(\nabla\delta B,\nabla\delta v)\|_{L_t^2(L^2)}\big)$$
$$\leqslant \|\tau^{-\frac{1}{2}}\delta B\|_{L_t^\infty(L^2)}\,\|A_2\|_{L_t^\infty(L^\infty)}\,\|\tau^{\frac{1}{2}}(\nabla B_2,\nabla v_2)\|_{L_t^2(L^\infty)}$$
$$+ \|B_1\|_{L_t^\infty(L^3)}\,\|\tau^{-\frac{1}{2}}\delta A\|_{L_t^\infty(L^6)}\,\|\tau^{\frac{1}{2}}(\nabla B_2,\nabla v_2)\|_{L_t^2(L^\infty)}$$
$$+ \|B_1\|_{L_t^2(L^\infty)}\,\|A_1\|_{L_t^\infty(L^\infty)}\,\|(\nabla\delta B,\nabla\delta v)\|_{L_t^\infty(L^2)}$$
$$\leqslant \varepsilon(t)\delta F(t).$$

Step 4 Estimate of $\|(\nabla\operatorname{div}\delta g_1,\ \nabla\operatorname{div}\delta g_2)\|_{L_t^2(L^2)}$.

By the chain rule (3.6), we get that

$$\|(\nabla\operatorname{div}\delta g_1,\nabla\operatorname{div}\delta g_2)\|_{L_t^2(L^2)}$$
$$= \|(\ \nabla\operatorname{div}[(Id-A_2)\delta v - \delta A v_1],\ \nabla\operatorname{div}[(Id-A_2)\delta B-\delta A B_1]\ )\|_{L_t^2(L^2)}$$
$$\leqslant \|(\ \nabla((Id-A_2)^{\mathrm{T}}:\nabla\delta v),\ \nabla((Id-A_2)^{\mathrm{T}}:\nabla\delta B)\ )\|_{L_t^2(L^2)}$$
$$+ \|(\ \nabla(\delta A^{\mathrm{T}}:\nabla v_1),\ \nabla(\delta A^{\mathrm{T}}:\nabla B_1)\ )\|_{L_t^2(L^2)}.$$

From (3.13), (3.15), (3.18) and Lemma 3.2, we get that

$$\|(\ \nabla(\delta A^{\mathrm{T}}:\nabla v_1),\ \nabla(\delta A^{\mathrm{T}}:\nabla B_1)\ )\|_{L_t^2(L^2)}$$
$$\leqslant C\big(\|\nabla\delta A(\nabla v_1,\nabla B_1)\|_{L_t^2(L^2)} + \|\delta A(\nabla^2 v_1,\nabla^2 B_1)\|_{L_t^2(L^2)}\big)$$
$$\leqslant C\big(\|\tau^{-\frac{1}{2}}\nabla\delta A\|_{L_t^\infty(L^2)}\|\tau^{\frac{1}{2}}(\nabla v_1,\nabla B_1)\|_{L_t^2(L^\infty)}$$
$$+ \|\tau^{-\frac{1}{2}+\frac{s-\frac{1}{2}}{2}}\int_0^\tau |\nabla\delta v|\,\mathrm{d}\tau'\|_{L_t^\infty(L^6)}\,\|\tau^{\frac{3-s}{2}}(\nabla^2 v_1,\nabla^2 B_1)\|_{L_t^2(L^3)}\big)$$
$$\leqslant \varepsilon(t)\delta F(t), \tag{3.23}$$

and

$$\|(\ \nabla((Id-A_2)^{\mathrm{T}}:\nabla\delta v),\ \nabla((Id-A_2)^{\mathrm{T}}:\nabla\delta B)\ )\|_{L_t^2(L^2)}$$

$$\leqslant C\|\nabla A_2\|_{L_t^\infty(L^3)}\|(\nabla\delta v,\nabla\delta B)\|_{L_t^2(L^6)}$$
$$+\Big\|\int_0^\tau |\nabla v_2|\,d\tau'\Big\|_{L_t^\infty(L^\infty)}\|(\nabla^2\delta v,\nabla^2\delta B)\|_{L_t^2(L^2)}$$
$$\leqslant \varepsilon(t)\delta F(t). \tag{3.24}$$

So, by (3.23) and (3.24), we obtain
$$\|(\nabla\operatorname{div}\delta g_1,\nabla\operatorname{div}\delta g_2)\|_{L_t^2(L^2)}\leqslant \varepsilon(t)\delta F(t).$$

Step 5 Estimate of $\|(\partial_t\delta g_2,\partial_t\delta g_2)\|_{L_t^2(L^2)}$.

$$\|(\partial_t\delta g_1,\partial_t\delta g_2)\|_{L_t^2(L^2)}$$
$$\leqslant \|(\partial_t((Id-A_2)\delta v),\partial_t((Id-A_2)\delta B))\|_{L_t^2(L^2)}+\|(\partial_t(\delta Av_1),\partial_t(\delta AB_1))\|_{L_t^2(L^2)}$$

By (3.13) and (3.15), we know that

$$\|(\partial_t(\delta Av_1),\partial_t(\delta AB_1))\|_{L_t^2(L^2)}$$
$$\leqslant C\Big(\|\nabla\delta v(v_1,B_1)\|_{L_t^2(L^2)}+\Big\|\int_0^\tau|\nabla\delta v|\,d\tau'\,|(\nabla v_1,\nabla v_2)|(v_1,B_1)\Big\|_{L_t^2(L^2)}$$
$$+\|\delta A(\partial_t v_1,\partial_t B_1)\|_{L_t^2(L^2)}\Big)$$
$$\triangleq J_1+J_2+J_3. \tag{3.25}$$

By Lemma 3.2 and (3.21), one has
$$J_1\leqslant \|\nabla\delta v\|_{L_t^\infty(L^2)}\|(v_1,B_1)\|_{L_t^2(L^\infty)}\leqslant \varepsilon(t)\delta F(t), \tag{3.26}$$

and
$$J_2\leqslant \Big\|\tau^{-\frac{1}{2}}\int_0^\tau|\nabla\delta v|\,d\tau'\Big\|_{L_t^\infty(L^6)}\|\tau^{\frac{1}{2}}(\nabla v_1,\nabla v_2)\|_{L_t^2(L^\infty)}\|(v_1,B_1)\|_{L_t^\infty(L^3)}$$
$$\leqslant \varepsilon(t)\delta F(t). \tag{3.27}$$

Also, we get by (3.16) and Lemma 3.2 that
$$J_3\leqslant \|\tau^{-\frac{3}{2}-s}\delta A\|_{L_t^\infty(L^6)}\,\|\tau^{\frac{3}{2}-s}(\partial_t v_1,\partial_t B_1)\|_{L_t^2(L^3)}\leqslant \varepsilon(t)\delta F(t). \tag{3.28}$$

By substituting (3.26)–(3.28) into (3.25), we have
$$\|(\partial_t(\delta Av_1),\partial_t(\delta AB_1))\|_{L_t^2(L^2)}\leqslant \varepsilon(t)\delta F(t). \tag{3.29}$$

Similarly, by (3.22) and Lemma 3.2, one has

$$\|(\partial_t((Id-A_2)\delta v),\ \partial_t((Id-A_2)\delta B))\|_{L^2_t(L^2)}$$
$$\leqslant \|\tau^{\frac{1}{2}}\partial_t A_2\|_{L^2_t(L^\infty)}\|\tau^{-\frac{1}{2}}(\delta v,\delta B)\|_{L^\infty_t(L^2)} + \|Id - A_2\|_{L^\infty_t(L^\infty)}\|(\partial_t\delta v,\partial_t\delta B)\|_{L^2_t(L^2)}$$
$$\leqslant \|\tau^{\frac{1}{2}}\nabla v_2\|_{L^2_t(L^\infty)}\|(\partial_t\delta v,\partial_t\delta B)\|_{L^2_t(L^2)} + \|\nabla v_2\|_{L^1_t(L^\infty)}\|(\partial_t\delta v,\partial_t\delta B)\|_{L^2_t(L^2)}$$
$$\leqslant \varepsilon(t)\delta F(t). \tag{3.30}$$

Then by (3.29) and (3.30), we obtain

$$\|(\partial_t\delta g_2,\ \partial_t\delta g_2)\|_{L^2_t(L^2)} \leqslant \varepsilon(t)\delta F(t).$$

This finishes the proof of Lemma 3.4.

4 The proof of Theorem 1.2

In this part, we are devoted to studying the decay of the solution by using the dual method to deal with the MHD system with discontinuous density.

Let (ρ, u, H) be the solution obtained in Theorem 1.1. For any given $T > 0$, let's introduce the adjoint system of the momentum equation and the magnetic equation of (1.1):

$$\begin{cases} \rho(\partial_t\phi + u\cdot\nabla\phi) + \Delta\phi + \nabla\widetilde{P} = H\cdot\nabla\psi & \text{in } [0,T]\times\mathbf{R}^3, \\ \partial_t\psi + u\cdot\nabla\psi + \Delta\psi = H\cdot\nabla\phi, \\ \text{div}\,\phi = \text{div}\,\psi = 0, \\ \phi(T,x) = u(T,x), \psi(T,x) = H(T,x). \end{cases}$$

Lemma 4.1 *For $0 \leqslant t \leqslant T$, and $i = 0, 1$, it holds that*

$$A_i(t) \leqslant C\|(u(T), H(T))\|^2_{L^2},$$

where

$$A_0(t) = \int (\rho|\phi(t,x)|^2 + |\psi(t,x)|^2)\,\mathrm{d}x + \int_t^T\int(|\nabla\phi|^2 + |\nabla\psi|^2)\,\mathrm{d}x\mathrm{d}\tau,$$
$$A_1(t) = (T-t)\int(|\nabla\phi(t,x)|^2 + |\nabla\psi(t,x)|^2)\,\mathrm{d}x$$
$$+ \int_t^T\int(T-\tau)(\rho|\phi_t|^2 + |\psi_t|^2 + |\nabla^2\phi|^2 + |\nabla^2\psi|^2)\,\mathrm{d}x\mathrm{d}\tau. \tag{4.1}$$

Proof Since the proof of (4.1) is similar to that of (2.5) and (2.13), here we omit it.

Taking the L^2 inner product of the momentum equation and the magnetic equation of (1.1) with ϕ, ψ, respectively, and using integration by parts, we obtain that

$$\int ((\rho u)(T,x)\phi(T,x), H(T,x)\psi(T,x)) \, dx$$
$$- \int ((\rho u)(0,x)\phi(0,x), H(0,x)\psi(0,x)) \, dx$$
$$= \int_0^T \int u(\rho\phi_t + \rho u \cdot \nabla\phi + \Delta\phi + \nabla\tilde{P}) \, dx d\tau + \int_0^T \int H \cdot \nabla H \phi \, dx d\tau$$
$$+ \int_0^T \int H(\psi_t + u \cdot \nabla\psi + \Delta\psi) \, dx d\tau + \int_0^T \int H \cdot \nabla u \psi \, dx d\tau$$
$$= \int_0^T \int u(H \cdot \nabla)\psi \, dx d\tau + \int_0^T \int H(H \cdot \nabla)\phi \, dx d\tau$$
$$+ \int_0^T \int H \cdot \nabla H \phi \, dx d\tau + \int_0^T \int H \cdot \nabla u \psi \, dx d\tau$$
$$= 0.$$

By the dual and Sobolev inequality, we get that for $\xi = 2\zeta(q)$,

$$\int (\rho(T,x)|u(T,x)|^2, |H(T,x)|^2) \, dx = \int ((\rho_0 u_0)(x)\phi(0,x), H_0(x)\psi(0,x)) \, dx$$
$$\leq \|(u_0, H_0)\|_{L^q} \|(\phi(0), \psi(0))\|_{L^{q'}}$$
$$\leq C\|(u_0, H_0)\|_{L^q} \|(\phi(0), \psi(0))\|_{\dot{H}^\xi}, \quad (4.2)$$

where $\dfrac{1}{q} + \dfrac{1}{q'} = 1$, $\xi = 3\left(\dfrac{1}{q} - \dfrac{1}{2}\right)$. In addition, by Lemma 4.1 and interpolation, we get

$$\|(\phi(t), \psi(t))\|_{\dot{H}^\xi} \leq C(T-t)^{-\frac{\xi}{2}} \|(u(T), H(T))\|_{L^2}. \quad (4.3)$$

From (4.2) and (4.3), one has

$$\|(u(T), H(T))\|_{L^2} \leq CT^{-\zeta(q)} \|(u_0, H_0)\|_{L^q}. \quad (4.4)$$

We know from the proof of (2.10) and (2.11) that

$$\frac{d}{dt} \int |(\nabla u(t,x), \nabla H(t,x))|^2 \, dx + c \int |(\nabla^2 u(t,x), \nabla^2 H(t,x))|^2 \, dx \leq 0, \quad (4.5)$$

where the constant $c > 0$. By (4.4) and interpolation, we have

$$\|(\nabla^2 u(t), \nabla^2 H(t))\|_{L^2} \geqslant c t^{\varsigma(q)} \|(\nabla u(t), \nabla H(t))\|_{L^2}^2. \qquad (4.6)$$

Define $U(t) = \int |(\nabla u(t,x), \nabla H(t,x))|^2 \, \mathrm{d}x$, then by (4.5) and (4.6), we obtain

$$\frac{\mathrm{d}}{\mathrm{d}t} U(t) + c t^{2\varsigma(q)} U(t)^2 \leqslant 0,$$

which ensures that for $t \geqslant 1$,

$$U(t) \leqslant C(1+t)^{-2\varsigma(q)-1}.$$

That is

$$\|(\nabla u(t), \nabla H(t))\|_{L^2} \leqslant C(1+t)^{-\varsigma(q)-\frac{1}{2}}. \qquad (4.7)$$

Then the proof of Theorem 1.2 is completed by (4.4) and (4.7).

Acknowledgements This work is supported by Postdoctoral Science Foundation of China under grant number 2017M620688 and by NNSF of China under grant number 11731014, 11571254 and 11601533.

References

[1] Abidi H, Gui G, Zhang P. On the decay and stability of global solutions to the 3D inhomogeneous Navier-Stokes equations [J]. Comm. Pure Appl. Math., 2011, 64: 0832–0881.

[2] Abidi H, Gui G, Zhang P. On the wellposedness of three-dimensional inhomogeneous Navier-Stokes equations in the critical spaces [J]. Arch. Rational Mech. Anal., 2012, 204: 189–230.

[3] Abidi H, Hmidi T. Résultats d'existence dans des espaces critiques pour le système de la MHD inhomogène [J]. Ann. Math. Blaise Pascal, 2007, 14: 103–148.

[4] Abidi H, Paicu M. Global existence for the magnetohydrodynamic system in critical spaces [J]. Proc. Roy. Soc. Edinburgh Sect. A, 2008, 138: 447–476.

[5] Bahouri H, Chemin J Y, Danchin R. Fourier Analysis and Nonlinear Partial Differential Equations [M]. Heidelberg, Springer, 2011.

[6] Cao C, Wu J. Two regularity criteria for the 3D MHD equations [J]. J. Differential Equations, 2010, 248: 2263–2274.

[7] Cao C, Wu J. Global regularity for the 2D MHD equations with mixed partial dissipation and magnetic diffusion [J]. Adv. Math., 2011, 226: 1803–1822.

[8] Chen F, Li Y, Xu H. Global solution to the 3D nonhomogeneous incompressible MHD equations with some large initial data [J]. Discrete Contin. Dyn. Syst., 2016, 36: 2945–2967.

[9]   Chen D, Zhang Z, Zhao W. Fujita-Kato theorem for the 3-D inhomogenous Navier-Stokes equations [J]. J. Differential Equations, 2016, 261: 738–761.
[10] Chen Q, Miao C, Zhang Z. The Beale-Kato-Majda criterion for the 3D magnetohydrodynamics equations [J]. Comm. Math. Phys., 2007, 275: 861–872.
[11] Chen Q, Tan Z, Wang Y. Strong solutions to the incompressible magnetohydrodynamic equations [J]. Math. Methods Appl. Sci., 2011, 34: 94–107.
[12] Danchin R. Density-dependent incompressible viscous fluids in critical spaces [J]. Proc. Roy. Soc. Edinburgh Sect. A, 2003, 133: 1311–1334.
[13] Danchin R, Mucha P B. A Lagrangian approach for the incompressible Navier-Stokes equations with variable density [J]. Comm. Pure Appl. Math., 2012, 65: 1458–1480.
[14] Danchin R, Mucha P B. Incompressible flows with piecewise constant density [J]. Arch. Ration. Mech. Anal., 2013, 207: 991–1023.
[15] Desjardins B, Le Bris C. Remarks on a nonhomogeneous model of magnetohydrodynamics [J]. Differential Integral Equations, 1998, 11: 377–394.
[16] Duvaut G, Lions J L. Inéquations en thermoélasticité et magnétohydrodynamique [J]. Arch. Rational Mech. Anal., 1972, 46: 241–279.
[17] Evans L C. Partial Differential Equations [M]. Grad. Stud. Math., vol. 19, AMS, Providence, RI, 1998.
[18] Gerbeau J F, Le Bris C. Existence of solution for a density-dependent magnetohydrodynamic equation [J]. Adv. Differential Equations, 1997, 2: 427–452.
[19] Gong H, Li J. Global existence of strong solutions to incompressible MHD [J]. Commun. Pure Appl. Anal., 2014, 13: 1337–1345.
[20] Gui G. Global well-posedness of the two-dimensional incompressible magnetohydrodynamics system with variable density and electrical conductivity [J]. J. Funct. Anal., 2014, 267: 1488–1539.
[21] He C, Xin Z. On the regularity of weak solutions to the magnetohydrodynamic equations [J]. J. Differential Equations, 2005, 213: 235–254.
[22] He C, Xin Z. Partial regularity of suitable weak solutions to the incompressible magnetohydrodynamic equations [J]. J. Funct. Anal., 2005, 227: 113–152.
[23] Hoff D. Global solutions of the Navier-Stokes equations for mutidimensional compressible flow with discontinuous initial data [J]. J. Differential Equations, 1995, 120: 215–254.
[24] Hoff D. Dynamics of singularity surfaces for compressible, viscous flows in two space dimensions [J]. Comm. Pure Appl. Math., 2002, 55: 1365–1407.
[25] Huang X, Li J, Xin Z. Global well-posedness of classical solutions with large oscillations and vacuum to the three-dimentional isentropic compressible Navier-Stokes equations [J]. Comm. Pure Appl. Math., 2012, 65:0549–0585.
[26] Huang X, Wang Y. Global strong solution to the 2D nonhomogeneous incompressible MHD system [J]. J. Differential Equations, 2013, 254: 511–527.
[27] Jia J, Peng J, Li K. On the decay and stability of global solutions to the 3D

inhomogenous MHD system [J]. Comm. Pure Appl. Anal., 2017, 16: 745–780.

[28] Kazhikhov A V. Solvability of the initial-boundary value problem for the equations of the motion of an inhomogeneous viscous incompressible fluid [J]. Dokl. Akad. Nauk SSSR, 1974, 216: 1008–1010.

[29] Lin F, Xu L, Zhang P. Global small solutions to 2-D incompressible MHD system [J]. J. Differential Equations, 2015, 259: 5440–5485.

[30] Lin F, Zhang P. Global small solutions to an MHD-type system: the three-dimensional case [J]. Comm. Pure Appl. Math., 2014, 67: 531–580.

[31] Lions P L. Mathematical Topics in Fluid Mechanics [M]. Vol. 1. Incompressible Models, Oxford Lecture Ser. Math. Appl., vol. 3, Oxford Sci. Publ., The Clarendon Press, New York, Oxford University Press, 1996.

[32] Paicu M, Zhang P, Zhang Z. Global unique solvability of inhomogeneous Navier-Stokes equations with boundary density [J]. Comm. Partial Differential Equations, 2013, 38: 1208–1234.

[33] Ren X, Wu J, Xiang Z, Zhang Z. Global existence and decay of smooth solution for the 2-D MHD equations without magentic diffusion [J]. J. Funct. Anal., 2014, 267: 503–541.

[34] Sermange M, Temam R. Some mathematical questions related to the MHD equations [J]. Comm. Pure Appl. Math., 1983, 36: 635–664.

[35] Xu H, Li Y, Zhai X. On the well-posedness of 2-D incompressible Navier-Stokes equations with variable viscosity in critical spaces [J]. J. Differential Equations, 2016, 260: 6604–6637.

[36] Zhai X, Li Y, Xu H. Global well-posedness for the 2-D nonhomogeneous incompressible MHD equations with large initial data [J]. Nonlinear Anal. Real World Appl., 2017, 33: 1–18.

[37] Zhai X, Li Y, Yan W. Global well-posedness for the 3-D incompressible inhomogeneous MHD system in the ciritical Besov spaces [J]. J. Math. Anal. Appl., 2015, 432: 179–195.

[38] Zhai X, Yin Z. Global well-posedness for the 3D incompressible inhomogeneous Navier-Stokes equations and MHD equations [J]. J. Differential Equations, 2017, 262: 1359–1412.